To our many students, fellows, and colleagues who have made
our careers in immunology a continued time of excitement and joy.
We hope that future generations of immunologists will find
the subject as rewarding as we have.

# ABOUT THE AUTHORS

**Left to right:** *Richard A. Goldsby, Barbara A. Osborne, and Thomas J. Kindt*

Thomas J. Kindt regularly consults on immunology and infectious disease issues for governmental and private organizations and served for many years as Director of Intramural Research for the National Institutes of Health's National Institute of Allergy and Infectious Disease, a position that placed him in daily contact with the cutting edge of experimental and clinical immunology. Tom holds an adjunct professorship with the University of New Mexico Department of Biology and belongs to the Regional Association of Medical and Biological Organizations in New Mexico.

Richard A. Goldsby teaches immunology to undergraduate and graduate students at Amherst College. His research interests include technologies for the generation of engineered and human antibodies in animal bioreactors. He has served on many occasions as a course director in the National Science Foundation Chautauqua Short Course Program, presenting current advances in immunology to college teachers.

Barbara A. Osborne, University of Massachusetts at Amherst, is a recognized contributor to the fast-moving area of programmed cell death and development of T-cell responses. A highly active researcher, Barbara also teaches immunology to undergraduate and graduate students.

Janis Kuby, who died in 1997, taught at San Francisco State University and the University of California at Berkeley. Professor Kuby was the originator of this textbook and author of the first three editions. Her expert teaching and writing skills made *Immunology* the best-selling text for the course, and her vision for the text as a way to combine cutting-edge content in an accessible, pedagogically rich format lives on in the new edition.

# KUBY
# IMMUNOLOGY

SIXTH EDITION

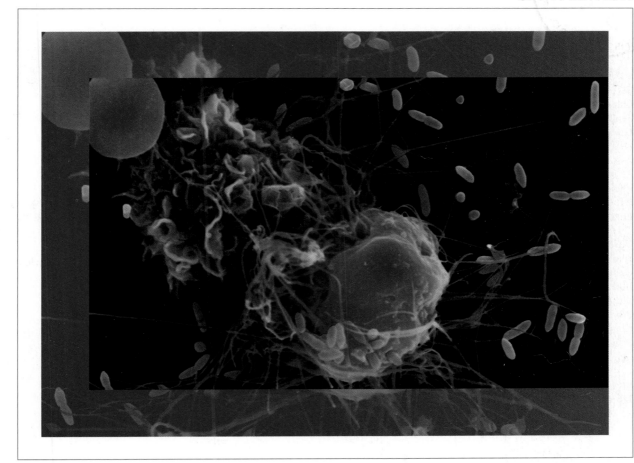

## Thomas J. Kindt
National Institutes of Health

## Richard A. Goldsby
Amherst College

## Barbara A. Osborne
University of Massachusetts at Amherst

W. H. Freeman and Company  |  New York

## About the Cover

The macrophage (blue) is a key player in innate immunity, sensing bacterial proteins through its pattern recognition receptors and then internalizing and digesting (phagocytosing) the invading bacteria (yellow). This encounter stimulates the macrophage to secrete soluble factors that attract other cells such as the monocyte (green) to the area of attack. (© 2004 Dennis Kunkel Microscopy, Inc.)

Publisher: Sara Tenney
Senior Acquisitions Editor: Kate Ahr
Director of Marketing: John Britch
Developmental Editors: Morgan Ryan, Matthew Tontonoz
Associate Project Manager: Hannah Thonet
Assistant Editor: Nick Tymoczko
Media Editors: Alysia Baker, Martin Batey
Photo Editor: Ted Szczepanski
Photo Researcher: Dena Digilio Betz
Text Designer: Marsha Cohen
Cover Designer: Vicki Tomaselli
Senior Project Editor: Georgia Lee Hadler
Copy Editor: Karen Taschek
Illustrations: Imagineering
Illustration Coordinator: Susan Timmins
Production Coordinator: Paul Rohloff
Composition: Techbooks
Printing and Binding: RR Donnelley

**Library of Congress Control Number:** 2006927337

ISBN-13: 978-0-7167-8590-3
ISBN-10: 0-7167-8590-0

Printed in the United States of America

First printing

W. H. Freeman and Company
41 Madison Avenue
New York, NY 10010
Houndmills, Basingstoke RG21 6XS, England
www.whfreeman.com

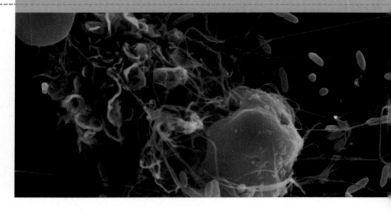

In the second edition of *Immunology*, Janis Kuby wrote, ". . . the continued growth of immunology is inevitable and challenges both the medical and academic community to stay current." Our goal with each new edition of *Immunology* is to present the knowledge to a new generation of scientists and medical professionals. We must provide newcomers with a broad view of the field of immunology. We must be up to date. And we must also introduce the experiments and the model systems upon which our knowledge of the immune system has been built.

## New Chapter 3—Innate Immunity

The means by which the workings of the innate immune system were deduced and the rapid progress made in learning about this arm of immunity rank among the most exciting developments in immunology since the last edition of this book. New Chapter 3, Innate Immunity, explores how:

- The activities of immune effectors like the pattern recognition receptors are integrated within the innate immune response.

- Cells, soluble antimicrobial molecules, and membrane-bound receptors collaborate to mount an instantaneous attack on infectious agents.

- The innate immune system acts not only as a first line of defense, but as an essential trigger for the adaptive immune system.

- Defects in components of the innate immune system often result in weakened or inadequate responses by the adaptive immune system.

## Emphasis on Clinical Relevance

An immune response that is either deficient or excessive can lead to dire consequences. It is critical that those interested in medical careers comprehend the working of this system. We cover a wide range of new diseases and infections throughout the book and we've updated our featured Clinical Focus essays and their corresponding end-of-chapter Study Questions. New to this edition are:

- Explanation of antigen cross-presentation as it pertains to immunity to viruses and other infectious agents (Chapter 8).

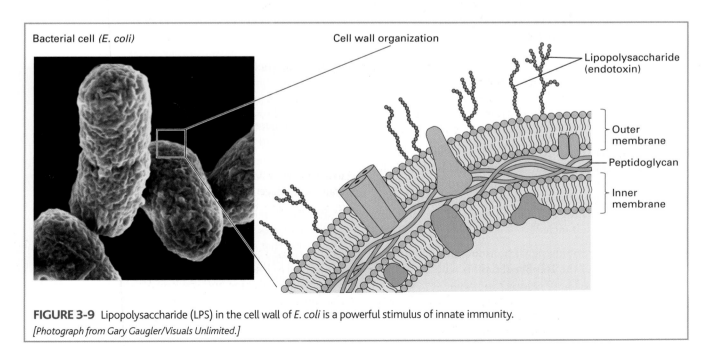

**FIGURE 3-9** Lipopolysaccharide (LPS) in the cell wall of *E. coli* is a powerful stimulus of innate immunity. *[Photograph from Gary Gaugler/Visuals Unlimited.]*

- Expanded discussion of cytokines and their role in inflammation and disease (Chapter 12 and throughout).

- Latest findings on the diversity of NK cell receptors and how their genetic variability influences disease susceptibility (Chapter 14).

- New Clinical Focus essay on the KIR/MHC influence on disease (Chapter 14).

- New coverage of central and peripheral tolerance and how they relate to autoimmune disease and to rejection of allografts (Chapter 16).

- New discussions on methods to alleviate the suffering from various autoimmune conditions such as arthritis, multiple sclerosis, lupus erythematosis, and Crohn's disease (Chapter 16).

- Coverage of the expanding clinical use of monoclonal antibodies as therapeutic agents (Chapters 4, 5, 6, 16, 17, and elsewhere).

- Expanded coverage of infectious diseases including the descriptions of major pathogen groups and the characteristic immune responses they provoke; updated material on influenza, including the avian strains and their threat to human populations; and why fungal disease has increased significantly due to the spread of AIDS and the increase of people taking drugs to counter autoimmune conditions (Chapter 18).

- Coverage of SARS, including discovery and assessment of how it jumped from animals to humans (Chapters 18 and 19).

- New data on consequences of AIDS (Chapter 20).

- Report on the linkage between human papilloma virus (HPV) and cervical cancer, the vaccine trials aimed at its prevention, and a new Clinical Focus essay further examining these findings (Chapter 21).

## Increased Coverage of Signaling

The past few years have seen a great increase in our knowledge of the events that occur after receptors bind their ligands. We now devote a section of the Introduction to the general topic of signal transduction, summarizing the general pattern of signaling and naming some of the more universal key players, and include:

- New section describing signal transduction subsequent to engagement of the Toll-like receptors with their ligands (Chapter 3).

- Expanded coverage of molecular interactions involved in cell migration and extravasation (Chapters 3 and 13).

- New detail on the signaling pathways that lead to maturity, differentiation, and activation of a variety of cell types. (Chapters 10 and 11).

- Expanded coverage of the γδ T-cell receptor, including a recent 3-D image revealing differences in how γδ and αβ T-cell receptors bind antigen (Chapters 9 and 10).

## New Techniques

The following discussions of important modern techniques have produced exciting new knowledge in the field of immunology:

- Description of the **plasmon surface resonance** technique and its application to basic questions in immunology (Chapter 6).

- Use of **labeled MHC-antigen tetramers** to mark membrane bound T-cell receptors (Chapter 14).

- Illustration of the power of **two-photon phase microscopy** to track cells within a lymph node (Chapters 11 and 22).

## Updated Organization

In consultation with numerous teachers of immunology, we have made the following organizational changes in the sixth edition to improve flow and avoid redundancies:

- Antibody and Antigen chapters combined (Chapter 4).

- Complement chapter moved earlier to follow antibody chapters (Chapter 7).

- MHC and Antigen Presentation chapters combined (Chapter 8).

For complete chapter-by-chapter details of the many other additions of new material and updated material in the sixth edition of *Immunology,* please visit

**www.whfreeman.com/immunology6e**

## Pedagogic Tools

### Overview Figures and Pedagogically Focused Illustrations

A number of concepts are especially crucial for students to grasp in developing a firm understanding of immunology. We employ consistent imagery to help the students master the material.

- Principal concepts are illustrated in **Overview Figures.** These figures summarize important ideas and processes in a way that written text alone cannot convey; they are often illustrated as "walkthrough" diagrams, which include more extensive and systematic labeling to assist in visualizing key processes.

- Consistent use of icons depicting various immune system cells and important membrane molecules to assist students in visualizing complex relationships. These iconic elements appear in a table as the frontispiece of the book, providing an easily accessible guide.

# OVERVIEW FIGURE 1-6:  Common Themes in Signal Transduction

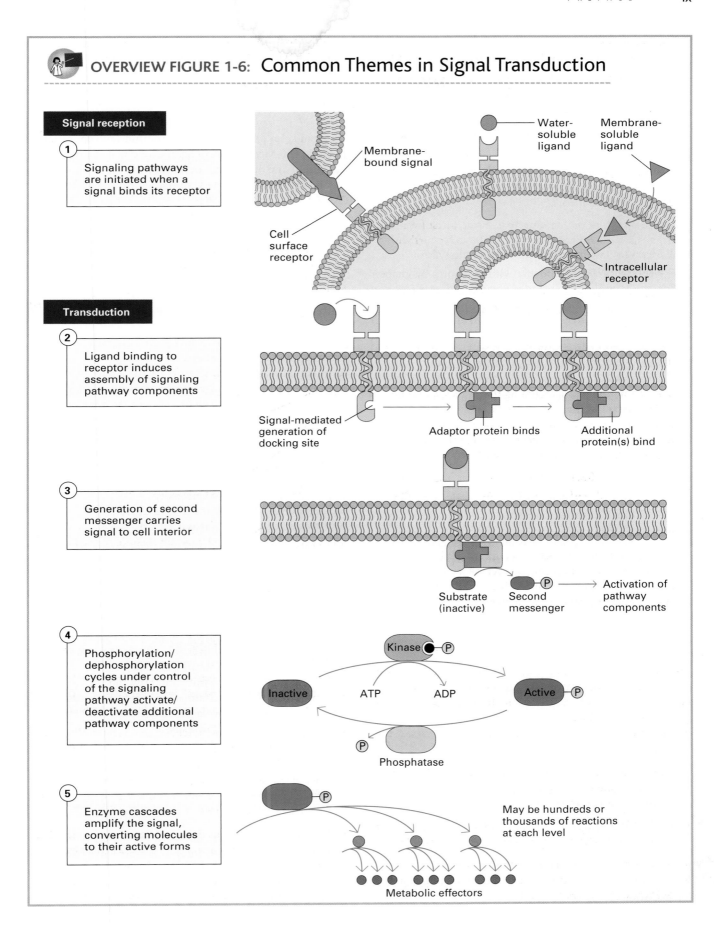

**Signal reception**

**1** Signaling pathways are initiated when a signal binds its receptor

- Water-soluble ligand
- Membrane-soluble ligand
- Membrane-bound signal
- Cell surface receptor
- Intracellular receptor

**Transduction**

**2** Ligand binding to receptor induces assembly of signaling pathway components

- Signal-mediated generation of docking site
- Adaptor protein binds
- Additional protein(s) bind

**3** Generation of second messenger carries signal to cell interior

- Substrate (inactive)
- Second messenger
- Activation of pathway components

**4** Phosphorylation/ dephosphorylation cycles under control of the signaling pathway activate/ deactivate additional pathway components

- Kinase — P
- Inactive
- ATP
- ADP
- Active — P
- P
- Phosphatase

**5** Enzyme cascades amplify the signal, converting molecules to their active forms

- P
- May be hundreds or thousands of reactions at each level
- Metabolic effectors

## Study Questions

Originated by Janis Kuby, the Study Questions in *Immunology* have proven a valuable resource for instructors and students alike. The sixth edition features new and revised questions for every chapter, including an all-new series of **Analyze the Data** questions, which employ modern literature and quantitative data, and challenge students to extrapolate information with the tools learned from the text. Questions are matched with extended, updated answers in the back matter.

## Media

### Web Site: www.whfreeman.com/immunology6e

The Web site provides students with a chapter-by-chapter online study guide that includes a variety of exercises and resources such as Clinical Case Studies. Students are directed to the Internet resources with icons listed at the end of each chapter.

- **Animations** of key concepts facilitate understanding of complex immunological processes by adding a dimension beyond the printed page.

- In **Molecular Visualizations** by David Marcey of California Lutheran University, enhanced and accessible dynamic guided tours of three-dimensional molecular models make the relationship between molecular structure and function easier to grasp.

- **Vocabulary Self-Tests** allow students to review and then quiz and grade themselves on definitions of key terms from the text.

- Students can access the **Useful Web Sites** mentioned in the textbook for further research.

### Instructor's Resource CD-ROM (ISBN: 0-7167-3223-8)

Available to adopters, this electronic resource now includes all text images in both JPEG and PowerPoint formats. (All text images are also available from the **Instructor's Resource Center** on the book's companion Web site.) In addition, the CD-ROM contains all animations from the Web site.

## Acknowledgments

We owe special thanks to several colleagues who assisted with complex revisions and who provided detailed reviews that led to major improvements in the text. These notable contributors include Drs. J. Donald Capra and Kendra White of the Oklahoma Medical Research Foundation, Dr. JoAnn Meerschaert of Saint Cloud State University, Dr. Jiri Mestecky of the University of Alabama at Birmingham, Dr. Jonathan Yewdell at NIAID, NIH, Dr. James Kindt of Emory University, Dr. Johnna Wesley at Brown University, and Dr. Eric Long at NIAID, NIH. We hope that the final product reflects the high quality of the input from these experts and from all those listed below who provided critical insights and guidance.

We would also like to express our gratitude and appreciation to Dr. JoAnn Meerschaert of Saint Cloud State University for her authorship of excellent new problems and to Dr. Stephen K. Chapes of Kansas State University for his creation of our "Analyze the Data" questions.

We thank the following reviewers for their comments and suggestions about the manuscript during preparation of this sixth edition. Their expertise and insights have contributed greatly to the book.

**Ruth D. Allen**, Indiana University-Purdue University Indianapolis
**Avery August**, The Pennsylvania State University
**Pamela J. Baker**, Bates College
**Kenneth J. Balazovich**, University of Michigan
**Cynthia L. Baldwin**, University of Massachusetts Amherst
**Scott R. Barnum**, University of Alabama, Birmingham
**Stephen H. Benedict**, University of Kansas
**Earl F. Bloch**, College of Medicine Howard University
**Lisa Borghesi**, University of Pittsburgh
**Laurent Brossay**, Brown University
**Jane Bruner**, California State University, Stanislaus
**James W. Campbell**, Rice University
**Stephen Keith Chapes**, Kansas State University
**Koteswara R. Chintalacharuvu**, UCLA
**Jeffrey R. Dawson**, Duke University, School of Medicine
**Janet M. Decker**, University of Arizona
**Michael Edidin**, The Johns Hopkins University
**Sherry D. Fleming**, Kansas State University
**Scott C. Garman**, University of Massachusetts Amherst
**Elizabeth Godrick**, Boston University
**Sandra O. Gollnick**, Roswell Park Cancer Institute
**Hans W. Heidner**, The University of Texas at San Antonio
**Vincent W. Hollis, Jr.**, Howard University
**W. Martin Kast**, University of Southern California
**Dennis J. Kitz**, Southern Illinois University, Edwardsville
**Katherine L. Knight**, Loyola University
**Paul M. Knopf**, Brown University
**Kay K. Lee-Fruman**, California State University, Long Beach
**Alan D. Levine**, Case Western Reserve University
**Judith Manning**, University of Wisconsin School of Medicine
**James A. Marsh**, Cornell University College of Veterinary Medicine
**John Martinko**, Southern Illinois University Carbondale
**Andrea M. Mastro**, The Pennsylvania State University
**Jennifer M. Mataraza**, Boston College
**Dennis W. McGee**, Binghamton University
**JoAnn Meerschaert**, Saint Cloud State University

**Jiri Mestecky**, University of Alabama, Birmingham
**Michael F. Minnick**, University of Montana
**Thomas W. Molitor**, University of Minnesota, College of Veterinary Medicine
**David M. Mosser**, University of Maryland
**Rita B. Moyes**, Texas A&M University
**Philip C. Nelson**, University of Pennsylvania
**Alma Moon Novotny**, Rice University
**Kim O'Neill**, Brigham Young University
**Luke O'Neill**, Trinity College, Dublin, Ireland
**Leonard D. Pearson**, Colorado State University
**Christopher A. Pennel**, University of Minnesota
**Wendy E. Raymond**, Williams College
**Robert C. Rickert**, University of California, San Diego
**Kenneth H. Roux**, Florida State University
**Abhineet Sheoran**, Tufts Cummings School of Veterinary Medicine
**Michail Sitkovsky**, Northeastern University

**Robert C. Sizemore**, Alcorn State University
**Gary Splitter**, University of Wisconsin, Madison
**Douglas A. Steeber**, University of Wisconsin, Milwaukee
**Lisa Steiner**, Massachusetts Institute of Technology
**Jeffrey L. Stott**, University of California, Davis School of Veterinary Medicine
**Denise G. Wingett**, Boise State University
**Jon Yewdell**, NIH-NIAID
**Kirk Ziegler**, Emory University School of Medicine

We would also like to thank our experienced and talented colleagues at W. H. Freeman and Company. Particular thanks are due to Kate Ahr, Georgia Lee Hadler, Karen Taschek, Vicki Tomaselli, Paul Rohloff, Susan Timmins, Ted Szczepanski, Hannah Thonet, and Nick Tymoczko. The execution of this work would not have been possible without the dogged determination of our developmental editor, Morgan Ryan, who helped us tell and illustrate the story of immunology.

# Contents in Brief

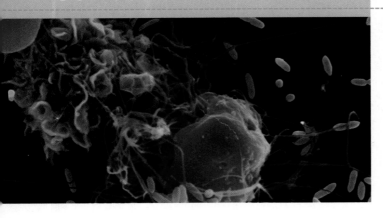

# Contents

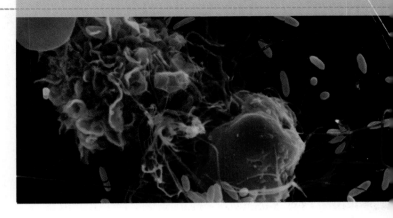

# Part III Immune Effector Mechanisms

# Part IV The Immune System in Health and Disease

# Overview of the Immune System

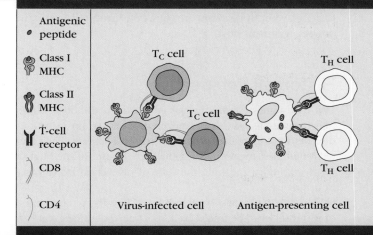

| | |
|---|---|
| ● | Antigenic peptide |
| | Class I MHC |
| | Class II MHC |
| Y | T-cell receptor |
| | CD8 |
| | CD4 |

T_C cell     T_H cell

T_C cell

T_C cell     T_H cell

Virus-infected cell     Antigen-presenting cell

*T-cell recognition of antigen-MHC.*

# chapter 1

- Historical Perspective
- Early Studies of Humoral and Cellular Immunity
- Theoretical Challenges
- Infection and Immunity
- Innate and Adaptive Immunity
- Immune Dysfunction and Its Consequences

*Innate immunity —*

---

**T**HE IMMUNE SYSTEM EVOLVED TO PROTECT MULTICELLULAR organisms from pathogens. Highly adaptable, it defends the body against invaders as diverse as the virus that causes polio and the flatworm that causes schistosomiasis. The immune system generates an enormous variety of cells and molecules capable of specifically recognizing and eliminating foreign invaders, all of which act together in a dynamic network.

Protection by the immune system can be divided into two related activities—recognition and response. Immune recognition is remarkable for its capacity to distinguish foreign invaders from self components. The immune system is able to recognize molecular patterns that characterize groups of common pathogens and deal with these in a rapid and decisive manner. It can even detect subtle chemical differences that distinguish one foreign pathogen from another. Above all, the system is able to discriminate between foreign molecules and the body's own cells and molecules (self-nonself discrimination). In addition, the system is able to recognize host cells that are altered and that may lead to cancer. Typically, recognition of a pathogen by the immune system triggers an **effector response** that eliminates or neutralizes the invader. The multiple components of the immune system are able to convert the initial recognition event into a variety of effector responses, each uniquely suited for eliminating a particular type of pathogen. Certain exposures induce a **memory response,** characterized by a more rapid and heightened immune reaction upon later attack. It is the remarkable property of memory that prevents us from catching some diseases a second time, and immunological memory is the foundation for vaccination, which is a means of "educating" the immune system to prepare it for later attacks.

Although reference is made to the immune *system,* it must be pointed out that there are *two* systems of immunity, innate immunity and adaptive immunity, which collaborate to protect the body. **Innate immunity** includes molecular and cellular mechanisms predeployed before an infection and poised to prevent or eliminate it. This highly effective first line of defense prevents most infections at the outset or eliminates them within hours of encounter with the innate immune system.

The recognition elements of the innate immune system precisely distinguish between self and pathogens, but they are not specialized to distinguish small differences in foreign molecules. A second form of immunity, known as **adaptive immunity,** develops in response to infection and adapts to recognize, eliminate, and then remember the invading pathogen. Adaptive immunity is contingent upon innate immunity and begins a few days after the initial infection. It provides a second, comprehensive line of defense that eliminates pathogens that evade the innate responses or persist in spite of them. An important consequence of adaptive immune response is memory. If the same, or a closely related, pathogen infects the body, memory cells provide the means for the adaptive immune system to make a rapid and often highly effective attack on the invading pathogen.

This chapter introduces the study of immunology from a historical perspective. The highly practical or applied aspects of immunology are outlined, emphasizing the role of vaccination in the development of immunology as a scientific field and as an important aspect of public health. A bird's-eye view

1

of the common pathogens to which we are exposed is given, with a broad overview of the processes, cells, and molecules that make up the innate and adaptive immune systems. Last, we describe circumstances in which the immune system fails to act or when it becomes an aggressor, turning its awesome powers against its host.

## Historical Perspective

The discipline of immunology grew out of the observation that individuals who had recovered from certain infectious diseases were thereafter protected from the disease. The Latin term *immunis,* meaning "exempt," is the source of the English word *immunity,* meaning the state of protection from infectious disease.

Perhaps the earliest written reference to the phenomenon of immunity can be traced back to Thucydides, the great historian of the Peloponnesian War. In describing a plague in Athens, he wrote in 430 BC that only those who had recovered from the plague could nurse the sick because they would not contract the disease a second time. Although early societies recognized the phenomenon of immunity, almost 2000 years passed before the concept was successfully converted into medically effective practice.

### Early vaccination studies led the way to immunology

The first recorded attempts to induce immunity deliberately were performed by the Chinese and Turks in the fifteenth century. They were attempting to prevent smallpox, a disease that is fatal in about 30% of cases and that leaves survivors disfigured for life (Figure 1-1). Reports suggest that the dried crusts derived from smallpox pustules were either inhaled into the nostrils or inserted into small cuts in the skin (a technique called *variolation*) in order to prevent this dreaded disease. In 1718, Lady Mary Wortley Montagu, the wife of the British ambassador in Constantinople, observed the positive effects of variolation on the native Turkish population and had the technique performed on her own children. The English physician Edward Jenner, in 1798, made a giant advance in the deliberate development of immunity. Intrigued by the fact that milkmaids who had contracted the mild disease cowpox were subsequently immune to the much more severe smallpox, Jenner reasoned that introducing fluid from a cowpox pustule into people (i.e., inoculating them) might protect them from smallpox. To test this idea, he inoculated an eight-year-old boy with fluid from a cowpox pustule and later intentionally infected the child with smallpox. As predicted, the child did not develop smallpox.

Jenner's technique of inoculating with cowpox to protect against smallpox spread quickly throughout Europe. However, it was nearly a hundred years before this technique was applied to other diseases. As so often happens in science, serendipity in combination with astute observation led to the

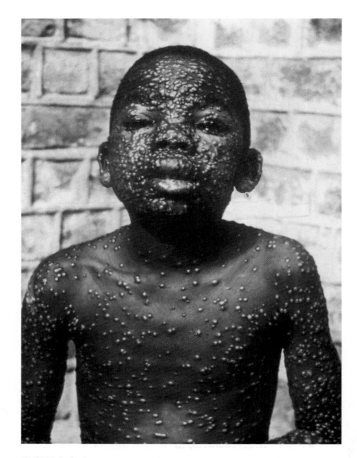

**FIGURE 1-1  African child with rash typical of smallpox on face, chest, and arms.** Smallpox, caused by the virus *Variola major,* has a 30% mortality rate. Survivors are often left with disfiguring scars. *[Centers for Disease Control.]*

next major advance in immunology, the induction of immunity to cholera. Louis Pasteur had succeeded in growing the bacterium thought to cause fowl cholera in culture, confirming its role when chickens injected with the cultured bacterium developed fatal cholera. After returning from a summer vacation, he injected some chickens with an old bacterial culture. The chickens became ill, but to Pasteur's surprise, they recovered. Pasteur then grew a fresh culture of the bacterium with the intention of injecting it into some fresh chickens. But as the story is told, his supply of chickens was limited, and therefore he used the previously injected chickens. Unexpectedly, the chickens were completely protected from the disease. Pasteur hypothesized and proved that aging had weakened the virulence of the pathogen and that such an attenuated strain might be administered to protect against the disease. He called this attenuated strain a **vaccine** (from the Latin *vacca,* meaning "cow"), in honor of Jenner's work with cowpox inoculation.

Pasteur extended these findings to other diseases, demonstrating that it was possible to **attenuate,** or weaken, a pathogen and administer the attenuated strain as a vaccine. In

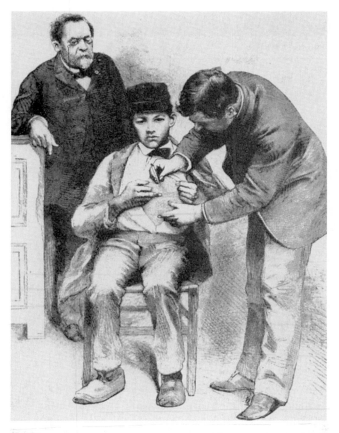

**FIGURE 1-2**  Wood engraving of Louis Pasteur watching Joseph Meister receive the rabies vaccine. *[From* Harper's Weekly *29:836; courtesy of the National Library of Medicine.]*

a now classic experiment performed in the small village of Pouilly-le-Fort in 1881, Pasteur first vaccinated one group of sheep with heat-attenuated anthrax bacteria *(Bacillus anthracis);* he then challenged the vaccinated sheep and some unvaccinated sheep with a virulent culture of the anthrax bacillus. All the vaccinated sheep lived, and all the unvaccinated animals died. These experiments marked the beginnings of the discipline of immunology. In 1885, Pasteur administered his first vaccine to a human, a young boy who had been bitten repeatedly by a rabid dog (Figure 1-2). The boy, Joseph Meister, was inoculated with a series of attenuated rabies virus preparations. He lived and later became a custodian at the eponymous Pasteur Institute.

## Vaccination is an ongoing, worldwide enterprise

The emergence of the study of immunology and the discovery of vaccines are tightly linked. The discovery, development, and appropriate use of vaccines remains a challenge to immunologists today.

In 1977, the last known case of naturally acquired smallpox was seen in Somalia. This dreaded disease was eradicated by universal application of a vaccine that did not differ greatly from that used by Jenner in the 1790s. A consequence of eradication is that universal vaccination becomes unnecessary; this is a tremendous benefit because the smallpox vaccine carried a slight degree of risk to persons vaccinated as well as to persons exposed to recent vaccinees. There is a darker side, however, to eradication and the end of universal vaccination. Over time, the number of people with no immunity to the disease will necessarily rise. Yet one day the disease may be reintroduced by unnatural means. In fact, smallpox is considered one of the most potent bioterrorism threats. In response, new and safer vaccines against smallpox are being developed.

A milestone in vaccinology comparable to smallpox eradication may be around the corner for paralytic polio, a crippling disease targeted for eradication in the near future. A campaign spearheaded by the World Health Organization has relied on massive immunization programs to carry out this objective. The project was slowed by resistance in certain areas based on rumors that immunization causes sterility in male children. The regional resurgence in polio cases in certain areas of Asia and Africa resulting from this resistance is a setback, but one that can be overcome by education and observation of the benefits of vaccination. The fact that polio is not a threat worldwide and has been eliminated in most countries is a triumph that should not be overshadowed by delays in the eradication program.

In the United States and other industrialized nations, vaccines have eliminated a host of childhood diseases that were considered a part of growing up just 50 years ago. Measles, mumps, whooping cough (pertussis), tetanus, diphtheria, and polio are extremely rare or nonexistent because of current vaccination practices (Table 1-1). One can barely estimate the savings to society resulting from the prevention of these diseases. Aside from suffering and mortality, the cost to treat these illnesses and their aftereffects or sequelae (such as paralysis, deafness, blindness, and mental retardation) is immense and dwarfs the costs of immunization.

For some diseases, immunization is the best, if not the only, effective defense. Because few antiviral drugs are now available, the major defense against influenza must be an effective vaccine. If an influenza pandemic recurs, as predicted by many experts, there will be a race between its spread and the manufacture and administration of an effective vaccine. At the time of this writing, much attention is being paid to an emerging strain of avian influenza. About 200 cases of infection in humans have been documented, of which about half were fatal. If this virus adapts to spread efficiently in humans, a major pandemic will result. Without a preventative vaccine, the devastation caused by the influenza pandemic of 1918, which left as many as 50 million dead, may be equaled or surpassed.

Despite the record of success for vaccines and our reliance on them, opponents to vaccination programs have made claims that vaccines do more harm than good and that childhood vaccination should be curtailed or even eliminated. There is no dispute that vaccines represent unique safety issues, since vaccines are given to people who

| TABLE 1-1 | Cases of selected infectious disease before and after the introduction of effective vaccines | | |
|---|---|---|---|
| | ANNUAL CASES/YR | | CASES IN 2004 |
| Disease | Prevaccine | Postvaccine | Reduction (%) |
| Smallpox | 48,164 | 0 | 100 |
| Diphtheria | 175,885 | 0 | 100 |
| Measles | 503,282 | 37 | 99.99 |
| Mumps | 152,209 | 236 | 99.85 |
| Pertussis (whooping cough) | 147,271 | 18,957 | 87.13 |
| Paralytic polio | 16,316 | 0 | 100 |
| Rubella (German measles) | 47,745 | 12 | 99.97 |
| Tetanus ("lockjaw") | 1,314 (deaths) | 26 (cases) | 98.02 |
| Invasive hemophilus influenzae | 20,000 | 172 | 99.14 |

SOURCE: Adapted from W. A. Orenstein et al., 2005. *Health Affairs* **24**:599.

are healthy. Furthermore, there is general agreement that vaccines must be regulated and the public must have access to clear and complete information about them. Although the claims of critics must be evaluated, many can be answered by careful and objective examination of records. A recent example is the claim that the mercury-containing preservative thimerosol, previously used in some vaccine preparations, causes autism and was responsible for recent increases in the incidence of the disorder, which is marked by intense self-absorption and inability to relate to others. Autism generally manifests itself between 1 and 2 years of age, the time window in which many vaccines are given (see Chapter 19). The Danish government maintains meticulous health records on its citizens, a source of data that throws revealing light on the putative causal connection between thimerosol and autism. The records indicated that the incidence of autism increased significantly since 1992 in Denmark. However, the records also showed that the use of thimerosol as a vaccine preservative had already been stopped completely in that country several years earlier. Data such as these make it difficult to link autism to the use of thimerosol and suggest the need for further research into the causes for the increase in autism.

Perhaps the greatest current challenge in vaccine development is the lack of effective vaccines for major killers such as malaria and AIDS. It is hoped that the immunologists of today, using the tools of molecular and cellular biology, genomics, and proteomics, will make inroads into preventing these diseases. A further issue in vaccines is the fact that millions of children in developing countries die from diseases that are fully preventable by available, safe vaccines. High manufacturing costs, instability of the products, and delivery problems keep these vaccines from those they could greatly benefit. This problem could be alleviated in many cases by development of future-generation vaccines that are inexpensive, heat stable, and administered without a needle.

## Early Studies of Humoral and Cellular Immunity

Pasteur proved that vaccination worked, but he did not understand how. The experimental work of Emil von Behring and Shibasaburo Kitasato in 1890 gave the first insights into the mechanism of immunity, earning von Behring the Nobel prize in medicine in 1901 (Table 1-2). Von Behring and Kitasato demonstrated that **serum** (the liquid, noncellular component recovered from coagulated blood) from animals previously immunized to diphtheria could transfer the immune state to unimmunized animals. In search of the protective agent, various researchers during the next decade demonstrated that an active component from immune serum could neutralize toxins, precipitate toxins, and agglutinate (clump) bacteria. In each case, the active agent was named for the activity it exhibited: antitoxin, precipitin, and agglutinin, respectively. Initially, different serum components were thought to be responsible for each activity, but during the 1930s, mainly through the efforts of Elvin Kabat, a fraction of serum first called gamma globulin (now **immunoglobulin**) was shown to be responsible for all these activities. The active molecules in the immunoglobulin fraction are called **antibodies.** (The terms *antibody* and *immunoglobulin* may be used interchangeably, but usually the term *antibody* is reserved for immunoglobulins with known specificity for an antigen.) Because immunity was mediated by antibodies contained in body fluids (known at the time as humors), it was called humoral immunity.

The observation of von Behring and Kitasato was applied to clinical practice. Prior to the advent of antibiotic therapy for infectious diseases, antisera, often prepared in horses, were given to patients with a variety of illnesses. Today there are therapies that rely on transfer of immunoglobulins, and with the development of monoclonal antibody technology, antibody therap

*Phagocytes - WBC ingesting microorganisms.*

| TABLE 1-2 | Nobel prizes for immunologic research | | |
|---|---|---|---|
| **Year** | **Recipient** | **Country** | **Research** |
| 1901 | Emil von Behring | Germany | Serum antitoxins |
| 1905 | Robert Koch | Germany | Cellular immunity to tuberculosis |
| 1908 | Elie Metchnikoff<br>Paul Ehrlich | Russia<br>Germany | Role of phagocytosis (Metchnikoff) and antitoxins (Ehrlich) in immunity |
| 1913 | Charles Richet | France | Anaphylaxis |
| 1919 | Jules Bordet | Belgium | Complement-mediated bacteriolysis |
| 1930 | Karl Landsteiner | United States | Discovery of human blood groups |
| 1951 | Max Theiler | South Africa | Development of yellow fever vaccine |
| 1957 | Daniel Bovet | Switzerland | Antihistamines |
| 1960 | F. Macfarlane Burnet<br>Peter Medawar | Australia<br>Great Britain | Discovery of acquired immunological tolerance |
| 1972 | Rodney R. Porter<br>Gerald M. Edelman | Great Britain<br>United States | Chemical structure of antibodies |
| 1977 | Rosalyn R. Yalow | United States | Development of radioimmunoassay |
| 1980 | George Snell<br>Jean Dausset<br>Baruj Benacerraf | United States<br>France<br>United States | Major histocompatibility complex |
| 1984 | Cesar Milstein<br>Georges E. Köhler<br>Niels K. Jerne | Great Britain<br>Germany<br>Denmark | Monoclonal antibodies<br><br>Immune regulatory theories |
| 1987 | Susumu Tonegawa | Japan | Gene rearrangement in antibody production |
| 1991 | E. Donnall Thomas<br>Joseph Murray | United States<br>United States | Transplantation immunology |
| 1996 | Peter C. Doherty<br>Rolf M. Zinkernagel | Australia<br>Switzerland | Role of major histocompatibility complex in antigen recognition by T cells |
| 2002 | Sydney Brenner<br>H. Robert Horvitz<br>J. E. Sulston | S. Africa<br>United States<br>Great Britain | Genetic regulation of organ development and cell death (apoptosis) |

is now a well-established commercial enterprise (see Clinical Focus in Chapter 4). Emergency use of sera containing antibodies against snake or scorpion venoms for bite victims is a common practice. Whereas a vaccine is said to engender **active immunity** in the host, transfer of antibody with a given specificity confers **passive immunity.** Newborn infants benefit from passive immunization conferred by the presence of maternal antibodies in their circulation. Passive immunity may be given as a preventative (prophylaxis) to those who anticipate exposure to a given disease or to those with compromised immunity.

Antisera to certain bacterial toxins such as diphtheria or tetanus can be administered to infected individuals to stop the disease caused by the toxin. A dramatic vignette of this application was the relay of dogsleds racing over 674 miles of frozen Alaskan landscape to bring diphtheria antitoxin from Fairbanks to the affected children in the icebound city of Nome during a diphtheria outbreak in 1925. The successful outcome is commemorated annually in the Iditarod dogsled race, and there is a statue of Balto, one of the lead dogs used for the final leg of the delivery, in New York's Central Park.

In 1883, even before the discovery that a serum component could transfer immunity, Elie Metchnikoff demonstrated that cells also contribute to the immune state of an animal. He observed that certain white blood cells, which he termed **phagocytes,** ingested (phagocytosed) microorganisms and other foreign material (Figure 1-3). Noting that these phagocytic cells were more active in animals that had been immunized, Metchnikoff hypothesized that cells, rather than serum components, were the major effector of immunity. The active phagocytic cells identified by Metchnikoff were likely blood monocytes and neutrophils (see Chapter 2).

## Theoretical Challenges

Whereas development of safe and effective vaccines and the use of passive immune therapy remain challenging problems in clinical (also called *translational*) research, the study of immunology also raised puzzling theoretical questions that occupied the scientific community.

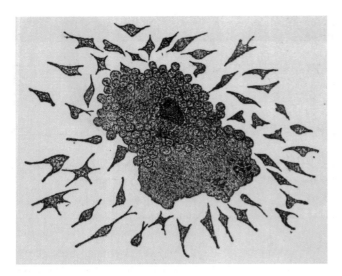

**FIGURE 1-3 Drawing by Elie Metchnikoff of phagocytic cells surrounding a foreign particle.** He first described and named the process of phagocytosis, or ingestion of foreign matter by white blood cells. *[By permission of The British Library:7616.h.19,* Lectures on the Comparative Pathology of Inflammation delivered at the Pasteur Institute in 1891, *translated by F. A. Starling and E. H. Starling, with plates by Il'ya Il'ich Mechnikov, 1893, p. 64, fig. 32.]*

One question that surfaced early involved the relative roles of cellular and humoral immunity. A controversy developed between those who held to the concept of humoral immunity and those who agreed with Metchnikoff's concept of **cell-mediated immunity.** It is now obvious that both are correct—the full immune response requires both cellular and humoral responses. Early studies of immune cells were hindered by the lack of genetically defined animal models and modern tissue culture techniques, whereas studies with serum took advantage of the ready availability of blood and established biochemical techniques to purify protein mediators of humoral immunity. Information about cellular immunity therefore lagged behind the unraveling of humoral immunity.

In a key experiment in the 1940s, Merrill Chase, working at The Rockefeller Institute, succeeded in transferring immunity against the tuberculosis organism by transferring white blood cells between guinea pigs. Until that point, attempts to develop an effective vaccine or antibody therapy against tuberculosis had met with failure. Thus, Chase's demonstration helped to rekindle interest in cellular immunity. With the emergence of improved cell culture techniques in the 1950s, the **lymphocyte** was identified as the cell responsible for both cellular and humoral immunity. Soon thereafter, experiments with chickens pioneered by Bruce Glick at Mississippi State University indicated that there are two types of lymphocytes: T lymphocytes, derived from thymus, mediated cellular immunity, and B lymphocytes, from the bursa of Fabricius (an outgrowth of the cloaca in birds), were involved in humoral immunity. The controversy about the roles of humoral and cellular immunity was resolved when the systems were shown to be intertwined. It became clear that both systems were necessary for the immune response.

One of the great enigmas confronting early immunologists was the specificity of the antibody molecule for foreign material, or **antigen** (the general term for a substance that binds with a specific antibody). Around 1900, Jules Bordet at the Pasteur Institute expanded the concept of immunity beyond infectious diseases by demonstrating that nonpathogenic substances, such as red blood cells from other species, could serve as antigens. Serum from an animal inoculated with noninfectious material would nevertheless react with the injected material in a specific manner. The work of Karl Landsteiner and those who followed him showed that injecting an animal with almost any organic chemical could induce production of antibodies that would bind specifically to the chemical. These studies demonstrated that antibodies have a capacity for an almost unlimited range of reactivity, including responses to compounds that had only recently been synthesized in the laboratory and had not previously existed in nature. In addition, it was shown that molecules differing in the smallest detail could be distinguished by their reactivity with different antibodies. Two major theories were proposed to account for this specificity: the selective theory and the instructional theory.

The earliest conception of the *selective theory* dates to Paul Ehrlich in 1900. In an attempt to explain the origin of serum antibody, Ehrlich proposed that cells in the blood expressed a variety of receptors, which he called "side-chain receptors," that could react with infectious agents and inactivate them. Borrowing a concept used by Emil Fischer in 1894 to explain the interaction between an enzyme and its substrate, Ehrlich proposed that binding of the receptor to an infectious agent was like the fit between a lock and key. Ehrlich suggested that interaction between an infectious agent and a cell-bound receptor would induce the cell to produce and release more receptors with the same specificity (Figure 1-4). According to Ehrlich's theory, the specificity of the receptor was determined in the host before its exposure to antigen, and the antigen selected the appropriate receptor. Ultimately, all aspects of Ehrlich's theory would be proven correct with the minor exception that the "receptor" exists as both a soluble antibody molecule and a cell-bound receptor; the soluble form is secreted rather than the bound form released.

In the 1930s and 1940s, the selective theory was challenged by various *instructional theories,* in which antigen played a central role in determining the specificity of the antibody molecule. According to the instructional theories, a particular antigen would serve as a template around which antibody would fold. The antibody molecule would thereby assume a configuration complementary to that of the antigen template. This concept was first postulated by Friedrich Breinl and Felix Haurowitz in about 1930 and redefined in the 1940s in terms of protein folding by Linus Pauling. The instructional theories were formally disproved in the 1960s, by which time information was emerging about the structure of protein, RNA, and DNA that would offer new insights into the vexing problem of how an individual could make antibodies against almost anything.

disease-causing org - pathogen → Attack is called pathogenesis

human pathogens can be grouped into major categories, shown below and in Figure 1-5.

| Major groups of human pathogens | Examples of diseases |
| --- | --- |
| Viruses | Polio, smallpox, influenza, measles, AIDS |
| Bacteria | Tuberculosis, tetanus, whooping cough |
| Fungi | Thrush, ringworm |
| Parasites | Malaria, leishmaniasis |

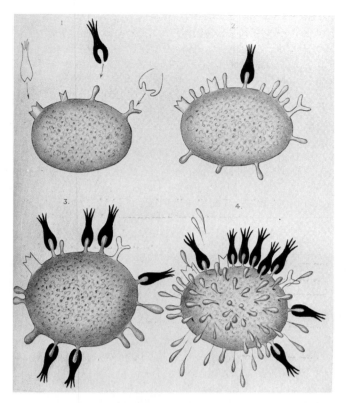

**FIGURE 1-4  Representation of Paul Ehrlich's side chain theory to explain antibody formation.** The cell is pluripotent in that it expresses a number of different receptors or side chains, all with different specificities; if an antigen encounters this cell and there is a good fit with one of its side chains, synthesis of that receptor is triggered and the receptor will be released. *[From Ehrlich's Croonian lecture of 1900 to the Royal Society.]*

In the 1950s, selective theories resurfaced as a result of new experimental data and, through the insights of Niels Jerne, David Talmadge, and F. Macfarlane Burnet, were refined into a theory that came to be known as the **clonal-selection** theory. According to this theory, an individual lymphocyte expresses membrane receptors that are specific for a distinct antigen. This unique receptor specificity is determined before the lymphocyte is exposed to the antigen. Binding of antigen to its specific receptor activates the cell, causing it to proliferate into a clone of cells that have the same immunologic specificity as the parent cell. The clonal-selection theory has been further refined and is now accepted as an underlying paradigm of modern immunology.

## Infection and Immunity

Along with the developing field of immunology grew the study of medical microbiology, which covered the identification of infectious agents and their modes of causing disease. Organisms causing disease are termed **pathogens,** and their means of attacking the host is called **pathogenesis.** The

Of major interest here are the means by which effective immunity to the pathogens can be achieved. An effective defense relies heavily on the nature of the individual microorganism. For example, because viruses require mammalian cells for their replication and multiplication, an effective defense strategy may involve recognizing and killing any cell that is virus infected before the life cycle of the virus can be completed. For organisms that replicate outside host cells, prompt recognition of these invaders by antibodies or soluble molecules can be used, followed by cellular and molecular immune mechanisms that eliminate the pathogen.

Some pathogens that pervade our environment cause no problem for healthy individuals because there is adequate pre-existing immunity. However, individuals with deficiencies in immune function may be susceptible to disease caused by these ubiquitous microbes. For example, the fungus *Candida albicans,* present nearly everywhere, causes no problems for most individuals. However, in those with lowered immunity, it can cause an irritating rash and a spreading infection in the mucosa lining the oral and vaginal cavities. The rash, called thrush, may be a first sign of immune dysfunction. If left unchecked, *C. albicans* can spread, causing systemic candidiasis, a life-threatening condition. Another example is the virus *Herpesvirus simplex,* which normally causes small lesions around the lips or the genitalia. In those with deficient immunity, these lesions may spread to cover large portions of the body. Such infections by ubiquitous microorganisms, often observed in cases of immune deficiency, are termed **opportunistic infections.** Several rarely seen opportunistic infections identified in early AIDS patients were the signal that the immune systems of these patients were seriously compromised.

For some pathogens known to cause serious disease, the means to achieve immunity are well documented and allow control of the disease. For example, tetanus (commonly called lockjaw) is caused by a common soil bacterium (*Clostridium tetani*), which acts through a toxin that attacks the nervous system (neurotoxin). Untreated, tetanus leads to death within a short time. There is an effective vaccine to tetanus, and should that fail, antibodies to the toxin can be administered to offset the potentially fatal disease. Prior to the availability of these prevention and treatment measures, anyone suffering a

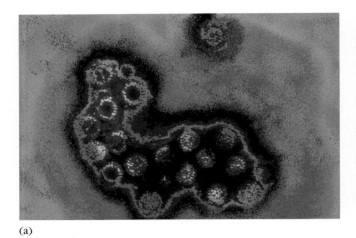

(a)

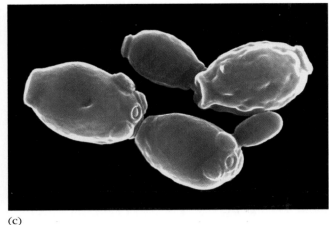

(c)

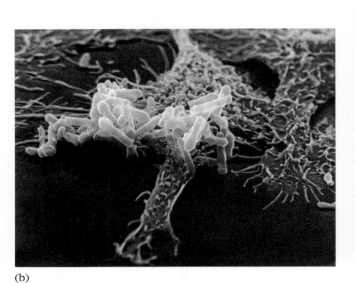

(b)

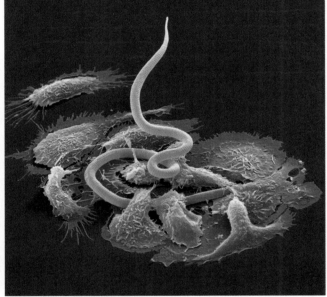

(d)

**FIGURE 1-5 Pathogens representing the major categories of microorganisms causing human disease.** (a) Viruses: Transmission electron micrograph of rotavirus, a major cause of infant diarrhea. Rotavirus accounts for approximately 1 million infant deaths per year in developing countries and hospitalization of about fifty thousand infants per year in the United States. *[VEM/Photo Researchers.]* (b) Bacteria: *Pseudomonas*, a soil bacterium that is an opportunistic pathogen for humans, being ingested by human macrophages. *[David M. Phillips/* *Photo Researchers.]* (c) Fungi: *Candida albicans*, a yeast inhabiting human mouth, throat, intestines, and genitourinary tract; *C. albicans* commonly causes an oral rash (thrush) or vaginitis in immunosuppressed individuals or in those taking antibiotics that kill normal bacterial flora. *[Stem Jems/Photo Researchers.]* (d) Parasites: The larval form of filaria, a parasitic worm, being attacked by macrophages. Approximately 120 million persons worldwide have some form of filariasis. *[Oliver Meckes/Nicole Ottawa/Eye of Science/Photo Researchers.]*

puncture wound by a dirty object, such as a rusty nail, was at risk for a fatal tetanus infection. In marked contrast to the success in controlling tetanus is the failure to devise immune strategies to control HIV infection, which leads to AIDS.

The immune system must deal with all types of pathogens and has evolved multiple strategies for dealing with invasion by a pathogen that has passed the first barriers of skin and mucosa. We will see that some of these strategies are poised to respond the instant a pathogen breaches the host's barriers; other defenses are crafted to act after infection is established.

## Innate and Adaptive Immunity

Immunity—the state of protection from infectious disease— has both a less specific and a more specific component. The less specific component, **innate immunity,** provides the first line of defense against infection. Most components of innate immunity are present before the onset of infection and constitute a set of disease-resistance mechanisms that are not specific to a particular pathogen, but include cellular and molecular components that recognize classes of molecules peculiar to frequently encountered pathogens.

The first hurdle for a pathogen involves breaching the barriers that protect the host. Obvious barriers include the skin and the mucosal membranes. The acidity of the stomach contents and of perspiration poses a further barrier to organisms unable to grow in acidic conditions. Enzymes such as lysozyme, which is present in tears, attack the cell walls of certain bacteria on contact. The importance of these barriers becomes obvious when they are breached. Bites of animals or insects puncture the skin and can introduce a number of diseases. Animal bites can communicate rabies or tetanus, whereas insects carry a host of diseases, including malaria from mosquitoes, plague from fleas, and Lyme disease from ticks. A dramatic example of barrier loss is seen in burn victims, who lose the protective skin at the burn site and must be treated aggressively with drugs to prevent bacterial and fungal infection. Beyond the primary barriers, innate immunity includes a host of cells, such as the phagocytes demonstrated by Metchnikoff, as well as antimicrobial compounds synthesized by the host that can recognize and neutralize invaders based on common molecular surface markers.

## Phagocytic cells are a barrier to infection

An important innate defense mechanism is the ingestion of extracellular particulate material by phagocytosis. Phagocytosis is one type of **endocytosis,** the general term for the uptake by a cell of material from its environment. In phagocytosis, a cell's plasma membrane expands around the particulate material, which may include whole pathogenic microorganisms. Most phagocytosis is conducted by specialized cells, such as blood monocytes, neutrophils, and tissue macrophages (see Chapter 2). Most cell types are capable of other forms of endocytosis, such as *receptor-mediated endocytosis,* in which extracellular molecules are internalized after binding to specific cellular receptors, and *pinocytosis,* the process by which cells take up fluid from the surrounding medium along with any molecules contained in it.

## Soluble molecules contribute to innate immunity

A variety of soluble factors contribute to innate immunity, among them the protein lysozyme, the interferon proteins, and components of the complement system (see Chapter 7). **Lysozyme,** a hydrolytic enzyme found in mucous secretions and in tears, is able to cleave the peptidoglycan layer of the bacterial cell wall. **Interferon** comprises a group of proteins produced by virus-infected cells. Among the many functions of the interferons is the ability to bind to nearby cells and induce a generalized antiviral state.

**Complement,** examined in detail in Chapter 7, includes a group of serum proteins that circulate in an inactive state. A variety of specific and nonspecific immunologic mechanisms can convert the inactive forms of complement proteins into an active state with the ability to damage the membranes of pathogenic organisms, either destroying the pathogens or

facilitating their clearance. Complement occupies a position that truly straddles the innate and adaptive immune systems, in that certain components may directly deal with pathogens, whereas others require prior binding of antibodies to activate its effector system. Reactions between complement molecules or fragments of complement molecules and cellular receptors trigger activation of cells of the innate or adaptive immune systems. Recent studies on **collectins** indicate that these proteins may kill certain bacteria directly by disrupting their lipid membranes or, alternatively, by aggregating the bacteria to enhance their susceptibility to phagocytosis.

Many of the molecules involved in innate immunity have the property of **pattern recognition,** the ability to recognize a given class of molecules. Because certain types of molecules are unique to microbes and never found in multicellular organisms, the ability to immediately recognize and combat invaders displaying such molecules is a strong feature of innate immunity. Molecules with pattern recognition ability may be soluble, like lysozyme and the complement components described above, or they may be cell-associated receptors, such as those designated the **Toll-like receptors (TLRs),** described in Chapter 3.

Should the invading pathogen breach the host's physical and chemical barriers, it may then be detected by pattern recognition molecules of the host and taken up by phagocytic cells, causing the system to react with an inflammatory response (see Chapters 3 and 13). This response concentrates elements of innate immunity at the inflammation site and may result in the marshaling of a specific immune response against the invader. The specific response called upon by the inflammation is the adaptive (sometimes called acquired) immune response.

In contrast to the broad reactivity of the innate immune system, which is uniform in all members of a species, the specific component, adaptive immunity, does not come into play until there is a recognized antigenic challenge to the organism. Adaptive immunity responds to the challenge with a high degree of specificity as well as the remarkable property of "memory." Typically, there is an adaptive immune response against an antigen within five or six days after the initial exposure to that antigen. Future exposure to the same antigen results in a memory response: the immune response to the second challenge occurs more quickly than the first, is stronger, and is often more effective in neutralizing and clearing the pathogen. The major agents of adaptive immunity are lymphocytes and the antibodies they produce. Table 1-3 compares innate and adaptive immunity.

Because adaptive immune responses require some time to marshal, innate immunity provides the first line of defense during the critical period just after the host's exposure to a pathogen. In general, most of the microorganisms encountered by a healthy individual are readily cleared within a few days by defense mechanisms of the innate immune system before they activate the adaptive immune system.

## Collaboration between innate and adaptive immunity increases immune responsiveness

It is important to appreciate that the innate and adaptive immune systems do not operate independently—they

| TABLE 1-3 | Comparison of innate and adaptive immunity | |
|---|---|---|
| | **Innate** | **Adaptive** |
| Response time | Hours | Days |
| Specificity | Limited and fixed | Highly diverse; improves during the course of immune response |
| Response to repeat infection | Identical to primary response | Much more rapid than primary response |
| Major components | Barriers (e.g., skin); phagocytes; pattern recognition molecules | Lymphocytes; antigen-specific receptors; antibodies |

function as a highly interactive and cooperative system, producing a combined response more effective than either branch could produce by itself. Certain cellular and molecular immune components play important roles in both types of immunity.

An example of cooperation is seen in the encounter between macrophages and microbes. Interactions between receptors on macrophages and microbial components generate soluble proteins that stimulate and direct adaptive immune responses, facilitating the participation of the adaptive immune system in the elimination of the pathogen. The soluble proteins are growth factor–like molecules known by the general name **cytokines.** The cytokines react with receptors on various cell types and signal the cell to perform functions such as synthesis of new factors or to undergo differentiation to a new cell type. A restricted class of cytokines have chemotactic activity and recruit specific cells to the site of the cell secreting that cytokine; these are called **chemokines.**

The type of intracellular communication mediated by cytokines is referred to by the general term **signaling.** Basically, signaling involves the reaction between a soluble molecule (ligand) and a cell membrane–bound molecule (receptor) or between membrane-bound molecules on two different cells. The interaction between a receptor and its ligand leads to metabolic adaptations in cells. There are an enormous number of different signal transduction pathways, all exhibiting common themes typical of these integrative processes:

- *Signal transduction begins when a signal binds to its receptor.* Receptors may be inside or outside the cell. Signals that cannot penetrate the cell membrane bind to receptors on the cell surface. This group includes water-soluble signaling molecules and membrane-bound ligands (MHC-peptide complexes, for example). Hydrophobic signals, such as steroids, can diffuse through the cell membrane and are bound by intracellular receptors.

- *Many signal transduction pathways involve the signal-induced assembly of pathway components.* Molecules

known as adaptor proteins bind specifically and simultaneously to two or more different molecules with roles in the signaling pathway, bringing them together and promoting their combined activity.

- *Signal reception often leads to the generation within the cell of a "second messenger,"* a molecule or ion that can diffuse to other sites in the cell and stimulate metabolic changes. Examples are cyclic nucleotides (cAMP, cGMP), calcium ion ($Ca^{2+}$), and membrane phospholipid derivatives such as diacylglycerol (DAG) and inositol triphosphate ($IP_3$).

- *Protein kinases and protein phosphatases are activated or inhibited.* Kinases catalyze the phosphorylation of target residues (tyrosine, serine, or threonine) of key signal transduction proteins. Phosphatases catalyze dephosphorylation, reversing the effect of kinases. These enzymes play essential roles in many signal transduction pathways of immunological interest.

- *Signals are amplified by enzyme cascades.* An enzyme in a signaling pathway, once activated, catalyzes many additional reactions. The enzyme may generate many molecules of the next component in the pathway or activate many copies of the next enzyme in the sequence. This greatly amplifies the signal at each step and offers opportunities to modulate the intensity of a signal.

These processes are schematically represented in Figure 1-6.

The activity that results from signal transduction may be synthesis and/or secretion of certain proteins, differentiation, or the initiation or cessation of specific functions. For example, stimulated macrophages secrete cytokines that can direct adaptive immune responses of lymphocytes against specific pathogens.

In a complementary fashion, the adaptive immune system produces signals and components that increase the effectiveness of innate responses. Some T cells, when they encounter appropriately presented antigen, synthesize and secrete cytokines that increase the ability of macrophages to kill the microbes they have ingested. Also, antibodies produced against an invader bind to the pathogen, marking it as a target for attack by macrophages or by complement proteins and serving as a potent activator of the attack.

A major difference between adaptive and innate immunity is the rapidity of the innate immune response, which utilizes a preexisting but limited repertoire of responding components. Adaptive immunity compensates for its slower onset by its ability to recognize a much wider repertoire of foreign substances and also by its ability to improve during a response, whereas innate immunity remains constant.

## Adaptive immunity is highly specific

Adaptive immunity is capable of recognizing and selectively eliminating specific foreign microorganisms and molecules (i.e., foreign antigens). Unlike innate immune responses,

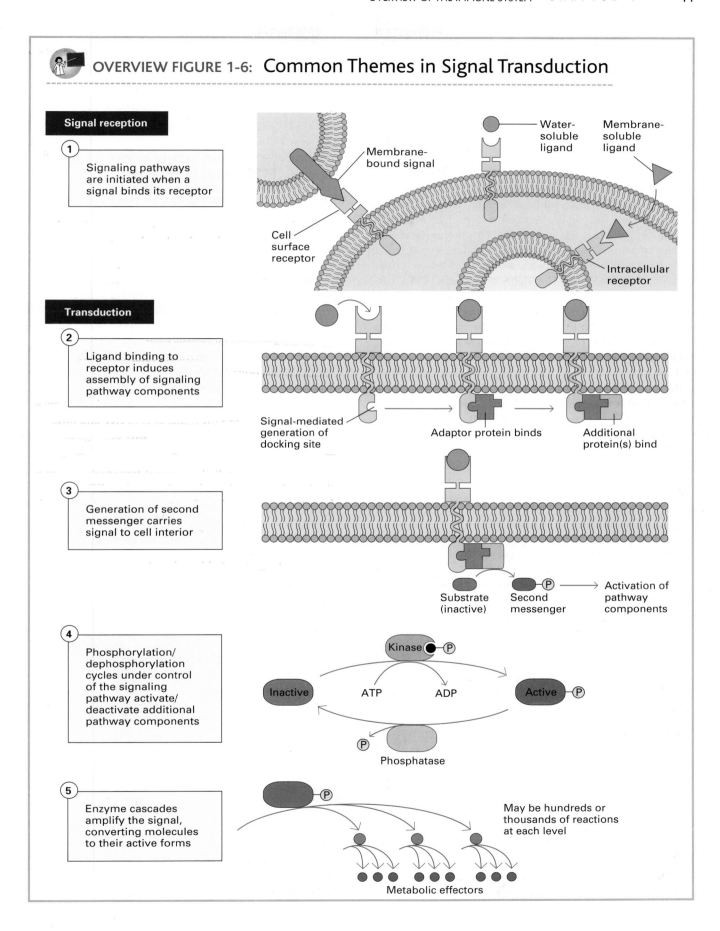

**OVERVIEW FIGURE 1-6:** Common Themes in Signal Transduction

**Signal reception**

**1** Signaling pathways are initiated when a signal binds its receptor

Membrane-bound signal

Cell surface receptor

Water-soluble ligand

Membrane-soluble ligand

Intracellular receptor

**Transduction**

**2** Ligand binding to receptor induces assembly of signaling pathway components

Signal-mediated generation of docking site

Adaptor protein binds

Additional protein(s) bind

**3** Generation of second messenger carries signal to cell interior

Substrate (inactive)

Second messenger

Activation of pathway components

**4** Phosphorylation/ dephosphorylation cycles under control of the signaling pathway activate/ deactivate additional pathway components

Kinase

ATP    ADP

Inactive

Active

Phosphatase

**5** Enzyme cascades amplify the signal, converting molecules to their active forms

May be hundreds or thousands of reactions at each level

Metabolic effectors

adaptive responses are not the same in all members of a species but are reactions to specific antigenic challenges. Adaptive immunity displays four characteristic attributes:

- Antigenic specificity
- Diversity
- Immunologic memory
- Self-nonself recognition

The **antigenic specificity** of the adaptive immune system permits it to distinguish subtle differences among antigens. Antibodies can distinguish between two protein molecules that differ in only a single amino acid. The immune system is capable of generating tremendous *diversity* in its recognition molecules, allowing it to recognize billions of unique structures on foreign antigens. This ability is in contrast to the pattern recognition molecules of the innate system, which recognize broad classes of organisms based on molecular structures present on them. The adaptive system can recognize a single type of organism and differentiate among those with minor genetic variations.

Once the adaptive immune system has recognized and responded to an antigen, it exhibits *immunologic memory;* that is, a second encounter with the same antigen induces a heightened state of immune reactivity. Because of this attribute, the immune system can confer lifelong immunity to many infectious agents after an initial encounter. Finally, the immune system normally responds only to foreign antigens, indicating that it is capable of *self-nonself recognition.* The ability of the immune system to distinguish self from nonself and respond only to nonself molecules is essential. As described below, failure of the ability to distinguish self from nonself leads to an inappropriate response to self components and can be fatal.

## Lymphocytes and antigen-presenting cells cooperate in adaptive immunity

An effective immune response involves two major groups of cells: *lymphocytes* and *antigen-presenting cells.* Lymphocytes are one of many types of white blood cells produced in the bone marrow by the process of hematopoiesis (see Chapter 2). Lymphocytes leave the bone marrow, circulate in the blood and lymphatic systems, and reside in various lymphoid organs. Because they produce and display antigen-binding cell surface receptors, lymphocytes mediate the defining immunologic attributes of specificity, diversity, memory, and self-nonself recognition. The two major populations of lymphocytes—**B lymphocytes (B cells)** and **T lymphocytes (T cells)**—are described briefly here and in greater detail in later chapters.

### B Lymphocytes

B lymphocytes mature in the bone marrow; on release, each expresses a unique antigen-binding receptor on its membrane (Figure 1-7). This antigen-binding or B-cell receptor

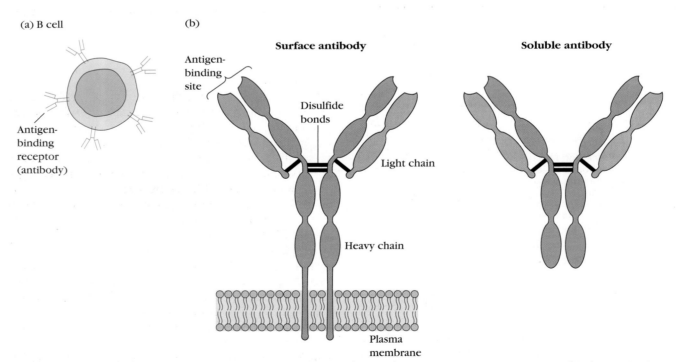

**(a) B cell**

**(b)**

Antigen-binding site

**Surface antibody**

**Soluble antibody**

Disulfide bonds

Light chain

Antigen-binding receptor (antibody)

Heavy chain

Plasma membrane

**FIGURE 1-7 B cell.** (a) The surfaces of B cells host about $10^5$ molecules of membrane-bound antibody per cell. All the antibody molecules on a given B cell have the same antigenic specificity and can interact directly with antigen. (b) Membrane-bound and soluble (secreted) antibody showing the heavy chains (blue) and light chains (red). Note that the soluble form lacks the sequences that bind the cell membrane. (The ovals in the diagrams represent characteristic protein folds called Ig domains, discussed in Chapter 4.)

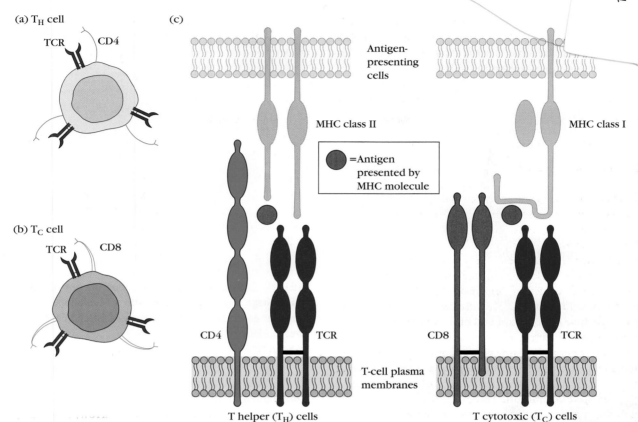

**FIGURE 1-8  T cells.** Key to the function of T cells is an antigen-binding molecule, the T-cell receptor (TCR). In general, T cells bearing CD4 (CD4⁺ cells) act as (a) helper cells and CD8⁺ cells act as (b) cytotoxic cells. (c) CD4⁺ cells recognize only antigen bound to class II MHC molecules on antigen-presenting cells. CD8⁺ cells recognize only antigen associated with class I MHC molecules. Both types of T cells express about $10^5$ identical molecules of the antigen-binding T-cell receptor per cell, all with the same antigenic specificity. (The ovals in the diagrams represent characteristic protein folds, discussed in Chapter 4.)

is a membrane-bound **antibody molecule** (Figure 1-7b, c). Antibodies are glycoproteins that consist of two identical polypeptides called the heavy chains and two shorter, identical polypeptides called the light chains. Each heavy chain is joined to a light chain by disulfide bonds, and the heavy/light chain pairs are linked together by additional disulfide bonds. The amino-terminal ends of the pairs of heavy and light chains form a site to which antigen binds. When a naive B cell (one that has not previously encountered antigen) first encounters the antigen that matches its membrane-bound antibody, the binding of the antigen to the antibody causes the cell to divide rapidly; its progeny differentiate into **memory B cells** and **effector B cells** called **plasma cells.** Memory B cells have a longer life span than naive cells, and they express the same membrane-bound antibody as their parent B cell. Plasma cells produce the antibody in a form that can be secreted (Figure 1-7b, right) and have little or no membrane-bound antibody. Although plasma cells live for only a few days, they secrete enormous amounts of antibody during this time. A single plasma cell can secrete hundreds to thousands of molecules of antibody per second. Secreted antibodies are the major effector molecules of humoral immunity.

## T Lymphocytes

T lymphocytes also arise in the bone marrow. Unlike B cells, which mature in the marrow, T cells migrate to the thymus gland to mature. Maturing T cells express a unique antigen-binding molecule, the **T-cell receptor (TCR),** on their membranes. There are two well-defined subpopulations of T cells: **T helper ($T_H$)** and **T cytotoxic ($T_C$) cells.** T helper (Figure 1-8a) and T cytotoxic (Figure 1-8b) cells can be distinguished from one another by the presence of either **CD4** or **CD8** membrane glycoproteins on their surfaces. T cells displaying CD4 generally function as $T_H$ cells, whereas those displaying CD8 generally function as $T_C$ cells (see Chapter 2). A recently characterized third type of T cell, called a **T regulatory ($T_{reg}$) cell,** carries CD4 on its surface but may be distinguished from $T_H$ and $T_C$ cells by cell surface markers associated with its stage of activation.

Unlike membrane-bound antibodies on B cells, which recognize free antigen, most T-cell receptors can recognize only antigen that is bound to cell membrane proteins called **major histocompatibility complex (MHC) molecules.** MHC molecules are polymorphic (genetically diverse) glycoproteins found on cell membranes (see Chapter 8). There are two major types of MHC molecules: class I MHC molecules, which are expressed by nearly all nucleated cells of vertebrate species, and Class II MHC molecules, which are expressed only by **antigen-presenting cells (APCs)** (Figure 1-8c). When a naive T cell encounters antigen combined with a MHC molecule on a cell, the T cell proliferates and differentiates into memory T cells and various effector T cells.

After a $T_H$ cell recognizes and interacts with an antigen–MHC class II molecule complex, the cell is activated—it undergoes a metabolic transformation and begins to secrete various cytokines. The secreted cytokines play an important role in activating B cells, $T_C$ cells, macrophages, and various other cells that participate in the immune response. Differences in the types of cytokines produced by activated $T_H$ cells result in different patterns of immune response. One possible response is the induction of a change in $T_C$ cells to form **cytotoxic T lymphocytes (CTLs)** that exhibit cell-killing or cytotoxic activity. CTLs have a vital function in monitoring the cells of the body and eliminating any that display antigen, such as virus-infected cells, tumor cells, and cells of a foreign tissue graft.

## Antigen-presenting cells interact with T cells

Activation of both the humoral and cell-mediated branches of the immune system requires cytokines produced by $T_H$ cells. It is essential that activation of $T_H$ cells themselves be carefully regulated, because a T-cell response directed against self components can have fatal autoimmune consequences. One safeguard against unregulated activation of $T_H$ cells is that the antigen receptors of $T_H$ cells can recognize only antigen that is displayed together with class II MHC molecules on the surface of antigen-presenting cells. These specialized cells, which include macrophages, B lymphocytes, and dendritic cells, are distinguished by two properties: (1) they express class II MHC molecules on their membranes, and (2) they can produce cytokines that cause $T_H$ cells to become activated.

Antigen-presenting cells first internalize antigen, by either phagocytosis or endocytosis, and then display a part of that antigen on their membrane bound to a class II MHC molecule. The $T_H$ cell interacts with the antigen–class II MHC molecule complex on the membrane of the antigen-presenting cell (Figure 1-9). The antigen-presenting cell then produces an additional signal leading to activation of the $T_H$ cell.

## Humoral and cellular immune responses exhibit different effector functions

As mentioned earlier, immune responses can be divided into humoral and cell-mediated responses. Humoral immunity refers to immunity that can be conferred on a nonimmune individual by administration of serum antibodies from an immune individual. In contrast, cell-mediated immunity can be transferred only by administration of T cells from an immune individual.

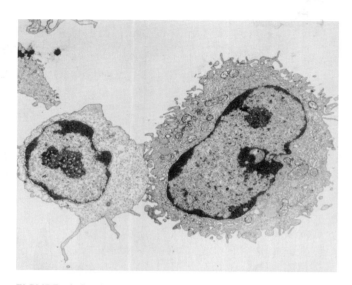

**FIGURE 1-9 Electron micrograph of an antigen-presenting macrophage *(right)* associating with a T lymphocyte.** *[From A. S. Rosenthal et al., 1982, in* Phagocytosis—Past and Future, *Academic Press, p. 239.]*

The humoral branch of the immune system is at work in the interaction of B cells with antigen and their subsequent proliferation and differentiation into antibody-secreting plasma cells (Figure 1-10). Antibody functions as the effector of the humoral response by binding to antigen and facilitating its elimination. Antigen coated with antibody is eliminated in several ways. For example, antibody can cross-link several antigens, forming clusters that are more readily ingested by phagocytic cells. Binding of antibody to antigen on a microorganism can also activate the complement system, resulting in lysis of the foreign organism. In addition, antibody can neutralize toxins or viral particles by coating them, which prevents them from binding to host cells.

Effector T cells generated in response to antigen are responsible for cell-mediated immunity (see Figure 1-10). Both activated $T_H$ cells and cytotoxic T lymphocytes serve as effector cells in cell-mediated immune reactions. Cytokines secreted by $T_H$ cells can activate various phagocytic cells, enabling them to phagocytose and kill microorganisms more effectively. This type of cell-mediated immune response is especially important in ridding the host of bacteria and protozoa contained by infected host cells. CTLs participate in cell-mediated immune reactions by killing altered self cells, including virus-infected cells and tumor cells.

## The antigen receptors of B and T lymphocytes are diverse

The antigenic specificity of each B cell is determined by the membrane-bound antigen-binding receptor (i.e., antibody) expressed by the cell. As a B cell matures in the bone marrow, its specificity is created by random rearrangements of a series of gene segments that encode the antibody molecule

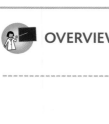

**OVERVIEW FIGURE 1-10:** The Humoral and Cell-Mediated Branches of the Immune System

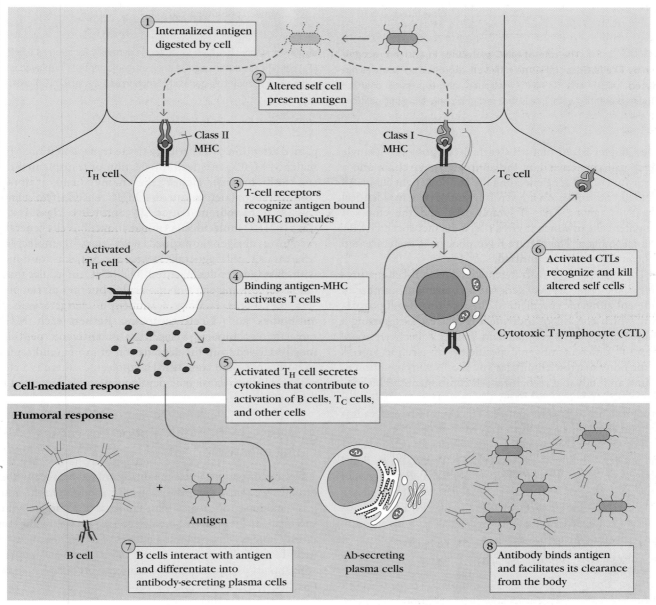

In the humoral response, B cells interact with antigen and then differentiate into antibody-secreting plasma cells. The secreted antibody binds to the antigen and facilitates its clearance from the body. In the cell-mediated response, various subpopulations of T cells recognize antigen presented on self cells. $T_H$ cells respond to antigen by producing cytokines. $T_C$ cells respond to antigen by developing into cytotoxic T lymphocytes (CTLs), which mediate killing of altered self cells (e.g., virus-infected cells).

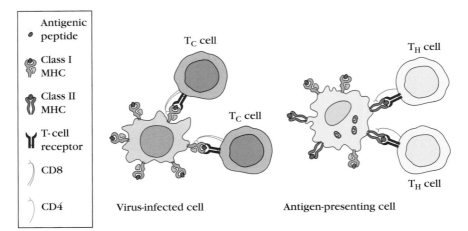

**FIGURE 1-11 The role of MHC molecules in antigen recognition by T cells.** Class I MHC molecules are expressed on nearly all nucleated cells. Class II MHC molecules are expressed only on antigen-presenting cells. T cells that recognize only antigenic peptides displayed with a class II MHC molecule generally function as T helper ($T_H$) cells. T cells that recognize only antigenic peptides displayed with a class I MHC molecule generally function as T cytotoxic ($T_C$) cells.

(see Chapter 5). At maturity, each B cell possesses a single functional gene encoding the antibody heavy chain and a single functional gene encoding the antibody light chain. All antibody molecules on a given B lymphocyte have identical specificity, giving each B lymphocyte, and the clone of daughter cells to which it gives rise, a distinct specificity for a single antigen. The mature B lymphocyte is therefore said to be **antigenically committed.**

The random gene rearrangements during B-cell maturation in the bone marrow generate an enormous number of different antigenic specificities. The resulting B-cell population, which consists of individual B cells each expressing a unique antibody, is estimated to exhibit collectively more than $10^{10}$ different antigenic specificities. A selection process in the bone marrow eliminates any B cells with membrane-bound antibody that recognizes self components. (This process is discussed in detail in Chapter 11.) The selection process helps to ensure that self-reactive antibodies (autoantibodies) are not propagated.

The attributes of specificity and diversity also characterize the antigen-binding T-cell receptor on T cells. As in B-cell maturation, the process of T-cell maturation includes random rearrangements of a series of gene segments that encode the cell's antigen-binding receptor (see Chapter 9). Each T lymphocyte cell expresses about $10^5$ receptors, all with identical specificity for antigen. The random rearrangement of the TCR genes is capable of generating on the order of $10^9$ unique antigenic specificities.

## The major histocompatibility complex molecules bind antigenic peptides

The major histocompatibility complex is a large genetic complex with multiple loci. This group of genes first gained attention as a barrier to tissue transplantation. Mismatched MHC genes lead to rejection of transplanted tissue and organs. Hence the name histo- (tissue) compatibility. Loci within the MHC encode two major classes of membrane-bound glycoproteins: **class I** and **class II MHC molecules.** As noted above, $T_H$ cells generally recognize antigen combined with class II molecules, whereas $T_C$ cells generally recognize antigen combined with class I molecules (Figure 1-11). MHC molecules function as antigen-recognition molecules, but they do not possess the fine specificity for antigen characteristic of antibodies and T-cell receptors. Rather, each MHC molecule can bind to a spectrum of **antigenic peptides** mostly derived from the degradation of protein molecules. In both class I and class II MHC molecules, there is a cleft within which the antigenic peptide sits and is presented to T lymphocytes (see Figure 1-11).

## Antigen selection of lymphocytes causes clonal expansion

A mature immunocompetent animal contains a large number of antigen-reactive clones of T and B lymphocytes; the antigenic specificity of each of these clones is determined by the specificity of the antigen-binding receptor on the membrane of the clone's lymphocytes. As noted above, the specificity of each T and B lymphocyte is determined before its contact with antigen by random gene rearrangements during maturation in the thymus or bone marrow.

The role of antigen becomes critical when it interacts with and activates mature, antigenically committed T and B lymphocytes, bringing about expansion of the population of cells with a given antigenic specificity. In this process of **clonal selection,** an antigen binds to a particular T or B cell

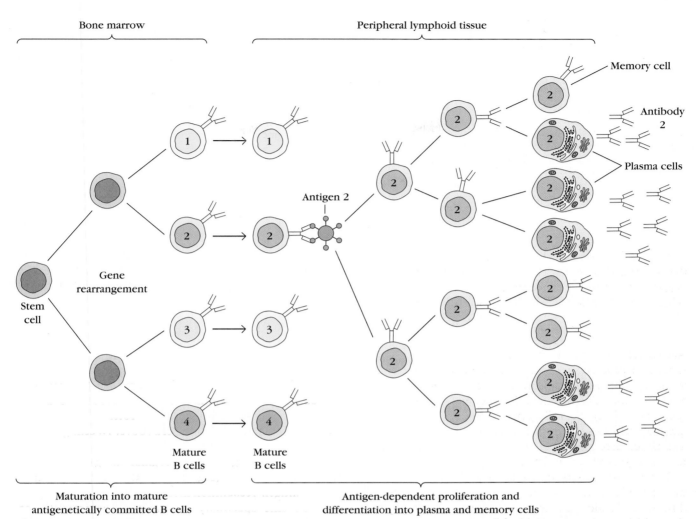

**FIGURE 1-12 Maturation and clonal selection of B lympho-cytes.** Maturation, which occurs in the absence of antigen, produces antigenically committed B cells, each of which expresses antibody with a single antigenic specificity (indicated by 1, 2, 3, and 4). Clonal selection occurs when an antigen binds to a B cell whose membrane-bound antibody molecules are specific for that antigen. Clonal expansion of an antigen-activated B cell (number 2 in this example) leads to a clone of memory B cells and effector B cells, called plasma cells; all cells in the expanded clone are specific for the original antigen. The plasma cells secrete antibody reactive with the activating antigen. Similar processes take place in the T-lymphocyte population, resulting in clones of memory T cells and effector T cells; the latter include activated $T_H$ cells, which secrete cytokines, and cytotoxic T lymphocytes (CTLs).

and stimulates it to divide repeatedly into a clone of cells with the same antigenic specificity as the original parent cell (Figure 1-12).

Clonal selection provides a framework for understanding the specificity and self-nonself recognition that is characteristic of adaptive immunity. Specificity is present because only lymphocytes whose receptors are specific for a given antigen will be clonally expanded and thus mobilized for an immune response. Self-nonself discrimination is accomplished by the elimination, during development, of lymphocytes bearing self-reactive receptors or by the functional suppression of these cells if they reach maturity.

Immunologic memory is another consequence of clonal selection. During clonal selection, the number of lymphocytes specific for a given antigen is greatly amplified. Moreover, many of these lymphocytes, referred to as memory cells, have a longer life span than the naive lymphocytes from which they arise. The initial encounter of a naive immunocompetent lymphocyte with an antigen induces a **primary response;** a later contact of the host with antigen will induce a more rapid and heightened **secondary response.** The amplified population of memory cells accounts for the rapidity and intensity that distinguishes a secondary response from the primary response.

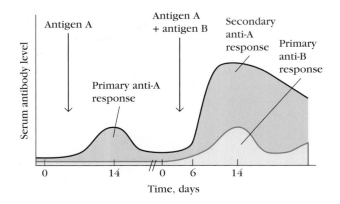

**FIGURE 1-13 Differences in the primary and secondary response to injected antigen (humoral response) reflect the phenomenon of immunologic memory.** When an animal is injected with an antigen, it produces a primary serum antibody response of low magnitude and short duration, peaking at about 10 to 17 days. A second immunization with the same antigen results in a secondary response that is greater in magnitude, peaks in less time (2–7 days), and lasts longer (months to years) than the primary response. Compare the secondary response to antigen A with the primary response to antigen B administered to the same mice (light blue shading).

In the humoral branch of the immune system, antigen induces the clonal proliferation of B lymphocytes into antibody-secreting plasma cells and memory B cells. As seen in Figure 1-13, the primary response has a lag of approximately 5 to 7 days before antibody levels start to rise. This lag is the time required for activation of naive B cells by antigen and $T_H$ cells and for the subsequent proliferation and differentiation of the activated B cells into plasma cells. Antibody levels peak in the primary response at about day 14 and then begin to drop off as the plasma cells begin to die. In the secondary response, the lag is much shorter (only 1–2 days), antibody levels are much higher, and they are sustained for much longer. The secondary response reflects the activity of the clonally expanded population of memory B cells. These memory cells respond to the antigen more rapidly than naive B cells; in addition, because there are many more memory cells than there were naive B cells for the primary response, more plasma cells are generated in the secondary response, and antibody levels are consequently 100- to 1000-fold higher.

In the cell-mediated branch of the immune system, the recognition of an antigen-MHC complex by a specific mature T lymphocyte induces clonal proliferation into various T cells with effector functions ($T_H$ cells and CTLs) and into memory T cells. As with humoral immune responses, secondary cell-mediated responses are faster and stronger than primary.

## Immune Dysfunction and Its Consequences

The above overview of innate and adaptive immunity depicts a multicomponent interactive system that protects the host from invasion by pathogens that cause infectious diseases and from altered cells that can reproduce in an uncontrolled manner to cause cancer. This overview would not be complete without mentioning that the immune system can, and does at times, function improperly. Sometimes the immune system fails to protect the host adequately because of a deficiency, and sometimes it overreacts or misdirects its activities to cause discomfort, debilitating disease, or even death. There are several common manifestations of immune dysfunction:

- Allergies and asthma
- Graft rejection and graft-versus-host disease
- Autoimmune disease
- Immunodeficiency

Allergies and asthma are results of inappropriate immune responses, often to common antigens such as plant pollen, food, or animal dander. The possibility that certain substances induced increased sensitivity (hypersensitivity) rather than protection was recognized in about 1902 by Charles Richet, who attempted to immunize dogs against the toxins of a type of jellyfish, *Physalia*. He and his colleague Paul Portier observed that dogs exposed to sublethal doses of the toxin reacted almost instantly, and fatally, to subsequent challenge with minute amounts of the toxin. Richet concluded that a successful immunization or vaccination results in *phylaxis,* or protection, whereas the opposite result may occur—**anaphylaxis**—in which exposure to antigen can result in a potentially lethal sensitivity to the antigen if the exposure is repeated. Richet received the Nobel prize in 1913 for his discovery of the anaphylactic response.

Fortunately, most allergic reactions in humans are not rapidly fatal. A specific allergic or anaphylactic response usually involves a type of antibody called IgE (for *immunoglobulin E*). Binding of IgE to its specific antigen (allergen) releases substances that cause irritation and inflammation. When an allergic individual is exposed to an allergen, symptoms may include sneezing, wheezing, and difficulty in breathing (asthma); dermatitis or skin eruptions (hives); and, in more extreme cases, strangulation due to blockage of airways by inflammation (Figure 1-14). A significant fraction of our health resources is expended to care for those suffering from allergies and asthma. The frequency of allergy and asthma in the United States places these complaints among the most common reasons for a visit to the doctor's office

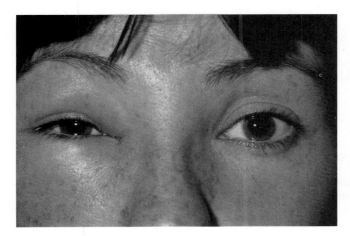

**FIGURE 1-14  Patient suffering swelling of the right eye from effects of allergic reaction to a bee sting.** Such hypersensitivity reactions result from sensitization caused by previous exposure to the bee venom. Bee stings may cause pain, redness, and swelling as shown here or may cause systemic anaphylactic reactions that result in death if not quickly treated. *[Dr. P. Marazzi/Photo Researchers.]*

or to the hospital emergency room (see Clinical Focus on page 20).

When the immune system encounters foreign cells or tissue, it responds strongly to rid the host of the invader. The same response may be raised against mutant host cells, including cancer cells. However, in some cases, the transplantation of cells or an organ from another individual, although viewed by the immune system as a foreign invasion, may be the only possible treatment for life-threatening disease. For example, it is estimated that more than 70,000 persons in the United States alone would benefit from a kidney transplant. The fact that the immune system will attack and reject any transplanted organ that it recognizes as nonself raises a formidable barrier to this potentially lifesaving treatment. An additional danger in transplantation is that any transplanted cells with immune function (for example, when bone marrow is transplanted to restore immune function) may view the new host as nonself and react against it. This reaction, which is termed graft-versus-host disease, can be fatal. The rejection reaction and graft-versus-host disease can be suppressed by drugs, but treatment with these drugs suppresses general immune function, so that the host is no longer protected by its own immune system and becomes susceptible to infectious diseases. Transplantation studies have played a major role in the development of immunology. A Nobel prize was awarded to Karl Landsteiner (mentioned above for his contributions to the concept of immune specificity) in 1930 for the discovery of the human ABO blood groups, a finding that allowed blood transfusions to be carried out safely. In 1980, G. Snell, J. Dausset, and B. Benacerraf were recognized for discovery of the major histocompatibility complex, and in 1991, E. D. Thomas and

J. Murray were awarded Nobel prizes for advances in transplantation immunity. Development of procedures that would allow a foreign organ to be accepted without suppressing immunity to all antigens remains a challenge for immunologists today.

In certain individuals, the immune system malfunctions by losing its sense of self and nonself, which permits an immune attack on the host. This condition, **autoimmunity,** can cause a number of chronic debilitating diseases. The symptoms of autoimmunity differ, depending on which tissues and organs are under attack. For example, multiple sclerosis is due to an autoimmune attack on a protein in nerve sheaths in the brain and central nervous system, Crohn's disease is an attack on intestinal tissues, and rheumatoid arthritis is an attack on joints of the hands, feet, arms, and legs. The genetic and environmental factors that trigger and sustain autoimmune disease are very active areas of immunologic research, as is the search for improved treatments.

If any components of innate or specific immunity are defective because of genetic abnormality or if any immune function is lost because of damage by chemical, physical, or biological agents, the host suffers from **immunodeficiency.** The severity of the immunodeficiency disease depends on the number of affected components. A common type of immunodeficiency in North America is a selective immunodeficiency in which only one type of immunoglobulin, IgA, is lacking; the symptoms may be an increase in certain types of infections or the deficiency may even go unnoticed. In contrast, a rarer immunodeficiency called **severe combined immunodeficiency (SCID),** which affects both B and T cells, may result in death from infection at an early age if untreated. Since the 1980s, the most common form of immunodeficiency has been acquired immune deficiency syndrome, or AIDS, which results from infection with the retrovirus human immunodeficiency virus, or HIV. In AIDS, T helper cells are destroyed by HIV, causing a collapse of the immune system. It is estimated that 40 million people worldwide suffer from this disease, which if not treated is usually fatal within 8 to 10 years after infection. Although certain treatments can now prolong the life of AIDS patients, there is no known cure for the disease.

This chapter has been a brief introduction to the immune system, and it has given a thumbnail sketch of how this complex system functions to protect the host from disease. The following chapters will examine the structure and function of the individual cells, organs, and molecules that make up this system. They will describe our current understanding of how the components of immunity interact and the experiments that allowed discovery of these mechanisms. Specific areas of applied immunology, such as immunity to infectious diseases, cancer, current vaccination practices, and the major types of immune dysfunction are the subject of later chapters.

## CLINICAL FOCUS
## Allergies and Asthma Are Serious Public Health Problems

**Although** the immune system serves to protect the host from infection and cancer, inappropriate responses of this system can lead to disease. Common among the results of immune dysfunction are allergies and asthma, both serious public health problems. Details of the mechanisms that underlie allergic and asthmatic responses to environmental antigens (or allergens) will be considered in Chapter 15. Simply stated, allergic reactions are responses to antigenic stimuli that result in immunity based mainly on the IgE class of immunoglobulin. Exposure to the antigen (or allergen) triggers an IgE-mediated release of molecules that cause symptoms ranging from sneezing and dermatitis to inflammation of the lungs in an asthmatic attack. The sequence of events in an allergic response is depicted in the accompanying figure.

The discomfort from common allergies such as plant pollen allergy (often called ragweed allergy) is short-lived, consisting of sneezing and runny nose—trivial in comparison to the effects of cancer, cardiac arrest, or life-threatening infections. A more serious allergic reaction is asthma, a chronic disease of the lungs in which inflammation, mediated by environmental antigens or infections, causes severe difficulty in breathing. According to 2002 statistics from the Centers for Disease Control, 20 million people suffer from asthma in the United States, and 12 million per year experience an asthma attack. About 5000 people die each year from asthma. In the past 20 years, the prevalence of asthma in the Western world has doubled.*

The importance of allergy as a public health problem is underscored by the fact that the annual numbers of doctor visits for hypertension, routine medical examinations, or normal pregnancy are each fewer than the number of visits for allergic conditions. In fact, the most common reason for a trip to a hospital emergency room is an asthma attack, accounting for one third of

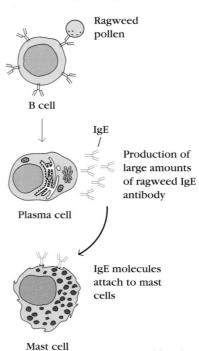

**First contact with an allergen (ragweed)**

Ragweed pollen

B cell

IgE

Production of large amounts of ragweed IgE antibody

Plasma cell

IgE molecules attach to mast cells

Mast cell

**Subsequent contact with allergen**

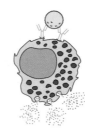

IgE-primed mast cell releases molecules that cause wheezing, sneezing, runny nose, watery eyes, and other symptoms

Sequence of events leading to an allergic response. When the antibody produced on contact with an allergen is IgE, this class of antibody reacts with a mast cell. Subsequent reaction of the antibody-binding site with the allergen triggers the mast cell to which the IgE is bound to secrete molecules that cause the allergic symptoms.

all visits. In addition to those treated in the ER, there were about 160,000 hospitalizations for asthma in the past year, with an average stay of 3 to 4 days.

Although all ages and races are affected, deaths from asthma are 3.5 times more common among African American children. The reasons for the increases in number of asthma cases and for the higher death rate in African American children remain unknown, although recent studies of genetic factors in allergic disease may have uncovered some clues (see Clinical Focus in Chapter 15).

An increasingly serious health problem is food allergy, especially to peanuts and tree nuts (almonds, cashews, and walnuts).[†] Approximately 3 million Americans are allergic to these foods, and they are the leading causes of fatal and near-fatal food allergic (anaphylactic) reactions. Although avoiding these foods can prevent harmful consequences, the ubiquitous use of peanut protein and other nut products in a variety of foods makes avoidance very difficult for the allergic individual. At least 50% of serious reactions are caused by accidental exposures to peanuts, tree nuts, or their products. This has led to controversial movements to ban peanuts from schools and airplanes.

Anaphylaxis generally occurs within an hour of ingesting the food allergen, and the most effective treatment is injection of the drug epinephrine. Those prone to anaphylactic attacks often carry injectable epinephrine to be used in case of exposure.

In addition to the suffering and anxiety caused by inappropriate immune responses or allergies to environmental antigens, there is a staggering cost—estimated to be almost $20 billion—in terms of lost work time for those affected and for caregivers. These costs well justify the extensive efforts by basic and clinical immunologists and allergists to relieve the suffering caused by these disorders.

*Holgate, S. T. 1999. The epidemic of allergy and asthma, *Nature* supp. to vol. 402, B2.
†Hughes, D. A., and C. Mills. 2001. Food allergy: A problem on the rise. *Biologist* (London) **48**:201.

## SUMMARY

- Immunity is the state of protection against foreign organisms or substances (antigens). Vertebrates have two types of immunity: innate and adaptive.

- Innate immunity constitutes a first line of defense, which includes barriers, phagocytic cells, and molecules that recognize certain classes of pathogens.

- Innate and adaptive immunity operate cooperatively; activation of the innate immune response produces signals that stimulate a subsequent adaptive immune response.

- Adaptive immune responses exhibit four immunologic attributes: specificity, diversity, memory, and self-nonself recognition.

- The high degree of specificity in adaptive immunity resides in molecules (antibodies and T-cell receptors) that recognize and bind specific antigens.

- Antibodies recognize and interact directly with antigen. T-cell receptors recognize only antigen combined with major histocompatibility complex (MHC) molecules.

- The two major subpopulations of T lymphocytes are the $CD4^+$ T helper ($T_H$) cells and $CD8^+$ T cytotoxic ($T_C$) cells, which give rise to cytotoxic T cells (CTLs).

- Dysfunctions of the immune system include common maladies such as allergies and asthma as well as immunodeficiency and autoimmunity.

## References

Burnet, F. M. 1959. *The Clonal Selection Theory of Acquired Immunity.* Cambridge University Press, Cambridge, MA.

Cohen, S. G., and M. Samter. 1992. *Excerpts from Classics in Allergy.* Symposia Foundation, Carlsbad, CA.

Desour, L. 1922. *Pasteur and His Work* (translated by A. F. and B. H. Wedd). T. Fisher Unwin, London.

Kimbrell, D. A., and B. Beutler. 2001. The evolution and genetics of innate immunity. *Nature Reviews Genetics* **2**:256.

Kindt, T. J., and J. D. Capra. 1984. *The Antibody Enigma.* Plenum Press, New York.

Landsteiner, K. 1947. *The Specificity of Serologic Reactions.* Harvard University Press, Cambridge, MA.

Medawar, P. B. 1958. *The Immunology of Transplantation: The Harvey Lectures, 1956–1957.* Academic Press, New York.

Metchnikoff, E. 1905. *Immunity in the Infectious Diseases.* Macmillan, New York.

O'Neill, A. J. 2005. Immunity's early warning system. *Scientific American* **292**:38.

Paul, W., ed. 2003. *Fundamental Immunology,* 5th ed. Lippincott Williams & Wilkins, Philadelphia.

Roitt, I. M., and P. J. Delves, eds. 1998. *An Encyclopedia of Immunology,* 2nd ed., vols. 1–4. Academic Press, London.

## Useful Web Sites

**http://www.aaaai.org/**

The American Academy of Allergy Asthma and Immunology site includes an extensive library of information about allergic diseases.

**http://www.aai.org**

The Web site of the American Association of Immunologists contains a good deal of information of interest to immunologists.

**http://www.ncbi.nlm.nih.gov/PubMed/**

PubMed, the National Library of Medicine database of more than 9 million publications, is the world's most comprehensive bibliographic database for biological and biomedical literature. It is also a highly user-friendly site.

## Study Questions

**CLINICAL FOCUS QUESTION** You have a young nephew who has developed a severe allergy to tree nuts. What precautions would you advise for him and for his parents? Should school officials be aware of this condition?

1. Why was Jenner's vaccine superior to previous methods for conferring resistance to smallpox?

2. Did the treatment for rabies used by Pasteur confer active or passive immunity to the rabies virus? Is there any way to test this?

3. Infants immediately after birth are often at risk for infection with group B streptococcus. A vaccine is proposed for administration to women of childbearing years. How can immunizing the mothers help the babies?

4. Indicate to which branch(es) of the immune system the following statements apply, using H for the humoral branch and CM for the cell-mediated branch. Some statements may apply to both branches.

   a. Involves class I MHC molecules
   b. Responds to viral infection
   c. Involves T helper cells
   d. Involves processed antigen
   e. Responds following an organ transplant
   f. Involves T cytotoxic cells
   g. Involves B cells
   h. Involves T cells
   i. Responds to extracellular bacterial infection
   j. Involves secreted antibody
   k. Kills virus-infected self cells

5. Adaptive immunity exhibits four characteristic attributes, which are mediated by lymphocytes. List these four attributes and briefly explain how they arise.

6. Name three features of a secondary immune response that distinguish it from a primary immune response.

7. Compare and contrast the four types of antigen-binding molecules used by the immune system—antibodies, T-cell

receptors, class I MHC molecules, and class II MHC molecules—in terms of the following characteristics:

a. Specificity for antigen
b. Cellular expression
c. Types of antigen recognized

8. Fill in the blanks in the following statements with the most appropriate terms:

a. _____, _____, and_____ all function as antigen-presenting cells.
b. Antigen-presenting cells deliver a _____ signal to _____ cells.
c. Only antigen-presenting cells express class _____ molecules, whereas nearly all cells express class _____ MHC molecules.
d. The scientific term that refers to white blood cells in general is _____.
e. The _____ arm of the immune system is so called because antibodies are generated in response to specific pathogens. Prior exposure to a pathogen is required for this part of the immune system to develop.
f. T cells must have coreceptors so they can efficiently bind to MHC molecules. The coreceptor for recognition of class I MHC is _____ and the coreceptor for recognition of class II is called _____.
g. The part of the antigen bound by an antibody is known as the _____.

9. The T cell is said to be class I restricted. What does this mean?

10. Innate and adaptive immunity act in cooperative and interdependent ways to protect the host. Discuss the collaboration of these two forms of immunity.

11. Give examples of mild and severe consequences of immune dysfunction. What is the most common cause of immunodeficiency throughout the world today?

12. Which of the following statements about how B and T cells recognize antigen are true?

a. B cells only recognize antigen presented by class I or class II MHC molecules.

b. Both cell types can recognize antigen in solution (without cells).
c. Both cell types recognize extracellular matrix-bound antigens.
d. T cells only recognize antigen presented by class I or class II MHC molecules.

13. For each of the following statements, indicate whether the statement is true or false. If you think the statement is false, explain why.

a. Booster shots are required because repeated exposure to an antigen builds a stronger immune response.
b. The gene for the T cell receptor must be cut and spliced together, deleting entire sections, before it can be transcribed.
c. Our bodies face the greatest onslaught from foreign invaders through our mucus membranes.
d. Increased production of antibody in the immune system is driven by the presence of antigen.
e. Antigen is bound directly by T cells.
f. Peptides are added to the binding cleft of class MHC I molecules in the cytosol.
g. In order for B cells to mature into plasma cells they need "help" from T cells.

14. Match the cell type with the receptor found on that cell.

| Cell Type | Receptor |
| --- | --- |
| a. Antigen presenting cell | 1. CD8$^+$ |
| b. B cell | 2. MHC |
| c. Helper T cell | 3. BCR |
| d. Cytotoxic T cell | 4. CD4$^+$ |

 ## Interactive Study

 **www.whfreeman.com/kuby**

SELF-TEST
Review of Key Terms

# chapter 2

# Cells and Organs of the Immune System

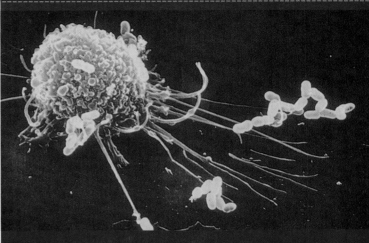

*Scanning electron micrograph of a macrophage. [L. Nilsson, © Boehringer Ingelheim International GmbH.]*

- Hematopoiesis
- Cells of the Immune System
- Organs of the Immune System
- Lymphoid Cells and Organs—Evolutionary Comparisons

THE MANY CELLS, ORGANS, AND TISSUES OF THE IMMUNE system are found throughout the body. They can be classified functionally into two main groups. The **primary lymphoid organs** provide appropriate microenvironments for the development and maturation of lymphocytes. The **secondary lymphoid organs** trap antigen, generally from nearby tissues or vascular spaces and are sites where mature lymphocytes can interact effectively with antigen. Blood vessels and lymphatic systems connect these organs, uniting them into a functional whole.

Carried within the blood and lymph and populating the lymphoid organs are various white blood cells, or **leukocytes,** that participate in the immune response. Of these cells, only the antigen-specific lymphocytes possess the attributes of diversity, specificity, memory, and self-nonself recognition, the hallmarks of an adaptive immune response. Other leukocytes also play important roles, some as antigen-presenting cells and others participating as effector cells in the elimination of antigen by phagocytosis or the secretion of immune effector molecules. Some leukocytes, especially T lymphocytes, secrete various protein molecules called cytokines. These molecules act as immunoregulatory hormones and play important roles in the coordination and regulation of immune responses. Here we describe the maturation of blood cells, the properties of the various immune system cells, and the functions of the lymphoid organs.

## Hematopoiesis

All blood cells arise from a type of cell called the **hematopoietic stem cell (HSC). Stem cells** are cells that can differentiate into other cell types. They are self-renewing, maintaining their population level by cell division. In humans, **hematopoiesis,** the formation and development of red and white blood cells, begins in the embryonic yolk sac during the first weeks of development. Yolk sac stem cells differentiate into primitive erythroid cells that contain embryonic hemoglobin. By the third month of gestation, hematopoietic stem cells have migrated from the yolk sac to the fetal liver

and subsequently colonize the spleen; these two organs have major roles in hematopoiesis from the third to the seventh months of gestation. After that, the differentiation of HSCs in the bone marrow becomes the major factor in hematopoiesis, and by birth there is little or no hematopoiesis in the liver and spleen. The migration of hematopoiesis during development from a series of early sites to bone marrow is shown in Figure 2-1.

It is remarkable that every functionally specialized, mature blood cell arises from an HSC. However, the study of hematopoietic stem cells is difficult for two reasons. They are scarce, normally fewer than one HSC per $5 \times 10^4$ cells in the bone marrow, and they are difficult to grow in vitro. As a result, the knowledge we have about how their proliferation and differentiation is regulated remains incomplete. By virtue of their capacity for self-renewal, hematopoietic stem cells are maintained at stable levels throughout adult life; however, when there is an increased demand for hematopoiesis, HSCs display an enormous proliferative capacity. This can be demonstrated in mice whose hematopoietic systems have been completely destroyed by a lethal dose of x-rays (950 rads[1]).

[1]One rad represents the absorption by an irradiated target of an amount of radiation corresponding to 100 ergs/gram of target.

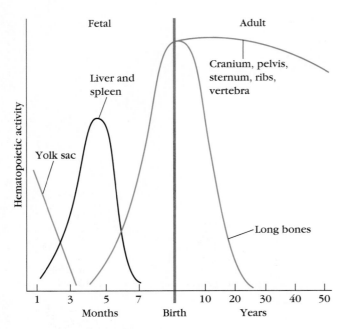

**FIGURE 2-1 Sites of hematopoiesis at various times during prenatal and postnatal human development.**

Such irradiated mice die within 10 days unless they are infused with normal bone marrow cells from a syngeneic (genetically identical) mouse. Although a normal mouse has $3 \times 10^8$ bone marrow cells, infusion of only $10^4$ to $10^5$ bone marrow cells (i.e., 0.01%–0.1% of the normal amount) from a donor is sufficient to completely restore the hematopoietic system, which demonstrates the enormous capacity of HSCs for self-renewal.

Early in hematopoiesis, a multipotent stem cell differentiates along one of two pathways, giving rise to either a **lymphoid progenitor cell** or a **myeloid progenitor cell** (Figure 2-2). **Progenitor cells** have lost the capacity for self-renewal and are committed to a particular cell lineage. Lymphoid progenitor cells give rise to B, T, and NK (natural killer) cells. Myeloid stem cells generate progenitors of red blood cells (erythrocytes), many of the various white blood cells (neutrophils, eosinophils, basophils, monocytes, mast cells, dendritic cells), and platelet-generating cells called megakaryocytes. In bone marrow, hematopoietic cells and their descendants grow, differentiate, and mature on a mesh-like scaffold of **stromal cells,** which include fat cells, endothelial cells, fibroblasts, and macrophages. Stromal cells influence the differentiation of hematopoietic stem cells by providing a **hematopoietic-inducing microenvironment (HIM)** consisting of a cellular matrix and factors that promote growth and differentiation. Many of these hematopoietic growth factors are soluble agents that arrive at their target cells by diffusion, whereas others are membrane-bound molecules on the surface of stromal cells that require cell-to-cell contact between the responding cells and the stromal cells.

During hematopoiesis, erythrocytes and many different kinds of white blood cells descend from the few hematopoietic stem cells in a complicated process that involves many steps traversing a hierarchy of precursor populations. Why has such a complex process evolved to generate blood cells? Over a lifetime, a person will produce about $10^{16}$ blood cells. This requires a great many cell divisions. John Dick has pointed out that cell division is error prone and provides an opportunity for the genome to suffer mutations, some of which may cause cancer. He has suggested that to make this potentially catastrophic event less likely, the blood-forming system is organized as an ingenious hierarchy in which most of the proliferation takes place within more differentiated precursors rather than in the hematopoietic stem cell population itself. These progressively differentiating precursors are not self-renewing, and the mature blood cells they form are either incapable of division or only divide under special circumstances. Consequently, the chance of generating cancer in HSCs and their immediate descendants is lowered, even if it is not reduced to zero.

## Hematopoiesis is regulated at the genetic level

The development of pluripotent hematopoietic stem cells into different cell types requires the expression of different sets of lineage-determining and lineage-specific genes at appropriate times and in the correct order. The proteins specified by these genes are critical components of regulatory networks that direct the differentiation of the stem cell and its descendants. Much of what we know about the dependence of hematopoiesis on a particular gene comes from studies of mice in which a gene has been inactivated or "knocked out" by targeted disruption, which blocks the production of the protein that it encodes (see Chapter 22). If mice fail to produce red cells or particular white blood cells when a gene is knocked out, we conclude that the protein specified by the gene is necessary for development of those cells. Knockout technology is one of the most powerful tools available for determining the roles of particular genes in a broad range of processes, and it has made important contributions to the identification of many genes that regulate hematopoiesis.

Although much remains to be learned, targeted disruption and other approaches have identified a number of transcription factors that play important roles in hematopoiesis (Table 2-1). Some of these transcription factors affect many different hematopoietic lineages, and others affect only a single lineage, such as the developmental pathway that leads to lymphocytes. One transcription factor that affects multiple lineages is GATA-2, a member of a family of transcription factors that recognize the tetranucleotide sequence GATA, a motif found in target genes. A functional **GATA-2 gene,** which specifies this transcription factor, is essential for the development of the lymphoid, erythroid, and myeloid lineages. As might be expected, animals in

## OVERVIEW FIGURE 2-2:  Hematopoiesis

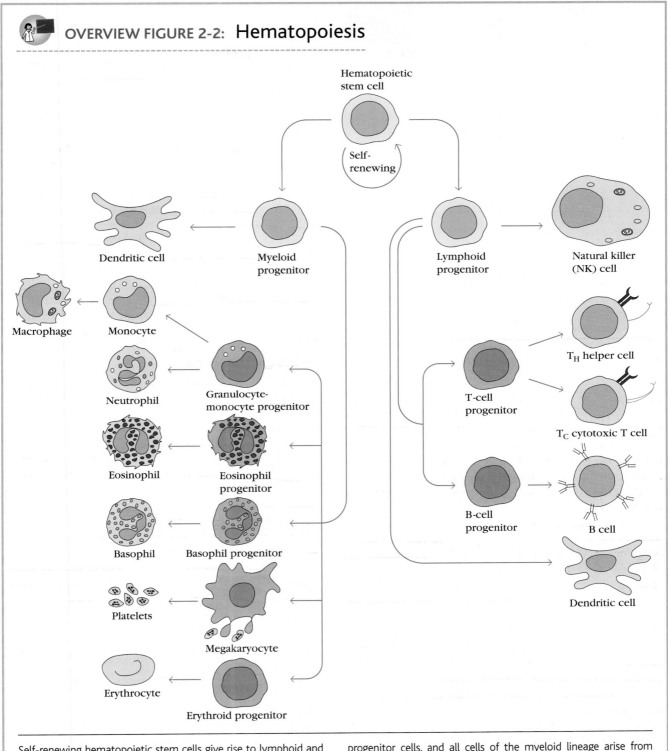

Self-renewing hematopoietic stem cells give rise to lymphoid and myeloid progenitors. All lymphoid cells descend from lymphoid progenitor cells, and all cells of the myeloid lineage arise from myeloid progenitors.

| TABLE 2-1 | Some transcription factors essential for hematopoietic lineages |
| --- | --- |
| Factor | Dependent lineage |
| GATA-1 | Erythroid |
| GATA-2 | Erythroid, myeloid, lymphoid |
| PU.1 | Erythroid (maturational stages), myeloid (later stages), lymphoid |
| Bmi-1 | All hematopoietic lineages |
| Ikaros | Lymphoid |
| Oct-2 | B lymphoid (differentiation of B cells into plasma cells) |

which this gene is disrupted die during embryonic development. In contrast to GATA-2, another transcription factor, **Ikaros,** is required only for the development of cells of the lymphoid lineage. Although Ikaros knockout mice do not produce significant numbers of B, T, and NK cells, their production of erythrocytes, granulocytes, and other cells of the myeloid lineage is unimpaired. Ikaros knockout mice survive embryonic development, but they are severely compromised immunologically and die of infections at an early age. Yet another transcriptional regulator, Bmi-1, is a transcriptional repressor that is a key determinant of the ability of HSCs to self-renew. When the gene for Bmi-1, which is highly expressed in HSCs of humans and mice, is knocked out, the Bmi-1 deficient mice die within two months of birth. The cause of death is the eventual failure of the bone marrow to generate red and white blood cells. This bone marrow failure was traced to a lack of self-renewal by HSCs.

## Hematopoietic homeostasis involves many factors

Hematopoiesis is a steady-state process in which mature blood cells are produced at the same rate at which they are lost. (The principal cause of blood cell loss is aging.) The average erythrocyte has a life span of 120 days before it is phagocytosed and digested by macrophages in the spleen. The various white blood cells have life spans ranging from a day, for neutrophils, to as long as 20 to 30 years for some T lymphocytes. To maintain steady-state levels, the average human being must produce an estimated $3.7 \times 10^{11}$ white blood cells per day. This massive system is regulated by complex mechanisms that affect all of the individual cell types, and ultimately, the number of cells in any hematopoietic lineage is set by a balance between the number of cells removed by cell death and the number that arise from division and differentiation. Any one or a combination of regulatory factors can affect rates of cell reproduction and differentiation. These factors can also determine whether a hematopoietic cell is induced to die.

## Programmed cell death is an essential homeostatic mechanism

**Programmed cell death,** an induced and ordered process in which the cell actively participates in bringing about its own demise, is a critical factor in the homeostatic regulation of many types of cell populations, including those of the hematopoietic system.

Cells undergoing programmed cell death often exhibit distinctive morphologic changes, collectively referred to as **apoptosis** (Figures 2-3, 2-4). These changes include a pronounced decrease in cell volume; modification of the cytoskeleton, which results in membrane blebbing; a condensation of the chromatin; and degradation of the DNA into fragments. Following these morphologic changes, an apoptotic cell sheds tiny membrane-bound apoptotic bodies containing intact organelles. Macrophages phagocytose apoptotic bodies and cells in the advanced stages of apoptosis. This ensures that their intracellular contents, including proteolytic and other lytic enzymes, cationic proteins, and oxidizing molecules, are not released into the surrounding tissue. As a consequence, apoptosis does not induce a local inflammatory response. Apoptosis differs markedly from **necrosis,** the changes associated with cell death arising from injury. In necrosis, injured cells swell and burst, releasing their contents and possibly triggering a damaging inflammatory response.

Each of the leukocytes produced by hematopoiesis has a characteristic life span and then dies by programmed cell death. In the adult human, for example, there are about $5 \times 10^{10}$ neutrophils in the circulation. These cells have a life span of only a day before programmed cell death is initiated. A stable number of these cells is maintained by constant neutrophil production. Programmed cell death also plays a role in maintaining proper numbers of hematopoietic progenitor cells for erythrocytes and various types of leukocytes. Beyond hematopoiesis, apoptosis is important in such immunological processes as tolerance and the killing of target cells by cytotoxic T cells or natural killer cells. Details of the mechanisms underlying apoptosis are emerging; they are fully discussed in Chapters 10 and 14.

The expression of several genes accompanies apoptosis in leukocytes and other cell types (Table 2-2). Some of the proteins specified by these genes induce apoptosis, others are critical as apoptosis progresses, and still others inhibit apoptosis. For example, apoptosis can be induced in thymocytes by radiation, but only if the protein p53 is present; many cell deaths are induced by signals from Fas, a molecule present on the surface of many cells; and proteases known as caspases take part in a cascade of reactions that lead to apoptosis. On the other hand, members of the bcl-2 (B-cell lymphoma 2) family of genes, bcl-2 and bcl-$X_L$, encode protein products that inhibit apoptosis. Interestingly, the first member of this gene family to be discovered, bcl-2, was found in studies that were concerned not with cell death but with the uncontrolled proliferation of B cells in a type of cancer

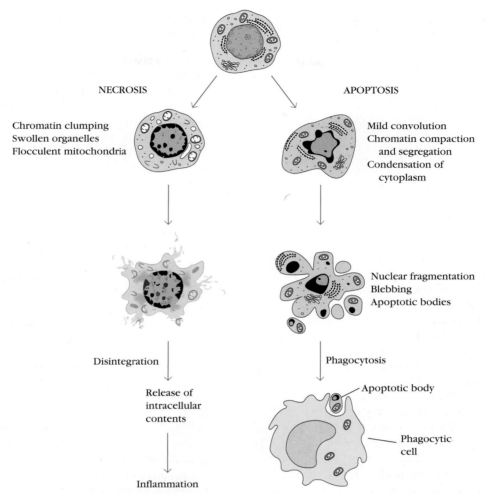

**FIGURE 2-3  Comparison of morphologic changes that occur in apoptosis and necrosis.**  Apoptosis, which results in the programmed cell death of hematopoietic cells, does not induce a local inflammatory response. In contrast, necrosis, the process that leads to death of injured cells, results in release of the cells' contents, which may induce a local inflammatory response.

known as B-cell lymphoma. In this case, the *bcl-2* gene was near the breakpoint of a chromosomal translocation in a human B-cell lymphoma. The translocation moved the *bcl-2* gene into the immunoglobulin heavy-chain locus, resulting in transcription of *bcl-2* along with the immunoglobulin gene, with consequent overproduction of the encoded Bcl-2 protein by the lymphoma cells. The resulting high levels of Bcl-2 are thought to help transform lymphoid cells into cancerous lymphoma cells by inhibiting the signals that would normally induce apoptotic cell death.

Bcl-2 levels have been found to play an important role in regulating the normal life span of various hematopoietic cell lineages, including lymphocytes. A normal adult has about five liters of blood, with about 2000 lymphocytes/mm$^3$, for a total of about $10^{11}$ circulating lymphocytes. During acute infection, the lymphocyte count increases fourfold or more, producing a circulating lymphocyte level of over $4 \times 10^{11}$. The immune system cannot sustain such a massive increase in cell numbers for an extended period, so the system must

eliminate unneeded activated lymphocytes once the antigenic threat has passed. Activated lymphocytes have been found to express lower levels of Bcl-2 and therefore are more susceptible to the induction of apoptotic death than are naive lymphocytes or memory cells. However, if the lymphocytes

| TABLE 2-2 | Genes that regulate apoptosis | |
|---|---|---|
| **Gene** | **Function** | **Role in apoptosis** |
| *bcl-2* | Prevents apoptosis | Inhibits |
| *bax* | Opposes *bcl-2* | Promotes |
| *bcl-X*$_L$ *(bcl-Long)* | Prevents apoptosis | Inhibits |
| *bcl-X*$_S$ *(bcl-Short)* | Opposes *bcl-X*$_L$ | Promotes |
| *caspase* (several different ones) | Protease | Promotes |
| *Fas* | Induces apoptosis | Initiates |

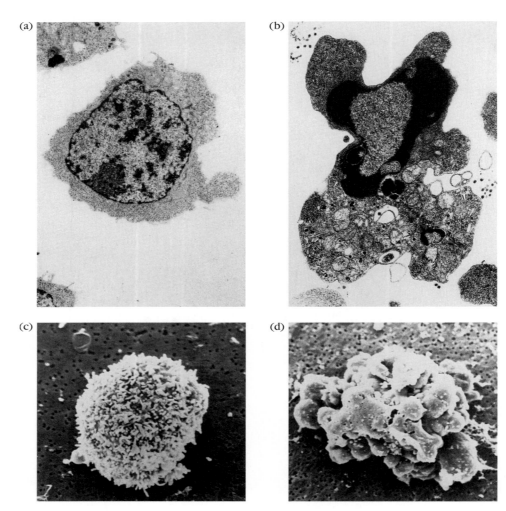

(a)  (b)

(c)  (d)

**FIGURE 2-4  Apoptosis.** Light micrographs of (a) normal thymocytes (developing T cells in the thymus) and (b) apoptotic thymocytes. Scanning electron micrographs of (c) normal and (d) apoptotic thymocytes. *[From B. A. Osborne and S. Smith, 1997, Journal of NIH Research 9:35; courtesy of B. A. Osborne, University of Massachusetts at Amherst.]*

continue to be activated by antigen, signals received during activation block the apoptotic signal. As antigen levels subside, the levels of the signals that block apoptosis diminish, and the lymphocytes begin to die by apoptosis.

## Hematopoietic stem cells can be enriched

Irv Weissman and colleagues developed a novel way of enriching the concentration of mouse hematopoietic stem cells, which normally constitute less than 0.05% of all bone marrow cells in mice. Their approach relied on the use of antibodies specific for molecules known as **differentiation antigens,** which are expressed only by particular cell types. They exposed bone marrow samples to antibodies that had been labeled with a fluorescent compound and were specific for the differentiation antigens expressed on the surface of mature red and white blood cells but not on stem cells (Figure 2-5). The labeled cells were then removed by flow cytometry with a fluorescence-activated cell sorter (see Chapter 6). After each sorting, the remaining cells were

assayed to determine the minimum number needed for restoration of hematopoiesis in a lethally x-irradiated mouse. As the pluripotent stem cells became relatively more numerous in the remaining population, fewer and fewer cells were needed to restore hematopoiesis in this system. Removal of the hematopoietic cells by selecting for the presence of differentiation antigens allowed a 50- to 200-fold enrichment of pluripotent stem cells. To further enrich the pluripotent stem cells, the remaining cells were incubated with various antibodies raised against cells likely to be in the early stages of hematopoiesis. One of these antibodies recognized a differentiation antigen called stem cell antigen 1 (Sca-1). Treatment with this antibody aided capture of undifferentiated stem cells and yielded a preparation so enriched in pluripotent stem cells that an aliquot containing only 30 to 100 cells routinely restored hematopoiesis in a lethally x-irradiated mouse, whereas more than $10^4$ nonenriched bone marrow cells were needed for restoration. Using a refinement of this approach, H. Nakauchi and his colleagues have devised procedures so effective that in one out

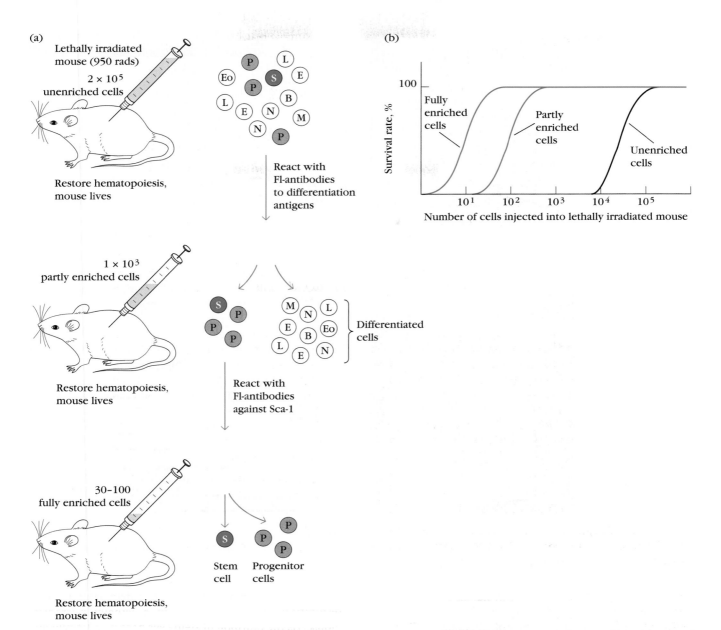

**FIGURE 2-5  Enrichment of the pluripotent stem cells from bone marrow.** (a) Differentiated hematopoietic cells (white) are removed by treatment with fluorescently labeled antibodies (Fl-antibodies) specific for membrane molecules expressed on differentiated lineages but absent from the undifferentiated stem cells (S) and progenitor cells (P). Treatment of the resulting partly enriched preparation with antibody specific for Sca-1, an early differentiation antigen, removed most of the progenitor cells. M = monocyte; B = basophil; N = neutrophil;

Eo = eosinophil; L = lymphocyte; E = erythrocyte. (b) Enrichment of stem cell preparations is measured by their ability to restore hematopoiesis in lethally irradiated mice. Only animals in which hematopoiesis occurs survive. Progressive enrichment of stem cells is indicated by the decrease in the number of injected cells needed to restore hematopoiesis. A total enrichment of about 1000-fold is possible by this procedure.

of five lethally irradiated mice, a single hematopoietic cell can restore both myeloid and lymphoid lineages (Table 2-3).

It has been found that CD34, a marker found on about 1% of hematopoietic cells, although not unique to stem cells is found on a small population of cells that includes stem cells. The administration of human cell populations suitably enriched for CD34$^+$ cells (the "+" indicates that the factor is

present on the cell membrane) can reconstitute a patient's entire hematopoietic system (see Clinical Focus on page 32).

A major tool in studies to identify and characterize the human hematopoietic stem cell is the use of **c m     mm      c   c** mice as an in vivo assay system for the presence and function of HSCs. SCID mice do not have B and T lymphocytes and are unable to mount

| TABLE 2-3 | Reconstitution of hematopoiesis by HSCs |
|---|---|
| Number of enriched HSCs | Number of mice reconstituted (%) |
| 1 | 9 of 41 (21.9) |
| 2 | 5 of 21 (23.8) |
| 5 | 9 of 17 (52.9) |
| 10 | 10 of 11 (90.9) |
| 20 | 4 of 4 (100) |

SOURCE: Adapted from M. Osawa et al., 1996, *Science* **273**:242.

| TABLE 2-4 | Normal adult blood cell counts | |
|---|---|---|
| Cell type | Cells/mm$^3$ | Total leukocytes (%) |
| Red blood cells | $5.0 \times 10^6$ | |
| Platelets | $2.5 \times 10^5$ | |
| Leukocytes | $7.3 \times 10^3$ | |
| Neutrophil | $3.7–5.1 \times 10^3$ | 50–70 |
| Lymphocyte | $1.5–3.0 \times 10^3$ | 20–40 |
| Monocyte | $1–4.4 \times 10^2$ | 1–6 |
| Eosinophil | $1–2.2 \times 10^2$ | 1–3 |
| Basophil | $<1.3 \times 10^2$ | <1 |

adaptive immune responses such as those that act in the normal rejection of foreign cells, tissues, and organs. Consequently, these animals do not reject transplanted human cell populations containing HSCs or tissues such as thymus and bone marrow. Immunodeficient mice have been useful surrogates or alternative hosts for in vivo studies of human stem cells. Implanted with fragments of human thymus and bone marrow, SCID mice support the differentiation of human hematopoietic stem cells into mature hematopoietic cells. This system has made possible the study of subpopulations of CD34$^+$ cells and the effect of human growth factors on the differentiation of various hematopoietic lineages.

## Cells of the Immune System

Lymphocytes bearing antigen receptors are the central cells of adaptive immunity and are responsible for its signature properties of diversity, specificity, and memory. Important as lymphocytes are, other types of white blood cells also play essential roles in adaptive immunity, presenting antigens, secreting cytokines, and engulfing and destroying microorganisms. Furthermore, as we will see in the next chapter, the innate immune system, which shares many cells with the adaptive immune system, plays an indispensable collaborative role in the induction of adaptive responses.

### Lymphoid cells

Lymphocytes constitute 20% to 40% of the body's white blood cells and 99% of the cells in the lymph (Table 2-4). There are approximately a trillion ($10^{12}$) lymphocytes in the human body. Lymphocytes circulate continuously in the blood and lymph and are capable of migrating into the tissue spaces and lymphoid organs, serving thereby as a bridge between parts of the immune system.

The lymphocytes can be broadly subdivided into three major populations—B cells, T cells, and natural killer cells—on the basis of function and cell membrane components. Key cells of adaptive immunity, B cells and T cells each bear

their own distinctive family of antigen receptors.

Natural killer cells are large, granular lymphocytes (*granular* refers to their grainy appearance under a microscope) that are part of the innate immune system and do not express the set of surface markers that characterize B or T cells. B and T lymphocytes that have not interacted with antigen—referred to as         or unprimed—are small, motile, nonphagocytic cells that cannot be distinguished from each other morphologically. In their unactivated state, they remain in the G$_0$ phase of the cell cycle. Known as small lymphocytes, these cells are only about 6 μm in diameter; their cytoplasm forms a barely discernible rim around the nucleus. Small lymphocytes have densely packed chromatin, few mitochondria, and a poorly developed endoplasmic reticulum and Golgi apparatus. The naive lymphocyte is generally thought to have a short life span. Under appropriate conditions, the interaction of small lymphocytes with antigen induces these cells to progress through the cell cycle from G$_0$ into G$_1$ and subsequently into S, G$_2$, and M (Figure 2-6a). As the cell cycle proceeds, lymphocytes enlarge into 15 μm-diameter cells called **m**         these cells have a higher cytoplasm-to-nucleus ratio and more organellar complexity than small lymphocytes (Figure 2-6b).

Lymphoblasts proliferate and eventually differentiate into **c    c** or into **m m    c** Effector cells function in various ways to eliminate antigen. These cells have life spans generally ranging from a few days to a few weeks. **m c** —the antibody-secreting effector cells of the B-cell lineage—have a characteristic cytoplasm that contains abundant, concentric layers of endoplasmic reticulum (to support their high rate of protein synthesis) and many Golgi vesicles (see Figure 2-6b). The effector cells of the T-cell lineage include the cytokine-secreting T helper cell (T$_H$ cell) and antigen-activated mature cells of the T cytotoxic lymphocyte (T$_C$ cell) lineage known as CTLs (cytotoxic T lymphocytes). Some of the progeny of B and T lymphoblasts differentiate into memory cells. The persistence of this population of cells is responsible for lifelong immunity to many pathogens. Memory cells look like small lymphocytes but can be distinguished from naive cells by the presence or absence of certain molecules on their cell membranes.

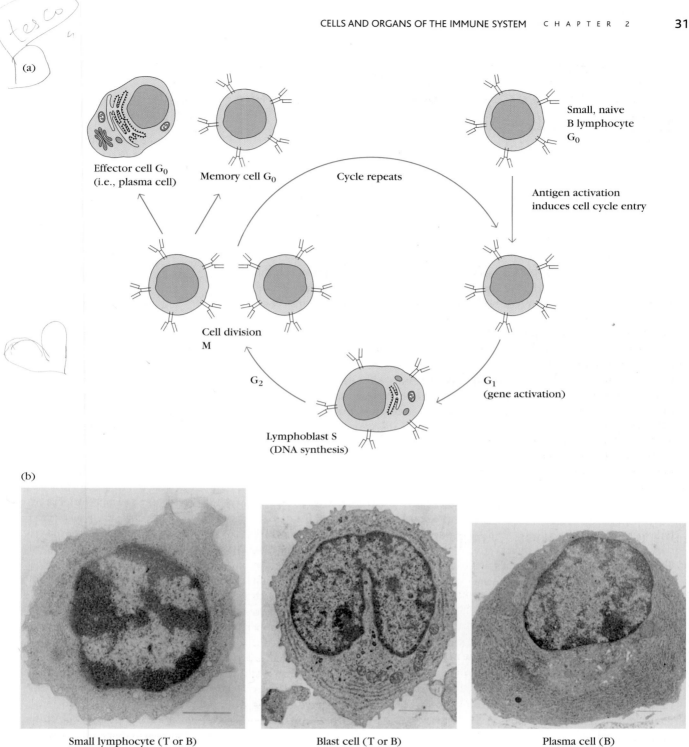

Small lymphocyte (T or B)
6 μm diameter

Blast cell (T or B)
15 μm diameter

Plasma cell (B)
15 μm diameter

**FIGURE 2-6 Fate of antigen-activated small lymphocytes.**
(a) A small resting (naive or unprimed) lymphocyte resides in the $G_0$ phase of the cell cycle. At this stage, B and T lymphocytes cannot be distinguished morphologically. After antigen activation, a B or T cell enters the cell cycle and enlarges into a lymphoblast, which undergoes several rounds of cell division and, eventually, generates effector cells and memory cells. Shown here are cells of the B-cell lineage. (b) Electron micrographs of a small lymphocyte *(left)* showing condensed chromatin indicative of a resting cell, an enlarged lymphoblast *(center)* showing decondensed chromatin, and a plasma cell *(right)* showing abundant endoplasmic reticulum arranged in concentric circles and a prominent nucleus that has been pushed to a characteristically eccentric position. The three cells are shown at different magnifications. *[Micrographs courtesy of Dr. J. R. Goodman, Dept. of Pediatrics, University of California at San Francisco.]*

## CLINICAL FOCUS
# Stem Cells—Clinical Uses and Potential

**Stem** cell transplantation holds great promise for the regeneration of diseased, damaged, or defective tissue. Hematopoietic stem cells are already used to restore hematopoietic function, and their use in the clinic is described below. However, rapid advances in stem cell research have raised the possibility that other stem cell types may soon be routinely employed for replacement of a variety of cells and tissues. Two properties of stem cells underlie their utility and promise. They have the capacity to give rise to lineages of differentiated cells, and they are self-renewing—each division of a stem cell creates at least one stem cell. If stem cells are classified according to their descent and developmental potential, four levels of stem cells can be recognized: totipotent, pluripotent, multipotent, and unipotent.

*Totipotent cells* can give rise to an entire organism. A fertilized egg, the zygote, is a totipotent cell. In humans, the initial divisions of the zygote and its descendants produce cells that are also totipotent. In fact, identical twins develop when totipotent cells separate and develop into genetically identical fetuses. *Pluripotent cells* arise from totipotent cells and can give rise to most but not all of the cell types necessary for fetal development. For example, human pluripotent stem cells can give rise to all of the cells of the body but cannot generate a placenta. Further differentiation of pluripotent stem cells leads to the formation of multipotent and unipotent stem cells. *Multipotent cells* can give rise to only a limited number of cell types, and *unipotent cells* can generate only the same cell type as themselves. Pluripotent cells, called *embryonic stem cells*, or simply *ES cells*, can be isolated from early embryos, and for many years it has been possible to grow mouse ES cells as cell lines in the laboratory. Strikingly, these cells can be induced to generate many different types of cells. Mouse ES cells have been shown to give rise to muscle cells, nerve cells, liver cells, pancreatic cells, and hematopoietic cells.

Recent advances have made it possible to grow lines of human pluripotent cells. This is a development of considerable importance to the understanding of human development, and it also has great therapeutic potential. In vitro studies of the factors that determine or influence the development of human pluripotent stem cells along specific developmental paths are providing considerable insight into how cells differentiate into specialized cell types. This research is driven in part by the great potential for using pluripotent stem cells to generate cells and tissues that could replace diseased or damaged tissue. Success in this endeavor would be a major advance because transplantation medicine now depends entirely on donated organs and tissues, yet the need far exceeds the number of donations, and the need is increasing. Success in deriving cells, tissues, and organs from pluripotent stem cells could provide skin replacement for burn patients, heart muscle cells for those with chronic heart disease, pancreatic islet cells for patients with diabetes, and neurons for the treatment of Parkinson's disease or Alzheimer's disease.

The transplantation of hematopoietic stem cells (HSCs) is an important therapy for patients whose hematopoietic systems must be replaced. It has three major applications:

- Providing a functional immune system to individuals with a genetically determined immunodeficiency, such as severe combined immunodeficiency (SCID).

- Replacing a defective hematopoietic system with a functional one to cure patients with life-threatening nonmalignant genetic disorders in hematopoiesis, such as sickle-cell anemia or thalassemia.

- Restoring the hematopoietic system of cancer patients after treatment with doses of chemotherapeutic agents and radiation so high that they destroy the system. These high-dose regimens can be much more effective at killing tumor cells than therapies using more conventional doses of cytotoxic agents. Stem cell transplantation makes it possible to recover from such drastic treatment. Also, certain cancers, such as some cases of acute myeloid leukemia, can be cured only by destroying the source of the leukemia cells, the patient's own hematopoietic system.

Hematopoietic stem cells have extraordinary powers of regeneration. Experiments in mice indicate that as few as one HSC can completely restore the erythroid population and the immune system. In humans, as little as 10% of a donor's total volume of bone marrow can provide enough HSCs to completely restore the recipient's hematopoietic system. Once injected into a vein, HSCs enter the circulation and find their own way to the bone marrow, where they begin the process of engraftment. There is no need for a surgeon to directly inject the cells into bones. In addition, HSCs can be preserved by freezing. This means that hematopoietic cells can be "banked." After collection, the cells are treated with a cryopreservative, frozen, and then stored for later use. When needed, the frozen preparation is thawed and infused into the patient, where it reconstitutes the hematopoietic system. This cell-freezing technology even makes it possible for individuals to store their own hematopoietic cells for transplantation to themselves at a later time. Currently, this procedure is used to allow cancer patients to donate cells before undergoing chemotherapy and radiation treatments, then using their own cells later to reconstitute their hematopoietic system. Hematopoietic stem cells are found in cell populations that display distinctive surface antigens. As discussed in the text, one of these antigens is CD34, which is present on only a small percentage (~1%) of the cells in adult bone marrow. An antibody specific for CD34 is used to select cells displaying this antigen, producing a population enriched in CD34$^+$ stem cells. Versions of this selection procedure have been used to enrich populations of stem cells from a variety of sources.

Transplantation of stem cell populations may be **autologous** (the recipient is also the donor), **syngeneic** (the donor is genetically identical, i.e., an identical twin of the recipient), or **allogeneic** (the donor and recipient are not genetically identical). In any transplantation procedure, genetic differences between donor and recipient can lead to immune-based rejection reactions. Aside from host rejection of transplanted tissue (host versus graft), lymphocytes conveyed to the recipient via the graft can attack the recipient's tissues, thereby causing **graft-versus-host disease (GVHD),** a life-threatening affliction. In order to suppress rejection reactions, powerful immunosuppressive drugs must be used. Unfortunately, these drugs have serious side effects, and immunosuppression increases the patient's risk of infection and susceptibility to tumors. Consequently, HSC transplantation has fewest complications when there is genetic identity between donor and recipient.

At one time, bone marrow transplantation was the only way to restore the hematopoietic system. However, the essential element of bone marrow transplantation is really stem cell transplantation. Fortunately, significant numbers of stem cells can be obtained from other tissues, such as peripheral blood and umbilical cord blood. These alternative sources of HSCs are attractive because the donor does not have to undergo anesthesia or the highly invasive procedure by which bone marrow is extracted. Many in the transplantation community believe that peripheral blood will replace marrow as the major source of hematopoietic stem cells for many applications. To obtain HSC-enriched preparations from peripheral blood, agents are used to induce increased numbers of circulating HSCs, and then the HSC-containing fraction is separated from the plasma and red blood cells in a process called leukopheresis. If necessary, further purification can be done to remove T cells and to enrich the CD34$^+$ population.

Umbilical cord blood already contains a significant number of hematopoietic stem cells. Furthermore, it is obtained from placental tissue (the "afterbirth"), which is normally discarded. Consequently, umbilical cord blood has become an attractive source of cells for HSC transplantation. Although HSCs from cord blood fail to engraft somewhat more often than do cells from peripheral blood, grafts of cord blood cells produce GVHD less frequently than marrow grafts, probably because cord blood has fewer mature T cells.

Beyond its current applications in cancer treatment, many researchers feel that autologous stem cell transplantation will be useful for **gene therapy,** the introduction of a normal gene to correct a disorder caused by a defective gene. Rapid advances in genetic engineering may soon make gene therapy a realistic treatment for genetic disorders of blood cells, and hematopoietic stem cells are attractive vehicles for such an approach. The therapy would entail removing a sample of hematopoietic stem cells from a patient, inserting a functional gene to compensate for the defective one, and then reinjecting the engineered stem cells into the donor. The advantage of using stem cells in gene therapy is that they are self-renewing. Consequently, at least in theory, patients would have to receive only a single injection of engineered stem cells. In contrast, gene therapy with engineered mature lymphocytes or other blood cells would require periodic injections because these cells are not capable of self-renewal.

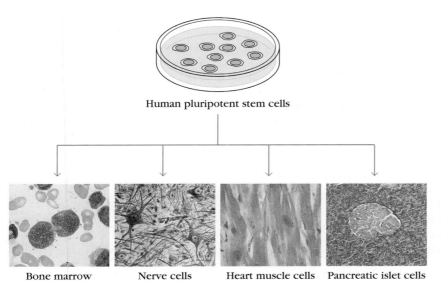

Human pluripotent stem cells

Bone marrow        Nerve cells        Heart muscle cells    Pancreatic islet cells

Human pluripotent stem cells can differentiate into a variety of different cell types, some of which are shown here. *[Adapted from Stem Cell Basics, NIH Web site http://stemcells. nih.gov/info/basics. Micrographs (left to right): Biophoto Associates/Science Source/Photo Researchers; Biophoto Associates/Photo Researchers; AFIP/Science Source/Photo Researchers; Astrid & Hanns-FriederMichler/Science PhotoLibrary/Photo Researchers.]*

| TABLE 2-5 | Common CD markers used to distinguish functional lymphocyte subpopulations | | | | | |
|-----------|---------|--------|----|----|----|

| | | | T cell | | |
|-----------|---------|--------|----|----|----|
| CD designation* | Function | B cell | $T_H$ | $T_C$ | NK cell |
| CD2 | Adhesion molecule; signal transduction | − | + | + | + |
| CD3 | Signal transduction element of T-cell receptor | − | + | + | − |
| CD4 | Adhesion molecule that binds to class II MHC molecules; signal transduction | − | + (usually) | − (usually) | − |
| CD5 | Unknown (subset) | − | − | + | + |
| CD8 | Adhesion molecule that binds to class I MHC molecules; signal transduction | − | − (usually) | + (usually) | + (variable) |
| CD16 (FcγRIII) | Low-affinity receptor for Fc region of IgG | − | − | − | + |
| CD21 (CR2) | Receptor for complement (C3d) and Epstein-Barr virus | + | − | − | − |
| CD28 | Receptor for costimulatory B7 molecule on antigen-presenting cells | − | + | + | − |
| CD32 (FcγRII) | Receptor for Fc region of IgG | + | − | − | − |
| CD35 (CR1) | Receptor for complement (C3b) | + | − | − | − |
| CD40 | Signal transduction | + | − | − | − |
| CD45 | Signal transduction | + | + | + | + |
| CD56 | Adhesion molecule | − | − | − | + |

*Synonyms are shown in parentheses.

Different lineages or maturational stages of lymphocytes can be distinguished by their expression of membrane molecules recognized by particular monoclonal antibodies (antibodies that are specific for a single epitope of an antigen; see Chapter 4 for a description of monoclonal antibodies). All of the monoclonal antibodies that react with a particular membrane molecule are grouped together as a **cluster of differentiation (CD).** Each new monoclonal antibody that recognizes a leukocyte membrane molecule is analyzed to determine if it falls within a recognized CD designation; if it does not, it is given a new CD designation reflecting a new membrane molecule. Although the CD nomenclature was originally developed for the membrane molecules of human leukocytes, the homologous membrane molecules of other species, such as mice, are commonly referred to by the same CD designations. Table 2-5 lists some common CD molecules (often referred to as CD markers) found on human lymphocytes. However, this is only a partial listing of the approximately 250 CD markers that have been described. A complete list and description of known CD markers is in Appendix 1.

The general characteristics and functions of B and T lymphocytes are reviewed briefly in the next sections. These central cells of the immune system will be examined much more fully in later chapters.

## B lymphocytes (B cells)

The B lymphocyte, also known as the B cell, derived its letter designation from its site of maturation, in the *b*ursa of Fabricius in birds; the name turned out to be apt, as *b*one marrow is its major site of maturation in a number of mammalian species, including humans and mice. Mature B cells are definitively distinguished from other lymphocytes and all other cells by their synthesis and display of membrane-bound immunoglobulin (antibody) molecules, which serve as receptors for antigen. Each of the approximately $1.5 \times 10^5$ molecules of antibody on the membrane of a single B cell has an identical binding site for antigen. When a naive B cell (one that has not previously encountered antigen) first encounters the antigen that matches its membrane-bound antibody, the binding of the antigen to the antibody causes the cell to divide rapidly; its progeny differentiate into **effector cells** called **plasma cells** and into **memory B cells.** Memory B cells have a longer life span than naive cells, and they express the same membrane-bound antibody as their parent B cell. Plasma cells, on the other hand, produce the antibody in a form that can be secreted and have little or no membrane-bound antibody. Plasma cells are end-stage cells and do not divide. Although some long-lived populations can be found in bone marrow, many die within 1 or 2 weeks. They are highly specialized for secretion of antibody, and a single cell is estimated to be capable of secreting from a few hundred to more than a thousand molecules of antibody per second.

## T lymphocytes (T cells)

T lymphocytes derive their letter designation from their site of maturation in the *t*hymus. During its maturation within the thymus, the T cell comes to express on its membrane a unique antigen-binding molecule called the **T-cell receptor.** Unlike membrane-bound antibodies on B cells, which can recognize antigen alone, T-cell receptors only recognize antigen that is bound to cell membrane proteins called **major histocompatibility complex (MHC) molecules.** MHC molecules that function in this recognition event, which is termed *antigen presentation,* are genetically diverse (polymorphic) glycoproteins found on cell membranes (see Chapter 8). There are two major types of MHC molecules: class I MHC molecules, expressed by nearly all nucleated cells of vertebrate species, and class II MHC molecules, which are expressed by only a few cell types that are specialized for antigen presentation. When a T cell recognizes antigen combined with an MHC molecule on a cell, under appropriate circumstances the T cell proliferates and differentiates into various effector T cells and memory T cells.

There are two well-defined subpopulations of T cells: **T helper ($T_H$)** and **T cytotoxic ($T_C$) cells;** recently a third T-cell subpopulation, **T regulatory ($T_{reg}$ ) cells**, has been characterized. T helper and T cytotoxic cells can be distinguished from one another by the presence of either CD4 or CD8 membrane glycoproteins on their surfaces. T cells displaying CD4 generally function as $T_H$ cells, whereas those displaying CD8 generally function as $T_C$ cells. Thus, the ratio of $T_H$ to $T_C$ cells in a sample can be approximated by assaying the number of CD4$^+$ and CD8$^+$ T cells. This ratio is approximately 2:1 in normal human peripheral blood, but it may be significantly altered by immunodeficiency diseases, autoimmune diseases, and other disorders. Following activation by interaction with appropriate antigen-MHC complexes, $T_H$ cells differentiate into effector cells that enable or "help" the activation of B cells, $T_C$ cells, macrophages, and various other cells that participate in the immune response. Alternatively, some $T_H$ cells differentiate into memory cells instead of effector cells.

Recognition of antigen-MHC complexes by a $T_C$ cell triggers its proliferation and differentiation into an effector cell called a cytotoxic T lymphocyte (CTL) or into a memory cell. The CTL has a vital function in monitoring the cells of the body and eliminating any that display foreign antigen complexed with class I MHC, such as virus-infected cells, tumor cells, and cells of a foreign tissue graft.

T regulatory cells are identified by the presence of both CD4 and CD25 on their membranes. However, unlike CD4-bearing T helper cells, $T_{reg}$ cells suppress immune responses—they are negative regulators of the immune system. Like $T_H$ and $T_C$ cells, members of the $T_{reg}$ subpopulation of T cells may be progenitors of memory cells.

## B- and T-cell populations comprise subpopulations of clones

All of the antigen receptors on a given B or T cell have identical structures and therefore have identical specificities for antigen. If a given lymphocyte divides to form two daughter cells, both daughters bear antigen receptors with antigen specificities identical to each other and to the parental cell from which they arose, and so will any descendants they produce. The resulting population of lymphocytes, all arising from the same founding lymphocyte, is a **clone.** At any given moment, a human or a mouse will contain tens of thousands, perhaps a hundred thousand, distinct T and B cell clones, each clone distinguished by its own distinctive and identical cohort of antigen receptors. Contact with antigen induces these cells to proliferate and differentiate. The products of this process include both effector cells and memory cells. The effector cells carry out specific functions while the memory cells *persist* in the host and on rechallenge with same antigen mediate a response that is both quicker and greater in magnitude. The first encounter is termed a primary response and the re-encounter a secondary response.

## Natural killer Cells

The body contains a small population of large, granular lymphocytes called natural killer cells that display cytotoxic activity against a wide range of tumor cells and against cells infected with some but not all viruses. An extraordinary feature of these cells, which constitute 5% to 10% of lymphocytes in human peripheral blood, is their ability to recognize tumor or virus-infected cells despite lacking antigen-specific receptors. Natural killer cells are part of the innate immune system, and most do not have T-cell receptors or immunoglobulin incorporated in their plasma membranes; in other words, they do not express the membrane molecules and receptors that distinguish T- and B-cell lineages. NK cells recognize potential target cells in two different ways. In some cases, an NK cell employs NK-cell receptors to distinguish abnormalities, such as a reduction in the display of class I MHC molecules or the unusual profile of surface antigens displayed by some tumor cells and cells infected by some viruses. In addition, some tumor cells and cells infected by certain viruses display antigens against which the immune system has raised an antibody response, so that antitumor or antiviral antibodies are bound to their surfaces. Because NK cells express a membrane receptor (CD16) for a specific region of the antibody molecule, they can attach to these antibodies and subsequently destroy the targeted cells. This is an example of a process known as **antibody-dependent cell-mediated cytotoxicity (ADCC).** The exact mechanism of NK-cell cytotoxicity, the focus of much current experimental study, is described further in Chapter 14.

Recognition has recently been growing of another cell type, the **NKT cell,** which has some of the characteristics of both T cells and NK cells. Like T cells, NKT cells have T cell receptors (TCRs). Unlike most T cells, the TCRs of NKT cells interact with MHC-like molecules called CD1 rather than with class I or class II MHC molecules. Like NK cells, they have variable levels of CD16 and other receptors typical of NK cells, and they can kill target cells. A population of activated NKT cells can rapidly secrete large amounts of the cytokines needed to support antibody production by B cells

(a) Monocyte

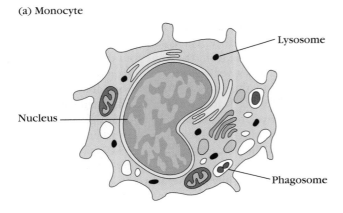

(b) Macrophage

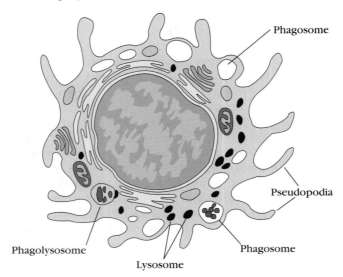

**FIGURE 2-7 Typical morphology of a monocyte (a) and a macrophage (b).** Macrophages are five- to tenfold larger than monocytes and contain more organelles, especially lysosomes.

as well as inflammation and the development and expansion of cytotoxic T cells. Determining the exact roles played by NKT cells in immunity is of great interest.

## Mononuclear phagocytes

The mononuclear phagocytic system consists of **monocytes** circulating in the blood and **macrophages** in the tissues (Figure 2-7). During hematopoiesis in the bone marrow, granulocyte-monocyte progenitor cells differentiate into promonocytes, which leave the bone marrow and enter the blood, where they further differentiate into mature monocytes. Monocytes circulate in the bloodstream for about 8 hours, during which time they enlarge; they then migrate into the tissues and differentiate into specific tissue macrophages.

Differentiation of a monocyte into a tissue macrophage involves a number of changes: the cell enlarges five- to tenfold; its intracellular organelles increase in both number and complexity; and it acquires increased phagocytic ability, produces higher levels of hydrolytic enzymes, and begins to secrete a variety of soluble factors. Macrophages are dispersed throughout the body. Some take up residence in particular tissues, becoming fixed macrophages, whereas others remain motile and are called free or wandering macrophages. Free macrophages travel by amoeboid movement throughout the tissues. Macrophage-like cells serve different functions in different tissues and are named according to their tissue location:

- **Intestinal macrophages** in the gut
- **Alveolar macrophages** in the lung
- **Histiocytes** in connective tissues
- **Kupffer cells** in the liver
- **Mesangial cells** in the kidney
- **Microglial cells** in the brain
- **Osteoclasts** in bone

Macrophages are activated by a variety of stimuli in the course of an immune response. Phagocytosis of particulate antigens or contact with receptors that sense molecules present in microbial pathogens often serves as an initial activating stimulus. However, macrophage activity can be further enhanced by cytokines secreted by activated $T_H$ cells and by mediators of the inflammatory response.

Activated macrophages are more effective than resting ones in eliminating potential pathogens for several reasons. They exhibit greater phagocytic activity, an increased ability to kill ingested microbes, increased secretion of inflammatory mediators, and an increased ability to activate T cells. In addition, activated macrophages, but not resting ones, secrete various cytotoxic proteins that help them eliminate a broad range of targets, including virus-infected cells, tumor cells, and intracellular bacteria. More will be said about the antimicrobial activities of macrophages in Chapter 3. Activated macrophages also express higher levels of class II MHC molecules, allowing them to function more effectively as antigen-presenting cells for $T_H$ cells. Thus, macrophages and $T_H$ cells facilitate each other's activation during the immune response.

## Phagocytosis is followed by digestion and presentation of antigen

Macrophages are capable of ingesting and digesting exogenous antigens, such as whole microorganisms and insoluble particles, and endogenous matter, such as injured or dead host cells, cellular debris, and activated clotting factors. Phagocytosis is initiated by the adherence of the antigen to the macrophage cell membrane. Complex antigens, such as

(a)

(b)

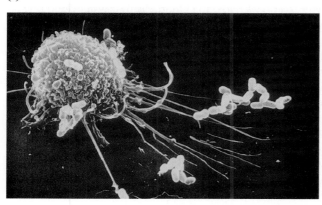

**FIGURE 2-8  Macrophages can ingest and degrade particulate antigens, including bacteria.** (a) Scanning electron micrograph of a macrophage. Note the long pseudopodia extending toward and making contact with bacterial cells, an early step in phagocytosis. (b) Phagocytosis and processing of exogenous antigen by macrophages.

Most of the products resulting from digestion of ingested material are exocytosed, but some peptide products may interact with class II MHC molecules, forming complexes that move to the cell surface, where they are presented to $T_H$ cells. *[(a) L. Nilsson, Boehringer Ingelheim International GmbH.]*

whole bacterial cells or viral particles, tend to adhere well and are readily phagocytosed; isolated proteins and encapsulated bacteria tend to adhere poorly and are less readily phagocytosed. Adherence induces membrane protrusions, called **pseudopodia,** to extend around the attached material (Figure 2-8a). Fusion of the pseudopodia encloses the material within a membrane-bound structure called a **phagosome,** which then enters the endocytic processing pathway (Figure 2-8b). In this pathway, a phagosome moves toward the cell interior, where it fuses with a **lysosome** to form a **phagolysosome.** Lysosomes contain a variety of hydrolytic enzymes that digest the ingested material. The digested contents of the phagolysosome are then eliminated in a process called **exocytosis** (see Figure 2-8b).

The macrophage membrane has receptors for certain classes of antibody. If an antigen (e.g., a bacterium) is coated with the appropriate antibody, the complex of antigen and antibody binds to antibody receptors on the macrophage membrane more readily than antigen alone and phagocytosis is enhanced. In one study, for example, the rate of phagocytosis of an antigen was 4000-fold higher in the presence of specific antibody to the antigen than in its absence. Thus, antibody functions as an **opsonin,** a molecule that binds to both antigen and phagocyte, thereby enhancing phagocytosis. The process by which opsonins render particulate antigens more susceptible to phagocytosis is called **opsonization.**

Although most of the antigen ingested by macrophages is degraded and eliminated, experiments with radiolabeled antigens have demonstrated the presence of antigen peptides on the macrophage membrane. As depicted in Figure 2-8b, phagocytosed antigen is digested within the endocytic processing pathway into peptides that associate with class II

MHC molecules; these peptide–class II MHC complexes then move to the macrophage membrane. This processing and presentation of antigen, examined in detail in Chapter 8, are critical to $T_H$-cell activation, a central event in the development of both humoral and cell-mediated immune responses. Finally, as discussed in Chapter 3, activated macrophages secrete regulatory proteins that are important for the development of immune responses.

## Granulocytic cells

The **granulocytes** are classified as neutrophils, eosinophils, or basophils on the basis of cellular morphology and cytoplasmic-staining characteristics (Figure 2-9). The **neutrophil** has a multilobed nucleus and a granulated cytoplasm that stains with both acid and basic dyes; it is often called a polymorphonuclear (PMN) leukocyte for its multilobed nucleus. The **eosinophil** has a bilobed nucleus and a granulated cytoplasm that stains with the acid dye eosin red (hence its name). The **basophil** has a lobed nucleus and heavily granulated cytoplasm that stains with the basic dye methylene blue. Both neutrophils and eosinophils are phagocytic, whereas basophils are not. Neutrophils, which constitute 50% to 70% of the circulating white blood cells, are much more numerous than eosinophils (1%–3%) or basophils (<1%).

### Neutrophils

Neutrophils are produced by hematopoiesis in the bone marrow. They are released into the peripheral blood and circulate for 7 to 10 hours before migrating into the tissues, where they have a life span of only a few days. In response to many types of infections, the bone marrow releases more than the usual

(a) Neutrophil

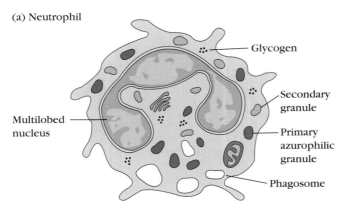

Multilobed nucleus

Glycogen

Secondary granule

Primary azurophilic granule

Phagosome

(b) Eosinophil

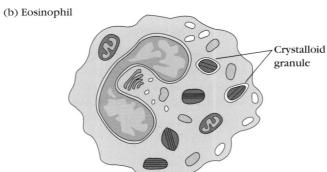

Crystalloid granule

(c) Basophil

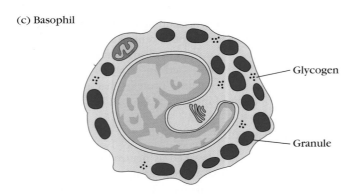

Glycogen

Granule

**FIGURE 2-9  Drawings showing typical morphology of granulocytes.** Note differences in the shape of the nucleus and in the number, color, and shape of the cytoplasmic granules.

number of neutrophils, and these cells generally are the first to arrive at a site of inflammation. The resulting transient increase in the number of circulating neutrophils, called **leukocytosis,** is used medically as an indication of infection.

Movement of circulating neutrophils into tissues, called **extravasation,** takes several steps: the cell first adheres to the vascular endothelium, then penetrates the gap between adjacent endothelial cells lining the vessel wall, and finally penetrates the vascular basement membrane, moving out into the tissue spaces. (This process is fully discussed in Chapter 3.) A number of substances generated in an inflammatory reaction serve as **chemotactic factors** that promote accumulation of neutrophils at an inflammatory site. Among these

chemotactic factors are some of the complement components, components of the blood-clotting system, and several cytokines secreted by activated $T_H$ cells and macrophages.

Like macrophages, neutrophils are phagocytes. Phagocytosis by neutrophils is similar to that described for macrophages, except that the lytic enzymes and bactericidal substances in neutrophils are contained within primary and secondary granules (Figure 2-9a). The larger, denser primary granules are a type of lysosome containing peroxidase, lysozyme, and various hydrolytic enzymes. The smaller secondary granules contain collagenase, lactoferrin, and lysozyme. Both primary and secondary granules fuse with phagosomes, whose contents are then digested and eliminated. Neutrophils generate a variety of antimicrobial agents, which will be examined in Chapter 3.

### Eosinophils

Eosinophils, like neutrophils, are motile phagocytic cells (Figure 2-9b) that can migrate from the blood into the tissue spaces. Their phagocytic role is significantly less important than that of neutrophils, and it is thought that they play a role in the defense against parasitic organisms by secreting the contents of eosinophilic granules, which may damage the parasite membrane.

### Basophils

Basophils are nonphagocytic granulocytes (Figure 2-9c) that arise by hematopoiesis and function by releasing pharmacologically active substances from their cytoplasmic granules. These substances play a major role in certain allergic responses.

## Mast cells

Mast cell precursors, which are formed in the bone marrow by hematopoiesis, are released into the blood as undifferentiated cells; they do not differentiate until they leave the blood and enter the tissues. Mast cells can be found in a wide variety of tissues, including the skin, connective tissues of various organs, and mucosal epithelial tissue of the respiratory, genitourinary, and digestive tracts. Like circulating basophils, these cells have large numbers of cytoplasmic granules that contain histamine and other pharmacologically active substances. Mast cells play an important role in the development of allergies (see Chapter 15).

## Dendritic cells

Identified in 1868 during a close anatomical study of skin by Paul Langerhans, dendritic cells were the first cells of the immune system discovered. The **dendritic cell (DC)** acquired its name because it is covered with long membranous extensions that resemble the dendrites of nerve cells. There are many types of dendritic cells, and at least four major categories are recognized: Langerhans DCs, interstitial DCs, monocyte-derived DCs, and plasmacytoid-derived DCs.

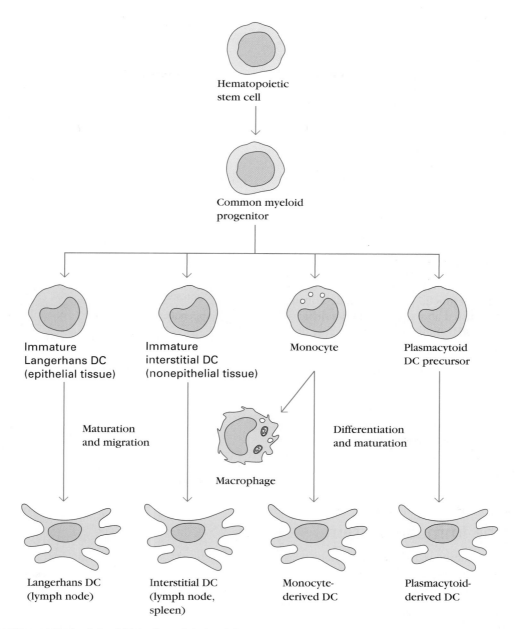

**FIGURE 2-10 Different kinds of dendritic cells and their origins.**

Each arises from hematopoietic stem cells via different pathways and in different locations (Figure 2-10). Langerhans DCs are in epidermal layers of skin, and interstitial DCs are present in the interstitial spaces of virtually all organs except the brain. As indicated by their name, monocyte-derived DCs arise from monocytes that have migrated from the bloodstream into tissues. From the tissues, they can move through lymph to lymph nodes or cross back into the bloodstream and use it as an avenue of transport to lymphoid tissue.

Plasmacytoid-derived DCs arise from plasmacytoid cells. They play roles in innate immune defense and as antigen-presenting cells. Although there are important differences in the functions and phenotypes of different varieties of DCs, they all share display of class I and II MHC molecules, and

the B7 family of costimulatory molecules, CD80 and CD86, are present. Dendritic cells also have CD40, a molecule that can influence T cell behavior by interaction with a complementary ligand on T cells.

Dendritic cells are versatile, occurring in many forms and performing the distinct functions of antigen capture in one location and antigen presentation in another. Outside lymph nodes, immature forms of these cells monitor the body for signs of invasion by pathogens and capture intruding or foreign antigens. They then migrate to lymph nodes, where they present the antigen to T cells. When acting as sentinels in the periphery, immature dendritic cells take on their cargo of antigen in three ways. They engulf it by phagocytosis, internalize it by receptor-mediated endocytosis, or imbibe it by pinocytosis. Indeed, immature dendritic cells pinocytose fluid volumes of 1000 to 1500 μm³ per hour, a volume that rivals that of the cell itself. Through a process of maturation, they shift from an antigen-capturing phenotype to one that supports presentation of antigen to T cells. In making the transition, some attributes are lost and others are gained. Lost is the capacity for phagocytosis and large-scale pinocytosis. Expression of class II MHC increases, which is necessary to present antigen to $T_H$ cells, and there is an increase in the production increases of costimulatory molecules that are essential for the activation of naive T cells. On maturation, dendritic cells abandon residency in peripheral tissues, enter the blood or lymphatic circulation, and migrate to regions of the lymphoid organs, where T cells reside, and present antigen.

## Follicular dendritic cells

**Follicular dendritic cells** do not arise in bone marrow and have completely different functions from those described for the dendritic cells discussed above. Follicular dendritic cells do not express class II MHC molecules and therefore do not function as antigen-presenting cells for $T_H$-cell activation. These dendritic cells were named for their exclusive location in organized structures of the lymph node called lymph follicles, which are rich in B cells. Although they do not express class II MHC molecules, follicular dendritic cells do express high levels of membrane receptors for antibody, which allows efficient binding of antigen-antibody complexes. As discussed in Chapter 11, the interaction of B cells with follicular dendritic cells is an important step in the maturation and diversification of B cells.

## Organs of the Immune System

A number of morphologically and functionally diverse organs and tissues contribute to the development of immune responses (Figure 2-11). These organs can be distinguished by function as the **primary** and **secondary lymphoid organs.** The thymus and bone marrow are the primary (or central) lymphoid organs, where maturation of lymphocytes takes place. The secondary (or peripheral) lymphoid organs include lymph nodes, the spleen, and various mucosa-associated lymphoid tissues (MALT) such as gut-associated lymphoid tissue (GALT). These organs provide sites for mature lymphocytes to interact with antigen. Once mature lymphocytes have been generated in the primary lymphoid organs, they circulate in the blood and **lymphatic system,** a network of vessels that collect fluid that has escaped into the tissues from capillaries of the circulatory system. Ultimately, this escaped fluid is returned to the blood.

## Primary lymphoid organs

Immature lymphocytes generated in hematopoiesis mature and become committed to a particular antigenic specificity within the primary lymphoid organs. Only after a lymphocyte has matured within a primary lymphoid organ is the

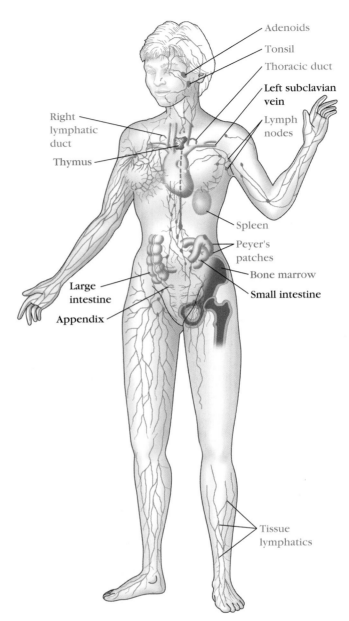

**FIGURE 2-11 The human lymphoid system.** The primary organs (bone marrow and thymus) are shown in red; secondary organs and tissues, in blue. These structurally and functionally diverse lymphoid organs and tissues are interconnected by the blood vessels (not shown) and lymphatic vessels (purple). Most of the body's lymphatics eventually drain into the thoracic duct, which empties into the left subclavian vein. However, the vessels draining the right arm and right side of the head converge to form the right lymphatic duct, which empties into the right subclavian vein (not shown). Bones containing marrow are part of the lymphoid system; normally samples of bone marrow are taken from the iliac crest or the sternum. *[Adapted from H. Lodish et al., 1995, Molecular Cell Biology, 3rd ed., Scientific American Books, New York.]*

cell **immunocompetent** (capable of mounting an immune response). T cells arise in the bone marrow and develop in the **thymus.** In many mammals—humans and mice, for example—B cells originate in **bone marrow.**

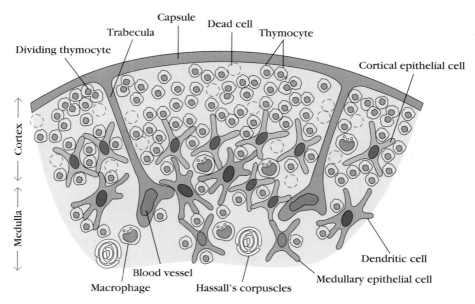

**FIGURE 2-12  Diagrammatic cross section of a portion of the thymus, showing several lobules separated by connective tissue strands (trabeculae).** The densely populated outer cortex contains many immature thymocytes (blue), which undergo rapid proliferation coupled with an enormous rate of cell death. The medulla is sparsely populated and contains thymocytes that are more mature. During their stay within the thymus, thymocytes interact with various stromal cells, including cortical epithelial cells (light red), medullary epithelial cells (tan), dendritic cells (purple), and macrophages (yellow). These cells produce regulatory factors and express high levels of class I and class II MHC molecules. Hassall's corpuscles, found in the medulla, contain concentric layers of degenerating epithelial cells. *[Adapted with permission from W. van Ewijk, 1991,* Annual Review of Immunology *9:591, © 1991 by Annual Reviews.]*

## Thymus

The thymus, site of T-cell development and maturation, is a flat, bilobed organ situated above the heart. Each lobe is surrounded by a capsule and divided into lobules, which are separated from each other by strands of connective tissue called trabeculae (Figure 2-12). Each lobule is organized into two compartments: the outer compartment, or **cortex,** is densely packed with immature T cells, called thymocytes, whereas the inner compartment, or **medulla,** is sparsely populated with thymocytes.

Both the cortex and the medulla of the thymus are crisscrossed by a three-dimensional stromal cell network composed of epithelial cells, dendritic cells, and macrophages, which make up the framework of the organ and contribute to the growth and maturation of thymocytes. Many of these stromal cells interact physically with the developing thymocytes. The function of the thymus is to generate and select a repertoire of T cells that will protect the body from infection. As thymocytes develop, an enormous diversity of T-cell receptors is generated by gene rearrangement (see Chapter 9), which produces some T cells with receptors capable of recognizing antigen-MHC complexes. However, most of the T-cell receptors produced by this random process are incapable of recognizing antigen-MHC complexes, and a small portion react with combinations of self antigen–MHC complexes. Using mechanisms that are discussed in Chapter 10, the thymus induces the death of those T cells that cannot recognize antigen-MHC complexes and those that react with self antigen–MHC strongly enough to pose a danger of causing autoimmune disease. More than 95% of all thymocytes die by apoptosis in the thymus without ever reaching maturity.

The role of the thymus in immune function can be studied in mice by examining the effects of neonatal thymectomy, a procedure in which the thymus is surgically removed from newborn mice. Thymectomized mice show a dramatic decrease in circulating lymphocytes of the T-cell lineage and an absence of cell-mediated immunity. Other evidence of the importance of the thymus comes from studies of a congenital birth defect in humans (**DiGeorge's syndrome**) and in certain mice (**nude mice**). In both cases, the thymus fails to develop, and there is an absence of circulating T cells and of cell-mediated immunity and an increase in infectious disease.

Thymic function is known to decline with age. The thymus reaches its maximal size at puberty and then atrophies, with a significant decrease in both cortical and medullary cells and an increase in the total fat content of the organ. Whereas the average weight of the thymus is 30 grams in human infants, its age-dependent involution leaves an organ with an average weight of only 3 grams in the elderly (Figure 2-13). The age-dependent loss in mass is accompanied by a decline in T-cell output. By age 35, the thymic generation of T cells has dropped to 20% of production in newborns, and by age 65, the output has fallen to only 2% of the newborn rate.

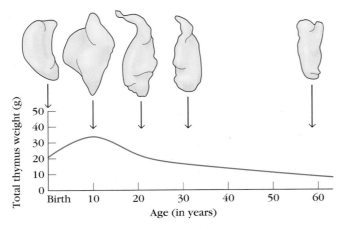

**FIGURE 2-13 Changes in the thymus with age.** The thymus decreases in size and cellularity after puberty.

A number of experiments have been designed to look at the effect of age on the immune function of the thymus. In one experiment, thymectomized adult mice were implanted with the thymus from either a 1-day-old or 33-month-old mouse. (For most laboratory mice, 33 months is very old.) Mice receiving the newborn thymus graft showed a significantly greater improvement in immune function than mice receiving the 33-month-old thymus.

## Bone Marrow

Bone marrow is a complex tissue that is the site of hematopoiesis and a fat depot. In fact, with the passage of time, fat eventually fills 50% or more of the marrow compartment of bone. Hematopoietic cells generated in bone marrow move through the walls of blood vessels and enter the bloodstream, which carries them out of the marrow and distributes these various cell types to the rest of the body.

In humans and mice, bone marrow is the site of B-cell origin and development. Arising from lymphoid progenitors, immature B cells proliferate and differentiate within the bone marrow, and stromal cells within the bone marrow interact directly with the B cells and secrete various cytokines that are required for development. Bone marrow B cells are the source of about 90% of the immunoglobulins IgG and IgA in plasma. Like thymic selection during T-cell maturation, a selection process within the bone marrow eliminates B cells that possess self-reactive antibody receptors. (This process is explained in Chapter 11.) Despite its key role in humans and mice, bone marrow is not the site of B-cell development in all species. In birds, a lymphoid organ associated with the gut called the bursa of Fabricius is the primary site of B-cell maturation. In mammals such as primates and rodents, there is no bursa or counterpart that serves as a primary lymphoid organ. In cattle and sheep, the primary lymphoid tissue hosting the maturation, proliferation, and diversification of B cells early in gestation is the fetal spleen. Later in gestation, this function is assumed by a patch of tissue embedded in the wall of the intestine called the ileal **Peyer's patch,** which contains a large number of B cells as well as T cells. The rabbit, too, uses gut-associated tissues, especially the appendix, as primary lymphoid tissue for important steps in the proliferation and diversification of B cells.

## Lymphatic System

As blood circulates under pressure, its fluid component (**plasma**) seeps through the thin walls of the capillaries into the surrounding tissue. In an adult, depending on size and activity, seepage can add up to 2.9 liters or more during a 24-hour period. This fluid, called **interstitial fluid,** permeates all tissues and bathes all cells. If this fluid was not returned to the circulation, edema, a progressive swelling, would result and eventually become life threatening. We are not afflicted with such catastrophic edema because much of the fluid is returned to the blood through the walls of venules. The remainder of the interstitial fluid enters a delicate network of thin-walled tubes called primary lymphatic vessels. The walls of the primary vessels consist of a single layer of loosely apposed endothelial cells (Figure 2-14). Even though the capillaries are closed ("blind"), the porous architecture of the primary vessels allows fluids and even cells to enter the lymphatic network. Within these vessels, the fluid, now called **lymph,** flows from the network of tiny tubes into a series of progressively larger collecting vessels called **lymphatic vessels** (Figure 2-14).

The largest lymphatic vessel, the **thoracic duct,** empties into the left subclavian vein. It collects lymph from all of the body except the right arm and right side of the head. Lymph from these areas is collected into the right lymphatic duct, which drains into the right subclavian vein (see Figure 2-11). In this way, the lymphatic system captures fluid lost from the blood and returns it to the blood, ensuring steady-state levels of fluid within the circulatory system. Lymphocytes, dendritic cells, macrophages, and other cells can also gain entry through the thin wall of loosely joined endothelial cells of the primary lymphatic vessels and join the flow of lymph (see Figure 2-14). The heart does not pump the lymph through the lymphatic system; instead, the slow, low-pressure flow of lymph is achieved as the lymph vessels are squeezed by movements of the body's muscles. A series of one-way valves along the lymphatic vessels ensures that lymph flows in only one direction.

When a foreign antigen gains entrance to the tissues, it is picked up by the lymphatic system (which drains all the tissues of the body) and is carried to various organized lymphoid tissues such as lymph nodes, which trap the foreign antigen. As lymph passes from the tissues to lymphatic vessels, it becomes progressively enriched in lymphocytes. Thus, the lymphatic system also serves as a means of transporting lymphocytes and antigen from the connective tissues to organized lymphoid tissues, where the lymphocytes can interact with the trapped antigen and undergo activation.

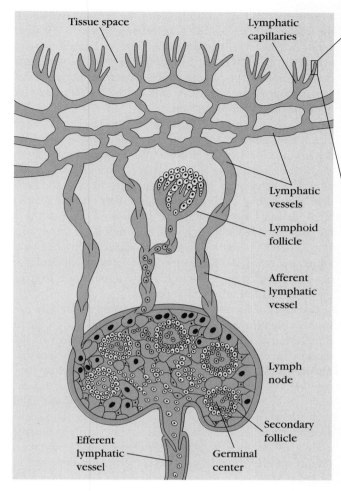

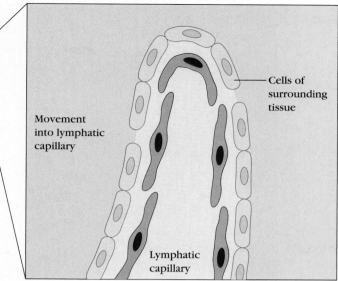

**FIGURE 2-14 Lymphatic vessels.** Interstitial fluid enters small closed ("blind") lymphatic capillaries by moving between the loosely joined flaps of the thin layer of endothelial cells that form the vessel wall. The fluid, now called lymph, is carried through progressively larger lymphatic vessels to regional lymph nodes. As lymph leaves the nodes, it is carried through larger efferent lymphatic vessels, which eventually drain into the circulatory system at the thoracic duct or right lymph duct (see Figure 2-11).

## Secondary lymphoid organs

Various types of organized lymphoid tissues are situated along the vessels of the lymphatic system. Some lymphoid tissue in the lung and lamina propria of the intestinal wall consists of diffuse collections of lymphocytes and macrophages. Other lymphoid tissue is organized into structures called lymphoid follicles, which consist of aggregates of lymphoid and nonlymphoid cells surrounded by a network of draining lymphatic capillaries. Until it is activated by antigen, a lymphoid follicle—called a **primary follicle**—comprises a network of follicular dendritic cells and small resting B cells. After an antigenic challenge, a primary follicle becomes a larger **secondary follicle**—a ring of concentrically packed B lymphocytes surrounding a center (the **germinal center**) in which one finds a focus of proliferating B lymphocytes and an area that contains nondividing B cells and some helper T cells interspersed with macrophages and follicular dendritic cells (Figure 2-15).

**Lymph nodes** and the **spleen** are the most highly organized of the secondary lymphoid organs; in addition to lymphoid follicles, they have additional distinct regions of T-cell and B-cell activity, and they are surrounded by a fibrous capsule. Less organized lymphoid tissue, collectively called

**mucosa-associated lymphoid tissue (MALT),** is found in various body sites. MALT includes Peyer's patches (in the small intestine), the tonsils, and the appendix as well as numerous lymphoid follicles within the lamina propria of the intestines and in the mucous membranes lining the upper airways, bronchi, and genitourinary tract.

### Lymph Nodes

Lymph nodes are the sites where immune responses are mounted to antigens in lymph. They are encapsulated bean-shaped structures containing a reticular network packed with lymphocytes, macrophages, and dendritic cells. Clustered at junctions of the lymphatic vessels, lymph nodes are the first organized lymphoid structure to encounter antigens that enter the tissue spaces. As lymph percolates through a node, any particulate antigen that is brought in with the lymph will be trapped by the cellular network of phagocytic cells and dendritic cells. The overall architecture of a lymph node supports an ideal microenvironment for lymphocytes to effectively encounter and respond to trapped antigens.

Morphologically, a lymph node can be divided into three roughly concentric regions: the cortex, the paracortex, and the medulla, each of which supports a distinct microenvironment

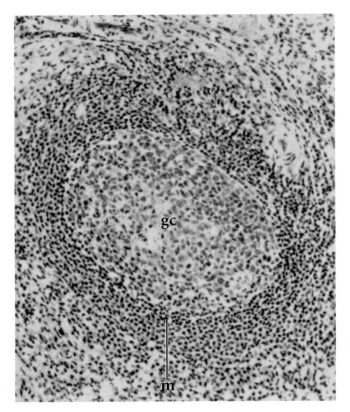

gc

m

**FIGURE 2-15** A secondary lymphoid follicle consisting of a large germinal center (gc) surrounded by a dense mantle (m) of small lymphocytes. *[From W. Bloom and D. W. Fawcett, 1975,* Textbook of Histology, *10th ed., © 1975 by W. B. Saunders Co.]*

(Figure 2-16a). The outermost layer, the **c**        contains lymphocytes (mostly B cells), macrophages, and follicular dendritic cells arranged in primary follicles. After antigenic challenge, the primary follicles enlarge into secondary follicles, each containing a germinal center. In children with B-cell deficiencies, the cortex lacks primary follicles and germinal centers. Beneath the cortex is the        **c**        which is populated largely by T lymphocytes and also contains dendritic cells that migrated from tissues to the node. These dendritic cells express high levels of class II MHC molecules, which are necessary for presenting antigen to $T_H$ cells. The **m**        is more sparsely populated with lymphoid lineage cells, and of those present, many are plasma cells actively secreting antibody molecules.

As antigen is carried into a regional node by the lymph, it is trapped, processed, and presented together with class II MHC molecules by dendritic cells in the paracortex, resulting in the activation of $T_H$ cells. The initial activation of B cells takes place within the T-cell-rich paracortex. Once activated, $T_H$ and B cells form small foci consisting largely of proliferating B cells at the edges of the paracortex. Some B cells within the foci differentiate into plasma cells secreting antibody. These foci reach maximum size within 4 to 6 days of antigen challenge. Within 4 to 7 days of antigen challenge,

a few B cells and $T_H$ cells migrate to the primary follicles of the cortex, where cellular interactions between follicular dendritic cells, B cells, and $T_H$ cells take place, leading to development of a secondary follicle with a central germinal center. Some of the plasma cells generated in the germinal center move to the medullary areas of the lymph node, and many migrate to bone marrow.

Afferent (incoming) lymphatic vessels pierce the capsule of a lymph node at numerous sites and empty lymph into the subcapsular sinus (Figure 2-16b). Lymph coming from the tissues percolates slowly inward through the cortex, paracortex, and medulla, allowing phagocytic cells and dendritic cells to trap any bacteria or particulate material carried by the lymph. After infection or the introduction of other antigens into the body, the lymph, leaving a node through its single efferent (outgoing) lymphatic vessel, is enriched with antibodies newly secreted by medullary plasma cells and also has an approximately 50-fold higher concentration of lymphocytes than the afferent lymph.

Some of the increase in the concentration of lymphocytes in the lymph leaving a node is due to lymphocyte proliferation within the node in response to antigen. Most of the increase, however, represents blood-borne lymphocytes that migrate into the node through the walls of postcapillary venules. These venules are lined with unusually plump endothelial cells that give them a thickened appearance, and they are called    **g**                    The HEVs are important because most of the lymphocytes that enter the node arrive by passing between the specialized endothelial cells of the HEV by extravasation, a mechanism discussed in Chapter 3. A portion of the lymphocytes leaving a lymph node have migrated across this endothelial layer and entered the node from the blood. Because antigenic stimulation within a node can increase this migration 10-fold, the number of lymphocytes in a node that is actively responding can increase greatly, and the node swells visibly. Factors released in lymph nodes during antigen stimulation are thought to facilitate this increased migration. Figure 2-16b summarizes the flow of lymph and lymphocytes through a lymph node.

## Spleen

The spleen, situated high in the left abdominal cavity, is a large, ovoid secondary lymphoid organ that plays a major role in mounting immune responses to antigens in the bloodstream. Whereas lymph nodes are specialized for trapping antigen from local tissues, the spleen specializes in filtering blood and trapping blood-borne antigens; thus, it can respond to systemic infections. Unlike the lymph nodes, the spleen is not supplied by lymphatic vessels. Instead, blood-borne antigens and lymphocytes are carried into the spleen through the splenic artery. Experiments with radioactively labeled lymphocytes show that more recirculating lymphocytes pass daily through the spleen than through all the lymph nodes combined.

The spleen is surrounded by a capsule from which a number of projections (trabeculae) extend into the interior to

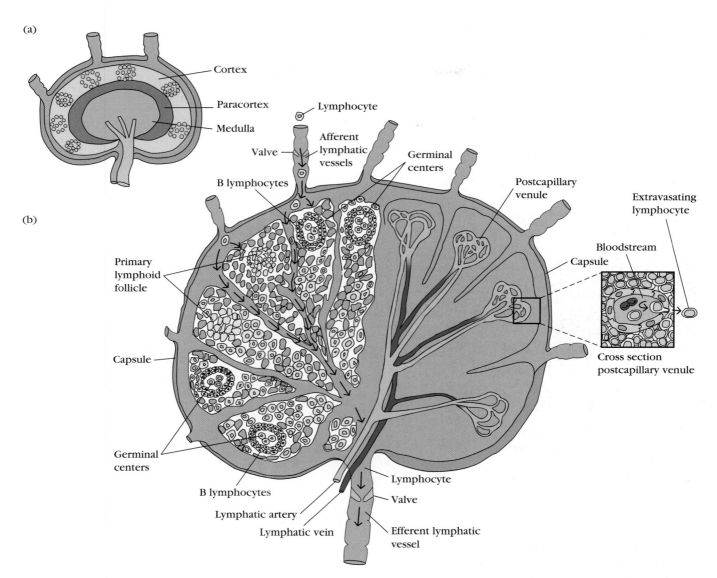

**FIGURE 2-16   Structure of a lymph node.** (a) The three layers of a lymph node support distinct microenvironments. (b) The left side depicts the arrangement of reticulum and lymphocytes within the various regions of a lymph node. Macrophages and dendritic cells, which trap antigen, are present in the cortex and paracortex. $T_H$ cells are concentrated in the paracortex; B cells are primarily in the cortex, within follicles and germinal centers. The medulla is populated largely by antibody-producing plasma cells. Lymphocytes circulating in the lymph are carried into the node by afferent lymphatic vessels; they either enter the reticular matrix of the node or pass through it and leave by the efferent lymphatic vessel. The right side of (b) depicts the lymphatic artery and vein and the postcapillary venules. Lymphocytes in the circulation can pass into the node from the postcapillary venules by a process called extravasation (*inset*).

form a compartmentalized structure. The compartments are of two types, the red pulp and white pulp, which are separated by a diffuse marginal zone (Figure 2-17). The splenic **red pulp** consists of a network of sinusoids populated by macrophages, numerous red blood cells, and few lymphocytes; it is the site where old and defective red blood cells are destroyed and removed. Many of the macrophages within the red pulp contain engulfed red blood cells or iron-containing pigments from degraded hemoglobin. The splenic **white pulp** surrounds the branches of the splenic artery, forming a **periarteriolar lymphoid sheath (PALS)** populated mainly by

T lymphocytes. Primary lymphoid follicles are attached to the PALS. These follicles are rich in B cells, and some of them contain germinal centers. The **marginal zone,** peripheral to the PALS, is populated by lymphocytes and macrophages.

Blood-borne antigens and lymphocytes enter the spleen through the splenic artery, which empties into the marginal zone. In the marginal zone, antigen is trapped by dendritic cells, which carry it to the PALS. Lymphocytes in the blood also enter sinuses in the marginal zone and migrate to the PALS.

The initial activation of B and T cells takes place in the T-cell-rich PALS. Here dendritic cells capture antigen and

(a)

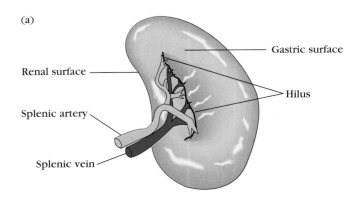

Gastric surface

Renal surface

Hilus

Splenic artery

Splenic vein

(b)

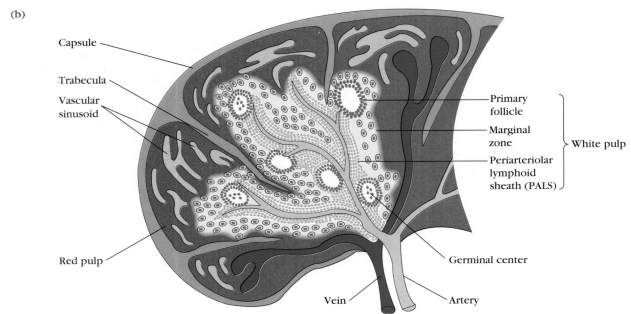

Capsule

Trabecula

Vascular sinusoid

Primary follicle

Marginal zone

Periarteriolar lymphoid sheath (PALS)

White pulp

Red pulp

Germinal center

Vein

Artery

**FIGURE 2-17 Structure of the spleen.** (a) The spleen, which is about 5 inches long in adults, is the largest secondary lymphoid organ. It is specialized for trapping blood-borne antigens. (b) Diagrammatic cross section of the spleen. The splenic artery pierces the capsule and divides into progressively smaller arterioles, ending in vascular sinusoids that drain back into the splenic vein. The erythrocyte-filled red pulp surrounds the sinusoids. The white pulp forms a sleeve, the periarteriolar lymphoid sheath (PALS), around the arterioles; this sheath contains numerous T cells. Closely associated with the PALS is the marginal zone, an area rich in B cells that contains lymphoid follicles that can develop into secondary follicles containing germinal centers.

present it combined with class II MHC molecules to $T_H$ cells. Once activated, these $T_H$ cells can then activate B cells. The activated B cells, together with some $T_H$ cells, then migrate to primary follicles in the marginal zone. On antigenic challenge, these primary follicles develop into characteristic secondary follicles containing germinal centers (like those in the lymph nodes), where rapidly dividing B cells and plasma cells are surrounded by dense clusters of concentrically arranged lymphocytes.

The severity of the consequences if the spleen is lost depends on the age at which splenectomy takes place. In children, splenectomy often leads to an increased incidence of bacterial sepsis caused primarily by *Streptococcus pneumoniae, Neisseria meningitidis,* and *Haemophilus influenzae.* Although fewer adverse effects are experienced by adults, splenectomy does lead to some increase in blood-borne bacterial infections (**bacteremia).**

### Mucosa-Associated Lymphoid Tissue

The mucous membranes lining the digestive, respiratory, and urogenital systems have a combined surface area of about 400 m$^2$ (nearly the size of a basketball court) and are the major sites of entry for most pathogens. These vulnerable membrane surfaces are defended by a group of organized lymphoid tissues mentioned earlier and known collectively as **mucosa-associated lymphoid tissue (MALT).** The secondary lymphoid tissue associated with the respiratory epithelium is designated **bronchus-associated lymphoid tissue (BALT),** and that associated with the epithelium of the digestive tract is collectively called **gut-associated**

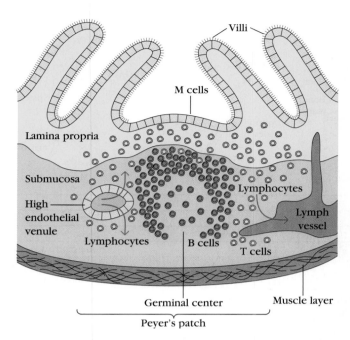

**FIGURE 2-18 Cross-sectional diagram of the mucous membrane lining the intestine, showing a Peyer's patch lymphoid nodule in the submucosa.** The intestinal lamina propria contains loose clusters of lymphoid cells and diffuse follicles.

**m** Structurally, mucosa-associated lymphoid tissue ranges from loose, barely organized clusters of lymphoid cells in the lamina propria of intestinal villi to well-organized structures such as Peyer's patches, which are found within the intestinal lining. MALT also includes the tonsils and the appendix. The functional importance of MALT in the body's defense is attested to by its large population of antibody-producing plasma cells, whose number far exceeds that of plasma cells in the spleen, lymph nodes, and bone marrow combined.

As shown in Figures 2-18 and 2-19, lymphoid cells are found in various regions within this tissue. The outer mucosal epithelial layer contains so-called **m**
**c** many of which are T cells. The lamina propria, which lies under the epithelial layer, contains large numbers of B cells, plasma cells, activated $T_H$ cells, and macrophages in loose clusters. Histologic sections have revealed more than 15,000 lymphoid follicles within the intestinal lamina propria of a healthy child. Peyer's patches, nodules of 30 to 40 lymphoid follicles, extend from the subepithelium to the muscle layers. Like lymphoid follicles in other sites, those that compose Peyer's patches can develop into secondary follicles with germinal centers.

The epithelial cells of mucous membranes play an important role in promoting the immune response by delivering small samples of foreign antigen from the lumina of the respiratory, digestive, and urogenital tracts to the underlying mucosa-associated lymphoid tissue. This antigen transport is carried out by specialized **c** The structure of the M cell is striking: they are flattened epithelial cells lacking the microvilli that characterize the rest of the mucous epithelium. M cells have a deep invagination, or pocket, in the basolateral plasma membrane, which is filled with a cluster of B cells, T cells, and macrophages (Figure 2-19a). Antigens in the intestinal lumen are endocytosed into vesicles that are transported from the luminal membrane (membrane bordering the intestinal lumen) to the underlying pocket membrane.

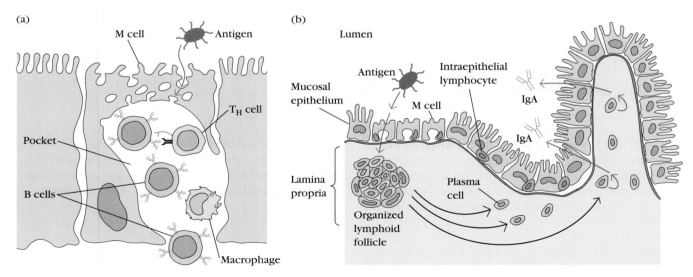

**FIGURE 2-19 Structure of M cells and production of IgA at inductive sites.** (a) M cells, situated in mucous membranes, endocytose antigen from the lumen of the digestive, respiratory, and urogenital tracts. The antigen is transported across the cell and released into the large basolateral pocket. (b) Antigen transported across the epithelial layer by M cells at an inductive site activates B cells in the underlying lymphoid follicles. The activated B cells differentiate into IgA-producing plasma cells, which migrate along the submucosa. The outer mucosal epithelial layer contains intraepithelial lymphocytes, of which many are T cells.

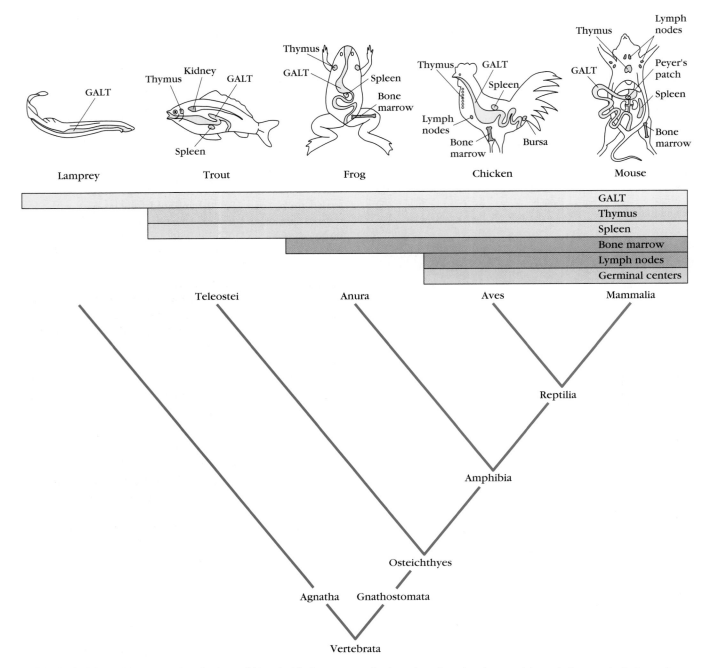

**FIGURE 2-20 Evolutionary distribution of lymphoid tissues.** The presence and location of lymphoid tissues in several major orders of vertebrates are shown. Although they are not shown in the diagram, cartilaginous fish such as sharks and rays have GALT, a thymus, and a spleen. Reptiles also have GALT, a thymus, and a spleen and may also have lymph nodes that participate in immunological reactions. The sites and nature of primary lymphoid tissues in reptiles are under investigation. *[Adapted from Dupasquier and M. Flajnik, 2004, in* Fundamental Immunology, *5th ed., W. E. Paul, ed., Lippincott-Raven, Philadelphia.]*

The vesicles then fuse with the pocket membrane, delivering the potentially response-activating antigens to the clusters of lymphocytes and antigen-presenting cells present, the most important of which are dendritic cells, contained within the pocket. Antigen transported across the mucous membrane by M cells ultimately leads to the activation of B cells that differentiate and then secrete IgA. This class of antibody is specialized for secretion and is an important tool used by the body to combat many types of infection at mucosal sites.

## Cutaneous-Associated Lymphoid Tissue

The skin is an important anatomic barrier to the external environment. It is the largest organ in the body and plays an important role in nonspecific (innate) defenses. The epidermal (outer) layer of the skin is composed largely of specialized epithelial cells called keratinocytes. These cells secrete a number of cytokines that may function to induce a local inflammatory reaction. Scattered among the epithelial-cell matrix of the epidermis are Langerhans cells, a type of dendritic cell, which internalize antigen by phagocytosis or endocytosis. As mentioned earlier, these Langerhans cells then undergo maturation and migrate from the epidermis to regional lymph nodes, where they function as potent activators of naive $T_H$ cells. In addition to Langerhans cells, the epidermis also contains so-called **intraepidermal lymphocytes,** which are mostly T cells. Some immunologists believe that they play a role in combating antigens that enter through the skin, a function for which they are well positioned. The underlying dermal layer of the skin also contains scattered T cells and macrophages. Most of these dermal T cells appear to be either previously activated cells or memory cells.

## Lymphoid Cells and Organs— Evolutionary Comparisons

In the next chapter, we will see that innate systems of immunity are found in vertebrates, invertebrates, and even in plants. Adaptive immunity, which depends on lymphocytes and is mediated by antibodies and T cells, only evolved in the subphylum Vertebrata. However, as shown in Figure 2-20, the kinds of lymphoid tissues seen in different orders of vertebrates differ dramatically.

As one considers the spectrum from the earliest vertebrates, the jawless fishes (Agnatha), to the birds and mammals, it is apparent that evolution has added new organs of immunity, such as lymph nodes and tissues (e.g., Peyer's patches), but has tended to retain those (the thymus, for example) evolved by earlier orders. Whereas all vertebrates have gut-associated lymphoid tissue (GALT) and most have some version of a spleen and thymus, not all have lymph nodes, and many do not generate lymphocytes in bone marrow. The differences seen at the level of organs and tissues are also reflected at the cellular level. So far, neither T nor B lymphocytes, nor other components of an adaptive immune system, have been found in the jawless fishes. In fact, only jawed vertebrates (Gnathosomata), of which the cartilaginous fish (sharks, rays) are the earliest example, have B and T lymphocytes and support adaptive immune responses.

## SUMMARY

- The humoral (antibody) and cell-mediated responses of the immune system result from the coordinated activities of many types of cells, organs, and tissues found throughout the body.
- Many of the body's cells, tissues, and organs arise from different stem cell populations. Leukocytes develop from a pluripotent hematopoietic stem cell during a highly regulated process called hematopoiesis.
- Apoptosis, a type of programmed cell death, is a key factor in regulating the levels of hematopoietic and other cell populations.
- There are three types of lymphoid cells: B cells, T cells, and natural killer (NK) cells. Only B and T cells are members of clonal populations distinguished by antigen receptors of unique specificity. B cells synthesize and display membrane antibody, and T cells synthesize and display T-cell receptors. Most NK cells do not synthesize antigen-specific receptors, although a small subpopulation of this group, NK-T cells, do synthesize and display a T cell receptor.
- Macrophages and neutrophils are specialized for the phagocytosis and degradation of antigens. Macrophages also have the capacity to present antigen to T cells.
- Immature forms of dendritic cells have the capacity to capture antigen in one location, undergo maturation, and migrate to another location, where they present antigen to $T_H$ cells. Dendritic cells are the major population of antigen-presenting cells.
- Primary lymphoid organs are the sites where lymphocytes develop and mature. T cells arise in the bone marrow and develop in the thymus; in humans and mice, B cells arise and develop in bone marrow.
- Secondary lymphoid organs provide sites where lymphocytes encounter antigen, become activated, and undergo clonal expansion and differentiation into effector cells.
- Vertebrate orders differ greatly in the kinds of lymphoid organs, tissues, and cells they possess. The most primitive, the jawless fish, lack B and T cells and cannot mount adaptive immune responses; jawed vertebrates have T and B cells, have adaptive immunity, and display an increasing variety of lymphoid tissues.

## References

Appelbaum, F. R. 1996. Hematopoietic stem cell transplantation. In *Scientific American Medicine.* D. Dale and D. Federman, eds. Scientific American Publishers, New York.

Banchereau J., et al. 2000. Immunobiology of dendritic cells. *Annual Review of Immunology* **18:**767.

Clevers, H. C., and R. Grosschedl. 1996. Transcriptional control of lymphoid development: lessons from gene targeting. *Immunology Today* **17:**336.

Cory, S. 1995. Regulation of lymphocyte survival by the BCL-2 gene family. *Annual Review of Immunology* **12:**513.

Ema, H., et al. 2005. Quantification of self-renewal capacity in single hematopoietic stem cells from normal and Lnk-deficient mice. *Developmental Cell* **6:**907.

Godfrey, D. I., MacDonald, H. R. Kronenberg, M., Smyth, M. J., and Van Kaer, L. (2004). NKT cells: What's in a name? *Nature Reviews Immunology* **4:**231.

Liu, Y. J. 2001. Dendritic cell subsets and lineages, and their functions in innate and adaptive immunity. *Cell* **106**:259.

Melchers, F., and A. Rolink. 1999. B-lymphocyte development and biology. In *Fundamental Immunology,* 4th ed. W. E. Paul, ed. Lippincott-Raven, Philadelphia, p. 183.

Nathan, C., and M. U. Shiloh. 2000. Reactive oxygen and nitrogen intermediates in the relationship between mammalian hosts and microbial pathogens. *Proceedings of the National Academy of Science* **97**:8841.

Pedersen, R. A. 1999. Embryonic stem cells for medicine. *Scientific American* **280**:68.

Osborne, B. A. 1996. Apoptosis and the maintenance of homeostasis in the immune system. *Current Opinion in Immunology* **8**:245.

Piccirillo, C.A., and E. M. Shevach. 2004. Naturally occurring CD4+CD25+ immunoregulatory T cells: central players in the arena of peripheral tolerance. *Seminars in Immunology* **16**:81.

Picker, L. J., and M. H. Siegelman. 1999. Lymphoid tissues and organs. In *Fundamental Immunology,* 4th ed. W. E. Paul, ed. Lippincott-Raven, Philadelphia, p. 145.

Rothenberg, E. V. 2000. Stepwise specification of lymphocyte developmental lineages. *Current Opin. Gen. Dev.* **10**:370.

Shizuru, J. A., R. S. Negrin, and I. L. Weissman. 2005. Hematopoietic stem and progenitor cells: clinical and preclinical regeneration of the hematolymphoid system. *Annual Review of Medicine* **56**:509.

Ward, A. C., et al. 2000. Regulation of granulopoiesis by transcription factors and cytokine signals. *Leukemia* **14**:973.

Weissman, I. L. 2000. Translating stem and progenitor cell biology to the clinic: barriers and opportunities. *Science* **287**:1442.

 ## Useful Web Sites

**http://www.ncbi.nlm.nih.gov/prow**
The PROW guides are authoritative short, structured reviews on proteins and protein families that bring together the most relevant information on each molecule into a single document of standardized format. Information on CD1 through 339 is available.

**http://stemcells.nih.gov/**
Links to information on stem cells, including basic and clinical information, and registries of available embryonic stem cells.

 ## Study Questions

**CLINICAL FOCUS QUESTION** The T and B cells that differentiate from hematopoietic stem cells recognize as self the bodies in which they differentiate. Suppose a woman donates HSCs to a genetically unrelated man whose hematopoietic system was totally destroyed by a combination of radiation and chemotherapy. Suppose further that, although most of the donor HSCs differentiate into hematopoietic cells, some differentiate into cells of the pancreas, liver, and heart. Decide which of the following outcomes is likely and justify your choice.

a. The T cells from the donor HSCs do not attack the pancreatic, heart, and liver cells that arose from donor cells but mount a GVH response against all of the other host cells.

b. The T cells from the donor HSCs mount a GVH response against all of the host cells.

c. The T cells from the donor HSCs attack the pancreatic, heart, and liver cells that arose from donor cells but fail to mount a GVH response against all of the other host cells.

d. The T cells from the donor HSCs do not attack the pancreatic, heart, and liver cells that arose from donor cells and fail to mount a GVH response against all of the other host cells.

1. Explain why each of the following statements is false.

   a. All $T_H$ cells express CD4 and recognize only antigen associated with class II MHC molecules.
   b. The pluripotent stem cell is one of the most abundant cell types in the bone marrow.
   c. Activation of macrophages increases their expression of class I MHC molecules, making the cells present antigen more effectively.
   d. Lymphoid follicles are present only in the spleen and lymph nodes.
   e. Infection has no influence on the rate of hematopoiesis.
   f. Follicular dendritic cells can process and present antigen to T lymphocytes.
   g. All lymphoid cells have antigen-specific receptors on their membrane.
   h. All vertebrates generate B lymphocytes in bone marrow.
   i. All vertebrates produce B or T lymphocytes, and most produce both.

2. For each of the following sets of cells, state the latest common progenitor cell that gives rise to both cell types.

   a. Dendritic cells and macrophages
   b. Monocytes and neutrophils
   c. $T_C$ cells and basophils
   d. Natural killer cells and B cells

3. List the primary lymphoid organs and summarize their functions in the immune response.

4. List the secondary lymphoid organs and summarize their functions in the immune response.

5. What are the two primary characteristics that distinguish hematopoietic stem cells and progenitor cells?

6. What are the two primary roles of the thymus?

7. What do nude mice and humans with DiGeorge's syndrome have in common?

8. At what age does the thymus reach its maximal size?

   a. During the first year of life
   b. Teenage years (puberty)
   c. Between 40 and 50 years of age
   d. After 70 years of age

9. Preparations enriched in hematopoietic stem cells are useful for research and clinical practice. In Weissman's method for enriching hematopoietic stem cells, why is it necessary to use lethally irradiated mice to demonstrate enrichment?

10. Explain the difference between a monocyte and a macrophage.

11. What effect would removal of the bursa of Fabricius (bursectomy) have on chickens?

12. Some microorganisms (e.g., *Neisseria gonorrhoeae, Mycobacterium tuberculosis,* and *Candida albicans*) are classified as intracellular pathogens. Define this term and explain why the immune response to these pathogens differs from that to other pathogens such as *Staphylococcus aureus* and *Streptococcus pneumoniae.*

13. Indicate whether each of the following statements about the spleen is true or false. If you think a statement is false, explain why.

    a. It filters antigens out of the blood.
    b. The marginal zone is rich in T cells, and the periarteriolar lymphoid sheath (PALS) is rich in B cells.
    c. It contains germinal centers.
    d. It functions to remove old and defective red blood cells.
    e. Lymphatic vessels draining the tissue spaces enter the spleen.
    f. Lymph node but not spleen function is affected by a knockout of the Ikaros gene

14. For each type of cell indicated (a–p), select the most appropriate description (1–16) listed below. Each description may be used once, more than once, or not at all.

    Cell Types

    a. _____ Common myeloid progenitor cells

    b. _____ Monocytes

    c. _____ Eosinophils

    d. _____ Dendritic cells

    e. _____ Natural killer (NK) cells

    f. _____ Kupffer cells

    g. _____ Lymphoid dendritic cell

    h. _____ Mast cells

    i. _____ Neutrophils

    j. _____ M cells

    k. _____ Bone marrow stromal cells

    l. _____ Lymphocytes

    m. _____ NK1-T cell

    n. _____ Microglial cell

    o. _____ Myeloid dendritic cell

    p. _____ Hematopoietic stem cell

Descriptions

(1) Major cell type presenting antigen to $T_H$ cells

(2) Phagocytic cell of the central nervous system

(3) Phagocytic cells important in the body's defense against parasitic organisms

(4) Macrophages found in the liver

(5) Give rise to red blood cells

(6) An antigen-presenting cell derived from monocytes that is not phagocytic

(7) Generally first cells to arrive at site of inflammation

(8) Secrete colony-stimulating factors (CSFs)

(9) Give rise to thymocytes

(10) Circulating blood cells that differentiate into macrophages in the tissues

(11) An antigen-presenting cell that arises from the same precursor as a T cell but not the same as a macrophage

(12) Cells that are important in sampling antigens of the intestinal lumen

(13) Nonphagocytic granulocytic cells that release various pharmacologically active substances

(14) White blood cells that migrate into the tissues and play an important role in the development of allergies

(15) Cells that sometimes recognize their targets with the aid of an antigen-specific cell-surface receptor and sometimes by mechanisms that resemble those of natural killer cells

(16) Members of a category of cells not found in jawless fishes

 **Interactive Study**

**www.whfreeman.com/kuby**

 SELF-TEST
Review of Key Terms

 ANIMATION
Cell Death
Cells and Organs of
the Immune System

# chapter 3

# Innate Immunity

**V**ERTEBRATES ARE PROTECTED BY TWO SYSTEMS OF immunity: innate and adaptive. **Innate immunity** consists of the defenses against infection that are ready for immediate activation prior to attack by a pathogen. The innate immune system includes physical, chemical, and cellular barriers. The main physical barriers are skin and mucous membranes. Chemical barriers include the acidity of the stomach contents and specialized soluble molecules that possess antimicrobial activity. The cellular line of innate defense comprises an array of cells with sensitive receptors that detect microbial products and instigate a counterattack. The response to invasion by a microbial agent of infection that overcomes the initial barriers of skin and mucous membranes is rapid, typically initiating within minutes of invasion.

Despite the multiple layers of the innate system, some pathogens may evade the innate defenses. On call is a second system called **adaptive immunity** (or **acquired immunity**), which is induced by exposure to microbes and counters infection with a specific response tailor-made to the attacking pathogen in the form of a large population of B and T lymphocytes that specifically recognize the invader. Raising an adaptive response takes time—as much as a week or more before it is fully effective. Adaptive immunity displays the phenomenon of immunological memory, and once triggered by a particular pathogen, future exposures elicit quicker and often more vigorous responses. Recognition of invaders is mediated by antibodies and T cell receptors, the "sensors" of adaptive immunity. These molecules are products of genes with an extraordinary feature: they undergo modification and diversification—genetic recombination—in the host to generate a unique, gigantic population of sentinels on the lookout for invaders. The processes of modification and generation of immunological diversity are discussed in Chapters 5 and 9.

Innate immunity is the most ancient defense of vertebrates against microbes; some form of innate immunity has been found in all multicellular plants and animals. Adaptive immunity evolved in jawed vertebrates and is a much more recent evolutionary invention than innate immunity. In vertebrates, adaptive immunity complements a well-developed system of innate immunity. Table 3-1 compares innate and adaptive immunity.

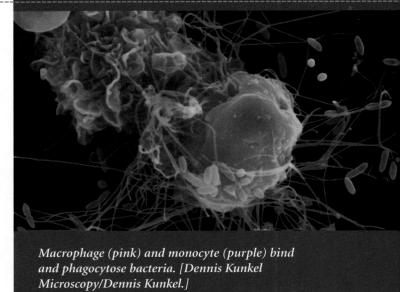

*Macrophage (pink) and monocyte (purple) bind and phagocytose bacteria. [Dennis Kunkel Microscopy/Dennis Kunkel.]*

- Anatomical Barriers
- Connections Between Innate and Adaptive Immunity
- Inflammation
- Soluble Molecules and Membrane-Associated Receptors
- Toll-like Receptors
- Cell Types of Innate Immunity
- Signal Transduction Pathways
- Ubiquity of Innate Immunity

A large and growing body of research has revealed that as innate and adaptive immunity have co-evolved, a high degree of interaction and interdependence has arisen between the two systems. In fact, if a pathogen completely evades the first line of defense, the innate immune system, the response of the adaptive immune system may be quite feeble. Recognition by the innate immune system sets the stage for an effective adaptive immune response.

This chapter describes the components of the innate immune system—physical and physiologic barriers, soluble chemical agents, and several types of cells and their receptors—and illustrates how they act together to defend against infection. We conclude with an overview of innate immunity across the phyla of animals and plants.

| TABLE 3-1 | Innate and adaptive immunity | |
|---|---|---|
| Attribute | Innate immunity | Adaptive immunity |
| Response time | Minutes/hours | Days |
| Specificity | Specific for molecules and molecular patterns associated with pathogens | Highly specific; discriminates even minor differences in molecular structure; details of microbial or nonmicrobial structure recognized with high specificity |
| Diversity | A limited number of germ line–encoded receptors | Highly diverse; a very large number of receptors arising from genetic recombination of receptor genes |
| Memory responses | None | Persistent memory, with faster response of greater magnitude on subsequent infection |
| Self/nonself discrimination | Perfect; no microbe-specific patterns in host | Very good; occasional failures of self/nonself discrimination result in autoimmune disease |
| Soluble components of blood or tissue fluids | Many antimicrobial peptides and proteins | Antibodies |
| Major cell types | Phagocytes (monocytes, macrophages, neutrophils), natural killer (NK) cells, dendritic cells | T cells, B cells, antigen-presenting cells |

## Anatomical Barriers

The most obvious components of innate immunity are the external barriers to microbial invasion—skin and mucous membranes, which include the mucosal epithelia that line the respiratory, gastrointestinal, and urogenital tracts and insulate the body's interior from the pathogens of the exterior world (Figure 3-1). The skin consists of two distinct layers: a thin outer layer, the **epidermis,** and a thicker layer, the **dermis.** The epidermis contains several tiers of tightly packed epithelial cells. The outer epidermal layer consists of mostly dead cells filled with a waterproofing protein called keratin. The dermis is composed of connective tissue and contains blood vessels, hair follicles, sebaceous glands, and sweat glands. Skin and epithelia provide a kind of living "saran wrap" that encases and protects the inner domains of the body from the outer world. But these anatomical barriers are more than just passive wrappers. They also mount active biochemical defenses by synthesizing and deploying peptides and proteins with antimicrobial activity. Among the many such agents produced by human skin, recent research has identified psoriasin, a small protein with potent antibacterial activity against *Escherichia coli*. This finding answered a long-standing question: Why is human skin resistant to colonization by *E. coli* despite constant exposure to it? As shown in Figure 3-2, incubation of *E. coli* on human skin for as little as 30 minutes specifically kills the bacteria. (Other bacterial species are generally less sensitive.) The capacity of skin and epithelia to produce a wide variety of antimicrobial agents is important because breaks in the skin from scratches, wounds, or abrasion provide routes of infection that would be readily exploited by pathogenic microbes if not defended by biochemical means. The skin may also be penetrated by biting insects (e.g., mosquitoes, mites, ticks,

fleas, and sand flies), which can introduce pathogenic organisms into the body as they feed. The protozoan that causes malaria, for example, is deposited in humans by mosquitoes when they take a blood meal, as is the virus causing West Nile fever. Similarly, the bubonic plague bacterium is spread by the bite of fleas, and the bacterium that causes Lyme disease is spread by the bite of ticks.

In place of skin, the alimentary, respiratory, and urogenital tracts and the eyes are lined by mucous membranes that consist of an outer epithelial layer and an underlying layer of connective tissue. Many pathogens enter the body by penetrating these membranes; opposing this entry are a number of nonspecific defense mechanisms. For example, saliva, tears, and mucous secretions wash away potential invaders and also contain antibacterial or antiviral substances. The viscous fluid called mucus, secreted by epithelial cells of mucous membranes, entraps foreign microorganisms. In the lower respiratory tract, the mucous membrane is covered by **cilia,** hairlike protrusions of the epithelial cell membrane. The synchronous movement of cilia propels mucus-entrapped microorganisms from these tracts. With every meal, we ingest huge numbers of microorganisms, but they must run a gauntlet of defenses that begins with the antimicrobial compounds in saliva and in the epithelia of the mouth and includes the hostile mix of acid and digestive enzymes found in the stomach. In addition to the array of biochemical and anatomical defenses, pathogenic microbes must compete for the body's resources with the many nonpathogenic organisms that colonize the mucosal surfaces. These *normal flora,* highly adapted to their internal environment, generally outcompete pathogens for attachment sites and necessary nutrients on epithelial surfaces.

Some organisms have evolved ways to evade the defenses of mucous membranes. For example, influenza virus (the agent

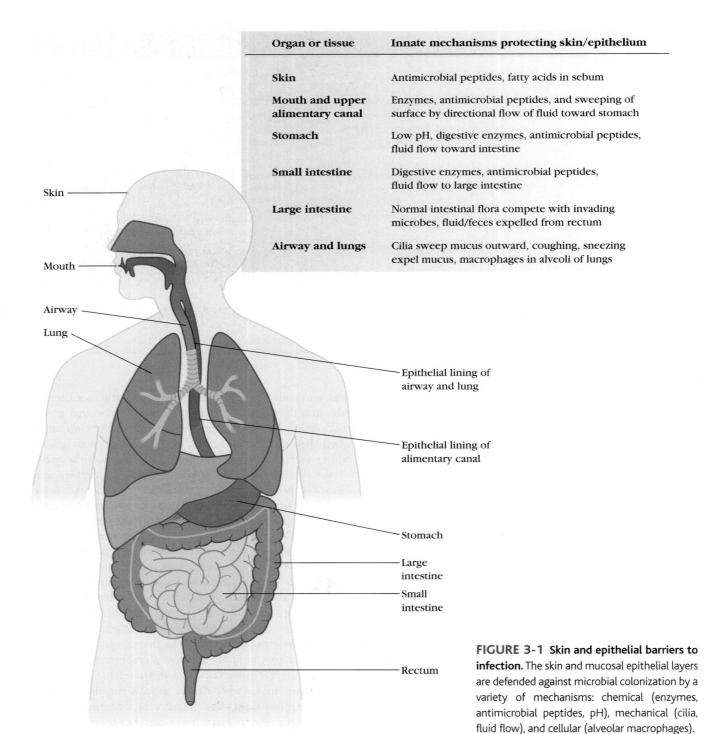

| Organ or tissue | Innate mechanisms protecting skin/epithelium |
| --- | --- |
| **Skin** | Antimicrobial peptides, fatty acids in sebum |
| **Mouth and upper alimentary canal** | Enzymes, antimicrobial peptides, and sweeping of surface by directional flow of fluid toward stomach |
| **Stomach** | Low pH, digestive enzymes, antimicrobial peptides, fluid flow toward intestine |
| **Small intestine** | Digestive enzymes, antimicrobial peptides, fluid flow to large intestine |
| **Large intestine** | Normal intestinal flora compete with invading microbes, fluid/feces expelled from rectum |
| **Airway and lungs** | Cilia sweep mucus outward, coughing, sneezing expel mucus, macrophages in alveoli of lungs |

Skin

Mouth

Airway

Lung

Epithelial lining of airway and lung

Epithelial lining of alimentary canal

Stomach

Large intestine

Small intestine

Rectum

**FIGURE 3-1 Skin and epithelial barriers to infection.** The skin and mucosal epithelial layers are defended against microbial colonization by a variety of mechanisms: chemical (enzymes, antimicrobial peptides, pH), mechanical (cilia, fluid flow), and cellular (alveolar macrophages).

that causes flu) has a surface molecule that enables it to attach firmly to cells in mucous membranes of the respiratory tract, preventing the virus from being swept out by the ciliated epithelial cells. The organism that causes gonorrhea has surface projections that bind to epithelial cells in the mucous membrane of the urogenital tract. Adherence of bacteria to mucous membranes is generally mediated by hairlike protrusions on the bacterium called **fimbriae** or **pili,**

which interact with certain glycoproteins or glycolipids only expressed by epithelial cells of the mucous membrane of particular tissues (Figure 3-3). For these reasons and others, some tissues are susceptible to invasion by particular pathogens, despite the general effectiveness of protective epithelial barriers. When this happens, the receptors of innate immunity play the essential roles of detecting the infection and triggering an effective defense against it.

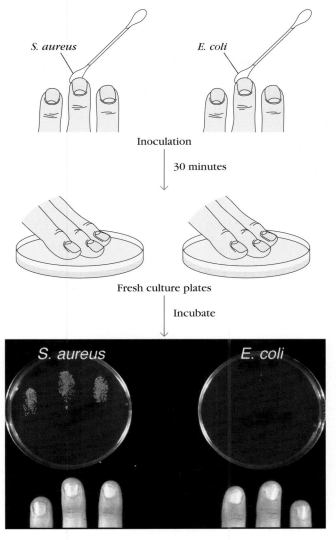

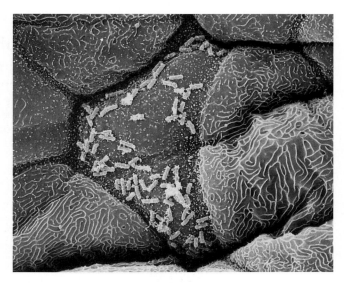

**FIGURE 3-3 Electron micrograph of rod-shaped *E. coli* bacteria adhering to surface of epithelial cells of the urinary tract.** *[From N. Sharon and H. Lis, 1993, Scientific American **268(1)**:85; courtesy of K. Fujita.]*

**FIGURE 3-2 Psoriasin prevents colonization of skin by *E. coli*.** Skin secretes psoriasin, an antimicrobial protein that kills *E. coli*. Fingertips of a healthy human were inoculated with *Staphylococcus aureus (S. aureus)* and *E. coli*. After 30 minutes, the fingertips were pressed on a nutrient agar plate and the number of colonies of *S. aureus* and *E. coli* determined. Almost all of the inoculated *E. coli* were killed; most of the *S. aureus* survived. *[Photograph courtesy of Nature Immunology; from R. Gläser et al., 2005, Nature Immunology **6**:57–64.]*

# Connections Between Innate and Adaptive Immunity

Once a pathogen breaches the nonspecific anatomical and physiologic barriers of the host, infection and disease may ensue. The immune system responds to invasion with two critical functions: sensors detect the invader, and an elaborate response mechanism attacks the invader. The first detection event of the immune response occurs when the invader interacts with soluble or membrane-bound molecules of the host capable of precisely discriminating between self (host) and nonself (pathogen). These molecular sensors recognize broad structural motifs that are highly conserved within a microbial species (and are usually necessary for survival) but are generally absent from the host. Because they recognize particular overall molecular patterns, such receptors are called **pattern recognition receptors (PRRs),** and when such patterns are found on pathogens, they are called **pathogen-associated molecular patterns (PAMPs).** The PAMPs recognized by PRRs include combinations of sugars, certain proteins, particular lipid-bearing molecules, and some nucleic acid motifs. The restriction of innate recognition to molecular patterns found on microbes focuses the innate system on entities that can cause infection rather than substances that are merely nonself, such as artificial hip joints. In contrast, antibodies and T cell receptors, the sensors of adaptive immunity, recognize finer details of molecular structure and can discriminate with exquisite specificity between antigens featuring only slight structural differences. Typically, the ability of pattern recognition receptors to distinguish between self and nonself is flawless, because the molecular pattern targeted by the receptor is produced only by the pathogen and never by the host. This contrasts sharply with the occasional recognition of self antigens by receptors of adaptive immunity, a potentially dangerous malfunction from which autoimmune diseases may arise.

Detection of pathogen-associated molecular patterns by soluble and membrane-bound mediators of innate immunity brings multiple components of immunity into play. The soluble mediators include initiators of the **complement system,** such as **mannose-binding lectin (MBL)** and **C-reactive protein (CRP).** If the pathogen bears PAMPs recognized by these mediators, the complement system (see Chapter 7) will

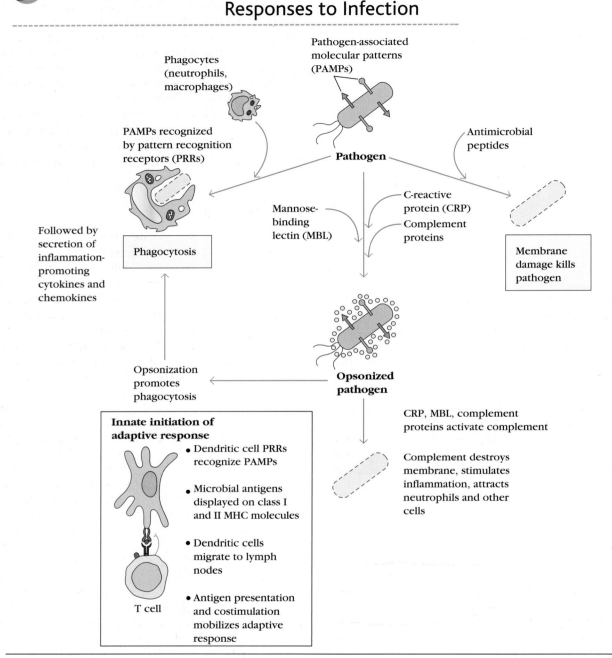

OVERVIEW FIGURE 3-4: Effectors of Innate Immune Responses to Infection

Microbial invasion brings many effectors of innate immunity into play. Entry of microbial invaders through lesions in epithelial barriers generates inflammatory signals and exposes the invaders to attack by different effector molecules and cells. Microbes recognized by C-reactive protein (CRP) or mannose-binding lectin (MBL) are bound by these opsonizing and complement-activating molecules. Some pathogens, such as zymosan-bearing fungi, can activate complement, which can cause direct lysis or opsonization, marking the pathogen for phagocytosis by neutrophils and macrophages. Inflammatory signals cause phagocytes such as macrophages and neutrophils to bind to the walls of blood vessels, extravasate, and move to the site of infection, where they phagocytose and kill infecting microorganisms. During the action of these cellular and molecular effectors, additional inflammatory signals are generated that intensify the response by bringing more phagocytes and soluble mediators (CRP, MBL, and complement) from the bloodstream to the site of infection. Dendritic cells internalize microbial components, mature, and present microbial peptides on MHC molecules. Dendritic cells then migrate through lymphatic vessels to nearby lymph nodes, where they present antigen to T cells. Antigen-activated T cells then initiate adaptive immune responses against the pathogen. Cytokines produced during innate immune responses also support and direct the adaptive immune responses to infection.

be activated. One part of the complement system is a collection of proteins that, when activated, form aggregates that punch holes in the cell membranes of targeted microbes, killing the cells by lysis. The complement system also includes serum glycoproteins that, when activated, promote uptake of microorganisms by phagocytes (opsonization). The complement system straddles the innate and adaptive immune systems: the activation cascade of complement, leading to opsonization or lysis of invaders, can be activated either by molecules that recognize PAMPs (innate) or by antibodies (adaptive) binding to specific foreign antigens. In addition, some of the byproducts of complement activation promote inflammation and thereby bring leukocytes to the site of infection, launching another layer of response.

The invaded tissue's immature dendritic cells and macrophages have a variety of receptors, including the most important group of innate receptors discovered to date: the Toll-like receptors (TLRs), which detect microbial products. Thus far 12 of these receptors have been described for the mouse and 11 for humans; each TLR reacts with a specific microbial product. These versatile receptors, which are described in detail in a later section, allow dendritic cells and macrophages to detect a broad spectrum of pathogens. Signals initiated at the TLRs of macrophages stimulate phagocytic activity and production of chemical agents that are toxic to phagocytosed microbes. Activated macrophages also secrete a class of molecules known collectively as **cytokines,** which are hormone- or growth-factor-like proteins that communicate via cell receptors to induce specific cell activities (see Chapter 12). Like hormones, the cytokines modify the behavior and physiology of target cells and tissues. For example, activated macrophages secrete cytokines such as interleukin-1 (IL-1), interleukin-6 (IL-6), and tumor necrosis factor alpha (TNF-$\alpha$), which induce and support inflammatory responses.

Immature dendritic cells at the site of infection internalize and process the antigen, mature, and then migrate to lymphoid tissue, where their presentation of antigen to T cells is the key step in initiating an adaptive immune response to pathogen invasion. This activity is therefore a bridge between the innate and adaptive systems of immunity. Dendritic cells also secrete a variety of cytokines that promote inflammation and help direct the host's adaptive immune response.

All of the innate immune functions occur early in the course of infection, prior to the generation of significant populations of pathogen-specific T cells and pathogen-specific antibodies from B cells. However, cytokines released by cells involved in the innate response affect the nature of subsequent adaptive immune responses to the infection. The major molecular and cellular effectors employed by the innate immune system to attack infection are summarized in Figure 3-4. In many cases, the innate immune system can defeat and clear an infection by itself. When it cannot, the pathogen faces a coordinated attack by the adaptive immune system. During the adaptive response, cytotoxic T cells detect and destroy pathogens lurking in host cells, and antibodies neutralize the capacity of the invader to infect other cells while increasing the likelihood that the invader will be phagocytosed by macrophages and neutrophils (antibody-mediated uptake, a type of opsonization). Antibodies also collaborate with the complement system to bring about the lysis of pathogenic microbes. After the infection is cleared, some of the B and T cells generated during the adaptive phase of the response will persist in the host for long periods in the form of memory T and memory B cells. Future infections by the pathogen will then be met by a ready reserve of lymphocytes specific for the pathogen and capable of mounting a rapid response.

This thumbnail sketch introduces the major players in the innate immune system and its linkage to the adaptive system. The rest of this chapter will describe the components and mechanisms of the innate immune system in more detail.

------------------------

# Inflammation

When pathogens breach the outer barriers of innate immunity—skin and mucous membranes—the attending infection or tissue injury can induce a complex cascade of events known as the **inflammatory response.** Inflammation may be acute, for example in response to tissue damage, or it may be chronic, leading to pathologic consequences such as arthritis and the wasting associated with certain cancers (see Chapter 13). The acute inflammatory response combats the early stages of infection and initializes processes that result in the repair of damaged tissue. The hallmarks of a localized inflammatory response were first described by the Romans almost 2000 years ago: swelling (Latin = *tumor*), redness (*rubor*), heat (*calor*) and pain (*dolor*). An additional mark of inflammation added in the second century by the physician Galen is loss of function (*functio laesa*). Within minutes after tissue injury, there is an increase in vascular diameter (vasodilation), resulting in a rise of blood volume in the area. Higher blood volume heats the tissue and causes it to redden. Vascular permeability also increases, leading to leakage of fluid from the blood vessels, particularly at postcapillary venules. This results in an accumulation of fluid (**edema**) that swells the tissue. Within a few hours, leukocytes adhere to endothelial cells in the inflamed region and pass through the walls of capillaries and into the tissue spaces, a process called **extravasation** (Figure 3-5). These leukocytes phagocytose invading pathogens and release molecular mediators that contribute to the inflammatory response and the recruitment and activation of effector cells.

Among the mediators are low molecular weight regulatory proteins of the cytokine family mentioned above. Cytokines are secreted by white blood cells and various other cells in the body in response to stimuli and play major roles in regulating the development and behavior of immune effector cells. **Chemokines** (see Chapter 13) are a major subgroup of cytokines whose signature activity is their capacity to act as **chemoattractants** (agents that cause cells to move toward higher concentrations of the agent). However, not all chemoattractants are chemokines. Other important chemoattractants are the byproducts of complement

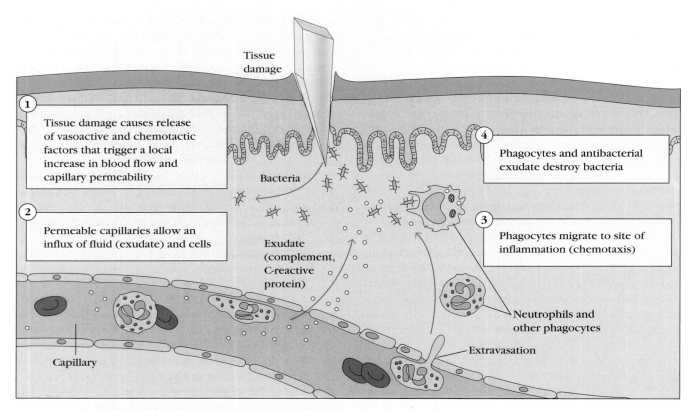

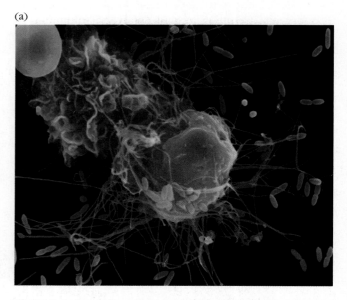

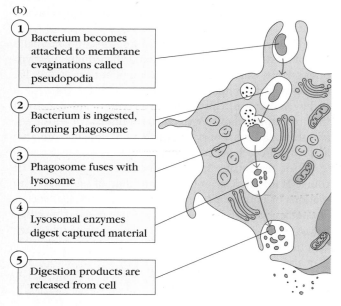

**FIGURE 3-5 Recruitment of macrophages and antimicrobial agents from the bloodstream.** Bacterial entry through wounds initiates an inflammatory response that brings antimicrobial substances and phagocytes (first neutrophils and then macrophages and monocytes) to the site of infection.

**FIGURE 3-6 (a) Electron micrograph of macrophage (center, pink) attacking *Escherichia coli* (green).** The bacteria are phagocytosed as described in part b, and breakdown products are secreted. The monocyte (purple, top left) has been recruited to the vicinity by soluble factors secreted by the macrophage. The red sphere is an erythrocyte. (b) Steps in the phagocytosis of a bacterium. *[Part a, Dennis Kunkel Microscopy/Dennis Kunkel.]*

(C5a, C3a) and various *N*-formyl peptides produced by the breakdown of bacterial proteins during an infection. Binding of chemokines or other chemoattractants to receptors on the membrane of neutrophil cells triggers an activating signal that induces a conformational change in a molecule of the neutrophil membrane called **integrin,** increasing its affinity for **intercellular adhesion molecules (ICAMs)** on the endothelium. Although there are many different chemoattractants, chemokines are the most important and versatile regulators of leukocyte traffic, selectively controlling the adhesion, chemotaxis, and activation of a variety of leukocyte subpopulations. Inflammatory chemokines are typically induced in response to infection or the response of cells to **proinflammatory** (inflammation-promoting) **cytokines.** Chemokines cause leukocytes to move into various tissue sites by inducing the adherence of these cells to the vascular endothelium lining the walls of blood vessels. After migrating into tissues, leukocytes move by chemotaxis toward the higher localized concentrations of chemokines at the site of infection. Thus, targeted phagocytes and effector lymphocyte populations are attracted to the focus of the inflammation.

A major role of the cells attracted to the inflamed site is phagocytosis of invading organisms. Elie Metchnikoff described the process of phagocytosis in the 1880s and ascribed to it a major role in immunity. He hypothesized that white blood cells engulfing and killing pathogens were the major effectors in immunity, more critical, in his judgment, than defenses mediated by serum components (antibodies). Metchnikoff was correct in assigning a critical role to the process of phagocytosis, and we now know that lack of this function leads to severe immunodeficiency. The general process of phagocytosis of bacteria is shown in Figure 3-6. The microorganism is engulfed and lysed within the macrophage, and the products of lysis are discharged. These products include molecules carrying PAMPs, which alert cell receptors to the presence of the pathogen. The assembly of leukocytes at sites of infection, orchestrated by chemokines, is an essential stage in the focused response to infection.

Finally, some signals generated at sites of inflammation are carried systemically to other parts of the body, where they induce changes that support the innate immune response (discussed below in the section describing the acute phase response).

## Leukocyte extravasation is a highly regulated, multistep process

The tightly regulated process of extravasation is responsible for the migration of leukocytes from the bloodstream to sites of infection. As an inflammatory response develops, various cytokines and other inflammatory mediators act on the endothelium of local blood vessels, inducing increased expression of **cell adhesion molecules (CAMs).** The affected epithelium is said to be inflamed or activated. Since neutrophils are generally the first cell type to bind to inflamed endothelium and extravasate into the tissues, we will focus on their entry, bearing in mind that other leukocytes use similar mechanisms. Extravasation presents the neutrophil with a number of formidable challenges. First, they must recognize the inflamed endothelium; they must adhere strongly so that they are not swept away by the flowing blood; and while clinging to the vessel wall, the neutrophils must penetrate the endothelial layer and gain access to the underlying tissue.

Neutrophil extravasation can be divided into four steps: (1) rolling, (2) activation by chemoattractant stimulus, (3) arrest and adhesion, and (4) transendothelial migration (Figure 3-7a). In the first step, neutrophils attach loosely to the endothelium by a low-affinity interaction between glycoproteins—mucins on the neutrophil, selectins on the endothelial cell (Figure 3-7b). In the absence of additional signals, the weak interactions tethering the neutrophil to the endothelial cell are quickly broken by shearing forces as the circulating blood sweeps past the cell. As regions of the neutrophil surface successively bind and break free, the neutrophil tumbles end over end along the surface of the endothelium.

As the neutrophil rolls along the endothelium, it may encounter chemokines or other chemoattractants that have arisen from the site of an inflammatory process. Subsequent interaction between integrins and ICAMs stabilizes adhesion of the neutrophil to the endothelial cell, enabling the cell to then force its way between cells of the endothelium.

# Soluble Molecules and Membrane-Associated Receptors

The innate immune system is multifaceted, utilizing a variety of soluble molecules as well as cell membrane–bound receptors as its effectors. Certain of the soluble molecules are produced at the site of infection or injury and act locally. These include antimicrobial peptides such as defensins and cathelicidins as well as the **interferons,** an important group of cytokines with antiviral action, discussed below and more fully in Chapter 12. Other soluble effectors are produced at distant sites and arrive at their target tissues via the bloodstream. Complement proteins and acute phase proteins fall into this group. The nature of these effectors and their contributions to host defense are discussed below.

## Antimicrobial peptides contribute to the innate defense against bacteria and fungi

Peptides with antimicrobial activity have been isolated from sources as diverse as humans, frogs, flies, nematodes, and several species of plants (Table 3-2). The early evolution and retention of this defensive strategy, coupled with the identification of more than 800 different antimicrobial peptides, testifies to their effectiveness. They range in size from six to 59 amino acid residues, and most are positively charged (cationic), like the magainins found in the skin of frogs and the defensins present in humans and other species. Human

(a) Rolling and extravasation

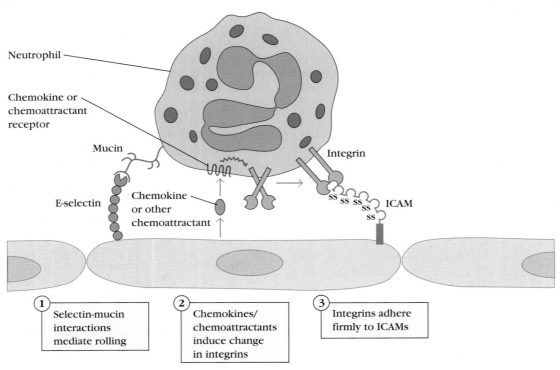

FIGURE 3-7 (a) Neutrophil rolling and extravasation. (b) Cell adhesion molecules and chemokines involved in neutrophil extravasation. Rolling is mediated by transient binding of selectins on the vascular endothelium to mucins on the neutrophil. Chemokines or other chemoattractants that bind to a specific receptor on the neutrophil activate a signal transduction pathway, resulting in a conformational change in integrin molecules that enables them to adhere firmly to intracellular adhesion molecules on the surface of endothelial cells.

defensins are cationic peptides, 29 to 35 amino acid residues in length, with six invariant cysteines that form two to three disulfide bonds, stabilizing relatively rigid three-dimensional structures. They kill a wide variety of bacteria, including *Staphylococcus aureus, Streptococcus pneumoniae, E. coli, Pseudomonas aeruginosa,* and *Hemophilus influenzae.* Neutrophils are rich sources of these peptides, but there are other sources as well: paneth cells secrete defensins into the intestine, and epithelial cells of the pancreas and kidney release defensins into the serum. Defensins kill microbes rapidly, typically within minutes. Even slow-acting antimicrobial peptides kill within 90 minutes.

Antimicrobial peptides often work by disrupting microbial membranes. How these compounds discriminate between microbial and host membranes is a matter of current research interest. Although membrane disruption is a major mechanism of action, antimicrobial peptides also produce a variety of intracellular effects, such as inhibiting the synthesis of DNA, RNA, or proteins, and activating antimicrobial enzymes that lyse components of the microbe. Antimicrobial peptides attack not just bacteria and fungi but viruses as well, having been shown to effectively target the lipoprotein envelope of some enveloped viruses such as influenza virus and some herpesviruses. The breadth of these activities, their

| TABLE 3-2 | Some antimicrobial peptides | |
|---|---|---|
| **Peptide** | **Typical producer species\*** | **Typical microbial activity\*** |
| Defensin family<br>α-Defensins | Human (found in paneth cells of intestine and in cytoplasmic granules of neutrophils) | Antibacterial |
| β-Defensins | Human (found in epithelia and other tissues) | Antibacterial |
| Cathelicidins | Human, bovine | Antibacterial |
| Magainins | Frog | Antibacterial; antifungal |
| Cercropins | Silk moth | Antibacterial |
| Drosomycin | Fruit fly | Antifungal |
| Spinigerin | Termite | Antibacterial; antifungal |

\*In many cases, production of the indicated antimicrobial peptide or family is not limited to the typical producer but is produced by many different species. Also, some members of the indicated peptide or family may have broader antimicrobial activity than the typical one indicated.

proven antimicrobial effectiveness, and the increasing emergence of resistance to existing antibiotics has stimulated investigation of the suitability of antimicrobial peptides for therapeutic use. However, questions remain about their in vivo stability, toxicity, and efficiency when administered in a clinical setting. Concerns have also been raised about the danger that bacteria might rapidly acquire resistance to these antimicrobials if they are used widely, undermining an essential tier of innate immunity against infection. For these reasons, antimicrobial peptides are not yet in clinical use.

## Proteins of the acute phase response contribute to innate immunity

During the 1920s and 1930s, before the introduction of antibiotics, much attention was given to controlling pneumococcal pneumonia. Researchers noted changes in the concentration of several serum proteins during the acute phase of the disease, the phase preceding recovery or death. The serum changes were collectively called the **acute phase response (APR)**, and the proteins whose concentrations rose or fell during the acute phase are still called **acute phase response proteins (APR proteins)**. The physiological significance of many APR proteins is still not understood, but we now know that some, such as components of the complement system and C-reactive protein, are part of the innate immune response to infection. The acute phase response (discussed fully in Chapter 13) is induced by signals that travel through the blood from sites of injury or infection. The liver is one of the major sites of APR protein synthesis, and the proinflammatory cytokines TNF-α, IL-1, and IL-6 are the major signals responsible for induction of the acute phase response. Production of these cytokines is one of the early responses of phagocytes, and the increase in the level of C-reactive protein and other acute phase proteins such as complement contribute to defense in several ways. C-reactive protein belongs to a family of pentameric proteins called

**pentraxins,** which bind ligands in a calcium-dependent reaction. Among the ligands recognized by CRP are a polysaccharide found on the surface of pneumococcal species and phosphorylcholine, which is present on the surface of many microbes. C-reactive protein bound to these ligands on the surface of a microbe promotes uptake by phagocytes and activates a complement-mediated attack on the microbe. (The Clinical Focus box on p. 66 discusses the link between CRP's role in inflammation and heart disease.) Mannose-binding lectin is an acute phase protein that recognizes mannose-containing molecular patterns found on microbes but not on vertebrate cells. Mannose-binding lectin, too, directs a complement attack on the microbes to which it binds.

## Innate immunity uses a variety of receptors to detect infection

A number of pattern recognition molecules have been identified; several examples appear in Table 3-3. The Toll-like receptors are perhaps the most important of these and are discussed below. Others are present in the bloodstream and tissue fluids as soluble, circulating proteins or bound to the membranes of macrophages, neutrophils, and dendritic cells. MBL and CRP, discussed above, are soluble pattern recognition receptors that bind to microbial surfaces, promoting phagocytosis or making the invader a likely target of complement-mediated lysis. Yet another soluble receptor of the innate immune system, lipopolysaccharide-binding protein (LBP), is an important part of the system that recognizes and signals a response to lipopolysaccharide, a component of the outer cell wall of gram-negative bacteria. NOD proteins (from *n*ucleotide-binding *o*ligomerization *d*omain) are the most recent group of receptors found to play roles in innate immunity. These proteins are cytosolic, and two members of this family, NOD1 and NOD2, recognize products derived from bacterial peptidoglycans. NOD1 binds to tripeptide products of peptidoglycan

| TABLE 3-3 | Receptors of the innate immune system | |
|---|---|---|
| **Receptor (location)** | **Target (source)** | **Effect of recognition** |
| Complement (bloodstream, tissue fluids) | Microbial cell wall components | Complement activation, opsonization, lysis |
| Mannose-binding lectin (MBL) (bloodstream, tissue fluids) | Mannose-containing microbial carbohydrates (cell walls) | Complement activation, opsonization |
| C-reactive protein (CRP) (bloodstream, tissue fluids) | Phosphatidylcholine, pneumococcal polysaccharide (microbial membranes) | Complement activation, opsonization |
| Lipopolysaccharide (LPS) receptor;* LPS-binding protein (LBP) (bloodstream, tissue fluids) | Bacterial lipopolysaccharide (gram-negative bacterial cell walls) | Delivery to cell membrane |
| Toll-like receptors (cell surface or internal compartments) | Microbial components not found in hosts | Induces innate responses |
| NOD[†] family receptors (intracellular) | Bacterial cell wall components | Induces innate responses |
| Scavenger receptors (SRs) (cell membrane) | Many targets; gram-positive and gram-negative bacteria, apoptotic host cells | Induces phagocytosis or endocytosis |

\* LPS is bound at the cell membrane by a complex of proteins that includes CD14, MD-2, and a TLR (usually TLR4).
[†] Nucleotide-binding oligomerization domain.

**FIGURE 3-8 Severe fungal infection in a fruit fly (color) with a disabling mutation in the signal-transduction pathway required for the synthesis of the antifungal peptide drosomycin.** *[Electron micrograph adapted from B. Lemaitre et al., 1996, Cell 86:973; courtesy of J. A. Hoffman, University of Strasbourg.]*

breakdown, and NOD2 recognizes muramyl dipeptide, derived from the degradation of peptidoglycan from gram-positive bacterial cell walls. Pattern recognition receptors found on the cell membrane include scavenger receptors (SRs) that are present on macrophages and many types of dendritic cells. SRs are involved in the binding and internalization of gram-positive and gram-negative bacteria as well as the phagocytosis of apoptotic host cells. The roles and mechanisms of scavenger receptors are under active investigation.

## Toll-like Receptors

The protein Toll first attracted attention during the 1980s, when researchers in Germany found that developing flies could not establish a proper dorsal-ventral axis without Toll. (*Toll,* referring to the mutant flies' bizarrely scrambled anatomy, means "weird" in German slang.) Toll is a trans-membrane signal receptor protein; related molecules with roles in innate immunity came to be known as **Toll-like receptors (TLRs).** Three recent discoveries ignited an explosion of knowledge about the central role of TLRs in innate immunity. The first observation came from the fruit fly. In 1996, Jules Hoffman and Bruno Lemaitre found that mutations in Toll, previously known to function in fly development, made flies highly susceptible to lethal infection with *Aspergillus fumigatus,* a fungus to which wild-type flies were immune (Figure 3-8). This landmark experiment convincingly demonstrated the importance of pathogen-triggered immune responses in a nonvertebrate organism. A year

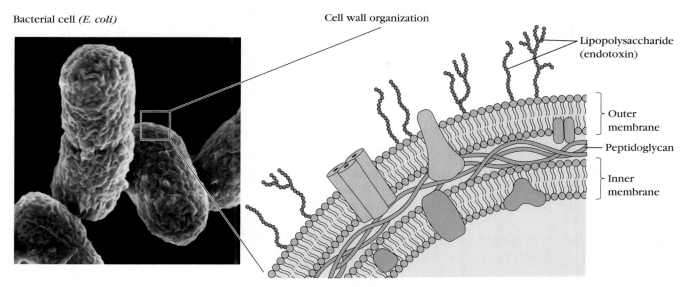

Bacterial cell *(E. coli)*

Cell wall organization

Lipopolysaccharide (endotoxin)

Outer membrane

Peptidoglycan

Inner membrane

**FIGURE 3-9 Lipopolysaccharide (LPS) in the cell wall of *E. coli*.** LPS is a powerful stimulus of innate immunity. *[Photograph from Gary Gaugler/Visuals Unlimited.]*

later, in 1997, Ruslan Medzhitov and Charles Janeway discovered that a certain human protein, identified by homology between its cytoplasmic domain and that of Toll, activated the expression of immune response genes when transfected into a human experimental cell line. This human protein was subsequently named TLR4. This was the first evidence that an immune response pathway was conserved between fruit flies and humans. In 1998, proof that TLRs are part of the normal immune physiology of mammals came from studies with mutant mice conducted in the laboratory of Bruce Beutler. Mice homozygous for the *lps* locus were resistant to lipolysaccharide (LPS), also known as endotoxin, which comes from the cell walls of gram-negative bacteria (Figure 3-9). In humans, a buildup of endotoxin from severe bacterial infection can cause septic shock, a life-threatening condition in which vital organs such as the brain, heart, kidney, and liver may fail. Each year, about 20,000 people die of septic shock caused by gram-negative infections, so it was striking that some mutant strains of mice were resistant to fatal doses of LPS. DNA sequencing revealed that the mouse *lps* gene encoded a mutant form of a Toll-like receptor, TLR4, which differed from the normal form by a single amino acid. This work provided an unequivocal demonstration that TLR4 is indispensable for the recognition of LPS and showed that TLRs do indeed play a role in normal immunophysiology. In a very few years, the work of many investigators has determined that there are many TLRs. So far, 11 have been found in humans and 12 in mice.

Toll-like receptors are membrane-spanning proteins that share a common structural element in their extracellular region, repeating segments of 24 to 29 amino acids containing the sequence xLxxLxLxx (x is any amino acid and L is leucine). These structural motifs are called leucine-rich repeats (**LRRs**; Figure 3-10). All TLRs contain several LRRs, and a subset of the LRRs make up the extracellular ligand-binding region of

the TLR. The intracellular domain of TLRs is called the TIR domain, from *Toll/IL-1 receptor*, referring to the similarity between the cytoplasmic domains of TLRs and the comparable region of a receptor for IL-1, an important regulatory

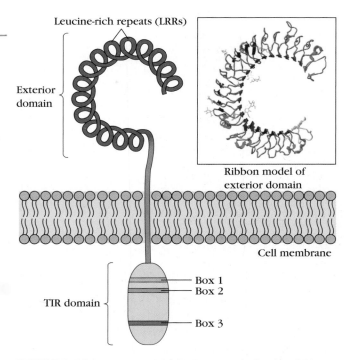

Leucine-rich repeats (LRRs)

Exterior domain

Ribbon model of exterior domain

Cell membrane

TIR domain

Box 1
Box 2

Box 3

**FIGURE 3-10 Structure of a Toll-like receptor (TLR).** Toll-like receptors have an exterior region that contains many leucine-rich repeats (LRRs), a membrane-spanning domain, and an interior domain called the TIR domain. The ligand-binding site of the TLR is found among the LRRs. The TIR domain interacts with the TIR domains of other members of the TLR signal-transduction pathway; three highly conserved sequences of amino acids called box 1, 2, and 3 are essential for this interaction and are characteristic features of TIR domains.

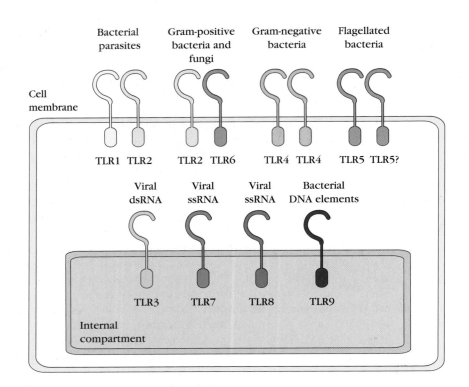

**FIGURE 3-11 Toll-like receptors and their ligands.** TLRs that interact with extracellular ligands reside in the plasma membrane; TLRs that bind ligands generated within the cell are localized to intracellular membranes. Some TLRs form dimers with other TLRs; TLR4 dimerizes with itself (and TLR5 might do so). Other TLRs may function as monomers, or dimeric partners may yet be discovered.

| TLRs | Ligands | Target microbes |
|------|---------|-----------------|
| TLR1 | Triacyl lipopeptides | Mycobacteria |
| TLR2 | Peptidoglycans<br>GPI-linked proteins<br>Lipoproteins<br>Zymosan | Gram-positive bacteria<br>Trypanosomes<br>Mycobacteria<br>Yeasts and other fungi |
| TLR3 | Double-stranded RNA (dsRNA) | Viruses |
| TLR4 | LPS<br>F-protein | Gram-negative bacteria<br>Respiratory syncytial virus (RSV) |
| TLR5 | Flagellin | Bacteria |
| TLR6 | Diacyl lipopeptides<br>Zymosan | Mycobacteria<br>Yeasts and fungi |
| TLR7 | Single-stranded RNA (ssRNA) | Viruses |
| TLR8 | Single-stranded RNA (ssRNA) | Viruses |
| TLR9 | CpG unmethylated dinucleotides<br>Dinucleotides<br>Herpesvirus infection | Bacterial DNA<br><br>Some herpesviruses |
| TLR10,11 | Unknown | Unknown |

molecule. As shown in Figure 3-10, TIR domains have three regions, highly conserved among all members of the TIR family, called boxes 1, 2, and 3, that serve as binding sites for intracellular proteins participating in the signaling pathways mediated by TLRs.

Functions have been determined for nine of the 11 TLRs present in humans. Strikingly, each TLR detects a distinct repertoire of highly conserved pathogen molecules. The complete set of TLRs present in a mouse or human can detect a broad variety of viruses, bacteria, fungi, and even some protozoa. The set of human TLRs whose functions and ligands have been determined appear in Figure 3-11. It is notable that the ligands that bind TLRs are indispensable components of pathogens: a virus could not function without its nucleic acid, gram-negative bacteria cannot be constructed without their LPS-containing walls, and fungi must incorporate the polysaccharide zymosan in their cell walls. Pathogens simply do not have the option of mutating to forms that lack the essential building blocks recognized by TLRs. As shown in Figure 3-11, TLRs that recognize extracellular ligands are found on the surface of cells, whereas those that recognize intracellular ligands, such as viral RNA or fragments of DNA from bacteria, are localized in intracellular compartments.

Several Toll-like receptors, TLRs 1, 2, 4, and 6, operate as dimers (in some cases, additional proteins are incorporated in the complexes formed). One of this set, TLR4, pairs with itself (forming a homodimer), and the others form complexes with another TLR (heterodimers). Partners have yet to be found for TLRs 3, 7, 8, and 9, which may act as monomers, and some data suggest that TLR5 may exist as a homodimer.

The pairing of TLRs affects their specificity. TLR2 coupled with TLR6 binds a wide variety of molecular classes found in microbes, including peptidoglycans, zymosans, and bacterial lipopeptides. When paired with TLR1, however, TLR2 recognizes bacterial lipoproteins and some characteristic surface proteins of parasites. TLR4 is the key receptor for most bacterial lipopolysaccharides. TLR5 recognizes flagellin, the major structural component of bacterial flagella. TLR3 recognizes the double-stranded RNA (dsRNA) that appears in cells after infection by RNA viruses, and viral single-stranded RNA (ssRNA) is the ligand for TLR8 and TLR7. Finally, TLR9 recognizes and initiates a response to the DNA sequence CpG (unmethylated cytosine linked to guanine). Unmethylated sequences such as this are abundant in microbial DNA and much less common in vertebrate DNA.

## Cell Types of Innate Immunity

Innate immune responses typically involve the participation of many different cell types. Key actors are neutrophils, macrophages, monocytes, natural killer cells, and dendritic cells.

The roles of the major cell types in innate immunity are highlighted in Figure 3-12.

## Neutrophils are specialized for phagocytosis and killing

Neutrophils are the first cells to migrate from the blood to sites of infection, and they arrive with a vast array of weapons to deploy against infecting agents. They are essential for the innate defense against bacteria and fungi. Although phagocytosis is the neutrophil's main weapon against invaders, other mechanisms contribute to the containment and elimination of pathogens. Neutrophils display several Toll-like receptors on their surfaces. TLR2 allows them to detect the peptidoglycans of gram-positive bacteria, and TLR4 detects the lipopolysaccharide present on the cell walls of gram-negative microbes. In addition to TLRs, there are other pattern recognition receptors on the surface of the neutrophil.

Although neutrophils are capable of direct recognition of pathogens, binding and phagocytosis improve dramatically when microbes are marked (opsonized) by the attachment of antibody, complement components, or both. Even in the absence of antigen-specific antibodies, the complement proteins in serum can deposit protein fragments on the surface of intruding pathogens that facilitate binding by neutrophils, followed by rapid phagocytosis.

In neutrophils, monocytes, and macrophages, two additional antimicrobial devices, oxidative and nonoxidative attack, contribute to a multipronged, coordinated, and highly effective defense. The oxidative arm employs reactive oxygen species (ROS) and reactive nitrogen species (RNS). The reactive oxygen species are generated by the **NADPH phagosome oxidase (phox)** enzyme complex. Phagocytosed microbes are internalized in vacuoles called phagosomes, in which reactive oxygen species are used as microbicides. The oxygen consumed by phagocytes to support ROS production by the phox enzyme is provided by a metabolic process known as the **respiratory burst**, during which oxygen

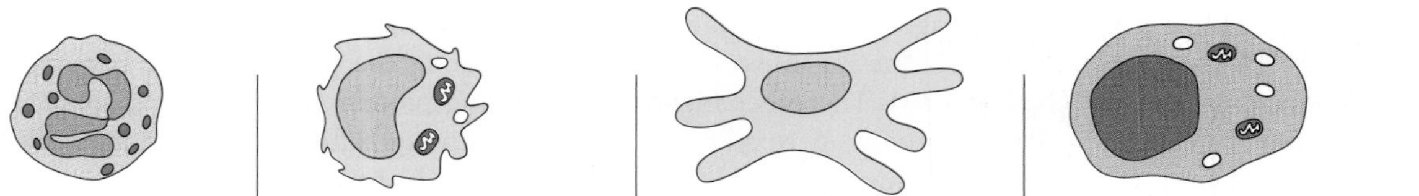

| Cell type | Neutrophils | Macrophages | Dendritic cells | Natural killer cells |
|---|---|---|---|---|
| **Function** | Phagocytosis<br>Reactive oxygen and nitrogen species<br>Antimicrobial peptides | Phagocytosis<br>Inflammatory mediators<br>Antigen presentation<br>Reactive oxygen and nitrogen species<br>Cytokines<br>Complement proteins | Antigen presentation<br>Costimulatory signals<br>Reactive oxygen species<br>Interferon<br>Cytokines | Lysis of viral-infected cells<br>Interferon<br>Macrophage activation |

**FIGURE 3-12 The major leukocytes of innate immunity.** Monocytes, not shown here, have many of the same capabilities as macrophages.

## CLINICAL FOCUS

# C-Reactive Protein Is a Key Marker of Cardiovascular Risk

**Cardiovascular** disease[1] is the leading killer in the United States and Europe, and only infectious disease causes more deaths worldwide. The most frequent cause of cardiovascular disease is atherosclerosis, the progressive accumulation of lipids and fibrous elements in the arteries. Atherosclerosis is a complex disease still far from completely understood. However, a growing body of evidence identifies inflammation as an important factor in the progression of atherosclerosis. A connection between inflammation, the immune system, and artery disease was first suggested by studies in which animals were fed an atherosclerosis-inducing diet and then the walls of the arteries were examined and compared with those of control animals. Light microscopy revealed that the arterial walls of the controls were free of leukocytes, whereas many were found to be firmly attached to the arterial walls of animals fed the atherosclerosis-inducing diet. This was surprising because arterial blood flow normally prevents firm adhesion of leukocytes to arterial walls. Further studies have shown

that leukocytes are important in the development of atherosclerotic plaques. In the initial stages of the disease, monocytes adhere to the arterial walls and migrate through the layer of endothelial cells and differentiate into macrophages. Scavenger receptors on the macrophages bind lipoprotein particles and internalize them, accumulating lipid droplets and assuming a "foamy" appearance. These foamy macrophages secrete proteolytic enzymes, reactive oxygen species (ROS), and cytokines. The proteases degrade the local extracellular matrix, which undergoes some remodeling during repair processes. The cytokines and ROS intensify inflammation, and more cells and lipids migrate into the newly forming plaque, leaving the artery narrowed and more susceptible to blockage. Blocked arteries in the heart are called myocardial infarctions. They choke off blood flow to regions of the heart, denying oxygen to the muscle served by the occluded vessel—a "heart attack." In a significant percentage of cases, the first heart attack is fatal. It is therefore advantageous to identify individuals at risk of a first heart

attack so that preventative therapies and lifestyle changes can be instituted.

The connection between inflammation and plaque formation has led researchers to examine inflammatory markers as predictors of cardiovascular events. A recent study of men and women measured the blood levels of several markers of inflammation, including interleukin-6 (IL-6), soluble tumor necrosis factor alpha receptors (TNF-$\alpha$), and the acute phase response protein CRP. In addition to the inflammatory markers just mentioned, the study also examined the more traditional risk factor, cholesterol, and its related markers LDL and HDL.[2] The risk associated with inflammatory markers was compared with the risk associated with cholesterol-related markers. Thorough medical histories were taken, and the data were adjusted for the risk associated with the following medical history and lifestyle factors known to increase the risk of heart disease:

- Parental history of coronary heart disease before age 60
- Excessive alcohol intake
- Smoking
- Obesity
- Inadequate physical activity
- Hypertension (high blood pressure)
- Diabetes

uptake by the cell increases severalfold. The ROS include a mix of superoxide anion ($\cdot O_2^-$), hydrogen peroxide ($H_2O_2$), and hypochlorous acid (HOCl), the active component of household bleach. The generation of ROS by neutrophils and macrophages is triggered by phagocytosis, which activates the NADPH phagosome oxidase. The enzyme complex then produces superoxide (Figure 3-13). The other highly toxic reactive oxygen species (hydrogen peroxide and hypochlorous acid) are generated from superoxide.

As shown in Figure 3-13, reaction of nitric oxide with superoxide generates reactive nitrogen species. Thus, the respiratory burst contributes to both ROS and RNS production. The importance to antimicrobial defense of NADPH phagosome oxidase and its products, ROS and RNS, is illustrated by the dramatically increased susceptibility to fungal and bacterial infection observed in patients afflicted with **chronic granulomatous disease,** which is caused by a defect in the ability of phox to generate oxidizing species.

Some pathogens, such as the yeast *Candida albicans* and the bacterium *S. aureus*, are not efficiently killed solely by oxida-

tive assault. The inclusion of nonoxidative defenses in the arsenal of neutrophils (and macrophages) greatly increases their defense against microbes. Nonoxidant defenses are deployed when neutrophil granules fuse with phagosomes, adding their cargo of antimicrobial peptides and proteins to the mix. Among the proteins is the bactericidal/permeability-increasing protein (BPI), a remarkable 55-kDa protein that binds with high affinity to LPS in the walls of gram-negative bacteria and causes damage to the bacterium's inner membrane. Other neutrophil granule agents include enzymes (proteases and lysozyme, for example) that cause the hydrolytic breakdown of essential structural components of microbes. The antimicrobial peptides include defensins and cathelicidins, cationic peptides with a broad range of antimicrobial activity.

## Macrophages deploy a number of antipathogen devices

Macrophages in a resting state are activated by a variety of stimuli. TLRs on the surfaces of macrophages recognize

The study group was followed for six to eight years and the number of nonfatal and fatal heart attacks recorded. Of the markers of inflammation studied, only CRP is associated with higher risk of coronary heart disease. Comparing the predictive value of CRP levels in patients with different ratios of (total cholesterol)/(HDL-bound cholesterol), this inflammatory marker correlates with increased risk. Specifically, the study found that high levels of CRP are a greater risk factor for patients with lower (total cholesterol)/(HDL-bound cholesterol) ratios than in patients with high ratios.

The past decade has seen the increasing use of a class of cholesterol-lowering drugs known as **statins.** These drugs inhibit cholesterol biosynthesis while also lowering inflammation. A recent study examined whether statins lowered CRP levels and whether patients with acute coronary syndromes who had lower CRP values as a result of statin therapy would have a lower risk of another heart attack than those patients who had higher CRP levels. The investigators found that administration of statins produced impressive reductions in CRP levels. The study also found that in those patients for whom statin treatment lowered CRP levels to values of 2 mg/liter or less, the rate of heart attack was significantly lower than in patients where values remained above this level, and there was a striking parallel between lower CRP and lower levels of LDL.

Evidence of a link between cardiovascular disease and inflammation has been accumulating for many years. In view of the established role of CRP as an agent of innate immunity and a mediator of inflammation, the finding that CRP levels are useful in evaluating the risk of heart attack strengthens this hypothesis. The finding that statin therapy, originally introduced to reduce cholesterol levels, also reduces the level of CRP is an unexpected benefit, one that supports the hypotheses linking inflammation to cardiovascular disease.

[1] Cardiovascular disease includes strokes and myocardial infarctions, or heart attacks.

[2] HDL, high-density lipoprotein, often imprecisely referred to as "good cholesterol" and LDL, low-density lipoprotein, often called "bad cholesterol," are complexes of cholesterol and protein. Elevated LDL is a risk factor for cardiovasular disease.

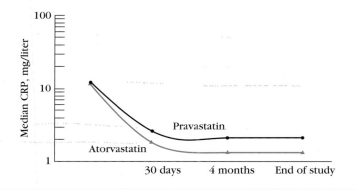

Statin treatment reduces serum levels of C-reactive protein (CRP). Subjects were tested with either of two statins, pravastatin or atorvastatin. Reduction of CRP levels were striking and long term. *[Adapted from P. M. Ridker et al., 2005, New England Journal of Medicine* **352:20.***]*

microbial components, such as LPS, peptidoglycans, and flagellins, and cytokine receptors detect cytokines released by other cells as part of the inflammatory response. On activation, macrophages exhibit greater phagocytic activity, increased ability to kill ingested microbes, and they secrete mediators of inflammation. They also express higher levels of class II MHC molecules, which present antigen to $T_H$ cells—another important point of collaboration between the innate and adaptive immune systems. Pathogens ingested by macrophages are efficiently killed in phagosomes by many of the same microbicidal agents used by neutrophils, with roles played by both reactive oxygen species and reactive nitrogen species (see Figure 3-13). An additional chemical weapon of macrophages and neutrophils has been well studied. Following activation, mediated by receptors such as TLRs or exposure to appropriate cytokines, phagocytes express high levels of **inducible nitric oxide synthetase (iNOS),** an enzyme that oxidizes L-arginine to yield L-citrulline and nitric oxide (NO).

$$\text{L-arginine} + O_2 + \text{NADPH} \xrightarrow{\text{iNOS}} \text{NO} + \text{L-citrulline} + \text{NADP}$$

The enzyme is called *inducible* NOS to distinguish it from other forms of the enzyme present in the body.

Nitric oxide has potent antimicrobial activity and can combine with superoxide to yield even more potent antimicrobial substances. Recent evidence indicates that nitric oxide and substances derived from it account for much of the antimicrobial activity of macrophages against bacteria, fungi, parasitic worms, and protozoa. This was impressively demonstrated using mice in which the genes encoding inducible nitric oxide reductase were knocked out. These mice lost much of their ability to control infections by such intracellular pathogens as *Mycobacterium tuberculosis*, the bacterium responsible for tuberculosis, and *Leishmania major*, the intracellular protozoan parasite that causes leishmaniasis.

Besides killing and clearing pathogens, macrophages also play a role in the coordination of other cells and tissues of the

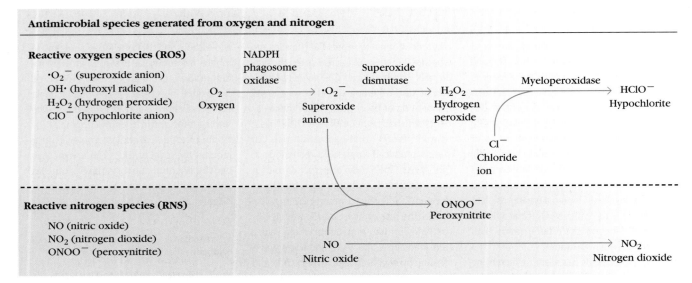

**FIGURE 3-13 Generation of antimicrobial reactive oxygen and reactive nitrogen species.** Within the confines of neutrophils and macrophages, several enzymes transform molecular oxygen into highly reactive oxygen species (ROS) that have antimicrobial activity. One of the products of this pathway, superoxide anion, can interact with a reactive nitrogen species (RNS) to produce peroxynitrite, another RNS. NO can also undergo oxidation to generate the RNS nitrogen dioxide.

immune and other supporting systems. They exert this influence by the secretion of a variety of cytokines, including IL-1, TNF-$\alpha$, and IL-6. These cytokines are particularly adept at the promotion of inflammatory responses, although each of these agents has a variety of effects. For example, IL-1 activates lymphocytes, and IL-1, IL-6, and TNF-$\alpha$ promote fever by affecting the thermoregulatory center in the hypothalamus. These cytokines also promote the acute phase response discussed earlier and in Chapter 13. In addition to cytokines, activated macrophages produce complement proteins that promote inflammation and assist in eliminating pathogens. Although the major site of synthesis of complement proteins is the liver, these proteins are also produced in macrophages and other cell types.

## NK cells are an important first line of defense against viruses and provide a key activating signal to other cells

**Natural killer (NK) cells** provide a first line of defense against many different viral infections. Using a system discussed in Chapter 14 that allows them to distinguish between infected and uninfected host cells, NK cells target and kill infected cells, which are potential sources of large numbers of additional infectious virus particles. NK cell-mediated lysis effectively eliminates the infection or holds it in check until days later, when the adaptive immune system engages the infection with virus-specific cytotoxic T cells and antibodies. However, some viral infections are probably cleared completely by innate mechanisms such as NK cells without any aid from adaptive immunity. Acti-

vated natural killer cells are also potent producers of a variety of cytokines that regulate other cells of the immune system and thereby shape and modify the body's ongoing and future defenses against the pathogen. It is notable that NK cells produce interferon-$\gamma$ and TNF-$\alpha$, two potent and versatile immunoregulatory cytokines. These two cytokines can stimulate the maturation of dendritic cells, the key coordinators of innate and adaptive immunity, discussed in the next section. Interferon-$\gamma$ is also a powerful mediator of macrophage activation and an important regulator of $T_H$ cell development, establishing a direct link between NK cells and the adaptive system.

## Dendritic cells engage pathogens and invoke adaptive immune responses by activating T cells

Dendritic cells provide a broader link between innate and adaptive immunity than the other cells of innate immunity by interacting with both $T_H$ cells *and* $T_C$ cells. Mature dendritic cells are able to activate both $T_H$ and $T_C$ cells because they are able to present exogenous antigens on either MHC I or MHC II and deliver strong costimulatory signals to the T cells. As agents of innate immunity, immature dendritic cells use a variety of PRRs, especially TLRs, to recognize pathogens. The recognition causes the activation of dendritic cells, which then undergo a maturation process that includes the increased production of MHC class II molecules and costimulatory molecules for T cell activation. (Like most nucleated cells, dendritic cells normally express class I MHC molecules.) Dendritic cells then migrate to lymphoid tissues, where they present antigen to both MHC class II-dependent T helper ($T_H$) cells and MHC class I-dependent T cytotoxic ($T_C$) cells.

The dendritic cell response is not limited to the vitally important role of communication between innate and adaptive immunity. These versatile cells also mount a direct assault on the pathogens they detect. Dendritic cells are capable of generating the reactive oxygen species and nitric oxide, and they have been reported to produce antimicrobial peptides as well. Pathogens that suffer phagocytosis by dendritic cells are therefore killed by many of the same agents used by macrophages. In addition, there is a subset of dendritic cells, plasmacytoid dendritic cells, that are potent producers of type I interferons, a family of antiviral cytokines that induce a state in virally infected cells and other nearby cells that is incompatible with viral replication. A critical activity of viruses is expression of their genomes in host cells. Engagement of TLRs on plasmacytoid dendritic cells with foreign nucleic acid triggers production of the type I interferons that block viral replication. Other subsets of dendritic cells produce interleukin-12, TNF-α, and IL-6, potent inducers of inflammation. One of this group, IL-12, plays a major role in shaping the T helper cell responses of adaptive immunity.

## Signal Transduction Pathways

Cell surface receptors receive the initial signals that activate complex innate immune system responses. The next step, transmission of signals to the cell interior, or **signal transduction,** is a universal theme in biological systems and an area of intense research in many fields beyond immunology. Response to signals requires three elements: the signal itself, a receptor, and a **signal transduction pathway** that connects the detector to effector mechanisms.

Signal ⟶ receptor ⟶
  signal transduction ⟶ effector mechanism

This general pathway is illustrated in Figure 1-6.

In the case of innate immunity, the signal will be a microbial product, the receptor will be a PRR on a leukocyte, and the signal will be transduced by the interactions of specific intracellular molecules. The effector mechanism—the action that takes place as a consequence of the signal—results in the clearance of the invading organism. Signaling and its consequences is a recurring theme in immunology. Some general features of signal tranduction pathways are outlined here, followed by the example of signal transduction through TLRs.

### TLR signaling is typical of signal transduction pathways

TLRs and their roles in innate immunity were discovered only recently, yet the major signal transduction pathways used by these receptors have already been worked out. Here we examine a signaling pathway (Figure 3-14) used by several TLRs, which can serve as an example for the signaling pathways of other receptors of innate immunity listed in Table 3-3, all of which follow a similar outline. The pathway discussed below results in the induction of many of the signature features of innate immunity, including the generation of inflammatory cytokines and chemokines, generation of antimicrobial peptides, and so on.

- *Initiation by interaction of signal with receptor:* Microbial products bind the extracellular portion of the TLR (see Figure 3-10). On the cytoplasmic side, a separate protein domain contains the highly conserved TIR structural motifs found in signaling molecules of animals and plants. The TIR domain offers binding sites for other components of the pathway.

- *Signal-induced assembly of pathway components/ involvement of an adaptor molecule:* Adaptor proteins, themselves containing TIR domains, interact with the TIR domain of TLRs. The most common adaptor protein for TLRs is MyD88, which promotes the association of two protein kinases, IRAK1 and IRAK4.

- *Protein kinase-mediated phosphorylation:* The protein kinase IRAK4, of the IRAK1:IRAK4 complex, phosphorylates its partner, IRAK1. The newly attached phosphate provides a docking site on IRAK1 for TRAF6, which binds and then dissociates in company with IRAK1 to form an intermediate IRAK1:TRAF6 complex. Another protein kinase, TAK1, joins this complex with several other proteins, resulting in the activation of the TAK1 kinase activity.

- *Initiation of an enzyme cascade:* TAK1 is pivotal in the pathway because its protein kinase activity allows it to perform the phosphorylation-mediated activation of two other signal transduction modules. One of these is the *m*itogen-*a*ctivated *p*rotein *k*inase (MAP kinase) pathway, and the other is the NFκB pathway (see below). MAP kinase pathways are signal-transducing enzyme cascades found in many cell types and conserved across a spectrum of eukaryotes from yeasts to humans. The end product of the cascade enters the nucleus and promotes phosphorylation of one or more transcription factors, which then affect the cell cycle or cellular differentiation.

TAK1 also phosphorylates the protein kinase IKK, which is the key step in activating the NFκB pathway. NFκB is a powerful transcription factor whose activity is inhibited by the unphosphorylated form of a cytoplasmic protein, IκB. NFκB bound to unphosphorylated IκB is sequestered in the cytoplasm. IKK phosphorylates IκB, causing the release of NFκB, which can then migrate to the nucleus.

NFκB in the nucleus initiates the transcription of many genes necessary for the effector functions of innate immunity. In vertebrates, NFκB-dependent pathways generate cytokines, adhesion molecules, and other effectors of the innate immune response. NFκB also plays a role in some key

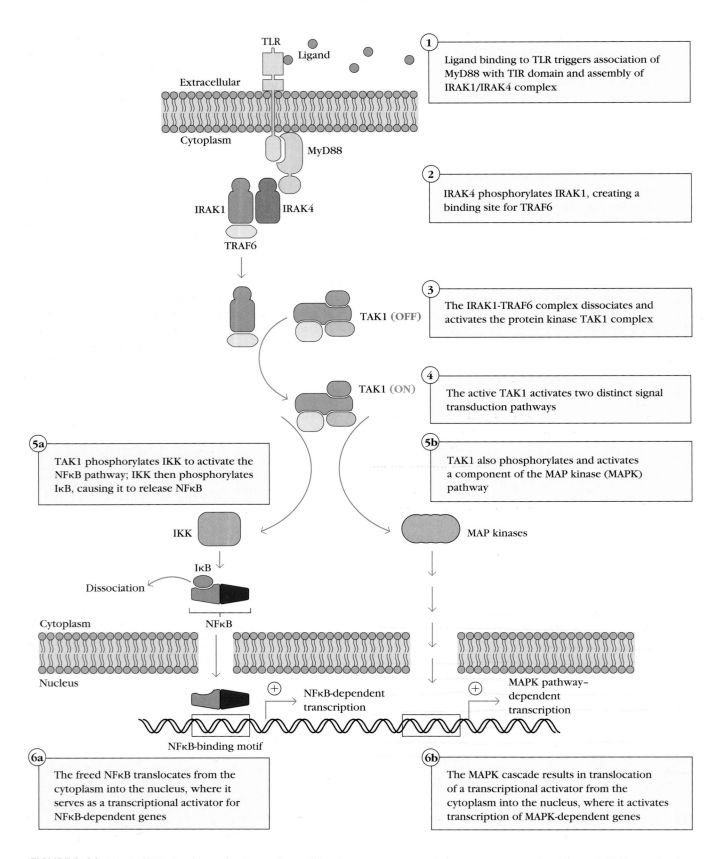

**FIGURE 3-14 A typical TLR signal transduction pathway.** Abbreviations: MyD88, myeloid differentiation primary-response protein 88; IRAK, IL-1R-associated kinase; IL-1R, interleukin-1 receptor; TRAF6, tumor-necrosis-factor-receptor-associated factor 6; TAK1, transforming-growth-factor-β-activated kinase 1; MAPK, mitogen-activated protein kinase; IκB, inhibitor of nuclear factor NFκB; IKK, IκB kinase.

signal transduction pathways of T and B cells and is therefore also important in adaptive immunity.

Activation of TLR signaling pathways has many effects. It promotes the expression of genes that contribute to inflammation, induces changes in antigen-presenting cells that make them more efficient at antigen presentation, and causes the synthesis and export of intercellular signaling molecules that affect the behavior of leukocytes and other cells. Engagement of TLRs can increase the phagocytic activity of macrophages and neutrophils and change their physiology in ways that increase their ability to kill and clear pathogens. In nonvertebrate systems, TLR signaling activates a variety of effective systems of immunity. Most, but not all, TLRs employ the signal transduction pathway schematized in Figure 3-14. TLR3 uses a pathway that is independent of MyD88, and TLR4 uses both the pathway described above and the MyD88-independent pathway employed by TLR3.

## Ubiquity of Innate Immunity

A determined search for antibodies, T cells, and B cells in organisms of the nonvertebrate phyla has failed to find any traces of these signature features of adaptive immunity. Yet despite their prominence in vertebrate immune systems, it would be a mistake to conclude that these extraordinary molecules and versatile cells are essential for immunity. The interior spaces of organisms as diverse as the sea squirt (a chordate without a backbone), fruit fly, and tomato do not contain unchecked microbial populations. Careful studies of these organisms and many other representatives of nonvertebrate phyla have found well-developed systems of innate

immunity. The accumulating evidence leads to the conclusion that some system of immunity protects all multicellular organisms from microbial infection and exploitation. The genome of the sea squirt, *Ciona intestinalis* (Figure 3-15a), encodes many of the genes associated with innate immunity, including those for complement-like lectins and Toll-like receptors. In fruit flies, a pathway involving a member of the NFκB family is triggered by gram-negative bacterial infections, leading to the production of diptericin, a potent antibacterial peptide. In addition to these pathways, *Drosophila* and other arthropods have a variety of other strategies of innate immunity, which include the activation of prophenoloxidase cascades that result in the deposition of melanin around invading organisms. The tomato, *Lycopersicon esculentum* (Figure 3-15b), like other plants, has evolved a repertoire of innate immune defenses to protect itself against infection. These include generation of oxidative bursts, raising of internal pH, localized death of infected regions, and the induction of a variety of proteins, including enzymes that can digest the walls of invading fungi (chitinases) or bacteria (α-1,3 glucanase). Plants also respond to infection by producing a wide variety of antimicrobial peptides, as well as small nonpeptide organic molecules, such as phytoalexins, that have antibiotic activity. Mutations that disrupt synthesis of phytoalexins result in loss of resistance to many plant pathogens. In some cases, the response of plants to pathogens even goes beyond a chemical assault to include an architectural response, in which the plant isolates cells in the infected area by strengthening the walls of surrounding cells. Table 3-4 compares the capabilities of immune systems in a wide range of multicellular organisms, both animals and plants.

(a)

(b)

**FIGURE 3-15 Innate immunity in species extending across kingdoms.** (a) Sea squirts, nonvertebrate chordates. (b) A member of the plant kingdom, the tomato. *[Part a, Gary Bell/Getty Images; part b, George Glod, SuperStock.]*

| TABLE 3-4 | Immunity in multicellular organisms |
|-----------|-------------------------------------|

| Taxonomic group | Innate immunity (nonspecific) | Adaptive immunity (specific) | Invasion-induced protective enzymes and enzyme cascades | Phagocytosis | Anti-microbial peptides | Pattern recognition receptors | Graft rejection | T and B cells | Anti-bodies |
|---|---|---|---|---|---|---|---|---|---|
| *Higher plants* | + | − | + | − | + | + | − | − | − |
| *Invertebrate animals* | | | | | | | | | |
| Porifera (sponges) | + | − | ? | + | ? | ? | + | − | − |
| Annelids (earthworms) | + | − | ? | + | ? | ? | + | − | − |
| Arthropods (insects, crustaceans) | + | − | + | + | + | + | ? | − | − |
| *Vertebrate animals* | | | | | | | | | |
| Elasmobranchs (cartilaginous fish; e.g., sharks, rays) | + | + | + | + | Equivalent agents | + | + | + | + |
| Teleost fish and bony fish (e.g., salmon, tuna) | + | + | + | + | Probable | + | + | + | + |
| Amphibians | + | + | + | + | + | + | + | + | + |
| Reptiles | + | + | + | + | ? | + | + | + | + |
| Birds | + | + | + | + | ? | + | + | + | + |
| Mammals | + | + | + | + | + | + | + | + | + |

KEY: + = definitive demonstration;   − = failure to demonstrate thus far; ? = presence or absence remains to be established.

SOURCES: M. J. Flajnik, K. Miller, and L. Du Pasquier, 2003, "Origin and Evolution of the Vertebrate Immune System," in *Fundamental Immunology,* 5th ed., W. E. Paul (ed.), Lippincott, Philadelphia; M. J. Flajnik and L. Du Pasquier, 2004, *Trends in Immunology* **25**:640.

## SUMMARY

- Two systems of immunity protect vertebrates: innate immunity, which is in place or ready for activation prior to infection, and adaptive immunity, which is induced by infection and requires days to weeks to respond.

- The receptors of innate immunity recognize pathogen-associated molecular patterns (PAMPs), which are molecular motifs found in microbes. In contrast, the receptors of adaptive immunity recognize specific details of molecular structure.

- The receptors of innate immunity are encoded in the host germ line, but the genes that encode antibodies and T cells, the signature receptors of adaptive immunity, are formed by a process of genetic recombination.

- Adaptive immune responses display memory, whereas innate responses do not.

- The skin and mucous membranes constitute an anatomical barrier that is highly effective in protecting against infection.

- Inflammation increases vascular permeability, allowing soluble mediators of defense such as complement, mannose-binding lectin (MBL), C-reactive protein (CRP), and later antibodies to reach the infected site. In addition, inflammation causes migration of phagocytes and antiviral cells by extravasation and chemotaxis to the focus of infection.

- Antimicrobial peptides are important effectors of innate immunity and have been found in a broad diversity of species. They kill a wide variety of microorganisms, often working by disrupting microbial membranes.

- Many cytokines are generated by the innate immune system. These cytokines include type 1 interferons that have antiviral effects and others such as TNF-$\alpha$ and interferon-$\gamma$ that exert powerful effects on other cells and organs.

- Certain cytokines induce an acute phase response, a process during which several antimicrobial proteins are released from the liver to the bloodstream. Among these proteins are MBL, CRP, and complement, which can act to kill microbes.

- The innate immune system employs pattern recognition receptors (PRRs) to detect infection. Toll-like receptors (TLRs) are an important category of PRRs; each TLR detects a distinct subset of pathogens, and the entire repertoire can detect a wide variety of viruses, bacteria, fungi, and protozoa.

■ Phagocytes use a variety of strategies to kill pathogens. These strategies include cytolytic proteins, antimicrobial peptides, and the generation of reactive oxygen species (ROS) and reactive nitrogen species (RNS).

■ Dendritic cells are a key cellular bridge between adaptive and innate immunity. Microbial components acquired during the innate response by dendritic cells are brought from the site of infection to lymph nodes, and microbial antigens are displayed on MHC molecules and presented to T cells, resulting in T cell activation and an adaptive immune response.

■ TLRs use signal-transduction pathways common to those found throughout the plant and animal kingdom. TLR signaling initiates events that enable cells to control and clear infections.

■ Innate immunity appeared early in the evolution of multicellular organisms and has been found in all multicellular plants and animals examined. Adaptive immunity is found only in vertebrates.

## References

Akira, S., and K. Takeda. 2004. Toll-like receptor signaling. *Nature Reviews Immunology* **4:** 499.

Basset, C., et al. 2003. Innate immunity and pathogen-host interaction. *Vaccine* **21:** s2/12.

Beutler, B., and E. T. Rietschel. 2003. Innate immune sensing and its roots: the story of endotoxin. *Nature Reviews Immunology* **3:** 169.

Bulet, P., R. Stocklin, and L. Menin. 2004. Antimicrobial peptides: from invertebrates to vertebrates. *Immunological Reviews* **198:** 169.

Fang, F. C. 2004. Antimicrobial reactive oxygen and nitrogen species: concepts and controversies. *Nature Reviews Microbiology* **2:** 820.

Iwasaki, A., and R. Medzhitov. 2004. Toll-like receptor control of the adaptive immune responses. *Nature Immunology* **5:** 987.

Lemaitre, B. 2004. The road to Toll. *Nature Reviews Immunology* **4:** 521.

Medzhitov, R., et al. 1997. A human homologue of the Toll protein signals activation of adaptive immunity. *Nature* **388:** 394.

O'Neill, A. J. 2005. Immunity's early warning system. *Scientific American* **292:** 38.

Pai, J. K., et al. 2004. Inflammatory markers and the risk of coronary heart disease in men and women. *New England Journal of Medicine* **351:** 2599.

Poltorak, A., et al. 1998. Defective LPS signaling in C3Hej and C57BL/10ScCr mice: mutations in TLR4 gene. *Science* **282:** 2085.

Ridiker, P. M., et al. 2005. C-reactive protein and outcomes after statin therapy. *New England Journal of Medicine* **352:** 20.

Ulevitch, R. J. 2004. Therapeutics targeting the innate immune system. *Nature Reviews Immunology* **4:** 512.

 Useful Web Sites

http://www.ncbi.nlm.nih.gov/PubMed/

PubMed, the National Library of Medicine database of more than 15 million publications, is the world's most comprehensive bibliographic database for biological and biomedical literature. It is also highly user friendly.

http://cpmcnet.columbia.edu/dept/curric-pathology/pathology/pathology/pathoatlas/GP_I_menu.html
Images are shown of the major inflammatory cells involved in acute and chronic inflammation, as well as examples of specific inflammatory diseases.

http://animaldiversity.ummz.umich.edu/site/index.html
The Animal Diversity Web (ADW), based at the University of Michigan, is an excellent and comprehensive database of animal classification as well as a source of information on animal natural history and distribution. Here is a place to find information about animals that are not humans or mice.

 Study Questions

**CLINICAL FOCUS QUESTION** Comment on the role of inflammatory processes in the development and progression of atherosclerosis. How might inflammation raise CRP levels?

1. Innate immunity collaborates with adaptive immunity to protect the host. Discuss this collaboration, naming key points of interaction between the two systems.

2. What are the hallmark characteristics of a localized inflammatory response? How do these characteristics contribute to the mounting of an effective innate immune response?

3. Use this list to complete the statements that follow. Some terms may be used more than once.

| | |
|---|---|
| Interferon | Antimicrobial peptides |
| NO | TLR8 |
| TLR2 | NADPH phagosome oxidase |
| TNF-α | Antibody |
| NK cells | PAMPs |
| TLR4 | $O_2$ |
| NOD | Complement |
| Class I MHC molecules | Class II MHC molecules |
| Phagocytosis | CRP |
| iNOS | T cell receptors |
| Costimulatory molecules | Innate immunity |
| Adaptive immunity | APR |
| PRRs | ROS |
| TLR7 | RNS |
| Dendritic cell | TLR9 |
| $T_H$ | $T_C$ |
| Arginine | Proinflammatory cytokines |
| MBL | NADPH |

a. Both _____, a cytokine, and _____ cells defend against viral infection.

b. The enzyme _____ uses the amino acids _____ and _____ to generate _____, an antimicrobial gas.

c. The enzyme _____ uses _____ to generate microbe-killing _____, which can combine with the antimicrobial gas _____ to produce _____, which are also antimicrobial.

d. During the innate response, a cell known as a _____ captures antigen and displays it on _____ and _____ for presentation to _____ and _____ cells when the _____ migrates from the tissues to a lymph node.

e. _____ and _____ are pattern recognition receptors that can activate complement and facilitate opsonization.

f. The _____ occurs when _____ such as _____ are generated by inflammation and arrive at the liver.

g. Some cells use _____ and _____ to detect RNA virus infections and _____ to detect infections by bacteria and some DNA viruses.

h. _____, the receptors of innate immunity, are encoded in the germ line, but _____ and _____, the signature receptors of adaptive immunity, are encoded by genes that arise by somatic recombination.

i. Dendritic cells, which have _____, can display antigen on both _____ and _____ and therefore are efficient activators of helper and cytotoxic T cells.

j. _____ is an intracellular _____ that detects bacterial cell wall components.

k. _____ are receptors of innate immunity that detect _____.

l. _____ detects gram-positive bacterial infections and _____ detects gram-negative infections.

m. Certain microbial cell wall components can activate _____, triggering opsonization and damage to the microbe's plasma membrane.

**For use with Question 7**

| Treatment or experimental modification | Induction of adaptive immunity | Leukocyte extravasation | Acute phase response | Induction of complement-mediated lysis | Induction of inflammation | ROS and RNS |
|---|---|---|---|---|---|---|
| a. Injection of antibodies that block integrin-ICAM interactions | | | | | | |
| b. Knockout of the gene for TLR4 | | | | | | |
| c. Antibodies that neutralize TNF-α and IL-1 | | | | | | |
| d. Mutation in phox enzyme | | | | | | |
| e. Mice with class II MHC gene knocked out | | | | | | |

Justifications for your predictions

a. _____

b. _____

c. _____

d. _____

e. _____

4. What were the two experimental observations that linked TLRs to innate immunity?

5. As adaptive immunity evolved in vertebrates, the more ancient system of innate immunity was retained. Can you think of any disadvantages to having a dual system of immunity? Would you argue that either system is more essential?

6. How might an arthropod, such as a cockroach or beetle, protect itself from fungal infection? Compare the immune responses of an arthropod to those of a human.

7. What effects would the treatments or experimental modifications shown in the table at the bottom of page 74 have on the indicated responses by a human or mouse host to the first infection with a gram-positive bacterium? Justify each of your answers.

**ANALYZE THE DATA** François and colleagues (*J. Immunol.* 2005, **174**:363) investigated the effects of Toll-like receptor (TLR) agonists on apoptosis of neutrophils, which make up the largest population of white blood cells and play the role of first responders against bacteria. Although neutrophils are short-lived, they play a critical role in innate immunity by targeting infected tissue, where they defend the host by making reactive oxygen species, proteolytic enzymes, and other antimicrobial products. François and coworkers incubated neutrophil cells with SN50, an inhibitor of NF-κB, or a control. Samples were then treated with TLR agonists and a measure of the inhibition of apoptosis was determined (part a in figure below). The investigators also examined a signal transduction pathway activated upon TLR ligand binding (part b in figure). Answer the following questions based on the data presented and what you have learned from reading this text book.

(a) Effect of TLR engagement on apoptosis

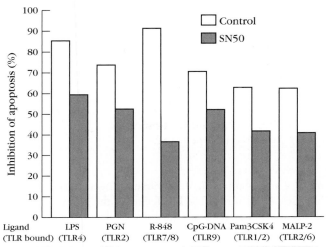

(b) Amount of phospho-IKK after TLR engagement, measured by flow cytometry

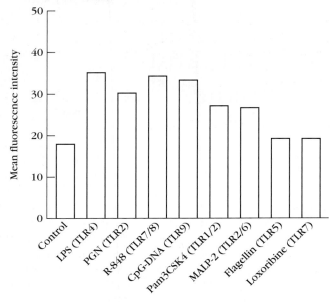

a. Does engagement of TLR receptors affect neutrophil life span? Why is this an advantage to the host? Why might this be a disadvantage?

b. If a mutant mouse has a defect in IKK (IκB kinase), would you predict a longer or shorter neutrophil half life? (*Hint*: See Figure 3–14).

c. What TLR molecules appear to use a different signal transduction pathway than TLR4? Assume that the agonists used in this study are appropriate. Explain your answer.

 **Interactive Study**

**www.whfreeman.com/kuby**

SELF-TEST
Review of Key Terms

ANIMATION
Leukocyte Extravasation

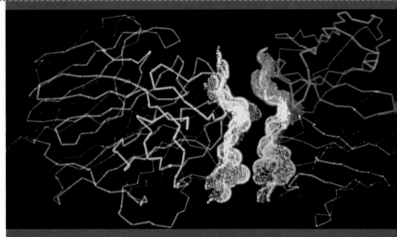

# chapter 4

# Antigens and Antibodies

THE HALLMARK MOLECULES OF THE ADAPTIVE IMMUNE system are the antibody and the T-cell receptor. Whereas components of innate immunity are programmed for pattern recognition and thus recognize features shared by groups of foreign molecules, the antibody and T-cell receptor molecules display a higher degree of specificity, recognizing specific **antigenic determinants** or **epitopes.** Epitopes are the immunologically active regions of an immunogen that bind to antigen-specific membrane receptors on lymphocytes or to secreted antibodies. Antibodies are epitope-binding proteins found in two forms, as either membrane-bound constituents of B cells or as soluble molecules secreted by plasma cells. Membrane-bound antibodies confer antigenic specificity on B cells; proliferation of antigen-specific B-cell clones is elicited by the interaction of membrane antibody with antigen. Secreted antibodies circulate in the blood, where they serve as the effectors of **humoral** immunity by searching out antigens and marking them for elimination. All antibodies share structural features, bind to antigen, and participate in a limited number of **effector functions.** The T-cell receptor expressed on the surface membrane of the T cell recognizes only processed antigen fragments that are complexed with MHC molecules. Although the present chapter emphasizes the antibody and the nature of B-cell antigen recognition, comparisons with T-cell antigen recognition will be made to develop the general topics of antigenicity and illustrate the contrast between T- and B-cell interactions with antigens. The molecular features of MHC molecules, the manner in which antigens are processed and bound to MHC molecules, and the nature of T-cell receptors that recognize antigen-MHC complexes will be covered in Chapters 8 and 9.

In general, the population of antibodies produced in response to a particular antigenic stimulus is heterogeneous. Most antigens are structurally complex, containing many different epitopes, and the immune system usually responds by producing antibodies to several of the epitopes on the antigen. In other words, several different B-cell clones are stimulated and proliferate. The output of the plasma cells of a single B-cell clone is a monoclonal antibody that specifically

*Complementarity between antibody and influenza virus protein (yellow). [Based on x-ray crytallography data collected by P. M. Colman and W. R. Tulip; from G. J. V. H. Nossal, 1993, Sci. Am. 269(3): 22.]*

- Immunogenicity Versus Antigenicity
- Epitopes
- Basic Structure of Antibodies
- Antibody Binding Site
- Antibody-Mediated Effector Functions
- Antibody Classes and Biological Activities
- Antigenic Determinants on Immunoglobulins
- The B-Cell Receptor
- The Immunoglobulin Superfamily
- Monoclonal Antibodies

binds a single antigenic determinant. Together the secreted products of all stimulated B-cell clones, the group of monoclonal antibodies, make up the polyclonal and heterogeneous serum antibody response to an immunizing antigen.

Members of the antibody protein family have common structural features and are known collectively as **immunoglobulins (Igs).** Yet despite their similarities, the members of this family perform an incredibly diverse set of binding functions as well as several distinct effector functions subsequent to antigen binding. Before exploring the structure and complexity of the antibody family, this chapter will discuss the nature of antigens and the concepts and features of immunogenic substances.

Antigens are defined specifically as molecules that interact with the immunoglobulin receptor of B cells (or with the T-cell receptor when complexed with MHC). The molecular properties of antigens and how these properties contribute to immune activation are central to our understanding of the immune system. This chapter describes the molecular features of antigens recognized by antibodies and explores the contribution made to immunogenicity by the biological system of the host.

## Immunogenicity Versus Antigenicity

Immunogenicity and antigenicity are related but distinct immunologic properties that are sometimes confused. **Immunogenicity** is the ability to induce a humoral and/or cell-mediated immune response:

B cells + antigen → effector B cells + memory B cells
↓
Plasma cell → secretes antibody

T cells + antigen → effector T cells + memory T cells
↓
CTLs, $T_H$, etc. → secretes cytokines and cytotoxic factors

Although a substance that induces a specific immune response is usually called an antigen, it is more appropriately called an **immunogen.**

**Antigenicity** is the ability to combine specifically with the final products of the above responses (i.e., secreted antibodies and/or surface receptors on T cells). Although all molecules that have the property of immunogenicity also have the property of antigenicity, the reverse is not true. Some small molecules, called **haptens,** are antigenic but incapable by themselves of inducing a specific immune response. In other words, they lack immunogenicity.

### Haptens are valuable research and diagnostic tools

The chemical coupling of a hapten to a large immunogenic protein, called a **carrier,** yields an immunogenic **hapten-carrier conjugate.** Animals immunized with such a conjugate produce antibodies specific for three types of antigenic determinant: (1) the hapten determinant, (2) unaltered epitopes on the carrier protein, and (3) new epitopes formed by regions of both the hapten and the carrier molecule in combination (Figure 4-1). By itself, a hapten cannot function as an immunogenic epitope. But when multiple molecules of a single hapten are coupled to a carrier protein (or even to a nonimmunogenic homopolymer), the hapten becomes accessible to the immune system and can function as an immunogen.

In addition to showing the broad range of epitopes that can be found on a single immunogen, the hapten-carrier

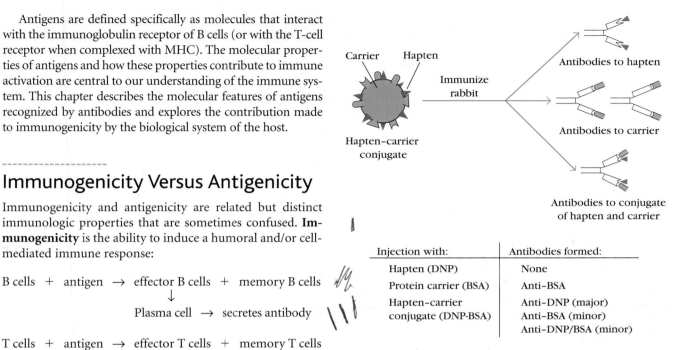

| Injection with: | Antibodies formed: |
|---|---|
| Hapten (DNP) | None |
| Protein carrier (BSA) | Anti–BSA |
| Hapten-carrier conjugate (DNP-BSA) | Anti–DNP (major) Anti–BSA (minor) Anti–DNP/BSA (minor) |

**FIGURE 4-1 A hapten-carrier conjugate is the immunogen in this illustration, and the hapten is an antigen that is not by itself immunogenic.** The immunogen contains multiple copies of the hapten—a small nonimmunogenic organic compound, in this case, dinitrophenol (DNP)—chemically linked to a large protein carrier such as bovine serum albumin (BSA). Immunization with DNP alone elicits no anti-DNP antibodies, but immunization with DNP-BSA elicits three types of antibodies. Of these, anti-DNP antibody is predominant, indicating that in this case the hapten is the immunodominant epitope or antigenic determinant, as it often is in such conjugates.

system provides immunologists with a tool to probe the effects of minor variations in chemical structures on immune specificity. The hapten can be a chemically defined determinant that is then subtly modified by chemical means to determine whether this impacts recognition by antibodies. A classic illustration of this strategy is seen in the pioneering work of Karl Landsteiner, who in the 1920s and 1930s created a simple, chemically defined system for studying the binding of an individual antibody. Landsteiner employed as haptens small organic molecules that are antigenic but *not* immunogenic. In his studies, Landsteiner immunized rabbits with a hapten-carrier conjugate and then tested the reactivity of the rabbits' immune sera with that hapten and with closely related haptens coupled to a different carrier protein. He was thus able to measure specifically the reaction of the antihapten antibodies in the immune serum and not that of antibodies to the original carrier epitopes. Landsteiner tested whether an antihapten antibody could bind to other haptens with slightly different chemical structures. If binding occurred, it was called a **cross-reaction.** By observing which hapten modifications prevented or permitted cross-reactions, Landsteiner was able to gain insight into the specificity of antigen-antibody interactions.

| TABLE 4-1 | Reactivity of antisera with various haptens | | | |
|---|---|---|---|---|
| | REACTIVITY WITH | | | |
| Antiserum against | Aminobenzene (aniline) | o-Aminobenzoic acid | m-Aminobenzoic acid | p-Aminobenzoic acid |
| Aminobenzene | + | 0 | 0 | 0 |
| o-Aminobenzoic acid | 0 | + | 0 | 0 |
| m-Aminobenzoic acid | 0 | 0 | + | 0 |
| p-Aminobenzoic acid | 0 | 0 | 0 | + |

KEY: 0 = no reactivity; + = strong reactivity

SOURCE: Based on K. Landsteiner, 1962, *The Specificity of Serologic Reactions*, Dover Press. Modified by J. Klein, 1982, *Immunology: The Science of Self-Nonself Discrimination*, Wiley.

Using various derivatives of aminobenzene as haptens, Landsteiner found that the overall configuration of a hapten plays a major role in determining whether it can react with a given antibody. For example, antiserum from rabbits immunized with aminobenzene or one of its carboxyl derivatives (o-aminobenzoic acid, m-aminobenzoic acid, or p-aminobenzoic acid) coupled to a carrier protein reacted only with the original immunizing hapten and did not cross-react with any of the other haptens (Table 4-1). In contrast, if the overall configuration of the hapten was kept the same and the hapten was modified in the para position with various nonionic derivatives, then the antisera showed various degrees of cross-reactivity. Landsteiner's work demonstrated both the specificity of the immune system for small structural variations on haptens and the enormous diversity of epitopes that the immune system is capable of recognizing.

Many biologically important substances, including drugs, peptide hormones, and steroid hormones, can function as haptens. Conjugates of these haptens with large protein carriers can be used to produce hapten-specific antibodies. These antibodies are useful for measuring the presence of various substances in the body. For example, the original home pregnancy test kit employed antihapten antibodies to determine whether a woman's urine contained human chorionic gonadotropin (HCG), which is a sign of pregnancy.

Although studies with hapten-carrier conjugates give a clear delineation between the concepts of antigenicity and immunogenicity, in practice, multiple factors must be considered to determine whether a substance encountered by the immune system will evoke a response. Immunogenicity depends not only on intrinsic properties of an antigen but also on a number of properties of the particular biological system that the antigen encounters and on the manner in which the immunogen is presented. The following sections describe the properties that most immunogens share and the contribution that the biological system makes to the expression of immunogenicity.

## Properties of the immunogen contribute to immunogenicity

Immunogenicity is determined in part by four properties of the immunogen: its foreignness, molecular size, chemical composition and complexity, and ability to be processed and presented with an MHC molecule on the surface of an antigen-presenting cell or altered self cell.

### Foreignness

To elicit an immune response, a molecule must be recognized as nonself by the biological system. The flip side of the capacity to recognize nonself is **tolerance** of self, a specific unresponsiveness to self antigens (see Chapter 16). Much of the ability to tolerate self antigens arises during lymphocyte development, when immature lymphocytes are exposed to self components. Cells that recognize self components during this process are inactivated. Survivors of the process are released. Antigens that have not been presented to immature lymphocytes during this critical period may be later recognized as nonself, or foreign, by the immune system. When an antigen is introduced into an organism, the degree of its immunogenicity depends on the degree of its foreignness. Generally, the greater the phylogenetic distance between two species, the greater the structural (and therefore the antigenic) disparity between their constituent molecules. For example, the common experimental antigen bovine serum albumin (BSA) is not immunogenic when injected into a cow but is strongly immunogenic when injected into a rabbit. Moreover, BSA would be expected to exhibit greater immunogenicity in a chicken than in a goat, which is more

– Lys – Ala – His – Gly – Lys – Lys – Val – Leu

Amino acid sequence
of polypeptide chain

**PRIMARY STRUCTURE**

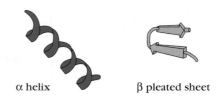

α helix          β pleated sheet

**SECONDARY STRUCTURE**

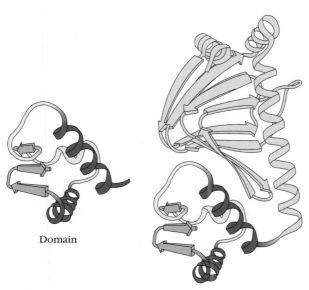

Domain

Monomeric polypeptide molecule

**TERTIARY STRUCTURE**

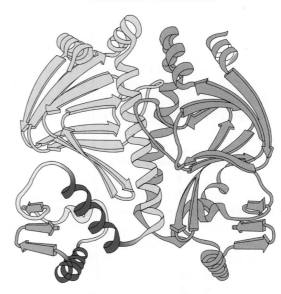

Dimeric protein molecule

**QUATERNARY STRUCTURE**

**FIGURE 4-2  The four levels of protein organizational structure.** The linear arrangement of amino acids constitutes the primary structure. Folding of parts of a polypeptide chain into regular structures (e.g., α helices and β pleated sheets) generates the secondary structure. Tertiary structure includes the folding of regions between secondary features to give the overall shape of the molecule or parts of it (domains) with specific functional properties. Quaternary structure results from the association of two or more polypeptide chains into a single polymeric protein molecule.

closely related to bovines. There are some exceptions to this rule. Some macromolecules (e.g., collagen and cytochrome *c*) have been highly conserved throughout evolution and therefore display very little immunogenicity across diverse species lines. Conversely, some self components (e.g., corneal tissue and sperm) are effectively sequestered from the immune system, so that if these tissues are injected even into the animal from which they originated, they will function as immunogens.

## Molecular Size

There is a correlation between the size of a macromolecule and its immunogenicity. The most active immunogens tend to have a molecular mass of >100,000 daltons (Da). Generally, substances with a molecular mass less than 5000 to 10,000 Da are poor immunogens, although a few substances with a molecular mass less than 1000 Da have proven to be immunogenic.

## Chemical Composition and Heterogeneity

Size and foreignness are not, by themselves, sufficient to make a molecule immunogenic; other properties are needed as well. For example, synthetic homopolymers (polymers composed of multiple copies of a single amino acid or sugar) tend to lack immunogenicity regardless of their size. Heteropolymers are usually more immunogenic than homopolymers. These studies show that chemical complexity contributes to immunogenicity. It is notable that all four levels of protein organization—primary, secondary, tertiary, and quaternary—contribute to the structural complexity of a protein and hence affect its immunogenicity (Figure 4-2).

Appropriately presented lipid antigens can induce B-cell responses. For example, lipids can serve as haptens attached to suitable carrier molecules, such as the proteins keyhole limpet hemocyanin (KLH) or bovine serum albumin (BSA). By immunizing with these lipid-protein conjugates, it is possible to obtain antibodies that are highly specific for the target lipids. Using this approach, scientists have raised antibodies against a wide variety of lipid molecules, including steroids, complex fatty-acid derivatives, and fat-soluble vitamins such as vitamin E. Such antibodies are of considerable practical importance—many clinical assays for the presence and amounts of medically important lipids are antibody based. For example, a determination of the levels of a complex group of lipids known as leukotrienes can be useful in evaluating asthma patients (see Chapter 15). Assays based on the use of antilipid antibodies allow the detection of picogram amounts of leukotriene $C_4$. Because antileukotriene $C_4$ has little or no reactivity with similar compounds, such as leukotriene $D_4$ or leukotriene $E_4$, it can be used to assay leukotriene $C_4$ in samples that contain this compound and a variety of other structurally related lipids. Another medically important application involves detection of prednisone, an immunosuppressive steroid often administered as part of a program to suppress the rejection of a transplanted organ. Maintaining adequate blood levels of this and other immunosuppressive drugs is important to a successful outcome of transplantation, and antibody-based immunoassays are routinely used to make these evaluations.

### Susceptibility to Antigen Processing and Presentation

The development of both humoral (antibody-mediated) and T-cell-mediated immune responses requires interaction of T cells with antigen that has been processed and presented together with MHC molecules. Large, insoluble, or aggregated macromolecules generally are more immunogenic than small, soluble ones because the larger molecules are more readily phagocytosed and processed. Macromolecules that cannot be degraded and presented with MHC molecules are poor immunogens. This can be illustrated with polymers of D-amino acids, which are stereoisomers of the naturally occurring L-amino acids. Because the degradative enzymes within antigen-presenting cells can degrade only proteins containing L-amino acids, polymers of D-amino acids cannot be processed and thus are poor immunogens.

## The biological system contributes to immunogenicity

Even if a macromolecule has the properties that contribute to immunogenicity, its ability to induce an immune response will depend on certain properties of the biological system that the antigen encounters. Factors contributing to immunogenicity include host genetic makeup, the manner in which the material is presented, and the use of agents (adjuvants) to enhance immunogenicity.

### Genotype of the Recipient Animal

A major factor determining immune responsiveness can be the **genotype** of the recipient. The genetic constitution (or genotype) of an immunized animal influences the type of immune response the animal manifests as well as the degree of the response. For example, Hugh McDevitt showed that two different inbred strains of mice responded very differently after exposure to a synthetic polypeptide immunogen. One strain produced high levels of serum antibody, the other low levels. When the two strains were crossed, the $F_1$ generation showed an intermediate response to the immunogen. By backcross analysis, the gene controlling immune respon-

siveness was mapped to a subregion of the array of genes called the major histocompatibility complex (MHC; see Chapter 8). Numerous experiments with simple defined immunogens have demonstrated genetic control of immune responsiveness, largely confined to genes within the MHC. These data indicate that MHC gene products, which function in the presentation of processed antigen to T cells, play a central role in determining the degree to which an animal responds to an immunogen.

The response of an animal to an antigen is also influenced by the genes that encode B-cell and T-cell receptors and by genes that encode various proteins involved in immune regulatory mechanisms. Genetic variability in all of these genes affects the immunogenicity of a given macromolecule in different animals. These genetic contributions to immunogenicity will be described more fully in later chapters.

### Immunogen Dosage and Route of Administration

Each experimental immunogen exhibits a particular dose-response curve, which is determined by measuring the immune response to different doses and different administration routes. An antibody response is measured by determining the level of antibody present in the serum of immunized animals. Evaluating T-cell responses is more complex but may be determined by evaluating the increase in the number of T cells bearing T-cell receptors that recognize the immunogen. Some combination of optimal dosage and route of administration will induce a peak immune response in a given animal.

An insufficient dose will not stimulate an immune response either because it fails to activate enough lymphocytes or because, in some cases, certain ranges of low doses can induce a state of immunologic unresponsiveness, or tolerance. Conversely, an excessively high dose can also induce tolerance. The immune response of mice to the purified pneumococcal capsular polysaccharide illustrates the importance of dose. A 0.5 mg dose of antigen fails to induce an immune response in mice, whereas a 1000-fold lower dose of the same antigen ($5 \times 10^{-4}$ mg) induces a humoral antibody response. A single dose of most experimental immunogens will not induce a strong response; rather, repeated administration over a period of weeks is usually required. Such repeated administrations, or **boosters,** increase the clonal proliferation of antigen-specific B cells or T cells and thus increase the lymphocyte populations specific for the immunogen.

Experimental immunogens are generally administered parenterally (*para,* around; *enteric,* gut)—that is, by routes other than orally. The following administration routes are common:

- Intravenous (iv): into a vein

- Intradermal (id): into the skin

- Subcutaneous (sc): beneath the skin

- Intramuscular (im): into a muscle

- Intraperitoneal (ip): into the peritoneal cavity

The administration route strongly influences which immune organs and cell populations will be involved in the response. Antigen administered intravenously is carried first to the spleen, whereas antigen administered subcutaneously moves first to local lymph nodes. Differences in the lymphoid cells that populate these organs may be reflected in the subsequent immune response.

### Adjuvants

**Adjuvants** (from the Latin *adjuvare,* to help) are substances that, when mixed with an antigen and injected with it, enhance the immunogenicity of that antigen. Adjuvants are often used in research and clinical settings to boost the immune response when an antigen has low immunogenicity or when only small amounts of an antigen are available. For example, the antibody response of mice to immunization with BSA can be increased fivefold or more if the BSA is administered with an adjuvant. Precisely how adjuvants augment the immune response is not entirely clear, but some of the known adjuvants (e.g., synthetic polyribonucleotides and bacterial lipopolysaccharides) are now recognized as ligands for the Toll-like receptors on dendritic cells and macrophages (see Chapter 3) and thus stimulate immune responses via activation of the innate immune system.

In general, adjuvants appear to exert one or more of the following effects:

- Antigen persistence is prolonged.

- Costimulatory signals are enhanced.

- Local inflammation is increased.

- The nonspecific proliferation of lymphocytes is stimulated.

Aluminum potassium sulfate (alum) is an adjuvant that prolongs the persistence of antigen and is the only adjuvant approved for general human use. When an antigen is mixed with alum, the salt precipitates the antigen. Injection of this alum precipitate results in a slower release of antigen from the injection site, so that the effective time of exposure to the antigen increases from a few days without adjuvant to several weeks with the adjuvant. The alum precipitate also increases the size of the antigen, thus increasing the likelihood of phagocytosis.

Water-in-oil adjuvants also prolong the persistence of antigen. A preparation known as **Freund's incomplete adjuvant** contains antigen in aqueous solution, mineral oil, and an emulsifying agent such as mannide monooleate, which disperses the oil into small droplets surrounding the antigen; the antigen is then released very slowly from the site of injection. This preparation is based on **Freund's complete adjuvant,** the first deliberately formulated highly effective adjuvant, developed by Jules Freund many years ago and containing heat-killed *Mycobacteria* as an additional ingredient. Muramyl dipeptide, a component of the mycobacterial cell wall, activates dendritic cells and macrophages, making Freund's complete adjuvant far more potent than the incomplete form. Activated dendritic cells and macrophages are more phagocytic than their unactivated counterparts and express higher levels of the molecules that trigger costimulation and enhancement of the T-cell immune response. Thus, antigen presentation and the requisite costimulatory signal are usually increased in the presence of adjuvant.

Alum and Freund's adjuvants also stimulate a local, chronic inflammatory response that attracts both phagocytes and lymphocytes. This infiltration of cells at the site of the adjuvant injection often results in formation of a dense, macrophage-rich mass of cells called a **granuloma.**

-----

# Epitopes

As mentioned above, immune cells do not interact with, or recognize, an entire immunogen molecule; instead, lymphocytes recognize discrete sites on the macromolecule called epitopes, or antigenic determinants. Studies with small antigens have revealed that B and T cells recognize different epitopes on the same antigenic molecule. For example, when mice were immunized with glucagon, a small human hormone of 29 amino acids, antibody was elicited to epitopes in the amino-terminal portion, whereas the T cells responded only to epitopes in the carboxyl-terminal portion.

Lymphocytes may interact with a complex antigen on several levels of antigen structure. An epitope on a protein antigen may involve elements of the primary, secondary, tertiary, and even quaternary structure of the protein (see Figure 4-2). In polysaccharides, branched chains are commonly present, and branch points may contribute to the conformation of epitopes. B cells recognize soluble antigen when it binds to antibody molecules in the B-cell membrane. Because B cells bind antigen that is free in solution, the epitopes they recognize tend to be highly accessible sites on the exposed surface of the immunogen. T-cell epitopes from proteins differ in that they are peptides usually derived from enzymatic digestion of pathogen proteins and are recognized by the T-cell receptor only when in complex with an MHC antigen. Thus, there is no requirement for solution accessibility such as with the B-cell epitope. Table 4-2 summarizes major differences in B- and T-cell antigenicity.

## B-cell epitopes have characteristic properties

*B-cell epitopes on native proteins generally are composed of hydrophilic amino acids on the protein surface that are topographically accessible to membrane-bound or free antibody.* A B-cell epitope must be accessible in order to be able to bind to an antibody; in general, protruding regions on the surface of the protein are the most likely to be recognized as epitopes, and these regions are usually composed of predominantly hydrophilic amino acids. Amino acid sequences that are hidden within the interior of a protein often consist of predominantly hydrophobic amino acids and cannot function as B-cell epitopes unless the protein is first denatured.

| TABLE 4-2 | Comparison of antigen recognition by T cells and B cells | |
|---|---|---|
| **Characteristic** | **B cells** | **T cells** |
| Interaction with antigen | Involves binary complex of membrane Ig and Ag | Involves ternary complex of T-cell receptor, Ag, and MHC molecule |
| Binding of soluble antigen | Yes | No |
| Involvement of MHC molecules | None required | Required to display processed antigen |
| Chemical nature of antigens | Protein, polysaccharide, lipid | Mostly proteins, but some lipids and glycolipids presented on MHC-like molecules |
| Epitope properties | Accessible, hydrophilic, mobile peptides containing sequential or nonsequential amino acids | Internal linear peptides produced by processing of antigen and bound to MHC molecules |

B-cell epitopes may be composed of sequential contiguous residues along the polypeptide chain or nonsequential residues from segments of the chain brought together by the folded conformation of an antigen. Most antibodies elicited by globular proteins bind to the protein only when it is in its native conformation. Because denaturation of such antigens usually changes the structure of their epitopes, antibodies to the native protein do not bind to the denatured protein.

In one series of experiments, five distinct **sequential epitopes,** each containing six to eight contiguous amino acids, were found in sperm whale myoglobin. Each of these epitopes is on the surface of the molecule at bends between the α-helical regions (Figure 4-3a). Sperm whale myoglobin also contains several **nonsequential epitopes,** or **conformational determinants.** The residues that constitute these epitopes are far apart in the primary amino acid sequence but close together in the tertiary structure of the molecule. Such epitopes are present only when the protein is in its native conformation. One well-characterized nonsequential epitope in hen egg-white lysozyme (HEL) is

## OVERVIEW FIGURE 4-3

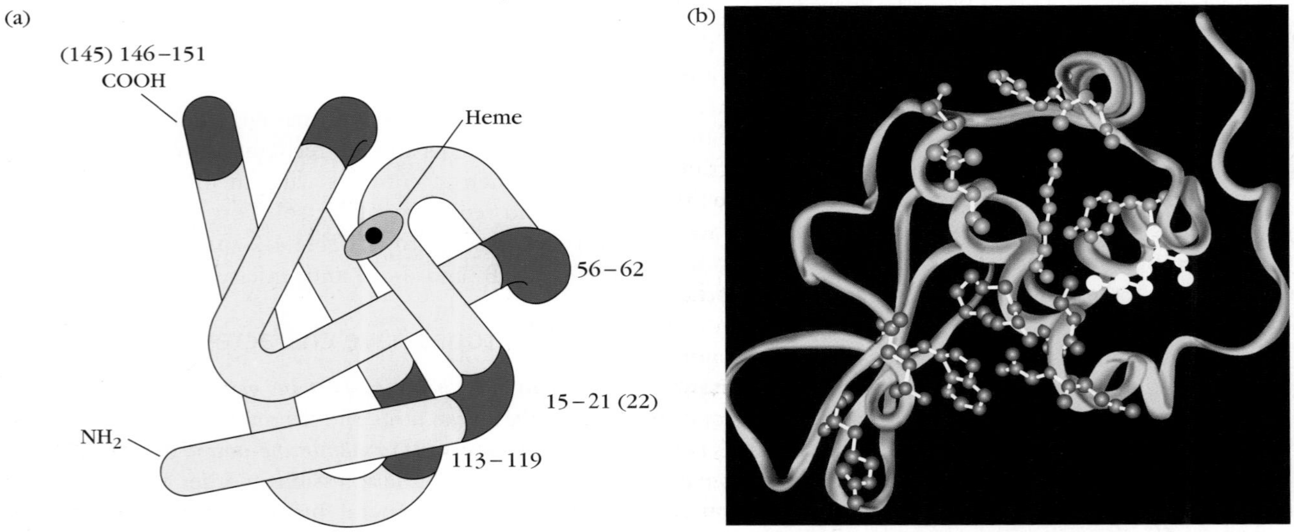

(a)

(b)

Protein antigens usually contain both sequential and nonsequential B-cell epitopes. (a) Diagram of sperm whale myoglobin showing locations of five sequential B-cell epitopes (blue). (b) Ribbon diagram of hen egg-white lysozyme showing residues that compose one nonsequential (conformational) epitope. Residues that contact antibody light chains, heavy chains, or both are shown in red, blue, and white, respectively. These residues are widely spaced in the amino acid sequence but are brought into proximity by folding of the protein. [Part (a) adapted from M. Z. Atassi and A. L. Kazim, 1978, Advances in Experimental Medicine and Biology **98**:9; part (b) from W. G. Laver et al., 1990, Cell **61**:554.]

(a) Hen egg-white lysosome

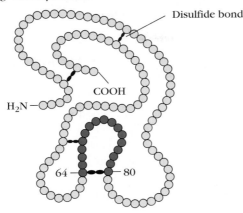

(b) Synthetic loop peptides

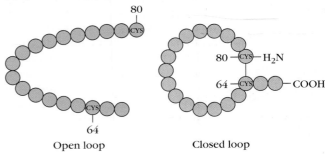

Open loop          Closed loop

(c) Inhibition of reaction between HEL loop and anti-loop antiserum

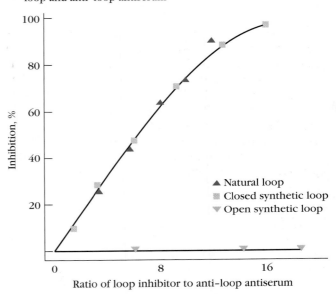

**FIGURE 4-4 Experimental demonstration that binding of antibody to conformational determinants in hen egg-white lysozyme (HEL) depends on maintenance of the tertiary structure of the epitopes by intrachain disulfide bonds.** (a) Diagram of HEL primary structure, in which circles represent amino acid residues. The loop (blue circles) formed by the disulfide bond between the cysteine residues at positions 64 and 80 constitutes one of the conformational determinants in HEL. (b) Synthetic open-loop and closed-loop peptides corresponding to the HEL loop epitope. (c) Inhibition of binding between HEL loop epitope and antiloop antiserum. Antiloop antiserum was first incubated with the natural loop sequence, the synthetic closed-loop peptide, or the synthetic open-loop peptide; the ability of the antiserum to bind the natural loop sequence then was measured. The absence of any inhibition by the open-loop peptide indicates that it does not bind to the anti-loop antiserum. [Adapted from D. Benjamin et al., 1984, Annual Review of Immunology **2**:67.]

shown in Figure 4-3b. Although the amino acid residues that compose this epitope of HEL are far apart in the primary amino acid sequence, they are brought together by the tertiary folding of the protein.

Sequential and nonsequential epitopes generally behave differently when a protein is denatured, fragmented, or reduced. For example, appropriate fragmentation of sperm whale myoglobin can yield five fragments, each retaining one sequential epitope, as demonstrated by the observation that antibody can bind to each fragment. On the other hand, nonsequential epitopes are generally eliminated when a protein is fragmented or its disulfide bonds are reduced. For example, HEL has four intrachain disulfide bonds, which determine the final protein conformation (Figure 4-4a). Many antibodies to HEL recognize several epitopes, and each of eight different epitopes has been recognized by a distinct antibody. Most of these epitopes are conformational determinants dependent on the overall structure of the protein. If the intrachain disulfide bonds of HEL are reduced with mercaptoethanol, the

nonsequential epitopes are lost; for this reason, antibody to native HEL does not bind to reduced HEL.

The inhibition experiment shown in Figure 4-4b and c nicely demonstrates this point. An antibody to a conformational determinant, in this example a peptide loop present in native HEL, was able to bind the epitope only if the disulfide bond that maintains the structure of the loop was intact. Information about the structural requirements of the antibody combining site was obtained by examining the ability of structural relatives of the natural antigen to bind to that antibody. If a structural relative has the critical epitopes present in the natural antigen, it will bind to the antibody combining site, thereby blocking its occupation by the natural antigen. In this inhibition assay, the ability of the closed loop (Figure 4-4b, right) to inhibit binding showed that the closed loop was sufficiently similar to HEL to be recognized by antibody to native HEL. Even though the open loop (Figure 4-4b, left) had the same sequence of amino acids as the closed loop, it lacked the epitopes recognized by the antibody and therefore was unable to block binding of HEL (4-4c).

*B-cell epitopes tend to be located in flexible regions of an immunogen and often display site mobility.* John A. Tainer and his colleagues analyzed the epitopes on a number of protein antigens (myohemerytherin, insulin, cytochrome *c*, myoglobin, and hemoglobin) by comparing the positions of the known B-cell epitopes with the mobility of the same residues. Their analysis revealed that the major antigenic determinants in these proteins generally were located in the most mobile regions. These investigators proposed that site mobility of epitopes maximizes complementarity with the antibody's binding site; more rigid epitopes appear to bind less effectively. However, because of the loss of entropy due to binding to a flexible site, the binding of antibody to a flexible epitope is generally of lower affinity than the binding of antibody to a rigid epitope.

*Complex proteins contain multiple overlapping B-cell epitopes, some of which are immunodominant.* For many years, it was dogma in immunology that each globular protein had a small number of epitopes, each confined to a highly accessible region and determined by the overall conformation of the protein. It has been shown more recently, however, that most of the surface of a globular protein is potentially immunogenic. This has been demonstrated by comparing the antigen-binding profiles of different monoclonal antibodies to various globular proteins. For example, when 64 different monoclonal antibodies to BSA were compared for their ability to bind to a panel of 10 different mammalian albumins, 25 different overlapping antigen-binding profiles emerged, suggesting that these 64 different antibodies recognized a minimum of 25 different epitopes on BSA.

The surface of a protein, then, presents a large number of potential antigenic sites. The subset of antigenic sites on a given protein that is recognized by the immune system of an animal is much smaller than the potential antigenic repertoire, and it varies from species to species and even among individual members of a given species. Within an animal, certain epitopes of an antigen are recognized as immunogenic; others are not. Furthermore, some epitopes, called **immunodominant,** induce a more pronounced immune response in a particular animal than other epitopes. It is highly likely that the intrinsic topographical properties of the epitope as well as the animal's regulatory mechanisms influence the immunodominance of epitopes.

## Basic Structure of Antibodies

Recognition of an immunogen by the surface antibodies of the B cell triggers proliferation and differentiation into memory B cells and plasma cells (see Chapter 11). Plasma cells secrete soluble antibody molecules with specificity for antigen that is identical to the surface receptor of the parent B cell. The following sections will describe the structural

features of the antibody molecules that allow them to fulfill their two major functions:

1. Binding foreign antigens encountered by the host

2. Mediating effector functions to neutralize or eliminate foreign invaders

It has been known since the late nineteenth century that antibodies reside in the blood serum. (In the mid-twentieth century, antibodies were shown to be present in other secreted body fluids: milk, tears, saliva, bile, etc.) Blood can be separated in a centrifuge into a fluid and a cellular fraction. The fluid fraction is the **plasma,** and the cellular fraction contains red blood cells, leukocytes, and platelets. Plasma contains all of the soluble small molecules and macromolecules of blood, including fibrin and other proteins required for the formation of blood clots. If the blood or plasma is allowed to clot, the fluid phase that remains is called **serum.** The first evidence that antibodies were contained in particular serum protein fractions came from a classic experiment by Arne Tiselius and Elvin A. Kabat in 1939. They immunized rabbits with the protein ovalbumin (the albumin of egg whites), prepared serum, and then divided the serum into two aliquots. Electrophoretic separation of one serum aliquot revealed four major fractions corresponding to serum albumin and the alpha ($\alpha$), beta ($\beta$), and gamma ($\gamma$) globulins. Prior to electrophoresis, the other serum aliquot was mixed with the immunizing antigen (ovalbumin), allowing formation of an immune precipitate

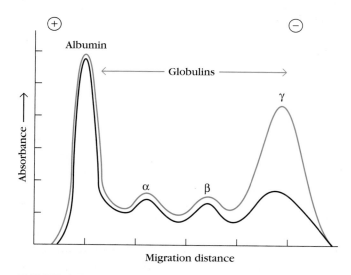

**FIGURE 4-5 Experimental demonstration that most antibodies are in the $\gamma$-globulin fraction of serum proteins.** After rabbits were immunized with ovalbumin (OVA), their antisera were pooled and electrophoresed, which separated the serum proteins according to their electric charge and mass. The blue line shows the electrophoretic pattern of untreated antiserum. The black line shows the pattern of antiserum that was first incubated with OVA to remove anti-OVA antibody and then subjected to electrophoresis. *[Adapted from A. Tiselius and E. A. Kabat, 1939,* Journal of Experimental Medicine **69:***119, with copyright permission of the Rockefeller University Press.]*

(complex of antigen and antibody) that was removed, leaving the remaining serum proteins, which did not react with the antigen. A comparison of the electrophoretic profiles of these two serum aliquots revealed that there was a significant drop in the γ-globulin peak in the aliquot that had been reacted with antigen (Figure 4-5). Thus, the γ-**globulin fraction** was identified as containing serum antibodies, which were called **immunoglobulins** to distinguish them from any other proteins that might be contained in the γ-globulin fraction. The early experiments of Kabat and Tiselius resolved serum proteins into three major nonalbumin fractions—α, β, and γ. We now know that although immunoglobulin G (IgG), the most abundant class of antibody molecules, is indeed mostly found in the γ-globulin fraction, significant amounts of it and other important classes of antibody molecules are found in the α and the β fractions of serum.

## Antibodies are heterodimers

Antibody molecules have a common structure of four peptide chains (Figure 4-6). This structure consists of two identical **light (L) chains,** polypeptides of about 22,000 Da, and two identical **heavy (H) chains,** larger polypeptides of around 55,000 Da or more. Each light chain is bound to a heavy chain by a disulfide bond and by noncovalent interactions such as salt linkages, hydrogen bonds, and hydrophobic interactions to form a heterodimer (H-L). Similar noncovalent interactions and disulfide bridges link the two identical heavy and light (H-L) chain combinations to each other to form the basic four-chain (H-L)$_2$ antibody structure, a dimer of dimers. As we will see, the exact number and precise positions of the disulfide bonds linking dimers differs among antibody classes and subclasses.

The first 110 or so amino acids of the amino-terminal region of a light or heavy chain varies greatly among antibodies of different antigen specificity. These segments of highly variable sequence are called *V regions:* $V_L$ in light chains and $V_H$ in heavy. All of the differences in specificity displayed by different antibodies can be traced to differences in the amino acid sequences of V regions. In fact, most of the differences among antibodies fall within areas of the V regions called *complementarity-determining regions (CDRs)*, and it is these CDRs, on both light and heavy chains, that constitute the antigen binding site of the antibody molecule. By contrast, within each particular class of antibody, far fewer differences are seen when one compares sequences throughout the rest of the molecule. The regions of relatively constant sequence beyond the variable regions have been dubbed C regions, $C_L$ in light chains and $C_H$ in heavy. Antibodies are glycoproteins; with few exceptions, the sites of attachment for carbohydrates are restricted to the constant region. We do not completely understand the role played by glycosylation of antibodies, but it probably increases the solubility of the molecules. Inappropriate glycosylation or its absence affects the rate

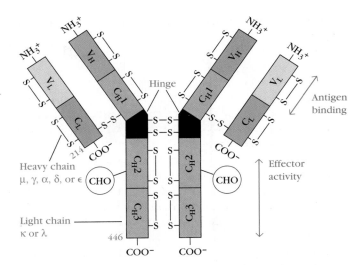

**FIGURE 4-6 Schematic diagram of structure of immunoglobulins derived from amino acid sequence analysis.** Each heavy and light chain in an immunoglobulin molecule contains an amino-terminal variable (V) region (aqua and tan, respectively) that consists of 100 to 110 amino acids and differs from one antibody to the next. The remainder of each chain in the molecule—the constant (C) regions (purple and red)—exhibits limited variation that defines the two light-chain subtypes and the five heavy-chain subclasses. Some heavy chains (γ, δ, and α) also contain a proline-rich hinge region (black). The amino-terminal portions, corresponding to the V regions, bind to antigen; effector functions are mediated by the carboxy-terminal domains. The μ and ε heavy chains, which lack a hinge region, contain an additional domain in the middle of the molecule. CHO denotes a carbohydrate group linked to the heavy chain.

at which antibodies are cleared from the serum and decreases the efficiency of interaction between antibody and other proteins with which it interacts.

## Chemical and enzymatic methods revealed basic antibody structure

Our knowledge of basic antibody structure was derived from a variety of experimental observations. When the γ-globulin fraction of serum is separated into high- and low-molecular-weight fractions, antibodies of around 150,000 Da, designated as immunoglobulin G (IgG), are found in the low-molecular-weight fraction. In a key experiment, brief digestion of IgG with the proteolytic enzyme papain produced three fragments, two of which were identical, with a third that was quite different (Figure 4-7). The two identical fragments (45,000 Da each) had antigen-binding activity and were called **Fab fragments** ("fragment, antigen binding"). The other fragment (50,000 Da) had no antigen-binding activity at all. Because it was found to crystallize during cold storage, it was called the **Fc fragment**

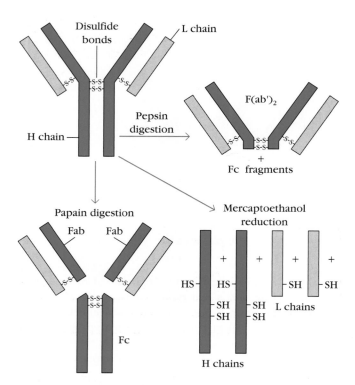

**FIGURE 4-7 Prototype structure of IgG, showing chain structure and interchain disulfide bonds.** The fragments produced by enzymatic digestion with pepsin or papain or by cleavage of the disulfide bonds with mercaptoethanol are indicated. Light (L) chains are in gray and heavy (H) chains in blue.

("fragment, crystallizable"). Digestion of IgG with a different proteolytic enzyme, pepsin, demonstrated again that the antigen-binding properties of an antibody can be separated from the rest of the molecule. Pepsin digestion generated a single 100,000 Da fragment, composed of two Fab-like subunits designated the **F(ab')₂ fragment,** which binds antigen. The Fc fragment was not recovered from pepsin treatment because it had been digested into numerous small peptides.

A key observation in deducing the multichain structure of IgG was made when the molecule was subjected to mercaptoethanol reduction and alkylation, a chemical treatment that irreversibly cleaves disulfide bonds but not peptide bonds. If the post-treatment sample is chromatographed on a column that separates molecules by size, it becomes clear that the intact 150,000 Da IgG molecule was, in fact, composed of subunits. Each IgG molecule contains two larger polypeptide chains of around 50,000 Da, designated as heavy (H) chains, and two smaller chains of about 22,000 Da, designated as light (L) chains (see Figure 4-7).

Antibodies themselves were used to determine how the enzyme digestion products—Fab, F(ab')₂, and Fc—were related to the heavy-chain and light-chain reduction products.

This question was answered by using antisera from goats that had been immunized with either the Fab fragments or the Fc fragments of rabbit IgG. The antibody to the Fab fragment could react with both the H and the L chains, whereas the antibody to the Fc fragment reacted only with the H chain. These observations led to the conclusion that the Fab fragment consists of a portion of a heavy plus an intact light chain and that Fc contains only heavy-chain components. From these results and those mentioned above the structure of IgG shown in Figure 4-6 was deduced. According to this model, the IgG molecule consists of two identical H chains and two identical L chains, which are linked by disulfide bridges.

The likely next step in developing a picture of antibody structure was determining their amino acid sequence, yet here a problem arose. The population of antibodies in the serum γ-globulin fraction consists of a heterogeneous spectrum of antibodies—many different antibodies each present in small amounts. Even if immunization is done with a hapten-carrier conjugate, the population of antibodies raised against the hapten is heterogeneous, because multiple antibodies recognize different epitopes of the hapten. Amino acid sequence analysis, however, requires a pure sample of the molecule under study. This obstacle to sequence analysis was overcome by the use of immunoglobulins from patients with **multiple myeloma,** a cancer of antibody-producing plasma cells. The plasma cells in a normal individual are end-stage cells that secrete a single molecular species of antibody for a limited period of time and then die. In contrast, a clone of plasma cells in an individual with multiple myeloma has escaped normal controls on their life span and proliferate in an unregulated way without requiring any activation by antigen to induce proliferation. Although such a cancerous plasma cell, called a **myeloma cell,** has been transformed, its protein-synthesizing machinery and secretory functions are not altered; thus, the cell continues to secrete molecularly homogeneous antibody. This antibody is indistinguishable from normal antibody molecules but is called **myeloma protein** to denote its source. In a patient afflicted with multiple myeloma, myeloma protein can account for 95% of the serum immunoglobulins. In many patients, the myeloma cells also secrete excessive amounts of light chains. These excess light chains were first discovered in the urine of myeloma patients and were named **Bence-Jones proteins** for their discoverer.

Multiple myeloma also occurs in other animals. In mice it can arise spontaneously, as it does in humans, or conditions favoring myeloma induction can be created by injecting mineral oil into the peritoneal cavity. The clones of malignant plasma cells that develop are called **plasmacytomas,** and many of these are designated MOPCs, denoting the *m*ineral-*o*il induction of *p*lasmacytoma *c*ells. A large number of mouse MOPC lines secreting different immunoglobulin classes is presently carried by the American Type-Culture Collection, a nonprofit repository of cell lines commonly used in research. The method for producing monoclonal antibodies of a desired specificity pioneered by Georges Köhler and Cesar Milstein (see below) makes use of the plasmacytoma lines. This hybridoma technology allows production of homogenous antibodies for structural and functional studies and for a variety of clinical applications.

## Light-chain sequences revealed constant and variable regions

When the amino acid sequences of several Bence-Jones proteins (light chains) from different individuals were compared, a striking pattern emerged. The amino-terminal half of the chain, consisting of 100 to 110 amino acids, was found to vary among different Bence-Jones proteins. This region was called the **variable (V) region.** The carboxyl-terminal halves of the molecules, called the **constant (C) region,** had one of two amino acid sequences. This led to the recognition that there were two light-chain types, **kappa (κ)** and **lambda (λ).** In humans, 60% of the light chains are kappa and 40% are lambda, whereas in mice, 95% of the light chains are kappa and only 5% are lambda. A normal antibody molecule contains only one light-chain type, either κ or λ, never both.

The amino acid sequences of λ light-chain constant regions show minor differences that are used to classify λ light chains into subtypes. In mice and in humans, there are four subtypes: λ1, λ2, λ3, and λ4. Amino acid substitutions at only a few positions are responsible for the subtype differences.

## There are five major classes of heavy chains

For heavy-chain sequence analysis, myeloma proteins were reduced with mercaptoethanol and alkylated, and the heavy chains were separated by gel filtration in a denaturing solvent. When the amino acid sequences of several myeloma protein heavy chains were compared, a pattern similar to that of the light chains emerged. The amino-terminal part of the chain, consisting of 100 to 110 amino acids, showed great sequence variation among myeloma heavy chains and was therefore called the variable (V) region. The remaining part of the protein revealed five basic sequence patterns, corresponding to five different heavy-chain constant (C) regions: μ, δ, γ, ε and α. Each of these five different heavy chains is called an **isotype.** The length of the constant regions is approximately 330 amino acids for δ, γ, and α and 440 amino acids for μ and ε. The heavy chains of a given antibody molecule determine the *class* of that antibody: IgM(μ), IgG(γ), IgA(α), IgD(δ), or IgE(ε). H chains of each class may pair with either κ or λ light chains. A single antibody molecule has two identical heavy chains and two identical light chains, $H_2L_2$, or a multiple $(H_2L_2)_n$ of this basic four-chain structure (Table 4-3).

Minor differences in the amino acid sequences of the α and γ heavy chains led to further classification of the heavy

| TABLE 4-3 | Chain composition of the five immunoglobulin classes in humans | | | |
|---|---|---|---|---|
| Class[*] | Heavy chain | Subclasses | Light chain | Molecular formula |
| IgG | γ | γ1, γ2, γ3, γ4 | κ or λ | $\gamma_2\kappa_2$ <br> $\gamma_2\lambda_2$ |
| IgM | μ | None | κ or λ | $(\mu_2\kappa_2)_n$ <br> $(\mu_2\lambda_2)_n$ <br> n = 1 or 5 |
| IgA | α | α1, α2 | κ or λ | $(\alpha_2\kappa_2)_n$ <br> $(\alpha_2\lambda_2)_n$ <br> n = 1, 2, 3, or 4 |
| IgE | ε | None | κ or λ | $\epsilon_2\kappa_2$ <br> $\epsilon_2\lambda_2$ |
| IgD | δ | None | κ or λ | $\delta_2\kappa_2$ <br> $\delta_2\lambda_2$ |

[*]See Figure 4-1 for general structures of five antibody classes.

chains into subisotypes that determine the subclass of antibody molecules they constitute. In humans, there are two subisotypes of α heavy chains—α1 and α2 (and thus two subclasses, IgA1 and IgA2)—and four subisotypes of γ heavy chains: γ1, γ2, γ3, and γ4 (and thus four subclasses, IgG1, IgG2, IgG3, and IgG4). In mice, there are four subisotypes, γ1, γ2a, γ2b, and γ3, and the corresponding subclasses.

## Immunoglobulins possess multiple domains based on the immunoglobulin fold

The overall structure of the immunoglobulin molecule is determined by the primary, secondary, tertiary, and quaternary organization of the protein. The primary structure—the sequence of amino acids—comprises the variable and constant regions of the heavy and light chains. The secondary structure is formed by the folding of the extended polypeptide chain upon itself into a series of antiparallel β pleated sheets (Figure 4-8). The chains are then folded into a tertiary structure of compact globular domains, which are connected to neighboring domains by stretches of the polypeptide chain between regions of β pleated sheet. Finally, the globular domains of adjacent heavy and light polypeptide chains interact in the quaternary structure (Figure 4-9), forming functional domains that enable the molecule to specifically bind antigen and at the same time perform a number of biological effector functions.

Careful analysis of the amino acid sequences of immunoglobulin heavy and light chains showed that both chains contain several homologous units of about 110 amino acid residues. Within each unit, termed a domain, an intrachain disulfide bond forms a loop of about 60 amino acids. Light chains contain one variable domain

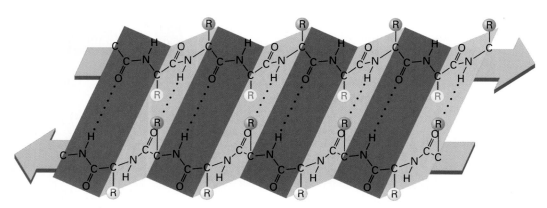

**FIGURE 4-8  Structural formula of a β pleated sheet containing two antiparallel β strands.** The structure is held together by hydrogen bonds between peptide bonds of neighboring stretches of polypeptide chains. The amino acid side groups (R) are arranged perpendicular to the plane of the sheet. [*Adapted from H. Lodish et al., 1995, Molecular Cell Biology, 4th ed., Scientific American Books, New York; reprinted by permission of W. H. Freeman and Company.*]

(a)

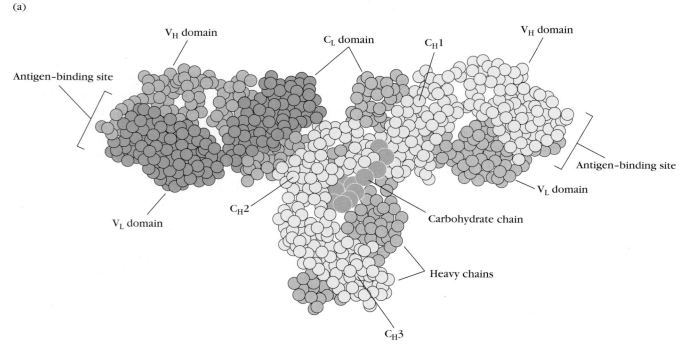

(b)

**FIGURE 4-9  Two representations of an intact antibody molecule. Interactions between domains in the separate chains of the immunoglobulin molecule are critical to its quaternary structure.** (a) Model of IgG molecule, based on x-ray crystallographic analysis, showing associations between domains in the separate chains of an immunoglobulin molecule that are critical to its quaternary structure. Each solid ball represents an amino acid residue; the larger tan balls are carbohydrate. Note that the $C_H2$ domains protrude because of the interior carbohydrate; this allows accessibility to other molecules such as complement components. The two light chains are shown in shades of red; the two heavy chains, in shades of blue. (b) Schematic diagram showing the interacting heavy- and light-chain domains. [*Part a from E. W. Silverton et al., 1977, Proceedings of the National Academy of Science U.S.A. **74**:5140.*]

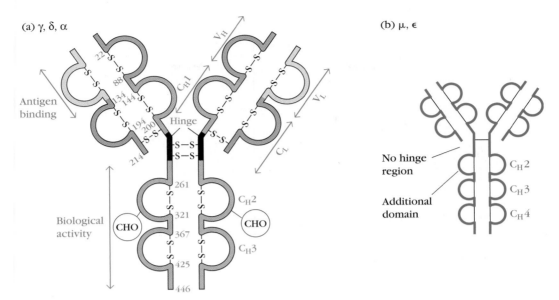

(a) γ, δ, α

(b) μ, ε

**FIGURE 4-10** **(a) Heavy and light chains are folded into domains, each containing about 110 amino acid residues and an intrachain disulfide bond that forms a loop of 60 amino acids.** The amino-terminal domains, corresponding to the V regions, bind to antigen; effector functions are mediated by the other domains. **(b)** The μ and ε heavy chains contain an additional domain that replaces the hinge region.

($V_L$) and one constant domain ($C_L$); heavy chains contain one variable domain ($V_H$) and either three or four constant domains ($C_H1$, $C_H2$, $C_H3$, and $C_H4$), depending on the antibody class (Figure 4-10).

X-ray crystallographic analysis revealed that immunoglobulin domains are folded into a characteristic compact structure called the **immunoglobulin fold.** This structure consists of a "sandwich" of two β pleated sheets, each containing antiparallel β strands of amino acids, which are connected by loops of various lengths (Figure 4-11). The β strands within a sheet are stabilized by hydrogen bonds between the amino (–NH) groups in one strand and carbonyl groups of an adjacent strand (see Figure 4-8). The β strands are characterized by alternating hydrophobic and hydrophilic amino acids whose side chains are arranged perpendicular to the plane of the sheet; the hydrophobic amino acids are oriented toward the interior of the sandwich, and the hydrophilic amino acids face outward. The two β sheets within an immunoglobulin fold are stabilized by the hydrophobic interactions between them and by a conserved disulfide bond.

Although variable and constant domains have a similar structure, there are subtle differences between them. The sequence of the V domain is slightly longer than the C domain and contains an extra pair of β strands within the β-sheet structure, as well as the extra loop sequence connecting this pair of β strands (see Figure 4-11).

The basic structure of the immunoglobulin fold contributes to the quaternary structure of immunoglobulins by facilitating noncovalent interactions between domains across the faces of the β sheets (see Figure 4-8). Interactions form links between identical domains (e.g., $C_H2/C_H2$, $C_H3/C_H3$, and $C_H4/C_H4$) and between nonidentical domains (e.g.,

$V_H/V_L$ and $C_H1/C_L$). The structure of the immunoglobulin fold also allows for variable lengths and sequences of amino acids that form the loops connecting the β strands. As the next section explains, some of the loop sequences of the $V_H$ and $V_L$ domains contain variable amino acids and constitute the antigen binding site of the molecule.

## Antibody Binding Site

Antibody molecules have two roles, antigen binding and mediation of effector functions. Antigen binding is accomplished by the amino-terminal portions and effector functions by the carboxyl-terminal regions. The structural features relating to these functions are discussed in later sections.

Detailed comparisons of the amino acid sequences of a large number of $V_L$ and $V_H$ domains revealed that the sequence variation is concentrated in a few discrete regions of these domains. The pattern of this variation is best summarized by a quantitative plot of the variability at each position of the polypeptide chain. The **variability** is defined as

$$\text{Variability} = \frac{\text{Number of different amino acids at a given position}}{\text{Frequency of the most common amino acid at a given position}}$$

Thus, if a comparison of the sequences of 100 heavy chains revealed that a serine was found in position 7 in 51 of the sequences (frequency 0.51), it would be the most common amino acid. If examination of the other 49 sequences showed that position 7 was occupied by either glutamine, histidine, proline, or tryptophan, the variability at that position would be 9.8 (5/0.51). Variability plots of $V_L$ and $V_H$ domains

(a)

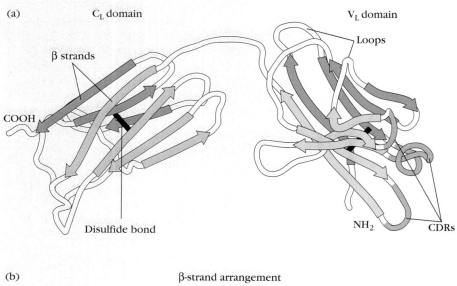

(b)                    β-strand arrangement

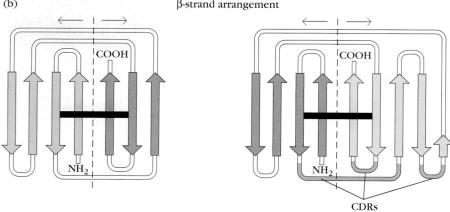

**FIGURE 4-11 (a) Diagram of an immunoglobulin light chain depicting the immunoglobulin-fold structure of its variable and constant domains.** The two β pleated sheets in each domain are held together by hydrophobic interactions and the conserved disulfide bond. The β strands that compose each sheet are shown in different colors. The amino acid sequences in three loops of each variable domain show considerable variation; these hypervariable regions (blue) make up the antigen binding site. Hypervariable regions are usually called CDRs (complementarity-determining regions). Heavy-chain domains have the same characteristic structure. (b) The β pleated sheets are opened out to reveal the relationship of the individual β strands and joining loops. Note that the variable domain contains two more β strands than the constant domain. *[Part a adapted from M. Schiffer et al., 1973, Biochemistry **12**:4620; reprinted with permission; part b adapted from A. F. Williams and A. N. Barclay, 1988, Annual Review of Immunology **6**:381.]*

of human antibodies show that maximum variation is seen in those portions of the sequence that correspond to the loops that join the β strands (Figure 4-12). These regions were originally called **hypervariable regions** in recognition of their high variability. Hypervariable regions form the antigen binding site of the antibody molecule. Because the antigen binding site is complementary to the structure of the epitope, these areas are now more widely called **complementarity-determining regions (CDRs).** The three heavy-chain and three light-chain CDR regions are located on the loops that connect the β strands of the $V_H$ and $V_L$ domains. The remainder of the $V_L$ and $V_H$ domains exhibit far less variation; these stretches are called the **framework regions (FRs).** The wide range of specificities exhibited by antibodies is due to variations in the length and amino acid sequence of the six CDRs that fall within the region corre-

sponding to Fab fragments. The framework region acts as a scaffold that supports these six loops. The three-dimensional structure of the framework regions of virtually all antibodies analyzed to date can be superimposed on one another; in contrast, the hypervariable loops (i.e., the CDRs) are essentially unique to each antibody.

## CDRs bind antigen

The finding that CDRs are the antigen-binding regions of antibodies has been confirmed directly by high-resolution x-ray crystallography of antigen-antibody complexes. Crystallographic analysis has been completed for many Fab fragments of monoclonal antibodies complexed either with large globular protein antigens or with smaller antigens, including carbohydrates, nucleic acids, peptides, and

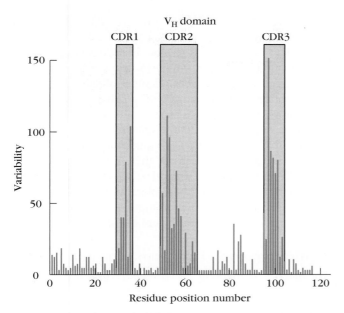

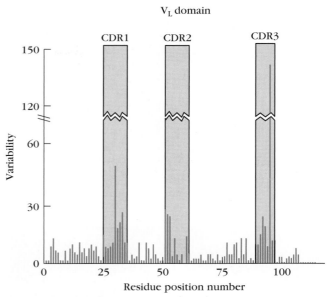

**FIGURE 4-12  Variability of amino acid residues in the V$_L$ and V$_H$ domains of human antibodies with different specificities.** Three hypervariable (HV) regions, also called complementarity-determining regions (CDRs), are present in both heavy- and light-chain V domains. As shown in Figure 4-11 *(right)*, the three HV regions in the light chain V domain are brought into proximity in the folded structure. The same is true of the heavy-chain V domain. *[Based on E. A. Kabat et al., 1977, Sequence of Immunoglobulin Chains, U.S. Dept. of Health, Education, and Welfare.]*

small haptens. In addition, complete structures have been obtained for several intact monoclonal antibodies. X-ray diffraction analysis of antibody-antigen complexes has shown that several CDRs may make contact with the antigen, and a number of complexes have been observed in which all six CDRs contact the antigen. In general, more residues in the heavy-chain CDRs appear to contact antigen than in the light-chain CDRs. In other words, the V$_H$ domain often contributes more to antigen binding than the V$_L$ domain. The dominant role of the heavy chain in antigen binding was demonstrated in a study in which a single heavy chain specific for a glycoprotein antigen of

the human immunodeficiency virus (HIV) was combined with various light chains of different antigenic specificity (see Chapter 20). All of the hybrid antibodies bound to the HIV glycoprotein antigen, indicating that the heavy chain alone was sufficient to confer specificity. However, one should not conclude that the light chain is largely irrelevant; in some antibody-antigen reactions, the light chain makes the more important contribution.

The nature of the contact surface between antibody and antigen is dramatically revealed by a computer-generated picture of the interaction between an influenza virus antigen and antibody (Figure 4-13). Contacts between a large globular

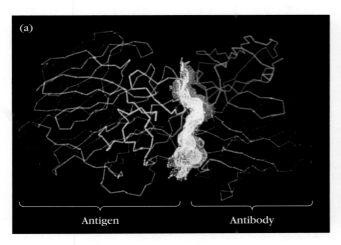

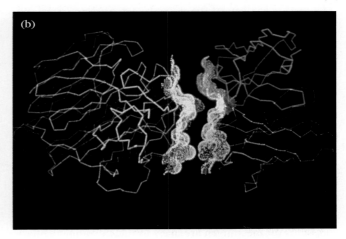

**FIGURE 4-13  Computer simulation of an interaction between antibody and influenza virus antigen, a globular protein.** (a) The antigen (yellow) is shown interacting with the antibody molecule; the variable region of the heavy chain is red, and the variable region of the light chain is blue. (b) The complementarity of the two molecules is revealed by separating the antigen from the antibody by 8 Å. *[Based on x-ray crystallography data collected by P. M. Colman and W. R. Tulip. From G. J. V. H. Nossal, 1993, Scientific American **269**(3):22.]*

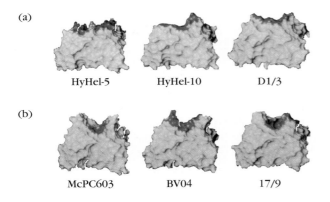

(a)

HyHel-5          HyHel-10          D1/3

(b)

McPC603          BV04          17/9

(c)

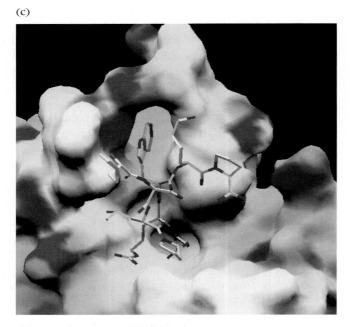

**FIGURE 4-14 Antibody binding sites.** In general, the binding sites for small molecules are deep pockets, whereas binding sites for large proteins are flatter, more undulating surfaces. (a, b) Models of the variable domains of six Fab fragments with their antigen-binding regions shown in purple. The three antibodies in (a) are specific for lysozyme, a large globular protein. The three antibodies in (b) are specific for smaller molecules or very small segments of macromolecules: McPC603 for phosphocholine; BV04 for a small segment of a single-stranded DNA molecule; and 17/9 for a peptide from hemagglutinin, an envelope protein of influenza virus. (c) Close-up of the complex between a small peptide derived from HIV protease and a Fab fragment from an antiprotease antibody. *[(a, b) From I. A. Wilson and R. L. Stanfield, 1993, Current Opinion in Structural Biology **3**:113; (c) from J. Lescar et al., 1997, Journal of Molecular Biology **267**:1207, courtesy of G. Bentley, Institute Pasteur.]*

that the surface areas of interaction are quite large, ranging from about 650 Å$^2$ to more than 900 Å$^2$. Within this area, some 15 to 22 amino acid residues in the antibody contact a similar number of residues in the protein antigen.

In Figure 4-14, we see that globular protein antigens and small peptide or hapten antigens interact with antibody in different ways. Typically, large areas of protein antigens are engaged by large areas of the antibody binding site (Figure 4-14a). Smaller antigens such as peptides occupy less space and can fit into pockets or clefts of the binding site (Figure 4-14b). The interaction of a peptide derived from HIV protease and a Fab fragment from an antiprotease antibody nicely illustrates the intimate complementarity between antibody and smaller antigen (Figure 4-14c).

## Conformational changes may be induced by antigen binding

As more x-ray crystallographic analyses of Fab fragments were completed, it became clear that in some cases, binding of antigen induces conformational changes in the antibody, antigen, or both. A striking example of conformational change is seen in the formation of the complex between the Fab fragment and its peptide epitope shown in Figure 4-15.

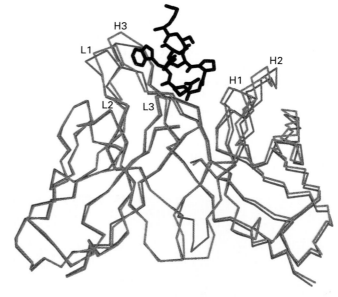

**FIGURE 4-15 Conformational change on binding of antigen to antibody.** Shown is the same complex as in Figure 4-14c, between a peptide derived from HIV protease and a Fab fragment from an antiprotease antibody, comparing the Fab structure before and after peptide binding. The red line shows the structure of the Fab fragment before it binds the peptide, and the blue line shows its structure when bound. There are significant conformational changes in the CDRs of the Fab on binding the antigen. These are especially pronounced in the light chain CDR1 (L1) and the heavy chain CDR3 (H3). *[From J. Lescar et al., 1997, Journal of Molecular Biology **267**:1207; courtesy of G. Bentley, Institute Pasteur.]*

protein antigen and antibody occur over a broad, often rather flat, undulating face. In the area of contact, protrusions or depressions on the antigen are likely to match complementary depressions or protrusions on the antibody binding site, formed by CDRs. In the case of the well-studied lysozyme/antilysozyme system, crystallographic studies have shown

Significant structural changes occur in the Fab on binding. The CDR1 region of the light chain moves as much as 1 Å, and the heavy chain CDR3 moves 2.7 Å. Thus, in addition to variability in the length and amino acid composition of the CDR loops, the ability of these loops to change conformation significantly on antigen binding enables antibodies to assume a shape more effectively complementary to that of their epitopes.

As already indicated, conformational changes following antigen binding need not be limited to the antibody. Although not shown in Figure 4-15, the conformation of the protease peptide bound to the Fab shows no structural similarity to the corresponding epitope in the native HIV protease. It has been suggested that the inhibition of protease activity by this anti-HIV protease antibody is a result of its distortion of the enzyme's native conformation.

## Constant-region domains

The immunoglobulin constant-region domains take part in various biological functions that are determined by the amino acid sequence of each domain.

### $C_H1$ and $C_L$ Domains

The $C_H1$ and $C_L$ domains serve to extend the Fab arms of the antibody molecule, thereby facilitating interaction with antigen and increasing the maximum rotation of the Fab arms. In addition, an interchain disulfide bond between these constant-region domains helps to hold the $V_H$ and $V_L$ domains together (see Figure 4-10). The $C_H1$ and $C_L$ domains may also contribute to antibody diversity by allowing more random associations between the $V_H$ and $V_L$ domains than would

occur if this association were driven by the $V_H/V_L$ interaction alone. These considerations have important implications for building a diverse repertoire of antibodies. As Chapter 5 will show, random rearrangements of the immunoglobulin genes generate unique $V_H$ and $V_L$ sequences for the heavy and light chains expressed by each B lymphocyte; association of the $V_H$ and $V_L$ sequences then generates a unique antigen binding site. The presence of $C_H1$ and $C_L$ domains appears to increase the number of stable $V_H$ and $V_L$ interactions that are possible, thus contributing to the overall diversity of antibody molecules that can be expressed by an animal.

### Hinge Region

The $\gamma$, $\delta$, and $\alpha$ heavy chains contain an extended peptide sequence between the $C_H1$ and $C_H2$ domains that has no homology with the other domains (see Figure 4-10). This region, called the **hinge region,** is rich in proline residues and is flexible, giving IgG, IgD, and IgA segmental flexibility. As a result, the two Fab arms can assume various angles to each other when antigen is bound. This flexibility of the hinge region can be visualized in electron micrographs of antigen-antibody complexes. For example, when a molecule containing two dinitrophenol (DNP) groups reacts with anti-DNP antibody and the complex is captured on a grid, negatively stained, and observed by electron microscopy, large complexes (e.g., dimers, trimers, tetramers) are seen. The angle between the arms of the Y-shaped antibody molecules differs in the different complexes, reflecting the flexibility of the hinge region (Figure 4-16).

Two prominent amino acids in the hinge region are proline and cysteine. The large number of proline residues in

**FIGURE 4-16 Experimental demonstration of the flexibility of the hinge region in antibody molecules.** (a) A hapten in which two dinitrophenyl (DNP) groups are tethered by a short connecting spacer group reacts with anti-DNP antibodies to form trimers, tetramers, and other larger antigen-antibody complexes. A trimer is shown schematically. (b) In an electron micrograph of a negatively stained preparation of these complexes, two triangular trimeric structures are clearly visible. The antibody protein stands out as a light structure against the electron-dense background. Because of the flexibility of the hinge region, the angle between the arms of the antibody molecules varies. *[Photograph from R. C. Valentine and N. M. Green, 1967, Journal of Molecular Biology 27:615; reprinted by permission of Academic Press Inc. (London) Ltd.]*

the hinge region gives it an extended polypeptide conformation, making it particularly vulnerable to cleavage by proteolytic enzymes; it is this region that is cleaved with papain or pepsin (see Figure 4-7). The cysteine residues form interchain disulfide bonds that hold the two heavy chains together. The number of interchain disulfide bonds in the hinge region varies considerably among different classes of antibodies and between species. Although $\mu$ and $\epsilon$ chains lack a hinge region, they have an additional domain of 110 amino acids ($C_H2/C_H2$) that has hingelike features.

### Other Constant-Region Domains

As noted already, the heavy chains in IgA, IgD, and IgG contain three constant-region domains and a hinge region, whereas the heavy chains in IgE and IgM contain four constant-region domains and no hinge region. The corresponding domains of the two groups are as follows:

| IgA, IgD, IgG | IgE, IgM |
|---|---|
| $C_H1/C_H1$ | $C_H1/C_H1$ |
| Hinge region | $C_H2/C_H2$ |
| $C_H2/C_H2$ | $C_H3/C_H3$ |
| $C_H3/C_H3$ | $C_H4/C_H4$ |

Although the $C_H2/C_H2$ domains in IgE and IgM occupy the same position in the polypeptide chains as the hinge region in the other classes of immunoglobulin, a function for this extra domain has not yet been determined.

X-ray crystallographic analyses have revealed that the two $C_H2$ domains of IgA, IgD, and IgG (and the $C_H3$ domains of IgE and IgM) are separated by oligosaccharide side chains that prevent contact between the appressed domains (see Figure 4-9b); as a result, these two globular domains are much more accessible than the others to the aqueous environment. This accessibility is one of the elements that contributes to the biological activity of these domains in the activation of complement components by IgG and IgM.

The carboxyl-terminal domain is designated $C_H3/C_H3$ in IgA, IgD, and IgG and $C_H4/C_H4$ in IgE and IgM. The five classes of antibody and their subclasses can be expressed either as **secreted immunoglobulin (sIg)** or as **membrane-bound immunoglobulin (mIg)**. The carboxyl-terminal domain in secreted immunoglobulin differs in both structure and function from the corresponding domain in membrane-bound immunoglobulin. Secreted immunoglobulins have a hydrophilic amino acid sequence of various length at the carboxyl-terminal end. The functions of this domain in the various classes of secreted antibody will be described later. In membrane-bound immunoglobulin, the carboxyl-terminal domain contains three regions:

- An extracellular hydrophilic "spacer" sequence composed of 26 amino acid residues

- A hydrophobic transmembrane sequence

- A short cytoplasmic tail

The length of the transmembrane sequence is constant among all immunoglobulin isotypes, whereas the lengths of the extracellular spacer sequence and the cytoplasmic tail vary.

B cells express different classes of mIg at different developmental stages. The immature B cell, called a pre–B cell, expresses only mIgM; later in maturation, mIgD appears and is co-expressed with IgM on the surface of mature B cells before they have been activated by antigen. A memory B cell can express mIgG, mIgA, or mIgE. Even when different classes are expressed sequentially on a single cell, the antigenic specificity of all the membrane antibody molecules expressed by a single cell is identical, so that each antibody molecule binds to the same epitope. The genetic mechanism that allows a single B cell to express multiple immunoglobulin isotypes all with the same antigenic specificity is described in Chapter 5.

## Antibody-Mediated Effector Functions

In addition to binding antigen, antibodies participate in a broad range of other biological activities. When considering the role of antibody in defending against disease, one must remember that antibodies generally do not kill or remove pathogens solely by binding to them. In order to be effective against pathogens, antibodies must not only recognize antigen but also invoke responses—effector functions—that result in removal of the antigen and death of the pathogen. Although variable regions of antibody are the sole agents of binding to antigen, the heavy-chain constant region ($C_H$) is responsible for a variety of collaborative interactions with other proteins, cells, and tissues that result in the effector functions of the humoral response.

Because these effector functions result from interactions between heavy-chain constant regions and other serum proteins or cell membrane receptors, not all classes of immunoglobulin have the same functional properties. An overview of four major effector functions mediated by domains of the constant region is presented here. A fifth function unique to IgE, the activation of mast cells, eosinophils, and basophils, will be described later.

### Opsonization is promoted by antibody

**Opsonization,** promotion of phagocytosis of antigens by macrophages and neutrophils, is an important factor in antibacterial defenses. Protein molecules called *Fc receptors (FcR)*, which can bind the constant region of Ig molecules, are present on the surfaces of macrophages and neutrophils as well as other cells not involved in phagocytosis. The binding of phagocyte Fc receptors by several antibody molecules complexed with the same target antigen, such as a bacterial cell, produces an interaction that se-cures the pathogen to the phagocyte membrane. This cross-linking of the FcR by binding to an array of antibody Fc regions initiates a signal-transduction pathway that re-sults in the phagocytosis of the antigen-antibody complex. Inside the phagocyte, the pathogen becomes the target of various destructive processes that include enzymatic di-gestion, oxidative damage, and the membrane-disrupting effects of antibacterial peptides.

## Antibodies activate complement

IgM and, in humans, most IgG subclasses can activate a col-lection of serum glycoproteins called the **complement** sys-tem. Complement includes a group of proteins that can perforate cell membranes. An important byproduct of the complement activation pathway is a protein fragment called C3b, which binds nonspecifically to cell- and antigen-antibody complexes near the site at which complement was activated. Many cell types—for example, red blood cells and macrophages—have receptors for C3b and so bind cells or complexes to which C3b has adhered. Binding of adherent C3b by macrophages leads to phagocytosis of the cells or molecular complexes attached to C3b. Binding of antigen-antibody complexes by the C3b receptors of a red blood cell allows the erythrocyte to deliver the complexes to liver or spleen, where resident macrophages remove them without destroying the red cell. The collaboration between antibody and the complement system is important for the inactiva-tion and removal of antigens and the killing of pathogens. The process of complement activation is described in detail in Chapter 7.

## Antibody-dependent cell-mediated cytotoxicity (ADCC) kills cells

The linking of antibody bound to target cells (e.g., virus-infected cells of the host) with the Fc receptors of a num-ber of cell types, particularly natural killer (NK) cells, can direct the cytotoxic activities of the effector cell against the target cell. In this process, called **antibody-dependent cell-mediated cytotoxicity (ADCC),** the antibody acts as a newly acquired receptor enabling the attacking cell to rec-ognize and kill the target cell. The phenomenon of ADCC is discussed in Chapter 14.

## Some antibodies can cross epithelial layers by transcytosis

The delivery of antibody to the mucosal surfaces of the respiratory, gastrointestinal, and urogenital tracts, as well as its export to breast milk, requires the movement of im-munoglobulin across epithelial layers, a process called **transcytosis.** The capacity to be transported depends on properties of the constant region. In humans and mice, IgA is the major antibody class that undergoes such trans-

cytosis, although IgM can also be tra ⌐ surfaces. Some mammalian species, s⌐ mice, also transfer significant amounts of mo⌐ of IgG from mother to fetus. Since maternal and fe⌐ culatory systems are separate, antibody must be trans-ported across the placental tissue that separates mother and fetus. In humans, this transfer takes place during the third trimester of gestation. The important consequence is that the developing fetus receives a sample of the mother's repertoire of antibody as a protective endowment against pathogens. As with the other effector functions described here, the capacity to undergo transplacental transport de-pends on properties of the constant region of the antibody molecule.

The transfer of IgG from mother to fetus is a form of **passive immunization,** which is the acquisition of immu-nity by receipt of preformed antibodies rather than by ac-tive production of antibodies after exposure to antigen. The ability to transfer immunity from one individual to another by the transfer of antibodies is the basis of antibody ther-apy, an important and widely practiced medical procedure (see Clinical Focus, page 98).

# Antibody Classes and Biological Activities

The various immunoglobulin isotypes and classes have been mentioned briefly already. Each class is distinguished by unique amino acid sequences in the heavy-chain constant region that confer class-specific structural and functional properties. In this section, the structure and effector func-tions of each class are described in more detail. The molecular properties and biological activities of the im-munoglobulin classes are summarized in Table 4-4. The structures of the five major classes are diagramed in Figure 4-17.

## Immunoglobulin G (IgG)

IgG, the most abundant class in serum, constitutes about 80% of the total serum immunoglobulin. The IgG molecule consists of two γ heavy chains and two κ or two λ light chains (see Figure 4-17a). There are four human IgG subclasses, distinguished by differences in γ-chain constant-region se-quence and numbered according to their decreasing average serum concentrations: IgG1, IgG2, IgG3, and IgG4 (see Table 4-3).

The amino acid sequences that distinguish the four IgG subclasses are encoded by different germ-line $C_H$ genes, whose DNA sequences are 90% to 95% homologous. The structural characteristics that distinguish these subclasses from one another are the size of the hinge region and the number and position of the interchain disulfide bonds be-tween the heavy chains (Figure 4-18). The subtle amino acid

| TABLE 4-4 | Properties and biological activities* of classes and subclasses of human serum immunoglobulins | | | | | | | | |
|---|---|---|---|---|---|---|---|---|---|
| | IgG1 | IgG2 | IgG3 | IgG4 | IgA1 | IgA2 | IgM‡ | IgE | IgD |
| Molecular weight† | 150,000 | 150,000 | 150,000 | 150,000 | 150,000 – 600,000 | 150,000 – 600,000 | 900,000 | 190,000 | 150,000 |
| Heavy-chain component | $\gamma 1$ | $\gamma 2$ | $\gamma 3$ | $\gamma 4$ | $\alpha 1$ | $\alpha 2$ | $\mu$ | $\epsilon$ | $\delta$ |
| Normal serum level (mg/ml) | 9 | 3 | 1 | 0.5 | 3.0 | 0.5 | 1.5 | 0.0003 | 0.03 |
| In vivo serum half-life (days) | 23 | 23 | 8 | 23 | 6 | 6 | 5 | 2.5 | 3 |
| Activates classical complement pathway | + | +/− | ++ | − | − | − | ++ | − | − |
| Crosses placenta | + | +/− | + | + | − | − | − | − | − |
| Present on membrane of mature B cells | − | − | − | − | − | − | + | − | + |
| Binds to Fc receptors of phagocytes | ++ | +/− | ++ | + | − | − | ? | − | − |
| Mucosal transport | − | − | − | − | ++ | ++ | + | − | − |
| Induces mast cell degranulation | − | − | − | − | − | − | − | + | − |

*Activity levels indicated as follows: ++ = high; + = moderate; +/− = minimal; − = none; ? = questionable.
†IgG, IgE, and IgD always exist as monomers; IgA can exist as a monomer, dimer, trimer, or tetramer. Membrane-bound IgM is a monomer, but secreted IgM in serum is a pentamer.
‡IgM is the first isotype produced by the neonate and during a primary immune response.

differences between subclasses of IgG affect the biological activity of the molecule:

- IgG1, IgG3, and IgG4 readily cross the placenta and play an important role in protecting the developing fetus.

- IgG3 is the most effective complement activator, followed by IgG1; IgG2 is less efficient, and IgG4 is not able to activate complement at all.

- IgG1 and IgG3 bind with high affinity to Fc receptors on phagocytic cells and thus mediate opsonization. IgG4 has an intermediate affinity for Fc receptors and IgG2 has an extremely low affinity.

## Immunoglobulin M (IgM)

IgM accounts for 5% to 10% of the total serum immunoglobulin, with an average serum concentration of 1.5 mg/ml. Monomeric IgM (180,000 Da) is expressed as membrane-bound antibody on B cells. IgM is secreted by plasma cells as a pentamer in which five monomeric units are held together by disulfide bonds that link their carboxyl-terminal heavy-chain domains ($C_\mu 4/C_\mu 4$) and their $C_\mu 3/C_\mu 3$ domains (see Figure 4-17e). The five monomeric subunits are arranged with their Fc regions in the center of the pentamer and the 10 antigen binding sites on the periphery of the molecule. Each pentamer contains an additional Fc-linked polypeptide called the J (joining) chain, which is disulfide-bonded to the carboxyl-terminal cysteine residue of two of the 10 $\mu$ chains. The J chain appears to be required for polymerization of the monomers to form pentameric IgM; it is added just before secretion of the pentamer.

IgM is the first immunoglobulin class produced in a primary response to an antigen, and it is also the first immunoglobulin to be synthesized by the neonate. Because of its pentameric structure with 10 antigen binding sites, serum IgM has a higher valency than the other isotypes. An IgM molecule can bind 10 small hapten molecules; however, because of steric hindrance, only five or fewer molecules of larger antigens can be bound simultaneously. Because of its high valency, pentameric IgM is more efficient than other isotypes in binding antigens with many

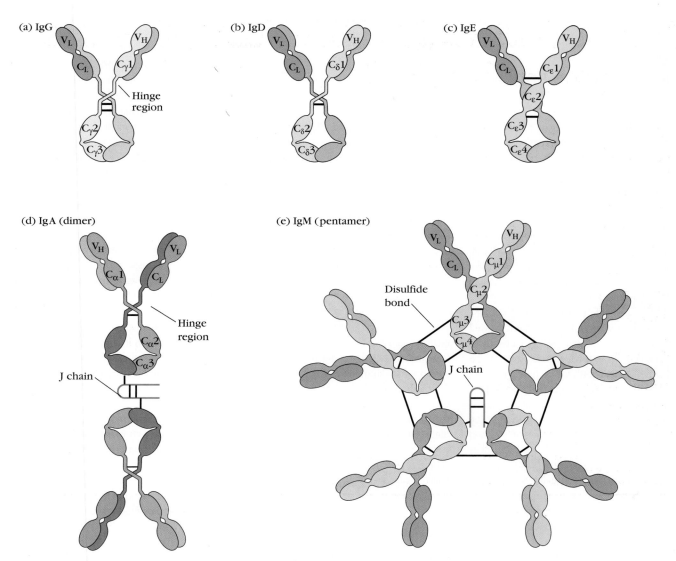

**FIGURE 4-17  General structures of the five major classes of secreted antibody.** Light chains are shown in shades of pink, disulfide bonds are indicated by thick black lines. Note that the IgG, IgA, and IgD heavy chains (blue, orange, and green, respectively) contain four domains and a hinge region, whereas the IgM and IgE heavy chains (purple and yellow, respectively) contain five domains but no hinge region. The polymeric forms of IgM and IgA contain a polypeptide, called the J chain, that is linked by two disulfide bonds to the Fc region in two different monomers. Serum IgM is always a pentamer; most serum IgA exists as a monomer, although dimers, trimers, and even tetramers are sometimes present. Not shown in these figures are intrachain disulfide bonds and disulfide bonds linking light and heavy chains (see Figure 4-6).

repeating epitopes such as viral particles and red blood cells (RBCs). For example, when RBCs are incubated with specific antibody, they clump together into large aggregates in a process called **agglutination.** It takes 100 to 1000 times more molecules of IgG than of IgM to achieve the same level of agglutination. A similar phenomenon occurs with viral particles: less IgM than IgG is required to neutralize viral infectivity. IgM is also more efficient than IgG at activating complement. Complement activation requires at least two Fc regions in close proximity, and the pentameric structure of a single molecule of IgM fulfills this requirement.

Because of its large size, IgM does not diffuse well and therefore is found in very low concentrations in the intercellular tissue fluids. The presence of the J chain allows IgM to bind to receptors on secretory cells, which transport it across epithelial linings to enter the external secretions that bathe mucosal surfaces. Although IgA is the major isotype found in these secretions, IgM plays an important accessory role as a secretory immunoglobulin.

## CLINICAL FOCUS
# Passive Antibody Therapy

**In 1890,** Emil Behring and Shibasaburo Kitasato reported an extraordinary experiment. After immunizing rabbits with tetanus and then collecting serum from these animals, they injected 0.2 ml of the immune serum into the abdominal cavity of six mice. Twenty-four hours later, they infected the treated animals and untreated controls with live, virulent tetanus bacteria. All of the control mice died within 48 hours of infection, whereas the treated mice not only survived but showed no effects of infection. This landmark experiment demonstrated two important points. One, it showed that following immunization, substances appeared in serum that could protect an animal against pathogens. Two, this work demonstrated that immunity could be passively acquired. Immunity could be transferred from one animal to another by taking serum from an immune animal and injecting it into a nonimmune one. These and subsequent experiments did not go unnoticed. Both men eventually received titles (Behring became von Behring and Kitasato became Baron Kitasato). A few years later, in 1901, von Behring was awarded the first Nobel Prize in Medicine.

These early observations and others paved the way for the introduction of passive immunization into clinical practice.

During the 1930s and 1940s, passive immunotherapy, the endowment of resistance to pathogens by transfer of the agent of immunity from an immunized donor to an unimmunized recipient, was used to prevent or modify the course of measles and hepatitis A. During subsequent years, clinical experience and advances in the technology of preparation of immunoglobulin for passive immunization have made this approach a standard medical practice. Passive immunization based on the transfer of antibodies is widely used in the treatment of immunodeficiency diseases and as a protection against anticipated exposure to infectious and toxic agents against which one does not have immunity.

Immunoglobulin for passive immunization is prepared from the pooled plasma of thousands of donors. In effect, recipients of these antibody preparations are receiving a sample of the antibodies produced by many people to a broad diversity of pathogens—a gram of intravenous immune globulin (IVIG) contains about $10^{18}$ molecules of antibody (mostly IgG) and may incorporate more than $10^7$ different antibody specificities. During the course of therapy, patients receive significant amounts of IVIG, usually 200 to 400 mg per kilogram of body weight. This means that an immunodeficient

patient weighing 70 kilograms would receive 14 to 28 grams of IVIG every 3 to 4 weeks. A product derived from the blood of such a large number of donors carries a risk of harboring pathogenic agents, particularly viruses. The risk is minimized by the processes used to produce intravenous immune globulin. The manufacture of IVIG involves treatment with solvents, such as ethanol, and the use of detergents that are highly effective in inactivating viruses such as HIV and hepatitis. In addition to removing or inactivating infectious agents, the production process must also eliminate aggregated immunoglobulin, because antibody aggregates can trigger massive activation of the complement pathway, leading to severe, even fatal anaphylaxis.

Passively administered antibody exerts its protective action in a number of ways. One of the most important is the recruitment of the complement pathway to the destruction or removal of a pathogen. In bacterial infections, antibody binding to bacterial surfaces promotes opsonization, the phagocytosis and killing of the invader by macrophages and neutrophils. Toxins and viruses can be bound and neutralized by antibody, even as the antibody marks the pathogen for removal from the body by phagocytes and by organs such as liver and kidneys. By the initiation of antibody-dependent cell-mediated cytotoxicity (ADCC), antibodies can also mediate the killing of target cells by cytotoxic cell populations such as natural killer cells.

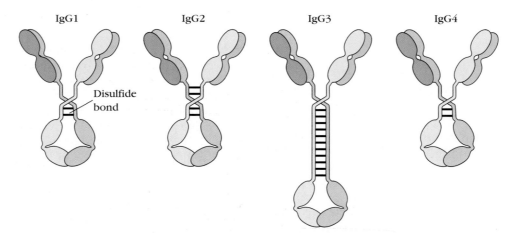

**FIGURE 4-18** **General structure of the four subclasses of human IgG, which differ in the number and arrangement of the interchain** disulfide bonds (thick black lines) linking the heavy chains. A notable feature of human IgG3 is its 11 interchain disulfide bonds.

## Immunoglobulin A (IgA)

Although IgA constitutes only 10% to 15% of the total immunoglobulin in serum, it is the predominant immunoglobulin class in external secretions such as breast milk, saliva, tears, and mucus of the bronchial, genitourinary, and digestive tracts. In serum, IgA exists primarily as a monomer, but polymeric forms (dimers, trimers, and some tetramers) are sometimes seen, all containing a J-chain polypeptide (see Figure 4-17d). The IgA of external secretions, called **secretory IgA,** consists of a dimer or tetramer, a J-chain polypeptide, and a polypeptide chain called **secretory component** (Figure 4-19a). As is explained below, secretory component is derived from the receptor that is

(a)  Structure of secretory IgA

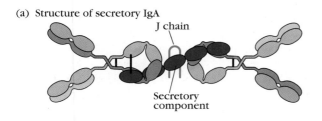

(b)  Formation of secretory IgA

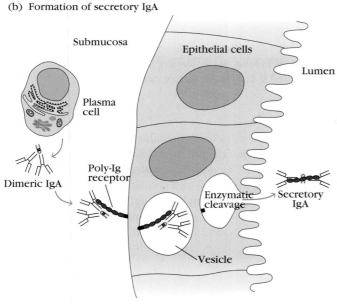

**FIGURE 4-19 Structure and formation of secretory IgA.** (a) Secretory IgA consists of at least two IgA molecules, which are covalently linked to each other through a J chain and are also covalently linked with the secretory component. The secretory component contains five Ig-like domains and is linked to dimeric IgA by a disulfide bond between its fifth domain and one of the IgA heavy chains. (b) Secretory IgA is formed during transport through mucous membrane epithelial cells. Dimeric IgA binds to a poly-Ig receptor on the basolateral membrane of an epithelial cell and is internalized by receptor-mediated endocytosis. After transport of the receptor-IgA complex to the luminal surface, the poly-Ig receptor is enzymatically cleaved, releasing the secretory component bound to the dimeric IgA.

responsible for transporting polymeric IgA across cell membranes. The J-chain polypeptide in IgA is identical to that found in pentameric IgM and serves a similar function in facilitating the polymerization of both serum IgA and secretory IgA. The secretory component is a polypeptide of 70,000 Da produced by epithelial cells of mucous membranes. It consists of five immunoglobulin-like domains that bind to the Fc region domains of the IgA dimer. This interaction is stabilized by a disulfide bond between the fifth domain of the secretory component and one of the chains of the dimeric IgA.

The daily production of secretory IgA is greater than that of any other immunoglobulin class. IgA-secreting plasma cells are concentrated along mucous membrane surfaces. Along the jejunum of the small intestine, for example, there are more than $2.5 \times 10^{10}$ IgA-secreting plasma cells—a number that surpasses the total plasma cell population of the bone marrow, lymph, and spleen combined! Every day, a human delivers from 5 g to 15 g of secretory IgA into mucous secretions.

The B cells that produce IgA preferentially migrate (home) to subepithelial tissue, where they differentiate into plasma cells that secrete IgA, which binds tightly to a receptor for polymeric immunoglobulin molecules (Figure 4-19b). This **poly-Ig receptor** is expressed on the basolateral surface of most mucosal epithelia (e.g., the lining of the digestive, respiratory, and genital tracts) and on glandular epithelia in the mammary, salivary, and lacrimal glands. After polymeric IgA binds to the poly-Ig receptor, the receptor-IgA complex is transported across the epithelial barrier to the lumen. Transport of the receptor-IgA complex involves receptor-mediated endocytosis of coated pits and directed transport of the vesicle across the epithelial cell to the luminal membrane, where the vesicle fuses with the plasma membrane. The poly-Ig receptor is then cleaved enzymatically from the membrane and becomes the secretory component, which is bound to and released together with polymeric IgA into the mucous secretions. The secretory component masks sites susceptible to protease cleavage in the hinge region of secretory IgA, allowing the polymeric molecule to exist longer in the protease-rich mucosal environment than would be possible otherwise. Pentameric IgM is also transported into mucous secretions by this mechanism, although it accounts for a much lower percentage of antibody in the mucous secretions than does IgA. The poly-Ig receptor interacts with the J chain of both polymeric IgA and IgM antibodies.

Secretory IgA serves an important effector function at mucous membrane surfaces, which are the main entry sites for most pathogenic organisms. Because it is polymeric, secretory IgA can cross-link large antigens with multiple epitopes. Binding of secretory IgA to bacterial and viral surface antigens prevents attachment of the pathogens to the mucosal cells, thus inhibiting viral infection and bacterial colonization. Complexes of secretory IgA and antigen are easily entrapped in mucus and then eliminated by the ciliated epithelial cells of the respiratory tract or by peristalsis of the gut. Secretory IgA has been shown to provide an important line of defense against bacteria such as *Salmonella, Vibrio cholerae,* and *Neisseria gonorrhoeae* and viruses such as polio, influenza, and reovirus.

Breast milk contains secretory IgA and many other molecules that help protect the newborn against infection during the first months of life. Because the immune system of infants is not fully functional, breast feeding plays an important role in maintaining the health of newborns.

## Immunoglobulin E (IgE)

The potent biological activity of IgE allowed it to be identified in serum despite its extremely low average serum concentration (0.3 μg/ml). IgE antibodies mediate the immediate hypersensitivity reactions that are responsible for the symptoms of hay fever, asthma, hives, and anaphylactic shock. The presence of a serum component responsible for allergic reactions was first demonstrated in 1921 by K. Prausnitz and H. Kustner, who injected serum from an allergic person intradermally into a nonallergic individual. When the appropriate antigen was later injected at the same site, a wheal and flare reaction (swelling and reddening) developed there. This reaction, called the **P-K reaction** (named for its originators, Prausnitz and Kustner), was the basis for the first biological assay for IgE activity.

Actual identification of IgE was accomplished by K. and T. Ishizaka in 1966. They obtained serum from an allergic individual and immunized rabbits with it to prepare anti-isotype antiserum. The rabbit antiserum was then allowed to react with each class of human antibody known at that time (i.e., IgG, IgA, IgM, and IgD). In this way, each of the known anti-isotype antibodies was precipitated and removed from the rabbit antiserum. What remained was an anti-isotype antibody specific for an unidentified class of antibody. This antibody turned out to completely block the P-K reaction. The new antibody was called IgE (in reference to the E antigen of ragweed pollen, which is a potent inducer of this class of antibody).

IgE binds to Fc receptors on the membranes of blood basophils and tissue mast cells. Cross-linkage of receptor-bound IgE molecules by antigen (allergen) induces basophils and mast cells to translocate their granules to the plasma membrane and release their contents to the extracellular environment, a process known as degranulation. As a result, a variety of pharmacologically active mediators are released and give rise to allergic manifestations (Figure 4-20). Localized mast cell degranulation induced by IgE also may release mediators that facilitate a buildup of various cells necessary for an ADCC antiparasitic defense (see Chapter 15).

## Immunoglobulin D (IgD)

IgD was first discovered when a patient developed a multiple myeloma whose myeloma protein failed to react with anti-isotype antisera against the then-known isotypes: IgA, IgM, and IgG. When rabbits were immunized with

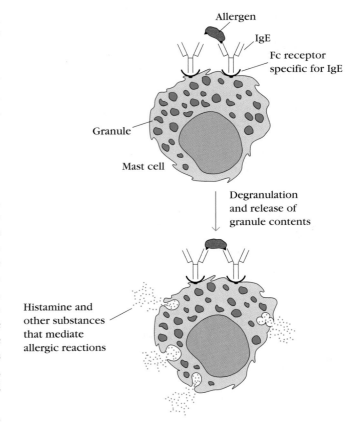

**FIGURE 4-20** Allergen cross-linkage of receptor-bound IgE on mast cells induces degranulation, causing release of substances (blue dots) that mediate allergic manifestations.

this myeloma protein, the resulting antisera were used to identify the same class of antibody at low levels in normal human serum. The new class, called IgD, has a serum concentration of 30 μg/ml and constitutes about 0.2% of the total immunoglobulin in serum. IgD, together with IgM, is the major membrane-bound immunoglobulin expressed by mature B cells, and its role in the physiology of B cells is under investigation. No biological effector function has been identified for IgD.

## Antigenic Determinants on Immunoglobulins

Immunoglobulin molecules can themselves, when injected into other animal species, function as potent immunogens to induce an antibody response. Such anti-Ig antibodies are powerful tools for the study of B-cell development and humoral immune responses. The antigenic determinants, or epitopes, on immunoglobulin molecules fall into three major categories: isotypic, allotypic, and idiotypic determinants, which are located in characteristic portions of the molecule (Figure 4-21).

(a) Isotypic determinants

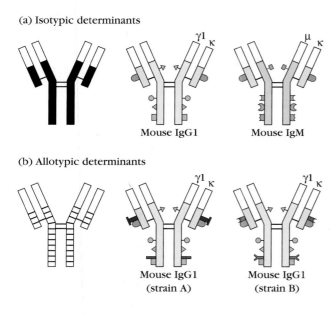

Mouse IgG1          Mouse IgM

(b) Allotypic determinants

Mouse IgG1          Mouse IgG1
(strain A)          (strain B)

(c) Idiotypic determinants

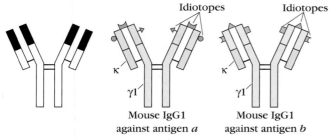

Mouse IgG1          Mouse IgG1
against antigen *a*    against antigen *b*

**FIGURE 4-21 Antigenic determinants of immunoglobulins. For each type of determinant, the general location of determinants within the antibody molecule is shown (*left*) and two examples are illustrated (*center* and *right*).** (a) Isotypic determinants are constant-region determinants that distinguish each Ig class and subclass within a species. (b) Allotypic determinants are subtle amino acid differences encoded by different alleles of isotype genes. Allotypic differences can be detected by comparing the same antibody class among different inbred strains. (c) Idiotypic determinants are generated by the conformation of the amino acid sequences of the heavy- and light-chain variable regions specific for each antigen. Each individual determinant is called an idiotope, and the sum of the individual idiotopes is the idiotype.

## Isotype

Isotypic determinants are constant-region determinants that collectively define each heavy-chain class and subclass and each light-chain type and subtype within a species (see Figure 4-21a). Each isotype is encoded by a separate constant-region gene, and all members of a species carry the same collection of constant-region genes (which may include multiple alleles). Within a species, each normal individual will express all isotypes in the serum. Different species inherit different constant-region genes and therefore express different isotypes. Thus, when an antibody

from one species is injected into another species, the isotypic determinants will be recognized as foreign, inducing an antibody response to the isotypic determinants on the foreign antibody. Anti-isotype antibody is routinely used for research purposes to determine the class or subclass of serum antibody produced during an immune response or to characterize the class of membrane-bound antibody present on B cells.

## Allotype

Although all members of a species inherit the same set of isotype genes, multiple alleles exist for some of the genes (see Figure 4-21b). These alleles encode subtle amino acid differences, called allotypic determinants, that occur in some, but not all, members of a species. The sum of the individual allotypic determinants displayed by an antibody determines its **allotype.** In humans, allotypes have been characterized for all four IgG subclasses, for one IgA subclass, and for the κ light chain. The γ-chain allotypes are referred to as Gm markers. At least 25 different Gm allotypes have been identified; they are designated by the class and subclass followed by the allele number, for example, G1m(1), G2m(23), G3m(11), G4m(4a). Of the two IgA subclasses, only the IgA2 subclass has allotypes, A2m(1) and A2m(2). The κ light chain has three allotypes, designated κm(1), κm(2), and κm(3). Each of these allotypic determinants represents differences in one to four amino acids that are encoded by different alleles of the same gene.

Antibody to allotypic determinants can be produced by injecting antibodies from one member of a species into another member of the same species who carries different allotypic determinants. Antibody to allotypic determinants sometimes is produced by a mother during pregnancy in response to paternal allotypic determinants on the fetal immunoglobulins. Antibodies to allotypic determinants can also arise from a blood transfusion.

## Idiotype

The unique amino acid sequence of the $V_H$ and $V_L$ domains of a given antibody can function not only as an antigen binding site but also as a set of antigenic determinants. The idiotypic determinants arise from the sequence of the heavy- and light-chain variable regions. Each individual antigenic determinant of the variable region is referred to as an **idiotope** (see Figure 4-21c). Each antibody will present multiple idiotopes, some of which are the actual antigen binding site, some comprising variable-region sequences outside of binding site. The sum of the individual idiotopes is called the **idiotype** of the antibody.

Because the antibodies produced by individual B cells derived from the same clone have identical variable-region sequences, they all have the same idiotype. Anti-idiotype antibody is produced by injecting antibodies that have minimal variation in their isotypes and allotypes, so that the idiotypic difference can be recognized. Often a homogeneous antibody such as myeloma protein or monoclonal antibody is used. Injection of such an antibody into a recipient who is genetically identical to the donor will result in the formation of anti-idiotype antibody to the idiotypic determinants.

## The B-Cell Receptor

Immunologists have long been puzzled about how membrane-bound Ig (mIg) on B cells mediates an activating signal after contact with an antigen. The dilemma is that all isotypes of mIg have very short cytoplasmic tails: the mIgM and mIgD cytoplasmic tails contain only three amino acids; the mIgA tail, 14 amino acids; and the mIgG and mIgE tails, 28 amino acids. In each case, the cytoplasmic tail is too short to be able to associate with intracellular signaling molecules (e.g., tyrosine kinases and G proteins).

The answer to this puzzle is that mIg does not constitute the entire antigen binding receptor on B cells. Rather, the **B-cell receptor (BCR)** is a transmembrane protein complex composed of mIg and disulfide-linked heterodimers called Ig-α/Ig-β. Molecules of this heterodimer associate with an mIg molecule to form a BCR (Figure 4-22). The Ig-α chain has a long cytoplasmic tail containing 61 amino acids; the tail of the Ig-β chain contains 48 amino acids. The tails in both Ig-α and Ig-β are long enough to interact with intracellular signaling molecules. Discovery of the Ig-α/Ig-β heterodimer by Michael Reth and his colleagues in the early 1990s has substantially furthered understanding of B-cell activation, which is discussed in detail in Chapter 11.

## Fc receptors bind to Fc regions of antibodies

Although the biosynthesis and surface expression of immunoglobulins is unique to the B-cell lineage, many cells feature membrane glycoproteins called **Fc receptors (FcR)** that have an affinity for the Fc portions of secreted antibody molecules. These receptors are essential for many of the biological functions of antibodies. Fc receptors are responsible for the movement of antibodies across cell membranes and the transfer of IgG from mother to fetus across the placenta. These receptors also allow passive acquisition of antibody by many cell types, including B and T lymphocytes, neutrophils, mast cells, eosinophils, macrophages, and natural killer cells. Consequently, Fc receptors provide a means by which antibodies—the products of the adaptive immune system—can recruit such key cellular elements of innate immunity as macrophages and natural killer cells. Engagement of antibody-bound antigens by the Fc receptors of macrophages or neutrophils provides an effective signal for the efficient phagocytosis (opsonization) of antigen-antibody complexes. In addition to triggering such effector functions as opsonization or ADCC, antigen-mediated cross-linking of FcR-bound antibodies can generate immunoregulatory signals that affect cell activation, induce

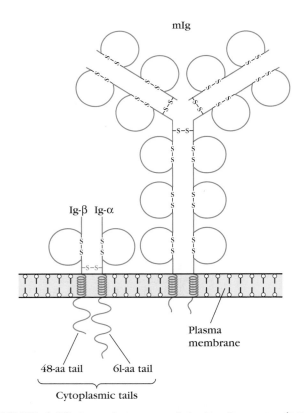

**FIGURE 4-22 General structure of the B-cell receptor (BCR).** This antigen-binding receptor is composed of membrane-bound immunoglobulin (mIg) and disulfide-linked heterodimers called Ig-a/Ig-b. Each heterodimer contains the immunoglobulin-fold structure and cytoplasmic tails much longer than those of mIg. As depicted, an mIg molecule is associated with one Ig-a/Ig-b heterodimer. [Adapted from A. D. Keegan and W. E. Paul, 1992, Immunology Today **13**:63, and M. Reth, 1992, Annual Review of Immunology **10**:97.]

differentiation, and, in some cases, downregulate cellular responses.

There are many different Fc receptors (Figure 4-23). The poly-Ig receptor is essential for the transport of polymeric immunoglobulins (polymeric IgA and to some extent, pentameric IgM) across epithelial surfaces. In humans, the **neonatal Fc receptor (FcR$_N$)** transfers IgGs from mother to fetus during gestation and also plays a role in the regulation of IgG serum levels. Fc receptors have been discovered for many of the Ig classes. Thus, there is an FcαR receptor that binds IgA, an FcεR that binds IgE (see also Figure 4-20), and several varieties of FcγR (RI, RII-A, RII-B1, RII-B2, RIII) capable of binding IgG and its subclasses are found in humans. In many cases, the cross-linking of these receptors by binding of antigen-antibody complexes results in the initiation of signal transduction cascades that result in such behaviors as phagocytosis or ADCC. The Fc receptor is often part of a signal-transducing complex that involves the participation of other accessory polypeptide chains. As shown in Figure 4-23, this may

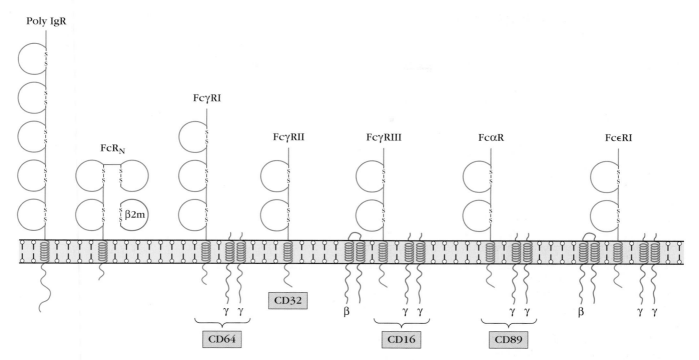

**FIGURE 4-23  The structure of a number of human Fc receptors.** The Fc-binding polypeptides are shown in blue and, where present, accessory signal-transducing polypeptides are shown in green. The loops in these structures represent portions of the molecule with the characteristic immunoglobulin-fold structure. These molecules appear on the plasma membrane of various cell types as cell surface antigens and, as indicated in the figure, many have been assigned CD designations (for clusters of differentiation, see Appendix 1). The FcγRII has three different forms, A, B1, and B2, differing in their intracellular regions. *[Adapted from M. Daeron, 1999, in* The Antibodies, *M. Zanetti and J. D. Capra, eds., vol. 5, p. 53, Harwood Academic Publishers, Amsterdam.]*

involve a pair of γ chains or, in the case of the IgE receptor, a more complex assemblage of two γ chains and a β chain. The association of an extracellular receptor with an intracellular signal-transducing unit was seen in the B-cell receptor (Figure 4-22) and is a central feature of the T-cell receptor complex (see Chapter 9).

## The Immunoglobulin Superfamily

The structures of the various immunoglobulin heavy and light chains described earlier share several features, suggesting that they have a common evolutionary ancestry. In particular, all heavy- and light-chain classes have the immunoglobulin-fold domain structure (see Figure 4-8). The presence of this characteristic structure in all immunoglobulin heavy and light chains suggests that the genes encoding them arose from a common primordial gene encoding a polypeptide of about 110 amino acids. Gene duplication and later divergence could then have generated the various heavy- and light-chain genes.

Large numbers of membrane proteins have been shown to possess one or more regions homologous to an immunoglobulin domain. Each of these membrane proteins is classified as a member of the **immunoglobulin**

**superfamily.** The term *superfamily* is used to denote proteins whose corresponding genes derived from a common primordial gene encoding the basic domain structure. These genes have evolved independently and do not share genetic linkage or function. The following proteins, in addition to the immunoglobulins themselves, are representative members of the immunoglobulin superfamily (Figure 4-24):

- Ig-α/Ig-β heterodimer, part of the B-cell receptor

- Poly-Ig receptor, which contributes the secretory component to secretory IgA and IgM

- T-cell receptor

- T-cell accessory proteins, including CD2, CD4, CD8, CD28, and the γ, δ, and ε chains of CD3

- Class I and class II MHC molecules

- $\beta_2$-microglobulin, an invariant protein associated with class I MHC molecules

- Various cell adhesion molecules, including VCAM-1, ICAM-1, ICAM-2, and LFA-3

- Platelet-derived growth factor

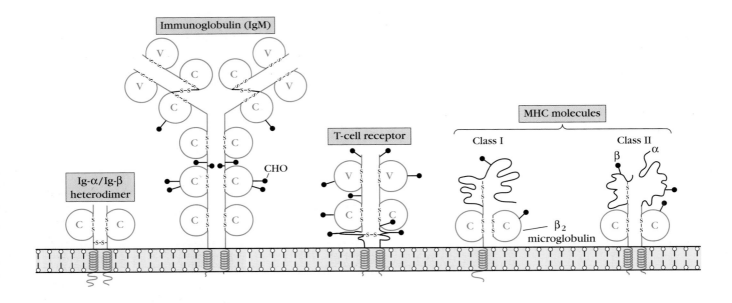

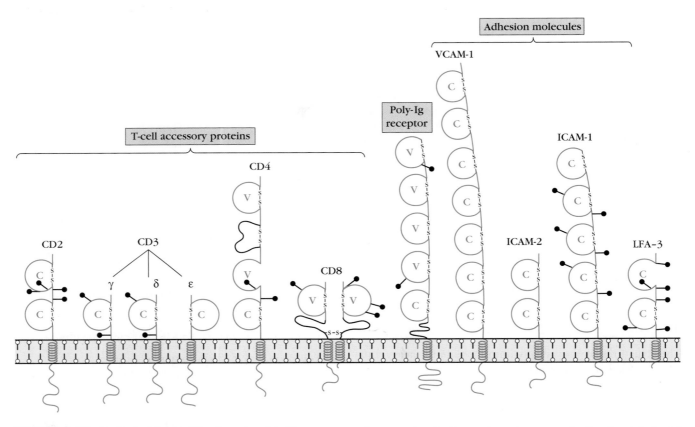

**FIGURE 4-24 Some members of the immunoglobulin super-family, a group of structurally related, usually membrane-bound glycoproteins.** In all cases shown here except for $\beta_2$-microglobulin, the carboxyl-terminal end of the molecule is anchored in the membrane.

Numerous other proteins, some of them discussed in other chapters, also belong to the immunoglobulin superfamily.

X-ray crystallographic analysis has not been accomplished for all members of the immunoglobulin superfamily. Nevertheless, the primary amino acid sequence of these proteins suggests that they all contain the typical immunoglobulin-fold domain. Specifically, all members of the immunoglobulin superfamily contain at least one or more stretches of about 110 amino acids, capable of arrangement into pleated sheets of antiparallel $\beta$ strands, usually with an invariant intrachain disulfide bond that closes a loop spanning 50 to 70 residues.

Most members of the immunoglobulin superfamily cannot bind antigen. Thus, the characteristic Ig-fold structure found in so many membrane proteins must have some function other than antigen binding. One possibility is that the immunoglobulin fold may facilitate interactions between membrane proteins. As described earlier, interactions can occur between the faces of β pleated sheets both of homologous immunoglobulin domains (e.g., $C_H2/C_H2$ interaction) and of nonhomologous domains (e.g., $V_H/V_L$ and $C_H1/C_L$ interactions).

## Monoclonal Antibodies

As noted earlier, most antigens offer multiple epitopes and therefore induce proliferation and differentiation of a variety of B-cell clones, each derived from a B cell that recognizes a particular epitope. The resulting serum antibodies are heterogeneous, comprising a mixture of antibodies, each specific for one epitope (Figure 4-25). Such a **polyclonal antibody** response facilitates the localization, phagocytosis, and complement-mediated lysis of antigen; it thus has clear advantages for the organism in vivo. Unfortunately, the antibody heterogeneity that increases immune protection in vivo often reduces the efficacy of an antiserum for various in vitro uses. For most research, diagnostic, and therapeutic purposes, **monoclonal antibodies,** derived from a single clone and thus specific for a single epitope, are preferable.

Direct biochemical purification of a monoclonal antibody from a polyclonal antibody preparation is not feasible. In 1975, Georges Köhler and Cesar Milstein devised a method for preparing monoclonal antibody, which quickly became one of immunology's key technologies. By fusing a normal activated, antibody-producing B cell with a myeloma cell (a cancerous plasma cell), they were able to generate a hybrid cell, called a **hybridoma,** that possessed the immortal-growth properties of the myeloma cell and secreted the antibody

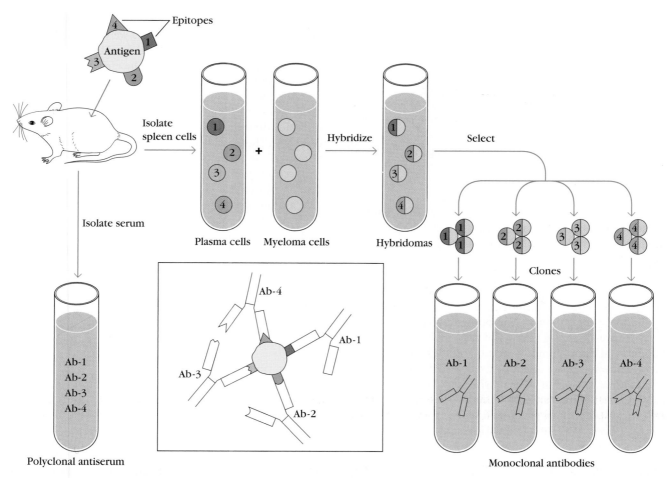

**FIGURE 4-25 The conventional polyclonal antiserum produced in response to a complex antigen contains a mixture of monoclonal antibodies, each specific for one of the four epitopes shown on the antigen (inset).** In contrast, a monoclonal antibody, which is derived from a single plasma cell, is specific for one epitope on a complex antigen. The outline of the basic method for obtaining a monoclonal antibody is illustrated here.

produced by the B cell (see Figure 4-25). The resulting clones of hybridoma cells, which secrete large quantities of monoclonal antibody, can be cultured indefinitely. The development of techniques for producing monoclonal antibodies, the details of which are discussed in Chapter 22, gave immunologists a powerful and versatile research tool. The significance of the work by Köhler and Milstein was acknowledged when each was awarded a Nobel prize.

## Monoclonal antibodies have important clinical uses

Monoclonal antibodies are proving to be very useful as diagnostic, imaging, and therapeutic reagents in clinical medicine. Initially, monoclonal antibodies were used primarily as in vitro diagnostic reagents. Among the many monoclonal antibody diagnostic reagents now available are products for detecting pregnancy, diagnosing numerous pathogenic microorganisms, measuring the blood levels of various drugs, matching histocompatibility antigens, and detecting antigens shed by certain tumors.

Radiolabeled monoclonal antibodies can also be used in vivo to detect or locate tumor antigens, permitting earlier diagnosis of some primary or metastatic tumors in patients. For example, monoclonal antibody to breast cancer cells is labeled with iodine-131 and introduced into the blood to detect the spread of a tumor to regional lymph nodes. This monoclonal imaging technique can reveal breast cancer metastases that would be undetected by other, less sensitive scanning techniques.

As discussed in the Clinical Focus on page 98, the use of antibodies as therapeutic agents has a long history. Recently the availability of monoclonal antibodies and the ability to "humanize" them by genetic-engineering techniques has given new vigor to this area. Early attempts to use mouse monoclonal antibodies to treat humans faced the possibility that there would be a strong reaction against these foreign proteins. Recent products are made in human systems or are genetically engineered (see Chapter 5) to incorporate V regions or CDRs of nonhuman antibodies into C regions and frameworks of human antibodies, thus minimizing the possibility of raising an immune response to them.

There are a growing number of approved therapeutic antibodies and over a hundred new ones in the drug development pathways. These account for several billion dollars a year in sales. The major areas in which antibody therapy has proven useful are cancer treatment and alleviation of arthritic disorders. The products Rituxan (generic name, rituximab) for treatment of non-Hodgkin's lymphoma and Remicade (generic name, infliximab) or Humira (adalimumab) for rheumatoid arthritis are among the most widely used. The generic names of the drugs reflect the type of antibody; for example, the suffix -umab denotes a human monoclonal antibody, whereas -imab denotes a chimeric antibody, one containing sequences from both humans and other species.

## Abzymes are monoclonal antibodies that catalyze reactions

The binding of an antibody to its antigen is similar in many ways to the binding of an enzyme to its substrate. In both cases the binding involves weak, noncovalent interactions and exhibits high specificity and often high affinity. What distinguishes an antibody-antigen interaction from an enzyme-substrate interaction is that the antibody does not alter the covalent bonds of the antigen, whereas the enzyme catalyzes a chemical change in its substrate. However, like enzymes, antibodies of appropriate specificity can stabilize the transition state of a bound substrate, thus reducing the activation energy for chemical modification of the substrate.

The similarities between antigen-antibody interactions and enzyme-substrate interactions raised the question of whether some antibodies could behave like enzymes and catalyze chemical reactions. To investigate this possibility, a hapten-carrier complex was synthesized in which the hapten structurally resembled the transition state of an ester undergoing hydrolysis. Spleen cells from mice immunized with this transition-state analogue were fused with myeloma cells to generate monoclonal antihapten monoclonal antibodies. When these monoclonal antibodies were incubated with an ester substrate, some of them accelerated hydrolysis by about 1000-fold; that is, they acted like the enzyme that normally catalyzes the substrate's hydrolysis. The catalytic activity of these antibodies was highly specific: they hydrolyzed only esters whose transition-state structure closely resembled the transition-state analogue used as a hapten in the immunizing conjugate. These catalytic antibodies have been called **abzymes** in reference to their dual role as antibody and enzyme.

A central goal of catalytic antibody research is the derivation of a battery of abzymes that cut peptide bonds at specific amino acid residues, much as restriction enzymes cut DNA at specific sites. Such abzymes would be invaluable tools in the structural and functional analysis of proteins. In addition, it may be possible to generate abzymes with the ability to dissolve blood clots or to cleave viral glycoproteins at specific sites, thus blocking viral infectivity. Unfortunately, catalytic antibodies that cleave the peptide bonds of proteins have been exceedingly difficult to derive. Much of the research currently being pursued in this field is devoted to the solution of this important but difficult problem.

------------------

## SUMMARY

- All immunogens are antigens, but not all antigens are immunogens. For example, haptens are small-molecule antigens, but they can induce an immune response (are immunogenic) only when conjugated with a large molecule.

- Immunogenicity is determined by many factors, including foreignness, molecular size, chemical composition, complexity, dose, susceptibility to antigen processing and presentation, the genotype of the recipient animal (in particular, its MHC genes), route of administration, and adjuvants.

- Antibodies recognize a wide range of antigenic determinants, or epitopes. Small antigens are often bound in narrow grooves or deep pockets of the antibody; larger antigens, such as globular proteins, interact with a larger, flatter complementary surface on the antibody molecule.

- An antibody molecule consists of two identical light chains and two identical heavy chains. Light chains are linked to heavy chains by disulfide bonds, and the heavy chains are linked to each other by disulfide bonds. Each antibody chain consists of an amino-terminal variable region and a carboxy-terminal constant region.

- In any given antibody molecule, the constant region contains one of five basic heavy-chain sequences, called isotypes, which determine antibody class ($\mu$, IgM; $\gamma$, IgG; $\delta$, IgD; $\alpha$, IgA; $\epsilon$, IgE) and one of two basic light-chain sequences ($\kappa$ or $\lambda$) called types.

- The five antibody classes have different effector functions, average serum concentrations, and half-lives.

- Each of the domains in the immunoglobulin molecule has a characteristic tertiary structure called the immunoglobulin fold. The presence of an immunoglobulin-fold domain also identifies many other nonantibody proteins as members of the immunoglobulin superfamily.

- Within the amino-terminal variable domain of each heavy and light chain are three complementarity-determining regions (CDRs). These polypeptide regions contribute the antigen binding site of an antibody, determining its specificity.

- Immunoglobulins are expressed in two forms: secreted antibody that is produced by plasma cells and membrane-bound antibody that associates with Ig-$\alpha$/Ig-$\beta$ heterodimers to form the B-cell antigen receptor present on the surface of B cells.

- The three major effector functions of antibodies are (1) opsonization, which promotes antigen phagocytosis by macrophages and neutrophils; (2) complement activation, which activates a pathway that leads to the generation of a collection of proteins that can perforate cell membranes; and (3) antibody-dependent cell-mediated cytotoxicity (ADCC), which can kill antibody-bound target cells.

- Unlike polyclonal antibodies, which arise from many B cell clones and have a heterogeneous collection of binding sites, a monoclonal antibody is derived from a single B-cell clone and has a single binding site.

## References

Berzofsky, J. A., and J. J. Berkower. 1999. Immunogenicity and antigen structure. In *Fundamental Immunology*, 4th ed., W. E. Paul, ed. Lippincott-Raven, Philadelphia.

Demotz, S., H. M. Grey, E. Appella, and A. Sette. 1989. Characterization of a naturally processed MHC class II-restricted T-cell determinant of hen egg lysozyme. *Nature* **342**:682.

Frazer, J. K., and J. D. Capra. 1999. Immunoglobulins: structure and function. In *Fundamental Immunology*, 4th ed. W. E. Paul, ed. Lippincott-Raven, Philadelphia.

Grey, H. M., A. Sette, and S. Buus. 1989. How T cells see antigen. *Scientific American* **261**(5):56.

*Immunology Today: The Immune Receptor Supplement*, 2nd ed. 1997. Elsevier Trends Journals, Cambridge, UK (ISSN 1365-1218).

Kindt, T. J., and J. D. Capra. 1984. *The Antibody Enigma*. Plenum Press, New York.

Köhler, G., and C. Milstein. 1975. Continuous cultures of fused cells secreting antibody of predefined specificity. *Nature* **256**:495.

Kraehenbuhl, J. P., and M. R. Neutra. 1992. Transepithelial transport and mucosal defence II: secretion of IgA. *Trends in Cell Biology* **2**:134.

Landsteiner, K. 1945. *The Specificity of Serological Reactions*. Harvard University Press, Cambridge, MA.

Laver, W. G., G. M. Air, R. G. Webster, and S. J. Smith-Gill. 1990. Epitopes on protein antigens: misconceptions and realities. *Cell* **61**:553.

Reth, M. 1995. The B-cell antigen receptor complex and coreceptor. *Immunology Today* **16**:310.

Stanfield, R. L., and I. A. Wilson. 1995. Protein-peptide interactions. *Current Opinion in Structural Biology* **5**:103.

Tainer, J. A., et al. 1985. The atomic mobility component of protein antigenicity. *Annual Review of Immunology* **3**:501.

Wedemayer, G. J., et al. 1997. Structural insights into the evolution of an antibody combining site. *Science* **276**:1665.

Wentworth, P., and K. Janda. 1998. Catalytic antibodies. *Current Opinion in Chemical Biology* **8**:138.

Wilson, I. A., and R. L. Stanfield. 1994. Antibody-antibody interactions: new structures and new conformational changes. *Current Opinion in Structural Biology* **4**:857.

 ## Useful Web Sites

http://www.umass.edu/microbio/rasmol/

RasMol is free software for visualizing molecular structures that can be run on Windows-based, Macintosh, or Unix PCs. With it one can view three-dimensional structures of many types of molecules, including proteins and nucleic acids.

http://www.expasy.ch/

This is the excellent and comprehensive Swiss Institute of Bioinformatics (SIB) Web site, which contains extensive information on protein structure. From it one can obtain protein sequences and three-dimensional structures of proteins as well as the versatile Swiss-PdbViewer software, which has several advanced capabilities not found in RasMol.

http://www.bioinf.org.uk/abs/

The Antibodies: Structure and Sequence Web site summarizes useful information on antibody structure and sequence. It provides general information on antibodies and crystal structures and links to other antibody-related information.

http://www.ncbi.nlm.nih.gov/Structure/

The Molecular Modeling Database (MMDB) contains three-dimensional structures determined by x-ray crystallography and NMR spectroscopy. The data for MMDB are obtained from the Protein Data Bank (PDB). The National Center for Biotechnology Information (NCBI) has structural data cross-linked to bibliographic information, to databases of protein and nucleic acid sequences, and to the NCBI animal taxonomy database. The NCBI has developed a 3-D structure viewer, Cn3D, for easy interactive visualization of molecular structures.

http://www.umass.edu/microbio/chime/pe/protexpl/frntdoor.htm

Protein Explorer is a molecular visualization program created by Eric Martz of the University of Massachusetts, Amherst, with the support of the National Science Foundation to make it easier for students, educators, and scientists to use interactive and dynamic molecular visualization techniques.

http://imgt.cines.fr

IMGT, the international ImMunoGeneTics database, created by Marie-Paule Lefranc, is a well-organized, powerful, and comprehensive information system that specializes in immunoglobulins, T-cell receptors and major histocompatibility complex (MHC) molecules of all vertebrate species.

## Study Questions

**CLINICAL FOCUS QUESTION** Two pharmaceutical companies make intravenous immune globulin (IVIG). Company A produces their product from pools of 100,000 donors drawn exclusively from the population of the United States. Company B makes their IVIG from pools of 60,000 donors drawn in equal numbers from North America, Europe, Brazil, and Japan.

a. Which product would you expect to have the broadest spectrum of pathogen reactivities? Why?

b. Assume the patients receiving the antibody will (1) never leave the U.S.A. or (2) travel extensively in many parts of the world. Which company's product would you choose for each of these patient groups? Justify your choices.

1. Indicate whether each of the following statements is true or false. If you think a statement is false, explain why.

a. Most antigens induce a response from more than one clone.

b. A large protein antigen generally can combine with many different antibody molecules.

c. A hapten can stimulate antibody formation but cannot combine with antibody molecules.

d. MHC genes play a major role in determining the degree of immune responsiveness to an antigen.

e. T-cell epitopes tend to be accessible amino acid residues that can combine with the T-cell receptor.

f. Many B-cell epitopes are nonsequential amino acids brought together by the tertiary conformation of a protein antigen.

g. All antigens are also immunogens.

h. Antibodies can bind hydrophilic or hydrophobic compounds, but T-cell receptors can only bind peptide-MHC complexes.

i. B-cell epitopes can be deduced with great accuracy from the primary structure of a protein.

2. For each pair of antigens listed below, indicate which is likely to be more immunogenic when injected into a rabbit. Explain your answer.

a. Native bovine serum albumin (BSA)
   Heat-denatured BSA
b. Hen egg-white lysozyme (HEL)
   Hen collagen
c. A protein with a molecular weight of 30,000
   A protein with a molecular weight of 150,000
d. BSA in Freund's complete adjuvant
   BSA in Freund's incomplete adjuvant

3. Indicate which of the following statements regarding haptens and carriers are true.

a. Haptens are large protein molecules such as BSA.

b. When a hapten-carrier complex containing multiple hapten molecules is injected into an animal, most of the induced antibodies are specific for the hapten.

c. Carriers are needed only if one wants to elicit a cell-mediated response.

d. It is necessary to immunize with a hapten-carrier complex in order to obtain antibodies directed against the hapten.

e. Carriers include small molecules such as dinitrophenol and penicillenic acid (derived from penicillin).

4. For each of the following statements, indicate whether it is true only of B-cell epitopes (B), only of T-cell epitopes (T), or both types of epitopes (BT) within a large antigen.

a. They almost always consist of a linear sequence of amino acid residues.

b. They generally are located in the interior of a protein antigen.

c. They generally are located on the surface of a protein antigen.

d. They lose their immunogenicity when a protein antigen is denatured by heat.

e. Immunodominant epitopes are determined in part by the MHC molecules expressed by an individual.

f. They generally arise from proteins.

g. Multiple different epitopes may occur in the same antigen.

h. Their immunogenicity may depend on the three-dimensional structure of the antigen.

i. The immune response to them may be enhanced by co-administration of Freund's complete adjuvant.

5. Indicate whether each of the following statements is true or false. If you think a statement is false, explain why.

a. A rabbit immunized with human IgG3 will produce antibody that reacts with all subclasses of IgG in humans.

b. All immunoglobulin molecules on the surface of a given B cell have the same idiotype.

c. All immunoglobulin molecules on the surface of a given B cell have the same isotype.

d. All myeloma protein molecules derived from a single myeloma clone have the same idiotype and allotype.

e. Although IgA is the major antibody species that undergoes transcytosis, polymeric IgM, but not monomeric IgA, can also undergo transcytosis.

f. The hypervariable regions make significant contact with the epitope.

g. IgG functions more effectively than IgM in bacterial agglutination.

h. Although monoclonal antibodies are often preferred for research and diagnostic purposes, both monoclonal and polyclonal antibodies can be highly specific.

i. All isotypes are normally found in each individual of a species.

j. The heavy-chain variable region ($V_H$) is twice as long as the light-chain variable region ($V_L$).

6. You are an energetic immunology student who has isolated protein X, which you believe is a new isotype of human immunoglobulin.

   a. What structural features must protein X possess in order to be classified as an immunoglobulin?

   b. You prepare rabbit antisera to whole human IgG, human κ chain, and human γ chain. Assuming protein X is, in fact, a new immunoglobulin isotype, to which of these antisera would it bind? Why?

   c. Devise an experimental procedure for preparing an antiserum that is specific for protein X.

7. According to the clonal-selection theory, all the immunoglobulin molecules on a single B cell have the same antigenic specificity. Explain why the presence of both IgM and IgD on the same B cell does not violate the unispecificity implied by clonal selection.

8. IgG, which contains γ heavy chains, developed much more recently during evolution than IgM, which contains μ heavy chains. Describe two advantages and two disadvantages that IgG has in comparison with IgM.

9. Although the five immunoglobulin isotypes share many common structural features, the differences in their structures affect their biological activities.

   a. Draw a schematic diagram of a typical IgG molecule and label each of the following parts: H chains, L chains, interchain disulfide bonds, intrachain disulfide bonds, hinge, Fab, Fc, and all the domains. Indicate which domains are involved in antigen binding.

   b. How would you have to modify the diagram of IgG to depict an IgA molecule isolated from saliva?

   c. How would you have to modify the diagram of IgG to depict serum IgM?

10. Fill out the accompanying table relating to the properties of IgG molecules and their various parts. Insert a ($+$) if the molecule or part exhibits the property, a ($-$) if it does not, and a ($+/-$) if it does so only weakly.

| Property | Whole IgG | H chain | L chain | Fab | F(ab')$_2$ | Fc |
|---|---|---|---|---|---|---|
| Binds antigen | | | | | | |
| Bivalent antigen binding | | | | | | |
| Binds to Fc receptors | | | | | | |
| Fixed complement in presence of antigen | | | | | | |
| Has V domains | | | | | | |
| Has C domains | | | | | | |

11. Because immunoglobulin molecules possess antigenic determinants, they themselves can function as immunogens, inducing formation of antibody. For each of the following immunization scenarios, indicate whether anti-immunoglobulin antibodies would be formed to isotypic (IS), allotypic (AL), or idiotypic (ID) determinants:

   a. Anti-DNP antibodies produced in a BALB/c mouse are injected into a C57BL/6 mouse.

   b. Anti-BGG monoclonal antibodies from a BALB/c mouse are injected into another BALB/c mouse.

   c. Anti-BGG antibodies produced in a BALB/c mouse are injected into a rabbit.

   d. Anti-DNP antibodies produced in a BALB/c mouse are injected into an outbred mouse.

   e. Anti-BGG antibodies produced in a BALB/c mouse are injected into the same mouse.

12. Write *yes* or *no* in the accompanying table to indicate whether the rabbit antisera listed at the top react with the mouse antibody components listed at the left.

| | γ chain | κ chain | IgG Fab fragment | IgG Fc fragment | J chain |
|---|---|---|---|---|---|
| Mouse γ chain | | | | | |
| Mouse κ chain | | | | | |
| Mouse IgM whole | | | | | |
| Mouse IgM Fc fragment | | | | | |

13. The characteristic structure of immunoglobulin domains, termed the immunoglobulin fold, also occurs in the numerous

membrane proteins belonging to the immunoglobulin superfamily.

a. Describe the typical features that define the immunoglobulin-fold domain structure.

b. Consider proteins that belong to the immunoglobulin superfamily. What do all of these proteins have in common? Describe two different Ig superfamily members that bind antigen. Identify four different Ig superfamily members that do not bind antigen.

14. Where are the CDR regions located on an antibody molecule and what are their functions?

15. The variation in amino acid sequence at each position in a polypeptide chain can be expressed by a quantity termed the *variability*.

a. What are the largest and smallest values of variability possible?

b. How would you expect the variability plot of an enzyme present in multiple mammalian species to differ from that of a group of human immunoglobulins?

16. An investigator wanted to make a rabbit antiserum specific for mouse IgG. She injected a rabbit with purified mouse IgG and obtained an antiserum that reacted strongly with mouse IgG. To her dismay, the antiserum also reacted with each of the other mouse isotypes. Explain why she got this result. How could she make the rabbit antiserum specific for mouse IgG?

17. You fuse spleen cells having a normal genotype for immunoglobulin heavy chains (H) and light chains (L) with three myeloma cell preparations differing in their immunoglobulin genotype as follows: (a) $H^+$, $L^+$; (b) $H^+$, $L^-$; and (c) $H^-$, $L^-$. For each hybridoma, predict how many unique antigen binding sites, composed of one H and one L chain, theoretically could be produced and show the chain structure of the possible antibody molecules. For each possible antibody molecule indicate whether the chains would originate from the spleen (S) or from the myeloma (M) fusion partner (e.g., $H_S L_S / H_m L_m$).

18. For each immunoglobulin isotype (a–e) select the description(s) listed below (1–12) that describe that isotype. Each description may be used once, more than once, or not at all; more than one description may apply to some isotypes.

*Isotypes*

a. _____ IgA    c. _____ IgE    e. _____ IgM

b. _____ IgD    d. _____ IgG

*Descriptions*

(1) Secreted form is a pentamer of the basic $H_2L_2$ unit
(2) Binds to Fc receptors on mast cells
(3) Multimeric forms have a J chain
(4) Present on the surface of mature, unprimed B cells
(5) The most abundant isotype in serum
(6) Major antibody in secretions such as saliva, tears, and breast milk
(7) Present on the surface of immature B cells
(8) The first serum antibody made in a primary immune response
(9) Plays an important role in immediate hypersensitivity
(10) Plays primary role in protecting against pathogens that invade through the gut or respiratory mucosa
(11) Multimeric forms may contain a secretory component
(12) Least abundant isotype in serum

19. You have just finished producing a monoclonal antibody to a tyrosine kinase-linked cell surface receptor. When the antibody binds the receptor expressed on your cells, you find that the cells are stimulated instead of inhibited. That is, the antibody fails to block the receptor and instead results in increased signaling from the receptor. Give a possible explanation for this result. How could you modify the antibody to eliminate the stimulation reaction, allowing you to determine if your antibody blocks ligand binding?

20. IgG can be transferred from mother to developing fetus, conferring immune protection against antigens the mother is exposed to in the third trimester. In what other ways can a mother provide immune system protection to her infant after birth?

 Interactive Study

**www.whfreeman.com/kuby**

 SELF-TEST
Review of Key Terms

 ANIMATION
Immunoglobulins
hCG Pregnancy Test

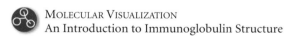 MOLECULAR VISUALIZATION
An Introduction to Immunoglobulin Structure

# chapter 5

# Organization and Expression of Immunoglobulin Genes

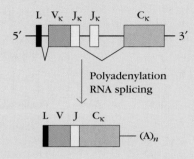

*Kappa light-chain genes are rearranged and spliced to generate the complete message.*

- Devising a Genetic Model Compatible with Ig Structure
- Multigene Organization of Ig Genes
- Variable-Region Gene Rearrangements
- Mechanism of Variable-Region DNA Rearrangements
- Generation of Antibody Diversity
- Class Switching Among Constant-Region Genes
- Expression of Ig Genes
- Synthesis, Assembly, and Secretion of Immunoglobulins
- Regulation of Ig-Gene Transcription
- Antibody Genes and Antibody Engineering

O NE OF THE MOST REMARKABLE FEATURES of the vertebrate immune system is its ability to respond to an apparently limitless array of foreign antigens. As immunoglobulin (Ig) sequence data accumulated, virtually every antibody molecule studied was found to contain a unique amino acid sequence in its variable region but only one of a limited number of invariant sequences in its constant region. The genetic basis for this combination of constancy and tremendous variation in a single protein molecule lies in the organization of the immunoglobulin genes.

In germ-line DNA, multiple gene segments encode portions of a single immunoglobulin heavy or light chain. These gene segments are carried in the germ cells but cannot be transcribed and translated into complete chains until they are rearranged into functional genes. During B-cell maturation in the bone marrow, certain of these gene segments are randomly shuffled by a dynamic genetic system capable of generating more than $10^6$ combinations. Subsequent processes increase the diversity of the repertoire of antibody binding sites to a very large number that exceeds $10^8$. During B-cell development, the maturation of a progenitor B cell progresses through an ordered sequence of Ig-gene rearrangements, coupled with modifications to the genes that contribute to the diversity of the final product. By the end of this process, a mature, immunocompetent B cell will contain coding sequences for one functional heavy-chain variable region and one light-chain variable region. Thus, an individual B cell arises with its specificity already determined. After antigenic stimulation of a mature B cell in peripheral lymphoid organs, further rearrangement of constant-region gene segments can generate changes in the isotype expressed, which produce changes in the biological effector functions of the immunoglobulin molecule without changing its specificity. Thus, mature B cells contain chromosomal DNA that is no longer identical to germ-line DNA. Although

we think of genomic DNA as a stable genetic blueprint, the lymphocyte cell lineage does not retain an intact copy of this blueprint. Genomic rearrangement is an essential feature of lymphocyte differentiation, and no other vertebrate cell type has been shown to undergo this process.

This chapter first describes the detailed organization of the immunoglobulin genes, the process of Ig-gene rearrangement, and the mechanisms by which the dynamic immunoglobulin genetic system generates more than $10^8$ different antigenic specificities. Then it describes the mechanism of class switching, the role of differential RNA processing in the expression of

## OVERVIEW FIGURE 5-1: B-Cell Development

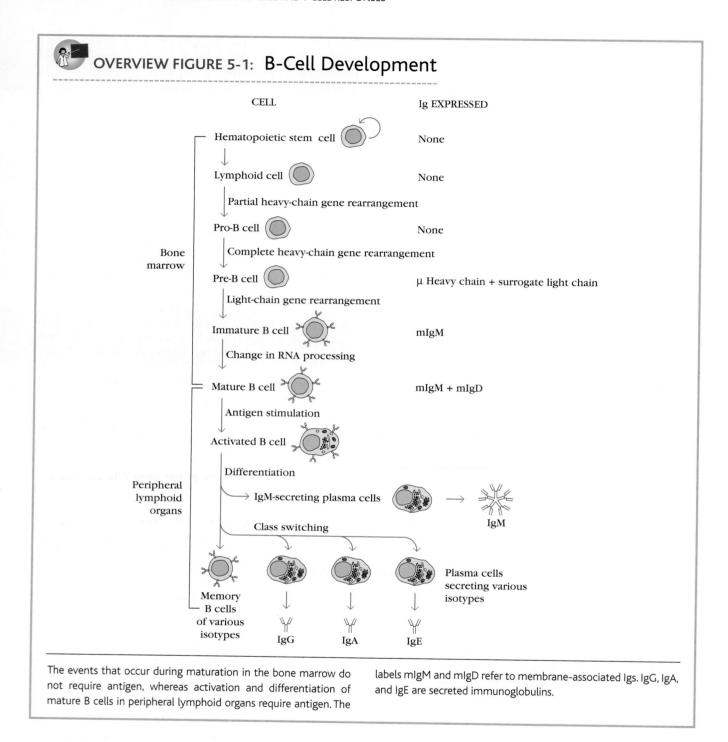

The events that occur during maturation in the bone marrow do not require antigen, whereas activation and differentiation of mature B cells in peripheral lymphoid organs require antigen. The labels mIgM and mIgD refer to membrane-associated Igs. IgG, IgA, and IgE are secreted immunoglobulins.

immunoglobulin genes, and the regulation of Ig-gene transcription. The chapter concludes with the application of our knowledge of the molecular biology of immunoglobulin genes to the engineering of antibody molecules for therapeutic and research applications. Chapter 11 covers in detail the entire process of B-cell development from the first gene rearrangements in progenitor B cells to final differentiation into memory B cells and antibody-secreting plasma cells. Figure 5-1 outlines the sequential stages in B-cell development, many of which result from critical rearrangements.

## Devising a Genetic Model Compatible with Ig Structure

The results of the immunoglobulin-sequence analysis described in Chapter 4 revealed a number of features of immunoglobulin structure that were difficult to reconcile with classic genetic models. Any viable model of the immunoglobulin genes had to account for the following properties of antibodies:

- The vast diversity of antibody specificities

- The presence in Ig heavy and light chains of a variable region at the amino-terminal end and a constant region at the carboxyl-terminal end

- The existence of isotypes with the same antigenic specificity, which result from the association of a given variable region with different heavy-chain constant regions

## Germ-line and somatic-variation models contended to explain antibody diversity

For several decades, immunologists sought to imagine a genetic mechanism that could explain the tremendous diversity of antibody structure. Two different sets of theories emerged. The **germ-line theories** maintained that the genome contributed by the germ cells, egg and sperm, contains a large repertoire of immunoglobulin genes; thus, these theories invoked no special genetic mechanisms to account for antibody diversity. They argued that the immense survival value of the immune system justified the dedication of a significant fraction of the genome to the coding of antibodies. In contrast, the **somatic-variation theories** maintained that the genome contains a relatively small number of immunoglobulin genes, from which a large number of antibody specificities are generated in the somatic cells by mutation or recombination.

As the amino acid sequences of more and more immunoglobulins were determined, it became clear that there must be mechanisms not only for generating antibody diversity but also for maintaining constancy. Whether diversity was generated by germ-line or by somatic mechanisms, a paradox remained: How could stability be maintained in the constant (C) region while some kind of diversifying mechanism generated the variable (V) region?

Neither the germ-line nor the somatic-variation proponents could offer a reasonable explanation for this central feature of immunoglobulin structure. Germ-line proponents found it difficult to envisage an evolutionary mechanism that could generate diversity in the variable part of the many heavy- and light-chain genes while preserving the constant region of each unchanged. Somatic-variation proponents found it difficult to conceive of a mechanism that could diversify the variable region of a single heavy- or light-chain gene in the somatic cells without allowing alteration in the amino acid sequence encoded by the constant region.

A third structural feature requiring an explanation emerged when amino acid sequencing of a human myeloma protein called Ti1 revealed that identical variable-region sequences were associated with both γ and μ heavy-chain constant regions. A similar phenomenon was observed in rabbits by C. Todd, who found that a particular allotypic marker in the heavy-chain variable region could be associated with α, γ, and μ heavy-chain constant regions. Considerable additional evidence has confirmed that a single variable-region sequence, defining a particular antigenic specificity, can be associated with multiple heavy-chain constant-region sequences; in other words, different classes, or isotypes, of antibody (e.g., IgG, IgM) can be expressed with identical variable-region sequences.

## Dreyer and Bennett proposed a revolutionary two-gene one-polypeptide model

In an attempt to develop a genetic model consistent with the known findings about the structure of immunoglobulins, W. Dreyer and J. Bennett suggested, in their classic theoretical paper of 1965, that two separate genes encode a single immunoglobulin heavy or light chain, one gene for the V region (variable region) and the other for the C region (constant region). They suggested that these two genes must somehow come together at the DNA level to form a continuous message that can be transcribed and translated into a single Ig heavy or light chain. At that time it was generally accepted that one gene encoded one polypeptide, so their proposal was a revolutionary one. Moreover, they proposed that hundreds or thousands of V-region genes were carried in the germ line, whereas only single copies of C-region class and subclass genes need exist.

The strength of this type of recombinational model (which combined elements of the germ-line and somatic-variation theories) was that it could account for those immunoglobulins in which a single V region was combined with various C regions. By postulating a single constant-region gene for each immunoglobulin class and subclass, the model also could account for the conservation of necessary biological effector functions while allowing for evolutionary diversification of variable-region genes.

At first, support for the Dreyer and Bennett hypothesis was indirect. Early studies of DNA hybridization kinetics using a radioactive constant-region DNA probe indicated that the probe hybridized with only one or two genes, confirming the model's prediction that only one or two copies of each constant-region class and subclass gene existed. However, indirect evidence was not enough to overcome stubborn resistance in the scientific community to the hypothesis of Dreyer and Bennett. The suggestion that two genes encoded a single polypeptide contradicted the existing one-gene one-polypeptide principle and was without precedent in any known biological system.

As so often is the case in science, theoretical and intellectual understanding of Ig-gene organization progressed ahead of the available methodology. Although the Dreyer and Bennett model provided a theoretical framework for reconciling the dilemma between Ig-sequence data and gene organization, actual validation of their hypothesis had to wait for several major technological advances in the field of molecular biology.

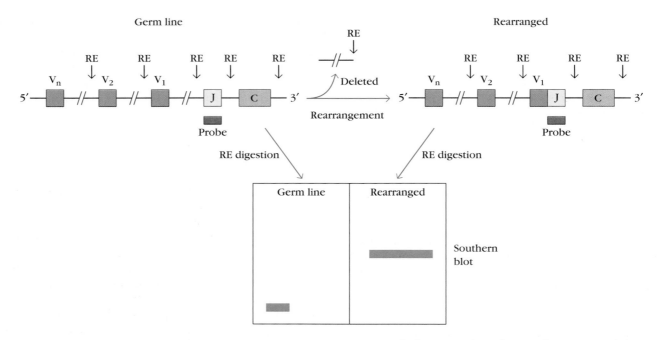

**FIGURE 5-2 Experimental basis for diagnosis of rearrangement at an immunoglobulin locus.** The number and size of restriction fragments generated by the treatment of DNA with a restriction enzyme are determined by the sequence of the DNA. The digestion of rearranged DNA with a restriction enzyme (RE) yields a pattern of restriction fragments that differ from those obtained by digestion of an unrearranged locus with the same RE. Typically, the fragments are analyzed by the technique of Southern blotting. In this example, a probe that includes a J gene segment is used to identify RE digestion fragments that include all or portions of this segment. As shown, rearrangement results in the deletion of a segment of germ-line DNA and the loss of the restriction sites that it includes. It also results in the joining of gene segments, in this case a V and a J segment, that are separated in the germ line. Consequently, fragments dependent on the presence of this segment for their generation are absent from the restriction-enzyme digest of DNA from the rearranged locus. Furthermore, rearranged DNA gives rise to novel fragments that are absent from digests of DNA in the germ-line configuration. This can be useful because both B cells and non–B cells have two immunoglobulin loci. One of these is rearranged, and the other is not. Consequently, unless a genetic accident has resulted in the loss of the germ-line locus, digestion of DNA from a myeloma or normal B-cell clone will produce a pattern of restriction that includes all of those in a germ-line digest plus any novel fragments that are generated from the change in DNA sequence that accompanies rearrangement. Note that only one of the several J gene segments present is shown.

## Tonegawa's bombshell—immunoglobulin genes rearrange

In 1976, S. Tonegawa and N. Hozumi found the first direct evidence that separate genes encode the V and C regions of immunoglobulins and that the genes are rearranged in the course of B-cell differentiation. Selecting DNA from embryonic cells and adult myeloma cells—cells at widely different stages of development—Tonegawa and Hozumi used various restriction endonucleases to generate DNA fragments. The fragments were then separated by size and analyzed for their ability to hybridize with a radiolabeled mRNA probe. Two separate restriction fragments from the embryonic DNA hybridized with the mRNA, whereas only a single restriction fragment of the adult myeloma DNA hybridized with the same probe. Tonegawa and Hozumi suggested that during differentiation of lymphocytes from the embryonic state to the fully differentiated plasma cell stage (represented in their system by the myeloma cells), the V and C genes undergo rearrangement. In the embryo, the V and C genes are separated by a large DNA segment that contains a restriction-endonuclease site; during differentiation, the V and C genes are brought closer together and the intervening DNA sequence is eliminated. This work changed the field of immunology, and in 1987, Tonegawa was awarded the Nobel prize.

The pioneering experiments of Tonegawa and Hozumi employed a tedious and time-consuming procedure that has since been replaced by the much more powerful approach of Southern-blot analysis. This method, now universally used to investigate the rearrangement of immunoglobulin genes, eliminates the need to elute the separated DNA restriction fragments from gel slices prior to analysis by hybridization with an immunoglobulin gene segment probe. Figure 5-2 shows the detection of rearrangement at the κ light-chain locus by comparing the fragments produced by digestion of DNA from a clone of B-lineage cells with the pattern obtained by digestion of non–B cells (e.g., sperm or liver cells). The rearrangement of a V gene deletes an extensive section of germ-line DNA, thereby creating differences between rearranged and unrearranged Ig loci in the distribution and number of restriction sites. This results in the generation of different restriction patterns by rearranged and unrearranged loci. Extensive application of this approach has demonstrated that the Dreyer and Bennett two-gene model—one gene encoding the variable region and another encoding the constant region—applies to both heavy- and light-chain genes.

## Multigene Organization of Ig Genes

As cloning and sequencing of the light- and heavy-chain DNA was accomplished, even greater complexity was revealed than had been predicted by Dreyer and Bennett. The κ and λ light chains and the heavy chains are encoded by separate multigene families situated on different chromosomes (Table 5-1). In germ-line DNA, each of these multigene families contains several coding sequences, called **gene segments,** separated by noncoding regions. During B-cell maturation, these gene segments are rearranged and brought together to form functional immunoglobulin genes.

### Each multigene family has distinct features

The κ and λ light-chain families contain **V, J,** and **C gene segments;** the rearranged VJ segments encode the variable region of the light chains. The heavy-chain family contains **V, D, J,** and **C gene segments;** the rearranged VDJ gene segments encode the variable region of the heavy chain. In each gene family, C gene segments encode the constant regions. Each V gene segment is preceded at its 5′ end by a small exon that encodes a short **signal** or **leader (L) peptide** that guides the heavy or light chain through the endoplasmic reticulum. The signal peptide is cleaved from the nascent light and heavy chains before assembly of the finished immunoglobulin molecule. Thus, amino acids encoded by this leader sequence do not appear in the finished immunoglobulin molecule.

### λ-Chain Multigene Family

The first evidence that the light-chain variable region was actually encoded by two gene segments appeared when Tonegawa cloned the germ-line DNA that encodes the variable region of mouse λ light chain and determined its complete nucleotide sequence. When the nucleotide sequence was compared with the known amino acid sequence of the λ-chain variable region, an unusual discrepancy was observed. Although the first 97 amino acids of the λ-chain variable region corresponded to the nucleotide codon sequence, the remaining 13 carboxyl-terminal amino acids of the protein's variable region did not. It turned out that many base pairs away, a separate, 39-bp gene segment, called J for *joining,* encoded the remaining 13 amino acids of the λ-chain variable region. Thus, a functional λ variable-region gene contains two coding segments—a 5′ V segment and a 3′ J segment—which are separated by a noncoding DNA sequence in unrearranged germ-line DNA.

The λ multigene family in the mouse germ line contains two $V_\lambda$ gene segments, three functional $J_\lambda$ gene segments, and two functional $C_\lambda$ gene segments (Figure 5-3a). In addition to functional versions of these genes, some forms are **pseudogenes,** defective genes that are incapable of encoding protein; such genes are indicated with the Greek letter psi ($\Psi$). Interestingly, $J_\lambda 4$'s constant region partner, $C_\lambda 4$, is a perfectly functional gene. The $V_\lambda$ and the three functional $J_\lambda$ gene segments encode the variable region of the light chain, and each of the three functional $C_\lambda$ gene segments encodes the constant region of one of the three λ-chain subtypes (λ1, λ2, and λ3). In humans, the lambda locus is more complex. There are 30 functional $V_\lambda$ gene segments, four functional $J_\lambda$ segments, and seven $C_\lambda$ segments, only four of which are functional. In addition to the functional gene segments, the human lambda complex contains many $V_\lambda$ and $J_\lambda$ pseudogenes.

### κ-Chain Multigene Family

The κ-chain multigene family in the mouse contains approximately 75 $V_\kappa$ gene segments, each with an adjacent leader sequence a short distance upstream (i.e., on the 5′ side). There are five $J_\kappa$ gene segments (one of which is a nonfunctional pseudogene) and a single $C_\kappa$ gene segment (Figure 5-3b). As in the λ multigene family, the $V_\kappa$ and $J_\kappa$ gene segments encode the variable region of the κ light chain, and the $C_\kappa$ gene segment encodes the constant region. Since there is only one $C_\kappa$ gene segment, there are no subtypes of κ light chains. Comparison of parts *a* and *b* of Figure 5-3 shows that the arrangement of the gene segments is quite different in the κ and λ gene families. The κ-chain multigene family in humans, which has an organization similar to that of the mouse, contains approximately 40 $V_\kappa$ gene segments, five $J_\kappa$ segments, and a single $C_\kappa$ segment.

### Heavy-chain multigene family

The organization of the immunoglobulin heavy-chain genes is similar to, but more complex than, that of the κ and λ light-chain genes (Figure 5-3c). An additional gene segment encodes part of the heavy-chain variable region. The existence of this gene segment was first proposed by Leroy Hood and his colleagues, who compared the heavy-chain variable-region amino acid sequence with the $V_H$ and $J_H$ nucleotide sequences.

| TABLE 5-1 | Chromosomal locations of immunoglobulin genes in human and mouse | |
|---|---|---|
| | **CHROMOSOME** | |
| **Gene** | **Human** | **Mouse** |
| λ Light chain | 22 | 16 |
| κ Light chain | 2 | 6 |
| Heavy chain | 14 | 12 |

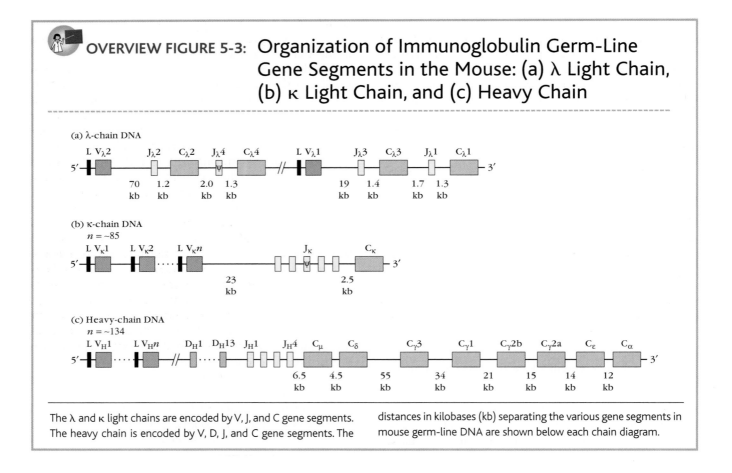

**OVERVIEW FIGURE 5-3:** Organization of Immunoglobulin Germ-Line Gene Segments in the Mouse: (a) λ Light Chain, (b) κ Light Chain, and (c) Heavy Chain

The λ and κ light chains are encoded by V, J, and C gene segments. The heavy chain is encoded by V, D, J, and C gene segments. The distances in kilobases (kb) separating the various gene segments in mouse germ-line DNA are shown below each chain diagram.

The $V_H$ gene segment was found to encode amino acids 1 to 94 and the $J_H$ segment was found to encode amino acids 98 to 113; however, neither of these gene segments carried the information to encode amino acids 95 to 97. When the nucleotide sequence was determined for a rearranged myeloma DNA and compared with the germ-line DNA sequence, an additional nucleotide sequence was observed between the $V_H$ and $J_H$ gene segments. This nucleotide sequence corresponded to amino acids 95 to 97 of the heavy chain.

From these results, Hood and his colleagues proposed that a third germ-line gene segment must join with the $V_H$ and $J_H$ gene segments to encode the entire variable region of the heavy chain. This gene segment, which encoded amino acids within the third complementarity-determining region (CDR3), was designated D for *diversity*, because of its contribution to the generation of antibody diversity. Tonegawa and his colleagues located the D gene segments within mouse germ-line DNA with a cDNA probe complementary to the D region, which hybridized with a stretch of DNA lying between the $V_H$ and $J_H$ gene segments.

The heavy-chain multigene family on human chromosome 14 has been shown by direct sequencing of DNA to contain 39 $V_H$ gene segments located upstream from a cluster of 23 functional $D_H$ gene segments. As with the light-chain genes, each $V_H$ gene segment is preceded by a leader sequence a short distance upstream. Downstream from the

$D_H$ gene segments are six functional $J_H$ gene segments, followed by a series of $C_H$ gene segments. Each $C_H$ gene segment encodes the constant region of an immunoglobulin heavy-chain isotype. The $C_H$ gene segments consist of coding exons and noncoding introns. Each exon encodes a separate domain of the heavy-chain constant region. A similar heavy-chain gene organization is found in the mouse.

The conservation of important biological effector functions of the antibody molecule is maintained by the limited number of heavy-chain constant-region genes. In humans and mice, the $C_H$ gene segments are arranged sequentially in the order $C_\mu$, $C_\delta$, $C_\gamma$, $C_\epsilon$, $C_\alpha$ (see Figure 5-3c). This sequential arrangement is no accident; it is generally related to the sequential expression of the immunoglobulin classes in the course of B-cell development and the initial IgM response of a B cell to its first encounter with an antigen.

## Variable-Region Gene Rearrangements

The preceding sections have shown that functional genes that encode immunoglobulin light and heavy chains are assembled by recombinational events at the DNA level. These events, and parallel events involving T-cell receptor genes, are the only known site-specific DNA rearrangements in vertebrates. Variable-region gene rearrangements occur in an ordered sequence during B-cell maturation in the bone marrow. The heavy-chain variable-region genes rearrange first, then the light-chain variable-region genes. At the end of this process, each B cell contains a single functional variable-region DNA sequence for its heavy chain and another for its light chain.

The process of variable-region gene rearrangement produces mature, immunocompetent B cells; each such cell is committed to produce antibody with a binding site encoded by the particular sequence of its rearranged V genes. As described later in this chapter, rearrangements of the heavy-chain constant-region genes, a process called class switching, will generate further changes in the immunoglobulin class (isotype) expressed by a B cell, but those changes will not affect the product's antigenic specificity.

The steps in variable-region gene rearrangement occur in an ordered sequence, but for the purposes of this discussion they can be considered as random events that result in the random determination of B-cell specificity. The order, mechanism, and consequences of these rearrangements are described in this section.

## Light-chain DNA undergoes V-J rearrangements

Expression of both κ and λ light chains requires rearrangement of the variable-region V and J gene segments. In humans, any of the functional $V_\lambda$ genes can combine with any of the four functional $J_\lambda$-$C_\lambda$ combinations. In the mouse, things are slightly more complicated. DNA rearrangement can join the $V_\lambda 1$ gene segment with either the $J_\lambda 1$ or the $J_\lambda 3$ gene segment, or the $V_\lambda 2$ gene segment can be joined with the $J_\lambda 2$ gene segment. In human or mouse κ light-chain DNA, any one of the $V_\kappa$ gene segments can be joined with any one of the functional $J_\lambda$ gene segments.

Rearranged κ and λ genes contain the following regions in order from the 5′ to 3′ end: a short leader (L) exon, a noncoding sequence (intron), a joined VJ gene segment, a second intron, and the constant region. Upstream from each leader gene segment is a promoter sequence. The rearranged light-chain sequence is transcribed by RNA polymerase from the L exon through the C segment to the stop signal, generating a light-chain primary RNA transcript (Figure 5-4). The introns in the primary transcript are

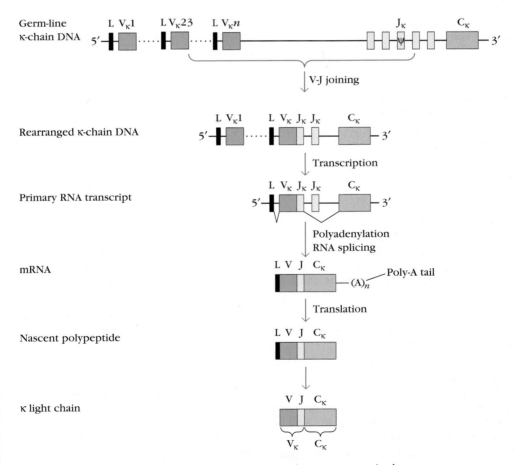

**FIGURE 5-4 Kappa light-chain gene rearrangement and RNA processing events required to generate a κ light-chain protein.** In this example, rearrangement joins $V_\kappa 23$ and $J_\kappa 4$.

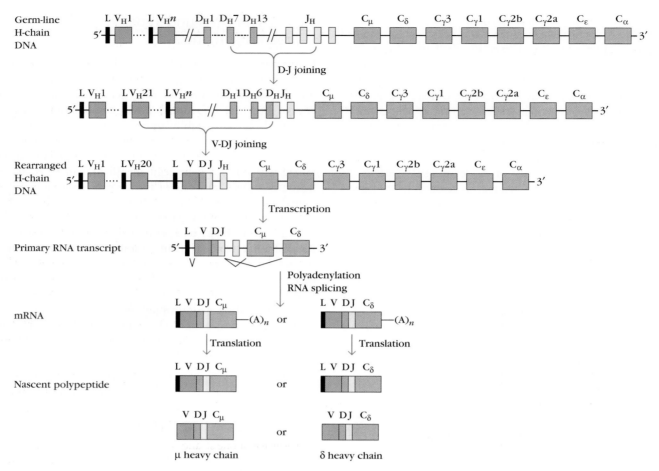

**FIGURE 5-5 Heavy-chain gene rearrangement and RNA processing events required to generate finished μ or δ heavy-chain protein.** Two DNA joinings are necessary to generate a functional heavy-chain gene: a $D_H$ to $J_H$ joining and a $V_H$ to $D_H J_H$ joining. In this example, $V_H 21$, $D_H 7$, and $J_H 3$ are joined. Expression of functional heavy-chain genes, although generally similar to expression of light-chain genes, involves differential RNA processing, which generates several different products, including μ or δ heavy chains. Each C gene is drawn as a single coding sequence; in reality, each is organized as a series of exons and introns.

removed by RNA-processing enzymes, and the resulting light-chain messenger RNA then exits from the nucleus. The light-chain mRNA binds to ribosomes and is translated into the light-chain protein. The leader sequence at the amino terminus targets the growing polypeptide chain to the lumen of the rough endoplasmic reticulum and is then cleaved, so it is not present in the finished light-chain protein product.

## Heavy-chain DNA undergoes V-D-J rearrangements

Generation of a functional immunoglobulin heavy-chain gene requires two separate rearrangement events within the variable region. As illustrated in Figure 5-5, a $D_H$ gene segment first joins to a $J_H$ segment; the resulting $D_H J_H$ segment then joins a $V_H$ segment to generate a $V_H D_H J_H$ unit that encodes the entire variable region. In heavy-chain DNA, variable-region rearrangement produces a rearranged gene consisting of the following sequences, starting from the 5' end: a short L exon, an intron, a joined VDJ segment, another intron, and a series of C gene segments. As with the light-chain genes, a promoter sequence is located a short distance upstream from each heavy-chain leader sequence.

Once heavy-chain gene rearrangement is accomplished, RNA polymerase can bind to the promoter sequence and transcribe the entire heavy-chain gene, including the introns. Initially, both $C_\mu$ and $C_\delta$ gene segments are transcribed. Differential polyadenylation and RNA splicing remove the introns and process the primary transcript to generate mRNA including either the $C_\mu$ or the $C_\delta$ transcript. These two mRNAs are then translated, and the leader peptide of the resulting nascent polypeptide is cleaved, generating finished μ and δ chains. The production of two different heavy-chain mRNAs allows a mature, immunocompetent B cell to express both IgM and IgD with identical antigenic specificity on its surface.

# Mechanism of Variable-Region DNA Rearrangements

Now that we've seen the results of variable-region gene rearrangements, let's examine in detail how this process occurs during the maturation of B cells.

## Recombination signal sequences direct recombination

The discovery of two closely related conserved sequences in variable-region germ-line DNA paved the way to fuller understanding of the mechanism of gene rearrangements. DNA sequencing studies revealed the presence of unique **recombination signal sequences (RSSs)** flanking each germ-line V, D, and J gene segment. One RSS is located 3′ to each V gene segment, 5′ to each J gene segment, and on both sides of each D gene segment. These sequences function as signals for the recombination process that rearranges the genes. Each RSS contains a conserved palindromic heptamer and a conserved AT-rich nonamer sequence separated by an intervening sequence of 12 or 23 base pairs (Figure 5-6a). The intervening 12- and 23-bp sequences correspond, respectively, to one and two turns of the DNA helix; for this reason the sequences are called **one-turn recombination signal sequences** and **two-turn signal sequences.** The $V_\kappa$ signal sequence has a one-turn spacer, and the $J_\kappa$ signal sequence has a two-turn spacer. In λ light-chain DNA, this order is reversed; that is, the $V_\lambda$ signal sequence has a two-turn spacer, and the $J_\lambda$ signal sequence has a one-turn spacer. In heavy-chain DNA, the signal sequences of the $V_H$ and $J_H$ gene segments have two-turn spacers; the signals on either side of the $D_H$ gene segment have one-turn spacers (Figure 5-6b). Signal sequences having a one-turn spacer can join only with sequences having a two-turn spacer (the so-called one-

turn/two-turn joining rule). This joining rule ensures, for example, that a $V_L$ segment joins only to a $J_L$ segment and not to another $V_L$ segment; the rule likewise ensures that $V_H$, $D_H$, and $J_H$ segments join in proper order and that segments of the same type do not join each other.

## Gene segments are joined by recombinases

V-(D)-J recombination, which takes place at the junctions between RSSs and coding sequences, is catalyzed by enzymes collectively called **V(D)J recombinase.** Identification of the enzymes that catalyze recombination of V, D, and J gene segments began in the late 1980s and is still ongoing. In 1990 David Schatz, Marjorie Oettinger, and David Baltimore first reported the identification of two **recombination-activating genes,** designated *RAG-1* and *RAG-2*, whose encoded proteins act synergistically and are required to mediate V-(D)-J joining. The RAG-1 and RAG-2 proteins are the only lymphoid-specific gene products that have been shown to be involved in V-(D)-J rearrangement.

The recombination of variable-region gene segments consists of the following steps, catalyzed by a system of recombinase enzymes (Figure 5-7):

1. Recognition of recombination signal sequences (RSSs) by recombinase enzymes, followed by synapsis in which two signal sequences and the adjacent coding sequences (gene segments) are brought into proximity

2. Cleavage of one strand of DNA by RAG-1 and RAG-2 at the junctures of the signal sequences and coding sequences

3. A reaction catalyzed by RAG-1 and RAG-2 in which the free 3′-OH group on the cut DNA strand attacks the phosphodiester bond linking the opposite strand to the signal sequence, simultaneously producing a hairpin structure at the cut end of the coding sequence and a flush, 5′-phosphorylated, double-strand break at the signal sequence

(a) Nucleotide sequence of RSSs

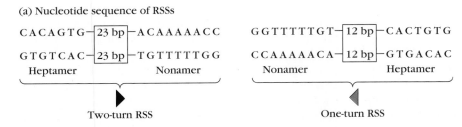

(b) Location of RSSs in germ-line immunoglobulin DNA

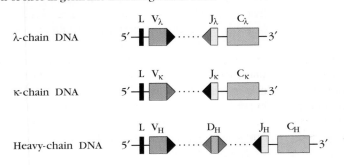

**FIGURE 5-6 Two conserved sequences in light-chain and heavy-chain DNA function as recombination signal sequences (RSSs).** (a) Both signal sequences consist of a conserved palindromic heptamer and conserved AT-rich nonamer; these are separated by nonconserved spacers of 12 or 23 base pairs. (b) The two types of RSS—designated one-turn RSS and two-turn RSS—have characteristic locations within λ-chain, κ-chain, and heavy-chain germ-line DNA. During DNA rearrangement, gene segments adjacent to the one-turn RSS can join only with segments adjacent to the two-turn RSS.

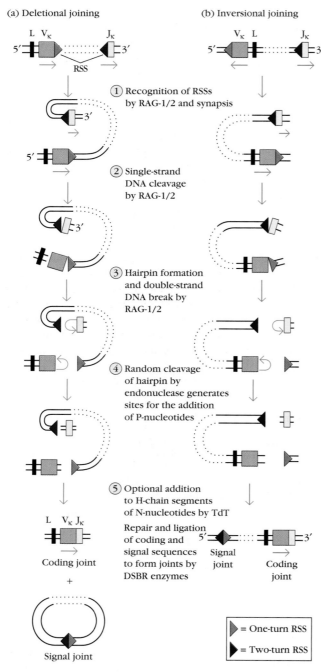

**(a) Deletional joining**

**(b) Inversional joining**

① Recognition of RSSs by RAG-1/2 and synapsis

② Single-strand DNA cleavage by RAG-1/2

③ Hairpin formation and double-strand DNA break by RAG-1/2

④ Random cleavage of hairpin by endonuclease generates sites for the addition of P-nucleotides

⑤ Optional addition to H-chain segments of N-nucleotides by TdT

Repair and ligation of coding and signal sequences to form joints by DSBR enzymes

Coding joint

Signal joint

Coding joint

Signal joint

▷ = One-turn RSS

▶ = Two-turn RSS

**FIGURE 5-7 Model depicting the general process of recombination of immunoglobulin gene segments is illustrated with V$_\kappa$ and J$_\kappa$.** (a) Deletional joining occurs when the gene segments to be joined have the same transcriptional orientation (indicated by horizontal blue arrows). This process yields two products: a rearranged VJ unit that includes the coding joint and a circular excision product consisting of the recombination signal sequences (RSSs), signal joint, and intervening DNA. (b) Inversional joining occurs when the gene segments have opposite transcriptional orientations. In this case, the RSSs, signal joint, and intervening DNA are retained, and the orientation of one of the joined segments is inverted. In both types of recombination, a few nucleotides may be deleted from or added to the cut ends of the coding sequences before they are rejoined.

4. Cutting of the hairpin to generate sites for the addition of **P-region nucleotides,** followed by the trimming of a few nucleotides from the coding sequence by a single-strand endonuclease

5. Addition of up to 15 nucleotides, called **N-region nucleotides,** at the cut ends of the V, D, and J coding sequences of the heavy chain by the enzyme terminal deoxynucleotidyl transferase

6. Repair and ligation to join the coding sequences and to join the signal sequences, catalyzed by normal double-strand break repair (DSBR) enzymes

Recombination results in the formation of a **coding joint,** falling between the coding sequences, and a **signal joint,** between the RSSs. The transcriptional orientation of the gene segments to be joined determines the fate of the signal joint and intervening DNA. When the two gene segments are in the same transcriptional orientation, joining results in deletion of the signal joint and intervening DNA as a circular excision product (Figure 5-8). Less frequently, the two gene segments have opposite orientations. In this case joining occurs by inversion of the DNA, resulting in the retention of both the coding joint and the signal joint (and intervening DNA) on the chromosome. In the human $\kappa$ locus, about half of the V$_\kappa$ gene segments are inverted with respect to J$_\kappa$ and their joining is thus by inversion.

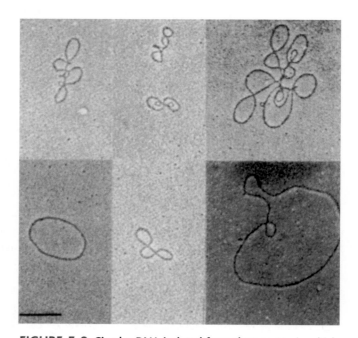

**FIGURE 5-8 Circular DNA isolated from thymocytes in which the DNA encoding the chains of the T-cell receptor (TCR) undergoes rearrangement in a process like that involving the immunoglobulin genes.** Isolation of this circular excision product is direct evidence for the mechanism of deletional joining shown in Figure 5-7. *[From K. Okazaki et al., 1987, Cell **49**:477.]*

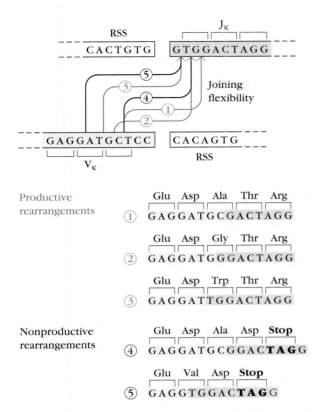

Productive rearrangements

|       | Glu | Asp | Ala | Thr | Arg |
| (1)   |     |     |     |     |     |
| GAGGATGCGACTAGG |

|       | Glu | Asp | Gly | Thr | Arg |
| (2)   |     |     |     |     |     |
| GAGGATGGGACTAGG |

|       | Glu | Asp | Trp | Thr | Arg |
| (3)   |     |     |     |     |     |
| GAGGATTGGACTAGG |

Nonproductive rearrangements

|       | Glu | Asp | Ala | Asp | **Stop** |
| (4)   |     |     |     |     |     |
| GAGGATGCGGAC**TAG**G |

|       | Glu | Val | Asp | **Stop** |
| (5)   |     |     |     |     |
| GAGGTGGAC**TAG**G |

**FIGURE 5-9 Junctional flexibility in the joining of immunoglobulin gene segments is illustrated with $V_\kappa$ and $J_\kappa$.** In-phase joining (arrows 1, 2, and 3) generates a productive rearrangement, which can be translated into protein. Out-of-phase joining (arrows 4 and 5) leads to a nonproductive rearrangement that contains stop codons and is not translated into protein.

## Ig-gene rearrangements may be productive or nonproductive

One of the striking features of gene-segment recombination is the diversity of the coding joints that are formed between any two gene segments. Although the double-strand DNA breaks that initiate V-(D)-J rearrangements are introduced precisely at the junctions of signal sequences and coding sequences, the subsequent joining of the coding sequences is imprecise. Junctional diversity at the V-J and V-D-J coding joints is generated by a number of mechanisms: variation in cutting of the hairpin to generate P-nucleotides, variation in trimming of the coding sequences, variation in N-nucleotide addition, and flexibility in joining the coding sequences. The introduction of randomness in the joining process helps generate antibody diversity by contributing to the hypervariability of the antigen-binding site. (This phenomenon is covered in more detail below in the section on generation of antibody diversity.)

Another consequence of imprecise joining is that gene segments may be joined out of phase, so that the triplet reading frame for translation is not preserved. In such a **nonproductive rearrangement,** the resulting VJ or VDJ unit is likely to contain numerous stop codons, which interrupt translation (Figure 5-9). When gene segments are joined in phase,

the reading frame is maintained. In such a **productive rearrangement,** the resulting VJ or VDJ unit can be translated in its entirety, yielding a complete antibody.

If one allele rearranges nonproductively, a B cell may still be able to rearrange the other allele productively. If an in-phase rearranged heavy-chain and light-chain gene are not produced, the B cell dies by apoptosis. It is estimated that only one in three attempts at $V_L$-$J_L$ joining, and one in three attempts at $V_H$-$D_H J_H$ joining, are productive. As a result, less than 1/9 (11%) of the early-stage pre–B cells in the bone marrow progress to maturity and leave the bone marrow as mature immunocompetent B cells.

## Allelic exclusion ensures a single antigenic specificity

B cells, like all somatic cells, are diploid and contain both maternal and paternal chromosomes. Even though a B cell is diploid, it expresses the rearranged heavy-chain genes from only one chromosome and the rearranged light-chain genes from only one chromosome. The process by which this is accomplished, called **allelic exclusion,** ensures that functional B cells never contain more than one $V_H D_H J_H$ and one $V_L J_L$ unit (Figure 5-10). This is, of course, essential for the antigenic specificity of the B cell, because the expression of both

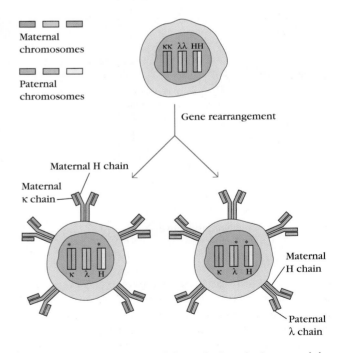

**FIGURE 5-10 Because of allelic exclusion, the immunoglobulin heavy- and light-chain genes of only one parental chromosome are expressed per cell.** This process ensures that B cells possess a single antigenic specificity. The allele selected for rearrangement is chosen randomly. Thus, the expressed immunoglobulin may contain one maternal and one paternal chain or both chains may derive from only one parent. Only B cells and T cells exhibit allelic exclusion. Asterisks (∗) indicate the expressed alleles.

alleles would render the B cell multispecific. The phenomenon of allelic exclusion suggests that once a productive $V_H$-$D_H$-$J_H$ rearrangement and a productive $V_L$-$J_L$ rearrangement have occurred, the recombination machinery is turned off so that the heavy- and light-chain genes on the homologous chromosomes are not expressed.

G. D. Yancopoulos and F. W. Alt have proposed a model to account for allelic exclusion (Figure 5-11). They suggest that once a productive rearrangement is attained, its encoded protein is expressed and the presence of this protein acts as a signal to prevent further gene rearrangement. According to their model, the presence of $\mu$ heavy chains signals the maturing B cell to turn off rearrangement of the other heavy-chain allele and to turn on rearrangement of the $\kappa$ light-chain genes. If a productive $\kappa$ rearrangement occurs, $\kappa$ light chains are produced and then pair with $\mu$ heavy chains to form a complete antibody molecule. The presence of this antibody then turns off further light-chain rearrangement. If $\kappa$ rearrangement is nonproductive for both $\kappa$ alleles,

rearrangement of the $\lambda$-chain genes begins. If neither $\lambda$ allele rearranges productively, the B cell presumably ceases to mature and soon dies by apoptosis.

Two studies with transgenic mice support the hypothesis that the protein products encoded by rearranged heavy- and light-chain genes regulate rearrangement of the remaining alleles. In one study, transgenic mice carrying a rearranged $\mu$ heavy-chain transgene were prepared. The $\mu$ transgene product was expressed by a large percentage of the B cells, and rearrangement of the endogenous immunoglobulin heavy-chain genes was blocked. Similarly, cells from a transgenic mouse carrying a $\kappa$ light-chain transgene did not rearrange the endogenous $\kappa$-chain genes when the $\kappa$ transgene was expressed and was associated with a heavy chain to form complete immunoglobulin. These studies suggest that expression of the heavy- and light-chain proteins may indeed prevent gene rearrangement of the remaining alleles and thus account for allelic exclusion.

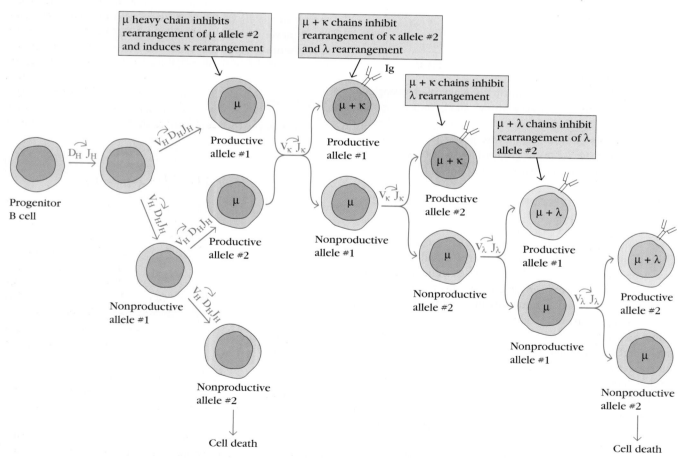

**FIGURE 5-11 Model to account for allelic exclusion.** Heavy-chain genes rearrange first, and once a productive heavy-chain gene rearrangement occurs, the $\mu$ protein product prevents rearrangement of the other heavy-chain allele and initiates light-chain gene rearrangement. In the mouse, rearrangement of $\kappa$ light-chain genes precedes rearrangement of the $\lambda$ genes, as shown here. In humans, either $\kappa$ or $\lambda$ rearrangement can proceed once a productive heavy-chain rearrangement has occurred. Formation of a complete immunoglobulin inhibits further light-chain gene rearrangement. If a nonproductive rearrangement occurs for one allele, then the cell attempts rearrangement of the other allele. [Adapted from G. D. Yancopoulos and F. W. Alt, 1986, Annual Review of Immunology **4**:339.]

# Generation of Antibody Diversity

As the organization of the immunoglobulin genes was deciphered, the sources of the vast diversity in the variable region began to emerge. The germ-line theory, mentioned earlier, argued that the entire variable-region repertoire is encoded in the germ line of the organism and is transmitted from parent to offspring through the germ cells (egg and sperm). The somatic-variation theory held that the germ line contains a limited number of variable genes, which are diversified in the somatic cells by mutational or recombinational events during development of the immune system. With the cloning and sequencing of the immunoglobulin genes, both models were partly vindicated. To date, seven means of antibody diversification have been identified in mice and humans:

- Multiple germ-line gene segments
- Combinatorial V-(D)-J joining
- Junctional flexibility
- P-region nucleotide addition (P-addition)
- N-region nucleotide addition (N-addition)
- Somatic hypermutation
- Combinatorial association of light and heavy chains

Although the exact contribution of each of these avenues of diversification to total antibody diversity is not known, they each contribute significantly to the immense number of distinct antibodies that the mammalian immune system is capable of generating.

## There are numerous germ-line V, D, and J gene segments

An inventory of functional V, D, and J gene segments in the germ-line DNA of one human reveals 48 $V_H$, 23 D, 6 $J_H$, 41 $V_\kappa$, 5 $J_\kappa$, 34 $V_\lambda$, and 5 $J_\lambda$ gene segments. In addition to these functional segments, there are many pseudogenes. It should be borne in mind that these numbers were largely derived from a landmark study that sequenced the DNA of the immunoglobulin loci of a single individual. The immunoglobulin loci of other individuals might contain slightly different numbers of particular types of gene segments. In the mouse, although the numbers are known with less precision than in the human, there appear to be about 85 $V_\kappa$ gene segments and 134 $V_H$ gene segments, 4 functional $J_H$, 4 functional $J_\kappa$, 3 functional $J_\lambda$, and an estimated 13 $D_H$ gene segments, but only three $V_\lambda$ gene segments. Although the number of germ-line genes found in either humans or mice is far fewer than predicted by early proponents of the germ-line model, multiple germ-line V, D, and J gene segments clearly do contribute to the diversity of the antigen-binding sites in antibodies.

## Combinatorial V-J and V-D-J joining generates diversity

The contribution of multiple germ-line gene segments to antibody diversity is magnified by the random rearrangement of these segments in somatic cells. It is possible to calculate how much diversity can be achieved by gene rearrangements (Table 5-2). In humans, the ability of any of the 48 $V_H$ gene segments to combine with any of the 23 $D_H$ segments and any of the 6 $J_H$ segments allows a considerable amount of heavy-chain gene diversity to be generated (48 × 23 × 6 = 6624 possible combinations). Similarly, 41 $V_\kappa$ gene segments randomly combining with 5 $J_\kappa$ segments has the potential of generating 205 possible combinations at the κ locus, whereas 34 $V_\lambda$ and 5 $J_\lambda$ gene segments allow up to 170 possible combinations at the human λ locus. It is important to realize that these are minimal calculations of potential diversity. Junctional flexibility and P- and N-nucleotide addition, as mentioned above, and, especially, somatic hypermutation, which will be described shortly, together make an enormous contribution to antibody diversity. Although it is not possible to make an exact calculation of their contribution, most workers in this field agree that they raise the potential for antibody combining-site diversity in humans to well over $10^{10}$. This does not mean that, at any given time, a single individual has a repertoire of $10^{10}$ different antibody combining sites. These very large numbers describe the set of possible variations, of which any individual carries a subset that is smaller by several orders of magnitude.

## Junctional flexibility adds diversity

The enormous diversity generated by means of V, D, and J combinations is further augmented by a phenomenon called **junctional flexibility.** As described above, recombination involves both the joining of recombination signal sequences to form a signal joint and the joining of coding sequences to form a coding joint (see Figure 5-7). Although the signal sequences are always joined precisely, joining of the coding sequences is often imprecise. In one study, for example, joining of the $V_\kappa 21$ and $J_\kappa 1$ coding sequences was analyzed in several pre–B cell lines. Sequence analysis of the signal and coding joints revealed the contrast in junctional precision (Figure 5-12).

As illustrated previously, junctional flexibility leads to many nonproductive rearrangements, but it also generates productive combinations that encode alternative amino acids at each coding joint (see Figure 5-9), thereby increasing antibody diversity. The amino acid sequence variation generated by junctional flexibility in the coding joints has been shown to fall within the third hypervariable region (CDR3) in immunoglobulin heavy-chain and light-chain DNA (Table 5-3). Since CDR3 often makes a major contribution to antigen binding by the antibody molecule, amino acid changes generated by junctional flexibility are important in the generation of antibody diversity.

**TABLE 5-2 Combinatorial antibody diversity in humans and mice**

| Multiple germ-line segments | Heavy chain | LIGHT CHAINS κ | LIGHT CHAINS λ |
|---|---|---|---|
| ESTIMATED NUMBER OF SEGMENTS IN HUMANS* | | | |
| V | 48 | 41 | 34 |
| D | 23 | 0 | 0 |
| J | 6 | 5 | 5 |
| Combinatorial V-D-J and V-J joining (possible number of combinations) | 48 × 23 × 6 = 6624 | 41 × 5 = 205 | 34 × 5 = 170 |
| Possible combinatorial associations of heavy and light chains† | | 6624 × (205 + 170) = $2.48 \times 10^6$ | + |
| ESTIMATED NUMBER OF SEGMENTS IN MICE* | | | |
| V | 134 | 85 | 2 |
| D | 13 | 0 | 0 |
| J | 4 | 4 | 3 |
| Combinatorial V-D-J and V-J joining (possible number of combinations) | 101 × 13 × 4 = 5252 | 85 × 4 = 340 | 2 × 3 = 6 |
| Possible combinatorial associations of heavy and light chains† | | 5252 × (340 + 6) = $1.82 \times 10^6$ | |

*These numbers have been determined from studies of single subjects; slight differences may be seen among different individuals. In the cases of both human and mouse, only the functional gene segments have been listed. The genome contains additional segments that are incapable of rearrangement or contain stop codons or both.

†Because of the diversity contributed by junctional flexibility, P-region nucleotide addition, N-region nucleotide addition, and somatic mutation, the actual potential exceeds these estimates by several orders of magnitude.

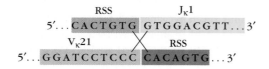

RSS
$J_\kappa 1$
5′. . . CACTGTG GTGGACGTT. . . 3′
$V_\kappa 21$
RSS
5′. . . GGATCCTCCC CACAGTG. . . 3′
Pre–B cell lines
Coding joints ($V_\kappa 21\ J_\kappa 1$)
Signal joints (RSS/RSS)
Cell line #1 5′-GGATCC GGACGTT-3′ 5′-CACTGTG CACAGTG-3′
Cell line #2 5′-GGATC TGGACGTT-3′ 5′-CACTGTG CACAGTG-3′
Cell line #3 5′-GGATCCTC GTGGACGTT-3′ 5′-CACTGTG CACAGTG-3′
Cell line #4 5′-GGATCCT TGGACGTT-3′ 5′-CACTGTG CACAGTG-3′

**FIGURE 5-12 Experimental evidence for junctional flexibility in immunoglobulin-gene rearrangement.** The nucleotide sequences flanking the coding joints between $V_\kappa 21$ and $J_\kappa 1$ and the corresponding signal joint sequences were determined in four pre–B cell lines. The sequence constancy in the signal joints contrasts with the sequence variability in the coding joints. Pink and yellow shading indicate nucleotides derived from $V_\kappa 21$ and $J_\kappa 1$, respectively, and purple and orange shading indicate nucleotides from the two RSSs.

**TABLE 5-3 Sources of sequence variation in complementarity-determining regions of immunoglobulin heavy- and light-chain genes**

| Source of variation | CDR1 | CDR2 | CDR3 |
|---|---|---|---|
| Sequence encoded by: | V segment | V segment | $V_L$-$J_L$ junction; $V_H$-$D_H$-$J_H$ junctions |
| Junctional flexibility | − | − | + |
| P-nucleotide addition | − | − | + |
| N-nucleotide addition* | − | − | + |
| Somatic hypermutation | + | + | + |

*N-nucleotide addition occurs only in heavy-chain DNA.

## P-addition adds diversity at palindromic sequences

As described earlier, after the initial single-strand DNA cleavage at the junction of a variable-region gene segment and attached signal sequence, the nucleotides at the end of the coding sequence turn back to form a hairpin structure (see Figure 5-7). This hairpin is later cleaved by an endonuclease. This second cleavage sometimes occurs at a position that leaves a short single strand at the end of the coding sequence. The subsequent addition of complementary nucleotides to this strand (**P-addition**) by repair enzymes generates a palindromic sequence in the coding joint, and so these nucleotides are called **P-nucleotides** (Figure 5-13a). Variation in the position at which the hairpin is cut thus leads to variation in the sequence of the coding joint.

## N-addition adds considerable diversity by addition of nucleotides

Variable-region coding joints in rearranged heavy-chain genes have been shown to contain short amino acid sequences that are not encoded by the germ-line V, D, or J gene segments. These amino acids are encoded by nucleotides added during the D-J and V to D-J joining process in a reaction catalyzed by a terminal deoxynucleotidyl transferase (TdT; Figure 5-13b). Evidence that TdT is responsible for the addition of these **N-nucleotides** comes from transfection studies in fibroblasts. When fibroblasts were transfected with the *RAG-1* and *RAG-2* genes, V-D-J rearrangement occurred, but no N-nucleotides were present in the coding joints. However, when the fibroblasts were also transfected with the gene encoding the enzyme TdT, V-D-J rearrangement was accompanied by the addition of N-nucleotides.

Up to 15 N-nucleotides can be added to both the $D_H$-$J_H$ and $V_H$-$D_H J_H$ joints. Thus, a complete heavy-chain variable region is encoded by a $V_H N D_H N J_H$ unit. The additional heavy-chain diversity generated by N-region nucleotide addition is quite large because N regions appear to consist of wholly random sequences. Since this diversity occurs at V-D-J coding joints, it is localized in CDR3 of the heavy-chain genes.

## Somatic hypermutation adds diversity in already-rearranged gene segments

All the antibody diversity described so far stems from mechanisms that operate during formation of specific variable regions by gene rearrangement. Additional antibody diversity is generated in rearranged variable-region gene units by a process called **somatic hypermutation.** As a result of somatic hypermutation, individual nucleotides in VJ or VDJ units are replaced with alternatives, thus potentially altering the specificity of the encoded immunoglobulins.

Normally, somatic hypermutation occurs only within germinal centers (see Chapter 11), structures that form in secondary lymphoid organs within a week or so of immunization with an antigen that activates a T-cell-dependent B-cell response. Somatic hypermutation is targeted to rearranged V regions located within a DNA sequence containing about 1500 nucleotides, which includes the whole of the VJ or VDJ segment. Somatic hypermutation occurs at a frequency approaching $10^{-3}$ per base pair per generation. This rate is at least a hundred thousand–fold higher (hence the name *hypermutation*) than the spontaneous mutation rate for other genes (about $10^{-8}$/bp/generation). Since the combined length of the H-chain and L-chain variable-region genes is about 600 bp, one expects that somatic hypermutation

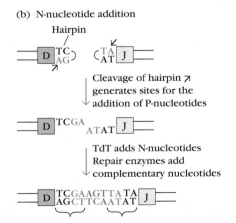

**(a) P-nucleotide addition**

**(b) N-nucleotide addition**

**FIGURE 5-13 P-nucleotide and N-nucleotide addition during joining.** (a) If cleavage of the hairpin intermediate yields a double-stranded end on the coding sequence, then P-nucleotide addition does not occur. In many cases, however, cleavage yields a single-stranded end. During subsequent repair, complementary nucleotides are added, called P-nucleotides, to produce palindromic sequences (indicated by brackets). In this example, four extra base pairs (blue) are present in the coding joint as the result of P-nucleotide addition. (b) Besides P-nucleotide addition, addition of random N-nucleotides (light red) by a terminal deoxynucleotidyl transferase (TdT) can occur during joining of heavy-chain coding sequences.

will introduce at least one mutation per every two cell divisions in the pair of $V_H$ and $V_L$ genes that encode an antibody.

During somatic hypermutation, most of the mutations are nucleotide substitutions rather than deletions or insertions. Somatic hypermutation introduces these substitutions in a largely but not completely random fashion. Certain nucleotide sequence motifs and palindromic sequences within $V_H$ and $V_L$ are especially susceptible to somatic hypermutation and are called "hot spots."

Somatic hypermutations occur throughout the VJ or VDJ segment, but in mature B cells they are clustered within the CDRs of the $V_H$ and $V_L$ sequences, where they are most likely to influence the overall affinity for antigen. Following exposure to antigen, those B cells with higher-affinity receptors will be preferentially selected for survival. The result of this differential selection is an increase in the antigen affinity of a population of B cells. The overall process, called **affinity**

**maturation,** takes place within germinal centers and is described more fully in Chapter 11.

In a now classic experiment, Claudia Berek and Cesar Milstein documented the reality and course of somatic hypermutation during an immune response to a hapten-carrier conjugate. These researchers were able to sequence mRNA that encoded antibodies raised against a hapten in response to primary, secondary, or tertiary immunization (first, second, or third exposure) with a hapten-carrier conjugate. The hapten they chose was 2-phenyl-5-oxazolone (phOx), coupled to a protein carrier. They chose this hapten because it had previously been shown that the majority of antibodies it induced were encoded by a single germ-line $V_H$ and $V_\kappa$ gene segment. Berek and Milstein immunized mice with the phOx-carrier conjugate and then used the mouse spleen cells to prepare hybridomas secreting monoclonal antibodies specific for the phOx hapten. The mRNA sequence for the H

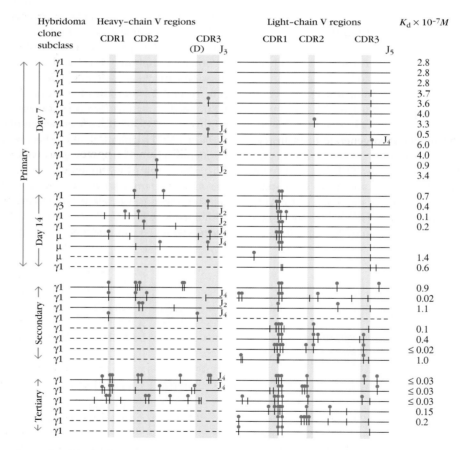

**FIGURE 5-14 Experimental evidence for somatic mutation in variable regions of immunoglobulin genes.** The diagram compares the mRNA sequences of heavy chains and of light chains from hybridomas specific for the phOx hapten. The horizontal solid lines represent the germ-line $V_H$ and $V_\kappa$ Ox-1 sequences; dashed lines represent sequences derived from other germ-line genes. Blue shading shows the areas where mutations clustered; the blue circles with vertical lines indicate locations of mutations that encode a different amino acid than the germ-line sequence. These data show that the

frequency of mutation (1) increases in the course of the primary response (day 7 vs. day 14) and (2) is higher after secondary and tertiary immunizations than after primary immunization. Moreover, the dissociation constant ($K_d$) of the anti-phOx antibodies decreases during the transition from the primary to tertiary response, indicating an increase in the overall affinity of the antibody. Note also that most of the mutations are clustered within CDR1 and CDR2 of both the heavy and the light chains. [Adapted from C. Berek and C. Milstein, 1987, Immunology Review **96**:23.]

chain and κ light chain of each hybridoma was then determined to identify deviations from the germ-line sequences.

The results of this experiment are depicted in Figure 5-14. Of the 12 hybridomas obtained from mice seven days after a primary immunization, all used a particular $V_H$, the $V_H$ Ox-1 gene segment, and all but one used the same $V_L$ gene segment, $V_\kappa$ Ox-1. Moreover, only a few mutations from the germ-line sequence were present in these hybridomas. By day 14 after primary immunization, analysis of eight hybridomas revealed that six continued to use the germ-line $V_H$ Ox-1 gene segment and all continued to use the $V_\kappa$ Ox-1 gene segment. Now, however, all of these hybridomas included one or more mutations from the germ-line sequence. Hybridomas analyzed from the secondary and tertiary responses showed a larger percentage utilizing germ-line $V_H$ gene segments other than the $V_H$ Ox-1 gene. In those hybridoma clones that utilized the $V_H$ Ox-1 and $V_\kappa$ Ox-1 gene segments, most of the mutations were clustered in the CDR1 and CDR2 hypervariable regions. The number of mutations in the anti-phOx hybridomas progressively increased following primary, secondary, and tertiary immunizations, as did the overall affinity of the antibodies for phOx.

## A final source of diversity is combinatorial association of heavy and light chains

In humans, there is the potential to generate 6624 heavy-chain genes and 375 light-chain genes as a result of variable-region gene rearrangements. If one assumes that any one of the possible heavy-chain and light-chain genes can occur randomly in the same cell, the potential number of heavy- and light-chain combinations is 2,484,000. This number is probably higher than the amount of combinatorial diversity actually generated in an individual, because it is not likely that all $V_H$ and $V_L$ will pair with each other. Furthermore, the recombination process is not completely random: not all $V_H$, D, or $V_L$ gene segments are used at the same frequency. Some are used often, others only occasionally, and still others almost never.

Although the number of different antibody binding sites the immune system can generate is difficult to calculate with precision, we know that it is quite high. Because the very large number of new sequences created by junctional flexibility, P-nucleotide addition, and N-nucleotide addition are within the third CDR, they are positioned to influence the structure of the antibody binding site. In addition to these sources of antibody diversity, the phenomenon of somatic hypermutation contributes enormously to the repertoire after antigen stimulation.

## Immunoglobulin gene diversification differs among species

In humans and mice, the primary repertoire of immunoglobulin genes is created somatically by the combinatorial rearrangement of an extensive library of germ-line V (D) and J gene segments. Furthermore, in mice and humans the generation of a primary repertoire occurs mostly in bone marrow. As will be fully discussed in Chapter 11, in most other vertebrates we see a departure from this pattern. In fact, bone marrow is not a site of B-cell development, and in birds, rabbits, sheep, cattle, and some other species, much of this key process takes place in gut-associated lymphoid tissues (GALT). Also, many species use mechanisms other than (or in addition to) combinatorial rearrangement of germ-line V (D) and J genes to generate a diversified primary repertoire of antibody genes. Two major differences are seen between many other species and humans and mice. First, some species use only one or a few rearrangements of germ-line V(D)J genes. This contrasts with the many rearrangements typical of humans and mice. Second, somatic hypermutation or another process called *gene conversion* is used in other species to extensively diversify the rearranged genes, whereas in humans and mice, the primary repertoire of rearranged genes is not further diversified by somatic mutation or gene conversion. During **gene conversion,** a special case of somatic mutation, portions of the sequence of one gene, the recipient, are modified to assume the corresponding sequence of a donor gene. The donor gene, whose sequence remains unchanged, acts a template for the conversion of a portion of the recipient sequence. In fact, gene conversion, which only occurs between genes that are very similar in sequence (>80% identical), is sometimes called **templated somatic mutation** because the donor acts as a template for the modification of a homologous region of the recipient to the sequence of the donor gene.

In chickens, the germ line contains only a single functional $V_L$, $V_H$, and (D)-J gene that can undergo rearrangement and large numbers of upstream $V_\lambda$ and $V_H$ pseudogenes that cannot rearrange (Figure 5-15). Following RAG-mediated rearrangement of their drastically limited repertoire of functional germ-line V-(D)-J genes, which yields a single V-(D)-J unit, chicken B cells migrate to an organ called the bursa of Fabricius, which is part of the GALT in birds. In specialized microenvironments of the bursa, the B cells undergo rapid proliferation and extensive gene-conversion-mediated diversification of their rearranged immunoglobulin genes. The large number of upstream pseudogenes act as donors for the conversion of short stretches of the rearranged V-(D)-J recipient. As shown in Figure 5-15, several pseudogenes can act as donors to a single rearranged V-(D)-J gene, modifying its germ-line sequence at several places. In addition to gene conversion, chickens also undergo some conventional somatic hypermutation.

Rabbits also rearrange very few heavy-chain $V_H$ genes and diversify the antibody repertoire by gene conversion and somatic hypermutation. In rabbits, as in birds, these processes take place in the GALT, particularly in specialized microenvironments of the appendix. Ruminants such as cattle and sheep diversify their antibody repertoires by somatic hypermutation in the ileal Peyer's patch, a type of GALT.

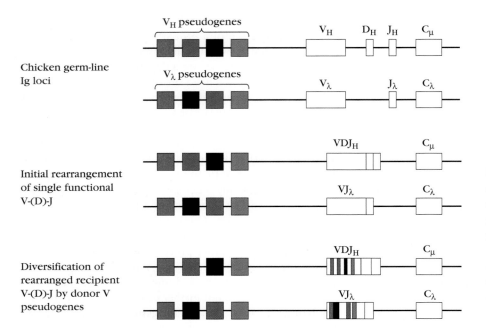

**FIGURE 5-15 Immunoglobulin diversification occurs by gene conversion in chickens.** In the chicken germ line, the single functional $V_H$ and $V_\lambda$ immunoglobulin genes are preceded by many pseudogenes. Rearrangement creates a single functional rearranged V-(D)-J. Gene conversion introduces diversity into the V segments of rearranged V-(D)-J genes using upstream V pseudogenes as a template.

# Class Switching Among Constant-Region Genes

After antigenic stimulation of a B cell, the heavy-chain DNA can undergo a further rearrangement in which the $V_H D_H J_H$ unit can combine with any $C_H$ gene segment. The exact mechanism of this process, called **class switching** or **isotype switching,** is unclear, but it involves DNA flanking sequences (called **switch regions**) located 2 to 3 kb upstream from each $C_H$ segment (except $C_\delta$). These switch regions, though rather large (2–10 kb), are composed of multiple copies of short repeats (GAGCT and TGGGG). One hypothesis is that a protein or system of proteins that constitute the switch recombinase recognizes these repeats and on binding carries out the DNA recombination that results in class switching. Intercellular regulatory proteins known as cytokines act as "switch factors" and play major roles in determining the particular immunoglobulin class that is expressed as a consequence of switching. Interleukin-4 (IL-4), for example, induces class switching from $C_\mu$ to $C_\gamma 1$ or $C_\epsilon$. In some cases, IL-4 has been observed to induce class switching in a successive manner: first from $C_\mu$ to $C_\gamma 1$ and then from $C_\gamma 1$ to $C_\epsilon$ (Figure 5-16). Examination of the DNA excision products produced during class switching from $C_\mu$ to $C_\gamma 1$ showed that a circular excision product containing $C_\mu$ together with the 5′ end of the γ1 switch region ($S_\gamma 1$) and the 3′ end of the μ switch region ($S_\mu$) was generated. Furthermore, the switch from $C_\gamma 1$ to $C_\epsilon$ produced circular excision products containing $C_\gamma 1$ together with portions of the μ, γ, and ε switch regions. Overall, class switching depends on the interplay of four elements: switch regions, a switch recombinase, the cytokine signals that dictate the isotype to which the

B cell switches, and an enzyme called *activation-induced cytidine deaminase (AID)*, whose critical role is discussed in the next section. A more complete description of the role of cytokines in class switching appears in Chapter 11.

## AID (activation-induced cytidine deaminase) mediates both somatic hypermutation and class switching

In this chapter, three different types of immunoglobulin gene modification have been presented for humans and mice: V-(D)-J recombination, somatic hypermutation, and class-switch recombination. (We also described gene conversion in other species.) The rearrangement and assembly of germ-line gene segments during V-(D)-J recombination forms a functional immunoglobulin gene. Somatic hypermutation introduces changes in the variable region of Ig genes that can affect the antigen binding properties of the antibodies they encode. Class-switch recombination allows a given V-(D)-J unit to be associated with different constant regions, which determine the effector function of the Ig molecule. The recombination-activating genes, *RAG1* and *RAG2,* are responsible for V-(D)-J recombination. Recent work has shown that one enzyme, **activation-induced cytidine deaminase (AID),** is the key mediator of somatic hypermutation, gene conversion, and class-switch recombination. AID belongs to an enzyme family called RNA editing enzymes. AID deaminates selected cytosines in certain mRNAs, changing the cytosines to uracils and thereby altering (editing) the protein-encoding instructions of the targeted messenger RNA. There are reports that this enzyme is also capable of directly modifying DNA by the deamination of cytosine, again resulting in formation of

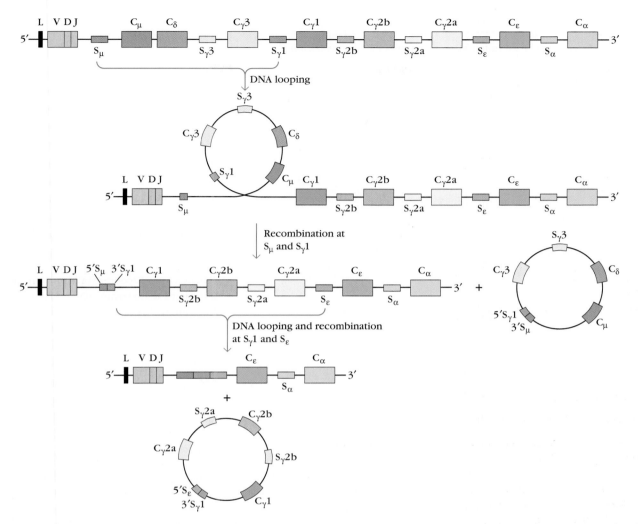

**FIGURE 5-16 Proposed mechanism for class switching induced by interleukin-4 in rearranged immunoglobulin heavy-chain genes.** A switch site is located upstream from each $C_H$ segment except $C_\delta$. Identification of the indicated circular excision products containing portions of the switch sites suggested that IL-4 induces sequential class switching from $C_\mu$ to $C_\gamma 1$ to $C_\varepsilon$.

uracil, which is not one of the usual four nucleotides in DNA. The site of deamination may then be repaired to form an A-T pair in place of the G-C or alternatively the uracil may be excised and replaced with any of the four DNA bases. Although the precise mechanism of AID action is under active investigation, its importance to both somatic hypermutation and class-switch recombination is quite clear (Figure 5-17a).

In a dramatic experiment performed at Japan's Kyoto University, Tasuku Honjo and his colleagues generated mice with the gene for AID knocked out. The ability of these mice (AID –/–) to conduct class switching and somatic hypermutation was compared with that of mice retaining a functional copy of the AID gene (AID +/–). The data in Figure 5-17b shows that as an immune response developed following successive immunizations, mice with a functional copy of the AID gene developed first IgM and then IgG antibodies against the target antigen. In

the AID knockout mice, however, only IgM antibodies were generated in response to immunization with the same antigen. To test for somatic hypermutation, the investigators examined the messenger RNA that encoded the heavy-chain variable regions of antibodies against the antigen from both (AID –/–) and (AID +/–) mice. They discovered that whereas (AID +/–) mice had normal levels of hypermutation, the AID knockout mice showed no somatic hypermutation at all. This surprising result demonstrated that the same protein, AID, was essential for both somatic hypermutation and class switching. In subsequent work by these investigators, the AID gene was introduced into fibroblasts, known to perform neither somatic hypermutation nor class switching. Both functions became active when AID was expressed in these nonlymphoid cells. This highly significant finding demonstrated that AID is the only factor needed to carry out either of these signature B-cell processes.

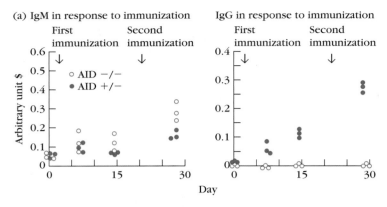

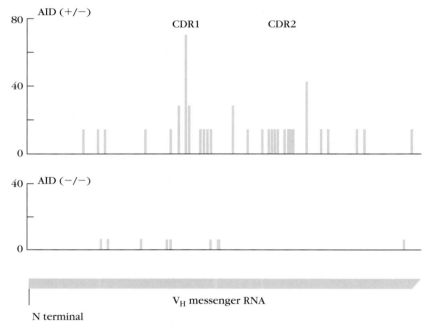

**FIGURE 5-17 Experimental demonstration of the role of the enzyme AID in class switching and somatic hypermutation.** (a) AID-expressing (+/–) and AID knockout (–/–) mice were immunized twice with a hapten-carrier conjugate and the antihapten antibody responses measured and plotted in arbitrary units. IgM responses were detected in both types of mice. Production of IgG, which requires class switching, occurred only in AID-expressing (+/–) mice. (b) Messenger RNA encoding the variable regions of antigen-reactive antibodies in immunized AID-expressing and AID knockout mice was sequenced and the position and frequency of mutations plotted. Many mutations are seen in the AID-expressing mice; only background levels of mutation are seen in the AID knockout mice. *[Adapted from M. Muramatsu et al., 2000, Cell* **102**:560.]

# Expression of Ig Genes

As is true for many genes, post-transcriptional processing of immunoglobulin primary transcripts is required to produce functional mRNAs (see Figures 5-4 and 5-5). The primary transcripts produced from rearranged heavy-chain and light-chain genes contain intervening DNA sequences that include noncoding introns and J gene segments not lost during V-(D)-J rearrangement. In addition, as noted earlier, the heavy-chain C-gene segments are organized as a series of coding exons and noncoding introns. Each exon of a $C_H$ gene segment corresponds to a constant-region domain or a hinge region of the heavy-chain polypeptide. The primary transcript must be processed to remove the intervening DNA sequences, and the remaining exons must be connected by a process called RNA splicing. Short, moderately conserved splice sequences, or splice sites, which are located at the intron-exon boundaries within a primary transcript, signal the positions at which splicing occurs. Processing of the primary transcript in the nucleus

removes each of these intervening sequences to yield the final mRNA product. The mRNA is then exported from the nucleus to be translated by ribosomes into complete H or L chains.

## Heavy-chain primary transcripts undergo differential RNA processing

Processing of an immunoglobulin heavy-chain primary transcript can yield several different mRNAs, which explains how a single B cell can produce secreted or membrane-bound forms of a particular immunoglobulin and simultaneously express IgM and IgD.

### Expression of Membrane or Secreted Immunoglobulin

As explained in Chapter 3, a particular immunoglobulin can exist in either membrane-bound or secreted form. The two forms differ in the amino acid sequence of the heavy-chain carboxyl-terminal domains ($C_H3/C_H3$ in IgA, IgD, and IgG and $C_H4/C_H4$ in IgE and IgM). The secreted form has a

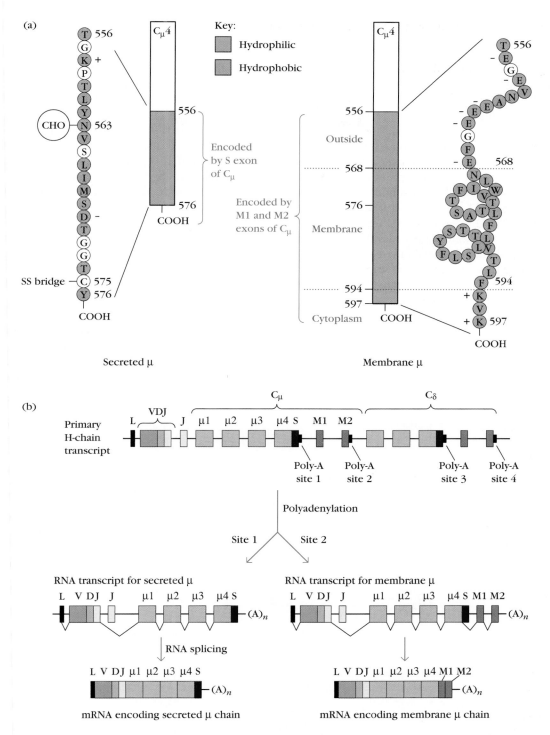

**FIGURE 5-18 Expression of secreted and membrane forms of the heavy chain by alternative RNA processing.** (a) Amino acid sequence of the carboxyl-terminal end of secreted and membrane μ heavy chains. Residues are indicated by the single-letter amino acid code. Hydrophilic and hydrophobic residues and regions are indicated by purple and orange, respectively, and charged amino acids are indicated with a + or −. The white regions of the sequences are identical in both forms. (b) Structure of the primary transcript of a rearranged heavy-chain gene showing the $C_\mu$ exons and poly-A sites. Polyadenylation of the primary transcript at either site 1 or site 2 and subsequent splicing (indicated by V-shaped lines) generates mRNAs encoding either secreted or membrane μ chains.

hydrophilic sequence of about 20 amino acids in the carboxyl-terminal domain; this is replaced in the membrane-bound form with a sequence of about 40 amino acids containing a hydrophilic segment that extends outside the

cell, a hydrophobic transmembrane segment, and a short hydrophilic segment at the carboxyl terminus that extends into the cytoplasm (Figure 5-18a). For some time, the existence of these two forms seemed inconsistent with the structure of germ-line heavy-chain DNA, which had been shown to contain a single $C_H$ gene segment corresponding to each class and subclass.

The resolution of this puzzle came from DNA sequencing of the $C_\mu$ gene segment, which consists of four exons ($C_\mu 1$, $C_\mu 2$, $C_\mu 3$, and $C_\mu 4$) corresponding to the four domains of the IgM molecule. The $C_\mu 4$ exon contains a nucleotide sequence (called S) at its 3′ end that encodes the hydrophilic sequence in the $C_H 4$ domain of secreted IgM. Two additional exons called M1 and M2 are located 1.8 kb downstream from the 3′ end of the $C_\mu 4$ exon. The M1 exon encodes the transmembrane segment, and M2 encodes the cytoplasmic segment of the $C_H 4$ domain in membrane-bound IgM. DNA sequencing revealed that all the $C_H$ gene segments have two additional downstream M1 and M2 exons that encode the transmembrane and cytoplasmic segments.

The primary transcript produced by transcription of a rearranged μ heavy-chain gene contains two polyadenylation signal sequences, or poly-A sites, in the $C_\mu$ segment. Site 1 is located at the 3′ end of the $C_\mu 4$ exon, and site 2 is at the 3′

end of the M2 exon (Figure 5-18b). If cleavage of the primary transcript and addition of the poly-A tail occurs at site 1, the M1 and M2 exons are lost. Excision of the introns and splicing of the remaining exons then produces mRNA encoding the secreted form of the heavy chain. If cleavage and polyadenylation of the primary transcript occurs instead at site 2, then a different pattern of splicing results. In this case, splicing removes the S sequence at the 3′ end of the $C_\mu 4$ exon, which encodes the hydrophilic carboxyl-terminal end of the secreted form and joins the remainder of the $C_\mu 4$ exon with the M1 and M2 exons, producing mRNA for the membrane form of the heavy chain.

Thus, differential processing of a common primary transcript determines whether the secreted or membrane form of an immunoglobulin will be produced. As noted previously, mature naive B cells produce only membrane-bound antibody, whereas differentiated plasma cells produce secreted antibodies. It remains to be determined precisely how naive B cells and plasma cells direct RNA processing preferentially toward the production of mRNA encoding one form or the other.

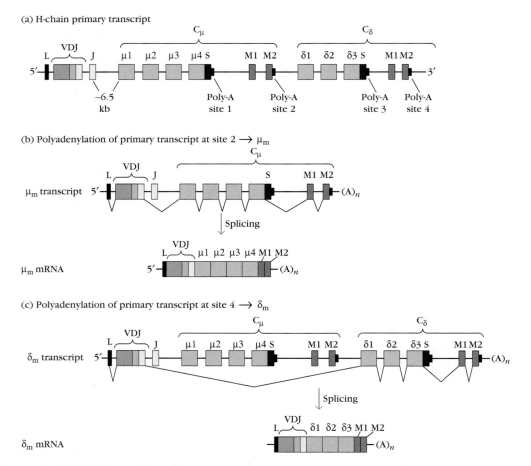

**FIGURE 5-19 Expression of membrane forms of μ and δ heavy chains by alternative RNA processing.** (a) Structure of rearranged heavy-chain gene showing $C_\mu$ and $C_\delta$ exons and poly-A sites. (b) Structure of $\mu_m$ transcript and $\mu_m$ mRNA resulting from polyadenyla-tion at site 2 and splicing. (c) Structure of $\delta_m$ transcript and $\delta_m$ mRNA resulting from polyadenylation at site 4 and splicing. Both processing pathways can proceed in any given B cell.

### Simultaneous Expression of IgM and IgD

Differential RNA processing also underlies the simultaneous expression of membrane-bound IgM and IgD by mature B cells. As mentioned already, transcription of rearranged heavy-chain genes in mature B cells produces primary transcripts containing both the $C_\mu$ and $C_\delta$ gene segments. The $C_\mu$ and $C_\delta$ gene segments are close together in the rearranged gene (only about 5 kb apart), and the lack of a switch site between them permits the entire $VDJC_\mu C_\delta$ region to be transcribed into a single primary RNA transcript about 15 kb long, which contains four poly-A sites (Figure 5-19). Sites 1 and 2 are associated with $C_\mu$, as described in the previous section; sites 3 and 4 are located at similar places in the $C_\delta$ gene segment. If the heavy-chain transcript is cleaved and polyadenylated at site 2 after the $C_\mu$ exons, then the mRNA will encode the membrane form of the $\mu$ heavy chain (Figure 5-19b); if polyadenylation is instead further downstream at site 4, after the $C_\delta$ exons, then RNA splicing will remove the intervening $C_\mu$ exons and produce mRNA encoding the membrane form of the $\delta$ heavy chain (Figure 5-19c).

Since the mature B cell expresses both IgM and IgD on its membrane, both processing pathways must occur simultaneously. Likewise, cleavage and polyadenylation of the primary heavy-chain transcript at poly-A site 1 or 3 in plasma cells and subsequent splicing will yield the secreted form of the $\mu$ or $\delta$ heavy chains, respectively (see Figure 5-18b).

## Synthesis, Assembly, and Secretion of Immunoglobulins

Antibody production poses unique problems related to the variability of the product and the amount that is produced. An extraordinarily diverse repertoire of antibodies is produced by the mechanisms of gene rearrangements, imprecise joining of V-(D)-J segments, addition of nucleotides to the cleaved ends of these segments, and, in many cases, somatic hypermutation, but an expensive byproduct of all this variability is the generation of a significant number of antibody genes that harbor premature stop codons or produce immunoglobulins that do not fold or assemble properly. A second problem arises from the scale of antibody production by plasma cells. In addition to the proteins required for all of their normal metabolic activities, plasma cells manufacture and secrete more than 1000 antibody molecules per cell per second. Simply moving the intracellular vesicles containing cargoes of antibody to the plasma membrane for discharge requires a remarkable feat of intracellular traffic control and coordination.

Like all proteins destined for incorporation into membranes or discharge to the extracellular environment, immunoglobulin polypeptides are synthesized in the rough endoplasmic reticulum (RER). Immunoglobulin heavy- and light-chain mRNAs are translated on separate polyribosomes of the RER (Figure 5-20). Newly synthesized chains contain an amino-terminal leader sequence, which serves to guide the chains into the lumen of the RER, where the signal sequence is then cleaved. The order of assembly of light (L) and heavy (H) chains varies among the immunoglobulin classes. In the case of IgM, the L and H chains assemble within the RER to form half-molecules, and then two half-molecules assemble to form the complete molecule. In the case of IgG, two H chains assemble, then an $H_2L$ intermediate is assembled, and finally the complete $H_2L_2$ molecule is formed. Enzymes present in the RER catalyze the formation of the interchain disulfide bonds necessary for assembly of immunoglobulin polypeptides, as well as the intrachain–S-S bonds that secure the folding of immunoglobulin domains. The glycosylation of antibody molecules is also carried out by enzymes of the RER.

Antibodies are directed to their destinations (either secreted or bound to the membrane) based on the presence of hydrophobic or hydrophilic carboxyl-terminal domains on their heavy chains, mentioned earlier. Membrane-bound antibodies contain the hydrophobic sequence, which becomes anchored in the membrane of a secretory vesicle. When the vesicle fuses with the plasma membrane, the antibody takes up residence in the plasma membrane (Figure 5-20, inset at top). Antibodies containing the hydrophilic sequence characteristic of secreted immunoglobulins are transported as free molecules in secretory vesicles and are released from the cell when these vesicles fuse with the plasma membrane.

The ER has quality control mechanisms (Figure 5-21), which ensure the export of only completely assembled antibody molecules and promote the destruction of incomplete or improperly folded Ig molecules. One mediator of these functions is BiP (immunoglobulin heavy chain *b*inding *p*rotein), which binds to incompletely assembled antibody molecules but dissociates from completely assembled ones. A specific sequence of amino acids in BiP (lysine-glutamate-aspartate-leucine) causes it to be retained in the endoplasmic reticulum, preventing incomplete antibody molecules to which it is bound from leaving the ER. Antibody molecules that are misfolded or retained in the ER eventually become labeled for degradation by attachment of a protein called ubiquitin. Proteins tagged with ubiquitin are translocated out of the nucleus and subjected to proteolysis by proteosomes, large multienzyme complexes that will be discussed more fully in Chapter 7.

## Regulation of Ig-Gene Transcription

Immunoglobulin genes are expressed only in B-lineage cells, and even within this lineage, the genes are expressed at different rates during different developmental stages. As with other eukaryotic genes, two major classes of cis regulatory sequences in DNA regulate transcription of immunoglobulin genes:

- **Promoters:** relatively short nucleotide sequences, extending about 200 bp upstream from the transcription

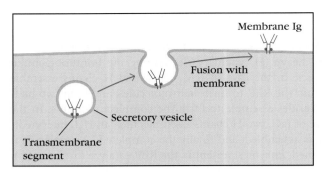

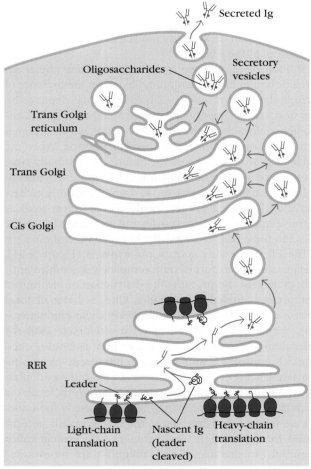

**FIGURE 5-20 Synthesis, assembly, and secretion of the immunoglobulin molecule.** The heavy and light chains are synthesized on separate polyribosomes (polysomes). The assembly of the chains, the formation of intrachain and interchain disulfide linkages, and the addition of carbohydrate all take place in the rough endoplasmic reticulum (RER). Vesicular transport brings the Ig to the Golgi, which it transits, departing in vesicles that fuse with the cell membrane. The main figure depicts the assembly of a secreted antibody. The inset depicts a membrane-bound antibody, which contains the carboxyl-terminal transmembrane segment. This form becomes anchored in the membrane of secretory vesicles and is retained in the cell membrane when the vesicles fuse with the cell membrane.

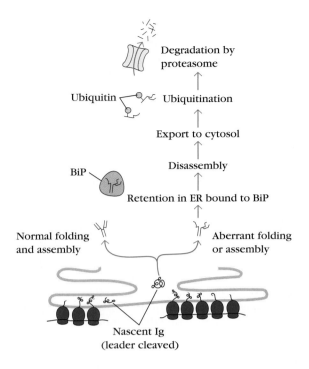

**FIGURE 5-21 Quality control during antibody synthesis.** Ig molecules that fail to fold or assemble properly remain bound to the chaperone protein BiP. Interaction with BiP and other factors causes the malformed antibody to be disassembled, exported from the ER, marked for degradation by conjugation with ubiquitin, and degraded by the proteosome.

initiation site, that promote initiation of RNA transcription in a specific direction

- **Enhancers:** nucleotide sequences situated some distance upstream or downstream from a gene that activate transcription from the promoter sequence in an orientation-independent manner

The locations of the regulatory elements in germ-line immunoglobulin DNA are shown in Figure 5-22. All of these regulatory elements have clusters of sequence motifs that can bind specifically to one or more nuclear proteins.

Each $V_H$ and $V_L$ gene segment has a promoter located just upstream from the leader sequence. In addition, the $J_\kappa$ cluster and each of the $D_H$ genes of the heavy-chain locus are preceded by promoters. Like other promoters, the immunoglobulin promoters contain a highly conserved AT-rich sequence called the TATA box, which serves as the binding site for a number of proteins necessary for the initiation of RNA transcription. The actual process of transcription is performed by RNA polymerase II, which starts transcribing DNA from the initiation site, located about 25 bp downstream of the TATA box. Ig promoters also contain an essential and conserved octamer that confers B-cell specificity on the promoter. The octamer binds two transcription factors, oct-1, found in many cell types, and oct-2, found only in B cells.

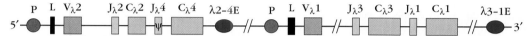

**FIGURE 5-22 Location of promoters (dark red) and enhancers (green) in mouse heavy-chain, κ light-chain, and λ light-chain germ-line DNA.** Variable-region DNA rearrangement moves an enhancer close enough to a promoter that the enhancer can activate transcription from the promoter. The promoters that precede the DH cluster, a number of the C genes, and the $J_\lambda$ cluster are omitted from this diagram for clarity.

While much remains to be learned about the function of enhancers, they offer binding sites for a number of proteins, many of which are transcription factors. A particularly important role is played by two proteins encoded by the *E2A* gene, which can undergo alternate splicing to generate two collaborating proteins, both of which bind to the μ and κ intronic enhancers. These proteins are essential for B-cell development; *E2A* knockout mice produce normal numbers of T cells but show a complete absence of B cells. Interestingly, transfection of these enhancer-binding proteins into a T-cell line resulted in a dramatic increase in the transcription of μ chain mRNA and even induced the T cell to undergo $D_H + J_H \rightarrow D_H J_H$ rearrangement.

One heavy-chain enhancer is located within the intron between the last (3′) J gene segment and the first (5′) C gene segment ($C_\mu$), which encodes the μ heavy chain. Because this heavy-chain enhancer ($E_\mu$) is located 5′ of the $S_\mu$ switch site near $C_\mu$, it can continue to function after class switching has occurred. Another heavy-chain enhancer ($3'_\alpha E$) has been detected 3′ of the $C_\alpha$ gene segment. One κ light-chain enhancer ($E_\kappa$) is located between the $J_\kappa$ segment and the $C_\kappa$ segment, and another enhancer ($3'_\kappa E$) is located 3′ of the $C_\kappa$ segment. The λ light-chain enhancers are located 3′ of $C_\lambda 4$ and 3′ of $C_\lambda 1$.

## DNA rearrangement greatly accelerates transcription

The promoters associated with the immunoglobulin V gene segments bind RNA polymerase II very weakly, and the variable-region enhancers in germ-line DNA are quite distant from the promoters (about 250–300 kb), too remote to significantly influence transcription. For this reason, the rate of transcription of $V_H$ and $V_L$ coding regions is negligible in unrearranged germ-line DNA. Variable-region gene rearrangement brings a promoter and enhancer within 2 kb of each other, close enough for the enhancer to influence transcription from the nearby promoter. As a result, the rate of transcription of a rearranged $V_L J_L$ or $V_H D_H J_H$ unit is as much as $10^4$ times the rate of transcription of unrearranged $V_L$ or $V_H$ segments. This effect was demonstrated directly in a study in which B cells transfected with rearranged heavy-chain genes from which the enhancer had been deleted did not transcribe the genes, whereas B cells transfected with similar genes that contained the enhancer transcribed the transfected genes at a high rate. These findings highlight the importance of enhancers in the normal transcription of immunoglobulin genes.

Genes that regulate cellular proliferation or inhibit apoptosis sometimes translocate to the immunoglobulin heavy- or light-chain loci. Here, under the influence of an immunoglobulin enhancer, the expression of these genes is significantly elevated, resulting in high levels of growth-promoting or apoptosis-inhibiting proteins. Translocations of the c-*myc* and *bcl*-2 oncogenes have each been associated with malignant B-cell lymphomas. The translocation of c-*myc* leads to constitutive expression of c-Myc and an aggressive, highly proliferative B-cell lymphoma called Burkitt's lymphoma. The translocation of *bcl*-2 leads to suspension of programmed cell death in B cells, resulting in follicular B-cell lymphoma. These cancer-promoting translocations are covered in greater detail in Chapter 21.

## Ig-gene expression is inhibited in T cells

As noted earlier, the site-specific mutation in B cells has a parallel in T cells. Germ-line DNA encoding the T-cell receptor (TCR) undergoes V-(D)-J rearrangement to generate functional TCR genes. Rearrangement of both immunoglobulin and TCR germ-line DNA occurs by similar recombination processes mediated by RAG-1 and RAG-2 and involving recombination signal sequences with one-turn or two-turn spacers (see Figure 5-7). Despite the similarity of the processes, complete Ig-gene rearrangement of H and L chains occurs only in B cells and complete TCR-gene rearrangement is limited to T cells.

Hitoshi Sakano and coworkers have obtained results suggesting that a sequence within the κ-chain 3′ enhancer (3′$_\kappa$E) serves to regulate the joining of V$_\kappa$ to J$_\kappa$ in B and T cells. When a sequence known as the PU.1 binding site within the 3′ κ-chain enhancer was mutated, these researchers found that V$_\kappa$-J$_\kappa$ joining occurred in T cells as well as B cells. They propose that binding of a protein expressed by T cells, but not B cells, to the unmutated κ-chain enhancer normally prevents V$_\kappa$-J$_\kappa$ joining in T cells. The identity of this DNA-binding protein in T cells remains to be determined. Similar processes may prevent rearrangement of heavy-chain and λ-chain DNA in T cells.

# Antibody Genes and Antibody Engineering

In many clinical applications, the exquisite specificity of a mouse monoclonal antibody would be useful. However, when mouse monoclonal antibodies are introduced into humans, they are recognized as foreign and evoke a *human antimouse antibody* (HAMA) response that quickly clears the mouse monoclonal antibody from the bloodstream and lowers the therapeutic efficiency of subsequent administration of the antibody. Circulating complexes of mouse and human antibodies can cause allergic reactions, and in some cases, the buildup of these complexes in organs such as the kidney can cause serious and even life-threatening reactions. Clearly, one way to avoid these undesirable reactions is to use human monoclonal antibodies for clinical applications. However, the use of hybridoma technology for the preparation of human monoclonal antibodies has been severely hampered by numerous technical problems, and the generation of hybridomas secreting human antibodies remains a major challenge. To overcome these difficulties, alternative methods of antibody engineering using recombinant DNA technology have been developed.

Knowledge of antibody gene sequences and antibody structure have been joined with techniques of molecular biology to make possible what Cesar Milstein, one of the inventors of monoclonal antibody technology, has called "man-made antibodies." It is now possible to design and construct genes that encode immunoglobulin molecules in which the variable regions come from one species, such as mouse, for which hybridomas with desired antigen specificity are readily generated, and the constant regions come from another, such as human. New genes have been created that link nucleotide sequences coding nonantibody proteins with sequences that encode antibody variable regions specific for particular antigens.

These molecular hybrids, or **chimeras,** can deliver powerful toxins to particular antigenic targets, such as tumor cells. By capturing a significant sample of all of the immunoglobulin heavy- and light-chain variable-region genes via incorporation into libraries of bacteriophages, it has been possible to achieve significant and useful reconstructions of the entire antibody repertoires of individuals. (The next few sections describe each of these types of antibody genetic engineering.) Finally, by replacement of the immunoglobulin loci of one species with that of another, animals of one species, such as mice or cattle, have been endowed with the capacity to respond to immunization by producing antibodies encoded by genetically transplanted Ig genes from humans.

## Chimeric and humanized monoclonal antibodies have potent clinical potential

One approach to engineering an antibody is to clone recombinant DNA containing the promoter, leader, and variable-region sequences from a mouse antibody gene and the constant-region exons from a human antibody gene (Figure 5-23). The antibody encoded by such a recombinant gene is a mouse-human chimera, commonly known as a **chimeric antibody.** Its antigenic specificity, which is determined by the variable region, is derived from mouse DNA; its isotype, which is determined by the constant region, is derived from human DNA. Because the constant regions of these chimeric antibodies are encoded by human genes, the antibodies have fewer mouse antigenic determinants and are far less immunogenic when administered to humans than mouse monoclonal antibodies (Figure 5-23a). One such chimeric human-mouse antibody has been used to treat patients with B-cell varieties of non-Hodgkin's lymphoma (see Clinical Focus, page 140). However, because the entire mouse V$_H$ and V$_L$ regions are retained in chimeric antibodies, they contain significant amounts of mouse Ig sequence, which can cause HAMA responses. Fortunately, it is possible to engineer antibodies in which all of the sequence is human except the CDRs, which are taken from mouse monoclonal antibody having the desired antigen specificity. These **humanized antibodies** are schematized in Figure 5-23b. Humanized antibodies retain the biological effector functions of human antibody and are more effective than mouse antibodies in triggering complement activation and Fc receptor-mediated processes, such as phagocytosis, in humans. However, they are less immunogenic in humans than chimeric mouse-human antibodies. Chimeric monoclonal antibodies can also be adapted to serve as immunotoxins. For example, the terminal constant-region domain in a tumor-specific antibody can be made highly radioactive by conjugation with a radioactive label such as yttrium-90, or it can be replaced or conjugated with a toxin such as diphtheria toxin (Figure 5-23c). These immunotoxins are selected for their ability to bind to tumor cells, making them highly specific as therapeutic reagents.

**Heteroconjugates,** or **bispecific antibodies,** are hybrids of two different antibody molecules (Figure 5-23d). They can be constructed by chemically cross-linking two different

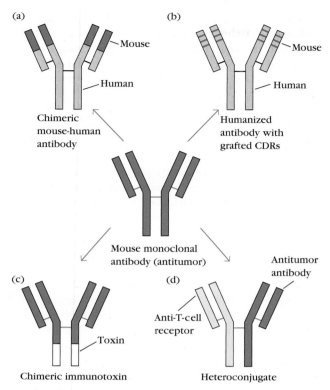

**FIGURE 5-23 Antibodies engineered by recombinant DNA technology.** (a) Chimeric mouse-human monoclonal antibody containing the $V_H$ and $V_L$ domains of a mouse monoclonal antibody (blue) and $C_L$ and $C_H$ domains of a human monoclonal antibody (gray). (b) A humanized monoclonal antibody containing only the CDRs of a mouse monoclonal antibody (blue bands) grafted into the framework regions of a human monoclonal antibody. (c) A chimeric monoclonal antibody in which the terminal Fc domain is replaced by toxin chains (white). (d) A heteroconjugate in which one half of the mouse antibody molecule is specific for a tumor antigen and the other half is specific for the CD3/T-cell receptor complex.

antibodies or by synthesizing them in hybridomas consisting of two different monoclonal-antibody-producing cell lines that have been fused. Both of these methods generate mixtures of monospecific and bispecific antibodies from which the desired bispecific molecule must be purified. A simpler and more elegant approach is to use genetic engineering to construct genes that encode molecules only with the two desired specificities. Several bispecific molecules have been designed in which one half of the antibody has specificity for a tumor cell and the other half has specificity for a surface molecule on an immune effector cell, such as an NK cell or a cytotoxic T lymphocyte (CTL), thereby promoting destruction of the tumor cell by the effector cell.

## Mice have been engineered with human immunoglobulin loci

Using knockout technology, it is possible to disable the capacity of a mouse to rearrange its germ-line heavy- and light-chain loci. These knockout mice make no mouse Ig, and because the development of B cells requires the generation of functional μ immunoglobulin, the mice make no B cells either. However, the ability to generate B cells and to make antibody can be restored by the introduction of functional mouse germ-line Ig loci into lines of Ig knockout mice. Some resourceful investigators have shown that the introduction of portions of human H and L germ-line loci can restore B-cell development as well and yield mice that produce human antibodies. However, complete germ-line Ig loci are very large stretches of DNA (the human H locus is 1.5 Mb and the human λ is 1 Mb), and technical difficulties prevented the introduction of the entire human H and L loci into mice. Only partial segments of the loci could be introduced, and consequently, not all of the elements of these large and complex loci may have been expressed. A powerful new approach to this problem emerged when Isao Ishida and his colleagues at Kirin Pharmaceutical in Japan generated a human artificial chromosome (HAC) bearing the entire human heavy-chain and λ-chain loci. They then introduced the HAC into embryonic stem cells derived from mouse H- and L-chain knockout mice and used the HAC-transgenic cells to derive lines of transgenic mice that respond to antigenic challenge by producing antigen-specific *human* antibodies (Figure 5-24). Immunizing these HAC-transgenic mice with human serum albumin induced a response consisting of human antihuman serum albumin antibodies.

One of the strengths of this method, as well as other approaches to transplanting human germ-line Ig loci into mice, is that these completely human antibodies are made in cells of the mouse B-cell lineage, from which antibody-secreting hybridomas are readily derived by cell fusion. This approach thus offers a solution to the problem of producing human monoclonal antibodies of any specificity desired. A further and unique advantage of the HAC approach is the facility with which it can be used to transfer entire human immunoglobulin loci into species other than mice. Recently, cloned HAC-transgenic cattle have been shown to produce human polyclonal antibodies. Because cattle have large blood volumes (~60 liters), harvest of their serum offers the potential to produce large amounts of human polyclonal antibodies. As discussed in the Clinical Focus of Chapter 4, human polyclonal antibodies have an important and growing role in the prevention and treatment of disease, and strategies to produce them in nonhuman hosts, economically and on a large scale, are of considerable biomedical interest.

## Phage display libraries allow the derivation of monoclonal antibodies without immunization

Another approach for generating monoclonal antibodies bypasses hybridoma technology completely. Beginning with a population of B cells, gene segments for the rearranged $V_H$ and $V_L$ domains of the B-cell antibodies are isolated. The result is a large and diverse library of $V_H$ and $V_L$ genes that

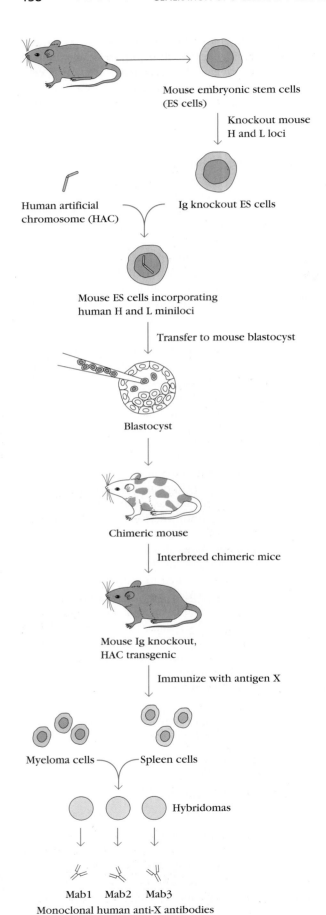

Mouse embryonic stem cells
(ES cells)

Knockout mouse
H and L loci

Ig knockout ES cells

Human artificial
chromosome (HAC)

Mouse ES cells incorporating
human H and L miniloci

Transfer to mouse blastocyst

Blastocyst

Chimeric mouse

Interbreed chimeric mice

Mouse Ig knockout,
HAC transgenic

Immunize with antigen X

Myeloma cells — Spleen cells

Hybridomas

Mab1   Mab2   Mab3
Monoclonal human anti-X antibodies

**FIGURE 5-24 Human antibody from mice bearing a human artificial chromosome (HAC) that includes entire human heavy- and light-chain loci.** A human artificial chromosome bearing the entire unrearranged human heavy-chain and λ light-chain loci was introduced into mouse embryonic stem (ES) cells in which the mouse heavy-chain and κ and λ light-chain loci had been knocked out. The modified ES cells were introduced into blastocysts, which were transferred to surrogate mothers and allowed to generate chimeric mice. Interbreeding of the chimeric mice produced mice that made human antibodies but not mouse Ig. Immunization of these mice allowed the generation of antigen-specific antiserum that contained only human antibodies or the generation of hybridomas that secreted antigen-specific human monoclonal antibodies.

represent the repertoire of rearranged V genes present in the source B-cell population. These $V_H$ and $V_L$ genes are then linked together, along with an intervening peptide linker sequence, to generate a collection of genes encoding polypeptides known as single-chain fragment variable (scFv) units (Figure 5-25a). In each scFv unit, a $V_H$ domain pairs with a $V_L$ to create a minimal antigen-binding unit. When scFvs are constructed for the creation of a phage library, any $V_L$ gene can join with any $V_H$ gene, generating an enormous diversity of antibody specificities; phage libraries containing more than $10^{11}$ unique members have been obtained. This level of diversity is comparable to the in vivo human repertoire, and these highly diverse libraries contain antibody binding sites specific for an exceedingly broad range of antigens.

The power of a large library of scFv unit genes is harnessed in a protocol called **phage display technology** (Figure 5-25b). Phages (short for bacteriophages) are bacterial viruses consisting of a genome and protein coat. Genetic engineering is used to join genes encoding scFv units with the gene that encodes one of the coat proteins expressed on the phage surface. The genetically transformed phage then produces altered coat proteins that display the scFv unit ($V_H$ and $V_L$ domains linked by a peptide) on the phage surface. Very large repertoires ($10^6$–$10^9$ different antibody binding sites) of scFv genes can be captured by cloning into phages, producing a phage display library.

Phage populations are amplified by introducing phage into *E. coli*. The progeny phage can then be screened for specificity by incubating the library with an immobilized antigen. Phages that do not bind antigen are washed away, leaving behind those that express scFv units that bind specifically to the antigen, along with some phage that bind to the plate nonspecifically. To further enrich for antigen-specific phage, infection of *E. coli* is repeated with recovered phages, followed by further rounds of selection. By repetitions of growth, antigen selection, and harvest, it is possible to isolate as few as 1 in $10^9$ antigen-specific phages.

Once phages bearing desirable binding sites are identified, genetic engineering techniques can be used to recover the $V_H$ and $V_L$ encoding regions from the phage genome, which are then grafted onto the genes encoding the immunoglobulin heavy- and light-chain constant regions. When the engineered heavy- and light-chain genes are expressed in cells, a monoclonal antibody specific for the desired antigen is produced. Thus, it is possible to completely bypass immunization and produce a useful antibody without resort to hybridoma technology. Furthermore, if the phage library is produced using human B cells as a source of $V_H$ and $V_L$ and these fragments are grafted to human H and L constant regions, the antibody produced is fully human. Indeed, this technology has been successfully used to generate a human monoclonal antibody against the inflammatory cytokine TNF-$\alpha$. This antibody was approved by the FDA and is now in clinical use for the treatment of rheumatoid arthritis.

(a)

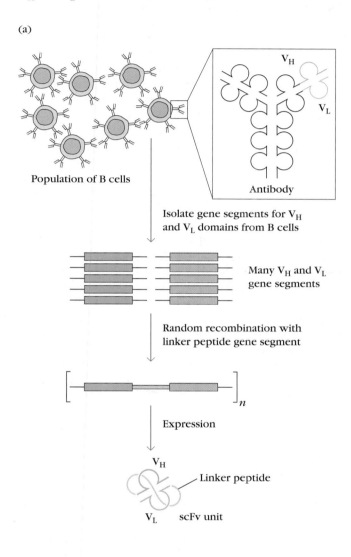

(b)

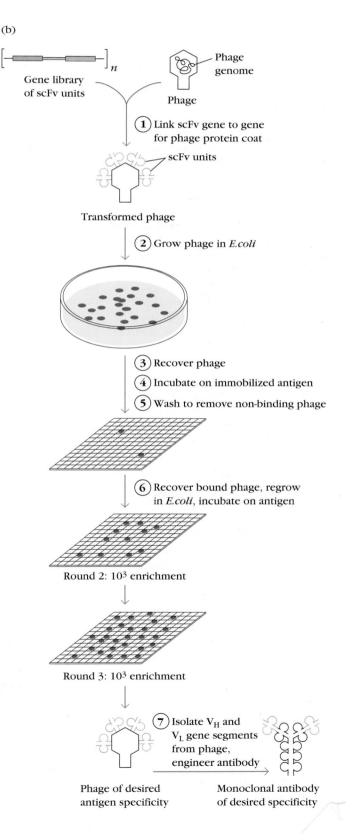

**FIGURE 5-25 Generation of phage libraries containing antibody binding sites.** (a) Derivation of scFv library. (b) The scFv library is cloned into phages that are used to infect *E. coli*. Growth of the phage-infected bacteria generates a phage library that is screened on antigen-coated plates for the presence of phage that bind to the desired antigen. Repeated cycles of binding-elution-regrowth results in the enrichment of the phage isolates for antigen specificity. Clones of antigen-specific phage can be isolated and the techniques of recombinant DNA technology used to graft the genes for $V_H$ and $V_L$ domains from selected phage onto the constant-region scaffolding of an antibody to engineer a monoclonal antibody of the desired antigen specificity. *[Adapted from J. Marks, 2004, Monoclonal antibodies from display libraries, in* Molecular Biology of B Cells, *T. Honjo, F. Alt, and M. Neuberger, eds., Elsevier, Amsterdam.]*

## CLINICAL FOCUS

# Therapy for Non-Hodgkin's Lymphoma and Other Diseases with Genetically Engineered Antibodies

**Lymphomas** are cancers of lymphatic tissue in which the tumor cells are of lymphocytic origin. There are two major forms of lymphoma: Hodgkin's lymphoma and non-Hodgkin's lymphoma. The less common form is Hodgkin's lymphoma, named for its discoverer, Thomas Hodgkin, an English physician. This unusually gifted early pathologist, who worked without the benefit of a microscope, recognized this condition in several patients and first described the anatomical features of the disease in 1832. Because many tissue specimens taken from patients Hodgkin suspected of harboring the disease were saved in the Gordon Museum of Guy's Hospital in London, it has been possible for later generations to judge the accuracy of his diagnoses. Hodgkin has fared well. Studies of these preserved tissues confirm that he was right in about 60% of the cases, a surprising achievement, considering the technology of the time. Actually, most lymphoma is non-Hodgkin's type and includes about 10 different types of disease. B-cell lymphomas are an important fraction of them.

For some years now, the major therapies directed against lymphomas have been radiation, chemotherapy, or a combination of both. Although these therapies benefit large numbers of patients by increasing survival, relapses after treatment are common, and many treated patients experience debilitating side effects. The side effects are an expected consequence of these therapies, because the agents used kill or severely damage a broad spectrum of normal cells as well as tumor cells. One of the holy grails of cancer treatment is the discovery of therapies that will affect only the tumor cells and completely spare normal cells. If particular types of cancer cells had antigens that were tumor specific, these antigens would be ideal targets for immune attack. Unfortunately, few such molecules are known.

However, a number of antigens are known that are restricted to the cell lineage in which the tumor originated and are expressed on the tumor cells.

Many cell-lineage-specific antigens have been identified for B lymphocytes and B lymphomas, including immunoglobulin, the hallmark of the B cell, and CD20, a membrane-bound phosphoprotein. CD20 has emerged as an attractive candidate for antibody-mediated immunotherapy because it is present on B lymphomas, and antibody-mediated cross-linking does not cause it to down-regulate or internalize. Indeed, some years ago, mouse monoclonal antibodies were raised against CD20, and one of these has formed the basis for an anti-B-cell lymphoma immunotherapy. This approach appears ready to take its place as an adjunct or alternative to radiation and chemotherapy. The development of this antitumor antibody is an excellent case study of the combined application of immunological insights and molecular biology to engineer a novel therapeutic agent.

The original anti-CD20 antibody was a mouse monoclonal antibody with murine $\gamma$ heavy chains and $\kappa$ light chains. The DNA sequences of the light- and heavy-chain variable regions of this antibody were amplified by polymerase chain reaction (PCR). Then a chimeric gene was created by replacing the CDR gene sequences of a human $\gamma$1 heavy chain with those from the murine heavy chain. In a similar maneuver, CDRs from the mouse $\kappa$ were ligated into a human $\kappa$ gene. The chimeric genes thus created were incorporated into vectors that permitted high levels of expression in mammalian cells. When an appropriate cell line was cotransfected with both of these constructs, it produced chimeric antibodies containing CDRs of mouse origin together with human variable-

region frameworks and constant regions. After purification, the biological activity of the antibody was evaluated, first in vitro and then in a primate animal model. The initial results were quite promising. The grafted human constant region supported effector functions such as the complement-mediated lysis or antibody-dependent cell-mediated cytotoxicity (ADCC) of human B lymphoid cells. Furthermore, weekly injections of the antibody into monkeys resulted in the rapid and sustained depletion of B cells from peripheral blood, lymph nodes, and even bone marrow. When the anti-CD20 antibody infusions were stopped, the differentiation of new B cells from progenitor populations allowed B-cell populations eventually to recover and approach normal levels. From these results, the hope grew that this immunologically active chimeric antibody could be used to clear entire B-cell populations, including B lymphoma cells, from the body in a way that spared other cell populations. This led to the trial of the antibody in human patients.

The human trials enrolled patients with B-cell lymphoma who had a relapse after chemotherapy or radiation treatment. These trials addressed three important issues: efficacy, safety, and immunogenicity. Although not all patients responded to treatment with anti-CD20, close to 50% exhibited full or partial remission. Thus, efficacy was demonstrated, because this level of response is comparable to the success rate with traditional approaches that employ highly cytotoxic drugs or radiation—it offers a truly alternative therapy. Side effects such as nausea, low blood pressure, and shortness of breath were seen in some patients (usually during or shortly after the initiation of therapy); these were, for the most part, not serious or life threatening. Consequently, treatment with the chimeric anti-CD20 appears safe. Patients who received the antibody have been observed closely for the appearance of human antimouse-Ig antibodies (HAMA) and for human anti-chimeric-antibody (HACA) responses. Such responses were not observed. Therefore, the antibody was not immunogenic. The absence of such responses demonstrates that antibodies can be genetically engineered to minimize, or even avoid, untoward immune reactions. Another reason for humanizing mouse antibodies is the very short half-life (a few hours) of mouse IgG antibodies in humans compared with the three-week half-lives of their human or humanized counterparts.

Antibody engineering has also contributed to the therapy of other malignancies, such as breast cancer, which is diagnosed in more than 180,000 American women each year. A little more than a quarter of all breast cancer patients have cancers that overexpress a growth factor receptor called HER2 (human epidermal growth factor receptor 2). Many tumors that overexpress HER2 grow faster and pose a more serious threat than those with normal levels of this protein on their sur-

face. A chimeric anti-HER2 monoclonal antibody in which all of the protein except the CDRs are of human origin was created by genetic engineering. Specifically, the DNA sequences for the heavy-chain and light-chain CDRs were taken from cloned mouse genes encoding an anti-HER2 monoclonal antibody. As in the anti-CD20 strategy described above, each of the mouse CDR gene segments were used to replace the corresponding human CDR gene segments in human genes encoding the human $IgG_1$ heavy chain and the human κ light chain. When this engineered antibody is used in combination with a chemotherapeutic drug, it is highly effective against metastatic breast cancer. The effects on patients who were given only a chemotherapeutic drug were compared with those for patients receiving both the chemotherapeutic drug and the

engineered anti-HER2 antibody. The combination anti-HER2/chemotherapy treatment showed significantly reduced rates of tumor progression, a higher percentage of responding patients, and a higher one-year survival rate. Treatment with Herceptin, as this engineered monoclonal antibody is called, has become part of the standard repertoire of breast cancer therapies.

The development of engineered and conventional monoclonal antibodies is one of the most active areas in the pharmaceutical industry. The table provides a partial compilation of monoclonal antibodies that have received approval from the Food and Drug Administration (FDA) for use in the treatment of human disease. Many more are in various stages of development and testing.

## Some monoclonal antibodies in clinical use

| Monoclonal antibody [mAB] (product name) | Nature of antibody | Target (antibody specificity) | Treatment for |
|---|---|---|---|
| Muromonab-CD3 (Orthoclone OKT3) | Mouse mAB | T cells (CD3, a T-cell antigen) | Acute rejection of liver, heart, and kidney transplants |
| Abciximab (ReoPro) | Human-mouse chimeric | Clotting receptor of platelets (GP IIb/IIIa) | Blood clotting during angioplasty and other cardiac procedures |
| Daclizumab (Zenapax) | Humanized mAB | Activated T cells (IL-2 receptor alpha subunit) | Acute rejection of kidney transplants |
| Inflixibmab (Remicade) | Human-mouse chimeric | Tumor necrosis factor (TNF), a mediator of inflammation (TNF) | Rheumatoid arthritis and Crohn's disease |
| Palivizumab (Synagis) | Humanized mAB | Respiratory syncytial virus (RSV) (F protein, a component of RSV) | RSV infection in children, particularly infants |
| Gemtuzumab (Mylotarg) | Humanized mAB | Many cells of the myeloid lineage (CD33, an adhesion molecule) | Acute myeloid leukemia (AML) |
| Alemtuzumab (Campath) | Humanized mAB | Many types of leukocytes (CD52 a cell surface antigen) | B-cell chronic lymphocytic leukemia |
| Trastuzumab (Herceptin) | Humanized mAB | An epidermal growth factor receptor (HER2 receptor) | HER2-receptor-positive advanced breast cancers |
| Rituximab (Rituxan) | Human-mouse chimeric | B cells (CD20, a B-cell surface antigen) | Relapsed or refractory non-Hodgkins lymphoma |
| Ibritumomab (Zevalin) | Mouse mAB | B cells (CD20, a B-cell surface antigen) | Relapsed or refractory non-Hodgkins lymphoma |

SOURCE: Adapted from P. Carter. 2001, Improving the efficacy of antibody-based cancer therapies, *Nature Reviews/Cancer* **1**:118.

## SUMMARY

- Immunoglobulin κ and λ light chains and heavy chains are encoded by three separate multigene families, each containing numerous gene segments and located on different chromosomes.

- Functional light-chain and heavy-chain genes are generated by random rearrangement of the variable-region gene segments in germ-line DNA.

- V(D)J joining is catalyzed by the recombinase activating genes, *RAG-1* and *RAG-2*, and the participation of other enzymes and proteins. The joining of segments is directed by recombination signal sequences (RSSs), conserved DNA sequences that flank each V, D, and J gene segment.

- Each recombination signal sequence contains a conserved heptamer sequence, a conserved nonamer sequence, and either a 12-bp (one-turn) or 23-bp (two-turn) spacer. During rearrangement, gene segments flanked by a one-turn spacer join only to segments flanked by a two-turn spacer, ensuring proper $V_L$-$J_L$ and $V_H$-$D_H$-$J_H$ joining.

- Immunoglobulin gene rearrangements occur in sequential order: heavy-chain rearrangements first, followed by light-chain rearrangements. Allelic exclusion is a consequence of the functional rearrangement of the immunoglobulin DNA of only one parental chromosome and is necessary to ensure that a mature B cell expresses immunoglobulin with a single antigenic specificity.

- The major sources of antibody diversity, which can generate greater than $10^{10}$ possible antibody combining sites, are random joining of multiple V, J, and D germ-line gene segments; random association of heavy and light chains; junctional flexibility; P-addition; N-addition; and somatic mutation. However, in some species, such as birds and ruminants, somatic diversification processes are major generators of primary repertoire diversity.

- Following immunization, somatic hypermutation is a major factor in generating antibodies with increased affinity for the immunizing antigen.

- After antigenic stimulation of mature B cells, class switching results in the expression of different classes of antibody (IgG, IgA, and IgE) with the same antigenic specificity.

- Activation-induced cytidine deaminase (AID) is essential for somatic hypermutation, gene conversion, and class switching.

- Differential RNA processing of the immunoglobulin heavy-chain primary transcript generates membrane-bound antibody in mature B cells, secreted antibody in plasma cells, and the simultaneous expression of IgM and IgD by mature B cells.

- Transcription of immunoglobulin genes is regulated by at least two types of DNA regulatory sequences: promoters and enhancers.

- Growing knowledge of the molecular biology of immunoglobulin genes has made it possible to engineer antibodies for research and therapy. The approaches include chimeric antibodies, phage-based libraries of Ig genes, and the introduction of complete human immunoglobulin loci into mice and cattle to generate human antibodies.

## References

Dreyer, W. J., and J. C. Bennett. 1965. The molecular basis of antibody formation. *Proceedings of the National Academy of Sciences U.S.A.* **54**:864.

Eichmann, K. 2005. *Köhler's Invention*. Birkhäuser, Boston.

Flajnik, M. F. 2002. Comparative analyses of immunoglobulin genes: Surprises and portents. *Nature Reviews Immunology* **2**:688.

Fugmann, S. D., et al. 2000. The RAG proteins and V(D)J recombination: Complexes, ends and transposition. *Annual Review of Immunology* **18**:495.

Hozumi, N., and S. Tonegawa. 1976. Evidence for somatic rearrangement of immunoglobulin genes coding for variable and constant regions. *Proceedings of the National Academy of Sciences U.S.A.* **73**:3628.

Hudson, P. J., and C. Souriau. 2003. Engineered antibodies. *Nature Medicine* **9**:129.

Maizels, N., and M. D. Scharff. 2004. Molecular mechanism of hypermutation. In *Molecular Biology of B Cells*. T. Honjo, F. Alt, and M. Neuberger, eds. Elsevier, Amsterdam.

Maloney, D. G., et al. 1997. IDEC-C2B8 (Rituximab) anti-CD20 monoclonal antibody therapy in patients with relapsed low-grade non-Hodgkin's lymphoma. *Blood* **90**:2188.

Marks, J. D. 2004. Monoclonal antibodies from display libraries. In *Molecular Biology of B Cells*. T. Honjo, F. W. Alt, and M. S. Neuberger, eds. Elsevier, Amsterdam.

Matsuda, F., et al. 1998. The complete nucleotide sequence of the human immunoglobulin heavy chain variable region locus. *Journal of Experimental Medicine* **188**:2151.

Max, E. E. 2004. Immunoglobulins: molecular genetics. In *Fundamental Immunology*, 5th ed., W. E. Paul, ed. Lippincott-Raven, Philadelphia.

Mostoslavsky, R., F. W. Alt, and K. Rajewsky. 2004. The lingering enigma of the allelic exclusion mechanism. *Cell* **118**:539.

Muramatsu, M., et al. 2000. Class switch recombination and hypermutation require activation-induced cytidine deaminase (AID), a potential RNA editing enzyme. *Cell* **102**:553.

Oettinger, M. A., et al. 1990. *RAG-1* and *RAG-2*, adjacent genes that synergistically activate V(D)J recombination. *Science* **248**:1517.

Tonegawa, S. 1983. Somatic generation of antibody diversity. *Nature* **302**:575.

 ## Useful Web Sites

**http://www.clinicaltrials.gov/**

This site furnishes updated information about federally and privately supported clinical trials using human volunteers. It gives information about the purpose, guidelines for recruitment of participants, locations, and phone numbers for more details. Detailed information on scores of clinical trials involving monoclonal antibodies can be found at this site.

## http://www.google.com

Select Google Groups, a free online community and discussion group service that offers a comprehensive archive of Usenet postings on many topics. At the Google Group site, typing in *antibody* or *phage display* delivers many sites with useful information on antibody engineering.

## http://imgt.cines.fr/

The IMGT site contains a collection of databases of genes relevant to the immune system. The IMGT/LIGM database houses sequences belonging to the immunoglobulin superfamily and of T-cell antigen receptor sequences.

 ## Study Questions

**CLINICAL FOCUS QUESTION** The Clinical Focus section includes a table of monoclonal antibodies approved for clinical use. Two, Rituxan and Zevalin, are used for the treatment of non-Hodgkin's lymphoma. Both target CD20, a B-cell surface antigen. Zevalin is chemically modified by attachment of radioactive isotopes (yttrium-90, a $\beta$ emitter, or indium-111, a high-energy $\gamma$ emitter) that lethally irradiate cells to which the monoclonal antibody binds. Early experiments found that Zevalin without a radioactive isotope attached was an ineffective therapeutic agent, whereas unlabeled Rituxan, a humanized mAB, was effective. Furthermore, Rituxan with a radioactive isotope attached was too toxic; Zevalin bearing the same isotope in equivalent amounts was far less toxic. Explain these results. (Hint: The longer a radioactive isotope stays in the body, the greater the dose of radiation absorbed by the body.)

1. Indicate whether each of the following statements is true or false. If you think a statement is false, explain why.

   a. $V_\kappa$ gene segments sometimes join to $C_\lambda$ gene segments.
   b. With the exception of a switch to IgD, immunoglobulin class switching is mediated by DNA rearrangements.
   c. Separate exons encode the transmembrane portion of each membrane immunoglobulin.
   d. Although each B cell carries two alleles encoding the immunoglobulin heavy and light chains, only one allele is expressed.
   e. Primary transcripts are processed into functional mRNA by removal of introns, capping, and addition of a poly-A tail.
   f. The primary transcript is an RNA complement of the coding strand of the DNA and includes both introns and exons.

2. Explain why a $V_H$ segment cannot join directly with a $J_H$ segment in heavy-chain gene rearrangement.

3. Considering only combinatorial joining of gene segments and association of light and heavy chains, how many different antibody molecules potentially could be generated from germ-line DNA containing 500 $V_L$ and 4 $J_L$ gene segments and 300 $V_H$, 15 $D_H$, and 4 $J_H$ gene segments?

4. For each incomplete statement below (a–g), select the phrase(s) that correctly completes the statement. More than one choice may be correct.

   a. Recombination of immunoglobulin gene segments serves to
      (1) promote Ig diversification
      (2) assemble a complete Ig coding sequence0
      (3) allow changes in coding information during B-cell maturation
      (4) increase the affinity of immunoglobulin for antibody
      (5) all of the above
   b. Somatic mutation of immunoglobulin genes accounts for
      (1) allelic exclusion
      (2) class switching from IgM to IgG
      (3) affinity maturation
      (4) all of the above
      (5) none of the above
   c. The frequency of somatic mutation in Ig genes is greatest during
      (1) differentiation of pre–B cells into mature B cells
      (2) differentiation of pre–T cells into mature T cells
      (3) generation of memory B cells
      (4) antibody secretion by plasma cells
      (5) none of the above
   d. Kappa and lambda light-chain genes
      (1) are located on the same chromosome
      (2) associate with only one type of heavy chain
      (3) can be expressed by the same B cell
      (4) all of the above
      (5) none of the above
   e. Generation of combinatorial diversity among immunoglobulins involves
      (1) mRNA splicing
      (2) DNA rearrangement
      (3) recombination signal sequences
      (4) one-turn/two-turn joining rule
      (5) switch sites
   f. A B cell becomes immunocompetent
      (1) following productive rearrangement of variable-region heavy-chain gene segments in germ-line DNA
      (2) following productive rearrangement of variable-region heavy-chain and light-chain gene segments in germ-line DNA
      (3) following class switching
      (4) during affinity maturation
      (5) following binding of $T_H$ cytokines to their receptors on the B cell
   g. The mechanism that permits immunoglobulins to be synthesized in either a membrane-bound or secreted form is
      (1) allelic exclusion
      (2) codominant expression
      (3) class switching
      (4) the one-turn/two-turn joining rule
      (5) differential RNA processing

5. What mechanisms generate the three hypervariable regions (complementarity-determining regions) of immunoglobulin heavy and light chains? Why is the third hypervariable region (CDR3) more variable than the other two (CDR1 and CDR2)?

6. You have been given a cloned myeloma cell line that secretes IgG with the molecular formula $\gamma_2\lambda_2$. Both the heavy and light chains in this cell line are encoded by genes derived from allele 1. Indicate the form(s) in which each of the genes listed below would occur in this cell line using the following symbols: G = germ-line form; R = productively rearranged form; NP = nonproductively rearranged form. State the reason for your choice in each case.

   a. Heavy-chain allele 1
   b. Heavy-chain allele 2
   c. κ-chain allele 1
   d. κ-chain allele 2
   e. λ-chain allele 1
   f. λ-chain allele 2

7. You have a B-cell lymphoma that has made nonproductive rearrangements for both heavy-chain alleles. What is the arrangement of its light-chain DNA? Why?

8. Indicate whether each of the class switches indicated below can occur (yes) or cannot occur (no).

   a. IgM to IgD
   b. IgM to IgA
   c. IgE to IgG
   d. IgA to IgG
   e. IgM to IgG

9. Describe one advantage and one disadvantage of N-nucleotide addition during the rearrangement of immunoglobulin heavy-chain gene segments.

10. X-ray crystallographic analyses of many antibody molecules bound to their respective antigens have revealed that the CDR3 of both the heavy and light chains make contact with the epitope. Moreover, sequence analyses reveal that the variability of CDR3 is greater than that of either CDR1 or CDR2. What mechanisms account for the greater diversity in CDR3?

11. How many chances does a developing B cell have to generate a functional immunoglobulin light-chain gene?

12. Match the terms below (a–h) to the description(s) that follow (1–11). Each description may be used once, more than once, or not at all; more than one description may apply to some terms.

   Terms
   a. _____ RAG-1 and RAG-2
   b. _____ Double-strand break repair (DSBR) enzymes
   c. _____ Coding joints
   d. _____ RSSs
   e. _____ P-nucleotides
   f. _____ N-nucleotides
   g. _____ Promoters
   h. _____ Enhancers

   Descriptions
   (1) Junctions between immunoglobulin gene segments formed during rearrangement
   (2) Source of diversity in antibody heavy chains
   (3) DNA regulatory sequences
   (4) Conserved DNA sequences, located adjacent to V, D, and J segments, that help direct gene rearrangement
   (5) Enzymes expressed in developing B cells
   (6) Enzymes expressed in mature B cells
   (7) Nucleotide sequences located close to each leader segment in immunoglobulin genes to which RNA polymerase binds
   (8) Product of endonuclease cleavage of hairpin intermediates in Ig-gene rearrangement
   (9) Enzymes that are defective in SCID mice
   (10) Nucleotide sequences that greatly increase the rate of transcription of rearranged immunoglobulin genes compared with germ-line DNA
   (11) Nucleotides added by TdT enzyme

13. Many B-cell lymphomas express surface immunoglobulin on their plasma membranes. It is possible to isolate this lymphoma antibody and make a high-affinity, highly specific mouse monoclonal anti-idiotype antibody against it. What steps should be taken to make this mouse monoclonal antibody most suitable for use in a patient? Is it highly likely that, once made, such an engineered antibody will be generally useful for lymphoma patients?

14. At which level, DNA or RNA, are each of the following activities determined?

   a. Membrane vs. secreted antibody
   b. IgM vs. IgD
   c. V-J joining
   d. IgA vs. IgE
   e. V-D-J joining

15. The tendency to become allergic can be inherited, based on a genetic predisposition favoring the production of IgE antibodies. However, the antigen that an individual is allergic to (the allergen) is not inherited. Knowing how antibody specificity is produced, explain this apparent contradiction.

 **Interactive Study**

www.whfreeman.com/kuby

SELF-TEST
Review of Key Terms

# chapter 6

# Antigen-Antibody Interactions: Principles and Applications

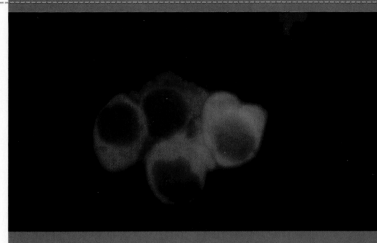

*Fluorescent antibody staining reveals intracellular immunoglobulin. [H. A. Schreuder et al., 1997, Nature 386:196; courtesy H. Schreuder, Hoechst Marion Roussel.]*

- Strength of Antigen-Antibody Interactions
- Cross-Reactivity
- Surface Plasmon Resonance (SPR)
- Precipitation Reactions
- Agglutination Reactions
- Radioimmunoassay
- Enzyme-Linked Immunosorbent Assay
- Western Blotting
- Immunoprecipitation
- Immunofluorescence
- Flow Cytometry and Fluorescence
- Alternatives to Antigen-Antibody Reactions
- Immunoelectron Microscopy

T HE ANTIGEN-ANTIBODY INTERACTION IS A BIOMOLECULAR association similar to an enzyme-substrate interaction, with an important distinction: it does not lead to an irreversible chemical alteration in either the antibody or the antigen. The association between an antibody and an antigen involves various noncovalent interactions between the antigenic determinant, or epitope, of the antigen and the variable-region ($V_H/V_L$) domain of the antibody molecule, particularly the hypervariable regions, or complementarity-determining regions (CDRs). The exquisite specificity of antigen-antibody interactions has led to the development of a variety of immunologic assays, which can be used to detect the presence of either antibody or antigen. Immunoassays have played vital roles in diagnosing diseases, monitoring the level of the humoral immune response, and identifying molecules of biological or medical interest. These assays differ in their speed and sensitivity; some are strictly qualitative, others are quantitative. This chapter examines the nature of the antigen-antibody interaction and various immunologic assays that measure or exploit this interaction.

## Strength of Antigen-Antibody Interactions

The noncovalent interactions that form the basis of antigen-antibody (Ag-Ab) binding include hydrogen bonds, ionic bonds, hydrophobic interactions, and van der Waals interactions (Figure 6-1). Because these interactions are individually weak (compared with a covalent bond), a large number of such interactions are required to form a strong Ag-Ab interaction. Furthermore, each of these noncovalent interactions operates over a very short distance, generally about $1 \times 10^{-7}$ mm (1 angstrom, Å); consequently, a strong Ag-Ab interaction depends on a very close fit between the antigen and antibody. Such fits require a high degree of complementarity between antigen and antibody, a requirement that underlies the exquisite specificity that characterizes antigen-antibody interactions.

## Antibody affinity is a quantitative measure of binding strength

The combined strength of the noncovalent interactions between a *single* antigen-binding site on an antibody and a

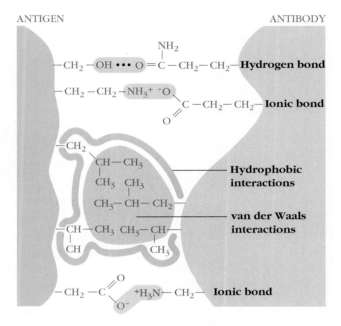

**FIGURE 6-1 The interaction between an antibody and an antigen depends on four types of noncovalent forces: (1) hydrogen bonds, in which a hydrogen atom is shared between two electronegative atoms; (2) ionic bonds between oppositely charged residues; (3) hydrophobic interactions, in which water forces hydrophobic groups together; and (4) van der Waals interactions between the outer electron clouds of two or more atoms.** In an aqueous environment, noncovalent interactions are extremely weak and depend on close complementarity of the shapes of antibody and antigen.

*single* epitope is the **affinity** of the antibody for that epitope. Low-affinity antibodies bind antigen weakly and tend to dissociate readily, whereas high-affinity antibodies bind antigen more tightly and remain bound longer. The association between a binding site on an antibody (Ab) with a monovalent antigen (Ag) can be described by the equation

$$Ag + Ab \underset{k_{-1}}{\overset{k_1}{\rightleftharpoons}} Ag\text{-}Ab$$

where $k_1$ is the forward (association) rate constant and $k_{-1}$ is the reverse (dissociation) rate constant. In biochemical terms, the ratio $k_1/k_{-1}$ is the **association constant, $K_a$** (i.e., $k_1/k_{-1} = K_a$), a measure of affinity. In immunology, $K_a$ is called the **affinity constant.** Because $K_a$ is the equilibrium constant for the reaction of a binding site of the antibody with antigen and since the molar concentration of a single binding site is equal to the antibody concentration, $K_a$ can be calculated from the ratio of the molar concentration of bound Ag-Ab complex to the molar concentrations of unbound antigen and antibody at equilibrium as follows:

$$K_a = \frac{[Ag\text{-}Ab]}{[Ab][Ag]}$$

The value of $K_a$ varies for different Ag-Ab complexes and depends on both $k_1$, which is expressed in units of liters/mole/second (L/mol/s), and $k_{-1}$, which is expressed in units of 1/second. For small haptens, the forward rate constant $k_1$ can be extremely high; in some cases, as high as $4 \times 10^8$ L/mol/s, approaching the theoretical upper limit of diffusion-limited reactions ($10^9$ L/mol/s). For larger protein antigens, however, $k_1$ is smaller, with values in the range of $10^5$ L/mol/s.

The rate at which bound antigen leaves an antibody's binding site (i.e., the dissociation rate constant, $k_{-1}$) is a major determinant of the antibody's affinity for an antigen. Table 6-1 illustrates the role of $k_{-1}$ in determining the association constant $K_a$ for several Ag-Ab interactions. For example, the $k_1$ for the DNP-L-lysine system is about one fifth that for the fluorescein system, but its $k_{-1}$ is 200 times greater; consequently, the affinity of the antifluorescein antibody $K_a$ for the fluorescein system is about 1000-fold higher than that of anti-DNP antibody. Low-affinity Ag-Ab complexes have $K_a$ values between $10^4$ and $10^5$ L/mol; high-affinity complexes can have $K_a$ values as high as $10^{11}$ L/mol.

For some purposes, the dissociation of the antigen-antibody complex is of interest:

$$Ag\text{-}Ab \rightleftharpoons Ab + Ag$$

The equilibrium constant for this dissociation reaction, $K_d$, the **dissociation constant,** is equal to the reciprocal of $K_a$:

$$K_d = \frac{[Ab][Ag]}{[Ab\text{-}Ag]} = 1/K_a$$

Very stable complexes have very low values of $K_d$; less stable ones have higher values. Even though $K_d$ is not the affinity

| TABLE 6-1 | Forward and reverse rate constants ($k_1$ and $k_{-1}$) and association and dissociation constants ($K_a$ and $K_d$) for three ligand-antibody interactions | | | | |
|---|---|---|---|---|---|
| Antibody | Ligand | $k_1$ | $k_{-1}$ | $K_a$ | $K_d$ |
| Anti-DNP | ε-DNP-L-lysine | $8 \times 10^7$ | 1 | $1 \times 10^8$ | $1 \times 10^{-8}$ |
| Anti-fluorescein | Fluorescein | $4 \times 10^8$ | $5 \times 10^{-3}$ | $1 \times 10^{11}$ | $1 \times 10^{-11}$ |
| Anti-bovine serum albumin (BSA) | Dansyl-BSA | $3 \times 10^5$ | $2 \times 10^{-3}$ | $1.7 \times 10^8$ | $5.9 \times 10^{-9}$ |

SOURCE: Adapted from H. N. Eisen, 1990, *Immunology,* 3rd ed., Harper & Row, Publishers.

(a)

Control: No antibody present
(ligand equilibrates on both sides equally)

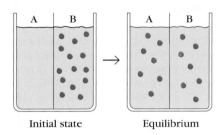

Initial state          Equilibrium

Experimental: Antibody in A
(at equilibrium more ligand in A due to Ab binding)

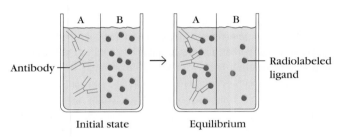

Antibody                                    Radiolabeled
                                            ligand

Initial state          Equilibrium

(b)

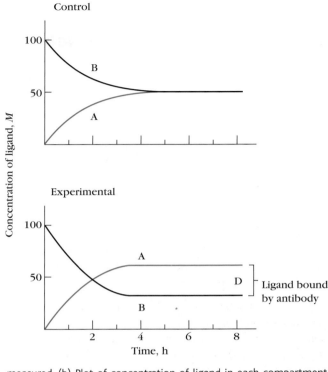

**FIGURE 6-2  Determination of antibody affinity by equilibrium dialysis.** (a) The dialysis chamber contains two compartments (A and B) separated by a semipermeable membrane. Antibody is added to one compartment and a radiolabeled ligand to another. At equilibrium, the concentration of radioactivity in both compartments is measured. (b) Plot of concentration of ligand in each compartment with time. At equilibrium, the difference in the concentration of radioactive ligand in the two compartments represents the amount of ligand bound to antibody.

constant, the affinity constant, $K_a$, can of course be readily calculated from $K_d$ as follows: $K_a = 1/K_d$.

The affinity constant, $K_a$, formerly determined by **equilibrium dialysis,** is now derived by newer methods, especially surface plasmon resonance (SPR), which is discussed in a later section. Because equilibrium dialysis illustrates important principles and remains for some immunologists the standard against which other methods are evaluated, it is described here. This procedure uses a dialysis chamber containing two equal compartments separated by a semipermeable membrane. Antibody is placed in one compartment, and a radioactively labeled ligand small enough to pass through the semipermeable membrane is placed in the other compartment (Figure 6-2). Suitable ligands include haptens, oligosaccharides, and oligopeptides. In the absence of antibody, ligand added to compartment B will equilibrate on both sides of the membrane (Figure 6-2a). In the presence of antibody, however, part of the labeled ligand will be bound to the antibody at equilibrium, trapping the ligand on the antibody side of the vessel, whereas unbound ligand will be equally distributed in both compartments. Thus, the total concentration of ligand will be greater in the compartment containing antibody (Figure 6-2b). The difference in the ligand concentration in the two compart-

ments represents the concentration of ligand bound to the antibody (i.e., the concentration of Ag-Ab complex). The higher the affinity of the antibody, the more ligand is bound.

Since the concentration of antibody in the equilibrium dialysis chamber is known, the concentration of antigen-antibody complexes can be determined by [Ab-Ag]/$n$, where $n$ is the number of binding sites per antibody molecule. The equilibrium constant or affinity constant, $K_a$, for a single binding site is:

$$K_a = [\text{Ab-Ag}]/[\text{Ab}][\text{Ag}] = \frac{r}{c(n-r)}$$

where $r$ equals the ratio of the concentration of bound ligand to total antibody concentration, $c$ is the concentration of free ligand, and $n$ remains the number of binding sites per antibody molecule. This expression can be rearranged to give the **Scatchard equation:**

$$\frac{r}{c} = K_a n - K_a r$$

Values for $r$ and $c$ can be obtained by repeating the equilibrium dialysis with the same concentration of antibody but with different concentrations of ligand. If $K_a$ is a constant, that is, if all the antibodies within the dialysis chamber have the same affinity for the ligand, then a Scatchard plot of $r/c$

(a) Homogeneous antibody

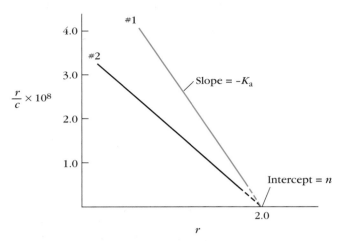

(b) Heterogeneous antibody

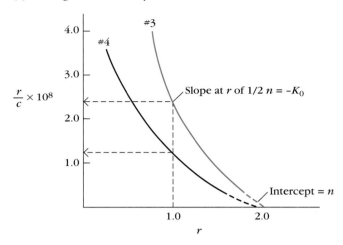

**FIGURE 6-3 Scatchard plots are based on repeated equilibrium dialyses with a constant concentration of antibody and varying concentration of ligand.** In these plots, $r$ equals moles of bound ligand/mole antibody and $c$ is the concentration of free ligand. From a Scatchard plot, both the equilibrium constant ($K_a$) and the number of binding sites per antibody molecule ($n$), or its valency, can be obtained. (a) If all antibodies have the same affinity, then a Scatchard plot yields a straight line with a slope of $-K_a$. The $x$ intercept is $n$, the valency of the antibody, which is 2 for IgG and other divalent Igs. For IgM, which is pentameric, $n = 10$, and for dimeric IgA, $n = 4$. In this graph, antibody #1 has a higher affinity than antibody #2. (b) If the antibody preparation is polyclonal and has a range of affinities, a Scatchard plot yields a curved line whose slope is constantly changing. The average affinity constant $K_0$ can be calculated by determining the value of $K_a$ when half of the binding sites are occupied (i.e., when $r = 1$ in this example). In this graph, antiserum #3 has a higher affinity ($K_0 = 2.4 \times 10^8$) than antiserum #4 ($K_0 = 1.25 \times 10^8$). Note that the curves shown in (a) and (b) are for divalent antibodies such as IgG.

versus $r$ will yield a straight line with a slope of $-K_a$ (Figure 6-3a). As the concentration of unbound ligand $c$ increases, $r/c$ approaches 0 and $r$ approaches $n$, the **valency,** equal to the number of binding sites per antibody molecule.

Most antibody preparations are polyclonal—a heterogeneous mixture of antibodies with a range of affinities is present. A Scatchard plot of heterogeneous antibody yields a curved line whose slope is constantly changing, reflecting this antibody heterogeneity (Figure 6-3b). With this type of Scatchard plot, it is possible to determine the average affinity constant, $K_0$, by determining the value of $K_a$ when half of the antigen-binding sites are filled. This is conveniently done by determining the slope of the curve at the point where half of the antigen binding sites are filled.

## Antibody avidity incorporates affinity of multiple binding sites

The affinity at one binding site does not always reflect the true strength of the antibody-antigen interaction. When complex antigens containing multiple, repeating antigenic determinants are mixed with antibodies containing multiple binding sites, the interaction of an antibody molecule with an antigen molecule at one site will increase the probability of reaction between those two molecules at a second site. The strength of such multiple interactions between a multivalent antibody and antigen is called the **avidity.** The avidity

is a better measure than the affinity for quantifying the binding capacity of an antibody within biological systems (e.g., the reaction of an antibody with antigenic determinants on a virus or bacterial cell). Avidity is the equilibrium constant, $K_{eq}$, for the interaction between the intact antibody and the antigen; affinity is the equilibrium constant for the reaction between an individual binding site of the antibody and the antigen. Under ideal conditions the avidity should be the product of the affinities of the individual binding sites of an antibody. For an IgG molecule, which has two binding sites with identical affinities, the avidity under ideal conditions would be:

$$\text{Avidity} = K_a \times K_a = (K_a)^2 = K_{eq} = \frac{[\text{Ab-Ag}]}{[\text{Ab}][\text{Ag}]}$$

This equation represents a theoretical maximum for avidity; actual avidities are always several orders of magnitude lower than the product of the affinities. The difference is due primarily to the geometry of Ab-Ag binding. The binding site of an antibody in contact with a single epitope of an antigen may be optimally oriented for the best fit, whereas binding to multiple epitopes on the target antigen may require somewhat strained geometry and less optimal binding interactions. Despite geometrical barriers to the achievement of maximum theoretical values of avidity, the avidity is higher than the affinity. Thus, high avidity can often compensate for low affinity. For example, the binding sites of secreted pentameric IgM molecules often have a significantly lower affinity than those of monomeric IgG, but the high avidity of IgM, resulting from its higher valence, enables it to bind antigen effectively. Indeed, in the case of an antigen with many closely spaced, repeating epitopes, such as those found on the surface of bacteria and other pathogens, an IgM with lower-affinity binding sites may bind more tightly than an IgG with sites of higher affinity.

## Cross-Reactivity

Although Ag-Ab reactions are highly specific, in some cases antibody elicited by one antigen can cross-react with an unrelated antigen. Such **cross-reactivity** occurs if two different antigens share an identical or very similar epitope. In the latter case, the antibody's affinity for the cross-reacting epitope is usually less than that for the original epitope.

Cross-reactivity is often observed among polysaccharide antigens that contain similar oligosaccharide residues. The **ABO blood-group antigens,** for example, are glycolipids expressed on red blood cells. Subtle differences in the terminal residues of the sugars attached to these cell surface glycolipids distinguish the A and B blood-group antigens. Although the mechanisms of tolerance prevent formation of antibodies against one's own blood group antigens, an individual lacking one or both of these antigens will have serum antibodies to the missing antigen(s), induced not by exposure to red blood cell antigens but by exposure to cross-reacting microbial antigens present on common intestinal bacteria. These microbial antigens induce the formation of antibodies in individuals lacking the similar blood-group antigens on their red blood cells. (In individuals possessing these antigens, complementary antibodies would be eliminated during the developmental stage, in which B cells making antibodies that recognize self epitopes are weeded out; see Chapter 16.) The blood-group antibodies, although elicited by microbial antigens, will cross-react with similar oligosaccharides on foreign red blood cells, providing the basis for blood-typing tests and accounting for the necessity of compatible blood types during blood transfusions. A type A individual has anti-B antibodies, a type B individual has anti-A, and a type O individual has anti-A and anti-B (Table 6-2).

| TABLE 6-2 | ABO blood types | |
|---|---|---|
| **Blood type** | **Antigens on RBCs** | **Serum antibodies** |
| A | A | Anti-B |
| B | B | Anti-A |
| AB | A and B | Neither |
| O | Neither | Anti-A and anti-B |

A number of viruses and bacteria have antigenic determinants identical or similar to normal host cell components. In some cases, these microbial antigens have been shown to elicit antibody that cross-reacts with the host cell components, resulting in a tissue-damaging autoimmune reaction. The bacterium *Streptococcus pyogenes,* for example, expresses cell wall proteins called M antigens. Antibodies produced to streptococcal M antigens have been shown to cross-react with several myocardial and skeletal muscle proteins and have been implicated in heart and kidney damage following streptococcal infections. The role of other cross-reacting antigens in the development of autoimmune diseases is discussed in Chapter 16.

Some vaccines also exhibit cross-reactivity. For instance, vaccinia virus, which causes cowpox, expresses cross-reacting epitopes with variola virus, the causative agent of smallpox. This cross-reactivity was the basis of Jenner's method of using vaccinia virus to induce immunity to smallpox, as mentioned in Chapter 1.

## Surface Plasmon Resonance (SPR)

Equilibrium dialysis methods for measuring antibody affinity have been superseded since the mid-1990s by **surface plasmon resonance (SPR),** which is far more sensitive, convenient, and rapid (Figure 6-4a). In addition to allowing measurements of affinity constants, SPR can be used to determine rates of antigen-antibody reactions and can even be adapted to measure concentrations of antibody. SPR works by detecting changes in the reflectance properties of the surface of an antigen-coated sensor when it binds antibody. Although the physics underlying SPR is rather sophisticated and best explored by consulting the reading reference (Rich and Myszka, 2003) and Web site listed at the end of the chapter, the method is straightforward and ingenious. A beam of polarized light is directed through a prism onto a thin gold film (coated with antigen on the opposite side) and reflected off the gold film toward a light-collecting sensor. However, at a unique angle, some incident light is absorbed by the gold layer, its energy transformed into charge waves called surface plasmons. A sharp dip in the reflected light intensity can be measured at that angle, called the resonant angle. The angle depends on the color of the light, the thickness and conductivity of the metal film, and the optical properties of the material close to the gold layer's surfaces. The SPR method takes advantage of the last of these factors: the binding of antibodies to the antigen attached to the chip has a strong enough effect to produce a detectable change in the resonant angle. The change has been found to be proportional to the number of bound antibodies. By measuring the rate at which the resonant angle changes during an antigen-antibody reaction, the rate of the antigen-antibody reaction can be determined (Figure 6-4b). Operationally, this is done by passing a solution with a known concentration of antibody

(a)

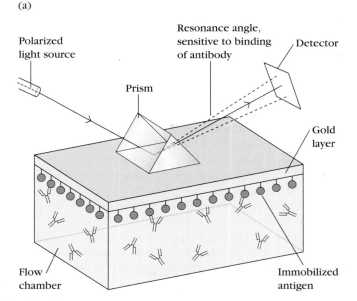

(b)

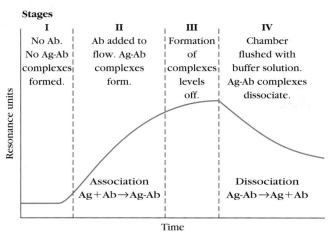

**FIGURE 6-4 Surface plasmon resonance (SPR).** (a) A buffer solution containing antibody is passed through a flow chamber, one wall of which contains a layer of immobilized antigen. As explained in the text, formation of antigen-antibody complexes on this layer causes a change in the resonant angle of a beam of polarized light against the back face of the layer. A sensitive detector records changes in the resonant angle as antigen-antibody complexes form. (b) Interpretation of a sensorgram. There are four stages in the plot of the detector response (expressed as resonance units, which represent a change of 0.0001 degree in the resonance angle) versus time. Stage I: Buffer is passed through the flow chamber. No Ag-Ab complexes are present, establishing a baseline. Stage II: Antibody is introduced into the flow and Ag-Ab complexes form. The ascending slope of this curve is proportional to the forward rate of the reaction. Stage III: The curve plateaus when all sites that can be bound at the prevailing antibody concentration are filled. The height of the plateau is directly proportional to the antibody concentration. Stage IV: The flow cell is flushed with buffer containing no antibody and the Ag-Ab complexes dissociate. The rate of dissociation is proportional to the slope of the dissociation curve. The ratio of the slopes, ascending over descending, equals $k_1/k_2 = k_a$.

over the antigen-coated chip. A plot of the changes measured during an SPR experiment versus time is called a *sensorgram*. In the course of an antigen-antibody reaction, the sensorgram plot rises until all of the sites capable of binding antibody (at a given concentration) have done so. Beyond that point the sensorgram plateaus. The data from these measurements can be used to calculate $k_1$, the association rate constant for the reaction:

$$\text{Ab} + \text{Ag} \xrightarrow{k_1} \text{Ab-Ag}$$

Once the plateau has been reached, solution containing no antibody can be passed through the chamber. Under these conditions, the antigen-antibody complexes dissociate, allowing calculation of the dissociation rate constant, $k_2$. Measurement of $k_1$ and $k_2$ allows determination of the affinity constant, $K_a$, since $K_a = k_1/k_2$.

Surface plasmon resonance can also be used to measure the concentration of antibody in a sample. The amount of antigen-antibody complex formed on the chip is represented by the height of the sensorgram, and this measure is directly proportional to the amount of antibody in the solution flowing through the chamber. By comparison to reference data, antibody concentrations can be determined.

## SPR can be used to characterize the epitope specificities of collections of antibodies

SPR has uses beyond measuring affinity or concentration of antibodies. Suppose one has two different monoclonal antibodies against an antigen such as gp120, the envelope protein of HIV. Do these antibodies bind to the same or different epitopes of this key protein? Surface plasmon resonance provides a powerful approach to answering this question. Antibodies that bind to different epitopes on an antigen give a characteristic sensorgram plot when added to the chamber serially (Figure 6-5a). Antibodies that bind to the same epitope compete with each other for binding sites, each blocking the binding of the other, demonstrated by the plots in Figure 6-5b.

In a variation of this procedure, one can use chemical synthesis or judicious enzymatic hydrolysis to generate a set of fragments of an antigen and then immobilize each of these fragments on a chip. Assuming that the antigen fragments have the same three-dimensional conformation as the native antigen, one can determine the reactivities (or lack thereof) of antibodies and the locations of the epitopes within the original antigen molecule can be isolated, a procedure is known as **epitope mapping.**

(a) $Ab_1$ and $Ab_2$ bind different epitopes of the same antigen.

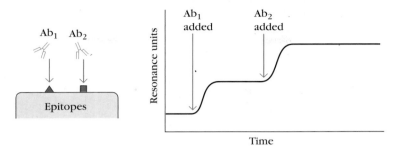

(b) $Ab_1$ and $Ab_2$ bind to the same epitope.

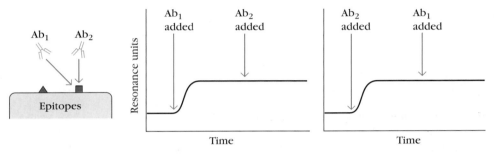

**FIGURE 6-5 SPR can be used to determine if different antibodies bind to the same or different epitopes.** (a) $Ab_1$ and $Ab_2$ recognize different epitopes of the antigen. Initial injection of $Ab_1$ causes a rise from baseline to a plateau; subsequent addition of $Ab_2$ causes a rise to a second, higher plateau. This indicates that $Ab_1$ and $Ab_2$ react with different epitopes on the antigen, for example, two spatially distinct regions of a protein. (b) $Ab_1$ and $Ab_2$ recognize the same epitope of the antigen. In the first sensorgram, addition of $Ab_1$ reveals a response, and subsequent injection of $Ab_2$ generates no additional response. In the second sensorgram, addition of $Ab_2$ first reveals a response, but subsequent addition of $Ab_1$ generates no additional response. This pattern indicates that $Ab_1$ and $Ab_2$ bind competitively, both recognizing the same epitope on the antigen; binding of one blocks subsequent binding of the other.

## Precipitation Reactions

Antibody and soluble antigen interacting in aqueous solution form a lattice that eventually develops into a visible precipitate. Antibodies that aggregate soluble antigens are called **precipitins.** Although formation of the soluble Ag-Ab complex occurs within minutes, formation of visible precipitate occurs more slowly, often taking a day or two to reach completion.

Formation of an Ag-Ab lattice depends on the valency of both the antibody and antigen:

- The antibody must be bivalent; a precipitate will not form with monovalent Fab fragments.

- The antigen must be either bivalent or polyvalent; that is, it must have at least two copies of the same epitope or have different epitopes that react with different antibodies present in polyclonal antisera.

Although various modifications of the precipitation reaction were at one time the major types of assay used in immunology, they have been largely replaced by methods that are faster and, because they are far more sensitive, require only very small quantities of antigen or antibody. Also, modern assay methods are not limited to antigen-antibody reactions that produce a precipitate. Table 6-3 presents a comparison of the *sensitivity*, or minimum amount of antibody detectable, of a number of immunoassays.

## Precipitation reactions in gels yield visible precipitin lines

Immune precipitates can form not only in solution but also in an agar matrix. When antigen and antibody diffuse toward one another in agar or when antibody is incorporated into the agar and antigen diffuses into the antibody-containing matrix, a visible line of precipitation will form. As in a precipitation reaction in fluid, visible precipitation occurs in the region of equivalence, whereas no visible precipitate forms in regions of antibody or antigen excess. Two types of *immunodiffusion reactions* can be used to determine relative concentrations of antibodies or antigens, to compare antigens, or to determine the relative purity of an antigen preparation. They are **radial immunodiffusion** (the Mancini method) and **double immunodiffusion** (the Ouchterlony method); both are carried out in a semisolid medium such as agar. In radial immunodiffusion, an antigen sample is placed in a well and allowed to diffuse into agar containing a suitable dilution of an antiserum. As the antigen diffuses into the agar, the region of

| Assay | Sensitivity*<br>($\mu$g antibody/ml) |
|---|---|
| **TABLE 6-3    Sensitivity of various immunoassays** | |
| Precipitation reaction in fluids | 20–200 |
| Precipitation reactions in gels | |
|    Mancini radial immunodiffusion | 10–50 |
|    Ouchterlony double immunodiffusion | 20–200 |
|    Immunoelectrophoresis | 20–200 |
|    Rocket electrophoresis | 2 |
| Agglutination reactions | |
|    Direct | 0.3 |
|    Passive agglutination | 0.006–0.06 |
|    Agglutination inhibition | 0.006–0.06 |
| Radioimmunoassay (RIA) | 0.0006–0.006 |
| Enzyme-linked immunosorbent assay (ELISA) | ~0.0001–0.01 |
| ELISA using chemiluminescence | ~0.00001–0.01† |
| Immunofluorescence | 1.0 |
| Flow cytometry | 0.006–0.06 |

*The sensitivity depends on the affinity of the antibody used for the assay as well as the epitope density and distribution on the antigen.

†Note that the sensitivity of chemiluminescence-based ELISA assays can be made to match that of RIA.

SOURCE: Updated and adapted from N. R. Rose et al., eds., 1997, *Manual of Clinical Laboratory Immunology,* 5th ed., American Society for Microbiology, Washington, DC.

RADIAL IMMUNODIFFUSION

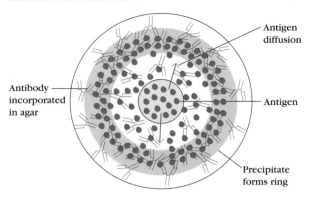

DOUBLE IMMUNODIFFUSION

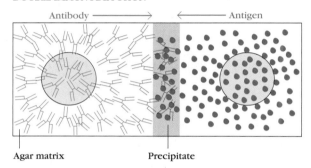

**FIGURE 6-6 Diagrammatic representation of radial immuno-diffusion (Mancini method) and double immunodiffusion (Ouchterlony method) in a gel.** In both cases, large insoluble complexes form in the agar in the zone of equivalence, visible as lines of precipitation (purple regions). Only the antigen (red) diffuses in radial immunodiffusion, whereas both the antibody (blue) and antigen (red) diffuse in double immunodiffusion.

equivalence is established and a ring of precipitation, a precipitin ring, forms around the well (Figure 6-6, upper panel). The area of the precipitin ring is proportional to the concentration of antigen. By comparing the area of the precipitin ring with a standard curve (obtained by measuring the precipitin areas of known concentrations of the antigen), the concentration of the antigen sample can be determined. In the Ouchterlony method, both antigen and antibody diffuse radially from wells toward each other, thereby establishing a concentration gradient. As equivalence is reached, a visible line of precipitation, a precipitin line, forms (Figure 6-6, lower panel).

## Immunoelectrophoresis combines electrophoresis and double immunodiffusion

In **immunoelectrophoresis,** the antigen mixture is first electrophoresed to separate its components by charge. Troughs are then cut into the agar gel parallel to the direction of the electric field, and antiserum is added to the troughs. Antibody and antigen then diffuse toward each other and produce lines of precipitation where they meet in appropriate

proportions (Figure 6-7). Immunoelectrophoresis is used in clinical laboratories to detect the presence or absence of proteins in the serum. A sample of serum is electrophoresed, and the individual serum components are identified with antisera specific for a given protein or immunoglobulin class. This technique is useful in determining whether a patient produces abnormally low amounts of one or more isotypes, characteristic of certain immunodeficiency diseases. It can also show whether a patient overproduces some serum protein, such as albumin, immunoglobulin, or transferrin. The immunoelectrophoretic patterns of serum from patients with multiple myeloma show a large amount of myeloma protein. Because immunoelectrophoresis is a strictly *qualitative* technique that only detects relatively high antibody concentrations (greater than several hundred $\mu$g/ml), its utility is limited to the detection of quantitative abnormalities only when the departure from normal is striking, as in immunodeficiency states and immunoproliferative disorders.

A related *quantitative* technique, **rocket electrophoresis,** does permit measurement of antigen levels. In rocket electrophoresis, a negatively charged antigen is electrophoresed

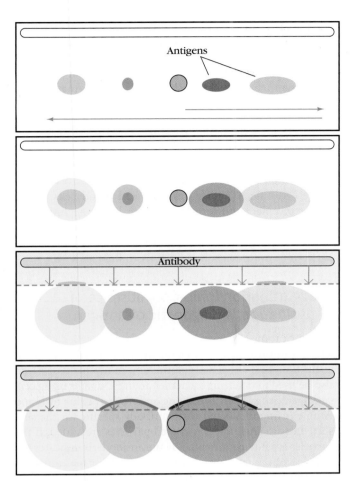

**FIGURE 6-7 Immunoelectrophoresis of an antigen mixture.** An antigen preparation (orange) is first electrophoresed, which separates the component antigens on the basis of charge. Antiserum (blue) is then added to troughs on one or both sides of the separated antigens and allowed to diffuse; in time, lines of precipitation (colored arcs) form where specific antibody and antigen interact.

in a gel containing antibody. The precipitate formed between antigen and antibody has the shape of a rocket, the height of which is proportional to the concentration of antigen in the well. One limitation of rocket electrophoresis is the need for the antigen to be negatively charged, which is a requirement for electrophoretic movement within the agar matrix. Some proteins, immunoglobulins, for example, are not sufficiently charged to be quantitatively analyzed by rocket electrophoresis, nor is it possible to measure the amounts of several antigens in a mixture at the same time.

## Agglutination Reactions

The interaction between antibody and a particulate antigen results in visible clumping called **agglutination.** Antibodies that produce such reactions are called **agglutinins.** Agglutination reactions are similar in principle to precipitation reactions;

they depend on the cross-linking of polyvalent antigens. Just as an excess of antibody inhibits precipitation reactions, such excess can also inhibit agglutination reactions; this inhibition is called the **prozone effect.** Because prozone effects are encountered in many types of immunoassays, understanding the basis of this phenomenon is of general importance.

Several mechanisms can cause the prozone effect. First, at high antibody concentrations, the number of antibody binding sites may greatly exceed the number of epitopes. As a result, most antibodies bind antigen only univalently instead of multivalently. Antibodies that bind univalently cannot cross-link one antigen to another. Prozone effects are readily diagnosed by performing the assay at a variety of antibody (or antigen) concentrations. As one dilutes a sample to an optimum antibody concentration, one sees higher levels of agglutination or whatever parameter is measured in the assay being used. When polyclonal antibodies are used, the prozone effect can occur for another reason. The antiserum may contain high concentrations of antibodies that bind to the antigen but do not induce agglutination; these antibodies, called **incomplete antibodies,** are often of the IgG class. At high concentrations of IgG, incomplete antibodies may occupy most of the antigenic sites, thus blocking access by IgM, which is a good agglutinin. This effect is not seen with agglutinating monoclonal antibodies. The lack of agglutinating activity of an incomplete antibody may be due to restricted flexibility in the hinge region, making it difficult for the antibody to assume the required angle for optimal cross-linking of epitopes on two or more particulate antigens. Alternatively, the density of epitope distribution or the location of some epitopes in deep pockets of a particulate antigen may make it difficult for the antibodies specific for these epitopes to agglutinate certain particulate antigens. When feasible, the solution to both of these problems is to try different antibodies that may react with other epitopes of the antigen that do not present these limitations.

## Hemagglutination is used in blood typing

Agglutination reactions (Figure 6-8) are routinely performed to type red blood cells (RBCs). With tens of millions of blood-typing determinations run each year, this is one of

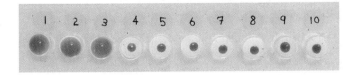

**FIGURE 6-8 Demonstration of hemagglutination using antibodies against sheep red blood cells (SRBCs).** The control tube (10) contains only SRBCs, which settle into a solid "button." The experimental tubes 1–9 contain a constant number of SRBCs plus serial twofold dilutions of anti-SRBC serum. The spread pattern in the experimental series indicates positive hemagglutination through tube 3. *[Louisiana State University Medical Center/MIP. Courtesy of Harriet C. W. Thompson.]*

the world's most frequently used immunoassays. In typing for the ABO antigens, RBCs are mixed on a slide with antisera to the A or B blood-group antigens. If the antigen is present on the cells, they agglutinate, forming a visible clump on the slide. As an additional check, the donor's blood is examined for the presence of antibodies against ABO antigens. If a donor does not contain antibody against his or her own ABO antigens, the cell and antibody determinations are concordant, and the cell typing is confirmed as correct. If the results are discordant and agglutination assays indicate the donor has antibodies that react with their own ABO antigens, then there is either an assay error (the cell or the antibody assay is incorrect) or the donor is making antibodies to self, a manifestation of autoimmunity (see Chapter 16). Samples that produce discordant results on retesting are not used for transfusion.

## Bacterial agglutination is used to diagnose infection

A bacterial infection often elicits the production of serum antibodies specific for surface antigens on the bacterial cells. The presence of such antibodies can be detected by bacterial agglutination reactions. Serum from a patient thought to be infected with a given bacterium is serially diluted in an array of tubes to which the bacteria is added. The last tube showing visible agglutination will reflect the serum antibody **titer** of the patient. The agglutinin titer is defined as the reciprocal of the greatest serum dilution that elicits a positive agglutination reaction. For example, if serial twofold dilutions of serum are prepared and if the dilution of 1/640 shows agglutination but the dilution of 1/1280 does not, then the agglutination titer of the patient's serum is 640. In some cases, serum can be diluted up to 1/50,000 and still show agglutination of bacteria.

The agglutinin titer of an antiserum can be used to diagnose a bacterial infection. Patients with typhoid fever, for example, show a significant rise in the agglutination titer to *Salmonella typhi*. Agglutination reactions also provide a way to type bacteria. For instance, different species of the bacterium *Salmonella* can be distinguished by agglutination reactions with a panel of typing antisera.

## Passive agglutination is useful with soluble antigens

The sensitivity and simplicity of agglutination reactions can be extended to soluble antigens by the technique of passive hemagglutination. In this technique, antigen-coated red blood cells are prepared by mixing a soluble antigen with red blood cells that have been treated with tannic acid or chromium chloride, both of which promote adsorption of the antigen to the surface of the cells. Serum containing antibody is serially diluted into microtiter plate wells, and the antigen-coated red blood cells are then added to each well; agglutination is assessed by the size of the characteristic

spread pattern of agglutinated red blood cells on the bottom of the well, like the pattern seen in agglutination reactions (see Figure 6-8).

In recent years, there has been a shift away from red blood cells to synthetic particles, such as latex beads, as matrices for agglutination reactions. Once the antigen has been coupled to latex beads, the preparation can be used immediately or stored. The use of synthetic beads offers the advantages of consistency, uniformity, and stability. Furthermore, agglutination reactions employing synthetic beads can be read rapidly, often within 3 to 5 minutes of mixing the beads with the test sample. Whether based on red blood cells or the more convenient and versatile synthetic beads, agglutination reactions are simple to perform, do not require expensive equipment, and can detect small amounts of antibody (concentrations as low as nanograms per milliliter).

## In agglutination inhibition, absence of agglutination is diagnostic of antigen

A modification of the agglutination reaction, called **agglutination inhibition,** provides a highly sensitive assay for small quantities of an antigen. Agglutination inhibition assays can also be used to determine whether an individual is using certain illicit drugs, such as cocaine or heroin. A urine or blood sample is first incubated with antibody specific for the suspected drug. Then antigen-coated particles are added. If the particles are not agglutinated by the antibody, it indicates the sample contained an antigen recognized by the antibody, suggesting that the individual was using the drug. One problem with these tests is that legal drugs having chemical structures similar to those of illicit drugs may cross-react with the antibody, giving a false-positive reaction. For this reason, positive reactions are confirmed by a nonimmunologic method.

Agglutination inhibition assays are widely used in clinical laboratories to determine whether an individual has been exposed to certain types of viruses that cause agglutination of red blood cells. If an individual's serum contains specific antiviral antibodies, then the antibodies will bind to the virus and interfere with hemagglutination by the virus. This technique is commonly used in premarital testing to determine the immune status of women with respect to rubella virus. The reciprocal of the last serum dilution to show inhibition of rubella hemagglutination is the titer of the serum. A titer greater than 10 (1:10 dilution) indicates that a woman is immune to rubella, whereas a titer of less than 10 is indicative of a lack of immunity and the need for immunization with the rubella vaccine.

## Radioimmunoassay

One of the most sensitive techniques for detecting antigen or antibody is **radioimmunoassay (RIA).** The technique was first developed in 1960 by two endocrinologists, S. A. Berson and Rosalyn Yalow, to determine levels of insulin–anti-insulin complexes in diabetics. Although their technique encountered some skepticism, it soon proved its value for measuring hormones, serum proteins, drugs, and vitamins at concentrations of 0.001 *micrograms* per milliliter or less. In 1977, some years after Berson's death, the significance of the technique was acknowledged by the award of a Nobel prize to Yalow.

The principle of RIA involves competitive binding of radiolabeled antigen and unlabeled antigen to a high-affinity antibody. The labeled antigen is mixed with antibody at a concentration that saturates the antigen-binding sites of the antibody. Then test samples of unlabeled antigen of unknown concentration are added in progressively larger amounts. The antibody does not distinguish labeled from unlabeled antigen, so the two kinds of antigen compete for available binding sites on the antibody. As the concentration of unlabeled antigen increases, more labeled antigen will be displaced from the binding sites. The decrease in the amount of radiolabeled antigen bound to specific antibody in the presence of the test sample is measured in order to determine the amount of antigen present in the test sample.

The antigen is generally labeled with a gamma-emitting isotope such as $^{125}$I, but beta-emitting isotopes such as tritium ($^3$H) are also routinely used as labels. The radiolabeled antigen is part of the assay mixture; the test sample may be a complex mixture, such as serum or other body fluids, that contains the unlabeled antigen. The first step in setting up an RIA is to determine the amount of antibody needed to bind 50% to 70% of a fixed quantity of radioactive antigen (Ag*) in the assay mixture. This ratio of antibody to Ag* is chosen to ensure that the number of epitopes presented by the labeled antigen always exceeds the total number of antibody binding sites. Consequently, unlabeled antigen added to the sample mixture will compete with radiolabeled antigen for the limited supply of antibody. Even a small amount of unlabeled antigen added to the assay mixture of labeled antigen and antibody will cause a decrease in the amount of radioactive antigen bound, and this decrease will be proportional to the amount of unlabeled antigen added. To determine the amount of labeled antigen bound, the Ag-Ab complex is precipitated to separate it from free antigen (antigen not bound to antibody), and the radioactivity in the precipitate is measured. A standard curve can be generated using unlabeled antigen samples of known concentration (in place of the test sample), and from this plot the amount of antigen in the test mixture may be precisely determined.

Several methods have been developed for separating bound antigen from free antigen in RIA. One method involves precipitating Ag-Ab complexes with a secondary anti-isotype antiserum. For example, if the Ag-Ab complex contains rabbit IgG antibody, then goat anti-rabbit IgG will bind to the rabbit IgG and precipitate the complex. Another method makes use of the fact that protein A

of *Staphylococcus aureus* has high affinity for IgG. If the Ag-Ab complex contains an IgG antibody, the complex can be precipitated by mixing with formalin-killed *S. aureus*. After removal of the complex by either of these methods, the amount of free labeled antigen remaining in the supernatant can be measured in a radiation counter; subtracting this value from the total amount of labeled antigen added yields the amount of labeled antigen bound.

Various solid-phase RIAs have been developed that make it easier to separate the Ag-Ab complex from the unbound antigen. In some cases, the antibody is covalently cross-linked to Sepharose beads. The amount of radiolabeled antigen bound to the beads can be measured after the beads have been centrifuged and washed. Alternatively, the antibody can be immobilized on polystyrene or polyvinylchloride wells and the amount of free labeled antigen in the supernatant can be determined in a radiation counter. In another approach, the antibody is immobilized on the walls of microtiter wells and the amount of bound antigen determined. Because the procedure requires only small amounts of sample and can be conducted in small 96-well microtiter plates (slightly larger than a 3 × 5 card), this procedure is well suited for determining the concentration of a particular antigen in large numbers of samples. For example, a microtiter RIA has been widely used to screen for the presence of the hepatitis B virus (Figure 6-9). RIA screening of donor blood has sharply reduced the incidence of hepatitis B infections in recipients of blood transfusions.

# Enzyme-Linked Immunosorbent Assay

**Enzyme-linked immunosorbent assay,** commonly known as **ELISA** (or EIA), is similar in principle to RIA but depends on an enzyme rather than a radioactive label. An enzyme conjugated with an antibody reacts with a colorless substrate to generate a colored reaction product. Such a substrate is called **a chromogenic substrate.** A number of enzymes have been employed for ELISA, including alkaline phosphatase, horseradish peroxidase, and β-galactosidase. These assays match the sensitivity of RIAs and have the advantage of being safer and less costly.

## There are numerous variants of ELISA

A number of variations of ELISA have been developed, allowing qualitative detection or quantitative measurement of either antigen or antibody. Each type of ELISA can be used qualitatively to detect the presence of antibody or antigen. Alternatively, a standard curve based on known concentrations of antibody or antigen is prepared, from which the unknown concentration of a sample can be determined.

(a)

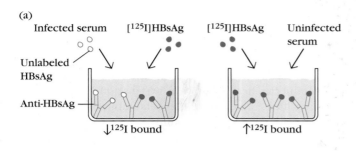

(b)

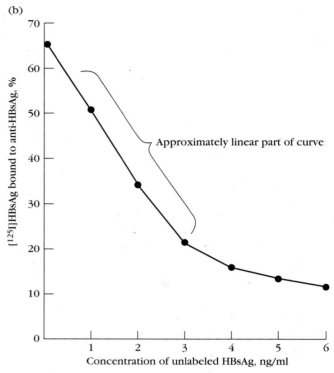

**FIGURE 6-9  A solid-phase radioimmunoassay (RIA) to detect hepatitis B virus in blood samples.** (a) Microtiter wells are coated with a constant amount of antibody specific for HBsAg, the surface antigen on hepatitis B virions. A serum sample and [$^{125}$I]HBsAg are then added. After incubation, the supernatant is removed and the radioactivity of the antigen-antibody complexes is measured. If the sample is infected, the amount of label bound will be less than in controls with uninfected serum. (b) A standard curve is obtained by adding increasing concentrations of unlabeled HBsAg to a fixed quantity of [$^{125}$I]HBsAg and specific antibody. From the plot of the percentage of labeled antigen bound versus the concentration of unlabeled antigen, the concentration of HBsAg in unknown serum samples can be determined by using the linear part of the curve.

## Indirect ELISA

Antibody can be detected or quantitatively determined with an indirect ELISA (Figure 6-10a). Serum or some other sample containing primary antibody (Ab$_1$) is added to an antigen-coated microtiter well and allowed to react with the antigen attached to the well. After any free Ab$_1$ is washed away, the presence of antibody bound to the antigen is detected by adding an enzyme-conjugated secondary antibody (Ab$_2$) that binds to the primary antibody. Any free Ab$_2$ is then washed away, and a substrate for the enzyme is added. The amount of colored reaction product that forms is measured by specialized spectrophotometric plate readers, which can measure the absorbance of all of the wells of a 96-well plate in seconds.

Indirect ELISA is the method of choice to detect the presence of serum antibodies against human immunodeficiency virus (HIV), the causative agent of AIDS. In this assay, recombinant envelope and core proteins of HIV are adsorbed as solid-phase antigens to microtiter wells. Individuals infected with HIV will produce serum antibodies to epitopes on these viral proteins. Generally, serum antibodies to HIV can be detected by indirect ELISA within 6 weeks of infection.

## Sandwich ELISA

Antigen can be detected or measured by a sandwich ELISA (Figure 6-10b). In this technique, the antibody (rather than the antigen) is immobilized on a microtiter well. A sample containing antigen is added and allowed to react with the immobilized antibody. After the well is washed, a second enzyme-linked antibody specific for a different epitope on the antigen is added and allowed to react with the bound antigen.

After any free second antibody is removed by washing, substrate is added, and the colored reaction product is measured.

## Competitive ELISA

Another variation for measuring amounts of antigen is competitive ELISA (Figure 6-10c). In this technique, antibody is first incubated in solution with a sample containing antigen. The antigen-antibody mixture is then added to an antigen-coated microtiter well. The more antigen present in the sample, the less free antibody will be available to bind to the antigen-coated well. Addition of an enzyme-conjugated secondary antibody (Ab$_2$) specific for the isotype of the primary antibody can be used to determine the amount of primary antibody bound to the well, as in an indirect ELISA. In the competitive assay, however, the higher the concentration of antigen in the original sample, the lower the absorbance.

## Chemiluminescence

Light produced by chemiluminescence during certain chemical reactions provides a convenient and highly sensitive alternative to absorbance measurements in the ELISA. In versions of the ELISA using chemiluminescence, a luxogenic (light-generating) substrate takes the place of the chromogenic substrate in conventional ELISA reactions. For example, oxidation of the compound luminol by H$_2$O$_2$ and the enzyme horseradish peroxidase (HRP) produces light:

$$\text{Ab-HRP} + \text{Ag} \longrightarrow \text{Ab-HRP-Ag} \xrightarrow{\text{luminol} + \text{H}_2\text{O}_2} \text{light}$$

The light produced during luxogenic reactions may be detected by its ability to expose photographic film. Quantitative

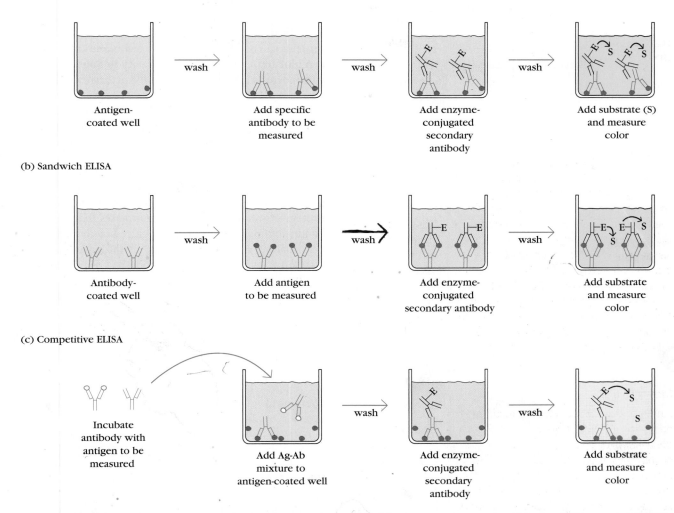

**FIGURE 6-10  Variations in the enzyme-linked immunosorbent assay (ELISA) technique allow determination of antibody or antigen.** Each assay can be used qualitatively or quantitatively by comparison with standard curves prepared with known concentrations of antibody or antigen. Antibody can be determined with an indirect ELISA (a), whereas antigen can be determined with a sandwich ELISA (b) or competitive ELISA (c). In the competitive ELISA, which is an inhibition-type assay, the concentration of antigen is inversely proportional to the color produced.

measurement of light emission can be made by use of a luminometer. The advantage of chemiluminescence assays over chromogenic ones is enhanced sensitivity. In general, the detection limit can be increased at least 10-fold by switching from a chromogenic to a luxogenic substrate and, with the addition of enhancing agents, more than 200-fold. In fact, under ideal conditions, as little as $5 \times 10^{-18}$ moles (5 attomoles) of target antigen have been detected.

## ELISPOT Assay

A modification of the ELISA called the ELISPOT assay allows the quantitative determination of the number of cells in a population that are producing antibodies specific for a given antigen or an antigen for which one has a specific antibody (Figure 6-11). In this approach, the plates are coated with either the antigen (capture antigen) recognized by the antibody of interest or with the antibody (capture antibody) specific for the antigen whose production is being assayed. A suspension of the cell population under investigation is then added to the coated plates and incubated. The cells settle onto the surface of the plate, and secreted molecules reactive with the capture molecules are bound in the vicinity of the secreting cells, producing a ring of antigen-antibody complexes around each cell that produces the molecule of interest. The plate is then washed and an enzyme-linked antibody specific for the secreted antigen or specific for the species (e.g., goat antirabbit) of the secreted antibody is added and allowed to bind. Subsequent development of the assay by addition of a suitable chromogenic or luxogenic substrate reveals the position of each antibody- or antigen-producing cell as a point of color or light.

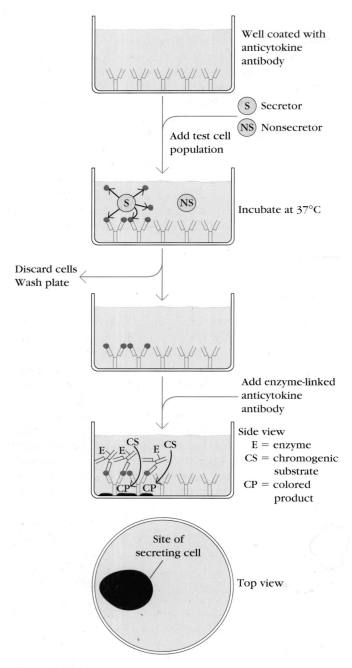

## Western Blotting

Identification of a specific protein in a complex mixture of proteins can be accomplished by a technique known as **Western blotting,** named for its similarity to Southern blotting, which detects DNA fragments, and Northern blotting, which detects mRNAs. In Western blotting, a protein mixture is electrophoretically separated on an **SDS-polyacrylamide gel (SDS-PAGE),** a slab gel infused with sodium dodecyl sulfate (SDS), a dissociating agent (Figure 6-12). The protein bands are transferred to a nitrocellulose membrane by electrophoresis, and the individual protein bands are identified by flooding the membrane with radiolabeled or enzyme-linked polyclonal or monoclonal antibody specific for the protein of interest. The Ag-Ab complexes that form on the band containing the protein recognized by the antibody can be visualized in a variety of ways. If the protein of interest was bound by a radioactive antibody, its position on the blot can be determined by exposing the membrane to a sheet of x-ray film, a procedure called autoradiography. However, the most generally used detection procedures employ enzyme-linked antibodies against the protein. After binding of the enzyme-antibody conjugate, addition of a chromogenic substrate that produces a highly colored and insoluble product causes the appearance of a colored band at the site of the target antigen. Even greater sensitivity can be achieved if a chemiluminescent compound with suitable enhancing agents is used to produce light at the antigen site.

Western blotting can also identify a specific antibody in a mixture. In this case, known antigens of well-defined molecular weight are separated by SDS-PAGE and blotted onto nitrocellulose. The separated bands of known antigens are then probed with the sample suspected of containing antibody specific for one or more of these antigens. Reaction of an antibody with a band is detected by using either radiolabeled or enzyme-linked secondary antibody that is specific for the species of the antibodies in the test sample. The most widely used application of this procedure is in confirmatory testing for HIV, where Western blotting is used to determine whether the patient has antibodies that react with one or more viral proteins.

## Immunoprecipitation

The immunoprecipitation technique has the advantage of allowing the isolation of the antigen of interest for further analysis. It also provides a sensitive assay for the presence of a particular antigen in a given cell or tissue type. An extract produced by disruption of cells or tissues is mixed with an antibody against the antigen of interest in order to form an antigen-antibody complex that will precipitate. However, if the antigen concentration is low (often the case in cell and tissue extracts), the assembly of antigen-antibody complexes into precipitates can take hours, even days, and it is difficult to isolate the small amount of immunoprecipitate that forms.

**FIGURE 6-11** In the ELISPOT assay, a well is coated with antibody against the antigen of interest, a cytokine in this example, and then a suspension of a cell population thought to contain some members synthesizing and secreting the cytokine is layered onto the bottom of the well and incubated. Most of the cytokine molecules secreted by a particular cell react with nearby well-bound antibodies. After the incubation period, the well is washed and an enzyme-labeled anticytokine antibody is added. After washing away unbound antibody, a chromogenic substrate that forms an insoluble colored product is added. The colored product (purple) precipitates and forms a spot only on the areas of the well where cytokine-secreting cells were deposited. By counting the number of colored spots, it is possible to determine how many cytokine-secreting cells were present in the added cell suspension.

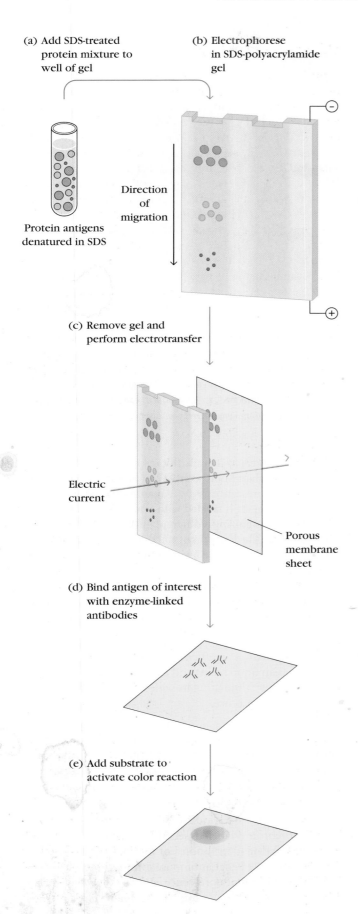

(a) Add SDS-treated protein mixture to well of gel

(b) Electrophorese in SDS-polyacrylamide gel

Protein antigens denatured in SDS

Direction of migration

(c) Remove gel and perform electrotransfer

Electric current

Porous membrane sheet

(d) Bind antigen of interest with enzyme-linked antibodies

(e) Add substrate to activate color reaction

**FIGURE 6-12 In Western blotting, a protein mixture is (a) treated with SDS, a strong denaturing detergent, (b) then separated by electrophoresis in an SDS polyacrylamide gel (SDS-PAGE), which separates the components according to their molecular weight; lower-molecular-weight components migrate farther than higher-molecular-weight ones.** (c) The gel is removed from the apparatus and applied to a protein-binding sheet of nitrocellulose or nylon, and the proteins in the gel are transferred to the sheet by the passage of an electric current. (d) Addition of enzyme-linked antibodies detects the antigen of interest, and (e) the position of the antibodies is visualized by means of an ELISA reaction that generates a highly colored insoluble product that is deposited at the site of the reaction. Alternatively, a chemiluminescent ELISA can be used to generate light that is readily detected by exposure of the blot to a piece of photographic film.

There are a number of ways to avoid these limitations. One is to attach the antibody to a solid support, such as a synthetic bead, which allows the antigen-antibody complex to be collected by centrifugation. Another is to add a secondary antibody specific for the primary antibody to bind the antigen-antibody complexes. If the secondary antibody is attached to a bead, the immune complexes can be collected by centrifugation. A particularly ingenious version of this procedure involves the coupling of the secondary antibody to magnetic beads. After the secondary antibody binds to the primary antibody, immunoprecipitates are collected by placing a magnet against the side of the tube (Figure 6-13).

When used in conjunction with biosynthetic radioisotope labeling, immunoprecipitation can also be used to determine whether a particular antigen is actually synthesized by a cell or tissue. Radiolabeling of proteins synthesized by cells of interest can be done by growing the cells in cell culture medium containing one or more radiolabeled amino acids. Generally, the amino acids used for this application are those most resistant to metabolic modification, such as leucine, cysteine, or methionine. After growth in the radioactive medium, the cells are lysed and subjected to a primary antibody specific for the antigen of interest. The Ag-Ab complex is collected by immunoprecipitation, washed free of unincorporated radiolabeled amino acid and other impurities, and then analyzed. The complex can be counted in a scintillation counter to obtain a quantitative determination of the amount of the protein synthesized. Further analysis often involves disruption of the complex, usually by use of SDS and heat, so that the identity of the immunoprecipitated antigen can be confirmed by checking that its molecular weight is that expected for the antigen of interest. This is done by separation of the disrupted complex by SDS-PAGE and subsequent autoradiography to determine the position of the radiolabeled antigen on the gel.

(a)                    (b)                    (c)

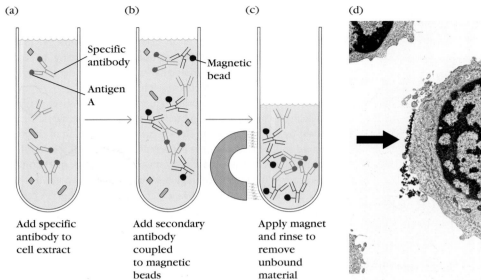

Add specific          Add secondary          Apply magnet
antibody to           antibody               and rinse to
cell extract          coupled                remove
                      to magnetic            unbound
                      beads                  material

(d)

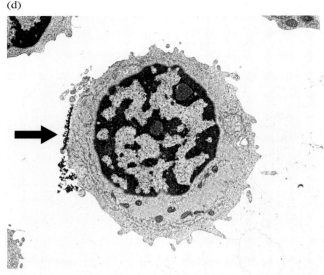

**FIGURE 6-13 Immunoprecipitates can be collected using magnetic beads coupled to a secondary antibody.** (a) Treatment of a cell extract containing antigen A (red) with a mouse anti-A antibody (blue) results in the formation of antigen-antibody complexes. (b) Addition of magnetic beads to which a rabbit antimouse antibody is linked binds the antigen-antibody complexes (and any unreacted mouse Ig).

(c) Placing a magnet against the side of the tube allows the rapid collection of the antigen-antibody complexes. After rinsing to remove any unbound material, the antigen-antibody complexes can be dissociated and the antigen studied. (d) An electron micrograph showing a cell with magnetic beads attached to its surface via antibodies. *[Part d, P. Groscurth, Institute of Anatomy, University of Zurich-Irchel.]*

## Immunofluorescence

In 1944, Albert Coons showed that antibodies could be labeled with molecules that have the property of fluorescence. Fluorescent molecules absorb light of one wavelength (excitation) and emit light of another wavelength (emission). If antibody molecules are tagged with a fluorescent dye, or **fluorochrome,** immune complexes containing these fluorescently labeled antibodies (FA) can be detected by colored light emission when excited by light of the appropriate wavelength. Antibody molecules bound to antigens in cells or tissue sections can similarly be visualized. The emitted light can be viewed with a fluorescence microscope, which is equipped with a UV light source. In this technique, known as **immunofluorescence,** fluorescent compounds such as fluorescein and rhodamine are in common use, but other highly fluorescent substances are also routinely employed, such as phycoerythrin, an intensely colored and highly fluorescent pigment obtained from algae. These molecules can be conjugated to the Fc region of an antibody molecule without affecting the specificity of the antibody. Each of the fluorochromes below absorbs light at one wavelength and emits light at a longer wavelength:

■ **Fluorescein,** an organic dye that is the most widely used label for immunofluorescence procedures, absorbs blue light (490 nm) and emits an intense yellow-green fluorescence (517 nm).

■ **Rhodamine,** another organic dye, absorbs in the yellow-green range (515 nm) and emits a deep red fluorescence

(546 nm). Because it emits fluorescence at a longer wavelength than fluorescein, it can be used in two-color immunofluorescence assays. An antibody specific to one determinant is labeled with fluorescein, and an antibody recognizing a different antigen is labeled with rhodamine. The location of the fluorescein-tagged antibody will be visible by its yellow-green color, easy to distinguish from the red color emitted where the rhodamine-tagged antibody has bound. By conjugating fluorescein to one antibody and rhodamine to another, one can, for example, visualize simultaneously two different cell membrane antigens on the same cell.

■ **Phycoerythrin** is an efficient absorber of light (~30-fold greater than fluorescein) and a brilliant emitter of red fluorescence, stimulating its wide use as a label for immunofluorescence.

Fluorescent-antibody staining of cell membrane molecules or tissue sections can be direct or indirect (Figure 6-14). In **direct staining,** the specific antibody (the primary antibody) is directly conjugated with fluorescein; in **indirect staining,** the primary antibody is unlabeled and is detected with an additional fluorochrome-labeled reagent. A number of reagents have been developed for indirect staining. The most common is a fluorochrome-labeled secondary antibody raised in one species against antibodies of another species, such as fluorescein-labeled goat antimouse immunoglobulin.

Indirect immunofluorescence staining has two advantages over direct staining. First, the primary antibody does not need to be conjugated with a fluorochrome. Because the

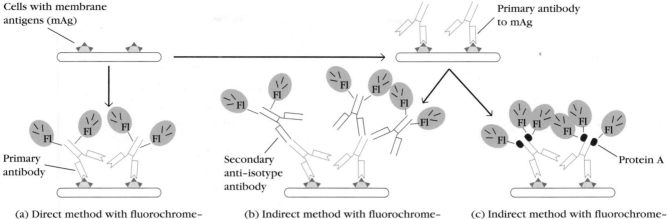

(a) Direct method with fluorochrome-labeled antibody to mAg

(b) Indirect method with fluorochrome-labeled anti–isotype antibody

(c) Indirect method with fluorochrome-labeled protein A

**FIGURE 6-14 Direct and indirect immunofluorescence staining of membrane antigen (mAg).** Cells are affixed to a microscope slide. In the direct method (a), cells are stained with anti-mAg antibody that is labeled with a fluorochrome (Fl). In the indirect methods (b and c), cells are first incubated with unlabeled anti-mAg antibody and then stained with a fluorochrome-labeled secondary reagent that binds to the primary antibody. Cells are viewed under a fluorescence microscope to see if they have been stained. (d) In this micrograph, antibody molecules bearing μ heavy chains are detected by indirect staining of cells with rhodamine-conjugated second antibody. *[Part d, H. A. Schreuder et al., 1997, Nature **386**:196, courtesy H. Schreuder, Hoechst Marion Roussel.]*

(d)

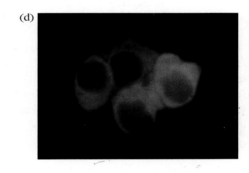

supply of primary antibody is often a limiting factor, indirect methods avoid the loss of antibody that usually occurs during the conjugation reaction. Second, indirect methods increase the sensitivity of staining because multiple molecules of the fluorochrome reagent bind to each primary antibody molecule, increasing the amount of light emitted at the location of each primary antibody molecule.

Immunofluorescence has been applied to identify a number of subpopulations of lymphocytes, notably the CD4+ and CD8+ T-cell subpopulations. The technique is also suitable for identifying bacterial species, detecting Ag-Ab complexes in autoimmune disease, detecting complement components in tissues, and localizing hormones and other cellular products stained in situ. Indeed, a major application of the fluorescent antibody technique is the localization of antigens in tissue sections or in subcellular compartments. Because it can be used to map the actual location of target antigens, fluorescence microscopy is a powerful tool for relating the molecular architecture of tissues and organs to their overall gross anatomy.

## Flow Cytometry and Fluorescence

The fluorescent antibody techniques described are extremely valuable qualitative tools, but they do not give quantitative data. This shortcoming was remedied by development of the flow cytometer, which was designed to automate the analysis and separation of cells stained with fluorescent antibody. The flow cytometer uses a laser beam and light detector to count single intact cells in suspension (Figure 6-15). Every time a cell passes the laser beam, light is deflected from the detector, and this interruption of the laser signal is recorded. Those cells having a fluorescently tagged antibody bound to their cell surface antigens are excited by the laser and emit light that is recorded by a second detector system located at a right angle to the laser beam. The simplest form of the instrument counts each cell as it passes the laser beam and records the level of fluorescence the cell emits; an attached computer generates plots of the number of cells as the ordinate and their fluorescence intensity as the abscissa. More sophisticated versions of the instrument are capable of sorting populations of cells into different containers according to their fluorescence profile. Use of the instrument to determine which and how many members of a cell population bind fluorescently labeled antibodies is called *analysis;* use of the instrument to place cells having different patterns of reactivity into different containers is called *cell sorting.*

The flow cytometer has multiple applications to clinical and research problems. A common clinical use is to determine the kind and number of white blood cells in blood samples. By treating appropriately processed blood samples with a fluorescently labeled antibody and performing flow cytometric analysis, one can obtain the following information:

- *How many cells express the target antigen as an absolute number and also as a percentage of cells passing the beam.* For example, if one uses a fluorescent antibody specific

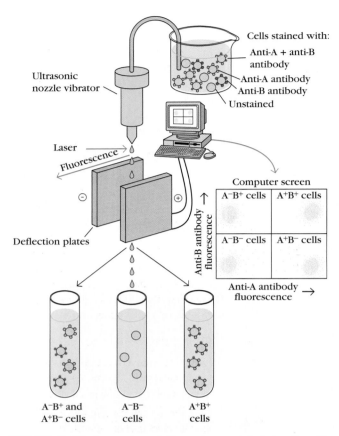

**FIGURE 6-15 Separation of fluorochrome-labeled cells with the flow cytometer.** In the example shown, a mixed cell population is stained with two antibodies, one specific for surface antigen A and the other specific for surface antigen B. The anti-A antibodies are labeled with fluorescein (green) and the anti-B antibodies with rhodamine (red). The stained cells are loaded into the sample chamber of the cytometer. The cells are expelled, one at a time, from a small vibrating nozzle that generates microdroplets, each containing no more than a single cell. As it leaves the nozzle, each droplet receives a small electric charge, and the computer that controls the flow cytometer can detect exactly when a drop generated by the nozzle passes through the beam of laser light that excites the fluorochrome. The intensity of the fluorescence emitted by each droplet that contains a cell is monitored by a detector and displayed on a computer screen. Because the computer tracks the position of each droplet, it is possible to determine when a particular droplet will arrive between the deflection plates. By applying a momentary charge to the deflection plates when a droplet is passing between them, it is possible to deflect the path of a particular droplet into one or another collecting vessel. This allows the sorting of a population of cells into subpopulations having different profiles of surface markers.

In the computer display, each dot represents a cell. Cells that fall into the lower left-hand panel have background levels of fluorescence and are judged not to have reacted with either antibody anti-A or anti-B. Those that appear in the upper-left panel reacted with anti-B but not anti-A, and those in the lower-right panel reacted with anti-A but not anti-B. The upper-right panel contains cells that react with both anti-A and anti-B. In the example shown here, the $A^-B^-$—and the $A^+B^+$—subpopulations have each been sorted into a separate tube. Staining with anti-A and anti-B fluorescent antibodies allows four subpopulations to be distinguished: $A^-B^-$, $A^+B^+$, $A^-B^+$, and $A^+B^-$.

for an antigen present on all T cells, it would be possible to determine the percentage of T cells in the total white blood cell population. Then, using the cell-sorting capabilities of the flow cytometer, it would be possible to isolate the T-cell fraction of the leukocyte population.

■ *The distribution of cells in a sample population according to antigen densities as determined by fluorescence intensity.* It is thus possible to obtain a measure of the distribution of antigen density within the population of cells that possess the antigen. This is a powerful feature of the instrument, since the same type of cell may express different levels of antigen depending on its developmental or physiological state.

■ *The size of cells.* This information is derived from analysis of the light-scattering properties of members of the cell population under examination.

Flow cytometry also makes it possible to analyze cell populations that have been labeled with two or even three different fluorescent antibodies. For example, if a blood sample is reacted with a fluorescein-tagged antibody specific for T cells and also with a phycoerythrin-tagged antibody specific for B cells, the percentages of B and T cells may be determined simultaneously with a single analysis. Numerous variations of such multicolor analyses are common, including experiments employing many "colors" at once.

Flow cytometry now occupies a key position in immunology and cell biology, and it has become an indispensable clinical tool as well. In many medical centers, the flow cytometer is one of the essential tools for the detection and classification of leukemias (see Clinical Focus). The choice of treatment for leukemia depends heavily on the cell types involved, and precise identification of the neoplastic cells is an essential part of clinical practice. Likewise, the rapid measurement of T-cell subpopulations, an important prognostic indicator in AIDS, is routinely done by flow-cytometric analysis. In this procedure, labeled monoclonal antibodies against the major T-cell subtypes bearing the CD4 and CD8 antigens are used to determine their ratios in the patient's blood. When the number of CD4 T cells falls below a certain level, the patient is at high risk for opportunistic infections.

## Alternatives to Antigen-Antibody Reactions

As a defense against host antibodies, some bacteria have evolved the ability to make proteins that bind to the Fc region of IgG molecules with high affinity ($K_a \sim 10^8$). One such molecule, known as **protein A,** is found in the cell walls of some strains of *Staphylococcus aureus,* and another, **protein G,** appears in the walls of group C and G *Streptococcus.* By cloning the genes for protein A and protein G and generating a hybrid of both, one can make a recombinant protein, known as **protein A/G,** that combines features of both. These molecules are useful because they bind IgG from many different species. Thus, they can be labeled with fluorochromes

## CLINICAL FOCUS
# Flow Cytometry and Leukemia Typing

**Leukemia** is the unchecked proliferation of an abnormal clone of hematopoietic cells. Typically, leukemic cells respond poorly or inappropriately to regulatory signals, display aberrant patterns of differentiation, or even fail to differentiate. Furthermore, they sometimes suppress the growth of normal lymphoid and myeloid cells. Leukemia can arise at any maturational stage of any one of the hematopoietic lineages. Lymphocytic leukemias display many characteristics of cells of the lymphoid lineage; another broad group, myelogenous leukemias, have attributes of members of the myeloid lineage. Aside from lineage, many leukemias can be classified as acute or chronic. Some examples are acute lymphocytic leukemia (ALL), the most common childhood leukemia; acute myelogenous leukemia (AML), found more often in adults than in children; and chronic lymphocytic leukemia (CLL), which is rarely seen in children but is the most common form of adult leukemia in the Western world. A fourth type, chronic myelogenous leukemia (CML), occurs much more often in older adults than in children.

The diagnosis of leukemia is based on detection of abnormal cells in the bloodstream and bone marrow. Designing the most appropriate therapy for the patient requires knowing which type of leukemia is present. In this regard, two of the important questions are: (1) What is the lineage of the abnormal cells, and, (2) What is their maturational stage? A variety of approaches, including cytologic examination of cell morphology and staining characteristics, immunophenotyping, and, in some cases, an analysis of gene rearrangements, are useful in answering these questions. One of the most powerful of these approaches is immunophenotyping, the determination of the profile of selected cell surface markers displayed by the leukemic cell. Although leukemia-specific antigens have not yet been found, profiles of expressed surface antigens often can establish cell lineage, and they are frequently helpful in determining the maturational stages present in leukemic cell populations. For example, an abnormal cell that displays surface immunoglobulin would be assigned to the B-cell lineage and its maturational stage would be that of a mature B cell. On the other hand, a cell that had cytoplasmic μ heavy chains, but no surface immunoglobulin, would be a B-lineage leukemic cell but at the maturational stage of a pre–B cell. The most efficient and precise technology for immunophenotyping uses flow cytometry and monoclonal antibodies. The availability of monoclonal antibodies specific for each of the scores of antigens found on various types and subtypes of hematopoietic cells has made it possible to identify patterns of antigen expression that are typical of cell lineages, maturational stages, and a number of different types of leukemia. Most cancer centers are equipped with flow cytometers that are capable of performing and interpreting the multiparameter analyses necessary to provide useful profiles of surface markers on tumor cell populations. Flow cytometric determination of immunophenotypes allows:

- Confirmation of diagnosis

- Diagnosis when no clear judgment can be made based on morphology or patterns of cytochemical staining

- Identification of aberrant antigen profiles that can help identify the return of leukemia after remission

- Improved prediction of the course of the disease

Distribution of selected markers on some leukemic cell types. Shown are typical surface antigen profiles found on many, but not all, ALLs and CLLs.

An ALL of the pre–B lineage
(the most commonly occurring ALL)

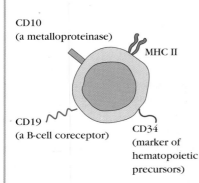

ALL of the T lineage

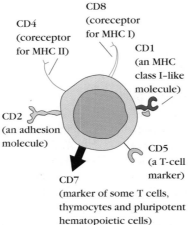

A B-lineage CLL

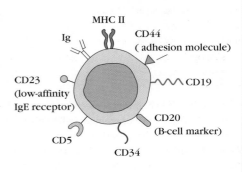

or radioactivity and used to detect IgG molecules in the antigen-antibody complexes formed during ELISA, RIA, or such fluorescence-based assays as flow cytometry or fluorescence microscopy. These bacterial IgG-binding proteins can also be used to make affinity columns for the isolation of IgG.

Egg whites contain a protein called *avidin* that binds biotin, a vitamin that is essential for fat synthesis. Avidin is believed to have evolved as a defense against marauding rodents that rob nests and eat the stolen eggs. The binding between avidin and biotin is extremely specific and of much higher affinity ($K_a \sim 10^{15}$) than any known antigen-antibody reaction. A bacterial protein called **streptavidin,** made by *Streptomyces avidinii,* has similarly high affinity and specificity. The extraordinary affinity and exquisite specificity of the interaction of these proteins with biotin is widely used in many immunological procedures. The primary or secondary antibody is labeled with biotin and allowed to react with the target antigen, and the unbound antibody is then washed away. Subsequently, streptavidin or avidin conjugated with an enzyme, fluorochrome, or radioactive label is used to detect the bound antibody.

## Immunoelectron Microscopy

The fine specificity of antibodies has made them powerful tools for visualizing specific intracellular tissue components by **immunoelectron microscopy.** In this technique, an electron-dense label is either conjugated to the Fc portion of a specific antibody for direct staining or conjugated to an anti-immunoglobulin reagent for indirect staining. A number of electron-dense labels have been employed, including *ferritin* and *colloidal gold.* The electron-dense label can be visualized with the electron microscope as small black dots. In the case of immunogold labeling, different antibodies can be conjugated with gold particles of different sizes, allowing identification of several antigens within a cell by the different sizes of the electron-dense gold particles attached to the antibodies (Figure 6-16).

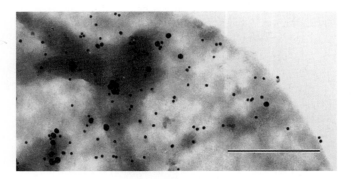

**FIGURE 6-16 An immunoelectronmicrograph of the surface of a B-cell lymphoma was stained with two antibodies: one against class II MHC molecules labeled with 30-nm gold particles and another against MHC class I molecules labeled with 15-nm gold particles.** The density of class I molecules exceeds that of class II on this cell. Bar = 500 nm. *[From A. Jenei et al., 1997, PNAS **94**:7269–7274; courtesy of A. Jenei and S. Damjanovich, University Medical School of Debrecen, Hungary.]*

## SUMMARY

- Antigen-antibody interactions depend on four types of noncovalent interactions: hydrogen bonds, ionic bonds, hydrophobic interactions, and van der Waals interactions.

- The affinity constant, which can be determined by Scatchard analysis, provides a quantitative measure of the strength of the interaction between an epitope of the antigen and a single binding site of an antibody. The avidity reflects the overall strength of the interactions between a multivalent antibody molecule and a multivalent antigen molecule at multiple sites.

- Surface plasmon resonance (SPR) allows the characterization of the affinity and kinetics of antigen-antibody interactions.

- The interaction of a soluble antigen and precipitating antibody in a liquid or gel medium forms an Ag-Ab precipitate. Electrophoresis can be combined with precipitation in gels in a technique called immunoelectrophoresis.

- The interaction between a particulate antigen and agglutinating antibody (agglutinin) produces visible clumping, or agglutination, that forms the basis of simple, rapid, and sensitive immunoassays.

- Radioimmunoassay (RIA) is a highly sensitive and quantitative procedure that utilizes radioactively labeled antigen or antibody.

- The enzyme-linked immunosorbent assay (ELISA) depends on an enzyme-substrate reaction that generates a colored reaction product. ELISA assays that employ chemiluminescence instead of a chromogenic reaction are the most sensitive immunoassays available.

- In Western blotting, a protein mixture is separated by electrophoresis; then the protein bands are electrophoretically transferred onto nitrocellulose and identified with labeled antibody or labeled antigen.

- Fluorescence microscopy using antibodies labeled with fluorescent molecules can be used to visualize antigen on or within cells.

- Flow cytometry provides an unusually powerful technology for the quantitative analysis and sorting of cell populations labeled with one or more fluorescent antibodies.

## References

Bierer, B., et al. 2005. *Current Protocols in Immunology.* Wiley, New York.

Harlow, E., and D. Lane. 1999. *Using Antibodies: A Laboratory Manual.* Cold Spring Harbor Laboratory Press, Cold Spring Harbor, NY.

Herzenberg, L. A., ed. 1996. *Weir's Handbook of Experimental Immunology,* 5th ed. Oxford, Blackwell Scientific Publications.

Malmqvist, M. 1993. Surface plasmon resonance for detection and measurement of antibody-antigen affinity and kinetics. *Current Opinion in Immunology* **5:**282.

Rich, R. L., and D. G. Myszka. 2003. Spying on HIV with SPR. *Trends in Microbology* **11**:124.

Rose, N. R., et al. 1997. *Manual of Clinical Laboratory Immunology.* American Society of Microbiology, Washington, DC.

Wild, D., ed. 2005. *The Immunoassay Handbook,* 3rd ed. Elsevier, Amsterdam.

 ## Useful Web Sites

**http://pathlabsofark.com/flowcyttests.html**

Explore the Pathology Laboratories of Arkansas to see what kinds of samples are taken from patients and what markers are used to evaluate lymphocyte populations by flow cytometry.

**http://jcsmr.anu.edu.au/facslab/protocol.html**

At the highly informative Australian Flow Cytometry Group Web site, one can find a carefully detailed and illustrated guide to the interpretation of flow cytometric analyses of clinical samples.

**http://www.rockefeller.edu/spectroscopy/ instruments_spr.php**

Excellent definition of the units of response expressed in surface plasmon resonance.

 ## Study Questions

**CLINICAL FOCUS QUESTION** Flow-cytometric analysis for the detection and measurement of subpopulations of leukocytes, including those of leukemia, is usually performed using monoclonal antibodies. Why?

1. Indicate whether each of the following statements is true or false. If you think a statement is false, explain why.

   a. Indirect immunofluorescence is a more sensitive technique than direct immunofluorescence.

   b. Most antigens induce a polyclonal response.

   c. A papain digest of anti-SRBC antibodies can agglutinate sheep red blood cells (SRBCs).

   d. A pepsin digest of anti-SRBC antibodies can agglutinate SRBCs.

   e. Indirect immunofluorescence can be performed using a Fab fragment as the primary, nonlabeled antibody.

   f. For precipitation to occur, both antigen and antibody must be multivalent.

   g. Analysis of a cell population by flow cytometry can simultaneously provide information on both the size distribution and antigen profile of cell populations containing several different cell types.

   h. ELISA tests using chemiluminescence are more sensitive than chromogenic ones and precipitation tests are more sensitive than agglutination tests.

   i. Western blotting and immunoprecipitation assays are useful quantitative assays for measuring the levels of proteins in cells or tissues.

   j. Assume antibody A and antibody B both react with an epitope C. Furthermore, assume that antibody A has a $K_a$ 5 times greater than that of antibody B. The strength of the monovalent reaction of antibody A with epitope C will always be greater than the avidity of antibody B for an antigen with multiple copies of epitope C.

2. You have obtained a preparation of purified bovine serum albumin (BSA) from normal bovine serum. To determine whether any other serum proteins remain in this preparation of BSA, you decide to use immunoelectrophoresis.

   a. What antigen would you use to prepare the antiserum needed to detect impurities in the BSA preparation?

   b. Assuming that the BSA preparation is pure, draw the immunoelectrophoretic pattern you would expect if the assay was performed with bovine serum in a well above a trough containing the antiserum you prepared in (a) and the BSA sample in a well below the trough as shown below:

3. The labels from four bottles (A, B, C, and D) of hapten-carrier conjugates were accidentally removed. However, it was known that each bottle contained either (1) hapten 1–carrier 1 (H1-C1), (2) hapten 1–carrier 2 (H1-C2), (3) hapten 2–carrier 1 (H2-C1), or (4) hapten 2–carrier 2 (H2-C2). Carrier 1 has a molecular weight of 60,000 daltons, and carrier 2 has a molecular weight of over 120,000 daltons. Assume you have an anti-H1 antibody and an anti-H-2 antibody and a molecular-weight marker that is 100,000 daltons. Use Western blotting to determine the contents of each bottle and show the Western blots you would expect from 1, 2, 3, and 4. Your answer should also tell which antibody or combination of antibodies was used to obtain each blot.

4. The concentration of a small amount (250 nanograms/ml) of hapten can be determined by which of the following assays: (a) ELISA (chromogenic), (b) Ouchterlony method, (c) RIA, (d) fluorescence microscopy, (e) flow cytometry, (f) immunoprecipitation, (g) immunoelectron microscopy, (h) ELISPOT assay, (i) chemiluminescent ELISA?

5. You have a myeloma protein, X, whose isotype is unknown and several other myeloma proteins of all known isotypes (e.g., IgG, IgM, IgA, and IgE).

   a. How could you produce isotype-specific antibodies that could be used to determine the isotype of myeloma protein, X?

   b. How could you use this anti-isotype antibody to measure the level of myeloma protein X in normal serum?

6. For each antigen or antibody listed below, indicate an appropriate assay method and the necessary test reagents. Keep in mind the sensitivity of the assay and the expected concentration of each protein.

   a. IgG in serum

   b. Insulin in serum

   c. IgE in serum

d. Complement component C3 on glomerular basement membrane

e. Anti-A antibodies to blood-group antigen A in serum

f. Horse meat contamination of hamburger

g. Syphilis spirochete in a smear from a chancre

7. Which of the following does *not* participate in the formation of antigen-antibody complexes?

a. Hydrophobic bonds

b. Covalent bonds

c. Electrostatic interactions

d. Hydrogen bonds

e. Van der Waals forces

8. Explain the difference between antibody affinity and antibody avidity. Which of these properties of an antibody better reflects its ability to contribute to the humoral immune response to invading bacteria?

9. You want to develop a sensitive immunoassay for a hormone that occurs in the blood at concentrations near $10^{-7}$ M. You are offered a choice of three different antisera whose affinities for the hormone have been determined by equilibrium dialysis. The results are shown below in the Scatchard plots.

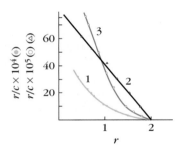

a. What is the value of $K_0$ for each antiserum?

b. What is the valence of each of the antibodies?

c. Which of the antisera might be a monoclonal antibody?

d. Which of the antisera would you use for your assay? Why?

10. In preparing a demonstration for her immunology class, an instructor purified IgG antibodies to sheep red blood cells (SRBCs) and digested some of the antibodies into Fab, Fc, and F(ab.)$_2$ fragments. She placed each preparation in a separate tube, labeled the tubes with a water-soluble marker, and left them in an ice bucket. When the instructor returned for her class period, she discovered that the labels had smeared and were unreadable. Determined to salvage the demonstration, she relabeled the tubes 1, 2, 3, and 4 and proceeded. Based on the test results described below, indi-

cate which preparation was contained in each tube and explain how you identified the contents.

a. The preparation in tube 1 agglutinated SRBCs but did not lyse them in the presence of complement.

b. The preparation in tube 2 did not agglutinate SRBCs or lyse them in the presence of complement. However, when this preparation was added to SRBCs before the addition of whole anti-SRBC, it prevented agglutination of the cells by the whole anti-SRBC antiserum.

c. The preparation in tube 3 agglutinated SRBCs and also lysed the cells in the presence of complement.

d. The preparation in tube 4 did not agglutinate or lyse SRBCs and did not inhibit agglutination of SRBCs by whole anti-SRBC antiserum.

11. Consider equation 1 and derive the form of the Scatchard equation that appears in equation 2.

1. $S + L = SL$
2. $B/F = K_a([S]_t - B)$

Where S = antibody binding sites; [S] = molar concentration of antibody binding sites; L = ligand (monovalent antigen); [L] = molar concentration of ligand; SL = site-ligand complex; [SL] = molar concentration of site ligand complex; B is substituted for [SL] and F for [L]. Hint: It will be helpful to begin by writing the law of mass action for the reaction shown in equation 1.

12. You have taken a job in a clinical lab over the summer to help pay off your student loans. You order a kit designed to detect Hantavirus antibodies in patient sera using indirect ELISA, but when it arrives there are no instructions, just bottles labeled "reagent A," "reagent B," and so forth. Several patients have been potentially exposed to Hantavirus, and you need to use this kit to determine if the patients have actually been exposed.

a. Which of the following would be included in the kit as reagents required to run this assay?

1. Positive control soluble antigen

2. Primary antibody to Hanta virus

3. Substrate

4. Enzyme-labeled secondary antibody to primary

5. Enzyme-labeled secondary antibody to virus

6. Antigen-coated plates

b. Using the kit, you obtain results for three patient sera (P1–P3 in the table below). What are the titers of the patient sera (give a titer for each of the three patients)? Which (if any) of these patients were exposed?

13. A new strain of flu has been discovered that has the potential to infect large numbers of people due to its high rate of transmission. To assess the severity of a wide-spread infection,

**For use with Question 12**

| Blank | − Control | + Control | | 1:1 | 1:10 | 1:100 | 1:1000 | 1:10,000 | 1:100,000 |
|---|---|---|---|---|---|---|---|---|---|
| 0.001 | 0.103 | 0.857 | P1 | 1.002 | 0.562 | 0.352 | 0.202 | 0.096 | 0.005 |
| 0.006 | 0.056 | 0.952 | P2 | 0.002 | 0.005 | 0.008 | 0.002 | 0.004 | 0.001 |
| 0.005 | 0.096 | 0.903 | P3 | 0.568 | 0.203 | 0.086 | 0.079 | 0.082 | 0.045 |

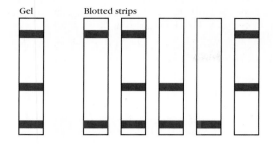

Gel            Blotted strips

you test whether average citizens already express antibodies that could cross-react with the flu antigens. First you run solublized surface antigens from the flu strain in lanes of an SDS-PAGE gel. One lane is then stained with the non-specific stain Coomassie blue (left in the figure above). The remaining lanes are electrophoretically transferred to nitrocellulose and cut into strips so that each lane can be incubated with normal volunteer sera as a primary antibody for Western blotting. The results of the experiment are shown below. From these results, would you expect a widespread health crisis from the new flu strain?

14. Your research advisor has just discovered a new receptor thought to be important in the development of Alzheimer's disease. Other lab members have cloned the receptor and transfected the gene into a cell line for further study. Your task is to perform cell sorting on the transfected cells to screen out nonexpressers and select a population of high-level expressers. You have available a rabbit antibody to the receptor, a control mouse antibody that binds to untransfected cells, fluorescein-conjugated goat anti-mouse antibodies, and rhodamine-conjugated donkey anti-rabbit antibodies.

a. Describe how you would set up the FACS experiment.

b. Draw a representative FACS histogram of the expected results with the $x$ axis showing the level of fluorescein and the $y$ axis showing the level of rhodamine. Include in your histogram where you would set the gates for cell separation.

15. What is the advantage of using an ELISPOT versus a standard sandwich ELISA? What are the similarities and differences between these two assays?

**ANALYZE THE DATA** Kanayama and colleagues (*J. Immunology* **169**:6865, 2002) made a transgenic mouse (QM mouse) whose major B-cell receptor (present on 80% of all B cells in the mouse) was made up of $V_H$ 17.2.25-encoded heavy chains and $\lambda 1$ or $\lambda 2$ light chains. The antigen receptor was specific for the hapten 4-hydroxy-3-nitrophenylacetyl (NP). These investigators estimated the relative affinity of these B cells for other NP-related haptens using competitive inhibition assays ($IC_{50} =$ concentration needed to compete for 50% of the binding of radiolabeled NP) presented in the table. They also immunized the footpads of QM mice with pNP-CGG (chicken globulin) and then bled the mice at various intervals (graphs). Answer the following questions based on the information presented and what you have learned from reading this book.

a. True or false: This idiotype (antigen receptor) is specific for the NP hapten; therefore, the Ig receptor will always have the highest affinity for NP compared to other NP-related haptens. Explain.

b. Did class switching occur in response to immunization? Explain your answer.

| DNP ligand | | $IC_{50}$ (M) |
|---|---|---|
| NIP | 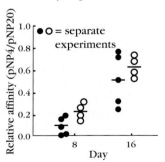 structure with $O_2N$, $HO$, $CH_2CO-$, $I$ | $2 \times 10^{-5}$ (15) |
| NP | structure with $O_2N$, $HO$, $CH_2CO-$ | $3 \times 10^{-4}$ (1) |
| mNP | structure with $O_2N$, $CH_2CO-$ | $3 \times 10^{-4}$ (1) |
| pNP | $O_2N$ — ring — $CH_2CO-$ | $6 \times 10^{-3}$ (0.05) |
| HP | $HO$ — ring — $CH_2CO-$ | No inhibition |

c. What two processes contributed to the outcome illustrated in graph b?

d. $\lambda$ light chains were expressed by the QM mouse B cells examined in this study. What can you say about the events that occurred in the pre–B-cell stage of B-cell differentiation that led to the production of antibody with $\lambda$ light chains?

e. Do B cells expressing IgM with the $V_H$ 17.2.25-encoded heavy chain and the $\lambda 1$ light chain have the same idiotype as B cells expressing IgG with $V_H$ 17.2.25-encoded heavy chain and the $\lambda 2$ light chain?

(a) Anti-IgG response to pNP over time

(b) Relative affinity of anti-pNP antibody response over time

Graph (a): $x$ axis Day (0, 8, 16), $y$ axis Relative concentration of IgG (measured by ELISA), scale 0.01 to 1.0. ● ○ = separate experiments

Graph (b): $x$ axis Day (8, 16), $y$ axis Relative affinity (pNP4/pNP20), scale 0.0 to 1.0. ● ○ = separate experiments

 **Interactive Study**

**www.whfreeman.com/kuby**

 SELF-TEST
Review of Key Terms

ANIMATION
hCG Pregnancy Test (ELISA)

# chapter 7

## The Complement System

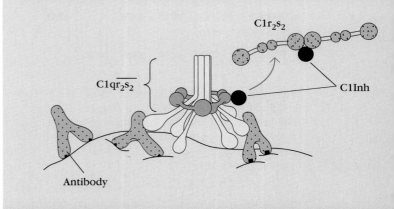

*Initiation of the classical complement pathway by antibody binding to a cell membrane and its inhibition by C1 inhibitor.*

- The Functions of Complement
- The Components of Complement
- Complement Activation
- Regulation of the Complement System
- Biological Consequences of Complement Activation
- Complement Deficiencies

THE COMPLEMENT SYSTEM IS THE MAJOR EFFECTOR OF THE humoral branch of the immune system. Research on complement began in the 1890s, when Jules Bordet at the Institut Pasteur in Paris showed that sheep antiserum to the bacterium *Vibrio cholerae* caused lysis of the bacteria and that heating the antiserum destroyed its bacteriolytic activity. Surprisingly, the ability to lyse the bacteria was restored to the heated serum by adding fresh serum that contained no antibodies directed against the bacterium and that was unable to kill the bacterium by itself. Bordet correctly reasoned that bacteriolytic activity requires two different substances: first, the specific antibacterial antibodies, which survive the heating process, and a second, heat-sensitive component responsible for the lytic activity. Bordet devised a simple test for the lytic activity, the easily detected lysis of antibody-coated red blood cells, called **hemolysis.** Paul Ehrlich in Berlin independently carried out similar experiments and coined the term *complement,* defining it as "the activity of blood serum that completes the action of antibody." In ensuing years, researchers discovered that the lytic action of complement was the result of interactions of a complex group of proteins. It was further shown that the results of complement activation range far beyond the originally observed antibody-mediated cell lysis and that complement plays a key role in both innate and adaptive immunity.

An important aspect of complement activity is its position in the innate immune system. Although the discovery of complement and most early studies were linked to the activity of complement following antibody binding, a major role for this system is the recognition and destruction of pathogens based on recognition of pathogen-associated molecular patterns, or PAMPs, rather than on antibody specificity. The activation of the complement cascade may be initiated by several proteins that circulate in normal serum. These molecules, termed **acute phase proteins,** possess pattern-recognition capacity and undergo changes in concentration during inflammation. All of the various actions of the complement system may thus be triggered not only by antibody, as shown originally by Bordet, but also by components of innate immunity.

This chapter describes the components of the complement system, their activation via three major pathways, the regulation of the complement system, the effector functions of various complement components, and the clinical consequences of complement deficiencies. The Clinical Focus describes the consequences of a defect in proteins that regulate complement activity.

## The Functions of Complement

Research on complement extends to include a plethora of soluble and cell-bound proteins. The biological activities of this system affect both innate and adaptive immunity and reach far beyond antibody-mediated lysis of bacteria and red blood cells. Structural comparisons of the proteins involved in complement pathways place the origin of this system in invertebrate organisms possessing the most

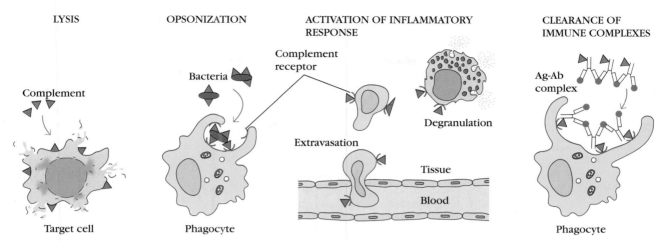

**FIGURE 7-1  The multiple activities of the complement system.** Serum complement proteins and membrane-bound complement receptors partake in a number of immune activities: lysis of foreign cells by antibody-dependent or antibody-independent pathways; opsonization or uptake of particulate antigens, including bacteria, by phagocytes; activation of inflammatory responses; and clearance of circulating immune complexes by cells in the liver and spleen. Soluble complement proteins are schematically indicated by a triangle and receptors by a semicircle; no attempt is made to differentiate among individual components of the complement system here.

rudimentary innate immune systems. Primitive multicellular organisms lacking adaptive immune components possess proteins related to the complement system. By contrast, interaction of cellular receptors with complement proteins can direct B-cell activities, which reveals a role for this system in the highly developed adaptive immune system. Thus, in vertebrate species, we have a system that straddles innate and adaptive immunity, contributing to each in a variety of ways.

After initial activation, the various complement components interact, in a highly regulated cascade, to carry out a number of basic functions (Figure 7-1), including

- Lysis of cells, bacteria, and viruses

- Opsonization, which promotes phagocytosis of particulate antigens

- Binding to specific complement receptors on cells of the immune system, triggering specific cell functions, inflammation, and secretion of immunoregulatory molecules

- Immune clearance, which removes immune complexes from the circulation and deposits them in the spleen and liver

## The Components of Complement

The soluble proteins and glycoproteins that constitute the complement system are synthesized mainly by liver hepatocytes, although significant amounts are also produced by blood monocytes, tissue macrophages, and epithelial cells of the gastrointestinal and genitourinary tracts. Complement components constitute 5% (by weight) of the serum globulin fraction. Most circulate in the serum in functionally inactive forms as proenzymes, or *zymogens,* which are inactive until proteolytic cleavage removes an inhibitory fragment and exposes the active site of the molecule. The complement reaction sequence starts with an enzyme cascade.

Complement components are designated by numerals (C1–C9), by letter symbols (e.g., factor D), or by trivial names (e.g., homologous restriction factor). Peptide fragments formed by activation of a component are denoted by small letters. In most cases, the smaller fragment resulting from cleavage of a component is designated "a" and the larger fragment designated "b" (e.g., C3a, C3b; note that C2 is an exception: C2a is the larger cleavage fragment). The larger fragments bind to the target near the site of activation, and the smaller fragments diffuse from the site and can initiate localized inflammatory responses by binding to specific receptors. The complement fragments interact with one another to form functional complexes. Those complexes that have enzymatic activity are designated by a bar over the number or symbol (e.g., $\overline{\text{C4b2a}}$, $\overline{\text{C3bBb}}$).

## Complement Activation

Figure 7-2 outlines the pathways of complement activation. The early steps, culminating in the formation of C5b, can occur by the **classical pathway,** the **alternative pathway,** or the **lectin pathway.** The final steps that lead to formation of the membrane-attack complex, or MAC, are identical in all three pathways.

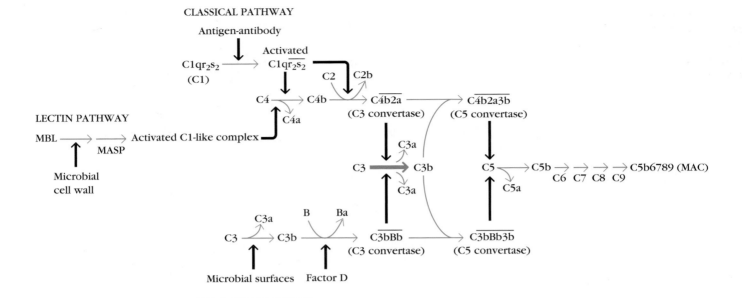

**FIGURE 7-2 Overview of the complement activation pathways.** The classical pathway is initiated when C1 binds to antigen-antibody complexes. The alternative pathway is initiated by binding of spontaneously generated C3b to activating surfaces such as microbial cell walls. The lectin pathway is initiated by binding of the serum protein MBL to the surface of a pathogen. All three pathways generate C3 and C5 convertases and bound C5b, which is converted into a membrane-attack complex (MAC) by a common sequence of terminal reactions.

## The classical pathway begins with antigen-antibody binding

Complement activation by the classical pathway commonly begins with the formation of soluble antigen-antibody complexes (immune complexes) or with the binding of antibody to antigen on a suitable target, such as a bacterial cell. IgM and certain subclasses of IgG (human IgG1, IgG2, and IgG3, but not IgG4) can activate the classical complement pathway. The initial stage of activation involves components C1, C2, C3, and C4, which are present in plasma in functionally inactive forms. Because the components were named in order of their discovery and before their functional roles had been determined, their numerical designations do not always reflect the order in which they react.

The formation of an antigen-antibody complex induces conformational changes in the Fc portion of the IgM molecule that expose a binding site for the C1 component of the complement system. C1 in serum is a macromolecular complex consisting of C1q and two molecules each of C1r and C1s, held together in a complex ($C1qr_2s_2$) stabilized by $Ca^{2+}$ ions. The C1q molecule is composed of 18 polypeptide chains that associate to form six collagen-like triple helical arms, the tips of which bind to exposed C1q-binding sites in the $C_H2$ domain of the antibody molecule (Figure 7-3). Each C1r and C1s monomer contains a catalytic domain and an interaction domain; the latter facilitates interaction with C1q or with each other.

Each C1 macromolecular complex must bind by its C1q globular heads to at least two Fc sites for a stable C1-antibody interaction to occur. When pentameric IgM is bound to antigen on a target surface, it assumes the so-called staple configuration, in which at least three binding sites for C1q are exposed. Circulating IgM, however, adopts a planar configuration in which the C1q-binding sites are not exposed (Figure 7-4) and therefore cannot activate the complement cascade. An IgG molecule, on the other hand, contains only a single C1q-binding site in the $C_H2$ domain of the Fc, so that firm C1q binding is achieved only when two IgG molecules are within 30 to 40 nm of each other on a target surface or in a complex, providing two attachment sites for C1q. This difference accounts for the observation that a single molecule of IgM bound to a red blood cell can activate the classical complement pathway and lyse the red blood cell, whereas some 1000 molecules of cell-bound IgG are required to ensure that two IgG molecules are close enough to each other on the cell surface to initiate C1q binding.

The intermediates in the classical activation pathway are depicted schematically in Figure 7-5. Binding of C1q to Fc-binding sites induces a conformational change in C1r that converts C1r to an active serine protease enzyme, $\overline{C1r}$, which then cleaves C1s to a similar active enzyme, $\overline{C1s}$. $\overline{C1s}$ has two substrates, C4 and C2. The C4 component is a glycoprotein containing three polypeptide chains α, β, and γ. C4 is activated when $\overline{C1s}$ hydrolyzes a small fragment (C4a) from the amino terminus of the α chain, exposing a binding

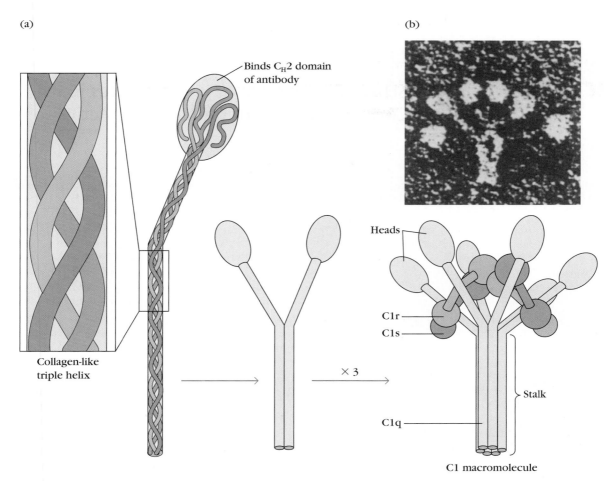

**FIGURE 7-3  Structure of the C1 macromolecular complex.**
(a) C1q molecule consists of 18 polypeptide chains in six collagen-like triple helices, each of which contains one A, B, and C chain, shown as blue, green, and red respectively. The head group of each triplet contains elements of all three chains. In the C1 macromolecule, C1q interacts with two molecules each of C1r and C1s. Each C1r and C1s monomer contains a catalytic domain with enzymatic activity and an interaction domain that facilitates binding with C1q or with each other. (b) Electron micrograph of C1q molecule showing stalk and six globular heads. *[Inset electron micrograph from H. R. Knobel et al., 1975, European Journal of Immunology **5**:78.]*

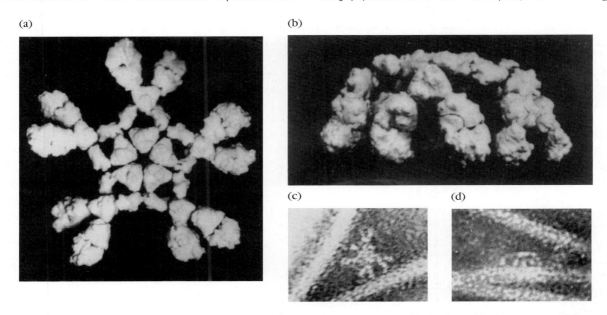

**FIGURE 7-4  Models of pentameric IgM in planar form (a) and "staple" form (b).** Several C1q-binding sites in the Fc region are accessible in the staple form, whereas none are exposed in the planar form. Electron micrographs of IgM antiflagellum antibody bound to flagella, showing the planar form (c) and staple form (d). *[From A. Feinstein et al., 1981, Monographs in Allergy, **17**:28, and 1981, Annals of the New York Academy of Sciences **190**:1104.]*

**OVERVIEW FIGURE 7-5:** Schematic Diagram of Intermediates in the Classical Pathway of Complement Activation

**1** C1q binds antigen-bound antibody. C1r activates autocatalytically and activates the second C1r; both activate C1s.

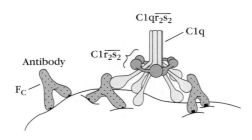

**2** C1s cleaves C4 and C2. Cleaving C4 exposes the binding site for C2. C4 binds the surface near C1 and C2 binds C4, forming C3 convertase.

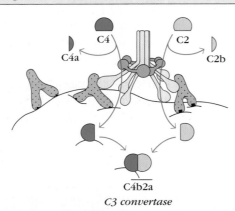

*C3 convertase*

**3** C3 convertase hydrolyzes many C3 molecules. Some combine with C3 convertase to form C5 convertase.

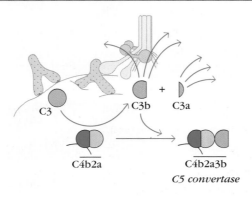

*C5 convertase*

**4** The C3b component of C5 convertase binds C5, permitting C4b2a to cleave C5.

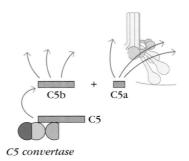

*C5 convertase*

**5** C5b binds C6, initiating the formation of the membrane-attack complex.

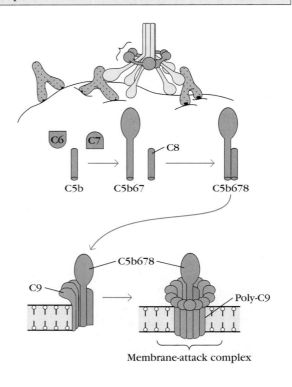

*Membrane-attack complex*

The completed membrane-attack complex (MAC, bottom right) forms a large pore in the membrane.

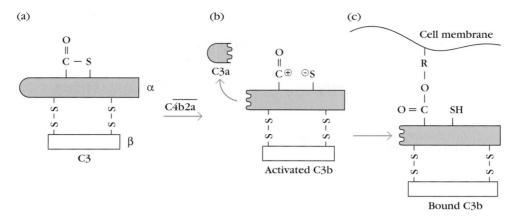

**FIGURE 7-6 Hydrolysis of C3 by C3 convertase $\overline{\text{C4b2a}}$.** (a) Native C3. (b) Activated C3 showing site of cleavage by $\overline{\text{C4b2a}}$, resulting in production of the C3a and C3b fragments. (c) A labile internal thioester bond in C3 is activated as C3b is formed, allowing the C3b fragment to bind to free hydroxyl or amino groups (R) on a cell membrane. Bound C3b exhibits various biological activities, including binding of C5 and binding to C3b receptors on phagocytic cells.

site on the larger fragment (C4b). The C4b fragment attaches to the target surface in the vicinity of C1, and the C2 proenzyme then attaches to the exposed binding site on C4b, where the C2 is then cleaved by the neighboring $\overline{\text{C1s}}$; the smaller fragment (C2b) diffuses away. The resulting $\overline{\text{C4b2a}}$ complex is called C3 convertase, referring to its role in converting C3 into an active form. The smaller fragment from C4 cleavage, C4a, is an anaphylatoxin, or mediator of inflammation, which does not participate directly in the lytic function of the complement cascade; the anaphylatoxins, which include the smaller fragments of C4, C3, and C5, are described below.

The native C3 component consists of two polypeptide chains, α and β. Hydrolysis of a short fragment (C3a) from the amino terminus of the α chain by the C3 convertase generates C3b (Figure 7-6). A single C3 convertase molecule can generate over 200 molecules of C3b, resulting in tremendous amplification at this step of the sequence. Some of the C3b binds to $\overline{\text{C4b2a}}$ to form a trimolecular complex, $\overline{\text{C4b2a3b}}$, called C5 convertase. The C3b component of this complex binds C5 and alters its conformation, allowing the $\overline{\text{C4b2a}}$ component to cleave C5 into C5a, which diffuses away, and C5b, which attaches to C6 and initiates formation of the membrane-attack complex in a sequence described later. Some of the C3b generated by C3 convertase activity does not associate with $\overline{\text{C4b2a}}$; instead it diffuses away and coats immune complexes and particulate antigens, functioning as an opsonin or promoter of phagocytosis. C3b may also bind directly to cell membranes.

## The alternative pathway is antibody-independent

The alternative pathway generates active products similar to those of the classical pathway, but it does so without the requirement of antigen-antibody complexes for initiation. Because no antibody is required, the alternative pathway is a component of the innate immune system. This major pathway of complement activation involves four serum proteins: C3, factor B, factor D, and properdin. The alternative pathway is initiated in most cases by cell surface constituents that are foreign to the host (Table 7-1). For example, both gram-negative and gram-positive bacteria

| TABLE 7-1 | Initiators of the alternative pathway of complement activation |
|---|---|
| **PATHOGENS AND PARTICLES OF MICROBIAL ORIGIN** ||
| Many strains of gram-negative bacteria ||
| Lipopolysaccharides from gram-negative bacteria ||
| Many strains of gram-positive bacteria ||
| Teichoic acid from gram-positive cell walls ||
| Fungal and yeast cell walls (zymosan) ||
| Some viruses and virus-infected cells ||
| Some tumor cells (Raji) ||
| Parasites (trypanosomes) ||
| **NONPATHOGENS** ||
| Human IgG, IgA, and IgE in complexes ||
| Rabbit and guinea pig IgG in complexes ||
| Cobra venom factor ||
| Heterologous erythrocytes (rabbit, mouse, chicken) ||
| Anionic polymers (dextran sulfate) ||
| Pure carbohydrates (agarose, inulin) ||

SOURCE: Adapted from M. K. Pangburn, 1986, in *Immunobiology of the Complement System*, G. Ross, ed., Academic Press, Orlando.

OVERVIEW FIGURE 7-7:    Schematic Diagram of Intermediates in the Formation of Bound C5b by the Alternative Pathway of Complement Activation

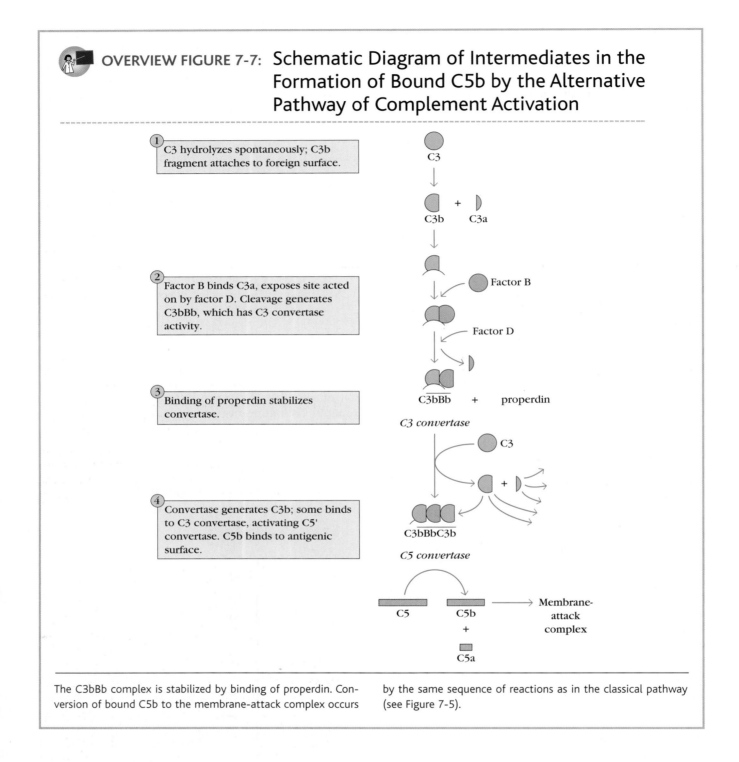

① C3 hydrolyzes spontaneously; C3b fragment attaches to foreign surface.

② Factor B binds C3a, exposes site acted on by factor D. Cleavage generates C3bBb, which has C3 convertase activity.

③ Binding of properdin stabilizes convertase.

④ Convertase generates C3b; some binds to C3 convertase, activating C5' convertase. C5b binds to antigenic surface.

The C3bBb complex is stabilized by binding of properdin. Conversion of bound C5b to the membrane-attack complex occurs by the same sequence of reactions as in the classical pathway (see Figure 7-5).

have cell wall constituents that can activate the alternative pathway. The intermediates in the alternative pathway for generating C5b are shown schematically in Figure 7-7.

In the classical pathway, C3 is rapidly cleaved to C3a and C3b by the enzymatic activity of the C3 convertase. In the alternative pathway, serum C3, which contains an unstable thioester bond, is subject to slow spontaneous hydrolysis to yield C3a and C3b. The C3b component can bind to for-

eign surface antigens (such as those on bacterial cells or viral particles) or even to the host's own cells (see Figure 7-6c). The membranes of most mammalian cells have a high level of sialic acid, which contributes to the rapid inactivation of bound C3b molecules on host cells; consequently, this binding rarely leads to further reactions on the host cell membrane. Because many foreign antigenic surfaces (e.g., bacterial cell walls, yeast cell walls, and certain viral envelopes) have only low levels of sialic acid, C3b bound to these surfaces remains active for a longer time. The C3b present on the surface of the foreign cells can bind another serum protein called factor B to form a complex stabilized by $Mg^{2+}$. Binding to C3b exposes a site on factor B that serves as the substrate for an enzymatically active serum protein called factor D. Factor D cleaves the C3b-bound factor B, releasing a small fragment (Ba) that diffuses away and generating C3b$\overline{\text{Bb}}$. The C3b$\overline{\text{Bb}}$ complex has C3 convertase activity and thus is analogous to the $\overline{\text{C4b2a}}$ complex in the classical pathway. The C3 convertase activity of C3b$\overline{\text{Bb}}$ has a limited half-life unless the serum protein properdin binds to it.

The C3b$\overline{\text{Bb}}$ generated in the alternative pathway can activate unhydrolyzed C3 to produce more C3b autocatalytically. As a result, the initial steps are repeated and amplified, with the result that more than $2 \times 10^6$ molecules of C3b can be deposited on an antigenic surface in less than 5 minutes. The C3 convertase activity of C3b$\overline{\text{Bb}}$ generates the C3bBb3b complex, which exhibits C5 convertase activity, analogous to the $\overline{\text{C4b2a3b}}$ complex in the classical pathway. The nonenzymatic C3b component binds C5, and the Bb component subsequently hydrolyzes the bound C5 to generate C5a and C5b (see Figure 7-7); the latter binds to the antigenic surface, commencing the final phase of the lytic cycle.

## The lectin pathway originates with host proteins binding microbial surfaces

Lectins are proteins that recognize and bind to specific carbohydrate targets. Because the lectin that activates complement binds to mannose residues, some researchers designate this the MBLectin pathway or mannan-binding lectin pathway. The lectin pathway, like the alternative pathway, does not depend on antibody for its activation. However, the mechanism is more like that of the classical pathway than the alternative pathway, because after initiation, it proceeds through the action of C4 and C2 to produce activated proteins of the complement system (see Figure 7-2).

The lectin pathway is activated by the binding of mannose-binding lectin (MBL) to mannose residues on glycoproteins or carbohydrates on the surface of microorganisms, including bacteria such as *Salmonella, Listeria,* and *Neisseria* strains, as well as the fungi *Cryptococcus neoformans* and *Candida albicans* and to some viruses, including HIV-1 and respiratory syncytial virus. Human cells normally have sialic acid residues covering the sugar groups recognized by MBL and are not a target for binding.

MBL, a member of the collectin family, is an acute phase protein, and its concentration increases during inflammatory responses. Its function in the complement pathway is similar to that of C1q, which it resembles in structure (see Figure 7-3). After MBL binds to the carbohydrate residues on the surface of a cell or pathogen, MBL-associated serine proteases, MASP-1 and MASP-2, bind to MBL. The active complex formed by this association causes cleavage and activation of C4 and C2. The MASP-1 and MASP-2 proteins

are structurally similar to C1r and C1s and mimic their activities. This means of activating the C2 to C4 components to form a C5 convertase without need for specific antibody binding represents an important innate defense mechanism comparable to the alternative pathway but utilizing the elements of the classical pathway except for the C1 proteins.

## The three complement pathways converge at the membrane-attack complex

The terminal sequence of complement activation involves C5b, C6, C7, C8, and C9, which interact sequentially to form a macromolecular structure called the **membrane-attack complex (MAC).** This complex forms a large channel through the membrane of the target cell, enabling ions and small molecules to diffuse freely across the membrane.

The end result of activating the classical, alternative, or lectin pathways is production of an active C5 convertase. This enzyme cleaves C5, which contains two protein chains, $\alpha$ and $\beta$. After binding of C5 to the nonenzymatic C3b component of the convertase, the amino terminus of the $\alpha$ chain is cleaved. This generates the small C5a fragment, which diffuses away, and the large C5b fragment, which binds to the surface of the target cell and provides a binding site for the subsequent components of the membrane-attack complex (see Figure 7-5, step 5). The C5b component is extremely labile and rapidly becomes inactive unless it is stabilized by the binding of C6.

Up to this point, all of the complement reactions take place on the hydrophilic surfaces of membranes or on immune complexes in the fluid phase. As C5b6 binds to C7, the resulting complex undergoes a structural transition that exposes

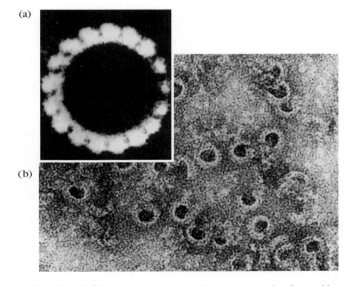

(a)

(b)

**FIGURE 7-8 (a) Photomicrograph of poly-C9 complex formed by in vitro polymerization of C9 and (b) complement-induced lesions on the membrane of a red blood cell.** These lesions result from formation of membrane-attack complexes. *[Part a from E. R. Podack, 1986, in Immunobiology of the Complement System, G. Ross, ed., Academic Press, Orlando; part b from J. Humphrey and R. Dourmashkin, 1969, Advances in Immunology 11:75.]*

hydrophobic regions, which serve as binding sites for membrane phospholipids. If the reaction occurs on a target cell membrane, the hydrophobic binding sites enable the C5b67 complex to insert into the phospholipid bilayer. If, however, the reaction occurs on an immune complex or other noncellular activating surface, then the hydrophobic binding sites cannot anchor the complex and it is released. Released C5b67 complexes can insert into the membrane of nearby cells and mediate "innocent bystander" lysis. Regulator proteins normally prevent this from occurring, but in certain diseases, cell and tissue damage may result from innocent bystander lysis. A hemolytic disorder resulting from deficiency in a regulatory protein is explained in the Clinical Focus on page 183, and an autoimmune process in which immune complexes mediate tissue damage will be considered in Chapter 15.

Binding of C8 to membrane-bound C5b67 induces a conformational change in C8, so that it too undergoes a structural transition, exposing a hydrophobic region, which interacts with the plasma membrane. The C5b678 complex creates a small pore, 10 Å in diameter; formation of this pore can lead to lysis of red blood cells but not of nucleated cells. The final step in formation of the MAC is the binding and polymerization of C9, a perforin-like molecule, to the C5b678 complex. As many as 10 to 17 molecules of C9 can be bound and polymerized by a single C5b678 complex. During polymerization, the C9 molecules undergo a transition, so that they too can insert into the membrane. The completed MAC, which has a tubular form and functional pore size of 70 to 100 Å, consists of a C5b678 complex surrounded by a poly-C9 complex (Figure 7-8). Since ions and small molecules can diffuse freely through the central channel of the MAC, the cell cannot maintain its osmotic stability and is killed by an influx of water and loss of electrolytes. The cascade of events for the three pathways of complement activation are summarized in Figure 7-9.

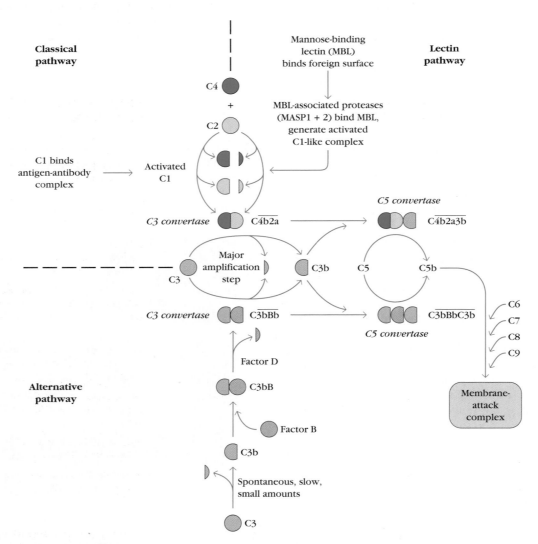

**FIGURE 7-9 Summary of the three pathways of complement activation showing initiation via either the classical mode, which begins with C1 binding antigen-antibody complex, or by the alternative or lectin modes, which are antibody-independent. All** pathways convert C3 to its active form, C3b, which participates in formation of a C5 convertase. With the formation of C5b, all three pathways converge to form the membrane-attack complex.

# Regulation of the Complement System

Because many elements of the complement system are capable of attacking host cells as well as foreign cells and microorganisms, a number of regulatory mechanisms have evolved to restrict complement activity to designated targets. One passive mechanism of regulation in all complement pathways is the inclusion of highly labile components that undergo spontaneous inactivation if they are not stabilized by reaction with other components. For example, the C3 convertase activity generated in the alternative pathway has an active half-life of only 5 minutes unless stabilized by reaction with properdin. Active regulation of complement activity is accomplished by a series of regulatory proteins that inactivate various complement components (Table 7-2). For example, the glycoprotein C1 inhibitor (C1Inh) can form a complex with $C1r_2s_2$, causing it to dissociate from C1q and preventing further activation of C4 or C2 (Figure 7-10a).

The reaction catalyzed by the C3 convertase enzymes of the classical, lectin, and alternative pathways is the major amplification step in complement activation, generating hundreds of molecules of C3b. The C3b generated by these enzymes has the potential to bind to nearby cells, mediating damage to the healthy cells by induction of the membrane-attack complex. Damage to normal host cells is minimized because C3b undergoes spontaneous hydrolysis by the time it has diffused 40 nm away from the $C\overline{4b2a}$ or $C3\overline{bBb}$ convertase enzymes, eliminating its ability to bind to its target site. The potential destruction of healthy host cells by C3b is further limited by a family of related proteins that regulate C3 convertase activity in the classical and alternative pathways. These regulatory proteins all contain repeating amino acid sequences (or motifs) of about 60 residues, termed *short consensus repeats* (SCRs). All these proteins are encoded at a single location on chromosome 1 in humans, known as the *regulators of complement activation* (RCA) gene cluster.

In the classical and lectin pathways, three structurally distinct RCA proteins act similarly to prevent assembly of C3

| TABLE 7-2 | Proteins that regulate the complement system | | |
|---|---|---|---|
| **Protein** | **Type of protein** | **Pathway affected** | **Immunologic function** |
| C1 inhibitor (C1Inh) | Soluble | Classical | Serine protease inhibitor: causes $C1r_2s_2$ to dissociate from C1q |
| C4b-binding protein (C4bBP)* | Soluble | Classical and lectin | Blocks formation of C3 convertase by binding C4b; cofactor for cleavage of C4b by factor I |
| Factor H* | Soluble | Alternative | Blocks formation of C3 convertase by binding C3b; cofactor for cleavage of C3b by factor I |
| Complement receptor type 1 (CR1 or CD35)* Membrane-cofactor protein (MCP or CD46)* | Membrane bound | Classical, alternative, and lectin | Block formation of C3 convertase by binding C4b or C3b; cofactor for factor I–catalyzed cleavage of C4b or C3b |
| Decay-accelerating factor (DAF or CD55)* | Membrane bound | Classical, alternative, and lectin | Accelerates dissociation of $C\overline{4b2a}$ and $C3\overline{bBb}$ (classical and alternative C3 convertases) |
| Factor I | Soluble | Classical, alternative, and lectin | Serine protease: cleaves C4b or C3b using C4bBP, CR1, factor H, DAE, or MCP as cofactor |
| S protein | Soluble | Terminal | Binds soluble C5b67 and prevents its insertion into cell membrane |
| Homologous restriction factor (HRF), also called membrane inhibitor of reactive lysis (MIRL or CD59)* | Membrane bound | Terminal | Bind to C5b678 on autologous cells, blocking binding of C9 |
| Anaphylatoxin inactivator | Soluble | Effector | Inactivates anaphylatoxin activity of C3a, C4a, and C5a by carboxypeptidase N-catalyzed removal of C-terminal Arg |

*An RCA (regulator of complement activation) protein. In humans, all RCA proteins are encoded on chromosome 1 and contain short consensus repeats.

OVERVIEW FIGURE 7-10: Regulation of the Complement System by Regulatory Proteins

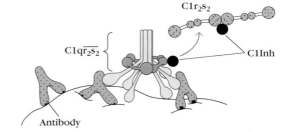

**Regulation of the Complement System**

(a) Before assembly of convertase activity

1. C1 inhibitor (C1Inh) binds $C1r_2s_2$, causing dissociation from C1q.

2. Association of C4b and C2a is blocked by binding C4b-binding protein (C4bBP), complement receptor type I, or membrane cofactor protein (MCP).

3. Inhibitor-bound C4b is cleaved by factor I.

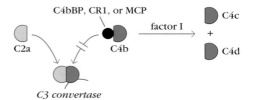

4. In alternative pathway, CR1, MCP, or factor H prevents binding of C3b and factor B.

5. Inhibitor-bound C3b is cleaved by factor I.

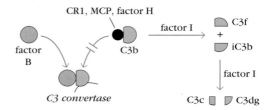

(b) After assembly of convertase

C3 convertases are dissociated by C4bBP, CR1, factor H, and decay-accelerating factor (DAF).

(c) Regulation at assembly of membrane-attack complex (MAC)

1. S protein prevents insertion of C5b67 MAC component into the membrane.

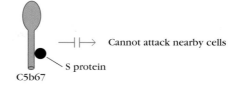

2. Homologous restriction factor (HRF) or membrane inhibitor of reactive lysis (MIRL or CD59) bind C5b678, preventing assembly of poly-C9 and blocking formation of MAC.

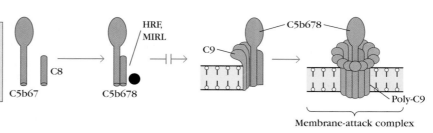

These proteins (shown in black) act at various points in the complement cascade.

convertase (Figure 7-10a(2)). These regulatory proteins include soluble C4b-binding protein (C4bBP) and two membrane-bound proteins, complement receptor type 1 (CR1) and membrane cofactor protein (MCP). Each of these regulatory proteins binds to C4b and prevents its association with C2a. Once C4bBP, CR1, or MCP is bound to C4b, another regulatory protein, factor I, cleaves the C4b into bound C4d and soluble C4c (Figure 7-10a(3)). A similar regulatory sequence operates to prevent assembly of the C3 convertase C3bBb in the alternative pathway. In this case, CR1, MCP, or a regulatory component called factor H binds to C3b and prevents its association with factor B (Figure 7-10a(4)). Once CR1, MCP, or factor H is bound to C3b, factor I cleaves the C3b into a bound iC3b fragment and a soluble C3f fragment. Further cleavage of iC3b by factor I releases C3c and leaves C3dg bound to the membrane (Figure 7-10a(5)). The molecular events involved in regulation of cell-bound C4b and C3b are depicted in Figure 7-11.

Several RCA proteins also act on the assembled C3 convertase, causing it to dissociate; these include the previously mentioned C4bBP, CR1, and factor H. In addition, decay-accelerating factor (DAF or CD55), which is a glycoprotein covalently anchored to a glycophospholipid membrane protein, has the ability to dissociate C3 convertase. The consequences of DAF deficiency are described in the Clinical Focus. Each of

these RCA proteins accelerates decay (dissociation) of C3 convertase by releasing the component with enzymatic activity (C2a or Bb) from the cell-bound component (C4b or C3b). Once dissociation of the C3 convertase occurs, factor I cleaves the remaining membrane-bound C4b or C3b component, irreversibly inactivating the convertase (Figure 7-10b).

Regulatory proteins also operate at the level of the membrane-attack complex. The potential release of the C5b67 complex poses a threat of innocent bystander lysis to healthy cells. A number of serum proteins counter this threat by binding to released C5b67 and preventing its insertion into the membrane of nearby cells. A serum protein called S protein can bind to C5b67, inducing a transition and thereby preventing insertion of C5b67 into the membrane of nearby cells (Figure 7-10c(1)).

Complement-mediated lysis of cells is more effective if the complement is from a species different from that of the cells being lysed. This phenomenon depends on a membrane protein that blocks MAC formation. This protein, present in many cell types, is called *homologous restriction factor (HRF)* or *membrane inhibitor of reactive lysis (MIRL or CD59)*. CD59 protects cells from nonspecific complement-mediated lysis by binding to C8, preventing assembly of poly-C9 and its insertion into the plasma membrane (Figure 7-10c(2)). However, this inhibition occurs only if the complement

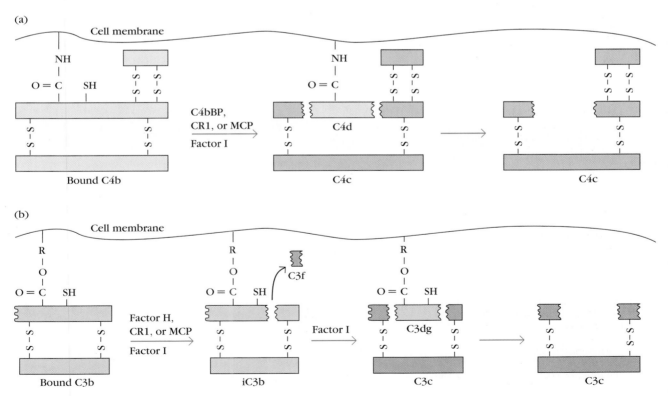

**FIGURE 7-11 Inactivation of bound C4b and C3b by regulatory proteins of the complement system.** (a) In the classical pathway, C4bBP (C4b-binding protein), CR1 (complement receptor type 1), or MCP (membrane cofactor protein) bind to C4b and act as cofactors for factor I–mediated cleavage of C4b. (b) In the alternative pathway, factor H, CR1, or MCP bind to C3b and act as cofactors for factor I–mediated cleavage of C3b. Free diffusible fragments are shown in dark shades; membrane-bound components in light shades.

components are from the same species as the target cells. For this reason, CD59 is said to display homologous restriction, for which it was formerly named.

## Biological Consequences of Complement Activation

Complement serves as an important mediator of the humoral response by amplifying the response and converting it into an effective defense mechanism to destroy invading microorganisms. The MAC mediates cell lysis, while other complement components or split products participate in the inflammatory response, opsonization of antigen, viral neutralization, and clearance of immune complexes (Table 7-3).

Many of the biological activities of the complement system depend on the binding of complement fragments to complement receptors, which are expressed by various cells. In addition, some complement receptors play an important role in regulating complement activity by binding biologically active complement components. The complement receptors and their primary ligands, which include various complement components and their proteolytic breakdown products, are listed in Table 7-4.

### The membrane-attack complex can lyse a broad spectrum of cells

The membrane-attack complex formed by complement activation can lyse gram-negative bacteria, parasites, viruses, erythrocytes, and nucleated cells. Because the alternative and lectin pathways of activation generally occur without an initial antigen-antibody interaction, these pathways serve as important innate immune defenses against infectious microorganisms. The requirement for an initial antigen-antibody reaction in the classical pathway supplements these nonspecific innate defenses with a more specific defense mechanism. In some instances, the requirement for antibody in the activating event may be supplied by so-called natural antibodies, which are raised against common components of ubiquitous microbes.

The importance of cell-mediated immunity in host defense against viral infections has been emphasized in numerous studies and will be discussed in later chapters. Nevertheless, antibody and complement do play a role in host defense against viruses and are often crucial in containing viral spread during acute infection and in protecting against reinfection. Most—perhaps all—enveloped viruses are susceptible to complement-mediated lysis. The viral envelope is largely derived from the plasma membrane of infected host cells and is therefore susceptible to pore formation by the membrane-attack complex. Among the pathogenic viruses susceptible to lysis by complement-mediated lysis are herpesviruses, orthomyxoviruses, such as those causing measles and mumps, paramyxoviruses, such as influenza, and retroviruses.

The complement system is generally quite effective in lysing gram-negative bacteria (Figure 7-12). However, some gram-negative bacteria and most gram-positive bacteria have mechanisms for evading complement-mediated damage (Table 7-5). For example, a few gram-negative bacteria can develop resistance to complement-mediated lysis in a manner that correlates with the virulence of the organism.

| TABLE 7-3 | Summary of biological effects mediated by complement products |
|---|---|
| **Effect** | **Complement product mediating*** |
| Cell lysis | C5b–9, the membrane-attack complex (MAC) |
| Inflammatory response | |
|   Degranulation of mast cells and basophils† | C3a, C4a, and C5a (anaphylatoxins) |
|   Degranulation of eosinophils | C3a, **C5a** |
|   Extravasation and chemotaxis of leukocytes at inflammatory site | C3a, **C5a**, C5b67 |
|   Aggregation of platelets | C3a, C5a |
|   Inhibition of monocyte/macrophage migration and induction of their spreading | Bb |
|   Release of neutrophils from bone marrow | C3c |
|   Release of hydrolytic enzymes from neutrophils | C5a |
|   Increased expression of complement receptors type 1 and 3 (CR1 and CR3) on neutrophils | C5a |
| Opsonization of particulate antigens, increasing their phagocytosis | **C3b**, C4b, iC3b |
| Viral neutralization | C3b, C5b–9 (MAC) |
| Solubilization and clearance of immune complexes | C3b |

*Boldfaced component is most important in mediating indicated effect.

†Degranulation leads to release of histamine and other mediators that induce contraction of smooth muscle and increased permeability of vessels.

| TABLE 7-4 | Complement-binding receptors | | |
|---|---|---|---|
| Receptor | Major ligands | Activity | Cellular distribution |
| CR1 (CD35) | C3b, C4b | Blocks formation of C3 convertase; binds immune complexes to cells | Erythrocytes, neutrophils, monocytes, macrophages, eosinophils, follicular dendritic cells, B cells, some T cells |
| CR2 (CD21) | C3d, C3dg,* iC3b | Part of B-cell coreceptor; binds Epstein-Barr virus | B cells, follicular dendritic cells, some T cells |
| CR3 (CD11b/18) CR4 (CD11c/18) | iC3b | Bind cell adhesion molecules on neutrophils, facilitating their extravasation; bind immune complexes, enhancing their phagocytosis | Monocytes, macrophages, neutrophils, natural killer cells, some T cells |
| C3a/C4a receptor | C3a, C4a | Induces degranulation of mast cells and basophils | Mast cells, basophils, granulocytes |
| C5a receptor | C5a | Induces degranulation of mast cells and basophils | Mast cells, basophils, granulocytes, monocytes, macrophages, platelets, endothelial cells |

*Cleavage of C3dg by serum proteases generates C3d and C3g.

In *Escherichia coli* and *Salmonella,* resistance to complement is associated with the smooth bacterial phenotype, which is characterized by the presence of long polysaccharide side chains in the cell wall lipopolysaccharide (LPS) component. It has been proposed that the increased LPS in the wall of resistant strains may prevent insertion of the MAC into the bacterial membrane, so that the complex is released from the bacterial cell rather than forming a pore. Strains of *Neisseria gonorrheae* resistant to complement-mediated killing have been associated with disseminated gonococcal infec-

tions (also known as gonococcal arthritis) in humans. Some evidence suggests that the membrane proteins of resistant *Neisseria* strains undergo noncovalent interactions with the MAC that prevent its insertion into the outer membrane of the bacterial cells. These examples of resistant gram-negative bacteria are the exception; most gram-negative bacteria are susceptible to complement-mediated lysis.

Gram-positive bacteria are generally resistant to complement-mediated lysis because the thick peptidoglycan layer in their cell wall prevents insertion of the MAC

(a)

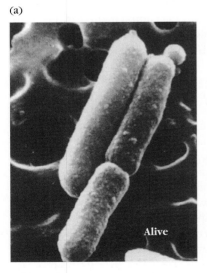

(b)

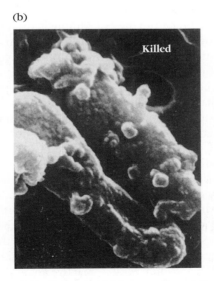

(c)

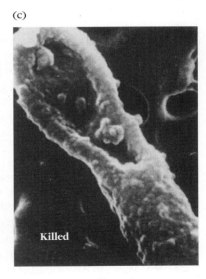

**FIGURE 7-12 Scanning electron micrographs of *E. coli* showing (a) intact cells and (b, c) cells killed by complement-mediated lysis.** Note membrane blebbing on lysed cells. *[From R. D. Schreiber et al., 1979,* Journal of Experimental Medicine ***149:**870.]*

| TABLE 7-5 | Microbial evasion of complement-mediated damage | | |
|---|---|---|---|
| Microbial component | Mechanism of evasion | | Examples |
| | GRAM-NEGATIVE BACTERIA | | |
| Long polysaccharide chains in cell wall LPS* | Side chains prevent insertion of MAC into bacterial membrane* | | Resistant strains of *E. coli* and *Salmonella* |
| Outer membrane protein | MAC interacts with membrane protein and fails to insert into bacterial membrane | | Resistant strains of *Neisseria gonorrhoeae* |
| Elastase | Anaphylatoxins C3a and C5a are inactivated by microbial elastase | | *Pseudomonas aeruginosa* |
| | GRAM-POSITIVE BACTERIA | | |
| Peptidoglycan layer of cell wall | Insertion of MAC into bacterial membrane is prevented by thick layer of peptidoglycan | | *Streptococcus* |
| Bacterial capsule | Capsule provides physical barrier between C3b deposited on bacterial membrane and CR1 on phagocytic cells* | | *Streptococcus pneumoniae* |
| | OTHER MICROBES | | |
| Proteins that mimic complement regulatory proteins | Protein present in various bacteria, viruses, fungi, and protozoans inhibit the complement cascade | | Vaccinia virus, herpes simplex, Epstein-Barr virus, *Trypanosoma cruzi*, Candida *albicans* |

*LPS = lipopolysaccharide; MAC = membrane-attack complex; CR1 = complement receptor type 1.

into the inner membrane. Although complement activation can occur on the cell membrane of encapsulated bacteria such as *Streptococcus pneumoniae,* the capsule prevents interaction between C3b deposited on the membrane and the CR1 on phagocytic cells. Some bacteria possess an elastase that inactivates C3a and C5a, preventing these split products from inducing an inflammatory response. In addition to these mechanisms of evasion, various bacteria, viruses, fungi, and protozoans contain proteins that can interrupt the complement cascade on their surfaces, thus mimicking the effects of the normal complement regulatory proteins C4bBP, CR1, and CD55 (DAF).

Lysis of nucleated cells requires formation of multiple membrane-attack complexes, whereas a single MAC can lyse a red blood cell. Many nucleated cells, including the majority of cancer cells, can endocytose the MAC. If the complex is removed soon enough, the cell can repair any membrane damage and restore its osmotic stability. An unfortunate consequence of this effect is that complement-mediated lysis by antibodies specific for tumor-cell antigens, which offers a potential weapon against cancer, may be rendered ineffective by endocytosis of the MAC (see Chapter 21).

## Cleavage products of complement components mediate inflammation

Although discussions of the complement cascade normally focus on its role in cell lysis, equally important functions are accomplished in the process of complement activation. Critically important are the various smaller fragments generated during formation of the MAC; these peptides play a decisive role in the development of an effective inflammatory response (see Table 7-3). The smaller fragments resulting from complement cleavage, C3a and C5a, called **anaphylatoxins,** bind to receptors on mast cells and blood basophils and induce degranulation, with release of histamine and other pharmacologically active mediators. The anaphylatoxins also induce smooth muscle contraction and increased vascular permeability. Activation of the complement system thus results in influxes of fluid that carries antibody and phagocytic cells to the site of antigen entry. The activities of these highly reactive anaphylatoxins are regulated by a serum protease called carboxypeptidase N, which cleaves an Arg residue from the C terminus of the molecules, yielding so-called *des-Arg* forms. The des-Arg form of C3a is completely

## CLINICAL FOCUS

# Paroxymal Nocturnal Hemoglobinuria: A Defect in Regulation of Complement Lysis

**Common** conditions associated with deficiency in the complement components include increased susceptibility to bacterial infections and systemic lupus erythematosus, which is related to the inability to clear immune complexes. Deficiency in the proteins that regulate complement activity can cause equally serious disorders. An example is paroxymal nocturnal hemoglobinuria, or PNH, which manifests as increased fragility of erythrocytes, leading to chronic hemolytic anemia, pancytopenia (loss of blood cells of all types), and venous thrombosis (formation of blood clots). The name PNH derives from the presence of hemoglobin in the urine, most commonly observed in the first urine passed after a night's sleep. The cause of PNH is a general defect in synthesis of cell surface proteins, which affects the expression of two regulators of complement, DAF or CD55 and MIRL or CD59.

CD55 and CD59 are cell surface proteins that function as inhibitors of complement-mediated cell lysis but act at different stages of the process. CD55 inhibits cell lysis by causing dissociation and inactivation of the C3 convertases of the classical, lectin, and alternative pathways (see Figure 7-10b). CD59 acts later in the pathway by binding to the C5b678 complex, which inhibits C9 binding and prevents formation of the pores that de-

stroy the cell under attack. Both proteins are expressed on erythrocytes as well as a number of other hematopoetic cell types. Deficiency in these proteins leads to highly increased sensitivity of host cells to the lytic effects of the host's complement activity. PNH, the clinical consequence of deficiency in CD55 and CD59, is a chronic disease with a mean survival time between 10 and 15 years. The most common causes of mortality in PNH are venous thrombosis affecting hepatic veins and progressive bone marrow failure.

A curious feature of this rare but serious disease is the fact that two different proteins are involved in its pathogenesis. The simultaneous occurrence of a genetic defect in both proteins would be rarer than the 1 in 100,000 incidence of PNH. In fact, neither protein is itself defective in PNH; the defect lies in a post-translational modification of the peptide anchors that bind the proteins to the cell membrane. Whereas most proteins that are expressed on the surface of cells have hydrophobic sequences that traverse the lipid bilayer in the cell membrane, some are bound by glycolipid anchors (glycosyl phosphatidylinositol, or GPI) attached to amino acid residues in the protein. If the ability to form GPI anchors is lacking, proteins that attach in this manner, such as

CD55 and CD59, will be absent from the cell surface.

The defect identified in PNH occurs early in the enzymatic pathway leading to formation of a GPI anchor and resides in the *pig-a* gene (phosphatidylinositol glycan complementation class A gene). Transfection of cells from PNH patients with an intact *pig-a* gene restores the resistance of the cells to host complement lysis. Examination of *pig-a* sequences in a population of PNH patients revealed a number of different defects in this X-linked gene, indicating a somatic rather than genetic origin for the defect. Because the defect affects numerous cells, remedy of the condition by gene therapy is not a practical treatment option.

A recent breakthrough in treatment of PNH was reported using a humanized monoclonal antibody (see Chapter 4) that targets complement component C5 and thus inhibits the terminal steps of the complement cascade and formation of the membrane-attack complex. This antibody, with the trade name Eculizumab, was infused into patients, who were then monitored for the loss of red blood cells through lysis. A dramatic improvement was seen in patients during a 12-week period of treatment with Eculizumab (see Clinical Focus Figure 1).

The features of PNH underscore the fact that the complement system is a powerful defender of the host but also a dangerous one. Complex systems of regulation are necessary to protect host cells from the activated complement complexes generated to lyse intruders.

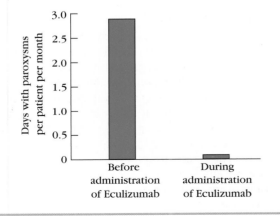

Treatment of PNH patients with Eculizumab relieves hemoglobinuria. The number of days with paroxysm (onset of attack) per patient per month in the month prior to treatment (left bar) and for a 12-week period of treatment with Eculizumab (right bar) is shown. *[From Hillmen, P., et al. 2004.* New England Journal of Medicine *350:6, 552–59.]*

inactive, whereas that of C5a retains a fraction of both its chemotactic activity and its ability to cause smooth muscle contraction.

C3a and C5a can each induce monocytes and neutrophils to adhere to vascular endothelial cells, extravasate through the endothelial lining of the capillary, and migrate toward the site of complement activation in the tissues. C5a is most potent in mediating these processes, effective in picomolar quantities. The role of complement in leukocyte chemotaxis is discussed more fully in Chapter 14.

## C3b and C4b binding facilitates opsonization

C3b is the major opsonin of the complement system, although C4b and iC3b also have opsonizing activity. The amplification that occurs with C3 activation results in a coating of C3b on immune complexes and particulate antigens. Phagocytic cells, as well as some other cells, express complement receptors (CR1, CR3, and CR4) that bind C3b, C4b, or iC3b (see Table 7-4). Antigen coated with C3b binds to cells bearing CR1. If the cell is a phagocyte (e.g., a neutrophil, monocyte, or macrophage), phagocytosis will be enhanced (Figure 7-13). Activation of phagocytic cells by various agents, including C5a anaphylatoxin, has been shown to increase the number of CR1s from 5000 on resting phagocytes to 50,000 on activated cells, greatly facilitating their phagocytosis of C3b-coated antigen.

Recent studies further indicate that complement fragment C3b acts as an adjuvant when coupled with protein antigens. C3b targets the antigen directly to the phagocyte, enhancing the initiation of antigen processing and accelerating specific antibody production.

## The complement system also neutralizes viral infectivity

For most viruses, the binding of serum antibody to the repeating subunits of viral structural proteins creates particulate immune complexes ideally suited for complement activation by the classical pathway. Some viruses (e.g., retroviruses, Epstein-Barr virus, Newcastle disease virus, and rubella virus) can activate the alternative, lectin, or even the classical pathway in the absence of antibody.

The complement system mediates viral neutralization by a number of mechanisms. Some degree of neutralization is achieved through the formation of larger viral aggregates, simply because these aggregates reduce the net number of infectious viral particles. Although antibody plays a role in the formation of viral aggregates, in vitro studies show that the C3b component facilitates aggregate formation in the presence of as little as two molecules of antibody per virion. For example, polyoma virus coated with antibody is neutralized when serum containing activated C3 is added.

The binding of antibody and/or complement to the surface of a viral particle creates a thick protein coating that can be visualized by electron microscopy (Figure 7-14). This coating neutralizes viral infectivity by blocking attachment to susceptible host cells. The deposits of antibody and complement on viral particles also facilitate binding of the viral particle to cells possessing Fc or type 1 complement receptors (CR1). In the case of phagocytic cells, such binding can be followed by phagocytosis and intracellular destruction of the ingested viral particle. Finally, complement is effective in lysing most, if not all, enveloped viruses, resulting in fragmentation of the envelope and disintegration of the nucleocapsid.

(a)

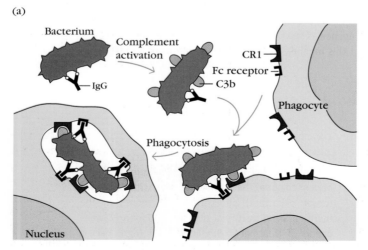

(b)

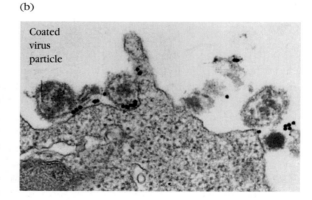

**FIGURE 7-13** **(a) Schematic representation of the roles of C3b and antibody in opsonization.** (b) Electron micrograph of Epstein-Barr virus coated with antibody and C3b and bound to the Fc and C3b receptor (CR1) on a B lymphocyte. *[Part b from N. R. Cooper and G. R. Nemerow, 1986, in* Immunobiology of the Complement System, *G. Ross, ed., Academic Press, Orlando.]*

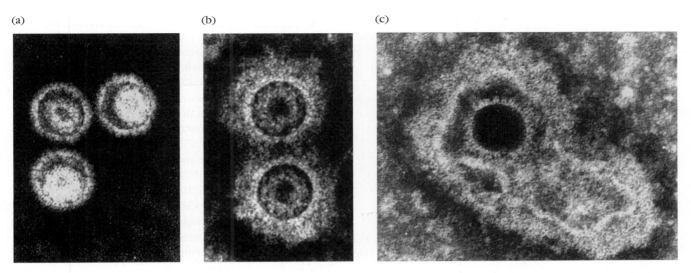

**FIGURE 7-14 Electron micrographs of negatively stained preparations of Epstein-Barr virus.** (a) Control without antibody. (b) Antibody-coated particles. (c) Particles coated with antibody and complement. *[From N. R. Cooper and G. R. Nemerow, 1986, in* Immunobiology of the Complement System, *G. Ross, ed., Academic Press, Orlando.]*

Viruses have developed a number of different strategies for evasion of complement activity. These strategies fall into three different categories:

1. *Interference with the binding of complement to antibody-antigen complexes.* This evasion strategy is utilized by herpesvirus, which synthesizes viral proteins that have Fc receptor activity; the binding of these proteins to nonspecific immunoglobulin blocks the binding of antiviral antibodies.

2. *Viral mimicry of mammalian complement regulators.* Vaccinia virus produces a protein that binds C4b and C3b, interfering with its activity, and that further acts as a cofactor for inhibitory factor I.

3. *Incorporation of cellular complement regulators in the virion.* HTLV-1 (human leukemia virus) incorporates significant levels of sialic acid in its viral envelope, masking the virus with a property of mammalian cells.

## The complement system clears immune complexes from circulation

The importance of the complement system in clearing immune complexes is seen in people with the autoimmune disease systemic lupus erythematosus (SLE). These individuals produce large quantities of immune complexes and suffer tissue damage as a result of complement-mediated lysis and the induction of type II or type III hypersensitivity (see Chapter 15). Although complement plays a significant role in the development of tissue damage in SLE, the paradoxical finding is that deficiencies in C1, C2, C4, and CR1 predispose an individual to SLE; indeed, 90% of individuals who completely lack C4 develop SLE. The complement deficiencies are thought to interfere with effective solubilization and clearance of immune complexes; as a result, these complexes persist, leading to tissue damage by the very system whose deficiency was to blame.

The coating of soluble immune complexes with C3b is thought to facilitate their binding to CR1 on erythrocytes. Although red blood cells express lower levels of CR1 ($\sim 5 \times 10^2$ per cell) than granulocytes do ($\sim 5 \times 10^4$ per cell), there are about $10^3$ red blood cells for every white blood cell; therefore, erythrocytes account for about 90% of the CR1 in the blood. For this reason, erythrocytes play an important role in binding C3b-coated immune complexes and carrying these complexes to the liver and spleen. In these organs, immune complexes are stripped from the red blood cells and are phagocytosed, thereby preventing their deposition in tissues (Figure 7-15). In SLE patients, deficiencies in C1, C2, and C4 each contribute to reduced levels of C3b on immune complexes and hence inhibit their clearance. The lower levels of CR1 expressed on the erythrocytes of SLE patients also may interfere with the proper binding and clearance of immune complexes.

## Complement Deficiencies

Genetic deficiencies have been described for each of the complement components. Homozygous deficiencies in any of the early components of the classical pathway (C1q, C1r, C1s, C4, and C2) result in similar symptoms, notably a marked increase in immune-complex diseases such as systemic lupus erythematosus, glomerulonephritis, and vasculitis. The effects of these deficiencies highlight the importance of the early complement reactions in generating C3b and the critical role of C3b in solubilization and clearance of immune complexes. In addition to immune-complex diseases, individuals with such complement deficiencies may suffer from recurrent

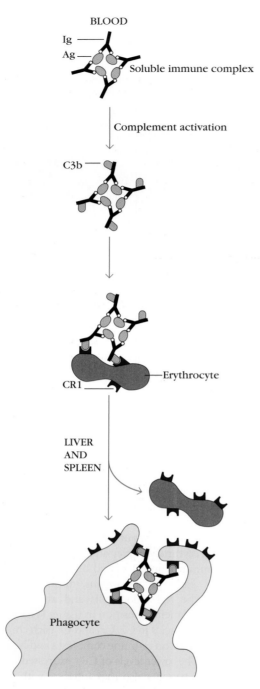

BLOOD

Ig

Ag

Soluble immune complex

Complement activation

C3b

Erythrocyte

CR1

LIVER
AND
SPLEEN

Phagocyte

**FIGURE 7-15 Clearance of circulating immune complexes by reaction with receptors for complement products on erythrocytes and removal of these complexes by receptors on macrophages in the liver and spleen.** Because erythrocytes have fewer receptors than macrophages, the latter can strip the complexes from the erythrocytes as they pass through the liver or spleen. Deficiency in this process can lead to renal damage due to accumulation of immune complexes.

infections by pyogenic (pus-forming) bacteria such as streptococci and staphylococci. These organisms are gram-positive and therefore resistant to the lytic effects of the MAC. Nevertheless, the early complement components ordinarily prevent recurrent infection by mediating a localized inflammatory response and opsonizing the bacteria. Deficiencies in factor D and properdin—early components of the alternative pathway—

appear to be associated with *Neisseria* infections but not with immune-complex disease. MBL deficiency has been shown to be relatively common and results in serious pyogenic infections in neonates and children. Children with MBL deficiency suffer increased respiratory tract infections. This deficiency is also found to be two to three times higher in SLE patients than in normal subjects, and certain mutant forms are prevalent in chronic carriers of hepatitis B.

People with C3 deficiencies have the most severe clinical manifestations, reflecting the central role of C3 in activation of C5 and formation of the MAC. The first person identified with a C3 deficiency was a child suffering from frequent severe bacterial infections and initially diagnosed as having agammaglobulinemia. After tests revealed normal immunoglobulin levels, a deficiency in C3 was discovered. This case highlights the critical function of the complement system in converting a humoral antibody response into an effective defense mechanism. The majority of people with C3 deficiency have recurrent bacterial infections and may have immune-complex diseases.

Levels of C4 vary considerably in the population, and people with lower levels may be subject to a higher incidence of autoimmune disease. The genes encoding C4 are in the major histocompatibility locus (see Chapter 8), and the number of C4 genes may vary from two to seven copies in an individual. Recent studies link the genetically determined lower levels of C4 to increased risk for SLE.

Individuals with homozygous deficiencies in the components involved in the MAC develop recurrent meningococcal and gonococcal infections caused by *Neisseria* species. In normal individuals, these gram-negative bacteria are generally susceptible to complement-mediated lysis or are cleared by the opsonizing activity of C3b. MAC-deficient individuals rarely have immune-complex disease, which suggests that they produce enough C3b to clear immune complexes. Interestingly, a deficiency in C9 results in no clinical symptoms, suggesting that the entire MAC is not always necessary for complement-mediated lysis.

Congenital deficiencies of complement regulatory proteins have also been reported. The C1 inhibitor (C1Inh) regulates activation of the classical pathway by preventing excessive C4 and C2 activation by C1. Deficiency of C1Inh is an autosomal dominant condition with a frequency of 1 in 1000. The deficiency gives rise to a condition called hereditary angioedema, which manifests clinically as localized edema of the tissue, often following trauma, but sometimes with no known cause. The edema can be in subcutaneous tissues or within the bowel, where it causes abdominal pain, or in the upper respiratory tract, where it causes obstruction of the airway that may be fatal.

Studies in humans and experimental animals with homozygous deficiencies in complement components are the major source of information on the role of individual complement components in immunity. These findings have been greatly extended by studies using knockout mice genetically engineered to lack expression of specific complement components. Investigations of in vivo complement activity in these animals has allowed dissection of the complex system of complement proteins and the assignment of precise biologic roles to each.

## SUMMARY

- The complement system comprises a group of serum proteins, many of which exist in inactive forms.

- Complement activation occurs by the classical, alternative, or lectin pathways, each of which is initiated differently.

- The three pathways converge in a common sequence of events that leads to generation of a molecular complex that causes cell lysis.

- The classical pathway is initiated by antibody binding to a cell target; reactions of IgM and certain IgG subclasses activate this pathway.

- Activation of the alternative and lectin pathways is antibody-independent. These pathways are initiated by reaction of complement proteins with surface molecules of microorganisms.

- In addition to its key role in cell lysis, the complement system mediates opsonization of bacteria, activation of inflammation, and clearance of immune complexes.

- Interactions of complement proteins and protein fragments with receptors on cells of the immune system control both innate and adaptive immune responses.

- Because of its ability to damage the host organism, the complement system requires complex passive and active regulatory mechanisms.

- Clinical consequences of inherited complement deficiencies range from increases in susceptibility to infection to tissue damage caused by immune complexes.

## References

Carroll, M. C. 2004. The complement system in regulation of adaptive immunity. *Nature Immunology* **5:**981.

Davis, A. E., III. 2005. The pathophysiology of hereditary angioedema. *Clinical Immunology* **114:**3.

Favoreel, H. W., et al. 2003. Virus complement evasion strategies. *Journal of General Virology* **84:**1.

Hillmen, P., et al. 2004. Effect of Eculizumab on hemolysis and transfusion requirements in patients with paroxysmal nocturnal hemoglobinuria. *New England Journal of Medicine* **350:**552.

Holmskov, U., S. Thiel, and J. C. Jensenius. 2003. Collectins and ficolins: Humoral lectins of the innate immune defense. *Annual Review of Immunology* **21:**547.

Kishore, U., et al. 2004. C1q and tumor necrosis factor superfamily: Modularity and versatility. *Trends in Immunology* **25:**551.

Lindahl, G., U. Sjobring, and E. Johnsson. 2000. Human complement regulators: A major target for pathogenic microorganisms. *Current Opinion in Immunology* **12:**44.

Manderson, A. P., M. Botto, and M. J. Walport. 2004. The role of complement in the development of systemic lupus erythematosus. *Annual Review of Immunology* **22:**431.

Matsumoto, M., et al. 1997. Abrogation of the alternative complement pathway by targeted deletion of murine factor B. *Proceedings of the National Academy of Sciences U.S.A.* **94:**8720.

Muller-Eberhard, H. J. 1988. Molecular organization and function of the complement system. *Annual Review of Biochemistry* **57:**321.

Nonaka, M., and F. Yohizaki. 2004. Primitive complement system of invertebrates. *Immunological Reviews* **198:**203.

Rautemaa, R., and S. Meri. 1999. Complement-resistance mechanisms of bacteria. *Microbes and Infection/Institut Pasteur* **1:**785.

Sloand, E. M., et al. 1998. Correction of the PNH defect by GPI-anchored protein transfer. *Blood* **92:**4439.

Turner, M. W. 2003. The role of mannose-binding lectin in health and disease. *Molecular Immunology* **40:**423.

Wen, L., J. P. Atkinson, and P. C. Gicias. 2004. Clinical and laboratory evaluation of complement deficiency. *Journal of Allergy and Clinical Immunology* **113:**585.

Yu, C. Y., and C. C. Whitacre. 2004. Sex, MHC and complement C4 in autoimmune diseases. *Trends in Immunology* **25:**694.

Zarkadis, I. K., D. Mastellos, and J. D. Lambris. 2001. Phylogenetic aspects of the complement system. *Developmental and Comparative Immunology* **25:**745.

 Useful Web Sites

http://www.complement-genetics.uni-mainz.de/

The Complement Genetics Homepage from the University of Mainz gives chromosomal locations and information on genetic deficiencies of complement proteins.

http://www.cehs.siu.edu/fix/medmicro/cfix.htm

A clever graphic representation of the basic assay for complement activity using red blood cell lysis, from D. Fix at the University of Southern Illinois, Carbondale.

 Study Questions

**CLINICAL FOCUS QUESTION** Explain why complement disorders involving regulatory components such as PNH may be more serious than deficiencies in the active complement components.

1. Indicate whether each of the following statements is true or false. If you think a statement is false, explain why.

   a. A single molecule of bound IgM can activate the C1q component of the classical complement pathway.

   b. C3a and C3b are fragments of C3.

   c. The C4 and C2 complement components are present in the serum in a functionally inactive proenzyme form.

   d. Nucleated cells tend to be more resistant to complement-mediated lysis than red blood cells.

   e. Enveloped viruses cannot be lysed by complement because their outer envelope is resistant to pore formation by the membrane-attack complex.

   f. C4-deficient individuals have difficulty eliminating immune complexes.

2. Explain why serum IgM cannot activate complement by itself.

3. Genetic deficiencies have been described in patients for all of the complement components except factor B. Particularly severe consequences result from a deficiency in C3. Describe the consequences of an absence of C3 for each of the following:

   a. Activation of the classical and alternate pathways

   b. Clearance of immune complexes

   c. Phagocytosis of infectious bacteria

   d. Presentation of antigenic peptides from infectious bacteria

4. Summarize the four major functions of the complement system.

5. Complement activation can occur via the classical, alternative, or lectin pathway.

   a. How do the three pathways differ in the substances that can initiate activation?

   b. Which portion of the overall activation sequence differs in the three pathways? Which portion is similar?

   c. How do the biological consequences of complement activation via these pathways differ?

6. Enucleated cells, such as red blood cells, are more susceptible to complement-mediated lysis than nucleated cells.

   a. Explain why the red blood cells of an individual are not normally destroyed as the result of innocent bystander lysis by complement.

   b. Under what conditions might complement cause lysis of an individual's own red blood cells?

7. Briefly explain the mechanism of action of the following complement regulatory proteins. Indicate which pathway(s) each protein regulates.

   a. C1 inhibitor (C1Inh)

   b. C4b-binding protein (C4bBP)

   c. Homologous restriction factor (HRF)

   d. Decay-accelerating factor (DAF)

   e. Factor H

   f. Membrane cofactor protein (MCP)

8. For each complement component(s) or reaction (a–l), select the most appropriate description listed (1–13). Each description may be used once, more than once, or not at all.

*Complement Component(s)/Reactions*

   a. _____ C3b

   b. _____ C1, C4, C2, and C3

   c. _____ C9

   d. _____ C3, factor B, and factor D

   e. _____ C1q

   f. _____ C4b2a3b

   g. _____ C5b, C6, C7, C8, and C9

   h. _____ C3 → C3a + C3b

   i. _____ C3a, C5a, and C5b67

   j. _____ C3a, C4a, and C5a

   k. _____ C4b2a

   l. _____ C3b + B → C3bBb + Ba

*Descriptions*

   (1) Reaction that produces major amplification during activation

   (2) Are early components of alternative pathway

   (3) Compose the membrane-attack complex

   (4) Mediates opsonization

   (5) Are early components of classical pathway

   (6) Has perforin-like activity

   (7) Binds to Fc region of antibodies

   (8) Have chemotactic activity

   (9) Has C3 convertase activity

   (10) Induce degranulation of mast cells (are anaphylatoxins)

   (11) Has C5 convertase activity

   (12) Reaction catalyzed by factor D

   (13) Reaction catalyzed by $C1qr_2s_2$

9. You have prepared knockout mice with mutations in the genes that encode various complement components. Each knockout strain cannot express one of the complement components listed across the top of the table below. Predict the effect of each mutation on the steps in complement activation and on the complement effector functions indicated in the table using the following symbols: NE = no effect; D = process/function decreased but not abolished; A = process/function abolished.

| | Component knocked out | | | | | | |
|---|---|---|---|---|---|---|---|
| | C1q | C4 | C3 | C5 | C6 | C9 | Factor B |
| **COMPLEMENT ACTIVATION** | | | | | | | |
| Formation of C3 convertase in classical pathway | | | | | | | |
| Formation of C3 convertase in alternative pathway | | | | | | | |
| Formation of C5 convertase in classical pathway | | | | | | | |
| Formation of C5 convertase in alternative pathway | | | | | | | |
| **EFFECTOR FUNCTIONS** | | | | | | | |
| C3b-mediated opsonization | | | | | | | |
| Neutrophil chemotaxis | | | | | | | |
| Cell lysis | | | | | | | |

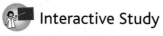

Interactive Study

**www.whfreeman.com/kuby**

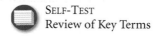

SELF-TEST
Review of Key Terms

# chapter 8

# The Major Histocompatibility Complex and Antigen Presentation

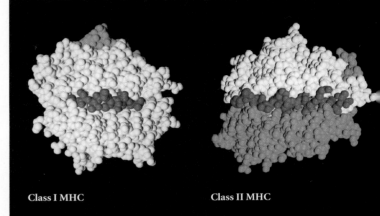

Class I MHC          Class II MHC

*Space-filling models of class I and class II MHC molecules showing bound peptides in red. [From D. A. Vignali and J. Strominger, 1994,* The Immunologist *2:112.]*

- General Organization and Inheritance of the MHC

- MHC Molecules and Genes

- Detailed Genomic Map of MHC Genes

- Cellular Expression of MHC Molecules

- Regulation of MHC Expression

- MHC and Disease Susceptibility

- MHC and Immune Responsiveness

- Self-MHC Restriction of T Cells

- Role of Antigen-Presenting Cells

- Evidence for Different Antigen-Processing and Presentation Pathways

- Endogenous Antigens: The Cytosolic Pathway

- Exogenous Antigens: The Endocytic Pathway

- Cross-Presentation of Exogenous Antigens

- Presentation of Nonpeptide Antigens

IN CONTRAST TO ANTIBODIES OR B-CELL RECEPTORS, WHICH CAN recognize an antigen alone, T-cell receptors only recognize antigen that has been processed and presented in the context of molecules encoded by the **major histocompatibility complex (MHC).** The MHC was first studied and so named as a genetic complex that influences the ability of an organism to accept or reject transplanted tissue from another member of the same species. Pioneering studies by R. Zinkernagel and P. Doherty, B. Benacerraf, and others made it clear that the molecules encoded by the MHC play a seminal role in determination of adaptive immune responses, and the particular set of MHC molecules expressed by an individual influences the repertoire of antigens to which that individual's $T_H$ and $T_C$ cells can respond. The MHC affects the response of an individual to antigens of infectious organisms, and it has therefore been implicated in the susceptibility to disease as well as the development of autoimmunity. The recent understanding that natural killer cells express receptors for MHC class I antigens and the fact that the receptor-MHC interaction may lead to inhibition or activation expands the known role of this gene family (see Chapter 14). The mechanisms by which this family of molecules exerts influence on the development of immunity to nearly all types of antigens has become a major theme in immunology and takes the study of the MHC far beyond transplantation biology.

The present chapter will explore briefly the history of the MHC and describe the genes and the molecules encoded within it. The association of certain MHC genes with disease susceptibility will be mentioned, as will the diversity of these genes in the human population. Emphasis is placed on how the MHC products produce the peptide-MHC complex derived from protein antigens (**antigen processing**) along with the way in which this molecular complex is transported to the membrane of the cell, where antigen is displayed for recognition by T cells (**antigen presentation**). In addition, the role of molecules that are not encoded within the MHC in the presentation of certain classes of antigen will be described.

# General Organization and Inheritance of the MHC

Every mammalian species studied to date possesses the tightly linked cluster of genes that constitute the MHC, whose products play roles in intercellular recognition and in discrimination between self and nonself. The MHC participates in the development of both humoral and cell-mediated immune responses. Studies of this gene cluster originated when it was found that the rejection of foreign tissue is the result of an immune response to cell surface molecules, now called **histocompatibility antigens.** In the mid-1930s, Peter Gorer, who was using inbred strains of mice to identify blood-group antigens, identified four groups of genes, designated I through IV, that encoded blood-cell antigens. Work carried out in the 1940s and 1950s by Gorer and George Snell established that antigens encoded by the genes in the group designated II took part in the rejection of transplanted tumors and other tissue. Snell called these genes "histocompatibility genes"; their current designation as histocompatibility-2 (H-2) genes was in reference to Gorer's group II blood-group antigens. Although Gorer died before

his contributions were recognized fully, Snell was awarded the Nobel prize in 1980 for this work.

## The MHC encodes three major classes of molecules

The major histocompatibility complex is a collection of genes arrayed within a long continuous stretch of DNA on chromosome 6 in humans and on chromosome 17 in mice. The MHC is referred to as the **HLA complex** in humans and as the **H-2 complex** in mice. Although the arrangement of genes is somewhat different in the two species, in both cases the MHC genes are organized into regions encoding three classes of molecules (Figure 8-1):

- **Class I MHC genes** encode glycoproteins expressed on the surface of nearly all nucleated cells; the major function of the class I gene products is presentation of peptide antigens to $T_C$ cells.

- **Class II MHC genes** encode glycoproteins expressed primarily on antigen-presenting cells (macrophages, dendritic cells, and B cells), where they present processed antigenic peptides to $T_H$ cells.

---

**OVERVIEW FIGURE 8-1:** Simplified Organization of the Major Histocompatibility Complex (MHC) in the Mouse and Human

Mouse H-2 complex

| Complex | H–2 | | | | | | |
|---|---|---|---|---|---|---|---|
| MHC class | I | II | | III | | I | |
| Region | K | IA | IE | S | | D | |
| Gene products | H–2K | IA αβ | IE αβ | C′ proteins | TNF-α TNF-β | H-2D | H-2L* |

*Not present in all haplotypes

Human HLA complex

| Complex | HLA | | | | | | | |
|---|---|---|---|---|---|---|---|---|
| MHC class | II | | | III | | I | | |
| Region | DP | DQ | DR | C4, C2, BF | | B | C | A |
| Gene products | DP αβ | DQ αβ | DR αβ | C′ proteins | TNF-α TNF-β | HLA-B | HLA-C | HLA-A |

The MHC is referred to as the H-2 complex in mice and as the HLA complex in humans. In both species, the MHC is organized into a number of regions encoding class I (pink), class II (blue), and class III (green) gene products. The class I and II gene products shown in this figure are considered to be the classical MHC molecules. The class III gene products include complement (C′) proteins and the tumor necrosis factors (TNF-α and TNF-β).

■ **Class III MHC genes** encode, in addition to other products, various secreted proteins that have immune functions, including components of the complement system and molecules involved in inflammation.

Class I MHC molecules encoded by the K and D regions in mice and by the A, B, and C loci in humans were the first discovered, and they are expressed in the widest range of cell types. These are referred to as *classical class I molecules*. Additional genes or groups of genes within the H-2 or HLA complexes also encode class I molecules; these genes are designated *nonclassical class I genes*. Expression of the nonclassical gene products is limited to certain specific cell types. Although functions are not known for all of these gene products, some may have highly specialized roles in immunity. For example, the expression of the class I HLA-G molecules on cytotrophoblasts at the fetal-maternal interface has been implicated in protection of the fetus from being recognized as foreign (this may occur when paternal antigens begin to appear) and from being rejected by maternal $T_C$ cells.

The two chains of the class II MHC molecules are encoded by the IA and IE regions in mice and by the DP, DQ, and DR regions in humans. The terminology is somewhat confusing, since the D region in mice encodes class I MHC molecules, whereas DR, DQ, and DP in humans refers to class II genes and molecules! As with the class I loci, additional class II molecules encoded within this region have specialized functions in the immune process.

The class I and II MHC molecules have common structural features, and both have roles in antigen processing and presentation. By contrast, the class III MHC region, which is flanked by the class I and II regions, encodes molecules that are critical to immune function but have little in common with class I or II molecules. Class III products include the complement components C4, C2, and factor B (encoded by *BF*; see Chapter 7),

and several inflammatory cytokines, including tumor necrosis factor (TNF).

## Allelic forms of MHC genes are inherited in linked groups called haplotypes

As described in more detail later, the loci constituting the MHC are highly **polymorphic;** that is, many alternative forms of the gene, or **alleles,** exist at each locus among the population. The genes of the MHC loci lie close together; for example, the recombination frequency within the H-2 complex (i.e., the frequency of chromosome crossover events during meiosis, indicative of the distance between given gene segments; crossover is more likely when genes are farther apart) is only 0.5%—crossover occurs only once in every 200 meiotic cycles. For this reason, most individuals inherit the alleles encoded by these closely linked loci as two sets, one from each parent. Each set of alleles is referred to as a **haplotype.** An individual inherits one haplotype from the mother and one haplotype from the father. In outbred populations, the offspring are generally heterozygous at many loci and will express both maternal and paternal MHC alleles. The alleles are *codominantly expressed;* that is, both maternal and paternal gene products are expressed in the same cells. If mice are inbred (that is, have identical alleles at all loci), each H-2 locus will be homozygous because the maternal and paternal haplotypes are identical, and all offspring therefore express identical haplotypes.

Certain inbred mouse strains have been designated as prototype strains, and the MHC haplotype expressed by these strains is designated by an arbitrary italic superscript (e.g., H-2$^a$, H-2$^b$). These designations refer to the entire set of inherited H-2 alleles within a strain without having to list each allele individually (Table 8-1). Different inbred strains may

| TABLE 8-1 | H-2 haplotypes of some mouse strains | | | | | | | |
|---|---|---|---|---|---|---|---|---|
| | | | | H-2 ALLELES | | | | |
| **Prototype strain** | **Other strains with the same haplotype** | **Haplotype** | **K** | **IA** | **IE** | **S** | **D** |
| CBA | AKR, C3H, B10.BR, C57BR | *k* | *k* | *k* | *k* | *k* | *k* |
| DBA/2 | BALB/c, NZB, SEA, YBR | *d* | *d* | *d* | *d* | *d* | *d* |
| C57BL/10 (B10) | C57BL/6, C57L, C3H.SW, LP, 129 | *b* | *b* | *b* | *b* | *b* | *b* |
| A | A/He, A/Sn, A/Wy, B10.A | *a* | *k* | *k* | *k* | *d* | *d* |
| B10.A (2R)* | | *b2* | *k* | *k* | *k* | *d* | *b* |
| B10.A (3R) | | *i3* | *b* | *b* | *k* | *d* | *d* |
| B10.A. (4R) | | *b4* | *k* | *k* | *b* | *b* | *b* |
| A.SW | B10.S, SJL | *s* | *s* | *s* | *s* | *s* | *s* |
| A.TL | | *t1* | *s* | *k* | *k* | *k* | *d* |
| DBA/1 | STOLI, B10.Q, BDP | *q* | *q* | *q* | *q* | *q* | *q* |

*The R designates a recombinant haplotype, in this case between the H-2ᵃ and H-2ᵇ types. Gene contribution from the *a* strain is shown in yellow and from the *b* strain in red.

## (a) Mating of inbred mouse strains with different MHC haplotypes

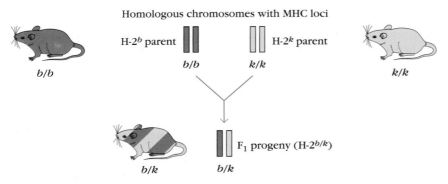

Homologous chromosomes with MHC loci

## (b) Skin transplantation between inbred mouse strains with same or different MHC haplotypes

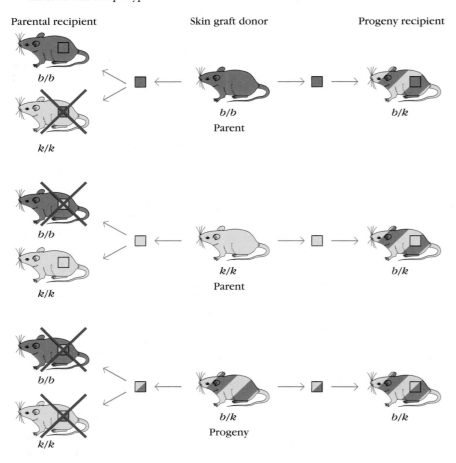

Parental recipient · Skin graft donor · Progeny recipient

**FIGURE 8-2 (a) Illustration of inheritance of MHC haplotypes in inbred mouse strains.** The letters *b/b* designate a mouse homozygous for the H-2$^b$ MHC haplotype, *k/k* homozygous for the H-2$^k$ haplotype, and *b/k* a heterozygote. Because the MHC loci are closely linked and inherited as a set, the MHC haplotype of F$_1$ progeny from the mating of two different inbred strains can be predicted easily. (b) Acceptance or rejection of skin grafts is controlled by the MHC type of the inbred mice. The progeny of the cross between two inbred strains with different MHC haplotypes (H-2$^b$ and H-2$^k$) will express both haplotypes (H-2$^{b/k}$) and will accept grafts from either parent and from one another. Neither parent strain will accept grafts from the offspring. (c) Inheritance of HLA haplotypes in a hypothetical human family. In humans, the paternal HLA haplotypes are arbitrarily designated A and B, maternal C and D. Note that a new haplotype, R (recombination), arises from recombination of maternal haplotypes. Because humans are an outbred species and there are many alleles at each HLA locus, the alleles constituting the haplotypes must be determined by typing parents and progeny. (d) The genes that make up each parental haplotype in the hypothetical family in (c) are shown along with the new haplotype (R) that arose from recombination of maternal haplotypes.

## (c) Inheritance of HLA haplotypes in a typical human family

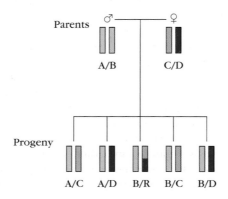

## (d) A new haplotype (R) arises from recombination of maternal haplotypes

| | | A | B | C | DR | DQ | DP |
|---|---|---|---|---|---|---|---|
| | | | | | HLA Alleles | | |
| Haplotypes | A | 1 | 7 | w3 | 2 | 1 | 1 |
| | B | 2 | 8 | w2 | 3 | 2 | 2 |
| | C | 3 | 44 | w4 | 4 | 1 | 3 |
| | D | 11 | 35 | w1 | 7 | 3 | 4 |
| | R | 3 | 44 | w4 | 7 | 3 | 4 |

have the same set of alleles, that is, the same MHC haplotype, as the prototype strain. For example, the CBA, AKR, and C3H strains all have the same MHC haplotype (H-$2^k$). The three strains differ, however, in genes outside the H-2 complex.

If two mice from inbred strains having different MHC haplotypes are bred to each other, the $F_1$ generation inherits haplotypes from both parental strains and therefore expresses both parental alleles at each MHC locus. For example, if an H-$2^b$ strain is crossed with an H-$2^k$, then the $F_1$ generation inherits both parental sets of alleles and is said to be H-$2^{b/k}$ (Figure 8-2a). Because such an $F_1$ generation expresses the MHC proteins of both parental strains on its cells, it is histocompatible with both strains and able to accept grafts from either parental strain (Figure 8-2b). However, neither of the inbred parental strains can accept a graft from the $F_1$ mice because half of the MHC molecules will be foreign to the parent.

The inheritance of HLA haplotypes from heterozygous human parents is illustrated in Figure 8-2c. In an outbred population, each individual is generally heterozygous at each locus. The human HLA complex is highly polymorphic, and multiple alleles of each class I and class II gene exist. However, as with mice, the human MHC loci are closely linked and usually inherited as a haplotype. When the father and mother have different haplotypes, as in the example shown (Figure 8-2c) there is a one-in-four chance that siblings will inherit the same paternal and maternal haplotypes and therefore be histocompatible with each other; none of the offspring will be histocompatible with the parents.

Although the rate of recombination by crossover is low within the HLA complex, it still contributes significantly to the diversity of the loci in human populations. Genetic recombination generates new allelic combinations (Figure 8-2d), and the high number of intervening generations since the appearance of humans as a species has allowed extensive recombination. As a result of recombination and other mechanisms for generating mutations, it is rare for any two unrelated individuals to have identical sets of HLA genes.

## Inbred mouse strains have aided the study of the MHC

Detailed analysis of the H-2 complex in mice was made possible by the development of congenic mouse strains. Inbred mouse strains are **syngeneic** or identical at all genetic loci. Two strains are **congenic** if they are genetically identical except at a single genetic locus or region. Any phenotypic differences that can be detected between congenic strains are related to the genetic region that distinguishes the strains. Congenic strains that are identical with each other except at the MHC can be produced by a series of crosses, backcrosses, and selections between two inbred strains that differ at the MHC. A frequently used congenic strain, designated B10.A, is derived from B10 mice (H-$2^b$) but has the H-$2^a$ haplotype. In several cases, recombinant haplotypes have been observed in congenic mice, allowing study of individual MHC genes

and their products. Examples of these are included in the list in Table 8-1. For example the B10.A (2R) strain has all MHC genes from the *a* haplotype except for the D region, which is derived from the H-$2^b$ parent.

Further insight into the function of the MHC products has been supplied by production of mouse strains with different genes knocked out. One of the earlier examples was a strain in which the gene for beta-2-microglobulin (which is expressed with class I molecules on the cell surface) was deleted, thus inhibiting most cell surface expression of class I MHC molecules. Although no class I antigens could be detected, the animals appeared normal and were fertile; however, they lacked CD8$^+$ cytotoxic T lymphocytes and had difficulty in resolving infections and in rejecting transplanted tumors. Class II knockouts lacked CD4$^+$ T lymphocytes in the spleen and lymph nodes and could not respond to T-dependent antigens (see Chapter 10). The respective roles for class I and class II antigens in development of immunity will be described below.

# MHC Molecules and Genes

Class I and class II MHC molecules are membrane-bound glycoproteins that are closely related in both structure and function. Both class I and class II MHC molecules have been isolated and purified, and the three-dimensional structures of their extracellular domains have been determined by x-ray crystallography. Both types of membrane glycoproteins function as highly specialized antigen-presenting molecules that form unusually stable complexes with peptide ligands, displaying them on the cell surface for recognition by T cells. In contrast, class III MHC molecules are a group of unrelated proteins that do not share structural similarity and common function with class I and II molecules. The class III molecules include complement proteins (Chapter 7).

## Class I molecules have a glycoprotein heavy chain and a small protein light chain

Class I MHC molecules contain a 45-kilodalton (kDa) α chain associated noncovalently with a 12-kDa **$β_2$-microglobulin** molecule (Figure 8-3). The α chain is a transmembrane glycoprotein encoded by polymorphic genes within the A, B, and C regions of the human HLA complex and within the K and D regions of the mouse H-2 complex (see Figure 8-1). $β_2$-Microglobulin is a protein encoded by a highly conserved gene located on a different chromosome. Association of the α chain with $β_2$-microglobulin is required for expression of class I molecules on cell membranes. The α chain is anchored in the plasma membrane by its hydrophobic transmembrane segment and hydrophilic cytoplasmic tail.

Structural analyses have revealed that the α chain of class I MHC molecules is organized into three external domains (α1, α2, and α3), each containing approximately 90 amino acids; a transmembrane domain of about 25 hydrophobic

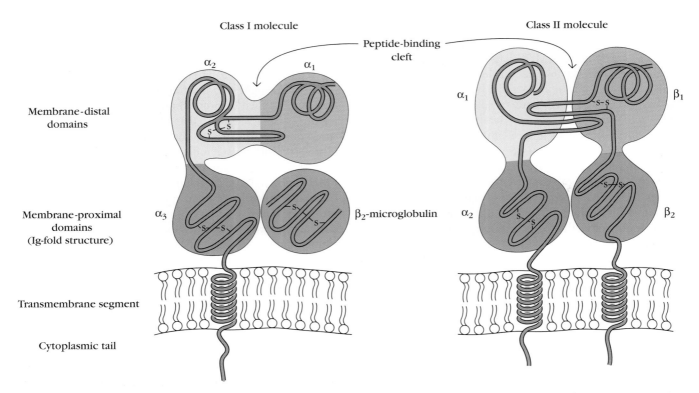

**FIGURE 8-3 Schematic diagrams of a class I and a class II MHC molecule showing the external domains, transmembrane segment, and cytoplasmic tail.** The peptide-binding cleft is formed by the membrane-distal domains in both class I and class II molecules. The membrane-proximal domains possess the basic immunoglobulin-fold structure; thus, class I and class II MHC molecules are classified as members of the immunoglobulin superfamily.

amino acids followed by a short stretch of charged (hydrophilic) amino acids; and a cytoplasmic anchor segment of 30 amino acids. $\beta_2$-Microglobulin is similar in size and organization to the $\alpha 3$ domain; it does not contain a transmembrane region and is noncovalently bound to the class I glycoprotein. Sequence data reveal homology between the $\alpha 3$ domain, $\beta_2$-microglobulin, and the constant-region domains in immunoglobulins. The enzyme papain cleaves the $\alpha$ chain just 13 residues proximal to its transmembrane domain, releasing the extracellular portion of the molecule, consisting of $\alpha 1$, $\alpha 2$, $\alpha 3$, and $\beta_2$-microglobulin. Purification and crystallization of the extracellular portion revealed two pairs of interacting domains: a membrane-distal pair made up of the $\alpha 1$ and $\alpha 2$ domains and a membrane-proximal pair composed of the $\alpha 3$ domain and $\beta_2$-microglobulin (Figure 8-4a).

The $\alpha 1$ and $\alpha 2$ domains interact to form a platform of eight antiparallel $\beta$ strands spanned by two long $\alpha$-helical regions. The structure forms a deep groove, or cleft, approximately $25 \text{ Å} \times 10 \text{ Å} \times 11 \text{ Å}$, with the long $\alpha$ helices as sides and the $\beta$ strands of the $\beta$ sheet as the bottom (Figure 8-4b). This *peptide-binding cleft* is located on the top surface of the class I MHC molecule, and it is large enough to bind a peptide of eight to 10 amino acids. The great surprise in the x-ray crystallographic analysis of class I molecules was the finding of small peptides in the cleft that had cocrystallized with the protein. These peptides are, in fact, processed antigen and self-peptides bound to the $\alpha 1$ and $\alpha 2$ domains in this deep groove.

The $\alpha 3$ domain and $\beta_2$-microglobulin are organized into two $\beta$ pleated sheets each formed by antiparallel $\beta$ strands of amino acids. As described in Chapter 4, this structure, known as the immunoglobulin fold, is characteristic of immunoglobulin domains. Because of this structural similarity, which is not surprising given the considerable sequence similarity with the immunoglobulin constant regions, class I MHC molecules and $\beta_2$-microglobulin are classified as members of the immunoglobulin superfamily (see Figure 4-23). The $\alpha 3$ domain appears to be highly conserved among class I MHC molecules and contains a sequence that interacts with the CD8 membrane molecule present on $T_C$ cells.

$\beta_2$-Microglobulin interacts extensively with the $\alpha 3$ domain and also interacts with amino acids of the $\alpha 1$ and $\alpha 2$ domains. The interaction of both $\beta_2$-microglobulin and a peptide with a class I $\alpha$ chain is essential for the class I molecule to reach its fully folded conformation. As described in detail below, assembly of class I molecules is believed to occur by the initial interaction of

(a)

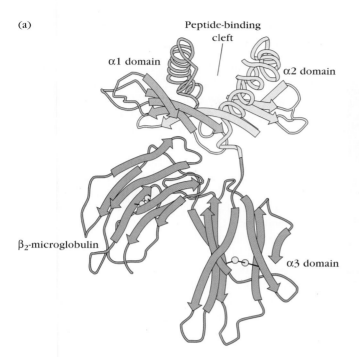

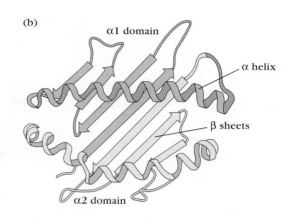

(b)

**FIGURE 8-4 Representations of the three-dimensional structure of the external domains of a human class I MHC molecule based on x-ray crystallographic analysis.** (a) Side view in which the β strands are depicted as thick arrows and the α helices as spiral ribbons. Disulfide bonds are shown as two interconnected spheres. The α1 and α2 domains interact to form the peptide-binding cleft.

Note the immunoglobulin-fold structure of the α3 domain and β$_2$-microglobulin. (b) The α1 and α2 domains as viewed from the top, showing the peptide-binding cleft, consisting of a base of antiparallel β strands and sides of α helices. This cleft in class I molecules can accommodate peptides containing eight to 10 residues.

β$_2$-microglobulin with the folding class I α chain. This metastable "empty" dimer is then stabilized by the binding of an appropriate peptide to form the native trimeric class I structure consisting of the class I α chain, β$_2$-microglobulin, and a peptide. This complete molecular complex is ultimately transported to the cell surface.

In vitro experiments indicate that in the absence of β$_2$-microglobulin, the class I MHC α chain is not expressed on the cell membrane. This is illustrated by Daudi tumor cells, which are unable to synthesize β$_2$-microglobulin. These tumor cells produce class I MHC α chains but do not express them on the membrane. However, if Daudi cells are transfected with a functional gene encoding β$_2$-microglobulin, class I molecules appear on the membrane.

## Class II molecules have two nonidentical glycoprotein chains

Class II MHC molecules contain two different polypeptide chains, a 33-kDa α chain and a 28-kDa β chain, which associate by noncovalent interactions (see Figure 8-3, right). Like class I α chains, class II MHC molecules are membrane-bound glycoproteins that contain external domains, a

transmembrane segment, and a cytoplasmic anchor segment. Each chain in a class II molecule contains two external domains: α1 and α2 domains in one chain and β1 and β2 domains in the other. The membrane-proximal α2 and β2 domains, like the membrane-proximal α3/β$_2$-microglobulin domains of class I MHC molecules, bear sequence similarity to the immunoglobulin-fold structure; for this reason, class II MHC molecules also are classified in the immunoglobulin superfamily. The membrane-distal portion of a class II molecule is composed of the α1 and β1 domains and forms the peptide-binding cleft for processed antigen.

X-ray crystallographic analysis reveals the similarity of class II and class I molecules, strikingly apparent when the molecules are surperimposed (Figure 8-5). The peptide-binding cleft of HLA-DR1, like that in class I molecules, is composed of a floor of eight antiparallel β strands and sides of antiparallel α helices. However, the class II molecule lacks the conserved residues in the class I molecule that bind to the terminal residues of short peptides and forms instead an open pocket; class I presents more of a socket, class II an open-ended groove. The functional consequences of these differences in fine structure will be explored below.

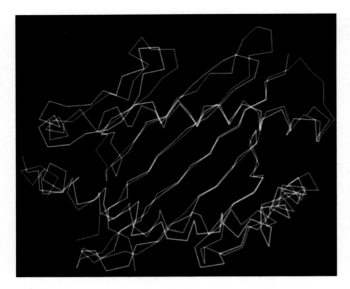

**FIGURE 8-5 The membrane-distal, peptide-binding cleft of a human class II MHC molecule, HLA-DR1 (blue), superimposed over the corresponding regions of a human class I MHC molecule, HLA-A2 (red).** *[From J. H. Brown et al., 1993, Nature **364**:33.]*

## The exon/intron arrangement of class I and II genes reflects their domain structure

Separate exons encode each region of the class I and II proteins (Figure 8-6). Each of the mouse and human class I genes has a 5′ leader exon encoding a short signal peptide followed by five or six exons encoding the α chain of the class I molecule (see Figure 8-6a). The signal peptide serves to facilitate insertion of the α chain into the endoplasmic reticulum and is removed by proteolytic enzymes in the endoplasmic reticulum after translation is completed. The next three exons encode the extracellular α1, α2, and α3 domains, and the following downstream exon encodes the transmembrane ($T_m$) region; finally, one or two 3′-terminal exons encode the cytoplasmic domains (C).

Like class I MHC genes, the class II genes are organized into a series of exons and introns mirroring the domain structure of the α and β chains (see Figure 8-6b). Both the α and the β genes encoding mouse and human class II MHC molecules have a leader exon, an α1 or β1 exon, an α2 or β2 exon, a transmembrane exon, and one or more cytoplasmic exons.

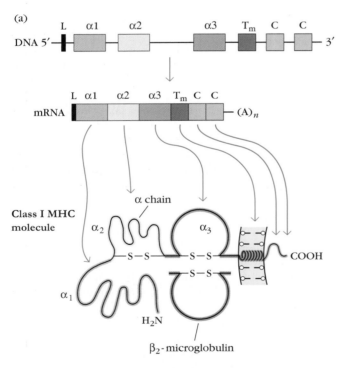

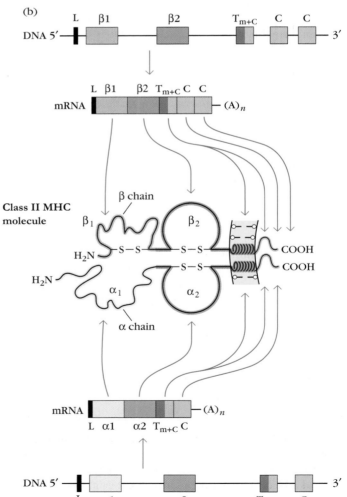

**FIGURE 8-6 Schematic diagram of (a) class I and (b) class II MHC genes, mRNA transcripts, and protein molecules.** There is correspondence between exons and the domains in the gene products; note that the mRNA transcripts are spliced to remove the intron sequences. Each exon, with the exception of the leader (L) exon, encodes a separate domain of the MHC molecule. The leader peptide is removed in a post-translational reaction before the molecule is expressed on the cell surface. The gene encoding $\beta_2$-microglobulin is located on a different chromosome. $T_m$ = transmembrane; C = cytoplasmic.

| TABLE 8-2 | Peptide binding by class I and class II MHC molecules | |
|---|---|---|
| | **Class I molecules** | **Class II molecules** |
| Peptide-binding domain | α1/α2 | α1/β1 |
| Nature of peptide-binding cleft | Closed at both ends | Open at both ends |
| General size of bound peptides | 8–10 amino acids | 13–18 amino acids |
| Peptide motifs involved in binding to MHC molecule | Anchor residues at both ends of peptide; generally hydrophobic carboxyl-terminal anchor | Anchor residues distributed along the length of the peptide |
| Nature of bound peptide | Extended structure in which both ends interact with MHC cleft but middle arches up away from MHC molecule | Extended structure that is held at a constant elevation above the floor of MHC cleft |

## Class I and II molecules exhibit polymorphism in the region that binds to peptides

Several hundred different allelic variants of class I and II MHC molecules have been identified in humans. Any one individual, however, expresses only a small number of these molecules—up to six different class I molecules and up to 12 different class II molecules. Yet this limited number of MHC molecules must be able to present an enormous array of different antigenic peptides to T cells, permitting the immune system to respond specifically to a wide variety of antigenic challenges. Thus, peptide binding by class I and II molecules does not exhibit the fine specificity characteristic of antigen binding by antibodies and T-cell receptors. Instead, a given MHC molecule can bind numerous different peptides, and some peptides can bind to several different MHC molecules. Because of this broad specificity, the binding between a peptide and an MHC molecule is often referred to as "promiscuous."

Given the similarities in the structures of the peptide-binding clefts in class I and II MHC molecules, it is not surprising that they exhibit some common peptide-binding features (Table 8-2). In both types of MHC molecules, peptide ligands are held in a largely extended conformation that runs the length of the cleft. The peptide-binding cleft in class I molecules is blocked at both ends, whereas the cleft is open in class II molecules (Figure 8-7). As a result of this difference, class I molecules bind peptides that typically contain eight to 10 amino acid residues, while the open groove of class II molecules accommodates slightly longer peptides of 13 to 18 amino acids. Another difference, explained in more detail below, is that class I binding requires that the peptide contain specific amino acid residues near the N and C termini; there is no such requirement for class II peptide binding.

The association of peptide and MHC molecule is very stable under physiologic conditions ($K_d$ values range from $\sim 10^{-6}$ to $10^{-10}$); thus, most of the MHC molecules expressed on the membrane of a cell will be associated with a peptide of self or nonself origin.

(a) **Class I MHC**

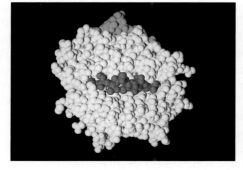

(b) **Class II MHC**

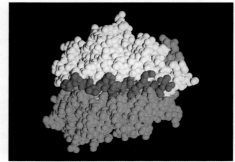

**FIGURE 8-7 MHC class I and class II molecules with bound peptides.** (a) Space-filling model of human class I molecule HLA-A2 (white) with peptide (red) from HIV reverse transcriptase (amino acid residues 309–317) in the binding groove. β$_2$-microglobulin is shown in blue. Residues above the peptide are from the α1 domain, those below from α2. (b) Space-filling model of human class II molecules HLA-DR1 with the DRα chain shown in white and the DRβ chain in blue. The peptide (red) in the binding groove is from influenza hemagglutinin (amino acid residues 306–318). *[From D. A. Vignali and J. Strominger, 1994, The Immunologist **2**:112.]*

## Class I MHC–Peptide Interaction

Class I MHC molecules bind peptides and present them to $CD8^+$ T cells. In general, these peptides are derived from endogenous intracellular proteins that are digested in the cytosol. The peptides are then transported from the cytosol into the cisternae of the endoplasmic reticulum, where they interact with class I MHC molecules. This process, known as the cytosolic or endogenous processing pathway, is discussed in detail below.

Each type of class I MHC molecule (K, D, and L in mice or A, B, and C in humans) binds a unique set of peptides. In addition, each allelic variant of a class I MHC molecule (e.g., $H-2K^k$ and $H-2K^d$) binds a distinct set of peptides. Because a single nucleated cell expresses about $10^5$ copies of each class I molecule, many different peptides will be expressed simultaneously on the surface of a nucleated cell by class I MHC molecules.

In a critical study of peptide binding by MHC molecules, peptides bound by two allelic variants of a class I MHC molecule were released chemically and analyzed by high-performance liquid chromatography (HPLC) mass spectrometry. More than 2000 distinct peptides were found among the peptide ligands released from these two class I MHC molecules. Since there are approximately $10^5$ copies of each class I allelic variant per cell, it is estimated that each of the 2000 distinct peptides is presented with a frequency of 100 to 4000 copies per cell. Evidence suggests that even a single peptide-MHC complex may be sufficient to target a cell for recognition and lysis by a cytotoxic T lymphocyte with a receptor specific for that target structure.

The bound peptides isolated from different class I molecules have been found to have two distinguishing features: they are eight to 10 amino acids in length, most commonly nine, and they contain specific amino acid residues that appear to be essential for binding to a particular MHC molecule. Binding studies have shown that nonameric peptides bind to class I molecules with a higher affinity than do peptides that are either longer or shorter, suggesting that this peptide length is most compatible with the closed-ended peptide-binding cleft in class I molecules. The ability of an individual class I MHC molecule to bind to a diverse spectrum of peptides is due to the presence of the same or similar amino acid residues at several defined positions along the peptides (Figure 8-8). Because these amino acid residues anchor the peptide into the groove of the MHC molecule, they are called *anchor residues*. The side chains of the anchor residues in the peptide are complementary with surface features of the binding cleft of the class I MHC molecule. The amino acid residues lining the binding sites vary among different class I allelic variants and determine the identity of the anchor residues that can interact with a given class I molecule.

All peptides examined to date that bind to class I molecules contain a carboxyl-terminal anchor. These anchors are generally hydrophobic residues (e.g., leucine, isoleucine), although a few charged amino acids have been

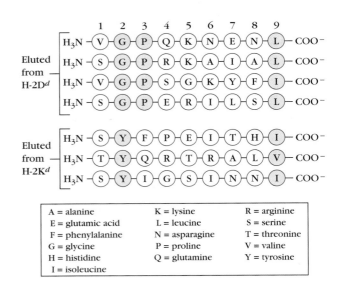

| A = alanine | K = lysine | R = arginine |
| E = glutamic acid | L = leucine | S = serine |
| F = phenylalanine | N = asparagine | T = threonine |
| G = glycine | P = proline | V = valine |
| H = histidine | Q = glutamine | Y = tyrosine |
| I = isoleucine | | |

**FIGURE 8-8 Examples of anchor residues (blue) in nonameric peptides eluted from two class I MHC molecules.** Anchor residues that interact with the class I MHC molecule tend to be hydrophobic amino acids. *[Data from V. H. Engelhard, 1994, Current Opinion in Immunology 6:13.]*

reported. Besides the anchor residue found at the carboxyl terminus, another anchor is often found at the second or second and third positions at the amino-terminal end of the peptide (see Figure 8-8). In general, any peptide of correct length that contains the same or similar anchor residues will bind to the same class I MHC molecule. The discovery of conserved anchor residues in peptides that bind to various class I MHC molecules may permit prediction of which peptides in a complex antigen will bind to a particular MHC molecule, based on the presence or absence of these motifs.

X-ray crystallographic analyses of peptide–class I MHC complexes have revealed how the peptide-binding cleft in a given MHC molecule can interact stably with a broad spectrum of different peptides. The anchor residues at both ends of the peptide are buried within the binding cleft, thereby holding the peptide firmly in place (see Figure 8-7). As noted already, nonameric peptides are bound preferentially; the main contacts between class I MHC molecules and peptides involve residue 2 at the amino-terminal end and residue 9 at the carboxyl terminus of the nonameric peptide. Between the anchors, the peptide arches away from the floor of the cleft in the middle (Figure 8-9), allowing peptides that are slightly longer or shorter to be accommodated. Amino acids that arch away from the MHC molecule are more exposed and presumably can interact more directly with the T-cell receptor.

## Class II MHC–Peptide Interaction

Class II MHC molecules bind peptides and present these peptides to $CD4^+$ T cells. Like class I molecules, molecules of class II can bind a variety of peptides. In general, these peptides are derived from exogenous proteins (either self or nonself), which are degraded within the endocytic processing

(a)

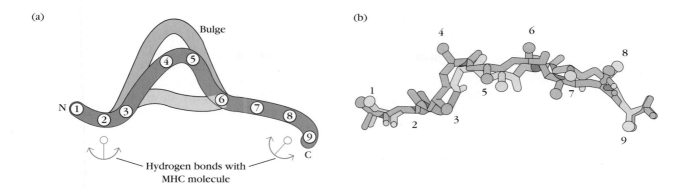

(b)

(c)

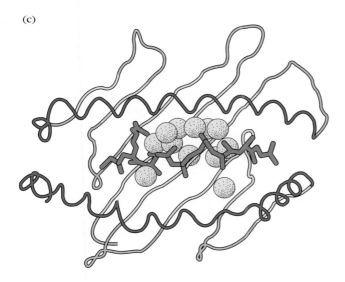

**FIGURE 8-9 Conformation of peptides bound to class I MHC molecules. (a) Schematic diagram of conformational difference in bound peptides of different lengths.** Longer peptides bulge in the middle, whereas shorter peptides are more extended. Contact with the MHC molecule is by hydrogen bonds to anchor residues 1/2 and 8/9. (b) Molecular model based on crystal structure of an influenza virus antigenic peptide (blue) and an endogenous peptide (purple) bound to a class I MHC molecule. Residues are identified by small numbers corresponding to those in part a. (c) Representation of α1 and α2 domains of HLA-B27 and a bound antigenic peptide based on x-ray crystallographic analysis of the cocrystallized peptide-HLA molecule. The peptide (purple) arches up away from the β strands forming the floor of the binding cleft and interacts with 12 water molecules (spheres). *[Part a adapted from P. Parham, 1992,* **Nature** *360:300, © 1992 Macmillan Magazines Limited; part b adapted from M. L. Silver et al., 1992,* **Nature** *360:367, © 1992 Macmillan Magazines Limited; part c adapted from D. R. Madden et al., 1992,* **Cell** *70:1035, reprinted by permission of Cell Press.]*

pathway (see below). Most of the peptides associated with class II MHC molecules are derived from self membrane-bound proteins or foreign proteins internalized by phagocytosis or by receptor-mediated endocytosis and then processed through the endocytic pathway. For instance, peptides derived from digestion of membrane-bound class I MHC molecules often are bound to class II MHC molecules.

Peptides recovered from class II MHC–peptide complexes generally contain 13 to 18 amino acid residues, somewhat longer than the nonameric peptides that most commonly bind to class I molecules. The peptide-binding cleft in class II molecules is open at both ends (see Figure 8-7b), allowing longer peptides to extend beyond the ends, like a long hot dog in a bun. Peptides bound to class II MHC molecules maintain a roughly constant elevation on the floor of the binding cleft, another feature that distinguishes peptide binding to class I and class II molecules.

Peptide-binding studies and structural data for class II molecules indicate that a central core of 13 amino acids determines the ability of a peptide to bind class II. Longer peptides may be accommodated within the class II cleft, but the binding characteristics are determined by the central 13 residues. The peptides that bind to a particular class II molecule often have internal conserved sequence motifs, but

unlike class I–binding peptides, they lack conserved anchor residues. Instead, hydrogen bonds between the backbone of the peptide and the class II molecule are distributed throughout the binding site rather than being clustered predominantly at the ends of the site as for class I–bound peptides. Peptides that bind to class II MHC molecules contain an internal sequence of seven to 10 amino acids that provide the major contact points. Generally, this sequence has an aromatic or hydrophobic residue at the amino terminus and three additional hydrophobic residues in the middle portion and carboxyl-terminal end of the peptide. In addition, over 30% of the peptides eluted from class II molecules contain a proline residue at position 2 and another cluster of prolines at the carboxyl-terminal end.

## Class I and class II molecules exhibit diversity within a species, and multiple forms occur in an individual

An enormous diversity is exhibited by the MHC molecules within a species and within individuals. This variability echoes the diversity of antibodies and T-cell receptors, but the source of diversity for MHC molecules is not the same. Antibodies and T-cell receptors are generated by several somatic processes, including gene rearrangement and somatic mutation of rearranged genes (see Table 5-2). Thus, the generation of T and B cell receptors is dynamic, changing over time within an individual. By contrast, the MHC molecules expressed by an individual are fixed in the genes and do not change over time. The diversity of the MHC within a species stems from polymorphism, the presence of multiple alleles at a given genetic locus within the species. Diversity of MHC molecules in an individual results not only from having different alleles of each gene but also from the presence of duplicated genes with similar or overlapping functions, not unlike the isotypes of immunoglobulins. Because it includes genes with similar, but not identical structure and function (for example, HLA-A, -B, and -C), the MHC may be said to be **polygenic.**

The MHC possesses an extraordinarily large number of different alleles at each locus and is one of the most polymorphic genetic complexes known in higher vertebrates. These alleles differ in their DNA sequences from one individual to another by 5% to 10%. The number of amino acid differences between MHC alleles can be quite significant, with up to 20 amino acid residues contributing to the unique structural nature of each allele. Analysis of human HLA class I genes has revealed, as of 2006, approximately 370 A alleles, 660 B alleles, and 190 C alleles. In mice, the polymorphism is similarly enormous. The human class II genes are also highly polymorphic, and, in some cases, different individuals have different numbers of genes. The number of HLA-DR beta-chain genes (DRB) may vary from two to nine in different haplotypes, and approximately 480 alleles of DRB genes have been reported. Interestingly, the DRA chain is highly conserved, with only three different alleles reported. Current estimates of actual polymorphism in the human MHC are probably on the low side because the most detailed data were obtained from populations of European descent. The fact that many non-European population groups cannot be typed using the MHC serologic typing reagents available indicates that the worldwide diversity of the MHC genes is far greater. Now that MHC genes can be sequenced directly, it is expected that many additional alleles will be detected.

This enormous polymorphism results in a tremendous diversity of MHC molecules within a species. Using the numbers given above for the allelic forms of human HLA-A, -B, and -C, we can calculate the theoretical number of combinations that can exist by multiplying $370 \times 660 \times 190$, yielding upward of 46 million different class I haplotypes possible in the population. If class II loci are considered, the five genes encoding the beta chain (DRB genes, B1 through B5) have 400, 1, 42, 13, and 18 alleles respectively, DQA1 and B1 contribute 28 and 62 alleles respectively, and DPB1 contributes 118 alleles; this allows approximately $8.0 \times 10^{11}$ different class II combinations. Because each haplotype contains both class I and class II genes, multiplication of the numbers gives a total of almost $4 \times 10^{19}$ possible combinations of these class I and II alleles.

## Linkage Disequilibrium

The calculation in the preceding paragraph leading to astronomical estimates of possible HLA haplotypes assumes completely random combinations of alleles. The actual diversity is known to be less, because certain allelic combinations occur more frequently in HLA haplotypes than predicted by random combination, a state referred to as *linkage disequilibrium.* Briefly, linkage disequilibrium is the difference between the frequency *observed* for a particular combination of alleles and that *expected* from the frequencies of the individual alleles. The expected frequency for the combination may be calculated by multiplying the frequencies of the two alleles. For example, if HLA-A1 occurs in 16% of individuals in a population (frequency = 0.16) and HLA-B8 in 9% of that group (frequency = 0.09), it is expected that about 1.4% of the group will have both alleles ($0.16 \times 0.09 = 0.014$). However, the data show that HLA-A1 and HLA-B8 are found together in 8.8% of individuals studied. This difference is a measure of the linkage disequilibrium between these alleles of class I MHC genes.

Several explanations have been advanced to explain linkage disequilibrium. The simplest is that too few generations have elapsed to allow the number of crossovers necessary to reach equilibrium among the alleles present in founders of the population. The haplotypes that are overrepresented in the population today would then reflect the combinations of alleles present in the founders. Alternatively, selective effects could result in the higher frequency of certain allelic combinations. For example, certain combinations of alleles might produce resistance to certain diseases, causing them to be selected for and therefore overrepresented. Alternatively, the underrepresented haplotypes might generate harmful effects, such as susceptibility to autoimmune disorders, and undergo negative selection. A third hypothesis is that crossovers are more frequent in certain DNA sequence regions, and the presence or absence of regions prone to crossover (hot spots) between alleles can dictate the frequency of allelic association. Data in support of this hypothesis were found in mouse-breeding studies that generated new recombinant H-2 types. The points of crossover in the new MHC haplotypes were not randomly distributed throughout the complex. Instead, the same regions of crossover were found in more than one recombinant haplotype. This suggests that hot spots of recombination do exist that would influence linkage disequilibrium in populations.

Despite linkage disequilibrium, enormous polymorphism still exists in the human MHC. This polymorphism has been generated by recombination, point mutation, and gene conversion, all of which contribute to the diversity of MHC genes within the population. Comparison of allelic gene sequences shows individual substitutions, which are evidence for point mutations within the genes, as well as clusters of variation within highly conserved regions, suggesting that gene conversion has taken place (see Chapter 5). The

(a)

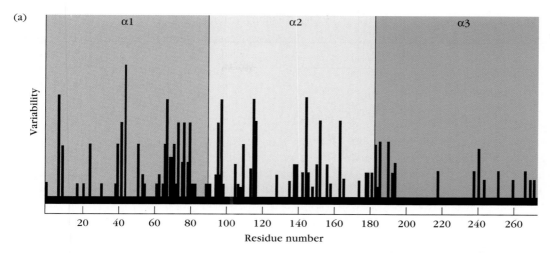

(b)

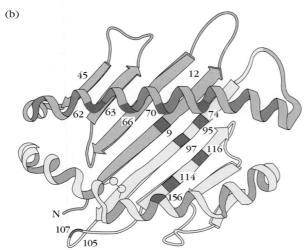

**FIGURE 8-10 (a) Plots of variability in the amino acid sequence of allelic class I MHC molecules in humans versus residue position.** In the external domains, most of the variable residues are in the membrane-distal α1 and α2 domains. (b) Location of polymorphic amino acid residues (red) in the α1/α2 domain of a human class I MHC molecule. *[Part a adapted from R. Sodoyer et al., 1984, EMBO Journal **3**:879, reprinted by permission of Oxford University Press; part b adapted, with permission, from P. Parham, 1989, Nature **342**:617, © 1989 Macmillan Magazines Limited.]*

high degree of diversity among the MHC loci of different individuals makes it very difficult to match donor and acceptor MHC types for successful organ transplants. The consequences of this major obstacle to the therapeutic use of transplantation are described in Chapter 17.

## Functional Relevance of MHC Polymorphism

Sequence divergence among alleles of the MHC within a species is very high, as great as the divergence observed for the genes encoding some enzymes across species. Also of interest is that the sequence variation among MHC molecules is not randomly distributed along the entire polypeptide chain but instead is clustered in short stretches, largely within the membrane-distal α1 and α2 domains of class I molecules (Figure 8-10a). Similar patterns of diversity are observed in the α1 and β1 domains of class II molecules.

Structural comparisons have located the polymorphic residues within the three-dimensional structure of the membrane-distal domains in class I and class II MHC molecules and have related allelic differences to functional

differences (Figure 8-10b). For example, of 17 amino acids previously shown to display significant polymorphism in the HLA-A2 molecule, 15 were shown by x-ray crystallographic analysis to be in the peptide-binding cleft of this molecule. The location of so many polymorphic amino acids within the binding site for processed antigen strongly suggests that allelic differences contribute to the observed differences in the ability of MHC molecules to interact with a given peptide ligand.

## Detailed Genomic Map of MHC Genes

The MHC spans some 2000 kb of mouse DNA and some 4000 kb of human DNA. The recently completed human genome sequence shows this region to be densely packed with genes, most of which have known functions. Our current understanding of the genomic organization of mouse and human MHC genes is diagrammed in Figure 8-11.

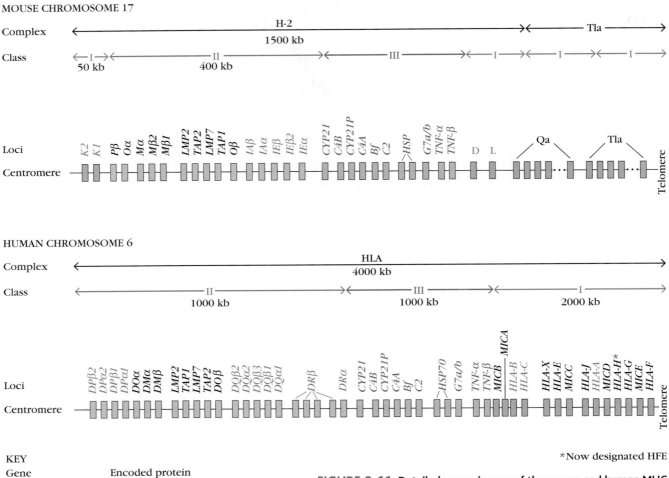

**FIGURE 8-11 Detailed genomic map of the mouse and human MHC, including genes encoding classical and nonclassical MHC molecules.** The class I MHC genes are colored red, MHC II genes are colored blue, and genes in MHC III are colored green. Classical class I genes are labeled in red, class II genes are labeled in blue, and the nonclassical MHC genes are labeled in black. The concept of classical and nonclassical does not apply to class III. The functions for certain proteins encoded by the nonclassical class I genes are known. In the mouse, there are nonclassical genes located downstream from Tla that are not shown.

**KEY**

| Gene | Encoded protein |
|---|---|
| C2, C4A, C4B, Bf | Complement components |
| CYP21, CYP21P | Steroid 21-hydroxylases |
| G7a/b | Valyl-tRNA synthetase |
| HSP | Heat-shock protein |
| LMP2, LMP7 | Proteasome-like subunits |
| TAP1, TAP2 | Peptide-transporter subunits |
| TNF-α, TNF-β | Tumor necrosis factors α and β |

*Now designated HFE

## The human class I region spans about 2000 kb at the telomeric end of the HLA complex

In humans, the class I MHC region is about 2000 kb long and contains approximately 20 genes. In mice, the class I MHC consists of two regions separated by the intervening class II and class III regions. Included within the class I region are the genes encoding the well-characterized classical class I MHC molecules designated HLA-A, HLA-B, and HLA-C in humans and in mice H-2K, H-2D, and H-2L (H-2L genes are present only in certain mouse haplotypes). Many nonclassical class I genes, identified by molecular mapping, are also present in both the mouse and human MHC. In mice, the nonclassical class I genes are located in three regions (*H-2Q, T,* and *M*) downstream from the H-2

complex (*M* is not shown in Figure 8-11). In humans, the nonclassical class I genes include the *HLA-E, HLA-F, HLA-G, HFE, HLA-J,* and *HLA-X* loci as well as a recently discovered family of genes called *MIC*, which includes *MICA* through *MICE*. Some of the nonclassical class I MHC genes are pseudogenes and do not encode a protein product, but others, such as *HLA-G* and *HFE*, encode class I–like products with highly specialized functions. The MIC family of class I genes has only 15% to 30% sequence identity to classical class I, and those designated as MICA are highly polymorphic. The MIC gene products are expressed at low levels in epithelial cells and are induced by heat or other stimuli that influence heat-shock proteins.

The functions of the nonclassical class I MHC molecules remain largely unknown, although a few studies suggest that some of these molecules, like the classical class I MHC molecules, may present peptides to T cells. One intriguing finding is that the mouse molecule encoded by the *H-2M* locus is able to bind a self peptide derived from a subunit of NADH dehydrogenase, an enzyme encoded by the mitochondrial genome. This particular self peptide contains an amino-terminal formylated methionine. What is interesting about this finding is that peptides derived from prokaryotic organisms often have formylated amino-terminal methionine residues. This *H-2M*–encoded class I molecule may thus be uniquely suited to present peptides from prokaryotic organisms that are able to grow intracellularly. *Listeria monocytogenes* is such an organism, and H-2m molecules present peptides from this bacterium.

## The class II MHC genes are located at the centromeric end of HLA

The class II MHC region contains the genes encoding the α and β chains of the classical class II MHC molecules designated HLA-DR, DP, and DQ in humans and H-2IA and IE in mice. Molecular mapping of the class II MHC has revealed multiple β-chain genes in some regions in both mice and humans, as well as multiple α-chain genes in humans (see Figure 8-11). In the human DR region, for example, there are three or four functional β-chain genes. All of the β-chain gene products can be expressed together with the α-chain gene product in a given cell, thereby increasing the number of different antigen-presenting molecules on the cell. Although the human DR region contains just one α-chain gene, the DP and DQ region each contains two.

Genes encoding nonclassical class II MHC molecules have also been identified in both humans and mice. In mice, several class II genes (*O*α, *O*β, *M*α, and *M*β) encode nonclassical MHC molecules that exhibit limited polymorphism and a different pattern of expression than the classical IA and IE class II molecules. In the human class II region, nonclassical genes designated *DM* and *DO* have been identified. The *DM* genes encode a class II–like molecule (HLA-DM) that facilitates the loading of antigenic peptides into the class II MHC molecules. Class II DO molecules, which are expressed only in the thymus and mature B cells, have been shown to serve as regulators of class II antigen processing. The functions of HLA-DM and HLA-DO are described below.

## Human MHC class III genes are between class I and II

The class III region of the MHC in humans and mice contains a heterogeneous collection of genes (see Figure 8-11). These genes encode several complement components, two steroid 21-hydroxylases, two heat-shock proteins, and two cytokines (TNF-α and TNF-β). Some of these class III MHC gene products play a role in certain diseases. For example, mutations in the genes encoding 21-hydroxylase have been linked to congenital adrenal hyperplasia. Interestingly, the presence of a linked class III gene cluster is conserved in all species with an MHC region.

## Cellular Expression of MHC Molecules

In general, the classical class I MHC molecules are expressed on most nucleated cells, but the level of expression differs among different cell types. The highest levels of class I molecules are expressed by lymphocytes, where they constitute approximately 1% of the total plasma membrane proteins, or some $5 \times 10^5$ molecules per cell. In contrast, fibroblasts, muscle cells, liver hepatocytes, and neural cells express very low levels of class I MHC molecules. The low level on liver cells may contribute to the considerable success of liver transplants by reducing the likelihood of graft rejection by $T_c$ cells of the recipient. A few cell types (e.g., neurons and sperm cells at certain stages of differentiation) appear to lack class I MHC molecules altogether.

As noted earlier, any particular MHC molecule can bind many different peptides. Since the MHC alleles are codominantly expressed, heterozygous individuals express on their cells the gene products encoded by both alleles at each MHC locus. An $F_1$ mouse, for example, expresses the K, D, and L from each parent (six different class I MHC molecules) on each of its nucleated cells (Figure 8-12). A similar situation

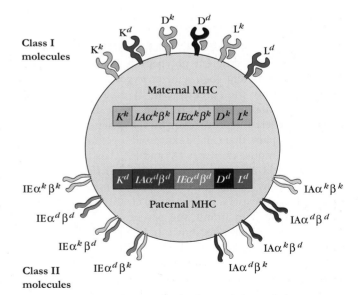

**FIGURE 8-12 Diagram illustrating various MHC molecules expressed on antigen-presenting cells of a heterozygous H-2$^{k/d}$ mouse.** Both the maternal and paternal MHC genes are expressed. Because the class II molecules are heterodimers, heterologous molecules containing one maternal-derived and one paternal-derived chain are produced. The β$_2$-microglobulin component of class I molecules (pink) is encoded by a gene on a separate chromosome and may be derived from either parent.

occurs in humans; that is, a heterozygous individual expresses the A, B, and C alleles from each parent (six different class I MHC molecules) on the membrane of each nucleated cell. The expression of so many class I MHC molecules allows each cell to display a large number of peptides in the peptide-binding clefts of its MHC molecules.

In normal, healthy cells, the class I molecules will display self peptides resulting from normal turnover of self proteins. In cells infected by a virus, viral peptides, as well as self peptides, will be displayed. A single virus-infected cell should be envisioned as having various class I molecules on its membrane, each displaying different sets of viral peptides. Because of individual allelic differences in the peptide-binding clefts of the class I MHC molecules, different individuals within a species will have the ability to bind different sets of viral peptides.

Unlike class I MHC molecules, class II molecules are expressed constitutively only by antigen-presenting cells, primarily macrophages, mature dendritic cells, and B cells; thymic epithelial cells and some other cell types can be induced to express class II molecules and to function as antigen-presenting cells under certain conditions and under stimulation of some cytokines. Among the various cell types that express class II MHC molecules, marked differences in expression have been observed. In some cases, class II expression depends on the cell's differentiation stage. For example, class II molecules cannot be detected on pre–B cells but are expressed constitutively on the membrane of mature B cells. Similarly, monocytes and macrophages express only low levels of class II molecules until they are activated by interaction with an antigen, after which the level of expression increases significantly.

Because each of the classical class II MHC molecules is composed of two different polypeptide chains, which are encoded by different loci, a heterozygous individual expresses not only the parental class II molecules but also molecules containing $\alpha$ and $\beta$ chains from different chromosomes. There is no restriction on the genetic origins of $\alpha$ and $\beta$ chain pairs that can be expressed together. For example, an H-2$^k$ mouse expresses IA$^k$ and IE$^k$ class II molecules; similarly, an H-2$^d$ mouse expresses IA$^d$ and IE$^d$ molecules. The F$_1$ progeny resulting from crosses of mice with these two haplotypes express four parental class II molecules and four molecules containing one parent's $\alpha$ chain and the other parent's $\beta$ chain (as shown in Figure 8-12). Since the human MHC contains three classical class II genes (*DP, DQ,* and *DR*), a heterozygous individual expresses six parental class II molecules and six molecules containing $\alpha$ and $\beta$ chain combinations from either parent. The number of different class II molecules expressed by an individual is increased further by the presence of multiple $\beta$-chain genes in mice and humans and in humans by multiple $\alpha$-chain genes. The diversity generated by these mechanisms presumably increases the number of different antigenic peptides that can be presented and thus is advantageous to the organism.

## Regulation of MHC Expression

Research on the regulatory mechanisms that control the differential expression of MHC genes in different cell types is still in its infancy, but much has been learned. The current availability of the complete genomic map of the MHC complex is expected to greatly accelerate the identification and investigation of coding and regulatory sequences, leading to new directions in research on how the system is controlled.

Both class I and class II MHC genes are flanked by 5′ promoter sequences, which bind sequence-specific transcription factors. The promoter motifs and the transcription factors that bind to these motifs have been identified for a number of MHC genes. Transcriptional regulation of the MHC is mediated by both positive and negative elements. For example, a class II MHC transcriptional activator called *CIITA* and another transcription factor called *RFX* have both been shown to bind to the promoter region of class II MHC genes. Defects in these transcription factors cause one form of *bare lymphocyte syndrome* (see the Clinical Focus on page 213). Patients with this disorder lack class II MHC molecules on their cells and as a result suffer a severe immunodeficiency due to the central role of class II MHC molecules in T-cell maturation and activation.

The expression of MHC molecules is also regulated by various cytokines. The interferons (alpha, beta, and gamma) and tumor necrosis factor have each been shown to increase expression of class I MHC molecules on cells. Interferon gamma (IFN-$\gamma$), for example, appears to induce the formation of a specific transcription factor that binds to the promoter sequence flanking the class I MHC genes. Binding of this transcription factor to the promoter sequence appears to coordinate the up-regulation of transcription of the genes encoding the class I $\alpha$ chain, $\beta_2$-microglobulin, and other proteins involved in antigen processing and presentation. IFN-$\gamma$ also has been shown to induce expression of the class II transcriptional activator (CIITA), thereby indirectly increasing expression of class II MHC molecules on a variety of cells, including non-antigen-presenting cells (e.g., skin keratinocytes, intestinal epithelial cells, vascular endothelium, placental cells, and pancreatic beta cells). Other cytokines influence MHC expression only in certain cell types; for example, IL-4 increases expression of class II molecules by resting B cells. Expression of class II molecules by B cells is down-regulated by IFN-$\gamma$; corticosteroids and prostaglandins also decrease expression of class II MHC molecules.

MHC expression on cell surfaces is decreased by infection with certain viruses, including human cytomegalovirus (CMV), hepatitis B virus (HBV), and adenovirus 12 (Ad12). In some cases, reduced expression of class I MHC molecules is due to decreased levels of a component needed for peptide transport or MHC class I assembly rather than decreased transcription. In cytomegalovirus infection, for example, a viral protein binds to $\beta_2$-microglobulin, preventing assembly of class I MHC molecules and their transport to the plasma membrane. Adenovirus 12 infection causes a pronounced decrease in transcription of the transporter genes (*TAP1* and *TAP2*). As described below, the TAP gene products play an important role in peptide transport from the cytoplasm into the rough endoplasmic reticulum. Blocking of *TAP* gene expression inhibits peptide transport; as a result, class I MHC molecules cannot assemble with $\beta_2$-microglobulin or be transported to the cell membrane. Decreased expression of class I MHC molecules, by whatever mechanism, is likely to help viruses evade the immune response by reducing the likelihood that virus-infected cells can display MHC-viral peptide complexes and become targets for CTL-mediated destruction. The Clinical Focus on page 213 describes how deficiencies in TAP are implicated in bare lymphocyte syndrome.

## MHC and Disease Susceptibility

Some HLA alleles occur at a much higher frequency in people suffering from certain diseases than in the general population. The diseases associated with particular MHC alleles include autoimmune disorders, certain viral diseases, disorders of the complement system, some neurologic disorders, and several different allergies. The association between an HLA allele and a given disease may be quantified by determining the frequency of that HLA allele expressed by individuals afflicted with the disease, then comparing these data with the frequency of the same allele in the general population. Such a comparison allows calculation of **relative risk (RR).**

$$RR = \frac{(Ag^+ / Ag^-) \text{ disease group}}{(Ag^+ / Ag^-) \text{ control group}}$$

A relative risk value of 1 means that the HLA allele is expressed with the same frequency in the patient and general populations, indicating that the allele confers no increased risk for the disease. A relative risk value substantially above 1 indicates an association between the HLA allele and the disease. For example, individuals with the HLA-B27 allele are 90 times more likely (relative risk of 90) to develop the autoimmune disease ankylosing spondylitis, an inflammatory disease of vertebral joints characterized by destruction of cartilage, than individuals who lack this HLA-B allele. Other disease associations with significantly high relative risk include HLA-DR2 and narcolepsy (RR of 130) and hereditary hemochromatosis and the haplotype A3/B14, with an RR of 90.

The existence of an association between an MHC allele and a disease should not be interpreted to imply that the expression of the allele has caused the disease—the relationship between MHC alleles and development of disease is complex. In the case of ankylosing spondylitis, for example, it has been suggested that because of the close linkage of the TNF-$\alpha$ and TNF-$\beta$ genes with the HLA-B locus, these cytokines may be involved in the destruction of cartilage. The

association of hemochromatosis and A3/B14 is likely due to mutations in HFE, which is linked to these class I genes. HFE encodes a membrane protein with a role in iron metabolism; its absence or malfunction causes the iron overload characteristic of hemochromatosis.

When the associations between MHC alleles and disease are weak, reflected by low relative risk values, it is likely that multiple genes influence susceptibility, of which only one is in the MHC. The genetic origins of several autoimmune diseases, such as multiple sclerosis (associated with DR2, with an RR of 5) and rheumatoid arthritis (associated with DR4, RR of 10) have been studied in depth. That these diseases are not inherited by simple Mendelian segregation of MHC alleles can be seen in identical twins: both inherit the MHC risk factor, but it is by no means certain that both will develop the disease. This finding suggests that multiple genetic and environmental factors have roles in the development of disease, especially autoimmune diseases, with the MHC playing an important but not exclusive role. An additional difficulty in associating a particular MHC product with disease is the genetic phenomenon of linkage disequilibrium, which was described above. The fact that some of the class I MHC alleles are in linkage disequilibrium with the class II MHC alleles makes their contribution to disease susceptibility appear more pronounced than it actually is. If, for example, DR4 contributes to risk of a disease and if it occurs frequently in combination with A3 because of linkage disequilibrium, then A3 would incorrectly appear to be associated with the disease. Improved genomic mapping techniques make it possible to analyze the linkage between the MHC and various diseases more fully and to assess the contributions from other loci.

A number of hypotheses have been offered to account for the role of the MHC in disease susceptibility. As noted earlier, allelic differences may yield differences in immune responsiveness arising from variation in the ability to present processed antigen or the ability of T cells to recognize presented antigen. Allelic forms of MHC genes may also encode molecules that are recognized as receptors by viruses or bacterial toxins. As will be explained in Chapter 15, in the genetic analysis of disease, one must consider the possibility that genes at multiple loci may be involved and that complex interactions among them may be needed to trigger disease.

Some evidence suggests that a reduction in MHC polymorphism within a species may predispose that species to infectious disease. Cheetahs and certain other wild cats, such as Florida panthers, that have been shown to be highly susceptible to viral disease have very limited MHC polymorphism. It is postulated that the present cheetah population (Figure 8-13) arose from a limited breeding stock, causing a loss of MHC diversity. The increased susceptibility of cheetahs to various viral diseases may result from a reduction in the number of different MHC molecules available to the species as a whole and a corresponding limitation on the range of processed antigens with which these MHC molecules can

**FIGURE 8-13 Cheetah female with two nearly full-grown cubs.** Polymorphism in MHC genes of the cheetah is very limited, presumably because of a bottleneck in breeding that occurred in the not-too-distant past. It is assumed that all cheetahs alive today are descendants of a very small breeding pool. [*Photograph taken in the Okavango Delta, Botswana, by T. J. Kindt.*]

interact. Thus, the high level of MHC polymorphism that has been observed in various species may provide the advantage of a broad range of antigen-presenting MHC molecules. Although some individuals within a species probably will not be able to develop an immune response to any given pathogen and therefore will be susceptible to infection by it, extreme polymorphism ensures that at least some members of a species will be able to respond and will be resistant. In this way, MHC diversity appears to protect a species from a wide range of infectious diseases.

## MHC and Immune Responsiveness

Early studies by Benacerraf in which guinea pigs were immunized with simple synthetic antigens were the first to show that the ability of an animal to mount an immune response, as measured by the production of serum antibodies, is determined by its MHC haplotype. Later experiments by H. McDevitt, M. Sela, and their colleagues used congenic and recombinant congenic mouse strains to map the control of *immune responsiveness* to class II MHC genes. In early reports, the genes responsible for this phenotype were designated *Ir* or immune response genes; retaining the initial *I*, mouse class II products are called IA and IE. We now know that the dependence of immune responsiveness on the class II MHC reflects the central role of class II MHC molecules in presenting antigen to $T_H$ cells.

Two explanations have been proposed to account for the variability in immune responsiveness observed among different haplotypes. According to the *determinant-selection model*, different class II MHC molecules differ in their ability to bind processed antigen. According to the alternative *holes-in-the-repertoire model*, T cells bearing receptors that recognize foreign antigens closely resembling self antigens

may be eliminated during thymic processing. Since the T-cell response to an antigen involves a trimolecular complex of the T cell's receptor, an antigenic peptide, and an MHC molecule (discussed in detail in Chapter 9), both models may be correct. That is, the absence of an MHC

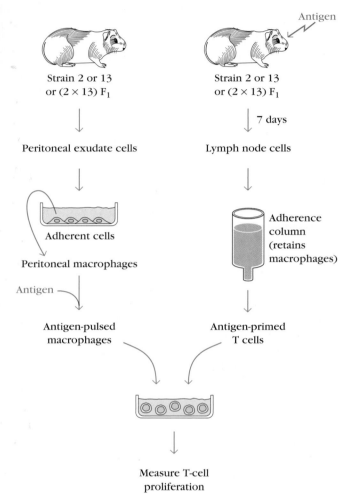

| Antigen-primed T cell | Antigen-pulsed macrophages | | |
|---|---|---|---|
| | Strain 2 | Strain 13 | $(2 \times 13)$ $F_1$ |
| Strain 2 | + | − | + |
| Strain 13 | − | + | + |
| $(2 \times 13)$ $F_1$ | + | + | + |

**FIGURE 8-14 Experimental demonstration of self-MHC restriction of $T_H$ cells.** Peritoneal exudate cells from strain 2, strain 13, or $(2 \times 13)$ $F_1$ guinea pigs were incubated in plastic petri dishes, allowing enrichment of macrophages, which are adherent cells. The peritoneal macrophages were then incubated with antigen. These "antigen-pulsed" macrophages were incubated in vitro with T cells from strain 2, strain 13, or $(2 \times 13)$ $F_1$ guinea pigs, and the degree of T-cell proliferation was assessed. The results indicated that $T_H$ cells could proliferate only in response to antigen presented by macrophages that shared MHC alleles. [*Adapted from A. Rosenthal and E. Shevach, 1974,* Journal of Experimental Medicine **138**:1194, *by copyright permission of the Rockefeller University Press.*]

molecule that can bind and present a given peptide, or the absence of T-cell receptors that can recognize a given peptide–MHC molecule complex, could result in the absence of immune responsiveness and so account for the observed relationship between MHC haplotype and immune responsiveness to exogenous antigens.

## Self-MHC Restriction of T Cells

In the 1970s a series of experiments were carried out to further explore the relationship between MHC and immune response. These investigations proved that both $CD4^+$ and $CD8^+$ T cells can recognize antigen only when presented by a self-MHC molecule, a constraint referred to as self-MHC restriction. A. Rosenthal and E. Shevach showed that antigen-specific proliferation of $T_H$ cells occurs only in response to antigen presented by macrophages of the same MHC haplotype as the T cells. In their experimental system, guinea pig macrophages from strain 2 were initially incubated with an antigen. After the "antigen-pulsed" macrophages had processed the antigen and presented it on their surface, they were mixed with T cells from the same strain (strain 2), a different strain (strain 13), or (2 × 13) $F_1$ animals, and the magnitude of T-cell proliferation in response to the antigen-pulsed macrophages was measured.

The results of these experiments, outlined in Figure 8-14, showed that strain-2 antigen-pulsed macrophages activated strain-2 and $F_1$ T cells but not strain-13 T cells. Similarly, strain-13 antigen-pulsed macrophages activated strain-13 and $F_1$ T cells but not strain-2 T cells. Subsequently, congenic and recombinant congenic strains of mice, which differed from each other only in selected regions of the H-2 complex, were used as the source of macrophages and T cells. These experiments confirmed that the $CD4^+$ $T_H$ cell is activated and proliferates only in the presence of antigen-pulsed macrophages that share class II MHC alleles. Thus, antigen recognition by the $CD4^+$ $T_H$ cell is *class II MHC restricted*.

In 1974 Zinkernagel and Doherty demonstrated the self-MHC restriction of $CD8^+$ T cells. In their experiments, mice were immunized with lymphocytic choriomeningitis (LCM) virus; several days later, the animals' spleen cells, which included $T_C$ cells specific for the virus, were isolated and incubated with LCM-infected target cells of the same or different haplotype (Figure 8-15). They found that the $T_C$ cells killed only syngeneic virus-infected target cells. Later studies with congenic and recombinant congenic strains showed that the $T_C$ cell and the virus-infected target cell must share class I molecules encoded by the K or D regions of the MHC. Thus, antigen recognition by $CD8^+$ $T_C$ cells is *class I MHC restricted*. In 1996, Doherty and Zinkernagel were awarded the Nobel prize for their major contribution to understanding the role of the MHC in cell-mediated immunity.

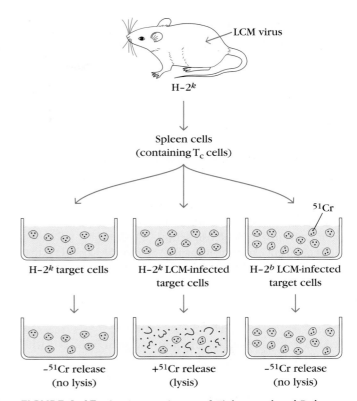

**FIGURE 8-15 Classic experiment of Zinkernagel and Doherty demonstrating that antigen recognition by $T_C$ cells exhibits MHC restriction.** H-$2^k$ mice were primed with the lymphocytic choriomeningitis (LCM) virus to induce cytotoxic T lymphocytes (CTLs) specific for the virus. Spleen cells from this LCM-primed mouse were then added to target cells of different H-2 haplotypes that were intracellularly labeled with $^{51}Cr$ (black dots) and either infected or not with the LCM virus. CTL-mediated killing of the target cells, as measured by the release of $^{51}Cr$ into the culture supernatant, occurred only if the target cells were infected with LCM and had the same MHC haplotype as the CTLs. [*Adapted from P. C. Doherty and R. M. Zinkernagel, 1975, Journal of Experimental Medicine **141**:502.*]

## Role of Antigen-Presenting Cells

As early as 1959, immunologists were confronted with data suggesting that T cells and B cells recognized antigen by different mechanisms. The dogma of the time, which persisted until the 1980s, was that cells of the immune system recognize the entire protein in its native conformation. However, experiments by P. G. H. Gell and Benacerraf demonstrated that, while both a primary-antibody and cell-mediated response were induced by a protein in its native conformation, a secondary antibody response (mediated by B cells) could be induced only by the native protein, whereas a secondary cell-mediated response could be induced by either the native or the denatured protein. These findings were viewed as an interesting enigma, but implications for antigen presentation were completely overlooked until the early 1980s.

## Processing of antigen is required for recognition by T cells

The results obtained by K. Ziegler and E. R. Unanue were among those that contradicted the prevailing dogma that antigen recognition by B and T cells was basically similar. These researchers observed that $T_H$-cell activation by bacterial protein antigens was prevented by treating the antigen-presenting cells with paraformaldehyde prior to antigen exposure. However, if the antigen-presenting cells were first allowed to ingest the antigen and were fixed with paraformaldehyde 1 to 3 hours later, $T_H$-cell activation still occurred (Figure 8-16a, b). During that interval of 1 to 3 hours, the antigen-presenting cells had processed the antigen and had displayed it on the membrane in a form able to activate T cells.

Subsequent experiments by R. P. Shimonkevitz showed that internalization and processing could be bypassed if antigen-presenting cells were exposed to peptide digests of an antigen instead of the native antigen (Figure 8-16c). In these experiments, antigen-presenting cells were treated with glutaraldehyde (this chemical, like paraformaldehyde, fixes the cell, rendering it metabolically inactive) and then incubated with native ovalbumin or with ovalbumin that had been subjected to partial enzymatic digestion. The digested ovalbumin was able to interact with the glutaraldehyde-fixed antigen-presenting cells, thereby activating ovalbumin-specific $T_H$ cells, whereas the native ovalbumin failed to do so. These results suggest that antigen processing involves the digestion of the protein into peptides that are recognized by the ovalbumin-specific $T_H$ cells.

At about the same time, several investigators, including W. Gerhard, A. Townsend, and their colleagues, began to identify the proteins of influenza virus that were recognized by $T_C$ cells. Contrary to their expectations, they found that internal proteins of the virus, such as polymerase and nucleocapsid proteins, were often recognized by $T_C$ cells better than the more exposed envelope proteins. Moreover, Townsend's work revealed that $T_C$ cells recognized short linear peptide sequences of the influenza protein. In fact, when noninfected target cells were incubated in vitro with synthetic peptides corresponding to sequences of internal influenza proteins, these cells could be recognized by $T_C$ cells and subsequently lysed just as well as target cells that had been infected with live influenza virus. These findings, along with those presented in Figure 8-16, suggest that antigen

EXPERIMENTAL CONDITIONS

T-CELL
ACTIVATION

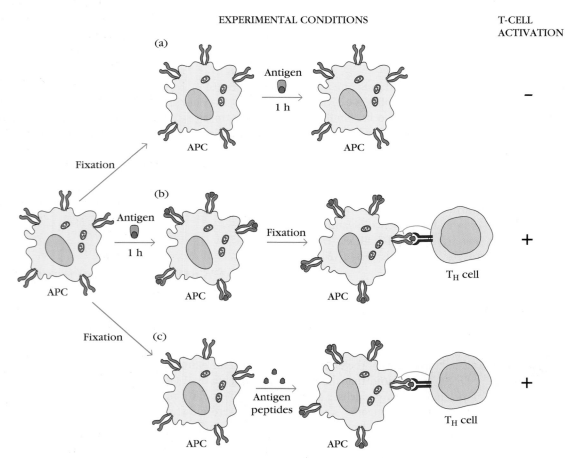

**FIGURE 8-16 Experimental demonstration that antigen processing is necessary for $T_H$-cell activation.** (a) When antigen-presenting cells (APCs) are fixed before exposure to antigen, they are unable to activate $T_H$ cells. (b) In contrast, APCs fixed at least 1 hour after antigen exposure can activate $T_H$ cells. (This simplified figure does not show costimulatory molecules needed for T-cell activation.) (c) When APCs are fixed before antigen exposure and incubated with peptide digests of the antigen (rather than native antigen), they also can activate $T_H$ cells. $T_H$-cell activation is determined by measuring a specific $T_H$-cell response (e.g., cytokine secretion).

processing is a metabolic process that digests proteins into peptides, which can then be displayed on the cell membrane together with a class I or class II MHC molecule.

## Most cells can present antigen with class I MHC; presentation with class II MHC is restricted to APCs

Since all cells expressing either class I or class II MHC molecules can present peptides to T cells, strictly speaking, they all could be designated as antigen-presenting cells. However, by convention, cells that display peptides associated with class I MHC molecules to $CD8^+$ $T_C$ cells are referred to as *target cells;* cells that display peptides associated with class II MHC molecules to $CD4^+$ $T_H$ cells are called **antigen-presenting cells (APCs).** This convention is followed throughout this text. It should be pointed out that in some cases APCs also present antigen in the context of the class I MHC molecules.

A variety of cells can function as antigen-presenting cells. Their distinguishing feature is their ability to express class II MHC molecules and to deliver a costimulatory signal. Three cell types are classified as *professional* antigen-presenting cells: dendritic cells, macrophages, and B lymphocytes. These cells differ from one another in their mechanisms of antigen uptake, in whether they constitutively express class II MHC molecules, and in their costimulatory activity:

- Dendritic cells are the most effective of the antigen-presenting cells. Because these cells constitutively express a high level of class II MHC molecules and have costimulatory activity, they can activate naive $T_H$ cells.

- Macrophages must be activated by phagocytosis of particulate antigens before they express class II MHC molecules or costimulatory membrane molecules such as B7.

- B cells constitutively express class II MHC molecules but must be activated before they express costimulatory molecules.

Several other cell types, classified as *nonprofessional* antigen-presenting cells, can be induced to express class II MHC molecules or a costimulatory signal (Table 8-3). Many of these cells function in antigen presentation only for short periods during a sustained inflammatory response.

Because nearly all nucleated cells express class I MHC molecules, virtually any nucleated cell is able to function as a target cell presenting endogenous antigens to $T_C$ cells. Most often, target cells are cells that have been infected by a virus or some other intracellular microorganism. However, altered self cells such as cancer cells, aging body cells, or allogeneic cells from a graft can also serve as targets.

## Evidence for Different Antigen-Processing and Presentation Pathways

The immune system uses different pathways to eliminate intracellular and extracellular antigens. As a general rule, endogenous antigens (those generated within the cell) are processed in the *cytosolic pathway* and presented on the membrane with class I MHC molecules, and exogenous antigens (those taken up by endocytosis) are processed in the *endocytic pathway* and presented on the membrane with class II MHC molecules (Figure 8-17).

Experiments carried out by L. A. Morrison and T. J. Braciale provided an excellent demonstration that the antigenic peptides presented by class I and class II MHC molecules are derived from different processing pathways. Using a T-cell clone that recognized an influenza virus antigen associated with a class I MHC molecule and another T cell clone that recognized the same virus antigen associated with a class II MHC molecule, they deduced general principles about the two pathways:

- Class I presentation requires synthesis of virus protein, as shown by the requirement that the target cell be infected by live virus and by the inhibition of class I presentation with an inhibitor (emetine) of protein synthesis.

- Class II presentation occurs with either live or replication-incompetent virus; protein synthesis inhibitors had no effect, indicating that new protein synthesis is not a necessary condition of class II presentation.

- Class II, but not class I, presentation is inhibited by treatment of the cells with an agent (chloroquine) that blocks endocytic processing within the cell.

These results support the distinction between the processing of exogenous and endogenous antigens, including the preferential, but not absolute, associations of exogenous antigens with class II MHC molecules and of endogenous antigens with class I MHC molecules. Association of viral antigen with class I MHC molecules required replication of the influenza virus and viral protein synthesis within the target cells; association with class II did not. These findings suggested that the peptides presented by class I and class II MHC molecules are trafficked through separate intracellular compartments; class I MHC molecules interact with peptides derived from cytosolic degradation of endogenously synthesized proteins, class II molecules with peptides derived from endocytic degradation of exogenous antigens. The next two sections examine these two pathways in detail.

| TABLE 8-3 | Antigen-presenting cells | |
|---|---|---|
| **Professional antigen-presenting cells** | **Nonprofessional antigen-presenting cells** | |
| Dendritic cells (several types) | Fibroblasts (skin) | Thymic epithelial cells |
| Macrophages | Glial cells (brain) | Thyroid epithelial cells |
| B cells | Pancreatic beta cells | Vascular endothelial cells |

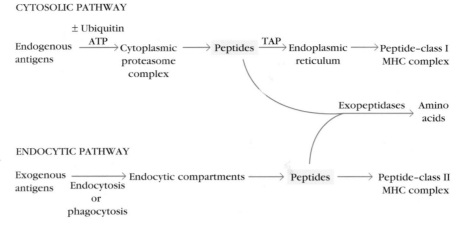

**CYTOSOLIC PATHWAY**

**ENDOCYTIC PATHWAY**

**FIGURE 8-17 Overview of cytosolic and endocytic pathways for processing antigen.** The proteasome complex contains enzymes that cleave peptide bonds, converting proteins into peptides. The antigenic peptides from proteasome cleavage and those from endocytic compartments associate with class I or class II MHC molecules, and the peptide-MHC complexes are then transported to the cell membrane. TAP (*transporter associated with antigen processing*) transports the peptides to the endoplasmic reticulum. It should be noted that the ultimate fate of most peptides in the cell is neither of these pathways but rather to be degraded completely into amino acids.

# Endogenous Antigens: The Cytosolic Pathway

In eukaryotic cells, protein levels are carefully regulated. Every protein is subject to continuous turnover and is degraded at a rate that is generally expressed in terms of its half-life. Some proteins (e.g., transcription factors, cyclins, and key metabolic enzymes) have very short half-lives; denatured, misfolded, or otherwise abnormal proteins also are degraded rapidly. Defective ribosomal products, or DRiPs,

are polypeptides that are synthesized with imperfections and constitute a large part of the products that are rapidly degraded. The average half-life for cellular proteins is about 2 days, but many are degraded within 10 minutes. The consequence of steady turnover of both normal and defective proteins is a deluge of degradation products within a cell. Most will be degraded to their constituent amino acids, but some persist in the cytosol as peptides, and the immune system samples these and presents some on the cell surface in association with class I MHC molecules. The pathway by which endogenous peptides are degraded for presentation

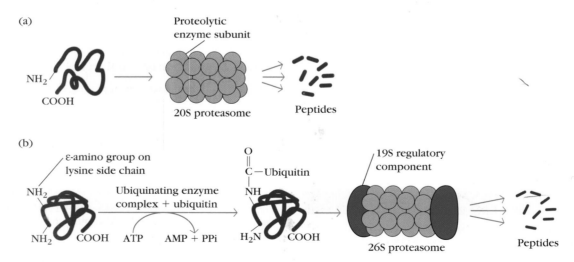

**FIGURE 8-18 Cytosolic proteolytic system for degradation of intracellular proteins.** (a) Degradation of misfolded proteins and defective ribosomal products occurs within the central channel of the proteasome, generating a variety of peptides. The 20S proteasome is a large cylindrical particle whose subunits catalyze cleavage of peptide bonds. (b) Intact proteins to be degraded are covalently linked to a small protein called ubiquitin. In the ubiquitination reaction, which requires ATP, an enzyme complex links several ubiquitin molecules to the ε-amino group of a lysine residue near the amino terminus of the protein. Ubiquitinated proteins are destined to be degraded by the 26S proteasome, consisting of the 20S proteasome and a 19S regulatory component that can attach at either end of the 20S unit.

with class I MHC molecules utilizes mechanisms similar to those involved in the normal turnover of intracellular proteins, but how particular peptides are selected remains unclear.

## Peptides for presentation are generated by protease complexes called proteasomes

Intracellular proteins are degraded into short peptides by a cytosolic proteolytic system present in all cells, the proteasome (Figure 8-18a). The large (20S) proteasome is composed of 14 subunits arrayed in a barrel-like structure of symmetrical rings. Some, but not all, of the subunits have protease activity. Entry to the proteasome occurs through narrow channels at each end.

Many proteins targeted for proteolysis have a small protein, called *ubiquitin,* attached to them (Figure 8-18b). Ubiquitin-protein conjugates can be degraded by a multifunctional protease complex consisting of a 20S proteasome to which a 19S regulatory component is added. The resulting 26S proteasome cleaves peptide bonds in an ATP-dependent process (Figure 8-18b). Degradation of ubiquitin-protein complexes is thought to occur within the central hollow of the proteasome.

Experimental evidence indicates that the immune system utilizes this general pathway of protein degradation to produce small peptides for presentation with class I MHC molecules. In addition to the standard 20S proteasomes resident in all cells, a distinct proteasome of the same size is present in cells with immune activity. This immunoproteasome, which can be induced by interferon γ or TNF α, is found in virus-infected cells, suggesting a role in the processing of viral proteins for class I presentation. The immunoproteasome turns over more rapidly than standard proteasomes, possibly because the increased level of protein degradation in their presence may have consequences beyond the targeting of virus-infected cells. It is possible that autoimmunity results from increased processing of self proteins in cells with high levels of immunoproteasomes.

## Peptides are transported from the cytosol to the rough endoplasmic reticulum

Insight into the role that peptide transport, the delivery of peptides to the MHC molecule, plays in the cytosolic processing pathway came from studies of cell lines with defects in peptide presentation by class I MHC molecules. One such mutant cell line, called RMA-S, expresses about 5% of the normal levels of class I MHC molecules on its membrane. Although RMA-S cells synthesize normal levels of class I α chains and β$_2$-microglobulin, few class I MHC complexes appear on the membrane. A clue to the mutation in the RMA-S cell line was the discovery by Townsend and his colleagues that "feeding" these cells peptides restored their level of membrane-associated class I MHC molecules to normal. These investigators suggested that peptides might be required to stabilize the interaction between the class I α chain and

β$_2$-microglobulin. The ability to restore expression of class I MHC molecules on the membrane by feeding the cells predigested peptides suggested that the RMA-S cell line might have a defect in peptide transport.

Subsequent experiments showed that the defect in the RMA-S cell line occurs in the protein that transports peptides from the cytoplasm to the RER, where class I molecules are synthesized. When RMA-S cells were transfected with a functional gene encoding the transporter protein, the cells began to express class I molecules on the membrane. The transporter protein, designated **TAP** (for **transporter associated with antigen processing**), is a membrane-spanning heterodimer consisting of two proteins: TAP1 and TAP2 (Figure 8-19a).

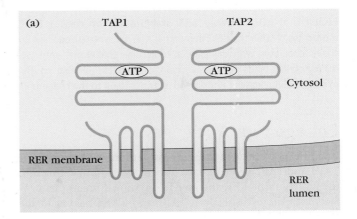

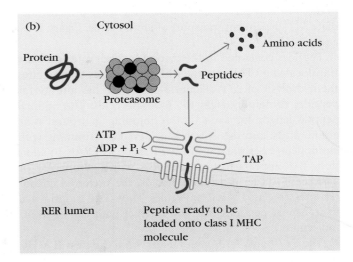

**FIGURE 8-19 TAP (transporter associated with antigen processing).** (a) Schematic diagram of TAP, a heterodimer anchored in the membrane of the rough endoplasmic reticulum (RER). The two chains are encoded by *TAP1* and *TAP2*. The cytosolic domain in each TAP subunit contains an ATP-binding site, and peptide transport depends on the hydrolysis of ATP. (b) In the cytosol, association of LMP2, LMP7, and LMP10 (black spheres) with a proteasome changes its catalytic specificity to favor production of peptides that bind to class I MHC molecules. These peptides are translocated by TAP into the RER lumen, where, in a process mediated by several other proteins, they will associate with class I MHC molecules.

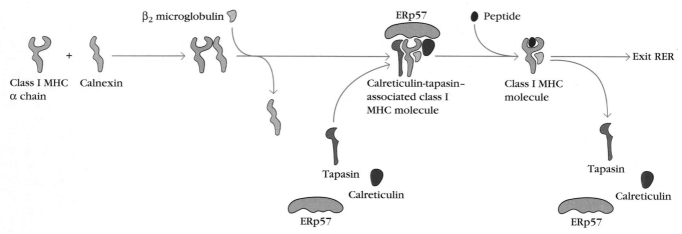

**FIGURE 8-20 Assembly and stabilization of class I MHC molecules.** Within the RER membrane, a newly synthesized class I α chain associates with calnexin, a molecular chaperone, until β₂-microglobulin binds to the α chain. Subsequent binding to β₂-microglobulin releases calnexin and allows binding to the chaperonin calreticulin and to tapasin, which is associated with the peptide transporter TAP. This association promotes binding of an antigenic peptide, which stabilizes the class I molecule–peptide complex, allowing its release from the RER.

In addition to their multiple transmembrane segments, the TAP1 and TAP2 proteins each have a domain projecting into the lumen of the RER and an ATP-binding domain that projects into the cytosol. Both TAP1 and TAP2 belong to the family of ATP-binding cassette proteins found in the membranes of many cells, including bacteria; these proteins mediate ATP-dependent transport of amino acids, sugars, ions, and peptides.

Peptides generated in the cytosol by the proteasome are translocated by TAP into the RER by a process that requires the hydrolysis of ATP (Figure 8-19b). TAP has affinity for peptides containing eight to 16 amino acids. The optimal peptide length for class I MHC binding is around nine amino acids, and this length is achieved by trimming with aminopeptidases present in the ER, such as ERAP. In addition, TAP appears to favor peptides with hydrophobic or basic carboxyl-terminal amino acids, the preferred anchor residues for class I MHC molecules. Thus, TAP is optimized to transport peptides that will interact with class I MHC molecules.

The *TAP1* and *TAP2* genes map within the class II MHC region, adjacent to the *LMP2* and *LMP7* genes (see Figure 8-11), and different allelic forms of these genes exist within the population. TAP deficiencies can lead to a disease syndrome that has aspects of both immunodeficiency and autoimmunity (see Clinical Focus).

## Peptides assemble with class I MHC aided by chaperone molecules

Like other proteins, the α chain and β₂-microglobulin components of the class I MHC molecule are synthesized on polysomes along the rough endoplasmic reticulum. Assembly of these components into a stable class I MHC molecular complex that can exit the RER requires the presence of a peptide in the binding groove of the class I molecule. The assembly process involves several steps and includes the participation of *molecular chaperones,* which facilitate the folding of polypeptides. The first molecular chaperone involved in class I MHC assembly is *calnexin,* a resident membrane protein of the endoplasmic reticulum. Calnexin associates with the free class I α chain and promotes its folding. When β₂-microglobulin binds to the α chain, calnexin is released and the class I molecule associates with the chaperone *calreticulin* and with *tapasin.* Tapasin (TAP-associated protein) brings the TAP transporter into proximity with the class I molecule and allows it to acquire an antigenic peptide (Figure 8-20). Tapasin may exist in a multimeric form of as many as four molecules but is shown as a monomer here for simplicity. An additional protein with enzymatic activity, ERp57, forms a disulfide bond to tapasin and noncovalently associates with calreticulin to stabilize their interaction and allow for the release of the MHC α chain and β₂-microglobulin after acquisition of peptide. The TAP protein (see Figure 8-19b) promotes peptide capture by the class I molecule before the peptides are exposed to the luminal environment of the RER.

Exoproteases in the ER will act on peptides not associated with class I MHC molecules. One ER aminopeptidase, ERAP1, mentioned above, removes the amino-terminal residue from peptides to achieve optimum binding size. ERAP1 has little affinity for peptides shorter than eight amino acids in length. Another ER aminopeptidase, ERAP2, can degrade peptides of any length and thus eliminates any that too are short for optimum binding to class I MHC molecules. ERAP activity on free peptides not bound to the

## CLINICAL FOCUS

# Deficiency in Transporters Associated with Antigen Processing (TAPs) Leads to a Diverse Disease Spectrum

**A relatively** rare condition known as bare lymphocyte syndrome, or BLS, has been recognized for more than 20 years. The lymphocytes in BLS patients express MHC molecules at below-normal levels and, in some cases, not at all. In type 1 BLS, a deficiency in MHC class I molecules exists; in type 2 BLS, expression of class II molecules is impaired. The pathogenesis of one type of BLS underscores the importance of the class I family of MHC molecules in their dual roles of preventing autoimmunity as well as defending against pathogens.

Defects that impair or preclude MHC gene transcription were found for some type 2 BLS cases, but in many instances the nature of the underlying defect is not known. A recent study has identified a group of patients with type 1 BLS due to defects in *TAP1* or *TAP2* genes. Manifestations of the TAP deficiency were consistent in this patient group and define a unique disease. As described in this chapter, TAP proteins are necessary for the loading of peptides onto class I molecules, a step that is essential for expression of class I MHC molecules on the cell surface. Lymphocytes in individuals with TAP deficiency express levels of class I molecules significantly lower than those of normal controls. Other cellular abnormalities include increased numbers of NK and $\gamma\delta$ T cells and decreased levels of CD8$^+$ $\alpha\beta$ T cells. As we will see, the disease manifestations are reasonably well explained by these deviations in the levels of certain cells involved in immune function.

In early life, the TAP-deficient individual suffers frequent bacterial infections of the upper respiratory tract and in the second decade begins to have chronic infec-tion of the lungs. It is thought that a post-nasal-drip syndrome common in younger patients promotes the bacterial lung infections in later life. Noteworthy is the absence of susceptibility to severe viral infection, which is common in immunodeficiencies with T-cell involvement (see Chapter 20). Bronchiectasis (dilation of the bronchial tubes) often occurs, and recurring infections can lead to lung damage that may be fatal. The most characteristic mark of the deficiency is the occurrence of necrotizing skin lesions on the extremities and the midface. These lesions ulcerate and may cause disfigurement. The skin lesions are probably due to activated NK cells and $\gamma\delta$ T cells; NK cells were isolated from biop-sied skin from several patients, supporting

this possibility. Normally, the activity of NK cells is limited through the action of killer-cell-inhibitory receptors (KIRs), which deliver a negative signal to the NK cell following interaction with class I molecules (see Chapter 14). The deficiency of class I molecules in TAP-related BLS patients explains the excessive activity of the NK cells. Activation of NK cells further explains the absence of severe viral infections, which are limited by NK and $\gamma\delta$ cells.

The best treatment for the characteristic lung infections appears to be antibiotics and intravenous immunoglobulin. Attempts to limit the skin disease by immunosuppressive regimens, such as steroid treatment or cytotoxic agents, can lead to exacerbation of lesions and is therefore contraindicated. Mutations in the promoter region of *TAP* that preclude expression of the gene were found for several patients, suggesting the possibility of gene therapy, but the cellular distribution of class I is so widespread that it is not clear what cells would need to be corrected to alleviate all symptoms.

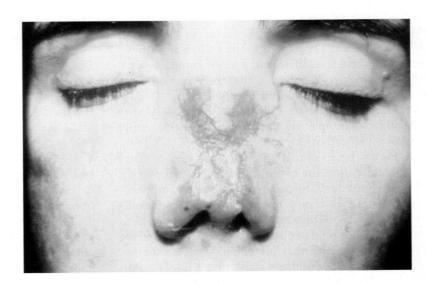

Necrotizing granulomatous lesions in the midface of a patient with TAP-deficiency syndrome. TAP deficiency leads to a condition with symptoms characteristic of autoimmunity, such as the skin lesions that appear on the extremities and the midface, as well as immunodeficiency that causes chronic sinusitis, leading to recurrent lung infection. *[From S. D. Gadola et al., 1999, Lancet **354**:1598, and 2000, Clinical and Experimental Immunology **121**:173.]*

class I molecules in the ER allows the compartment to eliminate peptides not suited to class I binding. As a consequence of productive peptide binding, the class I molecule displays increased stability and can dissociate from the complex with calreticulin, tapasin, and ERp57; exit from the RER; and proceed to the cell surface via the Golgi complex.

# Exogenous Antigens: The Endocytic Pathway

Antigen-presenting cells can internalize antigen by phagocytosis, endocytosis, or both. Macrophages and dendritic cells internalize antigen by both processes, whereas most other APCs are not phagocytic or are poorly phagocytic and therefore internalize exogenous antigen only by endocytosis (either receptor-mediated endocytosis or pinocytosis). B cells, for example, internalize antigen very effectively by receptor-mediated endocytosis using antigen-specific membrane antibody as the receptor.

## Peptides are generated from internalized molecules in endocytic vesicles

Once an antigen is internalized, it is degraded into peptides within compartments of the endocytic processing pathway. As the experiment shown in Figure 8-16 demonstrated, internalized antigen takes 1 to 3 hours to traverse the endocytic pathway and appear at the cell surface in the form of peptide–class II MHC complexes. The endocytic pathway appears to involve three increasingly acidic compartments: early endosomes (pH 6.0–6.5); late endosomes, or endolysosomes (pH 5.0–6.0); and lysosomes (pH 4.5–5.0). Internalized antigen moves from early to late endosomes and finally to lysosomes, encountering hydrolytic enzymes and a lower pH in each compartment (Figure 8-21). Lysosomes, for example, contain a unique collection of more than 40 acid-dependent hydrolases, including proteases, nucleases, glycosidases, lipases, phospholipases, and phosphatases. Within the compartments of the endocytic pathway, antigen is degraded into oligopeptides of about 13 to 18 residues, which bind to class II MHC molecules and are thus protected from further proteolysis. Because the hydrolytic enzymes are optimally active under acidic conditions (low pH), antigen processing can be inhibited by chemical agents that increase the pH of the compartments (e.g., chloroquine) as well as by protease inhibitors (e.g., leupeptin).

The mechanism by which internalized antigen moves from one endocytic compartment to the next has not been conclusively demonstrated. It has been suggested that early endosomes from the periphery move inward to become late endosomes and finally lysosomes. Alternatively, small trans-

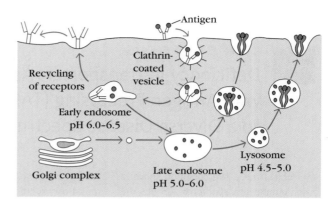

**FIGURE 8-21 Generation of antigenic peptides in the endocytic processing pathway.** Internalized exogenous antigen moves through several acidic compartments, in which it is degraded into peptides that ultimately associate with class II MHC molecules transported in vesicles from the Golgi complex. The cell shown here is a B cell, which internalizes antigen by receptor-mediated endocytosis, with the membrane-bound antibody functioning as an antigen-specific receptor.

port vesicles may carry antigens from one compartment to the next. Eventually the endocytic compartments, or portions of them, return to the cell periphery, where they fuse with the plasma membrane. In this way, the surface receptors are recycled.

## The invariant chain guides transport of class II MHC molecules to endocytic vesicles

Since antigen-presenting cells express both class I and class II MHC molecules, some mechanism must exist to prevent class II MHC molecules from binding to the same set of antigenic peptides as the class I molecules. When class II MHC molecule are synthesized within the RER, three pairs of class II αβ chains associate with a preassembled trimer of a protein called **invariant chain (Ii, CD74).** This trimeric protein interacts with the peptide-binding cleft of the class II molecules, preventing any endogenously derived peptides from binding to the cleft while the class II molecule is within the RER (Figure 8-22a). The invariant chain also appears to be involved in the folding of the class II α and β chains, their exit from the RER, and the subsequent routing of class II molecules to the endocytic processing pathway from the trans-Golgi network.

The role of the invariant chain in the routing of class II molecules has been demonstrated in transfection experiments with cells that lack the genes encoding class II MHC molecules and the invariant chain. Immunofluorescent labeling of such cells transfected only with class II MHC genes revealed class II molecules localized within the Golgi complex. However, in cells transfected with both the class II MHC genes and the invariant-chain gene, the class II

(a)

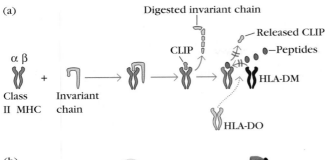

(b)

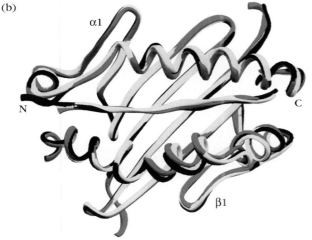

**FIGURE 8-22 (a) Assembly of class II MHC molecules.** Within the rough endoplasmic reticulum, a newly synthesized class II MHC molecule binds an invariant chain. The bound invariant chain prevents premature binding of peptides to the class II molecule and helps to direct the complex to endocytic compartments containing peptides derived from exogenous antigens. Digestion of the invariant chain leaves CLIP, a small fragment remaining in the binding groove of the class II MHC molecule. HLA-DM, a nonclassical MHC class II molecule expressed within endosomal compartments, mediates exchange of antigenic peptides for CLIP. The nonclassical class II molecule HLA-DO may act as a negative regulator of class II antigen processing by binding to HLA-DM and inhibiting its role in the dissociation of CLIP from class II molecules. (b) Comparison of three-dimensional structures showing the binding groove of HLA class II molecules ($\alpha$1 and $\beta$1) containing different antigenic peptides or CLIP. The red lines show DR4 complexed with collagen II peptide, yellow lines are DR1 with influenza hemagglutinin peptide, and blue lines are DR3 associated with CLIP. (N indicates the amino terminus and C the carboxyl terminus of the peptides.) No major differences in the structures of the class II molecules or in the conformation of the bound peptides are seen. This comparison shows that CLIP binds the class II molecule in a manner identical to that of antigenic peptides. *[Part b from Dessen et al., 1997, Immunity* **7:**473–481; *courtesy of Don Wiley, Harvard University.]*

## Peptides assemble with class II MHC molecules by displacing CLIP

Recent experiments indicate that most class II MHC–invariant chain complexes are transported from the RER, where they are formed, through the Golgi complex and trans-Golgi network, and then through the endocytic pathway, moving from early endosomes to late endosomes and finally to lysosomes. As the proteolytic activity increases in each successive compartment, the invariant chain is gradually degraded. However, a short fragment of the invariant chain termed *CLIP* (for *class II–associated invariant chain peptide*) remains bound to the class II molecule after the invariant chain has been cleaved within the endosomal compartment. CLIP physically occupies the peptide-binding groove of the class II MHC molecule, presumably preventing any premature binding of antigenic peptide (see Figure 8-22a).

A nonclassical class II MHC molecule called *HLA-DM* is required to catalyze the exchange of CLIP with antigenic peptides (see Figure 8-22a). Like other class II MHC molecules, HLA-DM is a heterodimer of $\alpha$ and $\beta$ chains. However, unlike other class II molecules, HLA-DM is not polymorphic and is not expressed at the cell membrane but is found predominantly within the endosomal compartment. The *DM*$\alpha$ and *DM*$\beta$ genes are located near the *TAP* and *LMP* genes in the MHC complex of humans, and DM is expressed in cells that express classical class II molecules.

The reaction between HLA-DM and the class II CLIP complex facilitating exchange of CLIP for another peptide is impaired in the presence of HLA-DO, which binds to HLA-DM and lessens the efficiency of the exchange reaction. HLA-DO, like HLA-DM, is a nonclassical and nonpolymorphic class II molecule that is also found in the MHC of other species. HLA-DO differs from HLA-DM in that it is expressed only by B cells and the thymus, and unlike other class II molecules, its expression is not induced by IFN-$\gamma$. An additional difference is that the genes encoding the $\alpha$ and the $\beta$ chains of HLA-DO are not adjacent in the MHC as are all other class II $\alpha$ and $\beta$ pairs (see Figure 8-11).

An HLA-DR3 molecule associated with CLIP was isolated from a cell line that did not express HLA-DM and was therefore defective in antigen processing. Superimposing the structure of HLA-DR3–CLIP on another DR molecule bound to antigenic peptide reveals that CLIP binds to class II in the same stable manner as the antigenic peptide (Figure 8-22b). The discovery of this stable complex in a cell with defective HLA-DM supports the argument that HLA-DM is required for the replacement of CLIP.

Although it certainly modulates the activity of HLA-DM, the precise role of HLA-DO remains obscure. One possibility is that it acts in the selection of peptides bound to class II MHC molecules in B cells. DO occurs in complex with DM in these cells, and this association continues in the endosomal compartments. Conditions of higher acidity weaken

molecules were localized in the cytoplasmic vesicular structures of the endocytic pathway. The invariant chain contains sorting signals in its cytoplasmic tail that direct the transport of the class II MHC complex from the trans-Golgi network to the endocytic compartments.

the association of DM/DO and increase the possibility of antigenic peptide binding despite the presence of DO. Such a pH-dependent interaction could lead to preferential selection of class II–bound peptides from lysosomal compartments in B cells compared with other APCs.

As with class I MHC molecules, peptide binding is required to maintain the structure and stability of class II MHC molecules. Once a peptide has bound, the peptide–class II complex is transported to the plasma membrane, where the neutral pH appears to enable the complex to assume a compact, stable form. Peptide is bound so strongly in this compact form that it is difficult to replace a class II–bound peptide on the membrane with another peptide at physiologic conditions.

Figure 8-23 recapitulates the endogenous pathway (left side) and compares it with the separate exogenous pathway

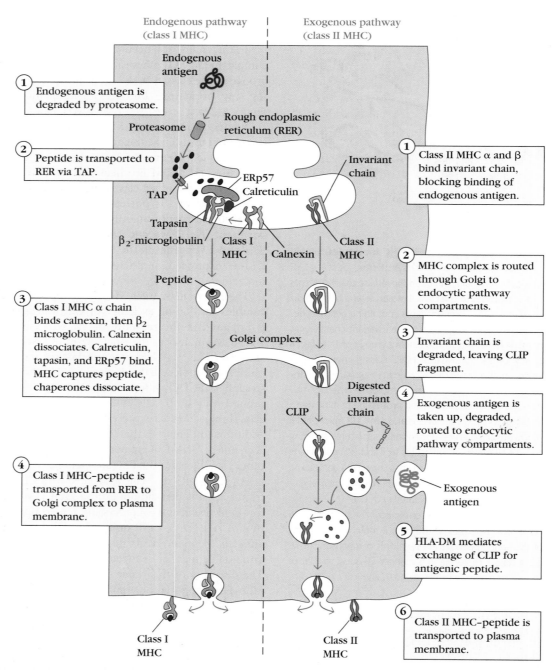

Endogenous pathway
(class I MHC)

Exogenous pathway
(class II MHC)

**① Endogenous antigen is degraded by proteasome.**

**② Peptide is transported to RER via TAP.**

**③ Class I MHC α chain binds calnexin, then β₂ microglobulin. Calnexin dissociates. Calreticulin, tapasin, and ERp57 bind. MHC captures peptide, chaperones dissociate.**

**④ Class I MHC-peptide is transported from RER to Golgi complex to plasma membrane.**

**① Class II MHC α and β bind invariant chain, blocking binding of endogenous antigen.**

**② MHC complex is routed through Golgi to endocytic pathway compartments.**

**③ Invariant chain is degraded, leaving CLIP fragment.**

**④ Exogenous antigen is taken up, degraded, routed to endocytic pathway compartments.**

**⑤ HLA-DM mediates exchange of CLIP for antigenic peptide.**

**⑥ Class II MHC-peptide is transported to plasma membrane.**

Endogenous antigen

Proteasome

Rough endoplasmic reticulum (RER)

Invariant chain

ERp57
Calreticulin

TAP

Tapasin

β₂-microglobulin    Class I MHC    Calnexin    Class II MHC

Peptide

Golgi complex

CLIP

Digested invariant chain

Exogenous antigen

Class I MHC

Class II MHC

**FIGURE 8-23 Separate antigen-presenting pathways are utilized for endogenous (green) and exogenous (red) antigens.** The mode of antigen entry into cells and the site of antigen processing determine whether antigenic peptides associate with class I MHC molecules in the rough endoplasmic reticulum or with class II molecules in endocytic compartments.

(right), considered above. Whether an antigenic peptide associates with class I or with class II molecules is dictated by the mode of entry into the cell, either exogenous or endogenous, and by the site of processing. However, we will see that these assignments are not absolute and that exogenous antigens may, in some APCs, be presented by class I antigens via a process designated cross-presentation.

## Cross-Presentation of Exogenous Antigens

In certain cases, APCs may present exogenous antigen to cytotoxic T cells in the context of class I MHC molecules. First reported by Michael Bevan and more recently described in detail by Peter Cresswell, the phenomenon of cross-presentation requires that internalized antigens that would normally be handled by the exogenous pathway leading to class II MHC presentation instead intersect with the endogenous pathway for class I peptide loading. These peptides

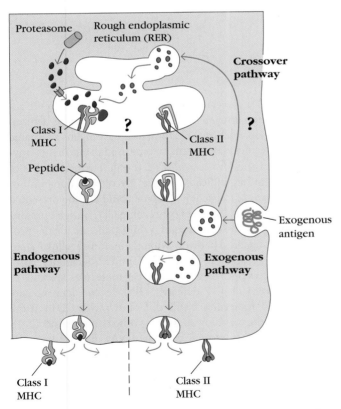

**FIGURE 8-24 Hypothetical mechanism for cross-presentation of exogenous antigen by MHC class I molecules.** This process occurs only in certain antigen-presenting cells and allows MHC class I binding of antigens that are acquired by phagocytic or endocytic mechanisms. The antigen may be internalized by any pathway and is processed to produce peptides suitable for class I binding. Question marks indicate uncertain routing in the pathway to class I MHC peptide loading.

are then presented (cross-presented) in the context of class I MHC molecules. Proposed mechanisms for cross-presentation include the exchange within an endosomal compartment of exogenous peptides for those already loaded onto class I molecules in the ER. It is also possible that exogenous peptides have access to the ER for the entire processing sequence. Currently unknown is whether the APCs capable of cross-presentation use this mechanism as an alternative to the normal presentation of endogenous antigen, or whether it is the exclusive pathway for class I MHC presentation in these cells. A hypothetical pathway for cross-presentation is shown in Figure 8-24; the exact means by which antigen achieves the crossover from the exogenous to the endogenous pathway remains undefined.

The clearest examples of cross-presentation are found in dendritic cells; it is not certain whether cross-presentation operates in any other APCs, although early reports indicate that macrophages may also have this capability. As discussed in Chapter 14, the ability of dendritic cells to cross-present antigens has great advantage for the host, in that it allows dendritic cells to capture viruses, process viral antigens, and generate CTLs that can attack viruses and virus-infected cells prior to general spread of the virus infection.

Although dendritic cells lack the specific host cell receptors exploited by viruses for cell entry, they can nevertheless capture many different viruses for processing and presentation. Perhaps the major question remaining for cross-presentation is how and where in the cell the two relatively well-defined pathways of antigen presentation merge to allow an exogenous antigen to be presented as an endogenous antigen. This is a highly active area of research and answers should be forthcoming.

## Presentation of Nonpeptide Antigens

To this point, the discussion of the presentation of antigens has been limited to peptide antigens and their presentation by classical class I and II MHC molecules. It is well known that nonprotein antigens are also recognized by T cells, and in the 1980s T-cell proliferation was detected in the presence of nonprotein antigens derived from infectious agents. More recent reports indicate that T cells expressing the $\gamma\delta$ TCR (T-cell receptors are dimers of either $\alpha\beta$ or $\gamma\delta$ chains) react with glycolipid antigens derived from bacteria such as *Mycobacterium tuberculosis*. These nonprotein antigens are presented by members of the CD1 family of nonclassical class I molecules.

The CD1 family of molecules associates with $\beta_2$-microglobulin and has a general structural similarity to class I MHC molecules. Five genes encode human CD1 molecules (*CD1A–E*, encoding the gene products CD1a–e. These genes are located not within the MHC but on chromosome 1 (Figure 8-25a). The genes are classified into two groups based

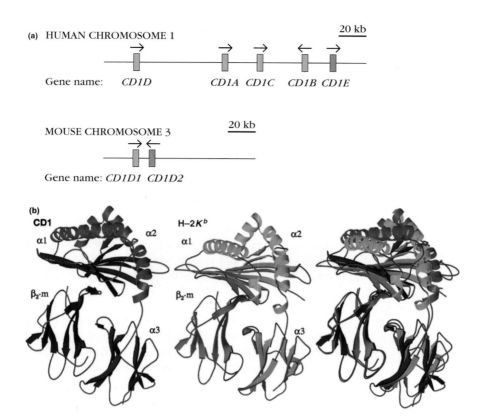

(a) HUMAN CHROMOSOME 1

20 kb

Gene name:   CD1D          CD1A CD1C   CD1B CD1E

MOUSE CHROMOSOME 3

20 kb

Gene name: CD1D1  CD1D2

(b)
CD1
α1      α2
β₂-m
α3

H–2K^b
α1      α2
β₂-m
α3

**FIGURE 8-25 The *CD1* family of genes and structure of a CD1 molecule.** (a) The genes encoding the CD1 family of molecules in human (top) and mouse (bottom). The genes are separated into two groups based on sequence identity: *CD1A, B, C*, and *E* are group 1, *CD1D* genes are group 2. The products of the pink genes have been characterized; functions have yet to be identified for products of gray genes. (b) Comparison of the crystal structures of mouse nonclassical CD1 and classical class I molecule H-2K^b. Note the differences in the antigen-binding grooves. [*Part b reprinted from* Trends in Immunology (*formerly* Immunology Today), *vol. 19, S. A. Porcelli and R. L. Modlin, The CD1 family of lipid antigen presenting molecules, pp. 362–68, 1998, with permission from Elsevier Science.*]

on sequence homology. Group 1 includes *CD1A, B, C,* and *E; CD1D* is in group 2. All mammalian species studied have CD1 genes, although the number varies. Rodents have only group 2 *CD1* genes, the counterpart of human *CD1D*, whereas rabbits, like humans, have five genes, including both group 1 and 2 types. Sequence identity of CD1 with classical class I molecules is considerably lower than the identity of the class I molecules with each other. Comparison of the three-dimensional structure of a mouse CD1 with the class I MHC molecule H-2K^b shows that the antigen-binding groove of the CD1 molecule is deeper and more voluminous than that of the classical class I molecule (Figure 8-25b).

Expression of CD1 molecules varies according to subset; *CD1D1* genes are expressed mainly in nonprofessional APCs and on certain B-cell subsets. The mouse CD1d1 is more widely distributed and found on T cells, B cells, dendritic cells, hepatocytes, and some epithelial cells. The *CD1A, B,* and *C* genes are expressed on immature thymocytes and professional APCs, mainly those of the dendritic type. *CD1C* gene expression is seen on B cells, whereas the *CD1A* and *B* products are not. *CD1* genes can be induced by exposure to certain cytokines such as GM-CSF or IL-3. The intracellular trafficking patterns of the CD1 molecules differ; for example, CD1a is found mostly in early endosomes or on the cell surface, CD1b and CD1d localize to late endosomes, and CD1c is found throughout the endocytic system.

Certain CD1 molecules are recognized by T cells in the absence of foreign antigens, and self restriction can be demonstrated in these reactions. Examination of antigens presented by CD1 molecules revealed them to be lipid components (mycolic acid) of the *M. tuberculosis* cell wall. Further studies of CD1 presentation indicated that a glycolipid (lipoarabinomannan) from *Mycobacterium leprae* could also be presented by these molecules. The data concerning CD1 antigen presentation point out the existence of a third pathway for the processing of antigens, a pathway with distinct intracellular steps that do not involve the molecules found to facilitate class I antigen processing. For example, CD1 molecules are able to process antigen in TAP-deficient cells. Recent data indicate that the CD1 molecules traffic differently, with CD1a at the surface or in the recycling endocytic compartments and CD1b and CD1d in the lysomal compartments. Exactly how the CD1 pathway complements or intersects the better-understood class I and class II pathways remains an open question. The T-cell types reactive to CD1 were first thought to be limited to T cells expressing the γδ TCR and lacking both CD4 and CD8 or to T cells with a CD1-specific TCR α chain, but recent reports indicate that a wider range of T-cell types will recognize CD1-presenting cells. Recent evidence indicates that natural killer T cells (NK T cells) recognize CD1d molecules presenting autologous antigen, as well as bacterial glycosphingolipids, suggesting a role in an innate-type immune response to certain bacteria. This activity is discussed in Chapter 14.

## SUMMARY

- The major histocompatibility complex (MHC) encodes class I and II molecules, which function in antigen presentation to T cells, and class III molecules, which have diverse functions.

- MHC genes are tightly linked and generally inherited as a unit from parents; these linked units are called haplotypes.

- MHC genes are polymorphic—many alleles exist for each gene—and polygenic—many different MHC genes exist.

- Class I MHC molecules consist of a large glycoprotein chain and $\beta_2$-microglobulin, a protein with a single domain.

- Class II MHC molecules are composed of two noncovalently associated glycoproteins, the $\alpha$ and $\beta$ chain, encoded by separate MHC genes.

- Class I molecules are expressed on most nucleated cells; class II molecules are restricted to B cells, macrophages, and dendritic cells.

- Detailed maps of the human and mouse MHC reveal the presence of genes involved in antigen processing, including proteasomes and transporters.

- MHC alleles influence immune responsiveness and the ability to present antigen as well as susceptibility to a number of diseases.

- In general, class I molecules present processed endogenous antigen to $CD8^+$ $T_C$ cells and class II molecules present processed exogenous antigen to $CD4^+$ $T_H$ cells.

- Endogenous antigens are degraded into peptides within the cytosol by proteasomes, assemble with class I molecules in the RER, and are presented on the membrane to $CD8^+$ $T_C$ cells.

- Exogenous antigens are internalized and degraded within the acidic endocytic compartments and subsequently combine with class II molecules for presentation to $CD4^+$ $T_H$ cells.

- Peptide binding to class II molecules involves replacing a fragment of invariant chain in the binding cleft by a process catalyzed by nonclassical MHC molecule HLA-DM.

- Exogenous peptide antigens in certain cell types may gain access to class I presentation pathways via phagosomes in a process called cross-presentation.

- Presentation of nonpeptide (lipid and glycolipid) antigens derived from bacteria involves the class I–like CD1 molecules.

## References

Ackerman, A. L., and P. Cresswell. 2004. Cellular mechanisms governing cross-presentation of exogenous antigens. *Nature Immunology* **5**:678.

Brigl, M., and M. B. Brenner. 2004. CD1: Antigen presentation and T cell function. *Annual Review of Immunology* **22**:817.

Brode, S., and P. A. Macary. 2004. Cross-presentation: Dendritic cells and macrophages bite off more than they can chew! *Immunology* **112**:345.

Brown, J. H., et al. 1993. Three-dimensional structure of the human class II histocompatibility antigen HLA-DR1. *Nature* **364**:33.

Doherty, P. C., and R. M. Zinkernagel. 1975. H-2 compatibility is required for T-cell mediated lysis of target cells infected with lymphocytic choriomeningitis virus. *Journal of Experimental Medicine* **141**:502.

Fahrer, A. M., et al. 2001. A genomic view of immunology. *Nature* **409**:836.

International Human Genome Sequencing Consortium. 2001. Initial sequencing and analysis of the human genome. *Nature* **409**:860.

Gadola, S. D., et al. 2000. TAP deficiency syndrome. *Clinical and Experimental Immunology* **121**:173.

Horton, R., et al. 2004. Gene map of the extended human MHC. *Nature Reviews Genetics* **5**:1038.

Kelley, J., et al. 2005. Comparative genomics of major histocompatibility complexes. *Immunogenetics* **56**:683.

Madden, D. R. 1995. The three-dimensional structure of peptide-MHC complexes. *Annual Review of Immunology* **13**:587.

Margulies, D. 1999. The major histocompatibility complex. In *Fundamental Immunology*, 4th ed. W. E. Paul, ed. Lippincott-Raven, Philadelphia.

Parham, P. 1999. Virtual reality in the MHC. *Immunological Reviews* **167**:5.

Rock, K. L., et al. 2004. Post-proteosomal antigen processing for major histocompatibility complex class I presentation. *Nature Immunology* **5**:670.

Rothenberg, B. E., and J. R. Voland. 1996. Beta 2 knockout mice develop parenchymal iron overload: A putative role for class I genes of the major histocompatibility complex in iron metabolism. *Proceedings of the National Academy of Science U.S.A.* **93**:1529.

Rouas-Freiss, N., et al. 1997. Direct evidence to support the role of HLA-G in protecting the fetus from maternal uterine natural killer cytolysis. *Proceedings of the National Academy of Science U.S.A.* **94**:11520.

Sugita, M., et al. 2004. New insights into pathways for CD1-mediated antigen presentation. *Current Opinion in Immunology* **16**:90.

## Useful Web Sites

http://www.bioscience.org/

This *Bioscience* site includes a section called Databases, which lists gene knockouts. Information is available on studies of the consequences of targeted disruption of MHC molecules and other component molecules, including $\beta_2$-microglobulin and the class II invariant chain.

http://www.bshi.org.uk/

The British Society for Histocompatibility and Immunogenetics home page contains information on tissue typing and transplantation and links to worldwide sites related to the MHC.

http://www.nature.com/nrg/journal/v5/n12/poster/MHCmap

A poster with an extended gene map of the human MHC from the Horton et al. reference cited above.

http://www.ebi.ac.uk/imgt/hla/

The International ImMunoGeneTics (IMGT) database section contains links to sites with information about HLA gene structure and genetics and up-to-date listings and sequences for all HLA alleles officially recognized by the World Health Organization HLA nomenclature committee.

 ## Study Questions

**CLINICAL FOCUS QUESTION** Patients with TAP deficiency have partial immunodeficiency as well as autoimmune manifestations. How do the profiles for patients' immune cells explain the partial immunodeficiency? Why is it difficult to design a gene therapy treatment for this disease, despite the fact that a single gene defect is implicated?

1. Indicate whether each of the following statements is true or false. If you think a statement is false, explain why.

    a. A monoclonal antibody specific for $\beta_2$-microglobulin can be used to detect both class I MHC K and D molecules on the surface of cells.

    b. Antigen-presenting cells express both class I and class II MHC molecules on their membranes.

    c. Class III MHC genes encode membrane-bound proteins.

    d. In outbred populations, an individual is more likely to be histocompatible with one of its parents than with its siblings.

    e. Class II MHC molecules typically bind to longer peptides than do class I molecules.

    f. All cells express class I MHC molecules.

    g. The majority of the peptides displayed by class I and class II MHC molecules on cells are derived from self proteins.

2. You cross a BALB/c ($H$-$2^d$) mouse with a CBA ($H$-$2^k$) mouse. What MHC molecules will the $F_1$ progeny express on (a) its liver cells and (b) its macrophages?

3. To carry out studies on the structure and function of the class I MHC molecule $K^b$ and the class II MHC molecule $IA^b$, you decide to transfect the genes encoding these proteins into a mouse fibroblast cell line (L cell) derived from the C3H strain ($H$-$2^k$). L cells do not normally function as antigen-presenting cells. In the following table, indicate which of the listed MHC molecules will ($+$) or will not ($-$) be expressed on the membrane of the transfected L cells.

| Transfected gene | MHC molecules expressed on the membrane of the transfected L cells | | | | | |
| --- | --- | --- | --- | --- | --- | --- |
| | $D^k$ | $D^b$ | $K^k$ | $K^b$ | $IA^k$ | $IA^b$ |
| None | | | | | | |
| $K^b$ | | | | | | |
| $IA\alpha^b$ | | | | | | |
| $IA\beta^b$ | | | | | | |
| $IA\alpha^b$ and $IA\beta^b$ | | | | | | |

4. The SJL mouse strain, which has the $H$-$2^s$ haplotype, has a deletion of the $IE\alpha$ locus.

    a. List the classical MHC molecules that are expressed on the membrane of macrophages from SJL mice.

    b. If the class II $IE\alpha$ and $IE\beta$ genes from an $H$-$2^k$ strain are transfected into SJL macrophages, what additional classical MHC molecules would be expressed on the transfected macrophages?

5. Draw diagrams illustrating the general structure, including the domains, of class I MHC molecules, class II MHC molecules, and membrane-bound antibody on B cells. Label each chain and the domains within it, the antigen-binding regions, and regions that have the immunoglobulin-fold structure.

6. One of the characteristic features of the MHC is the large number of different alleles at each locus.

    a. Where are most of the polymorphic amino acid residues located in MHC molecules? What is the significance of this location?

    b. How is MHC polymorphism thought to be generated?

7. As a student in an immunology laboratory class, you have been given spleen cells from a mouse immunized with the LCM virus (LCMV). You determine the antigen-specific functional activity of these cells with two different assays. In assay 1, the spleen cells are incubated with macrophages that have been briefly exposed to LCMV; the production of interleukin 2 (IL-2) is a positive response. In assay 2, the spleen cells are incubated with LCMV-infected target cells; lysis of the target cells represents a positive response in this assay. The results of the assays using macrophages and target cells of different haplotypes are presented in the table at the top of page 221. Note that the experiment has been set up in a way to exclude alloreactive responses (reactions against nonself MHC molecules).

    a. The activity of which cell population is detected in each of the two assays?

    b. The functional activity of which MHC molecules is detected in each of the two assays?

    c. From the results of this experiment, which MHC molecules are required, in addition to the LCM virus, for specific reactivity of the spleen cells in each of the two assays?

| Mouse strain used as source of macrophages and target cells | MHC haplotype of macrophages and virus-infected target cells | | | | Response of spleen cells | |
|---|---|---|---|---|---|---|
| | | | | | IL-2 production in response to LCMV-pulsed macrophages (assay 1) | Lysis of LCMV-infected cells (assay 2) |
| | K | IA | IE | D | | |
| C3H | k | k | k | k | + | − |
| BALB/c | d | d | d | d | − | + |
| (BALB/c x B10.A)F₁ | d/k | d/k | d/k | d/d | + | + |
| A.TL | s | k | k | d | + | + |
| B10.A (3R) | b | b | b | d | − | + |
| B10.A (4R) | k | k | — | b | + | − |

For use with Question 7

d. What additional experiments could you perform to un-ambiguously confirm the MHC molecules required for antigen-specific reactivity of the spleen cells?

e. Which of the mouse strains listed in the table above could have been the source of the immunized spleen cells tested in the functional assays? Give your reasons.

8. A $T_C$-cell clone recognizes a particular measles virus peptide when it is presented by H-2D$^b$. Another MHC molecule has a peptide-binding cleft identical to the one in H-2D$^b$ but differs from H-2D$^b$ at several other amino acids in the $\alpha 1\beta 1$ domain. Predict whether the second MHC molecule could present this measles virus peptide to the $T_C$-cell clone. Briefly explain your answer.

9. Human red blood cells are not nucleated and do not express any MHC molecules. Why is this property fortuitous for blood transfusions?

10. The hypothetical allelic combination *HLA-A99* and *HLA-B276* carries a relative risk of 200 for a rare, and yet un-named, disease that is fatal to pre-adolescent children.

a. Will every individual with *A99/B276* contract the disease?

b. Will everyone with the disease have the *A99/B276* combination?

c. How frequently will the *A99/B276* allelic combination be observed in the general population? Do you think that this combination will be more or less frequent than predicted by the frequency of the two individual alleles? Why?

11. Explain the difference between the terms *antigen-presenting cell* and *target cell*, as they are commonly used in immunology.

12. Define the following terms:

a. Self-MHC restriction

b. Antigen processing

c. Endogenous antigen

d. Exogenous antigen

13. For each of the following cell components or processes, indicate whether it is involved in the processing and presentation of exogenous antigens (EX), endogenous antigens (EN), or both (B). Briefly explain the function of each item.

a. _____ Class I MHC molecules

b. _____ Class II MHC molecules

c. _____ Invariant (Ii) chains

d. _____ Lysosomal hydrolases

e. _____ TAP1 and TAP2 proteins

f. _____ Transport of vesicles from the RER to the Golgi complex

g. _____ Proteasomes

h. _____ Phagocytosis or endocytosis

i. _____ Calnexin

j. _____ CLIP

k. _____ Tapasin

14. Antigen-presenting cells have been shown to present lysozyme peptide 46–61 together with the class II IA$^k$ molecule. When CD4$^+$ $T_H$ cells are incubated with APCs and native lysozyme or the synthetic lysozyme peptide 46–61, $T_H$-cell activation occurs.

a. If chloroquine is added to the incubation mixture, presentation of the native protein is inhibited, but the peptide continues to induce $T_H$-cell activation. Explain why this occurs.

b. If chloroquine addition is delayed for 3 hours, presentation of the native protein is not inhibited. Explain why this occurs.

15. Cells that can present antigen to $T_H$ cells have been classified into two groups—professional and nonprofessional APCs.

a. Name the three types of professional APCs. For each type, indicate whether it expresses class II MHC molecules and a costimulatory signal constitutively or must be activated before doing so.

b. Give three examples of nonprofessional APCs. When are these cells most likely to function in antigen presentation?

16. Predict whether $T_H$-cell proliferation or CTL-mediated cytolysis of target cells will occur with the following mixtures of cells. The CD4$^+$ $T_H$ cells are from lysozyme-primed mice, and the CD8$^+$ CTLs are from influenza-infected

mice. Use R to indicate a response and NR to indicate no response.

   a. _____ H-$2^k$ T$_H$ cells + lysozyme-pulsed H-$2^k$ macrophages

   b. _____ H-$2^k$ T$_H$ cells + lysozyme-pulsed H-$2^{b/k}$ macrophages

   c. _____ H-$2^k$ T$_H$ cells + lysozyme-primed H-$2^d$ macrophages

   d. _____ H-$2^k$ CTLs + influenza-infected H-$2^k$ macrophages

   e. _____ H-$2^k$ CTLs + influenza-infected H-$2^d$ macrophages

   f. _____ H-$2^d$ CTLs + influenza-infected H-$2^{d/k}$ macrophages

17. HLA-DM and HLA-DO are termed nonclassical MHC class II molecules. How do they differ from the classical MHC class II molecules? How do they differ from each other?

18. Molecules of the CD1 family were recently shown to present nonpeptide antigens.

   a. What is a major source of nonpeptide antigens?

   b. Why are CD1 molecules not classified as members of the MHC family even though they associate with $\beta_2$-microglobulin?

   c. What evidence suggests that the CD1 pathway is different from that utilized by classical class I MHC molecules?

19. A slide of macrophages was stained by immunofluorescence using a monoclonal antibody for the TAP1/TAP2 complex. Which of the following intracellular compartments would exhibit positive staining with this antibody?

   a. Cell surface

   b. Endoplasmic reticulum

   c. Golgi apparatus

   d. The cells would not be positive because the TAP1/TAP2 complex is not expressed in macrophages.

20. HLA determinants are used not only for tissue typing of transplant organs, but also as one set of markers in paternity testing. Given the following phenotypes, which of the potential fathers is most likely the actual biological father? Indicate why each could or could not be the biological father.

| | HLA-A | HLA-B | HLA-C |
|---|---|---|---|
| Offspring: | A3, 43 | B54, 59 | C5, 8 |
| Biological mother: | A3, 11 | B59, 78 | C8, 8 |
| Potential Father 1: | A3, 33 | B54, 27 | C5, 5 |
| Potential Father 2: | A11, 43 | B54, 27 | C5, 6 |
| Potential Father 3: | A11, 33 | B59, 26 | C6, 8 |

**ANALYZE THE DATA** Smith and colleagues (*J. Immunology* **169**: 3105, 2002) examined the ability of two peptides to bind two different allotypes, $L^d$ and $L^q$, of the mouse class I MHC molecule. They looked at (a) the binding of a murine cytomegalovirus peptide (MCMV; amino acid sequence YPHFMPTNL) and (b) a synthesized peptide, tum$^-$ P91A 14–22 (TQNHRALDL). Before and after pulsing the cells with the target peptides, the investigators measured the amount of peptide-free MHC molecules on the surface of cells expressing either $L^d$, $L^q$, or a mutant $L^d$ in which tryptophan had been mutated to arginine at amino acid 97 ($L^d$ W97R). The ability of the peptide to decrease the relative number of open forms on the cell surface reflects increased peptide binding to the class I MHC molecules. Answer the following questions based on the data and what you have learned from reading this book.

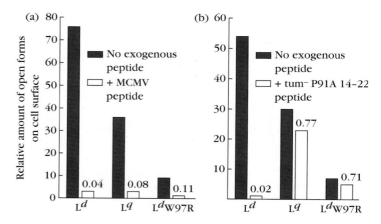

   a. Are there allotypic differences in the binding of peptide by native class I MHC molecules on the cell surface?

   b. Is there a difference in the binding of MCMV peptide to $L^d$ after a tryptophan (W) to arginine (R) mutation in the $L^d$ molecule at position 97? Explain your answer.

   c. Is there a difference in the binding of tum$^-$ P91A 14–22 peptide to $L^d$ after a tryptophan (W) to arginine (R) mutation in the $L^d$ molecule at position 97? Explain your answer.

   d. If you wanted to successfully induce a CD8$^+$ T-cell response against tum$^-$ P91A 14–22 peptide, would you inject an mouse that expresses $L^q$ or $L^d$? Explain your answer.

   e. The T cells generated against the MCMV peptide have a different specificity than the T cells generated against the tum$^-$ P91A 14–22 peptide when $L^d$ is the restricting MHC molecule. Explain how different peptides can bind the same $L^d$ molecule yet restrict/present peptide to T cells with different antigen specificities.

 **Interactive Study**

www.whfreeman.com/kuby

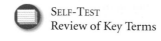 SELF-TEST
Review of Key Terms

# chapter 9

# T-Cell Receptor

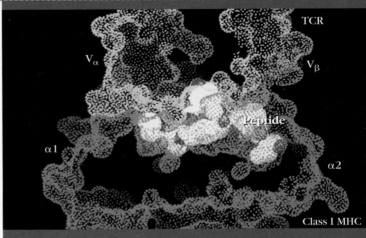

*Interaction between T-cell receptor (yellow) and Class I MHC molecule (blue) presenting bound peptide (white and red). [From D. N. Garboczi et al., 1996, Nature 384: 134–141, courtesy D. C. Wiley, Harvard University.]*

- Early Studies of the T-Cell Receptor
- αβ and γδ T-Cell Receptors: Structures and Roles
- Organization and Rearrangement of TCR Genes
- T-Cell Receptor Complex: TCR-CD3
- T-Cell Accessory Membrane Molecules
- Three-Dimensional Structures of TCR-Peptide-MHC Complexes
- Alloreactivity of T Cells

THE ANTIGEN-SPECIFIC NATURE OF T-CELL RESPONSES clearly implies that T cells possess an antigen-specific and clonally restricted receptor. However, the identity of this receptor remained unknown long after the B-cell receptor (immunoglobulin molecule) had been identified. Relevant experimental results were contradictory and difficult to conceptualize within a single model because the T-cell receptor (TCR) differs from the B-cell antigen-binding receptor in important ways. First, the T-cell receptor is membrane bound and does not appear in a soluble form as the B-cell receptor does; therefore, assessment of its structure by classic biochemical methods was complicated, and complex cellular assays were necessary to determine its specificity. Second, the antigen-binding interaction of T-cell receptors is weaker than that of antibodies, requiring more sensitive assays. Finally, most T-cell receptors are specific not for antigen alone but for antigen combined with a molecule encoded by the major histocompatibility complex (MHC). This property precludes purification of the T-cell receptor by simple antigen-binding techniques and adds complexity to any experimental system designed to investigate the receptor.

A combination of immunologic, biochemical, and molecular-biological manipulations has overcome these problems. The molecule responsible for T-cell specificity is a heterodimer composed of either α and β or γ and δ chains. The genomic organization of the T-cell receptor gene families and the means by which the diversity of the component chains is generated were found to resemble those of the B-cell receptor chains. Further, the T-cell receptor is associated on the membrane with a multicomponent signal-transducing complex, CD3, whose function is similar to that of the Ig-α/ Ig-β complex of the B-cell receptor.

Important new insights concerning T-cell receptors have been gained through recent studies, including new awareness of the major differences between cells bearing a receptor consisting of α and β chains and those with γ and δ chains. The αβ TCR, like the antibody, is characterized by its high degree of specificity and thus is considered a signature molecule of the adaptive immune system. By contrast, certain receptors on γδ T cells appear to recognize classes of antigens present on groups of pathogens and so function in a manner more consistent with innate immunity.

Structural investigations using X-ray crystallography have revealed differences in how αβ TCRs bind antigens presented by class I versus class II MHC molecules and have yielded insight into how the γδ TCRs recognize certain antigens in their native form. This chapter will explore the nature of the T-cell receptor molecules and the genes that encode them and discuss TCR recognition of foreign and self antigens.

## Early Studies of the T-Cell Receptor

By the early 1980s, investigators had learned much about T-cell function but were thwarted in their attempts to identify and isolate its antigen-binding receptor. The obvious parallels between the recognition functions of T cells and B cells stimulated a great deal of experimental effort to take advantage of the anticipated structural similarities between immunoglobulins and T-cell receptors. Reports published in the 1970s claimed discovery of immunoglobulin isotypes associated exclusively with T cells (IgT) and of antisera that recognize variable-region markers (idiotypes) common to antibodies and T-cell receptors with similar specificity. These experiments could not be reproduced and were proven to be incorrect when it was demonstrated that the T-cell receptor and immunoglobulins do not have common recognition elements and are encoded by entirely separate gene families. As the following sections will show, a sequence of well-designed experiments using the then-developing technologies of monoclonal antibody production and molecular biology was required to correctly answer questions about the structure of the T-cell receptor, the genes that encode it, and the manner in which it recognizes antigen.

## Classic experiments demonstrated the self-MHC restriction of the T-cell receptor

By the early 1970s, immunologists had learned to generate cytotoxic T lymphocytes (CTLs) specific for virus-infected target cells. For example, when mice were infected with lymphocytic choriomeningitis (LCM) virus, they would produce CTLs that could lyse LCM-infected target cells in vitro. Yet these same CTLs failed to bind free LCM virus or viral antigens. Why didn't the CTLs bind the virus or viral antigens directly as immunoglobulins did? The answer began to emerge in the classic experiments of R. M. Zinkernagel and P. C. Doherty in 1974 (see Figure 8-15). These studies demonstrated that antigen recognition by T cells is specific not for viral antigen alone but for antigen associated with an MHC molecule (Figure 9-1). T cells were shown to recognize antigen only when presented on the membrane of a cell by a

self-MHC molecule. This attribute, called *self-MHC restriction,* distinguishes recognition of antigen by T cells and B cells. In 1996, Zinkernagel and Doherty were awarded the Nobel prize for this work.

Two models were proposed to explain the MHC restriction of the T-cell receptor. The *dual-receptor model* envisioned a T cell with two separate receptors, one for antigen and one for class I or class II MHC molecules. The *altered-self model* proposed that a single receptor recognizes an alteration in self-MHC molecules induced by their association with foreign antigens. The debate between proponents of these two models was waged for a number of years until an elegant experiment by J. Kappler and P. Marrack demonstrated that specificity for both MHC and antigen resides in a single receptor. An overwhelming amount of structural and functional data have since been added in support of the altered-self model.

## T-cell receptors were isolated by using clonotypic antibodies

Identification and isolation of the T-cell receptor was accomplished by producing large numbers of monoclonal antibodies to various T-cell clones and then screening the antibodies to find one that was clone specific, or *clonotypic.* This approach assumes that, since the T-cell receptor is specific for both an antigen and an MHC molecule, there should be significant structural differences in the receptor from clone to clone; each T-cell clone should have an antigenic marker similar to the idiotype markers that characterize monoclonal antibodies. Using this approach, researchers in the early 1980s isolated the receptor and found that it was a heterodimer consisting of α and β chains.

When antisera were prepared using αβ heterodimers isolated from membranes of various T-cell clones, some antisera bound to αβ heterodimers from all the clones, whereas other antisera were clone specific. This finding suggested that the amino acid sequences of the TCR α and β chains, like those of the immunoglobulin heavy and light chains, have constant and variable regions identifiable by antibodies recognizing isotypic or idiotypic determinants. Later, a second type of TCR heterodimer consisting of δ and γ chains was identified. In human and mouse, the majority of T cells express the αβ heterodimer; the remaining T cells express the γδ heterodimer. As described below, the exact proportion of T cells expressing αβ or γδ TCRs differs by organ and species, but αβ T cells normally predominate. Because of the predominance of cells bearing the αβ TCR and the greater amount of information available for them, future discussions of the T-cell receptor will assume that αβ is meant if no specific reference to γδ is given.

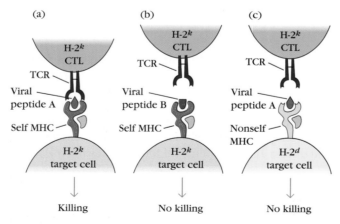

**FIGURE 9-1 Self-MHC restriction of the T-cell receptor (TCR).**
A particular TCR is specific for both an antigenic peptide and a self-MHC molecule. In this example, the H-2$^k$ CTL is specific for viral peptide A presented on an H-2$^k$ target cell (a). Antigen recognition does not occur when peptide B is displayed on an H-2$^k$ target cell (b) nor when peptide A is displayed on an H-2$^d$ target cell (c).

## The TCR β-chain gene was cloned by use of subtractive hybridization

In order to identify and isolate the TCR genes, S. M. Hedrick and M. M. Davis sought to isolate mRNA that encodes the α and β chains from a $T_H$-cell clone. This was no easy task because the receptor mRNA represents only a minor fraction of the total cellular mRNA. By contrast, in the plasma cell, immunoglobulin is a major secreted cell product, and mRNAs encoding the heavy and light chains are abundant and easy to purify.

The successful scheme of Hedrick and Davis assumed that the TCR mRNA—like the mRNAs that encode other integral membrane proteins—would be associated with membrane-bound polyribosomes rather than with free cytoplasmic ribosomes. They therefore isolated the membrane-bound polyribosomal mRNA from a $T_H$-cell clone and used reverse transcriptase to synthesize $^{32}$P-labeled cDNA probes (Figure 9-2). Because only 3% of lymphocyte mRNA is in the membrane-bound polyribosomal fraction, this step eliminated 97% of the cell mRNA.

Hedrick and Davis next used a technique called *DNA subtractive hybridization* to remove from their preparation the $^{32}$P-labeled cDNA that was not unique to T cells. Their rationale for this step was based on earlier measurements by Davis showing that 98% of the genes expressed in lymphocytes are common to B cells and T cells. The 2% of the expressed genes that is unique to T cells should include the genes encoding the T-cell receptor. Therefore, by hybridizing B-cell mRNA with their $T_H$-cell $^{32}$P-labeled cDNA, they were able to remove, or subtract, all the cDNA that was common to B cells and T cells. The unhybridized $^{32}$P-labeled cDNA remaining after this step presumably represented the expressed polyribosomal mRNA that was unique to the $T_H$-cell clone, including the mRNA encoding its T-cell receptor.

Cloning of the unhybridized $^{32}$P-labeled cDNA generated a library from which 10 different cDNA clones were identified. To determine which of these T-cell-specific cDNA clones represented the T-cell receptor, all were used as probes to look for genes that rearranged in mature T cells. This approach was based on the assumption that, since the αβ T-cell receptor appeared to have constant and variable regions, its genes should undergo DNA rearrangements like

those observed in the Ig genes of B cells. The two investigators tested DNA from T cells, B cells, liver cells, and macrophages by Southern-blot analysis using the 10 $^{32}$P-labeled cDNA probes to identify unique T-cell genomic DNA sequences. One clone showed bands indicating DNA rearrangement in T cells but not in the other cell types. This cDNA probe identified six different patterns for the DNA from six different mature T-cell lines (see Figure 9-2 inset, upper panel). These different patterns presumably represented rearranged TCR genes. Such results would be

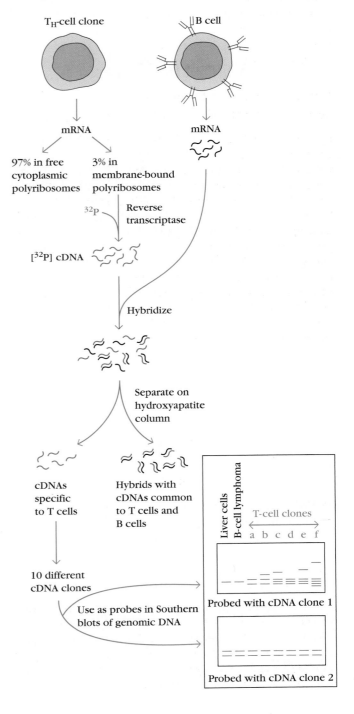

**FIGURE 9-2 Production and identification of a cDNA clone encoding the T-cell receptor.** The flowchart outlines the procedure used by S. M. Hedrick and M. M. Davis to obtain [$^{32}$P]cDNA clones corresponding to T-cell-specific mRNAs. The technique of DNA subtractive hybridization enabled them to isolate [$^{32}$P]cDNA unique to the T cell. The labeled $T_H$-cell cDNA clones were used as probes (*inset*) in Southern-blot analyses of genomic DNA from liver cells, B-lymphoma cells, and six different $T_H$-cell clones (a–f). Probing with cDNA clone 1 produced a distinct blot pattern for each T-cell clone, whereas probing with cDNA clone 2 did not. When it was assumed that liver cells and B cells contained unrearranged germ-line TCR DNA and that each of the T-cell clones contained different rearranged TCR genes, the results using cDNA clone 1 as the probe identified clone 1 as the T-cell receptor gene. The cDNA of clone 2 identified the gene for another T-cell membrane molecule encoded by DNA that does not undergo rearrangement. [*Based on S. Hedrick et al., 1984, Nature **308**:149.*]

expected if rearranged TCR genes occur only in mature T cells. The observation that each of the six T-cell lines showed different Southern-blot patterns was consistent with the predicted differences in TCR specificity in each T-cell line.

The cDNA clone 1 identified by the Southern-blot analyses shown in Figure 9-2 has all the hallmarks of a putative TCR gene: it represents a gene sequence that rearranges, is expressed as a membrane-bound protein, and is expressed only in T cells. This cDNA clone was found to encode the β chain of the T-cell receptor. Later, cDNA clones were identified encoding the α chain, the γ chain, and finally the δ chain. These findings opened the way to understanding the T-cell receptor and made possible subsequent structural and functional studies.

## αβ and γδ T-Cell Receptors: Structures and Roles

The domain structures of αβ and γδ TCR heterodimers are strikingly similar to those of the immunoglobulins; thus, they are classified as members of the immunoglobulin superfamily (see Figure 4-24). Each chain in a TCR has two domains containing an intrachain disulfide bond that spans 60 to 75 amino acids. The amino-terminal domain in both chains exhibits marked sequence variation, but the sequences of the remainder of each chain are conserved. Thus the TCR domains—one variable (V) and one constant (C)—are structurally homologous to the V and C domains of immunoglobulins, and the TCR molecule resembles a Fab fragment attached to the cell membrane instead of to the constant region of an Ig molecule (Figure 9-3). The TCR variable domains have three hypervariable regions, which appear to be equivalent to the complementarity-determining regions (CDRs) in immunoglobulin light and heavy chains.

In addition to the constant domain, each TCR chain contains a short connecting sequence, in which a cysteine residue forms a disulfide link with the other chain of the heterodimer. Following the connecting region is a transmembrane region of 21 or 22 amino acids, which anchors each chain in the plasma membrane. The transmembrane domains of both chains are unusual in that they contain positively charged amino acid residues. As will be discussed below, these residues promote interaction between the chains of the TCR heterodimer and chains of the signal-transducing CD3 complex. Finally, each TCR chain contains a short cytoplasmic tail of five to 12 amino acids at the carboxyl-terminal end.

αβ and γδ T-cell receptors were initially difficult to investigate because, like all transmembrane proteins, they are insoluble. This problem was circumvented by expressing modified forms of the protein in vitro that had been engineered to contain premature in-frame stop codons that preclude translation of the membrane-binding sequence that makes the molecule insoluble.

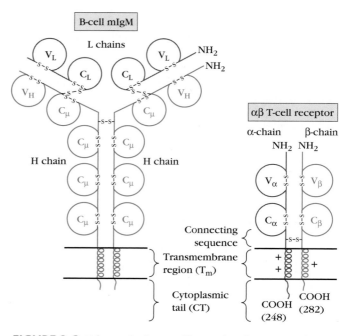

**FIGURE 9-3 Schematic diagram illustrating the structural similarity between the αβ T-cell receptor and membrane-bound IgM on B cells.** The TCR α and β chain each contains two domains with the immunoglobulin-fold structure. The amino-terminal domains ($V_\alpha$ and $V_\beta$) exhibit sequence variation and contain three hypervariable regions equivalent to the CDRs in antibodies. The sequence of the constant domains ($C_\alpha$ and $C_\beta$) does not vary. The two TCR chains are connected by a disulfide bond between their constant sequences; the IgM H chains are connected to each other by a disulfide bond in the hinge region of the H chain, and the L chains are connected to the H chains by disulfide links between the C termini of the L chains and the $C_\mu$ region. TCR molecules interact with CD3 via positively charged amino acid residues (indicated by +) in their transmembrane regions. Numbers indicate the length of the chains in the TCR molecule. Unlike the antibody molecule, which is bivalent, the TCR is monovalent.

The majority of circulating T cells in human and mouse express T-cell receptors encoded by the αβ genes. These receptors interact with peptide antigens processed and presented on the surface of antigen-presenting cells. Early indications that certain T cells reacted with *nonpeptide* antigens were puzzling until some light was shed on the problem when products of the CD1 family of genes were found to present nonpeptide antigens such as glycolipids and phospholipids. More recently, it has been found that certain γδ cells react with protein antigens that are neither processed nor presented in the context of MHC molecules.

Differences in the antigen-binding regions of TCR αβ and γδ were anticipated because of the different antigens they recognize, but no extreme dissimilarities were expected. However, the three-dimensional structure for a γδ receptor that reacts with a phospholipid antigen, reported by Tim Allison, Dave Garboczi, and their coworkers, reveals significant differences in the overall structures of the two receptor types, pointing to

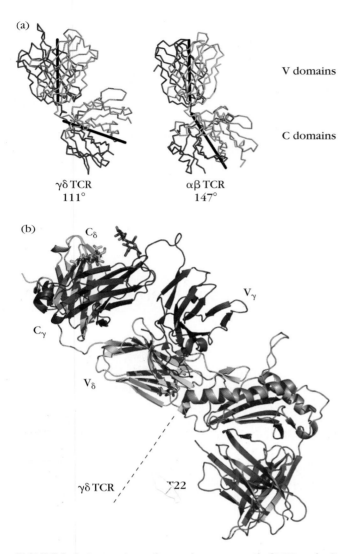

**FIGURE 9-4 Comparison of crystal structures of γδ TCR and αβ TCR.** (a) The human γδ molecule, showing the difference (highlighted with black lines) in the elbow angle compared to that of αβ. (b) The mouse G8 γδ TCR in complex with the nonclassical MHC molecule T22. This binding occurs independent of the peptide binding site of T22. Note how the binding involves mainly contact with CDR3 of the δ chain of the receptor. *[Part a from T. Allison et al., 2001, Nature **411**:820; part b from E. J. Adams et al., 2005, Science **308**:227.]*

contribute to differences in signaling mechanisms and in how the molecules interact with coreceptor molecules.

A more recent x-ray crystallography structure from Garcia and coworkers of a mouse γδ receptor in complex with a nonclassical MHC class I molecule termed T22 indicates that the γδ receptor binds directly to T22, with no requirement for the presence of a peptide antigen (Figure 9-4b). They further showed, using surface plasmon resonance (see Chapter 6), that binding between the γδ TCR and T22 was quite strong ($K_d = 0.10$ M). The structure of the complex revealed that the major contact between the antigen and the γδ TCR involves the CDR3 loop of the δ chain. In the interaction between αβ TCRs and antigen, CDR3 provides the primary contact with antigenic peptide presented by MHC molecules; the CDR1 and 2 loops interact with conserved surface features of the MHC molecule. In the γδ TCR-T22 interaction, CDR3 again provides the major contact with antigen, but the contact is not with a presented peptide within an MHC molecule, but with a site outside the peptide-binding region of the nonclassical MHC molecule (T22) itself. This recognition of the T22 molecule in the absence of a presented peptide suggests that the γδ receptor is more similar to pattern recognition receptors (PRRs; see Chapter 3) than to the αβ receptor.

Whereas the αβ T cells recognize antigen processed and presented in the context of an MHC molecule, an accumulating body of data indicates that γδ T cells do not require either MHC processing or presentation for antigen recognition. The precise functions of the γδ T cells is still unknown, and their role in immunity to foreign pathogens and possible role in autoimmunity remains to be learned. Recognition of antigens common to classes of microorganisms, as well as the ability of γδ TCRs to bind to nonclassical self-MHC molecules, as described above, suggest a rapid-response role more characteristic of innate than adaptive immunity.

The number of γδ T cells in circulation is small compared with cells that have αβ receptors, and the V gene segments of γδ receptors exhibit limited diversity. As seen from the data in Table 9-1, the majority of γδ cells lack both CD4 and CD8, and the majority express the same γδ-chain pair. In humans the predominant receptor expressed on circulating γδ cells recognizes a microbial phospholipid antigen, 3-formyl-1-butyl pyrophosphate, found on *Mycobacterium tuberculosis* and other bacteria and parasites. This specificity for frequently encountered pathogens supports speculation that γδ cells may function as an arm of the innate immune response, allowing rapid reactivity to certain antigens without the need for a processing step. Interestingly, the specificity of circulating γδ cells in the mouse and other species studied does not parallel that of humans, suggesting that the γδ response may be directed against pathogens commonly encountered by a given species. In certain types of infection, the percentage of γδ T cells in blood or in a specific organ may dramatically increase. Infections with a variety of bacteria, including the very common *M. tuberculosis* and *Hemophilus influenzae* as well as the malaria and leishmania parasites, lead to

possible differences in function. The receptor they studied was a dimer of γ and δ chains (γ9 and δ2), which are those most frequently expressed in human peripheral blood. A deep cleft on the surface of the molecule accommodates the microbial phospholipid for which the γδ receptor is specific. This antigen is recognized without MHC presentation.

The most striking feature of the structure is how it differs from the αβ receptor in the orientation of its V and C regions. The so-called elbow angle between the long axes of the V and C regions of γδ TCR is 111°; in the αβ TCR, the elbow angle is 147°, giving the molecules distinct shapes (Figure 9-4a). The full significance of this difference is not known, but it could

| TABLE 9-1 | Comparison of $\alpha\beta$ and $\gamma\delta$ T cells in peripheral blood | |
|---|---|---|
| Feature | $\alpha\beta$ T cells | $\gamma\delta$ T cells |
| Proportion of CD3$^+$ cells | 90–99% | 1–10% |
| TCR V gene germ-line repertoire | Large | Small |
| CD4/CD8 phenotype | | |
| CD4$^+$ | ~60% | <1% |
| CD8$^+$ | ~30% | ~30% |
| CD4$^+$ CD8$^+$ | <1% | <1% |
| CD4$^-$ CD8$^-$ | <1% | ~60% |
| MHC restriction | CD4$^+$: MHC class II | No MHC restriction |
| | CD8$^+$: MHC class I | |
| Ligands | MHC + peptide antigen | Phospholipid, intact protein |

SOURCE: D. Kabelitz et al., 1999, *Springer Seminars in Immunopathology* **21** (55): 36.

increased numbers of $\gamma\delta$ cells. The $\gamma\delta$ cells may kill infected cells or organisms using mechanisms similar to those of CTLs (granulysin and perforin; see Chapter 14). Additional data on $\gamma\delta$ cells implicate them in chronic autoimmune conditions, including lupus, myositis, and multiple sclerosis. Furthermore, data indicating that $\gamma\delta$ cells can secrete a spectrum of chemokines and cytokines suggest that they may play a regulatory role in recruiting $\alpha\beta$ T cells to the site of invasion by pathogens. The recruited $\alpha\beta$ T cells would presumably display a broad spectrum of receptors; those with the highest affinity would be selectively activated and amplified to deal with the pathogen.

The roles and mechanisms of $\gamma\delta$ T-cell function are an active research area, and much remains to be learned. Of particular interest is a more precise profile of the range of antigens that can be recognized by the $\gamma\delta$ receptor and whether it can recognize antigen in the context of nonclassical MHC molecules. Recent studies also suggest that $\gamma\delta$ cells may themselves serve as antigen-presenting cells (APCs), thus greatly expanding their functional potential.

## Organization and Rearrangement of TCR Genes

The genes that encode the $\alpha\beta$ and $\gamma\delta$ T-cell receptors are expressed only in cells of the T-cell lineage. The four TCR loci ($\alpha$, $\beta$, $\gamma$, and $\delta$) are organized in the germ line in a manner that is remarkably similar to the multigene organization of the immunoglobulin (Ig) genes (Figure 9-5). As in the case of Ig genes, functional TCR genes are produced by rearrangements of V and J segments in the $\alpha$-chain and $\gamma$-chain families and V, D, and J segments in the $\beta$-chain and $\delta$-chain families. In the mouse, the $\alpha$-, $\beta$-, and $\gamma$-chain gene segments are located on chromosomes 14, 6, and 13, respectively. The $\delta$-chain gene segments are located on chromosome 14 between

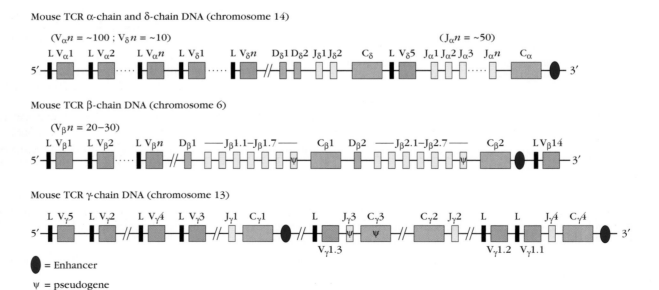

**FIGURE 9-5 Germ-line organization of the mouse TCR $\alpha$-, $\beta$-, $\gamma$-, and $\delta$-chain gene segments.** Each C gene segment is composed of a series of exons and introns, which are not shown. The organization of TCR gene segments in humans is similar, although the number of the various gene segments differs in some cases (see Table 9-2). The numbers of gene segments listed here are averages for different mouse strains and include pseudogenes. *[Adapted from D. Raulet, 1989, Annual Review of Immunology **7**:175, and M. Davis, 1990, Annual Review of Biochemistry **59**:475.]*

## TABLE 9-2 — TCR multigene families in humans

| Gene | Chromosome location | V | D | J | C |
|---|---|---|---|---|---|
| | | NO. OF GENE SEGMENTS* | | | |
| α Chain | 14 | 54 | | 61 | 1 |
| δ Chain† | 14 | 3 | 3 | 3 | 1 |
| β Chain‡ | 7 | 67 | 2 | 14 | 2 |
| γ Chain§ | 7 | 14 | | 5 | 2 |

*Not all gene segments listed here give rise to TCR products; pseudogenes are included in this list.

†The δ-chain gene segments are located between the $V_\alpha$ and $J_\alpha$ segments.

‡There are two repeats, each containing one $D_\beta$, six or seven $J_\beta$, and one $C_\beta$.

§There are two repeats, each containing two or three $J_\gamma$ and One $C_\gamma$.

SOURCE: Data from Immunogenetics Database, http://imgt.cines.fr.

the $V_\alpha$ and $J_\alpha$ segments. The location of the δ-chain gene family is significant: a productive rearrangement of the α-chain gene segments deletes $C_\delta$, so that, in a given T cell, the αβ TCR receptor cannot be co-expressed with the γδ receptor. This decisive lineage commitment is important in light of the different functions for αβ and γδ T cells.

Mouse germ-line DNA contains about 80 $V_\alpha$ and 40 $J_\alpha$ gene segments and a single $C_\alpha$ segment. The δ-chain gene family contains about 10 V gene segments, which are largely distinct from the $V_\alpha$ gene segments, although some sharing of V segments has been observed in rearranged α- and δ-chain genes. Two $D_\delta$ and two $J_\delta$ gene segments and one $C_\delta$ segment have also been identified. The β-chain gene family has 20 to 30 V gene segments and two almost identical repeats of D, J, and C segments, each repeat consisting of one $D_\beta$, six $J_\beta$, and one $C_\beta$. The γ-chain gene family consists of seven $V_\gamma$ segments and three different functional $J_\gamma$-$C_\gamma$ repeats. The organization of the TCR multigene families in humans is generally similar to that in mice, although the number of segments differs (Table 9-2). In all cases, the gene families may include members with mutations that render them nonfunctional (pseudogenes); these genes are normally included in the count of genes within the gene family unless *functional* genes are specified. For example, Table 9-3 includes only functional genes because pseudogenes do not contribute to diversity, whereas Figure 9-5 includes all genes in the germ line.

## TCR variable-region genes rearrange in a manner similar to antibody genes

The α chain, like the immunoglobulin L chain, is encoded by V, J, and C gene segments. The β chain, like the immunoglobulin

## TABLE 9-3 — Sources of possible diversity in mouse immunoglobulin and TCR genes

| Mechanism of diversity | IMMUNOGLOBULINS | | αβ T-CELL RECEPTOR | | γδ T-CELL RECEPTOR | |
|---|---|---|---|---|---|---|
| | H Chain | κ Chain | α Chain | β Chain | γ Chain | δ Chain |
| ESTIMATED NUMBER OF FUNCTIONAL GENE SEGMENTS* | | | | | | |
| V | 101 | 85 | 79 | 21 | 7 | 6 |
| D | 13 | 0 | 0 | 2 | 0 | 2 |
| J | 4 | 4 | 38 | 11 | 3 | 2 |
| POSSIBLE NUMBER OF COMBINATIONS† | | | | | | |
| Combinatorial V-J | $101 \times 13 \times 4$ | $85 \times 4$ | $79 \times 38$ | $21 \times 2 \times 11$ | $7 \times 3$ | $6 \times 2 \times 2$ |
| and V-D-J joining | $5.3 \times 10^3$ | $3.4 \times 10^2$ | $3.0 \times 10^3$ | $4.6 \times 10^2$ | 21 | 24 |
| Alternative joining | − | − | − | + | − | + |
| of D gene segments | | | | (some) | | (often) |
| Junctional flexibility | + | + | + | + | + | + |
| N-region nucleotide addition‡ | + | − | + | + | + | + |
| P-region nucleotide addition | + | + | + | + | + | + |
| Somatic mutation | + | + | − | − | − | − |
| Combinatorial association of chains | | + | | + | | + |

*Immunoglobulin data from Table 5-2; TCR data from Baum et al., 2004, *Nucleic Acids Research* **32**:D51.

†A plus sign (+) indicates mechanism makes a significant contribution to diversity but to an unknown extent. A minus sign (−) indicates mechanism does not operate.

‡See Figure 9-8d for theoretical number of combinations generated by N-region addition.

OVERVIEW FIGURE 9-6: Example of Gene Rearrangements That Yield a Functional Gene Encoding the αβ T-Cell Receptor

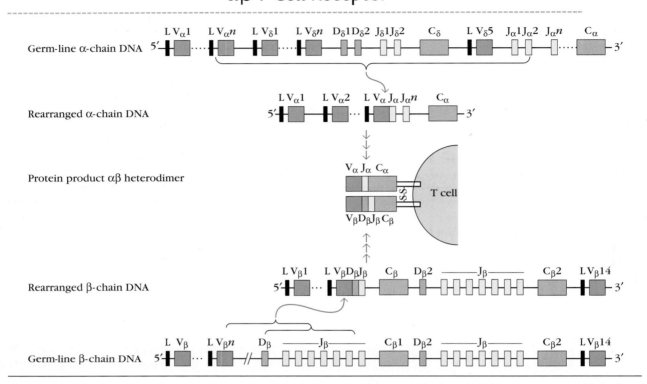

The α-chain DNA, analogous to immunoglobulin light-chain DNA, undergoes a variable-region $V_\alpha$-$J_\alpha$ joining. The β-chain DNA, analogous to immunoglobulin heavy-chain DNA, undergoes two variable-region joinings: first $D_\beta$ to $J_\beta$ and then $V_\beta$ to $D_\beta J_\beta$. Transcription of the rearranged genes yields primary transcripts, which are processed to give mRNAs encoding the α and β chains of the membrane-bound TCR. The leader sequence is cleaved from the nascent polypeptide chain and is not present in the finished protein. As no secreted TCR is produced, differential processing of the primary transcripts to produce membrane-bound and secreted forms does not occur. Although the β-chain DNA contains two C genes, the gene products of these two C genes exhibit no known functional differences. The C genes are composed of several exons and introns, which are not individually shown here (see Figure 9-7).

H chain, is encoded by V, D, J, and C gene segments. Rearrangement of the TCR α- and β-chain gene segments results in VJ joining for the α chain and VDJ joining for the β chain (Figure 9-6).

After transcription of the rearranged TCR genes, RNA processing, and translation, the α and β chains are expressed as a disulfide-linked heterodimer on the membrane of the T cell. Unlike immunoglobulins, which can be membrane bound or secreted, the αβ heterodimer is expressed only in a membrane-bound form; thus, no differential RNA processing to produce membrane and secreted forms is operative. Each TCR constant region includes a constant domain, a connecting sequence, a transmembrane sequence, and a cytoplasmic sequence.

The germ-line DNA encoding the TCR α- and β-chain constant regions is much simpler than the immunoglobulin heavy-chain germ-line DNA, which has multiple C gene segments encoding distinct isotypes with different effector functions. TCR α-chain DNA has only a single C gene segment; the β-chain DNA includes duplicated J and C gene segments. Protein products of the two groups differ by only a few amino acids and have no known functional differences.

## Mechanism of TCR DNA rearrangements

The mechanisms by which TCR germ-line DNA is rearranged to form functional receptor genes appear to be similar to the mechanisms of Ig-gene rearrangements. For example, conserved heptamer and nonamer recombination signal sequences (RSSs), containing either 12-bp (one-turn) or 23-bp (two-turn) spacer sequences, have been identified flanking each V, D, and J gene segment in TCR germ-line DNA (see Figure 5-6).

All of the TCR-gene rearrangements follow the one-turn/two-turn joining rule observed for the Ig genes, so recombination can occur only between the two different types of RSSs.

Like the pre–B cell, the pre–T cell expresses the recombination-activating genes (*RAG-1* and *RAG-2*). The RAG-1/2 recombinase enzyme recognizes the heptamer and nonamer recognition signals and catalyzes V-J and V-D-J joining during TCR-gene rearrangement by the same deletional or inversional mechanisms that occur in the Ig genes (see Figure 5-7). As described in Chapter 5 for the immunoglobulin genes, RAG-1/2 begins by introducing a nick on one DNA strand between the coding and signal sequences and proceeds to excise DNA loops formed in the process. Circular excision products thought to be generated by looping out and deletion during TCR-gene rearrangement have been identified in thymocytes (see Figure 5-8).

Studies with SCID mice, which lack functional T and B cells, provide evidence for the similarity in the mechanisms of Ig-gene and TCR-gene rearrangements. As explained in Chapter 20, SCID mice have a defect in a gene required for the repair of double-stranded DNA breaks. As a result of this defect, D and J gene segments are not joined during rearrangement of either Ig or TCR DNA (see Figure 5-10). This finding suggests that the same double-stranded break-repair enzymes are involved in V-D-J rearrangements in B cells and in T cells.

Although B cells and T cells use very similar mechanisms for variable-region gene rearrangements, complete rearrangement of Ig genes does not occur in T cells, and complete rearrangement of TCR genes does not occur in B cells. Not only is the recombinase enzyme system differently regulated in each lineage, but chromatin is uniquely reconfigured in B cells and T cells to allow the recombinase access to the appropriate specific antigen receptor genes. Rearrangement of the gene segments in both T and B cells creates a DNA sequence unique to that cell and its progeny. The large number of possible configurations of the rearranged genes makes this new sequence a marker that is specific for the cell clone. These unique DNA sequences have been used to aid in diagnosis and treatment of lymphoid leukemias and lymphomas, cancers that involve clonal proliferation of T or B cells (see Clinical Focus on page 232).

## Allelic exclusion of TCR genes

As mentioned above, the δ genes are located within the α-gene complex and are deleted by α-chain rearrangements. This event provides an irrevocable mode of exclusion for the δ genes located on the same chromosome as the rearranging α genes. Allelic exclusion of genes for the TCR α and β chains occurs as well, but exceptions have been observed.

The organization of the β-chain J and C gene segments into two duplicate clusters means that, if a nonproductive rearrangement occurs, the thymocyte can attempt a second rearrangement. This increases the likelihood of a productive V-J rearrangement for the β chain. Once a productive rearrangement occurs for one β-chain allele, the rearrangement of the other β allele is inhibited.

Exceptions to allelic exclusion are most often seen for the TCR α-chain genes. For example, analyses of T-cell clones that express a functional αβ T-cell receptor revealed a number of clones with productive rearrangements of both α-chain alleles. Furthermore, when an immature T-cell lymphoma that expressed a particular αβ T-cell receptor was subcloned, several subclones were obtained that expressed the same β-chain allele but an α-chain allele different from the one expressed by the original parent clone. Studies with transgenic mice also indicate that allelic exclusion is less stringent for TCR α-chain genes than for β-chain genes. Mice that carry a productively rearranged αβ-TCR transgene do not rearrange and express the endogenous β-chain genes. However, the endogenous α-chain genes sometimes are expressed at various levels in place of the already-rearranged α-chain transgene.

Since allelic exclusion is not complete for the TCR α chain, on rare occasions more than one α chain is expressed on the membrane of a given T cell. The obvious question is, how do the rare T cells that express two αβ T-cell receptors maintain a single antigen-binding specificity? One proposal suggests that when a T cell expresses two different αβ T-cell receptors, only one is likely to be self-MHC restricted and therefore functional.

## Rearranged TCR genes are assembled from V, J, and D gene segments

The general structure of rearranged TCR genes is shown in Figure 9-7. The variable regions of T-cell receptors are, of course, encoded by rearranged VDJ and VJ sequences. In

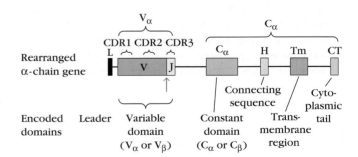

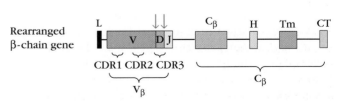

**FIGURE 9-7 Schematic diagram of rearranged αβ-TCR genes showing the exons that encode the various domains of the αβ T-cell receptor and approximate position of the CDRs.** Junctional diversity (vertical arrows) generates CDR3 (see Figure 9-8). The structures of the rearranged γ- and δ-chain genes are similar, although additional junctional diversity can occur in δ-chain genes.

## T-Cell Rearrangements as Markers for Cancerous Cells

**T-cell** cancers, which include leukemia and lymphoma, involve the uncontrolled proliferation of a clonal population of T cells. Successful treatment requires quick, certain, and detailed diagnosis in order to apply the most effective treatment. Once treatment is initiated, reliable tests are needed to determine whether the treatment regimen was successful. In principle, because T-cell cancers are clonal in nature, the cell population that is cancerous could be identified and monitored by the expression of its unique T-cell receptor molecules. Identification of the clone-specific TCR with a monoclonal antibody is rarely practical because it requires the tedious and lengthy preparation of a specific antibody directed against its variable region (an anti-idiotype antibody). Also, not all cancers will display a TCR molecule that is detectable by the antibody. An alternative means of identifying a clonal population of T cells is to look at their DNA rather than protein products. The pattern resulting from rearrangement of the TCR genes provides a unique marker for the cancerous T cell. Because rearrangement of the TCR genes in the T cells occurs before the product molecule is expressed, T cells in early stages of development can be detected. The unique gene fragments that result from TCR gene rearrangement can be detected by simple molecular-biological techniques and provide a true fingerprint for a clonal T-cell population.

DNA patterns that result from rearrangement of the genes in the TCR β region are used most frequently as markers. There are approximately 70 $V_\beta$ gene segments that can rearrange to one of two D-region gene segments and subsequently to one of 14 J gene segments (see Figures 9-6 and 9-8). Because each of the V-region genes is flanked by unique sequences, this process creates new DNA sequences that are unique to each cell that undergoes the rearrangement; these new sequences may be detected by Southern-blot techniques or by PCR (polymerase chain reaction). Since the entire sequence of the D, J, and C region of the TCR gene β complex is known, the appropriate probes and restriction enzymes are easily chosen for Southern blotting.

Detection of rearranged TCR DNA may be used as a diagnostic tool when abnormally enlarged lymph nodes persist; this condition could result either from inflammation due to chronic infection or from proliferation of a cancerous lymphoid cell. If inflammation is the cause, the cells would come from a variety of clones, and the DNA isolated from them would be a mixture of many different TCR sequences resulting from multiple rearrangements; no unique fragments would be detected. If the persistent enlargement of the nodes represents a clonal proliferation, there would be a single dominant DNA fragment, because the cancerous cells would all contain the same TCR DNA sequence produced by DNA rearrangement in the parent cell. Thus, the question of whether the observed enlargement was due to the cancerous growth of T cells could be answered by the presence of a single new gene fragment in the DNA from the otherwise heterogeneous cell population. Because Ig genes rearrange in the same fashion as the TCR genes, similar techniques use Ig probes to detect clonal B-cell populations by their unique DNA patterns. The technique therefore has value for a wide range of lymphoid cell cancers.

Although the detection of a unique DNA fragment resulting from rearranged TCR or Ig genes indicates clonal proliferation and possible malignancy of T or B cells, the absence of such a fragment does not rule out cancer of a population of lymphoid cells. The cell involved may not contain rearranged TCR or Ig genes that can be detected by the method or the probes used. Since the malignant clone may derive from γδ T cells, samples should be tested with δ and γ probes if negative with β.

If the DNA fragment test and other diagnostic criteria indicate that the patient has a lymphoid cell cancer, treatment by appropriate therapy would follow. The success of this treatment can be monitored by probing DNA from the patient for the unique sequence found in the cancerous cell. If the treatment regimen is successful, the number of cancerous cells will decline

---

TCR genes, combinatorial joining of α and β V gene segments appears to generate CDR1 and CDR2, whereas junctional flexibility and N-region nucleotide addition generate CDR3. Rearranged TCR genes also contain a short leader (L) exon upstream of the joined VJ or VDJ sequences. The amino acids encoded by the leader exon are cleaved as the nascent polypeptide enters the endoplasmic reticulum.

The constant region of each TCR chain is encoded by a C gene segment that has multiple exons (see Figure 9-7), corresponding to the structural domains in the protein (see Figure 9-3). The first exon in the C gene segment encodes most of the C domain of the corresponding chain. Next is a short exon that encodes the connecting sequence, followed by exons that encode the transmembrane region and the cytoplasmic tail.

## TCR Diversity is generated like antibody diversity but without somatic mutation

Although TCR germ-line DNA contains far fewer V gene segments than Ig germ-line DNA, several mechanisms that

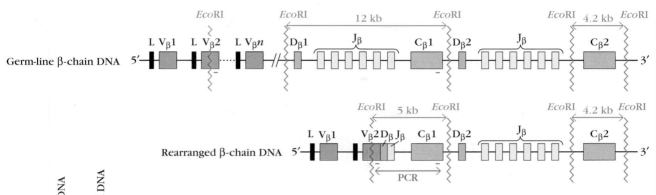

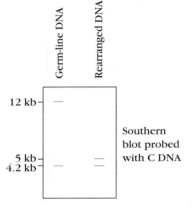

Southern blot probed with C DNA

Digestion of human TCR β-chain DNA in a germ-line (nonrearranged) configuration with *Eco*RI and then probing with a C-region sequence will detect the indicated C-containing fragments by Southern blotting. When the DNA has rearranged, a 5′ restriction site will be excised. Digestion with *Eco*RI will yield a different fragment unique to the specific $V_\beta$ and $J_\beta$ region gene segments incorporated into the rearranged gene, as indicated in this hypothetical example. The technique used for this analysis derives from that first used by S. M. Hedrick and his coworkers to detect unique TCR β genes in a series of mouse T-cell clones (see inset to Figure 9-2). For highly sensitive detection of the rearranged TCR sequence, the polymerase chain reaction (PCR) is used. The sequence of the 5′ primer (red bar) is based on a unique sequence in the ($V_\beta$) gene segment used by the cancerous clone ($V_\beta$2 in this example) and the 3′ primer (red bar) as a constant-region sequence. For DNA samples in which this V gene is not rearranged, the fragment will be absent because it is too large to be efficiently amplified and detected by Southern blotting.

greatly. If the number of cancerous cells falls below 1% or 2% of the total T-cell population, analysis by Southern blot may no longer detect the unique fragment. In this case, a more sensitive technique, PCR, may be used. (With PCR it is possible to amplify, or synthesize multiple copies of, a specific DNA sequence in a sample; primers can hybridize to the two ends of that specific sequence and direct a DNA polymerase to copy it; see Chapter 22 for details.) To detect a portion of the rearranged TCR DNA, amplification using a sequence from the rearranged V region as one primer and a sequence from the β-chain C region as the other primer will yield a rearranged TCR DNA fragment of predicted size in sufficient quantity to be detected by electrophoresis (see red arrow in the diagram). Recently, quantitative PCR methods have been used to follow patients who are in remission in order to make decisions about resuming treatment if the number of cancerous cells, as estimated by these techniques, has risen above a certain level. The presence of the rearranged DNA in the clonal population of T cells gives the clinician a valuable tool for diagnosing lymphoid cell cancer and for monitoring the progress of treatment.

operate during TCR gene rearrangements contribute to a high degree of diversity among T-cell receptors. Table 9-3 and Figure 9-8 compare the generation of diversity among antibody molecules and TCR molecules.

*Combinatorial joining* of variable-region gene segments generates a large number of random gene combinations for all the TCR chains, as it does for the Ig heavy- and light-chain genes. For example, in the mouse, the 79 functional $V_\alpha$ and 38 $J_\alpha$ gene segments can generate $3.0 \times 10^3$ possible VJ combinations for the TCR α chain. Similarly, 21 $V_\beta$, two $D_\beta$, and 11 $J_\beta$ gene segments can give $4.6 \times 10^2$ possible

combinations for the β chain. Although there are fewer TCR $V_\alpha$ and $V_\beta$ gene segments than immunoglobulin $V_H$ and $V_L$ segments, this difference is offset by the greater number of J segments in TCR germ-line DNA. If it is assumed that the antigen-binding specificity of a given T-cell receptor depends on the variable region in both chains, random association of $3 \times 10^3$ $V_\alpha$ combinations with $4.6 \times 10^2$ $V_\beta$ combinations can generate $1.4 \times 10^6$ possible combinations for the αβ T-cell receptor. Additional means to generate diversity in the TCR V genes are described below, making the 1.4 million combinations a minimum estimate.

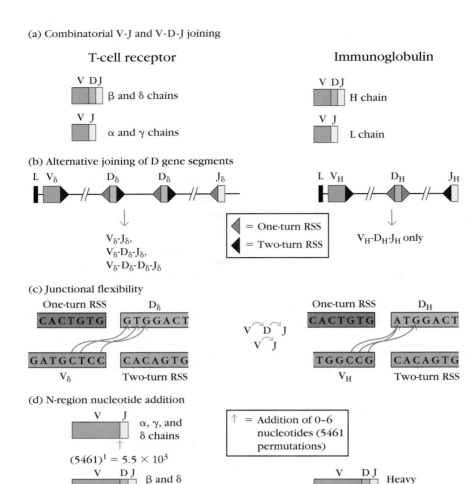

(a) Combinatorial V-J and V-D-J joining

T-cell receptor

V DJ  β and δ chains

V J  α and γ chains

Immunoglobulin

V DJ  H chain

V J  L chain

(b) Alternative joining of D gene segments

L $V_\delta$    $D_\delta$    $D_\delta$    $J_\delta$

$V_\delta$-$J_\delta$,
$V_\delta$-$D_\delta$-$J_\delta$,
$V_\delta$-$D_\delta$-$D_\delta$-$J_\delta$

◄ = One-turn RSS
◀ = Two-turn RSS

L $V_H$    $D_H$    $J_H$

$V_H$-$D_H$-$J_H$ only

(c) Junctional flexibility

One-turn RSS    $D_\delta$
CACTGTG    GTGGACT

GATGCTCC    CACAGTG
$V_\delta$    Two-turn RSS

V ⁀ D ⁀ J
V ⁀ J

One-turn RSS    $D_H$
CACTGTG    ATGGACT

TGGCCG    CACAGTG
$V_H$    Two-turn RSS

(d) N-region nucleotide addition

V    J    α, γ, and
         δ chains
$(5461)^1 = 5.5 \times 10^3$

V    D J    β and δ
           chains
$(5461)^2 = 3.0 \times 10^7$

V    D D J    δ chain

↑ = Addition of 0–6
nucleotides (5461
permutations)

V    D J    Heavy
           chain
$(5461)^2 = 3.0 \times 10^7$

**FIGURE 9-8 Comparison of mechanisms for generating diversity in TCR genes and immunoglobulin genes.** In addition to the mechanisms shown, P-region nucleotide addition occurs in both TCR and Ig genes, and somatic mutation occurs in Ig genes. Combinatorial association of the expressed chains generates additional diversity among both TCR and Ig molecules.

As illustrated in Figure 9-8b, the location of one-turn (12-bp) and two-turn (23-bp) recombination signal sequences (RSSs) in TCR β- and δ-chain DNA differs from that in Ig heavy-chain DNA. Because of the arrangement of the RSSs in TCR germ-line DNA, alternative joining of D gene segments can occur while the one-turn/two-turn joining rule is observed. Thus, it is possible for a $V_\beta$ gene segment to join directly with a $J_\beta$ or a $D_\beta$ gene segment, generating a $(VJ)_\beta$ or $(VDJ)_\beta$ unit.

Alternative joining of δ-chain gene segments generates similar units; in addition, one $D_\delta$ can join with another, yielding $(VDDJ)_\delta$ and, in humans, $(VDDDJ)_\delta$. This mechanism, which cannot operate in Ig heavy-chain DNA, generates considerable additional diversity in TCR genes.

The joining of gene segments during TCR-gene rearrangement exhibits **junctional flexibility.** As with the Ig genes, this flexibility can generate many nonproductive

rearrangements, but it also increases diversity by encoding several alternative amino acids at each junction (see Figure 9-8c). In both Ig and TCR genes, nucleotides may be added at the junctions between some gene segments during rearrangement (see Figure 5-13). Variation in endonuclease cleavage leads to the addition of further nucleotides that are palindromic. Such **P-region nucleotide addition** can occur in the genes encoding all the TCR and Ig chains. Addition of **N-region nucleotides,** catalyzed by a terminal deoxynucleotidyl transferase, generates additional junctional diversity. Whereas the addition of N-region nucleotides in immunoglobulins occurs only in the Ig heavy-chain genes, it occurs in the genes encoding all the TCR chains. As many as six nucleotides can be added by this mechanism at each junction, generating up to 5461 possible combinations, assuming random selection of nucleotides (see Figure 9-8d). Some of these combinations, however, lead to nonproductive rearrangements by inserting in-frame stop codons that prematurely terminate the TCR chain or by substituting amino acids that render the product nonfunctional. Although each junctional region in a TCR gene encodes only 10 to 20 amino acids, enormous diversity can be generated in these regions. Estimates suggest that the combined effects of P- and N-region nucleotide addition and joining flexibility can generate as many as $10^{13}$ possible amino acid sequences in the TCR junctional regions alone.

The mechanism by which TCR diversity is generated must allow the receptor to recognize a very large number of different processed antigens while restricting its MHC-recognition repertoire to a much smaller number of self-MHC molecules. TCR DNA has far fewer V gene segments than Ig DNA (see Table 9-3). Researchers have postulated that the smaller number of V gene segments in TCR DNA were selected to encode a limited number of CDR1 and CDR2 regions with affinity for regions of the $\alpha$ helices of MHC molecules. This attractive idea is unlikely to be correct in light of recent data on the structure of the TCR-peptide-MHC complex showing contact between peptide and CDR1 as well as CDR3. Therefore, the TCR residues that bind to peptide versus those that bind MHC are not confined solely to the highly variable CDR3 region.

In contrast to the limited diversity of CDR1 and CDR2, the CDR3 of the TCR has even greater diversity than that seen in immunoglobulins. Diversity in CDR3 is generated by junctional diversity in the joining of V, D, and J segments, joining of multiple D gene segments, and the introduction of P and N nucleotides at the V-D-J and V-J junctions (see Figure 9-7).

Unlike the Ig genes, the TCR genes do not appear to undergo somatic mutation. That is, the functional TCR genes generated by gene rearrangements during T-cell maturation in the thymus have the same sequences as those found in the mature peripheral T-cell population. The absence of somatic mutation in T cells ensures that T-cell specificity does not change after thymic selection and therefore reduces the possibility that random mutation might generate a self-reactive T cell. Although a few experiments have provided evidence for somatic mutation of receptor genes in T cells in the germinal center, this appears to be the exception and not the rule.

## T-Cell Receptor Complex: TCR-CD3

As explained in Chapter 4, membrane-bound immunoglobulin on B cells associates with another membrane protein, the Ig-$\alpha$/Ig-$\beta$ heterodimer, to form the B-cell antigen receptor (see Figure 4-22). Similarly, the T-cell receptor associates with **CD3,** forming the TCR-CD3 membrane complex. In both cases, the accessory molecule participates in signal transduction *after* interaction of a B or T cell with antigen; it does not influence interaction with antigen.

The first evidence suggesting that the T-cell receptor is associated with another membrane molecule came from experiments in which fluorescent antibody to the receptor was shown to cause aggregation of another membrane protein, designated CD3. Later experiments by J. P. Allison and L. Lanier using cross-linking reagents demonstrated that the two chains must be within 12 Å. Subsequent experiments demonstrated not only that CD3 is closely associated with the $\alpha\beta$ heterodimer but also that its expression is required for membrane expression of $\alpha\beta$ and $\gamma\delta$ T-cell receptors—each heterodimer forms a complex with CD3 on the T-cell membrane. Loss of the genes encoding either CD3 or the TCR chains results in loss of the entire molecular complex from the membrane.

CD3 is a complex of five invariant polypeptide chains that associate to form three dimers: a heterodimer of gamma and epsilon chains ($\gamma\epsilon$), a heterodimer of delta and epsilon chains ($\delta\epsilon$), and a homodimer of two zeta chains ($\zeta\zeta$) or a heterodimer of zeta and eta chains ($\zeta\eta$) (Figure 9-9a). The $\zeta$ and $\eta$ chains are encoded by the same gene but differ in their carboxyl-terminal ends because of differences in RNA splicing of the primary transcript. About 90% of the CD3 complexes examined to date incorporate the ($\zeta\zeta$) homodimer; the remainder have the ($\zeta\eta$) heterodimer. The T-cell receptor complex can thus be envisioned as four dimers: the $\alpha\beta$ or $\gamma\delta$ TCR heterodimer determines the ligand-binding specificity, whereas the CD3 dimers ($\gamma\epsilon$, $\delta\epsilon$, and $\zeta\zeta$ or $\zeta\eta$) are required for membrane expression of the T-cell receptor and for signal transduction.

The $\gamma$, $\delta$, and $\epsilon$ chains of CD3 are members of the immunoglobulin superfamily, each containing an immunoglobulin-like extracellular domain followed by a transmembrane region and a cytoplasmic domain of more than 40 amino acids. The $\zeta$ chain has a distinctly different structure, with a very short external region of only nine amino acids, a transmembrane region, and a long cytoplasmic tail containing 113 amino acids. The transmembrane regions of all the CD3 polypeptide chains contain a negatively charged amino acid residue (aspartic or glutamic acid) that interacts with one or two positively charged amino acids in the transmembrane region of each TCR chain. The components of the CD3 complex and the means by which they interact with the $\alpha\beta$ TCR are shown in Figure 9-9b.

The cytoplasmic tails of the CD3 chains contain a motif called the **immunoreceptor tyrosine-based activation motif (ITAM).** ITAMs are found in a number of other receptors, including the Ig-$\alpha$/Ig-$\beta$ heterodimer of the B-cell receptor complex and the Fc receptors for IgE and IgG. The ITAM sites have been shown to interact with tyrosine kinases and to play an important role in signal transduction. In CD3, the $\gamma$, $\delta$, and $\epsilon$ chains each contain a single ITAM, whereas the $\zeta$ and $\eta$ chains contain three copies (see Figure 9-9a). The function of CD3 in signal transduction is described more fully in Chapter 10.

(a)

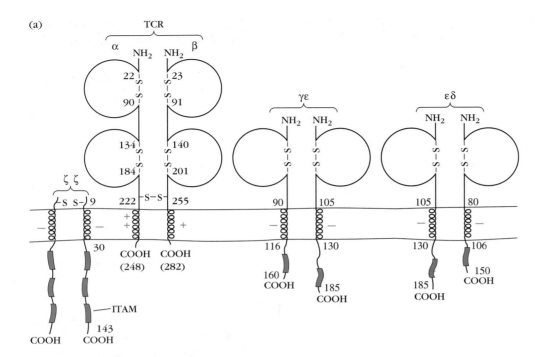

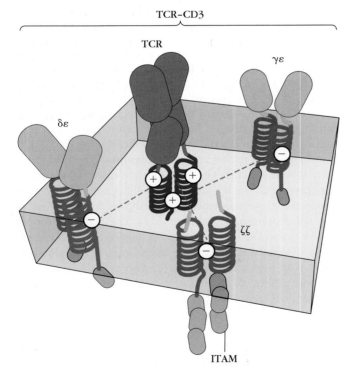

**FIGURE 9-9 Schematic diagram of the TCR-CD3 complex, which constitutes the T-cell antigen-binding receptor.** (a) Components of the CD3 complex include the ζζ homodimer (alternately, a ζη heterodimer) plus γε and δε heterodimers. The external domains of the γ, δ, and ε chains of CD3 consist of immunoglobulin folds, which facilitates their interaction with the T-cell receptor and each other. The long cytoplasmic tails of the CD3 chains contain a common sequence, the immunoreceptor tyrosine-based activation motif (ITAM), which functions in signal transduction; these sequences are shown as blue boxes. (b) Ionic interactions also may occur between the oppositely charged transmembrane regions in the TCR and CD3 chains. Proposed interactions between the CD3 components and the αβ TCR are shown. *[Adapted from M. E. Call and K. W. Wucherpfennig, 2004, Molecular Immunology **40**:1295.]*

# T-Cell Accessory Membrane Molecules

Although recognition of antigen-MHC complexes is mediated solely by the TCR-CD3 complex, various other membrane molecules play important accessory roles in antigen recognition and T-cell activation (Table 9-4). Some of these molecules strengthen the interaction between T cells and antigen-presenting cells or target cells, some act in signal transduction, and some do both.

## CD4 and CD8 coreceptors bind to conserved regions of MHC class II or I molecules

T cells can be subdivided into two populations according to their expression of CD4 or CD8 membrane molecules. As described in preceding chapters, CD4$^+$ T cells recognize

| TABLE 9-4 | Selected T-cell accessory molecules | | | | |
| --- | --- | --- | --- | --- |
| | | FUNCTION | | |
| Name | Ligand | Adhesion | Signal transduction | Member of Ig superfamily |
| CD4 | Class II MHC | + | + | + |
| CD8 | Class I MHC | + | + | + |
| CD2 (LFA-2) | CD58 (LFA-3) | + | + | + |
| LFA-1 (CD11a/CD18) | ICAM-1 (CD54) | + | ? | +/(−) |
| CD28 | B7 | ? | + | + |
| CTLA-4 | B7 | ? | + | − |
| CD45R | CD22 | + | + | + |
| CD5 | CD72 | ? | + | − |

antigen that is combined with class II MHC molecules and function largely as helper cells, whereas CD8$^+$ T cells recognize antigen that is combined with class I MHC molecules and function largely as cytotoxic cells. CD4 is a 55-kDa monomeric membrane glycoprotein that contains four extracellular immunoglobulin-like domains ($D_1$–$D_4$), a hydrophobic transmembrane region, and a long cytoplasmic tail (Figure 9-10) containing three serine residues that can be phosphorylated. CD8 generally takes the form of a disulfide-linked $\alpha\beta$ heterodimer or $\alpha\alpha$ homodimer. Both the $\alpha$ and $\beta$ chains of CD8 are small glycoproteins of approximately 30 to 38 kDa. Each chain consists of a single extracellular immunoglobulin-like domain, a stalk region, a hydrophobic transmembrane region, and a cytoplasmic tail (Figure 9-10) containing 25 to 27 residues, several of which can be phosphorylated.

CD4 and CD8 are classified as *coreceptors* based on their abilities to recognize the peptide-MHC complex and their roles in signal transduction. The extracellular domains of CD4 and CD8 bind to the conserved regions of MHC molecules on antigen-presenting cells (APCs) or target cells. Crystallographic studies of a complex composed of the class I MHC molecule HLA-A2, an antigenic peptide, and a CD8 $\alpha\alpha$ homodimer indicate that CD8 binds to class I molecules by contacting the MHC class I $\alpha2$ and $\alpha3$ domains as well as having some contact with $\beta_2$-microglobulin (Figure 9-11a). The orientation of the class I $\alpha3$ domain changes slightly on binding to CD8. This structure is consistent with a single MHC molecule binding to CD8; no evidence for the possibility of multimeric class I to CD8 complexes has been observed. Similar structural data document the mode by which CD4 binds to the class II molecule. The interaction between CD4 and MHC II involves contact of the membrane-distal domain of CD4 with a hydrophobic pocket formed by residues from the $\alpha2$ and $\beta2$ domains of MHC II (Figure 9-11b). CD4 facilitates signal transduction and T-cell activation of cells recognizing class II–peptide complexes.

Whether there are differences between the roles played by the CD4 and CD8 coreceptors remains open to speculation. Despite the similarities in structure, recall that the nature of the binding of a peptide to class I and class II molecules differs in that a class I molecule has a closed groove that binds a short peptide with a higher degree of specificity. Recent data indicate that the angle at which the TCR approaches the peptide MHC complex differs between class I and II. The differences in roles played by the CD4 and CD8 coreceptors may be due to these differences in binding requirements. As will be explained in Chapter 10, binding of the CD4 and

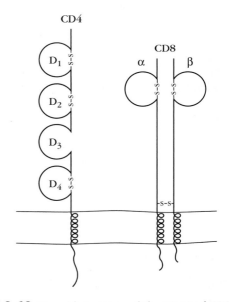

**FIGURE 9-10 General structure of the CD4 and CD8 coreceptors; the Ig like domains are shown as circles.** CD8 takes the form of an $\alpha\beta$ heterodimer or an $\alpha\alpha$ homodimer. The monomeric CD4 molecule contains four Ig-fold domains; each chain in the CD8 molecule contains one.

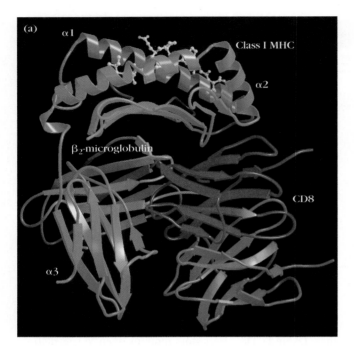

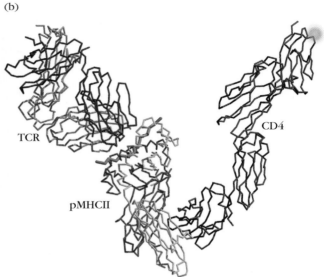

**FIGURE 9-11 Interactions of coreceptors with TCR and MHC molecules.** (a) Ribbon diagram showing three-dimensional structure of an HLA-A2 MHC class I molecule bound to a CD8 $\alpha\alpha$ homodimer. The HLA-A2 heavy chain is shown in green, $\beta_2$-microglobulin in gold, the CD8 $\alpha1$ in red, the CD8 $\alpha2$ in blue, and the bound peptide in white. A flexible loop of the $\alpha3$ domain (residues 223–229) is in contact with the two CD8 subunits. In this model, the right side of CD8 would be anchored in the T-cell membrane, and the lower left end of the class I MHC molecule (the $\alpha3$ domain) is attached to the surface of the target cell. (b) Interaction of extracellular domain of CD4 with the class II MHC peptide complex (pMHCII). *[Part a from Gao et al., 1997, Nature, **387**:630; part b from Wang et al., 2001, Proceedings of the National Academy of Sciences U.S.A., **98** (19): 10799.]*

CD8 molecules serves to transmit stimulatory signals to the T cells; the signal transduction properties of both CD4 and CD8 are mediated through their cytoplasmic domains.

## Affinity of TCR for peptide-MHC complexes is enhanced by coreceptors

The affinity of T-cell receptors for peptide-MHC complexes is low to moderate, with $K_d$ values ranging from $10^{-4}$ to $10^{-7}$ M. This level of affinity is weak compared with antigen-antibody interactions, which generally have $K_d$ values ranging from $10^{-6}$ to $10^{-10}$ M (Figure 9-12a). However, T-cell interactions do not depend solely on binding by the TCR; *cell-adhesion molecules* strengthen the bond between a T cell and an antigen-presenting cell or a target cell. Several accessory membrane molecules, including CD2, LFA-1, CD28, and CD45R, bind independently to other ligands on antigen-presenting cells or target cells (see Table 9-4 and Figure 9-12b). Once cell-to-cell contact has been made by the adhesion molecules, the T-cell receptor may scan the membrane for peptide-MHC complexes. During activation of a T cell by a particular peptide-MHC complex, there is a transient increase in the membrane expression of cell adhesion molecules, causing closer contact between the interacting cells, which allows cytokines or cytotoxic substances to be transferred more effectively. Soon after activation, the degree of adhesion declines and the T cell detaches from the antigen-presenting cell or target cell. Like CD4 and CD8, some of these other molecules also function as signal transducers. The importance of their role is shown by the ability of monoclonal antibodies specific for the binding sites of the cell adhesion molecules to block T-cell activation.

Recent data on the interaction between the extracellular portions of CD4 and the peptide–class II complex indicate that there is very weak affinity between them, and in some experiments no binding could be demonstrated. This finding is not consistent with the body of evidence showing the proximity of CD4 and class II molecules (see Figure 9-11). A satisfying explanation for this difference is shown in Figure 9-13, where a multistep interaction between CD4 and the TCR complex is proposed. The diagram indicates that the interaction between the TCR complex and the CD4 molecule is stabilized by dual interactions between the signal transduction molecule p56[lck] and both CD4 and the $\zeta\zeta$ homodimer of the CD3 complex. It further suggests that recruitment of molecules involved in signal transduction may be a major role for CD4 on the T-cell membrane.

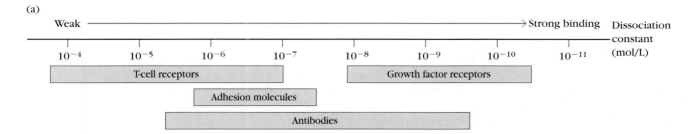

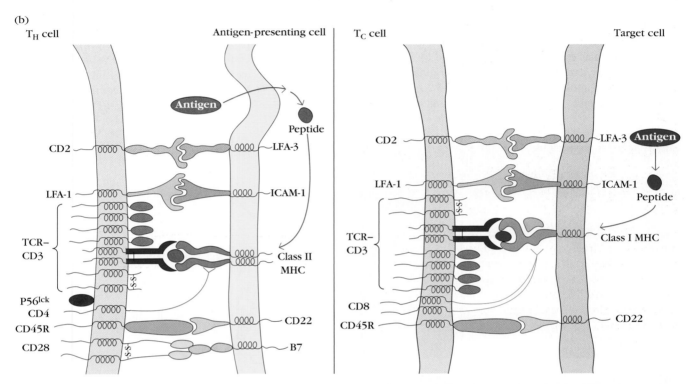

**FIGURE 9-12 Role of coreceptors in TCR binding affinity.** (a) Dissociation constants for various biologic systems. (b) Schematic diagram of the interactions between the T-cell receptor and the peptide-MHC complex and of various accessory molecules with their ligands on an antigen-presenting cell *(left)* or target cell *(right)*. Bind-

ing of the coreceptors CD4 and CD8 and the other accessory molecules to their ligands strengthens the bond between the interacting cells and/or facilitates the signal transduction that leads to activation of the T cell.

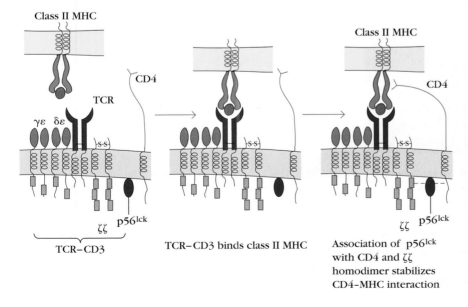

**FIGURE 9-13 The role of CD4 in stabilizing the interaction between T cell and APC is significantly enhanced by a two-point contact involving not only the extracellular portions of CD4 and class II MHC molecules (see Figure 9-11) but also the intracellular interaction of CD4 and the associated signal transduction molecule p56<sup>lck</sup> with the zeta chain of the CD3 complex.** [*Adapted from Davis et al., 2003, Nature Immunology **4:**217.*]

# Three-Dimensional Structures of TCR-Peptide-MHC Complexes

The interaction between the T-cell receptor and an antigen bound to an MHC molecule is central to both humoral and cell-mediated responses. The molecular elements of this interaction have now been described in detail by x-ray crystallography for TCR molecules binding to peptides bound to MHC class I and class II molecules. The three-dimensional structure of the trimolecular complex was first reported by Dave Garboczi, Don Wiley, and their colleagues. This original structure includes TCR α and β chains and an HLA-A2 molecule to which an antigenic peptide from the human retrovirus

HTLV-1 is bound. Comparisons of the TCR complexed with either class I or class II molecules suggest that there are differences in how the TCR contacts the MHC-peptide complex. These TCR structures differ significantly from those of the γδ receptor, which is relatively simple and may be bound to an antigen that does not require processing or to an MHC molecule independent of the bound peptide (see Figure 9-4).

From x-ray analysis, the αβ TCR-peptide-MHC complex consists of a single TCR molecule bound to a single MHC molecule and its peptide. The TCR contacts the MHC molecule through the TCR variable domains (Figure 9-14a, b). Viewing the MHC molecule with its bound peptide from above, we can see that the TCR is situated across it diagonally relative to the long axis of the peptide (Figure 9-14c). The CDR3 loops

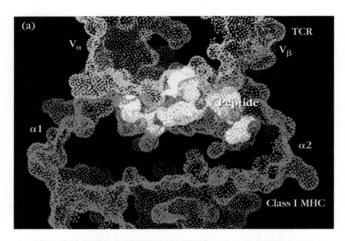

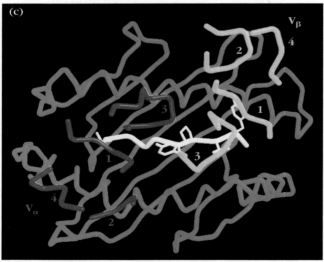

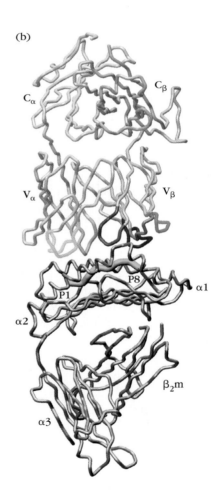

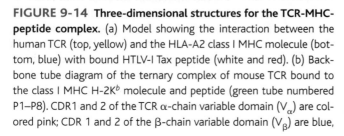

**FIGURE 9-14 Three-dimensional structures for the TCR-MHC-peptide complex.** (a) Model showing the interaction between the human TCR (top, yellow) and the HLA-A2 class I MHC molecule (bottom, blue) with bound HTLV-I Tax peptide (white and red). (b) Backbone tube diagram of the ternary complex of mouse TCR bound to the class I MHC H-2K$^b$ molecule and peptide (green tube numbered P1–P8). CDR1 and 2 of the TCR α-chain variable domain (V$_\alpha$) are colored pink; CDR 1 and 2 of the β-chain variable domain (V$_\beta$) are blue,

and the CDR3s of both chains are green. The HV4 of the β chain is orange. (c) The MHC molecule viewed from above, that is, from the top of part a, with the hypervariable loops (1–4) of the human TCR α (red) and β (yellow) variable chains superimposed on the Tax peptide (white) and the α1 and α2 domains of the HLA-A2 MHC class I molecule (blue). *[Parts a and c from D. N. Garboczi et al., 1996, Nature* **384:**134–41, courtesy of D. C. Wiley, Harvard University; part b from C. Garcia et al., 1996, Science **274:**209, courtesy of C. Garcia, Stanford University.]

of the TCR α and β chains meet in the center of the peptide, and the CDR1 loop of the TCR α chain is at the N terminus of the peptide, whereas CDR1 of the β chain is at the C terminus of the peptide. The CDR2 loops are in contact with the MHC molecule; CDR2α is over the α2 domain alpha helix and CDR2β over the α1 domain alpha helix (Figure 9-14c).

As predicted from data for immunoglobulins, the recognition of the peptide-MHC complex occurs through the variable loops in the TCR structure. CDR1 and CDR3 from both the TCR α and the TCR β chain contact the peptide and a large area of the MHC molecule. The peptide is buried more deeply in the MHC molecule than it is in the TCR (see Figure 9-14a, b), and the TCR molecule fits across the MHC molecule, contacting it through a flat surface of the TCR at the "high points" on the MHC molecule. The fact that the CDR1 region contacts both peptide and MHC indicates that regions other than CDR3 are involved in peptide binding.

## TCRs interact differently with class I and class II molecules

Can the conclusions drawn from the three-dimensional structure of TCR–peptide–class I complexes be extrapolated

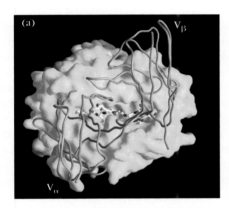

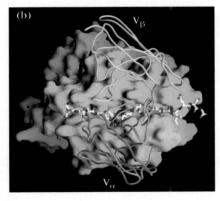

**FIGURE 9-15 Comparison of the interactions between αβ TCR and (a) class I MHC–peptide, and (b) class II MHC–peptide.** The TCR (wire diagram) is red in (a), blue-green in (b); the MHC molecules are shown as surface models; peptide is shown as ball and stick. *[From Reinherz et al., 1999,* Science **286**:1913.]*

to interactions of TCR with class II complexes? Ellis Reinherz and his colleagues resolved this question by analysis of a TCR molecule in complex with a mouse class II molecule and its specific antigen. Although the structures of the peptide-binding regions in class I and class II molecules are similar, in Chapter 8 we saw that there are differences in how these molecules accommodate bound peptide (see Figures 8-7a, b). A comparison of the interactions of a TCR with class I MHC–peptide and class II–peptide reveals a significant difference in the angle at which the TCR molecule sits on the MHC complexes (Figure 9-15). Also notable is the greater number of contact residues between TCR and class II MHC, which is consistent with the known higher affinity of interaction. However, it remains to be seen whether the evident difference in the number of contact points will exist in all class I and II structures.

## Alloreactivity of T Cells

The preceding sections have focused on the role of MHC molecules in the presentation of antigen to T cells and the interactions of TCRs with peptide-MHC complexes. However, as noted in Chapter 8, MHC molecules were first identified because of their role in rejection of foreign tissue. Graft rejection reactions result from the recognition by T cells of MHC molecules, which function as **histocompatibility antigens.** Because of the extreme polymorphism of the MHC, most individuals of the same species have unique sets of MHC molecules, or histocompatibility antigens, and are considered to be **allogeneic,** a term used to describe genetically different individuals of the same species (see Chapter 17). Therefore, T cells respond strongly to **allografts** (grafts from members of the same species), and MHC molecules are considered *alloantigens.* Generally, CD4⁺ T cells are alloreactive to class II alloantigens, and CD8⁺ T cells respond to class I alloantigens.

The alloreactivity of T cells is puzzling for two reasons. First, the ability of T cells to respond to allogeneic histocompatibility antigens alone appears to contradict all the evidence indicating that T cells can respond only to foreign antigen plus *self*-MHC molecules. In responding to allogeneic grafts, however, T cells recognize a *foreign* (nonself) MHC molecule. A second problem posed by the T-cell response to allogeneic MHC molecules is that the frequency of alloreactive T cells is quite high; it has been estimated that 1% to 5% of all T cells are reactive to a given alloantigen, which is higher than the normal frequency of T cells reactive with any particular foreign antigenic peptide plus self-MHC molecule. This high frequency of alloreactive T cells appears to contradict the basic tenet of clonal selection. If one T cell in 20 reacts with a given alloantigen and if one assumes there are on the order of 100 distinct H-2 haplotypes in mice, then there are not enough distinct T-cell specificities to cover all the unique H-2 alloantigens, let alone foreign antigens displayed by self-MHC molecules.

One possible and biologically satisfying explanation for the high frequency of alloreactive T cells is that a particular T-cell receptor specific for a foreign antigenic peptide plus a self-MHC molecule can also cross-react with certain allogeneic MHC molecules. In other words, if an allogeneic MHC molecule plus an allogeneic peptide structurally resembles a processed foreign peptide plus self-MHC molecule, the same T-cell receptor may recognize both peptide-MHC complexes. Since allogeneic cells express on the order of $10^5$ class I MHC molecules per cell, T cells bearing low-affinity cross-reactive receptors might be able to bind by virtue of the high density of membrane alloantigen. Foreign antigen, on the other hand, would be sparsely displayed on the membrane of an antigen-presenting cell or altered self cell associated with class I or class II MHC molecules, limiting responsiveness to only those T cells bearing high-affinity receptors.

Information relevant to mechanisms of alloreactive TCR binding was gained by J-B. Reiser and colleagues, who determined the structure of a mouse TCR complexed with an allogeneic class I molecule containing a bound octapeptide. This analysis revealed a structure similar to structures reported for TCR bound to class I self-MHC complexes, leading the authors to conclude that allogeneic recognition does not differ from recognition of self-MHC molecules.

The actual recognition of foreign MHC may be direct—T cells recognize allogeneic MHC molecules on foreign cells as if they were self-MHC molecules—or indirect, meaning T cells recognize foreign MHC after it is processed and fragments of it are presented bound to self MHC (Figure 9-16). The two options could give rise to a large number of combinations recognizable by the host T cells. In the indirect model, the absence of negative selection for the peptides contained in the foreign MHC molecules can contribute to the high frequency of alloreactive T cells. This condition, coupled with the differences in the structure of the exposed portions of the allogeneic MHC molecule, may account for the phenomenon of alloreactivity. An explanation for the large number of alloreactive cells can be found in the large number of potential antigens provided by the foreign MHC molecule plus the possible peptide antigens already bound by foreign MHC molecules at the time that they are processed for presentation.

Alloreactivity may also contribute insight to the general question of cross-reactivity, or "promiscuity," of the TCR, with consequences for our understanding of autoreactivity and the etiology of autoimmune disease. In an elegant study designed to measure the degree of TCR crossreactivity, Garcia and colleagues developed a decapeptide library in which substitutions of all 20 amino acids were introduced at each of the 10 amino acid residue positions. This combinatorial library was screened using a TCR known to recognize a peptide from myelin basic protein (MBP) in the context of a mouse class II MHC molecule. The three-dimensional structure of the complex had been resolved, and the TCR contact residues had previously been identified. These researchers found that peptides from the combinatorial library, which represented all possible substitutions at all locations in the peptide, did not react at all with the TCR when amino acids were substituted in regions of the peptide involved in anchoring to the MHC molecules. Furthermore, whereas a surprising number of substitutions in the other positions in the peptide resulted in some degree of recognition, it was uniformly much less than that seen for the original antigen. Substitutions at positions known to contact the TCR binding site generally had the most significant negative effect on affinity, leading the authors to conclude that the TCR is likely to bind with high affinity only to unique MHC-peptide combinations while binding with much less affinity to all variants, indicating that the danger of the TCR interacting promiscuously is quite low. As a consequence, TCR cross-reactivity is probably of biological importance only in exceptional circumstances, as in the case of self-reactivity leading to autoimmune disease.

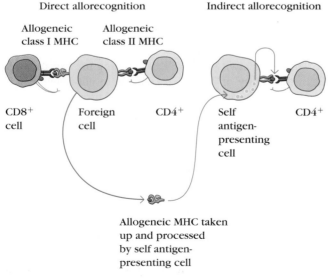

**FIGURE 9-16  Alloreactivity of T cells may result from direct or indirect recognition of alloantigens.** In the direct model, the nonself MHC molecule is recognized in its intact native form. Indirect recognition would result from peptides derived from the processed MHC molecule presented in the context of self-MHC molecules. *[Adapted from A. Whitelegg and L. D. Barber, 2004,* Tissue Antigens *63:101.]*

## SUMMARY

■ Most T-cell receptors, unlike antibodies, do not react with soluble antigen but rather with processed antigen bound to a self-MHC molecule.

■ T-cell receptors, first isolated by means of clonotypic monoclonal antibodies, are heterodimers consisting of an α and β chain or a γ and δ chain; certain γδ receptors recognize antigens not processed and presented with the MHC.

■ The membrane-bound T-cell receptor chains are organized into variable and constant domains. TCR domains are similar to those of immunoglobulins, and the V region has hypervariable regions.

■ TCR germ-line DNA is organized into multigene families corresponding to the α, β, γ, and δ chains. Each family contains multiple gene segments.

- The mechanisms that generate TCR diversity are generally similar to those that generate antibody diversity, although somatic mutation does not occur in TCR genes, as it does in immunoglobulin genes.

- The T-cell receptor is closely associated with CD3, a complex of polypeptide chains involved in signal transduction; CD3 is required for surface expression of the TCR.

- T cells express membrane molecules, including CD4, CD8, CD2, LFA-1, CD28, and CD45R, that play accessory roles in T-cell function or signal transduction.

- Formation of the ternary complex TCR-antigen-MHC requires binding of a peptide to the MHC molecule, then binding of the MHC–peptide complex by the T-cell receptor.

- Interactions between TCR and MHC class I/peptide differ from those with MHC class II/peptide in the contact points between the TCR and MHC molecules.

- The γδ T-cell receptor is distinguished from the αβ T-cell receptor by its ability to bind native antigens and by differences in the orientation of the variable and constant regions.

- In addition to reaction with self MHC plus foreign antigens, T cells respond to foreign MHC molecules, a reaction that leads to rejection of allogeneic grafts.

## References

Adams, E. J., Y-H. Chien, and K. C. Garcia. 2005. Structure of a γδ T cell receptor in complex with the nonclassical MHC T22. *Science* **308**:227.

Allison, T. J., et al. 2001. Structure of a human γδ T-cell antigen receptor. *Nature* **411**:820.

Baum, T-P., et al. 2004. IMGT/Geneinfo: Enhancing V(D)J recombination database accessibility. *Nucleic Acids Research* **32**:D51.

Call, M. E., and K. W. Wucherpfennig. 2004. Molecular mechanisms for the assembly of the T cell receptor-CD3 complex. *Molecular Immunology* **40**:1295.

Chen, Z. W., and N. L. Letvin. 2003. Vγ2Vδ2+ T cells and antimicrobial immune responses. *Microbes and Infection* **5**:491.

Davis, S. J., et al. 2003. The nature of molecular recognition by T cells. *Nature Immunology* **4**:217.

Gao, G. F., et al. 1997. Crystal structure of the complex between human CD8αα and HLA-A2. *Nature* **387**:630.

Garboczi, D. N., et al. 1996. Structure of the complex between human T-cell receptor, viral peptide, and HLA-A2. *Nature* **384**:134.

Garcia, K. C., et al. 1996. An αβ T-cell receptor structure at 2.5 Å and its orientation in the TCR-MHC complex. *Science* **274**:209.

Garcia, K. C., et al. 1998. T-cell receptor–peptide–MHC interactions: biological lessons from structural studies. *Current Opinions in Biotechnology* **9**:338.

Hodges, E., et al. 2003. Diagnostic role of tests for T cells receptor (TCR) genes. *Journal of Clinical Pathology* **56**:1.

Kabelitz, D., et al. 2000. Antigen recognition by γδ T lymphocytes. *International Archives of Allergy and Immunology* **122**:1.

Krogsgaard, M., and M. M. Davis. 2005. How T cells "see" antigen. *Nature Immunology.* **6**:239.

Maynard, J., et al. 2005. Structure of an autoimmune T-cell receptor complexed with class II peptide-MHC: insights into MHC bias and antigen specificity. *Immunity* **22**:81.

Reinherz, E., et al. 1999. The crystal structure of a T-cell receptor in complex with peptide and MHC class II. *Science* **286**:1913.

Reiser, J-B., et al. 2000. Crystal structure of a T-cell receptor bound to an allogeneic MHC molecule. *Nature Immunology* **1**:291.

Whitelegg, A., and L. D. Barber. 2003. The structural basis of T-cell recognition. *Tissue Antigens* **63**:101.

Xiong, Y., et al. 2001. T-cell receptor binding to a pMHCII ligand is kinetically distinct from and independent of CD4. *Journal of Biological Chemistry* **276**:5659.

Zinkernagel, R. M., and P. C. Doherty. 1974. Immunological surveillance against altered self-components by sensitized T lymphocytes in lymphocytic choriomeningitis. *Nature* **251**:547.

 ## Useful Web Sites

**http://imgt.cines.fr**

A comprehensive database of genetic information on TCRs, MHC molecules, and immunoglobulins from the International Immunogenetics Database, University of Montpellier, France.

**http://www.bioscience.org/knockout/tcrab.htm**

This location presents a brief summary of the effects of TCR knockouts.

**http://mif.dfci.harvard.edu/Tools/rankpep.html**

This server predicts which peptides will bind to class I and class II MHC molecules based on protein sequences or sequence alignments using Position Specific Scoring Matrices (PSSMs).

 ## Study Questions

**CLINICAL FOCUS QUESTION**   A patient presents with an enlarged lymph node, and a T-cell lymphoma is suspected. However, DNA sampled from biopsied tissue shows no evidence of a predominant gene rearrangement when probed with α and β TCR genes. What should be done next to rule out lymphocyte malignancy?

1. Indicate whether each of the following statements is true or false. If you think a statement is false, explain why.

   a. Monoclonal antibody specific for CD4 will coprecipitate the T-cell receptor along with CD4.

   b. Subtractive hybridization can be used to enrich for mRNA that is present in one cell type but absent in another cell type within the same species.

   c. Clonotypic monoclonal antibody was used to isolate the T-cell receptor.

   d. The T cell uses the same set of V, D, and J gene segments as the B cell but uses different C gene segments.

   e. The αβ TCR is bivalent and has two antigen-binding sites.

   f. Each αβ T cell expresses only one β-chain and one α-chain allele.

g. Mechanisms for generation of diversity of T-cell receptors are identical to those used by immunoglobulins.

h. The Ig-α/Ig-β heterodimer and CD3 serve analogous functions in the B-cell receptor and T-cell receptor, respectively.

2. What led Zinkernagel and Doherty to conclude that T-cell receptor recognition requires both antigen and MHC molecules?

3. Draw the basic structure of the αβ T-cell receptor and compare it with the basic structure of membrane-bound immunoglobulin.

4. Several membrane molecules, in addition to the T-cell receptor, are involved in antigen recognition and T-cell activation. Describe the properties and distinct functions of the following T-cell membrane molecules: (a) CD3, (b) CD4 and CD8, and (c) CD2.

5. Indicate whether each of the properties listed below applies to the T-cell receptor (TCR), B-cell immunoglobulin (Ig), or both (TCR/Ig).

a. _____ Is associated with CD3

b. _____ Is monovalent

c. _____ Exists in membrane-bound and secreted forms

d. _____ Contains domains with the immunoglobulin-fold structure

e. _____ Is MHC restricted

f. _____ Exhibits diversity generated by imprecise joining of gene segments

g. _____ Exhibits diversity generated by somatic mutation

6. A major obstacle to identifying and cloning TCR genes is the low level of TCR mRNA in T cells.

a. To overcome this obstacle, Hedrick and Davis made three important assumptions that proved to be correct. Describe each assumption and how it facilitated identification of the genes that encode the T-cell receptor.

b. Suppose instead that Hedrick and Davis wanted to identify the genes that encode IL-4. What changes in the three assumptions should they make?

7. Hedrick and Davis used the technique of subtractive hybridization to isolate cDNA clones that encode the T-cell receptor. You wish to use this technique to isolate cDNA clones that encode several gene products and have available clones of various cell types to use as the source of cDNA or mRNA for hybridization. For each gene product listed in the left column of the table below, select the most appropriate.

| Gene product | cDNA source | mRNA source |
|---|---|---|
| IL-2 | | |
| CD8 | | |
| J chain | | |
| IL-1 | | |
| CD3 | | |

cDNA and mRNA source clones are from the following cell types: $T_H1$ cell line (A); $T_H2$ cell line (B); $T_C$ cell line (C); macrophage (D); IgA-secreting myeloma cell (E); IgG-secreting myeloma cell (F); myeloid progenitor cell (G); and B-cell line (H). More than one cell type may be correct in some cases.

8. Mice from different inbred strains listed in the left column of the accompanying table were infected with LCM virus. Spleen cells derived from these LCM-infected mice were then tested for their ability to lyse LCM-infected $^{51}$Cr-labeled target cells from the strains listed across the top of the table. Indicate with (+) or (−) whether you would expect to see $^{51}$Cr released from the labeled target cells.

| Source of spleen cells from LCM-infected mice | Release of $^{51}$Cr from LCM-infected target cells | | | |
|---|---|---|---|---|
| | B10.D2 (H-$2^d$) | B10 (H-$2^b$) | B10.BR (H-$2^k$) | (BALB/c × B10) F$_1$ (H-$2^{b/d}$) |
| B10.D2 (H-$2^d$) | | | | |
| B10 (H-$2^b$) | | | | |
| BALB/c (H-$2^d$) | | | | |
| BALB/b (H-$2^b$) | | | | |

9. The γδ T-cell receptor differs from the αβ T-cell receptor in both structural and functional parameters. Describe how they are *similar* to each other and different from the B-cell antigen receptors.

10. A surprisingly large proportion of T cells are alloreactive. Explain how as many as one in 20 T cells can react to allografts.

11. A secretive and reclusive investigator solved a three-dimensional structure for a TCR molecule in complex with its antigen. The investigator disappeared from the laboratory abruptly without giving contact information but left behind a complete high-resolution structure. What features of the structure would you inspect to identify which type of TCR it is? How could you obtain information about the antigen and the presenting MHC molecule if one exists?

12. Mice with the gene for the TCR α chain knocked out thrive much better than those with a knockout of the CD3 ζ chain. Can you explain why?

13. Compare the mechanisms for generation of diversity in the αβ TCR and the BCR. Comment on both subunits for each receptor in your answer.

 **Interactive Study**

www.whfreeman.com/kuby

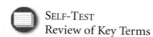 SELF-TEST
Review of Key Terms

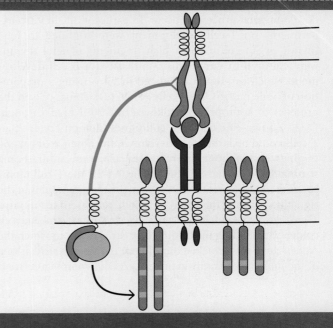

# chapter 10

# T-Cell Maturation, Activation, and Differentiation

THE ATTRIBUTE THAT DISTINGUISHES ANTIGEN RECOGNITION by most T cells from recognition by B cells is MHC restriction. In most cases, both the maturation of progenitor T cells in the thymus and the activation of mature T cells in the periphery are influenced by the involvement of MHC molecules. The potential antigenic diversity of the T-cell population is reduced during maturation by a selection process that allows only MHC-restricted and nonself-reactive T cells to mature. The final stages in the maturation of most T cells proceed along two different developmental pathways, which generate functionally distinct CD4$^+$ and CD8$^+$ subpopulations that exhibit class II and class I MHC restriction, respectively.

Activation of mature peripheral T cells begins with the interaction of the T-cell receptor (TCR) with an antigenic peptide displayed in the groove of an MHC molecule. Although the specificity of this interaction is governed by the TCR, its low affinity necessitates the involvement of coreceptors and other accessory membrane molecules that strengthen the TCR-antigen-MHC interaction and transduce the activating signal. Activation leads to the proliferation and differentiation of T cells into various types of effector cells and memory T cells. Because the vast majority of thymocytes and peripheral T cells express the αβ T-cell receptor rather than the γδ T-cell receptor, all references to the T-cell receptor in this chapter denote the αβ receptor unless otherwise indicated.

## T-Cell Maturation and the Thymus

Progenitor T cells begin to migrate to the thymus from the early sites of hematopoiesis at about day 11 of gestation in mice and in the eighth or ninth week of gestation in humans. In a manner similar to B-cell maturation in the bone marrow, T-cell maturation involves rearrangements of the germ-line TCR genes and the expression of various membrane markers. In the thymus, developing T cells, known as **thymocytes,**

*Engagement of TCR by peptide-MHC initiates signal transduction.*

- T-Cell Maturation and the Thymus
- Thymic Selection of the T-Cell Repertoire
- T-Cell Activation
- T-Cell Differentiation
- Cell Death and T-Cell Populations

proliferate and differentiate along developmental pathways that generate functionally distinct subpopulations of mature T cells.

As indicated in Chapter 2, the thymus occupies a central role in T-cell biology. Aside from being the main source of all T cells, it is where T cells diversify and are then shaped into an effective primary T-cell repertoire by an extraordinary pair of selection processes. One of these, **positive selection,** permits the survival of only those T cells whose TCRs are capable of recognizing self-MHC molecules. It is thus responsible for the creation of a self-MHC-restricted repertoire of T cells. The other selection process, **negative selection,** eliminates T cells that react too strongly with self MHC or with self MHC plus self peptides. Negative selection is an extremely important factor in generating a primary T-cell repertoire that is self-tolerant.

T-cell development begins with the arrival of small numbers of lymphoid precursors migrating from the blood into the thymus, where they proliferate, differentiate, and undergo selection processes that result in the development of mature T cells. Until recently, it was believed that the intact architecture of the thymus was a necessary condition for this developmental process to occur. In earlier years, in vitro experiments on T-cell development were confined to fetal thymic organ culture, in which fragments of fetal thymus were cultured in vitro. This contrasts with B-cell development, which can be recapitulated in vitro using a preparation of bone marrow stem cells grown on stromal cells in the presence of appropriate cytokines. In a 2002 breakthrough, J. C. Zuniga-Pfluker and colleagues demonstrated that T cells could be induced to develop in the absence of thymic fragments when bone marrow stem cells were cultured on a stromal cell line that expressed a ligand for the T-cell membrane receptor Notch. In retrospect, the requirement that the signaling pathway mediated by Notch be present for in vitro T-cell development had been suggested by several lines of evidence, including the finding that early T cells in which the *Notch1* gene was knocked out did not mature. (*Notch1* is one of the four Notch family members.) As shown in Figure 10-1,

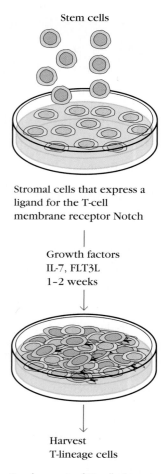

Stem cells

Stromal cells that express a
ligand for the T-cell
membrane receptor Notch

Growth factors
IL-7, FLT3L
1–2 weeks

Harvest
T-lineage cells

**FIGURE 10-1:** Development of T cells from hematopoietic stem cells on bone marrow stromal cells expressing the Notch ligand. *[Adapted from J. C. Zuniga-Pflucker, 2002, Nature Reviews Immunology 4:67–72.]*

growth of hematopoietic stem cells (HSCs) on stromal cells that express Notch ligand drives the development of these multipotent stem cells to the T lineage. In fact, forced expression of Notch in HSCs results in animals with T lymphocytes but no B lymphocytes. This result tells us that the expression of one protein, Notch, can redirect a potential B cell into the T lineage and firmly establishes Notch as a key protein in T-lineage specification.

As shown in Figure 10-2, when T-cell precursors arrive at the thymus, they do not express such signature surface markers of T cells as the T-cell receptor, the CD3 complex, or the coreceptors CD4 and CD8. In fact, these progenitor cells have not yet rearranged their TCR genes and do not express the proteins, such as RAG-1 and RAG-2, that are required for rearrangement. After arriving in the thymus, T-cell precursors enter the outer cortex and slowly proliferate. During approximately three weeks of development in the thymus, the differentiating T cells pass through a series of stages that are marked by characteristic changes in their cell surface phenotype. For example, thymocytes early in development lack detectable CD4 and CD8. Because these cells are $CD4^-CD8^-$, they are referred to as **double-negative (DN)** cells. In fact, DN T cells can be subdivided into four subsets (DN1–4) characterized by the presence or absence of cell surface molecules in addition to CD4 and CD8, such as c-Kit, the receptor for stem cell growth factor; CD44, an adhesion molecule; and CD25, the $\alpha$ chain of the IL-2 receptor. The cells that enter the thymus—DN1 cells—are capable of giving rise to all subsets of T cells and are phenotypically $c\text{-kit}^+$, $CD44^{high}$, and $CD25^-$. Once DN1 cells encounter the thymic environment, they begin to proliferate and express CD25, becoming $c\text{-kit}^+$, $CD44^{low}$, and $CD25^+$. These cells are called DN2 cells. During the critical DN2 stage of development, rearrangement of the genes for the TCR $\gamma$, $\delta$, and $\beta$ chains begins; however, the TCR $\alpha$ locus does not rearrange, presumably because the region of DNA encoding TCR$\alpha$ genes is densely compacted and not accessible to the recombinase machinery. As cells progress to DN3, the expression of both c-kit and CD44 is turned off and TCR$\gamma$, TCR$\delta$, and TCR$\beta$ rearrangements progress. Cells destined to become $\gamma\delta$ T cells (a small percentage ($<5\%$) of mature thymocytes) diverge at the transition between DN2 and DN3 and become mature $\gamma\delta$ T cells with very few changes in their surface phenotype. In mice, this thymocyte subpopulation can be detected by day 14 of gestation, reaches maximal numbers between days 17 and 18, and then declines until birth (Figure 10-3). Most DN2 cells are destined to give rise to $\alpha\beta$ T cells, and on assuming the DN3 phenotype ($c\text{-kit}^-$, $CD44^-$, $CD25^+$), the cells halt proliferation, and protein products of TCR$\beta$ rearrangements are detected in the cytoplasm of these cells. The newly synthesized $\beta$ chains combine with a 33-kDa glycoprotein known as the pre–T$\alpha$ chain and associate with the CD3 group to form a complex called the **pre-T-cell receptor** or **pre-TCR** (Figure 10-4). Of note, Notch proteins play a critical role at this point in T-cell development: cells that do not express Notch do not mature past this stage.

## OVERVIEW FIGURE 10-2: Development of T Cells in the Mouse

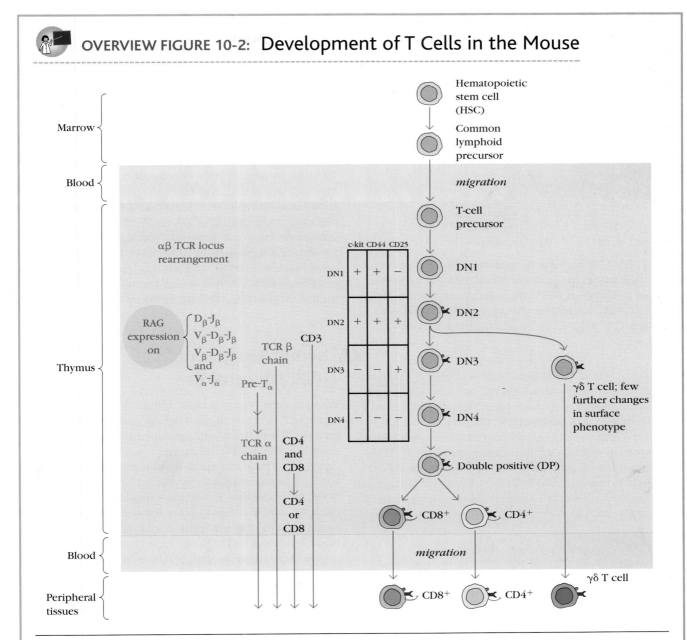

T-cell precursors arrive at the thymus from bone marrow via the bloodstream, undergo development to mature T cells, and are exported to the periphery, where they can undergo antigen-induced activation and differentiation into effector cells and memory cells. Each stage of development is characterized by stage-specific intracellular events and the display of distinctive cell surface markers.

Formation of the pre-TCR activates a signal transduction pathway that has several consequences:

- Indicates that a cell has made a productive TCR β-chain rearrangement and signals its further proliferation and maturation

- Suppresses further rearrangement of TCR β-chain genes, resulting in allelic exclusion

- Renders the cell permissive for rearrangement of the TCR α chain

- Induces developmental progression to the CD4$^+$8$^+$ *double-positive* state

After β-chain rearrangement is completed, the DN3 cells quickly progress to DN4, the level of CD25 falls, and both CD4 and CD8 coreceptors are expressed. This **double-positive (DP)**

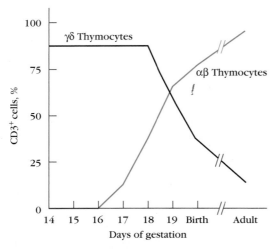

**FIGURE 10-3 Time course of appearance of γδ thymocytes and αβ thymocytes during mouse fetal development.** The graph shows the percentage of CD3⁺ cells in the thymus that are double negative (CD4⁻8⁻) and bear the γδ T-cell receptor (black) or are double positive (CD4⁺8⁺) and bear the αβ T-cell receptor (blue).

stage is a period of rapid proliferation. However, TCR α-chain gene rearrangement still has not occurred at this time. The rearrangement of α-chain genes does not begin until the double-positive thymocytes stop proliferating and RAG-2 protein levels increase. The proliferative phase prior to the rearrangement of the α chain increases the diversity of the T-cell repertoire by generating a clone of cells with a single TCR β-chain rearrangement. Each of the cells within this clone can then rearrange a different α-chain gene, thereby generating a much more diverse population than if the orig-

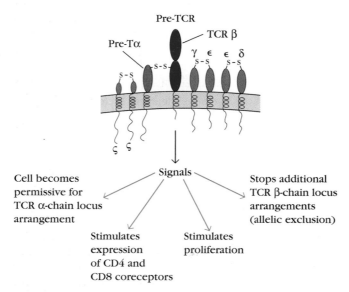

**FIGURE 10-4 Structure and activity of the pre-T-cell receptor (pre-TCR).** Binding of ligands yet to be identified to the pre-TCR generates intracellular signals that induce a variety of processes.

inal cell had first undergone rearrangement at both the β- and α-chain loci before it proliferated. In mice, the TCR α-chain genes are not expressed until day 16 or 17 of gestation; double-positive cells expressing both CD3 and the αβ T-cell receptor begin to appear at day 17 and reach maximal levels at about the time of birth (see Figure 10-3). The possession of a complete TCR enables DP thymocytes to undergo the rigors of positive and negative selection.

T-cell development is an expensive process for the host. An estimated 98% of all thymocytes do not mature—they die by apoptosis within the thymus either because they fail to make a productive TCR gene rearrangement or because they fail to survive thymic selection. Double-positive thymocytes that express the αβ TCR-CD3 complex and survive thymic selection develop into immature **single-positive CD4⁺** thymocytes or **single-positive CD8⁺** thymocytes. These single-positive cells undergo additional negative selection and migrate from the cortex to the medulla, where they pass from the thymus into the circulatory system.

## Thymic Selection of the T-Cell Repertoire

Random gene rearrangement within TCR germ-line DNA combined with junctional diversity can generate an enormous TCR repertoire, with an estimated potential diversity exceeding $10^{15}$ for the αβ receptor and $10^{18}$ for the γδ receptor. Gene products encoded by the rearranged TCR genes have no inherent affinity for foreign antigen plus a self-MHC molecule; they theoretically should be capable of recognizing soluble antigen (either foreign or self), self-MHC molecules, or antigen plus a nonself-MHC molecule. Nonetheless, the most distinctive property of mature T cells is that they recognize only foreign antigen combined with self-MHC molecules.

As noted, thymocytes undergo two selection processes in the thymus:

- Positive selection for thymocytes bearing receptors capable of binding self-MHC molecules, which results in **MHC restriction.** Cells that fail positive selection are eliminated within the thymus by apoptosis.

- Negative selection that eliminates thymocytes bearing high-affinity receptors for self-MHC molecules alone or self-antigen presented by self-MHC, which results in **self-tolerance.**

Both processes are necessary to generate mature T cells that are self-MHC restricted and self-tolerant. As noted already, some 98% or more of all thymocytes die by apoptosis within the thymus. The bulk of this high death rate appears to reflect a weeding out of thymocytes that fail positive selection because their receptors do not specifically recognize self-MHC molecules. These cells do not receive growth stimulation and die by a process known as death by neglect.

EXPERIMENT

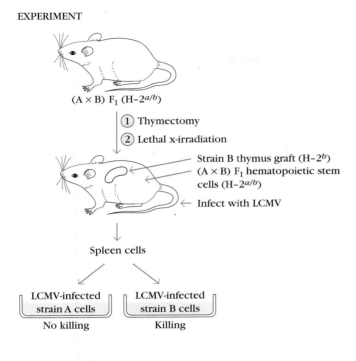

(A × B) F$_1$ (H-2$^{a/b}$)

① Thymectomy
② Lethal x-irradiation

Strain B thymus graft (H-2$^b$)
(A × B) F$_1$ hematopoietic stem
cells (H-2$^{a/b}$)
Infect with LCMV

Spleen cells

| LCMV-infected strain A cells | LCMV-infected strain B cells |
|---|---|
| No killing | Killing |

CONTROL

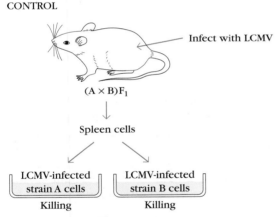

Infect with LCMV

(A × B)F$_1$

Spleen cells

| LCMV-infected strain A cells | LCMV-infected strain B cells |
|---|---|
| Killing | Killing |

**FIGURE 10-5 Experimental demonstration that the thymus selects for maturation only those T cells whose T-cell receptors recognize antigen presented on target cells with the haplotype of the thymus.** Thymectomized and lethally irradiated (A × B) F$_1$ mice were grafted with a strain-B thymus and reconstituted with (A × B) F$_1$ bone marrow cells. After infection with LCMV, the CTL cells were assayed for their ability to kill $^{51}$Cr-labeled strain A or strain B target cells infected with LCMV. Only strain B target cells were lysed, suggesting that the H-2$^b$ grafted thymus had selected for maturation only those T cells that could recognize antigen combined with H-2$^b$ MHC molecules.

Early evidence for the role of the thymus in selection of the T-cell repertoire came from chimeric mouse experiments by R. M. Zinkernagel and his colleagues (Figure 10-5). These researchers implanted thymectomized and irradiated (A × B) F$_1$ mice with a B-type thymus and then reconstituted the animal's immune system with an intravenous infu-

sion of F$_1$ bone marrow cells. To be certain that the thymus graft did not contain any mature T cells, it was irradiated before being transplanted. In such an experimental system, T-cell progenitors from the (A × B) F$_1$ bone marrow transplant mature within a thymus that expresses only B-haplotype MHC molecules on its stromal cells. Would these (A × B) F$_1$ T cells now be MHC restricted for the haplotype of the thymus? To answer this question, the chimeric mice were infected with lymphocytic choriomeningitis virus (LCMV) and the splenic T cells were then tested for their ability to kill LCMV-infected target cells from the strain A or strain B mice. As shown in Figure 10-5, when T$_C$ cells from the chimeric mice were tested on LCMV-infected target cells from strain A or strain B mice, they could lyse LCMV-infected target cells from strain B mice. These mice have the same MHC haplotype, B, as the implanted thymus. Thus, the MHC haplotype of the thymus in which T cells develop determines their MHC restriction.

Thymic stromal cells, including epithelial cells, macrophages, and dendritic cells, play essential roles in positive and negative selection. These cells express class I MHC molecules and can display high levels of class II MHC also. The interaction of immature thymocytes that express the TCR-CD3 complex with populations of thymic stromal cells results in positive and negative selection by mechanisms that are under intense investigation. First, we'll examine the details of each selection process and then review some experiments that provide insights into the operation of these processes.

## Positive selection ensures MHC restriction

Positive selection takes place in the cortical region of the thymus and involves the interaction of immature thymocytes with cortical epithelial cells (Figure 10-6). There is evidence that the T-cell receptors on thymocytes tend to cluster with MHC molecules on the cortical cells at sites of cell-cell contact. Some researchers have suggested that these interactions allow the immature thymocytes to receive a protective signal that prevents them from undergoing cell death; cells whose receptors are not able to bind MHC molecules would not interact with the thymic epithelial cells and consequently would not receive the protective signal, leading to their death by apoptosis.

During positive selection, the RAG-1, RAG-2, and TdT proteins required for gene rearrangement and modification (see Figure 5-7) continue to be expressed. Thus, each of the immature thymocytes in a clone expressing a given β chain has an opportunity to rearrange different TCR α-chain genes, and the resulting TCRs are then selected for self-MHC recognition. Only those cells whose αβ TCR heterodimer recognizes a self-MHC molecule are selected for survival. Consequently, the presence of more than one combination of αβ TCR chains among members of the clone is important because it increases the possibility that some members will "pass" the test for positive selection. Any cell that manages to rearrange an α chain that allows the resulting αβ TCR to recognize self-MHC will be spared; all

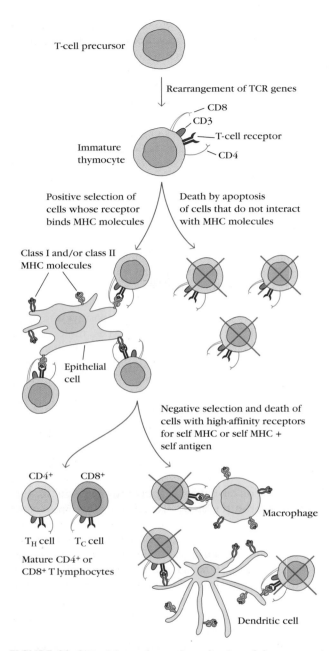

**FIGURE 10-6 Positive and negative selection of thymocytes in the thymus.** Thymic selection involves thymic stromal cells (epithelial cells, dendritic cells, and macrophages) and results in mature T cells that are both self-MHC restricted and self-tolerant.

members of the clone that fail to do so will die by apoptosis within 3 to 4 days.

## Negative selection ensures self-tolerance

The population of MHC-restricted thymocytes that survive positive selection includes cells with receptors having a range of affinities from low to high for self antigen presented by

self-MHC molecules. Thymocytes with high-affinity receptors are weeded out during negative selection via an interaction with thymic stromal cells. During negative selection, dendritic cells and macrophages bearing class I and class II MHC molecules interact with thymocytes bearing high-affinity receptors for self-antigen plus self-MHC molecules or for self-MHC molecules alone (see Figure 10-6). However, the precise details of the process are not yet known. Cells that experience negative selection are observed to undergo death by apoptosis. Tolerance to self antigens encountered in the thymus is thereby achieved by eliminating T cells that are reactive to these antigens.

## Experiments revealed the essential elements of positive and negative selection

Direct evidence that positive selection in the thymus requires binding of thymocytes to class I or class II MHC molecules came from experimental studies with knockout mice incapable of producing functional class I or class II MHC molecules (Table 10-1). Class I–deficient mice were found to have a normal distribution of double-negative, double-positive, and CD4$^+$ thymocytes but failed to produce CD8$^+$ thymocytes. Class II–deficient mice had double-negative, double-positive, and CD8$^+$ thymocytes but lacked CD4$^+$ thymocytes. Not surprisingly, the lymph nodes of these class II–deficient mice lacked CD4$^+$ T cells. Thus, the absence of class I or II MHC molecules prevents positive selection of CD8$^+$ or CD4$^+$ T cells, respectively.

Further experiments with transgenic mice provided additional evidence that interaction with MHC molecules plays a role in positive selection. In these experiments, rearranged αβ-TCR genes derived from a CD8$^+$ T-cell clone specific for influenza antigen plus H-2$^k$ class I MHC molecules were injected into fertilized eggs from two different mouse strains, one with the H-2$^k$ haplotype and one with the H-2$^d$ haplotype (Figure 10-7). Since the receptor transgenes were already

| TABLE 10-1 | Effect of class I or II MHC deficiency on thymocyte populations* | | |
|---|---|---|---|
| | | KNOCKOUT MICE | |
| Cell type | Control mice | Class I deficient | Class II deficient |
| CD4$^-$CD8$^-$ | + | + | + |
| CD4$^+$CD8$^+$ | + | + | + |
| CD4$^+$ | + | + | − |
| CD8$^+$ | + | − | + |

*Plus sign indicates normal distribution of indicated cell types in thymus. Minus sign indicates absence of cell type.

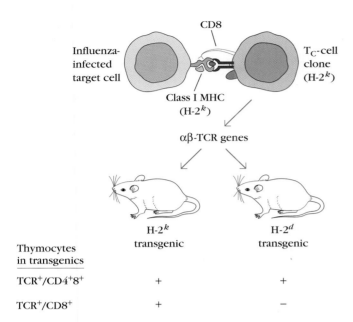

| Thymocytes in transgenics | H-2$^k$ transgenic | H-2$^d$ transgenic |
|---|---|---|
| TCR$^+$/CD4$^+$8$^+$ | + | + |
| TCR$^+$/CD8$^+$ | + | − |

**FIGURE 10-7 Effect of host haplotype on T-cell maturation in mice carrying transgenes encoding an H-2$^b$ class I–restricted T-cell receptor specific for influenza virus.** The presence of the rearranged TCR transgenes suppressed other gene rearrangements in the transgenics; therefore, most of the thymocytes in the transgenics expressed the αβ T-cell receptor encoded by the transgene. Immature double-positive thymocytes matured into CD8$^+$ T cells only in transgenics with the haplotype (H-2$^k$) corresponding to the MHC restriction of the TCR transgene.

rearranged, other TCR gene rearrangements were suppressed in the transgenic mice; therefore, a high percentage of the thymocytes in the transgenic mice expressed the T-cell receptor encoded by the transgene. Thymocytes expressing the TCR transgene were found to mature into CD8$^+$ T cells only in the transgenic mice with the H-2$^k$ class I MHC haplotype (i.e., the haplotype for which the transgene receptor was restricted). In transgenic mice with a different MHC haplotype (H-2$^d$), immature, double-positive thymocytes expressing the transgene were present, but these thymocytes failed to mature into CD8$^+$ T cells. These findings confirmed that interaction between T-cell receptors on immature thymocytes and self-MHC molecules is required for positive selection. In the absence of self-MHC molecules, as in the H-2$^d$ transgenic mice, positive selection and subsequent maturation do not occur.

Evidence for deletion of thymocytes reactive with self antigen plus MHC molecules comes from a number of experimental systems. In one system, thymocyte maturation was analyzed in transgenic mice bearing an αβ TCR transgene specific for the class I D$^b$ MHC molecule plus H-Y antigen, a small protein that is encoded on the Y chromosome and is therefore a self molecule only in male mice. In this experiment, the MHC haplotype of the transgenic mice was H-2$^b$, the same as the MHC restriction of the transgene-encoded receptor.

Therefore, any differences in the selection of thymocytes in male and female transgenics would be related to the presence or absence of H-Y antigen.

Analysis of thymocytes in the transgenic mice revealed that female mice contained thymocytes expressing the H-Y-specific TCR transgene, but male mice did not (Figure 10-8). In other words, H-Y-reactive thymocytes were self-reactive in the male mice and were eliminated. However, in the female transgenics, which did not express the H-Y antigen, these cells were not self-reactive and so were not eliminated. When thymocytes from these male transgenic mice were cultured in vitro with antigen-presenting cells expressing the H-Y antigen, the thymocytes were observed to undergo apoptosis, providing a striking example of negative selection.

## Some central issues in thymic selection remain unresolved

Although a great deal has been learned about the developmental processes that generate mature CD4$^+$ and CD8$^+$ T cells, some mysteries persist. Prominent among them is a paradox: If positive selection allows only thymocytes reactive with self-MHC molecules to survive and negative selection eliminates the self-MHC-reactive thymocytes, then no T cells would be allowed to mature. Since this is not the outcome of T-cell development, clearly, other factors operate to prevent these two MHC-dependent processes from eliminating the entire repertoire of MHC-restricted T cells.

Experimental evidence from fetal thymic organ culture (FTOC) has been helpful in resolving this puzzle. In this system, mouse thymic lobes are excised at a gestational age of day 16 and placed in culture. At this time, the lobes consist predominantly of CD4$^-$8$^-$ thymocytes. Because these immature, double-negative thymocytes continue to develop in the organ culture, thymic selection can be studied under conditions that permit a range of informative experiments. Particular use has been made of mice in which the peptide transporter, TAP-1, has been knocked out. In the absence of TAP-1, only low levels of MHC class I are expressed on thymic cells, and the development of CD8$^+$ thymocytes is blocked. However, when exogenous peptides are added to these organ cultures, then peptide-bearing class I MHC molecules appear on the surface of the thymic cells, and development of CD8$^+$ T cells is restored. Significantly, when a diverse peptide mixture is added, the extent of CD8$^+$ T-cell restoration is greater than when a single peptide is added. This indicates that the role of peptide is not simply to support stable MHC expression but also to be recognized itself in the selection process.

Two competing hypotheses attempt to explain the paradox of MHC-dependent positive and negative selection. The *avidity hypothesis* asserts that differences in the strength of the signals received by thymocytes undergoing positive and negative selection determine the outcome, with signal

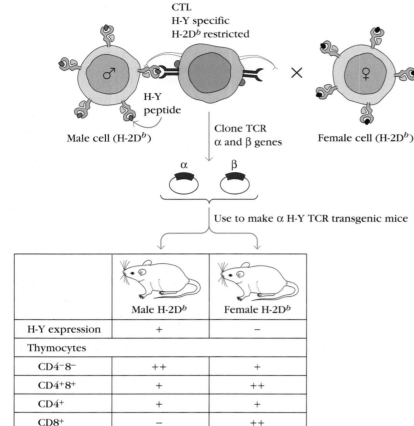

| | Male H-2D$^b$ | Female H-2D$^b$ |
|---|---|---|
| H-Y expression | + | − |
| Thymocytes | | |
| CD4$^-$8$^-$ | ++ | + |
| CD4$^+$8$^+$ | + | ++ |
| CD4$^+$ | + | + |
| CD8$^+$ | − | ++ |

**FIGURE 10-8 Experimental demonstration that negative selection of thymocytes requires both self antigen and self MHC.** In this experiment, H-2$^b$ male and female transgenics were prepared carrying TCR transgenes specific for H-Y antigen plus the D$^b$ molecule. This antigen is expressed only in males. FACS analysis of thymocytes from the transgenics showed that mature CD8$^+$ T cells expressing the transgene were absent in the male mice but present in the female mice, suggesting that thymocytes reactive with a self antigen (in this case, H-Y antigen in the male mice) are deleted during thymic selection. *[Adapted from H. von Boehmer and P. Kisielow, 1990,* Science **248***:1370.]*

strength dictated by the avidity of the TCR-MHC-peptide interaction. The *differential-signaling hypothesis* holds that the outcomes of selection are dictated by different signals rather than different strengths of the same signal.

The avidity hypothesis was tested with TAP-1 knockout mice transgenic for an αβ TCR that recognized an LCMV peptide–MHC complex. These mice were used to prepare fetal thymic organ cultures (Figure 10-9). The avidity of the TCR-MHC interaction was varied by the use of different concentrations of peptide. At low peptide concentrations, few MHC molecules bound peptide and the avidity of the TCR-MHC interaction was low. As peptide concentrations were raised, the number of peptide-MHC complexes displayed increased and so did the avidity of the interaction. In this experiment, very few CD8$^+$ cells appeared when peptide was not added, but even low concentrations of the relevant peptide resulted in the appearance of significant numbers of CD8$^+$ T cells bearing the transgenic TCR receptor. Increasing the peptide concentrations to an optimum range yielded the highest number of CD8$^+$ T cells. However, at higher concentrations of peptide, the numbers of CD8$^+$ T cells produced declined steeply. The results of these experiments show that positive and negative selection can be achieved with signals generated by the same peptide-MHC combination. No signal (no peptide) fails to support positive selection.

A weak signal (low peptide level) induces positive selection. However, too strong a signal (high peptide level) results in negative selection.

The differential-signaling model provides an alternative explanation for determining whether a T cell undergoes positive or negative selection. This model is a qualitative rather than a quantitative one, and it emphasizes the nature of the signal delivered by the TCR rather than its strength. At the core of this model is the observation that some MHC-peptide complexes can deliver only a weak or partly activating signal whereas others can deliver a complete signal. In this model, positive selection takes place when the TCRs of developing thymocytes encounter MHC-peptide complexes that deliver weak or partial signals to their receptors, and negative selection results when the signal is complete. At this point it is not possible to decide between the avidity model and the differential-signaling model; both have experimental support. It may be that in some cases, one of these mechanisms operates to the complete exclusion of the other. It is also possible that no single mechanism accounts for all the outcomes in the cellular interactions that take place in the thymus and more than one mechanism may play a significant role. Further work is required to complete our understanding of this matter.

The differential expression of the coreceptor CD8 also can affect thymic selection. In an experiment in which CD8

(a) Experimental procedure—fetal thymic organ culture (FTOC)

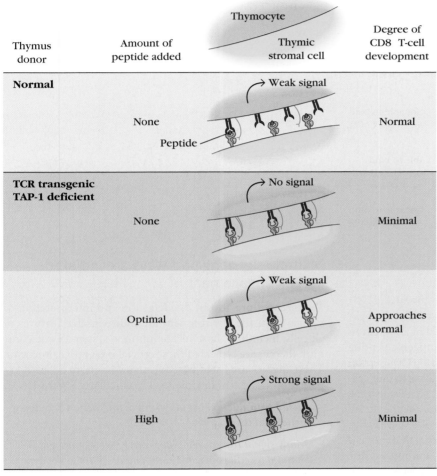

(b) Development of CD8 CD4⁻ MHC I-restricted cells

| Thymus donor | Amount of peptide added | Thymocyte / Thymic stromal cell | Degree of CD8 T-cell development |
|---|---|---|---|
| **Normal** | None | Weak signal | Normal |
| **TCR transgenic TAP-1 deficient** | None | No signal | Minimal |
| | Optimal | Weak signal | Approaches normal |
| | High | Strong signal | Minimal |

**FIGURE 10-9   Role of peptides in selection.** Thymuses harvested before their thymocyte populations have undergone positive and negative selection allow study of the development and selection of single positive (CD4⁺CD8⁻ and CD4⁻CD8⁺) T cells. (a) Outline of the experimental procedure for in vitro fetal thymic organ culture (FTOC). (b) The development and selection of CD8⁺CD4⁻ class I–restricted T cells depends on TCR peptide–MHC I interactions. TAP-1 knockout mice are unable to form peptide-MHC complexes unless peptide is added. The mice used in this study were transgenic for the α and β chains of a TCR that recognizes the added peptide bound to MHC I molecules of the TAP-1 knockout/TCR transgenic mice. Varying the amount of added peptide revealed that low concentrations of peptide, producing low avidity of binding, resulted in positive selection and nearly normal levels of CD4⁻CD8⁺ cells. High concentrations of peptide, producing high avidity of binding to the TCR, caused negative selection, and few CD4⁻CD8⁺ T cells appeared. *[Adapted from P. G. Ashton Rickardt et al., 1994,* Cell **25:**651.]

expression was artificially raised to twice its normal level, the concentration of mature CD8⁺ cells in the thymus was one thirteenth of the concentration in control mice bearing normal levels of CD8 on their surface. Since the interaction of T cells with class I MHC molecules is strengthened by participation of CD8, perhaps the increased expression of CD8 would increase the avidity of thymocytes for class I molecules, possibly making their negative selection more likely.

Another important open question in thymic selection is how double-positive thymocytes are directed to become either CD4⁺8⁻ or CD4⁻8⁺ T cells. Selection of CD4⁺8⁺ thymocytes gives rise to class I MHC–restricted CD8⁺ T cells and class II–restricted CD4⁺ T cells. Two models have been proposed to explain the transformation of a double-positive

precursor into one of two different single-positive lineages (Figure 10-10). The *instructive model* postulates that the multiple interactions between the TCR, CD8⁺, or CD4⁺ coreceptors and class I or class II MHC molecules instruct the cells to differentiate into either CD8⁺ or CD4⁺ single-positive cells, respectively. This model would predict that a class I MHC–specific TCR together with the CD8 coreceptor would generate a signal that is different from the signal induced by a class II MHC–specific TCR together with the CD4 coreceptor. The *stochastic model* suggests that CD4 or CD8 expression is switched off randomly with no relation to the specificity of the TCR. Only those thymocytes whose TCR and remaining coreceptor recognize the same class of MHC molecule will mature. At present, it is not possible to choose one model over the other.

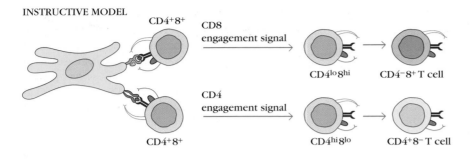

INSTRUCTIVE MODEL

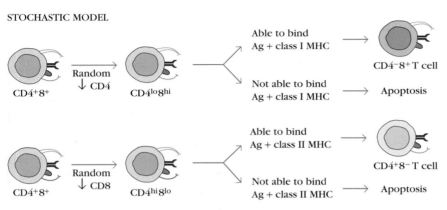

STOCHASTIC MODEL

**FIGURE 10-10 Proposed models for the role of the CD4 and CD8 coreceptors in thymic selection of double-positive thymocytes leading to single-positive T cells.** According to the instructive model, interaction of one coreceptor with MHC molecules on stromal cells results in down-regulation of the other coreceptor. According to the stochastic model, down-regulation of CD4 or CD8 is a random process.

# T-Cell Activation

The central event in the generation of both humoral and cell-mediated immune responses is the activation and clonal expansion of T cells. T-cell activation is initiated by interaction of the TCR-CD3 complex with a processed antigenic peptide bound to either a class I (CD8$^+$ cells) or class II (CD4$^+$ cells) MHC molecule on the surface of an antigen-presenting cell. This interaction and the resulting activating signals also involve various accessory membrane molecules on the T cell and the antigen-presenting cell. Interaction of a T cell with antigen initiates a cascade of biochemical events that induces the resting T cell to enter the cell cycle, proliferating and differentiating into memory cells or effector cells. Many of the gene products that appear on interaction with antigen can be grouped into one of three categories, depending on how early they can be detected after antigen recognition (Table 10-2):

- *Immediate genes,* expressed within half an hour of antigen recognition, encode a number of transcription factors, including c-Fos, c-Myc, c-Jun, NFAT, and NF-κB.

- *Early genes,* expressed within 1 to 2 hours of antigen recognition, encode IL-2, IL-2R (IL-2 receptor), IL-3, IL-6, IFN-γ, and numerous other proteins.

- *Late genes,* expressed more than 2 days after antigen recognition, encode various adhesion molecules.

These profound changes are the result of signal transduction pathways that are activated by the encounter between the TCR and MHC-peptide complexes. An overview of some of the basic strategies of cellular signaling will be useful background for appreciating the specific signaling pathways used by T cells.

## Multiple signaling pathways are initiated by TCR engagement

As discussed previously, the detection and interpretation of signals from the environment are indispensable features of all cells, including those of the immune system. Some common themes are typical of these crucial integrative processes. These themes are listed here for review; a more detailed explanation was given in Chapter 1, particularly Figure 1-6.

- *Signal transduction begins with the interaction between a signal and its receptor.*

- *Many signal transduction pathways involve the signal-induced assembly of some components of the pathway; these assemblies utilize adaptor proteins.*

- *Signal reception often leads to the generation within the cell of a "second messenger," a molecule or ion that can diffuse to other sites in the cell and evoke metabolic changes.*

- *Protein kinases and protein phosphatases are activated or inhibited.*

- *Signals are amplified by enzyme cascades.*

The events that link antigen recognition by the T-cell receptor to gene activation echo many of these themes. The key element in the initiation of T-cell activation is the recognition by the TCR of MHC-peptide complexes on antigen-presenting cells. This event catalyzes a series of intracellular events beginning at the inner surface of the plasma membrane and culminating in the nucleus, resulting in the transcription of genes that drive the cell cycle and/or differentiation of the T cell.

As described in Chapter 9, the TCR consists of a mostly extracellular ligand-binding unit, a predominantly intracellular

| TABLE 10-2 | Time course of gene expression by $T_H$ cells following interaction with antigen | | | |
|---|---|---|---|---|
| Gene product | Function | Time mRNA expression begins | Location | Ratio of activated to nonactivated cells |
| IMMEDIATE | | | | |
| c-Fos | Proto-oncogene; nuclear-binding protein | 15 min | Nucleus | >100 |
| c-Jun | Cellular oncogene; transcription factor | 15–20 min | Nucleus | ? |
| NFAT | Transcription factor | 20 min | Nucleus | 50 |
| c-Myc | Cellular oncogene | 30 min | Nucleus | 20 |
| NF-κB | Transcription factor | 30 min | Nucleus | >10 |
| EARLY | | | | |
| IFN-γ | Cytokine | 30 min | Secreted | >100 |
| IL-2 | Cytokine | 45 min | Secreted | >1000 |
| Insulin receptor | Hormone receptor | 1 h | Cell membrane | 3 |
| IL-3 | Cytokine | 1–2 h | Secreted | >100 |
| TGF-β | Cytokine | <2 h | Secreted | >10 |
| IL-2 receptor (p55) | Cytokine receptor | 2 h | Cell membrane | >50 |
| TNF-β | Cytokine | 1–3 h | Secreted | >100 |
| Cyclin | Cell cycle protein | 4–6 h | Cytoplasmic | >10 |
| IL-4 | Cytokine | <6 h | Secreted | >100 |
| IL-5 | Cytokine | <6 h | Secreted | >100 |
| IL-6 | Cytokine | <6 h | Secreted | >100 |
| c-Myb | Proto-oncogene | 16 h | Nucleus | 100 |
| GM-CSF | Cytokine | 20 h | Secreted | ? |
| LATE | | | | |
| HLA-DR | Class II MHC molecule | 3–5 days | Cell membrane | 10 |
| VLA-4 | Adhesion molecule | 4 days | Cell membrane | >100 |
| VLA-1, VLA-2, VLA-3, VLA-5 | Adhesion molecules | 7–14 days | Cell membrane | >100, ?, ?, ? |

SOURCE: Adapted from G. Crabtree, *Science* 243:357.

signaling unit, the CD3 complex, and the homodimer of ζ (zeta) chains. Experiments with knockout mice have shown that all of these components are essential for signal transduction. Two phases can be recognized in the antigen-mediated induction of T-cell responses: initiation and signal generation:

- *Initiation.* It is critical that T-cell signaling be regulated and activation occur as a result of specific antigen binding. In a resting T cell, p56[Lck], a protein tyrosine kinase essential for the initiation of TCR signaling, is sequestered from the TCR complex. The p56[Lck] kinase is usually found in *lipid rafts,* membrane microdomains rich in sphingomyelin, glycosphingolipids, and cholesterol. The unengaged TCR complex is excluded from these domains. On engagement

of MHC-peptide by the TCR, the TCR complex becomes associated with lipid rafts, followed by clustering with CD4 or CD8 coreceptors that bind to invariant regions of the MHC molecule (Figure 10-11). The tyrosine kinase p56[Lck], which is associated with the cytoplasmic tails of the coreceptors when the cell is in a resting state, is then brought close to the cytoplasmic tails of the TCR complex, where it phosphorylates the immunoreceptor tyrosine-based activation motifs (ITAMs; see Figure 9-9) of the CD3 component polypeptides. Phosphorylated tyrosines in the ITAMs of the zeta chain provide docking sites to which another protein tyrosine kinase called ZAP-70 attaches (step 2 in Figure 10-11) and becomes active. ZAP-70 then catalyzes the phosphorylation of a number of

## OVERVIEW FIGURE 10-11: TCR-Mediated Signaling

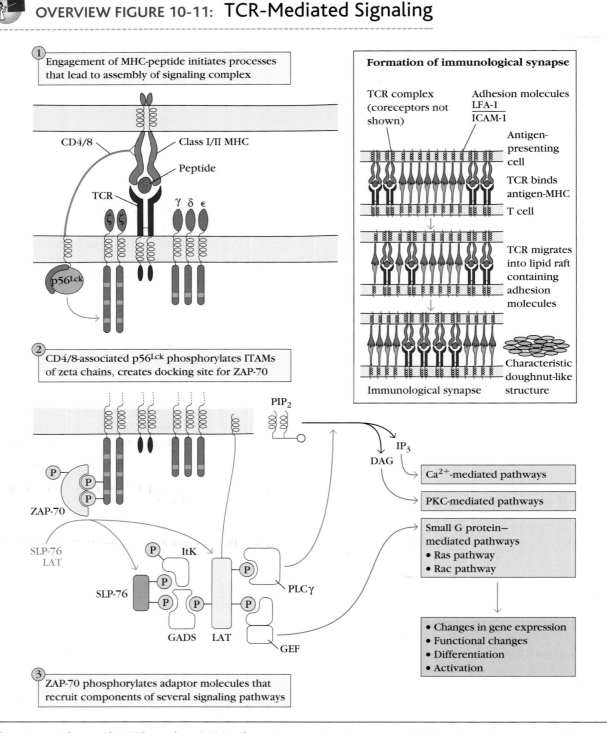

**1** Engagement of MHC-peptide initiates processes that lead to assembly of signaling complex

CD4/8

Class I/II MHC

Peptide

TCR

γ δ ε

ζ ζ

p56Lck

**Formation of immunological synapse**

TCR complex (coreceptors not shown)

Adhesion molecules
LFA-1
ICAM-1

Antigen-presenting cell

TCR binds antigen-MHC

T cell

TCR migrates into lipid raft containing adhesion molecules

Immunological synapse

Characteristic doughnut-like structure

**2** CD4/8-associated p56Lck phosphorylates ITAMs of zeta chains, creates docking site for ZAP-70

PIP₂

IP₃

DAG

Ca²⁺-mediated pathways

PKC-mediated pathways

ZAP-70

SLP-76
LAT

ItK

SLP-76

GADS    LAT

PLCγ

GEF

Small G protein–mediated pathways
• Ras pathway
• Rac pathway

• Changes in gene expression
• Functional changes
• Differentiation
• Activation

**3** ZAP-70 phosphorylates adaptor molecules that recruit components of several signaling pathways

TCR engagement by peptide-MHC complexes initiates the assembly of a signaling complex. An early step is the p56Lck-mediated phosphorylation of ITAMs on the zeta (ζ) chains of the TCR complex, creating docking sites to which the protein kinase ZAP-70 attaches and becomes activated by phosphorylation. A series of ZAP-70-catalyzed protein phosphorylations make possible the generation of a variety of signals. (Abbreviations: DAG = diacyl-glycerol; GADS = Grb2-like adaptor downstream of Shc; GEF = guanine nucleotide exchange factor; ITAM = immunoreceptor tyrosine-based activation motif; ItK = inducible T-cell kinase; IP₃ = inositol 1,4,5-triphosphate; LAT = linker of activated T cells; PIP₂ = phosphoinositol biphosphate; PLCγ = phospholipase C γ; SLP-76 = SH₂-containing leukocyte-specific protein of 76 kDa; ZAP-70 = zeta-associated protein of 70 kDa.)

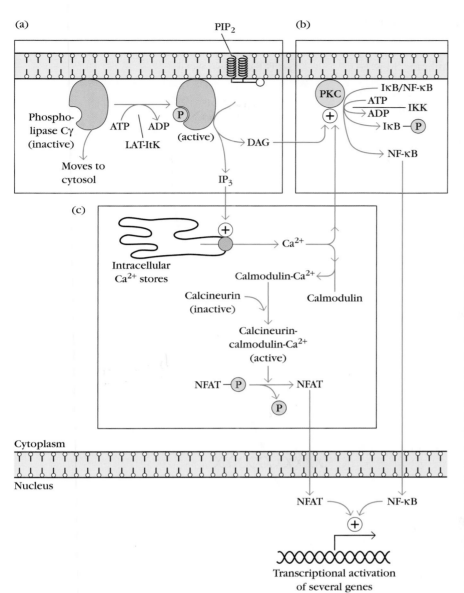

**FIGURE 10-12 Signal transduction pathways associated with T-cell activation.** (a) Phospholipase Cγ (PLCγ) is activated by phosphorylation. Active PLCγ hydrolyzes a phospholipid component of the plasma membrane to generate the second messengers, DAG and $IP_3$. (b) Protein kinase C (PKC) is activated by DAG and $Ca^{2+}$. Among the numerous effects of PKC is activation of IKK, which phosphorylates IkB, a cytoplasmic protein that binds the transcription factor NF-κB and prevents it from entering the nucleus. Phosphorylation of IkB releases NF-κB, which then translocates into the nucleus. (c) $Ca^{2+}$-dependent activation of calcineurin. Calcineurin is a $Ca^{2+}$/calmodulin-dependent phosphatase. $IP_3$ mediates the release of $Ca^{2+}$ from the endoplasmic reticulum. $Ca^{2+}$ binds the protein calmodulin, which then associates with and activates the $Ca^{2+}$/calmodulin-dependent phosphatase calcineurin. Active calcineurin removes a phosphate group from NFAT, which allows this transcription factor to translocate into the nucleus.

membrane-associated adaptor molecules, including Vav-1, Shc, LAT, and SLP-76 (step 3), which act as scaffolds for the recruitment of mediator molecules associated with several intracellular signal transduction pathways. One of these pathways involves a form of the enzyme phospholipase C (PLCγ), which anchors to an adaptor molecule, is activated by phosphorylation, and cleaves a membrane phospholipid to generate second messengers. This pathway also results in the activation of a transcription factor, nuclear factor kappa B, or NF-κB. Another pathway activates a G protein.

■ *Generation of multiple intracellular signals.* Many signaling pathways are activated as a consequence of the steps that occur in the initiation phase, as shown at bottom right in Figure 10-11 and described below.

A feature of the initiation phase of TCR signaling is the formation of a supramolecular structure known as the *immunological synapse*, or IS. The IS, so named by analogy with the synapses between neurons, is a dynamic and highly organized structure that forms where T cells contact antigen-presenting cells. It is thought that entry of the TCR into lipid rafts is a critical feature of IS formation. The central region of the IS is highly enriched with TCR/CD3 complexes as well as other intracellular signaling proteins associated with the TCR complex, whereas the outer, doughnut-like ring that surrounds this central region contains adhesion molecules such as LFA-1 (Figure 10-11, inset). Although the formation of an IS is not required for the initiation phase of TCR signaling, sustained T-cell activation is known to be more effective if an IS forms.

We will consider several of the signaling pathways recruited by T-cell activation, but the overall process is quite complex, and many of the details will not be presented here. The review articles listed at the end of the chapter provide extensive coverage of this very active research area.

### Phospholipase Cγ (PLCγ)

PLCγ is activated by phosphorylation and gains access to its substrate by binding to the membrane-associated adaptor protein LAT in association with inducible T-cell kinase (ItK) (Figure 10-12a). PLCγ hydrolyzes phosphoinositol bisphosphate,

PIP$_2$, a phospholipid component of the membrane, to generate inositol 1,4,5-triphosphate (IP$_3$) and diacylglycerol (DAG). IP$_3$ causes a rapid release of Ca$^{2+}$ from the endoplasmic reticulum and opens Ca$^{2+}$ channels in the cell membrane (Figure 10-12c). DAG activates protein kinase C, a multifunctional kinase that phosphorylates many different targets (Figure 10-12b).

## Ca$^{2+}$

Calcium ion is involved in an unusually broad range of processes, including vision, muscle contraction, and many others. It is an essential element in many T-cell responses, and calcium release from the endoplasmic reticulum ultimately results in the phosphorylation of an important transcription factor, NFAT, which culminates in the transport of NFAT from the cytoplasm into the nucleus (Figure 10-12c). In the nucleus, NFAT supports the transcription of genes required for the expression of the T-cell-growth-promoting cytokines IL-2, IL-4, and others.

## Protein Kinase C (PKC)

The production of DAG by PLCγ results in the activation of protein kinase C (PKC). Activation leads to the translocation of PKC to lipid rafts, where the enzyme initiates a cascade of events leading to the activation of the transcription factor NF-κB (Figure 10-12b).

## Nuclear Factor Kappa B (NF-κB)

Nuclear factor kappa B is an important transcription factor induced by a variety of signals in many cell types, including cells of the immune system. However, the pathway that leads to activation of NF-κB in T cells is novel. Activation of PKC by PLCγ leads to the assembly of a membrane-bound complex of proteins that includes CARMA1, BCL-10, and MALT1. This complex, once assembled, drives the activation of a multiprotein enzyme known as IKK (inhibitor of κB kinase). IKK, in turn, phosphorylates inhibitor of κB or IκB (Figure 10-12b). IκB normally binds NF-κB, retaining it in the cytosol, but on phosphorylation of IκB by IKK, NF-κB is released and migrates to the nucleus, where it activates gene transcription. IL-2 is a critical downstream target of NF-κB.

## The Ras/MAP Kinase Pathway

Ras is a pivotal component of signal transduction pathways found in many cell types and is evolutionarily conserved across a spectrum of eukaryotes from yeast to humans. Ras is a small G protein whose activation by GTP initiates a cascade of protein kinases known as the mitogen-activated protein kinase (MAP kinase) pathway. Phosphorylation of the end product of this cascade, MAP kinase (also called ERK), allows it to activate Elk, a transcription factor necessary for the expression of Fos (Figure 10-13). Phosphorylation of Fos by MAP kinase allows it to associate with Jun to form AP-1, which is an essential transcription factor for T-cell activation. AP-1 also regulates the transcription of IL-2.

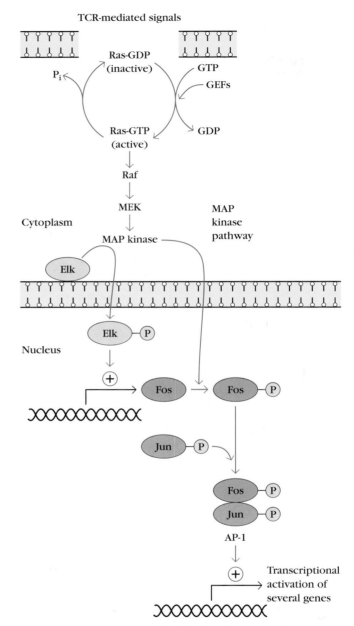

**FIGURE 10-13    Activation of the small G protein Ras.** Signals from the T-cell receptor result in activation of Ras via the action of specific guanine nucleotide exchange factors (GEFs) that catalyze the exchange of GDP for GTP. Active Ras causes a cascade of reactions that result in the increased production of the transcription factor Fos. Following their phosphorylation, Fos and Jun dimerize to yield the transcription factor AP-1. Note that all these pathways have important effects other than the specific examples shown in the figure.

## How many TCR complexes must be engaged to trigger T-cell activation?

The question of how many ligands a T cell must recognize before activation occurs has been addressed by many groups. Until recent developments in single-cell imaging, these calculations were determined by indirect and imprecise means.

Mark Davis and colleagues at Stanford Medical School approached this question using an acutely sensitive, two-part microscopic visualization technique. Antigen-presenting cells were treated briefly with peptide containing a terminal biotin molecule. The number of peptides bound to MHC could then be determined by adding a streptavidin-phycoerythrin conjugate, which binds the biotin of the biotinylated peptide and can be induced to emit light for subsequent imaging. T cells specific for the biotinylated peptide-MHC complex were loaded with fura-2, a dye used to visualize free intracellular calcium, which would serve as an indicator of T-cell activation. The T cells were then added to the antigen-presenting cells. Using this extremely sensitive technique, these researchers could determine (1) if a T cell was activated and (2) how many molecules of peptide were present at the point of contact between the T cell and the antigen-presenting cell—in other words, how many TCR-MHC-biotinylated peptide complexes had formed. The investigators determined that even a single peptide results in $Ca^{2+}$ release and that maximal $Ca^{2+}$ release was achieved in $CD4^+$ T cells when as few as 10 TCR complexes were engaged. Similar results were obtained with $CD8^+$ T cells.

## Costimulatory signals are required for full T-cell activation

T-cell activation requires the dynamic interaction of the multiple membrane-associated molecules described above, but these interactions by themselves are not sufficient to fully activate naive T cells. Naive T cells require more than one signal for activation and subsequent proliferation into effector cells:

- *Signal 1,* the initial signal, is generated by interaction of an antigenic peptide with the TCR-CD3 complex.

- A subsequent antigen-nonspecific costimulatory signal, *signal 2,* is provided primarily by interactions between CD28 on the T cell and members of the B7 family on the antigen-presenting cell (Figure 10-14).

There are two related forms of B7: B7-1 and B7-2. These molecules are members of the immunoglobulin superfamily and have a similar organization of extracellular domains but markedly different cytosolic domains. Both B7 molecules are constitutively expressed on dendritic cells and induced on activated macrophages and activated B cells. The ligands for B7 molecules are CD28 and CTLA-4 (also known as CD152), both of which are expressed on the T-cell membrane as disulfide-linked homodimers; like B7, they are members of the immunoglobulin superfamily (see Figure 10-14). Although CD28 and CTLA-4 are structurally similar glycoproteins, they act antagonistically. Signaling through CD28 delivers a positive costimulatory signal to the T cell; signaling through CTLA-4 is inhibitory and down-regulates the activation of the T cell. CD28 is expressed by both resting and activated T cells, but CTLA-4 is virtually undetectable on resting cells. Typically, engagement of the TCR causes the induction of CTLA-4 expression, and CTLA-4 is readily detectable within

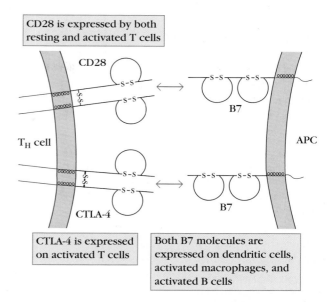

**FIGURE 10-14 T$_H$-cell activation requires a costimulatory signal provided by antigen-presenting cells (APCs).** Interaction of B7 family members on APCs with CD28 delivers the costimulatory signal. Engagement of the closely related CTLA-4 molecule with B7 produces an inhibitory signal. All of these molecules contain at least one immunoglobulin-like domain and thus belong to the immunoglobulin superfamily. *[Adapted from P. S. Linsley and J. A. Ledbetter, 1993,* Annual Review of Immunology *11:191.]*

24 hours of stimulation, with maximal expression within 2 to 3 days post-stimulation. Even though the peak surface levels of CTLA-4 are lower than those of CD28, it still competes favorably for B7 molecules because it has a significantly higher affinity for these molecules than CD28 does. Interestingly, the level of CTLA-4 expression is increased by CD28-generated costimulatory signals. This provides regulatory braking via CTLA-4 in proportion to the acceleration received from CD28. Some of the importance of CTLA-4 in the regulation of lymphocyte activation and proliferation is revealed by experiments with CTLA-4 knockout mice. T cells in these mice proliferate massively, which leads to lymphadenopathy (greatly enlarged lymph nodes), splenomegaly (enlarged spleen), and death at 3 to 4 weeks after birth. Clearly, the production of inhibitory signals by engagement of CTLA-4 is important in lymphocyte homeostasis.

## Clonal anergy ensues if a costimulatory signal is absent

T-cell recognition of an antigenic peptide-MHC complex sometimes results in a state of nonresponsiveness called **clonal anergy,** marked by the inability of cells to proliferate in response to engagement with a peptide-MHC complex. Whether clonal expansion or clonal anergy ensues is determined by the presence or absence of a costimulatory signal (signal 2), such as that produced by interaction of CD28 on T cells with B7 on antigen-presenting cells. Experiments with cultured cells show

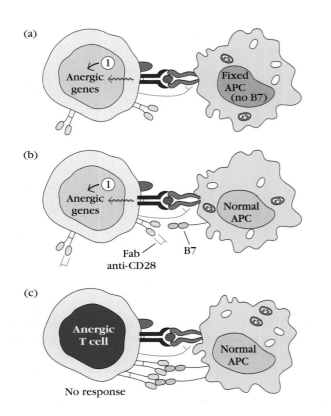

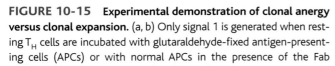

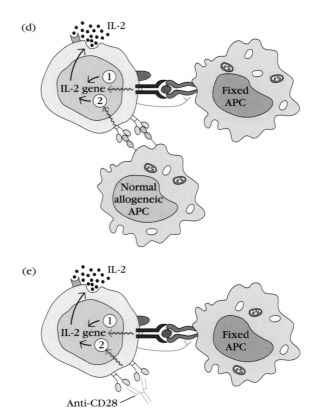

**FIGURE 10-15   Experimental demonstration of clonal anergy versus clonal expansion.** (a, b) Only signal 1 is generated when resting $T_H$ cells are incubated with glutaraldehyde-fixed antigen-presenting cells (APCs) or with normal APCs in the presence of the Fab portion of anti-CD28. (c) The resulting anergic T cells cannot respond to normal APCs. (d, e) In the presence of normal allogeneic APCs or anti-CD28, both of which produce the costimulatory signal 2, T cells are activated by fixed APCs.

that, if a resting T cell receives the TCR-mediated signal (signal 1) in the absence of a suitable costimulatory signal, then the T cell will become anergic. Specifically, if resting T cells are incubated with glutaraldehyde-fixed APCs, which do not express B7 (Figure 10-15a), the fixed APCs are able to present peptides together with class II MHC molecules, thereby providing signal 1, but they are unable to provide the necessary costimulatory signal 2. In the absence of a costimulatory signal, there is minimal production of cytokines, especially of IL-2. Anergy can also be induced by incubating T cells with normal APCs in the presence of the Fab portion of anti-CD28, which blocks the interaction of CD28 with B7 (Figure 10-15b).

Two different control experiments demonstrate that fixed APCs bearing appropriate peptide-MHC complexes can deliver an effective signal mediated by T-cell receptors. In one experiment, T cells are incubated both with fixed APCs bearing peptide-MHC complexes recognized by the TCR of the T cells and with normal APCs, which express B7 (Figure 10-15d). The fixed APCs engage the TCRs of the T cells, and the B7 molecules on the surface of the normal APCs cross-link the CD28 of the T cell. These T cells thus receive both signals and undergo activation. The addition of bivalent anti-CD28 to mixtures of fixed APCs and T cells also provides effective costimulation by cross-linking CD28 (Figure 10-15e).

There is good evidence that both $CD4^+$ and $CD8^+$ T cells can be anergized, but most studies of anergy have been conducted in $CD4^+$ $T_H$ cells. A T cell can be anergized in several ways. In addition to the lack of costimulatory signals, signal 2, the appearance of CTLA-4 in a cell that has received both signal 1 and signal 2 can block costimulation and functionally result in anergy. Although the existence of anergy has been established, the precise biochemical pathways that regulate this state of nonresponsiveness are not well understood. However, during the past few years experimental systems using microarray analysis (see Chapter 22) have identified several key molecular components of anergic T cells, which include several ubiquitin ligases that appear to target key components of the TCR signaling pathway for degradation by the proteasome.

## Superantigens induce T-cell activation by binding the TCR and MHC II simultaneously

Superantigens are viral or bacterial proteins that bind simultaneously to the $V_\beta$ domain of a T-cell receptor and to the $\alpha$ chain of a class II MHC molecule. Both exogenous and endogenous superantigens have been identified. Cross-linkage of a T-cell receptor and class II MHC molecule by either type of superantigen produces an activating signal that induces T-cell activation and proliferation (Figure 10-16).

*Exogenous* superantigens are soluble proteins secreted by bacteria. Among them are a variety of **exotoxins** secreted by

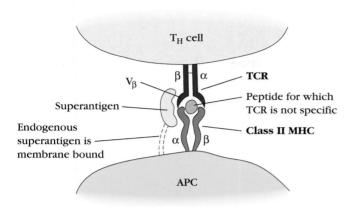

**FIGURE 10-16   Superantigen-mediated cross-linkage of T-cell receptor and class II MHC molecules.** A superantigen binds to all TCRs bearing a particular $V_\beta$ sequence regardless of their antigenic specificity. Exogenous superantigens are soluble secreted bacterial proteins, including various exotoxins. Endogenous superantigens are membrane-embedded proteins produced by certain viruses; they include Mls antigens encoded by mouse mammary tumor virus.

Gram-positive bacteria, such as staphylococcal enterotoxins, toxic shock syndrome toxin, and exfoliative dermatitis toxin. Each of these exogenous superantigens binds particular $V_\beta$ sequences in T-cell receptors (Table 10-3) and cross-links the TCR to a class II MHC molecule.

*Endogenous* superantigens are cell membrane proteins encoded by certain viruses that infect mammalian cells. One group, encoded by mouse mammary tumor virus (MTV), can integrate into the DNA of certain inbred mouse strains; after integration, retroviral proteins are expressed on the membrane of the infected cells. These viral proteins, called **minor lymphocyte-stimulating (Mls) determinants,** bind particular $V_\beta$ sequences in T-cell receptors and cross-link the TCR to a class II MHC molecule. Four Mls superantigens, originating in different MTV strains, have been identified.

Because superantigens bind outside the TCR antigen-binding cleft, any T cell expressing a particular $V_\beta$ sequence will be activated by a corresponding superantigen. Hence, the activation is polyclonal and can affect a significant percentage (5% is not unusual) of the total $T_H$ population. The massive activations that follow cross-linkage by a superantigen result in overproduction of $T_H$-cell cytokines, leading to systemic toxicity. The food poisoning induced by staphylococcal enterotoxins and the toxic shock induced by toxic shock syndrome toxin are two examples of the consequences of cytokine overproduction induced by superantigens.

Superantigens can also influence T-cell maturation in the thymus. A superantigen present in the thymus during thymic processing will induce the negative selection of all thymocytes bearing a TCR $V_\beta$ domain corresponding to the superantigen specificity. Such massive deletion can be caused by exogeneous or endogenous superantigens and is characterized by the absence of all T cells whose receptors possess $V_\beta$ domains targeted by the superantigen.

## T-Cell Differentiation

$CD4^+$ and $CD8^+$ T cells leave the thymus and enter the circulation as resting cells in the $G_0$ stage of the cell cycle. There are about twice as many $CD4^+$ T cells as $CD8^+$ T cells in the

| TABLE 10-3 | Exogenous superantigens and their $V_\beta$ specificity | | |
|---|---|---|---|
| | | $V_\beta$ SPECIFICITY | |
| **Superantigen** | **Disease*** | **Mouse** | **Human** |
| Staphylococcal enterotoxins | | | |
| SEA | Food poisoning | 1, 3, 10, 11, 12, 17 | nd |
| SEB | Food poisoning | 3, 8.1, 8.2, 8.3 | 3, 12, 14, 15, 17, 20 |
| SEC1 | Food poisoning | 7, 8.2, 8.3, 11 | 12 |
| SEC2 | Food poisoning | 8.2, 10 | 12, 13, 14, 15, 17, 20 |
| SEC3 | Food poisoning | 7, 8.2 | 5, 12 |
| SED | Food poisoning | 3, 7, 8.3, 11, 17 | 5, 12 |
| SEE | Food poisoning | 11, 15, 17 | 5.1, 6.1–6.3, 8, 18 |
| Toxic shock syndrome toxin (TSST1) | Toxic shock syndrome | 15, 16 | 2 |
| Exfoliative dermatitis toxin (ExFT) | Scalded skin syndrome | 10, 11, 15 | 2 |
| Mycoplasma arthritidis supernatant (MAS) | Arthritis, shock | 6, 8.1–8.3 | nd |
| Streptococcal pyrogenic exotoxins (SPE-A, B, C, D) | Rheumatic fever, shock | nd | nd |

*Disease results from infection by bacteria that produce the indicated superantigens.

periphery. T cells that have not yet encountered antigen (naive T cells) are characterized by condensed chromatin, very little cytoplasm, and little transcriptional activity. Naive T cells continually recirculate between the blood and lymph systems. During recirculation, naive T cells reside in secondary lymphoid tissues such as lymph nodes. If a naive cell does not encounter antigen in a lymph node, it exits through the efferent lymphatics, ultimately draining into the thoracic duct and rejoining the blood. It is estimated that each naive T cell recirculates from the blood to the lymph nodes and back again every 12 to 24 hours. Because only about 1 in $10^5$ naive T cells is specific for any given antigen, this large-scale recirculation increases the chances that a naive T cell will encounter appropriate antigen.

## Activated T cells generate effector and memory T cells

If a naive T cell recognizes an antigen-MHC complex on an appropriate antigen-presenting cell or target cell, it will be activated, initiating a *primary response*. About 48 hours after activation, the naive T cell enlarges into a blast cell and begins undergoing repeated rounds of cell division. As described earlier, activation depends on a signal induced by engagement of the TCR complex and a costimulatory signal induced by the CD28-B7 interaction (see Figure 10-16). These signals trigger entry of the T cell into the $G_1$ phase of the cell cycle and, at the same time, induce transcription of the gene for IL-2 and the α chain of the high-affinity IL-2 receptor (CD25). In addition, the costimulatory signal increases the half-life of the IL-2 mRNA. The increase in IL-2 transcription, together with stabilization of the IL-2 mRNA, increases IL-2 production by 100-fold in the activated T cell. Secretion of IL-2 and its subsequent binding to the high-affinity IL-2 receptor induces the activated naive T cell to proliferate and differentiate (Figure 10-17). T cells activated in this way divide 2 to 3 times per day for 4 to 5 days, generating a clone of progeny cells, which differentiate into memory or effector T-cell populations.

The various *effector T cells* carry out specialized functions such as cytokine secretion and B-cell help (activated CD4$^+$ T$_H$ cells) and cytotoxic killing activity (CD8$^+$ CTLs). The generation and activity of CTL cells are described in detail in Chapter 14. Effector cells are derived from both naive and memory cells after antigen activation. Effector cells are short-lived cells, whose life spans range from a few days to a few weeks. The effector and naive populations express different cell membrane molecules, which contribute to different recirculation patterns.

As described in more detail in Chapter 12, CD4$^+$ effector T cells form two subpopulations distinguished by the different panels of cytokines they secrete. One population, called the **T$_H$1 subset,** secretes IL-2, IFN-γ, and TNF-β. The T$_H$1 subset is responsible for classic cell-mediated functions, such as delayed-type hypersensitivity and the activation of cytotoxic T lymphocytes. The other subset, called the **T$_H$2 subset,** secretes IL-4, IL-5, IL-6, and IL-10. This subset functions more effectively as a helper for B-cell activation.

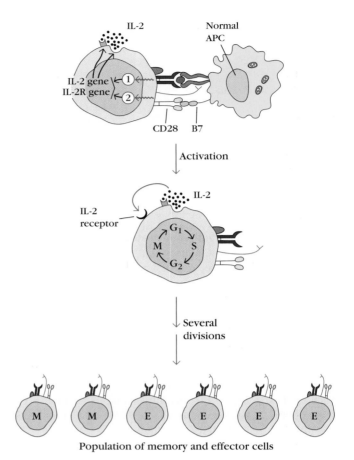

**FIGURE 10-17** Activation of a T$_H$ cell by both signal 1 and costimulatory signal 2 up-regulates expression of IL-2 and the high-affinity IL-2 receptor, leading to the entry of the T cell into the cell cycle and several rounds of proliferation. Some of the cells differentiate into effector cells, others into memory cells.

The *memory T-cell* population is derived both from naive T cells after they have encountered antigen and from effector cells after antigenic activation and differentiation. Memory T cells are antigen-generated, generally long-lived, quiescent cells that respond with heightened reactivity to a subsequent challenge with the same antigen, generating a *secondary response.* An expanded population of memory T cells appears to remain long after the population of effector T cells has declined. In general, memory T cells express many of the same cell surface markers as effector T cells; no cell surface markers definitively identify them as memory cells.

Like naive T cells, most memory T cells are resting cells in the $G_0$ stage of the cell cycle, but they appear to have less stringent requirements for activation than do naive T cells. For example, naive T$_H$ cells are activated almost exclusively by dendritic cells, whereas memory T$_H$ cells can be activated by macrophages, dendritic cells, and B cells. It is thought that the expression of high levels of numerous adhesion molecules by memory T$_H$ cells enables these cells to adhere to a broad spectrum of antigen-presenting cells. Memory cells also display recirculation patterns that differ from those of naive or effector T cells.

## A CD4⁺CD25⁺ subpopulation of T cells negatively regulates immune responses

During the early 1970s, investigators first described T-cell populations that could suppress immune responses. These cells were called suppressor T cells ($T_s$) and were believed to be CD8⁺ T cells. However, the cellular and molecular basis of the observed suppression remained obscure, and eventually great doubt was cast on the existence of CD8⁺ suppressor T cells. Recent research has shown that there are indeed T cells that suppress immune responses. Unexpectedly, these cells have turned out to be CD4⁺ rather than CD8⁺ T cells. Within the population of CD4⁺CD25⁺ T cells are regulatory T cells ($T_{reg}$ cells) that can inhibit the proliferation of other T-cell populations in vitro. Animal studies show that members of the CD4⁺CD25⁺ population inhibit the development of autoimmune diseases such as experimentally induced inflammatory bowel disease, experimental allergic encephalitis, and autoimmune diabetes. The suppression by these regulatory cells is antigen specific because it depends on activation through the T-cell receptor. Cell contact between the suppressing cells and their targets is required. If the regulatory cells are activated by antigen but separated from their targets by a permeable barrier, no suppression occurs. The existence of regulatory T cells that specifically suppress immune responses has clinical implications. The depletion or inhibition of regulatory T cells followed by immunization may enhance immune responses to conventional vaccines. In this regard, some have suggested that elimination of T cells that suppress responses to tumor antigens may facilitate the development of antitumor immunity. Conversely, increasing the suppressive activity of regulatory T-cell populations could be useful in the treatment of allergic or autoimmune diseases. The ability to increase the activity of regulatory T-cell populations might also be useful in suppressing organ and tissue rejection. Investigators doing future work on this regulatory cell population will seek deeper insights into the mechanisms by which members of CD4⁺CD25⁺ T-cell populations regulate immune responses. They will also make determined efforts to discover ways the activities of these populations can be increased to diminish unwanted immune responses and decreased to promote desirable ones.

## Antigen-presenting cells have characteristic costimulatory properties

Only professional antigen-presenting cells (dendritic cells, macrophages, and B cells) are able to present antigen together with class II MHC molecules and deliver the costimulatory signal necessary for complete T-cell activation that leads to proliferation and differentiation. The principal costimulatory molecules expressed on antigen-presenting cells are the glycoproteins B7-1 (CD80) and B7-2 (CD86) (see Figure 10-14). The professional antigen-presenting cells differ both in their ability to display antigen and to deliver the costimulatory signal (Figure 10-18).

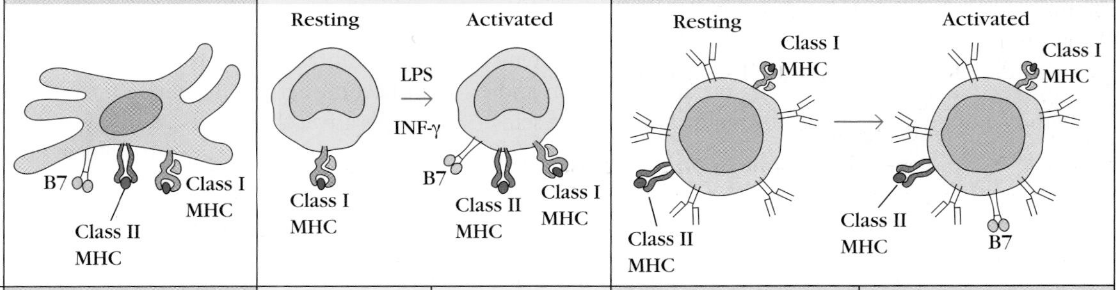

|  | Dendritic cell | Macrophage | | B lymphocyte | |
|---|---|---|---|---|---|
|  |  | Resting | Activated | Resting | Activated |
| Antigen uptake | Endocytosis phagocytosis (by Langerhans cells) | Phagocytosis | Phagocytosis | Receptor-mediated endocytosis | Receptor-mediated endocytosis |
| Class II MHC expression | Constitutive (+++) | Inducible (−) | Inducible (++) | Constitutive (++) | Constitutive (+++) |
| Costimulatory activity | Constitutive B7 (+++) | Inducible B7 (−) | Inducible B7 (++) | Inducible B7 (−) | Inducible B7 (++) |
| T-cell activation | Naive T cells Effector T cells Memory T cells | (−) | Effector T cells Memory T cells | Effector T cells Memory T cells | Naive T cells Effector T cells Memory T cells |

**FIGURE 10-18 Differences in the properties of professional antigen-presenting cells affect their ability to present antigen and induce T-cell activation.** Note that activation of effector and memory T cells does not require the costimulatory B7 molecule.

Dendritic cells constitutively express high levels of class I and class II MHC molecules as well as high levels of B7-1 and B7-2. For this reason, dendritic cells are very potent activators of naive, memory, and effector T cells. In contrast, all other professional APCs require activation for expression of costimulatory B7 molecules on their membranes; consequently, resting macrophages are not able to activate naive T cells and are poor activators of memory and effector T cells. Macrophages can be activated by phagocytosis of bacteria or by bacterial products such as LPS or by IFN-$\gamma$, a $T_H1$-derived cytokine. Activated macrophages up-regulate their expression of class II MHC molecules and costimulatory B7 molecules. Thus, activated macrophages are common activators of memory and effector T cells, but their effectiveness in activating naive T cells is considered minimal.

B cells also serve as antigen-presenting cells in T-cell activation. Resting B cells express class II MHC molecules but fail to express costimulatory B7 molecules. Consequently, resting B cells cannot activate naive T cells, although they can activate the effector and memory T-cell populations. On activation, B cells up-regulate their expression of class II MHC molecules and begin expressing B7. These activated B cells can now activate naive T cells as well as the memory and effector populations.

## Cell Death and T-Cell Populations

Cell death is an important feature of development in all multicellular organisms. During fetal life, it is used to mold and sculpt, removing unnecessary cells to provide shape and form. It also is an important feature of lymphocyte homeostasis, returning T- and B-cell populations to their appropriate levels after bursts of antigen-induced proliferation. Apoptosis also plays a crucial role in the deletion of potentially autoreactive thymocytes during negative selection and in the removal of developing T cells unable to recognize self-MHC molecules (failure to undergo positive selection).

Although the induction of apoptosis involves different signals depending on the cell types involved, the actual death of the cell is a highly conserved process among vertebrates and invertebrates. For example, T cells may be induced to die by many different signals, including the withdrawal of growth factor, treatment with glucocorticoids, or TCR signaling. Each of these signals engages unique signaling pathways, but in all cases, the actual execution of the cell involves the activation of a specialized set of proteases known as *caspases*. The role of these proteases was first revealed by studies of developmentally programmed cell deaths in the nematode *Caenorhabditis elegans*, where the death of cells was shown to be totally dependent on the activity of a gene that encoded a cysteine protease with specificity for aspartic acid residues. We now know that mammals have at least 14 cysteine proteases or caspases, and all cell deaths require the activity of at least a subset of these molecules. We also know that essentially every cell in the body produces caspase proteins, suggesting that every cell has the potential to initiate its own death.

Cells protect themselves from apoptotic death under normal circumstances by keeping caspases in an inactive form within a cell. On reception of the appropriate death signal, certain caspases are activated by proteolytic cleavage and then activate other caspases in turn, leading to the activation of *effector caspases*. This catalytic cascade culminates in cell death. Although it is not well understood how caspase activation directly results in apoptotic death of the cell, presumably it is through the cleavage of critical targets necessary for cell survival.

T cells use two different pathways to activate caspases (Figure 10-19). In peripheral T cells, antigen stimulation results in proliferation of the stimulated T cell and production of several cytokines, including IL-2. On activation, T cells increase the expression of two key cell surface proteins involved in T-cell death, Fas and Fas ligand (FasL). When Fas binds its ligand, FasL, FADD (Fas-associated protein with death domain) is recruited and binds to Fas, followed by the recruitment of procaspase 8, an inactive form of caspase 8. The association of FADD with procaspase 8 results in the proteolytic cleavage of procaspase 8 to its active form; caspase 8 then initiates a proteolytic cascade that leads to the death of the cell.

Other than in the thymus, most of the TCR-mediated apoptosis of mature T cells is mediated by the Fas pathway. Repeated or persistent stimulation of peripheral T cells results in the co-expression of both Fas and Fas ligand, followed by the apoptotic death of the cell. The Fas/FasL-mediated death of T cells as a consequence of activation is called *activation-induced cell death* (AICD) and is a major homeostatic mechanism, regulating the size of the pool of T cells and removing T cells that respond to stimulation by self antigens.

The importance of Fas and FasL in the removal of activated T cells is underscored by *lpr/lpr* mice, a naturally occurring mutation that results in nonfunctional Fas. When T cells become activated in these mice, the Fas/FasL pathway is not operative; the T cells continue to proliferate, producing IL-2 and maintaining an activated state. These mice spontaneously develop autoimmune disease, have excessive numbers of T cells, and clearly demonstrate the consequences of a failure to delete activated T cells. An additional mutation, *gld/gld*, is also informative. These mice lack functional FasL and display abnormalities similar to those found in the *lpr/lpr* mice. Recently, humans with defects in Fas have been reported. As expected, these individuals display characteristics of autoimmune disease (see Clinical Focus).

Fas and FasL are members of a family of related receptor/ligands including tumor necrosis factor (TNF) and its ligand, TNFR (tumor necrosis factor receptor). Like Fas and FasL, membrane-bound TNFR interacts with TNF to induce apoptosis. Also similar to Fas/FasL-induced apoptosis, TNF/TNFR-induced death is the result of the activation of caspase 8 followed by the activation of effector caspases such as caspase 3.

In addition to the activation of apoptosis through death receptor proteins such as Fas and TNFR, T cells can die through other pathways that do not activate procaspase 8. For example, negative selection in the thymus induces the

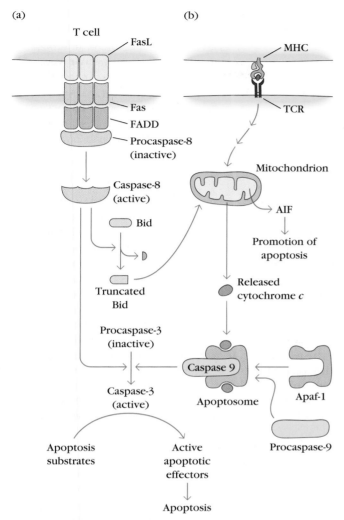

**FIGURE 10-19  Two pathways to apoptosis in T cells.** (a) Activated peripheral T cells are induced to express high levels of Fas and FasL. FasL induces the trimerization of Fas on a neighboring cell. FasL can also engage Fas on the same cell, resulting in a self-induced death signal. Trimerization of Fas leads to the recruitment of FADD, which leads in turn to the cleavage of associated molecules of procaspase 8 to form active caspase 8. Caspase 8 cleaves procaspase 3, producing active caspase 3, which results in the death of the cell. Caspase 8 can also cleave Bid to a truncated form that can activate the mitochondrial death pathway. (b) Other signals, such as the engagement of the TCR by peptide-MHC complexes on an APC, result in the activation of the mitochondrial death pathway. A key feature of this pathway is the release of AIF (apoptosis-inducing factor) and cytochrome *c* from the inner mitochondrial membrane into the cytosol. Cytochrome *c* interacts with Apaf-1 and subsequently with procaspase 9 to form the active apoptosome. The apoptosome initiates the cleavage of procaspase 3, producing active caspase 3, which initiates the execution phase of apoptosis by proteolysis of substances whose cleavage commits the cell to apoptosis. *[Adapted in part from S. H. Kaufmann and M. O. Hengartner, 2001,* Trends in Cell Biology **11**:526.*]*

apoptotic death of developing T cells via a signaling pathway that originates at the TCR. We still do not completely understand why some signals through the TCR induce positive selection and others induce negative selection, but we know that the strength of the signal plays a critical role. A strong, negatively selecting signal induces a route to apoptosis in which mitochondria play a central role. In mitochondrially dependent apoptotic pathways, cytochrome *c*, which normally resides in the inner mitochondrial membrane, leaks into the cytosol. Cytochrome *c* binds to a protein known as Apaf-1 (apoptotic protease-activating factor 1) and undergoes an ATP-dependent conformational change and oligomerization. Binding of the oligomeric form of Apaf-1 to procaspase 9 results in its transformation to active caspase 9. The complex of cytochrome *c*/Apaf-1/caspase 9, called the *apoptosome,* proteolytically cleaves procaspase 3, generating active caspase 3, which initiates a cascade of reactions that kills the cell (see Figure 10-19). Finally, mitochondria also release another molecule, AIF (apoptosis-inducing factor), which plays a role in the induction of cell death.

Cell death induced by Fas/FasL is swift, with rapid activation of the caspase cascade leading to cell death in 2 to 4 hours. On the other hand, TCR-induced negative selection appears to be a more circuitous pathway, requiring the activation of several processes, including mitochondrial membrane failure, the release of cytochrome *c*, and the formation of the apoptosome before caspases become involved. Consequently, TCR-mediated negative selection can take as long as 8 to 10 hours.

An important feature in the mitochondrially induced cell death pathway is the regulatory role played by Bcl-2 family members. Bcl-2 and Bcl-XL both reside in the mitochondrial membrane. These proteins are strong inhibitors of apoptosis, and although it is not clear how they inhibit cell death, one hypothesis is that they somehow regulate the release of cytochrome *c* from the mitochondria. There are at least three groups of Bcl-2 family members. Group I members are anti-apoptotic and include Bcl-2 and Bcl-XL. Group II and group III members are pro-apoptotic and include Bax and Bak in group II and Bid and Bim in group III. There is clear evidence that levels of anti-apoptotic Bcl-2 family members play an important role in regulating apoptosis in lymphocytes. Bcl-2 family members dimerize, and the anti-apoptotic group members may control apoptosis by dimerizing with pro-apoptotic members, blocking their activity. As indicated in Figure 10-19, cleavage of Bid, catalyzed by caspase 8 generated by the Fas pathway, can turn on the mitochondrial death pathway. Thus, signals initiated through Fas can also involve the mitochondrial pathway.

Although a lymphocyte can be signaled to die in several ways, all of these pathways to cell death converge on the activation of caspases. This part of the cell death pathway, the execution phase, is common to almost all death pathways known in both vertebrates and invertebrates, demonstrating that apoptosis is an ancient process that has been conserved throughout evolution.

## CLINICAL FOCUS
# Failure of Apoptosis Causes Defective Lymphocyte Homeostasis

**The** maintenance of appropriate numbers of various types of lymphocytes is extremely important to an effective immune system. One of the most important elements in this regulation is apoptosis mediated by the Fas/FasL ligand system. The following excerpts from medical histories show what can happen when this key regulatory mechanism fails.

**Patient A:** A woman, now 43, has had a long history of immunologic imbalances and other medical problems. By age 2, she was diagnosed with Canale-Smith syndrome (CSS), a severe enlargement of such lymphoid tissues as lymph nodes (lymphadenopathy) and spleen (splenomegaly). Biopsy of lymph nodes showed that, in common with many other CSS patients, she had greatly increased numbers of lymphocytes. She had reduced numbers of platelets (thrombocytopenia) and, because her red blood cells were being lysed, she was anemic (hemolytic anemia). The reduction in numbers of platelets and the lysis of red blood cells could be traced to the action of circulating antibodies that reacted with these host components. At age 21, she was diagnosed with grossly enlarged pelvic lymph nodes that had to be removed. Ten years later, she was again found to have an enlarged abdominal mass, which on surgical removal turned out to be a half-pound lymph node aggregate. She has continued to have mild lymphadenopathy and, typical of these patients, the lymphocyte populations of enlarged nodes had elevated numbers of T cells (87% as opposed to a normal range of 48%–67% T cells). Examination of these cells by flow cytometry and fluorescent antibody staining revealed an excess of double-negative T cells (Clinical Focus Figure 1). Also, like many patients with Canale-Smith syn-

drome, she has had cancer: breast cancer at age 22 and skin cancer at ages 22 and 41.

**Patient B:** A man who was eventually diagnosed with Canale-Smith syndrome had severe lymphadenopathy and splenomegaly as an infant and child. He was treated from age 4 to age 12 with corticosteroids and the immunosuppressive drug mercaptopurine. These appeared to help, and the swelling of lymphoid tissues became milder during adolescence and adulthood. At age 42, he died of liver cancer.

**Patient C:** An 8-year-old boy, the son of patient B, was also afflicted with Canale-Smith syndrome and showed elevated T-cell counts and severe lymphadenopathy at the age of seven months. At age 2 his spleen became so enlarged that it had to be removed. He also developed hemolytic

anemia and thrombocytopenia. However, although he continued to have elevated T-cell counts, the severity of his hemolytic anemia and thrombocytopenia have so far been controlled by treatment with methotrexate, a DNA-synthesis-inhibiting drug used for immunosuppression and cancer chemotherapy.

Recognition of the serious consequences of a failure to regulate the number of lymphocytes, as exemplified by these case histories, emerged from detailed study of several children whose enlarged lymphoid tissues attracted medical attention. In each of these cases of Canale-Smith syndrome, examination revealed grossly enlarged lymph nodes that were 1 to 2 cm in girth and sometimes large enough to distort the local anatomy. In four of a group of five children who were studied intensively, the spleens were so massive that they had to be removed.

Even though the clinical picture in Canale-Smith syndrome can vary from person to person, with some individuals suffer-

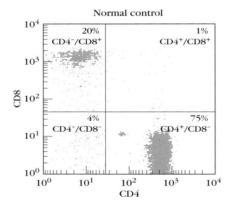
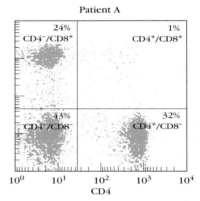

Flow-cytometric analysis of T cells in the blood of patient A and a control subject. The relative staining by an anti-CD8 antibody is shown on the *y* axis and the relative staining by an anti-CD4 antibody appears on the *x* axis. Mature T cells are either CD4$^+$ or CD8$^+$. Although almost all of the T cells in the control subject are CD4$^+$ or CD8$^+$, the CSS patient shows high numbers of double-negative T cells (43%), which express neither CD4 nor CD8. The percentage of each category of T cells is indicated in the quadrants. *[Adapted from M. D. Drappa et al., 1996, New England Journal of Medicine 335:1643.]*

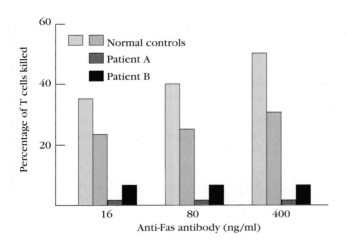

Fas-mediated killing takes place when Fas is cross-linked by FasL, its normal ligand, or by treatment with anti-Fas antibody, which artificially cross-links Fas molecules. This experiment shows the reduction in numbers of T cells after induction of apoptosis in T cells from patients and controls by cross-linking Fas with increasing amounts of an anti-Fas monoclonal antibody. T cells from the Canale-Smith patients (A and B) are resistant to Fas-mediated death. *[Adapted from M. D. Drappa et al., 1996, New England Journal of Medicine 335:1643.]*

ing severe chronic affliction and others only sporadic episodes of illness, there is a common feature: a failure of activated lymphocytes to undergo Fas-mediated apoptosis. Isolation and sequencing of Fas genes from a number of patients and more than 100 controls reveals that CSS patients are heterozygous ($fas^{+/-}$) at the *fas* locus and thus carry one copy of a defective *fas* gene. A comparison of Fas-mediated cell death in T cells from normal controls who do not carry mutant Fas genes with death induced in T cells from CSS patients shows a marked defect in Fas-induced death (Clinical Focus Figure 2). Characterization of the Fas genes so far seen in CSS patients reveals that they have mutations in or around the region encoding the death-inducing domain (the "death domain") of this protein (Clinical Focus Figure 3). Such mutations result in the production of Fas protein that lacks biological activity but still competes with normal Fas molecules for interactions with essential components of the Fas-mediated

death pathway. Other mutations have been found in the extracellular domain of Fas, often associated with milder forms of CSS or no disease at all.

A number of research groups have conducted detailed clinical studies of CSS patients, and the following general observations have been made:

- The cell populations of the blood and lymphoid tissues of CSS patients show dramatic elevations (fivefold to as much as 20-fold) in the numbers of lymphocytes of all sorts, including T cells, B cells, and NK cells.

- Most of the patients have elevated levels of one or more classes of immunoglobulin (hyper-gammaglobulinemia).

- Immune hyperactivity is responsible for such autoimmune phenomena as the production of autoantibodies against red blood cells, resulting in hemolytic anemia, and a depression in platelet counts due to the activity of antiplatelet autoantibodies.

These observations establish the importance of the death-mediated regulation of lymphocyte populations in lymphocyte homeostasis. Such cell death is necessary because the immune response to antigen results in a sudden and dramatic increase in the populations of responding clones of lymphocytes and temporarily distorts the representation of these clones in the repertoire. In the absence of cell death, the periodic stimulation of lymphocytes that occurs in the normal course of life would result in progressively increasing, and ultimately unsustainable, lymphocyte levels. As the Canale-Smith syndrome demonstrates, without the essential culling of lymphocytes by apoptosis, severe and life-threatening disease can result.

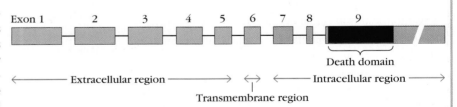

Map of *fas* locus. The *fas* gene is composed of nine exons separated by eight introns. Exons 1 through 5 encode the extracellular part of the protein, exon 6 encodes the transmembrane region, and exons 7 through 9 encode the intracellular region of the molecule. Much of exon 9 is responsible for encoding the critical death domain. *[Adapted from G. H. Fisher et al., 1995, Cell 81:935.]*

## SUMMARY

- Progenitor T cells from the bone marrow enter the thymus and rearrange their TCR genes. In most cases, these thymocytes rearrange $\alpha\beta$ TCR genes and become $\alpha\beta$ T cells. A small minority rearrange $\gamma\delta$ TCR genes and become $\gamma\delta$ T cells.

- The earliest thymocytes lack detectable CD4 and CD8 and are referred to as double-negative cells. During development, the majority of double-negative thymocytes develop into $CD4^+CD8^-$ $\alpha\beta$ T cells or $CD4^-CD8^+$ $\alpha\beta$ T cells.

- Positive selection in the thymus eliminates T cells unable to recognize self MHC and is the basis of MHC restriction. Negative selection eliminates thymocytes bearing high-affinity receptors for self-MHC molecules alone or self antigen plus self MHC and produces self-tolerance.

- T-cell activation is initiated by interaction of the TCR-CD3 complex with a peptide-MHC complex on an antigen-presenting cell. Activation also requires the activity of accessory molecules, including the coreceptors CD4 and CD8. Many different intracellular signal transduction pathways are activated by the engagement of the TCR.

- T cells that express CD4 recognize antigen combined with a class II MHC molecule and generally function as $T_H$ cells; T cells that express CD8 recognize antigen combined with a class I MHC molecule and generally function as $T_C$ cells.

- In addition to the signals mediated by the T-cell receptor and its associated accessory molecules (signal 1), activation of the $T_H$ cell requires a costimulatory signal (signal 2) provided by the antigen-presenting cell. The costimulatory signal is commonly induced by interaction between molecules of the B7 family on the membrane of the APC with CD28 on the $T_H$ cell. Engagement of CTLA-4, a close relative of CD28, by B7 inhibits T-cell activation.

- TCR engagement with antigenic peptide-MHC may induce activation or clonal anergy. The presence or absence of the costimulatory signal (signal 2) determines whether activation results in clonal expansion or clonal anergy.

- Naive T cells are resting cells $(G_0)$ that have not encountered antigen. Activation of naive cells leads to the generation of effector and memory T cells. Memory T cells, which are more easily activated than naive cells, are responsible for secondary responses. Effector cells are short-lived and perform helper, cytotoxic, or delayed-type hypersensitivity functions.

- The T-cell repertoire is shaped by apoptosis in the thymus and periphery.

## References

Ashton-Rickardt, P. G., et al. 1994. Evidence for a differential avidity model of T-cell selection in the thymus. *Cell* **74:**577.

Bosselut, R. 2004. CD4/CD8-lineage differentiation in the thymus: from nuclear effectors to membrane signals. *Nature Reviews Immunology* **4:**529–40.

Davis, D. M., and M. L. Dustin. 2004. What is the importance of the immunological synapse? *Trends in Immunology* **25:**323–27.

Drappa, M. D., et al. 1996. *Fas* gene mutations in the Canale-Smith syndrome, an inherited lymphoproliferative disorder associated with autoimmunity. *New England Journal of Medicine* **335:**1643.

Grakoui, A., et al. 1999. The immunological synapse: a molecular machine controlling T cell activation. *Science* **285:**221–27.

Hayday, A. 2000. $\gamma\delta$ Cells: a right time and right place for a conserved third way of protection. *Annual Review of Immunology* **18:**1975.

Herman, A., J. W. Kappler, P. Marrack, and A. M. Pullen. 1991. Superantigens: mechanism of T-cell stimulation and role in immune responses. *Annual Review of Immunology* **9:**745.

Huang, Y., and R. L. Wange. 2004. T cell receptor signaling: beyond complex complexes. *Journal of Biological Chemistry* **279:**28827–30.

Irvine, D. J., M. A. Purbhoo, M. Krogsgaard, and M. M. Davis. 2002. Direct observation of ligand recognition by T cells. *Nature* **419:**845–49.

Macian, F. 2005. NFAT proteins: key regulators of T-cell development and function. *Nature Review of Immunology* **5:**472–84.

Myung, P. S., N. J. Boerthe, and G. A. Koretzky. 2000. Adapter proteins in lymphocyte antigen-receptor signaling. *Current Opinion in Immunology* **12:**256.

Palmer, E. 2003. Negative selection—clearing out the bad apples from the T-cell repertoire. *Nature Reviews Immunology* **3:**383–91.

Rothenberg, E. V., and T. Taghon, T. 2005. Molecular genetics of T cell development. *Annual Review of Immunology* **23:**601–49.

Rothenberg, E. V., M. A. Yui, and J. C. Telfer. 2002. T cell developmental biology. In *Fundamental Immunology,* 5th ed., W. Paul, ed. Lippincott Williams and Wilkins, Philadelphia.

Salomon, B., and J. A. Bluestone. 2001. Complexities of CD28/B7: CTLA-4 costimulatory pathways in autoimmunity and transplantation. *Annual Review of Immunology* **19:**225.

Vaishnaw, A. K., et al. 1999. The molecular basis for apoptotic defects in patients with CD95 (Fas/Apo-1) mutations. *Journal of Clinical Investigation* **103:**355.

Weil, R., and A. Israel. 2004. T-cell-receptor- and B-cell-receptor-mediated activation of NF-kappaB in lymphocytes. *Current Opinion in Immunology* **16:**374–81.

Zuniga-Pflucker, J. C. 2004. T-cell development made simple. *Nature Review of Immunology* **4:**67–72.

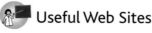 **Useful Web Sites**

**http://www.ncbi.nlm.nih.gov/Omim/**

The Online Mendelian Inheritance in Man Web site contains a subsite that features 10 different inherited diseases associated with defects in the TCR complex or associated proteins.

http://www.ultranet.com/~jkimball/BiologyPages/A/Apoptosis.html

http://www.ultranet.com/~jkimball/BiologyPages/B/B_and_Tcells.html

These subsites of John Kimball's Biology Pages Web site provide a clear introduction to T-cell biology and a good basic discussion of apoptosis.

http://www.bioscience.org/knockout/knochome.htm

In the Frontiers in Bioscience Database of Gene Knockouts, one can find information on the effects of knockouts of many genes involved in the development and function of T cells.

 ## Study Questions

**CLINICAL FOCUS QUESTION**   Over a period of several years, a group of children and adolescents are regularly dosed with compound X, a lifesaving drug. However, in addition to its beneficial effects, it was found that this drug interferes with Fas-mediated signaling.

a. What clinical manifestations of this side effect of compound X might be seen in these patients?

b. If white blood cells from an affected patient are stained with a fluorescein-labeled anti-CD4 and a phycoerythrin-labeled anti-CD8 antibody, what might be seen in the flow-cytometric analysis of these cells? What pattern would be expected if the same procedure were performed on white blood cells from a healthy control?

1. You have a CD8$^+$ CTL clone (from an H-2$^k$ mouse) that has a T-cell receptor specific for the H-Y antigen. You clone the αβ TCR genes from this cloned cell line and use them to prepare transgenic mice with the H-2$^k$ or H-2$^d$ haplotype.

a. How can you distinguish the immature thymocytes from the mature CD8$^+$ thymocytes in the transgenic mice?

b. For each transgenic mouse listed in the table below, indicate with (+) or (−) whether the mouse would have immature double-positive and mature CD8$^+$ thymocytes bearing the transgenic T-cell receptor.

c. Explain your answers for the H-2$^k$ transgenics.

d. Explain your answers for the H-2$^d$ transgenics.

| Transgenic mouse | Immature thymocytes | Mature CD8$^+$ thymocytes |
|---|---|---|
| H-2$^k$ female | | |
| H-2$^k$ male | | |
| H-2$^d$ female | | |
| H-2$^d$ male | | |

2. Cyclosporin and FK506 are powerful immunosuppressive drugs given to transplant recipients. Both drugs prevent the formation of a complex between calcineurin and Ca$^{2+}$/calmodulin. Explain why these compounds suppress T-cell-mediated aspects of transplant rejection. (Hint: See Figure 10-12.)

3. The antigenic activation of T cells leads to the release or induction of various nuclear factors that stimulate gene transcription.

a. What transcription factors that support proliferation of activated T cells are present in the cytoplasm of resting T cells in inactive forms?

b. Once in the nucleus, what might these transcription factors do?

4. You have fluorescein-labeled anti-CD4 and phycoerythrin-labeled anti-CD8. You use these antibodies to stain thymocytes and lymph-node cells from normal mice and from RAG-1 knockout mice. In the forms below, draw the FACS plots that you would expect.

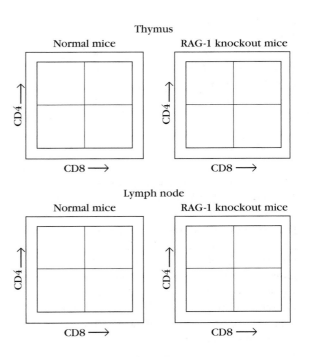

5. In order to demonstrate positive thymic selection experimentally, researchers analyzed the thymocytes from normal H-2$^b$ mice, which have a deletion of the class II *IE* gene, and from H-2$^b$ mice in which the class II *IA* gene had been knocked out.

a. What MHC molecules would you find on antigen-presenting cells from the normal H-2$^b$ mice?

b. What MHC molecules would you find on antigen-presenting cells from the IA knockout H-2$^b$ mice?

c. Would you expect to find CD4$^+$ T cells, CD8$^+$ T cells, or both in each type of mouse? Why?

6. In his classic chimeric mouse experiments, Zinkernagel took bone marrow from mouse 1 and a thymus from mouse 2 and transplanted them into mouse 3, which was thymectomized and lethally irradiated. He then challenged the reconstituted mouse with LCMV and removed its spleen cells. These spleen cells were incubated with LCMV-infected target cells with different MHC haplotypes, and the lysis of the target cells was monitored. The results of two such experiments using H-2$^b$ strain C57BL/6 mice and H-2$^d$ strain BALB/c mice are shown in the table at the top of page 270.

| Experiment | Bone marrow donor | Thymectomized x-irradiated recipient | Lysis of LCM-infected target cells | | |
|---|---|---|---|---|---|
| | | | $H\text{-}2^d$ | $H\text{-}2^k$ | $H\text{-}2^b$ |
| A | C57BL/6 × BALB/c | C57BL/6 × BALB/c | + | − | − |
| B | C57BL/6 × BALB/c | C57BL/6 × BALB/c | − | − | + |

a. What was the haplotype of the thymus-donor strain in experiment A and experiment B?

b. Why were the $H\text{-}2^b$ target cells not lysed in experiment A but were lysed in experiment B?

c. Why were the $H\text{-}2^k$ target cells not lysed in either experiment?

7. Fill in the blank(s) in each statement below (a–k) with the most appropriate term(s) from the box below. Terms may be used once, more than once, or not at all.

| protein phosphatase(s) | CD8 | class I MHC | CD45 |
|---|---|---|---|
| protein kinase(s) | CD4 | class II MHC | B7 |
| CD28 | IL-2 | IL-6 | CTLA-4 |

a. p56$^{\text{Lck}}$ and ZAP-70 are _____.

b. _____ is a T-cell membrane protein that has cytosolic domains with phosphatase activity.

c. Dendritic cells express _____ constitutively, whereas B cells must be activated before they express this membrane molecule.

d. Activation of $T_H$ cells results in secretion of _____ and expression of its receptor, leading to proliferation and differentiation.

e. The costimulatory signal needed for complete T-cell activation is triggered by interaction of _____ on the T cell and _____ on the APC.

f. Knockout mice lacking class I MHC molecules fail to produce thymocytes bearing _____.

g. Macrophages must be activated before they express _____ molecules and _____ molecules.

h. T cells bearing _____ are absent from the lymph nodes of knockout mice lacking class II MHC molecules.

i. $PIP_2$ is split by a _____ to yield DAG and $IP_3$.

j. In activated T cells, DAG activates a _____, which acts to generate the transcription factor NF-κB.

k. _____ stimulates and _____ inhibits T-cell activation when engaged by _____ or on antigen-presenting cells.

8. You wish to determine the percentage of various types of thymocytes in a sample of cells from mouse thymus using the indirect immunofluorescence method.

a. You first stain the sample with goat anti-CD3 (primary antibody) and then with rabbit FITC-labeled antigoat Ig (secondary antibody), which emits a green color. Analysis of the stained sample by flow cytometry indicates that 70% of the cells are stained. Based on this result, how many of the thymus cells in your sample are expressing antigen-binding receptors on their surface? Would all be expressing the same type of receptor? Explain your answer. What are the remaining unstained cells likely to be?

b. You then separate the CD3$^+$ cells with the fluorescence-activated cell sorter (FACS) and restain them. In this case, the primary antibody is hamster anti-CD4 and the secondary antibody is rabbit PE-labeled antihamster Ig, which emits a red color. Analysis of the stained CD3$^+$ cells shows that 80% of them are stained. From this result, can you determine how many $T_C$ cells are present in this sample? If yes, then how many $T_C$ cells are there? If no, what additional experiment would you perform to determine the number of $T_C$ cells that are present?

9. Many of the effects of engaging the TCR with MHC-peptide can be duplicated by the administration of ionomycin plus a phorbol ester. Ionomycin is a $Ca^{2+}$ ionophore, a compound that allows calcium ions in the medium to cross the plasma membrane and enter the cell. Phorbol esters are analogues of diacylglycerol (DAG). Why does the administration of phorbol and calcium ionophores mimic many effects of TCR engagement?

10. What effects on cell death would you expect to observe in mice carrying the following genetic modifications? Justify your answers.

a. Mice that are transgenic for BCL-2 and overexpress this protein.

b. Mice in which caspase 8 has been knocked out.

c. Mice in which caspase 3 has been knocked out.

11. Several basic themes of signal transduction were identified and discussed in this chapter. What are these themes? Consider the signal transduction processes of T-cell activation and provide an example for each of the five themes discussed.

12. Whereas the majority of T cell in our bodies express an αβ TCR, up to 5% of T cells express the γδ TCR instead. Explain the difference in antigen recognition between these two cell types.

13. T cells can be found that do not become stimulated when presented with their cognate antigen, whereas other T cells can be found that act as effector cells without having been presented with their cognate antigen. Explain how such T-cell activation/lack of activation can occur. Compare the signaling events that lead to these responses in your answer.

 Interactive Study

www.whfreeman.com/kuby

 SELF-TEST
Review of Key Terms

 ANIMATION
Cell Death

# chapter 11

# B-Cell Generation, Activation, and Differentiation

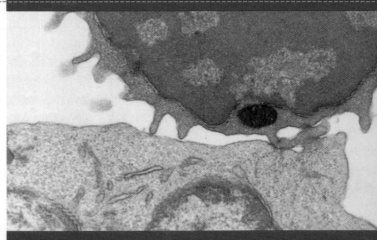

*Initial contact between B and T cells. [From V. M. Sanders et al., 1986, J. Immunol. 137:2395.]*

- B-Cell Maturation

- B-Cell Activation and Proliferation

- The Humoral Response

- In Vivo Sites for Induction of Humoral Responses

- Germinal Centers and Antigen-Induced B-Cell Differentiation

- Regulation of the Immune Effector Response

THE DEVELOPMENTAL PROCESS THAT RESULTS IN production of plasma cells and memory B cells can be divided into three broad stages: generation of mature, immunocompetent B cells (maturation), activation of mature B cells when they interact with antigen, and differentiation of activated B cells into plasma cells and memory B cells. In many vertebrates, including humans and mice, the bone marrow generates B cells. This process is an orderly sequence of Ig-gene rearrangements, which progresses in the absence of antigen. This is the antigen-independent phase of B-cell development.

An immature B cell bearing IgM in its membrane leaves the bone marrow and matures to express both membrane-bound IgM and IgD (mIgM and mIgD) with a single antigenic specificity. These **naive** B cells, which have not encountered antigen, circulate in the blood and lymph and are carried to the secondary lymphoid organs, most notably the spleen and lymph nodes (see Chapter 2). If a B cell is activated by interaction with an antigen recognized by its membrane-bound antibody, the cell proliferates (clonal expansion) and differentiates to generate a population of antibody-secreting plasma cells and memory B cells. As a consequence of activation, some B cells will undergo **affinity maturation,** the progressive increase in the average affinity of the antibodies produced as the response to activation proceeds. Many will also undergo **class switching,** the change in the isotype of the antibody produced by the B cell from μ to γ, α, or ε. Since B-cell activation and differentiation in the periphery requires antigen, this stage constitutes the antigen-dependent phase of B-cell development.

Many B cells are produced in the bone marrow throughout life, but very few of these cells mature. In mice, the size of the recirculating pool of B cells is about $2 \times 10^9$ cells. Most of these cells circulate as naive B cells, which have short life spans (half-lives of less than 3 days to about 8 weeks) if they fail to encounter antigen or lose in the competition with other B cells for residence in a supportive lymphoid environment. Given that the immune system is able to generate a total antibody diversity that is well above $10^{10}$, only a small fraction of this potential repertoire is displayed at any time by membrane immunoglobulin on recirculating B cells. Some aspects of B-cell developmental processes have been described in previous chapters. The overall pathway, beginning with the earliest distinctive B-lineage cell, is described in sequence in this chapter. Figure 11-1 presents an overview of the major events during B-cell development in humans and mice. Studies of a wide variety of vertebrates have shown, however, a surprising diversity of pathways and outcomes in B-cell development, some of which will be highlighted.

## B-Cell Maturation

The generation of mature B cells first occurs during embryonic stages and continues throughout life. Before birth, the yolk sac, fetal liver, and fetal bone marrow are the major sites of B-cell maturation; after birth, generation of mature B cells occurs in the bone marrow.

OVERVIEW FIGURE 11-1: **B-Cell Development**

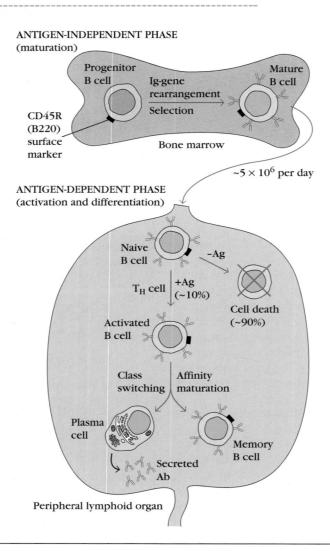

During the antigen-independent maturation phase, immature B cells expressing membrane IgM are generated in the bone marrow. These cells enter the bloodstream and develop into mature naive B cells that express both mIgM and mIgD. Only about 10% of the pool of potential B cells reaches maturity and exits the bone marrow. Naive B cells in the periphery die unless they encounter solu-ble protein antigen and activated $T_H$ cells. Once activated, B cells proliferate within secondary lymphoid organs. Those bearing high-affinity mIg differentiate into plasma cells and memory B cells, which may express different isotypes because of class switching. The numbers cited in the figure refer to B-cell development in the mouse, but the overall principles apply to humans as well.

## Progenitor B cells proliferate in bone marrow

B-cell development begins as lymphoid precursor cells differentiate into the earliest distinctive B-lineage cell—the **progenitor B cell (pro–B cell)**—which expresses a transmembrane tyrosine phosphatase called CD45R (sometimes called B220 in mice). Proliferation and differentiation of pro–B cells into **precursor B cells (pre–B cells)** requires the microenvironment provided by the bone marrow stromal cells. If pro–B cells are removed from the bone marrow and cultured in vitro, they will not progress to more mature B-cell stages unless stromal cells are present. The stromal cells play two important roles: they interact directly with pro–B and pre–B cells, and they secrete various cytokines, notably IL-7, that support the developmental process.

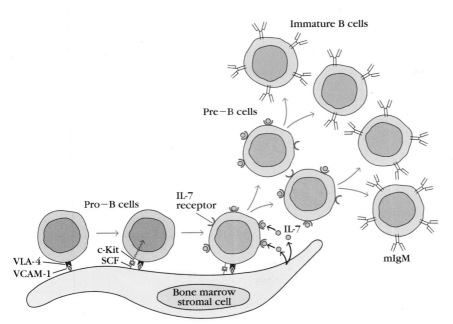

**FIGURE 11-2 Bone marrow stromal cells are required for maturation of progenitor B cells.** Pro–B cells bind to stromal cells by means of an interaction between VCAM-1 on the stromal cell and VLA-4 on the pro–B cell. This interaction promotes the binding of c-Kit on the pro–B cell to stem cell factor (SCF) on the stromal cell, which triggers a signal, mediated by the tyrosine kinase activity of c-Kit, that stimulates the pro–B cell to express receptors for IL-7. IL-7 released from the stromal cell then binds to the IL-7 receptors, inducing the pro–B cell to mature into a pre–B cell. Proliferation and differentiation eventually produces immature B cells.

At the earliest developmental stage, pro–B cells require direct contact with stromal cells in the bone marrow. This interaction is mediated by several cell adhesion molecules (CAMs), including VLA-4 on the pro–B cell and its ligand, VCAM-1, on the stromal cell (Figure 11-2). After initial contact is made, a receptor on the pro–B cell called c-Kit interacts with a stromal cell surface molecule known as stem cell factor (SCF). This interaction activates the thyrosine kinase activity of c-Kit, and the pro–B cell begins to divide and differentiate into a pre–B cell. The IL-7 secreted by the stromal cells binds the IL-7 receptor on the pre–B cell and drives the maturation process, eventually inducing down-regulation of the adhesion molecules on the pre–B cells so that the proliferating cells can detach from the stromal cells. At this stage, pre–B cell development no longer requires contact with stromal cells but continues to require IL-7.

## Ig-gene rearrangement produces immature B cells

B-cell maturation depends on rearrangement of the immunoglobulin DNA in the lymphoid stem cells. The mechanisms of Ig-gene rearrangement were described in Chapter 5. First to occur in the pro–B cell stage is a heavy-chain $D_H$-to-$J_H$ gene rearrangement, followed by a $V_H$-to-$D_H J_H$ rearrangement (Figure 11-3). If the first heavy-chain rearrangement is not productive, then $V_H$-$D_H$-$J_H$ rearrangement continues on the other chromosome. On completion of heavy-chain rearrangement, the cell is classified as a pre–B cell. Continued development of a pre–B cell into an immature B cell requires a productive light-chain gene rearrangement. Because of allelic exclusion, only one light-chain isotype is expressed on the membrane of a B cell. Comple-

tion of a productive light-chain rearrangement commits the now immature B cell to a particular antigenic specificity determined by the cell's heavy-chain VDJ sequence and light-chain VJ sequence. Immature B cells express mIgM (membrane IgM) on the cell surface along with Ig-α and Ig-β, thus forming a B-cell receptor (BCR) fully capable of signaling the engagement of antigen.

As would be expected, the recombinase enzymes RAG-1 and RAG-2, which are required for both heavy-chain and light-chain gene rearrangements, are expressed during the pro-B and pre-B-cell stages (see Figure 11-3). The enzyme terminal deoxyribonucleotidyl transferase (TdT), which catalyzes insertion of N-nucleotides at the $D_H$-$J_H$ and $V_H$-$D_H$-$J_H$ coding joints, is active during the pro-B-cell stage and ceases to be active early in the pre-B-cell stage. Because TdT expression is turned off during the part of the pre-B-cell stage when light-chain rearrangement occurs, N-nucleotides are not usually found in the $V_L$-$J_L$ coding joints.

The bone marrow phase of B-cell development culminates in the production of an IgM-bearing immature B cell. At this stage of development the B cell is not fully functional, and engagement of the BCR by antigen induces death or unresponsiveness (anergy) rather than division and differentiation. Full maturation is signaled by the co-expression of IgD and IgM on the membrane. This progression involves a change in RNA processing of the heavy-chain primary transcript to permit production of two mRNAs, one encoding the membrane form of the μ chain and the other encoding the membrane form of the δ chain. Although IgD is a characteristic cell surface marker of mature naive B cells, its function is not clear. However, since immunoglobulin δ knockout mice have essentially normal numbers of fully functional B cells, IgD is not essential to either B-cell development or antigen responsiveness.

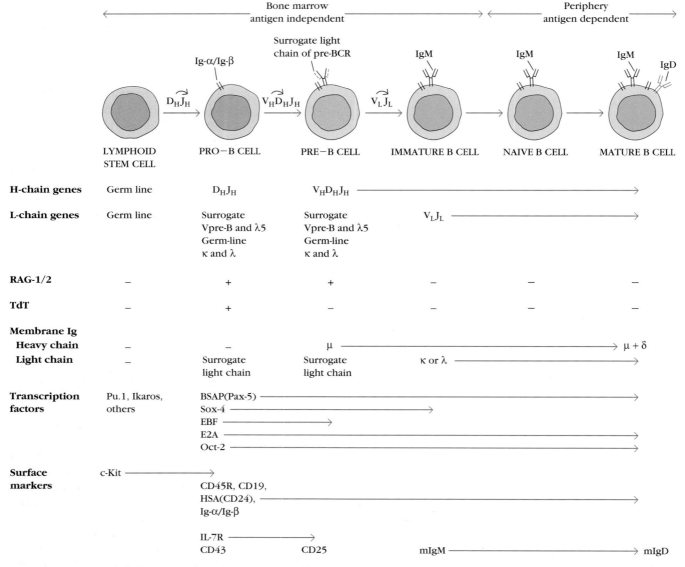

**FIGURE 11-3 Sequence of events and characteristics of the stages in B-cell maturation in the bone marrow.** The pre–B cell expresses a membrane immunoglobulin consisting of a heavy (H) chain and surrogate light chains, Vpre-B and λ5. Changes in the RNA processing of heavy-chain transcripts following the pre-B-cell stage lead to synthesis of both membrane-bound IgM and IgD by mature B cells. RAG-1/2 = two enzymes encoded by recombination-activating genes; TdT = terminal deoxyribonucleotidyl transferase. A number of B-cell-associated transcription factors are important at various stages of B-cell development; some are indicated here.

## The pre-B-cell receptor is essential for B-cell development

As we saw in Chapter 10, during one stage in T-cell development, the β chain of the T-cell receptor associates with pre-Tα to form the pre-T-cell receptor (see Figure 10-2). A parallel situation occurs during B-cell development. In the pre–B cell, the membrane μ chain is associated with the **surrogate light chain,** a complex consisting of two proteins: a V-like sequence called **Vpre-B** and a C-like sequence called λ5, which associate noncovalently to form a light-chain-like structure.

The membrane-bound complex of μ heavy chain and surrogate light chain appears on the pre–B cell associated with the Ig-α/Ig-β heterodimer to form the pre-B-cell receptor (Figure 11-4). Only pre–B cells that are able to express membrane-bound μ heavy chains in association with surrogate light chains are able to proceed along the maturation pathway.

Some researchers speculate that the pre-B-cell receptor transmits signals essential for the continued progression of B-cell development and for the prevention of $V_H$ to $D_H J_H$ rearrangement of the other heavy-chain allele, thus leading to allelic exclusion. Following the establishment of an effective pre-B-cell receptor, each pre–B cell undergoes multiple cell divisions, perhaps six to eight, producing as many as 256 descendants. Each of these progeny pre–B cells may then

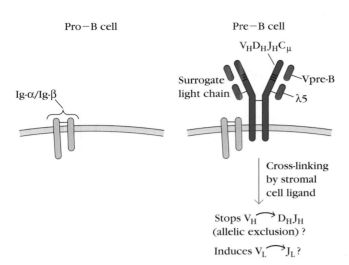

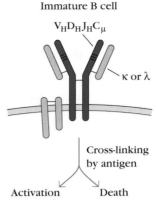

**FIGURE 11-4** **Schematic diagram of sequential expression of membrane immunoglobulin and surrogate light chain at different stages of B-cell differentiation in the bone marrow.** The pre-B-cell receptor contains a surrogate light chain consisting of a Vpre-B polypeptide and a λ5 polypeptide, which are noncovalently associated. The immature B cell no longer expresses the surrogate light chain and instead expresses the κ or λ light chain together with the μ heavy chain.

rearrange different light-chain gene segments, thereby increasing the overall diversity of the antibody repertoire.

The critical role of the pre-B-cell receptor was demonstrated with knockout mice in which the gene encoding the λ5 protein of the receptor was disrupted. B-cell development in these mice was shown to be blocked at the pre–B stage, which suggests that a signal generated through the receptor is necessary for pre–B cells to proceed to the immature B-cell stage.

## Knockout experiments identified essential transcription factors

As described in Chapter 2, many different transcription factors act in the development of hematopoietic cells. Nearly a dozen of them have so far been shown to play roles in B-cell development. Experiments in which particular transcription factors are knocked out by gene disruption have shown that four such factors, **E2A, early B-cell factor (EBF), B-cell-specific activator protein (BSAP),** and **Sox-4,** are particularly important for B-cell development (see Figure 11-3). All of these transcription factors affect development at an early stage; some of them are active at later stages also. Mice whose B cells lack E2A do not express RAG-1, are unable to make $D_H J_H$ rearrangements, and fail to express λ5, a critical component of the surrogate light chain. A similar pattern is seen in EBF-deficient mice. These findings point to important roles for both of these transcription factors in early B-cell development, and they may play essential roles in the early stages of commitment to the B-cell lineage. Knocking out the **Pax-5** gene, whose product is the transcription factor BSAP, also results in the arrest of B-cell development at an early stage. Binding sites for BSAP are found in the promoter regions of a number of B-cell-specific genes, including *Vpre-B* and λ5, the promoter for *RAG-2,* in a number of Ig switch regions and in the Ig heavy-chain enhancer. These findings indicate that BSAP plays a role beyond the early stages of B-cell development. BSAP is also expressed in the central nervous system, and its absence results in severe defects in midbrain development. Although the exact site of

action of the Sox-4 transcription factor is not known, it is required for early stages of B-cell activation.

## Cell-surface markers identify developmental stages

The developmental progression from progenitor to mature B cell is typified by a changing pattern of surface markers (see Figure 11-3). At the pro–B stage, the cells do not display the heavy or light chains of antibody, but they do express CD45R, a protein tyrosine phosphatase found on leukocytes, and the signal-transducing molecules Ig-α/Ig-β, which are found in association with the membrane forms of antibody in later stages of B-cell development. Also expressed on the surfaces of pro–B cells are CD19 (part of the B-cell coreceptor), CD43 (leukosialin), and CD24, a molecule also known as heat-stable antigen (HSA). At this stage, c-Kit, a receptor for a growth-promoting ligand present on stromal cells, is also found on the surface of pro–B cells. As cells progress from the pro–B to the pre–B stage, they express many of the same markers that were present during the pro–B stage; however, they cease to express c-Kit and CD43 and begin to express CD25, the α chain of the IL-2 receptor. The display of the pre-B-cell receptor (pre-BCR) is a salient feature of the pre-B-cell stage. After rearrangement of the light chain, surface immunoglobulin containing both heavy and light chains appears, and the cells, now classified as immature B cells, lose the pre-BCR and no longer express CD25. Monoclonal antibodies are available that can recognize all of these antigenic markers, making it possible to recognize and isolate the various stages of B-cell development by the techniques of immunohistology and flow cytometry described in Chapter 6.

## B-1 B cells are a self-renewing B-cell subset

A subset of B cells, called B-1 B cells, arises before the major group of B cells (sometimes called B-2 cells because of their later appearance) in humans and mice. In humans and mice, B-1 B cells make up only about 5% of the total B-cell

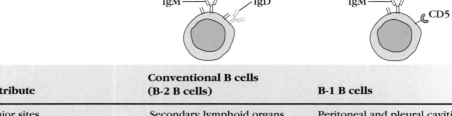

| Attribute | Conventional B cells (B-2 B cells) | B-1 B cells |
| --- | --- | --- |
| Major sites | Secondary lymphoid organs | Peritoneal and pleural cavities |
| Source of new B cells | From precursors in bone marrow | Self-renewing (division of existing B-1 cells) |
| V-region diversity | Highly diverse | Restricted diversity |
| Somatic hypermutation | Yes | No |
| Requirements for T-cell help | Yes | No |
| Isotypes produced | High levels of IgG | High levels of IgM |
| Response to carbohydrate antigens | Possibly | Definitely |
| Response to protein antigens | Definitely | Possibly |
| Memory | Yes | Very little or none |
| Surface IgD on mature B cells | Present on naive B cells | Little or none |

**FIGURE 11-5** **Comparison of conventional B cells and B-1 B cells.** Conventional B cells, the major B-cell population, are sometimes called B-2 cells because they develop after B-1 cells.

population. (In some species, such as rabbits and cattle, cells with many of the characteristics of B-1 B cells are the major B-cell population.) B-1 B cells arise from stem cells during fetal life and express surface Ig, the signature marker of B cells. However, most B-1 B cells bear surface IgM instead of IgG and have little or no IgD. Unlike naive B-2 B cells, the B-1 B cell population is self-renewing and can generate naive B-1 cells. B-1 B cells are also distinguished by the display of CD5, a marker usually associated with T cells. However, CD5 is not an indispensable component of the B-1 lineage. It does not appear on the B-1 cells of rats, and mice that lack a functional CD5 gene still produce B-1 cells. In animals such as humans and mice, where B-2 cells are the major B-cell population, B-1 cells are minor populations in such secondary tissues as lymph nodes and spleen. Despite their scarcity in many lymphoid sites, they are the major B-cell population of the peritoneal and pleural cavities.

In addition to self-renewal and the expression of CD5, other features set B-1 B cells apart from the B-2 B cells of humans and mice. The repertoire of V regions in B-1 B cells is much more restricted than that of the dominant B-2 population. Also, the antibodies produced by B-1 B cells generally bind antigen with lower affinity than the antibodies produced by the B-2 population. Studies of the B-1 population shows that these cells are more likely to be activated in response to carbohydrate antigen than to protein. They do not require T-cell help and exhibit little proclivity to differentiate into memory cells. Class switching is not common, and as mentioned above, most B-1 cells are IgM bearing. This

population also shows little or no somatic hypermutation of Ig genes. Consequently, there is no affinity maturation, which may be one of the reasons antibodies produced by a high proportion of B-1 cells exhibit low affinity for their antigens. Despite their lower affinity, many of the antibodies produced by B-1 cells are multispecific; that is, they can bind to many different antigens, including some present on pathogens. Some immunologists have suggested that the multispecific antibodies produced by this subset of B cells contribute to the innate immunity of the host. The properties of B-1 and B-2 cells are summarized in Figure 11-5.

## Self-reactive B cells are selected against in bone marrow

Mouse bone marrow is estimated to produce about $5 \times 10^7$ B cells per day, but only $5 \times 10^6$ (or about 10%) are actually recruited into the recirculating B-cell pool. This means that 90% of the B cells produced each day die without ever leaving the bone marrow. Some of this loss is attributable to **negative selection** and subsequent elimination (**clonal deletion**) of immature B cells that express auto-antibodies against self antigens in the bone marrow.

It has long been established that the cross-linkage of mIgM on immature B cells, demonstrated experimentally by treating immature B cells with antibody against the μ constant region, can cause the cells to die by apoptosis within the bone marrow. A similar process is thought to occur in vivo when immature B cells that express self-reactive mIgM

(a) H-2$^{d/k}$ transgenics

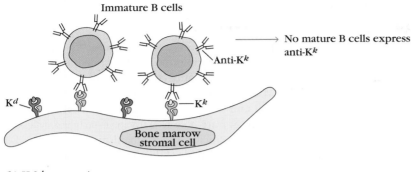

Immature B cells

No mature B cells express anti-K$^k$

Anti-K$^k$

K$^d$

K$^k$

Bone marrow stromal cell

(b) H-2$^d$ transgenics

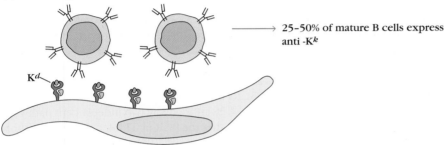

25–50% of mature B cells express anti -K$^k$

K$^d$

(c) H-2$^{d/k}$ transgenics

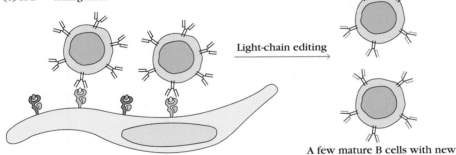

Light-chain editing

A few mature B cells with new light chains no longer bind K$^k$

**FIGURE 11-6 Experimental evidence for negative selection (clonal deletion) of self-reactive B cells during maturation in the bone marrow.** The presence or absence of mature peripheral B cells expressing a transgene-encoded IgM against the H-2 class I molecule K$^k$ was determined in H-2$^{d/k}$ mice (a) and H-2$^d$ mice (b). In the H-2$^{d/k}$ transgenics, the immature B cells recognized the self-antigen K$^k$ and were deleted by negative selection. In the H-2$^d$ transgenics, the immature B cells did not bind to a self antigen and consequently went on to mature, so that 25% to 50% of the splenic B cells expressed the transgene-encoded anti-K$^k$ as membrane Ig. More detailed analysis of the H-2$^{d/k}$ transgenics revealed a few peripheral B cells that expressed the transgene-encoded μ chain but a different light chain (c). Apparently, a few immature B cells underwent light-chain editing, so they no longer bound the K$^k$ molecule and consequently escaped negative selection. *[Adapted from D. A. Nemazee and K. Burki, 1989,* Nature **337***:562; S. L. Tiegs et al., 1993,* Journal of Experimental Medicine **177***:1009.]*

bind to self antigens in the bone marrow. For example, D. A. Nemazee and K. Burki introduced a transgene encoding the heavy and light chains of an IgM antibody specific for K$^k$, an H-2$^k$ class I MHC molecule, into H-2$^d$ and H-2$^{d/k}$ mice (Figure 11-6a, b). Since class I MHC molecules are expressed on the membrane of all nucleated cells, the endogenous H-2$^k$ and H-2$^d$ class I MHC molecules would be present on bone mar-

row stromal cells in the transgenic mice. In the H-2$^d$ mice, which do not express K$^k$, most of the mature, peripheral B cells expressed the transgene-encoded anti-K$^k$ both as a membrane antibody and as secreted antibody. In contrast, in the H-2$^{d/k}$ mice, which express K$^k$, no mature, peripheral B cells expressed the transgene-encoded antibody to H-2$^k$ (Table 11-1). These results suggest that there is negative

| TABLE 11-1 | Expression of transgene encoding IgM antibody to H-2$^k$ class I MHC molecules | | |
|---|---|---|---|
| | | EXPRESSION OF TRANSGENE | |
| Experimental animal | Number of animals tested | As membrane Ab | As secreted Ab (μg/ml) |
| Nontransgenics | 13 | (−) | <0.3 |
| H-2$^d$ transgenics | 7 | (+) | 93.0 |
| H-2$^{d/k}$ transgenics | 6 | (−) | <0.3 |
| SOURCE: Adapted from D. A. Nemazee and K. Burki, 1989, *Nature* **337**:562. | | | |

selection against any immature B cells expressing auto-antibodies on their membranes because these antibodies react with self antigen (e.g., the K$^k$ molecule in H-2$^{d/k}$ transgenics) present on stromal cells, leading to cross-linking of the antibodies and subsequent death of the immature B cells.

## Self-reactive B cells may be rescued by editing of light-chain genes

Later work using the transgenic system described by Nemazee and Burki showed that negative selection of immature B cells does not always result in their immediate deletion (Figure 11-6c). Instead, maturation of the self-reactive cell is arrested while the B cell "edits" the light-chain gene of its receptor. In this case, the H-2$^{d/k}$ transgenics produced a few mature B cells that expressed mIgM containing the μ chain encoded in the transgene but different, endogenous light chains. When some immature B cells bind a self antigen, maturation is arrested and the cells up-regulate *RAG-1* and *RAG-2* expression and begin further rearrangement of their endogenous light-chain DNA. Some of these cells succeed in replacing the light chain of the self-antigen-reactive antibody with a different light chain encoded by other light-chain gene segments. As a result, these cells will begin to express an "edited" mIgM with a different light chain. If the use of a different light chain results in a BCR that is not self-reactive, these cells escape negative selection and leave the bone marrow. Recent evidence suggests that receptor editing is the major mechanism of tolerance to antigen expressed in bone marrow.

# B-Cell Activation and Proliferation

After export of B cells from the bone marrow, activation, proliferation, and differentiation occur in the periphery in response to antigen. Antigen-driven activation and clonal selection of naive B cells leads to generation of plasma cells and memory B cells. In the absence of antigen-induced activation, naive B cells in the periphery have a short life span, dying within a few weeks by apoptosis (see Figure 11-1).

## Thymus-dependent and thymus-independent antigen have different requirements for response

Observations in nude mice and in patients born without a thymus led to the realization that depending on the nature of the antigen, there are two routes from B-cell activation to the production of antibody. Investigators found that although responses to most antigens required a thymus (**thymus-dependent antigens**), a few types of antigens (**thymus-independent antigens**) induced antibody even in the absence of a thymus. The B-cell response to thymus-dependent (TD) antigens requires direct contact with T$_H$ cells, not simply exposure to T$_H$-derived cytokines. Antigens that can activate B cells in the absence of this kind of direct participation by T$_H$

cells, thymus-independent (TI) antigens, are divided into types 1 and 2, and they activate B cells by different mechanisms. Some bacterial cell wall components, including lipopolysaccharide, function as *type 1 thymus-independent (TI-1) antigens*. *Type 2 thymus-independent (TI-2) antigens* are highly repetitive molecules such as polymeric proteins (e.g., bacterial flagellin) or bacterial cell wall polysaccharides with repeating polysaccharide units.

Most TI-1 antigens are polyclonal B-cell activators (**mitogens**); that is, they are able to activate B cells regardless of their antigenic specificity. The mechanism by which the T-1 antigen **lipopolysaccharide (LPS)**, a major component of the cell walls of gram-negative bacteria, induces antibody production independent of the presence of T cells is well understood. LPS interacts with two different receptors on B cells. One is the Toll-like receptor TLR4, an important receptor of the innate immune system (see Chapter 3). The other is the B-cell receptor. Only a few members of the B-cell population bear BCRs specific for LPS, but all of them have TLR4. Those B cells with BCRs that recognize LPS are stimulated to divide and secrete anti-LPS antibody by two independent pathways, originating with either the BCR or TLR4. Other B cells (those whose BCRs do not recognize LPS) are induced to divide and differentiate into antibody-secreting cells by the interaction of LPS with TLR4. The activation of many different B-cell clones by LPS produces a highly diverse collection of antibodies, some of which may interact with the invading gram-negative bacteria and bring about its neutralization.

The T-cell independence of TI-1 antigens is nicely illustrated by experiments using nude mice, which lack a thymus and have very few T cells. If a population of nude mice is divided into two groups, A and B, and group A is injected with T cells from normal mice, then both groups are immunized with a TI-1 antigen, both groups of animals make very similar levels of antibody against the immunizing antigen. The fact that the response in athymic mice is not greatly augmented by the injection of T cells demonstrates that TI-1 antigens are truly T cell independent.

TI-2 antigens activate B cells by extensively cross-linking the mIg receptor. TI-2 antigens differ from TI-1 antigens in three important respects. First, they are not B-cell mitogens and so do not act as polyclonal activators. Second, TI-1 antigens will activate both mature and immature B cells, whereas TI-2 antigens activate mature B cells but inactivate immature B cells. Third, although the B-cell response to TI-2 antigens does not require direct involvement of T$_H$ cells, cytokines derived from T$_H$ cells are required for efficient B-cell proliferation and for class switching to isotypes other than IgM. This can be shown by comparing the effect of TI-2 antigens in mice made T cell–deficient in various ways. In nude mice, which lack thymus-derived T cells but do contain a few T cells that probably arise in the intestine, TI-2 antigens do elicit B-cell responses. The partial dependence of TI-2 antigens on small populations of T cells is demonstrated by experiments showing that the administration of a few T cells to these T cell-deficient mice restores their ability to elicit B-cell responses to TI-2 antigens.

| TABLE 11-2 | Properties of thymus-dependent and thymus-independent antigens | | |
|---|---|---|---|
| | | TI antigens | |
| Property | TD antigens | Type 1 | Type 2 |
| Chemical nature | Soluble protein | Bacterial cell-wall components (e.g., LPS) | Polymeric protein antigens; capsular polysaccharides |
| Humoral response | | | |
|   Isotype switching | Yes | No | Limited |
|   Affinity maturation | Yes | No | No |
|   Immunologic memory | Yes | No | No |
|   Polyclonal activation | No | Yes (high doses) | No |

The humoral response to thymus-independent antigens is different from the response to thymus-dependent antigens (Table 11-2). The response to TI antigens is generally weaker, no memory cells are formed, and IgM is the predominant antibody secreted, reflecting a low level of class switching. These differences highlight the important role played by $T_H$ cells in generating memory B cells, affinity maturation, and class switching to other isotypes.

## Two types of signals drive B cells into and through the cell cycle

Naive, or resting, B cells are nondividing cells in the $G_0$ stage of the cell cycle. Activation drives the resting cell into the cell cycle, progressing through $G_1$ into the S phase, in which DNA is replicated. The transition from $G_1$ to S is a critical restriction point in the cell cycle. Once a cell has reached S, it completes the cell cycle, moving through $G_2$ and into mitosis (M). The activation of a B cell to proceed through the cell cycle requires not one, but two distinct sets of signaling events: the first set is collectively designated *signal 1* and the second set, *signal 2*. These signaling events are generated by different pathways in response to TI or TD antigens, but both pathways include signals generated when multivalent antigen binds and cross-links mIg (Figure 11-7). Once the B

cell has been stimulated by signal 1, signal 2, which may include cytokines and possibly other ligands, results in full activation, leading to cell division and differentiation into memory cells or antibody-secreting plasma cells.

## Transduction of activating signals involves Ig-α/Ig-β heterodimers

For many years, immunologists questioned how engagement of the Ig receptor by antigen could activate intracellular signaling pathways. All isotypes of mIg have very short cytoplasmic tails. Both mIgM and mIgD on B cells extend into the cytoplasm by only three amino acids, the mIgA tail consists of 14 amino acids, and the mIgG and mIgE tails contains 28 amino acids. In each case, the cytoplasmic tail is too short to be able to generate a signal by associating with intracellular signaling molecules, such as tyrosine kinases and G proteins. The discovery that membrane Ig is associated with the disulfide-linked heterodimer Ig-α/Ig-β, forming the **B-cell receptor (BCR),** solved this long-standing puzzle. Although it was originally thought that two Ig-α/Ig-β heterodimers associated with one mIg to form the B-cell receptor, careful biochemical analysis has shown that only one Ig-α/Ig-β heterodimer associates with a single mIg molecule to form the receptor complex (Figure 11-8). Thus, the BCR is functionally divided into the ligand-binding immunoglobulin molecule and the signal-transducing Ig-α/Ig-β heterodimer. A similar functional division marks the pre-BCR, which transduces signals via a complex consisting of an Ig-α/Ig-β heterodimer and μ heavy chains combined with the surrogate light chain (see Figure 11-4). The Ig-α chain has a long cytoplasmic tail containing 61 amino acids; the tail of the Ig-β chain contains 48 amino acids. The cytoplasmic tails of both Ig-α and Ig-β contain the 18-residue motif termed the **immunoreceptor *tyrosine-based activation *motif** (ITAM; Figure 11-8), which is also present in several molecules of the T-cell-receptor complex (see Figure 10-11). Interactions with the cytoplasmic tails of Ig-α/Ig-β transduce the stimulus produced by cross-linking of mIg molecules into an effective intracellular signal. This pattern of different members of a multimeric receptor assuming the functions of ligand binding and signal transduction is also

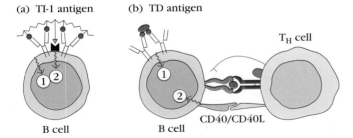

(a) TI-1 antigen     (b) TD antigen

$T_H$ cell

CD40/CD40L

B cell       B cell

**FIGURE 11-7 An effective signal for B-cell activation involves two distinct signals induced by membrane events.** Binding of a type 1 thymus-independent (TI-1) antigen to a B cell provides both signals. A thymus-dependent (TD) antigen provides signal 1 by cross-linking mIg, but a separate interaction between CD40 on the B cell and CD40L on an activated $T_H$ cell is required to generate signal 2.

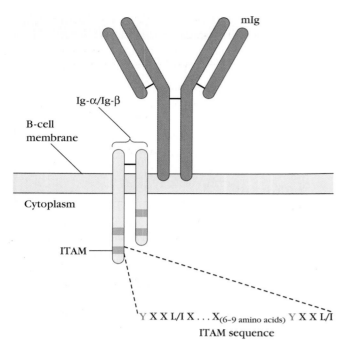

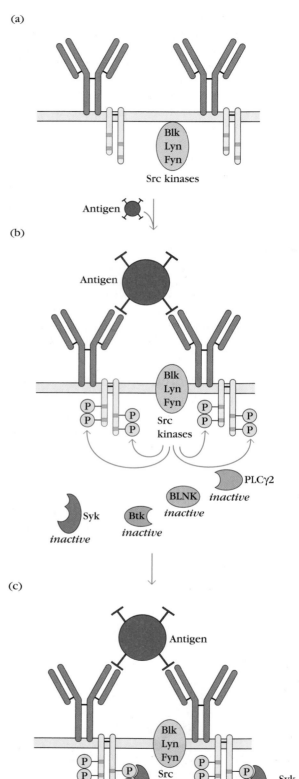

(a)

(b)

(c)

**FIGURE 11-8 The B-cell receptor.** The BCR is composed of a membrane immunoglobulin molecule associated with a disulfide-linked heterodimer of Ig-α and Ig-β. Antigen is recognized by the extracellular portion of the complex, and signal transduction is initiated by the cytoplasmic tails of Ig-α and Ig-β. Phosphorylation of tyrosines in the ITAMs (*immunoreceptor tyrosine-based activation motifs*) of Ig-α and Ig-β forms docking sites for the assembly of multimolecular signal transduction complexes. The highly conserved sequence motif of ITAMs is shown in the one-letter amino acid code, with tyrosines (Y) in red; L/I indicates that either leucine or isoleucine appears in that position, and X indicates any amino acid.

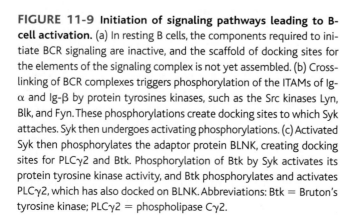

**FIGURE 11-9 Initiation of signaling pathways leading to B-cell activation.** (a) In resting B cells, the components required to initiate BCR signaling are inactive, and the scaffold of docking sites for the elements of the signaling complex is not yet assembled. (b) Cross-linking of BCR complexes triggers phosphorylation of the ITAMs of Ig-α and Ig-β by protein tyrosines kinases, such as the Src kinases Lyn, Blk, and Fyn. These phosphorylations create docking sites to which Syk attaches. Syk then undergoes activating phosphorylations. (c) Activated Syk then phosphorylates the adaptor protein BLNK, creating docking sites for PLCγ2 and Btk. Phosphorylation of Btk by Syk activates its protein tyrosine kinase activity, and Btk phosphorylates and activates PLCγ2, which has also docked on BLNK. Abbreviations: Btk = Bruton's tyrosine kinase; PLCγ2 = phospholipase Cγ2.

seen in receptors for the Fc regions of particular Ig classes (FcεRI for IgE; FcγRIIA, FcγRIIC, FcγRIIIA for IgG). In these receptors, and in the BCR and TCR, ligand binding and signal transduction are functionally compartmentalized, with the ligand-binding portions of these complexes (mIg in the case of the BCR) on the surface of the cell and the signal-transducing portion mostly or wholly within the cell. As is true of the TCR, protein tyrosine kinases (PTKs) play central roles in signaling from all of these receptors. Furthermore, like the TCR, the BCR itself has no PTK activity; this activity is acquired by recruitment of a number of different kinases to the cytoplasmic tails of the signal.

## B-cell signaling is initiated by antigen binding and induces many signal transduction pathways

The antigen-mediated cross-linking of BCRs initiates signal transduction processes that result in B-cell activation. As shown in Figure 11-9, binding of antigen results in phosphorylation of tyrosines within the ITAMs of the Ig-α and Ig-β chains of the BCR. Phosphorylation of these chains by receptor-associated PTKs is among the earliest events in B-cell activation, and it plays a key role in creating docking sites for other molecules that are essential for the initial stages of B-cell activation. These molecules include the critical PTK, Syk, and an adapter protein known as BLNK (B-cell-linker protein), which provides docking sites necessary for other proteins to join the signaling complex. Once BLNK has been phosphorylated by Syk, this adapter protein is joined by Bruton's tyrosine kinase (Btk) and phospholipase Cγ2 (PLCγ2). This step is essential for the phospholipase C–dependent activation of early calcium signaling and the initiation of protein kinase C (PKC)–dependent pathways. As shown in Figure 11-9, association of Btk and PLCγ2 with the adaptor protein BLNK allows Btk to phosphorylate PLCγ2, activating it. However, Btk must be phosphorylated by Syk before it can activate PLCγ2. The association of Btk with BLNK brings it close enough to Syk to allow this critical B-cell enzyme to perform the activating phosphorylation of Btk.

These initial stages set in motion a sequence of events that induce the many different signaling cascades that are necessary to activate the B cell. As shown in Figure 11-10, these signaling cascades include small G protein pathways, calcium-mediated pathways, and pathways mediated by protein kinase C, including the pathway that leads to production of the versatile transcription factor NF-κB. All of these pathways are also engaged during T-cell activation. Among the many similarities between BCR and TCR signaling, the following are notable:

- *Compartmentalization of function within receptor subunits:* Both the B-cell and T-cell pathways begin with antigen receptors that are composed of an antigen-binding and a signaling unit. The antigen-

binding unit confers specificity but has cytoplasmic tails too short to transduce signals to the cytoplasm of the cell. The signaling unit has long cytoplasmic tails that are the signal transducers of the receptor complex.

- *Activation by membrane-associated Src protein tyrosine kinases:* The receptor-associated PTKs (p56$^{Lck}$ in T cells and Lyn, Blk, and Fyn in B cells) catalyze phosphorylations during the early stages of signal transduction that are essential to the formation of a functional receptor signaling complex.

- *Assembly of a large signaling complex with protein tyrosine kinase activity:* The phosphorylated tyrosines in the ITAMs of the BCR and TCR provide docking sites for the molecules that endow these receptors with PTK activity: ZAP-70 in T cells and Syk in B cells. Adaptor molecules, BLNK in B cells and LAT and SLP-76 in T cells, play similar roles in providing a scaffold for assembly of essential elements of the signaling complex.

- *Recruitment of other signal transduction pathways:* Activation of B and T cells requires the coordinated activities of many different signal transduction pathways. The activation of these lymphocytes is *not* the consequence of a single signal transduction pathway.

- *Changes in gene expression:* One of the important outcomes of signal transduction processes set in motion by engagement of the BCR or the TCR is the generation or translocation to the nucleus of active transcription factors that stimulate or inhibit the transcription of specific genes.

Failures in signal transduction can have severe consequences for the immune system. The Clinical Focus on X-linked agammaglobulinemia (page 284) describes the effect of defective signal transduction on the development of B cells.

## The B-cell–coreceptor complex can enhance B-cell responses and CD22 can inhibit them

Stimulation through antigen receptors can be modified significantly by signals through coreceptors. Recall that costimulation through CD28 is an essential feature of effective positive stimulation of T lymphocytes, whereas signaling through CTLA-4 inhibits the response of the T cell. In B cells a component of the B-cell membrane, called the **B-cell coreceptor,** provides stimulatory modifying signals and another membrane protein, CD22, provides inhibitory signals. The B-cell coreceptor is a complex of three proteins: CD19, CR2 (CD21), and TAPA-1 (CD81) (Figure 11-11). CD19 is the key member of this complex and has a long cytoplasmic tail that provides docking sites for molecules that augment signals

**Some of the Many Signal-Transduction Pathways Activated by the BCR**

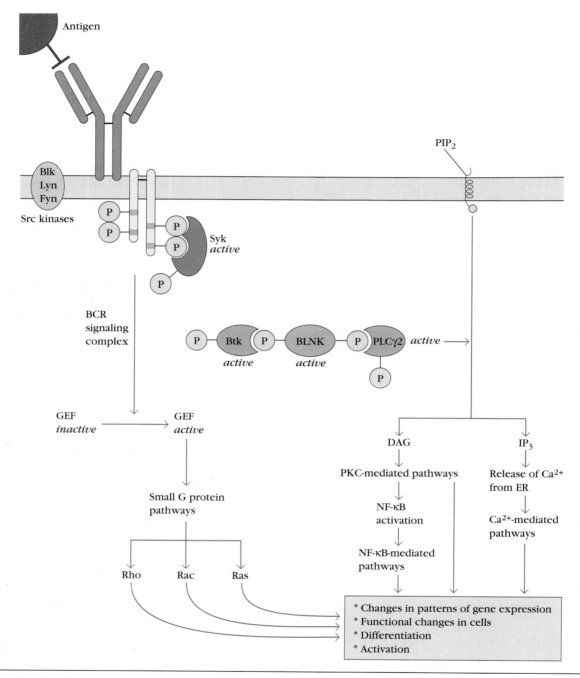

A series of phosphorylations initiated by Blk, Lyn, and Fyn lead to the generation of the BCR signaling complex. The adaptor protein BLNK facilitates the phosphorylation of PLCγ₂ by forming a complex that includes PLCγ₂ and Btk. The activation of PLCγ₂ leads to the hydrolysis of PIP₂, a membrane phospholipid, and produces the second messengers DAG and IP₃. DAG and Ca²⁺, released by the action of IP₃, collaboratively activate protein kinase C (PKC), which induces additional signal transduction pathways, including one that produces NF-κB. The activated receptor complex also generates signals that activate small G protein pathways. The result of activating all of these diverse pathways is changes in patterns of gene expression and functional changes in the cell that lead to its activation and differentiation. BLNK = B-cell-linker protein; Blk, Btk, Lyn, Fyn, and Syk are all protein tyrosine kinases; DAG = diacylglycerol; GEFs = guanine nucleotide exchange factors; IP₃ = inositol 1,4,5-triphosphate; PKC = protein kinase C; PLCγ₂ = phospholipase Cγ₂; Rho, Rac, and Ras are all small G proteins.

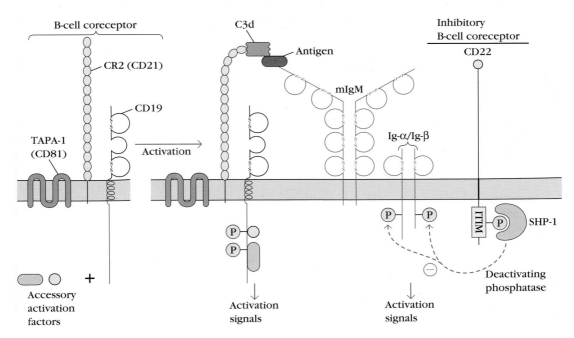

**FIGURE 11-11 The B-cell coreceptor is a complex of three cell membrane molecules: TAPA-1 (CD81), CR2 (CD21), and CD19.** Phosphorylation and association of the CD19 component of this complex with the BCR provides additional signals for B-cell activation. One of the mechanisms facilitating association of the stimulatory coreceptor complex with the BCR is binding of the CR2 component to complement-derived C3d that clings to antigen captured by the mIg component of the BCR. CD22 contains an ITIM (*im*munoreceptor *t*yrosine phosphate *i*nhibitory *m*otif) that contains a docking site for SHP-1, a tyrosine phosphatase that delivers negative signals by dephosphorylating tyrosines of the BCR.

delivered by the BCR complex. The CR2 component is a receptor of C3d, a breakdown product of the complement system, which is an important effector mechanism for destroying invaders (see Chapter 7); note that the involvement of C3d in the pathway for coreceptor activity reveals different arms of the immune system interacting with one another. CR2 also functions as a receptor for a membrane molecule and the transmembrane protein TAPA-1. As shown in Figure 11-11, the CR2 component of the coreceptor complex binds to complement-coated antigen that has been captured by the mIg on the B cell. This cross-links the coreceptor to the BCR and allows the CD19 component of the coreceptor to interact with the Ig-α/Ig-β component of the BCR. CD19 contains six tyrosine residues in its long cytoplasmic tail and is a major substrate of the protein tyrosine kinase activity that is mediated by cross-linkage of the BCR. Phosphorylation of CD19 permits it to bind a number of signaling molecules, including the protein tyrosine kinase Lyn.

The delivery of these signaling molecules to the BCR complex contributes to the activation process, and the coreceptor complex serves to amplify the activating signal transmitted through the BCR. In one experimental in vitro system, for example, $10^4$ molecules of mIgM had to be engaged by antigen for B-cell activation to occur when the coreceptor was not involved, but when CD19/CD2/TAPA-1 coreceptor was cross-linked to the BCR, only $10^2$ molecules of mIgM had to be engaged for B-cell activation. Another striking experiment highlights the role played by the B-cell

coreceptor. Mice were immunized with either unmodified lysozyme or a hybrid protein in which genetic engineering was used to join hen's egg lysozyme to C3d. The fusion protein, bearing two or three copies of C3d, produced antilysozyme responses that were 1000 to 10,000 times greater than those to lysozyme alone. Perhaps coreceptor phenomena such as these explain how naive B cells that often express mIg with low affinity for antigen are able to respond to low concentrations of antigen in a primary response. Such responses, even though initially of low affinity, can play a significant role in the ultimate generation of high-affinity antibody. As described later in this chapter, response to an antigen can lead to affinity maturation, resulting in a higher average affinity of the B-cell population. Finally, two experimental observations indicate that the CD19 component of the B-cell coreceptor can play a role independent of CR2, the complement receptor. First, in normal mice, artificially cross-linking the BCR with anti-CD19 antibodies results in the stimulation of some of the signal transduction pathways characteristic of B-cell activation. Second, B-cell activation in mice in which CD19 has been knocked out is significantly reduced, and CD19 knockout mice exhibit a greatly diminished antibody response to most antigens.

In addition to the stimulatory coreceptor, CD19, another molecule, CD22, which is constitutively associated with the B-cell receptor in resting B cells, delivers a negative signal that makes activation of B cells more difficult (see Figure 11-11). Activation of B cells results in the phosphorylation of an

## CLINICAL FOCUS

# X-Linked Agammaglobulinemia: A Failure in Signal Transduction and B-Cell Development

**X-linked** agammaglobulinemia is a genetically determined immunodeficiency disease characterized by the inability to synthesize all classes of antibody. It was discovered in 1952 by O. C. Bruton in what is still regarded as an outstanding example of research in clinical immunology. Bruton's investigation involved a young boy who had mumps three times and experienced 19 different episodes of serious bacterial infections during a period of just over 4 years. Because pneumococcus bacteria were isolated from the child's blood during 10 of the episodes of bacterial infection, attempts were made to induce immunity to pneumococcus by immunization with pneumococcus vaccine. The failure of these efforts to induce antibody responses prompted Bruton to determine whether the patient could mount antibody responses when challenged with other antigens. Surprisingly, immunization with diphtheria and typhoid vaccine preparations did not raise humoral responses in this patient. Electrophoretic analysis of the patient's serum revealed that although normal amounts of albumin and other typical serum proteins were present, gamma globulin, the major antibody fraction of serum, was absent. Having traced the immunodeficiency to a lack of antibody, Bruton tried a bold new treatment. He administered monthly doses of human immune serum globulin. The patient's experience of a 14-month period free of bacterial sepsis established the usefulness of immunoglobulin replacement for the treatment of immunodeficiency.

Though initially called Bruton's agammaglobulinemia, this hereditary immunodeficiency disease was renamed X-linked agammaglobulinemia, or X-LA, after the discovery that the gene responsible lies on the X chromosome. The disease has the following clinical features:

- Because this defect is X-linked, almost all afflicted individuals are male.

- Signs of immunodeficiency may appear as early as 9 months after birth, when the supply of maternal antibody acquired in utero has decreased below protective levels.

- There is a high frequency of infection by *Streptococcus pneumoniae* and *Haemophilus influenzae;* bacterial pneumonia, sinusitis, meningitis, or septicemia are often seen in these patients.

- Although infection by many viruses is no more severe in these patients than in normal individuals, long-term antiviral immunity is usually not induced.

- Analysis by fluorescence microscopy or flow cytometry shows few or no mature B cells in the blood.

Studies of this disease at the cellular and molecular level provided insights into the workings of the immune system. A scarcity of B cells in the periphery explained the inability of X-LA patients to make antibody. Studies of the cell populations in bone marrow traced the lack of B cells to failures in B-cell development. The samples displayed a ratio of pro–B cells to pre–B cells 10 times normal, suggesting inhibition of the transition from the pro- to the pre-B-cell stage. The presence of very few mature B cells in the marrow indicated a more profound blockade in the development of B cells from pre–B cells.

In the early 1990s, the gene responsible for X-LA was cloned. The normal counterpart of this gene encodes a protein tyrosine kinase that has been named Bruton's tyrosine kinase (Btk) in honor of the resourceful and insightful physician who discovered X-LA and devised a treatment for it. Parallel studies in mice have shown that the absence of Btk causes a syndrome known as xid, an immunodeficiency disease that is essentially identical to its human counterpart, X-LA. Btk has turned out to play important roles in B-cell signaling. For example, cross-linking of the B-cell receptor results in the phosphorylation of a tyrosine residue in the catalytic domain of Btk. This activates the protein-tyrosine-kinase activity of Btk, which then phosphorylates phospholipase C$\gamma_2$ (PLC$\gamma_2$); in vitro studies of cell cultures in which Btk has been knocked out show compromised PLC$\gamma_2$ activation. Once activated, PLC$\gamma_2$ hydrolyzes membrane phospholipids, liberating the potent second messengers IP$_3$ and DAG. As mentioned earlier, IP$_3$ causes a rise in intracellular Ca$^{2+}$, and DAG is an activator of protein kinase C (PKC). Thus, Btk plays a pivotal role in activating a network of intracellular signals vital to the function of mature B cells and earlier members of the B-cell lineage. Research has shown that it belongs to a family of PTKs known as Tec kinases; its counterpart in T cells is Itk. The insights gained from studies of X-LA, xid, and Btk are impressive examples of how the study of pathological states can clarify the workings of normal cells.

*immunoreceptor tyrosine inhibitory motif* (ITIM) in the cytoplasmic tail of CD22. This allows a tyrosine phosphatase to bind to the ITIMs of CD22 and strip activating phosphates from the tyrosines of neighboring signaling complexes. Recall that tyrosine phosphorylation is a key feature in B-cell activation. CD22 functions to deactivate B cells and plays a role in the negative regulation of B cells. Levels of B-cell activation are somewhat elevated in CD22 knockout mice, and aged CD22 KO mice have increased levels of autoimmunity, showing the importance of this negative regulation.

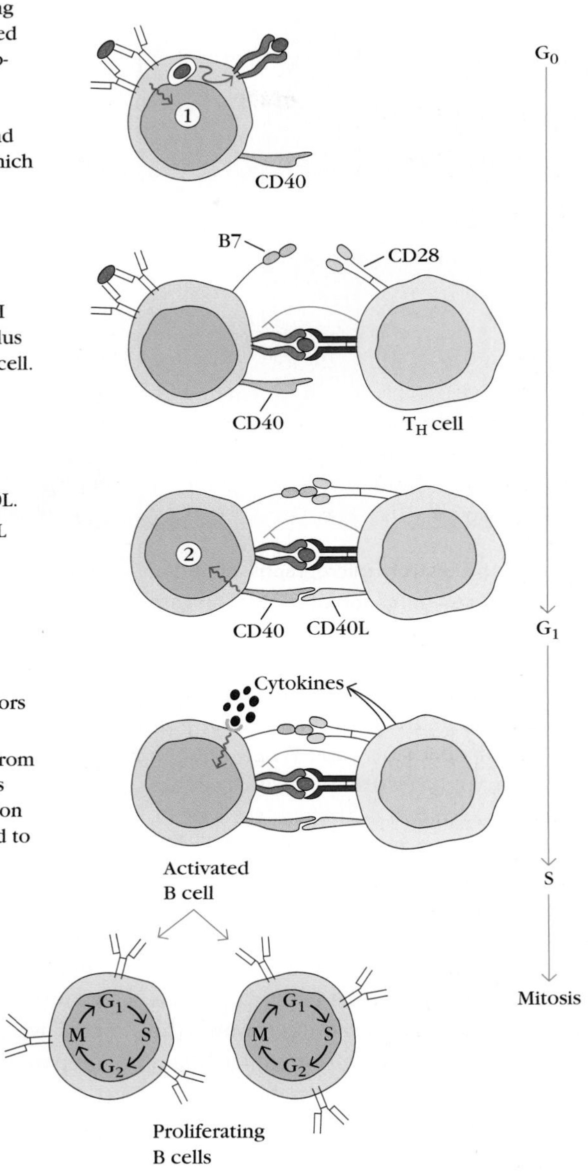

(a) Antigen cross-links mIg, generating signal ①, which leads to increased expression of class II MHC and costimulatory B7. Antigen-antibody complexes are internalized by receptor-mediated endocytosis and degraded to peptides, some of which are bound by class II MHC and presented on the membrane as peptide–MHC complexes.

(b) $T_H$ cell recognizes antigen–class II MHC on B-cell membrane. This plus costimulatory signal activates $T_H$ cell.

(c) 1. $T_H$ cell begins to express CD40L.

2. Interaction of CD40 and CD40L provides signal ②.

3. B7-CD28 interactions provide costimulation to the $T_H$ cell.

(d) 1. B cell begins to express receptors for various cytokines.

2. Binding of cytokines released from $T_H$ cell in a directed fashion sends signals that support the progression of the B cell to DNA synthesis and to differentiation.

FIGURE 11-12 Sequence of events in B-cell activation by a thymus-dependent antigen. The cell cycle phase of the interacting B cell is indicated on the right.

## $T_H$ cells play essential roles in most B-cell responses

As noted already, activation of B cells by soluble protein antigens requires the involvement of $T_H$ cells. Binding of antigen to B-cell mIg does not itself induce proliferation and differentiation to effector cells without additional interaction with membrane molecules on the $T_H$ cell and the presence of appropriate cytokines. Figure 11-12 outlines the probable sequence of events in B-cell activation by a thymus-dependent (TD) antigen. This process is considerably more complex than activation induced by thymus-independent (TI) antigens.

### Formation of T-B Conjugates

After binding of antigen by mIg on B cells, the antigen is internalized by receptor-mediated endocytosis and processed within the endocytic pathway into peptides. Antigen binding also initiates signaling through the BCR that induces the B cell to up-regulate the expression of a number of cell-membrane molecules, including class II MHC molecules and the costimulatory ligand B7 (see Figure 11-12a, b). Increased expression of both of these membrane proteins enhances the ability of the B cell to function as an antigen-presenting cell in $T_H$-cell activation. B cells could be regarded as helping their helpers because the antigenic peptides produced within the endocytic processing pathway associate with class II MHC molecules and are presented on the B-cell membrane to the $T_H$ cell, inducing its activation. It generally takes 30 to 60 minutes after internalization of antigen for processed antigenic peptides to be displayed on the B-cell membrane in association with class II MHC molecules.

Because a B cell recognizes and internalizes antigen specifically, by way of its membrane-bound Ig, a B cell is able

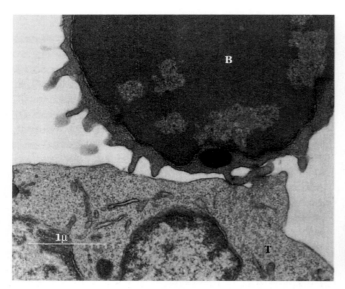
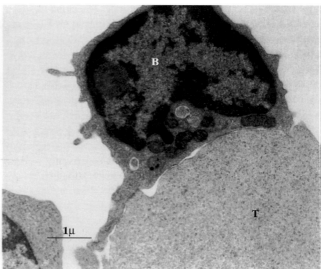

**FIGURE 11-13 Transmission electron micrographs of initial contact between a T cell and B cell (left) and of a T-B conjugate (right).** Note the broad area of membrane contact between the cells after formation of the conjugate. [From V. M. Sanders et al., 1986, Journal of Immunology **137**:2395.]

to present antigen to $T_H$ cells at antigen concentrations that are 100 to 10,000 times lower than the level required for presentation by macrophages or dendritic cells. When antigen concentrations are high, macrophages and dendritic cells are effective antigen-presenting cells, but as antigen levels drop, B cells take over as the major presenter of antigen to $T_H$ cells.

Once a $T_H$ cell recognizes a processed antigenic peptide displayed by a class II MHC molecule on the membrane of a B cell, the two cells interact to form a T-B conjugate (Figure 11-13). Micrographs of T-B conjugates reveal that in $T_H$ cells engaged in antigen-specific conjugates, the Golgi apparatus and microtubular-organizing center migrate toward the junction with the B cell. This structural adjustment facilitates the release of cytokines toward the antigen-specific B cell.

## Contact-Dependent Help Mediated by CD40/CD40L Interaction

Formation of a T-B conjugate not only leads to the directional release of $T_H$-cell cytokines but also to the up-regulation of CD40L (CD154), a $T_H$-cell membrane protein that then interacts with CD40 on B cells to provide an essential signal for T-cell-dependent B-cell activation. CD40 belongs to the tumor necrosis factor (TNF) family of cell surface proteins and soluble cytokines, which regulate cell proliferation and programmed cell death by apoptosis. CD40L belongs to the TNF receptor (TNFR) family. Interaction of CD40L with CD40 on the B cell delivers a signal (signal 2) to the B cell that, in concert with the signal generated by mIg cross-linkage (signal 1), drives the B cell into $G_1$ (see Figure 11-12c). The signals from CD40 are transduced by a number of intracellular signaling pathways, ultimately resulting in changes in gene expression. Studies have shown that although CD40 does not have kinase activity, cross-linking of CD40 by its T-cell-borne ligand, CD40L, is followed by the activation of protein tyrosine kinases

such as Lyn and Syk. Cross-linking of CD40 by CD40L also results in the activation of phospholipase C and the subsequent generation of the second messengers $IP_3$ and DAG and the activation of a number of transcription factors. Finally, cross-linking of CD40 results in its interaction with members of the TNFR-associated factor (TRAF) family. A consequence of this interaction is the activation of the transcription factor NF-κB.

Several lines of evidence have identified the CD40/CD40L interaction as the mediator of contact-dependent help. The role of an inducible $T_H$-cell membrane protein in B-cell activation was first revealed by experiments in which naive B cells were incubated with antigen and plasma membranes prepared from either activated or resting $T_H$-cell clones. Only the membranes from the activated $T_H$ cells induced B-cell proliferation, suggesting that one or more molecules expressed on the membrane of an activated $T_H$ cell engage receptors on the B cell to provide contact-dependent help. Furthermore, when antigen-stimulated B cells are treated with anti-CD40 monoclonal antibodies in the absence of $T_H$ cells, they become activated and proliferate. Thus, engagement of CD40, whether by antibodies to CD40 or by CD40L, is critical in providing signal 2 to the B cell. If appropriate cytokines are also added to this experimental system, then the proliferating B cells will differentiate into plasma cells. Conversely, antibodies to CD40L have been shown to block B-cell activation by blocking the CD40/CD40L interaction.

## Signals Provided by $T_H$-Cell Cytokines

Although B cells stimulated with membrane proteins from activated $T_H$ cells are able to proliferate, they fail to differentiate unless cytokines are also present; this finding suggests that both a membrane-contact signal and cytokine signals are necessary to induce B-cell proliferation and differentiation. Confocal microscopy of T-B conjugates reveals that the antigen-specific interaction between a $T_H$ and a B cell induces the formation of an immunological synapse (see Figure 10-11) accompanied by a redistribution of $T_H$-cell membrane proteins and cytoskeletal elements that results in the polarized release of cytokines toward the interacting B cell.

Once activated, the B cell begins to express membrane receptors for various cytokines, including IL-2, IL-4, and IL-5. These receptors then bind the cytokines produced by the interacting $T_H$ cell. The signals produced by these cytokine-receptor interactions support B-cell proliferation and can induce differentiation into plasma cells and memory B cells, class switching, and affinity maturation. Each of these events is described in a later section.

## Mature self-reactive B cells can be negatively selected in the periphery

Because some self antigens do not have access to the bone marrow, B cells expressing mIgM specific for such antigens cannot be eliminated by the negative-selection process in the bone marrow described earlier. To avoid autoimmune responses from such mature self-reactive B cells, some process for deleting them or rendering them inactive must occur in peripheral lymphoid tissue.

A transgenic system developed by C. Goodnow and his coworkers has helped to clarify the process of negative selection of mature B cells in the periphery. Goodnow's experimental system included two groups of transgenic mice (Figure 11-14a). One group carried a hen egg-white lysozyme (HEL) transgene linked to a metallothionine promoter, which placed transcription of the HEL gene under the control of zinc levels. The other group of transgenic mice carried rearranged immunoglobulin heavy- and light-chain transgenes encoding anti-HEL antibody; in normal mice, the frequency of HEL-reactive B cells is on the order of 1 in $10^3$, but in these transgenic mice the rearranged anti-HEL transgene is expressed by 60% to 90% of the mature peripheral B cells. Goodnow mated the two groups of transgenics to produce "double-transgenic" offspring carrying both the HEL and anti-HEL transgenes. Goodnow then asked what effect HEL, which is expressed in the periphery but not in the bone marrow, would have on the development of B cells expressing the anti-HEL transgene.

(a)

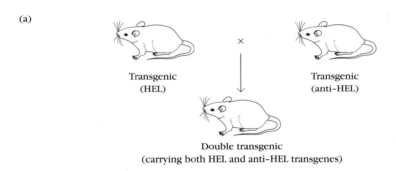

(b)

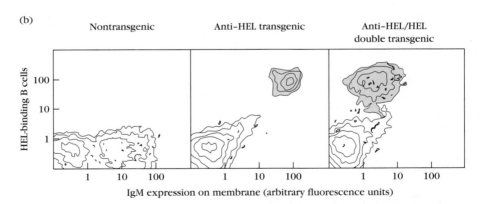

**FIGURE 11-14 Goodnow's experimental system for demonstrating clonal anergy in mature peripheral B cells.** (a) Production of double-transgenic mice carrying transgenes encoding HEL (hen egg-white lysozyme) and anti-HEL antibody. (b) Flow-cytometric analysis of peripheral B cells that bind HEL compared with membrane IgM levels. The number of B cells binding HEL was measured by determining how many cells bound fluorescently labeled HEL. Levels of membrane IgM were determined by incubating the cells with anti-mouse IgM antibody labeled with a fluorescent label different from that used to label HEL. Measurement of the fluorescence emitted from this label indicated the level of membrane IgM expressed by the B cells. The nontransgenics (*left*) had many B cells that expressed high levels of surface IgM but almost no B cells that bound HEL above the background level of 1. Both anti-HEL transgenics (*middle*) and anti-HEL/HEL double transgenics (*right*) had large numbers of B cells that bound HEL (blue), although the level of membrane IgM was about 20-fold lower in the double transgenics. The data in Table 11-3 indicate that the B cells expressing anti-HEL in the double transgenics cannot mount a humoral response to HEL.

| TABLE 11-3 | Expression of anti-HEL transgene by mature peripheral B cells in single- and double-transgenic mice | | | | |
|---|---|---|---|---|---|
| Experimental group | HEL level | Membrane anti-HEL | Anti-HEL PFC/spleen[*] | Anti-HEL serum titer[*] |
| Anti-HEL single transgenics | None | + | High | High |
| Anti-HEL/HEL double transgenics (group 1) | $10^{-9}$ M | + | Low | Low |

[*]Experimental animals were immunized with hen egg-white lysozyme (HEL). Several days later, hemolytic plaque assays for the number of plasma cells secreting anti-HEL antibody were performed and the serum anti-HEL titers were determined. PFC = plaque-forming cells.

SOURCE: Adapted from C. C. Goodnow, 1992, *Annual Review of Immunology* 10:489.

This approach has yielded several interesting findings on negative selection of B cells (Table 11-3). Double-transgenic mice expressing high levels of HEL ($10^{-9}$ M) continued to have mature, peripheral B cells bearing anti-HEL membrane antibody, but these B cells were functionally nonresponsive; that is, they were anergic. The flow-cytometric analysis of B cells from double-transgenic mice showed that, while large numbers of anergic anti-HEL cells were present, they expressed IgM at levels about 20-fold lower than anti-HEL single transgenics (Figure 11-14b). Further study demonstrated that the double transgenics had both surface IgM and IgD, indicating that the anergy was induced in mature rather than immature B cells. When these mice were given an immunizing dose of HEL, few anti-HEL plasma cells were induced and the serum anti-HEL titer was low.

To study what would happen if a class I MHC self antigen were expressed only in the periphery, Nemazee and Burki modified the transgenic system used in the experiments on negative selection in the bone marrow described previously (see Figure 11-6a). They first produced a transgene consisting of the class I $K^b$ gene linked to a liver-specific promoter so that the class I $K^b$ molecule could be expressed only in the liver. Transgenic mice expressing an anti-$K^b$ antibody on their B cells also were produced, and the two groups of transgenic mice were then mated (Figure 11-15a). In the resulting double-transgenic mice, the immature B cells expressing anti-$K^b$ mIgM would not encounter class I $K^b$ molecules in the bone marrow. Flow-cytometric analysis of the B cells in the double transgenics showed that immature B cells expressing the transgene-encoded anti-$K^b$ cells were

present in the bone marrow but not in the peripheral lymphoid organs (Figure 11-15b). In the previous experiments of Nemazee and Burki, the class I MHC self antigen (H-$2^k$) was expressed on all nucleated cells, and immature B cells expressing the transgene-encoded antibody to this class I molecule were selected against and deleted in the bone

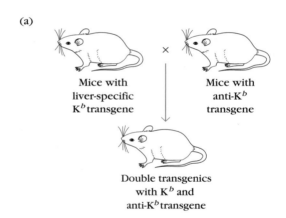

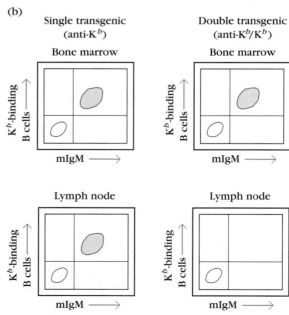

FIGURE 11-15 Experimental demonstration of clonal deletion of self-reactive mature peripheral B cells by Nemazee and Burki. (a) Production of double-transgenic mice expressing the class I $K^b$ molecule and anti-$K^b$ antibody. Because the $K^b$ transgene contained a liver-specific promoter, $K^b$ was not expressed in the bone marrow of the transgenics. (b) Flow-cytometric analysis of bone marrow and peripheral (lymph node) B cells for $K^b$ binding versus membrane IgM (mIgM). In the double transgenics, B cells expressing anti-$K^b$ (blue) were present in the bone marrow but were absent in the lymph nodes, indicating that mature self-reactive B cells were deleted in the periphery.

marrow (see Figure 11-6a). In their second system, however, the class I self antigen ($K^b$) was expressed only in the liver, so that negative selection and deletion occurred at the mature B-cell stage in the periphery.

# The Humoral Response

In this section we will consider the differences between the primary and secondary humoral response and the use of hapten-carrier conjugates in studying the humoral response.

## Primary and secondary responses differ significantly

The kinetics and other characteristics of the humoral response differ considerably depending on whether the humoral response results from activation of naive lymphocytes (primary response) or memory lymphocytes (secondary response). In both cases, activation leads to production of secreted antibodies of various isotypes, which differ in their ability to mediate specific effector functions (see Table 4-4).

The first contact of an exogenous antigen with an individual generates a <u>primary humoral response</u>, characterized by the production of antibody-secreting plasma cells and memory B cells. The kinetics of the primary response, as measured by serum antibody level, depends on the nature of the antigen, the route of antigen administration, the presence or absence of adjuvants, and the species or strain being immunized.

In all cases, however, a primary response to antigen is characterized by a lag phase, during which naive B cells undergo clonal selection, subsequent clonal expansion, and differentiation into memory cells or plasma cells (Figure 11-16). The lag phase is followed by a logarithmic increase in serum antibody level, which reaches a peak, plateaus for a variable time, and then declines. The duration of the lag phase varies with the nature of the antigen. Immunization of mice with an antigen such as sheep red blood cells (SRBCs) typically results in a lag phase of 4 to 5 days before antibody is reliably detected in serum, and peak serum antibody levels are attained by around day 7 to 10. For soluble protein antigens, the lag phase is a little longer, often lasting about a week, and peak serum titers do not occur until around 14 days. During a primary humoral response, IgM is secreted initially, often followed by a switch to an increasing proportion of IgG.

The memory B cells formed during a primary response stop dividing and enter the $G_0$ phase of the cell cycle. These cells have variable life spans, with some persisting for the life of the individual. The capacity to develop a secondary humoral response (see Figure 11-14) depends on the existence of this population of memory B cells as well as

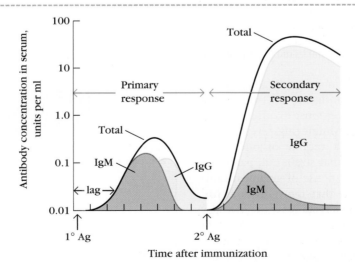

 **OVERVIEW FIGURE 11-16:** **Concentration and Isotype of Serum Antibody Following Primary (1°) and Secondary (2°) Immunization with Antigen**

The antibody concentrations are plotted on a logarithmic scale. The time units are not specified because the kinetics differ somewhat with type of antigen, administration route, presence or absence of adjuvant, and the species or strain of animal.

| TABLE 11-4 | Comparison of primary and secondary antibody responses | |
|---|---|---|
| Property | Primary response | Secondary response |
| Responding B cell | Naive B cell | Memory B cell |
| Lag period following antigen administration | Generally 4–7 days | Generally 1–3 days |
| Time of peak response | 7–10 days | 3–5 days |
| Magnitude of peak antibody response | Varies depending on antigen | Generally 100–1000 times higher than primary response |
| Isotype produced | IgM predominates early in the response | IgG predominates |
| Antigens | Thymus dependent and thymus independent | Thymus dependent |
| Antibody affinity | Lower | Higher |

memory T cells. Activation of memory cells by antigen results in a secondary antibody response that can be distinguished from the primary response in several ways (Table 11-4). The secondary response has a shorter lag period, reaches a greater magnitude, and lasts longer. The secondary response is also characterized by secretion of antibody with a higher affinity for the antigen, and isotypes other than IgM predominate.

A major factor in the more rapid onset and greater magnitude of secondary responses is the fact that the population of memory B cells specific for a given antigen is considerably larger than the population of corresponding naive B cells. Furthermore, memory cells are more easily activated than naive B cells. The processes of affinity maturation and class switching are responsible for the higher affinity and different isotypes exhibited in a secondary response. The higher levels of antibody coupled with the overall higher affinity provide an effective host defense against reinfection. The change in isotype provides antibodies whose effector functions are particularly suited to a given pathogen.

The existence of long-lived memory B cells accounts for a phenomenon called "original antigenic sin," which was first observed when the antibody response to influenza vaccines was monitored in adults. Monitoring revealed that immunization with an influenza vaccine of one strain elicited an antibody response to that strain but, paradoxically, also elicited an antibody response of greater magnitude to another influenza strain that the individual had been exposed to during childhood. It was as if the memory of the first antigen exposure had left a lifelong imprint on the immune system. This phenomenon can be explained by the presence of a memory-cell population, elicited by the influenza strain encountered in childhood, that is activated by cross-reacting epitopes on the vaccine strain encountered later. This process then generates a secondary response, characterized by antibodies with higher affinity for the earlier viral strain.

## T helper cells play a critical role in the humoral response to hapten-carrier conjugates

When animals are immunized with small organic compounds (haptens) conjugated with large proteins (carriers), the conjugate induces a humoral immune response consisting of antibodies both to hapten epitopes and to epitopes on the carrier protein. Hapten-carrier conjugates provided immunologists with an ideal system for studying cellular interactions of the humoral response, and such studies demonstrated that the generation of a humoral antibody response requires recognition of the antigen by both $T_H$ cells and B cells, each recognizing different epitopes on the same antigen. A variety of different hapten-carrier conjugates have been used in immunologic research (Table 11-5).

One of the earliest findings with hapten-carrier conjugates was that a hapten had to be chemically coupled to a larger carrier molecule to induce a humoral response to

| TABLE 11-5 | Common hapten-carrier conjugates used in immunologic research | |
|---|---|---|
| Hapten-carrier abbreviation | Hapten | Carrier protein |
| DNP-BGG | Dinitrophenol | Bovine gamma globulin |
| TNP-BSA | Trinitrophenyl | Bovine serum albumin |
| NIP-KLH | 5-Nitrophenyl acetic acid | Keyhole limpet hemocyanin |
| ARS-OVA | Azophenylarsonate | Ovalbumin |
| LAC-HGG | Phenyllactoside | Human gamma globulin |

the hapten. If an animal was immunized with both hapten and carrier separately, very little or no hapten-specific antibody was generated. A second important observation was that, in order to generate a secondary antibody response to a hapten, the animal had to be again immunized with the same hapten-carrier conjugate used for the primary immunization. If the secondary immunization was with the same hapten but conjugated to a different, unrelated carrier, no secondary antihapten response occurred. This phenomenon, called the **carrier effect,** could be circumvented by priming the animal separately with the unrelated carrier.

Similar experiments conducted with a cell transfer system showed that cells immunized against the hapten and cells immunized against the carrier were distinct populations. In these studies, one mouse was primed with the DNP-BSA conjugate and another was primed with the unrelated carrier bovine gamma globulin (BGG), which was not conjugated to the hapten. In one experiment, spleen cells from both mice were mixed and injected into a lethally irradiated syngeneic recipient. When this mouse was challenged with DNP conjugated to the unrelated carrier BGG, a secondary antihapten response to DNP occurred (Figure 11-17a). In a second experiment, spleen cells from the BGG-immunized mice were treated with anti-T-cell antiserum (anti-Thy-1) and complement to lyse the T cells. When this T-cell-depleted sample was mixed with the DNP-BSA–primed spleen cells and injected into an irradiated mouse, no secondary antihapten response was observed on immunizing with DNP-BGG (Figure 11-17b). However, similar treatment of the DNP-BSA–primed spleen cells with anti-Thy-1 and complement did not abolish the secondary antihapten response to DNP-BGG (Figure 11-17c). Later experiments, in which antisera were used to specifically deplete $CD4^+$ or $CD8^+$ T cells, showed that the $CD4^+$ T-cell subpopulation was responsible for the carrier effect. These experiments demonstrate that the response of hapten-primed B cells to the hapten-carrier conjugate requires the presence of carrier-primed $CD4^+$ $T_H$ cells specific for carrier epitopes. Keep in mind that the

B-cell response is not limited to the hapten determinant; in fact, some B cells do react to epitopes on the carrier. However, the assay can be conducted in such a way as to detect only antihapten responses.

The experiments with hapten-carrier conjugates revealed that both $T_H$ cells and B cells must recognize antigenic determinants on the same molecule for B-cell activation to occur. This feature of the T- and B-cell interaction in the humoral response is called associative, or linked, recognition. The conclusions drawn from hapten-carrier experiments apply to the humoral response to antigens in general and support the requirement for T-cell help in B-cell activation described earlier in this chapter.

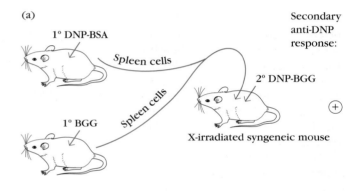

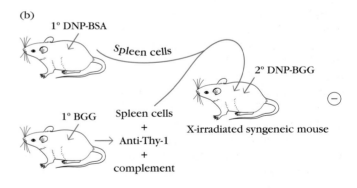

**FIGURE 11-17 Cell-transfer experiments demonstrating that hapten-primed and carrier-primed cells are separate populations.** (a) X-irradiated syngeneic mice reconstituted with spleen cells from both DNP-BSA–primed mice and BGG-primed mice and challenged with DNP-BGG generated a secondary anti-DNP response. (b) Removal of T cells from the BGG-primed spleen cells, by treatment with anti-Thy-1 antiserum, abolished the secondary anti-DNP response. (c) Removal of T cells from the DNP-BSA–primed spleen cells had no effect on the secondary response to DNP. These experiments show that carrier-primed cells are T cells and hapten-primed cells are B cells.

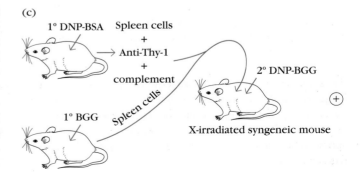

# In Vivo Sites for Induction of Humoral Responses

In vivo activation and differentiation of B cells occurs in defined anatomic sites whose structures place certain restrictions on the kinds of cellular interactions that can take place. When an antigen is introduced into the body, it becomes concentrated in various peripheral lymphoid organs. Blood-borne antigen is filtered by the spleen, whereas antigen from tissue spaces drained by the lymphatic system is filtered by regional lymph nodes or lymph nodules. The following description focuses on the generation of the humoral response in lymph nodes.

A lymph node is an extremely efficient filter capable of trapping more than 90% of any antigen carried into it by the afferent lymphatics. Antigen or antigen-antibody complexes enter the lymph nodes either alone or associated with antigen-transporting cells (e.g., Langerhans cells or dendritic cells) and macrophages. As antigen percolates through the cellular architecture of a node, it will encounter one of three types of antigen-presenting cells: dendritic cells in the paracortex, macrophages scattered throughout the node, or specialized follicular dendritic cells in the follicles and germinal centers. Antigenic challenge leading to a humoral immune response involves a complex series of events, which take place in distinct microenvironments within a lymph node (Figure 11-18). Slightly different pathways may operate during a primary and secondary response because much of the tissue antigen is complexed with circulating antibody in a secondary response.

The development of intravital microscopy (see two-photon microscopy in Chapter 22) has made it possible to observe live cells in real time within the interior of living tissues. Recently, investigators have used this technique to study the interactions of T and B cells within lymph nodes. They injected fluorescently labeled T and B cells from mice immunized with hen egg-white lysozyme into unimmunized mice. These recipients were then immunized with HEL, and the response of the labeled T and B cells to the antigen and to each other were videotaped inside the lymph node. Observations made 30 hours after immunization showed antigen-specific T and B cells approaching each other and interacting to form conjugates (Figure 11-19). These conjugates formed quickly (<1 minute) and lasted from 10 minutes to an hour or more. In contrast, the lifetime of conjugates between labeled T and B cells from unimmunized animals was of much shorter duration (<10 minutes). The antigen-specific T and B cells within the lymph node were observed to migrate through their respective areas of the lymph node (B cells through the B-cell-rich cortex, T cells through the T-enriched paracortex) and to arrive at the boundary between these areas, where they interacted with each other to form T-B conjugates.

Once antigen-mediated B-cell activation takes place, small foci of proliferating B cells form at the edges of the T-cell-rich zone. These B cells differentiate into plasma cells secreting IgM and IgG isotypes. Most of the antibody produced during a primary response comes from plasma cells in these foci. A similar sequence of events takes place in the spleen, where initial B-cell activation takes place in the T-cell-rich periarterial lymphatic sheath, PALS (see Figure 2-17).

A few days after the formation of foci within lymph nodes, a few activated B cells, together with a few $T_H$ cells, are thought to migrate from the foci to primary follicles. These follicles then develop into secondary follicles, which provide a specialized microenvironment favorable for interactions between B cells, activated $T_H$ cells, and follicular dendritic cells. Although they share the highly branched morphology of professional antigen-presenting dendritic cells derived from bone marrow, follicular dendritic cells do not arise in bone marrow, do not express class II MHC molecules, and do not present antigen to CD4$^+$ T cells. Follicular dendritic cells have long extensions, along which are arrayed Fc receptors and complement receptors. These receptors allow follicular dendritic cells to retain and present antigen-antibody complexes for long periods, even months, on the surface of the cell. Activated B cells, together with some activated $T_H$ cells, may migrate toward the center of the secondary follicle, forming a germinal center.

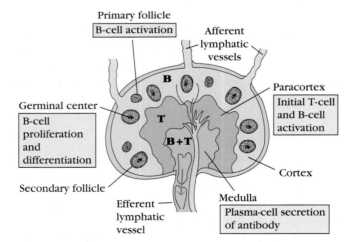

**FIGURE 11-18 Schematic diagram of a peripheral lymph node showing anatomic sites at which various steps in B-cell activation, proliferation, and differentiation occur.** The cortex is rich in B cells and the paracortex in T cells; both B and T cells are present in large numbers in the medulla. A secondary follicle contains the follicular mantle and a germinal center.

# Germinal Centers and Antigen-Induced B-Cell Differentiation

Germinal centers arise within 7 to 10 days after initial exposure to a thymus-dependent antigen. Three important B-cell differentiation events take place in germinal centers: affinity

(a) In vivo formation of T-B conjugate

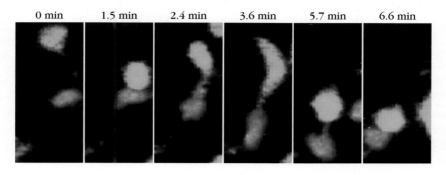

(b) In vivo movement of T-B conjugate

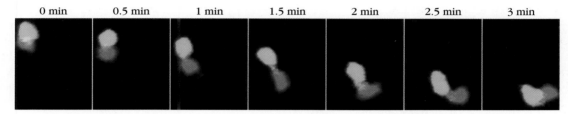

**FIGURE 11-19 In vivo dynamics of the interaction between antigen-engaged B cell and helper T cell.** (a) Two-photon microscopy was used to observe the encounter of living $T_H$ cells (green) and B cells (red) inside a lymph node. A T-B cell conjugate forms within a minute and remains intact for several minutes thereafter.

(b) Observation of live T-B conjugates within a lymph node shows that they are not static but motile. In most observations the B-cell member of the conjugate leads the migration. *[From T. Okada et al., 2005, Public Library of Science Biology 3:1047.]*

maturation, class switching, and formation of plasma cells and memory B cells. Affinity maturation and memory-cell formation require germinal centers. However some class switching and a significant amount of plasma-cell formation can occur outside germinal centers. During the first stage of germinal-center formation, activated B cells undergo intense proliferation. These proliferating B cells, known as centroblasts, appear in human germinal centers as a well-defined **dark zone** (Figure 11-20). Centroblasts are distinguished by their large size, expanded cytoplasm, diffuse chromatin, and absence or near absence of surface Ig. Centroblasts eventually give rise to centrocytes, which are small, nondividing B cells that now express membrane Ig. The centrocytes move from the dark zone into a region containing follicular dendritic cells called the **light zone,** where some centrocytes make contact with antigen displayed as antigen-antibody complexes on the surface of follicular dendritic cells.

## Affinity maturation is the result of repeated mutation and selection

The average affinity of the antibodies produced during the course of the humoral response increases remarkably during the process of affinity maturation, briefly described in Chapter 5, an effect first noticed by H. N. Eisen and G. W. Siskind when they immunized rabbits with the hapten-carrier complex DNP-BGG. The affinity of the serum anti-DNP antibodies produced in response to the antigen was then

measured at 2, 5, and 8 weeks after immunization. The average affinity of the anti-DNP antibodies increased about 140-fold from 2 weeks to 8 weeks. Subsequent work has shown that affinity maturation is mainly the result of somatic hypermutation.

### The Role of Somatic Hypermutation

Monitoring of antibody genes during an immune response shows that extensive mutation of the Ig genes that respond to the infection takes place in B cells within germinal centers. A direct demonstration that germinal centers are the sites of somatic hypermutation comes from the work of G. Kelsoe and his colleagues. These workers compared the mutation frequencies in B cells isolated from germinal centers with those from areas of intense B-cell activation outside the germinal centers. To do so, they prepared thin sections of spleen tissue from animals immunized with the hapten 4-hydroxy-3-nitrophenylacetyl (NP) conjugated with chicken gamma globulin as a carrier. This system is convenient because the initial response to this hapten is dominated by a particular heavy-chain gene rearrangement and the use of a γ light chain (in mice, >95% of antibodies bear κ light chains). Consequently, antibodies against the idiotype of this antibody can be used to readily distinguish responding B cells. Using antibodies to the idiotype and immunohistological staining techniques, these workers identified B cells bearing anti-NP antibody in germinal centers and nongerminal-center foci of B-cell activation present in thin sections cut from the spleens

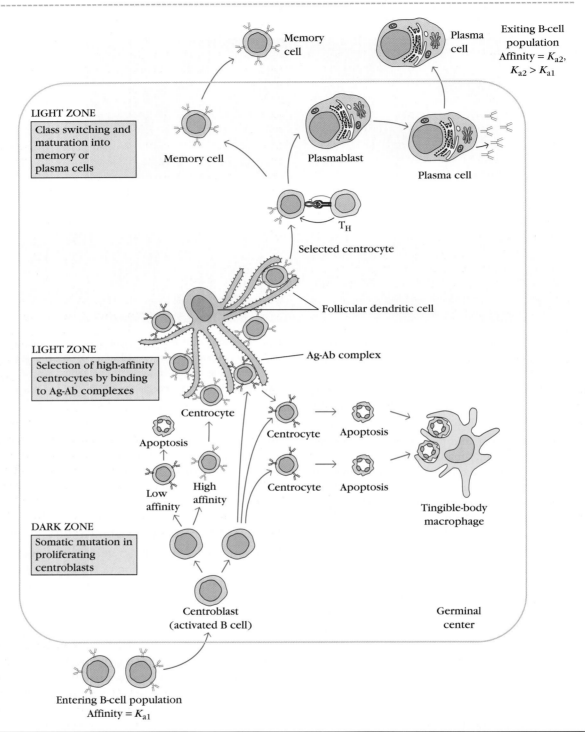

**LIGHT ZONE**

Class switching and maturation into memory or plasma cells

Memory cell

Memory cell

Plasmablast

Plasma cell

Plasma cell

Exiting B-cell population Affinity = $K_{a2}$, $K_{a2} > K_{a1}$

Selected centrocyte

$T_H$

Follicular dendritic cell

**LIGHT ZONE**

Selection of high-affinity centrocytes by binding to Ag-Ab complexes

Ag-Ab complex

Centrocyte

Apoptosis

Centrocyte

Apoptosis

Centrocyte

Apoptosis

Centrocyte

Low affinity

High affinity

Tingible-body macrophage

**DARK ZONE**

Somatic mutation in proliferating centroblasts

Centroblast (activated B cell)

Germinal center

Entering B-cell population Affinity = $K_{a1}$

Antigen-stimulated B cells migrate into germinal centers, where they reduce expression of surface Ig and undergo rapid cell division and mutation of rearranged immunoglobulin V-region genes within the dark zone. Subsequently, division stops and the B cells migrate to the light zone and increase their expression of surface Ig. At this stage they are called centrocytes. Within the light zone centrocytes must interact with follicular dendritic cells and T helper cells to survive. Follicular dendritic cells bind antigen-antibody complexes along their long extensions, and the centrocytes must compete with one an-

other to bind antigen. B cells bearing high-affinity membrane immunoglobulin (antibodies shown in blue) are most likely to compete successfully. Those that fail this antigen-mediated selection (antibodies shown in black) die by apoptosis. B cells that pass antigen selection and receive a second survival signal from $T_H$ cells differentiate into either memory B cells or antibody-secreting plasma cells. The encounter with $T_H$ cells may also induce class switching. A major outcome of the germinal center is to generate higher-affinity B cells ($K_{a2}$) from lower-affinity B cells ($K_{a1}$).

of recently immunized mice. They isolated these B cells by microdissection, used PCR to amplify the immunoglobulin genes of each individual cell, and then cloned and sequenced the immunoglobulin genes. Many mutations were found in the immunoglobulin genes obtained from B cells in germinal centers but few in the genes obtained from activated B cells in nongerminal-center foci. When the mutated sequences of the collection of B cells from germinal centers were examined, the sequences of many of the cells were similar enough to indicate common descent from the same precursor cell. Detailed analysis of the sequences allowed these workers to build genealogic trees in which one could clearly see the descent of progeny from progenitors by progressive somatic hypermutation.

The introduction of point mutations, deletions, and insertions into the rearranged immunoglobulin genes is strikingly focused. Figure 11-21 shows that the overwhelming majority of these mutations occur in a region that extends from about 0.5 kb 5′ to about 1.5 kb 3′ of the V(D)J segments of rearranged immunoglobulin genes. Although the hypermutation process delivers mutations throughout the V region, antigen-driven selection results in the eventual emergence of immunoglobulin genes in which the majority of the mutations lie within the three complementarity-determining regions (CDRs). It has been estimated that the mutation rate during somatic mutation is approximately $10^{-3}$/base pair/division, which is a millionfold greater than the normal mutation rate for other genes in humans or mice. Since the heavy- and light-chain V(D)J segments total about 700 base pairs, this rate of mutation means that, for every two cell divisions it undergoes, a centroblast will acquire a mutation in either the heavy- or light-chain variable regions. The extremely high rates and precise targeting of somatic hypermutation are remarkable features that are unique to the immune system. The enzyme activation-induced cytidine deaminase (AID), discussed in Chapter 5, is an important part of the molecular basis of this extraordinary process, but a complete understanding remains a challenge in immunology, including questions about how this process is primarily targeted to the variable regions of rearranged immunoglobulin genes.

Because somatic mutation occurs randomly, it will generate a few cells with receptors of higher affinity and many cells with receptors of unchanged or lower affinity for a particular antigen. Therefore, selection is needed to derive a population of cells that has increased affinity. The germinal center is the site of selection. As explained below, B cells that have high-affinity receptors for the antigen are likely to be positively selected and leave the germinal center; those with low affinity are likely to undergo negative selection and die in the germinal center.

### The Role of Selection

Somatic hypermutation of heavy- and light-chain variable-region genes occurs when centroblasts proliferate in the dark zone of the germinal center. Selection takes place in the light zone, among the nondividing centrocyte population. The most important factor influencing selection is the ability of the membrane Ig molecules on the centrocyte to recognize and bind antigen displayed by follicular dendritic cells (FDCs). Because the surfaces of FDCs are richly endowed with both Fc receptors and complement receptors, antigen complexed with antibody or antigen that has been bound by C3 fragments generated during complement activation (see Chapter 7) can bind to FDCs by antibody or C3 bridges. A centrocyte whose membrane Ig binds and undergoes cross-linking by FDC-bound antigen receives a signal that is essential for its survival. Those that fail to receive such signals die. However, centrocytes must compete for the small

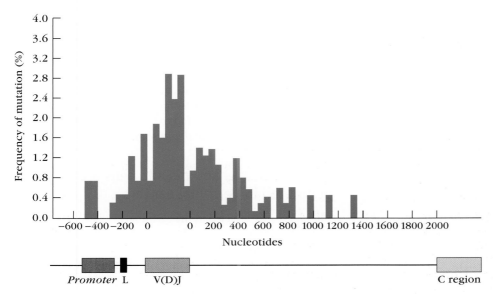

**FIGURE 11-21 The frequency of somatic hypermutation decreases with the distance from the rearranged V(D)J gene.** Experimental measurement of the mutation frequency shows that few if any mutations are seen upstream of the promoter of the rearranged gene. Mutations do not extend into the portion of the gene encoding the constant region because there are no mutations at positions more than about 1.5 kb 3′ of the rearranged gene. [Adapted from P. Gearhart, in Fundamental Immunology, 3rd ed., 1993, Lippincott Williams and Wilkins, Philadelphia, p. 877.]

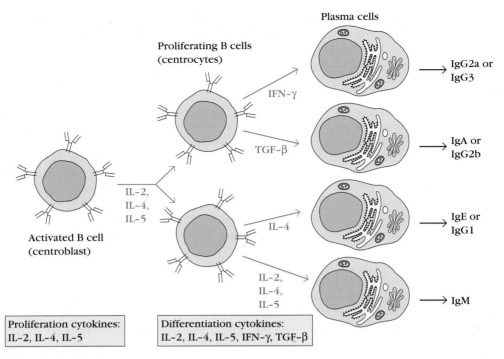

**FIGURE 11-22 The interactions of numerous cytokines with B cells generate signals required for proliferation and class switching during the differentiation of B cells into plasma cells.** Binding of the proliferation cytokines, which are released by activated $T_H$ cells, provides the signal needed for proliferation of activated B cells. Similar or identical effects may be mediated by cytokines beyond the ones shown. Class switching in the response to thymus-dependent antigens also requires the CD40/CD40L interaction, which is not shown here.

amounts of antigen present on FDCs. Because the amount of antigen is limited, centrocytes with receptors of high affinity are more likely to be successful in binding antigen than those of lower affinity (see Figure 11-20).

While antigen binding is necessary for centrocyte survival, it is not sufficient. A centrocyte must also receive signals generated by interaction with a CD4$^+$ $T_H$ cell to survive. An indispensable feature of this interaction is the engagement of CD40 on the B cell (centrocyte) by CD40L on the helper T cell. It is also necessary that processed antigen on class II MHC molecules of the B cell interact with the TCR of the collaborating $T_H$ cell. Centrocytes that fail to receive either the $T_H$-cell or the antigen-membrane Ig signal undergo apoptosis in the germinal center. Indeed, one of the striking characteristics of the germinal center is the extensive cell death by apoptosis that takes place there. This is clearly evident from the presence of condensed chromatin fragments, indicative of apoptosis, in tingible-body macrophages, an unusual type of macrophage that removes cells from lymphoid tissues by phagocytosis.

### Class Switching

Antibodies perform two important activities: the specific binding to an antigen, which is determined by the $V_H$ and $V_L$ domains, and participation in various biological effector functions, which is determined by the isotype of the heavy-chain constant domain. As described in Chapter 5, class switching allows any given $V_H$ domain to associate with the constant region of any isotype. This enables antibody specificity to remain constant while the biological effector activities of the molecule vary. A number of cytokines affect the decision of what Ig class is chosen when an IgM-bearing cell undergoes the class switch (Figure 11-22). As noted earlier, the humoral response to thymus-dependent antigens is marked by extensive class switching to isotypes other than IgM, whereas the antibody response to thymus-independent antigens is dominated by IgM. In the case of thymus-dependent antigens, membrane interaction between CD40 on the B cell and CD40L on the $T_H$ cell is essential for the induction of class switching. The importance of the CD40/CD40L interaction is illustrated by **X-linked hyper-IgM syndrome,** an immunodeficiency disorder in which $T_H$ cells fail to express CD40L (see Chapter 20). Patients with this disorder produce IgM but not other isotypes. Such patients fail to generate memory-cell populations, fail to form germinal centers, and their antibodies do not undergo somatic hypermutation.

## Memory B cells and plasma cells are generated in germinal centers

After B cells are selected in the germinal center for those bearing mIg that binds with high affinity to antigen displayed on follicular dendritic cells, some B cells differentiate into plasma cells and others become memory B cells (see Figure 11-20). Although germinal centers are important sites of plasma-cell

| TABLE 11-6 | Comparison of naive and memory B cells | |
|---|---|---|
| Property | Naive B cell | Memory B cell |
| Membrane markers Immunoglobulin Complement receptor | IgM, IgD Low | IgM, IgD(?), IgG, IgA, IgE High |
| Anatomic location | Spleen | Bone marrow, lymph node, spleen |
| Life span | Short-lived | May be long-lived |
| Recirculation | Yes | Yes |
| Receptor affinity | Lower average affinity | Higher average affinity due to affinity maturation* |
| Adhesion molecules | Low ICAM-1 | High ICAM-1 |

*Affinity maturation results from somatic mutation during proliferation of centroblasts and subsequent antigen selection of centrocytes bearing high-affinity mIg.

generation, these Ig-secreting cells are formed in other sites as well. Plasma cells generally lack detectable membrane-bound immunoglobulin and instead synthesize high levels of secreted antibody (at rates as high as 1000 molecules of Ig per cell per second). Differentiation of mature B cells into plasma cells requires a change in RNA processing so that the secreted form of the heavy chain rather than the membrane form is synthesized. In addition, the rate of transcription of heavy- and light-chain genes is significantly greater in plasma cells than in less-differentiated B cells. Several authors have suggested that the increased transcription by plasma cells might be explained by the synthesis of higher levels of transcription factors that bind to immunoglobulin enhancers. Some mechanism also must coordinate the increase in transcription of heavy-chain and light-chain genes, even though these genes are on different chromosomes.

As indicated above, B cells that survive selection in the light zone of germinal centers also differentiate into memory cells. Some properties of naive and memory B cells are summarized in Table 11-6. Except for membrane-bound immunoglobulins, few membrane molecules have been identified that distinguish naive B cells from memory B cells. Naive B cells express only IgM and IgD; as a consequence of class switching, however, memory B cells express additional isotypes, including IgG, IgA, and IgE.

## Regulation of the Immune Effector Response

On encountering an antigen, the immune system can either develop an immune response or enter a state of unresponsiveness called tolerance. The development of immunity or tolerance, both of which involve specific recognition of antigen by antigen-reactive T or B cells, must be carefully regulated since an inappropriate response—whether it is immunity to self antigens or tolerance to a potential pathogen—can have serious and possibly life-threatening consequences.

Regulation of the immune response takes place in both the humoral and the cell-mediated branch. Every time an antigen is introduced, important regulatory decisions determine the branch of the immune system to be activated, the intensity of the response, and its duration. Chapter 12 describes the importance of the cytokines to the orchestration of appropriate immune responses. In addition to cytokines, other regulatory mechanisms may play important immunoregulatory roles. Greater knowledge about these regulatory events, which are still not well understood, may allow the deliberate manipulation of immune responses, selectively up-regulating desirable responses and down-regulating undesirable ones.

## Different antigens can compete with each other

The immunologic history of an animal influences the quality and quantity of its immune response. A naive animal responds to antigen challenges very differently from a previously primed animal. Previous encounter with an antigen may have rendered the animal tolerant to the antigen or may have resulted in the formation of memory cells. In some cases, the presence of a competing antigen can regulate the immune response to an unrelated antigen. This antigenic competition is illustrated by injecting mice with a competing antigen a day or two before immunization with a test antigen. For example, the response to horse red blood cells (HRBCs) is severely reduced by prior immunization with sheep red blood cells (SRBCs) and vice versa (Table 11-7). Although antigenic competition is a well-established phenomenon, its molecular and cellular basis is not understood.

## The presence of antibody can suppress the response to antigen

Like many biochemical reactants, antibody exerts feedback inhibition on its own production. Because of antibody-mediated suppression, certain vaccines (e.g., those for measles

| TABLE 11-7 | Antigenic competition between sheep and horse RBCs | | |
|---|---|---|---|
| IMMUNIZING ANTIGEN | | HEMOLYTIC PLAQUE ASSAY (DAY 8) | |
| Ag1 (day 0) | Ag2 (day 3) | Test Ag | PFC/10$^{6*}$ spleen cells |
| None | HRBC | HRBC | 205 |
| SRBC | HRBC | HRBC | 13 |
| None | SRBC | SRBC | 626 |
| HRBC | SRBC | SRBC | 78 |

*PFC = plaque-forming cells.

and mumps) are not administered to infants before the age of 1 year. The level of naturally acquired maternal IgG, which the fetus acquires by transplacental transfer, remains high for about 6 months after birth. If an infant is immunized with measles or mumps vaccine while this maternal antibody is still present, the humoral response is low and the production of memory cells is inadequate to confer long-lasting immunity. If an animal is immunized with a specific antigen and is injected with pre-formed antibody to that same antigen just before or within a few days after antigen priming, the immune response to the antigen may be reduced as much as 100-fold.

Antibody-mediated suppression can be explained in two ways. The first explanation is that the circulating antibody competes with antigen-reactive B cells for antigen, inhibiting the clonal expansion of the B cells. The second explanation is that binding of antigen-antibody complexes by Fc receptors on B cells reduces signaling by the B-cell receptor complex.

As the antibody response proceeds, antibody feedback produces inhibition of the response. As more secreted IgG molecules become involved in antigen-antibody complexes, the Ig portions of these complexes become bound to Fcγ receptors present on the B-cell membrane, and the antigen of the complex binds the Ig of B-cell receptors. This cross-linking brings Fcγ receptors into close association with activated B-cell receptor complexes, allowing phosphatases bound to the cytoplasmic tails of the Fc receptor to dephosphorylate sites in the BCR complex that must remain phosphorylated for B-cell activation to be maintained. As a consequence, the activity of the B cell is progressively down-regulated as the amount of IgG bound to antigen increases. Evidence for such competition between passively administered antibody and antigen-reactive B cells comes from studies in which it took over 10 times more low-affinity anti-DNP antibody than high-affinity anti-DNP antibody to induce comparable suppression. In addition, the competition for antigen between passively administered antibody and antigen-reactive

B cells drives the B-cell response toward higher-affinity antibody. Only the high-affinity antigen-reactive cells can compete successfully with the passively administered antibody for the available antigen.

## SUMMARY

- B cells develop in bone marrow and undergo antigen-induced activation and differentiation in the periphery. Activated B cells can give rise to antibody-secreting plasma cells or memory B cells.

- During B-cell development, sequential Ig-gene rearrangements transform a pro–B cell into an immature B cell expressing mIgM with a single antigenic specificity. Further development yields mature naive B cells, still of a single specificity, expressing both mIgM and mIgD.

- When a self-reactive BCR is expressed in the bone marrow, negative selection occurs and the self-reactive cells are deleted by apoptosis or undergo receptor editing to produce non-self-reactive mIg. B cells reactive with self antigens encountered in the periphery are rendered anergic.

- In the periphery, the antigen-induced activation and differentiation of mature B cells generates an antibody response. For most antigens, this response requires T$_H$ cells; these are thymus-dependent (TD) responses. Responses to some antigens, such as certain bacterial cell wall products (e.g., LPS) and polymeric molecules with repeating epitopes, do not require T$_H$ cells and are thymus-independent (TI) antigens.

- B-cell activation is the consequence of signal transduction processes triggered by engagement of the B-cell receptor, ultimately leading to many changes in the cell, including changes in the expression of specific genes.

- B- and T-cell activation share many parallels, including compartmentalization of function within receptor subunits, activation by membrane-associated protein tyrosine kinases, assembly of large signaling complexes with protein tyrosine kinase activity, and recruitment of several signal transduction pathways.

- The B-cell coreceptor can intensify the activating signal resulting from cross-linkage of mIg; this may be particularly important during the primary response to low concentrations of antigen.

- The cell membrane–associated molecule CD22 can act as a negative regulator of B-cell activation. Protein tyrosine phosphatases docked to the ITIMs of the cytoplasmic tail of CD22 deactivate the BCR-associated signaling complexes by removing phosphates added by activating protein tyrosine kinases.

- Activation induced by TD antigens requires contact-dependent help delivered by the interaction between CD40 on B cells and CD40L on activated T$_H$ cells. The CD40/CD40L interaction is essential for B-cell survival, the formation of germinal centers, the generation of memory-cell populations, and somatic hypermutation.

■ The properties of the primary and secondary antibody responses differ. The primary response has a long lag period, and IgM is the first antibody class produced, followed by a gradual switch to other classes. The secondary response has a shorter lag time, and the response lasts longer. IgG and other isotypes are the main products generated in the secondary response rather than IgM, and the average affinity of antibody produced is higher.

■ Germinal centers, sites of somatic hypermutation of rearranged immunoglobulin genes, form within a week or so of exposure to a TD antigen. Germinal centers are the sites of affinity maturation, formation of memory B cells, class switching, and formation of plasma cells.

■ Humoral immune responses can be inhibited by antibody feedback, a process in which the binding of antigen-antibody complexes by Fcγ receptors on B cells inhibits BCR signaling.

## References

Berek, C. 1999. Affinity maturation. In *Fundamental Immunology,* 4th ed. W. E. Paul, ed. Lippincott-Raven, Philadelphia and New York.

Berland, R., and H. H. Wortis. 2002. Origins and functions of B-1 cells with notes on the role of CD5. *Annual Review of Immunology* **20:**253.

Bromburg, J. S. 2004. The beginnings of T-B collaboration. *Journal of Immunology* **173:**7.

Bruton, O. C. 1952. Agammaglobulinemia. *Pediatrics* **9:**722.

Burrows, P. D., et al. 2004. The development of human B lymphocytes. In *Molecular Biology of B Cells.* T. Honjo, F. W. Alt, and M. Neuberger, eds. Elsevier Academic Press, Amsterdam.

Dal Porto, J. M., et al. 2004. B cell antigen receptor signaling 101. *Molecular Immunology* **41:**599.

Jacob, J., G. Kelsoe, K. Rajewsky, and U. Weiss. 1991. Intraclonal generation of antibody mutants in germinal centres. *Nature* **354:**389.

Manis, J. P., M. Tian, and F. W. Alt. 2002. Mechanism and control of class-switch recombination. *Trends in Immunology* **23:**31.

Matthias, P. and A. G. Rolink, A. G. 2005. Transcriptional networks in developing and mature B cells. *Nature Reviews Immunology* **5:**497.

Matsuuchi, L., and M. R. Gold. 2001. New views of BCR structure and organization. *Current Opinion in Immunology* **13:**270.

Melchers, F., and A. Rolink. 1999. B-lymphocyte development and biology. In *Fundamental Immunology,* 4th ed. W. E. Paul, ed. Lippincott-Raven, Philadelphia and New York.

Meffre, E., R. Casellas, and M. C. Nussenzweig. 2000. Antibody regulation of B-cell development. *Nature Immunology* **1:**379.

Nitschke, L., and D. T. Fearon. 2004. Regulation of antigen receptor signaling by the co-receptors, CD19 and CD22. In *Molecular Biology of B Cells.* T. Honjo, F. W. Alt, and M. Neuberger, eds. Elsevier Academic Press, Amsterdam.

Okada, T., et al. 2005. Antigen-engaged B cells undergo chemotaxis toward the T zone and form motile conjugates with helper T cells. *Public Library of Science Biology* **3:**1047.

Papavasiliou, F. N., and D. G. Schatz. 2002. Somatic hypermutation of immunoglobulin genes: merging mechanisms for genetic diversity. *Cell* **109:**S35.

 ## Useful Web Sites

**http://www.ncbi.nlm.nih.gov/Omim/**

**http://www.ncbi.nlm.nih.gov/htbinpost/Omim/getmim**

The Online Mendelian Inheritance in Man Web site contains a subsite that lists more than a dozen different inherited diseases associated with B-cell defects.

**http://www.bioscience.org/knockout/knochome.htm**

The Frontiers in Bioscience Database of Gene Knockouts features information on the effects of knockouts of many genes important to the development and function of B cells.

**http://stke.sciencemag.org/**

An excellent and comprehensive offering from *Science,* covering signal transduction pathways. The database includes articles, reviews, and experimental protocols.

 ## Study Questions

**CLINICAL FOCUS QUESTION** Patients with X-linked agammaglobulinemia are subject to infection by a broad variety of pathogens. Suppose you have three sources of highly purified human immunoglobulin (HuIg) for the treatment of patients with X-linked agammaglobulinemia. The human Ig from all three sources is equally free of disease-causing agents and is equally well tolerated by recipients, but the number of donors whose blood was pooled for the preparation of each source differs widely: 100 individuals for source A, 1000 for source B, and 60,000 for source C. Which would you choose and what is the basis of your choice?

1. Indicate whether each of the following statements concerning B-cell maturation is true or false. If you think a statement is false, explain why.

   a. Heavy chain $V_H$-$D_H$-$J_H$ rearrangement begins in the pre-B-cell stage.

   b. Immature B cells express membrane IgM and IgD.

   c. The enzyme terminal deoxyribonucleotidyl transferase (TdT) is active in the pre-B-cell stage.

   d. The surrogate light chain is expressed by pre–B cells.

   e. Self-reactive B cells can be rescued from negative selection by the expression of a different light chain.

   f. In order to develop into immature B cells, pre–B cells must interact directly with bone marrow stromal cells.

g. Most of the B cells generated every day never leave the bone marrow as mature B cells.

2. You have fluorescein (Fl)-labeled antibody to the μ heavy chain and a rhodamine (Rh)-labeled antibody to the δ heavy chain. Describe the fluorescent antibody staining pattern of the following B-cell maturational stages, assuming that you can visualize both membrane and cytoplasmic staining: (a) progenitor B cell (pro–B cell), (b) precursor B cell (pre–B cell), (c) immature B cell, (d) mature B cell, and (e) plasma cell before any class switching has occurred.

3. Describe the general structure and probable function of the B-cell–coreceptor complex.

4. In the Goodnow experiment demonstrating clonal anergy of B cells, two types of transgenic mice were compared: single transgenics carrying a transgene-encoded antibody against hen egg-white lysozyme (HEL) and double transgenics carrying the anti-HEL transgene and a HEL transgene linked to the zinc-activated metallothionine promoter.

   a. In both the single and double transgenics, 60% to 90% of the B cells expressed anti-HEL membrane-bound antibody. Explain why.

   b. How could you show that the membrane antibody on these B cells is specific for HEL and how could you determine its isotype?

   c. Why was the metallothionine promoter used in constructing the HEL transgene?

   d. Design an experiment to prove that the B cells, not the $T_H$ cells, from the double transgenics were anergic.

5. Discuss the origin of the competence and progression signals required for activation and proliferation of B cells induced by (a) soluble protein antigens and (b) bacterial lipopolysaccharide (LPS).

6. Fill in the blank(s) in each statement below (a–h) with the most appropriate term(s) from the following list. Terms may be used more than once or not at all.

   dark zone        centroblasts              memory B cells
   light zone       centrocytes               plasmablasts
   paracortex       follicular dendritic cells   $T_H$ cells
   cortex           medulla

   a. Most centrocytes die by apoptosis in the _____.

   b. Initial activation of naive B cells induced by thymus-dependent antigens occurs within the _____ of lymph nodes.

   c. _____ are rapidly dividing B cells located in the _____ of germinal centers.

   d. _____ expressing high-affinity mIg interact with antigen captured by _____ in the light zone.

   e. Class switching occurs in the _____ and requires direct contact between B cells and _____.

   f. Centrocytes expressing mIg specific for a self-antigen present in the bone marrow are subjected to negative selection in the _____.

   g. Within lymph nodes, plasma cells are found primarily in the _____ of secondary follicles.

h. Somatic hypermutation, which occurs in proliferating _____, is critical to affinity maturation.

7. Activation and differentiation of B cells in response to thymus-dependent (TD) antigens requires $T_H$ cells, whereas the B-cell response to thymus-independent (TI) antigens does not.

   a. Discuss the differences in the structure of TD, TI-1, and TI-2 antigens and the characteristics of the humoral responses induced by them.

   b. Binding of which classes of antigen to mIg provides an effective competence signal for B-cell activation?

8. B-cell-activating signals must be transduced to the cell interior to influence developmental processes. Yet the cytoplasmic tails of all isotypes of mIg on B cells are too short to function in signal transduction.

   a. How do naive B cells transduce the signal induced by cross-linkage of mIg by antigen?

   b. Describe the general result of signal transduction in B cells during antigen-induced activation and differentiation.

9. In some of their experiments, Nemazee and Burki mated mice carrying a transgene encoding $K^b$, a class I MHC molecule, linked to a liver-specific promoter with mice carrying a transgene encoding antibody against $K^b$. In the resulting double transgenics, $K^b$-binding B cells were found in the bone marrow but not in lymph nodes. In contrast, the anti-$K^b$ single transgenics had $K^b$-binding B cells in both the bone marrow and lymph nodes.

   a. Was the haplotype of the mice that received the transgenes H-$2^b$ or some other haplotype?

   b. Why was the $K^b$ transgene linked to a liver-specific promoter in these experiments?

   c. What do these results suggest about the induction of B-cell tolerance to self antigens?

10. Indicate whether each of the following statements is true or false. If you believe a statement is false, explain why.

    a. Cytokines can determine which branch of the immune system is activated.

    b. Immunization with a hapten-carrier conjugate results in production of antibodies to both hapten and carrier epitopes.

    c. All the antibodies secreted by a single plasma cell have the same idiotype and isotype.

    d. If mice are immunized with HRBCs and then are immunized a day later with SRBCs, the antibody response to the SRBCs will be much higher than that achieved in control mice immunized only with SRBCs.

11. Four mice are immunized with antigen under the conditions listed below (a–d). In each case, indicate whether the induced serum antibodies will have high affinity or low affinity and whether they will be largely IgM or IgG.

    a. A primary response to a low antigen dose

    b. A secondary response to a low antigen dose

    c. A primary response to a high antigen dose

    d. A secondary response to a high antigen dose

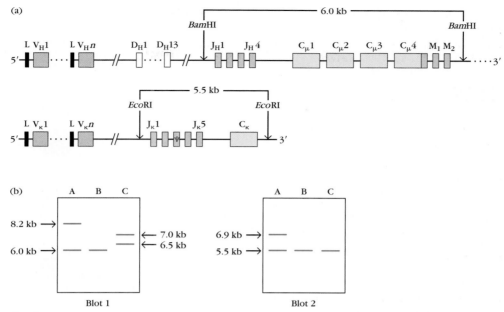

For use with **Question 12**

12. DNA was isolated from three sources: liver cells, pre–B lymphoma cells, and IgM-secreting myeloma cells. Each DNA sample was digested separately with the restriction enzymes *Bam*HI and *Eco*RI, which cleave germ-line heavy-chain and κ light-chain DNA as indicated in part *a* of the figure above. The digested samples were analyzed by Southern blotting using a radiolabeled $C_\mu 1$ probe with the *Bam*HI digests (blot 1) and a radiolabeled $C_\kappa$ probe with the *Eco*RI digests (blot 2). The blot patterns are illustrated in part *b* of the figure.

    From this information, which DNA sample (designated A, B, or C) was isolated from (a) liver cells, (b) pre–B lymphoma cells, and (c) IgM-secreting plasma cells? Explain your assignments.

13. For each of the following statements, indicate whether you think the statement is true or false. If you think a statement is false, explain why.

    a. CD21 is not only a complement receptor, it is an essential part of the B-cell receptor complex.

    b. Signal transduction through the BCR complex results in transport of AP-1 to the cytosol.

    c. Booster shots are required because repeated exposure to an antigen builds a stronger immune response.

    d. IgA is made against TI antigens.

    e. B cells mature into effector cells in the thymus.

    f. After a plasma cell secretes large amounts of antibody and little free antigen is present, the cell stops secreting antibody because its BCR is no longer stimulated.

    g. TI-1 antigens are substances such as LPS that can activate B cells nonspecifically at high concentrations.

14. For each of the following statements, fill in the blank with an appropriate word or phrase.

    a. X-linked hyper-IgM syndrome results from a defect in CD40 ligand on T cells, which affects B cells by blocking _____ .

    b. The cytokine _____ induces B cells to undergo class switching to IgE.

    c. The surrogate light chain of the pre–BCR is composed of _____ and _____ .

    d. Mature _____ produce large amounts of soluble IgG.

 **Interactive Study**

**www.whfreeman.com/kuby**

 SELF-TEST
Review of Key Terms

ANIMATION
Signal Transduction

# chapter 12

# Cytokines

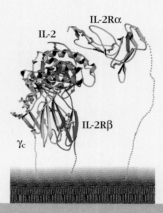

IL-2 Receptor with Bound IL-2 Ligand. [From X. Wang, M. Rickert, and K. C. Garcia, 2005, Science 310:1159.]

- Properties of Cytokines
- Cytokine Receptors
- Cytokine Antagonists
- Cytokine Secretion by $T_H1$ and $T_H2$ Subsets
- Cytokine-Related Diseases
- Cytokine-Based Therapies
- Cytokines in Hematopoiesis

T HE DEVELOPMENT OF AN EFFECTIVE IMMUNE RESPONSE involves lymphoid cells, inflammatory cells, and hematopoietic cells. The complex interactions among these cells are mediated by a group of proteins collectively designated **cytokines** to denote their role in cell-to-cell communication (*cyto-*, "cell," from the Greek *kinein*, "to move"). Cytokines are low-molecular-weight regulatory proteins or glycoproteins secreted by white blood cells and various other cells in the body in response to a number of stimuli. These proteins assist in regulating the development of immune effector cells, and some cytokines possess direct effector functions of their own.

The term *cytokine* encompasses those cytokines secreted by lymphocytes, substances formerly known as **lymphokines,** and those secreted by monocytes and macrophages, substances formerly known as **monokines.** Although these other two terms continue to be used, they are misleading because secretion of many lymphokines and monokines is not limited to lymphocytes and monocytes as these terms imply but extends to a broad spectrum of cells and types. For this reason, the more inclusive term *cytokine* is preferred.

Many cytokines are referred to as **interleukins,** a name indicating that they are secreted by some leukocytes and act on other leukocytes. Interleukins designated IL-1 through IL-29 have been described. There is reason to suppose that additional cytokines will be discovered and that the interleukin group will expand further. Some cytokines are known by common names, including the tumor necrosis factors and the interferons, although some recently described interferons such as members of the interferon λ family are also given an IL designation. Yet another subgroup of cytokines are the **chemokines,** a group of low-molecular-weight cytokines that affect chemotaxis and other aspects of leukocyte behavior. These molecules, which play an important role in the inflammatory response, were introduced in Chapter 3 and are more fully described in Chapter 13.

This chapter focuses on the biological activity of cytokines, the structure of cytokines and their receptors, signal transduction by cytokine receptors, the role of cytokine abnormalities in the pathogenesis of certain diseases, and therapeutic uses of cytokines or their receptors. The important role of cytokines in the inflammatory response is described in Chapter 13.

## Properties of Cytokines

Cytokines bind to specific receptors on the membrane of target cells, triggering signal transduction pathways that ultimately alter gene expression in the target cells (Figure 12-1a). The susceptibility of the target cell to a particular cytokine is determined by the presence of specific membrane receptors. We will see below that the cytokine receptors may be made up from several different chains and that the receptor for a given cytokine may exist in various combinations of these component chains on the cell surface. These chain combinations vary in the affinity with which they bind the cytokine and may also vary in their ability to initiate a signal-transducing pathway after binding. In general, the cytokines and their fully assembled receptors exhibit very high affinity for each other and deliver intracellular signals. Dissociation constants for cytokines and their optimum receptors range from $10^{-10}$ to $10^{-12}$ M. Because their affinities are so

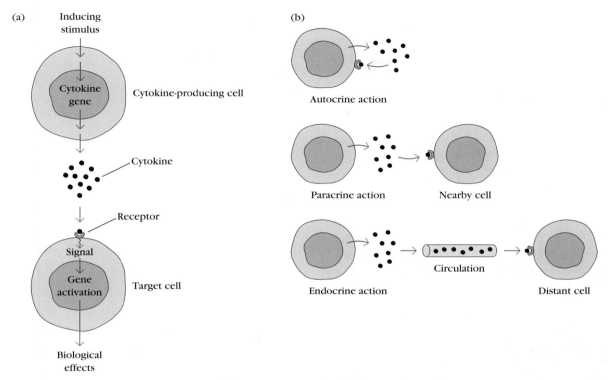

**FIGURE 12-1  (a) Overview of the induction and function of cytokines.** (b) Most cytokines exhibit autocrine and/or paracrine action; fewer exhibit endocrine action.

high, cytokines can mediate biological effects at picomolar concentrations.

A particular cytokine may bind to receptors on the membrane of the same cell that secreted it, exerting **autocrine** action; it may bind to receptors on a target cell in close proximity to the producer cell, exerting **paracrine** action; in a few cases, it may bind to target cells in distant parts of the body, exerting **endocrine** action (Figure 12-1b). Cytokines regulate the intensity and duration of the immune response by stimulating or inhibiting the activation, proliferation, and/or differentiation of various cells and by regulating the secretion of antibodies or other cytokines. As described later, binding of a given cytokine to responsive target cells generally stimulates increased expression of cytokine receptors and secretion of other cytokines, which affect other target cells in turn. Thus, the cytokines secreted by even a small number of lymphocytes activated by antigen can influence the activity of numerous cells involved in the immune response. For example, cytokines produced by activated $T_H$ cells can influence the activity of B cells, $T_C$ cells, natural killer cells, macrophages, granulocytes, and hematopoietic stem cells, thereby mobilizing an entire network of interacting cells.

Cytokines exhibit the attributes of pleiotropy, redundancy, synergy, antagonism, and cascade induction, which permit them to regulate cellular activity in a coordinated, interactive way (Figure 12-2). A given cytokine that has different biological effects on different target cells has a pleiotropic action. Two or more cytokines that mediate similar functions are said to be redundant; redundancy makes it difficult to ascribe a particular activity to a single cytokine. Cytokine synergism occurs when the combined effect of two cytokines on cellular activity is greater than the additive effects of the individual cytokines. In some cases, cytokines exhibit antagonism; that is, the effects of one cytokine inhibit or offset the effects of another cytokine. Cascade induction occurs when the action of one cytokine on a target cell induces that cell to produce one or more other cytokines, which in turn may induce other target cells to produce other cytokines.

Because cytokines share many properties with hormones and growth factors, the distinction between these three classes of mediators is often blurred. All three are secreted soluble factors that elicit their biological effects at picomolar concentrations by binding to receptors on target cells. However, some distinctions in their general features set them apart. For example, the three mediators differ in their modes of expression: growth factors tend to be produced constitutively, whereas cytokines and hormones are secreted in response to discrete stimuli, and secretion is short-lived, generally ranging from a few hours to a few days. Unlike hormones, which generally act at long range in an endocrine fashion, most cytokines act over a short distance in an autocrine or paracrine fashion. In addition, most hormones are produced by specialized glands and tend to have a unique action on one or a few types of target cell. In contrast, cytokines are often produced by and bind to a variety of cells.

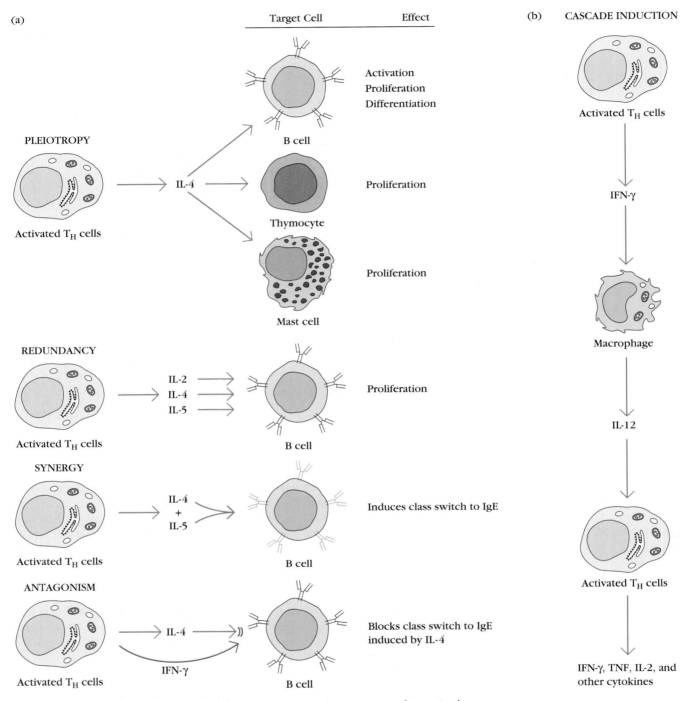

**FIGURE 12-2** Cytokine attributes of (a) pleiotropy, redundancy, synergy (synergism), antagonism, and (b) cascade induction.

The activity of cytokines was first recognized in the mid-1960s, when supernatants derived from in vitro cultures of lymphocytes were found to contain soluble factors that could regulate proliferation, differentiation, and maturation of immune system cells, including those from hosts differing in their genetic makeup. Soon after, it was discovered that production of these factors by cultured lymphocytes was induced by activation with antigen or with nonspecific mitogens. Biochemical isolation and purification of cytokines

was hampered because of their low concentration in culture supernatants and the absence of well-defined assay systems for individual cytokines. A great advance was made with the development of gene-cloning techniques during the 1970s and 1980s, which made it possible to produce pure cytokines by expressing the protein from cloned genes. The discovery of cell lines whose growth depended on the presence of a particular cytokine provided researchers with the first simple assay systems. The derivation of monoclonal antibodies

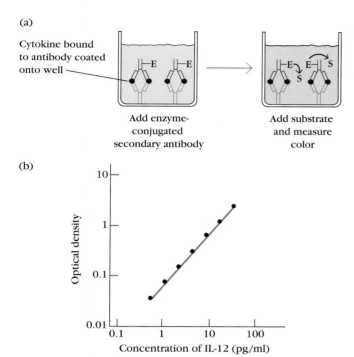

(a)

Cytokine bound to antibody coated onto well

Add enzyme-conjugated secondary antibody

Add substrate and measure color

(b)

**FIGURE 12-3 ELISA assay of a cytokine.** (a) The sample containing the cytokine of interest is captured by specific antibody (blue) coated onto wells of a microtiter plate. A second specific antibody (blue), conjugated to an enzyme (E) such as horseradish peroxidase, forms a sandwich with the captured cytokine, immobilizing the enzyme in the microtiter well. A chromogenic substrate (S) is added, and the enzyme generates a color whose intensity is proportional to the amount of cytokine bound to the captured antibody. The optical density of this color produced by the unknown is compared with values on an appropriately determined standard curve. (b) The standard curve shown here is for human interleukin-12 (IL-12). It is clear that this assay is sufficiently sensitive to detect as little as 1 picogram of IL-12. [Part b courtesy of R&D Systems.]

specific for each of the more important cytokines has made it possible to develop rapid quantitative immunoassays for each of them (Figure 12-3). In addition to the ELISA depicted here, specific anticytokine antibodies are used for intracellular staining of cells by FACS analysis and in the ELISPOT assay, which measures secretion of a specific cytokine by cells in a population (see Figure 6-11).

## Cytokines belong to four families

Recent years have seen an explosion of data about new cytokines and new receptors. Researchers will continue to amass information in this area, discovering many subtle features of the cytokines and their multiple activities and interactions. It is important to emphasize that however long the list of molecules becomes, all share common properties, and nearly all of the recently described cytokines act in concert with others. Advances in genomic analysis make it possible to search the genome for proteins that share a variety of properties. Thus, we find that there are seven members of the IL-1 group, as opposed to the

two previously recognized. The exact relevance of this to immune function remains to be learned and is beyond the scope of this discussion. The purpose here is not to provide an exhaustive list of cytokines (see Appendix II) but rather to give general principles that should apply as new cytokines are discovered.

Once the genes encoding various cytokines were cloned, sufficient quantities of purified preparations became available for detailed studies of their structure and function. These studies revealed common features. Cytokines generally have a molecular mass of less than 30 kDa. Structural studies have shown that the cytokines characterized so far belong to one of four groups: the hematopoietin family, the interferon family, the chemokine family, or the tumor necrosis factor family.

The structures of two members of the hematopoietin family, IL-2 and IL-4, are depicted in Figure 12-4. Although the

(a) Interleukin 2

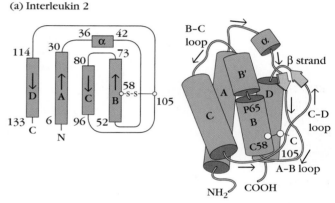

(b) Interleukin 4

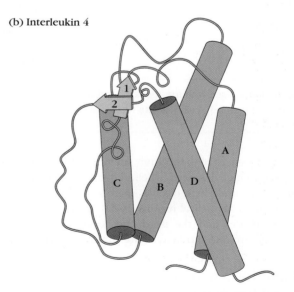

**FIGURE 12-4 Several representations of structures in the hematopoietin family.** (a) *Left:* Topographical representation of the primary structure of IL-2 showing α-helical regions (α and A–D) and connecting chains of the molecule. *Right:* Proposed three-dimensional model of IL-2. (b) Ribbon model of IL-4 deduced from x-ray crystallographic analysis of the molecule. In *a* and *b* the α helices are shown in red and the β sheets in blue. The structures of other cytokines belonging to the hematopoietin family are thought to be generally similar. [Part b from J. L. Boulay and W. E. Paul, 1993, Current Biology **3:**573.]

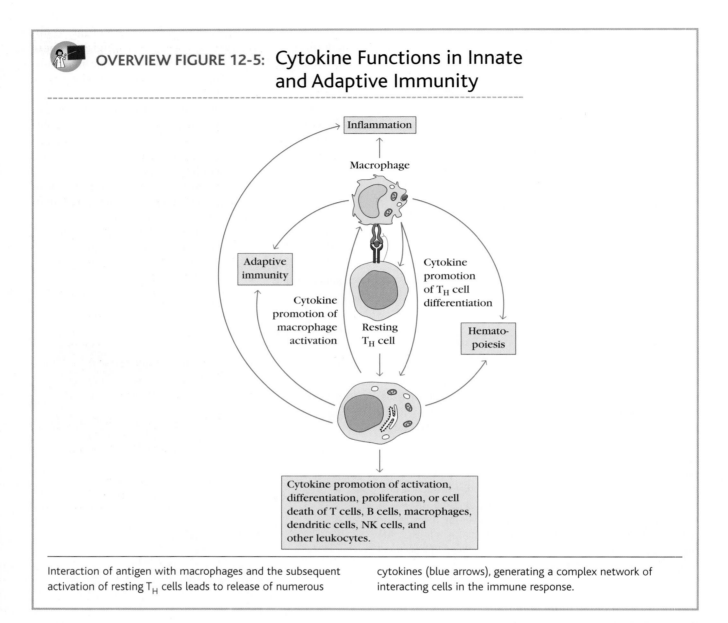

**OVERVIEW FIGURE 12-5:** Cytokine Functions in Innate and Adaptive Immunity

Interaction of antigen with macrophages and the subsequent activation of resting T$_H$ cells leads to release of numerous cytokines (blue arrows), generating a complex network of interacting cells in the immune response.

amino acid sequences of these family members differ considerably, all have a high degree of α-helical structure and little or no β-sheet structure. The molecules share a similar polypeptide fold, with four α-helical regions (A–D) in which the first and second helices and the third and fourth helices run roughly parallel to one another and are connected by loops.

## Cytokines have numerous biological functions

Although a variety of cells can secrete cytokines, the principal producers are T$_H$ cells, dendritic cells, and macrophages. Cytokines released from these cell types activate an entire network of interacting cells (Figure 12-5). Among the numerous physiologic responses that require cytokine involvement are development of cellular and humoral immune responses, induction of the inflammatory response, regulation of hematopoiesis, control of cellular proliferation and differentiation, and the healing of wounds. Although the immune response to a specific antigen may include the production of cytokines, it is important to remember that cytokines act in an antigen-nonspecific manner. That is, they affect whatever cells they encounter that bear appropriate receptors and are in a physiological state that allows them to respond.

Altogether, the total number of proteins with cytokine activity probably exceeds 200, and research continues to uncover new ones. Table 12-1 summarizes the activities of some commonly encountered cytokines and places them within functional groups. An expanded list of cytokines can be found in Appendix II. It should be kept in mind that most of the listed functions have been identified from analysis of the effects of recombinant cytokines, often added to in vitro systems at nonphysiologic concentrations. In vivo, however, cytokines rarely, if ever, act alone. Instead, a target cell is exposed to a

| TABLE 12-1 | Functional groups of selected cytokines* | | |
|---|---|---|---|
| **Cytokine†** | **Secreted by‡** | **Targets and effects** | |
| SOME CYTOKINES OF INNATE IMMUNITY | | | |
| Interleukin 1 (IL-1) | Monocytes, macrophages, endothelial cells, epithelial cells | Vasculature (inflammation); hypothalamus (fever); liver (induction of acute phase proteins) | |
| Tumor necrosis factor-α (TNF-α) | Macrophages | Vasculature (inflammation); liver (induction of acute phase proteins); loss of muscle, body fat (cachexia); induction of death in many cell types; neutrophil activation | |
| Interleukin 12 (IL-12) | Macrophages, dendritic cells | NK cells; influences adaptive immunity (promotes $T_H1$ subset) | |
| Interleukin 6 (IL-6) | Macrophages, endothelial cells | Liver (induces acute phase proteins); influences adaptive immunity (proliferation and antibody secretion of B cell lineage) | |
| Interferon α (IFN-α) (this is a family of molecules) | Macrophages | Induces an antiviral state in most nucleated cells; increases MHC class I expression; activates NK cells | |
| Interferon β (IFN-β) | Fibroblasts | Induces an antiviral state in most nucleated cells; increases MHC class I expression; activates NK cells | |
| SOME CYTOKINES OF ADAPTIVE IMMUNITY | | | |
| Interleukin 2 (IL-2) | T cells | T-cell proliferation; can promote AICD. NK cell activation and proliferation; B-cell proliferation | |
| Interleukin 4 (IL-4) | $T_H2$ cells, mast cells | Promotes $T_H2$ differentiation; isotype switch to IgE | |
| Interleukin 5 (IL-5) | $T_H2$ cells | Eosinophil activation and generation | |
| Transforming growth factor β (TGF-β) | T cells, macrophages, other cell types | Inhibits T-cell proliferation and effector functions; inhibits B-cell proliferation; promotes isotype switch to IgA; inhibits macrophages | |
| Interferon γ (IFN-γ) | $T_H1$ cells, $CD8^+$ cells, NK cells | Activates macrophages; increases expression MHC class I and class II molecules; increases antigen presentation | |

*Many cytokines play roles in more than one functional category.
†Only the major cell types providing cytokines for the indicated activity are listed; other cell types may also have the capacity to synthesize the given cytokine.
‡Also note that activated cells generally secrete greater amounts of cytokine than unactivated cells.

milieu containing a mixture of cytokines whose combined synergistic or antagonistic effects can have very different consequences. In addition, cytokines often induce the synthesis of other cytokines, resulting in cascades of activity.

The nonspecificity of cytokines appears to conflict with the established specificity of the immune system. What keeps cytokines from activating cells in a nonspecific fashion during the immune response? One way that specificity is maintained is by careful regulation of the expression of cytokine receptors on cells. Often, cytokine receptors are expressed on a cell only after that cell has interacted with antigen, limiting cytokine response to antigen-activated lymphocytes. Specificity may also be maintained if cytokine secretion occurs only when the cytokine-producing cell interacts directly with the target cell, thus ensuring that effective concentrations of the cytokine occur only in the vicinity of the intended target. In the case of the $T_H$ cell, a major producer of cytokines, cellular interaction occurs when the T-cell receptor recognizes an antigen-MHC complex on an appropriate antigen-presenting cell, such as a macrophage, dendritic cell, or B lymphocyte. The concentration of cytokines secreted at the junction of these interacting cells reaches high enough local concentrations to affect the target APC, but not more distant cells. In addition, the half-life of cytokines in the bloodstream or other extracellular fluids into which they are secreted is usually very short, ensuring that they act for only a limited period and thus over a short distance.

## Cytokine Receptors

As noted, cytokines exert their biological effects by binding to specific receptors expressed on the membrane of responsive target cells. Many types of cells express these receptors and are susceptible to the action of cytokines. Biochemical characterization of cytokine receptors initially progressed at a very slow pace because the levels of these receptors on the membrane of responsive cells are quite low. As with the cytokines themselves, cloning of the genes encoding cytokine receptors has led to rapid advances in the identification and characterization of these receptors.

## Cytokine receptors fall within five families

Receptors for the various cytokines are quite diverse structurally, but almost all belong to one of five families of receptor proteins (Figure 12-6):

- Immunoglobulin superfamily receptors

- Class I cytokine receptor family (also known as the hematopoietin receptor family)

- Class II cytokine receptor family (also known as the interferon receptor family)

- TNF receptor family

- Chemokine receptor family

The immunoglobulin superfamily includes the receptor for IL-1. Two forms of IL-1, called IL-1α and IL-1β, have been well characterized. Although these two proteins are only about 30% identical in sequence, they appear to be identical in function in that they bind to the same receptors and mediate the same responses. Two different receptors for IL-1 are known, both members of the Ig superfamily of proteins. The type 1 IL-1R is expressed on many cell types, whereas type 2 IL-1R is limited to B cells. The binding of IL-1 to the type 2 receptor does not initiate responses typical of IL-1, suggesting that type 2 IL-1R is a decoy meant to remove IL-1, thereby preventing its binding to type 1 receptors. We encountered the type 1 receptor for IL-1 in Chapter 3, when Toll-like receptors were introduced The cytoplasmic domain of TLRs, the TIR, or Toll/IL-1 receptor, is similar to the comparable region in IL-1R (see Figure 3-10). IL-1 plays a key role in inflammation and is designated as a proinflammatory cytokine. IL-18 is related to IL-1, uses the same receptor family, and has a similar function.

A majority of the cytokine-binding receptors that function in the immune and hematopoietic systems belong to the class I cytokine receptor family, also called the hematopoietin receptor family. The members of this family have conserved amino acid sequence motifs in the extracellular domain consisting of four positionally conserved cysteine residues (CCCC) and a conserved sequence of tryptophan-serine-(any amino acid)-tryptophan-serine (WSXWS, where X is a nonconserved amino acid). The receptors for all the cytokines classified as hematopoietins belong to the class I cytokine receptor family. IL-2 and about 13 other interleukins, as well

**FIGURE 12-6 Schematic diagrams showing the structural features that define the four types of receptor proteins to which most cytokines bind.** The receptors for most of the interleukins belong to the class I cytokine receptor family. C refers to conserved cysteine.

| RECEPTOR FAMILY | LIGANDS | |
|---|---|---|
| **(a) Immunoglobulin superfamily receptors** 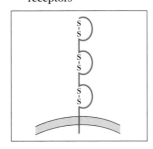 | IL-1<br>M-CSF<br>C-Kit<br>IL-18 | |
| **(b) Class I cytokine receptors (hematopoietin)** 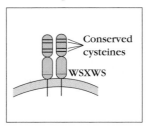 | IL-2<br>IL-3<br>IL-4<br>IL-5<br>IL-6<br>IL-7<br>IL-9<br>IL-11<br>IL-12<br>IL-13<br>IL-15 | IL-21<br>IL-23<br>IL-27<br>GM-CSF<br>G-CSF<br>OSM<br>LIF<br>CNTF<br>Growth hormone<br>Prolactin |
| **(c) Class II cytokine receptors (interferon)** 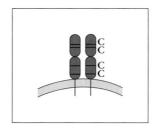 | IFN-α<br>IFN-β<br>IFN-γ<br>IL-10<br>IL-19<br>IL-20<br>IL-22<br>IL-24<br>IL-26<br>IL-28<br>IL-29 | |
| **(d) TNF receptors** 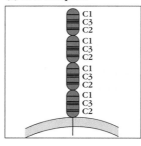 | TNF-α<br>TNF-β<br>CD27L<br>CD30L<br>CD40L<br>Nerve growth factor (NGF)<br>FAS | |
| **(e) Chemokine receptors** 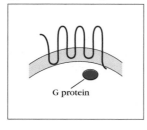 | IL-8<br>RANTES<br>MIP-1<br>PF4<br>MCAF<br>NAP-2 | |

as several colony-stimulating factors and growth hormones, bind to the receptors of this family. Structural data for most receptors of the type I group are available, with the exception of the recently described IL-27R, which has one chain in common with other receptors of the family and another chain that has not yet been identified.

The class II cytokine receptors (sometimes called the interferon receptor family) possess the conserved CCCC motif but lack the WSXWS motif present in class I cytokine receptors. Initially, only interferons α, β, and γ were thought to be ligands for these receptors. But recent work has shown that the family consists of 12 receptor chains that in their various assortments bind no fewer than 27 different cytokines, including six members of the IL-10 family, 17 type I interferons, one type II interferon, and three members of the recently described interferon λ family, including IL-28a, IL-28b, and IL-29.

A feature common to most of the class I and II cytokine receptor families is multiple subunits, often including one subunit that binds specific cytokine molecules and another that mediates signal transduction. These two functions are not always confined to one subunit or the other. In all cases studied to date, however, engagement of class I and II cytokine receptors induces tyrosine phosphorylation of the receptor through the activity of protein tyrosine kinases closely associated with the cytosolic domain of the receptors.

Chemokine receptors are structurally distinct from the other cytokine receptors and will be described in Chapter 13.

## Subfamilies of class I cytokine receptors have signaling subunits in common

Several subfamilies of class I cytokine receptors have been identified, with all the receptors in a subfamily having an identical signal-transducing subunit. Figure 12-7 shows schematically the members of three receptor subfamilies, named after GM-CSF, IL-2, and IL-6.

The sharing of signal-transducing subunits among receptors explains the redundancy and antagonism exhibited by some cytokines. Consider the GM-CSF receptor subfamily, which includes the receptors for IL-3, IL-5, and GM-CSF (Figure 12-7a). Each of these cytokines binds with low affinity to a unique cytokine-specific receptor protein, the α subunit of a dimeric receptor. All three low-affinity subunits can

**FIGURE 12-7  Schematic diagrams of three subfamilies of class I cytokine receptors.** All members of a subfamily have a common signal-transducing subunit (blue) but a unique cytokine-specific subunit. In addition to the conserved cysteines (double black lines) and WSXWS motifs (red lines) that characterize class I cytokine receptors, immunoglobulin-like domains are present in some of these receptors. CNTF = ciliary neurotrophic factor; LIF/OSM = leukemia inhibitory factor/oncostatin. *[Adapted from K. Sugamura et al., 1996, Annual Review of Immunology **14**:179.]*

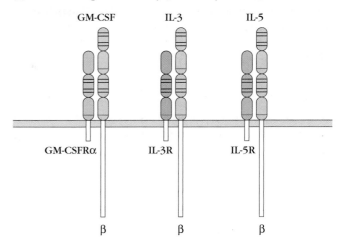

(a) GM-CSF receptor subfamily (common β subunit)

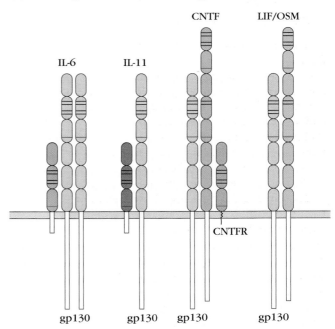

(b) IL-6 Receptor subfamily (common gp130 subunit)

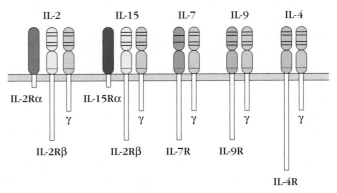

(c) IL-2 receptor subfamily (common γ subunit)

**FIGURE 12-8 Interactions between cytokine-specific subunits and a common signal-transducing subunit of cytokine receptors.** (a) Schematic diagram of the low-affinity and high-affinity receptors for IL-3, IL-5, and GM-CSF. The cytokine-specific subunits exhibit low-affinity binding and cannot transduce an activation signal. Noncovalent association of each subunit with a common β subunit yields a high-affinity dimeric receptor that can transduce a signal across the membrane. (b) Association of cytokine-specific subunits with a common signaling unit, the β subunit, allows the generation of identical signals by the different cytokine receptors shown. (c) Competition of ligand-binding chains of different receptors for a common subunit can produce antagonistic effects between cytokines. Here binding of IL-3 by α subunits of the IL-3 receptor allows them to outcompete α chains of the GM-CSF receptor for β subunits. *[Part a adapted from T. Kishimoto et al., 1992, Science **258**:593.]*

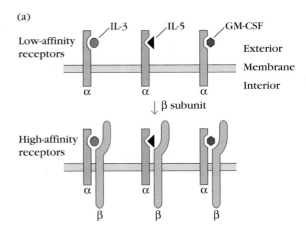

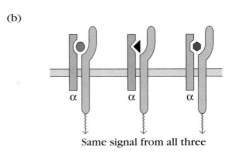

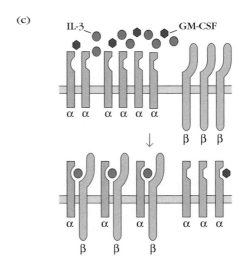

associate noncovalently with a common signal-transducing β subunit. The resulting dimeric receptor exhibits increased affinity for the cytokine and is capable of transducing a signal across the membrane after the cytokine binds (Figure 12-8a). Interestingly, IL-3, IL-5, and GM-CSF exhibit redundant activities. IL-3 and GM-CSF both act on hematopoietic stem cells and progenitor cells, activate monocytes, and induce megakaryocyte differentiation, and all three of these cytokines induce eosinophil proliferation and basophil degranulation with release of histamine.

Since the receptors for IL-3, IL-5, and GM-CSF share a common signal-transducing β subunit, each of these cytokines would be expected to transduce a similar activation signal, accounting for the redundancy among their biological effects (Figure 12-8b). In fact, all three cytokines induce the same patterns of protein phosphorylation, yet IL-3 and GM-CSF exhibit antagonism: binding of IL-3 is inhibited by GM-CSF, and binding of GM-CSF is inhibited by IL-3. This antagonism is caused by competition for a limited number of β subunits available to associate with the cytokine-specific α subunits of the dimeric receptors (Figure 12-8c).

A similar situation is found among the IL-6 receptor subfamily, which includes the receptors for IL-6, IL-11, leukemia inhibitory factor (LIF), oncostatin M (OSM), and ciliary neurotrophic factor (CNTF) (see Figure 12-7b). In this case, a common signal-transducing subunit called gp130 associates with one or two different cytokine-specific subunits. LIF and OSM, which share certain structural features, both bind to the same α subunit. As expected, the cytokines that bind to receptors in this subfamily display overlapping biological activities: IL-6, OSM, and LIF induce synthesis of acute-phase proteins by liver hepatocytes and differentiation of myeloid leukemia cells into macrophages; IL-6, LIF, and CNTF affect neuronal development; and IL-6, IL-11, and OSM stimulate megakaryocyte maturation and platelet production. The presence of gp130 in all receptors of the IL-6 subfamily explains their common signaling pathways as well as the binding competition for limited gp130 molecules that is observed among these cytokines.

A third signal-transducing subunit defines the IL-2 receptor subfamily, which includes receptors for IL-2, IL-4, IL-7, IL-9, IL-15, and IL-21 (see Figure 12-7c). The IL-2 and the IL-15 receptors are heterotrimers, consisting of a cytokine-specific α chain and two chains—β and γ—responsible for signal transduction. The IL-2 receptor γ chain functions as the signal-transducing subunit in the other receptors in this subfamily, which are all dimers. Recently, it has been shown that congenital **X-linked severe combined immunodeficiency (XSCID)** results from a defect in the γ-chain gene, which maps to the X chromosome. The immunodeficiency observed in this disorder, which includes loss of T-cell and NK-cell activity, is due to the loss of all the cytokine functions mediated by the IL-2 subfamily receptors. A variety of SCID in which T cells are defective but NK cells are normal is caused by genetic defects in the IL-7 receptor (see Chapter 20 for a complete discussion of SCID).

## IL-2R is the most thoroughly studied cytokine receptor

Because of the central role of IL-2 and its receptor in the clonal proliferation of T cells, the IL-2 receptor has received intensive study. As noted in the previous section, the complete trimeric receptor comprises three distinct subunits—the α, β, and γ chains. The β and γ chains belong to the class I cytokine receptor family, containing the characteristic CCCC and WSXWS motifs, whereas the α chain has a quite different structure and is not a member of this receptor family (see Figure 12-7c).

The IL-2 receptor occurs in three forms that exhibit different affinities for IL-2: the low-affinity monomeric IL-2Rα, the intermediate-affinity dimeric IL-2Rβγ, and the high-affinity trimeric IL-2Rαβγ (Figure 12-9a). A recent x-ray crystallographic structure for the high-affinity trimeric form of the IL-2 receptor with an IL-2 molecule in its binding site reveals that IL-2 binds in a pocket formed by the β and γ chains (Figure 12-9b). Important additional contacts with the

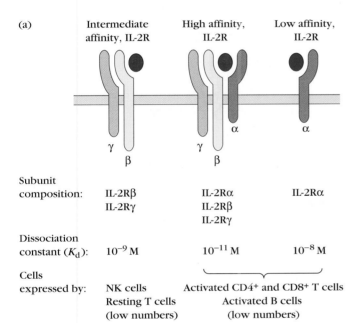

(a)

| | Intermediate affinity, IL-2R | High affinity, IL-2R | Low affinity, IL-2R |
|---|---|---|---|
| Subunit composition: | IL-2Rβ IL-2Rγ | IL-2Rα IL-2Rβ IL-2Rγ | IL-2Rα |
| Dissociation constant ($K_d$): | $10^{-9}$ M | $10^{-11}$ M | $10^{-8}$ M |
| Cells expressed by: | NK cells Resting T cells (low numbers) | Activated CD4+ and CD8+ T cells Activated B cells (low numbers) | |

**FIGURE 12-9  Comparison of the three forms of the IL-2 receptor.** (a) Schematic of the three forms of the receptor and listing of dissociation constants and properties for each. Signal transduction is mediated by the β and γ chains, but all three chains are required for high-affinity binding of IL-2. (b) Three-dimensional structure of the three-chain form of the IL-2 receptor with bound IL-2 (views rotated by 90°). Note that the α chain completes the pocket to which IL-2 binds, accounting for the higher affinity of the trimeric form. *[From X. Wang, M. Rickert, and K. C. Garcia, 2005, Science **310**:1159.]*

(b)

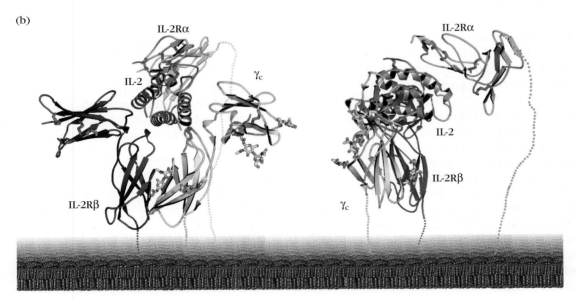

IL-2 ligand are contributed when the α chain is present, accounting for the higher affinity of binding by the trimer. Binding of IL-2 to the receptor including only the β and γ chains occurs at an intermediate affinity, and this form of the receptor is capable of signal transduction.

Because the IL-2Rα chain is expressed only by activated T cells, it is sometimes referred to as the TAC (T-cell activation) antigen or CD25; recall that we encountered CD25 earlier as a characterizing surface marker in the maturation of T cells (see Figure 10-2). A monoclonal antibody (anti-TAC or anti-CD25) that binds to the 55-kDa α chain is often used to identify IL-2Rα on cells. The presence of high levels of CD25 in mature T cells generally indicates that the cells are in a state of activation and, as discussed in Chapter 2 and in more detail in Chapter 16, may also indicate that cells of the $T_{reg}$ subset ($CD4^+$ cells with high levels of CD25) are present. The expression of the three chains of the IL-2 receptor varies among different cell types: the γ chain appears to be constitutively expressed on most lymphoid cells, whereas expression of the α and β chains is more restricted and is markedly enhanced after antigen has activated resting lymphocytes. This restricted expression ensures that only antigen-activated $CD4^+$ and $CD8^+$ T cells will express the high-affinity IL-2 receptor and proliferate in response to physiologic levels of IL-2. Activated T cells express approximately $5 \times 10^3$ high-affinity receptors and 10 times as many low-affinity receptors. NK cells express the β and γ subunits constitutively, accounting for their ability to bind IL-2 with an intermediate affinity and to be activated by IL-2.

## Cytokine receptors initiate signaling

Although some important cytokine receptors lie outside the class I and class II families, the majority are included within these two families. As noted, class I and class II cytokine receptors lack signaling motifs (for example, intrinsic tyrosine kinase domains). Yet early observations demonstrated that one of the first events after the interaction of a cytokine with one of these receptors is a series of protein tyrosine phosphorylations. These results were initially puzzling but were explained when a unifying model emerged from studies of the molecular events triggered by binding of interferon gamma (IFN-γ) to its receptor, a member of the class II family.

IFN-γ was originally discovered because of its ability to induce cells to block or inhibit the replication of a wide variety of viruses. Antiviral activity is a property it shares with IFN-α and IFN-β. However, unlike these other interferons, IFN-γ plays a central role in many immunoregulatory processes, including the regulation of mononuclear phagocytes, B-cell switching to certain IgG classes, and the support or inhibition of the development of $T_H$-cell subsets. The discovery of the major signaling pathway invoked by interaction of IFN-γ with its receptor led to the realization that signal transduction through most, if not all, class I and class II cytokine receptors involves the following steps,

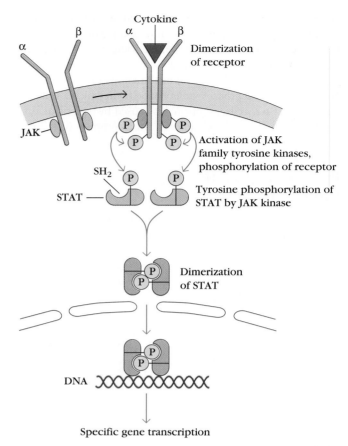

**FIGURE 12-10  General model of signal transduction mediated by most class I and class II cytokine receptors.** Binding of a cytokine induces dimerization of the receptor subunits, which leads to the activation of receptor subunit–associated JAK tyrosine kinases by reciprocal phosphorylation. Subsequently, the activated JAKs phosphorylate various tyrosine residues, resulting in the creation of docking sites for STATs on the receptor and the activation of the one or more STAT transcription factors. The phosphorylated STATs dimerize and translocate to the nucleus, where they activate transcription of specific genes.

which are the basis of a unifying cytokine-signaling model (Figure 12-10).

- *The cytokine receptor is composed of separate subunits,* one chain required mainly for cytokine binding and signal transduction and another chain necessary for signaling but often with only a minor role in binding.

- *Different inactive protein tyrosine kinases are associated with different subunits of the receptor.* The α chain of the receptor is associated with a novel family of protein tyrosine kinases, the Janus kinase (JAK)* family. The

*The Roman god Janus had two faces. Kinases of the Janus family have two functional sites: a binding site that associates with the cytokine receptor subunit and a catalytic site that, when activated, has protein tyrosine kinase activity. Some biochemists, wearied by the multitude of different protein kinases that have been discovered, claim JAK means *just another kinase*.

association of the JAK and the receptor subunit occurs spontaneously and does not require the binding of a cytokine. However, in the absence of a cytokine, JAKs lack protein tyrosine kinase activity.

■ *Cytokine binding induces the association of the two separate cytokine receptor subunits and activation of the receptor-associated JAKs.* The ability of IFN-γ, which binds to a class II cytokine receptor, to bring about the association of the ligand-binding chains of its receptor has been directly demonstrated by x-ray crystallographic studies, as shown in Figure 12-11.

■ *Activated JAKs create docking sites for the STAT transcription factors by phosphorylation of specific tyrosine residues on cytokine receptor subunits.* Once receptor-associated JAKs are activated, they phosphorylate specific tyrosines in the receptor subunits of the complex. Members of a family of transcription factors known as **STATs (signal transducers and activators of transcription)** bind to these phosphorylated tyrosine residues. Specific STATs play essential roles in the signaling pathways of a wide variety of cytokines (Table 12-2). The binding of STATs to receptor subunits

| TABLE 12-2 | STAT and JAK interaction with selected cytokine receptors during signal transduction | | |
|---|---|---|---|
| **Cytokine receptor** | **JAK** | | **STAT** |
| IFN-γ | JAK1 and JAK2 | | Stat1 |
| IFN-α/β | JAK1 and Tyk-2* | | Stat2 |
| IL-2 | JAK1 and JAK3 | | Stat5 |
| IL-3 | JAK2 | | Stat5 |
| IL-4 | JAK1 and JAK3 | | Stat6 |
| IL-6 | JAK1 (and sometimes others) | | Stat3 |
| IL-10 | JAK1 and Tyk-2 | | Stat3 |
| IL-12 | JAK2 and Tyk-2 | | Stat4 |

*Despite its name, Tyk-2 is also a Janus kinase.

SOURCE: Adapted from E. A. Bach, M. Aguet, and R. D. Schreiber, 1997, *Annual Review of Immunology* **15**:563.

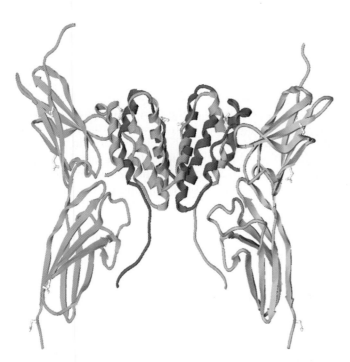

**FIGURE 12-11 The complex between IFN-γ and the ligand-binding chains of its receptor.** This model is based on the x-ray crystallographic analysis of a crystalline complex of interferon-γ (dark and light purple) bound to ligand-binding α chains of the receptor (green and yellow). Note that IFN-γ is shown in its native dimeric form; each member of the dimer engages the α chain of an IFN-γ receptor, thereby bringing about receptor dimerization and signal transduction. *[From M. R. Walter et al., 1995,* Nature ***376**:230, courtesy M. Walter, University of Alabama.]*

is mediated by the joining of the SH$_2$ domain on the STAT with the docking site created by the JAK-mediated phosphorylation of a particular tyrosine on receptor subunits.

■ *After undergoing JAK-mediated phosphorylation, STAT transcription factors translocate from receptor docking sites at the membrane to the nucleus, where they initiate the transcription of specific genes.* While docked to receptor subunits, STATs undergo JAK-catalyzed phosphorylation of a key tyrosine. This is followed by the dissociation of the STATs from the receptor subunits and their dimerization. The STAT dimers then translocate into the nucleus and induce the expression of genes containing appropriate regulatory sequences in their promoter regions.

In addition to IFN-γ, a number of other class I and class II ligands have been shown to cause dimerization of their receptors. An important element of cytokine specificity derives from the exquisite specificity of the match between cytokines and their receptors. Another aspect of cytokine specificity is that each particular cytokine (or group of redundant cytokines) induces transcription of a specific subset of genes in a given cell type; the resulting gene products then mediate the various effects typical of that cytokine. The specificity of cytokine effects is then traceable to three factors. First, particular cytokine receptors start particular JAK-STAT pathways. Second, the transcriptional activity of activated STATs is specific because a particular STAT homodimer or heterodimer will only recognize certain sequence motifs and thus can interact only with the promoters of certain genes. Third, only those target genes whose expression is permitted by a particular cell type can be activated within that variety of cell. That is, in any given cell type only a subset of the potential target genes of a particular STAT may be permitted expression. For example, IL-4 induces one set of genes in T cells, another in B cells, and yet a third in eosinophils.

It is worth mentioning that IL-1 does not signal via the JAK-STAT pathway but utilizes a kinase designated IL-1 receptor-associated kinase, or IRAK. The IRAK proteins are also utilized by TLRs for signal transduction (see Figure 3-14).

## Cytokine Antagonists

A number of proteins that inhibit the biological activity of cytokines have been reported. These proteins act in one of two ways: either they bind directly to a cytokine receptor but fail to activate the cell, or they bind directly to a cytokine, inhibiting its activity. The best-characterized inhibitor is the IL-1 receptor antagonist (IL-1Ra), which binds to the IL-1 receptor but has no activity. Binding of IL-1Ra to the IL-1 receptor blocks binding of both IL-1α and IL-1β, thus accounting for its antagonistic properties. Production of IL-1Ra has been thought by some to play a role in regulating the intensity of the inflammatory response. It has been cloned and is currently being investigated as a potential treatment for chronic inflammatory diseases.

Cytokine inhibitors are found in the bloodstream and extracellular fluid. These soluble antagonists arise from enzymatic cleavage of the extracellular domain of cytokine receptors. Among the soluble cytokine receptors that have been detected are those for IL-2, -4, -6, and -7; IFN-γ and -α; TNF-β; and LIF. Of these, the soluble IL-2 receptor (sIL-2R), which is released in chronic T-cell activation, is the best characterized. A segment containing the amino-terminal 192 amino acids of the α subunit is released by proteolytic cleavage, forming a 45-kDa soluble IL-2 receptor. The shed receptor can bind IL-2 and prevent its interaction with the membrane-bound IL-2 receptor. The presence of sIL-2R has been used as a clinical marker of chronic T-cell activation and is observed in a number of diseases, including autoimmunity, transplant rejection, and AIDS.

Some viruses have developed strategies to thwart cytokine activity. The evolution of such anticytokine strategies by microbial pathogens is biological evidence of the importance of cytokines in organizing and promoting effective antimicrobial immune responses. Among the various anticytokine strategies used by viruses are

- Cytokine homologs
- Soluble cytokine-binding proteins
- Homologs of cytokine receptors
- Interference with intracellular signaling
- Interference with cytokine secretion
- Induction of cytokine inhibitors in the host cell

| TABLE 12-3 | Viral mimics of cytokines and cytokine receptors |
|---|---|
| **Virus** | **Products** |
| Leporipoxvirus (a myxoma virus) | Soluble IFN-γ receptor |
| Several poxviruses | Soluble IFN-γ receptor |
| Vaccinia, smallpox virus | Soluble IL-1β receptor |
| Epstein-Barr | IL-10 homolog |
| Human herpesvirus-8 | IL-6 homolog; also homologs of the chemokines MIP-I and MIP-II |
| Cytomegalovirus | Three different chemokine receptor homologs, one of which binds three different soluble chemokines (RANTES, MCP-1, and MIP-1α) |

Epstein-Barr virus (EBV) produces an IL-10–like molecule (viral IL-10 or vIL-10) that binds to the IL-10 receptor and, like cellular IL-10, suppresses $T_H1$-type cell-mediated responses (see the next section), which are effective against many intracellular parasites such as viruses. Molecules produced by viruses that mimic cytokines allow the virus to manipulate the immune response in ways that aid the survival of the pathogen. This is an interesting and powerful adaptation some viruses have undergone in their continuing struggle to overcome the formidable barrier of host immunity. EBV also produces an inducer of IL-1Ra, the host antagonist of IL-1. The poxviruses have been shown to encode a soluble TNF-binding protein and a soluble IL-1-binding protein. Since both TNF and IL-1 exhibit a broad spectrum of activities in the inflammatory response, these soluble cytokine-binding proteins may prohibit or diminish the inflammatory effects of the cytokines, thereby conferring on the virus a selective advantage. Table 12-3 lists a number of viral products that inhibit cytokines and their activities.

## Cytokine Secretion by $T_H1$ and $T_H2$ Subsets

The immune response to a particular pathogen must induce an appropriate set of effector functions that can eliminate the disease agent or its toxic products from the host. For example, the neutralization of a soluble bacterial toxin requires antibodies, whereas the response to an intracellular virus or to a bacterial cell requires cell-mediated cytotoxicity or delayed-type hypersensitivity. A large body of evidence implicates differences in cytokine-secretion patterns among $T_H$-cell subsets as determinants of the type of immune response made to a particular antigenic challenge.

CD4$^+$ T$_H$ cells exert most of their helper functions through secreted cytokines, which either act on the cells that produce them in an autocrine fashion or modulate the responses of other cells through paracrine pathways. Although CD8$^+$ CTLs also secrete cytokines, their array of cytokines generally is more restricted than that of CD4$^+$ T$_H$ cells. As briefly discussed in Chapter 10, two CD4$^+$ T$_H$-cell subpopulations designated T$_H$1 and T$_H$2 can be distinguished in vitro by the cytokines they secrete. Both subsets secrete IL-3 and GM-CSF but differ in the other cytokines they produce (Table 12-4). T$_H$1 and T$_H$2 cells are characterized by the following functional differences:

- The T$_H$1 subset is responsible for many cell-mediated functions, such as delayed-type hypersensitivity and activation of T$_C$ cells, and for the production of opsonization-promoting IgG antibodies; that is, antibodies that bind to the high-affinity Fc receptors of phagocytes and interact with the complement system.

| TABLE 12-4 | Cytokine secretion and principal functions of mouse T$_H$1 and T$_H$2 subsets | |
|---|---|---|
| | **T$_H$1** | **T$_H$2** |
| **CYTOKINE SECRETION** | | |
| IL-2 | + | − |
| IFN-γ | ++ | − |
| TNF-β | ++ | − |
| GM-CSF | ++ | + |
| IL-3 | ++ | ++ |
| IL-4 | − | ++ |
| IL-5 | − | ++ |
| IL-10 | − | ++ |
| IL-13 | − | ++ |
| **FUNCTIONS** | | |
| Help for total antibody production | + | ++ |
| Help for IgE production | − | ++ |
| Help for IgG2a production | ++ | + |
| Eosinophil and mast-cell production | − | ++ |
| Macrophage activation | ++ | − |
| Delayed-type hypersensitivity | ++ | − |
| T$_C$-cell activation | ++ | − |

SOURCE: Adapted from F. Powrie and R. L. Coffman, 1993, *Immunology Today* 14:270.

This subset is also associated with the promotion of excessive inflammation and tissue injury.

- The T$_H$2 subset stimulates eosinophil activation and differentiation, provides help to B cells, and promotes the production of relatively large amounts of IgM, IgE, and noncomplement-activating IgG isotypes. The T$_H$2 subset also supports allergic reactions.

The differences in the cytokines secreted by T$_H$1 and T$_H$2 cells determine the different biological functions of these two subsets. A defining cytokine of the T$_H$1 subset, IFN-γ, activates macrophages, stimulating these cells to increase microbicidal activity, up-regulate the level of class II MHC, and secrete cytokines such as IL-12, which induces T$_H$ cells to differentiate into the T$_H$1 subset. IFN-γ secretion by T$_H$1 cells also induces antibody class switching to IgG classes (such as IgG2a in the mouse) that support phagocytosis and fixation of complement. TNF-β and IFN-γ are cytokines that mediate inflammation, and their secretion is part of the contribution of T$_H$1 cells to inflammatory phenomena such as delayed hypersensitivity (see Chapter 15). T$_H$1 cells produce IL-2 and IFN-γ cytokines that promote the differentiation of fully cytotoxic T$_C$ cells from CD8$^+$ precursors. This pattern of cytokine production makes the T$_H$1 subset particularly suited to respond to viral infections and intracellular pathogens. Finally, IFN-γ inhibits the expansion of the T$_H$2 population.

The secretion of IL-4 and IL-5 by cells of the T$_H$2 subset induces production of IgE and supports eosinophil-mediated attack on helminth (roundworm) infections. IL-4 promotes a pattern of class switching that produces IgG that does not activate the complement pathway (IgG1 in mice, for example). IL-4 also increases the extent to which B cells switch from IgM to IgE. This effect on IgE production meshes with eosinophil differentiation and activation by IL-5, because eosinophils are richly endowed with Fcε receptors, which bind IgE. Typically, roundworm infections induce T$_H$2 responses and evoke anti-roundworm IgE antibody. The antibody bound to the worm cross-links to the Fc receptors of eosinophils, thus forming an antigen-specific bridge between the worm and the eosinophils. The attack of the eosinophil on the worm is triggered by the cross-linking of the Fcε-bound IgE. Along with these beneficial actions of IgE, it is also the Ig class responsible for allergy. Finally, IL-4 and IL-10 suppress the expansion of T$_H$1-cell populations.

Because the T$_H$1 and T$_H$2 subsets were originally identified in long-term in vitro cultures of cloned T-cell lines, some researchers doubted that they represented true in vivo subpopulations. They suggested instead that these subsets might represent different maturational stages of a single lineage. Also, the initial failure to locate either subset in humans led some to believe that T$_H$1, T$_H$2, and other subsets of T helper cells did not occur in this species. Further research corrected these views. In many in vivo systems, the full commitment of populations of T cells to either the T$_H$1 or T$_H$2 phenotype often signals the end point of a chronic infection or allergy. Hence, it was difficult to find clear $T_H1$ or $T_H2$ subsets in studies employing healthy human subjects, who would not be at this stage of a response. Experiments with transgenic mice demonstrated conclusively that $T_H1$ and $T_H2$ cells arise independently. Furthermore, it was possible to demonstrate $T_H1$ or $T_H2$ populations in T cells isolated from humans during chronic infectious disease or chronic episodes of allergy. It is also important to emphasize that many helper T cells do not show either a $T_H1$ or a $T_H2$ profile; individual cells have shown striking heterogeneity in the $T_H$-cell population.

Numerous reports of studies in both mice and humans now document that the in vivo outcome of the immune response can be critically influenced by the relative levels of $T_H1$-like or $T_H2$-like activity. Typically, the $T_H1$ profile of cytokines is higher in response to intracellular pathogens, and the $T_H2$ profile is higher in allergic diseases and helminthic infections.

## The development of $T_H1$ and $T_H2$ subsets is determined by the cytokine environment

The cytokine environment in which antigen-primed $T_H$ cells differentiate determines the subset that develops (Figure 12-12). In particular, IL-4 is essential for the development of a $T_H2$ response, and IFN-$\gamma$, IL-12, and IL-18 all are important in the physiology of the development of $T_H1$ cells. $T_H1$ development is critically dependent on IFN-$\gamma$, which induces a number of changes, including the up-regulation of IL-12 production by macrophages and dendritic cells and the activation of the IL-12 receptor on activated T cells, which it accomplishes by up-regulating expression of the $\beta$ chain of the IL-12 receptor. At the beginning of an immune response, IFN-$\gamma$ is generated by stimulation of T cells and can also come from activated NK cells. The source of IL-12, one of the key mediators of $T_H1$ differentiation, is typically macrophages or dendritic cells activated by an encounter with intracellular bacteria or other intracellular parasites or with bacterial products such as LPS. Yet another cytokine, IL-18, promotes proliferation and IFN-$\gamma$ production by both developing and fully differentiated $T_H1$ cells and by NK cells. Thus, a regulatory network of cytokines positively controls the generation of $T_H1$ cells. The critical role played by each of these cytokines and their receptors has been demonstrated in experiments in which either the cytokine or its receptor has been knocked out. These mice fail to generate populations of $T_H1$ cells.

Recent studies have uncovered two new cytokines of the IL-12 family, IL-23 and IL-27, with important implications for the development of $T_H1$ cells. IL-23 has one chain in common with IL-12 (IL-12 p40), and the two chains of the IL-27 heterodimer are related to those of IL-12. IL-23 and IL-27 both share functions with IL-12; all are involved in differentiation to the $T_H1$ subset, and some activities tradi-

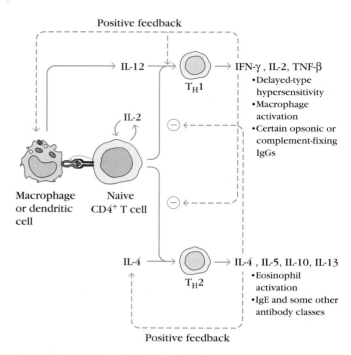

**FIGURE 12-12 Cytokine-mediated generation and cross-regulation of $T_H$ subsets.** An antigen-activated naive CD4$^+$ T cell produces IL-2 and proliferates. If it proliferates in an IL-12 dominated environment, it generates a population of $T_H1$ cells that secretes a characteristic profile of cytokines, including IFN-$\gamma$. A positive feedback loop is established when IFN-$\gamma$ secreted by the expanding $T_H1$ population stimulates dendritic cells or macrophages to produce more IL-12. If the environment is dominated by IL-4, a $T_H2$ population emerges and secretes a profile of cytokines that promotes eosinophil activation and the synthesis of certain antibody classes. Key cytokines produced by each subset positively regulate the subset that produces it and negatively regulate the other subset. *[Adapted from J. Rengarajan, S. Szabo, and L. Glimcher, 2000,* Immunology Today **21**:479.]

tionally attributed to IL-12 may be the result of IL-23 acting alone or in synergy with IL-12.

Just as $T_H1$ cells require IL-12 and IFN-$\gamma$, the generation of $T_H2$ cells depends critically on IL-4. Exposing naive helper cells to IL-4 at the beginning of an immune response causes them to differentiate into $T_H2$ cells. In fact, this influence of IL-4 is predominant in directing $T_H$ cells to the $T_H2$ route. Provided a threshold level of IL-4, $T_H2$ development is greatly favored over $T_H1$ even if IFN-$\gamma$ and IL-12 are present. The critical role of signals from IL-4 in $T_H2$ development is shown by the observation that knocking out the gene that encodes IL-4 prevents the development of this T-cell subset. Additional evidence supporting the central role of IL-4 comes from an experiment that interrupted the IL-4 signal transduction pathway. Like so many other cytokines, IL-4 uses a pathway that involves JAK and STAT proteins. The Stat6 transcription factor is the one activated in signaling by IL-4. Consequently, in mice in which the gene for Stat6 has been disrupted (Stat6 knockouts), IL-4-mediated processes are severely inhibited or absent. The observation that Stat6 knockout mice have very few $T_H2$ cells confirms the importance of IL-4 for the differentiation of this subset. In a similar fashion, disruption of Stat1, which is activated by IFN-$\gamma$, interferes with development of $T_H1$ cells.

## Cytokine profiles are cross-regulated

The critical cytokines produced by $T_H1$ and $T_H2$ subsets have two characteristic effects on subset development. First, they promote the growth of the subset that produces them; second, they inhibit the development and activity of the opposite subset, an effect known as *cross-regulation* (see Figure 12-12). For instance, IFN-$\gamma$ (secreted by the $T_H1$ subset) preferentially inhibits proliferation of the $T_H2$ subset, and IL-4 and IL-10 (secreted by the $T_H2$ subset) down-regulate secretion by both macrophages and dendritic cells of IL-12, one of the critical cytokines for $T_H1$ differentiation. Similarly, these cytokines have opposing effects on target cells other than $T_H$ subsets. In mice, where the $T_H1$ and $T_H2$ subsets have been studied most extensively, IFN-$\gamma$ secreted by the $T_H1$ subset promotes IgG2a production by B cells but inhibits IgG1 and IgE production. On the other hand, IL-4 secreted by the $T_H2$ subset promotes production of IgG1 and IgE and suppresses production of IgG2a. Thus, different antibody profiles result from the $T_H1$ or $T_H2$ response. The phenomenon of cross-regulation explains the observation that there is often an inverse relationship between antibody production and cell-mediated immunity; that is, when antibody production is high, cell-mediated immunity is low and vice versa. Furthermore, recent research has shown that IL-4 and IFN-$\gamma$ make members of the T-cell subset that release them less responsive to the cytokine that directs differentiation of the other T-cell subset. Thus, IL-4 enhances $T_H2$ cell development by making $T_H$ cells less susceptible to the cytokine signals that cause these cells to enter a differentiation pathway that would lead to $T_H1$ development. But as explained below, IFN-$\gamma$ up-regulates the expression of a key regulatory molecule that favors the differentiation and activity of $T_H1$ cells.

A third CD4$^+$ T-cell subset exercises regulatory control of T-cell responses and can limit autoimmune T-cell activity. This subset of T cells, designated $T_{reg}$ cells, is characterized by expression of Il-4, IL-10, and TGF-$\beta$ and exercises its suppressive effect through contact with target T cells. Details of the development pathway leading to $T_{reg}$ cells are a subject of current study.

Recent work has given insight into the molecular basis for the cytokine-mediated cross-regulation by which one subset promotes its own expansion and development while inhibiting the development of the opposite subset. Two transcription factors, T-Bet and GATA-3, are key elements in determining subset commitment and cross-regulation. The

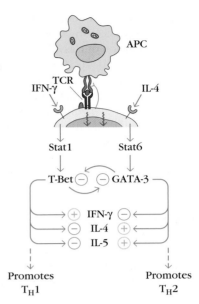

**FIGURE 12-13  Cross-regulation at the intracellular level.** Signals through the TCR and cytokine receptors determine whether the cell will produce the $T_H1$-promoting transcription factor, T-Bet, or the $T_H2$-promoting transcription factor, GATA-3. Experimental evidence supports a model in which exposure of cells bearing receptors for IFN-$\gamma$ to IFN-$\gamma$ induces the formation of T-Bet, which up-regulates the synthesis of IFN-$\gamma$ and represses the expression of GATA-3. Exposure of IL-4 R–bearing cells to IL-4 induces the formation of GATA-3, which up-regulates the synthesis of IL-4 and IL-5 but represses the expression of T-Bet. [Adapted from J. Rengarajan, S. Szabo, and L. Glimcher, 2000, Immunology Today **21**:479.]

expression of T-Bet drives cells to differentiate into $T_H1$ cells and suppresses their differentiation along the $T_H2$ pathway. Expression of GATA-3 does the opposite, promoting the development of naive T cells into $T_H2$ cells while suppressing their differentiation into $T_H1$ cells. As shown in Figure 12-13, the expression of T-Bet versus GATA-3 is determined by the cytokines IFN-$\gamma$ and IL-4. In the presence of IFN-$\gamma$, T cells up-regulate the expression of T-Bet and down-regulate GATA-3. This IFN-$\gamma$ receptor/Stat1–dependent process shifts the cytokine profile to the production of IFN-$\gamma$, the signature cytokine of $T_H1$ cells, and other cytokines typical of the $T_H1$ set. But in a process that involves the IL-4 receptor and Stat6, IL-4 induces the cell to produce IL-4 and other $T_H2$ cytokines. Further study has revealed that the up-regulation of T-Bet represses the expression of GATA-3. Similarly, expression of GATA-3 down-regulates T-Bet. Consequently, cytokine signals that induce one of these transcription factors set in motion a chain of events that represses the other. At the intracellular level, the differentiation of a T cell along the $T_H1$ pathway prevents its development of $T_H2$ characteristics and vice versa.

The cross-regulation of $T_H1$ cells by IL-10 secreted from $T_H2$ cells is not a direct inhibition of the $T_H1$ cells; instead, IL-10 acts on monocytes and macrophages, interfering with their ability to activate the $T_H1$ subset. This interference is thought to result from the demonstrated ability of IL-10 to down-regulate the expression of class II MHC molecules on these antigen-presenting cells. IL-10 has other potent immunosuppressant effects on the monocyte-macrophage lineage, such as suppressing the production of nitric oxide and other bactericidal metabolites involved in the destruction of pathogens and suppressing the production of various inflammatory mediators; for example, IL-1, IL-6, IL-8, GM-CSF, G-CSF, and TNF-γ. These suppressive effects on the macrophage serve to further diminish the biologic consequences of $T_H1$ activation.

## The $T_H1/T_H2$ balance determines disease outcomes

The progression of some diseases may depend on the balance between the $T_H1$ and $T_H2$ subsets. In humans, a well-studied example of this phenomenon is leprosy, which is caused by *Mycobacterium leprae*, an intracellular pathogen that can survive within the phagosomes of macrophages. Leprosy is not a single clinical entity; rather, the disease presents as a spectrum of clinical responses, with two major forms of disease, tuberculoid and lepromatous, at each end of the spectrum. In **tuberculoid leprosy,** a cell-mediated immune response forms granulomas, resulting in the destruction of most of the mycobacteria, so that only a few organisms remain in the tissues. Although skin and peripheral nerves are damaged, tuberculoid leprosy progresses slowly, and patients usually survive. In **lepromatous leprosy,** the cell-mediated response is depressed and humoral antibodies are formed instead, sometimes resulting in high levels of immunoglobulin (hyper-gamma-globulinemia). The mycobacteria are widely disseminated in macrophages, often reaching numbers as high as $10^{10}$ per gram of tissue. Lepromatous leprosy progresses into disseminated infection of the bone and cartilage with extensive nerve and tissue damage.

The development of lepromatous or tuberculoid leprosy depends on the balance of $T_H1$ and $T_H2$ cells (Figure 12-14). In tuberculoid leprosy, the immune response is characterized by a $T_H1$-type response with delayed-type hypersensitivity and a cytokine profile consisting of high levels of IL-2, IFN-γ, and TNF-β. In lepromatous leprosy, there is a $T_H2$-type immune response, with high levels of IL-4, IL-5, and IL-10. This cytokine profile explains the diminished cell-mediated immunity and increased production of serum antibody in lepromatous leprosy.

Evidence also exists for changes in $T_H$-subset activity in AIDS. Early in the disease, $T_H1$ activity is high, but as AIDS progresses, some workers have suggested, a shift may occur from a $T_H1$-like to a $T_H2$-like response. In addition, some pathogens may influence the activity of the $T_H$ subsets. The Epstein-Barr virus, for example, produces a homolog of human IL-10 called vIL-10, which has IL-10-like activity.

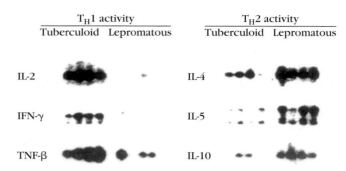

**FIGURE 12-14 Correlation between type of leprosy and relative $T_H1$ or $T_H2$ activity.** Messenger RNA isolated from lesions from tuberculoid and lepromatous leprosy patients was analyzed by Southern blotting using the cytokine probes indicated. Cytokines produced by $T_H1$ cells predominate in the tuberculoid patients, whereas cytokines produced by $T_H2$ cells predominate in the lepromatous patients. *[From P. A. Sieling and R. L. Modlin, 1994, Immunobiology* **191:378.]**

This viral mimic, like cellular IL-10, tends to suppress $T_H1$ activity by cross-regulation. Some researchers have speculated that vIL-10 may reduce the cell-mediated response to the Epstein-Barr virus, thus conferring a survival advantage on the virus.

## Cytokine-Related Diseases

Defects in the complex regulatory networks governing the expression of cytokines and cytokine receptors have been implicated in a number of diseases. Genetic defects in cytokines, their receptors, or the molecules involved in signal transduction following receptor-cytokine interaction lead to immunodeficiencies such as severe combined immunodeficiency (SCID), described in Chapter 20. Other defects in the cytokine network can cause inability to defend against specific families of pathogens. For example, people with a defective receptor for INF-γ are susceptible to mycobacterial infections that rarely occur in the normal population. In addition to the diseases rooted in genetic defects in cytokine activity, a number of disease conditions result from overexpression or underexpression of cytokines or cytokine receptors. Several examples of these diseases are given below, followed by an account of therapies aimed at preventing the potential harm caused by cytokine activity.

### Septic shock is common and potentially lethal

Despite the widespread use of antibiotics, bacterial infections remain a major cause of septic shock, which may develop a few hours after infection by certain gram-negative bacteria, including *E. coli, Klebsiella pneumoniae, Pseudomonas aeruginosa, Enterobacter aerogenes,* and *Neisseria meningitidis.* The symptoms of bacterial septic shock, which is often fatal, include a drop in blood pressure, fever, diarrhea, and widespread blood clotting in various organs.

Bacterial septic shock apparently develops because bacterial cell wall **endotoxins** bind TLRs on dendritic cells and macrophages, causing them to overproduce IL-1 and TNF-α to levels that cause septic shock. In one study, for example, higher levels of TNF-α were found in patients who died of meningitis than in those who recovered. Furthermore, a condition resembling bacterial septic shock can be produced by injecting mice with recombinant TNF-α in the absence of gram-negative bacterial infection. Several studies offer hope that neutralization of TNF-α or IL-1 activity with monoclonal antibodies or antagonists will prevent fatal shock from developing. In one study, monoclonal antibody to TNF-α protected animals from endotoxin-induced shock. In another study, injection of a recombinant IL-1 receptor antagonist (IL-1Ra), which prevents binding of IL-1 to the IL-1 receptor, resulted in a threefold reduction in mortality. However, neutralization of TNF-α does not reverse the progression of septic shock in all cases, and antibodies against TNF-α give little benefit to patients with advanced disease. Recent studies in which the cytokine profiles of patients with septic shock were followed over time shed some light on this apparent paradox.

Bacterial septic shock is one of the conditions that fall under the general heading of sepsis, which may be caused not only by bacterial infection but also by trauma, injury, ischemia (decrease in blood supply to an organ or a tissue), and certain cancers. Sepsis is the most common cause of death in U.S. hospital intensive-care units and accounts for 9.3% of total deaths in the United States. It is the third leading cause of death in developed countries. A common feature of sepsis, whatever the cause, is an overwhelming production of proinflammatory cytokines such as TNF-α and IL-1β. The cytokine imbalance often causes very abnormal body temperature and respiratory rate and high white blood cell counts, followed by capillary leakage, tissue injury, and lethal organ failure.

The increases in TNF-α and IL-1 occur rapidly in early sepsis, so neutralizing these cytokines is most beneficial early in the process. In animal experiments, early intervention can prevent sepsis altogether. However, approximately 24 hours following onset of sepsis, the levels of TNF-α and IL-1 fall dramatically, and other factors become more important. Cytokines critical in the later stages may include IL-6, MIF, and IL-8. Sepsis remains an area of intense investigation, and clarification of the process involved in bacterial septic shock and other forms of sepsis can be expected to lead to therapies for this major killer.

## Bacterial toxic shock is caused by superantigens

A variety of microorganisms produce toxins that act as **superantigens.** As described in Chapter 10, superantigens bind simultaneously to a class II MHC molecule and to the $V_\beta$ domain of the T-cell receptor, activating all T cells bearing a particular $V_\beta$ domain (see Figure 10-16). Because of their unique binding ability, superantigens can activate large numbers of T cells irrespective of their antigenic specificity.

Although less than 0.01% of T cells respond to a given conventional antigen, 5% or more of the T-cell population can respond to a given superantigen. The large proportion of T cells responsive to a particular superantigen results from the limited number of TCR $V_\beta$ genes carried in the germ line. Mice, for example, have about 20 $V_\beta$ genes. If one assumes that each $V_\beta$ gene is expressed with equal frequency, then each superantigen would be expected to interact with one in 20 T cells, or 5% of the total T-cell population.

Bacterial superantigens have been implicated as the causative agent of several diseases, such as bacterial toxic shock and food poisoning. Included among these bacterial superantigens are several enterotoxins, exfoliating toxins, and toxic shock syndrome toxin (TSST1) from *Staphylococcus aureus;* pyrogenic exotoxins from *Streptococcus pyrogenes;* and *Mycoplasma arthritidis* supernatant (MAS). The large number of T cells activated by these superantigens results in excessive production of cytokines. The toxic shock syndrome toxin, for example, has been shown to induce extremely high levels of TNF-α and IL-1. As in bacterial septic shock, these elevated concentrations of cytokines can induce systemic reactions that include fever, widespread blood clotting, and shock.

## Cytokine activity is implicated in lymphoid and myeloid cancers

Abnormalities in the production of cytokines or their receptors have been associated with some types of cancer. For example, abnormally high levels of IL-6 are secreted by cardiac myxoma cells (a benign heart tumor), myeloma and plasmacytoma cells, and cervical and bladder cancer cells. In myeloma and plasmacytoma cells, IL-6 appears to operate in an autocrine manner to stimulate cell proliferation. When monoclonal antibodies to IL-6 are added to in vitro cultures of myeloma cells, their growth is inhibited. In contrast, transgenic mice that express high levels of IL-6 have been found to exhibit a massive, fatal plasma-cell proliferation, called plasmacytosis. Although these plasma cells are not malignant, the high rate of plasma-cell proliferation possibly contributes to the development of cancer.

## Chagas's disease is caused by a parasite

The protozoan *Trypanosoma cruzi* is the causative agent of Chagas's disease, which is characterized by severe immune suppression. The ability of *T. cruzi* to mediate immune suppression can be observed by culturing peripheral-blood T cells in the presence and absence of *T. cruzi* and then evaluating their immune reactivity. Antigen, mitogen, or anti-CD3 monoclonal antibody normally can activate peripheral T cells, but in the presence of *T. cruzi*, T cells are not activated by any of these agents. The defect in these lymphocytes has been traced to a dramatic reduction in the expression of the 55-kDa α subunit of the IL-2 receptor (CD25). As noted earlier, the high-affinity IL-2 receptor contains α, β, and γ subunits. The α subunit is specific for cytokine binding (see Figure 12-9). Coculturing of T cells with *T. cruzi* and subsequent staining with fluorescein-labeled anti-CD25, which binds to the α subunit, revealed a 90% decrease in the level of the α subunit.

Although the mechanism by which *T. cruzi* suppresses expression of the α subunit is still unknown, the suppression can be induced across a filter that prevents contact between the lymphocytes and protozoa. This finding suggests that a diffusible factor mediates suppression. Such a factor, once isolated, might have numerous clinical applications for regulating the level of activated T cells in leukemias and autoimmune diseases.

## Cytokine-Based Therapies

The availability of purified cloned cytokines, monoclonal antibodies directed against cytokines, and soluble cytokine receptors offers the prospect of specific clinical therapies to modulate the immune response. A few cytokine-based therapies—notably, interferons (see Clinical Focus on page 322); colony-stimulating factors, such as GM-CSF; and blockers of TNF activity—have proven to be therapeutically useful in certain diseases. Despite its great potential, however, cytokine-based therapy has not been successful in all applications. As discussed above, the use of blockers for TNF-α is not a universal remedy for septic shock because septic shock is not mediated solely by TNF-α. By contrast, a soluble TNF-α receptor (Enbrel) and monoclonal antibodies against TNF-α (Remicade and Humira) have been used to treat rheumatoid arthritis in more than a million patients. These anti-TNF-α drugs reduce proinflammatory cytokine cascades and the resulting inflammatory response. For those suffering from rheumatoid arthritis, the drugs stop pain, stiffness, and joint swelling and promote healing and tissue repair. But this therapy carries an increased risk of infection, especially tuberculosis and pneumonia, because of reduced cytokine activity. The risk of lymphoma also increases for long-term users of TNF-α blockers.

Problems have arisen in adapting cytokines for safe, routine medical use. One problem is maintaining effective dose levels over a clinically significant period. During an immune response, interacting cells produce sufficiently high concentrations of cytokines in the vicinity of target cells, but achieving such local concentrations when cytokines must be administered systemically for clinical treatment is difficult. In addition, cytokines often have a very short half-life, so continuous administration may be required. For example, recombinant human IL-2 has a half-life of only 7 to 10 minutes when administered intravenously. Finally, cytokines are extremely potent biological response modifiers, and they can cause unpredictable and undesirable side effects. The side effects from administration of recombinant IL-2, for example, range from mild, such as fever,

(a) Suppression of $T_H$-cell proliferation and $T_C$-cell activation

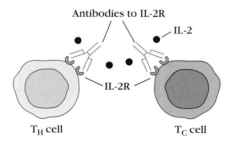

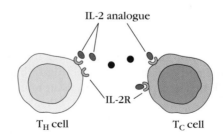

(b) Destruction of activated $T_H$ cells

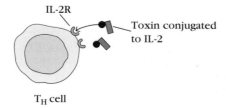

**FIGURE 12-15 Experimental cytokine-related therapeutic agents offer the prospect of selectively modulating the immune response.** (a) The anti-IL-2R monoclonal antibody binds to the cytokine receptor (IL-2R) on the cell surface, thereby preventing interaction of the cytokine with its receptor. (b) Conjugation of a toxin with a cytokine results in destruction of cells expressing the cytokine receptor.

chills, diarrhea, and weight gain, to serious, such as anemia, thrombocytopenia, shock, respiratory distress, and coma. Despite these difficulties, a multinational trial of recombinant IL-2 (Proleukin) has shown preliminary success in restoring CD4$^+$ T-cell levels in patients with AIDS (see Chapter 20). The cytokine is administered subcutaneously twice daily for 5 days every 8 weeks to patients who are simultaneously being given anti-retroviral drugs. Increased levels of CD4$^+$ T cells were sustained for longer periods in those given IL-2 compared to those given anti-retroviral therapy alone. The IL-2 group also had increases in a CD8$^+$ cell subset that had antiviral activity but did not show cytotoxicity against HIV-infected cells. Side effects of the IL-2 administered in this regimen were minimal, mostly flu-like symptoms and soreness at the injection site.

Other approaches to the medical use of cytokines include cytokine receptor blockade and the use of cytokine analogues and cytokine toxin conjugates. For example, proliferation of activated T$_H$ cells and activation of T$_C$ cells can be blocked by anti-CD25, the monoclonal antibody that binds to the α subunit of the high-affinity IL-2 receptor (Figure 12-15a). Administration of anti-CD25 has prolonged the survival rate for heart transplants in rats. Anti-CD25 has also been used for adult T-cell leukemia, which may result from infection with the human retrovirus HTLV-1 and is characterized by proliferation of a T-cell population that expresses IL-2Ra in abnormally high amounts. Similar results have been obtained with IL-2 analogues that retain their ability to bind the IL-2 receptor but lack the biological activity of IL-2 (see Figure 12-15a). Such analogues have been produced by site-directed mutagenesis of cloned IL-2 genes. Finally, cytokines conjugated to various toxins, such as the β chain of diphtheria toxin, have been shown to diminish rejection of kidney and heart transplants in animals. Conjugates containing IL-2 selectively bind to and kill activated T$_H$ cells (Figure 12-15b).

The use of cytokines in clinical medicine holds great promise, and efforts to develop safe and effective cytokine-related strategies continue, particularly in areas such as inflammation, cancer therapy, and modification of the immune response during organ transplantation, infectious disease, and allergy.

## Cytokines in Hematopoiesis

Early work in Australia and Israel demonstrated that soluble factors can support the growth and differentiation of red and white blood cells. The first of these soluble factors to be characterized, erythropoietin, was isolated from the urine of anemic patients and shown to support the development of red blood cells. Subsequently, many cytokines have been shown to play essential roles in hematopoiesis (Table 12-5). During hematopoiesis, cytokines act as developmental

| TABLE 12-5 | Hematopoietic cytokines | |
|---|---|---|
| **Hematopoietic growth factor** | **Sites of production** | **Main functions** |
| Erythropoietin | Kidney, liver | Erythrocyte production |
| G-CSF | Endothelial cells, fibroblasts, macrophages | Neutrophil production |
| Thrombopoietin | Liver, kidney | Platelet production |
| M-CSF | Fibroblasts, endothelial cells, macrophages | Macrophage and osteoclast production |
| SCF/c-*kit* ligand | Bone marrow stromal cells, constitutively | Stem cell, progenitor cells survival/division; mast cell differentiation |
| Flt-3 ligand | Fibroblasts, endothelial cells | Early progenitor cell expansion; pre–B cells |
| GM-CSF | T cells (T$_H$1 and T$_H$2), macrophages, mast cells | Macrophage, granulocyte production; dendritic cell maturation and activation |
| IL-3 | T cells (T$_H$1 and T$_H$2), macrophages | Stem cells and myeloid progenitor cell growth; mast cells |
| IL-5 | Activated helper, T cells—T$_H$2 response only | Eosinophil production; murine B-cell growth |
| IL-6 | Activated T cells, monocytes, fibroblasts, endothelial cells | Progenitor-cell stimulation; platelet production; immunoglobulin production in B cells |
| IL-7 | Bone marrow and lymphoid stromal cells | T-cell survival |
| IL-11 | Bone marrow stromal cells and IL-1-stimulated fibroblasts | Growth factor for megakaryocytes |

G-CSF, granulocyte colony–stimulating factor; GM-CSF, granulocyte-macrophage colony-stimulating factor; IL, interleukin; M-CSF, macrophage colony–stimulating factor; SCF, stem cell factor. Adapted from D. Thomas and A. Lopez, 2001, *Encyclopedia of Life Sciences:* Haematopoietic growth factors, Nature Publishing Group.

## CLINICAL FOCUS
# Therapy with Interferons

**Interferons** are an extraordinary group of proteins whose antiviral activity led to their discovery almost 50 years ago. Subsequent studies showed that interferons have other effects, including the capacity to induce cell differentiation, to inhibit proliferation by some cell types, to inhibit angiogenesis, and to function in various immunoregulatory roles. Their effects on the immune system are important and dramatic. Interferons induce increases in the expression of class I and class II MHC molecules and augment NK-cell activity. Increased class I expression increases the display of antigen to $CD8^+$ cells, a class that includes most of the $T_C$ population. This enhanced display of antigen not only makes the antigen-presenting cells more effective in inducing cytotoxic T-cell populations, it also makes them better targets for attack by $T_C$ cells. In addition to up-regulating class I MHC expression of many cell types, IFN-$\gamma$ increases the expression of class II MHC molecules on such antigen-presenting cells as macrophages and dendritic cells. This makes them better presenters of antigen to $T_H$ cells. IFN-$\gamma$ is also a potent inducer of macrophage activation and general promoter of inflammatory responses. Cloning of the genes that encode all three types of interferon, IFN-$\alpha$, IFN-$\beta$, and IFN-$\gamma$, has made it possible for the biotechnology industry to produce large amounts of all of these interferons at costs that make their clinical use practical. Some clinical uses of each type of interferon are as follows:

- IFN-$\alpha$ (also known by its trade names Roferon and Intron A) has been used for the treatment of hepatitis C and hepatitis B. It has also found a number of different applications in cancer therapy. A type of B-cell leukemia known as hairy-cell leukemia (because the cells are covered with fine, hairlike cytoplasmic projections) responds well to IFN-$\alpha$. Chronic myelogenous leukemia, a disease characterized by increased numbers of granulocytes, begins with a slowly developing chronic phase that changes to an accelerated phase and terminates in a blast phase, which is usually resistant to treatment. IFN-$\alpha$ is an effective treatment for this leukemia in the chronic phase (70% response rates have been reported), and some patients (as many as 20% in some studies) undergo complete remission. Kaposi's sarcoma, the cancer most often seen in American AIDS patients, also responds to treatment with IFN-$\alpha$, and there are reports of a trend toward longer survival and fewer opportunistic infections in patients treated with this agent. Most of the effects mentioned above have been obtained in clinical studies that used IFN-$\alpha$ alone, but certain applications such as hepatitis C therapy commonly use it with an antiviral drug such as ribavirin. The clearance time of INF-$\alpha$ is lengthened by treatment with polyethylene glycol (PEG), and this form, called pegylated interferon, or Pegasys, is now frequently used.

- IFN-$\beta$ has emerged as the first drug capable of producing clinical improvement in multiple sclerosis (MS). Young adults are the primary target of this autoimmune neurologic disease, in which nerves in the central nervous system (CNS) undergo demyelination. This results in progressive neurologic dysfunction, which leads to significant and, in many cases, severe disability. This disease is often characterized by periods of nonprogression and remission alternating with periods of relapse. Treatment with IFN-$\beta$ provides longer periods of remission and reduces the severity of relapses. Furthermore, magnetic-resonance-imaging studies of CNS damage in treated and untreated patients revealed that MS-induced damage was less severe in a group of IFN-$\beta$-treated patients than in untreated ones.

- IFN-$\gamma$ has been used, with varying degrees of success, to treat a variety of malignancies that include non-Hodgkin's lymphoma, cutaneous T-cell lymphoma, and multiple myeloma. A more successful clinical application of IFN-$\gamma$ in the clinic is in the treatment of the hereditary immunodeficiency chronic granulomatous disease (CGD; see Chapter 20). CGD features a serious impairment of the ability of phagocytic cells to kill ingested microbes, and patients with CGD suffer recurring infections by a number of bacteria (*Staphylococcus aureus*, *Klebsiella*, *Pseudomonas,* and others) and fungi such as *Aspergillus* and *Candida*. Before interferon therapy, standard treatment for the disease included attempts to avoid infection, aggressive administration of antibiotics, and surgical drainage of abscesses. A failure to generate microbicidal oxidants ($H_2O_2$, superoxide, and others) is the basis of CGD, and the administration of IFN-$\gamma$ significantly reverses this defect. Therapy of CGD patients with IFN-$\gamma$ significantly reduces the incidence of infections. Also, the

---

signals that direct commitment of progenitor cells into and through particular lineages. As shown in Figure 12-16, a myeloid progenitor in the presence of erythropoietin would proceed down a pathway that leads to the production of erythrocytes; suitable concentrations of a group of cytokines including IL-3, GM-CSF, IL-1, and IL-6 will cause it to enter differentiation pathways that lead to the generation of monocytes, neutrophils, and other leukocytes of the myeloid group. The participation of leukocytes in immune responses often results in their death and removal. How-

infections that are contracted are less severe and the average number of days spent by patients in the hospital goes down.

■ IFN-γ has also been shown to be effective in the treatment of osteopetrosis (*not* osteoporosis), a life-threatening congenital disorder characterized by overgrowth of bone that results in blindness and deafness. Another problem presented by this disease is that the buildup of bone

reduces the amount of space available for bone marrow, and the decrease in hematopoiesis results in fewer red blood cells and anemia. The decreased generation of white blood cells causes an increased susceptibility to infection.

The use of interferons in clinical practice is likely to expand as more is learned about their effects in combination with other therapeutic agents. Although interferons, in common with other cytokines, are powerful

modifiers of biological responses, the side effects accompanying their use are fortunately much milder. Typical side effects include flu-like symptoms, such as headache, fever, chills, and fatigue. These symptoms can largely be managed with acetaminophen (Tylenol) and diminish in intensity during continued treatment. Although interferon toxicity is usually not severe, serious manifestations such as anemia and depressed platelet and white-blood-cell counts have been seen.

**Cytokine-based therapies in clinical use**

| Agent | Nature of agent | Clinical application |
|---|---|---|
| Enbrel | Chimeric TNF-receptor/IgG constant region | Rheumatoid arthritis |
| Remicade or Humira | Monoclonal antibody against TNF-α receptor | Rheumatoid arthritis<br>Crohn's disease |
| Roferon | Interferon α-2a* | Hepatitis B<br>Hairy-cell leukemia<br>Kaposi's sarcoma |
| Intron A | Interferon α-2b | Hepatitis C†<br>Melanoma |
| Betaseron | Interferon β–1b | Multiple sclerosis |
| Avonex | Interferon β–1a | Multiple sclerosis |
| Actimmune | Interferon γ–1β | Chronic granulomatous disease (CGD)<br>Osteopetrosis |
| Neupogen | G-CSF (hematopoietic cytokine) | Stimulates production of neutrophils<br>Reduction of infection in cancer patients treated with chemotherapy, AIDS patients |
| Leukine | GM-CSF (hematopoietic cytokine) | Stimulates production of myeloid cells after bone marrow transplantation |
| Neumega or Neulasta | Interleukin-11 (IL-11), a hematopoietic cytokine | Stimulates production of platelets |
| Epogen | Erythopoietin (hematopoietic cytokine) | Stimulates red-blood-cell production |

*Interferon α-2a is also licensed for veterinary use to combat feline leukemia.

†Normally used in combination with an antiviral drug (ribavirin) for hepatitis C treatment.

ever, both adaptive and innate immune responses generate cytokines that stimulate and support the production of leukocytes. The steps at which a number of cytokines participate in hematopoiesis is shown in Figure 12-16. Hematopoetic cytokines that stimulate production of neu-

trophils (G-CSF), myeloid cells (GM-CSF), platelets (IL-11), and red blood cells (erythropoietin) have all been used in clinical applications, most often as supportive therapy for patients with immunodeficiency resulting from a genetic defect or from cancer chemotherapy (see Clinical Focus).

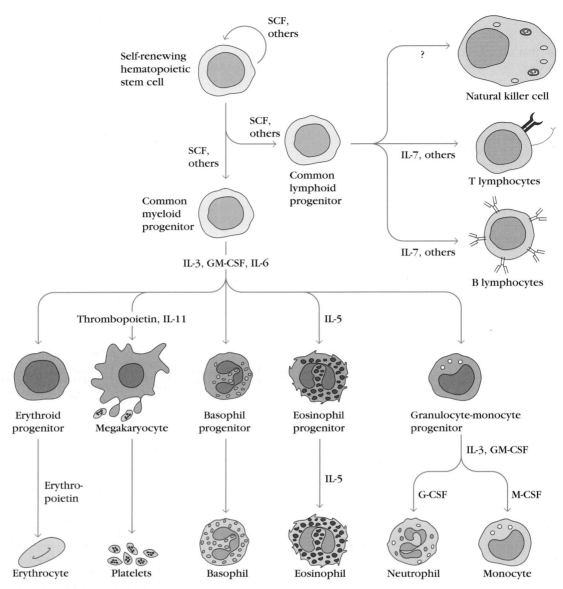

**FIGURE 12-16 Hematopoietic cytokines and hematopoiesis.** A variety of cytokines are involved in supporting the growth and directing the differentiation of hematopoietic cells. Note that additional factors may be required for some of the developmental pathways shown in the diagram.

## SUMMARY

- Cytokines are low-molecular-weight proteins that are produced and secreted by a variety of cell types. They play major roles in the induction and regulation of the cellular interactions involving cells of the immune, inflammatory, and hematopoietic systems.

- The biological activities of cytokines exhibit pleiotropy, redundancy, synergy, antagonism, and, in some instances, cascade induction.

- Of the more than 200 different cytokines, most fall into one of the following families: hematopoietins, interferons, chemokines, and tumor necrosis factors.

- Cytokines act by binding to cytokine receptors, most of which can be classified as immunoglobulin superfamily receptors, class I and class II cytokine receptors, members of the TNF receptor family, and chemokine receptors.

- A cytokine can act only on a cell that expresses a receptor for it. The activity of particular cytokines is directed to specific cells by regulation of the cell's profile of cytokine receptors.

- Cytokine-induced multimerization of class I and class II cytokine receptors activates a JAK/STAT signal transduction pathway.

- Antigen stimulation of $T_H$ cells in the presence of certain cytokines can lead to the generation of subpopulations of helper T cells known as $T_H1$ and $T_H2$. Each subset displays characterisic and different profiles of cytokine secretion.

- The cytokine profile of $T_H1$ cells supports immune responses that involve the marshaling of phagocytes, CTLs, and NK cells to eliminate intracellular pathogens. $T_H2$ cells produce cytokines that support production of particular immunoglobulin isotypes and IgE-mediated responses.

- Therapies based on cytokines and cytokine receptors have entered clinical practice.

## References

Abbas, A., K. M. Murphy, and A. Sher. 1996. Functional diversity of helper T lymphocytes. *Nature* **383**:787.

Bach, E. A., M. Aguet, and R. D. Schreiber. 1998. The IFN-γ receptor: a paradigm for cytokine receptor signaling. *Annual Review of Immunology* **15**:563.

Darnell, J. E., Jr. 1997. STATs and gene regulation. *Science* **5332**: 1630–35.

Feldman, M., and L. Steinman. 2005. Design of effective immunotherapy for human autoimmunity. *Nature* **435**:612.

Fitzgerald, K. A., et al. 2001. *The Cytokine Facts Book,* 2nd ed. Academic Press, New York.

Flynn, J. L., and J. Chan. 2001. Immunology of tuberculosis. *Annual Review of Immunology* **19**:93–129.

Gadina, M., et al. 2001. Signaling by type I and II cytokine receptors: ten years after. *Current Opinion in Immunology* **3**:363–73.

Jaeckel, E., et al. 2001. Treatment of acute hepatitis C with interferon α-2b. *New England Journal of Medicine* **345**:1452–57.

Johnson, D. C., and G. McFadden. 2002. Viral immune evasion. In *Immunology of Infectious Diseases.* S. H. E. Kaufman, A. Sher, and R. Ahmed, eds. ASM Press, Washington, p. 357.

Kotenko, S. V., and J. A. Langer. 2004. Full house: 12 receptors for 27 cytokines. *International Immunopharmacology* **4**:593.

Kovanem, P. E., and W. J. Leonard. 2004. Cytokines and immunodeficiency diseases: critical roles of the γc-dependent cytokines interleukins 2, 4, 7, 9, 15, and 21, and their signaling pathways. *Immunological Reviews* **202**:67.

Mossman, T. R., et al. 1986. Two types of murine helper T cell clone. I. Definition according to profiles of lymphokine activities and secreted proteins. *Journal of Immunology* **136**:2348.

Pestka, S., C. D. Krause, and M. R. Walter. 2004. Interferons, interferon-like cytokines and their receptors. *Immunological Reviews* **202**:8–32.

Szabo, S. J., et al. 2000. A novel transcription factor, T-bet, directs $T_H1$ lineage commitment. *Cell* **100**:655–69.

Szabo, Susanne J., Brandon M. Sullivan, Stanford L. Peng, and Laurie H. Glimcher. 2003. Molecular mechanisms regulating $T_H1$ immune responses. *Annual Review of Immunology* **21**:713.

Ulloa, L., and K. J. Tracey. 2005. The "cytokine profile": a code for sepsis. *Trends in Molecular Medicine* **11**:56.

Vandenbroeck, K., et al. 2004. Inhibiting cytokines of the interleukin-12 family: recent advances and novel challenges. *Journal of Pharmacy and Pharmacology* **56**:145.

Walter, M. R., et al. 1995. Crystal structure of a complex between interferon-γ and its soluble high-affinity receptor. *Nature* **376**:230.

Wang, X., Rickert, M., and K. C. Garcia. 2005. Structure of the quaternary complex of interleukin-2 with its α, β, and γc receptors. *Science* **310**:1159.

 ## Useful Web Sites

http://www.rndsystems.com/

The cytokine minireviews found at R&D Systems' Web site provide extensive, detailed, well-referenced, and often strikingly illustrated reviews of many cytokines and their receptors.

http://www.ncbi.nlm.nih.gov:80/LocusLink/index.html

Entrez Gene provides access to sequence and descriptive information about genetic loci of cytokines and other proteins. It also references papers discussing the basic biology (function and structure) of the gene or protein of interest.

 ## Study Questions

**CLINICAL FOCUS QUESTION** Cytokines are proving to be powerful drugs, but their use is accompanied by side effects that can be harmful to patients. What are some of the side effects produced by Actimmune, Roferon, and interferon beta? (Hint: Manufacturers' Web sites often provide detailed information on the side effects of drugs the manufacturers produce.)

1. Indicate whether each of the following statements is true or false. If you think a statement is false, explain why.

   a. The high-affinity IL-2 receptor consists of two transmembrane proteins.

   b. The anti-TAC monoclonal antibody recognizes the IL-1 receptor on T cells.

   c. All cytokine-binding receptors contain two or three subunits.

   d. Expression of the β subunit of the IL-2 receptor is indicative of T-cell activation.

   e. Some cytokine receptors possess domains with tyrosine kinase activity that function in signal transduction.

   f. All members of each subfamily of the class I cytokine (hematopoietin) receptors share a common signal-transducing subunit.

2. When IL-2 is secreted by one T cell in a peripheral lymphoid organ, do all the T cells in the vicinity proliferate in response to the IL-2 or only some of them? Explain.

3. Briefly describe the similarities and differences among cytokines, growth factors, and hormones.

4. Indicate which subunit(s) of the IL-2 receptor are expressed by the following types of cells:

   a. _____ Resting T cells

   b. _____ Activated T cells

   c. _____ Activated T cells + cyclosporin A

   d. _____ Resting $T_C$ cells

   e. _____ CTLs

   f. _____ NK cells

5. Superantigens have been implicated in several diseases and have been useful as research tools.

   a. What properties of superantigens distinguish them from conventional antigens?

   b. By what mechanism are bacterial superantigens thought to cause symptoms associated with food poisoning and toxic shock syndrome?

   c. Does the activity of superantigens exhibit MHC restriction?

6. IL-3, IL-5, and GM-CSF exhibit considerable redundancy in their effects. What structural feature of the receptors for these cytokines might explain this redundancy?

7. Considerable evidence indicates the existence of two $T_H$-cell subsets, differing in the pattern of cytokines they secrete.

   a. What type of immune response is mediated by the $T_H1$ subset? What type of antigen challenge is likely to induce a $T_H1$-mediated response?

   b. What type of immune response is mediated by the $T_H2$ subset? What type of antigen challenge is likely to induce a $T_H2$-mediated response?

8. Indicate whether each of the following statements is true or false. If you think a statement is false, explain why.

   a. IL-6 stimulates the liver to produce acute phase proteins which shut down the immune response after viral infections.

   b. Cytokines are expressed at relatively low concentrations and are very powerful, thus they can only act on cells nearby to the cell that secretes them.

   c. Whether or not a cell will respond to a cytokine depends on which subunits of the receptor are expressed and on the level of expression of the receptor on the cell surface.

   d. $T_H1$ cells release cytokines that result in an increased production of eosinophils and mast cells in the bone marrow.

   e. IL-4 inhibits differentiation of T0 cells to $T_H1$ cells.

   f. Activation of T cells results in secretion of IL-1 and its receptor on the T-cell surface.

9. The following figure shows the result of flow cytometry of human blood cells. The cells were stained with FITC-conjugated rabbit anti-human IL-2 receptor α subunit (y axis) and PE-conjugated mouse anti-human IL-2 receptor γ subunit (x-axis). Which quadrant shows cells expressing the medium affinity receptor?

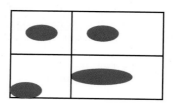

10. For each of the following cytokines, indicate which of the effects would result from the production of the cytokine.

    | Cytokines | Effects |
    | --- | --- |
    | a. IL-1 | 1. Activation of macrophages |
    | b. IL-4 | 2. Increased adhesiveness of the vasculature |
    | c. IL-7 | |
    | d. IL-5 | 3. Promotes $T_H1$ development |
    | e. Erythropoietin | 4. Promotes $T_H2$ development |
    | f. Interferon γ | 5. Production of eosinophils |
    | g. IL-25 | 6. Production of RBC |
    | | 7. Induction of acute phase proteins |
    | | 8. Isotype switching to IgE |
    | | 9. Development of lymphocytes |

11. Fill in the blank(s) in each statement.

    a. The cytokine _____ induces B cells to undergo class switching to IgE.

    b. _____ are chemotactic cytokines.

    c. The signal transduction cascade started by chemokines binding to their receptors involves _____.

    d. _____ is important for development of lymphocytes during hematopoiesis.

    e. If a cell expresses all three subunits of the IL-2 receptor (IL-2R), it will respond to _____ concentrations of GM-CSF.

    f. _____ is a cytokine you would expect to find increased in a person with a viral infection (not necessarily an enveloped virus).

12. Which of the following statements are not true regarding $T_H1$ cells?

    a. They are involved in delayed-type hypersensitivity.

    b. They activate $T_C$ cells.

    c. They promote inflammation.

    d. They promote allergic reactions by promoting class switching to IgE.

13. Which of the following is true regarding IFN-γ?

    a. It promotes differentiation of $T_C$ cells.

    b. It increases antimicrobial activity of macrophages.

    c. It inhibits the differentiation of T0 into $T_H2$.

    d. It helps B cells produce IgG subtypes that are good at opsonization.

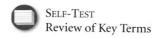 **Interactive Study**

www.whfreeman.com/kuby

SELF-TEST
Review of Key Terms

# Leukocyte Activation and Migration

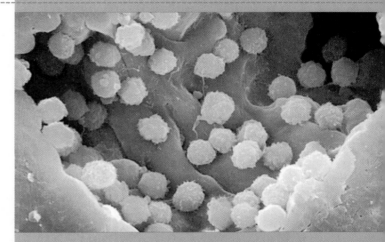

*Lymphocytes attached to the surface of a high-endothelial venule. [From S.D. Rosen and L.M. Stoolman, 1987,* Vertebrate Lectins, *Van Nostrand Reinhold.]*

- Cell Adhesion Molecules
- Chemokines
- Leukocyte Extravasation—The Multistep Paradigm
- Lymphocyte Recirculation
- Lymphocyte Extravasation
- Other Mediators of Inflammation
- The Inflammatory Process
- Anti-inflammatory Agents

**W**HEN INVADERS INFECT OUR BODIES, THE CELLS OF the immune system rally a defense. The cells of the innate immune system arrive first on the scene to mount an initial defense, with the adaptive immune response providing longer-term protection. Leukocytes continuously monitor our bodies by traversing the circulatory system. When an infection is detected, the cells cross the blood barrier and travel to the site of infection.

Inflammation is a complex response to local injury or other trauma; it is characterized by redness, heat, swelling, and pain, as described in Chapter 3. Inflammation involves various immune system cells and numerous mediators. Assembling and regulating **inflammatory responses** would be impossible without the controlled migration of leukocyte populations. We have discussed how lymphocytes become activated in response to antigens in Chapters 10 and 11. This chapter covers the molecules and processes that play a role in leukocyte migration, various molecules that mediate inflammation, and the characteristic physiologic changes that accompany inflammatory responses.

## Cell Adhesion Molecules

The tissues of our bodies are held together by molecular interactions between the cells. The immune system is unique in that the cells can traverse from one part of the body to another. The same molecules that hold our tissues together, **cell adhesion molecules (CAMs),** can be used by leukocytes to interact with tissue cells.

The vascular endothelium serves as an important "gatekeeper," regulating the movement of blood-borne molecules and leukocytes into the tissues. For circulating leukocytes to enter inflamed tissue or peripheral lymphoid organs, the cells must adhere to and pass between the endothelial cells lining the walls of blood vessels, a process called **extravasation.** Endothelial cells express leukocyte-specific CAMs. Some of these membrane proteins are expressed constitutively; others are expressed only in response to local concentrations of cytokines produced during an inflammatory response. Recircu-

lating lymphocytes, monocytes, and granulocytes bear receptors that bind to CAMs on the vascular endothelium, enabling these cells to extravasate into the tissues.

In addition to their role in leukocyte adhesion to vascular endothelial cells, CAMs on leukocytes serve to increase the strength of the functional interactions between cells of the immune system. Various adhesion molecules have been shown to contribute to the interactions between $T_H$ cells and APCs, $T_H$ and B cells, and CTLs and target cells. A number of endothelial and leukocyte CAMs have been cloned and characterized, providing new details about the extravasation process. Most of these CAMs belong to four families of proteins: the selectin family, the mucin-like family, the integrin family, and the immunoglobulin (Ig) superfamily (Figure 13-1).

**(a) General structure of CAM families**

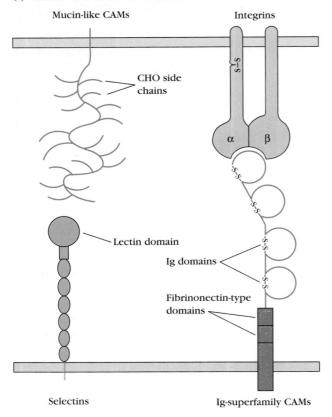

**(b) Selected CAMs belonging to each family**

| Mucin-like CAMs: | Selectins: |
|---|---|
| GlyCAM-1 | L-selectin |
| CD34 | P-selectin |
| PSGL-1 | E-selectin |
| MAdCAM-1 | |

| Ig-superfamily CAMs: | Integrins: |
|---|---|
| ICAM-1, -2, -3 | α4β1 (VLA-4, LPAM-2) |
| VCAM-1 | α4β7 (LPAM-1) |
| LFA-2 (CD2) | α6β1 (VLA-6) |
| LFA-3 (CD58) | αLβ2 (LFA-1) |
| MAdCAM-1 | αMβ2 (Mac-1) |
| | αXβ2 (CR4, p150/95) |

**FIGURE 13-1 Schematic diagrams depicting the general structures of the four families of cell adhesion molecules (a) and a list of representative molecules in each family (b).** The lectin domain in selectins interacts primarily with carbohydrate (CHO) moieties on mucin-like molecules. Both component chains in integrin molecules contribute to the binding site, which interacts with an Ig domain in CAMs belonging to the Ig superfamily. MAdCAM-1 contains both mucin-like and Ig-like domains and can bind to both selectins and integrins.

SELECTINS The **selectin** family of membrane glycoproteins has an extracellular lectin-like domain that enables these molecules to bind to specific carbohydrate groups. Selectins interact primarily with a sialylated carbohydrate moiety called sialyl-Lewis$^x$, which is often linked to mucin-like molecules. The selectin family includes three molecules, L-, E-, and P-selectin, also called CD62L, CD62E, and CD62P. Most circulating leukocytes express L-selectin, whereas E-selectin and P-selectin are expressed on vascular endothelial cells during an inflammatory response. P-selectin is stored within Weibel-Palade bodies, a type of granule within the endothelial cell. On activation of the endothelial cell, the granule fuses with the plasma membrane, resulting in the expression of P-selectin on the cell surface. Expression of E-selectin requires the synthesis of new proteins and occurs after the endothelium has been stimulated with pro-inflammatory cytokines. Selectin molecules are responsible for the initial stickiness of leukocytes to vascular endothelium.

MUCINS The **mucins** are a group of serine- and threonine-rich proteins that are heavily glycosylated. Their extended structure allows them to present the sialyl-Lewis$^x$ and other sulfated carbohydrate moieties as binding sites for selectins. For example, L-selectin (CD62L) on leukocytes binds CD34 and GlyCAM-1 on endothelial cells. Another mucin-like molecule (PSGL-1) found on neutrophils interacts with E- and P-selectin expressed on inflamed endothelium.

INTEGRINS The **integrins** are heterodimeric proteins consisting of an α and a β chain that are noncovalently associated at the cell surface. Most integrins bind extracellular matrix molecules and provide cell matrix interactions throughout the body. Some subfamilies (based on common subunit usage) bind cell surface adhesion molecules and are involved in cell-cell interactions. Leukocytes express a specific subfamily of integrins known as the β2-integrins (or CD18-integrins) as well as several integrins that are expressed on other cell types. Integrins containing the β2 (CD18) subunit bind to members of the Ig superfamily as well as proteins associated with the inflammatory response (Table 13-1). The combination of integrins expressed on a given cell type allows these cells to bind to different CAMs expressed on the surface of the vascular endothelium. As described later, some integrins must be activated before they can bind with high affinity to their ligands. Clustering of integrins on the cell surface also increases the likelihood of effective binding and plays a role in leukocyte migration. The importance of integrin molecules in leukocyte extravasation is demonstrated by **leukocyte adhesion deficiency (LAD)** type 1, an autosomal recessive disease described later in this chapter (see the Clinical Focus on page 343). It is characterized by recurrent bacterial infections and impaired healing of wounds. A similar deficiency is seen in individuals with impaired expression of selectins and has been coined LAD type 2.

| TABLE 13-1 | Some interactions between cell adhesion molecules implicated in leukocyte extravasation* | | | |
|---|---|---|---|---|
| **Receptor on cells** | **Expression** | **Ligands** | **Extravasation step[†]** | |
| PSGL-1 | Neutrophils | P-selectin (CD62P), Sialyl-Lewis$^x$ | Rolling/tethering | |
| L-selectin (CD62L) | Leukocytes | GlyCAM-1, CD34, MAdCAM-1 | Rolling/tethering | |
| CD11a/CD18 ($\alpha$L$\beta$2, LFA-1) | Leukocytes | ICAM-1, 2 | Tight adhesion | |
| CD11b/CD18 (Mac-1) | Monocytes, neutrophils, macrophages | ICAM-1, iC3b, fibrinogen | Tight adhesion | |
| CD49d/CD29 ($\alpha$4$\beta$1, VLA-4) | Lymphocytes, monocytes | VCAM-1, fibronectin | Tight adhesion | |
| $\alpha$4$\beta$7 | Lymphocytes | VCAM-1, MAdCAM-1, Fibronectin | Adhesion | |

*Most endothelial and leukocyte CAMs belong to four groups of proteins, as shown in Figure 13-1. In general, molecules in the integrin family bind to Ig-superfamily CAMs and molecules in the selectin family bind to mucin-like CAMs.
†See Figures 13-3a and 13-8 for an illustration of steps in the extravasation process.

**IG-SUPERFAMILY CAMS (ICAMS)** Several adhesion molecules contain a variable number of immunoglobulin-like domains and thus are classified as members of the immunoglobulin superfamily. Included in this group are ICAM-1 (CD54), ICAM-2 (CD102), ICAM-3 (CD50), and VCAM (CD106), which are expressed on vascular endothelial cells and bind to various integrin molecules. An interesting cell adhesion molecule called MAdCAM-1 has both Ig-like domains and mucin-like domains. This molecule is expressed on mucosal endothelium and directs lymphocyte entry into mucosa. It binds to integrin $\alpha$4$\beta$7 (LPAM) via its immunoglobulin-like domain and to L-selectin (CD62L) via its mucin-like domain. Platelet-endothelial cell adhesion molecule-1 (PECAM-1, CD31) is found on the surface of leukocytes (neutrophils, monocytes, and a subset of T lymphocytes) and within the junctional complex of endothelial cells. It exhibits *homotypic binding*—the binding of a PECAM-1 molecule on one cell to a PECAM-1 molecule on another cell. Junctional cell adhesion molecule-1 (JAM-1, CD321) is also located within the endothelial junctional complex. JAM-1 can interact with JAM-1 molecules and with CD11a/CD18 to play a role in transendothelial migration. Homotypic adhesion among Ig-superfamily members can be found among other cells types, such as L1 and NCAM on neural cells.

## Chemokines

Chemokines are a superfamily of small polypeptides, most of which contain 90 to 130 amino acid residues. They selectively control the adhesion, chemotaxis, and activation of many types of leukocyte populations and subpopulations. Consequently, they are major regulators of leukocyte traffic. Some chemokines are primarily involved in inflammatory processes; others are constitutively expressed and play important homeostatic or developmental roles. "Housekeeping" chemokines are produced in lymphoid organs and tissues or in nonlymphoid sites such as skin, where they direct normal trafficking of lymphocytes, such as the targeting to their proper destination of leukocytes newly generated by hematopoiesis and arriving from bone marrow. The thymus constitutively expresses chemokines, and normal B-cell lymphopoiesis is also dependent on appropriate chemokine expression.

Chemokine-mediated effects are not limited to the immune system. Mice that lack either the chemokine CXCL12 (also called SDF-1) or its receptor (Table 13-2) show major defects in the development of the brain and the heart. Members of the chemokine family have also been shown to play regulatory roles in the development of blood vessels (angiogenesis) and wound healing.

The inflammatory chemokines are typically induced in response to infection. Contact with pathogens or the action of proinflammatory cytokines, such as TNF-$\alpha$, up-regulate the expression of inflammatory cytokines at sites of developing inflammation. Chemokines cause leukocytes to move into various tissue sites by inducing the adherence of these cells to the vascular endothelium. After migrating into tissues, leukocytes travel in the direction of increasing localized concentrations of chemokines, resulting in the targeted recruitment of phagocytes and effector lymphocyte populations to inflammatory sites. The assembly of leukocytes at sites of infection, orchestrated by chemokines, is an essential part of mounting an appropriately focused response to infection.

The chemokine family consists of at least 43 members, with additional variation contributed by alternate RNA splicing pathways during transcription, post-translational processing, and isoforms (see Table 13-2). The chemokines possess four conserved cysteine residues, and based on the position of two of the four invariant cysteine residues, almost all fall into one of two distinctive subgroups:

- **CC subgroup** chemokines, in which the conserved cysteines are contiguous

- **CXC subgroup** chemokines, in which the conserved cysteines are separated by some other amino acid (X)

| TABLE 13-2 | Human chemokines and their receptors* |
|---|---|
| **Chemokine receptors** | **Chemokines bound by receptor** |
| CXC SUBGROUP | |
| CXCR1 | IL-8 (CXCL 8), GCP-2 (CXCL6) |
| CXCR2 | IL-8 (CXCL8), Gro-α (CXCL1), Gro-β (CXCL2), Gro-γ (CXCL3), NAP-2 (CXCL7), ENA-78 (CXCL5) |
| CXCR3 | IP-10 (CXCL10), Mig (CXCL9), I-TAC (CXCL11) |
| CXCR3b | PF4 (CXCL4), IP-10 (CXCL10), Mig (CXCL9), I-TAC |
| CXCR4 (CXCL11) | SDF-1 (CXCL12) |
| CXCR5 | BCA-1 (CXCL13) |
| CXCR6 | CXCL16-transmembrane cytokine |
| CC SUBGROUP | |
| CCR1 | MIP-1α (CCL3), RANTES (CCL5), MCP-2 (CCL8) MCP3(CCL7), MRP2 (mCCL9), MCP4 (hCCL13), HCCL1 (CCL14), HCC2, (hCCL15), HCC4 (CCL16), MPIF1 (hCCL23) |
| CCR2 | MCP-1 (CCL2), MCP-2 (CCL8), MCP-3 (CCL7), MCP4 (hCCL13), MCP5 (mCCL12),  HCC4 (CC16) |
| CCR3 | Eotaxin (CCL11), RANTES (CCL5),  MCP-2 (CCL8), MCP-3 (CCL7), MCP-4 (hCCL13), Eotaxin-2 (hCCL24), HCC2 (hCCL15), Eotaxin-3 (hCCL26), MEC (CCL28) |
| CCR4 | TARC (CCL17), MDC (CCL22) |
| CCR5 | MIP-1α (CCL3), RANTES (CCL5),  MIP-1β (CCL4), MCP2 (CCL8), Eotaxin (CCL11), HCCI (CCL14), HCC4 (CCL16) |
| CCR6 | MIP3α (LARC, CCL20) |
| CCR7 | ELC (MIP3β, CCL19), SLC (CCL21) |
| CCR8 | 1-309 (CCL1) |
| CCR9 | TECK (CCL25) |
| CCR10 | CTAK (CCL27), MEC (CCL28) |
| BOTH CC AND CXC SUBGROUPS | |
| DARC (the Duffy antigen of RBCs) | Binds to a number of CC and CXC chemokines |
| C SUBGROUP | |
| XCR1 | Lymphotactin (XCL1), SCM1b (hXCL2) |
| CX3C SUBGROUP | |
| CX3CR1 | Fractalkine-transmembrane chemokine (CX3CL1) |

*This table lists most known chemokine receptors but not all chemokines. The full names for a number of the chemokines abbreviated in the table are as follows: ELC (Ebl1 ligand chemokine); ENA-78 (epithelial-cell-derived neutrophil-activating protein); GCP-2 (granulocyte chemotactic protein 2); Gro-α, -β, -γ (growth-related oncogene α, β, γ); MCP-1, -2, -3, or -4 (monocyte chemoattractant protein 1, 2, 3, or 4); Mig (monokine induced by interferon γ); MIP-1α, -1β, or -5 (macrophage inflammatory protein 1α, 1β, or 5); NAP-2 (neutrophil-activating protein 2); RANTES (regulated on activation, normal T-cell expressed and secreted); TARC (thymus-and activation-regulated chemokine).

SOURCE: Adapted from Nelson and Krensky, 1998, *Current Opinion Immunology* 10:265; Baggiolini, 1998, *Nature* 392: 565; and Cyster, 2005, *Annual Review of Immunology* 23: 127.

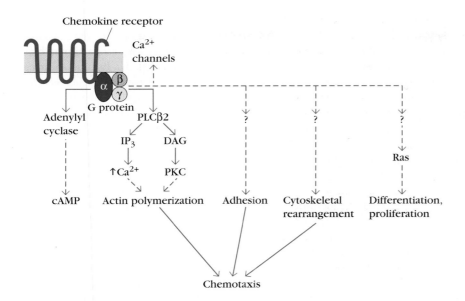

**FIGURE 13-2 Chemokines signal through receptors coupled with heterotrimeric large G proteins.** Binding of a chemokine to its receptor activates many signal transduction pathways, resulting in a variety of modifications in the physiology of the target cell. If the signal transduction pathway is not known or is incompletely worked out, dashed lines and question marks are used here to represent probable pathways. [Adapted from Premack et al., 1996, Nature Medicine **2**:1174, and Rot and von Adrian, 2004, Annual Review of Immunology **22**:891.]

The exceptions are CX3CL1, which has three amino acids between the conserved cysteines, and XCL1 and 2, which lack two of the four conserved cysteines.

Chemokine action is mediated by receptors whose polypeptide chains traverse the membrane seven times. The receptors are members of the G-protein–linked family of receptors and are grouped according to the type of chemokine(s) they bind. The CC receptors (CCRs) recognize CC chemokines, and CXC receptors (CXCRs) recognize CXC chemokines. As with cytokines, the interaction between chemokines and their receptors is both strong ($K_d > 10^{-9}$) and highly specific. However, as Table 13-2 shows, most receptors bind more than one chemokine. For example, CXCR2 recognizes at least six different chemokines, and many chemokines can bind to more than one receptor.

When a receptor binds an appropriate chemokine, it activates heterotrimeric large G proteins, initiating a signal transduction process that generates such potent second messengers

as cyclic AMP (cAMP), $IP_3$, $Ca^{2+}$, and activated small G proteins (Figure 13-2). Dramatic changes are effected by the chemokine-initiated activation of these signal transduction pathways. Within seconds, the addition of an appropriate chemokine to leukocytes can cause abrupt and extensive changes in shape, the promotion of greater adhesiveness to endothelial walls by activation of leukocyte integrins, and the generation of microbicidal oxygen radicals in phagocytes, as well as the release of granular contents, including the release of proteases from neutrophils and macrophages, histamine from basophils, and cytotoxic proteins from eosinophils.

## Chemokine-receptor profiles mediate leukocyte activity

Among major populations of human leukocytes, neutrophils express CXCR1, -2, and -4; eosinophils have CCR1 and CCR3 (Figure 13-3). Although resting naive T cells display CCR7, it

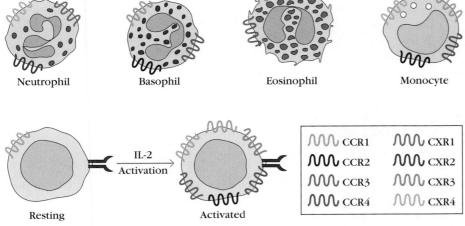

**FIGURE 13-3 Patterns of expression of some principal chemokine receptors on different classes of human leukocytes.** So far, the greatest variety of chemokine receptors has been observed on activated T lymphocytes. [Adapted from M. Baggiolini, 1998, Nature **392**:565.]

is not present on activated T cells, which may have CCR1, -4, -5, and -8, CXCR3, and possibly others. Clearly, a cell can respond to a chemokine only if it possesses a receptor that recognizes it. Consequently, differences in the expression of chemokine receptors by leukocytes coupled with the production of distinctive profiles of chemokines by destination tissues and sites provide rich opportunities for the differential regulation of activities of different leukocyte populations. Indeed, differences in patterns of chemokine-receptor expression occur within leukocyte populations. Recall that $T_H1$ and $T_H2$ subsets of $T_H$ cells can be distinguished by their different patterns of cytokine production. These subsets also display different profiles of chemokine receptors. $T_H2$ cells express CCR4 and -8, and a number of other receptors not expressed by $T_H1$ cells. On the other hand, $T_H1$ cells express CCR1 and -5, and CXCR3, whereas most $T_H2$ cells do not.

# Leukocyte Extravasation— The Multistep Paradigm

As an inflammatory response develops, various cytokines and other inflammatory mediators act on the local blood vessels, inducing increased expression of endothelial CAMs. The vascular endothelium is then said to be activated, or **inflamed.** Leukocytes then extravasate into the tissue and migrate to the site of infection. To accomplish this, leukocytes must recognize the inflamed endothelium and adhere strongly enough so that they are not swept away by the flowing blood. The bound leukocytes must then penetrate the endothelial layer and migrate into the underlying tissue.

Leukocyte extravasation can be divided into four steps: (1) rolling, mediated by selectins; (2) activation by chemoattractant stimulus; (3) arrest and adhesion, mediated by integrins binding to Ig-family members; and (4) transendothelial migration (Figure 13-4). In the first step, leukocytes attach loosely to the endothelium by a low-affinity selectin-carbohydrate interaction. During an inflammatory response, cytokines and other mediators act on the local endothelium, inducing expression of adhesion molecules of the selectin family. These

E- and P-selectin molecules bind to mucin-like cell adhesion molecules on the leukocyte membrane or with the sialylated lactosaminoglycan called sialyl Lewis$^x$. This interaction tethers the leukocyte briefly to the endothelial cell, but the shearing force of the circulating blood soon detaches the cell. Selectin molecules on another endothelial cell again tether the leukocyte; this process is repeated so that the cell tumbles end over end along the endothelium, a type of binding called *rolling.*

The process of rolling slows the cell long enough to allow interactions between chemokines (see Table 13-2) presented on the surface of the endothelium and receptors on the leukocyte. The various chemokines expressed on the endothelium and the repertoire of chemokine receptors on the leukocyte provide a degree of specificity to the recruitment of white blood cells to the infection site. Signal transduction events caused by the binding of a chemokine to the chemokine receptor on the surface of the leukocyte result in change of conformation and clustering of integrins on the leukocyte. This allows the cell to bind more tightly to the endothelium, lessening the likelihood that it will be carried away by the flowing blood. The leukocyte then squeezes in between two neighboring endothelial cells without disrupting the integrity of the endothelial barrier. It accomplishes this by homotypic binding of platelet-endothelial-cell adhesion molecule-1 (PECAM-1; CD31) on the leukocyte with PECAM-1 on the endothelium. PECAM-1 is normally found within the endothelial junction in a homotypic interaction. Thus, when the leukocyte PECAM-1 binds to endothelial PECAM-1, the junctional integrity is maintained. CD11a/CD18 on leukocytes also binds to JAM-1, another adhesion molecule located within the endothelial tight junction, to mediate egress from the bloodstream. Other integrins binding to matrix proteins within the basement membrane and extracellular matrix allow the leukocytes to follow a gradient of chemoattractants to the site of infection.

**NEUTROPHIL EXTRAVASATION** Neutrophils are generally the first cell type to bind to inflamed endothelium and extravasate into the tissues. Neutrophils do not bind to resting, non-inflamed endothelium and do not exit the bloodstream when no infection is present due to the lack of E- and P-selectin expressed on the surface of resting endothelium (Figure 13-5a). When

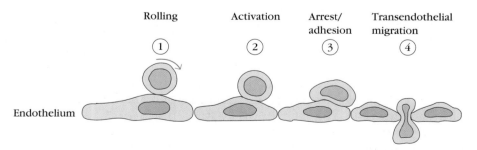

**FIGURE 13-4 The four sequential but overlapping steps in leukocyte extravasation.** (1) Initial rolling is mediated by binding of selectin molecules to sialylated carbohydrate moieties on mucin-like CAMs. (2) Chemokines then bind to G-protein–linked receptors on the leukocyte, triggering an activating signal. This signal induces a

conformational change in the integrin molecules (3), enabling them to adhere firmly to Ig-superfamily molecules on the endothelium. (4) Leukocytes traverse the tight endothelial junction and subsequently migrate into the underlying tissue.

(a) Resting endothelium

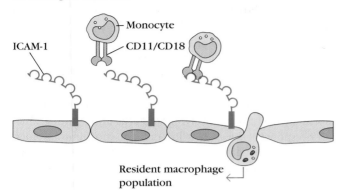

Resident macrophage population

(b)

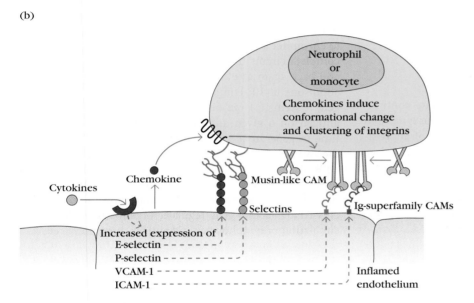

**FIGURE 13-5 Steps in extravasation of neutrophils and monocytes.** (a) Resting endothelium displays few adhesion molecules. Subpopulations of monocytes can bind via CD11/CD18 to ICAM to replenish tissue macrophage population. Little if any neutrophil emigration occurs. (b) Inflamed endothelium increases the level of several adhesion molecules, allowing transmigration of neutrophils and monocytes. See text for details.

endothelial cells become inflamed, P-selectin is released from the Weibel-Palade bodies and becomes available at the cell surface for binding to neutrophils. Neutrophils express L-selectin and mucin-like PSGL-1 to mediate the rolling on inflamed endothelium. As the neutrophil rolls, it is activated by various chemoattractants; these are either permanent features of the endothelial cell surface or secreted locally by cells involved in the inflammatory response. Two chemokines involved in the activation process are interleukin 8 (IL-8) and macrophage inflammatory protein 1β (MIP-1β). Binding of these chemoattractants to receptors on the neutrophil membrane triggers an activating signal mediated by G proteins associated with the receptor. This signal induces a conformational change in the integrin molecules in the neutrophil membrane, increasing their affinity for the Ig-superfamily adhesion molecules (ICAMs) on the endothelium. Neutrophils express CD11a/CD18 and CD11b/CD18 (LFA-1 and MAC-1), which convert to their active, more adhesive conformation when the neutrophil is stimulated by chemokines. Resting endothelium expresses low levels of ICAM-1 (CD54), which supports adhesion with activated CD11a/CD18 and CD11b/CD18 without the requirement of time-consuming synthesis of new Ig-family members.

Subsequently, the neutrophil migrates through the vessel wall into the tissues. Gradients of additional chemoattractants then guide the neutrophils to the site of infection. Complement split product C5a, bacterial peptides containing *N*-formyl peptides, and leukotrienes bind to specific receptors on the neutrophil, contributing to the directed migration and activation of these cells.

MONOCYTE EXTRAVASATION Monocytes from the peripheral blood home to an infection much later than neutrophils. This is because of the time it takes for the inflamed endothelium to increase expression of VCAM-1 (CD106) and ICAM-1 (CD54) (Figure 13-5b). Resting endothelium in humans expresses little or no VCAM-1, the ligand for integrin α4β1 (CD49d/CD29, VLA-4). A subpopulation of monocytes migrates at low levels into uninflamed tissue to replenish the population of resident macrophages and dendritic cells using a CD11/CD18-ICAM–dependent interaction. The chemokine CXCL14 is thought to facilitate this constitutive migration. Once ICAM-1 and VCAM-1 expression is elevated at the cell surface, peripheral blood monocytes adhere more effectively and migrate into the inflamed tissues. Rolling is mediated by L-selectin (CD62L), and integrins are activated by monocyte-specific chemoattractants, primarily monocyte-chemoattractant protein-1 (MCP-1, CCL2). As with neutrophils, transendothelial migration is mediated by PECAM-1 to PECAM-1 interactions and possibly by additional interactions involving JAMs.

Monocytes are lured to the area of infection by chemoattractants such as bacterial peptide fragments and complement fragments. The receptors expressed by monocytes for complement fragments include CR3, consisting of the integrins CD11b/CD18 (αMβ2, MAC-1) and CR4 (CD11c/CD18 (αXβ2, p150.95). Thus, the complement receptors contribute to the targeting of monocytes to the site of infection. Within the target tissue, monocytes can differentiate into macrophages, which become activated within the tissue as described in Chapter 3.

## Lymphocyte Recirculation

Unlike other leukocytes, lymphocytes recirculate continuously through the blood and lymph to the various lymphoid organs (Figure 13-6). After a brief transit time of approxi-

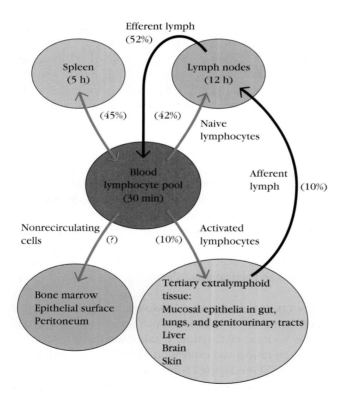

**FIGURE 13-6 Lymphocyte recirculation routes.** The percentage of the lymphocyte pool that circulates to various sites and the average transit times in the major sites are indicated. Lymphocytes migrate from the blood into lymph nodes through specialized areas in postcapillary venules called high-endothelial venules (HEVs). Although most lymphocytes circulate, some sites appear to contain lymphocytes that do not. *[Adapted from A. Ager, 1994, Trends in Cell Biology 4:326.]*

mately 30 minutes in the bloodstream, nearly 45% of all lymphocytes are carried from the blood directly to the spleen, where they reside for approximately 5 hours. Almost equal numbers (42%) of lymphocytes exit from the blood into various peripheral lymph nodes, where they reside for about 12 hours. A smaller number of lymphocytes (10%) migrate to tertiary extralymphoid tissues by crossing between endothelial cells that line the capillaries. These tissues normally host few, if any, lymphoid cells until an inflammatory response triggers their import. The most immunologically active tertiary extralymphoid tissues are those that interface with the external environment, such as the skin and various mucosal epithelia of the gastrointestinal, pulmonary, and genitourinary tracts.

The process of continual lymphocyte recirculation allows maximal numbers of antigenically committed lymphocytes to encounter antigen. An individual lymphocyte may make a complete circuit from the blood to the tissues and lymph and back again as often as 1 to 2 times per day. Since only about one in $10^5$ lymphocytes recognizes a particular antigen, it seems a large number of T or B cells must contact antigen on a given antigen-presenting cell within a short time to generate a specific immune response. The odds of the small percentage of lymphocytes committed to a given antigen actually making contact with that antigen when it is present are elevated by the extensive recirculation of lymphocytes. The likelihood of such contacts is profoundly increased by factors that regulate, organize, and direct the circulation of lymphocytes and antigen-presenting cells.

## Lymphocyte Extravasation

Various subsets of lymphocytes exhibit directed extravasation at inflammatory sites and secondary lymphoid organs. The recirculation of lymphocytes is thus carefully controlled to ensure that appropriate populations of B and T cells are recruited into different tissues. As with neutrophils, extravasation of lymphocytes involves interactions among a number of cell adhesion molecules (see Table 13-1). The overall process is similar to neutrophil extravasation and comprises the same four stages of rolling, activation, arrest and adhesion, and, finally, transendothelial migration.

### High-endothelial venules are sites of lymphocyte extravasation

Some regions of vascular endothelium in postcapillary venules of various lymphoid organs are composed of specialized cells with a plump, cuboidal ("high") shape; such regions are called **high-endothelial venules**, or **HEVs** (Figure 13-7a, b). Their cells contrast sharply in appearance with the flattened endothelial cells that line the rest of the capillary. Each of the secondary lymphoid organs, with the exception of the spleen, contains HEVs. When frozen sections of lymph nodes, Peyer's patches, or tonsils are incubated with

(a)

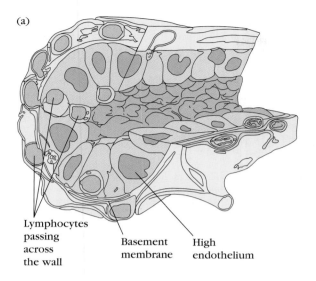

Lymphocytes passing across the wall

Basement membrane

High endothelium

(b)

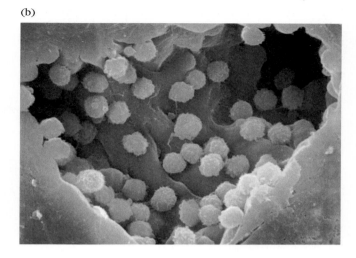

(c)

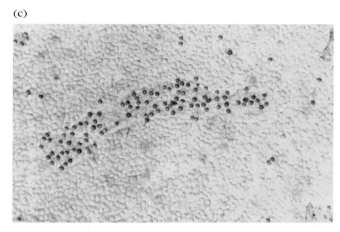

**FIGURE 13-7 (a) Schematic cross-sectional diagram of a lymph node postcapillary venule with high endothelium.** Lymphocytes are shown in various stages of attachment to the HEV and in migration across the wall into the cortex of the node. (b) Scanning electron micrograph showing numerous lymphocytes bound to the surface of a high-endothelial venule. (c) Micrograph of frozen sections of lymphoid tissue. Some 85% of the lymphocytes (darkly stained) are bound to HEVs (in cross section), which constitute only 1% to 2% of the total area of the tissue section. *[Part a adapted from A. O. Anderson and N. D. Anderson, 1981, in* Cellular Functions in Immunity and Inflammation, *J. J. Oppenheim et al., eds., Elsevier North-Holland; part b from S. D. Rosen and L. M. Stoolman, 1987,* Vertebrate Lectins, *Van Nostrand Reinhold; part c from S. D. Rosen, 1989,* Current Opinion in Cell Biology ***1:913.]***

lymphocytes and washed to remove unbound cells, over 85% of the bound cells are found adhering to HEVs, even though HEVs account for only 1% to 2% of the total area of the frozen section (Figure 13-7c).

It has been estimated that as many as $1.4 \times 10^4$ lymphocytes extravasate every second through HEVs into a single lymph node. The development and maintenance of HEVs in lymphoid organs is influenced by cytokines produced in response to antigen capture. For example, HEVs fail to develop in animals raised in a germ-free environment. The role of antigenic activation of lymphocytes in the maintenance of HEVs has been demonstrated by surgically blocking the afferent lymphatic vasculature to a node, so that antigen entry to the node is blocked. Within a short period, the HEVs show impaired function and eventually revert to a more flattened morphology.

High-endothelial venules express a variety of cell adhesion molecules. Like other vascular endothelial cells, HEVs express CAMs of the selectin family (E- and P-selectin), the mucin-like family (GlyCAM-1 and CD34), and the immunoglobulin superfamily (ICAM-1, ICAM-2, ICAM-3, VCAM-1, and MAdCAM-1). Some adhesion molecules that are distributed in a tissue-specific manner have been called **vascular addressins** (**VAs**) because they serve to direct the extravasation of different populations of recirculating lymphocytes to particular lymphoid organs.

Figure 13-8 depicts the typical interactions in extravasation of naive T cells across HEVs into lymph nodes. The first step is usually a selectin-carbohydrate interaction similar to that seen with neutrophil adhesion. Naive lymphocytes initially bind to HEVs through L-selectin, which serves as a homing receptor that directs the lymphocytes to particular tissues expressing a corresponding mucin-like vascular addressin such as CD34 or GlyCAM-1. Lymphocyte rolling is less pronounced than that of neutrophils. Although the initial selectin-carbohydrate interaction is quite weak, the slow rate of blood flow in postcapillary venules, particularly in regions of HEVs, reduces the likelihood that the shear force of the flowing blood will dislodge the tethered lymphocyte.

In the second step, an integrin-activating stimulus is mediated by chemokines that are either localized on the endothelial surface or secreted locally. The thick glycocalyx covering of the HEVs may function to retain these soluble chemoattractant factors on the HEVs. Chemokine binding

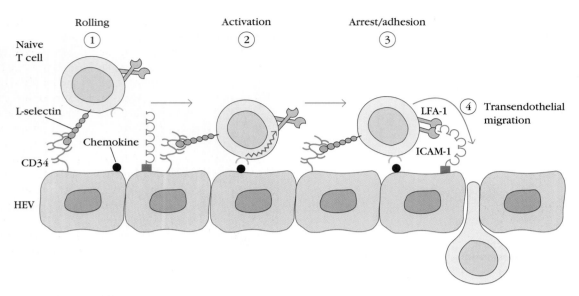

**FIGURE 13-8 Steps in extravasation of a naive T cell through a high-endothelial venule into a lymph node.** Extravasation of lymphocytes includes the same basic steps as neutrophil extravasation, but some of the cell adhesion molecules differ. Activation of the integrin LFA-1, induced by chemokine binding to the lymphocyte, leads to firm adhesion followed by migration between the endothelial cells into the tissue.

to G-protein–coupled receptors on the lymphocyte leads to activation of integrin molecules on the membrane, as occurs in neutrophil extravasation. Once activated, the integrin molecules interact with Ig-superfamily adhesion molecules (e.g., ICAM-1), allowing the lymphocyte to adhere firmly to the endothelium. The molecular mechanisms involved in the final step, transendothelial migration, are thought to involve junctional adhesion molecules PECAM-1 (CD31) and JAM-1 (CD321).

## Lymphocyte homing is directed by receptor profiles and signals

The general process of lymphocyte extravasation is similar to neutrophil extravasation. An important feature distinguishing the two processes is that different subsets of lymphocytes migrate differentially into different tissues. This process is called **trafficking,** or **homing.** The different trafficking patterns of lymphocyte subsets are mediated by unique combinations of adhesion molecules and chemokines; receptors that direct the circulation of various populations of lymphocytes to particular lymphoid and inflammatory tissues are called **homing receptors.**

## Naive lymphocytes recirculate to secondary lymphoid tissue

A naive lymphocyte is not able to mount an immune response until it has been activated to become an effector cell. Activation of a naive cell occurs in specialized microenvironments within secondary lymphoid tissue (e.g., peripheral lymph nodes, Peyer's patches, tonsils, and spleen). Within these microenvironments, dendritic cells capture antigen and present it to the naive lymphocyte, resulting in activation of the lymphocyte. Naive cells do not exhibit a preference for a particular type of secondary lymphoid tissue but instead circulate indiscriminately to secondary lymphoid tissue throughout the body.

The initial attachment of naive lymphocytes to HEVs is generally mediated by the binding of L-selectin to adhesion molecules such as GlyCAM-1 and CD34 on HEVs (Figure 13-8a). Recently, it has been found that CD11a/CD18 (LFA-1, $\alpha L\beta 2$) and CD49d/CD18 (VLA-4, $\alpha 4\beta 1$) also mediate rolling in their low-affinity conformation (prior to activation). The chemokines CCL21, CCL19, and CXCL12 promote tight adhesion of the naive T cells to the HEV by both a switch to the high-affinity conformation and clustering of CD11a/CD18 (LFA-1) and CD49d/CD18 (VLA-4). In addition to these chemokines, CXCL13 activates integrins on naive B cells. Other chemokines attract the lymphocyte subpopulations to their zones within the lymph nodes. The trafficking pattern of naive cells is designed to keep these cells constantly recirculating through secondary lymphoid tissue, whose primary function is to trap blood-borne or tissue-borne antigen.

Once naive lymphocytes encounter antigen trapped in a secondary lymphoid tissue, they become activated and enlarge into lymphoblasts. Activation takes about 48 hours, and during this time the blast cells are retained in the paracortical region of the secondary lymphoid tissue. In this phase, called the shutdown phase, the antigen-specific lymphocytes cannot be detected in the circulation (Figure 13-9). Rapid proliferation and differentiation of naive cells occurs during the shutdown phase. The effector and memory cells

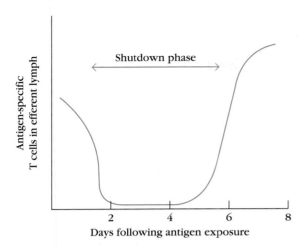

**FIGURE 13-9 T-cell activation in the paracortical region of a lymph node results in the brief loss of lymphocyte recirculation.** During this shutdown phase, antigen-specific T cells cannot be detected leaving the node in the efferent lymph.

that are generated by this process (described in detail in Chapter 11) then leave the lymphoid tissue and begin to recirculate.

## Effector and memory lymphocytes adopt different trafficking patterns

The trafficking patterns of effector and memory lymphocytes differ from those of naive lymphocytes. Effector cells tend to home to regions of infection by recognizing inflamed vascular endothelium and chemoattractant molecules that are generated during the inflammatory response. Memory lymphocytes, on the other hand, home selectively to the type of tissue in which they first encountered antigen. Presumably this ensures that a particular memory cell will return to

the tissue where it is most likely to re-encounter a subsequent threat by the antigen it recognizes.

Effector and memory cells express increased levels of certain cell adhesion molecules, such as LFA-1, that interact with ligands present on tertiary extralymphoid tissue (such as skin and mucosal epithelia) and at sites of inflammation, allowing effector and memory cells to enter these sites. Inflamed endothelium expresses a number of adhesion molecules, including E- and P-selectin and the Ig-superfamily molecules VCAM-1 and ICAM-1, that bind to the receptors expressed at high levels on memory and effector cells.

Unlike naive lymphocytes, subsets of the memory and effector populations exhibit tissue-selective homing behavior. Such tissue specificity is imparted by the repertoire of chemokine receptors expressed on the memory or effector cell surface and the chemokine presented on the endothelium. In addition, some tissues display unique sets of adhesion molecules that help select for the effector subset. For example, a mucosal homing subset of memory/effector cells has high levels of the integrins α4β7 (LPAM-1) and CD11a/CD18 (LFA-1, αLβ2), which bind to MAdCAM and various ICAMs on intestinal lamina propria venules (Figure 13-10a). However, these cells avoid direction to secondary lymphoid tissues because they have low levels of the L-selectin that would facilitate their entry into secondary lymphoid tissue. Instead, these cells express the chemokine receptor CCR9, which binds to CCL25 in the small intestine. IgA-secreting B cells are recruited into gut tissue via CCL25 and CCL28. A second subset of memory/effector cells displays preferential homing to the skin. This subset also expresses low levels of L-selectin but displays high levels of cutaneous lymphocyte antigen (CLA) and CD11a/CD18, which bind to E-selectin and ICAMs on dermal venules of the skin (Figure 13-10b). CCL17, CCL27, and CCL1 help in the recruitment of these T cells. Although effector and memory cells that express reduced levels of L-selectin do not tend to home through

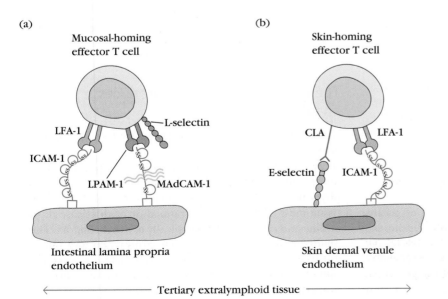

(a) Mucosal-homing effector T cell

LFA-1    L-selectin

ICAM-1

LPAM-1    MAdCAM-1

Intestinal lamina propria endothelium

(b) Skin-homing effector T cell

CLA    LFA-1

E-selectin    ICAM-1

Skin dermal venule endothelium

←———————— Tertiary extralymphoid tissue ————————→

**FIGURE 13-10 Examples of homing receptors and vascular addressins involved in selective trafficking of naive and effector T cells.** Various subsets of effector T cells express high levels of particular homing receptors that allow them to home to endothelium in particular tertiary extralymphoid tissues. The initial interactions in homing of effector T cells to (a) mucosal and (b) skin sites are illustrated.

HEVs into peripheral lymph nodes, they can enter peripheral lymph nodes through the afferent lymphatic vessels.

## Other Mediators of Inflammation

In addition to chemokines, a variety of other mediators released by cells of the innate and acquired immune systems trigger or enhance specific aspects of the inflammatory response. They are released by tissue mast cells, blood platelets, and a variety of leukocytes, including neutrophils, monocytes/macrophages, eosinophils, basophils, and lymphocytes. In addition to these sources, plasma contains four interconnected mediator-producing systems: the kinin system, the clotting system, the fibrinolytic system, and the complement system. The first three systems share a common intermediate, Hageman factor (factor XII), as illustrated in Figure 13-11. When tissue damage occurs, these four systems are activated to form a web of interacting processes that generate a number of mediators of inflammation.

### The kinin system is activated by tissue injury

The kinin system is an enzymatic cascade that begins when a plasma clotting factor, called Hageman factor (factor XII), is activated following tissue injury. The activated Hageman factor then activates prekallikrein to form kallikrein, which cleaves kininogen to produce **bradykinin** (see Figure 13-11). This potent peptide increases vascular permeability, causes

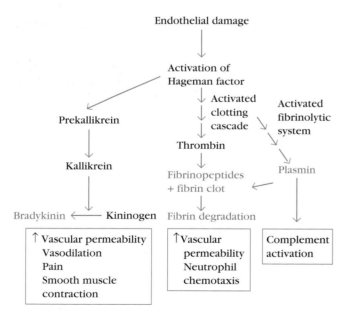

**FIGURE 13-11 Tissue damage induces formation of plasma enzyme mediators by the kinin system, the clotting system, and the fibrinolytic system.** These mediators cause vascular changes, among the earliest signs of inflammation, and various other effects. Plasmin not only degrades fibrin clots but also activates the classical complement pathway.

vasodilation, induces pain, and triggers contraction of smooth muscle. Kallikrein also acts directly on the complement system by cleaving C5 into C5a and C5b.

### The clotting system yields fibrin-generated mediators of inflammation

Another enzymatic cascade that is triggered by damage to blood vessels yields large quantities of thrombin. Initiation of an inflammatory response also triggers the clotting system through an interaction between P-selectin and PSGL-1, accompanied by the release of tissue factor from activated monocytes. Thrombin acts on soluble fibrinogen in tissue fluid or plasma to produce insoluble strands of **fibrin** and **fibrinopeptides.** The insoluble fibrin strands crisscross one another to form a **clot,** which serves as a barrier to the spread of infection. The clotting system is triggered very rapidly after tissue injury to prevent bleeding and limit the spread of invading pathogens into the bloodstream. The fibrinopeptides act as inflammatory mediators, inducing increased vascular permeability and neutrophil chemotaxis. Activated platelets release CD40L, which leads to increased production of proinflammatory cytokines, IL-6 and IL-8, and increased expression of adhesion molecules. The integrin CD11b/CD18 (MAC-1) binds two components of the clotting system, factor X and fibrinogen. Binding of factor X to CD11b/CD18 increases the activity of factor X, thereby promoting coagulation.

### The fibrinolytic system yields plasmin-generated mediators of inflammation

Removal of the fibrin clot from the injured tissue is achieved by the fibrinolytic system. The end product of this pathway is the enzyme **plasmin,** the activated form of plasminogen. Urokinase-type plasminogen activator (uPA) and its receptor, u-PAR, also seem to be involved in leukocyte binding to endothelium and the extracellular matrix. Plasmin, a potent proteolytic enzyme, breaks down fibrin clots into degradation products that are chemotactic for neutrophils. Plasmin also contributes to the inflammatory response by activating the classical complement pathway.

### The complement system produces anaphylatoxins

Activation of the complement system by both classical and alternative pathways results in the formation of a number of complement split products that serve as important mediators of inflammation (see Chapter 7). Binding of the **anaphylatoxins** (C3a and C5a) to receptors on the membrane of tissue mast cells induces degranulation with release of histamine and other pharmacologically active mediators. These mediators induce smooth muscle contraction and increase vascular permeability. C3a, C5a, and C5b67 act together to induce monocytes and neutrophils to adhere to vascular endothelial cells, extravasate through the endothelial lining of

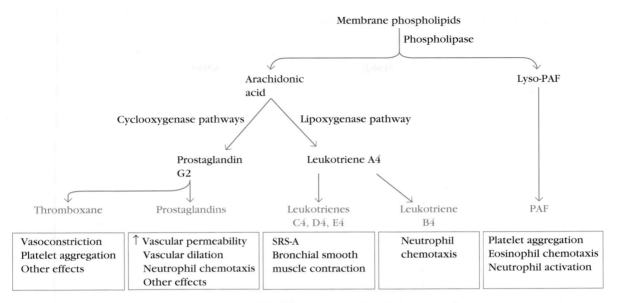

**FIGURE 13-12** The breakdown of membrane phospholipids generates important mediators of inflammation, including thromboxane, prostaglandins, leukotrienes, and platelet-activating factor (PAF).

the capillary, and migrate toward the site of complement activation in the tissues. Activation of the complement system thus results in influxes of fluid that carry antibody and phagocytic cells to the site of antigen entry.

## Some lipids act as inflammatory mediators

Following membrane perturbations, phospholipids in the membrane of several cell types (e.g., macrophages, monocytes, neutrophils, and mast cells) are degraded into arachidonic acid and lyso–platelet-activating factor (Figure 13-12). The latter is subsequently converted into platelet-activating factor (PAF), which causes platelet activation and has many inflammatory effects, including eosinophil chemotaxis and the activation and degranulation of neutrophils and eosinophils.

Metabolism of arachidonic acid by the cyclooxygenase pathway produces **prostaglandins** and **thromboxanes.** Different prostaglandins are produced by different cells: monocytes and macrophages produce large quantities of PGE2 and PGF2; neutrophils produce moderate amounts of PGE2; mast cells produce PGD2. Prostaglandins have diverse physiological effects, including increased vascular permeability, increased vascular dilation, and induction of neutrophil chemotaxis. The thromboxanes cause platelet aggregation and constriction of blood vessels.

Arachidonic acid is also metabolized by the lipoxygenase pathway to yield the four **leukotrienes:** LTB4, LTC4, LTD4, and LTE4. Three of these (LTC4, LTD4, and LTE4) together make up what was formerly called **slow-reacting substance of anaphylaxis (SRS-A);** these mediators induce smooth muscle contraction. LTB4 is a potent chemoattractant of neutrophils. The leukotrienes are produced by a variety of cells, including monocytes, macrophages, and mast cells.

## Some cytokines are important inflammatory mediators

A number of cytokines play a significant role in the development of an acute or chronic inflammatory response. IL-1, IL-6, TNF-α, IL-12, and many chemokines exhibit redundant and pleiotropic effects that together contribute to the inflammatory response. Some of the effects mediated by IL-1, IL-6, and TNF-α are listed in Table 13-3. In addition,

| TABLE 13-3 | Redundant and pleiotropic effects of IL-1, TNF-α, and IL-6 | | |
|---|---|---|---|
| **Effect** | **IL-1** | **TNF-α** | **IL-6** |
| Pyrogenic (fever inducing) | + | + | + |
| Synthesis of acute-phase proteins by liver | + | + | + |
| Increased vascular permeability | + | + | + |
| Increased adhesion molecules on vascular endothelium | + | + | − |
| Fibroblast proliferation | + | + | − |
| Platelet production | + | − | + |
| Chemokine induction (e.g., IL-8) | + | + | − |
| Induction of IL-6 | + | + | − |
| T-cell activation | + | + | + |
| B-cell activation | + | + | + |
| Increased immunoglobulin synthesis | − | − | + |

IFN-γ contributes to the inflammatory response, acting later in the acute response and making an important contribution to chronic inflammation by attracting and activating macrophages. IL-12 induces the differentiation of the proinflammatory T$_H$1 subset. The role of several of these inflammatory cytokines in the development of acute and chronic inflammation will be described more fully in the next section.

# The Inflammatory Process

Inflammation is a physiologic response to a variety of stimuli such as infections and tissue injury. In general, an acute inflammatory response has a rapid onset and lasts a short while. Acute inflammation is generally accompanied by a systemic reaction known as the acute-phase response, which is characterized by a rapid change in the levels of several plasma proteins. In some diseases, persistent immune activation can result in chronic inflammation, which often has pathologic consequences.

## Neutrophils play an early and important role in inflammation

In the early stages of an inflammatory response, the predominant cell type infiltrating the tissue is the neutrophil. Neutrophil infiltration into the tissue peaks within the first 6 hours of an inflammatory response, with production of neutrophils in the bone marrow increasing to meet this need. A normal adult produces more than $10^{10}$ neutrophils per day, but during a period of acute inflammation, neutrophil production may increase as much as 10-fold.

The neutrophils leave the bone marrow and circulate within the blood. In response to mediators of acute inflammation, vascular endothelial cells increase their expression of E- and P-selectin. Thrombin and histamine induce increased expression of P-selectin; cytokines such as IL-1 or TNF-α induce increased expression of E-selectin. The circulating neutrophils express mucins such as PSGL-1 or the tetrasaccharides sialyl Lewis$^a$ and sialyl Lewis$^x$, which bind to E- and P-selectin.

As described earlier, this binding mediates the attachment or tethering of neutrophils to the vascular endothelium, allowing the cells to roll in the direction of the blood flow. During this time, chemokines such as IL-8 or other chemoattractants act on the neutrophils, triggering a G-protein–mediated activating signal that leads to a conformational change in the integrin adhesion molecules, resulting in neutrophil adhesion and subsequent transendothelial migration (see Figure 13-5).

Once in tissues, the activated neutrophils also express increased levels of receptors for chemoattractants and consequently migrate up a gradient of the chemoattractant. Among the inflammatory mediators that are chemotactic for neutrophils are several chemokines, complement split products (C3a, C5a, and C5b67), fibrinopeptides, prostaglandins, and leukotrienes, as well as some molecules released by microorganisms, such as formyl methionyl peptides. Activated neutrophils express increased levels of Fc receptors for antibody and receptors for complement, enabling these cells to bind more effectively to antibody- or complement-coated pathogens, thus increasing phagocytosis.

The activating signal also stimulates metabolic pathways to a respiratory burst, which produces **reactive oxygen intermediates** and **reactive nitrogen intermediates** (see Chapter 3). Release of some of these reactive intermediates and the release of mediators from neutrophil primary and secondary granules (proteases, phospholipases, elastases, and collagenases) play an important role in killing various pathogens. These substances also contribute to the tissue damage that can result from an inflammatory response. The accumulation of dead cells and microorganisms, together with accumulated fluid and various proteins, makes up what is known as pus.

## Inflammatory responses may be localized or systemic

The inflammatory response provides early protection following infection or tissue injury by restricting the tissue damage to the affected site. The acute inflammatory response involves both localized and systemic responses, summarized in Chapter 3.

### Localized Inflammatory Response

The hallmarks of a localized acute inflammatory response, first described almost 2000 years ago, are swelling *(tumor)*, redness *(rubor)*, heat *(calor)*, pain *(dolor)*, and loss of function. Within minutes after tissue injury, there is an increase in vascular diameter (vasodilation), resulting in an increase in the volume of blood in the area and a reduction in the flow of blood. The increased blood volume heats the tissue and causes it to redden. Vascular permeability also increases, leading to leakage of fluid from the blood vessels, particularly at postcapillary venules. This results in an accumulation of fluid (**edema**) in the tissue and, in some instances, extravasation of leukocytes, contributing to the swelling and redness in the area. When fluid exudes from the bloodstream, the kinin, clotting, and fibrinolytic systems are activated (see Figure 13-11). Many of the vascular changes that occur early in a local response are due to the direct effects of plasma enzyme mediators such as bradykinin and fibrinopeptides, which induce vasodilation and increased vascular permeability. Some of the vascular changes are due to the indirect effects of the complement anaphylatoxins (C3a and C5a), which induce local mast cell degranulation with release of histamine. Histamine is a potent mediator of inflammation, causing vasodilation and smooth muscle contraction. The prostaglandins can also contribute to the vasodilation and increased vascular permeability associated with the acute inflammatory response.

Within a few hours of the onset of these vascular changes, neutrophils adhere to the endothelial cells and migrate out of

## OVERVIEW FIGURE 13-13: Cells and Mediators Involved in a Local Acute Inflammatory Response

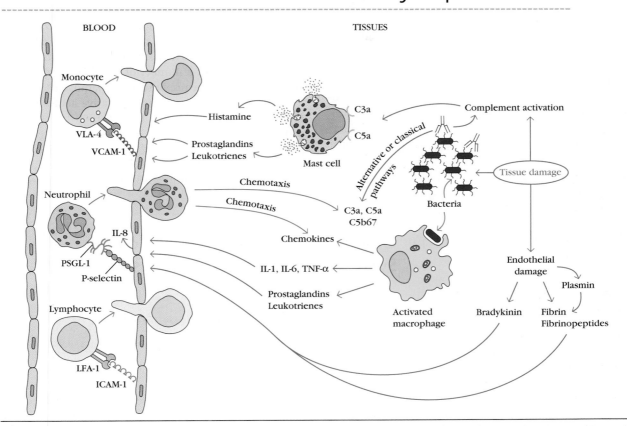

Tissue damage leads to the formation of complement products that act as opsonins, anaphylatoxins, and chemotactic agents. Bradykinin and fibrinopeptides induced by endothelial damage mediate vascular changes. Neutrophils generally are the first leukocytes to migrate into the tissue, followed by monocytes and lymphocytes. Only some of the interactions involved in the extravasation of leukocytes are depicted.

the blood into the tissue spaces (Figure 13-13). These neutrophils phagocytose invading pathogens and release mediators that contribute to the inflammatory response. Among the mediators are the macrophage inflammatory proteins (MIP-1α and MIP-1β), chemokines that attract macrophages to the site of inflammation. Macrophages arrive about 5 to 6 hours after an inflammatory response begins. These macrophages are activated cells that exhibit increased phagocytosis and increased release of mediators and cytokines that contribute to the inflammatory response.

Activated tissue macrophages secrete three cytokines (IL-1, IL-6, and TNF-α) that induce many of the localized and systemic changes observed in the acute inflammatory response (see Table 13-3). All three cytokines act locally, inducing coagulation and an increase in vascular permeability. Both TNF-α and IL-1 induce increased expression of adhesion molecules on vascular endothelial cells. For example, TNF-α stimulates expression of E-selectin, an endothelial adhesion molecule that selectively binds adhesion molecules on neutrophils. IL-1 induces increased expression of ICAM-1 and VCAM-1, which bind to integrins on lymphocytes and monocytes. Circulating neutrophils, monocytes, and lymphocytes recognize these adhesion molecules on the walls of blood vessels, adhere, and then move through the vessel wall into the tissue spaces. IL-1 and TNF-α also act on macrophages and endothelial cells to induce production of the chemokines that contribute to the influx of neutrophils by increasing their adhesion to vascular endothelial cells and by acting as potent chemotactic factors. In addition, IFN-γ and TNF-α activate macrophages and neutrophils, promoting increased phagocytic activity and increased release of lytic enzymes into the tissue spaces.

A local acute inflammatory response can occur without the overt involvement of the immune system. Often, however, cytokines released at the site of inflammation facilitate both the adherence of immune system cells to vascular endothelial cells and their migration through the vessel wall into the tissue spaces. The result is an influx of lymphocytes, neutrophils, monocytes, eosinophils, basophils, and mast cells to the site of tissue damage, where these cells participate in clearance of the antigen and healing of the tissue.

The duration and intensity of the local acute inflammatory response must be carefully regulated to control tissue damage and facilitate the tissue-repair mechanisms that are necessary for healing. TGF-β has been shown to play an important role in limiting the inflammatory response. It also promotes accumulation and proliferation of fibroblasts and the deposition of an extracellular matrix that is required for proper tissue repair.

Clearly, the processes of leukocyte adhesion are of great importance in the inflammatory response. A failure of proper leukocyte adhesion can result in disease, as exemplified by leukocyte adhesion deficiency (see Clinical Focus).

## Systemic Acute-Phase Response

The local inflammatory response is accompanied by a systemic response known as the **acute-phase response** (Figure 13-14). This response is marked by the induction of fever, increased synthesis of hormones such as ACTH and hydrocortisone, increased production of white blood cells (leukocytosis), and production of a large number of **acute-phase proteins** in the liver. The increase in body temperature inhibits the growth of a number of pathogens and appears to enhance the immune response to the pathogen.

C-reactive protein is a representative acute-phase protein whose serum level increases 1000-fold during an acute-phase response. It is composed of five identical polypeptides held together by noncovalent interactions. C-reactive protein binds to a wide variety of microorganisms and activates complement, resulting in deposition of

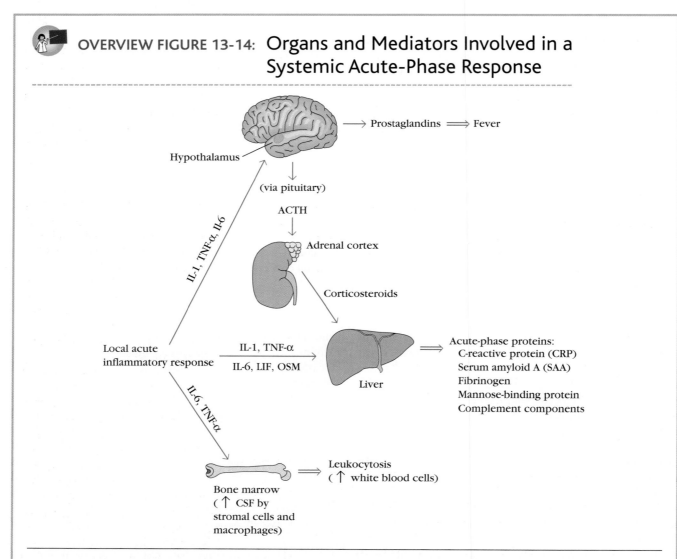

**OVERVIEW FIGURE 13-14:** Organs and Mediators Involved in a Systemic Acute-Phase Response

Prostaglandins ⟹ Fever

Hypothalamus

(via pituitary)

ACTH

IL-1, TNF-α, IL-6

Adrenal cortex

Corticosteroids

Local acute inflammatory response

IL-1, TNF-α
IL-6, LIF, OSM

Liver

Acute-phase proteins:
C-reactive protein (CRP)
Serum amyloid A (SAA)
Fibrinogen
Mannose-binding protein
Complement components

IL-6, TNF-α

Bone marrow
( ↑ CSF by stromal cells and macrophages)

Leukocytosis
( ↑ white blood cells)

IL-1, IL-6, and TNF-α, which are produced by activate macrophages at the site of inflammation, are particularly important in mediating acute-phase effects. LIF = leukemia inhibitory factor; OSM = oncostatin M.

## CLINICAL FOCUS

# Leukocyte Adhesion Deficiency (LAD) in Humans and Cattle

**The immune** system uses inflammation to assemble the components of an effective response and focus these resources at the site of infection. Inflammation is complex, involving vasodilation, increased vascular permeability, exudation of plasma proteins, and a gathering of inflammatory cells. Chemoattractants are key elements in calling leukocytes to sites of inflammation. These include chemokines such as IL-8, monocyte chemoattractant protein 1 (MCP-1), macrophage inflammatory protein 1 (MIP-1), and peptide fragments, such as C5a, generated during complement fixation.

Chemoattractants signal passing leukocytes to adhere tightly to the vascular surface, and, using adhesive interactions for traction, these cells push their way between endothelial cells and gain entry into the surrounding tissue. Once in, they are guided by gradients of chemoattractants to the sites of the inflammatory responses and become participants in the process. The key players in the adhesive interactions that are central to adhesion and extravasation are heterodimeric integrin molecules on the surface of the migrating leukocytes. There are a number of integrins, among which are LFA-1 (composed of CD11a and CD18); Mac-1, also called CR3 (components: CD11b and CD18); and p150/95 or CR4 (components: CD11c and CD18). When leukocytes encounter the appropriate chemokine or other chemoattractant, their complement of membrane integrin molecules undergoes a conformational change that transforms them from a slightly adhesive to a highly adhesive state.

In 1979, a paper titled "Delayed Separation of the Umbilical Cord, Widespread Infections, and Defective Neutrophil Mobility" appeared in *Lancet,* a British medical journal. This was the first in a series of reports that have appeared over the years describing patients afflicted with a rare autosomal recessive disease in which the first indication is often omphalitis, a swelling and reddening around the stalk of the umbilical cord. Although no more susceptible to virus infections than normal controls, those afflicted with this disorder suffer recurrent and often chronic bacterial infections, and sites where one would expect to find pus are instead pus free. This observation is particularly striking because the patients are not deficient in granulocytes; in fact, they typically have greatly elevated numbers of granulocytes in the circulation. Detailed immunological workups of these patients showed that Ig levels were in the normal range and that they had nearly normal B-, T-, and NK-cell function. However, examination of leukocyte migration in response to tissue damage revealed the root cause of the disease in these patients.

One method of evaluating leukocyte migration involves gently scraping the skin from a small area of the arm; the cell populations that move into the abraded area are then sampled by capturing some of those cells on a glass coverslip placed onto the wounded skin. A series of glass coverslips is sequentially placed, incubated, and removed over a period of several hours. Typically, each coverslip is left in place for two hours, and the procedure is repeated four times over an eight-hour period. Examination of the coverslips under a microscope reveals whether leukocytes have adhered to the coverslips. In normal individuals, the response of the immune system to tissue injury is to deliver leukocytes to the damaged area, and one finds these cells on the coverslips. However, in the patients described here, the coverslips were largely negative for leukocytes. Examination of white blood cells in these patients revealed an absence of CD18, an essential component of a number of integrins. A key element in the migration of leukocytes is integrin-mediated cell adhesion, and these patients suffer from an inability of their leukocytes to undergo adhesion-dependent migration into sites of inflammation. Hence, this syndrome has been named leukocyte adhesion deficiency (LAD).

Bacterial infections in these patients can be treated with antibiotics, but they recur. Furthermore, there are antibiotic-resistant strains of many pathogenic bacteria, and LAD patients must live under this microbial sword of Damocles, never knowing when the lifesaving thread of antibiotics will fail. If a suitable bone marrow donor can be found (almost always a close relative), however, there is a curative strategy. The LAD patient's hematopoietic system is destroyed, perhaps by treatment with cytotoxic chemicals, and then bone marrow transplantation is performed. If successful, this procedure provides the patient with leukocytes that have normal levels of functional integrin and display the full range of migratory capacities.

This disease is not limited to humans. A strikingly similar version known as bovine leukocyte adhesion disease (BLAD) occurs in cattle. The cause of BLAD in these animals is identical to the cause of LAD in human patients—the lack of a functional integrin subunit. What is different in some dairy herds is the incidence of the disease: though rare in humans, it can occur at economically important frequencies in cattle. This is a consequence of the high degree of inbreeding that exists in populations of dairy cattle. Typically, dairy herds are sired by the artificial insemination of semen from very few bulls. As a consequence of this practice, by the 1980s, almost one in 20 dairy bulls could be traced back to a single Holstein bull who happened to be heterozygous for BLAD. Such a high frequency of this recessive trait in the sire population dramatically raised the frequency of this disease in dairy herds. During the early 1990s, in some countries, the incidence of the BLAD gene was as high as 10% in a number of dairy herds. The gene for bovine CD18 has been cloned, which has allowed the design of a PCR-based assay for the aberrant forms of this gene. It is now possible to routinely screen sires and recipients for the BLAD allele. As a result, bulls that are carriers of the BLAD gene have been identified and eliminated from the breeding pool. This has led to a dramatic reduction in the frequency of new BLAD cases as well as in the overall frequency of the BLAD allele in dairy herd populations.

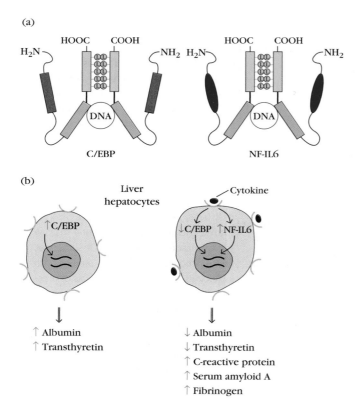

**FIGURE 13-15 Comparison of the structure and function of C/EBP and NF-IL6.** (a) Both transcription factors are dimeric proteins containing a leucine-zipper domain (orange) and a basic DNA-binding domain (blue). (b) C/EBP is expressed constitutively in liver hepatocytes and promotes transcription of albumin and transthyretin genes. During an inflammatory response, binding of IL-1, IL-6, TNF-α, LIF, or OSM to receptors on liver hepatocytes induces production of NF-IL6, which promotes transcription of the genes encoding various acute-phase proteins. Concurrently, C/EBP levels decrease and the levels of albumin and transthyretin consequently decrease.

the opsonin C3b on the surface of microorganisms. Phagocytic cells, which express C3b receptors, can then readily phagocytose the C3b-coated microorganisms.

Many systemic acute-phase effects are due to the combined action of IL-1, TNF-α, and IL-6 (see Figure 13-14). Each of these cytokines acts on the hypothalamus to induce a fever response. Within 12 to 24 hours of the onset of an acute-phase inflammatory response, increased levels of IL-1, TNF-α, and IL-6 (as well as leukemia inhibitory factor, LIF, and oncostatin M, OSM) induce production of acute-phase proteins by hepatocytes. TNF-α also acts on vascular endothelial cells and macrophages to induce secretion of colony-stimulating factors (M-CSF, G-CSF, and GM-CSF). These CSFs stimulate hematopoiesis, resulting in transient increases in the number of white blood cells needed to fight the infection.

The redundancy in the ability of at least five cytokines (TNF-α, IL-1, IL-6, LIF, and OSM) to induce production of acute-phase proteins by the liver results from the induction of a common transcription factor, NF-IL6, after each of these cytokines interacts with its receptor. Amino acid sequencing of cloned NF-IL6 revealed that it has a high degree of sequence identity with C/EBP, a liver-specific transcription factor (Figure 13-15a). Both NF-IL6 and C/EBP contain a leucine-zipper domain and a basic DNA-binding domain, and both proteins bind to the same nucleotide sequence in the promoter or enhancer of the genes encoding various liver proteins. C/EBP, which stimulates production of albumin and transthyretin, is expressed constitutively by hepatocytes. As an inflammatory response develops and the cytokines interact with their respective receptors on liver hepatocytes, expression of NF-IL6 increases and that of C/EBP decreases (Figure 13-15b). The inverse relationship between these two transcription factors accounts for the observation that serum levels of proteins such as albumin and transthyretin decline while those of acute-phase proteins increase during an inflammatory response.

## Chronic inflammation develops when antigen persists

Some microorganisms are able to evade clearance by the immune system, for example, by possessing cell wall components that enable them to resist phagocytosis. Such organisms often induce a chronic inflammatory response, resulting in significant tissue damage. Chronic inflammation also occurs in a number of autoimmune diseases in which self antigens continually activate T cells. Finally, chronic inflammation also contributes to the tissue damage and wasting associated with many types of cancer.

The accumulation and activation of macrophages is the hallmark of chronic inflammation. Cytokines released by the chronically activated macrophages also stimulate fibroblast proliferation and collagen production. A type of scar tissue develops at sites of chronic inflammation by a process called **fibrosis,** a wound-healing reaction that can interfere with normal tissue function. Chronic inflammation may also lead to formation of a **granuloma,** a tumor-like mass consisting of a central area of activated macrophages surrounded by activated lymphocytes. The center of the granuloma often contains multinucleated giant cells formed by the fusion of activated macrophages. These giant cells typically are surrounded by large modified macrophages that resemble epithelial cells and therefore are called epithelioid cells.

## Roles of IFN-γ and TNF-α in chronic inflammation

Two cytokines in particular, IFN-γ and TNF-α, play a central role in the development of chronic inflammation. $T_H1$ cells, NK cells, and $T_C$ cells release IFN-γ, while activated macrophages secrete TNF-α.

Members of the interferon family of glycoproteins (IFN-α and IFN-β) are released from virus-infected cells and confer

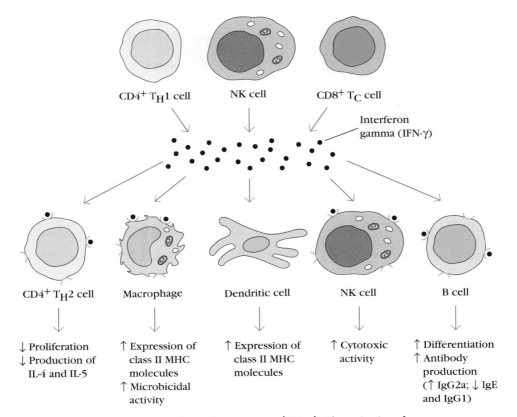

**FIGURE 13-16  Summary of pleiotropic activity of interferon gamma (IFN-γ).**  The activation of macrophages induced by IFN-γ plays a critical role in chronic inflammation. This cytokine is secreted by $T_H1$ cells, NK cells, and $T_C$ cells and acts on numerous cell types.  *[Adapted from Research News, 1993, Science **259**:1693.]*

antiviral protection on neighboring cells. Exactly which interferon is produced depends on the type of cell infected. IFN-α is produced by leukocytes; IFN-β, often called fibroblast interferon, is made largely by fibroblasts. IFN-γ is produced exclusively by T cells and NK cells. However, IFN-γ has a number of pleiotropic activities that distinguish it from IFN-α and IFN-β and contribute to the inflammatory response (Figure 13-16). One of the most striking effects of IFN-γ is its ability to activate macrophages. In their activated states, macrophages exhibit increased expression of class II MHC molecules, increased cytokine production, and increased microbicidal activity, making them more effective in antigen presentation and killing of intracellular microbial pathogens. In a chronic inflammatory response, however, the large numbers of activated macrophages release various hydrolytic enzymes and reactive oxygen and nitrogen intermediates, which are responsible for much of the damage to surrounding tissue.

One of the principal cytokines secreted by activated macrophages is TNF-α. The activity of this cytokine was first observed around the turn of the century by the surgeon William Coley. He noted that when cancer patients developed certain bacterial infections, the tumors would become necrotic. In the hope of finding a cure for cancer, Coley began to inject cancer patients with supernatants derived from various bacterial cultures. These culture supernatants,

called "Coley's toxins," did induce hemorrhagic necrosis in the tumors but had numerous undesirable side effects, making them unsuitable for cancer therapy. Decades later, the active component of Coley's toxin was shown to be a lipopolysaccharide (endotoxin) component of the bacterial cell wall. This endotoxin does not itself induce tumor necrosis but instead induces macrophages to produce TNF-α. This cytokine has a direct cytotoxic effect on some tumor cells but not on normal cells (Figure 13-17a). However, immunotherapeutic approaches using TNF-α for the treatment of cancer have been disappointing.

Several lines of evidence indicate that TNF-α also contributes to much of the tissue wasting that characterizes chronic inflammation. For example, mice carrying a TNF-α transgene become severely wasted (Figure 13-17b). In studies by A. Cerami and coworkers, rabbits were found to lose nearly half of their body mass within 2 months of being infected with trypanosomes. These workers subsequently discovered that a macrophage-derived factor was responsible for the profound wasting; they called the factor cachetin. Cloning of the genes for TNF-α and cachetin revealed that they were the same protein.

Activation of macrophages by IFN-α promotes increased transcription of the TNF-α gene and increases the stability of TNF-α mRNA. Both effects result in increased

(a)         Treated                    Untreated

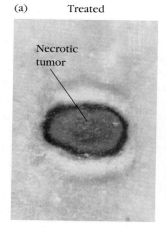
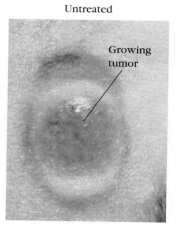

Necrotic tumor

Growing tumor

(b)

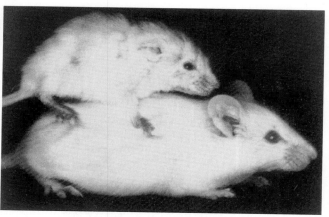

**FIGURE 13-17 Biological activities of TNF-α.** (a) A cancerous tumor in a mouse injected with endotoxin *(left)* shows hemorrhagic necrosis, unlike a tumor in an untreated mouse *(right)*. Endotoxin induces the production of TNF-α, which then acts to destroy the tumor.

(b) Transgenic mouse *(top)* bearing a TNF-α transgene becomes anorectic and severely wasted. Normal mouse is shown on the bottom. *[Part a from L. J. Old, 1988,* Scientific American ***258**:59; part b from B. Beutler, 1993,* Hospital Practice *(April 15):45.]*

TNF-α production. TNF-α acts synergistically with IFN-γ to initiate a chronic inflammatory response. Both cytokines together induce much greater increases in ICAM-1, E-selectin, and class I MHC molecules than either cytokine alone. The increase in intercellular adhesion molecules facilitates the recruitment of large numbers of cells in a chronic inflammatory response.

## HEV-like structures appear in chronic inflammatory disease

Recent studies suggest that regions of plump endothelial cells resembling HEVs appear along the vasculature in tertiary extralymphoid sites of chronic infection. These HEV-like regions, which appear to be sites of lymphocyte extravasation into the inflamed tissue, express several mucins (e.g., Gly-CAM-1, MAdCAM-1, and CD34) that are often displayed on normal HEVs. Several cytokines associated with chronic inflammation, notably IFN-γ and TNF-α, may play a role in the induction of HEV-like regions along the vasculature.

These HEV-like regions have been observed in a number of chronic inflammatory diseases in humans, including rheumatoid arthritis, Crohn's disease, ulcerative colitis, Graves' disease, Hashimoto's thyroiditis, and diabetes mellitus (Table 13-4). Development of this HEV-like vasculature is likely to facilitate a large-scale influx of leukocytes, contributing to chronic inflammation. These observations suggest that an effective approach for treating chronic inflammatory diseases may be to try to control the development of these HEV-like regions.

| TABLE 13-4 | Chronic inflammatory diseases associated with HEV-like vasculature | | |
|---|---|---|---|
| **Disease** | **Affected organ** | **Plump endothelium** | **Mucin-like CAMs on endothelium*** |
| Crohn's disease | Gut | + | + |
| Diabetes mellitus | Pancreas | + | + |
| Graves' disease | Thyroid | + | + |
| Hashimoto's thyroiditis | Thyroid | + | + |
| Rheumatoid arthritis | Synovium | + | + |
| Ulcerative colitis | Gut | + | + |

*Includes GlyCAM-1, MAdCAM-1, and CD34.

SOURCE: Adapted from J. P. Girard and T. A. Springer, 1995, *Immunology. Today* **16**:449.

# Anti-inflammatory Agents

Although development of an effective inflammatory response can play an important role in the body's defense, the response can sometimes be detrimental. Allergies, autoimmune diseases, microbial infections, transplants, and burns may initiate a chronic inflammatory response. Various therapeutic approaches are available for reducing long-term inflammatory responses and thus the complications associated with them.

## Antibody therapies reduce leukocyte extravasation

Because leukocyte extravasation is an integral part of the inflammatory response, one approach for reducing inflammation is to impede this process. Theoretically, leukocyte extravasation can be reduced by blocking the activity of various adhesion molecules with antibodies. In animal models, for example, antibodies to the integrin LFA-1 have been used to reduce neutrophil buildup in inflammatory tissue. Antibodies to ICAM-1 have also been used, with some success, in preventing the tissue necrosis associated with burns and in reducing the likelihood of kidney graft rejection in animal models. The results with antibodies specific for these adhesins have been so encouraging that a combination of antibodies (anti-ICAM-1 and anti-LFA-1) was used in clinical trials on human kidney transplant patients. A combination of two anti-adhesins had to be used because failure to block both LFA-1 and ICAM-1 results in rejection. Clinical trials have shown clear improvement in patients with relapsing multiple sclerosis, Crohn's disease, and rheumatoid arthritis when treated with antibodies that block the integrin α4 subunit of VLA-4 (CD49d), and severe psoriasis can be moderated when treated with antibodies that block the integrin αL subunit of LFA-1 (CD11a). However, the roles of these integrins in the immune system extend beyond leukocyte extravasation, and concerns have been raised about the risk versus the benefit when their activities are blocked. Engagement of VLA-4 (CD49d/CD29) by some of these antibodies results in the formation of a $T_H1$-type response, which would not be a favorable outcome in a disease already dominated by $T_H1$-type cells. In another trial, three patients receiving anti-α4 antibodies for multiple sclerosis or Crohn's disease developed virally linked leukoencephalopathy, possibly because decreased migration of immune cells into sites of the nervous system allowed the virus to spread. Before the blockade of adhesion molecules can achieve broader usage, a deeper understanding must be acquired of the side effects related to interfering with adhesion molecule function.

## Corticosteroids are powerful anti-inflammatory drugs

The corticosteroids, which are cholesterol derivatives, include prednisone, prednisolone, and methylprednisolone. These potent anti-inflammatory agents exert various effects that result in a reduction in the numbers and activity of immune system cells. They are regularly used in anti-inflammatory therapy.

Corticosteroid treatment causes a decrease in the number of circulating lymphocytes as a result either of steroid-induced lysis of lymphocytes (lympholysis) or of alterations in lymphocyte circulation patterns. Some species (e.g., hamster, mouse, rat, and rabbit) are particularly sensitive to corticosteroid-induced lympholysis. In these animals, corticosteroid treatment at dosages as low as $10^{-7}$ M causes such widespread lympholysis that the weight of the thymus is reduced by 90%; the spleen and lymph nodes also shrink visibly. Immature thymocytes in these species appear to be particularly sensitive to corticosteroid-mediated killing. In rodents, corticosteroids induce programmed cell death of immature thymocytes, whereas mature thymocytes are resistant to this activity. Within 2 hours following in vitro incubation with corticosteroids, immature thymocytes begin to show the characteristic morphology of apoptosis, and 90% of the chromatin is degraded into the characteristic nucleosome ladder by 24 hours after treatment. The steps involved in the induction of apoptosis by corticosteroids remain to be deter-

mined. In humans, guinea pigs, and monkeys, corticosteroids do not induce apoptosis but instead affect lymphocyte circulation patterns, causing a decrease in thymic weight and a marked decrease in the number of circulating lymphocytes.

Like other steroid hormones, the corticosteroids are lipophilic and thus can cross the plasma membrane and bind to receptors in the cytosol. The resulting receptor-hormone complexes are transported to the nucleus, where they bind to specific regulatory DNA sequences, regulating transcription up or down. The corticosteroids have been shown to induce increased transcription of the NF-κB inhibitor (IκB). Binding of this inhibitor to NF-κB in the cytosol prevents the translocation of NF-κB into the nucleus and consequently prevents NF-κB activation of a number of genes, including genes involved in T-cell activation and cytokine production.

Corticosteroids also reduce both the phagocytic and the killing ability of macrophages and neutrophils, and this effect may contribute to their anti-inflammatory action. In addition, they reduce chemotaxis, so that fewer inflammatory cells are attracted to the site of $T_H$-cell activation. In the presence of corticosteroids, expression of class II MHC molecules and IL-1 production by macrophages is dramatically reduced; such reductions would be expected to lead to corresponding reductions in $T_H$-cell activation. Finally, corticosteroids also stabilize the lysosomal membranes of participating leukocytes, so that decreased levels of lysosomal enzymes are released at the site of inflammation.

## NSAIDs combat pain and inflammation

Since the time of Hippocrates, extracts of willow bark have been used for relief of pain. The active ingredient, salicylate, which is found in aspirin, is just one of many nonsteroidal anti-inflammatory drugs (NSAIDs). NSAIDs are the most frequently used medication for treating pain and inflammation. Clinically, NSAIDs have been shown to be effective for treatment of many acute and chronic inflammatory reactions. The major mechanism by which these drugs exert anti-inflammatory effects is by inhibiting the cyclooxygenase pathway that produces prostaglandins and thromboxanes from arachidonic acid. The reduction in prostaglandin production limits the increase in vascular permeability and neutrophil chemotaxis in the inflammatory response. As shown in Figure 13-18, the cyclooxygenase pathway is mediated by two enzymes, cyclooxygenase 1 and cyclooxygenase 2 (Cox-1 and Cox-2).

Although NSAIDs such as aspirin, Tylenol, ibuprofen, Naproxen, and others are routinely prescribed for the treatment of ailments as diverse as arthritis, sprains, tissue injury, and back pain, the duration of their use is limited by gastrointestinal side effects that include unease and abdominal pain and in more serious cases bleeding or perforation of the stomach or upper GI tract. Investigation of the mechanism of NSAIDs has provided a basis for the beneficial and deleterious effects of many NSAIDs. Studies have shown that although most NSAIDs inhibit both Cox-1 and Cox-2, it is the inhibition of Cox-2 that is responsible for their anti-inflammatory effects. On the other hand, inhibition of Cox-1 by these agents

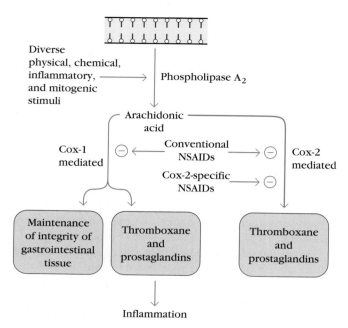

**FIGURE 13-18  Inhibition of cyclooxygenase 1 and 2 by NSAIDs.** A variety of agents trigger the release of arachidonic acid from the cell membrane by the action of phospholipase $A_2$. The subsequent action of Cox-1 and Cox-2 initiates the conversion of arachidonic acid to a variety of lipid mediators of inflammation and many other processes. Many NSAIDs inhibit both enzymatic pathways, but those with greater specificity for the Cox-2 arm produce anti-inflammatory effects with fewer side effects. *[Adapted from G. A. FitzGerald and C. N. Patrono, 2001, New England Journal of Medicine **345**:433.]*

causes damage to the GI tract but does not have significant anti-inflammatory benefits. This realization led to the design and development of a new generation of NSAIDs that specifically inhibit Cox-2 but have little effect on Cox-1 activity. The action of these highly targeted drugs is shown in Figure 13-18.

Although these new drugs were originally targeted to patients at risk for gastrointestinal side effects, usage spread much more widely. Subsequently, clinical data on patients using the newer Cox-2 specific inhibitors have shown an unacceptable increase in cardiovascular side effects, including heart attacks and strokes. The Cox-2 inhibitor Vioxx, targeted for those prone to gastrointestinal side effects, was voluntarily removed from the market in 2004 by Merck due to the concern over increased cardiovascular events. Currently Cox-2-specific inhibitors are recommended for use only as a second- or third-line agent.

## SUMMARY

- Lymphocytes undergo constant recirculation between the blood, lymph, lymphoid organs, and tertiary extralymphoid tissues, increasing the chances that the small number of lymphocytes specific for a given antigen (about 1 in $10^5$ cells) will actually encounter that antigen.

- Migration of leukocytes into inflamed tissue or into lymphoid organs requires interaction between cell adhesion molecules (CAMs) on the vascular endothelium and those on the circulating cells.

- Most CAMs fall into one of four protein families: the selectins, the mucin-like family, the integrins, or the Ig superfamily. Selectins and mucin-like CAMs interact with each other, and members of each family are expressed on both leukocytes and endothelial cells. Integrins, expressed on leukocytes, interact with Ig-superfamily CAMs, expressed on endothelial cells.

- Extravasation of both neutrophils and lymphocytes involves four steps: rolling, activation, arrest/adhesion, and transendothelial migration. Neutrophils are generally the first cell type to move from the bloodstream into inflammatory sites.

- Unlike neutrophils, various lymphocyte populations exhibit differential extravasation into various tissues. Homing receptors on lymphocytes interact with tissue-specific adhesion molecules, called vascular addressins, on high-endothelial venules (HEVs) in lymphoid organs and on the endothelium in tertiary extralymphoid tissues.

- Naive lymphocytes home to secondary lymphoid organs, extravasating across HEVs, whereas effector lymphocytes selectively home to inflamed vascular endothelium.

- Inflammation is a physiologic response to a variety of stimuli such as tissue injury and infection. An acute inflammatory response involves both localized and systemic effects. The localized response begins when tissue and endothelial damage induces formation of plasma enzyme mediators that lead to vasodilation and increased vascular permeability.

- Several types of mediators play a role in the inflammatory response. Chemokines act as chemoattractants and activating molecules during leukocyte extravasation. Plasma enzyme mediators include bradykinin and fibrinopeptides, which increase vascular permeability; plasmin is a proteolytic enzyme that degrades fibrin clots into chemotactic products and activates complement; and various complement products act as anaphylatoxins, opsonins, and chemotactic molecules for neutrophils and monocytes. Lipid inflammatory mediators include thromboxanes, prostaglandins, leukotrienes, and platelet-activating factor. Three cytokines, IL-1, IL-6, and TNF-$\alpha$, mediate many of the local and systemic features of the acute inflammatory response

- Activation of tissue macrophages and degranulation of mast cells lead to release of numerous inflammatory mediators, some of which induce the acute-phase response, which includes fever, leukocytosis, and production of corticosteroids and acute-phase proteins.

- A chronic inflammatory response may accompany allergies, autoimmune diseases, microbial infections, transplants, and burns. Drug-based therapies employing corticosteroids and a variety of nonsteroidal anti-inflammatory drugs (NSAIDs) are the most commonly used medications for pain and inflammation.

## References

Butcher, E., and L. J. Picker. 1996. Lymphocyte homing and homeostasis. *Science* **272**:60.

Cyster, J. G. 2005. Chemokines, sphingosine-1-phosphate, and cell migration in secondary lymphoid organs. *Annual Review of Immunology* **23**:127.

FitzGerald, G. A., and C. Patrono. 2002. The coxibs, selective inhibitors of cyclooxygenase-2. *New England Journal of Medicine* **345**:433.

Gabay, C., and I. Kushner. 1999. Acute-phase proteins and other systemic responses to inflammation. *New England Journal of Medicine* **340**:448.

González-Amaro, R., M. Mittelbrunn, and F. Sánchez-Madrid. 2005. Therapeutic anti-integrin ($\alpha$4 and $\alpha$L) monoclonal antibodies: two-edged swords? *Immunology* **116**:289.

Imhof, B. A., and M. Aurrand-Lions. 2004. Adhesion mechanisms regulating the migration of monocytes. *Nature Reviews Immunology* **4**:432.

Kim, C. H. 2005. The greater chemotactic network for lymphocyte trafficking: chemokines and beyond. *Current Opinion in Hematology* **12**:298.

Kuijpers, T. W., et al. 1997. Leukocyte adhesion deficiency type 1 (LAD-1)/variant: a novel immunodeficiency syndrome characterized by dysfunctional beta2 integrins. *Journal of Clinical Investigation* **100**:1725.

Kunkel, E. J., and E. C. Butcher. 2002. Chemokines and the tissue-specific migration of lymphocytes. *Immunity* **16**:1.

Pribila, J. T., A. C. Quale, K. L. Mueller, and Y. Shimizu. 2004. Integrins and T cell-mediated immunity. *Annual Review of Immunology* **22**:157.

Rosen, S. D. 2004. Ligands for L-selectin: homing, inflammation, and beyond. *Annual Review of Immunology* **22**:129.

Shuster, D. E., et al. 1992. Identification and prevalence of a genetic defect that causes leukocyte adhesion deficiency in Holstein cattle. *Proceedings of the National Academy of Sciences U.S.A.* **89**:9225.

Springer, T. A. 1994. Traffic signals for lymphocyte recirculation and leukocyte emigration: the multistep paradigm. *Cell* **76**:301.

Steel, D. M., and A. S. Whitehead. 1994. The major acute phase reactants: C-reactive protein, serum amyloid P component and serum amyloid A protein. *Immunology Today* **15**:81.

Stein, J. V., and C. Nombela-Arrieta. 2005. Chemokine control of lymphocyte trafficking: a general review. *Immunology* **116**:1.

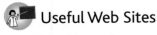

## Useful Web Sites

**http://cytokine.rndsystems.com**

The Cytokine Mini-Reviews Section of the R&D Systems site contains extensive, detailed, and well-illustrated reviews of many chemokines and chemokine receptors.

**http://www.ncbi.nlm.nih.gov/Omim/**

Online Mendelian Inheritance in Man is a catalog of human genes and genetic disorders. It contains pictures of and references to many diseases, including LAD.

 Study Questions

**CLINICAL FOCUS QUESTION** Why does a defect in CD18 result in an increased vulnerability to bacterial infection? Please address, as precisely as you can, the cell biology of cell migration.

1. Indicate whether each of the following statements is true or false. If you think a statement is false, explain why.

   a. Chemokines are chemoattractants for lymphocytes but not other leukocytes.
   b. $\beta$2-(CD18)-containing integrins are expressed on both leukocytes and endothelial cells.
   c. Leukocyte extravasation involves multiple interactions between cell adhesion molecules.
   d. Most secondary lymphoid organs contain high-endothelial venules (HEVs).
   e. Mucin-like CAMs interact with selectins.
   f. An acute inflammatory response involves only localized effects in the region of tissue injury or infection.
   g. MAdCAM-1 is an endothelial adhesion molecule that binds to L-selectin and to several integrins.
   h. Granuloma formation is a common symptom of local inflammation.

2. Various inflammatory mediators induce expression of ICAMs on a wide variety of tissues. What effect might this induction have on the localization of immune cells?

3. Extravasation of neutrophils and of lymphocytes occurs by generally similar mechanisms, although some differences distinguish the two processes.

   a. List in order the four basic steps in leukocyte extravasation.
   b. At which sites are neutrophils most likely to extravasate? Why?
   c. Different lymphocyte subpopulations migrate preferentially into different tissues, a process called homing (or trafficking). Discuss the roles of the three types of molecules that permit homing of lymphocytes.

4. Which three cytokines secreted by activated macrophages play a major role in mediating the localized and systemic effects associated with an acute inflammatory response?

5. An effective inflammatory response requires differentiation and proliferation of various nonlymphoid white blood cells. Explain how hematopoiesis in the bone marrow is induced by tissue injury or local infection.

6. For each pair of molecules listed below, indicate whether the molecules interact during the first, second, third, or fourth step in neutrophil extravasation at an inflammatory site. Use N to indicate any molecules that do not interact.

   a. _____ Chemokine and L-selectin
   b. _____ E-selectin and mucin-like CAM
   c. _____ IL-8 and E-selectin
   d. _____ Ig-superfamily CAM and integrin

e. _____ ICAM and chemokine
f. _____ Chemokine and G-protein–coupled receptor
g. _____ ICAM and integrin

7. Discuss the main effects of IFN-γ and TNF-α during a chronic inflammatory response.

8. Five cytokines (IL-1, IL-6, TNF-α, LIF, and OSM) induce production of C-reactive protein and other acute-phase proteins by hepatocytes. Briefly explain how these different cytokines can exert the same effect on hepatocytes.

9. For each inflammation-related term (a–h), select the descriptions listed below (1–11) that apply. Each description may be used once, more than once, or not at all; more than one description may apply to some terms.

Terms

a. _____ Tertiary extralymphoid tissue
b. _____ P- and E-selectin
c. _____ Prostaglandins
d. _____ Nonsteroidal anti-inflammatory drugs
e. _____ ICAM-1, -2, -3
f. _____ MAdCAM
g. _____ Bradykinin
h. _____ Inflamed endothelium

Descriptions

(1) Bind to sialylated carbohydrate moieties
(2) Inhibit cyclooxygenase pathway
(3) Induce expression of NF-κB inhibitor
(4) Has both Ig domains and mucin-like domains
(5) Region of vascular endothelium found in postcapillary venules
(6) Expressed by inflamed endothelium
(7) Exhibits HEV-like vasculature in chronic inflammation
(8) Belong to Ig superfamily of CAMs
(9) Exhibits increased expression of CAMs
(10) Increase vascular permeability and induce fever
(11) Induce fever

10. Predict the functional consequence(s) for the immune system in knockout mice lacking the following adhesion molecules.

a. MAdCAM-1
b. L-selectin
c. The β2 integrin subunit

11. Chemotaxis is one way that immune cells are targeted to specific areas. Which of the following are correct matches for a chemoattractant/chemokine and a cell type?

a. CXC chemokines, such as IL-8, attract neutrophils.
b. CC chemokines, such as MIP-1α, attract monocytes.
c. Complement component C7 attracts eosinophils.
d. Complement component C5a attracts monocytes and neutrophils.

12. Leukocyte adhesion deficiency type 1 is characterized by a mutation in a protein needed for neutrophils to leave the bloodstream (extravasation) to fight infections. Patients with this condition usually don't live past childhood because they cannot fight off bacterial infections. What approaches might be taken to treat this disease?

13. Why do neutrophils arrive at a site of infection before monocytes when both circulate within the bloodstream?

14. Mechanisms to heal tissue damage, such as the kinin system and the clotting system, increase the inflammatory response. Describe the interplay between these systems. What is the benefit of such interactions?

**ANALYZE THE DATA** Collin and colleagues. (2004, *J. Leukoc. Biol.* **76**:961) investigated the role of 5′-lipoxygenase (5-LOX) in the induction of organ failure by treating knock-out mice (5-$LOX^{-/-}$) with lipopolysaccharide. Answer the following questions based on the data in the figure below and what you have learned from reading this book.

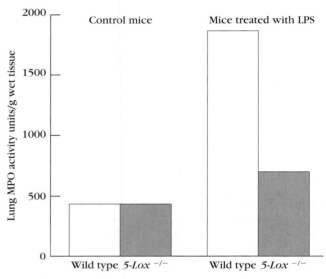

Myeloperoxidase (MPO) activity in lung tissue in wild-type and 5-$LOX^{-/-}$ mice treated with lipopolysaccharide (LPS).

a. The investigators used LPS as a model for what kind of infection?

b. Myeloperoxidase is an enzyme expressed by neutrophils and used to assay for the presence of neutrophils in tissues. The investigators measured MPO to quantify the recruitment of inflammatory cells into the lung. Does 5′-lipoxygenase increase or diminish inflammation? Explain your answer.

c. What chemotactic factor was most likely affected in the 5-$LOX^{-/-}$ knockout.

d. Do the data in the figure suggest that other recruitment mechanisms were active in this system?

 ## Interactive Study

www.whfreeman.com/kuby

SELF-TEST
Review of Key Terms

ANIMATION
Leukocyte Extravasation

# chapter 14

# Cell-Mediated Cytotoxic Responses

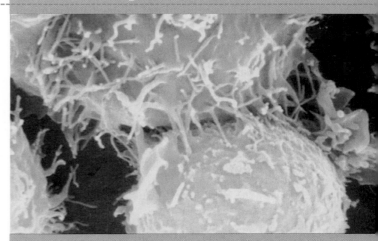

*Big CTL attacks little tumor cell. [From J. D. E. Young and Z. A. Cohn, 1988. Scientific American 258(1):38.]*

- Effector Responses
- General Properties of Effector T Cells
- Cytotoxic T Cells
- Natural Killer Cells
- NKT Cells
- Antibody-Dependent Cell-Mediated Cytotoxicity
- Experimental Assessment of Cell-Mediated Cytotoxicity

T HE CELL-MEDIATED AND HUMORAL BRANCHES OF the immune system assume different (though somewhat redundant) roles in protecting the host. The effectors of the humoral branch are antibodies, highly specific molecules on the surfaces of cells and in the extracellular spaces that can bind and neutralize antigens. The primary domain of antibody protection lies outside the cell. If antibodies were the only agents of immunity, pathogens that managed to evade them and colonize the intracellular environment would escape the immune system. This is not the case. A principal role of cell-mediated immunity is to detect and eliminate cells that harbor intracellular pathogens. Cell-mediated immunity also can recognize and eliminate cells, such as tumor cells, that have undergone genetic modifications, which often lead to the expression of antigens not typical of normal cells.

Both antigen-specific and -nonspecific cells can contribute to the cell-mediated immune response. Antigen-specific cells include CD8$^+$ cytotoxic T lymphocytes (T$_C$ cells or CTLs) and cytokine-secreting CD4$^+$ T$_H$ cells that mediate delayed-type hypersensitivity (DTH). The discussion of DTH reactions and the role of CD4$^+$ T cells in their orchestration appears in Chapter 15. Nonspecific cells include NK cells and nonlymphoid cell types such as macrophages, neutrophils, and eosinophils. Recent attention has focused on a previously unrecognized cell type with characteristics of both antigen-specific T cells and the nonspecific NK cells; this hybrid has been named the NKT cell. Although much remains to be learned about NKT cells, they are implicated in both antibacterial and antitumor immunity.

The activities of both specific and nonspecific cytotoxic components of immunity depend on effective local concentrations of various cytokines. T cells, NK cells, dendritic cells, and macrophages are the most important sources of the cytokines that organize and support cell-mediated toxicity. Activation of cells to carry out killer functions requires cooperation of different cell types. Finally, although humoral and cell-mediated immunity have many distinctive features, they are not completely independent. Cells such as macrophages, NK cells, neutrophils, and eosinophils can use antibodies as receptors to recognize and target cells for

killing. Also, chemotactic peptides generated by the activation of complement in response to antigen-antibody complexes can contribute to assembling the cell types required for a cell-mediated response.

In the preceding chapters, various aspects of the humoral and cell-mediated effector responses have been described. This chapter emphasizes the cytotoxic effector mechanisms mediated by T$_C$ cells, NK cells, and NKT cells, antibody-dependent cell-mediated cytotoxicity (ADCC), and the experimental assay of cytotoxicity.

## Effector Responses

The importance of cell-mediated immunity becomes evident when the system is defective. Children with DiGeorge syndrome, who are born without a thymus and therefore lack the T-cell component of the cell-mediated immune system, generally are able to cope with extracellular bacterial infections, but they cannot effectively eliminate intracellular pathogens. Their lack of functional cell-mediated immunity results in repeated infections with viruses, intracellular bacteria, and fungi. The severity of the cell-mediated immunodeficiency in these children is such that even the attenuated virus present in a vaccine, capable of only limited growth in normal individuals, can produce life-threatening infections.

Cell-mediated immune responses can be divided into two major categories according to the different effector populations that are mobilized. One group comprises effector cells that have direct cytotoxic activity. These effectors eliminate foreign cells and altered self cells, including virus-infected and tumor cells, by mounting a cytotoxic reaction that lyses their target. The various cytotoxic effector cells can be grouped into two general categories: one comprises antigen-specific cytotoxic T lymphocytes (CTLs) and nonspecific cells, such as natural killer (NK) cells and macrophages. The target cells to which these effectors are directed include allogeneic cells, malignant cells, virus-infected cells, and cells that have been stressed by heat or trauma. The other group is a subpopulation of effector CD4$^+$ T cells that mediates delayed-type hypersensitivity reactions (see Chapter 15). The next section reviews the general properties of effector T cells and how they differ from naive T cells.

## General Properties of Effector T Cells

The three types of effector T cells—CD4$^+$ T$_H$1 and T$_H$2 cells, and CD8$^+$ CTLs—exhibit several properties that set them apart from naive helper and cytotoxic T cells (Table 14-1). In particular, effector cells are characterized by their less stringent activation requirements, increased expression of cell adhesion molecules, and production of both membrane-bound and soluble effector molecules.

| TABLE 14-1 | Comparison of naive and effector T cells | |
|---|---|---|
| **Property** | **Naive T cells** | **Effector T cells** |
| Costimulatory signal (CD28-B7 interaction) | Required for activation | Not required for activation |
| CD45 isoform | CD45RA | CD45RO |
| Cell adhesion molecules (CD2 and LFA-1) | Low | High |
| Trafficking patterns | HEVs* in secondary lymphoid tissue | Tertiary lymphoid tissues; inflammatory sites |

*HEV = high-endothelial venules, sites in blood vessel used by lymphocytes for extravasation

## The activation requirements of T cells differ

T cells at different stages of differentiation may respond with different efficiencies to signals mediated by the T-cell receptor and may consequently require different levels of a second set of costimulatory signals. As described in Chapter 10, activation of naive T cells and their subsequent proliferation and differentiation into effector T cells require both a primary signal, delivered when the TCR complex and CD4 or CD8 coreceptor interact with a foreign peptide–MHC molecule complex, and a costimulatory signal, delivered by interaction between particular membrane molecules on the T cell and the antigen-presenting cell. In contrast, antigen-experienced effector cells and memory cells (as opposed to naive T cells) are able to respond to TCR-mediated signals with little if any costimulation.

The mechanisms underlying the different activation requirements of naive and activated T cells is an area of continuing research, but some clues have been found. One clue is that many populations of naive and effector T cells express different isoforms of CD45, designated CD45RA and CD45RO, which are produced by alternative splicing of the RNA transcript of the CD45 gene. Both of these membrane molecules mediate TCR signal transduction by catalyzing dephosphorylation of a tyrosine residue on the protein tyrosine kinases Lck and Fyn, activating these kinases and triggering the subsequent steps in T-cell activation (see Figures 10-10 and 10-11). Effector T cells express the CD45RO isoform, which associates much better with the TCR complex and its coreceptors, CD4 and CD8, than the CD45RA isoform expressed by naive T cells. Memory T cells have both isoforms, but CD45RO is predominant. As a result, effector and memory T cells are more sensitive to TCR-mediated activation by a peptide-MHC complex. They also have less stringent requirements for costimulatory signals and therefore are able to respond to peptide-MHC complexes displayed on target cells or antigen-presenting cells that lack the costimulatory B7 molecules.

## Cell adhesion molecules facilitate TCR-mediated interactions

CD2 and the integrin LFA-1 are cell adhesion molecules on the surfaces of T cells that bind, respectively, to LFA-3 and ICAMs (*inter*cellular *a*dhesion *m*olecules) on antigen-presenting cells and various target cells (see Figure 9-13). The levels of CD2 and LFA-1 are twofold to fourfold higher on effector T cells than on naive T cells, enabling the effector T cells to bind more effectively to antigen-presenting cells and to various target cells that express low levels of ICAMs or LFA-3.

As Chapter 9 showed, the initial interaction of an effector T cell with an antigen-presenting cell or target cell is relatively weak, allowing the TCR to scan the membrane for specific peptides presented by self-MHC molecules. If no peptide-MHC complex is recognized by the effector cell, it will disengage from the APC or target cell. Recognition of a peptide-MHC complex by the TCR, however, produces a signal that increases the affinity of LFA-1 for ICAMs on the

| TABLE 14-2 | Effector molecules produced by effector T cells | |
| --- | --- | --- |
| Cell type | Soluble effectors | Membrane-bound effectors |
| CTL | Cytotoxins (perforins and granzymes), IFN-γ, TNF-β | Fas ligand (FASL) |
| T$_H$1 | IL-2, IL-3, TNF-β, IFN-γ, GM-CSF (high) | Tumor necrosis factor β (TNF-β) |
| T$_H$2 | IL-3, IL-4, IL-5, IL-6, IL-10, IL-13, GM-CSF (low) | CD40 ligand |

APC or target-cell membrane, prolonging the interaction between the cells. For example, T$_H$1 effector cells remain bound to macrophages that display peptide–class II MHC complexes, T$_H$2 effector cells remain bound to B cells that display peptide–class II MHC complexes, and CTL effector cells bind tightly to virus-infected target cells that display peptide–class I MHC complexes.

## Effector T cells express a variety of effector molecules

Effector T cells express certain effector molecules, both membrane bound and soluble, that are not expressed by naive T cells (Table 14-2). The membrane-bound molecules belong to the tumor necrosis factor (TNF) family of membrane proteins and include the Fas ligand (FASL) on CD8$^+$ CTLs, TNF-β on T$_H$1 cells, and the CD40 ligand (CD40L or CD154) on T$_H$2 cells. Each of the effector T-cell populations also secretes distinct panels of soluble effector molecules. CTLs secrete cytotoxins (perforins and granzymes) as well as two cytokines, IFN-γ and TNF-β. As described in Chapter 12, the T$_H$1 and T$_H$2 subsets secrete largely nonoverlapping sets of cytokines.

Each of these membrane-bound and secreted molecules plays an important role in various T-cell effector functions. The Fas ligand, perforins, and granzymes, for example, mediate target-cell destruction by the CTL; membrane-bound TNF-β and soluble IFN-γ and GM-CSF promote macrophage activation by the T$_H$1 cell; and the membrane-bound CD40L and soluble IL-4, IL-5, and IL-6 all play a role in B-cell activation by the T$_H$2 cell.

## Cytotoxic T Cells

Cytotoxic T lymphocytes, or CTLs, are generated by immune activation of T cytotoxic (T$_C$) cells. These effector cells have lytic capability and are critical in the recognition and elimination of altered self cells (e.g., virus-infected cells and tumor cells) and genetically different cells in graft rejection reactions. In general, CTLs are CD8$^+$ and are therefore class I MHC restricted. Since virtually all nucleated cells in the body express class I MHC molecules, a CTL can recognize and eliminate almost any cell in the body that displays the specific antigen recognized by that CTL in the context of a class I MHC molecule.

The CTL-mediated immune response can be divided into two phases, reflecting different aspects of the response. In the first phase, naive T$_C$ cells undergo activation and differentiation into functional effector CTLs. In the second phase, effector CTLs recognize antigen–class I MHC complexes on specific target cells, which leads them to destroy the target cells.

## Effector CTLs are generated from CTL precursors

Naive T$_C$ cells are incapable of killing target cells and are therefore referred to as CTL precursors (CTL-Ps) to denote their functionally immature state. Only after a CTL-P has been activated will the cell differentiate into a functional CTL with cytotoxic activity. The threshold to activate CTLs from CTL-Ps is high compared to that of other effector T cells and appears to require at least three sequential signals (Figure 14-1):

- An antigen-specific signal transmitted by the TCR complex upon recognition of a peptide–class I MHC molecule complex on a "licensed" APC (*licensing* is explained below)

- A costimulatory signal transmitted by the CD28-B7 interaction of the CTL-P and the licensed APC

- A signal induced by the interaction of IL-2 with the high-affinity IL-2 receptor, resulting in proliferation and differentiation of the antigen-activated CTL-P into effector CTLs

Clearly activation of a CTL-P results from its recognition of antigen on an APC presented in the context of a class I MHC molecule, but what other contributing factors are required is not known. Convincing evidence is emerging that the APC must first acquire the ability to activate CTL-Ps via prior interactions in a process called *licensing* (Figure 14-1). Licensing may take place through interaction between the APC (most likely a dendritic cell) and a T$_H$1 cell via antigen processed in the context of a class II MHC molecule. Licensing also requires a costimulatory interaction between CD40 on the dendritic cell and CD40L on the T$_H$1 cell. This was demonstrated by the fact that an anti-CD40 monoclonal antibody has the same effect on the T$_H$1 cell as CD40L in experimental situations where the activation of CTL-P by dendritic cells was measured. Licensing may also occur in some cases through interaction of a Toll-like receptor (TLR) on the APC with a microbial product. The precise role of T$_H$1 cells in the generation of CTLs from naive CTL-Ps is not completely defined, and it is unlikely that a T$_H$1 cell and CTL-P interact directly. However, IL-2 and costimulation are important in the transformation of naive CTL-Ps into effector cells, and recent data suggest that T$_H$1 cells are essential in the provision of these requirements.

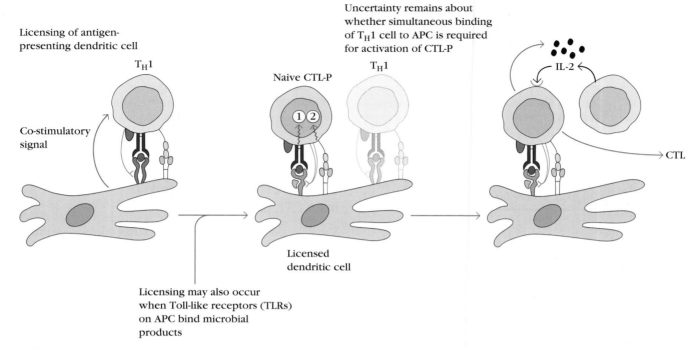

**FIGURE 14-1 Generation of effector CTLs.** The activation or licensing of a dendritic cell to give it the ability to activate a CTL-P may require prior or simultaneous interaction with a CD4$^+$ T$_H$1 cell via a class II–presented antigen and costimulation by CD40-CD40L. On interaction with antigen–class I MHC complexes and costimulator molecules on an appropriately licensed APC, CTL-Ps begin to express IL-2 receptors (IL-2R) and lesser amounts of IL-2. The CTL-Ps then become CTLs with the ability to kill target cells.

Once licensed, the dendritic cell interacts with the CTL-P via class I–presented antigen to effect its activation to a CTL. It is not yet known whether the interactions between the dendritic cell and the two different T cells must be concurrent, with each interacting simultaneously with presented antigens on the same dendritic cell, or if the dendritic cell licensed by a T$_H$1 cell interaction retains its capacity to activate CTL-Ps for some period after the T$_H$1 cell disengages. If the latter were true, then the specific T$_H$1 cells could move on to license other dendritic cells, amplifying the activation response. Figure 14-1 shows the interaction of T$_H$1 cells with a dendritic cell that is then licensed and can activate CTL-Ps to become CTLs when they interact with class I antigen on the dendritic cell.

The stringent requirement that both T$_H$1 and T$_C$ cells recognize antigen before the T$_C$ cell is activated to become a CTL provides a safeguard against inappropriate self-reactivity by cytotoxic cells. Recall that the phenomenon of antigen cross-presentation (see Chapter 8), most likely confined to dendritic cells, allows presentation of exogenous antigens by both class I and class II MHC molecules. This could allow simultaneous recognition, by both CD8$^+$ and CD4$^+$ cells, of the same antigen presented in the context of class I and class II on a dendritic cell. A further relevant observation is that dendritic cells take up, and can be infected by, a variety of viruses, including those for which they do not express specific receptors. This quality makes the dendritic cell a potential reservoir of viral antigens to which both CTL and T$_H$1 cells may be directed and allows generation of CTLs against viruses with strict tissue-specific patterns of infection that may not otherwise encounter CTLs until the infection is well established.

There are clearly discernible differences in the CD8$^+$ T cell before and after conversion from a CTL-P to CTL. Unactivated CTL-Ps do not express IL-2 or IL-2 receptors, do not proliferate, and do not display cytotoxic activity. Activation induces a CTL-P to begin expressing the IL-2 receptor and to a lesser extent IL-2, the principal cytokine required for proliferation and differentiation of activated CTL-Ps into effector CTLs. Memory CTL-Ps, which have lower activation requirements than naive cells (Figure 14-2), require less IL-2 to be activated than do naive CTL-Ps.

In general, though, most activated CTL-Ps require additional IL-2 produced by T$_H$1 cells to proliferate and differentiate into effector CTLs. In IL-2 knockout mice, the absence of IL-2 has been shown to abolish CTL-mediated cytotoxicity. The fact that the IL-2 receptor is not expressed until after a CTL-P has been activated by antigen plus a class I MHC molecule favors the clonal expansion and acquisition of cytotoxicity by only the antigen-specific CTL-Ps.

After clearance of antigen, the level of IL-2 declines, which induces T$_H$1 cells and CTLs to undergo programmed cell death by apoptosis. In this way, the immune response is rapidly terminated, lessening the likelihood of nonspecific tissue damage from the inflammatory response.

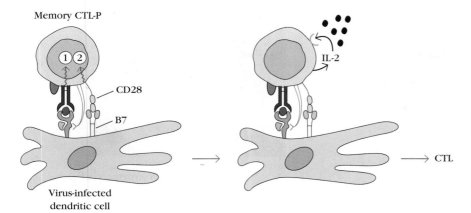

Memory CTL-P

CD28

B7

IL-2

CTL

Virus-infected
dendritic cell

**FIGURE 14-2 Proliferation of memory CTL-Ps may not require help from T$_H$ cells.** Antigen-activated memory CTL-Ps appear to secrete sufficient IL-2 to stimulate their own proliferation and differentiation into effector CTLs. They also may not require the CD28-B7 costimulatory signal for activation.

## CD8$^+$ CTLs can be tracked with MHC tetramer technology

MHC tetramers are laboratory-generated complexes of four MHC class I molecules bound to a specific peptide and linked to a fluorescent molecule (Figure 14-3). A given MHC tetramer–peptide complex binds only CD8$^+$ T cells that have TCRs specific for the particular peptide-MHC complex that makes up the tetramer. Thus, when a particular tetramer is added to a cell population containing T cells (recovered, for example, from spleen or lymph nodes), cells that bear TCRs specific for the tetramer become fluorescently labeled. Using flow cytometry, it is then possible to determine the proportion of cells in a population that have TCRs specific for a particular antigen by counting the number of fluorescently labeled cells in a cell population. This very sensitive approach can detect antigen-specific T cells even when their frequency in the CD8$^+$ population is as low as 0.1%.

In addition, one can directly measure the increase in antigen-specific CD8$^+$ T cells in response to exposure to pathogens such as viruses or cancer-associated antigens. In a related application, researchers infected mice with vesicular stomatitis virus (VSV) and systematically examined the distribution of CD8$^+$ cells specific for a VSV-derived peptide-MHC complex throughout the entire body. This study demonstrated that during acute infection with VSV, the distribution of VSV-specific CD8$^+$ cells is far from uniform (Figure 14-4); large populations of antigen-specific cells are not limited to the lymphoid system but can be found in the liver and kidney too.

## CTLs kill cells in two ways

The effector phase of a CTL-mediated response involves a carefully orchestrated sequence of events that begins with the binding of the target cell by the attacking cell (Figure 14-5). Long-term cultures of CTL clones have been used to identify many of the membrane molecules and membrane events involved in this process. As described below, studies with mouse strains carrying mutations that affect the ability of CTLs to induce death have led to the identification of the molecules responsible.

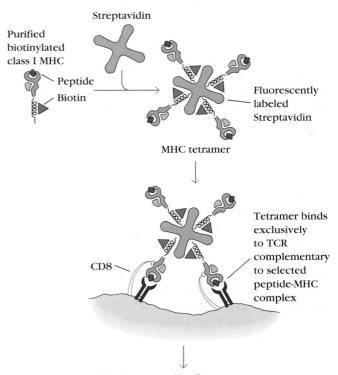

Streptavidin

Purified
biotinylated
class I MHC

Peptide

Biotin

Fluorescently
labeled
Streptavidin

MHC tetramer

CD8

Tetramer binds
exclusively
to TCR
complementary
to selected
peptide-MHC
complex

Signal measured by flow cytometer

**FIGURE 14-3 MHC tetramers.** A homogeneous population of peptide-bound class I MHC molecules (HLA-A1 bound to an HIV-derived peptide, for example) is conjugated to biotin and mixed with fluorescently labeled Streptavidin. Four biotinylated MHC-peptide complexes bind to the high-affinity binding sites of Streptavidin to form a tetramer. Addition of the tetramer to a population of T cells results in exclusive binding of the fluorescent tetramer to those CD8$^+$ T cells with TCRs complementary to the peptide-MHC complexes of the tetramer. This results in the labeling of the subpopulation of T cells that are specific for the target antigen, making them readily detectable by flow cytometry. *[Adapted in part from P. Klenerman, V. Cerundolo, and P. R. Dunbar, 2002,* Nature Reviews Immunology *2:264.]*

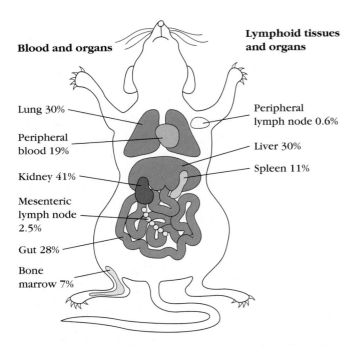

**Blood and organs**

**Lymphoid tissues and organs**

Lung 30%

Peripheral lymph node 0.6%

Peripheral blood 19%

Liver 30%

Kidney 41%

Spleen 11%

Mesenteric lymph node 2.5%

Gut 28%

Bone marrow 7%

**FIGURE 14-4 Localizing antigen-specific CD8⁺ T-cell populations in vivo.** Mice were infected with vesicular stomatitis virus (VSV), and during the course of the acute stage of the infection, cell populations were isolated from the tissues indicated in the figure and incubated with tetramers containing VSV-peptide–MHC complexes. Flow-cytometric analysis allowed determination of the percentages of CD8⁺ T cells that were VSV specific in each of the populations examined. *[Adapted from P. Klenerman, V. Cerundolo, and P. R. Dunbar, 2002, Nature Reviews Immunology 2:269.]*

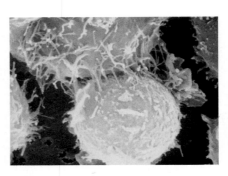

**FIGURE 14-5 Scanning electron micrograph of tumor-cell attack by a CTL.** The CTL (top) makes contact with a smaller tumor cell. *[From J. D. E. Young and Z. A. Cohn, 1988, Scientific American 258(1):38.]*

The primary events in CTL-mediated death are conjugate formation, membrane attack, CTL dissociation, and target-cell destruction (Figure 14-6). When antigen-specific CTLs are incubated with appropriate target cells, the two cell types interact and undergo conjugate formation. Formation of a CTL–target cell conjugate is followed within several minutes by a $Ca^{2+}$-dependent, energy-requiring step in which the CTL programs the target cell for death. The CTL then dissociates from the target cell and goes on to bind another target cell. Within a variable period (up to a few hours) after CTL dissociation, the target cell dies by apoptosis.

The process begins when the TCR-CD3 membrane complex on a CTL recognizes antigen in association with class I MHC molecules on a target cell. After this antigen-specific recognition, the integrin receptor LFA-1 on the CTL membrane binds to ICAMs on the target-cell membrane, resulting in the formation of a conjugate. Antigen-mediated

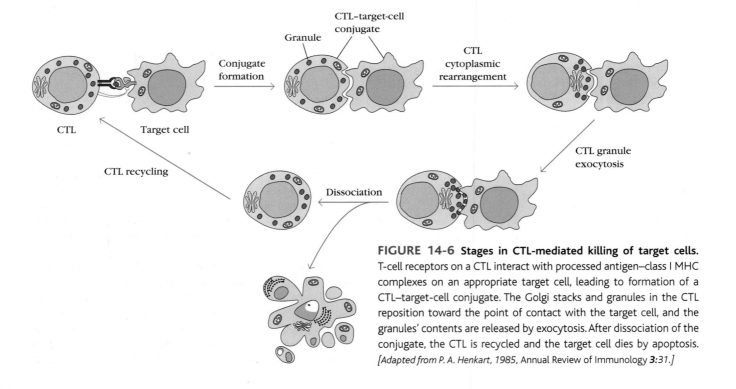

**FIGURE 14-6 Stages in CTL-mediated killing of target cells.** T-cell receptors on a CTL interact with processed antigen–class I MHC complexes on an appropriate target cell, leading to formation of a CTL–target-cell conjugate. The Golgi stacks and granules in the CTL reposition toward the point of contact with the target cell, and the granules' contents are released by exocytosis. After dissociation of the conjugate, the CTL is recycled and the target cell dies by apoptosis. *[Adapted from P. A. Henkart, 1985, Annual Review of Immunology 3:31.]*

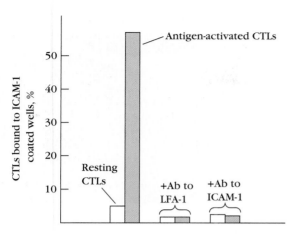

**FIGURE 14-7 Effect of antigen activation on the ability of CTLs to bind to the intercellular cell adhesion molecule ICAM-1.** Resting mouse CTLs were first incubated with anti-CD3 antibodies. Cross-linkage of CD3 molecules on the CTL membrane by anti-CD3 has the same activating effect as interaction with antigen–class I MHC complexes on a target cell. Adhesion was assayed by binding radiolabeled CTLs to microwells coated with ICAM-1. Antigen activation increased CTL binding to ICAM-1 more than 10-fold. The presence of excess monoclonal antibody to LFA-1 or ICAM-1 in the microwell abolished binding, demonstrating that both molecules are necessary for adhesion. *[Based on M. L. Dustin and T. A. Springer, 1989, Nature* **341:***619.]*

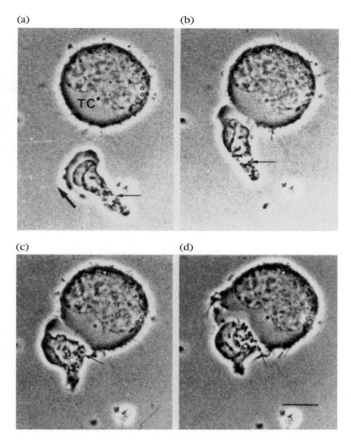

**FIGURE 14-8 Formation of a conjugate between a CTL and a target cell and reorientation of CTL cytoplasmic granules as recorded by time-lapse photography.** (a) A motile mouse CTL (thin arrow) approaches an appropriate target cell (TC). The thick arrow indicates direction of movement. (b) Initial contact of the CTL and target cell has occurred. (c) Within 2 minutes of initial contact, the membrane-contact region has broadened and the rearrangement of dark cytoplasmic granules within the CTL (thin arrow) is under way. (d) Further movement of dark granules toward the target cell is evident 10 minutes after initial contact. *[From J. R. Yanelli et al., 1986,* Journal of Immunology *136:377.]*

CTL activation converts LFA-1 from a low-affinity state to a high-affinity state (Figure 14-7). Because of this phenomenon, CTLs adhere to and form conjugates only with appropriate target cells that display antigenic peptides associated with class I MHC molecules. LFA-1 persists in the high-affinity state for only 5 to 10 minutes after antigen-mediated activation, and then it returns to the low-affinity state. This downshift in LFA-1 affinity may facilitate dissociation of the CTL from the target cell.

Electron microscopy of cultured CTL clones reveals the presence of intracellular electron-dense storage granules. These granules have been isolated by fractionation and shown to mediate target-cell damage by themselves. Analysis of their contents revealed 65-kDa monomers of a pore-forming protein called **perforin** and several serine proteases called **granzymes** (or **fragmentins**). CTL-Ps lack cytoplasmic granules and perforin; on activation, cytoplasmic granules appear, bearing newly expressed perforin monomers.

Immediately after formation of a CTL–target-cell conjugate, the Golgi stacks and storage granules reposition within the cytoplasm of the CTL to concentrate near the junction with the target cell (Figure 14-8). Evidence suggests that perforin monomers and the granzyme proteases are then released from the granules by exocytosis into the space at the junction between the two cells. As the perforin monomers contact the target-cell membrane, they undergo a conformational change, exposing an amphipathic domain that inserts into the target-cell membrane; the monomers then polymerize (in the presence of $Ca^{2+}$) to form cylindrical pores with an internal diameter of 5 to 20 nm (Figure 14-9a). A large number of perforin pores are visible on the target-cell membrane in the region of conjugate formation (Figure 14-9b). Interestingly, perforin exhibits some sequence homology with the terminal C9 component of the complement system, and the membrane pores formed by perforin are similar to those observed in complement-mediated lysis. The importance of perforin to CTL-mediated killing is demonstrated by perforin-deficient knockout mice, which are unable to eliminate lymphocytic choriomeningitis virus (LCMV) even though they mount a significant CD8$^+$ immune response to the virus.

Pore formation in the cell membrane of the target is one way that perforin mediates granzyme entry; another is the perforin-assisted pathway. Many target cells have a molecule known as the mannose 6-phosphate receptor on their surface that also binds to granzyme B. Granzyme B/mannose 6-phosphate receptor complexes are internalized and appear inside vesicles. Perforin internalized at the same time then

(a)

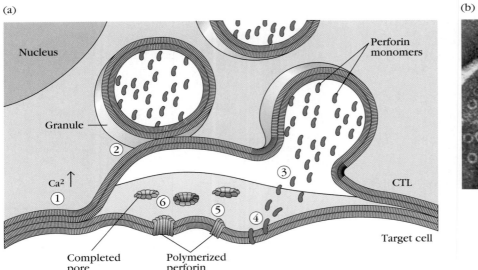

(b)

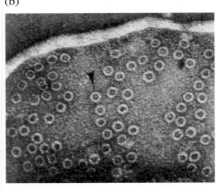

**FIGURE 14-9 CTL-mediated pore formation in target-cell membrane.** (a) In this model, a rise in intracellular $Ca^{2+}$ triggered by CTL–target-cell interaction (1) induces exocytosis, in which the granules fuse with the CTL cell membrane (2) and release monomeric perforin into the small space between the two cells (3). The released perforin monomers undergo a $Ca^{2+}$-induced conformational change that allows them to insert into the target-cell membrane (4). In the presence of $Ca^{2+}$, the monomers polymerize within the membrane (5), forming cylindrical pores (6). (b) Electron micrograph of perforin pores on the surface of a rabbit erythrocyte target cell. [Part a adapted from J. D. E. Young and Z. A. Cohn, 1988, Scientific American **258**(1):38; part b from E. R. Podack and G. Dennert, 1983, Nature **301**:442.]

forms pores that release granzyme B from the vesicle into the cytoplasm of the target cell.

Once it enters the cytoplasm of the target cell, granzyme B initiates a cascade of reactions that result in the fragmentation of the target-cell DNA into oligomers of 200 bp; this type of DNA fragmentation is typical of apoptosis. Granzymes, which are proteases, do not directly mediate DNA fragmentation. Rather, they activate an apoptotic pathway within the target cell. This apoptotic process does not require mRNA or protein synthesis in either the CTL or the target cell. Within 5 minutes of CTL contact, target cells begin to exhibit DNA fragmentation. Interestingly, viral DNA within infected target cells has also been shown to be fragmented during this process. This observation shows that CTL-mediated killing not only kills virus-infected cells but can also destroy the viral DNA in those cells. It has been suggested that the rapid onset of DNA fragmentation after CTL contact may prevent continued viral replication and assembly in the period before the target cell is destroyed.

Some potent CTL lines have been shown to lack perforin and granzymes. In these cases, cytotoxicity is mediated by Fas. As described in Chapter 10, this transmembrane protein, which is a member of the TNF-receptor family, can deliver a death signal when cross-linked by its natural ligand, a member of the tumor necrosis factor family called Fas ligand (see Figure 10-19). Fas ligand (FasL) is found on the membrane of CTLs, and the interaction of FasL with Fas on a target cell triggers apoptosis.

Key insight into the role of perforin and the Fas-FasL system in CTL-mediated cytolysis came from experiments with mutant mice. These experiments used two types of mutant mice: the perforin knockout mice mentioned above and a strain of mice known as *lpr* (Figure 14-10). Mice that are homozygous for the *lpr* mutation express little or no Fas, and consequently, cells from these mice cannot be killed by interaction with Fas ligand. If lymphocytes from normal H-2$^b$ mice are incubated with killed cells from H-2$^k$ mice, anti-H-2$^k$ CTLs are generated. These H-2$^b$ CTLs will kill target cells from normal H-2$^k$ mice or from H-2$^k$ animals that are homozygous for the *lpr* mutation. Incubation of H-2$^b$ cells of perforin knockout mice with killed cells from H-2$^k$ mice resulted in CTLs that killed wild-type target cells but failed to induce lysis in target cells from H-2$^k$ mice homozygous for the *lpr* mutation.

The results of these experiments, taken together with other studies, allowed the investigators to make the following interpretation. CTLs raised from normal mice can kill target cells by a perforin-mediated mechanism, by a mechanism involving engagement of target-cell Fas with Fas ligand displayed on the CTL membrane, or, in some cases, perhaps, by a combination of both mechanisms. Such CTLs can kill target cells that lack membrane Fas by using the perforin mechanism alone. On the other hand, CTLs from perforin knockout mice can kill only by the Fas-FasL mechanism. Consequently, CTLs from perforin knockout mice can kill Fas-bearing normal target cells but not *lpr* cells, which lack Fas. These workers also concluded that all of the CTL-mediated killing they observed could be traced to the action of perforin-dependent killing, Fas-mediated killing, or a combination of the two. No other mechanism was detected.

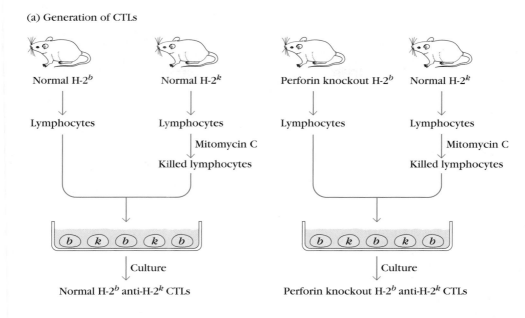

**FIGURE 14-10 Experimental demonstration that CTLs use Fas and perforin pathways.** (a) Generation of CTLs. Lymphocytes were harvested from mice of H-2$^b$ and H-2$^k$ MHC haplotypes. H-2$^k$ haplotype cells were killed by treatment with mitomycin C and cocultured with H-2$^b$ haplotype cells to stimulate the generation of H-2$^k$ CTLs. If the H-2$^b$ lymphocytes were derived from normal mice, they gave rise to CTLs that had both perforin and Fas ligand. If the CTLs were raised by stimulation of lymphocytes from perforin knockout (KO) mice, they expressed Fas ligand but not perforin. (b) Interaction of CTLs with Fas$^+$ and Fas$^-$ targets. Normal H-2$^b$ anti-H-2$^k$ CTLs that express both Fas ligand and perforin kill normal H-2$^k$ target cells and H-2$^k$ *lpr* mutant cells, which do not express Fas. In contrast, H-2$^b$ anti-H-2$^k$ CTLs from perforin KO mice kill Fas$^+$ normal cells by engagement of Fas with Fas ligand but are unable to kill the *lpr* cells, which lack Fas.

This experiment and others show that two mechanisms are responsible for initiating all CTL-mediated apoptotic death of target cells:

■ Directional delivery of cytotoxic proteins (perforin and granzymes) that are released from CTLs and enter target cells

■ Interaction of the membrane-bound Fas ligand on CTLs with the Fas receptor on the surface of target cells

Either of these initiating events results in the activation of a signaling pathway that culminates in the death of the target cell by apoptosis (Figure 14-11). A feature of cell death by apoptosis is the involvement of the **caspase** family of cysteine proteases, which cleave after an aspartic acid residue. The name *caspase* incorporates all of these elements (*c*ysteine, *asp*artate, prote*ase*). Normally, caspases are present in the cell as inactive proenzymes—procaspases—that require proteolytic cleavage for conversion to the active forms. More than a dozen different

caspases have been found, each with its own specificity. Cleavage of a procaspase produces an active initiator caspase, which cleaves other procaspases, thereby activating their proteolytic activity. The end result is the systematic and orderly disassembly of the cell that is the hallmark of apoptosis.

CTLs use granzymes and Fas ligand to initiate caspase cascades in their targets. The granzymes introduced into the target cell from the CTL mediate proteolytic events that activate an initiator caspase. Similarly, the engagement of Fas on a target cell by Fas ligand on the CTL causes the activation of an initiator caspase in the target cell. Fas is associated with a protein known as FADD (*Fas-associated protein with death domain), which in turn associates with a procaspase form of caspase 8. On Fas cross-linking, procaspase 8 is converted to caspase 8 and initiates an apoptotic caspase cascade. The end result of both the perforin/granzyme and Fas-mediated pathways is the activation of dormant death pathways that are present in the target cell. As one immunologist has so aptly put it, CTLs don't so much kill target cells as persuade them to commit suicide.

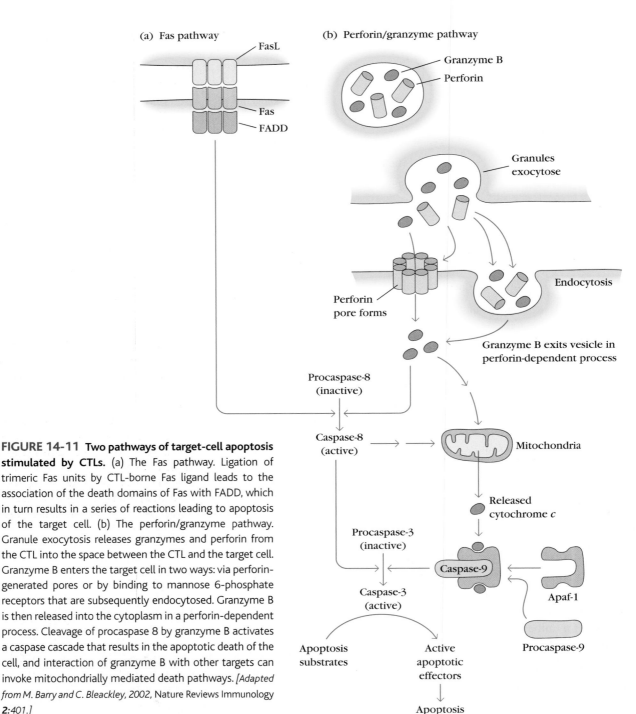

(a) Fas pathway

FasL

Fas

FADD

(b) Perforin/granzyme pathway

Granzyme B

Perforin

Granules exocytose

Endocytosis

Perforin pore forms

Granzyme B exits vesicle in perforin-dependent process

Procaspase-8 (inactive)

Caspase-8 (active)

Mitochondria

Released cytochrome *c*

Procaspase-3 (inactive)

Caspase-3 (active)

Caspase-9

Apaf-1

Procaspase-9

Apoptosis substrates

Active apoptotic effectors

Apoptosis

**FIGURE 14-11 Two pathways of target-cell apoptosis stimulated by CTLs.** (a) The Fas pathway. Ligation of trimeric Fas units by CTL-borne Fas ligand leads to the association of the death domains of Fas with FADD, which in turn results in a series of reactions leading to apoptosis of the target cell. (b) The perforin/granzyme pathway. Granule exocytosis releases granzymes and perforin from the CTL into the space between the CTL and the target cell. Granzyme B enters the target cell in two ways: via perforin-generated pores or by binding to mannose 6-phosphate receptors that are subsequently endocytosed. Granzyme B is then released into the cytoplasm in a perforin-dependent process. Cleavage of procaspase 8 by granzyme B activates a caspase cascade that results in the apoptotic death of the cell, and interaction of granzyme B with other targets can invoke mitochondrially mediated death pathways. *[Adapted from M. Barry and C. Bleackley, 2002,* Nature Reviews Immunology **2:**401.]

## Natural Killer Cells

Natural killer cells were discovered quite by accident when immunologists were measuring in vitro activity of tumor-specific cells taken from mice with tumors. Normal unimmunized mice and mice with unrelated tumors served as negative controls. Much to the consternation of the investigators, the controls showed significant lysis of the tumor cells too. Characterization of this nonspecific tumor-cell killing revealed that a population of large granular lymphocytes was responsible. The cells, which were named natural

killer (NK) cells for their nonspecific cytotoxicity, make up 5% to 10% of the circulating lymphocyte population. These cells are involved in immune defenses against viruses, other intracellular pathogens, and tumors. Because NK cells produce a number of immunologically important cytokines, they play contributing roles in immune regulation and influence both innate and adaptive immunity. In particular, IFN-γ production by NK cells can affect the participation of macrophages in innate immunity by activation of their phagocytic and microbicidal activities. IFN-γ derived from NK cells can influence the $T_H1$ versus $T_H2$ commitment of

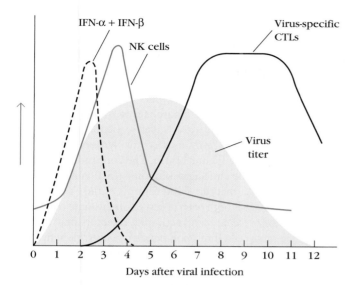

**FIGURE 14-12 Time course of viral infection.** IFN-α and IFN-β (dashed curve) are released from virus-infected cells soon after infection. These cytokines stimulate the NK cells, quickly leading to a rise in the NK-cell population (blue curve) from the basal level. NK cells help contain the infection during the period required for generation of CTLs (black curve). Once the CTL population reaches a peak, the virus titer (blue area) rapidly decreases.

helper T-cell populations by its inhibitory effects on $T_H2$ expansion and stimulate $T_H1$ development via induction of IL-12 by macrophages and dendritic cells.

NK cells are involved in the early response to infection with certain viruses and intracellular bacteria. NK activity is stimulated by IFN-α, IFN-β, and IL-12. In the course of a viral infection, the level of these cytokines rapidly rises, followed closely by a wave of NK cells that peaks in about 3 days (Figure 14-12). NK cells are the first line of defense against virus infection, controlling viral replication during the time required for activation, proliferation, and differentiation of CTL-P cells into functional CTLs at about day 7. The importance of NK cells in defense against viral infections is illustrated by the case of a young woman who completely lacked these cells. Even though this patient had normal T- and B-cell counts, she suffered severe varicella virus infections and a life-threatening cytomegalovirus infection.

## NK cells and T cells share some features

NK cells are lymphoid cells derived from bone marrow that share a common early progenitor with T cells, but their detailed lineage remains to be worked out. They express some membrane markers that are found on monocytes and granulocytes as well as some that are typical of T cells. Different NK cells express different sets of membrane molecules. It is not known whether this heterogeneity reflects subpopulations of NK cells or different stages in their activation or maturation. Among the membrane molecules expressed by NK cells are CD2, the 75-kDa β subunit of the IL-2 receptor, and, on almost all NK cells, CD16 (or FcγRIII), a receptor for the Fc

region of IgG. Cell depletion with monoclonal anti-CD16 antibody removes almost all NK-cell activity from peripheral blood. As will be discussed below, the number and type of activating and inhibitory receptors varies on NK cells.

Despite some similarities of NK cells to T lymphocytes, they do not develop exclusively in the thymus. Nude mice, which lack a thymus and have few or no T cells, have functional NK-cell populations. Unlike T cells and B cells, NK cells do not undergo rearrangement of receptor genes: NK cells develop in mice in which the recombinase genes *RAG-1* or *RAG-2* have been knocked out. Also, although no T or B cells are found in SCID mice, functional populations of NK cells exist. The power of NK cells and other protective mechanisms of innate immunity to protect animals totally lacking in adaptive immunity is nicely illustrated by the family of *RAG-1* knockout mice shown in Figure 14-13.

## Killing by NK cells is similar to CTL-mediated killing

Natural killer cells appear to kill tumor cells and virus-infected cells by processes similar to those employed by CTLs. NK cells bear FasL on their surface and readily induce death in Fas-bearing target cells. The cytoplasm of NK cells has numerous granules containing perforin and granzymes. Unlike CTLs, which must be activated before granules appear, NK cells are constitutively cytotoxic and always have large granules in their cytoplasm. After an NK cell adheres to a target cell, degranulation occurs, with release of perforin and granzymes at the junction between the interacting cells. Perforin and granzymes are thought to play the same roles in NK-mediated killing of target cells by apoptosis as they do in the CTL-mediated killing process.

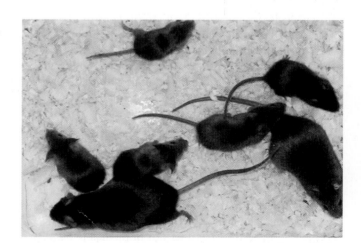

**FIGURE 14-13 Family of *RAG-1* knockout mice.** These mice have no adaptive immunity because they lack T and B cells. However, NK cells and other mechanisms of innate immunity provide sufficient protection against infection that, if maintained in clean conditions, these mice can reproduce and raise healthy offspring. However, they are more susceptible to infection than normal mice and have reduced life spans. *[From the laboratory of R. A. Goldsby.]*

Despite these similarities, NK cells differ from CTLs in several significant ways. First, NK cells do not express antigen-specific T-cell receptors or CD3. In addition, recognition of target cells by NK cells is not MHC restricted; that is, in many cases the same levels of NK-cell activity are observed with syngeneic and allogeneic tumor cells. Moreover, although prior priming enhances CTL activity, NK-cell activity does not increase after a second injection with the same tumor cells. In other words, the NK-cell response generates no immunologic memory.

## NK cells have both activation and inhibition receptors

Because NK cells do not express antigen-specific receptors, the mechanism by which these cells recognize altered self cells and distinguish them from normal body cells baffled immunologists for years. NK cells turned out to employ two different categories of receptors: one that delivers inhibition signals to NK cells and another that delivers activation signals. Initially, researchers thought that just two receptors were involved, one that activated and another that inhibited NK cells—the so-called *two-receptor model*. Current theory, the *opposing signals model*, holds that there are many different cell surface receptors for activation signals and a number of different inhibitory ones. NK cells distinguish healthy cells from infected or cancerous ones through a balance between activating signals and inhibitory signals. Additional NK-activating signals can be delivered by soluble factors, including cytokines such as $\alpha$ and $\beta$ interferons, TNF-$\alpha$, IL-12, and IL-15.

The receptors of NK cells fall into two general categories based on their structural characteristics: lectin-like and immunoglobulin-like receptors. As discussed in Chapter 7, lectins are a group of proteins that bind to specific carbohydrates; the lectin-like NK-cell receptors are so named because of their structural similarity to proteins of this group. Despite this similarity, most of the lectin-like NK-cell receptors bind proteins rather than carbohydrates. The second group of receptors, members of the immunoglobulin super-family, include the *killer-cell immunoglobulin-like receptors* (KIR), which bind to HLA-B or HLA-C molecules. There are also other Ig-like inhibitory receptors, such as ILT/LIR, which bind most class I molecules. Because the first discovered KIR molecules were inhibitory receptors, KIRs were initially called killer-cell *inhibitory* receptors. When subsequent studies revealed that both the KIR group and the lectin-like group of receptors included inhibitory and activating receptors, the designation was changed to killer *immunoglobulin-like* receptor. The extracellular structure of the NK-cell receptor (Ig or lectin-like) does not immediately identify it as inhibitory or activating because there are activating and inhibitory receptors that fall into both structural groups.

To further complicate the matter, some receptors may be either activating or inhibitory; the cytoplasmic regions of some NK receptors with identical extracellular structures have different intracellular domains and therefore different signaling properties. For example, the lectin-like receptor CD94:NKG2 has two forms, CD94:NKG2A and CD94:NKG2C. Although both bind similar ligands, the A form recruits phosphatases and delivers inhibitory signals, whereas the C form associates with an adaptor molecule and delivers activating signals. The intracellular sequences of the NK receptors that activate are ITAMs (*i*mmunoreceptor *t*yrosine phosphate *a*ctivation *m*otif, see Figure 11-8), whereas the intracellular sequences of inhibitory receptors are ITIMs (*i*mmunoreceptor *t*yrosine phosphate *i*nhibitory *m*otif; see Figure 11-11). The confusion caused by this dual role for the same ligand-binding reaction is somewhat relieved by the fact that the A and C forms of the CD94/NKG2 receptors have never both been observed on the same cell. However, inhibitory receptors and activating receptors may be present on the same cells. Inhibition signals can override activation signals, providing important safeguards for normal host cells.

The exact nature of the membrane-bound receptors on NK cells that produce activation is not completely clear. Antibody cross-linking of many molecules found on the surface of NK cells can activate these cells artificially, but the natural ligands for some of these putative activation receptors (ARs) are not known.

Some of the candidate ARs are members of a class of carbohydrate-binding proteins known as **C-type lectins,** so named because they have calcium-dependent carbohydrate recognition domains. In particular, NKG2D has emerged as an important receptor for activation of NK-cell function. NKG2D acts through a signaling cascade similar to that initiated by CD28 in T cells. The ligands for NKG2D include MIC-A, MIC-B, and the ULPB family of proteins in humans and H60, Mult1, and the Rae-1 family of proteins in mice. The expression of NKG2D ligands is often induced on cells undergoing stress, such as DNA damage or infection. NKG2D appears to be adapted to participate in a rapid innate immune response to a number of pathological conditions. However, a contribution of NKG2D to autoimmune responses has also been reported.

In addition to lectins, other molecules on NK cells might be involved in activation, including CD2 (the receptor for the adhesion molecule LFA-3), CD244 (also called 2B4, the receptor for CD48), and the Fc$\gamma$III receptor, CD16. Although CD16 is responsible for antibody-mediated recognition and killing of target cells by NK cells, it is probably not involved in non-antibody-dependent killing. In addition to the molecules already mentioned, three additional proteins, NKp30, NKp44, and NKp46, appear to play significant roles in the activation of human NK cells.

The target-cell ligands recognized by most of the activating receptors of NK cells remain unidentified, and even some that have been identified appear to function ambiguously as both activators and inhibitors. For example, the CD94/NKG2A and C pair of receptors mentioned above can transduce activating or inhibitory signals in response to binding of the same ligand. The best-characterized activation ligands are the HLA-encoded MIC-A and MIC-B molecules. These are nonpolymorphic MHC class I–like molecules (although they do not associate with β-2 microglobulin) that are recognized by the lectin-like receptor NKG2D. The MIC-A and -B proteins are inducible and are expressed on cells that have been stressed by infection, heat, or trauma. When the lectin-like NKG2D receptor binds to these stress-induced proteins, the response includes recruitment of cytotoxic granules and release of cytokines, leading to the death of the targeted cell. Other activating receptors present on most NK cells include NKp30, NKp44, and NKp46. These structurally related Ig-like molecules are involved in killing of tumor cells, but the identities of the tumor-cell ligands they recognize are not known.

NK-cell inhibitory ligands are better characterized than activating ligands. Clues to the sources of inhibitory signals came from studies of the killing of tumor cells and virus-infected cells by NK cells. Researchers determined that the preferential killing of mouse tumor cells compared with normal cells correlated with a lack of expression of MHC molecules by the tumor cells. Experiments with human cells showed that NK cells lysed a B-cell line that was MHC deficient as a result of transformation by Epstein-Barr virus. After this cell line was transfected with human HLA genes, which caused it to express high levels of MHC molecules, NK cells failed to lyse it. These observations led to the idea that NK cells target for killing those cells that have aberrant MHC expression. Since many virus-infected and tumor cells exhibit diminished MHC expression, this model, called the "missing self model," made good physiological sense. Support for this proposal has come from the discovery of receptors on NK cells that produce inhibitory signals when they recognize MHC molecules on potential target cells. These inhibitory receptors on the NK cell then prevent NK-cell killing, proliferation, and cytokine release.

In humans, most inhibitory receptors are Ig-like molecules. An exception is the lectin-like inhibitory receptor CD94/NKG2A, a disulfide-bonded heterodimer made up of two glycoproteins, CD94 and a member of the NKG2 family. The CD94/NKG2A receptors recognize HLA-E on potential target cells. Because HLA-E is not transported to the surface of a cell unless it has bound a peptide derived from HLA-A, HLA-B, or HLA-C, the amount of HLA-E on the surface serves as an indicator of the overall level of class I MHC biosynthesis in the cells. These inhibitory CD94/NKG2A receptors recognize the surface HLA-E and send inhibitory signals to the NK cell, with the net result that killing of potential target cells is inhibited if they are expressing adequate levels of class I.

KIR inhibitory receptors, which show considerable diversity, are generally specific for a single polymorphic product of a particular HLA locus or for a limited number of related HLA molecules. Unlike antibodies in B cells and TCRs in T cells, which have the property of allelic exclusion, NK cells are not limited to expressing a single inhibitory KIR but may express several, each specific for a different MHC molecule or for a set of closely related MHC molecules. For example, individual clones of human NK cells expressing a CD94/NKG2A receptor and as many as six different KIR receptors have been found.

Because signals from inhibitory receptors can counteract signals from activating receptors, a negative signal from any

inhibitory receptor, whether of the CD94/NKG2A or KIR type, can block the lysis of target cells by NK cells. Thus, cells expressing normal levels of unaltered MHC class I molecules tend to escape all forms of NK-cell-mediated killing. Surprisingly, the KIR family appears to have evolved extremely rapidly. Functional KIR receptors found in the primate line do not exist in rodents. Mice use a different family of receptors, the lectin-like Ly49 family, to achieve the function of KIRs, namely inhibition of NK cells through binding to MHC class I molecules on healthy cells. Functional Ly49 receptors do not exist in humans.

In the opposing-signals model of NK-cell regulation that is emerging from studies of NK cells (Figure 14-14), activating receptors engage ligands (most of which are not known) on the surface of a targeted tumorous, virus-infected, or otherwise stressed cell. Recognition of these determinants by activating receptors would signal NK cells to kill the target cells. Killing signals can be overridden by signals from inhibitory receptors. As we have already seen, the inhibitory receptors provide a signal that decisively overrides activation signals when these inhibitory receptors detect normal levels of MHC class I expression on potential target cells. This prevents the death of the target cell and also NK-cell proliferation and the induction of secretion of cytokines such as IFN-γ and TNF-α. The overall consequence of the opposing-signals model is to spare cells that

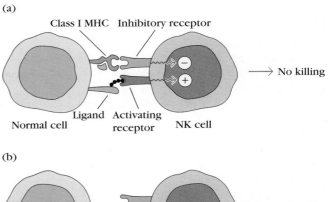

(a)

Class I MHC   Inhibitory receptor

Normal cell
Ligand   Activating receptor   NK cell
⟶ No killing

(b)

Virus-infected cell
( ↓ class I MHC)
Ligand   Activating receptor   NK cell
⟶ Killing

**FIGURE 14-14 Opposing-signals model of how cytotoxic activity of NK cells is restricted to altered self cells.** An activating receptor on NK cells interacts with its ligand on normal and altered self cells, inducing an activation signal that results in killing. However, engagement of inhibitory NK-cell receptors such as inhibitory KIRs and CD94/NKG2 by class I MHC molecules delivers an inhibition signal that counteracts the activation signal. Expression of class I molecules on normal cells thus prevents their destruction by NK cells. Because class I expression is often decreased on altered self cells, the killing signal predominates, leading to their destruction.

## CLINICAL FOCUS

# MHC-KIR Gene Combinations Influence Health

**NK cells** are an important component of innate immunity: these cells kill host cells that are altered by viral infection, stress, or transformation into a tumor cell. In addition, activated NK cells express cytokines that promote adaptive immune responses. The NK cell acts through two types of receptors: activating receptors that initiate killing of cells recognized as aberrant and inhibitory receptors that abort the killer function when they recognize certain host-cell surface markers such as class I MHC molecules. The balance between these opposing activities is critical to the well-being of the host. A deficiency of activating events can lead to cancer, viral infection, and the accumulation of damaged, non-functional cells. Defects in inhibitory activity can lead to the killing of normal host cells. The NK-cell receptors mediating these activities fall into two structural families: the lectin-like receptors and the killer-cell immunoglobulin-like receptors, or KIRs. Both structural groups include activating and inhibitory receptors.

Whereas the lectin-like receptors are highly conserved, showing little variation within a species, the KIR family of NK receptors is a diverse group of proteins. KIRs are encoded within a region on human chromosome 19 termed the *leukocyte receptor complex*. The number of KIR genes in a given haplotype ranges from nine to 14 members, and the individual KIR genes are themselves polymorphic. The KIR A haplotype (the most common haplotype of Caucasoid humans) is shown in the figure, with the nine encoded proteins identified by function. Note that there are two activating and four inhibitory receptors within this KIR haplotype.

Although ligands for most of the activating receptors remain to be discovered, ligands for the inhibitory receptors are known. Most are allelic forms of class I MHC molecules. Given that the KIR molecules are themselves variable in a population and that they recognize proteins encoded within the highly diverse MHC complex (see Chapter 8), an interesting genetic question arises: Since the MHC genes and KIR genes segregate independently, could some combinations of MHC and KIR genes result in the absence of required inhibitory MHC encoded ligands in the host, leading to defects in the inhibition of activation?

This possibility was examined in a number of studies, which found that certain

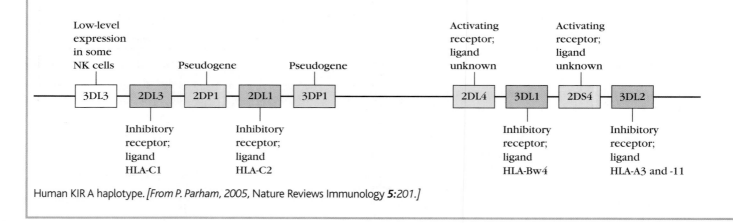

Human KIR A haplotype. *[From P. Parham, 2005, Nature Reviews Immunology 5:201.]*

express critical indicators of normal self, the MHC class I molecules, and to kill cells that lack indicators of self: normal levels of class I MHC are absent.

## NKT Cells

The above discussion covers the CTL, an important component of adaptive immunity that expresses an antigen-specific TCR, and the NK cell, an innate immune component bearing KIR and lectin-like receptors, which are analogous to pattern recognition receptors. Recently a third type of cell has been identified with characteristics common to both the CTL and the NK cell. This cell type, designated the NKT cell to reflect its hybrid quality, has TCR complexes on its surface but few other qualities in common with T lymphocytes. The NKT cell is considered part of the innate immune system, based on several properties common to innate responses:

- The T-cell receptor on the human NKT cell is invariant, with the TCRα and TCRβ chains encoded by specific gene segments ($V_\alpha 24$-$J_\alpha 18$ and $V_\beta 11$, respectively) within the germ-line DNA; the cells expressing this αβ TCR combination are sometimes referred to as iNKT or invariant NKT cells.

KIR-MHC combinations do influence disease progression and also are associated with reproductive problems such as frequent spontaneous abortions or preeclampsia (the lay term for the condition is toxemia; it is characterized by dangerous hypertension and fluid retention during pregnancy). The table lists some of the more common conditions associated with the various genetic combinations. Note that some KIR-MHC combinations appear to slow the progression of certain infections, whereas the same infection may be worsened by other combinations. Human papillomavirus-induced cervical neoplasia (see Clinical Focus, Chapter 21) is an example of a condition in which disease progression is increased by some combinations of KIR and MHC and decreased by others. The autoimmune conditions type I diabetes and psoriatic arthritis show increased progression in subjects with KIR-MHC combinations predicted to give less inhibition of NK activity.

The mechanisms underlying the link between different KIR-MHC combinations and disease and fertility problems remain a matter of speculation. For example, current research seeks to determine whether specific conditions are caused by too much or too little inhibition of NK-cell activity arising from KIR-MHC combinations. The table lists some early findings for several diseases.

Historically, KIR-MHC combinations have likely played roles in general fitness, the relative reproductive vigor of various genetic types, and even survival during epidemics. As we discover more about the specific mechanisms by which the products of these gene families influence human health, we can expect to learn why natural selection favored certain combinations.

## Disease associations with combinations of KIR and HLA genes

| Disease | KIR | HLA | Disease progression | Proposed contribution by KIRs |
|---|---|---|---|---|
| AIDS | *3DS1*<br>*3DS1* homozygous | *HLA-Bw4^{Ile80}*<br>No *HLA-Bw4^{Ile80}* | Decreased<br>Increased | Less inhibition<br>More inhibition |
| HCV infection | *2DL3* homozygous | *HLA-C1* homozygous | Decreased | Less inhibition |
| Cervical neoplasia (HPV induced) | *3DS1*<br>No *3DS1* | *HLA-C1* homozygous and no *HLA-Bw4*<br>*HLA-C2* and/or *HLA-Bw4* | Increased<br><br>Decreased | Less inhibition<br><br>More inhibition |
| Malignant melanoma | *2DL2* and/or *2DL3* | *HLA-C1* | Increased | More inhibition |
| Psoriatic arthritis | *2DS1* and/or *2DS2* | *HLA-C1* homozygous or *HLA-C2* homozygous | Increased | Less inhibition |
| Type 1 diabetes | *2DS2* | *HLA-C1* and no *HLA-C2*, no *HLA-Bw4* | Increased | Less inhibition |
| Preeclampsia | *2DL1* with fewer *2DS* (mother) | *HLA-C2* (fetus) | Increased | More inhibition |

SOURCE: S. Rajagopalan and E. Long, 2005, Understanding how combinations of HLA and KIR genes influence disease. *Journal of Experimental Medicine* **201**:1025.

- The TCR on NKT cells does not recognize MHC-bound peptides but rather a glycolipid presented by the nonpolymorphic CD1d molecule (see Figure 8-25).

- NKT cells do not form memory cells.

- NKT cells do not express a number of markers characteristic of T lymphocytes but do express those characteristic of NK cells.

The exact role of NKT cells in immunity remains to be defined. One clue is that development of NKT cells in the thymus requires a lysosomal glycosphingolipid, iGb3 (isoglobotrihexosylceramide): similar glycosyl ceramides are present in many bacterial species, and since these activate the NKT cell response, NKT cells may be involved in antibacterial immunity. Experiments show that mice lacking NKT cells mount a deficient response to certain low-dose bacterial infections with *Sphingomonas* or *Ehrlichia*. High-dose *Sphingomonas* infection leads to sepsis and death in wild-type mice, but those lacking NKT cells survive this challenge, suggesting that the NKT cells respond to the bacterial antigens with cytokine expression. (See Chapter 12 for a description of the role of proinflammatory cytokines in the onset of sepsis.) Other data implicating the NKT cell in immunity to tumors suggest that NKT cells recognize lipid antigens specific to tumor cells.

# Antibody-Dependent Cell-Mediated Cytotoxicity

A number of cells that have cytotoxic potential express membrane receptors for the Fc region of the antibody molecule. When antibody is specifically bound to a target cell, cells bearing the Fc receptor can bind to the antibody Fc region and thus to the target cells, subsequently causing lysis of the target cell. Although these cytotoxic cells are nonspecific for antigen, the specificity of the antibody directs them to specific target cells. This type of cytotoxicity is referred to as **antibody-dependent cell-mediated cytotoxicity (ADCC).**

Among the cells that can mediate ADCC are NK cells, macrophages, monocytes, neutrophils, and eosinophils. Antibody-dependent cell-mediated killing of cells infected with the measles virus can be observed in vitro by adding anti-measles antibody together with macrophages to a culture of measles-infected cells. Similarly, cell-mediated killing of helminths, such as schistosomes or blood flukes, can be observed in vitro by incubating larvae (schistosomules) with antibody to the schistosomules together with eosinophils.

Target-cell killing by ADCC appears to involve a number of different cytotoxic mechanisms but not complement-mediated lysis (Figure 14-15). When macrophages, neutrophils, or eosinophils bind to a target cell by way of the Fc receptor, they become more active metabolically; as a result, the level of lytic enzymes in their cytoplasmic lysosomes or granules increases. Release of these lytic enzymes at the site of the Fc-mediated contact may result in damage to the target cell. In addition, activated monocytes, macrophages, and NK cells have been shown to secrete tumor necrosis factor (TNF), which may have a cytotoxic effect on the bound target cell. Since both NK cells and eosinophils contain perforin in cytoplasmic granules, their target-cell killing also may involve perforin-mediated membrane damage similar to the mechanism described for CTL-mediated cytotoxicity.

# Experimental Assessment of Cell-Mediated Cytotoxicity

Three experimental systems have been particularly useful for measuring the activation and effector phases of cell-mediated cytotoxic responses:

- The *mixed-lymphocyte reaction (MLR)* is an in vitro system for assaying $T_H$-cell proliferation in a cell-mediated response.

- *Cell-mediated lympholysis (CML)* is an in vitro assay of effector cytotoxic function.

- The *graft-versus-host (GVH) reaction* in experimental animals provides an in vivo system for studying cell-mediated cytotoxicity.

## Coculturing T cells with foreign cells stimulates MLR

During the 1960s, early in the history of modern cellular immunology, immunologists observed that when rat lymphocytes were cultured on a monolayer of mouse fibroblast cells, the rat lymphocytes proliferated and destroyed the mouse fibroblasts. In 1970, several groups discovered that functional CTLs could also be generated by coculturing allogeneic spleen cells in a system termed the **mixed-lymphocyte reaction (MLR).** The T lymphocytes in an MLR undergo extensive blast transformation and cell proliferation. The degree of proliferation can be assessed by adding [³H]-thymidine to the culture medium and monitoring uptake of label into DNA in the course of repeated cell divisions.

Both populations of allogeneic T lymphocytes proliferate in an MLR unless one population is rendered unresponsive by treatment with mitomycin C or lethal x-irradiation (Figure 14-16). In the latter system, called a one-way MLR, the unresponsive population provides stimulator cells that express alloantigens foreign to the responder T cells. Within 24 to 48 hours, the responder T cells begin dividing in response to the alloantigens of the stimulator cells, and by 72 to 96 hours, an expanding population of functional CTLs is generated. With this experimental system, functional CTLs can be generated entirely in vitro, after which their activity can be assessed with various effector assays.

The significant role of $T_H$ cells in the one-way MLR can be demonstrated by use of antibodies to the $T_H$-cell membrane marker CD4. In a one-way MLR, responder $T_H$ cells recognize allogeneic class II MHC molecules on the stimulator cells and proliferate in response to these differences. Removal

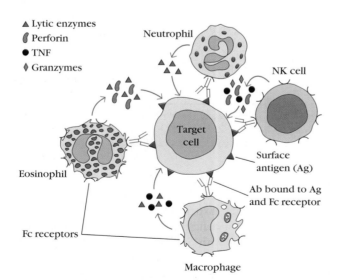

▲ Lytic enzymes
🍳 Perforin
● TNF
♦ Granzymes

Neutrophil

NK cell

Target cell

Surface antigen (Ag)

Ab bound to Ag and Fc receptor

Eosinophil

Fc receptors

Macrophage

**FIGURE 14-15 Antibody-dependent cell-mediated cytotoxicity (ADCC).** Nonspecific cytotoxic cells are directed to specific target cells by binding to the Fc region of antibody bound to surface antigens on the target cells. Various substances (e.g., lytic enzymes, TNF, perforin, granzymes) secreted by the nonspecific cytotoxic cells then mediate target-cell destruction.

(a)

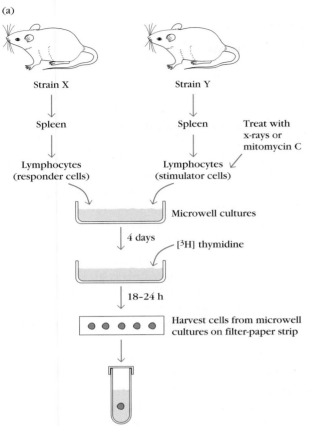

(b)

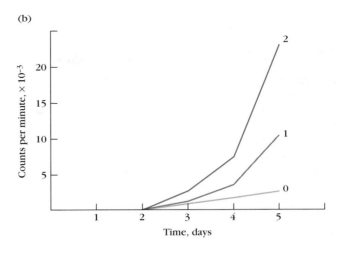

**FIGURE 14-16 One-way mixed-lymphocyte reaction (MLR).**
(a) This assay measures the proliferation of lymphocytes from one strain (responder cells) in response to allogeneic cells that have been x-irradiated or treated with mitomycin C to prevent proliferation (stimulator cells). The amount of [³H]-thymidine incorporated into the DNA is directly proportional to the extent of responder-cell proliferation.

(b) The amount of [³H]-thymidine uptake in a one-way MLR depends on the degree of differences in class II MHC molecules between the stimulator and responder cells. Curve 0 = no class II MHC differences; curve 1 = one class II MHC difference; curve 2 = two class II MHC differences. These results demonstrate that the greater the class II MHC differences, the greater the proliferation of responder cells.

of the $CD4^+$ $T_H$ cells from the responder population with anti-CD4 plus complement abolishes the MLR and prevents generation of CTLs. In addition to $T_H$ cells, accessory cells such as dendritic cells are necessary for the MLR to proceed. When adherent cells are removed from the stimulator population, the proliferative response in the MLR is abolished and functional CTLs are no longer generated. The function of these accessory cells is to activate the class II MHC–restricted $T_H$ cells, whose proliferation is measured in the MLR. In the absence of $T_H$-cell activation, there is no proliferation.

## CTL activity can be demonstrated by CML

Development of the **cell-mediated lympholysis (CML)** assay was a major experimental advance that contributed to understanding of the mechanism of target-cell killing by CTLs. In this assay, suitable target cells are labeled intracellularly with chromium-51 ($^{51}Cr$) by incubating the target cells with $Na_2{}^{51}CrO_4$. After the $^{51}Cr$ diffuses into a cell, it binds to cytoplasmic proteins, reducing passive diffusion of the label out of the cell. When specifically activated CTLs are incubated for

1 to 4 hours with such labeled target cells, the cells lyse and the $^{51}Cr$ is released. The amount of $^{51}Cr$ released correlates directly with the number of target cells lysed by the CTLs. By means of this assay, the specificity of CTLs for allogeneic cells, tumor cells, virus-infected cells, and chemically modified cells has been demonstrated (Figure 14-17).

The T cells responsible for CML were identified by selectively depleting different T-cell subpopulations by means of antibody-plus-complement lysis. In general, the activity of CTLs exhibits class I MHC restriction; that is, they can kill only target cells that present antigen associated with syngeneic class I MHC molecules. Occasionally, however, class II–restricted $CD4^+$ T cells have been shown to function as CTLs.

## The GVH reaction is an indication of cell-mediated cytotoxicity

The **graft-versus-host (GVH) reaction** develops when immunocompetent lymphocytes are injected into an allogeneic recipient whose immune system is compromised. Because the donor and recipient are not genetically identical, the

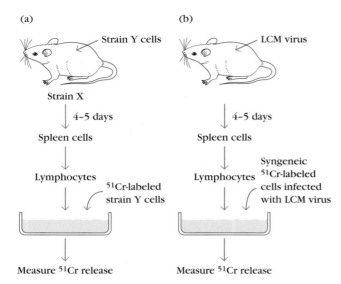

**FIGURE 14-17  In vitro cell-mediated lympholysis (CML) assay.**
This assay can measure the activity of cytotoxic T lymphocytes (CTLs) against allogeneic cells (a) or virus-infected cells (b). In both cases the release of ⁵¹Cr into the supernatant indicates the presence of CTLs that can lyse the target cells.

grafted lymphocytes begin to attack the host, and the host's compromised state prevents an immune response against the graft. In humans, GVH reactions often develop after transplantation of bone marrow into patients who have had radiation exposure or who have leukemia, immunodeficiency diseases, or autoimmune anemias. The clinical manifestations of the GVH reaction include diarrhea, skin lesions, jaundice, spleen enlargement, and death. Epithelial cells of the skin and gastrointestinal tract often become necrotic, causing the skin and intestinal lining to be sloughed.

Experimentally, GVH reactions develop when immunocompetent lymphocytes are transferred into an allogeneic neonatal or x-irradiated animal. The recipients, especially neonatal ones, often exhibit weight loss. The grafted lymphocytes generally are carried to a number of organs, including the spleen, where they begin to proliferate in response to the allogeneic MHC antigens of the host. This proliferation induces an influx of host cells and results in visible spleen enlargement, or splenomegaly. The intensity of a GVH reaction can be assessed by calculating the *spleen index* as follows:

$$\text{Spleen index} = \frac{\text{weight of experimental spleen/total body weight}}{\text{weight of control spleen/total body weight}}$$

A spleen index of 1.3 or greater is considered to be indicative of a positive GVH reaction. Spleen enlargement results from proliferation of both CD4⁺ and CD8⁺ T-cell populations. NK cells also have been shown to play a role in the GVH reaction, and these cells may contribute to some of the skin lesions and intestinal wall damage observed.

## SUMMARY

- The cell-mediated branch of the immune system involves two types of antigen-specific effector cells: cytotoxic T lymphocytes (CTLs) and CD4⁺ T cells. Compared with naive $T_H$ and $T_C$ cells, the effector cells are more easily activated, express higher levels of cell adhesion molecules, exhibit different trafficking patterns, and produce both soluble and membrane effector molecules.

- The first phase of the CTL-mediated immune response involves the activation and differentiation of $T_C$ cells, called CTL precursors (CTL-Ps). Details of the activation process, such as the involvement of $T_H1$ cells, remain unknown.

- Antigen-specific CD8⁺ populations can be identified and tracked by labeling with MHC tetramers.

- The second phase of the CTL-mediated response involves several steps: TCR-MHC mediated recognition of target cells, formation of CTL–target-cell conjugates, repositioning of CTL cytoplasmic granules toward the target cell, granule release, formation of pores in the target-cell membrane, dissociation of the CTL from the target, and death of the target cell.

- CTLs induce cell death via two mechanisms: the perforin/granzyme pathway and the Fas/FasL pathway.

- Various nonspecific (non-MHC-restricted) cytotoxic cells (NK cells, neutrophils, eosinophils, macrophages) can also kill target cells. Many of these cells bind to the Fc region of antibody on target cells and subsequently release lytic enzymes, perforin, or TNF, which damage the target-cell membrane, a process called antibody-dependent cell-mediated cytotoxicity (ADCC).

- NK cells mediate lysis of tumor cells and virus-infected cells by perforin-induced pore formation, a mechanism similar to one employed by CTLs.

- NK-cell receptors fall into two structural groups: the lectin-like receptors and the Ig-like receptors. Activating or inhibitory receptors fall into both groups. On engagement of their ligands, the NK-cell receptors transduce signals via ITAMs (activating) or ITIMs (inhibiting).

- The expression of relatively high levels of class I MHC molecules on normal cells protects them against NK-cell-mediated killing. NK-cell killing is regulated by the balance between positive signals generated by the engagement of activating receptors and negative signals from inhibitory receptors.

- NKT cells have characteristics common to both T lymphocytes and NK cells; most express an invariant TCR and markers common to NK cells.

## References

Behrens, G., et al. 2004. Helper T cells, dendritic cells and CTL immunity. *Immunology and Cell Biology* **82**:84.

Bossi, G., and G. M. Griffiths. 2005. CTL secretory lysosomes: biosynthesis and secretion of a harmful organelle. *Seminars in Immunology* **17**:87.

Haddad, E., et al. 1995. Treatment of Chediak-Higashi syndrome by allogeneic bone marrow transplantation: report of 10 cases. *Blood* **11**:3328.

Klenerman, P., et al. 2002. Tracking T cells with tetramers: new tales from new tools. *Nature Reviews Immunology* **2**:263.

Lekstrom-Himes, J. A., and J. I. Gallin. 2000. Advances in immunology: immunodeficiency diseases caused by defects in phagocytes. *New England Journal of Medicine* **343**:1703.

Parham, P. 2005. MHC class I molecules and KIRs in human history: health and survival. *Nature Reviews Immunology* **5**:201.

Rajagopolan, S., and E. Long. 2005. Understanding how combinations of HLA and KIR genes influence disease. *Journal of Experimental Medicine* **201**:1025.

Russell, J. H., and T. J. Ley. 2002. Lymphocyte-mediated cytotoxicity. *Annual Review of Immunology* **20**:370.

Van Kaer, L., and S. Joyce. 2005. Innate immunity: NKT cells in the spotlight. *Current Biology* **15**:R430.

Waterhouse, N. J., et al. 2004. Cytotoxic lymphocytes: instigators of dramatic target cell death. *Biochemical Pharmacology* **68**:1033.

## Useful Web Sites

**http://www.cellsalive.com/ctl.htm**

This Cells Alive subsite has a time-lapse video of cytotoxic T lymphocytes (CTLs) recognizing, attacking, and killing a much larger influenza-infected target.

## Study Questions

**CLINICAL FOCUS QUESTION** One inherited combination of KIR and MHC genes leads to increased susceptibility to a form of arthritis. Would you expect this to be caused by increased activation of NK activity or decreased activation? Could an increase in susceptibility to diabetes be explained using the same logic? Explain.

1. Indicate whether each of the following statements is true or false. If you believe a statement is false, explain why.

   a. Cytokines can regulate which branch of the immune system is activated.
   b. Both CTLs and NK cells release perforin after interacting with target cells.
   c. Antigen activation of naive CTL-Ps requires a costimulatory signal delivered by interaction of CD28 and B7.
   d. CTLs use a single mechanism to kill target cells.
   e. The secretion of certain critical cytokines is the basis of the role played by T cells in DTH reactions.

2. You have a monoclonal antibody specific for LFA-1. You perform CML assays of a CTL clone, using target cells for which the clone is specific, in the presence and absence of this antibody. Predict the relative amounts of $^{51}$Cr released in the two assays. Explain your answer.

3. You decide to coculture lymphocytes from the strains listed in the following table in order to observe the mixed-lymphocyte reaction (MLR). In each case, indicate which lymphocyte population(s) you would expect to proliferate.

| Population 1 | Population 2 | Proliferation |
|---|---|---|
| C57BL/6 (H-2$^b$) | CBA (H-2$^k$) | |
| C57BL/6 (H-2$^b$) | CBA (H-2$^k$) Mitomycin C–treated | |
| C57BL/6 (H-2$^b$) | (CBA × C57BL/6) F1 (H-2$^{k/b}$) | |
| C57BL/6 (H-2$^b$) | C57L (H-2$^b$) | |

4. In the mixed-lymphocyte reaction (MLR), the uptake of [$^3$H] thymidine is often used to assess cell proliferation.

   a. Which cell type proliferates in the MLR?
   b. How could you prove the identity of the proliferating cell?
   c. Explain why production of IL-2 also can be used to assess cell proliferation in the MLR.

5. Indicate whether each of the properties listed below is exhibited by T$_H$ cells, CTLs, both T$_H$ cells and CTLs, or neither cell type.

   a. _____ Can make IFN-γ
   b. _____ Can make IL-2
   c. _____ Is class I MHC restricted
   d. _____ Expresses CD8
   e. _____ Is required for B-cell activation
   f. _____ Is cytotoxic for target cells
   g. _____ Is the main proliferating cell in an MLR
   h. _____ Is the effector cell in a CML assay
   i. _____ Is class II MHC restricted
   j. _____ Expresses CD4
   k. _____ Expresses CD3
   l. _____ Adheres to target cells by LFA-1
   m. _____ Can express the IL-2 receptor
   n. _____ Expresses the αβ T-cell receptor
   o. _____ Is the principal target of HIV
   p. _____ Responds to soluble antigens alone
   q. _____ Produces perforin
   r. _____ Expresses the CD40 ligand on its surface

6. Mice from several different inbred strains were infected with LCM virus, and several days later their spleen cells were

| Source of primed spleen cells | $^{51}$Cr release from LCM-infected target cells | | | |
|---|---|---|---|---|
| | B10.D2 (H-2$^d$) | B10 (H-2$^b$) | B10.BR (H-2$^k$) | (BALB/c × B10) F1 (H-2$^{b/d}$) |
| B10.D2 (H-2$^d$) | | | | |
| B10 (H-2$^b$) | | | | |
| BALB/c (H-2$^d$) | | | | |
| BALB/c × B10 (H-2$^{b/d}$) | | | | |

isolated. The ability of the primed spleen cells to lyse LCM-infected, $^{51}$Cr-labeled target cells from various strains was determined. In the table at the bottom of page 369, indicate with a (+) or (−) whether the spleen cells listed in the left column would cause $^{51}$Cr release from the target cells listed in the headings across the top of the table.

7. A mouse is infected with influenza virus. How could you assess whether the mouse has $T_H$ and $T_C$ cells specific for influenza?

8. Explain why NK cells from a given host will kill many types of virus-infected cells but do not kill normal cells from that host.

9. Consider the following genetically altered mice and predict the outcome of the indicated procedures. H-2$^d$ mice in which both perforin and Fas ligand have been knocked out are immunized with LCM virus. One week after immunization, T cells from these mice are harvested and tested for cytotoxicity on the following:

   a. Target cells from normal LCM-infected H-2$^b$ mice
   b. Target cells from normal H-2$^d$ mice
   c. Target cells from H-2$^d$ mice in which both perforin and Fas have been knocked out
   d. Target cells from LCM-infected normal H-2$^d$ mice
   e. Target cells from H-2$^d$ mice in which both perforin and FasL have been knocked out

10. You wish to determine the levels of class I–restricted T cells in an HIV-infected individual that are specific for a peptide generated from gp120, a component of the virus. Assume that you know the HLA type of the subject. What method would you use and how would you perform the analysis? Be as specific as you can.

11. Indicate whether each of the following statements regarding Fas-mediated programmed cell death is true or false. If you believe a statement is false, explain why.

   a. It ultimately results in death of the cell.
   b. The interaction of Fas ligand and its receptor results in the recruitment of adapter proteins in the target cell.
   c. The caspase cascade results in the cleavage of proteins within the target cell.
   d. Fas ligand binding with Fas receptor stimulates a specific G protein.
   e. Caspase-8 may be activated by FADD or granzymes.
   f. FasL (Fas ligand) is expressed on the target cell.

12. NK cells do not express TCR molecules, yet they bind to class 1 MHC molecules on potential target cells.

   a. Explain how NK cells lacking TCRs can recognize infected cells.
   b. What is the mechanism used by NK cells to kill target cells?
   c. From what precursor cells do NK cells arise?

13. ADCC depends on the production of antibodies to recognize target cells. Describe the consequence if antibodies were developed against antigens expressed on normal cells as well as infected cells.

**ANALYZE THE DATA** E. Jäger and colleagues isolated an antigen from a tumor in a cancer patient who expressed HLA-A2. To characterize the cell-mediated immune response of the patient to this tumor antigen, the investigators generated a series

of peptide fragments from the tumor antigen and pulsed T2 cells with these peptides. They then measured the patient CTL response against each of these targets. Answer the following questions based on the data in the table and what you have learned from this book.

   a. Which epitope(s) of the tumor antigen were recognized by T cells as determined by these investigators? Explain your answer.
   b. Immunologists have identified anchor residues on HLA-A2 molecules that are the most important for antigen presentation. What amino acids are most likely to bind HLA-A2 anchor residues if these amino acids must be conserved? Explain your answer.
   c. Some of the peptides inspected by these investigators are poorly immunogenic. One explanation is that less immunogenic peptides may bind anchor residues ineffectively. Propose a different but reasonable hypothesis to explain why some of the peptides are poor immunogens.
   d. The investigators pulsed T2 cells with peptides and used CTLs from the patient's peripheral blood to perform CTL assays. What HLA-A molecule is expressed by T2 cells? Explain your answer.
   e. In what way is peptide 2 an unusual epitope?

| Peptide fragments derived from antigen isolated from tumor | | | |
|---|---|---|---|
| Number | Peptide sequence | Position | Specific lysis* |
| 1 | SLAQDAPPLPV | 108–118 | 0 |
| 2 | SLLMWITQCFL | 157–167 | 55 |
| 3 | QLSISSCLQQL | 146–156 | 1 |
| 4 | QLQLSISSCL | 144–153 | 1 |
| 5 | LLMWITQCFL | 158–167 | 15 |
| 6 | RLTAADHRQL | 136–145 | 5 |
| 7 | FTVSGNILTI | 126–135 | 1 |
| 8 | ITQCFLPVFL | 162–171 | 1 |
| 9 | SLAQDAPPL | 108–116 | 3 |
| 10 | PLPVPGVLL | 115–123 | 1 |
| 11 | WITQCFLPV | 161–169 | 5 |
| 12 | SLLMWITQC | 157–165 | 78 |
| 13 | RLLEFYLAM | 86–94 | 7 |
| 14 | SISSCLQQL | 148–156 | 6 |
| 15 | LMWITQCFL | 159–167 | 4 |
| 16 | QLQLSISSC | 144–152 | 1 |
| 17 | CLQQLSLLM | 152–160 | 1 |
| 18 | QLSLLMWIT | 155–163 | 84 |
| 19 | NILTIRLTA | 131–139 | 2 |
| 20 | GVLLKEFTV | 120–128 | 3 |
| 21 | ILTIRLTAA | 132–140 | 12 |
| 22 | TVSGNILTI | 127–135 | 4 |
| 23 | GTGGSTGDA | 7–15 | 9 |
| 24 | ATPMEAELA | 97–105 | 1 |
| 25 | FTVSGNILT | 126–134 | 1 |
| 26 | LTAADHRQL | 137–145 | 9 |

*Specific lysis is a relative measure of lytic activity.

## Interactive Study

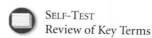

www.whfreeman.com/kuby

SELF-TEST
Review of Key Terms

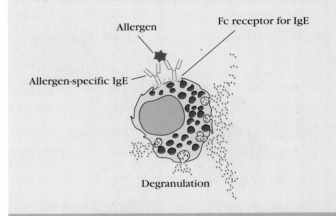

# chapter 15

# Hypersensitivity Reactions

Allergen

Fc receptor for IgE

Allergen-specific IgE

Degranulation

*IgE-mediated or type I hypersensitivity.*

- Gell and Coombs Classification

- IgE-Mediated (Type I) Hypersensitivity

- Antibody-Mediated Cytotoxic (Type II) Hypersensitivity

- Immune Complex–Mediated (Type III) Hypersensitivity

- Type IV or Delayed-Type Hypersensitivity (DTH)

**A**N IMMUNE RESPONSE MOBILIZES A BATTERY OF effector molecules that act to remove antigen by various mechanisms described in previous chapters. Generally, these effector molecules induce a localized inflammatory response that eliminates antigen without extensively damaging the host's tissue. Under certain circumstances, however, the inflammatory response can have deleterious effects, resulting in significant tissue injury, serious disease, or even death. Such an inappropriate and damaging immune response is termed **hypersensitivity.** Although the word *hypersensitivity* implies an increased response, the immune response is not always heightened but may instead be an inappropriate response to an antigen.

That the immune system could respond inappropriately to antigenic challenge was recognized early in the twentieth century. Two French scientists, Paul Portier and Charles Richet, investigated the problem of bathers in the Mediterranean reacting violently to the stings of Portuguese man-of-war jellyfish. Portier and Richet concluded that the localized reaction of the bathers was the result of toxins. To counteract this reaction, the scientists experimented with the use of isolated jellyfish toxins as vaccines. Their first attempts met with disastrous results. Portier and Richet injected dogs with the purified toxins, followed later by a booster of toxins. Instead of reacting to the booster by producing antibodies against the toxins, the dogs immediately reacted with vomiting, diarrhea, asphyxia, and death. Clearly this was an instance where the sensitized animals "overreacted" to the antigen. Richet coined the term *anaphylaxis,* loosely translated from Greek to mean the opposite of *prophylaxis,* to describe this overreaction. Richet was subsequently awarded the Nobel Prize in Physiology or Medicine in 1913 for the discovery of anaphylaxis.

Hypersensitivity reactions may develop in the course of either a humoral or a cell-mediated immune response. Anaphylactic reactions initiated by antibody or antigen-antibody complexes are referred to as **immediate hypersensitivity,** because the symptoms are manifest within minutes or hours after a sensitized recipient encounters antigen. **Delayed-type hypersensitivity** (DTH) is so named in recognition of the delay of symptoms until days after exposure. This chapter examines the mechanisms and consequences of the four primary types of hypersensitivity reactions.

## Gell and Coombs Classification

Hypersensitivity reactions can be distinguished by the type of immune response and differences in the effector molecules generated in the course of the reaction. In immediate hypersensitivity reactions, different antibody isotypes induce specific immune effector molecules. IgE antibodies, for example, induce mast-cell degranulation with release of histamine and other biologically active molecules. In contrast, IgG and IgM antibodies induce hypersensitivity reactions by activating complement. The effector molecules in the complement reactions are the membrane-attack complex and the anaphylactic complement split products C3a and C5a. In delayed-type hypersensitivity reactions, the effector molecules are various cytokines secreted by activated $T_H$ cells or $T_C$ cells themselves.

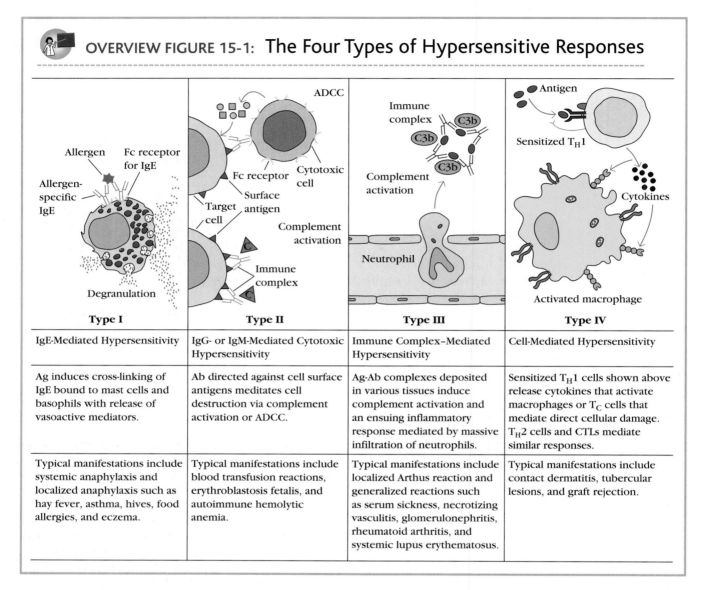

**OVERVIEW FIGURE 15-1:** The Four Types of Hypersensitive Responses

| Type I | Type II | Type III | Type IV |
|---|---|---|---|
| IgE-Mediated Hypersensitivity | IgG- or IgM-Mediated Cytotoxic Hypersensitivity | Immune Complex–Mediated Hypersensitivity | Cell-Mediated Hypersensitivity |
| Ag induces cross-linking of IgE bound to mast cells and basophils with release of vasoactive mediators. | Ab directed against cell surface antigens mediates cell destruction via complement activation or ADCC. | Ag-Ab complexes deposited in various tissues induce complement activation and an ensuing inflammatory response mediated by massive infiltration of neutrophils. | Sensitized $T_H1$ cells shown above release cytokines that activate macrophages or $T_C$ cells that mediate direct cellular damage. $T_H2$ cells and CTLs mediate similar responses. |
| Typical manifestations include systemic anaphylaxis and localized anaphylaxis such as hay fever, asthma, hives, food allergies, and eczema. | Typical manifestations include blood transfusion reactions, erythroblastosis fetalis, and autoimmune hemolytic anemia. | Typical manifestations include localized Arthus reaction and generalized reactions such as serum sickness, necrotizing vasculitis, glomerulonephritis, rheumatoid arthritis, and systemic lupus erythematosus. | Typical manifestations include contact dermatitis, tubercular lesions, and graft rejection. |

As it became clear that different immune mechanisms give rise to hypersensitivity reactions, P. G. H. Gell and R. R. A. Coombs proposed a classification scheme in which hypersensitivity reactions are divided into four types. Three types of hypersensitivity occur within the humoral branch and are mediated by antibody or antigen-antibody complexes: IgE-mediated (type I), antibody-mediated (type II), and immune complex–mediated (type III). A fourth type of hypersensitivity depends on activation of T cells within the cell-mediated branch and is termed delayed-type hypersensitivity, or DTH (type IV). Each type involves distinct immune mechanisms, cells, and mediator molecules (Figure 15-1). The Gell and Coombs classification scheme has served an important function in identifying the differences in the pathogenic immune mechanisms and the tissue damage characteristic of the various hypersensitivity reactions. However, it is important to point out that the complex mix of humoral and cell-mediated immune reactions and effector mechanisms often present in clinical hypersensitivity disorders blur the boundaries between the four categories.

# IgE-Mediated (Type I) Hypersensitivity

A type I hypersensitivity reaction is induced by certain types of antigens referred to as **allergens** and has all the hallmarks of a normal humoral response. That is, an allergen induces a humoral antibody response by the same mechanisms as described in Chapter 11 for other soluble antigens, resulting in the generation of antibody-secreting plasma cells and memory cells. What distinguishes a type I hypersensitive response from a normal humoral response is that the plasma cells secrete IgE (see Chapter 4) in response to the activation of allergen-specific $T_H2$ cells. This class of antibody binds with high affinity to **Fc receptors** on the surface of tissue

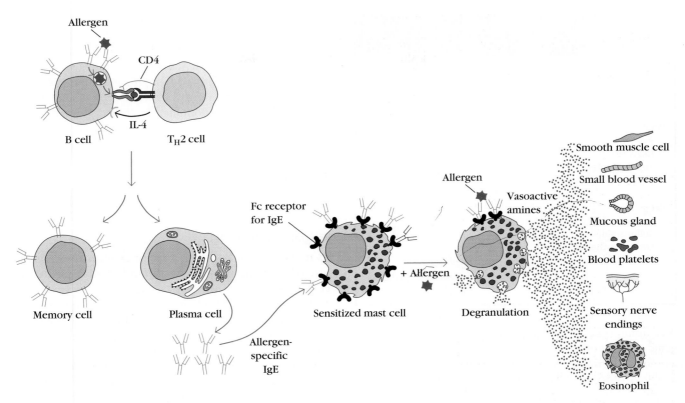

**FIGURE 15-2 General mechanism underlying an immediate type I hypersensitivity reaction.** Exposure to an allergen activates $T_H2$ cells that stimulate B cells to form IgE-secreting plasma cells. The secreted IgE molecules bind to IgE-specific Fc receptors (FcεRI) on mast cells and blood basophils. (Many molecules of IgE with various specificities can bind to the FcεRI.) Second exposure to the allergen leads to cross-linking of the bound IgE, triggering the release of pharmacologically active mediators (vasoactive amines) from mast cells and basophils. The mediators cause smooth muscle contraction, increased vascular permeability, and vasodilation.

mast cells and blood basophils; mast cells and basophils coated by IgE are said to be sensitized. A later exposure to the same allergen cross-links the membrane-bound IgE on sensitized mast cells and basophils, causing **degranulation** of these cells (Figure 15-2). The pharmacologically active mediators released from the granules act on the surrounding tissues. The principal effects—vasodilation and smooth muscle contraction—may be either systemic or localized, depending on the extent of mediator release. Today, the term **allergy** has come to be used interchangeably with type I hypersensitivity.

## There are several common components of type I reactions

Figure 15-2 depicts the general components that are critical to the development of type I hypersensitivity reactions. In this section, we will consider these common components first; further sections will then describe the mechanism of degranulation, biologically active mediators, clinical manifestations, and therapeutic approaches in type I reactions.

### Allergens

The majority of humans mount significant IgE responses only as a defense against parasitic infections. After an individual has been exposed to a parasite, serum IgE levels increase and remain high until the parasite is successfully cleared from the body. Some people, however, may have an abnormality called **atopy,** a hereditary predisposition to the development of immediate hypersensitivity reactions against common environmental antigens. The IgE regulatory defects suffered by atopic individuals allow nonparasitic antigens to stimulate inappropriate IgE production, leading to tissue-damaging type I hypersensitivity. The term *allergen* refers specifically to nonparasitic antigens capable of stimulating type I hypersensitive responses in allergic individuals on repeated exposure.

The abnormal IgE response of atopic individuals has a strong genetic component—it often runs in families. Atopic individuals have abnormally high levels of circulating IgE and elevated numbers of circulating eosinophils. These individuals are more susceptible to allergies such as hay fever, eczema, and asthma. The genetic propensity to atopic responses has been mapped to several candidate loci by gene linkage analyses. One locus, on chromosome 5, is linked to a region that encodes a variety of cytokines, including IL-3, IL-4, IL-5, IL-9, IL-13, and granulocyte-monocyte colony-stimulating factor (GM-CSF), all of which promote the production of IgE. Another locus, on chromosome 11, is linked to a region that encodes the β chain of the high-affinity IgE receptor. IgE responses have also been linked to MHC genes

| TABLE 15-1 | Common allergens associated with type I hypersensitivity |
|---|---|

| | |
|---|---|
| Proteins | Foods |
|    Foreign serum |    Nuts |
|    Vaccines |    Seafood |
| |    Eggs |
| Plant pollens |    Peas, beans |
|    Rye grass |    Milk |
|    Ragweed | |
|    Timothy grass | Insect products |
|    Birch trees |    Bee venom |
| |    Wasp venom |
| Drugs |    Ant venom |
|    Penicillin |    Cockroach calyx |
|    Sulfonamides |    Dust mites |
|    Local anesthetics | |
|    Salicylates | Mold spores |
| | Animal hair and dander |
| | Latex |

on chromosome 6. It is known that inherited atopy is multigenic and that other loci probably also are involved. Indeed, as more information from the Human Genome Project is analyzed, other candidate genes may be revealed.

Most allergic IgE responses occur on the surfaces of mucous membranes in response to allergens that enter the body by either inhalation or ingestion. Of the common allergens listed in Table 15-1, few have been purified and characterized. Those that have include the allergens from rye grass pollen, ragweed pollen, codfish, birch pollen, timothy grass pollen, bee venom, and latex. (In recent years, hypersensitivity to latex allergens in gloves has been increasing among health-care workers and is an example of an occupational allergy.) Each of these allergens has been shown to be a multiantigenic system that contains a number of allergenic components. Ragweed pollen, a major allergen in the United States, is a case in point. It has been reported that a square mile of ragweed yields 16 tons of pollen in a single season. All regions of the United States are plagued by ragweed pollen as well as pollen from trees indigenous to the region. The pollen particles are inhaled, and enzymes in the mucous secretions dissolve their tough outer wall, releasing the allergenic substances. Chemical fractionation of ragweed has revealed a variety of substances, most of which are not allergenic but are capable of eliciting an IgM or IgG response. Of the five fractions that are allergenic (able to induce an IgE response), two evoke allergenic reactions in about 95% of ragweed-sensitive individuals and are called major allergens; these are designated the E and K fractions. The other three, named Ra3, Ra4, and Ra5, are minor allergens that induce an allergic response in only 20% to 30% of sensitive subjects.

Why do some allergens (e.g., ragweed pollen) induce potent $T_H2$ reactions and allergic responses, whereas other equally abundant allergens (e.g., nettle pollen) do not? No single physicochemical property seems to distinguish the highly allergic E and K fractions of ragweed from the less allergenic Ra3, Ra4, and Ra5 fractions and from the nonallergenic

fractions. Rather, allergens as a group appear to possess diverse properties. Some allergens, including foreign serum and egg albumin, are potent antigens; others, such as plant pollens, are weak antigens. Although most allergens are small, soluble, glycosylated proteins or protein-bound substances having a molecular weight between 15,000 and 40,000, attempts to identify a dominant biological property shared by allergens have failed. It appears that allergenicity is most likely related to the ability to evoke a $T_H2$-mediated immune response and so results from a complex series of interactions involving not only the allergen but also the dose, the sensitizing route, sometimes an adjuvant, and, most important, the genetics of the recipient.

### Reaginic Antibody (IgE)

As described in Chapter 4, a factor in human serum that reacted with allergens was first found by K. Prausnitz and H. Kustner in 1921. These investigators injected serum from an allergic individual intradermally into a nonallergic test subject. A later injection of the allergen at the site of the serum injection caused reddening and swelling at the site. Although the specificity of this reaction (called the P-K reaction) for a given allergen led to the hypothesis that reaginic, or "skin-fixing," antibodies caused the reaction, this was not proved until the existence of the antibody class responsible for allergic reactions was revealed. Experiments conducted by K. and T. Ishizaka in the mid-1960s demonstrated that the biological activity of reaginic antibody in a P-K test could be neutralized by rabbit antiserum against whole atopic human sera but not by rabbit antiserum specific for the four human immunoglobulin classes known at that time (IgA, IgG, IgM, and IgD) (Table 15-2). In addition, when rabbits were immunized with sera from ragweed-sensitive individuals, the rabbit antiserum could inhibit (neutralize) a positive ragweed P-K test even after precipitation of rabbit antibodies specific for the human IgG, IgA, IgM, and IgD isotypes. The Ishizakas called this new isotype IgE in reference to the E antigen of ragweed that they used to characterize it.

In normal individuals, the level of IgE in serum is the lowest of all antibody classes, falling within the range of 0.1 to 0.4 $\mu$g/ml; atopic individuals often have 10 times the normal concentration of IgE in their circulation. The low levels of serum IgE made physiochemical studies difficult, and it was not until the discovery of an IgE myeloma by S. G. O. Johansson and H. Bennich in 1967 that extensive chemical analysis of IgE could be undertaken. IgE was found to be composed of two heavy $\varepsilon$ and two light chains with a combined molecular weight of 190,000. The higher molecular weight compared to IgG (150,000) is due to the presence of an additional constant-region domain (see Figure 4-13). This additional domain ($C_\varepsilon2$), which replaces the hinge region in IgG, contributes to an altered conformation of the Fc portion of the molecule that enables it to bind to glycoprotein receptors on the surface of basophils and mast cells. IgE from allergic subjects differs from that of other antibody isotypes in the limited selection of $V_H$ genes that contribute to its variable region and the pattern of somatic hypermutation in its $\varepsilon$ heavy chains. The unusual

| TABLE 15-2 | Identification of IgE based on reactivity of atopic serum in P-K test | | |
|---|---|:---:|:---:|
| Serum | Treatment | Allergen added | P-K reaction at skin site |
| Atopic | None | − | − |
| Atopic | None | + | + |
| Nonatopic | None | + | − |
| Atopic | Rabbit antiserum to human atopic serum[*] | + | − |
| Atopic | Rabbit antiserum to human IgM, IgG, IgA, and IgD[†] | + | + |

[*]Serum from an atopic individual was injected into rabbits to produce antiserum against human atopic serum. When this antiserum was reacted with human atopic serum, it neutralized the P-K reaction.

[†]Serum from an atopic individual was reacted with rabbit antiserum to the known classes of human antibody (IgM, IgA, IgG, and IgD) to remove these isotypes from the atopic serum. The treated atopic serum continued to give a positive P-K reaction, indicating that a new immunoglobulin isotype was responsible for this reactivity.

SOURCE: Based on K. Ishizaka and T. Ishizaka, 1967, *Journal of Immunology* **99**:1187.

$V_H$ repertoire of these IgE molecules is most likely selected for by the complex properties of the allergen. Although the half-life of IgE in the serum is only 2 to 3 days, IgE that has been bound to its receptor on mast cells and basophils is stable in that state for a number of weeks. The persistence of IgE on the mast-cell surface can be attributed to the local production of IgE in the mucosal tissue in allergic disease and the strong association of IgE with its receptor.

## Mast Cells and Basophils

The cells that bind IgE were identified by incubating human leukocytes and tissue cells with either $^{125}$I-labeled IgE myeloma protein or $^{125}$I-labeled anti-IgE. In both cases, autoradiography revealed that the labeled probe bound with high affinity to blood basophils and tissue mast cells. Basophils are granulocytes that circulate in the blood of most vertebrates; in humans, they account for 0.5% to 1.0% of the circulating white blood cells. In contrast to tissue mast cells, circulating basophils must be recruited into tissues at the sites of inflammation. Their granulated cytoplasm stains with basic dyes; hence the name basophil. Electron microscopy reveals a multilobed nucleus, few mitochondria, numerous glycogen granules, and electron-dense membrane-bound granules scattered throughout the cytoplasm that contain pharmacologically active mediators (see Figure 2-9c).

Mast-cell precursors are formed in the bone marrow during hematopoiesis and are carried to virtually all vascularized peripheral tissues, where they differentiate into mature cells. Mast cells are found throughout connective tissue, particularly near blood and lymphatic vessels and nerves. Some tissues, including the skin and mucous membrane surfaces of the respiratory and gastrointestinal tracts, contain high concentrations of mast cells; skin, for example, contains 10,000 mast cells per mm$^3$. Electron micrographs of mast cells reveal numerous membrane-bound granules distributed throughout the cytoplasm, which, like those in basophils, contain pharmacologically active mediators (Figure 15-3). After activation, these mediators are released from the granules, resulting in the clinical manifestations of the type I hypersensitivity reaction.

(a)    (b)    (c)

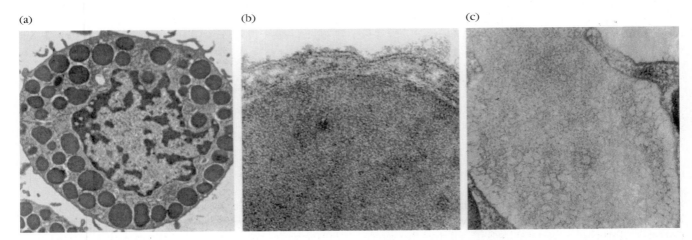

**FIGURE 15-3  Mast cell.** (a) Electron micrograph of a typical mast cell reveals numerous electron-dense membrane-bound granules prior to degranulation. (b) Close-up of intact granule underlying the plasma membrane of a mast cell. (c) Granule releasing its contents (toward top left) during degranulation. *[From S. Burwen and B. Satir, 1977, Journal of Cell Biology **73**:662.]*

Mast-cell populations in different anatomic sites differ significantly in the types and amounts of allergic mediators they contain and in their sensitivity to activating stimuli and cytokines. The ability of mast cells to adapt various phenotypes in response to their microenvironment is called mast-cell heterogeneity. Mast cells also secrete a large variety of cytokines that affect a broad spectrum of physiologic, immunologic, and pathologic processes (see Table 12-1).

## IgE-Binding Fc Receptors

The reaginic activity of IgE depends on its ability to bind to a receptor specific for the Fc region of the ε heavy chain. Two classes of FcεR have been identified, the high-affinity FcεRI and the low-affinity FcεRII, which are expressed by different cell types and differ by 1000-fold in their affinity for IgE.

**High affinity receptor (FCεRI)**  Mast cells and basophils constitutively express FcεRI, which binds IgE with exceptionally high affinity ($K_d = 1$–$2 \times 10^{-9}$ M). The high affinity of this receptor enables it to bind IgE despite the low serum concentration of IgE ($1 \times 10^{-7}$ M). The number of FcεRI molecules on human basophils is quite high, between 40,000 and 90,000. Several other cell types have also been shown to express FcεRI, including eosinophils, Langerhans cells, monocytes, and platelets, although at much lower levels.

The FcεRI receptor on mast cells and basophils contains four polypeptide chains: an α and β chain and two identical disulfide-linked γ chains (Figure 15-4a). Another form of the receptor lacking the β chain is found in monocytes and platelets. The external region of the α chain contains two domains of 90 amino acids that are homologous with the immunoglobulin-fold structure, placing the molecule in the immunoglobulin superfamily (see Figure 4-23). One FcεRI interacts with both $C_H3$ domains of one IgE molecule via the two Ig-like domains of the receptor α chain. This interaction was determined from the solved crystal structure of the α chain of FcεRI bound to the IgE Fc. Additional studies indicate that the $C_H2$ domains of IgE may help to stabilize the interaction with FcεRI, either directly or indirectly. The β chain spans the plasma membrane four times and is thought to link the α chain to the γ homodimer. The disulfide-linked γ chains extend a considerable distance into the cytoplasm and have the main responsibility for intracellular signal transduction. Each β and γ chain has a conserved sequence in its cytosolic domain known as an immunoreceptor tyrosine-based activation motif (ITAM). The β chain contributes to the formation of the signaling complex; the γ chain is homologous to the ξ chains of the T-cell receptor complex (see Figure 10-11) and the Ig-α/Ig-β chains associated with membrane immunoglobulin on B cells (see Figure 11-7). The ITAM motif on these receptors interacts with protein tyrosine

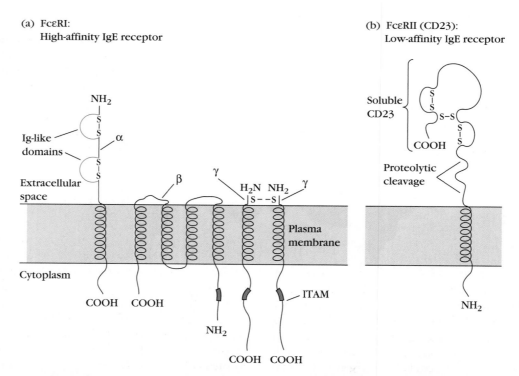

(a) FcεRI: High-affinity IgE receptor

(b) FcεRII (CD23): Low-affinity IgE receptor

**FIGURE 15-4  Schematic diagrams of the high-affinity FcεRI and low-affinity FcεRII receptors that bind the Fc region of IgE.** (a) FcεRI consists of an α chain that binds IgE, a β chain that participates in signaling, and two disulfide-linked γ chains that are the most important component in signal transduction. The β and γ chains contain a cytoplasmic ITAM, a motif also present in the Ig-α/Ig-β heterodimer of the B-cell receptor and in the T-cell receptor complex. (b) The single-chain FcεRII is unusual because it is oriented in the membrane with its $NH_2$-terminus directed toward the cell interior and its COOH-terminus directed toward the extracellular space.

kinases to transduce an activating signal to the cell. Allergen-mediated cross-linkage of the bound IgE results in aggregation of FcεRI and rapid tyrosine phosphorylation, which initiates the process of mast-cell degranulation. The role of FcεRI in anaphylaxis is confirmed by experiments conducted in mice that lack an FcεRI α chain. These mice have normal levels of mast cells but are resistant to localized and systemic anaphylaxis.

**Low-affinity receptor (FCεRII)** The other IgE receptor, designated FcεRII (or CD23), binds to the $C_H3$ domains of IgE and has a much lower affinity for IgE ($K_d = 1 \times 10^{-6}$ M) than does FcεRI (Figure 15-4b). CD23 has a single membrane-spanning region followed by an extracellular C-type lectin domain and is structurally unrelated to FcεRI. Two isoforms of CD23 have been identified, differing only slightly in the N-terminal cytoplasmic domain: CD23a is found on activated B cells, whereas CD23 is induced on various cell types by the cytokine IL-4. CD23 appears to play a variety of roles in regulating the intensity of the IgE response. Allergen cross-linkage of IgE bound to CD23 has been shown to activate B cells, alveolar macrophages, and eosinophils. When this receptor is blocked with monoclonal antibodies, IgE secretion by B cells is diminished. A soluble form of FcεRII (or sCD23), which is generated by autoproteolysis of the membrane receptor, has been shown to enhance IgE production by B cells. Interestingly, atopic individuals have higher levels of CD23 on their lymphocytes and macrophages and higher levels of sCD23 in their serum than do nonatopic individuals.

## IgE cross-linkage initiates degranulation

The biochemical events that mediate degranulation of mast cells and blood basophils have many features in common. For simplicity, in this section we give a general overview of mast-cell degranulation mechanisms without specifying the slight differences between mast cells and basophils. Although mast-cell degranulation generally is initiated by allergen cross-linkage of bound IgE, a number of non-IgE-mediated stimuli can also initiate the process, including the anaphylatoxins (C3a and C5a), various drugs, and even other mast-cell receptors. Here we focus on the biochemical events that follow allergen cross-linkage of bound IgE.

IgE-mediated degranulation begins when an allergen cross-links IgE that is bound (fixed) to the Fc receptor on the surface of a mast cell or basophil. In itself, the binding of IgE to FcεRI apparently has no effect on a target cell. It is only after allergen cross-links the fixed IgE-receptor complex that degranulation proceeds. The importance of cross-linkage is indicated by the finding that monovalent allergens, which cannot cross-link the fixed IgE, also do not initiate degranulation.

Experiments have revealed that the essential step in degranulation is cross-linkage of two or more FcεRI molecules—with or without bound IgE. Although cross-linkage is normally effected by the interaction of fixed IgE with divalent or multiva-

(a) Allergen cross-linkage of cell-bound IgE

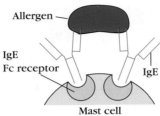

(c) Chemical cross-linkage of IgE

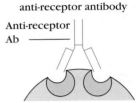

(b) Antibody cross-linkage of IgE

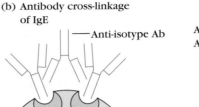

(d) Cross-linkage of IgE receptors by anti-receptor antibody

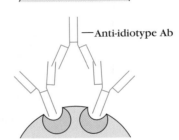

(e) Enhanced Ca²⁺ influx by ionophore that increases membrane permeability to Ca²⁺

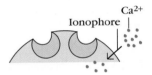

**FIGURE 15-5 Schematic diagrams of mechanisms that can initiate degranulation of mast cells.** Note that mechanisms (b) and (c) do not require allergen, mechanisms (d) and (e) require neither allergen nor IgE, and mechanism (e) does not even require receptor cross-linkage.

lent allergen, it can also be effected by a variety of experimental means that bypass the need for allergen and in some cases even for IgE (Figure 15-5).

## Intracellular events trigger mast-cell degranulation

The intracellular signaling events that ultimately result in mast-cell degranulation are multifaceted, involving cooperation among various protein and lipid kinases and phosphatases and rearrangements of the cytoskeleton. The Src family protein tyrosine kinase Lyn is associated with the cytoplasmic domain of the β chain of FcεRI in mast cells regardless of receptor cross-linkage. Cross-linkage of FcεRI receptors activates Lyn, which phosphorylates the tyrosines in the ITAMs of the β and γ chains, initiating a complex series of further phosphorylation reactions involving the tyrosine kinases Lyn, Syk, and Fyn and including the phosphorylation of phospholipase C (PLC). These phosphorylation events induce the production of a number of second messengers, including inositol 1,4,5-trisphosphate ($IP_3$)

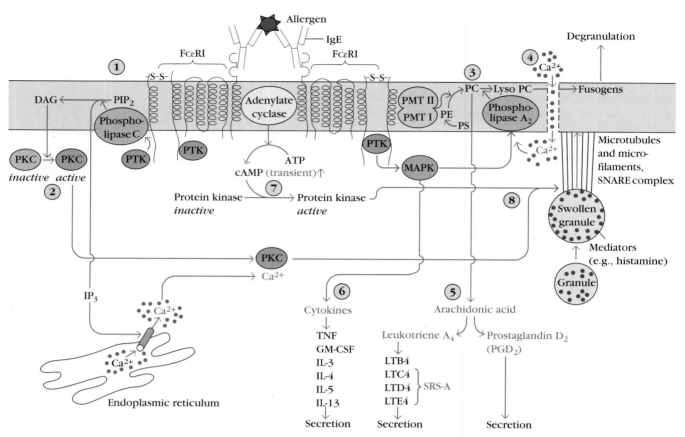

**FIGURE 15-6 Diagrammatic overview of biochemical events in mast-cell activation and degranulation.** Allergen cross-linkage of bound IgE results in FcεRI aggregation and activation of protein tyrosine kinase (PTK). (1) PTK then phosphorylates phospholipase C, which converts phosphatidylinositol-4,5 bisphosphate (PIP$_2$) into diacylglycerol (DAG) and inositol trisphosphate (IP$_3$). (2) DAG activates protein kinase C (PKC), which with Ca$^{2+}$ is necessary for microtubular assembly and the fusion of the granules with the plasma membrane. IP$_3$ is a potent mobilizer of intracellular Ca$^{2+}$ stores. (3) Cross-linkage of FcεRI also activates an enzyme that converts phosphatidylserine (PS) into phosphatidylethanolamine (PE). Eventually, PE is methylated to form phosphatidylcholine (PC) by the phospholipid methyl transferase enzymes I and II (PMT I and II). (4) The accumulation of PC on the exterior surface of the plasma membrane causes an increase in membrane fluidity and facilitates the formation of Ca$^{2+}$ channels. The resulting influx of Ca$^{2+}$ and PTK-activated mitogen-activated protein kinase (MAPK) activates phospholipase A$_2$, which promotes the breakdown of PC into lysophosphatidylcholine (lyso PC) and arachidonic acid. (5) Arachidonic acid is converted into potent mediators: the leukotrienes (slow-reactive substance of anaphylaxis, SRS-A) and prostaglandin D$_2$. (6) Activated MAPK also induces the secretion of cytokines by increasing transcription of cytokine genes. (7) FcεRI cross-linkage also activates the membrane adenylate cyclase, leading to a transient increase of cAMP within 15 seconds. A later drop in cAMP levels is mediated by protein kinase and is required for degranulation to proceed. (8) cAMP-dependent protein kinases are thought to phosphorylate the granule membrane proteins, thereby changing the permeability of the granules to water and Ca$^{2+}$. The consequent swelling of the granules and the formation of soluble N-ethylmaleimide attachment receptor (SNARE) protein complexes, facilitates fusion with the plasma membrane and release of the mediators.

and diacylglycerol (DAG), which mediate the process of degranulation. IP$_3$ increases intracellular Ca$^{2+}$ levels, and DAG, together with Ca$^{2+}$, activates protein kinase C (PKC; Figure 15-6). The increase in intracellular Ca$^{2+}$ concentration and activated PKC leads to degranulation. (A similar signaling sequence leads to the activation of T cells; see Figure 10-12).

Within 15 seconds after cross-linkage of FcεRI, methylation of various membrane phospholipids is observed, resulting in an increase in membrane fluidity and the formation of Ca$^{2+}$ channels. An increase of Ca$^{2+}$ reaches a peak within

2 minutes of FcεRI cross-linkage (Figure 15-7). This increase is due both to the uptake of extracellular Ca$^{2+}$ and to a release of Ca$^{2+}$ from intracellular stores in the endoplasmic reticulum (see Figure 15-6). The increase of Ca$^{2+}$ promotes the assembly of microtubules and the contraction of microfilaments, both of which are necessary for the movement of granules to the plasma membrane. SNARE (soluble N-ethylmaleimide attachment receptor) proteins, which are present on both the granule and plasma membrane, play a role in granule fusion with the membrane of the cell and exocytosis of the primary mediators associated with mast-cell

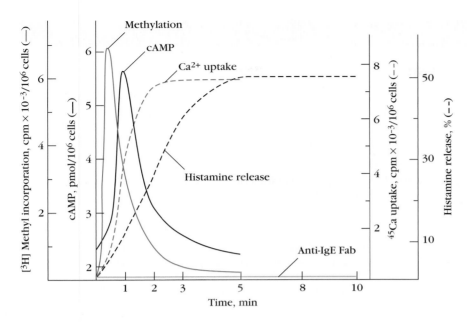

**FIGURE 15-7 Kinetics of major biochemical events that follow cross-linkage of bound IgE on cultured human basophils with F(ab')₂ fragments of anti-IgE.** Curves are shown for phospholipid methylation (solid blue), cAMP production (solid black), Ca²⁺ influx (dashed blue), and histamine release (dashed black). In control experiments with anti–IgE Fab fragments, no significant changes were observed. *[Adapted from T. Ishizaka et al., 1985, International Archives of Allergy and Applied Immunology **77:137**.]*

degranulation. In addition, the Ca²⁺ increase, along with the induction of a mitogen-activated protein kinase (MAPK), results in both cytokine production and the activation of the enzyme phospholipase A₂ (PLA₂). PLA₂ hydrolyzes membrane phospholipids, leading to the formation of arachidonic acid, which is converted into two classes of potent lipid mediators: **prostaglandins** and **leukotrienes** (see Figure 15-6). The importance of the Ca²⁺ increase in mast-cell degranulation is highlighted by the use of drugs, such as disodium cromoglycate (cromolyn sodium), that block this influx as a treatment for allergies.

Concomitant with phospholipid methylation and Ca²⁺ increase, there is a transient increase in the activity of membrane-bound adenylate cyclase, with a rapid peak of its reaction product, cyclic adenosine monophosphate (cAMP), reached about 1 minute after cross-linkage of FcεRI (see Figure 15-7). The effects of cAMP are exerted through the activation of cAMP-dependent protein kinases, which phosphorylate proteins on the granule membrane, thereby increasing the permeability of the granules to water and Ca²⁺ (see Figure 15-6). The consequent swelling of the granules facilitates their fusion with the plasma membrane, releasing their contents. The increase in cAMP is transient and is followed by a drop in cAMP to levels below baseline (see Figure 15-7). This drop in cAMP appears to be necessary for degranulation to proceed; when cAMP levels are increased by certain drugs, the degranulation process is blocked. Several of these drugs are given to treat allergic disorders and are considered later in this section.

## Several pharmacologic agents mediate type I reactions

The clinical manifestations of type I hypersensitivity reactions are related to the biological effects of the mediators released during mast-cell or basophil degranulation. These mediators are pharmacologically active agents that act on local tissues as well as on populations of secondary effector cells, including eosinophils, neutrophils, T lymphocytes, monocytes, and platelets. The mediators thus serve as an amplifying terminal effector mechanism, much as the complement system serves as an amplifier and effector of an antigen-antibody interaction. When generated in response to parasitic infection, these mediators initiate beneficial defense processes, including vasodilation and increased vascular permeability, which brings an influx of plasma and inflammatory cells to attack the pathogen. On the other hand, mediator release induced by inappropriate antigens, such as allergens, results in unnecessary increases in vascular permeability and inflammation whose detrimental effects far outweigh any beneficial effect.

The mediators can be classified as either primary or secondary (Table 15-3). The primary mediators are pre-made before degranulation and are stored in the granules. The most significant primary mediators are histamine, proteases, eosinophil chemotactic factor, neutrophil chemotactic factor, and heparin. The secondary mediators are either synthesized after target-cell activation or released by the breakdown of membrane phospholipids during the degranulation process. The secondary mediators include platelet-activating factor, leukotrienes, prostaglandins, bradykinins, and various cytokines and chemokines. The differing manifestations of type I hypersensitivity in different species or different tissues partly reflect variations in the primary and secondary mediators present. The main biological effects of several of these mediators are described briefly in the next sections.

### Histamine

Histamine, which is formed by decarboxylation of the amino acid histidine, is a major component of mast-cell granules, accounting for about 10% of granule weight.

| TABLE 15-3 | Principal mediators involved in type I hypersensitivity |
|------------|---------------------------------------------------------|
| **Mediator** | **Effects** |
| | PRIMARY |
| Histamine, heparin | Increased vascular permeability; smooth muscle contraction |
| Serotonin (rodents) | Increased vascular permeability; smooth muscle contraction |
| Eosinophil chemotactic factor (ECF-A) | Eosinophil chemotaxis |
| Neutrophil chemotactic factor (NCF-A) | Neutrophil chemotaxis |
| Proteases (tryptase, chymase) | Bronchial mucus secretion; degradation of blood vessel basement membrane; generation of complement split products |
| | SECONDARY |
| Platelet-activating factor | Platelet aggregation and degranulation; contraction of pulmonary smooth muscles |
| Leukotrienes (slow reactive substance of anaphylaxis, SRS-A) | Increased vascular permeability; contraction of pulmonary smooth muscles |
| Prostaglandins | Vasodilation; contraction of pulmonary smooth muscles; platelet aggregation |
| Bradykinin | Increased vascular permeability; smooth muscle contraction |
| Cytokines | |
| IL-1 and TNF-$\alpha$ | Systemic anaphylaxis; increased expression of CAMs on venular endothelial cells |
| IL-4 and IL-13 | Increased IgE production |
| IL-3, IL-5, IL-6, IL-10, TGF-$\beta$, and GM-CSF | Various effects (see Table 12-1) |

Because it is stored—preformed—in the granules, its biological effects are observed within minutes of mast-cell activation. Once released from mast cells, histamine initially binds to specific receptors on various target cells. Four types of histamine receptors—designated $H_1$, $H_2$, $H_3$, and $H_4$—have been identified; these receptors have different tissue distributions and mediate different effects when they bind histamine. Serotonin is present in the mast cells of rodents and has effects similar to histamine.

Most of the biologic effects of histamine in allergic reactions are mediated by the binding of histamine to $H_1$ receptors. This binding induces contraction of intestinal and bronchial smooth muscles, increased permeability of venules, and increased mucus secretion by goblet cells. Interaction of histamine with $H_2$ receptors increases vasopermeability (due to contraction of endothelial cells) and vasodilation (by relaxing the smooth muscle of blood vessels), stimulates exocrine glands, and increases the release of acid in the stomach. Binding of histamine to $H_2$ receptors on mast cells and basophils suppresses degranulation; thus, histamine exerts negative feedback on the release of mediators.

### Leukotrienes and Prostaglandins

As secondary mediators, the leukotrienes and prostaglandins are not formed until the mast cell undergoes degranulation and the enzymatic breakdown of phospholipids in the plasma membrane. An ensuing enzymatic cascade generates the prostaglandins and the leukotrienes (see Figure 15-6). Therefore, the biological effects of these mediators take longer to become apparent. Their effects are more pro-nounced and longer lasting, however, than those of histamine. The leukotrienes mediate bronchoconstriction, increased vascular permeability, and mucus production. Prostaglandin $D_2$ causes bronchoconstriction.

The contraction of human bronchial and tracheal smooth muscles appears at first to be mediated by histamine, but within 30 to 60 seconds, further contraction is mediated by the leukotrienes and prostaglandins. Active at nanomole levels, the leukotrienes are as much as 1000 times more potent as bronchoconstrictors than histamine is, and they are also more potent stimulators of vascular permeability and mucus secretion. In humans, the leukotrienes are thought to contribute to the prolonged bronchospasm and buildup of mucus seen in asthmatics.

### Cytokines

Adding to the complexity of the type I reaction is the variety of cytokines released from mast cells and basophils. Some of these may contribute to the clinical manifestations of type I hypersensitivity. Human mast cells secrete IL-3, IL-4, IL-5, IL-6, IL-10, IL-13, GM-CSF, and TNF-$\alpha$, among others. These cytokines alter the local microenvironment, eventually leading to the recruitment of inflammatory cells such as neutrophils and eosinophils. IL-4 and IL-13 stimulate a $T_H2$ response and thus increase IgE production by B cells. IL-5 is especially important in the recruitment and activation of eosinophils. The high concentrations of TNF-$\alpha$ secreted by mast cells may contribute to shock in systemic anaphylaxis. (This effect may parallel the role of TNF-$\alpha$ in bacterial septic shock and toxic shock syndrome described in Chapter 12.)

## Type I reactions can be systemic or localized

The clinical manifestations of type I reactions can range from life-threatening conditions, such as systemic anaphylaxis and severe asthma, to localized reactions, such as hay fever and eczema, which are merely annoyances.

### Systemic Anaphylaxis

Systemic anaphylaxis is a shock-like and often fatal state, the onset of which occurs within minutes of a type I hypersensitivity reaction, usually initiated by an allergen introduced directly into the bloodstream or absorbed from the gut or skin. This was the response observed by Portier and Richet in dogs after antigenic challenge by jellyfish toxin. Systemic anaphylaxis can be induced in a variety of experimental animals and is seen occasionally in humans. Each species exhibits characteristic symptoms, which reflect differences in the distribution of mast cells and in the biologically active contents of their granules. The animal model of choice for studying systemic anaphylaxis has been the guinea pig. Anaphylaxis can be induced in guinea pigs with ease, and its symptoms closely parallel those observed in humans.

Active sensitization in guinea pigs is induced by a single injection of a foreign protein such as egg albumin. After an incubation period of about 2 weeks, the animal is usually challenged with an intravenous injection of the same protein. Within 1 minute, the animal becomes restless, its respiration becomes labored, and its blood pressure drops, a state called **anaphylactic shock.** As the smooth muscles of the gastrointestinal tract and bladder contract, the guinea pig defecates and urinates. Finally, bronchiole constriction results in death by asphyxiation within 2 to 4 minutes of the injection. These events all stem from the systemic vasodilation and smooth muscle contraction brought on by mediators released in the course of the reaction. Postmortem examination reveals that massive edema, shock, and bronchiole constriction are the major causes of death.

Systemic anaphylaxis in humans is characterized by a similar sequence of events. A wide range of antigens have been shown to trigger this reaction in susceptible humans, including the venom from bee, wasp, hornet, and ant stings; drugs, such as penicillin, insulin, and antitoxins; and seafood and nuts. If not treated quickly, these reactions can be fatal. Epinephrine is the drug of choice for treating systemic anaphylactic reactions. Epinephrine counteracts the effects of mediators such as histamine and the leukotrienes by relaxing the smooth muscles and reducing vascular permeability. Epinephrine also improves cardiac output, which is necessary to prevent vascular collapse during an anaphylactic reaction. In addition, epinephrine increases cAMP levels in the mast cell, blocking further degranulation.

### Localized Hypersensitivity Reactions (Atopy)

In localized hypersensitivity reactions, the reaction is limited to a specific target tissue or organ, often involving epithelial surfaces at the site of allergen entry. The tendency to manifest localized hypersensitivity reactions is inherited and is called *atopy*. Atopic allergies, which afflict at least 20% of the population in developed countries, include a wide range of IgE-mediated disorders, including allergic rhinitis (hay fever), asthma, atopic dermatitis (eczema), and food allergies.

**Allergic rhinitis** The most common atopic disorder, affecting 10% of the U.S. population, is allergic rhinitis, commonly known as hay fever. This results from the inhalation of common airborne allergens and subsequent reaction with sensitized mast cells in the conjunctivae and nasal mucosa, inducing the release of pharmacologically active mediators from mast cells; these mediators then cause localized vasodilation and increased capillary permeability. The symptoms include watery exudation of the conjunctivae, nasal mucosa, and upper respiratory tract as well as sneezing and coughing.

**Asthma** Another common manifestation of organ-specific hypersensitivity reactions is asthma. In some cases, airborne or blood-borne allergens, such as pollens, dust, fumes, insect products, or viral antigens, trigger an asthmatic attack (allergic asthma); in other cases, an asthmatic attack can be induced by exercise or cold, apparently independently of allergen stimulation (intrinsic asthma). Like hay fever, asthma is triggered by degranulation of mast cells with release of mediators, but instead of occurring in the nasal mucosa, the reaction develops in the lower respiratory tract. The resulting contraction of the bronchial smooth muscles leads to bronchoconstriction. Airway edema, mucus secretion, and inflammation contribute to the bronchial constriction and to airway obstruction. Asthmatic patients may have abnormal levels of receptors for neuropeptides. For example, asthmatic patients have been reported to have increased expression of receptors for substance P, a peptide that contracts smooth muscles, and decreased expression of receptors for vasoactive intestinal peptide, which relaxes smooth muscles.

Most clinicians view asthma as primarily an inflammatory disease. The asthmatic response can be divided into early and late responses (Figure 15-8). The early response occurs within minutes of allergen exposure and primarily involves histamine, leukotrienes ($LTC_4$), and prostaglandin ($PGD_2$). The effects of these mediators lead to bronchoconstriction, vasodilation, and some buildup of mucus. The late response occurs hours later and involves additional mediators, including IL-4, IL-5, IL-16, TNF-$\alpha$, eosinophil chemotactic factor (ECF), and platelet-activating factor (PAF). The overall effects of these mediators is to induce expression of adhesion molecules on endothelial cells and to recruit inflammatory cells, including eosinophils and neutrophils, into the bronchial tissue.

The neutrophils and eosinophils are capable of causing significant tissue injury by releasing toxic enzymes, oxygen radicals, and cytokines. These events lead to occlusion of the bronchial lumen with mucus, proteins, and cellular

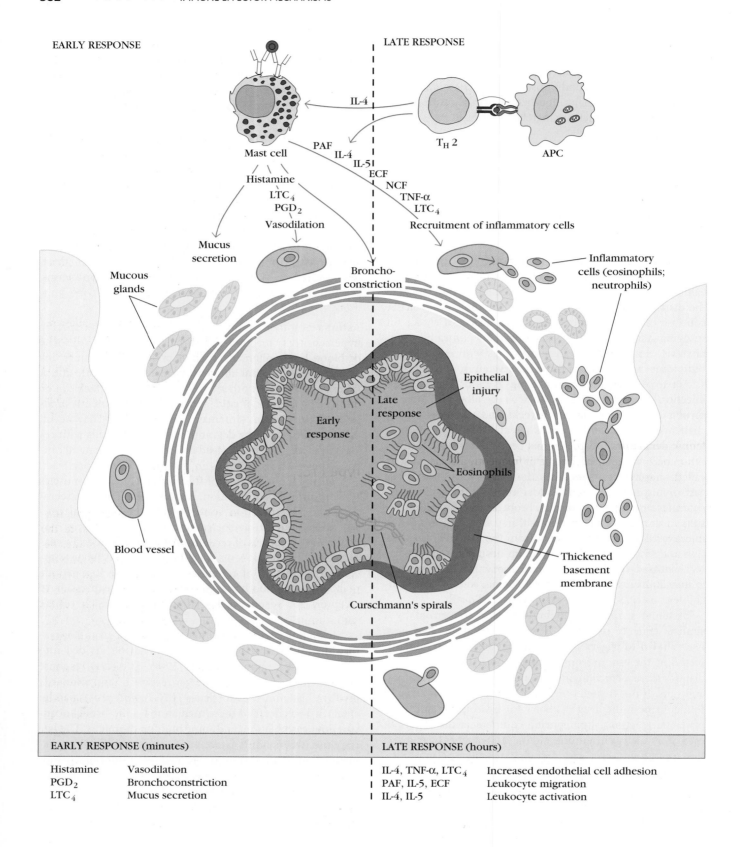

**FIGURE 15-8 The early and late inflammatory responses in asthma.** The immune cells involved in the early and late responses are represented at the top. The effects of various mediators on an airway, represented in cross section, are illustrated in the center.

debris; sloughing of the epithelium; thickening of the basement membrane; fluid buildup (edema); and hypertrophy of the bronchial smooth muscles. A mucus plug often forms and adheres to the bronchial wall. The mucus plug contains clusters of detached epithelial-cell fragments, eosinophils, some neutrophils, and spirals of bronchial tissue known as Curschmann's spirals. Asthma is increasing in prevalence in the United States, particularly among children in inner-city environments (see Clinical Focus on page 384).

**Food allergies** Various foods induce localized hypersensitivity reactions in allergic individuals. Allergen cross-linking of IgE on mast cells along the upper or lower gastrointestinal tract can induce localized smooth muscle contraction and vasodilation and thus such symptoms as vomiting or diarrhea. Mast-cell degranulation along the gut can increase the permeability of mucous membranes, so that the allergen enters the bloodstream. Various symptoms can ensue, depending on where the allergen is deposited. For example, some individuals develop asthmatic attacks after ingesting certain foods. Others develop atopic urticaria, commonly known as hives, when a food allergen is carried to sensitized mast cells in the skin, causing swollen (edematous) red (erythematous) eruptions; this is the wheal-and-flare response.

**Atopic dermatitis** Atopic dermatitis (allergic eczema) is an inflammatory disease of skin that is frequently associated with a family history of atopy. The disease is observed most frequently in young children, often developing during infancy. Serum IgE levels are often elevated. The allergic individual develops skin eruptions that are erythematous, and if there is an accompanying bacterial infection, the eruptions will be filled with pus. Unlike a delayed-type hypersensitivity reaction, which involves $T_H1$ cells, the skin lesions in atopic dermatitis contain $T_H2$ cells and an increased number of eosinophils.

## Late-phase reactions induce localized inflammatory reactions

As a type I hypersensitivity reaction begins to subside, mediators released during the course of the reaction often induce localized inflammation called the late-phase reaction. Distinct from the late response seen in asthma, the late-phase reaction begins to develop 4 to 6 hours after the initial type I reaction and persists for 1 to 2 days. The reaction is characterized by infiltration of neutrophils, eosinophils, macrophages, $T_H2$ cells, and basophils. The localized late-phase response also may be mediated partly by cytokines released from mast cells. Both TNF-$\alpha$ and IL-1 increase the expression of cell adhesion molecules on venular endothelial cells, thus facilitating the buildup of neutrophils, eosinophils, and $T_H2$ cells that characterizes the late-phase response.

Eosinophils play a principal role in the late-phase reaction, accounting for some 30% of the cells that accumulate. Eosinophil chemotactic factor, released by mast cells during the initial reaction, attracts large numbers of eosinophils to the affected site. Various cytokines released at the site, including IL-3, IL-5, and GM-CSF, contribute to the growth and differentiation of the eosinophils. Eosinophils express Fc receptors for IgG and IgE isotypes and bind directly to antibody-coated allergen. Much as in mast-cell degranulation, binding of antibody-coated antigen activates eosinophils, leading to their degranulation and release of inflammatory mediators, including leukotrienes, major basic protein, platelet activation factor, eosinophil cationic protein (ECP), and eosinophil-derived neurotoxin. The release of these eosinophil-derived mediators may play a protective role in parasitic infections. However, in response to allergens, these mediators contribute to extensive tissue damage in the late-phase reaction. The influx of eosinophils in the late-phase response has been shown to contribute to the chronic inflammation of the bronchial mucosa that characterizes persistent asthma.

Neutrophils are another major participant in late-phase reactions, accounting for another 30% of the inflammatory cells. Neutrophils are attracted to the area of a type I reaction by neutrophil chemotactic factor, released from degranulating mast cells. In addition, a variety of cytokines released at the site, including IL-8, have been shown to activate neutrophils, resulting in release of their granule contents, including lytic enzymes, platelet-activating factor, and leukotrienes.

## Type I responses are regulated by many factors

As noted earlier, the antigen dose, mode of antigen presentation, and genetic constitution of an animal influence the level of the IgE response induced by an antigen (i.e., its allergenicity). A genetic component influences susceptibility to type I hypersensitivity reactions in humans, and several candidate genes have been identified, as discussed earlier. If both parents are allergic, there is a 50% chance that a child will also be allergic; when only one parent is allergic, there is a 30% chance that a child will manifest some kind of type I reaction.

The effect of antigen dosage on the IgE response is illustrated by immunization of BDF1 mice. Repeated low doses of an appropriate antigen induce a persistent IgE response in these mice, but higher antigen doses result in transient IgE production and a shift toward IgG. The mode of antigen presentation also influences the development of the IgE response. For example, immunization of Lewis-strain rats with keyhole limpet hemocyanin (KLH) plus aluminum hydroxide gel or *Bordetella pertussis* as an adjuvant induces a strong IgE response, whereas injection of KLH with complete Freund's adjuvant produces a largely IgG response. Infection of mice with the nematode *Nippostrongylus brasiliensis* (Nb), like certain adjuvants, preferentially induces an IgE response. For example, Nb-infected mice, when challenged with an unrelated antigen, develop higher levels of IgE than do control mice not infected with Nb. The relative levels of

## CLINICAL FOCUS
# The Genetics of Asthma

**Asthma** affects almost 10% of the population of the United States. For reasons that are still unclear, the incidence of asthma recently has increased dramatically in developed countries. Even more alarming is that the severity of the disease also appears to be increasing. The increase in asthma mortality is highest among children, and in the United States the mortality is highest among African American children of the inner city. In 2003, 9.1 million children had asthma and more than 200 of them died of the disease. These statistics are increasing each year. In addition to its human costs, asthma imposes high financial costs on society. During 2004, the cost for the treatment of asthma in the United States was more than $16 billion.

Asthma is commonly defined as an inflammatory disease of the airway, and it is characterized by bronchial hyperresponsiveness, resulting in airway obstruction and wheezing. Atopic individuals, those with a predisposition to the type I hypersensitive response, are most suscepti-

ble to the development of bronchial hyperresponsiveness and asthma, but only 10% to 30% of atopic individuals actually develop asthma. The evidence that asthma has a genetic component originally was derived from family studies, in which it was estimated that the relative contribution of genetic factors to atopy and asthma is 40% to 60%. Although genetic factors are important, further studies have indicated that environmental factors also play a large role. In addition, asthma is a complex genetic disease, controlled by several genes, so that susceptibility to it is likely to involve the interaction of multiple genetic and environmental factors.

How do we determine which genes contribute to a complex multigenic disease such as this? One approach is the candidate-gene approach, in which a hypothesis suggests that a particular gene or set of genes may have some relation to the disease. After such a gene has been identified, families with apparent predisposition to the disease are examined for

polymorphic alleles of the gene in question. Comparing family members who do or do not have the disease allows correlation between a particular allele and the presence of the disease. The problem with this approach is its bias toward identification of genes already suspected to play a role in the disease, which precludes identification of new genes.

A good example of the use of the candidate-gene approach is the identification of a region on chromosome 5, region 5q31–33, that appears to be linked to the development of asthma. Using a candidate-gene approach, this region was investigated because it includes a cluster of cytokine genes, among them the genes that encode IL-3, -4, -5, -9, and -13 as well as the gene that encodes granulocyte-macrophage colony-stimulating factor. *IL-4* is thought to be a good candidate gene since it induces the Ig class switch to IgE. Several groups of investigators have examined this region in different populations and concluded that there is a polymorphism associated with predisposition to asthma that maps to the promoter region of *IL-4*. In addition, two alleles of *IL-9* associated with atopy have been identified.

---

the $T_H1$ and $T_H2$ subsets are also key to the regulation of type I hypersensitive responses. $T_H1$ cells reduce the response, whereas $T_H2$ cells enhance it. Cytokines secreted by $T_H2$ cells—the most important are IL-4, IL-5, IL-9, and IL-13—stimulate the type I response in several ways. IL-4 and IL-13 enhance class switching to IgE and regulates the clonal expansion of IgE-committed B cells; IL-4 and IL-9 enhance mast-cell production; and IL-5 and IL-9 enhance eosinophil maturation, activation, and accumulation. In contrast, $T_H1$ cells produce IFN-$\gamma$, which inhibits the type I response.

The pivotal role of IL-4 in the regulation of the type I response was demonstrated in experiments by W. E. Paul and coworkers. When these researchers activated normal, unprimed B cells in vitro with the bacterial endotoxin lipopolysaccharide (LPS), only 2% of the cells expressed membrane IgG1 and only 0.05% expressed membrane IgE. However, when unprimed B cells were incubated with LPS plus IL-4, the percentage of cells expressing IgG1 increased to 40% to 50%, and the percentage expressing IgE increased to 15% to 25%. In an attempt to determine whether IL-4

plays a role in regulating IgE production in vivo, Paul primed Nb-infected mice with the harmless antigen TNP-KLH in the presence and absence of monoclonal antibody to IL-4. The antibody to IL-4 reduced the production of IgE specific for TNP-KLH in these Nb-infected mice to 1% of the level in control animals.

Further support for the role of IL-4 in the IgE response comes from the experiments of K. Rajewsky and coworkers with IL-4 knockout mice. These IL-4–deficient mice were unable to mount an IgE response to helminthic antigens. Furthermore, increased levels of CD4$^+$ $T_H2$ cells and increased levels of IL-4 have been detected in atopic individuals. When allergen-specific CD4$^+$ T cells from atopic individuals are cloned and added to an autologous B-cell culture, the B cells synthesize IgE, whereas allergen-specific CD4$^+$ T cells from nonatopic individuals do not support IgE production.

In contrast to IL-4, IFN-$\gamma$ decreases IgE production, suggesting that the balance of IL-4 and IFN-$\gamma$ may determine the amount of IgE produced (Figure 15-9). Since IFN-$\gamma$ is secreted by the $T_H1$ subset and IL-4 by the $T_H2$ subset, the

Recent positional cloning experiments have identified three genes associated with asthma: *ADAM33*, which encodes an enzyme expressed in bronchial smooth muscle, and *PHF11* and *DPP10*, which are both involved in the induction of $T_H2$ and IgE inflammation.

Another approach to identifying genes associated with a particular disease is a random genomic search. With this method, the entire genome is scanned for polymorphisms associated with the disease in question. Using the random genomic approach, P. Lympany and colleagues identified a linkage between a polymorphism on chromosome 11—more specifically, region 11q13—associated with atopy in British families. This region maps to the vicinity of the β subunit of the FcεRIβ. This association is exciting since IgE is so important in mediating type I reactions. Additional genome-wide (and candidate gene) studies have confirmed the association of polymorphisms of the gene for the β subunit of FcεRIβ with atopy and with the development of asthma and atopic dermatitis. However, this association, like the others, will require more careful analysis since some studies do not confirm linkage of 11q with atopy and asthma.

More recently, a large genome-wide screen for loci linked to asthma susceptibility was conducted in ethnically diverse populations that included Caucasians, Hispanics, and African Americans. This study was published by the U.S. Collaborative Study on the Genetics of Asthma, a large collaborative group from medical centers throughout the United States, and identified many candidate loci associated with asthma. One locus on chromosome 5 coincided with the already identified region at 5q31–33. However, this locus was associated with asthma in Caucasians but not in Hispanics or African Americans. Similarly, some loci appeared to have a high correlation with asthma in Hispanics only, and other loci were identified as unique to African Americans.

Another interesting conclusion was that the association between chromosome 11q and atopy did not appear to be correlated with asthma. This could indicate that asthma and atopy have different molecular bases. More important, it suggests that genetic linkage to atopy should not be confused with genetic linkage to asthma. Overall, this study identified several genes linked to asthma and found that the number and relative importance of these genes

may differ among ethnic groups. This suggests that genetic differences as well as differences in environment may be the underlying basis of the differences observed in the prevalence as well as the severity of the disease among ethnic groups in the United States.

It is well documented that a higher-than-average percentage of African American inner-city children have serious complications with asthma. This has raised the question whether African Americans have a genetic predisposition for asthma. A 1997 report from D. L. Rosenstreich and colleagues indicates an important environmental linkage to asthma in the inner city. This group assessed the role of allergies to the cockroach in the development of asthma; they found that a combination of cockroach allergy and exposure to high levels of cockroach allergen can help explain the high frequency of asthma-related health problems in inner-city children. These data also point to defects in the public-health systems in large cities. Clearly, a concerted effort by public agencies to eradicate insect infestations would benefit the health of those who live in inner-city communities.

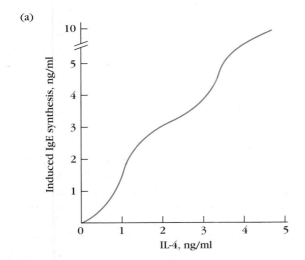
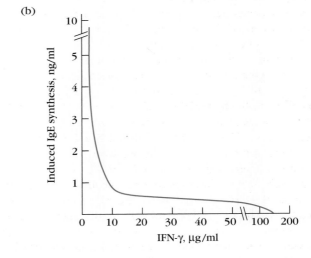
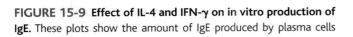

**FIGURE 15-9 Effect of IL-4 and IFN-γ on in vitro production of IgE.** These plots show the amount of IgE produced by plasma cells cultured in the presence of various concentrations of IL-4 (a) or IFN-γ (b). *[Adapted from G. Del Prete, 1988, Journal of Immunology 140:4193.]*

relative activity of these subsets may influence an individual's response to allergens. According to this proposal, atopic and nonatopic individuals would exhibit qualitatively different type I responses to an allergen: the response in atopic individuals would involve the $T_H2$ subset and result in production of IgE; the response in nonatopic individuals would involve the $T_H1$ subset and result in production of IgM or IgG. To test this hypothesis, allergen-specific T cells were cloned from atopic and nonatopic individuals. The cloned T cells from the atopic individuals were predominantly of the $T_H2$ phenotype (secreting IL-4), whereas the cloned T cells from nonatopic individuals were predominantly of the $T_H1$ phenotype (secreting IFN-$\gamma$). Needless to say, there is keen interest in down-regulating IL-4 as a possible treatment for allergic individuals.

## Several clinical methods are used to detect type I hypersensitivity reactions

Type I hypersensitivity is commonly identified and assessed by skin testing. Small amounts of potential allergens are introduced at specific skin sites either by intradermal injection or by superficial scratching. A number of allergens can be applied to sites on the forearm or back of an individual at one time. If a person is allergic to the allergen, local mast cells degranulate and the release of histamine and other mediators produces a wheal and flare within 30 minutes (Figure 15-10). The advantage of skin testing is that it is relatively inexpensive and allows screening of a large number of aller-

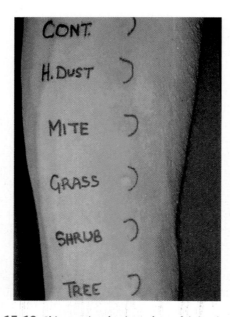

**FIGURE 15-10 Skin testing by intradermal injection of allergens into the forearm.** In this individual, a wheal-and-flare response developed within a few minutes at the site where grass was injected, indicating that the individual is allergic to grass. *[From L. M. Lichtenstein, 1993, Scientific American **269**(2):117. Used with permission.]*

gens at one time. The disadvantage of skin testing is that in rare cases it sensitizes the allergic individual to new allergens and even less commonly may even induce systemic anaphylactic shock. A few individuals also manifest a late-phase reaction, which comes 4 to 6 hours after testing and sometimes lasts for up to 24 hours. As noted already, eosinophils accumulate during a late-phase reaction, and release of eosinophil-granule contents contributes to the tissue damage in a late-phase reaction site.

Another method of assessing type I hypersensitivity is to determine the serum level of total IgE antibody by the radioimmunosorbent test (RIST). This highly sensitive technique, based on the radioimmunoassay, can detect nanomolar levels of total IgE. The patient's serum is reacted with agarose beads or paper disks coated with rabbit anti-IgE. After the beads or disks are washed, $^{125}$I-labeled rabbit anti-IgE is added. The radioactivity of the beads or disks, measured with a gamma counter, is proportional to the level of IgE in the patient's serum (Figure 15-11a).

The similar radioallergosorbent test (RAST) detects the serum level of IgE specific for a given allergen. The allergen is coupled to beads or disks, the patient's serum is added, and unbound antibody is washed away. The amount of specific IgE bound to the solid-phase allergen is then measured by adding $^{125}$I-labeled rabbit anti-IgE, washing the beads, and counting the bound radioactivity (Figure 15-11b). Newer adaptations of the RIST and RAST assays circumvent the use of radioactivity by using alternative labels, such as fluorescent markers, to detect IgE.

## Type I hypersensitivities can be controlled medically

The obvious first step in controlling type I hypersensitivities is to avoid contact with known allergens. Often the removal of house pets, dust-control measures, or avoidance of offending foods can eliminate a type I response. Elimination of inhalant allergens (such as pollens) is a physical impossibility, however, and other means of intervention must be pursued.

Immunotherapy with repeated injections of increasing doses of allergens (**hyposensitization**) has been known for some time to reduce the severity of type I reactions, or even eliminate them completely, in a significant number of individuals suffering from allergic rhinitis. Such repeated introduction of allergen by subcutaneous injections appears to cause a shift toward IgG production or to induce T-cell–mediated suppression (possibly by a shift to the $T_H1$ subset and IFN-$\gamma$ production) that turns off the IgE response (Figure 15-12). In this situation, the IgG antibody is referred to as *blocking antibody* because it competes for the allergen, binds to it, and forms a complex that can be removed by phagocytosis; as a result, the allergen is not available to cross-link the fixed IgE on the mast-cell membranes, and allergic symptoms decrease. The role of $T_{reg}$ cells in hyposensitization is being evaluated, as these cells may play

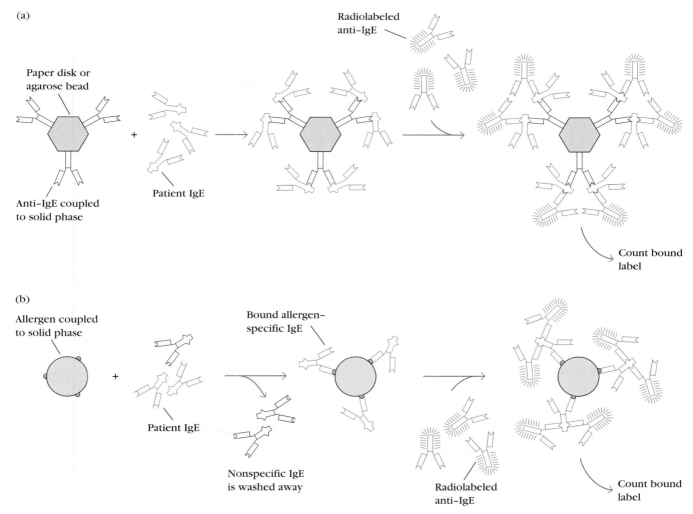

(a)

Radiolabeled anti–IgE

Paper disk or agarose bead

Anti–IgE coupled to solid phase

Patient IgE

Count bound label

(b)

Allergen coupled to solid phase

Bound allergen-specific IgE

Patient IgE

Nonspecific IgE is washed away

Radiolabeled anti–IgE

Count bound label

**FIGURE 15-11 Procedures for assessing type I hypersensitivity.** (a) Radioimmunosorbent test (RIST) can quantify nanogram amounts of total serum IgE. (b) Radioallergosorbent test (RAST) can quantify nanogram amounts of serum IgE specific for a particular allergen.

a role in blocking allergic responses. Newer hyposensitization protocols that incorporate peptides derived from the allergens, DNA vaccines, and adjuvants are also being evaluated.

Another form of immunotherapy is the use of humanized monoclonal anti-IgE. These antibodies bind to IgE, but only if IgE is not already bound to FcεRI; binding the latter would lead to histamine release. These antibodies selectively interact only with IgE that is not bound to FcεRI by binding to the same site on IgE that makes contact with FcεRI. These antibodies are humanized by the genetic engineering of the genes encoding the H and L chains; mouse framework regions are replaced with human framework sequences and the end result is a mouse-human chimeric monoclonal that is not likely to be recognized as foreign by the human immune system. When injected into people suffering from allergy, these antibodies can bind free IgE as well as down-regulate IgE production in B cells and the expression of FcεRI by mast cells and basophils. This results in lower serum IgE concentration,

which, in turn, reduces the sensitivity of mast cells and basophils. This form of immunotherapy is useful in treating many forms of allergies, especially crippling food allergies. Omalizumab was the first anti-IgE antibody to be approved in the United States (in 2003) for the treatment of allergic asthma. Although omalizumab therapy dramatically decreases IgE levels in asthma and rhinitis patients, the drug has met with only modest results in terms of quality of life, which must be weighed against the cost of the regimen.

Another approach for treating allergies stems from the finding that soluble antigens tend to induce a state of anergy by activating T cells in the absence of the necessary costimulatory signal (see Figure 10-15). Presumably, a soluble antigen is internalized by endocytosis, processed, and presented with class II MHC molecules but fails to up-regulate expression of the requisite costimulatory ligand (B7) on antigen-presenting cells. Other methods aimed at blocking the action of cytokines (such as anti-IL-4 and anti-IL-5) and cytokine

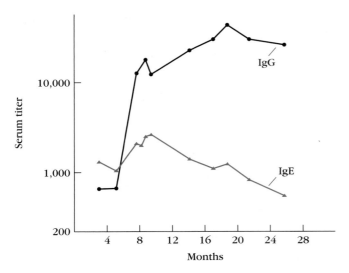

**FIGURE 15-12 Hyposensitization treatment of type I allergy.** Injection of ragweed antigen periodically for 2 years into a ragweed-sensitive individual induced a gradual decrease in IgE levels and a dramatic increase in IgG. Both antibodies were measured by a radioimmunoassay. *[Adapted from K. Ishizaka and T. Ishizaka, 1973, in* Asthma Physiology, Immunopharmacology and Treatment, *K. F. Austen and L. M. Lichtenstein, eds., Academic Press, London.]*

receptors and the use of anti-sense oligonucleotides are also being investigated.

Knowledge of the mechanism of mast-cell degranulation and the mediators involved in type I reactions opened the way to drug therapy for allergies. Antihistamines have been the most useful drugs for symptoms of allergic rhinitis. These drugs act by binding to the histamine receptors on target cells and blocking the binding of histamine. The $H_1$ receptors are blocked by the classical antihistamines; the second-generation $H_1$ antihistamines do not react with cholinergic receptors and thus have no sedating side effects.

| TABLE 15-4 | Mechanism of action of some drugs used to treat type I hypersensitivity |
|---|---|
| **Drug** | **Action** |
| Antihistamines | Block $H_1$ and $H_2$ receptors on target cells |
| Cromolyn sodium | Blocks $Ca^{2+}$ influx into mast cells |
| Theophylline | Prolongs high cAMP levels in mast cells by inhibiting phosphodiesterase, which cleaves cAMP to 5'-AMP[*] |
| Epinephrine (adrenaline) | Stimulates cAMP production by binding to β-adrenergic receptors on mast cells[*] |
| Cortisone | Reduces histamine levels by blocking conversion of histidine to histamine and stimulates mast-cell production of cAMP[*] |

[*]Although cAMP rises transiently during mast-cell activation, degranulation is prevented if cAMP levels remain high.

Newer allergy drugs block the activities of leukotrienes and prostaglandins rather than histamine.

Several drugs block release of allergic mediators by interfering with various biochemical steps in mast-cell activation and degranulation (Table 15-4). Disodium cromoglycate (cromolyn sodium) prevents $Ca^{2+}$ influx into mast cells. Theophylline, which is commonly administered to asthmatics orally or through inhalers, blocks phosphodiesterase, which catalyzes the breakdown of cAMP to 5'-AMP. The resulting prolonged increase in cAMP levels blocks degranulation. A number of drugs stimulate the β-adrenergic system by stimulating β-adrenergic receptors. As mentioned earlier, epinephrine (also known as adrenaline) is commonly administered during anaphylactic shock. It acts by binding to β-adrenergic receptors on bronchial smooth muscles and mast cells, elevating the cAMP levels within these cells. The increased levels of cAMP promote relaxation of the bronchial muscles and decreased mast-cell degranulation. A number of epinephrine analogues have been developed that bind to selected β-adrenergic receptors and induce cAMP increases with fewer side effects than are seen with epinephrine. Cortisone and various other anti-inflammatory drugs also have been used to reduce the inflammation seen in some type I reactions.

# Antibody-Mediated Cytotoxic (Type II) Hypersensitivity

Type II hypersensitivity reactions involve antibody-mediated destruction of cells. Antibody bound to a cell surface antigen can activate the complement system, creating pores in the membrane of a foreign cell (see Figure 7-8), or it can mediate cell destruction by antibody-dependent cell-mediated cytotoxicity (ADCC). In this process, cytotoxic cells with Fc receptors bind to the Fc region of antibodies on target cells and promote killing of the cells (see Figure 14-12). Antibody bound to a foreign cell also can serve as an opsonin, enabling phagocytic cells with Fc or C3b receptors to bind and phagocytose the antibody-coated cell (see Figure 7-13).

In this section we examine three examples of type II hypersensitivity reactions. Certain autoimmune diseases involve auto-antibody-mediated cellular destruction by type II mechanisms. These diseases are described in Chapter 16.

## Transfusion reactions are type II reactions

A large number of proteins and glycoproteins on the membrane of red blood cells are encoded by different genes, each of which has a number of alternative alleles. An individual possessing one allelic form of a blood-group antigen can recognize other allelic forms on transfused blood as foreign and mount an antibody response. In some cases, the antibodies have already been induced by natural exposure to similar antigenic determinants on a variety of microorganisms present in the normal flora of the gut. This is the case with the ABO blood-group antigens (Figure 15-13a).

(a)

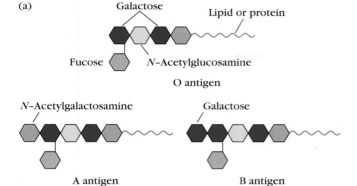

**FIGURE 15-13  ABO blood group.** (a) Structure of terminal sugars, which constitute the distinguishing epitopes, in the A, B, and O blood antigens. (b) ABO genotypes and corresponding phenotypes, agglutinins, and isohemagglutinins.

(b)

| Genotype | Blood-group phenotype | Antigens on erythrocytes *(agglutinins)* | Serum antibodies *(isohemagglutinins)* |
|---|---|---|---|
| AA or AO | A | A | Anti-B |
| BB or BO | B | B | Anti-A |
| AB | AB | A and B | None |
| OO | O | None | Anti-A and anti-B |

Antibodies to the A, B, and O antigens, called isohemagglutinins, are usually of the IgM class. An individual with blood type A, for example, recognizes B-like epitopes on intestinal microorganisms and produces isohemagglutinins to the B-like epitopes. This same individual does not respond to A-like epitopes on the same intestinal microorganisms because these A-like epitopes are too similar to self and a state of self-tolerance to these epitopes should exist (Figure 15-13b). If a type A individual is transfused with blood containing type B cells, a **transfusion reaction** occurs in which the anti-B isohemagglutinins bind to the B blood cells and mediate their destruction by means of complement-mediated lysis. Antibodies to other blood-group antigens may result from repeated blood transfusions because minor allelic differences in these antigens can stimulate antibody production. These antibodies are usually of the IgG class.

The clinical manifestations of transfusion reactions result from massive intravascular hemolysis of the transfused red blood cells by antibody plus complement. These manifestations may be either immediate or delayed. Reactions that begin immediately are most commonly associated with ABO blood-group incompatibilities, which lead to complement-mediated lysis triggered by the IgM isohemagglutinins. Within hours, free hemoglobin can be detected in the plasma; it is filtered through the kidneys, resulting in hemoglobinuria. Some of the hemoglobin gets converted to bilirubin, which at high levels is toxic. Typical symptoms include fever, chills, nausea, clotting within blood vessels, pain in the lower back, and hemoglobin in the urine. Treatment involves prompt termination of the transfusion and maintenance of urine flow with a diuretic, because the accumulation of hemoglobin in the kidney can cause acute tubular necrosis.

Delayed hemolytic transfusion reactions generally occur in individuals who have received repeated transfusions of ABO-compatible blood that is incompatible for other blood-group antigens. The reactions develop between 2 and 6 days after transfusion, reflecting the secondary nature of these reactions. The transfused blood induces clonal selection and production of IgG against a variety of blood-group membrane antigens, most commonly Rh, Kidd, Kell, and Duffy. The predominant isotype involved in these reactions is IgG, which is less effective than IgM in activating complement. For this reason, complement-mediated lysis of the transfused red blood cells is incomplete, and many of the transfused cells are destroyed at extravascular sites by agglutination, opsonization, and subsequent phagocytosis by macrophages. Symptoms include fever, low hemoglobin, increased bilirubin, mild jaundice, and anemia. Free hemoglobin is usually not detected in the plasma or urine in these reactions because red blood cell destruction occurs in extravascular sites.

## Hemolytic disease of the newborn is caused by type II reactions

Hemolytic disease of the newborn develops when maternal IgG antibodies specific for fetal blood-group antigens cross the placenta and destroy fetal red blood cells. The consequences of such transfer can be minor, serious, or lethal. Severe hemolytic disease of the newborn, called **erythroblastosis fetalis,** most commonly develops when an Rh⁺ fetus expresses an **Rh antigen** on its blood cells that the Rh⁻ mother does not express. During pregnancy, fetal red blood cells are separated from the mother's circulation by a layer of cells in the placenta called the trophoblast. During her first pregnancy with an Rh⁺ fetus, an Rh⁻ woman is usually not exposed to enough fetal red blood cells to activate

DEVELOPMENT OF ERYTHROBLASTOSIS FETALIS (WITHOUT RHOGAM)                    PREVENTION (WITH RHOGAM)

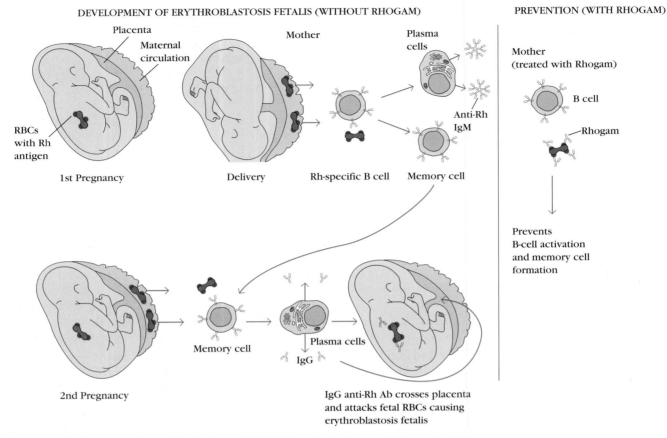

**FIGURE 15-14** Development of erythroblastosis fetalis (hemolytic disease of the newborn) caused when an Rh⁻ mother carries an Rh⁺ fetus *(left)* and effect of treatment with anti-Rh antibody, or Rhogam *(right)*.

her Rh-specific B cells. At the time of delivery, however, separation of the placenta from the uterine wall allows larger amounts of fetal umbilical cord blood to enter the mother's circulation. These fetal red blood cells activate Rh-specific B cells, resulting in production of Rh-specific plasma cells and memory B cells in the mother. The secreted IgM antibody clears the Rh⁺ fetal red cells from the mother's circulation, but the memory cells remain, a threat to any subsequent pregnancy with an Rh⁺ fetus. Activation of these memory cells in a subsequent pregnancy results in the formation of IgG anti-Rh antibodies, which cross the placenta and damage the fetal red blood cells (Figure 15-14). Mild to severe anemia can develop in the fetus, sometimes with fatal consequences. In addition, conversion of hemoglobin to bilirubin can present an additional threat to the newborn because the lipid-soluble bilirubin may accumulate in the brain and cause brain damage.

Hemolytic disease of the newborn caused by Rh incompatibility in a subsequent pregnancy can be almost entirely prevented by administering antibodies against the Rh antigen to the mother at around 28 weeks of pregnancy and within 24 to 48 hours after the first delivery. Anti-Rh antibodies are also administered to pregnant women after amniocentesis. These antibodies, marketed as **Rhogam,** bind to any fetal red blood

cells that may have entered the mother's circulation and facilitate their clearance before B-cell activation and ensuing memory-cell production can take place. In a subsequent pregnancy with an Rh⁺ fetus, a mother who has been treated with Rhogam is unlikely to produce IgG anti-Rh antibodies; thus, the fetus is protected from the damage that would occur when these antibodies crossed the placenta.

The development of hemolytic disease of the newborn caused by Rh incompatibility can be detected by testing maternal serum at intervals during pregnancy for antibodies to the Rh antigen. A rise in the titer of these antibodies as pregnancy progresses indicates that the mother has been exposed to Rh antigens and is producing increasing amounts of antibody. The presence of maternal IgG on the surface of fetal red blood cells can be detected by a Coombs test. Isolated fetal red blood cells are incubated with the Coombs reagent, goat antibody to human IgG antibody. If maternal IgG is bound to the fetal red blood cells, the cells agglutinate with the Coombs reagent.

If hemolytic disease caused by Rh incompatibility is detected during pregnancy, the treatment depends on the severity of the reaction. For a severe reaction, the fetus can be given an intrauterine blood-exchange transfusion to replace fetal

Rh$^+$ red blood cells with Rh$^-$ cells. These transfusions are given every 10 to 21 days until delivery. In less severe cases, a blood-exchange transfusion is not given until after birth, primarily to remove bilirubin; the infant is also exposed to low levels of UV light to break down the bilirubin and prevent cerebral damage. The mother can also be treated during the pregnancy by **plasmapheresis.** In this procedure, a cell separation machine is used to separate the mother's blood into two fractions, cells and plasma. The plasma containing the anti-Rh antibody is discarded, and the cells are reinfused into the mother in an albumin or fresh-plasma solution.

The majority of cases (65%) of hemolytic disease of the newborn have minor consequences and are caused by ABO blood-group incompatibility between the mother and fetus. Type A or B fetuses carried by type O mothers most commonly develop these reactions. A type O mother is most likely to develop IgG antibody to the A or B blood-group antigens either through natural exposure or through exposure to fetal blood-group A or B antigens in successive pregnancies. Usually the fetal anemia resulting from this incompatibility is mild; the major clinical manifestation is a slight elevation of bilirubin, with jaundice. Depending on the severity of the anemia and jaundice, a blood-exchange transfusion may be required in these infants. In general, the reaction is mild, however, and exposure of the infant to low levels of UV light is enough to break down the bilirubin and avoid cerebral damage.

## Drug-induced hemolytic anemia is a type II response

Certain antibiotics (e.g., penicillin, cephalosporin, and streptomycin) can adsorb nonspecifically to proteins on red blood cell membranes, forming a complex similar to a hapten-carrier complex. In some patients, such drug-protein complexes induce formation of antibodies, which then bind to the adsorbed drug on red blood cells, inducing complement-mediated lysis and thus progressive anemia. When the drug is withdrawn, the hemolytic anemia disappears. Penicillin is notable in that it can induce all four types of hypersensitivity with various clinical manifestations (Table 15-5).

| TABLE 15-5 | Penicillin-induced hypersensitive reactions | |
|---|---|---|
| **Type of reaction** | **Antibody or lymphocytes induced** | **Clinical manifestations** |
| I | IgE | Urticaria, systemic anaphylaxis |
| II | IgM, IgG | Hemolytic anemia |
| III | IgG | Serum sickness, glomerulonephritis |
| IV | T$_H$1 cells | Contact dermatitis |

# Immune Complex–Mediated (Type III) Hypersensitivity

The reaction of antibody with antigen generates immune complexes. Generally, this complexing of antigen with antibody facilitates the clearance of antigen by phagocytic cells and red blood cells (see Figure 7-15). In some cases, however, large amounts of immune complexes can lead to tissue-damaging type III hypersensitivity reactions. The magnitude of the reaction depends on the quantity of immune complexes as well as their distribution within the body. When the complexes are deposited in tissue very near the site of antigen entry, a localized reaction develops. When the complexes are formed in the blood, a reaction can develop wherever the complexes are deposited. In particular, complex deposition is frequently observed on blood vessel walls, in the synovial membrane of joints, on the glomerular basement membrane of the kidney, and on the choroid plexus of the brain. The deposition of these complexes initiates a reaction that results in the recruitment of neutrophils to the site. The tissue there is injured as a consequence of granular release from the neutrophil.

Type III hypersensitivity reactions develop when immune complexes activate the complement system's array of immune effector molecules (see Figure 7-2). As explained in Chapter 7, the C3a and C5a complement split products are anaphylatoxins that cause localized mast-cell degranulation and consequent increase in local vascular permeability. C3a, C5a, and C5b67 are also chemotactic factors for neutrophils, which can accumulate in large numbers at the site of immune-complex deposition. Larger immune complexes are deposited on the basement membrane of blood vessel walls or kidney glomeruli, whereas smaller complexes may pass through the basement membrane and be deposited in the subepithelium. The type of lesion that results depends on the site of deposition of the complexes.

The binding of the immune complexes to Fc and complement receptors on leukocytes leads to activation of an inflammatory response. Much of the tissue damage in type III reactions stems from release of lytic enzymes by neutrophils as they attempt to phagocytose immune complexes. The C3b complement component acts as an opsonin, coating immune complexes. A neutrophil binds to a C3b-coated immune complex by means of the type I complement receptor, which is specific for C3b. Because the complex is deposited on the basement membrane surface, phagocytosis is impeded, so that lytic enzymes are released during the unsuccessful attempts of the neutrophil to ingest the adhering immune complex. Further activation of the membrane-attack mechanism of the complement system can also contribute to the destruction of tissue. In addition, the activation of complement can induce aggregation of platelets, and the resulting release of clotting factors can lead to formation of microthrombi.

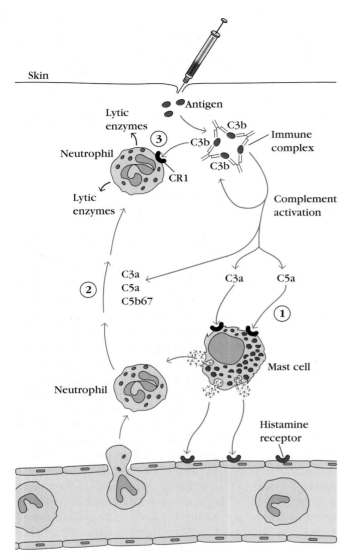

**FIGURE 15-15  Development of a localized Arthus reaction (type III hypersensitivity reaction).** Complement activation initiated by immune complexes (classical pathway) produces complement intermediates that (1) mediate mast-cell degranulation, (2) chemotactically attract neutrophils, and (3) stimulate release of lytic enzymes from neutrophils trying to phagocytose $C_3b$-coated immune complexes.

## Type III reactions can be localized

Injection of an antigen intradermally or subcutaneously into an animal that has high levels of circulating antibody specific for that antigen leads to formation of localized immune complexes, which mediate an acute Arthus reaction within 4 to 8 hours (Figure 15-15). Microscopic examination of the tissue reveals neutrophils adhering to the vascular endothelium and then migrating into the tissues at the site of immune-complex deposition. As the reaction develops, localized tissue and vascular damage results in an accumulation of fluid (edema) and red blood cells (erythema) at the site. The severity of the reaction can vary from mild swelling and redness to tissue necrosis.

After an insect bite, a sensitive individual may have a rapid, localized type I reaction at the site. Often, some 4 to 8 hours later, a typical Arthus reaction also develops at the site, with pronounced erythema and edema. Intrapulmonary Arthus-type reactions induced by bacterial spores, fungi, or dried fecal proteins can also cause pneumonitis or alveolitis. These reactions are known by a variety of common names reflecting the source of the antigen. For example, "farmer's lung" develops after inhalation of thermophilic actinomycetes from moldy hay, and "pigeon fancier's disease" results from inhalation of a serum protein in dust derived from dried pigeon feces.

## Type III reactions can also be generalized

When large amounts of antigen enter the bloodstream and bind to antibody, circulating immune complexes can form. If antigen is in significant excess compared to antibody, the complexes that form are smaller; because these are not easily cleared by the phagocytic cells, they can cause tissue-damaging type III reactions at various sites. Historically, generalized type III reactions were often observed after the administration of antitoxins containing foreign serum, such as horse antitetanus or anti-diphtheria serum. In such cases, the recipient of a foreign antiserum develops antibodies specific for the foreign serum proteins; these antibodies then form circulating immune complexes with the foreign serum antigens. Typically, within days or weeks after exposure to foreign serum antigens, an individual begins to manifest a combination of symptoms that are called **serum sickness** (Figure 15-16).

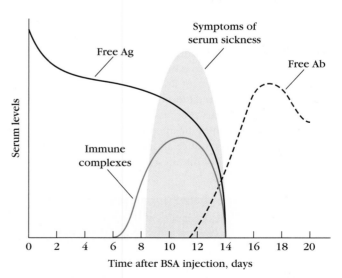

**FIGURE 15-16  Correlation between formation of immune complexes and development of symptoms of serum sickness.** A large dose of antigen (BSA) was injected into a rabbit at day 0. As antibody formed, it complexed with the antigen and was deposited in the kidneys, joints, and capillaries. The symptoms of serum sickness (light blue area) corresponded to the peak in immune-complex formation. As the immune complexes were cleared, free circulating antibody (dashed black curve) was detected and the symptoms of serum sickness subsided. *[Based on F. G. Germuth, Jr., 1953,* Journal of Experimental Medicine **97**:257.]

These symptoms include fever, weakness, generalized vasculitis (rashes) with edema and erythema, lymphadenopathy, arthritis, and sometimes glomerulonephritis. The precise manifestations of serum sickness depend on the quantity of immune complexes formed as well as the overall size of the complexes, which determine the site of their deposition. As mentioned above, the sites of deposition vary, but in general, complexes accumulate in tissues where filtration of plasma occurs. This explains the high incidence of glomerulonephritis (complex deposition in the kidney) and vasculitis (deposition in the arteries) and arthritis (deposition in the synovial joints) caused by serum sickness. Serum sickness is most commonly seen today after antibiotic treatment (not to be confused with drug allergy) and certain vaccinations; symptoms develop 7 to 21 days after initial exposure and can develop in 1 to 3 days on subsequent exposure to the same antigen.

Throughout the 1980s, the possible use of monoclonal antibodies as "magic bullets" to treat cancer and a variety of other diseases created much excitement. However, patients who received therapeutic monoclonal antibodies from mice produced their own antibodies against the foreign antibodies, a reaction called the human anti-mouse antibody (HAMA) response, and developed serum sickness–like symptoms. We know now that the injection of the mouse antibodies caused a generalized type III reaction; the therapeutic antibodies were cleared before they could even reach their pathogenic target. To avoid the HAMA response, current therapeutic antibodies are humanized or genetically engineered not to be recognized as foreign, as described earlier for anti-IgE therapy in allergy.

Formation of circulating immune complexes contributes to the pathogenesis of a number of conditions other than serum sickness, including the following:

- *Autoimmune diseases*
  Systemic lupus erythematosus
  Rheumatoid arthritis
  Goodpasture's syndrome
- *Drug reactions*
  Allergies to penicillin and sulfonamides
- *Infectious diseases*
  Poststreptococcal glomerulonephritis
  Meningitis
  Hepatitis
  Mononucleosis
  Malaria
  Trypanosomiasis

Complexes of antibody with various bacterial, viral, and parasitic antigens have been shown to induce a variety of type III hypersensitivity reactions, including skin rashes, arthritic symptoms, and glomerulonephritis. Poststreptococcal glomerulonephritis, for example, develops when circulating complexes of antibody and streptococcal antigens

are deposited in the kidney and damage the glomeruli. A number of autoimmune diseases stem from circulating complexes of antibody with self proteins, with glycoproteins, or even with DNA. In systemic lupus erythematosus, complexes of DNA and anti-DNA antibodies accumulate in synovial membranes, causing arthritic symptoms, or accumulate on the basement membrane of the kidney, causing progressive kidney damage.

# Type IV or Delayed-Type Hypersensitivity (DTH)

When some subpopulations of activated $T_H$ cells encounter certain types of antigens, they secrete cytokines that induce a localized inflammatory reaction called **delayed-type hypersensitivity (DTH)**. The reaction is characterized by large influxes of nonspecific inflammatory cells, in particular, macrophages. This type of reaction was first described in 1890 by Robert Koch, who observed that individuals infected with *Mycobacterium tuberculosis* developed a localized inflammatory response when injected intradermally with a filtrate derived from a mycobacterial culture. He called this localized skin reaction a "tuberculin reaction." Later, as it became apparent that a variety of other antigens could induce this response (Table 15-6), its name was changed to delayed-type or type IV hypersensitivity in reference to the delayed onset of the reaction and to the tissue damage (hypersensitivity) that is often associated with it. The term *hypersensitivity* is somewhat misleading, since it suggests that a DTH response is always detrimental. Although in some cases a DTH response does cause extensive tissue damage and is in itself pathologic, in many cases tissue damage is limited, and the response plays an important role in defense against intracellular pathogens and contact antigens. The hallmarks of a type IV reaction are the delay required for the reaction to develop and the recruitment of macrophages as opposed to neutrophils, as found in a type III reaction. Macrophages are the major component of the infiltrate that surrounds the site of inflammation.

| TABLE 15-6 | Intracellular pathogens and contact antigens that induce delayed-type (type IV) hypersensitivity |
|---|---|
| **Intracellular bacteria** | **Intracellular viruses** |
| *Mycobacterium tuberculosis* | Herpes simplex virus |
| *Mycobacterium leprae* | Variola (smallpox) |
| *Listeria monocytogenes* | Measles virus |
| *Brucella abortus* | |
| **Intracellular fungi** | **Contact antigens** |
| *Pneumocystis carinii* | Picrylchloride |
| *Candida albicans* | Hair dyes |
| *Histoplasma capsulatum* | Nickel salts |
| *Cryptococcus neoformans* | Poison ivy |
| **Intracellular parasites** | Poison oak |
| *Leishmania* sp. | |

## OVERVIEW FIGURE 15-17:  The DTH Response

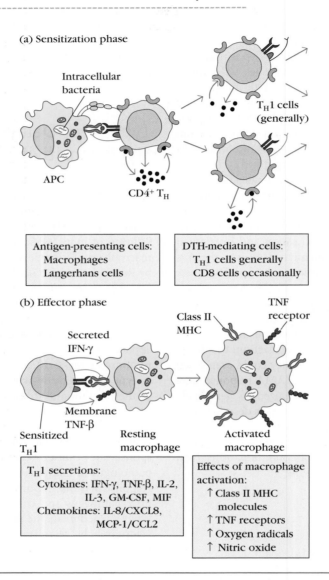

**(a) Sensitization phase**

Intracellular bacteria

APC

CD4⁺ T_H

$T_H1$ cells (generally)

| Antigen-presenting cells: | DTH-mediating cells: |
| --- | --- |
| Macrophages | $T_H1$ cells generally |
| Langerhans cells | CD8 cells occasionally |

**(b) Effector phase**

Secreted IFN-γ

Class II MHC

TNF receptor

Membrane TNF-β

Sensitized $T_H1$

Resting macrophage

Activated macrophage

| $T_H1$ secretions: | Effects of macrophage activation: |
| --- | --- |
| Cytokines: IFN-γ, TNF-β, IL-2, IL-3, GM-CSF, MIF | ↑ Class II MHC molecules |
| Chemokines: IL-8/CXCL8, MCP-1/CCL2 | ↑ TNF receptors |
| | ↑ Oxygen radicals |
| | ↑ Nitric oxide |

(a) In the sensitization phase after initial contact with antigen (e.g., peptides derived from intracellular bacteria), T_H cells proliferate and differentiate into T_H1 cells. Cytokines secreted by these T cells are indicated by the black balls. (b) In the effector phase after subsequent exposure of sensitized T_H1 cells to antigen, the T_H1 cells secrete a variety of cytokines and chemokines. These factors attract and activate macrophages and other nonspecific inflammatory cells. Activated macrophages are more effective in presenting antigen, thus perpetuating the DTH response, and function as the primary effector cells in this reaction.

## There are several phases of the DTH response

The development of the DTH response begins with an initial sensitization phase of 1 to 2 weeks after primary contact with an antigen. During this period, T_H cells are activated and clonally expanded by antigen presented together with the requisite class II MHC molecule on an appropriate antigen-presenting cell (Figure 15-17a).

A variety of antigen-presenting cells have been shown to be involved in the activation of a DTH response, including Langerhans cells and macrophages. Langerhans cells are dendritic cells found in the epidermis. These cells are thought to pick up antigen that enters through the skin and transport it to regional lymph nodes, where T cells are activated by the antigen. In some species, including humans, the vascular endothelial cells express class II MHC molecules and also function as antigen-presenting cells in the development of the DTH response. Generally, the T cells activated during the sensitization phase are CD4$^+$, primarily of the $T_H1$ subtype, but in a few cases CD8$^+$ cells have also been shown to induce a DTH response. The activated T cells previously were called $T_{DTH}$ cells to denote their function in the DTH response, although in reality they are simply a subset of activated $T_H1$ cells (or, in some cases, $T_C$ cells).

A subsequent exposure to the antigen induces the effector phase of the DTH response (see Figure 15-17b). In the effector phase, $T_H1$ cells secrete a variety of cytokines that recruit and activate macrophages and other nonspecific inflammatory cells. A DTH response normally does not become apparent until an average of 24 hours after the second contact with the antigen; the response generally peaks 48 to 72 hours after second contact. The delayed onset of this response reflects the time required for the cytokines to induce localized influxes of macrophages and their activation. Once a DTH response begins, a complex interplay of nonspecific cells and mediators is set in motion that can result in tremendous amplification. By the time the DTH response is fully developed, only about 5% of the participating cells are antigen-specific $T_H1$ cells; the remainder are macrophages and other nonspecific cells.

Macrophages are the principal effector cells of the DTH response. Cytokines elaborated by $T_H1$ cells induce blood monocytes to adhere to vascular endothelial cells and migrate from the blood into the surrounding tissues. During this process the monocytes differentiate into activated macrophages. As described in Chapter 2, activated macrophages exhibit increased levels of phagocytosis and an increased ability to kill microorganisms through various cytotoxic mediators. In addition, activated macrophages express increased levels of class II MHC molecules and cell adhesion molecules and therefore function more effectively as antigen-presenting cells.

The influx and activation of macrophages in the DTH response is important in host defense against parasites and bacteria that live within cells, where circulating antibodies cannot reach them. The heightened phagocytic activity and the buildup of lytic enzymes from macrophages in the area of infection lead to nonspecific destruction of cells and thus of the intracellular pathogen. Generally, the pathogen is cleared rapidly with little tissue damage. However, in some cases, especially if the antigen is not easily cleared, a prolonged DTH response can itself become destructive to the host as the intense inflammatory response develops into a visible granulomatous reaction. A granuloma develops when continuous activation of macrophages induces the macrophages to adhere closely to one another, assuming an epithelioid shape and sometimes fusing to form multinucleated giant cells (Figure 15-18). These giant cells displace the normal tissue cells, forming palpable nodules, and release high concentrations of lytic enzymes, which destroy surrounding tissue. In these cases, the response can

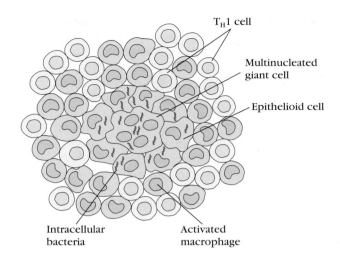

**FIGURE 15-18  A prolonged DTH response can lead to formation of a granuloma, a nodule-like mass.** Lytic enzymes released from activated macrophages in a granuloma can cause extensive tissue damage.

damage blood vessels and lead to extensive tissue necrosis. The response to *Mycobacterium tuberculosis* illustrates the double-edged nature of the DTH response. Immunity to this intracellular bacterium involves a DTH response in which activated macrophages wall off the organism in the lung and contain it within a granuloma-type lesion called a tubercle. Often, however, the concentrated release of lytic enzymes from the activated macrophages within tubercles damages lung tissue. Some examples of truly hypersensitive conditions, in which tissue damage far outweighs any beneficial effects, are described in Chapter 16.

## Numerous cytokines participate in the DTH reaction

Among the cytokines produced by $T_H1$ cells are a number that attract and activate macrophages to the site of infection. IL-3 and GM-CSF induce localized hematopoiesis of the granulocyte-monocyte lineage. IFN-$\gamma$ and TNF-$\beta$ (together with macrophage-derived TNF-$\alpha$ and IL-1) act on nearby endothelial cells, inducing a number of changes that facilitate extravasation of monocytes and other nonspecific inflammatory cells. Circulating neutrophils and monocytes adhere to the adhesion molecules displayed on the vascular endothelial cells and extravasate into the tissue spaces. Neutrophils appear early in the reaction, peaking by about 6 hours and then declining in numbers. The monocyte infiltration occurs between 24 and 48 hours after antigen exposure.

As the monocytes enter the tissues to become macrophages, they are chemotactically drawn to the site of the DTH response by chemokines such as monocyte chemotactic protein (MCP-1/CCL2). A cytokine called migration-inhibition factor (MIF) inhibits macrophages from migrating beyond the site of a DTH reaction. As macrophages accumulate at the site of a DTH reaction, they are activated by cytokines, particularly IFN-γ and membrane-bound TNF-β produced by $T_H1$ cells. As noted earlier, macrophages become more effective as antigen-presenting cells on activation. Thus, the activated macrophages can efficiently mediate activation of more $T_H1$ cells, primarily through the secretion of IL-12, which induces the $T_H1$ cell response and powerfully triggers IFN-γ production while suppressing the development of $T_H2$ cells. Another cytokine made by macrophages, IL-18, works together with IL-12 to induce even larger amounts of IFN-γ by $T_H1$ cells. The activated $T_H1$ cells in turn recruit and activate even more macrophages. This self-perpetuating response, however, is a double-edged sword, with a fine line existing between a beneficial, protective response and a detrimental response characterized by extensive tissue damage.

A report of experiments with knockout mice that could not produce IFN-γ demonstrated the importance of this cytokine in the DTH response. When these knockout mice were infected with an attenuated strain of *Mycobacterium bovis* known as BCG (Bacille Calmette Guérin), nearly all the animals died within 60 days, whereas wild-type mice survived (Figure 15-19). Macrophages from the IFN-γ knockout mice were shown to have reduced levels of class II MHC molecules and of bactericidal metabolites such as nitric oxide and superoxide anion.

IL-17 is another cytokine that acts as a potent mediator in delayed-type reactions by increasing chemokine production in various tissues to recruit monocytes and neutrophils to the site of inflammation, similar to the action of IFN-γ. IL-17 is produced by $T_H$ cells, although it is still controversial whether IL-17 is made by a subset of $T_H1$ cells or by a completely new and separate lineage of $T_H$ cell produced directly from naive CD4$^+$ T cells, called "$T_H$-17 cells" by the proponents of the latter theory. What is known is that the cells tentatively identified as $T_H$-17 cells are induced by IL-23 (versus the induction of $T_H1$ by IL-12) and produce a distinct cytokine profile, including IL-17 (versus IFN-γ for $T_H1$), that is responsible for destructive tissue damage in DTH reactions.

## The DTH reaction is detected with a skin test

The presence of a DTH reaction can be measured experimentally by injecting antigen intradermally into an animal and observing whether a characteristic skin lesion develops at the injection site. A positive skin-test reaction indicates that the individual has a population of sensitized $T_H1$ cells specific for the test antigen. For example, to determine whether an individual has been exposed to *M. tuberculosis*, PPD, a protein derived from the cell wall of this mycobacterium, is injected intradermally. Development of a red, slightly swollen, firm lesion at the site between 48 and 72 hours later indicates previous exposure. The skin lesion results from intense infiltration of cells to the site of injection during a DTH reaction; 80% to 90% of these cells are macrophages. Note, however, that a positive test does not allow one to conclude whether the exposure was to a pathogenic form of *M. tuberculosis* or to a vaccine form received through immunization, which is performed in some parts of the world.

## Contact dermatitis is a type of DTH response

Many contact dermatitis reactions, including the responses to formaldehyde, trinitrophenol, nickel, turpentine, and active agents in various cosmetics and hair dyes, poison oak, and poison ivy, are mediated by $T_H1$ cells. Most of these substances are small molecules that can complex with skin proteins. This complex is internalized by antigen-presenting cells in the skin (e.g., Langerhans cells), then processed and presented together with class II MHC molecules, causing activation of sensitized $T_H1$ cells. In the reaction to poison oak, for example, a pentadecacatechol compound from the leaves of the plant forms a complex with skin proteins. When $T_H$ cells react with this compound appropriately displayed by local antigen-presenting cells, they differentiate into sensitized $T_H1$ cells. A subsequent exposure to pentadecacatechol will elicit activation of $T_H1$ cells and induce cytokine production (Figure 15-20). Approximately 48 to 72 hours after the second exposure, the secreted cytokines cause macrophages to accumulate at the site. Activation of these macrophages and release of lytic enzymes result in the redness and pustules that characterize a reaction to poison oak.

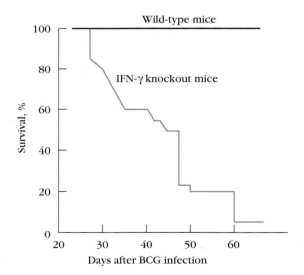

**FIGURE 15-19 Experimental demonstration of the role of IFN-γ in host defense against intracellular pathogens.** Knockout mice were produced by introducing a targeted mutation in the gene encoding IFN-γ. The mice were then infected with 10$^7$ colony-forming units of attenuated *Mycobacterium bovis* (BCG) and their survival monitored. *[Adapted from D. K. Dalton et al., 1993,* Science ***259**:1739.]*

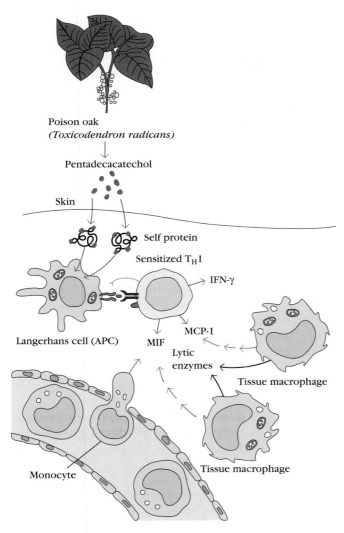

Poison oak
(*Toxicodendron radicans*)

Pentadecacatechol

Skin

Self protein

Sensitized $T_H1$

IFN-γ

MCP-1

Langerhans cell (APC)        MIF

Lytic
enzymes

Tissue macrophage

Tissue macrophage

Monocyte

**FIGURE 15-20 Development of delayed-type hypersensitivity reaction after a second exposure to poison oak.** Cytokines such as IFN-γ, monocyte chemotactic protein (MCP-1), and migration-inhibition factor (MIF) released from sensitized $T_H1$ cells mediate this reaction. Tissue damage results from lytic enzymes released from activated macrophages.

## SUMMARY

- Hypersensitivity reactions are inflammatory reactions within the humoral or cell-mediated branches of the immune system that lead to extensive tissue damage or even death. The four types of hypersensitivity reaction generate characteristic effector molecules and clinical manifestations.

- A type I hypersensitivity reaction is mediated by IgE antibodies, whose Fc region binds to receptors on mast cells or blood basophils. Cross-linkage of the fixed IgE by allergen leads to mast-cell or basophil degranulation with release of pharmacologically active mediators. The principal effects

of these mediators are smooth muscle contraction and vasodilation. Clinical manifestations of type I reactions include potentially life-threatening systemic anaphylaxis and localized responses such as hay fever and asthma.

- A type II hypersensitivity reaction occurs when antibody reacts with antigenic determinants present on the surface of cells, leading to cell damage or death through complement-mediated lysis or antibody-dependent cell-mediated cytotoxicity (ADCC). Transfusion reactions and hemolytic disease of the newborn are type II reactions.

- A type III hypersensitivity reaction is mediated by the formation of immune complexes and the ensuing activation of complement. Complement split products serve as immune effector molecules that elicit localized vasodilation and chemotactically attract neutrophils. Deposition of immune complexes near the site of antigen entry can induce an Arthus reaction, in which lytic enzymes released by the accumulated neutrophils and the complement membrane-attack complex cause localized tissue damage.

- A type IV hypersensitivity reaction involves the cell-mediated branch of the immune system. Antigen activation of sensitized $T_H1$ cells induces release of various cytokines that cause macrophages to accumulate and become activated. The net effect of the activation of macrophages is to release lytic enzymes that cause localized tissue damage.

## References

American Lung Association, Epidemiology & Statistics Unit, Research and Program Services. May 2005. Trends in asthma morbidity and mortality. http://www.lungusa.org/.

Aubry, J. P., et al. 1992. CD21 is a ligand for CD23 and regulates IgE production. *Nature* **358**:505.

Barnes, K. C., and D. G. Marsh. 1998. The genetics and complexity of allergy and asthma. *Immunology Today* **19**:325.

Borish, L. 1999. Genetics of allergy and asthma. *Annals of Allergy, Asthma, and Immunology* **82**:413.

Breiteneder, H., and O. Scheiner. 1998. Molecular and immunological characteristics of latex allergens. *International Archives of Allergy and Immunology* **116**:83.

Busse, W., and W. Neaville. 2001. Anti-immunoglobulin E for the treatment of allergic disease. *Current Opinion in Allergy and Immunology* **1**:105.

Chang, T. W. 2000. The pharmacological basis of anti-IgE therapy. *Nature Biotechnology* **18**:157.

Cohn, L., J. A. Elias, and G. L. Chupp. 2004. Asthma: mechanisms of disease persistence and progression. *Annual Review of Immunology* **22**:789.

Galli, S. J., et al. 2005. Mast cells as "tunable" effector and immunoregulatory cells: recent advances. *Annual Review of Immunology* **23**:749.

Gould, H. J., et al. 2003. The biology of IgE and the basis of allergic disease. *Annual Review of Immunology* **21**:579.

Jonkers, R. E., and J. S. van der Zee. 2005. Anti-IgE and other new immunomodulation-based therapies for allergic asthma. *Netherlands Journal of Medicine* **63**:121.

Kuhn, R., K. Rajewsky, and W. Muller. 1991. Generation and analysis of interleukin-4 deficient mice. *Science* **254**:707.

Marsh, D. G., et al. The Collaborative Study on the Genetics of Asthma (CSGA). 1997. A genome-wide search for asthma susceptibility loci in ethnically diverse populations. *Nature Genetics* **15**:389.

Nakae, S., et al. 2002. Antigen-specific T cell sensitization is impaired in IL-17-deficient mice, causing suppression of allergic cellular and humoral responses. *Immunity* **17**:375.

Novak, N., S. Kraft, and T. Bieber. 2001. IgE receptors. *Current Opinion in Immunology* **13**:721.

Oda, T., et al. 2000. Molecular cloning and characterization of a novel type of histamine receptor preferentially expressed in leukocytes. *Journal of Biological Chemistry* **275**:36781.

Ono, S. J. 2000. Molecular genetics of allergic diseases. *Annual Review of Immunology* **18**:347.

Park, H., et al. 2005. A distinct lineage of CD4 T cells regulates tissue inflammation by producing interleukin 17. *Nature Immunology* **6**:1133.

Romagnani, S. 2001. T-cell responses in allergy and asthma. *Current Opinion in Allergy and Clinical Immunology* **1**:73.

Rosenstreich, D. L., et al. 1997. The role of cockroach allergy and exposure to cockroach allergen in causing morbidity among inner-city children with asthma. *New England Journal of Medicine* **336**:1356.

Szabo, S. J., et al. 2003. Molecular mechanisms regulating $T_H1$ immune responses. *Annual Review of Immunology* **21**:713.

Ulrich, B., and J. Rivera. 2004. The ins and outs of IgE-dependent mast-cell exocytosis. *Trends in Immunology* **25**:266.

Van Eerdewegh, P., et al. 2002. Association of the ADAM33 gene with asthma and bronchial hyper-responsiveness. *Nature* **418**:426.

Young, R. P., et al. 1992. House dust mite sensitivity: interaction of genetics and allergen dosage. *Clinical and Experimental Allergy* **22**:205.

 Useful Web Sites

**http://www.niaid.nih.gov/**

The National Institute of Allergy and Infectious Diseases home page. NIAID is the NIH institute that sponsors research in infectious diseases. The NIAID Web site provides a number of links to other relevant sites.

**http://www.acaai.org/**

A site maintained by the American College of Allergy, Asthma & Immunology. An excellent source of patient information about many allergies. This site contains many valuable links.

**http://www.aaaai.org/**

The American Academy of Allergy, Asthma and Immunology Web site. A good site for exploring the many aspects of asthma.

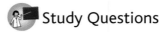 Study Questions

**CLINICAL FOCUS QUESTION** Discuss why IL-4 and FcεRIb are excellent candidate genes involved in the genetic susceptibility to asthma.

1. Indicate whether each of the following statements is true or false. If you think a statement is false, explain why.

   a. Mice infected with *Nippostrongylus brasiliensis* exhibit decreased production of IgE.

   b. IL-4 decreases IgE production by B cells.

   c. The initial step in the process of mast-cell degranulation is cross-linking of Fc receptors.

   d. Antihistamines are effective for the treatment of type III hypersensitivity.

   e. Most pollen allergens contain a single allergenic component.

   f. Babies can acquire IgE-mediated allergies by passive transfer of maternal antibody.

   g. Transfusion reactions are a manifestation of type II hypersensitivity.

   h. IgE is found attached to mast cells even if antigen is not present.

   i. Both eosinophils and mast cells are important mediators of allergic responses.

   j. $T_H1$ cells release cytokines that result in an increased production of eosinophils and mast cells in the bone marrow.

   k. Eosinophils play a role in the late-phase response in asthma.

   l. The wheal–and-flare response is typical of an early phase response in type I hypersensitivity.

   m. Smoke causes inflammation in the lung tissues.

   n. Allergy shots are thought to work by increasing the $T_H2$ response in allergic individuals.

2. In an immunology laboratory exercise, you are studying the response of mice injected intradermally with complete antibodies to the IgE Fc receptor (FcεR1) or with Fab fragments of such antibodies.

   a. Predict the response expected with each type of antibody.

   b. Would the responses observed depend on whether the mice were allergic? Explain.

3. Serum sickness can result when an individual is given a large dose of antiserum such as a mouse antitoxin to snake venom. How could you take advantage of recent technological advances to produce an antitoxin that would not produce serum sickness in patients who receive it?

4. What immunologic mechanisms most likely account for a person's developing each of the following reactions after an insect bite?

   a. Within 1 to 2 min after being bitten, swelling and redness appear at the site and then disappear by 1 h.

   b. 6 to 8 h later, swelling and redness again appear and persist for 24 h.

   c. 72 h later, the tissue becomes inflamed, and tissue necrosis follows.

5. Indicate which type(s) of hypersensitivity reaction (I–IV) apply to the following characteristics. Each characteristic can apply to one, or more than one, type.

   a. Is an important defense against intracellular pathogens.

   b. Can be induced by penicillin.

   c. Involves histamine as an important mediator.

   d. Can be induced by poison oak in sensitive individuals.

   e. Can lead to symptoms of asthma.

   f. Occurs as result of mismatched blood transfusion.

   g. Systemic form of reaction is treated with epinephrine.

   h. Can be induced by pollens and certain foods in sensitive individuals.

   i. May involve cell destruction by antibody-dependent cell-mediated cytotoxicity.

   j. One form of clinical manifestation is prevented by Rhogam.

   k. Localized form characterized by wheal-and-flare reaction.

6. In the table below, indicate whether each immunologic event listed does (+) or does not (−) occur in each type of hypersensitive response.

| Immunologic event | Hypersensitivity | | | |
|---|---|---|---|---|
| | Type I | Type II | Type III | Type IV |
| IgE-mediated degranulation of mast cells | | | | |
| Lysis of antibody-coated blood cells by complement | | | | |
| Tissue destruction in response to poison oak | | | | |
| C3a- and C5a–mediated mast-cell degranulation | | | | |
| Chemotaxis of neutrophils | | | | |
| Chemotaxis of eosinophils | | | | |
| Activation of macrophages by IFN-γ | | | | |
| Deposition of antigen-antibody complexes on basement membranes of capillaries | | | | |
| Sudden death due to vascular collapse (shock) shortly after injection or ingestion of antigen | | | | |

7. Describe the type II hypersensitivity reaction that can occur in an Rh⁺ infant of an Rh⁻ mother.

8. Type III hypersensitivities are characterized by immune complex deposition. Describe the clinical consequences of immune complex deposition.

9. When allergen binds to IgE attached to FcεRI on mast cells, the receptors cross-link, resulting in a signal transduction cascade leading to degranulation. For each of the following components of the signal transduction cascade, connect each of the production/activation events on the left to their cellular consequences on the right.

*Production or activation*

a. Increased intracellular calcium

b. Adenylate cyclase

c. Phospholipid methyl transferase enzymes I and II

d. Phospholipase C

e. Protein tyrosine kinase

f. Phosphatidylcholine

g. Diacylglycerol

*Cellular consequences*

1. Swelling of granules and fusion with plasma membrane

2. Formation of phoshatidyl choline

3. Production of diacyl glycerol and inositol triphosphate

4. Activation of protein kinase C

5. Formation of cyclic AMP

6. Increase in membrane fluidity

7. Activation of phospholipase C

10. Describe the difference between the early and late phases of asthma.

11. Fill in the blanks with the proper terms: A mother that is Rh⁻ is given _____ when she is carrying an Rh⁺ baby. If she develops a titer to Rh⁻ antigen, the baby could be born with _____.

**ANALYZE THE DATA** Edward Mitre and colleagues (2004, *J. Immunol.* **172**:2439) investigated the roles of basophils in filarial infections and host hypersensitivity responses. Helminth infections often pose challenges to host resistance. They looked at the amount of histamine released by the basophils in whole blood of normal and infected individuals (part a in figure on p 400). They also looked at the dose of antigen that induced histamine release (part b in the figure).

a. What type of hypersensitivity reaction is manifested by people that are infected?

b. Why is histamine used to measure this response? If you didn't have an essay to measure histamine, how could you measure a host response after BmAg administration?

c. Describe what immunoglobulin isotype you would isolate from the surface of the basophils if you analyzed the cells.

d. The data in part b show that basophils respond to antigen in a dose-dependent fashion. What occurs at the basophil surface to induce the response? What occurs intracellularly in response to that process?

e. Explain why you would or would not expect eosinophils to be found in filaria-infected individuals.

f. Does a filarial infection induce a humoral response? Explain your answer.

g. Mitre and coworkers found a correlation between the concentration of IL-4 secreted and the amount of IgE in the serum of infected patients. What kind of helper T cell is involved in this response? Why would this response suppress an IFN-γ secretion?

(a) Maximum amount of histamine released by basophil cells as percentage of total histamine released

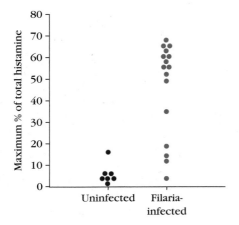

(b) Average histamine release versus dosgae of *Brugia malayi* antigen (BmAg)

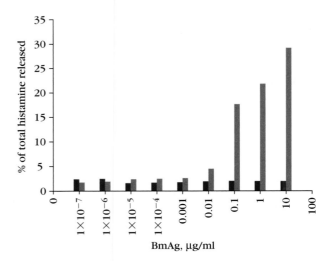

**Interactive Study**

www.whfreeman.com/kuby

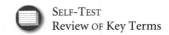

SELF-TEST
Review OF Key Terms

# Tolerance and Autoimmunity

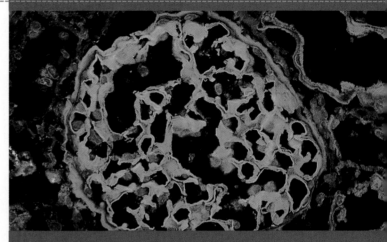

E
ARLY IN THE LAST CENTURY, PAUL EHRLICH REALIZED that the immune system could go awry and, instead of reacting only against foreign antigens, could focus its attack on self antigens. This condition, which he termed "horror autotoxicus," can result in a number of chronic and acute diseases, including rheumatoid arthritis, multiple sclerosis, lupus erythematosis, and certain types of diabetes. Simply stated, these diseases result from failure of the host's humoral and cellular immune systems to distinguish self from nonself, resulting in attack on self cells and organs by auto-antibodies and self-reactive T cells. A number of mechanisms exist to protect an individual from potentially self-reactive lymphocytes; these are given the general term **tolerance.** A primary mechanism termed **central tolerance** deletes T- or B-cell clones before the cells are allowed to mature if they possess receptors that recognize self antigens with greater than a low threshold affinity. Central tolerance occurs in the primary lymphoid organs, bone marrow, and thymus (Figure 16-1a). Because central tolerance is not perfect and some self-reactive lymphocytes find their way into the secondary lymphoid tissues, there are additional safeguards to limit their activity. These backup precautions include **peripheral tolerance,** which renders lymphocytes in secondary lymphoid tissues inactive or anergic (Figure 16-1b). The possibility of damage from self-reactive lymphocytes is further limited by the life span of activated lymphocytes, which is regulated by programs that induce cell death (apoptosis) on receipt of signals. Despite this layered system of regulation, self-reactive clones of T or B cells are occasionally activated, generating humoral or cell-mediated responses against self antigens. Such an inappropriate response of the immune system against self components is termed **autoimmunity.** Autoimmune reactions can cause serious damage to cells and organs, sometimes with fatal consequences.

In some cases the damage to self cells or organs is caused by antibodies; in other cases, T cells are the culprit. For example, a common form of autoimmunity is tissue injury by mechanisms similar to type II hypersensitivity reactions. As we saw in Chapter 15, type II hypersensitivity reactions

- Establishment and Maintenance of Tolerance

- Organ-Specific Autoimmune Diseases

- Systemic Autoimmune Diseases

- Animal Models for Autoimmune Diseases

- Evidence Implicating the CD4$^+$ T Cell, MHC, and TCR in Autoimmunity

- Proposed Mechanisms for Induction of Autoimmunity

- Treatment of Autoimmune Diseases

involve antibody-mediated destruction of cells. Autoimmune hemolytic anemia is an excellent example of such an autoimmune disease. In this disease, antigens on red blood cells are recognized by auto-antibodies, which results in the destruction of the blood cells, which in turn results in anemia. Auto-antibodies are also the major offender in Hashimoto's thyroiditis, in which antibodies reactive with tissue-specific antigens such as thyroid peroxidase and thyroglobulin cause severe tissue destruction. Other autoimmune diseases that involve auto-antibodies are listed in Table 16-1.

Many autoimmune diseases are characterized by tissue destruction mediated directly by T cells. A well-known example is rheumatoid arthritis, in which self-reactive T cells attack the tissue in joints, causing an inflammatory response that results in swelling and tissue destruction. Other

(a) Central tolerance

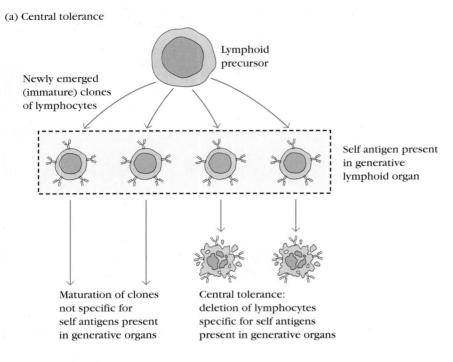

Lymphoid precursor

Newly emerged (immature) clones of lymphocytes

Self antigen present in generative lymphoid organ

Maturation of clones not specific for self antigens present in generative organs

Central tolerance: deletion of lymphocytes specific for self antigens present in generative organs

(b) Peripheral tolerance

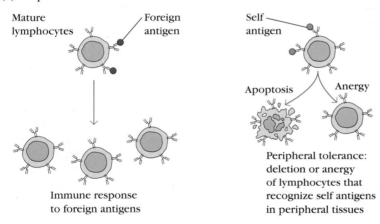

Mature lymphocytes

Foreign antigen

Self antigen

Apoptosis    Anergy

Immune response to foreign antigens

Peripheral tolerance: deletion or anergy of lymphocytes that recognize self antigens in peripheral tissues

**FIGURE 16-1  Central and peripheral tolerance.** (a) Central tolerance is established by deletion of lymphocytes possessing receptors that react with antigens in the primary lymphoid organs: thymus for T cells and bone marrow for B cells. (b) Peripheral tolerance involves deleting or rendering anergic lymphocytes that possess receptors that react with self antigens. This process occurs in secondary lymphoid organs.

examples include insulin-dependent diabetes mellitus and multiple sclerosis (see Table 16-1).

In this chapter we first describe the general mechanisms for limiting autoimmune reactivity by maintaining tolerance to self antigens. Common human autoimmune diseases resulting from failures of these mechanisms are described. These can be divided into two broad categories: organ specific and systemic autoimmune disease (see Table 16-1). Such diseases affect 5% to 7% of the human population, often causing chronic debilitating illnesses. Several experimental animal models used to study autoimmunity and various mechanisms that may contribute to induction of autoimmune reactions are also described, as well as current and experimental therapies for treating autoimmune diseases.

# Establishment and Maintenance of Tolerance

The several layers of protection imposed by the immune system to prevent the reaction of its cells and antibodies with host components and the onset of autoimmune disease are known under the general heading of **tolerance,** defined as a state of unresponsiveness to an antigen. The mechanisms mediating this unresponsiveness can vary. Under normal circumstances, encounter of the immune system with an antigen leads to an immune response, but presenting the antigen in some alternative form may lead to tolerance or the unresponsiveness of the immune system

| TABLE 16-1 | Some autoimmune diseases in humans | |
|---|---|---|
| **Disease** | **Self antigen** | **Immune response** |
| ORGAN-SPECIFIC AUTOIMMUNE DISEASES | | |
| Addison's disease | Adrenal cells | Auto-antibodies |
| Autoimmune hemolytic anemia | RBC membrane proteins | Auto-antibodies |
| Goodpasture's syndrome | Renal and lung basement membranes | Auto-antibodies |
| Graves' disease | Thyroid-stimulating hormone receptor | Auto-antibody (stimulating) |
| Hashimoto's thyroiditis | Thyroid proteins and cells | $T_H1$ cells, auto-antibodies |
| Idiopathic thrombocyopenia purpura | Platelet membrane proteins | Auto-antibodies |
| Insulin-dependent diabetes mellitus | Pancreatic beta cells | $T_H1$ cells, auto-antibodies |
| Myasthenia gravis | Acetylcholine receptors | Auto-antibody (blocking) |
| Myocardial infarction | Heart | Auto-antibodies |
| Pernicious anemia | Gastric parietal cells; intrinsic factor | Auto-antibody |
| Poststreptococcal glomerulonephritis | Kidney | Antigen-antibody complexes |
| Spontaneous infertility | Sperm | Auto-antibodies |
| SYSTEMIC AUTOIMMUNE DISEASES | | |
| Ankylosing spondylitis | Vertebrae | Immune complexes |
| Multiple sclerosis | Brain or white matter | $T_H1$ cells and $T_C$ cells, auto-antibodies |
| Rheumatoid arthritis | Connective tissue, IgG | Auto-antibodies, immune complexes |
| Scleroderma | Nuclei, heart, lungs, gastrointestinal tract, kidney | Auto-antibodies |
| Sjögren's syndrome | Salivary gland, liver, kidney, thyroid | Auto-antibodies |
| Systemic lupus erythematosus (SLE) | DNA, nuclear protein, RBC and platelet membranes | Auto-antibodies, immune complexes |

(Figure 16-2). Antigens that induce tolerance are called **tolerogens** rather than *immunogens*. The same chemical compound can be both an immunogen and a tolerogen, depending on how it is presented to the immune system. For example, an antigen presented to T cells without appropriate costimulation results in a form of tolerance known as anergy, whereas the very same antigen presented with costimulatory molecules present can become a potent immunogen. Factors promoting tolerance rather than stimulation of the immune system by a given antigen include the following:

- High doses of antigen

- Persistence of antigen in host

- Intravenous or oral introduction

- Absence of adjuvants

- Low levels of costimulators

It is well documented that antigens introduced orally may result in tolerance whereas the same antigen given as an intradermal or subcutaneous injection can be immunogenic. Notably, tolerance is antigen specific: the inactivation of an immune response due to tolerance does not result in general immune suppression, but rather is specific for the tolerogenic antigen.

In the 1960s, researchers believed that all self-reactive lymphocytes were eliminated during their development in the bone marrow and thymus and that a failure to eliminate these lymphocytes led to autoimmune consequences. More recent experimental evidence has countered that belief. Normal, healthy individuals have been shown to possess mature, recirculating, self-reactive lymphocytes. Since the presence of these self-reactive lymphocytes in the periphery does not inevitably result in autoimmune reactions, their activity must be regulated in normal individuals through other mechanisms. Means to maintain tolerance include induction of cell death or cell anergy and limitations on the activity of T cells by means of $T_{reg}$ cells.

## Central tolerance limits development of autoreactive T and B cells

The dominant mechanism in maintaining tolerance is the deletion during early maturation of lymphocyte clones that may react with self components. Consider the mechanisms that generate diversity in T-cell or B-cell receptors. As discussed in Chapters 5 and 9, the genetic rearrangements that give rise to a functional TCR or Ig occur through a process whereby any V-region gene can associate with any D or J gene segment. This means that the generation of V regions

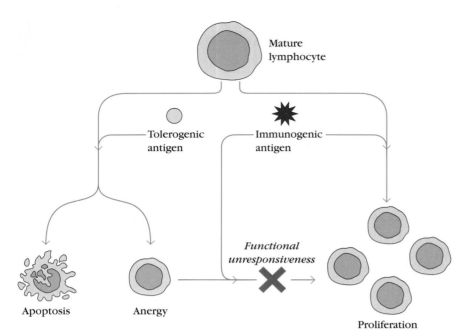

**FIGURE 16-2 An antigen may be immunogenic or tolerogenic, depending on a variety of factors that include dose and avenue of exposure.** Tolerogenic encounters with antigens result in either deletion of the reactive lymphocytes, usually by apoptosis, or induction of a state of anergy. Immunogenic exposure results in activation and proliferation of the reactive lymphocyte.

Mature lymphocyte

Tolerogenic antigen

Immunogenic antigen

*Functional unresponsiveness*

Apoptosis

Anergy

Proliferation

that react with self antigen is theoretically possible. If this were allowed to occur frequently, such TCR or Ig receptors could produce mature functional T or B cells that recognize self, and autoimmune disease would ensue. Alternatively, the receptors on those clones that react with self may be altered or *edited*, reducing their affinity for self antigens below a critical threshold that would lead to disease. As mentioned in Chapters 10 and 11, this so-called central tolerance process works to eliminate autoreactive B cells in the bone marrow and autoreactive T cells in the thymus. Although our understanding of the precise molecular mechanisms that mediate central tolerance in T and B cells is not complete, it is known that B and T cells undergo a developmentally regulated event known as negative selection, resulting in the induction of death in cells that carry potentially autoreactive TCR or Ig receptors.

Once the nature of V(D)J rearrangements was understood, it was logical to imagine that a process such as tolerance should exist to eliminate developing self-reactive T and B cells. One of the classic experiments demonstrating that self-reactive lymphocytes are removed or inactivated after encounter with self antigen is the elegant work of C. C. Goodnow and colleagues in 1991. Mice that expressed a transgenic immunoglobulin specific for hen egg-white lysozyme (HEL) were mated to transgenic mice that express HEL (Figure 16-3a). When anti-HEL$^+$ mice were mated to HEL$^+$ animals, B cells developing in the offspring encountered HEL in the bone marrow. Goodnow and colleagues noted the F1 mice lacked mature B cells expressing anti-HEL. Subsequent experiments from several laboratories demonstrated that, for the most part, autoreactive developing B cells are deleted in the bone marrow by the induction of apoptosis. David Nemazee and colleagues extended these observations and demonstrated that some

developing B cells can undergo a process they termed **receptor editing.** In cells undergoing receptor editing, the antigen-specific V region is "edited," with a different V-region gene segment switched for the autoreactive V gene segment via V(D)J recombination (Editing occurs most often within the $V_L$ rather than the $V_H$ region.) Receptor editing, as well as clonal deletion or apoptosis, is recognized as one of the mechanisms that lead to central tolerance in developing B cells. By a similar mechanism, T cells developing in the thymus that have too high an affinity for self antigen are deleted, primarily through the induction of apoptosis.

## Peripheral tolerance regulates autoreactive cells in circulation

Another important observation from the experiments with anti-HEL transgenics was that if HEL was expressed on the membrane of cells, it induced clonal deletion of all immature B cells with anti-HEL Ig. However, if HEL was secreted and detected as a soluble protein, the B cells matured and exited the bone marrow and were found in the periphery. These cells, however, were unresponsive to HEL antigen and existed in a state known as anergy. **Anergy** can be defined as unresponsiveness to antigenic stimulus. As shown by the HEL studies and by numerous other examples, central tolerance is not a foolproof process and does not completely eliminate all possible self-reactive lymphocytes because (1) not all self antigens are expressed in the central lymphoid organs where negative selection occurs, and (2) there is a threshold requirement for affinity to self antigens before clonal deletion is triggered, allowing some weakly self-reactive clones to survive the weeding-out process.

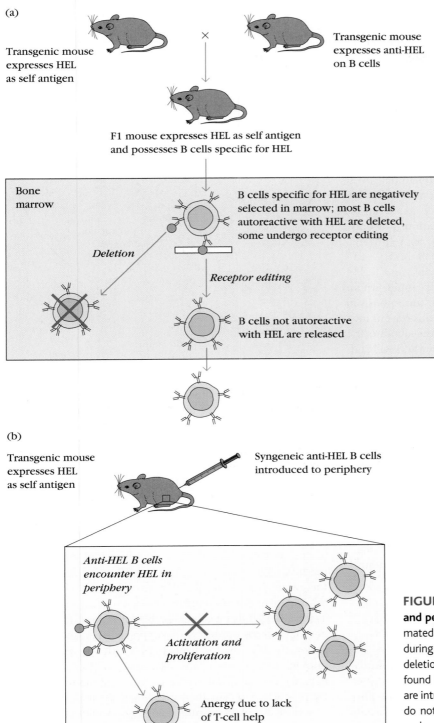

(a)

Transgenic mouse
expresses HEL
as self antigen

×

Transgenic mouse
expresses anti-HEL
on B cells

F1 mouse expresses HEL as self antigen
and possesses B cells specific for HEL

Bone
marrow

B cells specific for HEL are negatively
selected in marrow; most B cells
autoreactive with HEL are deleted,
some undergo receptor editing

*Deletion*

*Receptor editing*

B cells not autoreactive
with HEL are released

(b)

Transgenic mouse
expresses HEL
as self antigen

Syngeneic anti-HEL B cells
introduced to periphery

*Anti-HEL B cells
encounter HEL in
periphery*

*Activation and
proliferation*

Anergy due to lack
of T-cell help

**FIGURE 16-3 Experiments demonstrating central and peripheral tolerance.** (a) When anti-HEL⁺ mice are mated to HEL⁺ mice, the B cells of F1 mice encounter HEL during negative selection in the bone marrow, followed by deletion or receptor editing. HEL-reactive B cells are not found in the periphery of these mice. (b) Anti-HEL B cells are introduced into a mouse that expresses HEL. The B cells do not migrate to lymphoid follicles in spleen or lymph node and do not differentiate into plasma cells, instead becoming anergic due to lack of T-cell help.

In some cases, self-reactive T or B cells can escape deletion in the thymus or bone marrow and appear in the periphery. A form of tolerance called **peripheral tolerance** inactivates these cells. Peripheral tolerance can be defined as the inactivation of self-reactive T cells or B cells in the periphery, rendering them incapable of responding to self. As with central tolerance, the existence of peripheral tolerance was predicted in advance of the necessary experimental evidence. Again, when the anti-HEL Ig transgenics and HEL-expressing mouse models described above are used, HEL presented to mature anti-HEL B cells in peripheral tissues inactivates these B cells, and they never migrate to lymphoid follicles in spleen or lymph node (Figure 16-3b). Recall that B cells become antibody-secreting plasma cells after maturation and selection in lymphoid follicles/germinal centers. An important feature to appreciate in these experiments is that T cells that recognize HEL in the HEL-expressing mouse are deleted before maturation and release due to central tolerance. The B cells can recognize the antigen, but there is no subsequent help from T cells. These experiments demonstrated that when mature B cells encounter soluble antigen in the absence of T-cell help, they become unresponsive or anergic and never migrate to germinal centers.

Peripheral T-cell tolerance has been demonstrated by a variety of experimental strategies, providing evidence for the mechanisms that mediate peripheral tolerance. As we saw in Chapter 10, T cells, to become activated, require not only that the TCR recognize antigen presented on self-MHC molecules but also that costimulatory signals be present. Early experiments by M. K. Jenkins, D. Mueller, and R. H. Schwartz showed that in vitro, CD4$^+$ T-cell clones, when stimulated solely through the TCR, become unresponsive; they used the term **clonal anergy** to describe this state of unresponsiveness. Subsequent data from several laboratories showed that the interaction between CD28 on the T cell and B7 on the antigen-presenting cell provided the necessary costimulatory signals required for T-cell activation.

The understanding that CD28/B7 signals provide essential costimulatory signals for T-cell activation led to a careful examination of costimulation, revealing the existence of inhibitory receptors such as CTLA-4. CTLA-4, like CD28, binds to B7, but instead of providing activating signals, CTLA-4 inhibits T-cell activation. In fact, CTLA-4 expression is induced after T cells are activated, ensuring control and regulation of T cell activation. The role of CTLA-4 in tolerance was appreciated when the gene encoding this molecule was deleted. Mice lacking CTLA-4 display massive proliferation of lymphocytes and autoimmune disease, suggesting an important role for this molecule in maintaining peripheral tolerance.

## Regulatory T cells are a component of peripheral tolerance

Peripheral tolerance may also be induced by regulatory T cells (T$_{reg}$ cells). Acting in secondary lymphoid tissues and at sites of inflammation, T$_{reg}$ cells down-regulate autoimmune processes. Recall that T$_{reg}$ cells are a unique subset of CD4$^+$ T cells that express high levels of the IL-2R α chain (CD25). T$_{reg}$ cells have been shown to arise from a subset of T cells expressing receptors with intermediate affinity for self antigens in the thymus (Figure 16-4). Certain of these cells up-regulate the transcription factor Foxp3 and then develop into T$_{reg}$ cells capable of suppressing reaction to self antigens. The ability of T$_{reg}$ cells to suppress an immune response was demonstrated by early experiments in nonobese diabetic (NOD) mice and BB rats (BioBreeding rats, the first animal model of spontaneous autoimmune type I diabetes), two strains prone

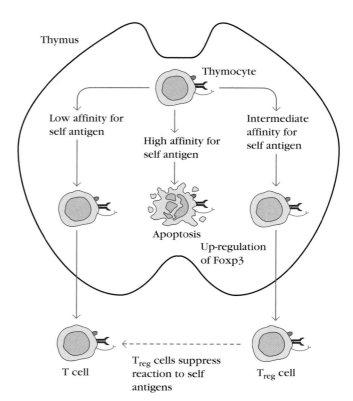

**FIGURE 16-4  T regulatory (T$_{reg}$) cells are generated from thymocytes during negative selection in the thymus.** Thymocytes with high affinity for self antigens are deleted at this stage; thymocytes with low affinity are positively selected and released. Thymocytes with intermediate affinity for antigens encountered in the thymus up-regulate the transcription factor Foxp3 and become T$_{reg}$ cells, which serve to keep self-reactive T-cell responses in check. *[Adapted from M. Kronenberg and A. Y. Rudensky, 2005,* Nature **435**:598.]

to the development of autoimmunity-based diabetes. The onset of diabetes in NOD mice and BB rats is delayed when these animals are injected with normal CD4$^+$ T cells from histocompatible donors. Further characterization of the CD4$^+$ T cells revealed that a subset notable for expressing high levels of CD25 was responsible for the suppression of diabetes in these animal models. The mechanisms by which T$_{reg}$ cells suppress immune responses is an area of intense investigation, but it is apparent that suppression is regulated, at least in part, through the production of cytokines, including IL-10 and TGFβ.

Cell death plays an important role in maintaining both central and peripheral tolerance. This is evidenced by the development of systemic autoimmune diseases in naturally occurring mutations of either the death receptor, Fas, or Fas ligand (FasL) in mice. As discussed in Chapter 10, activated T cells express increased levels of Fas and FasL. In both B and T cells, engagement of Fas by FasL induces a rapid apoptotic death known as activation-induced cell death (AICD). Mice that carry inactivating mutations in Fas *(lpr/lpr)* or FasL *(gld/gld)* are not able to engage the AICD pathway and develop autoimmune disease early in life.

## Antigen sequestration is a means to protect self antigens

In addition to the various mechanisms of central and peripheral tolerance described above, an effective means to avoid self-reactivity is sequestration or compartmentation of antigens such that they do not encounter reactive lymphocytes under normal circumstances. If the antigen is never exposed to immune cells, there is no possibility of reactivity. However, a consequence of sequestration is that the antigen is never encountered by developing lymphocytes, and no tolerance to the sequestered antigen is established. If barriers between immune cells and the sequestered antigens are breached by inoculation with the antigen, trauma, or use of certain chemicals leading to access and subsequent reactivity, the antigen will be seen as foreign because it was not previously encountered. For example, breaching the blood-brain barrier can cause reaction against components of the central nervous system.

## Failure of tolerance leads to autoimmunity

Simply stated, autoimmune disease is caused by failure of the tolerance processes to protect the host from the action of self-reactive lymphocytes. It is clear that tolerance should act to prevent autoimmune diseases. Convincing evidence that this is true comes from naturally occurring mutations that block tolerance; autoimmunity ensues. The link between Fas, FasL, and autoimmunity was reinforced when patients with a rare inherited form of autoimmune disease known as autoimmune lymphoproliferative syndrome, or ALPS, were found to carry mutations in Fas. Similar to observations in *lpr/lpr* mice, these patients have severe autoimmune disease affecting multiple organs.

A human autoimmune disease known as autoimmune polyendocrinopathy-candidiasis-ectodermal dystrophy (APECED) has been studied for some years because it is known to be caused by a single autosomal locus and displays a recessive inheritance pattern. Because it is the only known human autoimmune condition that is inherited in Mendelian fashion, it provided geneticists with an excellent model to decipher genetic components of autoimmunity. APECED is characterized by multiple autoimmune endocrinopathies, chronic mucocutaneous candidiasis, and ectodermal dystrophies. The causative gene lesion for APECED was isolated recently by two independent groups using a traditional positional cloning strategy (in which a gene's location is identified, then its function determined). The APECED gene encodes a novel protein known as AIRE. Expressed primarily in thymus, pancreas, and adrenal cortex, AIRE apparently regulates the presentation of peripheral-tissue self antigens on thymic medullary epithelial cells, thus linking it to the development of central tolerance. Quite recently, AIRE −/− mice have become available. Like humans with APECED, mice lacking AIRE are highly susceptible to the development of multi-organ autoimmune disease. How

AIRE regulates the expression of self antigens in the thymus is not well understood and is an area of intense investigation.

Another human autoimmune disorder, immune dysregulation, polyendocrinopathy, enteropathy, X-linked syndrome (IPEX), has contributed to our understanding of tolerance. IPEX is a fatal disorder known to share several features in common with a naturally occurring mutation in mice known as *scurfy*. Genetic analysis of both IPEX patients and *scurfy* mice led to the discovery that both conditions mapped to mutations in the Foxp3 gene. Foxp3 is a transcription factor now known to be required for the formation of $CD4^+/CD25^+$ regulatory T cells; mutations in Foxp3 lead to early and multifocal autoimmune disease. Thus, IPEX and *scurfy* appear to be caused by an inability of $T_{reg}$ cells to regulate immune responses.

The genetic bases of human autoimmune diseases are the subject of widespread investigation, and in most cases multiple genes have been implicated, making assignment of the exact defect a difficult task. In the next section we describe some of the more common autoimmune diseases, classified according to pathogenic mechanisms.

# Organ-Specific Autoimmune Diseases

In an organ-specific autoimmune disease, the immune response is directed to a target antigen unique to a single organ or gland, so that the manifestations are largely limited to that organ. The cells of the target organs may be damaged directly by humoral or cell-mediated effector mechanisms. Alternatively, the antibodies may overstimulate or block the normal function of the target organ.

## Some autoimmune diseases are mediated by direct cellular damage

Autoimmune diseases involving direct cellular damage occur when lymphocytes or antibodies bind to cell membrane antigens, causing cellular lysis and/or an inflammatory response in the affected organ. Gradually, the damaged cellular structure is replaced by connective tissue (fibrosis), and the function of the organ declines. In this section we briefly describe a few examples of this type of autoimmune disease.

### Hashimoto's Thyroiditis

In Hashimoto's thyroiditis, which is most frequently seen in middle-aged women, an individual produces auto-antibodies and sensitized $T_H1$ cells specific for thyroid antigens. An attending delayed-type hypersensitivity (DTH) response is characterized by an intense infiltration of the thyroid gland by lymphocytes, macrophages, and plasma cells, which form lymphocytic follicles and germinal centers (Figure 16-5). The ensuing inflammatory response causes a goiter, or visible enlargement of the thyroid gland, a physiological response to hypothyroidism (decreased production of thyroid hormones).

(a)

(b)

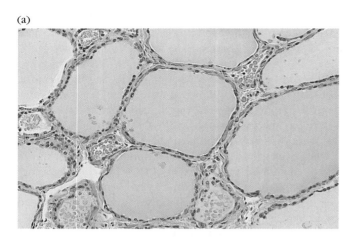
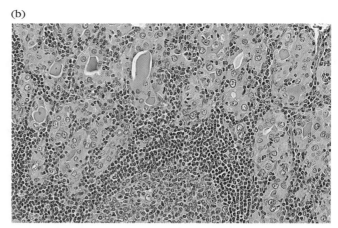

FIGURE 16-5 Photomicrographs of (a) normal thyroid gland showing a follicle lined by cuboidal follicular epithelial cells and (b) gland in Hashimoto's thyroiditis showing intense lymphocyte infiltration. *[From Web Path, courtesy of E. C. Klatt, University of Utah.]*

Hypothyroidism is caused when antibodies are formed to a number of thyroid proteins, including thyroglobulin and thyroid peroxidase, both of which are involved in the uptake of iodine. Binding of the auto-antibodies to these proteins interferes with iodine uptake, leading to hypothyroidism.

## Autoimmune Anemias

Autoimmune anemias include pernicious anemia, autoimmune hemolytic anemia, and drug-induced hemolytic anemia. Pernicious anemia is caused by auto-antibodies to intrinsic factor, a membrane-bound intestinal protein on gastric parietal cells. Intrinsic factor facilitates uptake of vitamin $B_{12}$ from the small intestine. Binding of the auto-antibody to intrinsic factor blocks the intrinsic factor–mediated absorption of vitamin $B_{12}$. In the absence of sufficient vitamin $B_{12}$, which is necessary for proper hematopoiesis, the number of functional mature red blood cells decreases below normal. Pernicious anemia is treated with injections of vitamin $B_{12}$, thus circumventing the defect in its absorption.

An individual with autoimmune hemolytic anemia makes auto-antibody to red–blood-cell antigens, triggering complement-mediated lysis or antibody-mediated opsonization and phagocytosis of the red blood cells. One form of autoimmune anemia is drug induced: when certain drugs such as penicillin or the anti-hypertensive agent methyldopa interact with red blood cells, the cells become antigenic. The immunodiagnostic test for autoimmune hemolytic anemias generally involves a Coombs test, in which the red cells are incubated with an anti–human IgG antiserum. If IgG auto-antibodies are present on the red cells, the cells are agglutinated by the antiserum.

## Goodpasture's Syndrome

In **Goodpasture's syndrome,** auto-antibodies specific for certain basement membrane antigens bind to the basement membranes of the kidney glomeruli and the alveoli of the lungs. Subsequent complement activation leads to direct cellular damage and an ensuing inflammatory response mediated by a buildup of complement split products. Damage to the glomerular and alveolar basement membranes leads to progressive kidney damage and pulmonary hemorrhage. Death may ensue within several months of the onset of symptoms. Biopsies from patients with Goodpasture's syndrome stained with fluorescent-labeled anti-IgG and anti-C3b reveal linear deposits of IgG and C3b along the basement membranes (Figure 16-6).

## Insulin-Dependent Diabetes Mellitus

A disease afflicting 0.2% of the population, **insulin-dependent diabetes mellitus (IDDM)** is caused by an autoimmune attack on the pancreas. The attack is directed

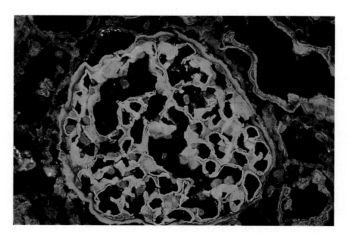

FIGURE 16-6 Fluorescent anti-IgG staining of a kidney biopsy from a patient with Goodpasture's syndrome reveals linear deposits of auto-antibody along the basement membrane. *[From Web Path, courtesy of E. C. Klatt, University of Utah.]*

(a)

(b)

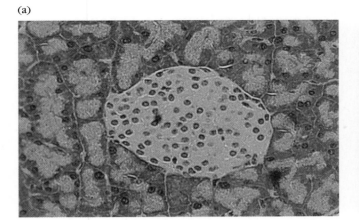
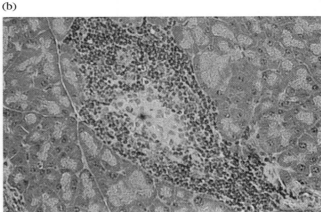

**FIGURE 16-7** Photomicrographs of an islet of Langerhans in (a) pancreas from a normal mouse and (b) pancreas from a mouse with a disease resembling insulin-dependent diabetes mellitus. Note the lymphocyte infiltration into the islet (insulitis) in (b). *[From M. A. Atkinson and N. K. Maclaren, 1990, Scientific American **263**(1):62.]*

against specialized insulin-producing cells (beta cells) that are located in spherical clusters, called the islets of Langerhans, scattered throughout the pancreas. The autoimmune attack destroys beta cells, resulting in decreased production of insulin and consequently increased levels of blood glucose. Several factors are important in the destruction of beta cells. First, activated CTLs migrate into an islet and begin to attack the insulin-producing cells. Local cytokine production during this response includes IFN-γ, TNF-α, and IL-1. Auto-antibody production can also be a contributing factor in IDDM. The first CTL infiltration and activation of macrophages, frequently referred to as insulitis (Figure 16-7), is followed by cytokine release and the presence of auto-antibodies, which leads to a cell-mediated DTH response. The subsequent beta-cell destruction is thought to be mediated by cytokines released during the DTH response and by lytic enzymes released from the activated macrophages. Auto-antibodies to beta cells may contribute to cell destruction by facilitating either antibody-mediated complement lysis or antibody-dependent cell-mediated cytotoxicity (ADCC).

The abnormalities in glucose metabolism that are caused by the destruction of islet beta cells result in serious metabolic problems that include ketoacidosis and increased urine production. The late stages of the disease are often characterized by atherosclerotic vascular lesions—which in turn cause gangrene of the extremities due to impeded vascular flow—renal failure, and blindness. If untreated, death can result. The most common therapy for diabetes is daily administration of insulin. This is quite helpful in managing the disease, but because sporadic doses are not the same as metabolically regulated continuous and controlled release of the hormone, periodically injected doses of insulin do not totally alleviate the problems caused by the disease. Another complicating feature of diabetes is that the disorder can go undetected for several years, allowing irreparable loss of pancreatic tissue to occur

before treatment begins. Improved techniques for transplantation of purified islet cells show promise for the treatment of IDDM (see Figure 17-12).

## Some autoimmune diseases are mediated by stimulating or blocking auto-antibodies

In some autoimmune diseases, antibodies act as agonists, binding to hormone receptors in lieu of the normal ligand and stimulating inappropriate activity. This usually leads to an overproduction of mediators or an increase in cell growth. Conversely, auto-antibodies may act as antagonists, binding hormone receptors and thereby blocking receptor function. This generally causes impaired secretion of mediators and gradual atrophy of the affected organ.

### Graves' Disease

The production of thyroid hormones is carefully regulated by thyroid-stimulating hormone (TSH), which is produced by the pituitary gland. Binding of TSH to a receptor on thyroid cells activates adenylate cyclase and stimulates the synthesis of two thyroid hormones, thyroxine and triiodothyronine. A patient with **Graves' disease** produces auto-antibodies that bind the receptor for TSH and mimic the normal action of TSH, activating adenylate cyclase and resulting in production of the thyroid hormones. Unlike TSH, however, the auto-antibodies are not regulated, and consequently they overstimulate the thyroid. For this reason these auto-antibodies are called long-acting thyroid-stimulating (LATS) antibodies (Figure 16-8).

### Myasthenia gravis

**Myasthenia gravis** is the prototype autoimmune disease mediated by blocking antibodies. A patient with this disease produces auto-antibodies that bind the acetylcholine

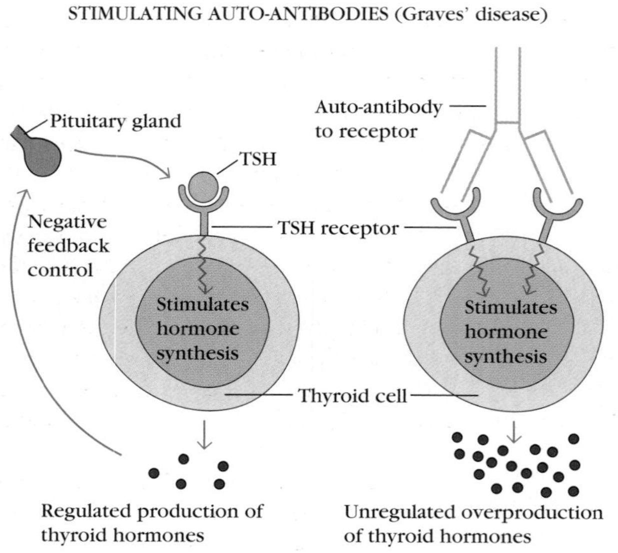

STIMULATING AUTO-ANTIBODIES (Graves' disease)

**FIGURE 16-8** In Graves' disease, binding of auto-antibodies to the receptor for thyroid-stimulating hormone (TSH) induces unregulated activation of the thyroid, leading to overproduction of the thyroid hormones (burgundy dots).

receptors on the motor end plates of muscles, blocking the normal binding of acetylcholine and also inducing complement-mediated lysis of the cells. The result is a progressive weakening of the skeletal muscles (Figure 16-9). Ultimately, the antibodies cause the destruction of the cells bearing the receptors. The early signs of this disease include drooping eyelids and inability to retract the corners of the mouth, which gives the appearance of snarling. Without treatment, progressive weakening of the muscles can lead to severe impairment of eating as well as problems with movement. However, with appropriate treatment, this disease can be managed quite well and afflicted individuals can lead a normal life.

# Systemic Autoimmune Diseases

In systemic autoimmune diseases, the response is directed toward a broad range of target antigens and involves a number of organs and tissues. These diseases reflect a general defect in immune regulation that results in hyperactive T cells and B cells. Tissue damage is widespread, both from cell-mediated immune responses and from direct cellular damage caused by auto-antibodies or by accumulation of immune complexes.

## Systemic lupus erythematosus attacks many tissues

One of the best examples of a systemic autoimmune disease is **systemic lupus erythematosus (SLE),** which typically appears in women between 20 and 40 years of age; the ratio of female to male patients is 10:1. SLE is characterized by fever, weakness, arthritis, skin rashes, pleurisy, and kidney dysfunction (Figure 16-10). Lupus is more frequent in African American and Hispanic women than in Caucasians, although it is not known why this is so. Affected individuals may produce auto-antibodies to a vast array of tissue antigens, such as DNA, histones, RBCs, platelets, leukocytes, and clotting factors; interaction of these auto-antibodies with their specific antigens produces various symptoms. Auto-antibody specific for RBCs and platelets, for example, can lead to complement-mediated lysis, resulting in hemolytic anemia and thrombocytopenia, respectively. When immune complexes of auto-antibodies with various nuclear antigens are deposited along the walls of small blood vessels, a type III hypersensitivity reaction develops. The complexes activate the complement system and generate membrane-attack complexes and complement split products that damage the wall of the blood vessel, resulting in vasculitis and glomerulonephritis.

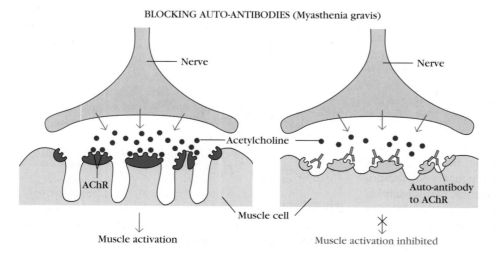

BLOCKING AUTO-ANTIBODIES (Myasthenia gravis)

**FIGURE 16-9** In myasthenia gravis, binding of auto-antibodies to the acetylcholine receptor (AChR; *right*) blocks the normal binding of acetylcholine (burgundy dots) and subsequent muscle activation *(left)*. In addition, the anti-AChR auto-antibody activates complement, which damages the muscle end plate; the number of acetylcholine receptors declines as the disease progresses.

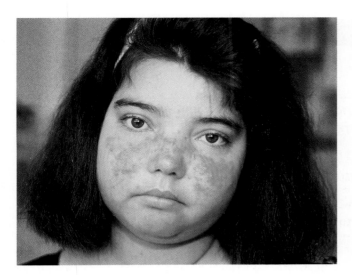

**FIGURE 16-10** Characteristic "butterfly" rash over the cheeks of a girl with systemic lupus erythematosus. *[From L. Steinman, 1993, Scientific American **269**(3):80.]*

Excessive complement activation in patients with severe SLE produces elevated serum levels of the complement split products C3a and C5a, which may be three to four times higher than normal. C5a induces increased expression of the type 3 complement receptor (CR3) on neutrophils, facilitating neutrophil aggregation and attachment to the vascular endothelium. As neutrophils attach to small blood vessels, the number of circulating neutrophils declines (neutropenia) and various occlusions of the small blood vessels develop (vasculitis). These occlusions can lead to widespread tissue damage.

Laboratory diagnosis of SLE focuses on the characteristic antinuclear antibodies, which are directed against double-stranded or single-stranded DNA, nucleoprotein, histones, and nucleolar RNA. Indirect immunofluorescent staining with serum from SLE patients produces various characteristic nucleus-staining patterns.

## Multiple sclerosis attacks the central nervous system

**Multiple sclerosis (MS)** is the most common cause of neurologic disability associated with disease in Western countries. The symptoms may be mild, such as numbness in the limbs, or severe, such as paralysis or loss of vision. Most people with MS are diagnosed between the ages of 20 and 40. Individuals with this disease produce autoreactive T cells that participate in the formation of inflammatory lesions along the myelin sheath of nerve fibers. The cerebrospinal fluid of patients with active MS contains activated T lymphocytes, which infiltrate the brain tissue and cause characteristic inflammatory lesions, destroying the myelin. Since myelin functions to insulate the nerve fibers, a breakdown in the myelin sheath leads to numerous neurologic dysfunctions.

Epidemiological studies indicate that MS is most common in the Northern Hemisphere and, interestingly, in the United States. Populations who live north of the 37th parallel have a prevalence of 110 to 140 cases per 100,000, whereas those who live south of the 37th parallel show a prevalence of 57 to 78 per 100,000. And individuals from south of the 37th parallel who move north assume a new risk if the move occurs before 15 years of age. These provocative data suggest that there is an environmental component affecting the risk of contracting MS. This is not the entire story, however, since genetic influences are also important. Whereas the average person in the United States has about one chance in 1000 of developing MS, close relatives of people with MS, such as children or siblings, have a one chance in 50 to 100 of developing MS. The identical twin of a person with MS has a one in three chance of developing the disease. These data point strongly to the genetic component of the disease. And MS affects women two to three times more frequently than men (see Clinical Focus).

The cause of MS, like most autoimmune diseases, is not well understood. However, there are some indications that infection by certain viruses may predispose a person to MS. Some viruses can cause demyelinating diseases, and it is tempting to speculate that viral infection plays a significant role in MS, but at present no definitive data implicate a particular virus.

## Rheumatoid arthritis attacks joints

**Rheumatoid arthritis** is a common autoimmune disorder, most often affecting women from 40 to 60 years old. The major symptom is chronic inflammation of the joints, although the hematologic, cardiovascular, and respiratory systems are also frequently affected. Many individuals with rheumatoid arthritis produce a group of auto-antibodies called **rheumatoid factors** that are reactive with determinants in the Fc region of IgG. The classic rheumatoid factor is an IgM antibody with that reactivity. Such auto-antibodies bind to normal circulating IgG, forming IgM-IgG complexes that are deposited in the joints. These immune complexes can activate the complement cascade, resulting in a type III hypersensitive reaction, which leads to chronic inflammation of the joints.

# Animal Models for Autoimmune Diseases

Animal models for autoimmune diseases have contributed valuable insights into the mechanism of autoimmunity, to our understanding of autoimmunity in humans, and to potential treatments. Autoimmunity develops spontaneously in certain inbred strains of animals and can also be induced by certain experimental manipulations (Table 16-2).

| TABLE 16-2 | Experimental animal models of autoimmune diseases | | |
|---|---|---|---|
| Animal model | Possible human disease counterpart | Inducing antigen | Disease transferred by T cells |
| SPONTANEOUS AUTOIMMUNE DISEASES | | | |
| Nonobese diabetic (NOD) mouse | Insulin-dependent diabetes mellitus (IDDM) | Unknown | Yes |
| (NZB × NZW) F$_1$ mouse | Systemic lupus erythematosus (SLE) | Unknown | Yes |
| Obese-strain chicken | Hashimoto's thyroiditis | Thyroglobulin | Yes |
| EXPERIMENTALLY INDUCED AUTOIMMUNE DISEASES* | | | |
| Experimental autoimmune myasthenia gravis (EAMG) | Myasthenia gravis | Acetylcholine receptor | Yes |
| Experimental autoimmune encephalomyelitis (EAE) | Multiple sclerosis (MS) | Myelin basic protein (MBP); proteolipid protein (PLP) | Yes |
| Autoimmune arthritis (AA) | Rheumatoid arthritis | *M. tuberculosis* (proteoglycans) | Yes |
| Experimental autoimmune thyroiditis (EAT) | Hashimoto's thyroiditis | Thyroglobulin | Yes |

*These diseases can be induced by injecting appropriate animals with the indicated antigen in complete Freund's adjuvant. Except for autoimmune arthritis, the antigens used correspond to the self antigens associated with the human disease counterpart. Rheumatoid arthritis involves reaction to proteoglycans, which are self antigens associated with connective tissue.

## Autoimmunity can develop spontaneously in animals

A number of autoimmune diseases that develop spontaneously in animals exhibit important clinical and pathologic similarities to certain autoimmune diseases in humans. Certain inbred mouse strains have been particularly valuable models for illuminating the immunologic defects involved in the development of autoimmunity.

New Zealand Black (NZB) mice and F$_1$ hybrids of NZB and New Zealand White (NZW) mice spontaneously develop autoimmune diseases that closely resemble systemic lupus erythematosus. NZB mice spontaneously develop autoimmune hemolytic anemia between 2 and 4 months of age, at which time various auto-antibodies can be detected, including antibodies to erythrocytes, nuclear proteins, DNA, and T lymphocytes. F$_1$ hybrid animals develop glomerulonephritis from immune-complex deposits in the kidney and die prematurely by 18 months. As in human SLE, the incidence of autoimmunity in the (NZB × NZW)F$_1$ hybrids is greater in females.

An accelerated and severe form of systemic autoimmune disease resembling SLE develops in a mouse strain called MRL/*lpr/lpr*. These mice are homozygous for the gene *lpr*, mentioned earlier, which has been identified as a defective *fas* gene. The *fas* gene product is a cell surface protein belonging to the TNF family of cysteine-rich membrane receptors (see Figure 12-6d). When the normal Fas protein

interacts with its ligand, it transduces a signal that leads to apoptotic death of the Fas-bearing cells. This mechanism may operate in destruction of target cells by some CTLs. Fas is also known to be essential in the death of hyperactivated peripheral CD4$^+$ cells. Normally, when mature peripheral T cells become activated, they are induced to express both Fas antigen and Fas ligand. When Fas-bearing cells come into contact with a neighboring activated cell bearing Fas ligand, the Fas-bearing cell is induced to die (see Figure 10-19). It is also possible that Fas ligand can engage Fas from the same cell, inducing a cellular suicide. In the absence of Fas, mature peripheral T cells do not die, and these activated cells continue to proliferate and produce cytokines that result in grossly enlarged lymph nodes and spleen. Defects in *fas* expression similar to that found in the *lpr* mouse are observed in humans, and such defects can have severe consequences. However, there is no link between *fas* expression and SLE in humans, which suggests that the *lpr* mouse may not be a true model for SLE.

Another important animal model is the nonobese diabetic (NOD) mouse, which spontaneously develops a form of diabetes that resembles human insulin-dependent diabetes mellitus (IDDM). Like the human disease, the NOD mouse disease begins with lymphocytic infiltration into the islets of the pancreas. Also, as in IDDM, there is a strong association between certain MHC alleles and the development of diabetes in these mice. Experiments have shown that T cells from diabetic mice can transfer diabetes to nondiabetic recipients. For example, when the immune systems of normal mice are destroyed by lethal doses of x-rays and then reconstituted with an injection of bone marrow cells from NOD mice, the reconstituted mice develop diabetes. Conversely, when the immune systems of still healthy NOD mice are destroyed by x-irradiation and then reconstituted with normal bone marrow cells, the NOD mice do not develop diabetes. Various studies have demonstrated a pivotal role for CD4$^+$ T cells in the NOD mouse, and recent evidence implicates the T$_H$1 subset in disease development.

Several other spontaneous autoimmune diseases have been discovered in animals that have served as models for similar human diseases. Among these are *Obese*-strain chickens, which develop both humoral and cell-mediated reactivity to thyroglobulin resembling that seen in Hashimoto's thyroiditis.

## Autoimmunity can be induced experimentally in animals

Autoimmune dysfunctions similar to certain human autoimmune diseases can be induced experimentally in some animals (see Table 16-2). One of the first such animal models was discovered serendipitously in 1973 when rabbits were immunized with acetylcholine receptors purified from electric eels. The animals soon developed muscular weakness similar to that seen in myasthenia gravis. This experimental autoimmune myasthenia gravis (EAMG) was shown to result when antibodies raised against the foreign (and by cross-reaction, the host) acetylcholine receptor blocked muscle stimulation by acetylcholine in the synapse. Within a year, this animal model had proved its value with the discovery that auto-antibodies to the acetylcholine receptor were the cause of myasthenia gravis in humans.

Experimental autoimmune encephalomyelitis (EAE) is one of the best-studied animal models of autoimmune disease. EAE is mediated solely by T cells and can be induced in a variety of species by immunization with myelin basic protein (MBP) or proteolipid protein (PLP) in complete Freud's adjuvant (Figure 16-11). Within 2 to 3 weeks the animals develop cellular infiltration of the myelin sheaths of the central nervous system, resulting in demyelination and paralysis. Most of the animals die, but others have milder symptoms, and some animals develop a chronic form of the disease that resembles chronic relapsing and remitting MS in humans. Those that recover are resistant to the development of disease from a subsequent injection of MBP and adjuvant.

The mouse EAE model provides a system for testing treatments for human MS. For example, because MBP- or PLP-specific T-cell clones are found in the periphery, it is assumed that these clones must have escaped negative selection in the thymus. Recent mouse experiments have suggested that orally administered MBP may make these antigen-specific peripheral T-cell clones self-tolerant. These studies have paved the way for clinical trials in MS patients.

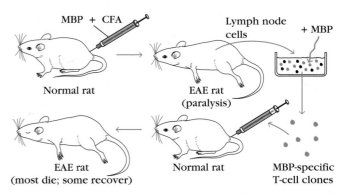

**FIGURE 16-11 Experimental autoimmune encephalomyelitis (EAE) can be induced in rats by injecting them with myelin basic protein (MBP) in complete Freud's adjuvant (CFA).** MBP-specific T-cell clones can be generated by culturing lymph node cells from EAE rats with MBP. When these T cells are injected into normal animals, most develop EAE and die, although a few recover.

Experimental autoimmune thyroiditis (EAT) can be induced in a number of animals by immunizing with thyroglobulin in complete Freud's adjuvant. Both humoral antibodies and T$_H$1 cells directed against the thyroglobulin develop, resulting in thyroid inflammation. EAT appears to best mimic Hashimoto's thyroiditis. In contrast to both EAE and EAT, which are induced by immunizing with self antigens, autoimmune arthritis (AA) is induced by immunizing rats with *Mycobacterium tuberculosis* in complete Freud's adjuvant. These animals develop an arthritis whose features are similar to those of rheumatoid arthritis in humans.

## Evidence Implicating the CD4$^+$ T Cell, MHC, and TCR in Autoimmunity

The inappropriate response to self antigens that characterizes all autoimmune diseases can involve either the humoral or cell-mediated branches of the immune system. Identifying the defects underlying human autoimmune diseases has been difficult; more success has been achieved in characterizing the immune defects in the various animal models. Each of the animal models has implicated the CD4$^+$ T cell as the primary mediator of autoimmune disease. For example, the evidence is quite strong that, in mice, EAE is caused by CD4$^+$ T$_H$1 cells specific for the immunizing antigen. The disease can be transferred from one animal into another by T cells from animals immunized with either MBP or PLP or by cloned T-cell lines from such animals. It has also been shown that disease can be prevented by treating animals with anti-CD4 antibodies. These data are compelling evidence for the involvement of CD4 in the establishment of EAE.

T-cell recognition of antigen, of course, involves a trimolecular complex of the T-cell receptor, an MHC molecule, and an antigenic peptide (see Figure 9-14). Thus, an individual susceptible to autoimmunity must possess MHC molecules and T-cell receptors capable of binding self-antigens.

## CD4$^+$ T cells and T$_H$1/T$_H$2 balance plays an important role in autoimmunity in some animal models

Autoimmune T-cell clones have been obtained from all of the animal models listed in Table 16-2 by culturing lymphocytes from the autoimmune animals in the presence of various T-cell growth factors and by inducing proliferation of specific autoimmune clones with the various autoantigens. For example, when lymph node cells from EAE rats are cultured in vitro with myelin basic protein (MBP), clones of activated T cells emerge. When sufficient numbers of these MBP-specific T-cell clones are injected intravenously into normal syngeneic animals, the cells cross the blood-brain barrier and induce demyelination; EAE develops very quickly, usually within 5 days (see Figure 16-11).

A similar experimental protocol has been used to isolate T-cell clones specific for thyroglobulin and for *M. tuberculosis* from EAT and AA animals, respectively. In each case, the T-cell clone induces the experimental autoimmune disease in normal animals. Examination of these T cells has revealed that they bear the CD4 membrane marker. In a number of animal models for autoimmune diseases, it has been possible to reverse the autoimmunity by depleting the T-cell population with antibody directed against CD4. For example, weekly injections of anti-CD4 monoclonal antibody abolished the autoimmune symptoms in (NZB × NZW) F$_1$ mice and in mice with EAE.

Most cases of organ-specific autoimmune disease develop as a consequence of self-reactive CD4$^+$ T cells. Analysis of these cells has revealed that the T$_H$1/T$_H$2 balance can affect whether autoimmunity develops. T$_H$1 cells have been implicated in the development of autoimmunity, whereas in a number of cases, T$_H$2 cells not only protect against the induction of disease but also against progression of established disease. In EAE, for example, immunohistologic studies revealed the presence of T$_H$1 cytokines (IL-2, TNF-α, and IFN-γ) in the central nervous system tissues at the height of the disease. In addition, the MBP-specific CD4$^+$ T-cell clones generated from animals with EAE, as shown in Figure 16-11, can be separated into T$_H$1 and T$_H$2 clones. Experiments have shown that only the T$_H$1 clones transfer EAE to normal healthy mice, whereas the T$_H$2 clones not only do not transfer EAE to normal healthy mice but also protect the mice against induction of EAE by subsequent immunization with MBP plus adjuvant.

Experiments that assessed the role of various cytokines or cytokine inhibitors on the development of EAE have provided further evidence for the different roles of T$_H$1 and T$_H$2 cells in autoimmunity. When mice were injected with IL-4 at the time of immunization with MBP plus adjuvant, the development of EAE was inhibited, whereas administration of

IL-12 had the opposite effect, promoting the development of EAE. As noted in Chapter 12, IL-4 promotes development of T$_H$2 cells, and IFN-γ, in addition to other cytokines such as IL-12, promotes development of T$_H$1 cells (see Figure 12-12). Thus, the observed effects of IL-4 and IL-12 on EAE development are consistent with a role for T$_H$1 cells in the genesis of autoimmunity.

## Autoimmunity can be associated with the MHC or with particular T-cell receptors

Several types of studies have supported an association between expression of a particular MHC allele and susceptibility to autoimmunity, an issue covered in detail in Chapter 8. The strongest association between an HLA allele and an autoimmune disease is seen in ankylosing spondylitis, an inflammatory disease of vertebral joints. Individuals who have *HLA-B27* have a 90 times greater likelihood of developing ankylosing spondylitis than individuals with a different *HLA-B* allele. However, the existence of such an association should not be interpreted to imply that the expression of a particular MHC allele has caused the disease, because the relationship between MHC alleles and development of autoimmune disease is complex. It is interesting to note that, unlike many other autoimmune diseases, 90% of the cases of ankylosing spondylitis are male.

The presence of T-cell receptors containing particular V$_α$ and V$_β$ domains also has been linked to a number of autoimmune diseases, including experimental EAE and its human counterpart, multiple sclerosis. In one approach, T cells specific for various encephalitogenic peptides of MBP were cloned and their T-cell receptors analyzed. For example, T-cell clones were obtained from PL/J mice by culturing their T cells with the acetylated amino-terminal nonapeptide of MBP presented in association with a class II IA$^u$ MHC molecule. Analysis of the T-cell receptors on these clones revealed a restricted repertoire of V$_α$ and V$_β$ domains: 100% of the T-cell clones expressed V$_α$ 4.3, and 80% of the T-cell clones expressed V$_β$ 8.2. In human autoimmune diseases, evidence for restricted TCR expression has been obtained for both multiple sclerosis and myasthenia gravis. The preferential expression of TCR variable-region genes in these autoimmune T-cell clones suggests that a single epitope might induce the clonal expansion of a small number of pathogenic T cells.

## Proposed Mechanisms for Induction of Autoimmunity

A variety of mechanisms have been proposed to account for the T-cell-mediated generation of autoimmune diseases (Figure 16-12). Evidence exists for each of these mechanisms, and it is likely that autoimmunity does not develop from a single event but rather from a number of different events.

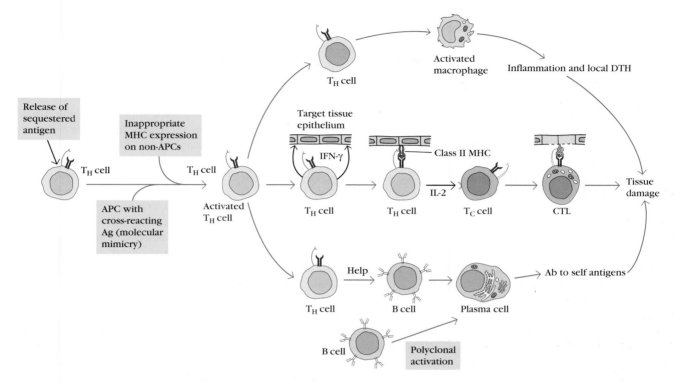

**FIGURE 16-12 Proposed mechanisms for inducing autoimmune responses.** Normal thymic selection appears to generate some self-reactive $T_H$ cells; abnormalities in this process may generate even more self-reactive $T_H$ cells. Activation of these self-reactive T cells in various ways, as well as polyclonal activation of B cells, is thought to induce an autoimmune response, in this case resulting in tissue damage. In all likelihood, several mechanisms are involved in each autoimmune disease. *[Adapted from V. Kumar et al., 1989,* Annual Review of Immunology ***7***:657.]*

In addition, susceptibility to many autoimmune diseases differs between the two sexes. As noted earlier, Hashimoto's thyroiditis, systemic lupus erythematosus, multiple sclerosis, rheumatoid arthritis, and scleroderma preferentially affect women. Factors that have been proposed to account for this preferential susceptibility, such as hormonal differences between the sexes and the potential effects of fetal cells in the maternal circulation during pregnancy, are discussed in the Clinical Focus on page 416.

## Release of sequestered antigens can induce autoimmune disease

As discussed in Chapter 10, the induction of self-tolerance in T cells results from exposure of immature thymocytes to self antigens and the subsequent clonal deletion of those that are self-reactive. As mentioned above, any tissue antigens that are sequestered from the circulation are not seen by the developing T cells in the thymus and will not induce self-tolerance. Exposure of mature T cells to such normally sequestered antigens at a later time might result in their activation.

Myelin basic protein (MBP) is an example of an antigen normally sequestered from the immune system, in this case by the blood-brain barrier. In the EAE model, animals are injected directly with MBP, together with adjuvant, under conditions that maximize immune exposure. In this type of animal model, the immune system is exposed to sequestered self antigens under nonphysiologic conditions; however, trauma to tissues following either an accident or a viral or bacterial infection might also release sequestered antigens into the circulation. A few tissue antigens are known to fall into this category. For example, sperm arise late in development and are sequestered from the circulation. However, after a vasectomy, some sperm antigens are released into the circulation and can induce auto-antibody formation in some men. Similarly, the release of lens protein after eye damage or of heart muscle antigens after myocardial infarction has been shown to lead on occasion to the formation of auto-antibodies.

Data indicate that injection of normally sequestered antigens directly into the thymus can reverse the development of tissue-specific autoimmune disease in animal models. For instance, intrathymic injection of pancreatic islet beta cells prevented development of autoimmunity in NOD mice. Moreover, EAE was prevented in susceptible rats by prior injection of MBP directly into the thymus. In these experiments, exposure of immature T cells to self antigens that normally are not present in the thymus presumably led to tolerance to these antigens.

# Why Are Women More Susceptible Than Men to Autoimmunity? Gender Differences in Autoimmune Disease

**Of** the nearly 9 million individuals in the United States with autoimmune disease, approximately 6.7 million are women. This predisposition to autoimmunity is more apparent in some diseases than others. For example, the female-to-male ratio of individuals who suffer from diseases such as multiple sclerosis (MS) or rheumatoid arthritis (RA) is approximately two or three females to one male, and there are nine women for every one man afflicted with systemic lupus erythematosus (SLE). However, these statistics do not tell the entire story since in some diseases, MS, for example, the severity of the disease can be worse in men than in women. That women are more susceptible to autoimmune disease has been recognized for many years, but the reasons for this increased risk are not entirely understood. Some possible explanations follow.

Although it may seem unlikely, considerable evidence suggests there are significant gender differences in immune responses in both humans and mice. Immunization studies in both species suggest that females produce a higher titer of antibodies than males. In fact, females in general tend to mount more vigorous immune responses. In humans, this is particularly apparent in young females. Women tend to have higher levels of CD4+ T cells and significantly higher levels of serum IgM.

In mice, whose gender differences are easier to study, a large body of literature documents gender differences in immune responses. Female mice are much more likely than male mice to develop $T_H1$ responses and, in infections for which proinflammatory $T_H1$ responses are beneficial, are more likely to be resistant to the infection. An excellent example is infection by viruses such as vesicular stomatitis virus (VSV), herpes simplex virus (HSV), and Theiler's murine encephalomyelitis virus (TMEV). Clearance of these viruses is enhanced by $T_H1$ responses. In some cases, however, a pro-inflammatory response can be deleterious. For example, a $T_H1$ response to lymphocytic choriomeningitis virus (LCMV) correlates with more severe disease and significant pathology. Thus, female mice are more likely to succumb to infection with LCMV. The fact that gender is important in LCMV infection is underscored by experiments demonstrating that castrated male mice behave immunologically like females and are more likely to succumb to infection than their uncastrated male littermates.

Another disease in which gender plays a role is infection by coxsackie virus type B-3 (CVB-3), an etiological agent of immune myocarditis. Male mice are much more susceptible to this disease than females. CVB-3 induces a predominant $T_H1$ response in males, whereas females, contrary to the situations described above, respond by mounting a protective $T_H2$ response. The response by females can be altered by injecting them with testosterone, which makes them susceptible to the disease. In addition, the male response can be altered by injecting males with estradiol, making them resistant to the virus. Possibly, based on these data in mice, basic differences exist between men and women in their responses to pathogens. However, the gender differences observed in mice may not extend to human populations.

How do these gender differences arise? That estradiol or testosterone can alter the outcome of infection by CVB-3 suggests a critical role for sex hormones. In humans, estrogen on its own seems not to play a significant role in the etiology of either RA or MS but may be important in SLE. This is suggested by data indicating that estrogen can stimulate auto-antibody production in SLE-prone mice and these effects can be modulated by an anti-estrogenic compound. Such data imply that, at least in mice, estrogen is capable of triggering SLE-like autoimmunity. In addition, androgens such as testosterone clearly play an important role in some autoimmune diseases. Female NOD mice are much more susceptible to spontaneous diabetes, and castration significantly increases the susceptibility of male NOD mice. Female SJL mice are more likely to be susceptible to EAE, a mouse MS-like disease. This indicates that testosterone may be effective in ameliorating some autoimmune responses and so may be protective against several autoimmune diseases, including MS, diabetes, SLE, and Sjögren's syndrome.

## Molecular mimicry may contribute to autoimmune disease

The notion that microbial or viral agents might play a role in autoimmunity is attractive for several reasons. It is well accepted that migrant human populations acquire the diseases of the area to which they move and that the incidence of autoimmunity has increased dramatically as populations have become more mobile. This, coupled with the fact that a number of viruses and bacteria have been shown to possess antigenic determinants that are identical or similar to normal host-cell components, led Michael Oldstone to propose that a pathogen may express a protein epitope that resembles a particular self component in conformation or primary sequence. Such molecular mimicry appears in a wide variety of organisms (Table 16-3). In one study, 600 different monoclonal antibodies specific for 11 different viruses were tested to evaluate their reactivity with normal tissue antigens. More

Why do sex steroids affect immune responses? This is not well understood, but it is likely that these hormones, which circulate throughout the body, alter immune responses by altering patterns of gene expression. The sex steroids, a highly lipophilic group of compounds, function by passing through the cell membrane and binding a cytoplasmic receptor. Each hormone has a cognate receptor, and binding of hormone to receptor leads to the activation or, in some instances, repression of gene expression. This is mediated by the binding of the receptor-hormone complex receptor to a specific DNA sequence. Thus, estrogen enters a cell, binds to the estrogen receptor, and induces the binding of the estrogen receptor to a specific DNA sequence, which in turn results in the modulation of transcription. So in cells that contain hormone receptors, sex hormones can regulate gene expression, and it is highly likely that sex steroids play an important role in the immune system through their receptors. Whether various cells of the immune system contain hormone receptors is not known; to understand how sex hormones mediate immune responses, we must determine which cells express which hormone receptors.

Hormonal effects on immune responses may not be limited to steroidal sex hormones. Prolactin, a hormone that is expressed in higher levels in women than in men, is not a member of the lipophilic sex steroid family that includes estrogen, progesterone, and testosterone. However, prolactin secretion (by the anterior pituitary) is stimulated by estrogen, explaining the higher levels of prolactin in women and the very high levels observed during pregnancy. Prolactin can have a profound influence on immune responses, as demonstrated in mice by removal of the anterior pituitary: this results in severe immunosuppression, which can be entirely reversed by treatment with exogenous prolactin. The presence of prolactin receptors on peripheral T and B cells in humans is further evidence that this hormone may play a role in regulating immune responses. In fact, some evidence suggests that prolactin may tend to turn cells toward $T_H1$-dominated immune responses.

Pregnancy may give us a clue to how sex plays a role in regulating immune response. Although women normally mount a normal response to foreign antigens, during pregnancy, it is critical that the mother tolerate the fetus, which is, in fact, a foreign graft. This makes it very likely that the female immune system undergoes important modifications during pregnancy. Recall that women normally tend to mount more $T_H1$-like responses than $T_H2$ responses. During pregnancy, however, women mount more $T_H2$-like responses. It is thought that pregnancy-associated levels of sex steroids may promote an anti-inflammatory environment. In this regard, it is notable that diseases enhanced by $T_H2$-like responses, such as SLE, which has a strong antibody-mediated component, can be exacerbated during pregnancy, whereas diseases that involve inflammatory responses, such as RA and MS, sometimes are ameliorated in pregnant women.

Another effect of pregnancy is the presence of fetal cells in the maternal circulation. Fetal cells can persist in the maternal circulation for decades, so these long-lived fetal cells may play a significant role in the development of autoimmune disease. Furthermore, the exchange of cells during pregnancy is bidirectional (cells of the mother may also appear in the fetal circulatory system), so that the presence of the mother's cells in the male circulation could be a contributing factor in autoimmune disease.

In summary, women and men differ significantly in their ability to mount an immune response. Women mount more robust immune responses, and these responses tend to be more $T_H1$-like. Estrogen has been reported to be immunostimulatory; this may be due in part to the ability of the hormone to regulate specific gene expression through the estrogen receptor. Furthermore, the incidence of autoimmune diseases is sharply higher in women than in men. These observations have generated the compelling hypothesis that the tendency of females to mount more $T_H1$-like responses may explain some of the differences in susceptibility to autoimmunity. Since this type of response is pro-inflammatory, it may enhance the development of autoimmunity. Whether the bias toward a $T_H1$ response is due to differences in sex steroids between males and females is less certain, but in the next several years, experiments that explore this idea are likely to be pursued vigorously.

SOURCE: The data discussed in this Clinical Focus are from a letter to *Science* (C. C. Whitacre, S. C. Reingold, and P. A. O'Looney, 1999, *Science* 283:1277) from the Task Force on Gender, MS, and Autoimmunity, a group convened by the National Multiple Sclerosis Society to begin a dialog on issues of gender and autoimmune disease.

than 3% of the virus-specific antibodies tested also bound to normal tissue, suggesting that molecular mimicry is a fairly common phenomenon.

Molecular mimicry has been suggested as one mechanism that leads to autoimmunity. One of the best examples of this type of autoimmune reaction is post-rabies encephalitis, which used to develop in some individuals who had received the rabies vaccine. In the past, the rabies virus was grown in rabbit brain-cell cultures, and preparations of the vaccine included antigens derived from the rabbit brain cells. In a vaccinated person, these rabbit brain-cell antigens could induce formation of antibodies and activated T cells, which could cross-react with the recipient's own brain cells, leading to encephalitis. Cross-reacting antibodies are also thought to be the cause of heart damage in rheumatic fever, which can sometimes develop after a *Streptococcus* infection. In this case, the antibodies are to streptococcal antigens, but they cross-react with the heart muscle.

| TABLE 16-3 | Molecular mimicry between proteins of infectious organisms and human host proteins |
|---|---|

| Protein* | Sequence† |
|---|---|
| Human cytomegalovirus IE2<br>HLA-DR molecule | 79 P D P L G R P D E D<br>60 V T E L G R P D A E |
| Poliovirus VP2<br>Acetylcholine receptor | 70 S T T K E S R G T T<br>176 T V I K E S R G T K |
| Papilloma virus E2<br>Insulin receptor | 76 S L H L E S L K D S<br>66 V Y G L E S L K D L |
| Rabies virus glycoprotein<br>Insulin receptor | 147 T K E S L V I I S<br>764 N K E S L V I S E |
| *Klebsiella pneumoniae* nitrogenase<br>HLA-B27 molecule | 186 S R Q T D R E D E<br>70 K A Q T D R E D L |
| Adenovirus 12 E1B<br>α-Gliadin | 384 L R R G M F R P S Q C N<br>206 L G Q G S F R P S Q Q N |
| Human immunodeficiency<br>virus p24<br>Human IgG constant region | 160 G V E T T T P S<br>466 G V E T T T P S |
| Measles virus P3<br>Corticotropin | 13 L E C I R A L K<br>18 L E C I R A C K |
| Measles virus P3<br>Myelin basic protein | 31 E I S D N L G Q E<br>61 E I S F K L G Q E |

*In each pair, the human protein is listed second. The proteins in each pair have been shown to exhibit immunologic cross-reactivity.

†Amino acids are indicated by a single-letter code. Identical residues are shown in blue. Numbers indicate amino acid position in the intact protein.

SOURCE: Adapted from M. B. A. Oldstone, 1987, *Cell* **50**:819.

## There is evidence for mimicry between MBP and viral peptides

Since the encephalitogenic MBP peptides are known, the extent to which they are mimicked by proteins from other organisms can be assessed. For example, one MBP peptide (amino acid residues 61–69) is highly homologous with a peptide in the P3 protein of the measles virus (see Table 16-3). In one study, the sequence of another encephalitogenic MBP peptide (66–75) was compared with the known sequences of a large number of viral proteins. This computer analysis revealed sequence homologies between this MBP peptide and a number of peptides from animal viruses, including influenza, polyoma, adenovirus, Rous sarcoma, Abelson leukemia, poliomyelitis, Epstein-Barr, and hepatitis B viruses.

One peptide from the polymerase enzyme of the hepatitis B virus was particularly striking, exhibiting 60% homology with a sequence in the encephalitogenic MBP peptide. To test the hypothesis that molecular mimicry can generate autoimmunity, rabbits were immunized with this hepatitis B virus peptide. The peptide was shown to induce both the formation of antibody and the proliferation of T cells that cross-reacted with MBP; in addition, central nervous system tissue from the immunized rabbits showed cellular infiltration characteristic of EAE.

These findings suggest that infection with certain viruses expressing epitopes that mimic sequestered self components, such as myelin basic protein, may induce autoimmunity to those components. Susceptibility to this type of autoimmunity may also be influenced by the MHC haplotype of the individual, since certain class I and class II MHC molecules may be more effective than others in presenting the homologous peptide for T-cell activation.

Another particularly compelling example of molecular mimicry comes from studies of herpes stromal keratinitis (HSK). In these studies, investigators showed that prior infection of mice with herpes simplex virus type 1 leads to HSK, an autoimmune-like disease in which T cells specific for a particular viral peptide attack corneal tissue, causing blindness. These data demonstrated very clearly that a particular epitope of HSV-1 is responsible for the disease and that mutant strains of HSV-1 lacking this epitope do not cause HSK. The data provide strong evidence for molecular mimicry in the development of a particular autoimmune disease.

## Inappropriate expression of class II MHC molecules can sensitize autoreactive T cells

The pancreatic beta cells of individuals with insulin-dependent diabetes mellitus (IDDM) express high levels of both class I and class II MHC molecules, whereas healthy beta cells express lower levels of class I and do not express class II at all. Similarly, thyroid acinar cells from people with Graves' disease have been shown to express class II MHC molecules on their membranes. This inappropriate expression of class II MHC molecules, which are normally expressed only on antigen-presenting cells, may serve to sensitize $T_H$ cells to peptides derived from the beta cells or thyroid cells, allowing activation of B cells or $T_C$ cells or sensitization of $T_H1$ cells against self antigens.

Other evidence suggests that certain agents can induce some cells that should not express class II MHC molecules to express them. For example, the T-cell mitogen phytohemagglutinin (PHA) has been shown to induce thyroid cells to express class II molecules. In vitro studies reveal that IFN-γ also induces increases in class II MHC molecules on a wide variety of cells, including pancreatic beta cells, intestinal epithelial cells, melanoma cells, and thyroid acinar cells. It was hypothesized that trauma or viral infection in an organ may induce a localized inflammatory response and thus increase concentrations of IFN-γ in the affected organ. If IFN-γ induces class II MHC expression on non-antigen-presenting cells, inappropriate $T_H$-cell activation might follow, with autoimmune consequences. It is noteworthy that SLE patients with active disease have higher serum titers of IFN-γ than patients with inactive disease. These data suggest that the increase in IFN-γ in these patients may lead to inappropriate expression of class II MHC molecules and thus to T-cell activation against a variety of autoantigens.

An interesting transgenic mouse system implicates IFN-γ and inappropriate class II MHC expression in autoimmunity.

(a)

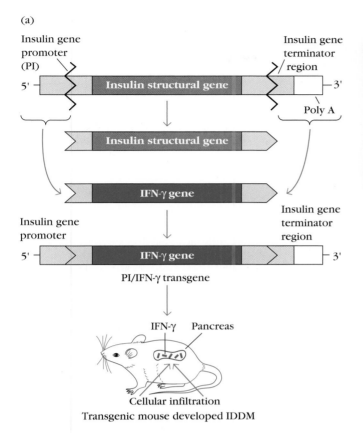

Transgenic mouse developed IDDM

(b)

**FIGURE 16-13  Insulin-dependent diabetes mellitus (IDDM) in transgenic mice.** (a) Production of transgenic mice containing an IFN-γ transgene linked to the insulin promoter (PI). The transgenics, which expressed the PI/IFN-γ transgene only in the pancreas, developed symptoms characteristic of IDDM. (b) Pancreatic islets of Langerhans from a normal BALB/c mouse *(left)* and from PI/IFN-γ transgenics at 3 weeks *(right)* showing infiltration of inflammatory cells. *[Part b from N. Sarvetnick, 1988, Cell* **52**:*773.]*

In this system, an IFN-γ transgene was genetically engineered with the insulin promoter, so that the transgenic mice secreted IFN-γ from their pancreatic beta cells (Figure 16-13a). Since IFN-γ up-regulates class II MHC expression, these transgenic mice also expressed class II MHC molecules on their pancreatic beta cells. The mice developed diabetes, which was associated with cellular infiltration of lymphocytes and inflammatory cells like the infiltration seen in autoimmune NOD mice and in patients with insulin-dependent diabetes mellitus (Figure 16-13b).

Although inappropriate class II MHC expression on pancreatic beta cells may be involved in the autoimmune reaction in these transgenic mice, other factors may also play a role. For example, IFN-γ is known to induce production of several other cytokines, including IL-1 and TNF. Therefore, the development of autoimmunity in this transgenic system may involve antigen presentation by class II MHC molecules on pancreatic beta cells together with a costimulatory signal, such as IL-1, that may activate self-reactive T cells. Some evidence also suggests that IL-1, IFN-γ, and TNF may directly impair the secretory function of human beta cells.

### Polyclonal B-cell activation can lead to autoimmune disease

A number of viruses and bacteria can induce nonspecific polyclonal B-cell activation. Gram-negative bacteria, cytomegalovirus, and Epstein-Barr virus (EBV) are all known to be such polyclonal activators, inducing the proliferation of numerous clones of B cells that express IgM in the absence of $T_H$ cells. If B cells reactive to self antigens are activated by this mechanism, auto-antibodies can appear. For instance, during infectious mononucleosis, which is caused by EBV, a variety of auto-antibodies are produced, including auto-antibodies reactive to T and B cells, rheumatoid factors, and antinuclear antibodies. Similarly, lymphocytes from patients with SLE produce large quantities of IgM in culture, suggesting that they have been polyclonally activated. Many AIDS patients also show high levels of nonspecific antibody and auto-antibodies to RBCs and platelets. These patients are often co-infected with other viruses such as EBV and cytomegalovirus, which may induce the polyclonal B-cell activation that results in auto-antibody production.

## Treatment of Autoimmune Diseases

Ideally, treatment for autoimmune diseases should be aimed at reducing only the autoimmune response while leaving the rest of the immune system intact. To date, this ideal has not been reached.

Current therapies for autoimmune diseases are not cures but merely palliatives, aimed at reducing symptoms to provide the patient with an acceptable quality of life. Immunosuppressive drugs (e.g., corticosteroids, azathioprine, and cyclophosphamide) are often given with the intent of slowing proliferation of lymphocytes. By depressing the immune response in general, such drugs can reduce the severity of autoimmune symptoms. The general reduction in immune responsiveness, however, puts the patient at greater risk for infection or the development of cancer. A somewhat more selective approach employs **cyclosporin A** or FK506 to treat autoimmunity. These agents block signal transduction mediated by the T-cell receptor; thus, they inhibit only antigen-activated T cells while sparing nonactivated ones. For the most part, these treatments provide nonspecific suppression of the immune system and thus do not distinguish between a pathologic autoimmune response and a protective immune response.

Another therapeutic approach that has produced positive results in some cases of myasthenia gravis is removal of the thymus. Because patients with this disease often have thymic abnormalities (e.g., thymic hyperplasia or thymomas), adult thymectomy often increases the likelihood of remission of symptoms. Patients with Graves' disease, myasthenia gravis, rheumatoid arthritis, or systemic lupus erythematosus may experience short-term benefit from plasmapheresis. In this process, plasma is removed from a patient's blood by continuous-flow centrifugation. The blood cells are then resuspended in a suitable medium and returned to the patient. Plasmapheresis has been beneficial to patients with autoimmune diseases involving antigen-antibody complexes, which are removed with the plasma. Removal of the complexes, although only temporary, can result in a short-term reduction in symptoms.

## Treatment of human autoimmune disease poses special challenges

Studies with experimental autoimmune animal models have provided evidence that it is possible to arrest the development of autoimmunity. Monoclonal antibodies have been used successfully to treat autoimmune disease in several animal models. For example, a high percentage of (NZB × NZW) F₁ mice given weekly injections of high doses of monoclonal antibody specific for the CD4 membrane molecule recovered from their autoimmune lupus-like symptoms (Figure 16-14). Similar positive results were observed in NOD mice, in which treatment with an anti-CD4 monoclonal antibody led to disappearance of the lymphocytic infiltration and diabetic symptoms.

Because anti-CD4 monoclonal antibodies block or deplete all $T_H$ cells, regardless of their specificity, they can threaten the overall immune responsiveness of the recipient. For this and other reasons, few of the treatments applied in model systems have been successful in treating human disease. One major limitation is identifying the disease at a stage that is susceptible to treatment. Animal models of induced autoimmunity where a triggering antigen is introduced or those in which genetic defects predictably lead to disease at a defined developmental stage allow capture of

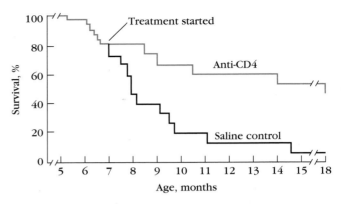

**FIGURE 16-14  Weekly injections of anti-CD4 monoclonal antibody into (NZB × NZW) F₁ mice exhibiting autoimmune lupus-like symptoms significantly increased their survival rate.** *[Adapted from D. Wofsy, 1988, Progress in Allergy **45**:106.]*

early events in the development of the disease. Inhibitory measures applied at these initial stages prevent further development of disease and give insight into disease processes. In contrast, human autoimmune disease is usually detected by the appearance of signs of advanced damage; the condition has progressed beyond the stage at which the inhibitory measures are effective. For example, diabetes is usually not diagnosed in humans until problems are evident, such as high blood glucose levels and evidence of insulin deficiency. The early events in the disease are asymptomatic and pass unnoticed, so opportunities for intervention are lost.

An additional problem is that animal models are not perfect predictors of human response. Although anti-CD4 antibodies successfully reversed MS and arthritis in animal models, human trials of this treatment have to date shown no efficacy. In addition to the problems cited above, a possible reason for this failure is that anti-CD4 may interfere with the activity of CD4⁺/CD25⁺ regulatory T cells, preventing their control of autoimmune activity.

## Inflammation is a target for treatment of autoimmunity

Chronic inflammation is a hallmark of debilitating autoimmunity, making pro-inflammatory processes a target for intervention. Recall that inflammation has several stages, including cytokine release and recruitment of leukocytes. Recruitment of cells to the site of inflammation involves extravasation (see Chapter 3) made possible by adherence of leukocytes to vessel walls, mediated by proteins of the integrin family. A blocker of α4β1 integrin showed initial promise in the treatment of MS, but the occurrence of untreatable infection in some test subjects caused the drug to be withdrawn. A more acceptable risk-benefit is seen with blockers of TNF-α. The drugs Enbrel, Remicade, and Humira are products that target TNF-α and are widely used for rheumatoid arthritis, psoriasis, Crohn's disease, and ankylosing spondylitis. A number of other strategies to reduce or prevent inflammation are being studied. An IL-1 receptor antagonist is approved for treatment of rheumatoid arthritis, as are antibodies directed against the IL-6 receptor and IL-15.

The class of drugs designated *statins,* used by millions to reduce cholesterol levels, were recently shown to lower serum levels of C-reactive protein, an acute-phase protein and indicator of inflammation (see the Clinical Focus in Chapter 3). Early-stage trials of statins for treatment of rheumatoid arthritis and MS have shown encouraging results. The use of drugs that are already approved for human use and for which extensive safety data are available is a tremendous advantage, considering that 95% of agents that undergo human trials fail at some point in the approval process.

Another agent considered for treatment of autoimmunity is the monoclonal antibody Rituxan, which kills B cells by targeting the surface marker CD20. Rituxan is currently approved for treatment of B-cell non-Hodgkin's lymphoma. Because antibodies play a major role in certain autoimmune diseases such as MS, rheumatoid arthritis, and SLE, depletion of B cells may have benefit. Rituxan has been tested for efficacy in these conditions, and preliminary results indicate success in control of rheumatoid arthritis.

## Activated T cells are a possible therapeutic target

A major disadvantage of immunosuppressive therapies for autoimmunity is that they render the host susceptible to infection by lowering immunity in general. One way to remedy this disadvantage is to block antigen-activated $T_H$ cells only, since these cells are important mediators of autoimmunity. Researchers have experimented with the use of monoclonal antibody directed against the α subunit of the high-affinity IL-2 receptor, which is expressed only by antigen-activated $T_H$ cells. Because the IL-2R α subunit (CD25) is expressed at higher levels on autoimmune T cells, monoclonal antibody to the α subunit (anti-CD25) might preferentially block autoreactive T cells. This approach was tested in adult rats injected with activated MBP-specific T cells in the presence or absence of anti-CD25. All the control rats died of EAE, whereas six of the nine treated with anti-CD25 had no symptoms and the symptoms in the other three were mild. A possible drawback to the use of this therapy is its potential negative effect on $T_{reg}$ cells. which express high levels of IL-2R α.

The association of autoimmune disease with specific TCR chain components in a number of animal models has prompted researchers to see if blockage of the preferred receptors with monoclonal antibody might be therapeutic. Injection of PL/J mice with monoclonal antibody specific for the $V_β$ 8.2 T-cell receptor prevented induction of EAE by MBP in adjuvant. Even more promising was the finding that the $V_β$ 8.2 monoclonal antibody could also reverse the symptoms of autoimmunity in mice manifesting induced EAE (Figure 16-15) and that these mice manifested long-term remission. Clearly, the use of anti-TCR antibodies as a

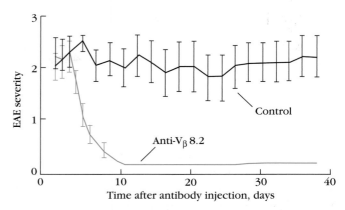

**FIGURE 16-15  Injection of monoclonal antibody to the $V_β$ 8.2 T-cell receptor into PL/J mice exhibiting EAE symptoms produced nearly complete remission of symptoms.** EAE was induced by injecting mice with MBP-specific T-cell clones. EAE severity scale: 3 = total paralysis of lower limbs; 2 = partial paralysis of lower limbs; 1 = limb tail; 0 = normal (no symptoms). *[Adapted from H. Acha-Orbea et al., 1989, Annual Review of Immunology 7:371.]*

treatment for human autoimmune diseases presents exciting possibilities.

Similarly, the association of various MHC alleles with autoimmunity, as well as the evidence for increased or inappropriate MHC expression in some autoimmune diseases, offers the possibility that monoclonal antibodies against appropriate MHC molecules might retard development of autoimmunity. Moreover, since antigen-presenting cells express many different class II MHC molecules, it should theoretically be possible to selectively block an MHC molecule that is associated with autoimmunity while sparing the others. In one study, injecting mice with monoclonal antibodies to class II MHC molecules before injecting MBP blocked the development of EAE. If, instead, the antibody was given after the injection of MBP, development of EAE was delayed but not prevented. In nonhuman primates, monoclonal antibodies to HLA-DR and HLA-DQ have been shown to reverse EAE.

## Oral antigens can induce tolerance

When antigens are administered orally, they tend to induce the state of immunologic unresponsiveness or tolerance. For example, as mentioned earlier, mice fed MBP do not develop EAE after subsequent injection of MBP. This finding led to a double-blind pilot trial in which 30 individuals with multiple sclerosis were fed either a placebo or 300 mg of bovine myelin every day for a year. The results of this study revealed that T cells specific for MBP were reduced in the myelin-fed group; MS symptoms seemed to be reduced in the male recipients (although the reduction fell short of statistical significance) but not in the female recipients. The results of oral tolerance induction are more promising in mice than in humans. However, the human clinical trials are in the early stages, and the peptides used so far and the doses may not be the most effective. Since the animal studies show the promise of this approach, more clinical trials will likely be conducted over the next few years.

## SUMMARY

- A major task of the immune system is to distinguish self from nonself. Failure to do so results in immune attacks against cells and organs of the host with the possible onset of autoimmune disease.

- Mechanisms to prevent self-reactivity, termed tolerance, operate at several levels. Central tolerance serves to delete self-reactive T or B lymphocytes; peripheral tolerance inactivates self-reactive lymphocytes that survive the initial screening process.

- Human autoimmune diseases can be divided into organ-specific and systemic diseases. The organ-specific diseases involve an autoimmune response directed primarily against a single organ or gland. The systemic diseases are directed against a broad spectrum of tissues.

- There are both spontaneous and experimental animal models for autoimmune diseases. Spontaneous autoimmune diseases result from genetic defects, whereas experimental animal models have been developed by immunizing animals with self antigens in the presence of adjuvant.

- Studies with experimental autoimmune animal models have revealed a central role for CD4$^+$ T$_H$ cells in the development of autoimmunity. T-cell clones can be isolated that induce the autoimmune disease in normal animals. The MHC haplotype of the experimental animal determines the ability to present various autoantigens to T$_H$ cells.

- The relative number of T$_H$1 and T$_H$2 cells appears to play a pivotal role in determining whether autoimmunity develops: T$_H$1 cells promote the development of autoimmunity, whereas T$_H$2 cells appear to block development and progression of autoimmune disease.

- A variety of mechanisms have been proposed for induction of autoimmunity, including release of sequestered antigens, molecular mimicry, and inappropriate class II MHC expression on cells. Evidence exists for each of these mechanisms, reflecting the many different pathways leading to autoimmune reactions.

- Current therapies for autoimmune diseases include treatment with immunosuppressive drugs, thymectomy, and plasmapheresis for diseases involving immune complexes. Blockers of TNF-α show success in controlling rheumatoid arthritis, Crohn's disease, and psoriasis.

## References

Anderson, M. S., et al. 2005. The cellular mechanism of Aire control of T cell tolerance. *Immunity* 23:227.

Benoist, C., and D. Matis. 2001. Autoimmunity provoked by infection: how good is the case for T cell epitope mimicry? *Nature Immunology* 2:797.

Feldman, M., and L. Steinman. 2005. Design of effective immunotherapy for human autoimmunity. *Nature* 435:612.

Goodnow, C. C., et al. 2005. Cellular and genetic mechanisms of self tolerance and autoimmunity. *Nature* 435:590.

Hafler, D. A., et al. 2005. Multiple sclerosis. *Immunological Reviews* 204:208.

Hausmann, S., and K. W. Wucherpfennig. 1997. Activation of autoreactive T cells by peptides from human pathogens. *Current Opinion in Immunology* 9:831.

Hogquist, K. A., T. A. Baldwin, and S. C. Jameson. 2005. Central tolerance: learning self-control in the thymus. *Nature Reviews Immunology* 5:772.

Horwitz, M. S., et al. 1998. Diabetes induced by Coxsackie virus: initiation by bystander damage and not molecular mimicry. *Nature Medicine* 4:781.

Jacobi, A. M., and B. Diamond. 2005. Balancing diversity and tolerance: lessons from patients with systemic lupus erythematosus. *Journal of Experimental Medicine* 202:341.

Levin, M. C., et al. 2002. Autoimmunity due to molecular mimicry as a cause of neurological disease. *Nature Medicine* 8:509.

Liblau, R. S., S. M. Singer, and H. O. McDevitt. 1995. T$_H$1 and T$_H$2 CD4$^+$ T cells in the pathogenesis of organ-specific autoimmune diseases. *Immunology Today* 16:34.

Lockshin, M. D. 1998. Why women? *Journal of the American Medical Women's Association* 53:4.

Mathis, D. 2005. *Lymphocyte Tolerance: Central Is Central.* Harvey Lectures, Series 99. Wiley-Liss, Hoboken, NJ, p. 95.

O'Garra, A., L. Steinman, and K. Gijbels. 1997. CD4$^+$ T-cell subsets in autoimmunity. *Current Opinion in Immunology* 9:872.

Paust, S., and H. Cantor. 2005. Regulatory T cells and autoimmune disease. *Immunological Reviews* 204:195.

Rioux, J. D., and A. K. Abbas. 2005. Paths to understanding the genetic basis of autoimmune disease. *Nature* 435:584.

Sakaguchi, S. 2004. Naturally arising CD4+ regulatory T cells for immunologic self-tolerance and negative control of immune responses. *Annual Review of Immunology* 22:531.

Silverstein, A. M., and N. R. Rose. 1997. On the mystique of the immunological self. *Immunological Reviews* 159:197.

Streilein, J. W., M. R. Dana, and B. R. Ksander. 1997. Immunity causing blindness: five different paths to herpes stromal keratitis. *Immunology Today* 18:443.

Vyse, T. J., and B. L. Kotzin. 1998. Genetic susceptibility to systemic lupus erythematosus. *Annual Review of Immunology* 16:261.

Weiner, H. L., et al. 1993. Double-blind pilot trial of oral tolerization with myelin antigens in multiple sclerosis. *Science* 259:1321.

Wucherpfennig, K. W. 2001. Mechanisms for the induction of autoimmunity by infectious agents. *Journal of Clinical Investigation* 108:1097.

Zhao, Z-S, et al. 1998. Molecular mimicry by herpes simplex virus type-1: autoimmune disease after viral infection. *Science* 279:1344.

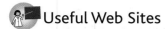

## Useful Web Sites

### http://www.lupus.org/index.html

The site for the Lupus Foundation of America contains valuable information for patients and family members as well as current information about research in this area.

### http://www.nih.gov/niams/

Home page for the National Institute for Arthritis and Musculoskeletal and Skin Diseases. This site contains links to other arthritis sites.

### http://www.niddk.nih.gov/

Home page for the National Institute for Diabetes and Digestive and Kidney Diseases. This site contains an exhaustive list of links to other diabetes health-related sites.

## Study Questions

**CLINICAL FOCUS QUESTION** What are some of the possible reasons why females are more susceptible to autoimmune diseases than males?

1. Explain why all self-reactive lymphocytes are not eliminated in the thymus or bone marrow. How are the surviving self reactors prevented from harming the host?

2. Why is tolerance critical to the normal functioning of the immune system?

3. What is the importance of receptor editing to B-cell tolerance?

4. For each of the following autoimmune diseases (a–l), select the most appropriate characteristic (1–12) listed below.

   *Disease*

   a. _____ Experimental autoimmune encephalitis (EAE)
   b. _____ Goodpasture's syndrome
   c. _____ Graves' disease
   d. _____ Systemic lupus erythematosus (SLE)
   e. _____ Insulin-dependent diabetes mellitus (IDDM)
   f. _____ Rheumatoid arthritis
   g. _____ Hashimoto's thyroiditis
   h. _____ Experimental autoimmune myasthenia gravis (EAMG)
   i. _____ Myasthenia gravis
   j. _____ Pernicious anemia
   k. _____ Multiple sclerosis
   l. _____ Autoimmune hemolytic anemia

   *Characteristics*

   (1) Auto-antibodies to intrinsic factor block vitamin $B_{12}$ absorption
   (2) Auto-antibodies to acetylcholine receptor
   (3) $T_H1$-cell reaction to thyroid antigens
   (4) Auto-antibodies to RBC antigens
   (5) T-cell response to myelin
   (6) Induced by injection of myelin basic protein plus complete Freund's adjuvant
   (7) Auto-antibody to IgG
   (8) Auto-antibodies to basement membrane
   (9) Auto-antibodies to DNA and DNA-associated protein
   (10) Auto-antibodies to receptor for thyroid-stimulating hormone
   (11) Induced by injection of acetylcholine receptors
   (12) $T_H1$-cell response to pancreatic beta cells

5. Experimental autoimmune encephalitis (EAE) has proved to be a useful animal model of autoimmune disorders.

   a. Describe how this animal model is made.
   b. What is unusual about the animals that recover from EAE?
   c. How has this animal model indicated a role for T cells in the development of autoimmunity?

6. Molecular mimicry is one mechanism proposed to account for the development of autoimmunity. How has induction of EAE with myelin basic protein contributed to the understanding of molecular mimicry in autoimmune disease?

7. Describe at least three different mechanisms by which a localized viral infection might contribute to the development of an organ-specific autoimmune disease.

8. Transgenic mice expressing the IFN-γ transgene linked to the insulin promoter developed diabetes.

   a. Why was the insulin promoter used?
   b. What is the evidence that the diabetes in these mice is due to autoimmune damage?
   c. What is unusual about MHC expression in this system?
   d. How might this system mimic events that might be caused by a localized viral infection in the pancreas?

9. Monoclonal antibodies have been administered for therapy in various autoimmune animal models. Which monoclonal antibodies have been used and what is the rationale for these approaches?

10. Indicate whether each of the following statements is true or false. If you think a statement is false, explain why.

   a. $T_H1$ cells have been associated with development of autoimmunity.
   b. Immunization of mice with IL-12 prevents induction of EAE by injection of myelin basic protein plus adjuvant.
   c. The presence of the *HLA B27* allele is diagnostic for ankylosing spondylitis, an autoimmune disease affecting the vertebrae.
   d. Individuals with pernicious anemia produce antibodies to intrinsic factor.
   e. A defect in the gene encoding Fas can reduce programmed cell death by apoptosis.

11. For each of the following autoimmune disorders (a–d), indicate which of the following treatments (1–5) may be appropriate:

   *Disease*

   a. Hashimoto's thyroiditis
   b. Systemic lupus erythematosis
   c. Grave's disease
   d. Myasthenia gravis

*Treatment*

1. Cylosporin A
2. Thymectomy
3. Plasmapharesis
4. Kidney transplant
5. Thyroid hormones

12. As discussed in Chapter 7, deficiencies in complement components C1q, C1r, C1s, C4, or C2 can contribute to the development of systemic lupus erythematosis, However, patients with SLE exhibit excessive complement activation. Explain how both of these findings can be true.

13. Which of the following are examples of mechanisms for the development of autoimmunity? For each possibility, give an example.

    a. Polyclonal B cell activation
    b. Tissue damage
    c. Viral infection
    d. Increased expression of TCR molecules
    e. Increased expression of class II MHC molecules

14. The figure below is a double immunodiffusion assay (Ouchterlony method) after several days' incubation, in which a patient was tested for autoimmune disease. The well in the center was loaded with a solution containing a mixture of nuclear antigens (nucleosomes and spliceosomes). The patient's sera was diluted at several different dilution ratios and placed in the outer wells (dilution ratios indicated in wells). A positive control (mouse anti-human histones) was placed in one well and an irrelevant control (normal human sera) was placed in another.

    a. Based on the assay, which of the following statements is true regarding this patient?

        1. The patient is normal because it is normal to express antibodies against nuclear proteins in your serum.
        2. The patient is probably male and is suffering from Goodpasture's syndrome.
        3. The patient is suffering from severe allergies.
        4. The patient may be suffering from SLE and there is a higher chance the patient is female.

    b. What was the titer of this patient?

    c. Which of the following is an experiment you could perform to verify the findings shown in the above figure?

        1. Direct ELISA with plates coated with nuclear antigens.
        2. Immunofluorescence assay with tissue sections or permeablized human cells.
        3. Western blots using nuclear proteins run on gels, sera used as primary Ab.

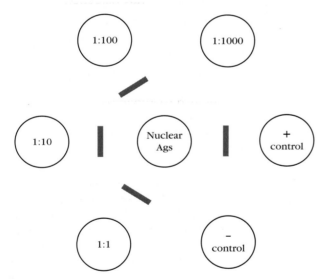

## Interactive Study

**www.whfreeman.com/kuby**

SELF-TEST
Review OF Key Terms

# chapter 17

# Transplantation Immunology

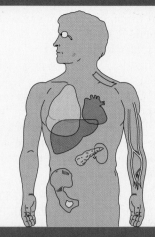

T RANSPLANTATION, AS THE TERM IS USED IN IMMUNOLOGY, refers to the act of transferring cells, tissues, or organs from one site to another. Many diseases can be cured by implantation of a healthy organ, tissue, or cells (a graft) from one individual (the donor) to another in need of the transplant (the recipient or host). The development of surgical techniques that allow the facile reimplantation of organs has removed one barrier to successful transplantation, but others remain. One is the lack of organs for transplantation. Although a supply of organs is provided by accident victims and, in some cases, living donors, more patients are in need of transplants than there are organs available. The seriousness of the donor-organ shortage is reflected in the fact that, as of April 2006, an estimated 91,700 patients in the United States were on the waiting list for an organ transplantation. The majority of those on the list (~70%) require a kidney; at present, the waiting period for this organ averages over 800 days. Although the lack of organs for transplantation is a serious issue, the most formidable barrier to making transplantation a routine medical treatment is the immune system. The immune system has evolved elaborate and effective mechanisms to protect the organism from attack by foreign agents, and these same mechanisms cause rejection of grafts from anyone who is not genetically identical to the recipient.

Alexis Carrel reported the first systematic study of transplantation in 1908; he interchanged both kidneys in a series of nine cats. Some of those receiving kidneys from other cats maintained urinary output for up to 25 days. Although all the cats eventually died, the experiment established that a transplanted organ could carry out its normal function in the recipient. The first human kidney transplant, attempted in 1935 by a Russian surgeon, failed because there was a mismatch of blood types between donor and recipient. This incompatibility caused almost immediate rejection of the kidney, and the patient died without establishing renal function. The rapid immune response experienced here, termed hyperacute rejection, is mediated by antibodies and will be described in this chapter. In 1954 a team headed by Joseph Murray at the Peter Bent Brigham Hospital in Boston performed the first successful human kidney transplant between identical twins. Today kidney, pancreas, heart, lung, liver, bone marrow, and cornea transplantations are performed among nonidentical individuals with ever-increasing frequency and with increasing rates of success at least for their short-term survival.

A variety of immunosuppressive agents can delay or prevent rejection of the transplanted organ, including drugs and specific antibodies developed to diminish the immunologic attack on grafts, but the majority of these agents have an overall immunosuppressive effect, and their long-term use has deleterious side effects for the recipient. New methods of inducing specific tolerance to the graft without suppressing other immune responses are being developed that promise longer survival of transplants without compromise of host immunity. This chapter describes the mechanisms underlying graft rejection, various procedures that are currently used to prolong graft survival, some newer experimental techniques that may find use in the future, and a summary of the current status of transplantation as a clinical tool. The Clinical Focus examines the use of organs from nonhuman species (xenotransplants) to circumvent the shortage of organs available for patients in need of them.

# Immunologic Basis of Graft Rejection

The degree of immune response to a graft varies with the type of graft. The following terms are used to denote different types of transplants:

- **Autograft** is self tissue transferred from one body site to another in the same individual. Transferring healthy skin to a burned area in burn patients and use of healthy blood vessels to replace blocked coronary arteries are examples of frequently used autografts.

- **Isograft** is tissue transferred between genetically identical individuals. In inbred strains of mice, an isograft can be performed from one mouse to another syngeneic mouse. In humans, an isograft can be performed between genetically identical (monozygotic) twins.

- **Allograft** is tissue transferred between genetically different members of the same species. In mice, an allograft is performed by transferring tissue or an organ from one strain to another. In humans, organ grafts from one individual to another are allografts unless the donor and recipient are identical twins.

- **Xenograft** is tissue transferred between different species (e.g., the graft of a baboon heart into a human). Because of significant shortages of donated organs, raising animals for the specific purpose of serving as organ donors for humans is under serious consideration.

Autografts and isografts are usually accepted, owing to the genetic identity between graft and host (Figure 17-1a). Because an allograft is genetically dissimilar to the host, it is often recognized as foreign by the immune system and is rejected. Obviously, xenografts exhibit the greatest genetic disparity and therefore engender a vigorous graft rejection.

## Allograft rejection displays specificity and memory

The rate of allograft rejection varies according to the tissue involved. In general, skin grafts are rejected faster than other tissues such as kidney or heart. Despite these time differences, the immune response culminating in graft rejection always displays the attributes of specificity and memory. If

an inbred mouse of strain A is grafted with skin from strain B, primary graft rejection, known as first-set rejection, occurs (Figure 17-1b). The skin first becomes revascularized between days 3 and 7; as the reaction develops, the vascularized transplant becomes infiltrated with lymphocytes, monocytes, neutrophils, and other inflammatory cells. There is decreased vascularization of the transplanted tissue by 7 to 10 days, visible necrosis by 10 days, and complete rejection by 12 to 14 days.

Immunologic memory is demonstrated when a second strain B graft is transferred to a previously grafted strain A mouse. In this case, a graft rejection reaction develops more quickly, with complete rejection occurring within 5 to 6 days; this secondary response is designated second-set rejection (Figure 17-1c). The specificity of second-set rejection can be demonstrated by grafting an unrelated strain C graft at the same time as the second strain B graft. Rejection of the strain C graft proceeds according to first-set rejection kinetics, whereas the strain B graft is rejected in an accelerated second-set fashion.

## T cells play a key role in allograft rejection

In the early 1950s, Avrion Mitchison showed in adoptive-transfer experiments (in which host lymphocytes are killed with x-rays and donor immune cells are introduced) that lymphocytes, but not serum antibody, could transfer allograft immunity. Later studies implicated T cells in allograft rejection. For example, nude mice, which lack a thymus and consequently lack functional T cells, were found to be incapable of allograft rejection; indeed, these mice even accept xenografts. In other studies, T cells derived from an allograft-primed mouse were shown to transfer second-set allograft rejection to an unprimed syngeneic recipient as long as that recipient was grafted with the same allogeneic tissue (Figure 17-2).

Analysis of the T-cell subpopulations involved in allograft rejection has implicated both CD4$^+$ and CD8$^+$ populations. In one study, mice were injected with monoclonal antibodies to deplete one or both types of T cells and then the rate of graft rejection was measured. As shown in Figure 17-3, removal of the CD8$^+$ population alone had no effect on graft survival, and the graft was rejected at the same rate as in control mice (15 days). Removal of the CD4$^+$ T-cell population alone prolonged graft survival from 15 days to 30 days. However, removal of both the CD4$^+$ and the CD8$^+$ T cells resulted in long-term survival (up to 60 days) of the allografts. This study indicated that both CD4$^+$ and CD8$^+$ T cells participated in rejection and that the collaboration of both subpopulations resulted in more pronounced graft rejection. These data are supported by a recent study of CD4 and CD8 gene expression of T cells infiltrating human kidney allografts. A study of RNA expression indicates that both CD4 and CD8 were overexpressed in T cells present in both acute and chronic rejection. Increased levels of INF-$\gamma$ but not of IL-4 in the infiltrating T cells placed them in the $T_H1$ subset.

**OVERVIEW FIGURE 17-1:    Schematic Diagrams of the Process of Graft Acceptance and Rejection**

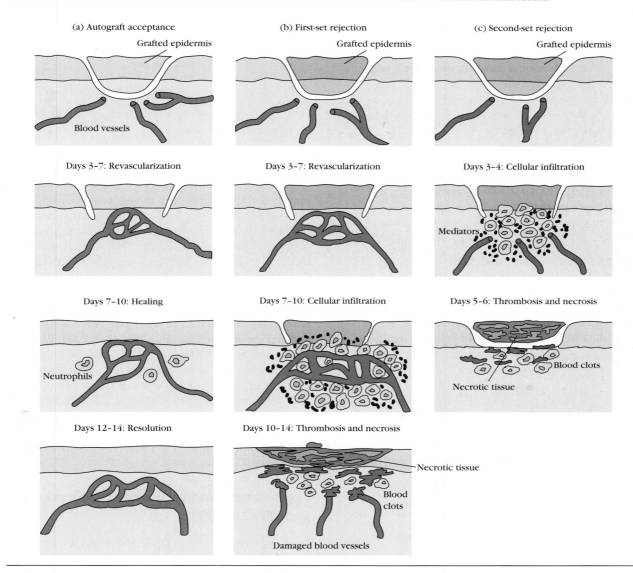

(a) Autograft acceptance

Grafted epidermis

Blood vessels

Days 3–7: Revascularization

Days 7–10: Healing

Neutrophils

Days 12–14: Resolution

(b) First-set rejection

Grafted epidermis

Days 3–7: Revascularization

Days 7–10: Cellular infiltration

Days 10–14: Thrombosis and necrosis

Necrotic tissue

Blood clots

Damaged blood vessels

(c) Second-set rejection

Grafted epidermis

Days 3–4: Cellular infiltration

Mediators

Days 5–6: Thrombosis and necrosis

Blood clots

Necrotic tissue

(a) Acceptance of an autograft is completed within 12 to 14 days. (b) First-set rejection of an allograft begins 7 to 10 days after grafting, with full rejection occurring by 10 to 14 days. (c) Second-set rejection of an allograft begins within 3 to 4 days, with full rejection by 5 to 6 days. The cellular infiltrate that invades an allograft (b, c) contains lymphocytes, phagocytes, and other inflammatory cells.

The role of dendritic cells in rejection or tolerance of an allograft is the subject of increasing interest because of their immunostimulatory capacity and their role in the induction of tolerance. As discussed in Chapter 8, dendritic cells can present exogenous antigens in the context of class I MHC molecules via the process of cross-presentation, giving CD8$^+$ T cells the opportunity to recognize alloantigens as part of the rejection process. Studies with mice have shown that inhibition of dendritic cells can aid graft acceptance, presumably by interfering with the presentation of antigen. On the other hand, pretreatment with donor dendritic cells can promote tolerance and has been shown to prolong the acceptance of both heart and pancreas transplants in mouse experiments.

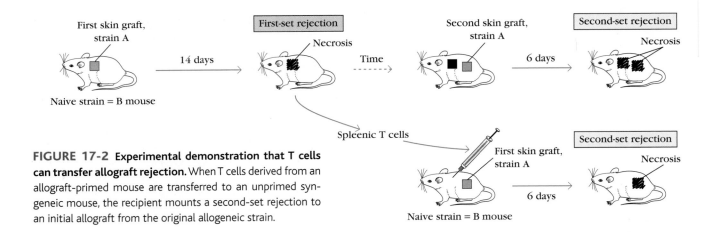

**FIGURE 17-2 Experimental demonstration that T cells can transfer allograft rejection.** When T cells derived from an allograft-primed mouse are transferred to an unprimed syngeneic mouse, the recipient mounts a second-set rejection to an initial allograft from the original allogeneic strain.

## Similar antigenic profiles foster allograft acceptance

Tissues that are antigenically similar are said to be **histocompatible;** such tissues do not induce an immunologic response that leads to tissue rejection. Tissues that display significant antigenic differences are *histoincompatible* and induce an immune response that leads to tissue rejection. The various antigens that determine histocompatibility are encoded by more than 40 different loci, but the loci responsible for the most vigorous allograft rejection reactions are located within the **major histocompatibility complex (MHC).** The organization of the MHC—called the H-2 complex in mice and the HLA complex in humans—was described in Chapter 8 (see Figure 8-1). Because the MHC loci are closely linked, they are usually inherited as a complete set, called a **haplotype,** from each parent.

Within an inbred strain of mice, all animals are homozygous at each MHC locus. When mice from two different

inbred strains, with haplotypes *b* and *k,* for example, are mated, the F$_1$ progeny each inherit one maternal and one paternal haplotype (see Figure 8-2). These F$_1$ offspring have the MHC type *b/k* and can accept grafts from either parent. Neither of the parental strains, however, can accept grafts from the F$_1$ offspring because each parent lacks one of the F$_1$ haplotypes. MHC inheritance in outbred populations is more complex because the high degree of polymorphism exhibited at each MHC locus gives a high probability of heterozygosity at most loci. In matings between members of an outbred species, there is only a 25% chance that any two offspring will inherit identical MHC haplotypes unless the parents share one or more haplotypes. Therefore, for purposes of organ or bone marrow grafts, it can be assumed that there is a 25% chance of MHC identity between siblings. With parent-to-child grafts, the donor and recipient will always have one haplotype in common but are nearly always mismatched for the haplotype inherited from the other parent.

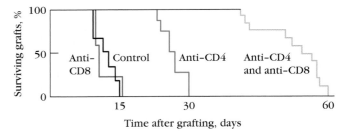

**FIGURE 17-3 The role of CD4$^+$ and CD8$^+$ T cells in allograft rejection is demonstrated by the curves showing survival times of skin grafts between mice mismatched at the MHC.** Animals in which the CD8$^+$ T cells were removed by treatment with an anti-CD8 monoclonal antibody (red) showed little difference from untreated control mice (black). Treatment with monoclonal anti-CD4 (blue) improved graft survival significantly, and treatment with both anti-CD4 and anti-CD8 antibody prolonged graft survival most dramatically (green). *[Adapted from S. P. Cobbold et al., 1986, Nature **323**:165.]*

## Graft donors and recipients are typed for RBC and MHC antigens

Since differences in blood group and major histocompatibility antigens are responsible for the most intense graft rejection reactions, various tissue-typing procedures to identify these antigens have been developed to screen potential donor and recipient cells. Initially, donor and recipient are screened for ABO blood-group compatibility. The blood-group antigens are expressed on RBCs, epithelial cells, and endothelial cells. Antibodies produced in the recipient to any of these antigens that are present on transplanted tissue will induce antibody-mediated complement lysis of the incompatible donor cells.

HLA typing of potential donors and a recipient can be accomplished with a microcytotoxicity test (Figure 17-4a, b). In this test, white blood cells from the potential donors and recipient are distributed into a series of wells on a microtiter plate, and then antibodies specific for various class I and

(a)

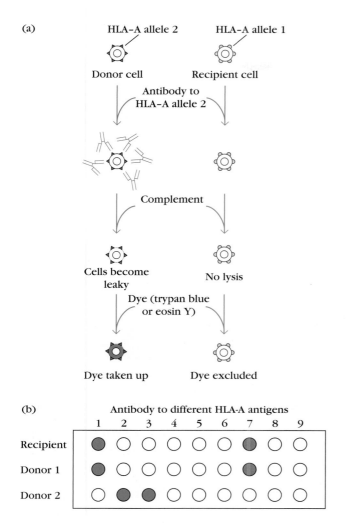

**FIGURE 17-4 Typing procedures for HLA antigens.** (a, b) HLA typing by microcytotoxicity. (a) White blood cells from potential donors and the recipient are added to separate wells of a microtiter plate. The example depicts the reaction of donor and recipient cells with a single antibody directed against an HLA-A antigen. The reaction sequence shows that if the antigen is present on the lymphocytes, addition of complement will cause them to become porous and unable to exclude the added dye. (b) Because cells express numerous HLA antigens, they are tested separately with a battery of antibodies specific for various HLA-A antigens. Here donor 1 shares HLA-A antigens recognized by antisera in wells 1 and 7 with the recipient, whereas donor 2 has none of the HLA-A antigens in common with the recipient. (c) Mixed lymphocyte reaction to determine identity of class II HLA antigens between a potential donor and recipient. Lymphocytes from the donor are irradiated or treated with mitomycin C to prevent cell division and then added to cells from the recipient. If the class II antigens on the two cell populations are different, the recipient cells will divide rapidly and take up large quantities of radioactive nucleotides into the newly synthesized nuclear DNA. The amount of radioactive nucleotide uptake is roughly proportional to the MHC class II differences between the donor and recipient lymphocytes.

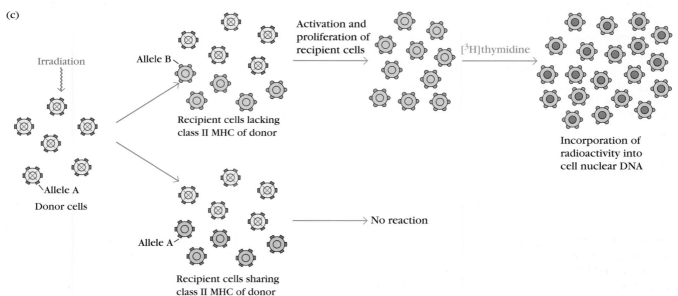

class II MHC alleles are added to different wells. After incubation, complement is added to the wells, and cytotoxicity is assessed by the uptake or exclusion of various dyes (e.g., trypan blue or eosin Y) by the cells. If the white blood cells express the MHC allele for which a particular monoclonal antibody is specific, then the cells will be lysed on addition of complement, and these dead cells will take up a dye such as trypan blue. HLA typing based on antibody-mediated microcytotoxicity can thus indicate the presence or absence of various MHC alleles.

Even when a fully HLA-compatible donor is not available, transplantation may be successful. In this situation, a one-way mixed-lymphocyte reaction (MLR) can be used to quantify the degree of class II MHC compatibility between potential donors and a recipient (Figure 17-4c). Lymphocytes from a potential donor that have been x-irradiated or treated with mitomycin C serve as the stimulator cells, and lymphocytes from the recipient serve as responder cells. Proliferation of the recipient T cells, which indicates T-cell activation, is measured by the uptake of [³H]thymidine into cellular DNA. The greater the class II MHC differences between the donor and recipient cells, the more [³H]thymidine uptake will be observed in an MLR assay. Intense proliferation of the recipient lymphocytes indicates a poor prognosis for graft survival. The advantage of the MLR over microcytotoxicity typing is that it gives a better indication of the degree of $T_H$-cell activation generated in response to the class II MHC antigens of the potential graft. The disadvantage of the MLR is that it takes several days to run the assay. If the potential donor is a cadaver, for example, it is not possible to wait for the results of the MLR because the organ must be used soon after removal from the cadaver. In that case, the microcytotoxicity test, which can be performed within a few hours, is the indicated protocol for tissue typing.

The importance of MHC matching for acceptance of allografts is confirmed by data gathered from recipients of kidney transplants. The data in Figure 17-5 reveal that survival of kidney grafts depends primarily on donor-recipient matching of the HLA class II antigens. Matching or mismatching of the class I antigens has a lesser effect on graft survival unless there is also mismatching of the class II antigens. A two-year survival rate of 90% is seen for kidney transplants in which one or two class I HLA loci are mismatched, whereas transplanted kidneys with differences in the class II MHC have only a 70% chance of surviving two years. Those with greater numbers of mismatches have a very low survival rate at one year after transplant. As described below, HLA matching is most important for kidney and bone marrow transplants; liver and heart transplants may survive with greater mismatching.

Data on the impact of HLA matching is confirmed by a recent survey of European renal transplant patients. The new data indicate that 36-month survival rates increased to almost 90% in the cohort transplanted between 1996 and 2000 compared to a 50% success in transplants carried out between 1966 and 1970. The increase is attributed mainly to

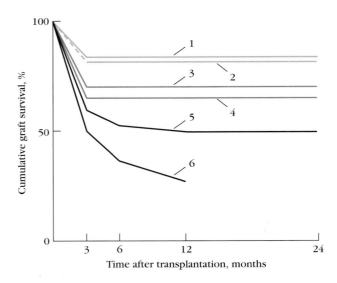

| Curve no. | HLA mismatches (no.) | |
| --- | --- | --- |
| | Class I | Class II |
| 1 | 0 | 0 |
| 2 | 1 or 2 | 0 |
| 3 | 3 or 4 | 0 |
| 4 | 0 | 1 or 2 |
| 5 | 1 or 2 | 1 or 2 |
| 6 | 3 or 4 | 1 or 2 |

**FIGURE 17-5 The effect of HLA class I and class II antigen matching on survival of kidney grafts.** Mismatching of one or two class I (HLA-A or HLA-B) antigens has little effect on graft survival. A single class II difference (line 4) has the same effect as three or four differences in class I antigens (line 3). When both class I and class II antigens are mismatched, rejection is accelerated. *[Adapted from T. Moen et al., 1980, New England Journal of Medicine **303**:850.]*

better immunosuppressive procedures. Within the recent groups of transplanted patients, there remained marked advantages for those transplanted with an organ from a donor with a close match of HLA types. In addition to a 5% to 10% increase in survival rate for the matched recipients, there were further advantages in a lowered incidence of skin cancer and other adverse effects. Importantly, if the transplant fails for any reason and a second transplant is required, recipients of closely matching grafts are favored because they will not have preexisting antibodies to multiple HLA types, which can hinder acceptance of a subsequent graft.

Current understanding of the killer-cell immunoglobulin-like receptors (KIRs) on the NK cell (see Chapter 14) suggests that absence of a class I antigen recognized by certain KIR molecules could lead to killing of the foreign cell. Rejection was observed in experimental bone marrow transplants where the class I molecule recognized by the recipient NK-inhibitory receptor is absent on donor cells. The effects of such class I mismatching on solid organ grafts may be less marked.

MHC identity of donor and host is not the sole factor determining tissue acceptance. When tissue is transplanted between genetically different individuals, even if their MHC antigens are identical, the transplanted tissue can be rejected because of differences at various **minor histocompatibility loci.** As described in Chapter 9, the major histocompatibility antigens are recognized directly by $T_H$ and $T_C$ cells, a phenomenon termed *alloreactivity.* In contrast, minor histocompatibility antigens are recognized only when they are presented in the context of self-MHC molecules. The tissue rejection induced by minor histocompatibility differences is usually less vigorous than that induced when the histocompatibility differences are more pronounced. Still, reaction to these minor tissue differences often results in graft rejection. For this reason, successful transplantation even between HLA-identical individuals requires some degree of immune suppression.

## Cell-mediated graft rejection occurs in two stages

Graft rejection is caused principally by a cell-mediated immune response to alloantigens (primarily, MHC molecules) expressed on cells of the graft. Both delayed-type hypersensitive and cell-mediated cytotoxicity reactions have been implicated. The process of graft rejection can be divided into two stages: (1) a sensitization phase, in which antigen-reactive lymphocytes of the recipient proliferate in response to alloantigens on the graft, and (2) an effector stage, in which immune destruction of the graft takes place.

### Sensitization Stage

During the sensitization phase, $CD4^+$ and $CD8^+$ T cells recognize alloantigens expressed on cells of the foreign graft and proliferate in response. Both major and minor histocompatibility alloantigens can be recognized. In general, the response to minor histocompatibility antigens is weak, although the combined response to several minor differences can sometimes be quite vigorous. The response to major histocompatibility antigens involves recognition of both the donor MHC molecule and an associated peptide ligand in the cleft of the MHC molecule. The peptides present in the groove of allogeneic class I MHC molecules are derived from proteins synthesized within the allogeneic cell. The peptides present in the groove of allogeneic class II MHC molecules are generally proteins taken up and processed through the endocytic pathway of the allogeneic antigen-presenting cell.

A host $T_H$ cell becomes activated when it interacts with an antigen-presenting cell (APC) that both expresses an appropriate antigenic ligand–MHC molecule complex and provides the requisite costimulatory signal. Depending on the tissue, different populations of cells within a graft may function as APCs. Because dendritic cells are found in most tissues and because they constitutively express high levels of class II MHC molecules, dendritic cells generally serve as the major APC in grafts. APCs of host origin can also migrate into a graft and endocytose the foreign alloantigens (both major and minor histocompatibility molecules) and present them as processed peptides together with self-MHC molecules. As mentioned above, the property of cross-presentation by dendritic cells allows them to present endocytic antigens in the context of class I MHC molecules to $CD8^+$ cells, which can then react to the allograft.

In some organ and tissue grafts (e.g., grafts of kidney, thymus, and pancreatic islets), a population of donor APCs has been shown to migrate from the graft to the regional lymph nodes. These APCs are dendritic cells, which express high levels of class II MHC molecules (together with normal levels of class I MHC molecules) and are widespread in mammalian tissues, with the chief exception of the brain. Because these APCs express the allogeneic MHC antigens of the donor graft, they are recognized as foreign and therefore can stimulate immune activation of T lymphocytes in the lymph node. In some experimental situations, these donor APCs have been shown to induce tolerance to their surface antigens by deletion of thymic T-cell populations with receptors specific for them. Consistent with the notion that exposure to donor cells can induce tolerance are data showing that blood transfusions from the donor prior to transplantation can aid acceptance of the graft. The concept that tolerance can be induced by using donor cells to create a chimeric host has been studied in various animal models but for numerous reasons has not yet found use in human transplantation.

Donor dendritic cells functioning as APCs are not the only cells involved in immune stimulation. In fact, they do not seem to play any role in skin graft rejection. Other cell types that have been implicated in alloantigen presentation to the immune system include Langerhans cells and endothelial cells lining the blood vessels. Both of these cell types express class I and class II MHC antigens.

Recognition of the alloantigens expressed on the cells of a graft induces vigorous T-cell proliferation in the host. This proliferation can be demonstrated in vitro in a mixed-lymphocyte reaction (see Figure 17-4c). Both dendritic cells and vascular endothelial cells from an allogeneic graft induce host T-cell proliferation. The major proliferating cell is the $CD4^+$ T cell, which recognizes class II alloantigens either directly or as alloantigen peptides presented by host antigen-presenting cells. This amplified population of activated $T_H$ cells is thought to play a central role in inducing the various effector mechanisms of allograft rejection. As discussed in Chapter 9, a relatively high percentage of T cells react to alloantigens.

### Effector Stage

A variety of effector mechanisms participate in allograft rejection (Figure 17-6). The most common are cell-mediated reactions involving delayed-type hypersensitivity and CTL-mediated cytotoxicity; less common mechanisms are antibody-plus-complement lysis and destruction by antibody-dependent cell-mediated cytotoxicity (ADCC). The hallmark

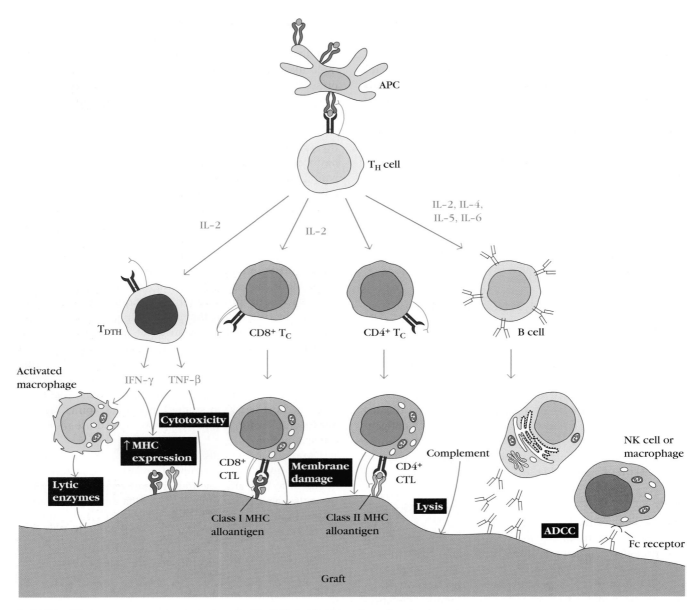

**FIGURE 17-6 Effector mechanisms (purple blocks) involved in allograft rejection.** The generation or activity of various effector cells depends directly or indirectly on cytokines (blue) secreted by activated T$_H$ cells. ADCC = antibody-dependent cell-mediated cytotoxicity.

of graft rejection involving cell-mediated reactions is an influx of T cells and macrophages into the graft. Histologically, the infiltration in many cases resembles that seen during a delayed-type hypersensitive response, in which cytokines produced by T cells promote macrophage infiltration (see Figure 14-15). Recognition by host CD8$^+$ cells of either foreign class I alloantigens on the graft or alloantigenic peptides cross-presented in the context of class I MHC by dendritic cells can lead to CTL-mediated killing (see Figure 9-16).

In each of these effector mechanisms, cytokines secreted by T$_H$ cells play a central role (see Figure 17-6). For example, IL-2, IFN-$\gamma$, and TNF-$\beta$ have each been shown to be

important mediators of graft rejection. IL-2 promotes T-cell proliferation and is generally necessary for the production of effector CTLs (see Figure 14-1). IFN-$\gamma$ is central to the development of a delayed-type hypersensitivity response, promoting the influx of macrophages into the graft and their subsequent activation into more destructive cells. TNF-$\beta$ has been shown to have a direct cytotoxic effect on the cells of a graft. A number of cytokines promote graft rejection by inducing expression of class I or class II MHC molecules on graft cells. The interferons ($\alpha$, $\beta$, and $\gamma$), TNF-$\alpha$, and TNF-$\beta$ all increase class I MHC expression, and IFN-$\gamma$ increases class II MHC expression as well.

During a rejection episode, the levels of these cytokines increase, inducing a variety of cell types within the graft to express class I or class II MHC molecules. In rat cardiac allografts, for example, dendritic cells are initially the only cells that express class II MHC molecules. However, as an allograft reaction begins, localized production of IFN-γ in the graft induces vascular endothelial cells and myocytes to express class II MHC molecules as well, making these cells targets for CTL attack.

A complete description of the effect of cytokines on the allograft must include some mention of those that promote tolerance of the foreign tissue. Whereas IL-12 and IL-15 are exclusively proinflammatory, supporting pathways leading to allograft rejection, IL-2 may have a pleiotropic effect. It promotes rejection by supporting expansion of T effector cells, but it also primes effector cells for deletion by apoptosis and supports the proliferation of T regulatory cells. IL-10 and TGF-β promote tolerance by mediating $T_{reg}$ function.

# Clinical Manifestations of Graft Rejection

Graft rejection reactions have various time courses, depending on the type of tissue or organ grafted and the immune response involved. Hyperacute rejection reactions occur within the first 24 hours after transplantation, acute rejection reactions usually begin in the first few weeks after transplantation, and chronic rejection reactions can occur from months to years after transplantation.

## Preexisting recipient antibodies mediate hyperacute rejection

In rare instances, a transplant is rejected so quickly that the grafted tissue never becomes vascularized. These hyperacute reactions are caused by preexisting host serum antibodies specific for antigens of the graft. The antigen-antibody complexes that form activate the complement system, resulting in an intense infiltration of neutrophils into the grafted tissue. The ensuing inflammatory reaction causes obstructing blood clots within the capillaries, preventing vascularization of the graft (Figure 17-7).

Several mechanisms can account for the presence of preexisting antibodies specific for allogeneic MHC antigens. Recipients of repeated blood transfusions sometimes develop significant levels of antibodies to MHC antigens expressed on white blood cells present in the transfused blood. If some of these MHC antigens are the same as those on a subsequent graft, then the antibodies can react with the graft, inducing a hyperacute rejection reaction. With repeated pregnancies, women are exposed to the paternal

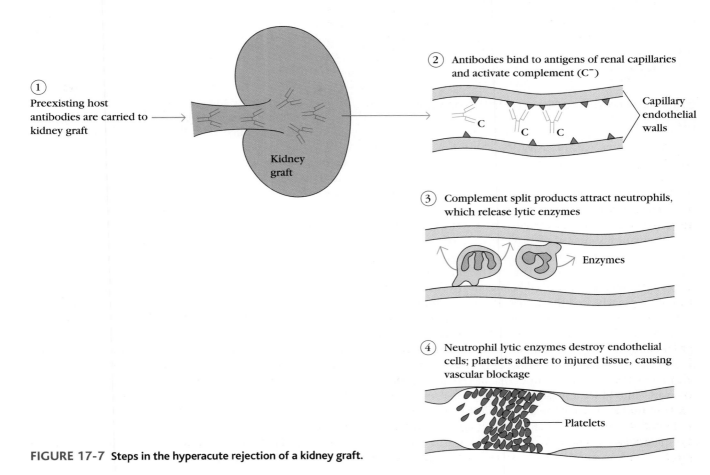

**FIGURE 17-7   Steps in the hyperacute rejection of a kidney graft.**

alloantigens of the fetus and may develop antibodies to these antigens. Finally, individuals who have had a previous graft sometimes have high levels of antibodies to the allogeneic MHC antigens of that graft.

In some cases, the preexisting antibodies participating in hyperacute graft rejection may be specific for blood-group antigens in the graft. If tissue typing and ABO blood-group typing are performed prior to transplantation, these preexisting antibodies can be detected and grafts that would result in hyperacute rejection can be avoided. Xenotransplants are often rejected in a hyperacute manner because of antibodies to cellular antigens of the donor species that are not present in the recipient species. Such an antigen is discussed in the Clinical Focus on the opposite page.

In addition to the hyperacute rejection mediated by preexisting antibodies, there is a less frequent form of rejection termed *accelerated rejection,* caused by antibodies that are produced immediately after transplantation.

## Acute rejection is mediated by T-cell responses

Cell-mediated allograft rejection manifests as an acute rejection of the graft beginning about 10 days after transplantation (see Figure 17-1b). Histopathologic examination reveals a massive infiltration of macrophages and lymphocytes at the site of tissue destruction, suggestive of $T_H$-cell activation and proliferation. Acute graft rejection is effected by the mechanisms described previously (see Figure 17-6).

## Chronic rejection occurs months or years post-transplant

Chronic rejection reactions develop months or years after acute rejection reactions have subsided. The mechanisms of chronic rejection include both humoral and cell-mediated responses by the recipient. Although the use of immunosuppressive drugs and the application of tissue-typing methods to obtain optimum match of donor and recipient have dramatically increased survival of allografts during the first years after engraftment, little progress has been made in long-term survival. The use of immunosuppressive drugs, which are described below, greatly increases the short-term survival of the transplant, but chronic rejection is not prevented in most cases. Data for rejection of kidney transplants since 1975 indicate an increase from 40% to almost 90% in 1-year survival of grafts. However, in the same period, long-term survival has risen only slightly; as in 1975, only about 50% of transplanted kidneys are still functioning at 10 years after transplant. Chronic rejection reactions are difficult to manage with immunosuppressive drugs and may necessitate another transplantation.

# General Immunosuppressive Therapy

Allogeneic transplantation requires some degree of immunosuppression if the transplant is to survive. Most of the immunosuppressive treatments that have been developed have the disadvantage of being nonspecific; that is, they result in generalized immunosuppression of responses to all antigens, not just those of the allograft, which places the recipient at increased risk of infection and lymphoid cancers. In addition, many immunosuppressive measures are aimed at slowing the proliferation of activated lymphocytes. This treatment affects any rapidly dividing nonimmune cells (e.g., epithelial cells of the gut or bone marrow hematopoietic stem cells), leading to serious or even life-threatening complications. Patients on long-term immunosuppressive therapy are at increased risk of cancer, hypertension, and metabolic bone disease.

## Mitotic inhibitors thwart T-cell proliferation

In 1959 Robert Schwartz and William Dameshek reported that treatment with 6-mercaptopurine suppressed immune responses in animal models. Joseph Murray and colleagues then screened a number of chemical analogues of 6-mercaptopurine for use in human transplantation and found one, azathioprine, that when used in combination with corticosteroids increased survival of allografts in a dramatic fashion. Murray received a Nobel prize in 1991 for this clinical advance, and the developers of the drug, Gertrude Elion and George Hitchings, received the Nobel prize in 1987.

Azathioprine (Imuran), a potent mitotic inhibitor, is often given just before and after transplantation to diminish T-cell proliferation in response to the alloantigens of the graft. Azathioprine acts on cells in the S phase of the cell cycle to block synthesis of inosinic acid, which is a precursor of the purines, adenylic and guanylic acid. Both B-cell and T-cell proliferation is diminished in the presence of azathioprine. Functional immune assays such as the MLR, CML, and skin test show a significant decline after azathioprine treatment, indicating an overall decrease in T-cell numbers.

Other mitotic inhibitors that are sometimes used in conjunction with immunosuppressive agents are cyclophosphamide, mycophenolate mofetil, and methotrexate. Cyclophosphamide is an alkylating agent that inserts into the DNA helix and becomes cross-linked, leading to disruption of the DNA chain. It is especially effective against rapidly dividing cells and is therefore sometimes given at the time of grafting to block T-cell proliferation. Mycophenolate mofetil is derived from the mold penicillium and blocks purine synthesis. Methotrexate acts as a folic-acid antagonist to block purine biosynthesis. The fact that the mitotic inhibitors act on all rapidly dividing cells and not specifically on those involved in the immune response against the allograft

# CLINICAL FOCUS

## Is There a Clinical Future for Xenotransplantation?

**Unless** organ donations increase drastically, most of the 91,700 U.S. patients on the waiting list for a transplant in April 2006 will not receive one. The majority need a kidney, but in 2005 only 16,477 kidneys were transplanted. A solution to this shortfall is to utilize animal organs. Some argue that xenografts bring the risk of introducing pathogenic retroviruses into the human population; others object based on ethical grounds relating to animal rights. Nevertheless, the use of pigs to supply organs for humans is under serious consideration. Pigs breed rapidly, have large litters, can be housed in pathogen-free environments, and share considerable anatomic and physiologic similarity with humans. In fact, pigs have served as donors of cardiac valves for humans for years. Primates are more closely related to humans than pigs are, but the availability of large primates as transplant donors is and will continue to be extremely limited.

Balancing the advantages of pig donors are serious difficulties. For example, if a pig kidney were implanted into a human by techniques standard for human transplants, it would likely fail in a rapid and dramatic fashion due to hyperacute rejection. This antibody-mediated rejection is due to the presence on the pig cells (and cells of most mammals other than humans and the highest nonhuman primates) of a disaccharide antigen (galactosyl-1,3-α-galactose) that is not present on human cells. The presence of this antigen on many microorganisms means that nearly everyone has been exposed to it and has formed antibodies against it. The preexisting antibodies react with pig cells, which are then lysed rapidly by comple-

ment. The absence of human regulators of complement activity on the pig cells, including human decay accelerating factor (DAF) and human membrane cofactor protein (MCP), intensifies the complement lysis cycle (see Chapter 7 for descriptions of DAF and MCP).

How can this major obstacle be circumvented? Being tested are strategies for absorbing the antibodies from the circulation on solid supports and using soluble gal-gal disaccharides to block antibody reactions. A more elegant solution involves genetically engineered pigs in which the gene for the enzyme that synthesizes α-1,3-galactosyltransferase has been knocked out. These GalT-KO pigs have been used as heart or kidney donors for baboons in experimental systems. K. Kuwaki and colleagues transplanted GalT-KO pig hearts into baboons immunosuppressed with antithymocyte globulin and an anti-CD154 monoclonal antibody and then maintained with commonly used drugs. They obtained a mean survival time of 92 days, and one GalT-KO pig heart transplant survived in a baboon for 179 days. K. Yamada and coworkers demonstrated kidney function in recipients of GalT-KO pig kidneys for up to 83 days. This study used a regimen involving simultaneous thymus transplant in an attempt to establish tolerance in the baboon recipients. Although these results are not conclusive, the availability of organs and the promising results of experiments aimed at establishing an optimal regimen prior to human use are quite encouraging for the use of xenotransplantation in a clinical setting.

Solving the immediate rejection reaction by eliminating a specific antigen may not prevent all antibody-mediated rejec-

tion. Certainly other antigenic differences to which human recipients have antibodies will be present in some if not all donor-recipient pairs. However, any antibody attack on the pig cells may be blunted if human DAF (CD55) is present on the targeted cell to dampen the complement reaction. The lack of human DAF in pig cells is remedied by producing transgenic pigs that express this protein. Addition of human complement regulators to the pig represents a more universal solution by conferring resistance on donor cells that might become a target of lysis by human complement.

Even if all issues of antigenic difference were resolved, an additional concern remains. Pig endogenous retroviruses introduced into humans as a result of xenotransplantation could cause disease. Opponents of xenotransplantation raise the specter of another HIV-type epidemic resulting from human infection by a new animal retrovirus. Recently, a Boston-based company announced development of pigs free of endogenous pig retroviruses, reducing the possibility of this bleak outcome.

Will we see the use of pig kidneys in humans in the near future? The increasing demand for organs is driving the commercial development of colonies of pigs suitable to become organ donors. Although kidneys are the most sought-after organ at present, other organs and cells from the specially bred and engineered animals will find use if they are proven to be safe and effective. A statement issued in 2000 from the American Society of Transplantation and the American Society of Transplant Surgeons endorses the use of xenotransplants if certain conditions are met, including the demonstration of feasibility in a nonhuman primate model, proven benefit to the patient, and lack of infectious disease risk. Although certain barriers remain to the clinical use of xenotransplants, serious efforts are in motion to overcome them.

can lead to deleterious side reactions by thwarting division of other functional cells in the body. More recent developments of cyclosporine and related inhibitors of cell metabolic processes replaced azathioprine as a first-line drug for transplantation.

## Corticosteroids suppress inflammation

As described in Chapter 16, corticosteroids, such as prednisone and dexamethasone, are potent anti-inflammatory agents that exert their effects at many levels of the immune response. These drugs are often given to transplant recipients together with other immunosuppressants to prevent acute episodes of graft rejection.

## Certain fungal metabolites are immunosuppressants

More specific immunosuppression became possible with the use of cyclosporin A (CsA), FK506 (tacrolimus), and rapamycin (sirolimus), which are fungal metabolites with immunosuppressive properties. Although chemically unrelated, CsA and FK506 have similar actions. Both drugs block activation of resting T cells by inhibiting the transcription of genes encoding IL-2 and the high-affinity IL-2 receptor (IL-2Rα), which are essential for activation. CsA and FK506 exert this effect by binding to cytoplasmic proteins called immunophilins, forming a complex that blocks the phosphatase activity of calcineurin. This prevents the formation and nuclear translocation of the cytoplasmic subunit NFATc and its subsequent assembly into NFAT, a DNA-binding protein necessary for transcription of the genes encoding a number of molecules important to T-cell activation (see Figure 10-12). Rapamycin is structurally similar to FK506 and also binds to an immunophilin. However, the rapamycin-immunophilin complex does not inhibit calcineurin activity; instead, it blocks the proliferation and differentiation of activated $T_H$ cells in the $G_1$ phase of the cell cycle. All three drugs, by inhibiting $T_H$-cell proliferation and thus $T_H$-cell cytokine expression, reduce the subsequent activation of various effector populations involved in graft rejection, including $T_H$ cells, $T_C$ cells, NK cells, macrophages, and B cells.

The profound immunosuppressive properties of these three agents have made them a mainstay of heart, liver, kidney, and bone marrow transplantation. Cyclosporin A has been shown to prolong graft survival in kidney, liver, heart, and heart-lung transplants. In one study of 209 kidney transplants from cadaver donors, the 1-year survival rate was 64% among recipients receiving other immunosuppressive treatments and 80% among those receiving cyclosporin A. Similar results have been obtained with liver transplants (Figure 17-8). Despite these impressive results, CsA does have some negative side effects, the most notable of which is toxicity to the kidneys. Acute nephrotoxicity is quite common, in some cases progressing to chronic nephrotoxicity

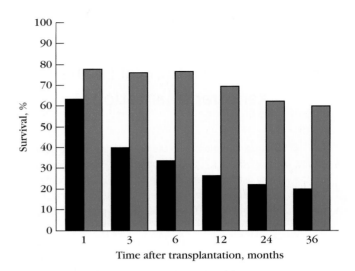

**FIGURE 17-8 Comparison of the survival rate of liver transplants in 84 patients who were immunosuppressed with azathioprine and corticosteroids (black) with the survival rate in 55 patients who were immunosuppressed with cyclosporin A and corticosteroids (blue).** [Adapted from S. M. Sabesin and J. W. Williams, 1987, Hospital Practice **15**(July):75.]

and drug-induced kidney failure. FK506 and rapamycin are 10 to 100 times more potent as immune suppressants than CsA and therefore can be administered at lower doses and with fewer side effects than CsA.

## Total lymphoid irradiation eliminates lymphocytes

Because lymphocytes are extremely sensitive to x-rays, x-irradiation can be used to eliminate them in the transplant recipient just before grafting. In total lymphoid x-irradiation, the recipient receives multiple x-ray exposures to the thymus, spleen, and lymph nodes before the transplant surgery. The typical protocol is daily x-irradiation treatments of about 200 rads per day for several weeks until a total of 3400 rads has been administered. The recipient is grafted in this immunosuppressed state. Because the bone marrow is not x-irradiated, lymphoid stem cells proliferate and renew the population of recirculating lymphocytes. These newly formed lymphocytes appear to be more tolerant to the antigens of the graft.

# Specific Immunosuppressive Therapy

In addition to harmful side effects peculiar to the various immunosuppressive treatments described above, a major limitation common to all is that they lack specificity, thus producing a more or less generalized immunosuppression and increasing the recipient's risk for infection. What is needed ideally is an antigen-specific immunosuppressant that reduces the immune response to the alloantigens of the graft while preserving the recipient's ability to respond to other foreign antigens. Although this goal has not yet been achieved in human transplants, recent successes in animal experiments indicate that it may be possible. Specific immunosuppression to allografts has been achieved in animal experiments using antibodies or soluble ligands reactive with cell surface molecules.

## Antibodies can suppress graft rejection responses

The use of antibodies directed against various surface molecules on cells of the immune system can successfully suppress T-cell activity in general, or the antibodies can target the activity of subpopulations of specific cells. Results from studies with animal models suggest further that certain monoclonals may be used to suppress only T cells that are activated. Successes with animal models and trials with humans give reason to believe that two types of strategies involving antibodies to suppress rejection are suitable for clinical use. Antibodies may be used to deplete the recipient of a certain broad or specific cell population; alternatively, they may be used to block costimulatory signals. In the latter case, a state of anergy is induced in those T cells that react to antigens present on the allograft.

Antithymocyte globulin prepared in animals has been used to deplete T cells in graft recipients prior to the transplant. An additional strategy to deplete immune cells involves use of a monoclonal antibody to the CD3 molecule of the TCR complex. Injection of such monoclonal antibodies results in a rapid depletion of mature T cells from the circulation. This depletion appears to be caused by binding of antibody-coated T cells to Fc receptors on phagocytic cells, which then phagocytose and clear the T cells from the circulation. In a further refinement of this strategy, a cytotoxic agent such as diphtheria toxin is coupled with the antibody. The cell with which the antibody reacts internalizes the toxin, causing its death. Another depletion strategy used to increase graft survival uses monoclonal antibodies specific for the high-affinity IL-2 receptor (CD25; the monoclonal antibody is anti-CD25). Since the high-affinity IL-2 receptor is expressed only on activated T cells, exposure to anti-CD25 after the graft specifically blocks proliferation of T cells activated in response to the alloantigens of the graft. More recent studies have used a monoclonal antibody directed against CD20 to deplete mature B cells, suppressing antibody-mediated rejection responses.

Monoclonal antibody therapy, which was initially employed to deplete T cells in graft recipients, also has been used to treat donors' bone marrow before it is transplanted. Such treatment is designed to deplete the immunocompetent T cells in the bone marrow transplant; these are the cells that react with the recipient tissues, causing graft-versus-host disease (described below). Monoclonal antibodies with isotypes that activate the complement system are most effective in all cell-depletion strategies.

The CD3 receptor and the high-affinity IL-2 receptor are targets present on all activated T cells; molecules present on particular T-cell subpopulations may also be targeted for immunosuppressive therapy. For example, a monoclonal antibody to CD4 has been shown to prolong graft survival. In one study, monkeys were given a single large dose of anti-CD4 just before they received a kidney transplant. Graft survival in the treated animals was markedly increased over that in untreated control animals. Interestingly, the anti-CD4 did not reduce the CD4$^+$ T-cell count but instead appeared to induce the T cells to enter an immunosuppressed state. This is an example of a nondepleting antibody.

Other targets for monoclonal antibody therapy are the cell surface adhesion molecules. Simultaneous treatment with monoclonal antibodies to the adhesion molecules ICAM-1 and LFA-1 for 6 days after transplantation has permitted indefinite survival of cardiac grafts between allogeneic mice. However, when either monoclonal antibody was administered alone, the cardiac transplant was rejected. The requirement that both monoclonal antibodies be given at the same time probably reflects redundancy of the adhesion molecules: LFA-1 is known to bind to ICAM-2 in addition to ICAM-1, and ICAM-1 is known to bind to Mac-1 (CD11b/CD18) and CD43 in addition to LFA-1. Only when all possible pairings among these adhesins are blocked at the same time is adhesion and signal transduction through these adhesins blocked.

A practical difficulty with using antibodies to prolong graft survival in humans is that they are generally of nonhuman origin. Early recipients of mouse monoclonals antibodies frequently developed an antibody response to the mouse antibody, rapidly clearing it from the body. This limitation has been overcome by the construction of human monoclonal antibodies and mouse-human chimeric antibodies (see Figure 5-24 and Clinical Focus in Chapter 5).

Because cytokines appear to play an important role in allograft rejection, another strategy for prolonging graft survival is to inject animals with monoclonal antibodies specific for the implicated cytokines, particularly TNF-α, IFN-γ, and IL-2. Monoclonal antibodies to TNF-α have been shown to prolong bone marrow transplants in mice and to reduce the incidence of graft-versus-host disease. Monoclonal antibodies to IFN-γ and to IL-2 have each been reported in some cases to prolong cardiac transplants in rats.

A number of new avenues to prevent rejection reactions are under consideration. A recent report by a consortium of investigators at Pfizer, Stanford, and the National Institutes of Health showed that an inhibitor of Janus kinase 3 (JAK3) promoted acceptance of kidney grafts in monkeys. This JAK3 inhibitor, designated CP-690550, can be administered orally and to date has revealed none of the deleterious side effects associated with the commonly used immunosuppressive drugs. In vitro experiments show that the inhibitor prevents IL-2 phosphorylation of JAK3 and Stat5. Because the

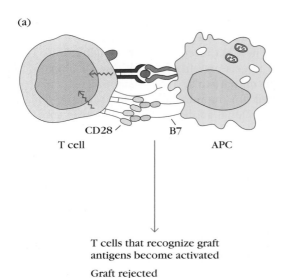

(a)

CD28    B7

T cell    APC

T cells that recognize graft
antigens become activated

Graft rejected

(b)

CTLA-4Ig

T cells that recognize graft antigens
lack costimulation and become anergic

Graft survives

**FIGURE 17-9 Blocking costimulatory signals at the time of transplantation can cause anergy instead of activation of the T cells reactive against the graft.** T-cell activation requires both the interaction of the TCR with its ligand and the reaction of costimulatory receptors with their ligands (a). In (b), contact between one of the costimulatory receptors, CD28 on the T cell, and its ligand, B7 on the APC, is blocked by reaction of B7 with the soluble ligand CTLA-4Ig. The CTLA-4 is coupled to an Ig H chain, which slows its clearance from the circulation. This process specifically suppresses graft rejection without inhibiting the immune response to other antigens.

compound has little effect on other related kinases, it is not predicted to inhibit general hematopoietic processes and should not cause disorders such as anemia, thrombocytopenia, or lymphopenia.

## Blocking costimulatory signals can induce anergy

As described in Chapter 10, $T_H$-cell activation requires a costimulatory signal in addition to the signal mediated by the T-cell receptor. The interaction between the B7 molecule on the membrane of antigen-presenting cells and the CD28 or CTLA-4 molecule on T cells provides one such signal (see Figure 10-14). Lacking a costimulatory signal, antigen-activated T cells become anergic (see Figure 10-15). CD28 is expressed on both resting and activated T cells and binds B7 with a moderate affinity; CTLA-4 is expressed at much lower levels and only on activated T cells but binds B7 with a 20-fold higher affinity. A second pair of costimulatory molecules required for T-cell activation are CD40, which is present on the APC, and CD40 ligand (CD154), which is present on the T cell.

D. J. Lenschow, J. A. Bluestone, and colleagues demonstrated that blocking the B7-mediated costimulatory signal with a soluble form of CTLA-4 after transplantation would cause the host's T cells directed against the grafted tissue to become anergic, thus enabling the tissue to survive. In their experiment, human pancreatic islets were transplanted into mice injected with CTLA-4Ig, a soluble fusion protein consisting of the extracellular domains of CTLA-4 and the constant region of the IgG1 heavy chain. Including the IgG1

heavy-chain constant region increases the half-life of the soluble fusion protein. The xenogeneic graft exhibited long-term survival in treated mice but was quickly rejected in untreated controls. The fact that the soluble form of the CTLA-4 receptor was able to block the rejection of the human tissue transplant in the recipient mice is evidence that blocking costimulatory signals in vivo is a viable strategy (Figure 17-9).

These exciting results were extended to transplantation of kidneys mismatched for class I and class II antigens in monkeys by Allan Kirk, David Harlan, and their colleagues. The recipients were treated for about 4 weeks after transplantation with either CTLA-4Ig, a monoclonal antibody directed against CD40L, or both in combination. Untreated control animals rejected the mismatched kidneys within 5 to 8 days; those treated with a single agent retained their grafts for 20 to 98 days. However, the animals given both reagents showed no evidence of rejection at 150 days after transplantation. This suppression of allograft rejection did not lead to a state of general immunosuppression; peripheral T-cell counts remained normal and other immune functions were present, including mixed lymphocyte reactivity between donor and recipients. If the procedures developed for monkeys can be successfully adapted to human use, these procedures could revolutionize clinical transplantation. The ability to block allograft rejection without general immunosuppression and without the deleterious side effects of suppressive drugs would enable recipients to lead normal lives.

The agents used in clinical transplantation are summarized in Figure 17-10, along with the sites of action for these drugs.

Transplant tissue

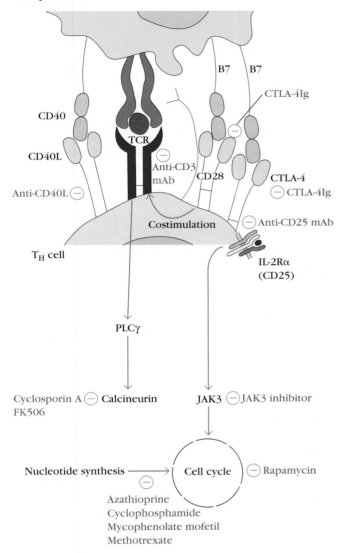

**FIGURE 17-10 Sites of action for various agents used in clinical transplantation.**

---

# Immune Tolerance to Allografts

In some instances an allograft may be accepted without the use of immunosuppressive measures. Obviously, with tissues that lack alloantigens, such as cartilage or heart valves, no immunologic barrier to transplantation exists. Sometimes, however, the strong predicted response to an allograft does not occur. An allograft may be accepted in one of two ways: when cells or tissue are grafted to a so-called privileged site that is sequestered from immune system surveillance or when a state of tolerance has been induced biologically, usually by previous exposure to the antigens of the donor in a manner that causes immune tolerance rather than sensitization in the recipient. Each of these exceptions is considered below.

## Privileged sites accept antigenic mismatches

In immunologically privileged sites, an allograft can be placed without engendering a rejection reaction. These sites include the anterior chamber of the eye, cornea, uterus, testes, and brain. The cheek pouch of the Syrian hamster is a privileged site used in experiments. Each of these sites is characterized by an absence of lymphatic vessels and in some cases by an absence of blood vessels as well. Consequently, the alloantigens of the graft are not able to sensitize the recipient's lymphocytes, and the graft has an increased likelihood of acceptance even when HLA antigens are not matched.

The privileged status of the cornea has allowed corneal transplants to be highly successful. The brain is an immunologically privileged site because the blood-brain barrier prevents the entry or exit of many molecules, including antibodies. The successful transplantation of allogeneic pancreatic islet cells into the thymus in a rat model of diabetes suggests that the thymus may also be an immunologically privileged site.

Immunologically privileged sites fail to induce an immune response because they are effectively sequestered from the cells of the immune system. This suggests the possibility of physically sequestering grafted cells. In one study, pancreatic islet cells were encapsulated in semipermeable membranes (fabricated from an acrylic copolymer) and then transplanted into diabetic mice. The islet cells survived and produced insulin. The transplanted cells were not rejected, because the recipient's immune cells could not penetrate the membrane. This novel transplant method enabled the diabetic mice to produce normal levels of insulin and may have application for treatment of human diabetics.

## Early exposure to alloantigens can induce specific tolerance

In 1945, Ray Owen reported that nonidentical twins in cattle retained the ability to accept cells or tissue from the genetically distinct sibling throughout their lives, unlike nonidentical twins of other mammalian species. A shared placenta in cattle allows free circulation of cells from one twin to the other throughout the embryonic period. Although the twins may have inherited distinct paternal and maternal antigens, they do not recognize those of their placental partner as foreign and can accept grafts from them.

Experimental support for the notion that tolerance comes from exposure of the developing organism to alloantigens came from mouse experiments. If neonates of mouse strain A are injected with cells from strain C, they will accept grafts from C strain as adults. Immunocompetence of the injected A strain mice and specificity of the tolerance is shown by the fact that they reject grafts from other strains as rapidly as their untreated littermates. Although no human experimental data demonstrate such specific tolerance, anecdotal data suggests that it may operate in humans as well.

There are examples in which allografts mismatched at a single HLA locus are accepted with little or no immune suppression. In cases where the mismatched antigen is expressed by the mother, but not inherited by the offspring, there is the possibility that perinatal exposure induced subsequent tolerance to this antigen. Because human maternal cells do not normally cross the placental barrier, such specific tolerance to noninherited maternal antigens would be an exception rather than a commonplace event.

Methods to exploit tolerance processes to allow acceptance of allografts have been studied in animal models, but there has been little application to human transplantation. Whether introduction of donor bone marrow to induce a state of mixed chimerism or manipulation of $T_{reg}$ cells to promote specific tolerance will be feasible in human trans-

plantation is difficult to predict but a focus of considerable contemporary research.

## Clinical Transplantation

For a number of illnesses, a transplant is the only means of therapy. Figure 17-11 summarizes the major organ and cell transplants being performed at the present time. In addition, certain combinations of organs, such as heart and lung or kidney and pancreas, are being transplanted simultaneously with increasing frequency. Since the first kidney transplant

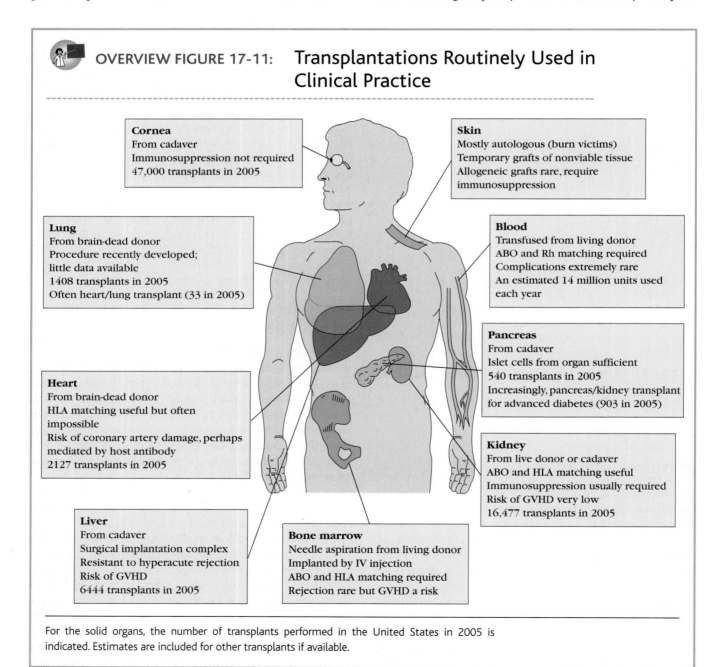

**OVERVIEW FIGURE 17-11:    Transplantations Routinely Used in Clinical Practice**

**Cornea**
From cadaver
Immunosuppression not required
47,000 transplants in 2005

**Skin**
Mostly autologous (burn victims)
Temporary grafts of nonviable tissue
Allogeneic grafts rare, require immunosuppression

**Lung**
From brain-dead donor
Procedure recently developed; little data available
1408 transplants in 2005
Often heart/lung transplant (33 in 2005)

**Blood**
Transfused from living donor
ABO and Rh matching required
Complications extremely rare
An estimated 14 million units used each year

**Pancreas**
From cadaver
Islet cells from organ sufficient
540 transplants in 2005
Increasingly, pancreas/kidney transplant for advanced diabetes (903 in 2005)

**Heart**
From brain-dead donor
HLA matching useful but often impossible
Risk of coronary artery damage, perhaps mediated by host antibody
2127 transplants in 2005

**Kidney**
From live donor or cadaver
ABO and HLA matching useful
Immunosuppression usually required
Risk of GVHD very low
16,477 transplants in 2005

**Liver**
From cadaver
Surgical implantation complex
Resistant to hyperacute rejection
Risk of GVHD
6444 transplants in 2005

**Bone marrow**
Needle aspiration from living donor
Implanted by IV injection
ABO and HLA matching required
Rejection rare but GVHD a risk

For the solid organs, the number of transplants performed in the United States in 2005 is indicated. Estimates are included for other transplants if available.

was performed in the 1950s, approximately 470,000 kidneys have been transplanted worldwide. The next most frequently transplanted solid organ is the liver (78,000), followed by the heart (55,000) and, more distantly, by the lung (10,000) and pancreas (3500). It is estimated that at the end of 2005 over 150,000 people were living in the United States with solid organ transplants. Although the clinical results of transplantation of various cells, tissues, and organs in humans have improved considerably in the past few years, major obstacles to the use of this treatment exist. As explained above, the use of immunosuppressive drugs greatly increases the short-term survival of the transplant, but medical problems arise from use of these drugs, and chronic rejection eventually prevails in most cases. The need for additional transplants after rejection exacerbates the shortage of organs that is a major obstacle to the widespread use of transplantation. Several of the organ systems for which transplantation is a common treatment are considered below. The frequency with which a given organ or tissue is transplanted depends on a number of factors:

- Clinical situations in which transplantation is indicated

- Availability of tissue or organs

- Difficulty of performing transplantation and caring for post-transplantation patients

- Specific factors that aid or hinder acceptance of the particular transplant

The urgency of the transplantation may depend on the affected organ. In the case of the heart, lung, and liver, few alternative procedures can keep the patient alive when these organs cease to function. Although dialysis is routinely used to maintain a patient awaiting a kidney transplant, this involves a difficult and strict regimen, and a number of patients voluntarily withdraw from treatment. Research on artificial organs continues, but there are no reports of universal long-term successes.

## The most commonly transplanted organ is the kidney

As mentioned above, the most commonly transplanted organ is the kidney; in 2005, there were 16,477 kidney transplants performed in the United States. Major factors contributing to this number are the numerous clinical indications for kidney transplantation. Many common diseases, such as diabetes and various types of nephritis, result in kidney failure that can be alleviated by transplantation. With respect to availability, kidneys can be obtained not only from cadavers but also from living relatives or volunteers, because it is possible to donate a kidney and live a normal life with the remaining kidney. In 2005, 6562 of the 16,477 kidneys transplanted in the United States came from living donors. Surgical procedures for transplantation are straightforward; technically, the kidney is simpler to re-implant than the liver or heart. Because many kidney transplants have

been done, patient-care procedures and immunosuppressive measures have been worked out in detail. Matching of blood and histocompatibility groups is advantageous in kidney transplantation because the organ is heavily vascularized, but the kidney presents no special problems that promote rejection or graft-versus-host disease (GVHD), as the bone marrow or liver do (discussed below).

Two major problems faced by patients waiting for a kidney are the short supply of available organs and the increasing number of sensitized recipients. The latter problem stems from rejection of a first transplant, which then sensitizes the individual and leads to the formation of antibodies and activation of cellular mechanisms directed against kidney antigens. Any subsequent graft containing antigens in common with the first would be quickly rejected. Therefore, detailed tissue typing procedures must be used to ascertain that the patient has no antibodies or active cellular mechanisms directed against the potential donor's kidney. In many cases, patients can never again find a match after one or two rejection episodes. It is almost always necessary to maintain kidney transplant patients on some form of immunosuppression, usually for their entire lives. Unfortunately, this gives rise to complications, including risks of cancer and infection as well as other side effects such as hypertension and metabolic bone disease.

## Bone marrow transplants are used for leukemia, anemia, and immunodeficiency

After the kidney, bone marrow is the most frequent transplant. Since the early 1980s, bone marrow transplantation has been increasingly adopted as a therapy for a number of malignant and nonmalignant hematologic diseases, including leukemia, lymphoma, aplastic anemia, thalassemia major, and immunodeficiency diseases, especially severe combined immunodeficiency, or SCID (see Chapter 20). The bone marrow, which is obtained from a living donor by multiple needle aspirations, consists of erythroid, myeloid, monocytoid, megakaryocytic, and lymphocytic lineages. The graft, usually about $10^9$ cells per kilogram of host body weight, is injected intravenously into the recipient. The first successful bone marrow transplantations were performed between identical twins. However, development of the tissue-typing procedures described earlier now makes it possible to identify allogeneic donors who have HLA antigens identical or near identical to those of the recipients. Although the supply of bone marrow for transplantation is not a problem, finding a matched donor may present an obstacle.

In the usual procedure, the recipient of a bone marrow transplant is immunologically suppressed before grafting. Leukemia patients, for example, are often treated with cyclophosphamide and total-body irradiation to kill all cancerous cells. The immune-suppressed state of the recipient makes graft rejection rare; however, because the donor bone marrow contains immunocompetent cells, the graft may reject the host, causing **graft-versus-host disease (GVHD).** GVHD affects 50% to 70% of bone marrow transplant patients; it develops as donor T cells recognize alloantigens on the host cells. The activation and proliferation of these T cells and the subsequent production of cytokines generate inflammatory reactions in the skin, gastrointestinal tract, and liver. In severe cases, GVHD can result in generalized erythroderma of the skin, gastrointestinal hemorrhage, and liver failure.

Various treatments are used to prevent GVHD in bone marrow transplantation. The transplant recipient is usually placed on a regimen of immunosuppressive drugs, often including cyclosporin A and methotrexate, in order to inhibit the immune responses of the donor cells. In another approach, the donor bone marrow is treated with anti-T-cell antisera or monoclonal antibodies specific for T cells before transplantation, thereby depleting the offending T cells. Complete T-cell depletion from donor bone marrow, however, increases the likelihood that the marrow will be rejected, and so the usual procedure now is a partial T-cell depletion. Apparently, a low level of donor T-cell activity, which results in a low-level GVHD, is actually beneficial because the donor cells kill any host T cells that survive the immunosuppression treatment. This prevents residual recipient cells from becoming sensitized and causing rejection of the graft. In leukemia patients, low-level GVHD also seems to result in destruction of host leukemic cells, thus making it less likely for the leukemia to recur.

## Heart transplantation is a challenging operation

Perhaps the most dramatic form of transplantation is that of the heart; once the damaged heart has been removed, the patient must be kept alive by wholly artificial means until the transplanted heart is in place and beating. Heart-lung machines are available to circulate and aerate the patient's blood after the heart is removed. The donor's heart must be maintained in such a manner that it will begin beating when it is placed in the recipient. It has been found that a human heart can be kept viable for a limited period in ice-cold buffer solutions that effectively short circuit the electric impulses that control the rhythmic beating, which could damage the isolated organ. The surgical methods for implanting a heart have been available for a number of years. The first heart transplant was carried out in South Africa by Dr. Christian Barnard, in 1964. Since then, the 1-year survival rate for transplantation of the heart has become greater than 80%. In 2005, 2127 heart transplants were performed in the United States, and about 3000 people are on waiting lists to receive one. An issue peculiar to heart transplantation has been a new type of atherosclerotic disease in the coronary arteries of the implanted organ. There is some possibility that host antibodies mediate injury to the vessels in the donated heart.

Although a heart transplant may greatly benefit patients with various types of heart disease or damage, obviously the number of available hearts is strictly limited. Accident victims who are declared brain dead but have an intact circulatory system and a functioning heart are the normal source of

these organs. HLA matching is desirable but not often possible, because of the limited supply of hearts and the urgency of the procedure.

## Lung transplants are on the increase

In recent years, lung transplantation, either by itself or in conjunction with heart transplantation, has been used to treat diseases such as cystic fibrosis and emphysema or acute damage to the lungs such as that caused by smoke inhalation. In 2005, 1408 lung and 33 heart-lung transplants were performed. First-year survival rate for lung transplants is reported at about 60%.

## Liver transplants treat congenital defects and damage from viral or chemical agents

The liver is a large organ that performs a number of functions related to clearance and detoxification of chemical and biological substances. Liver malfunction can be caused by damage to the organ from viral diseases such as hepatitis or by exposure to harmful chemicals, as in chronic alcoholism. Damage to the liver may correct itself when the damaged tissue regenerates after the causative injurious agent is cleared. If the liver tissue does not regenerate, damage may be fatal. The majority of liver transplants are used as a therapy for congenital abnormalities of the liver. Because the liver is large and has a complicated circulation, re-implantation of the liver initially posed a technical problem. Techniques have been developed to overcome this major surgical challenge, and the recent 1-year survival rate has risen to approximately 65%. In 2005, 6444 livers were transplanted in the United States. Increasingly, a liver from a single donor may be split and given to two recipients; normally, a child will receive the smaller portion and an adult the larger. Starting in 1998, the number of live donors giving a fraction of their liver to a patient in need (usually a close relative) climbed from less than a hundred to a few hundred per year.

The immunology of liver transplantation is interesting because the organ appears to resist rejection by hyperacute antibody-mediated mechanisms. It has been shown that even transplantation across blood-group barriers, which would be expected to trigger hyperacute rejection, can be successful in the short term. However, leukocytes within the donor organ together with anti-blood-group antibodies can mediate antibody-dependent hemolysis of recipient red blood cells if there is a mismatch of the blood groups. In addition, manifestations of GVHD have occurred in liver transplants even when donor and recipient are blood-group compatible. These reactions are obviously caused by donor lymphocytes carried by the transplanted liver.

## Transplantation of pancreatic cells offers a cure for diabetes mellitus

One of the more common diseases in the United States is diabetes mellitus. This disease is caused by malfunction of insulin-producing islet cells in the pancreas. Transplantation

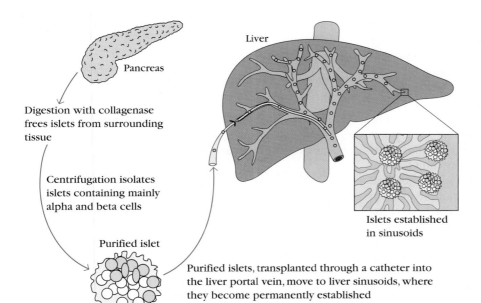

**FIGURE 17-12 Procedures used to harvest and implant pancreatic islet cells.** The pancreas is digested with collagenase to free the islets from surrounding tissue. The islets are then purified by gradient centrifugation and infused through a catheter into the portal vein of the liver, where they home to the liver sinusoids.

Pancreas

Digestion with collagenase frees islets from surrounding tissue

Centrifugation isolates islets containing mainly alpha and beta cells

Purified islet

Liver

Islets established in sinusoids

Purified islets, transplanted through a catheter into the liver portal vein, move to liver sinusoids, where they become permanently established

of a pancreas could provide the appropriately regulated levels of insulin necessary to make the diabetic individual normal. Recently, 1-year success rates for pancreas transplantation of about 55% have been reported. Transplantation of the complete pancreas is not necessary to restore the function needed to produce insulin in a controlled fashion; transplantation of the islet cells alone could restore function.

A multicenter study was recently launched to devise procedures for islet cell isolation and transplantation in order to treat insulin-dependent diabetes. The general procedure, shown in Figure 17-12, involves harvesting islet cells and perfusing them into the liver, where they become permanently established in the liver sinusoids. Initial results indicate that 53% of recipients are insulin-independent after transplant, some for periods up to 2 years. About 17% of the recipients withdrew or were withdrawn from the study, and the remainder of participants still require some exogenous insulin; these individuals are candidates for additional transplants. Several factors favor survival of functioning pancreatic cells, the most important being the condition of the islet cells used for implantation.

Kidney failure is a frequent complication of advanced diabetes, occurring in about 30% of diabetics, indicating that simultaneous kidney and pancreas transplants may be required. In 2005, there were 540 pancreas transplants and 903 simultaneous kidney-pancreas transplants. Whether it is better to carry out simultaneous kidney-pancreas transplants or to transplant separately remains an issue to be resolved on a case-by-case basis. However, the value of a kidney transplant for diabetics is reflected in the fact that the mean survival time for a type 1 diabetic on dialysis is 3.5 years compared to a 72% 8-year survival rate for recipients of a live kidney transplant.

## Skin grafts are used to treat burn victims

Most skin transplantation in humans is done with autologous tissue. However, in cases of severe burn, grafts of for-

eign skin thawed from frozen deposits in tissue banks may be used. These grafts generally act as biologic dressings, because the cellular elements are no longer viable and the graft does not grow in the new host; the grafts are left in place for several days but are regularly replaced. True allogeneic skin grafting using fresh viable donor skin has been undertaken in some cases, but rejection must be prevented by the use of immunosuppressive therapy. This is not desirable because a major problem with burn victims is the high risk of infection, and immunosuppressive therapy accentuates this risk.

The above list of common transplants is by no means all-inclusive and is expected to grow in future years. For example, intracerebral neural cell grafts have restored functionality in victims of Parkinson's disease. In studies conducted thus far, the source of neural donor cells was human embryos; the possibility of using those from other animal species is being tested. The use of cord-blood cells for leukemia patients is advocated, and repositories for storing umbilical cord blood are increasing. Advantages include less stringent matching criteria than for bone marrow transplants along with ease of acquisition from a repository. A disadvantage for adult recipients is the relatively low number of hematopoietic progenitor cells in cord blood.

## Xenotransplantation may be the answer to the shortage of donor organs

Although the immune system represents a formidable barrier to the use of transplantation, significant progress has been made in overcoming this obstacle. However, there has not been comparable progress in solving the complex problem of finding organs for those who need them. The insufficient supply of available organs means that a significant percentage of patients die while waiting for a transplant. There is about a 6% mortality rate for those awaiting a kidney and 14% for patients in need of a heart. The need for an alternative source of donor organs has focused attention on xenotransplantation. The larger nonhuman primates (chimpanzees and baboons) have served as the main transplant donors, and, as discussed in the Clinical Focus, genetically altered pigs may be raised as a source of organs for human use.

The earliest transplants of chimpanzee kidneys into humans date to 1964. Since that time, sporadic attempts at kidney, heart, liver, and bone marrow transplantation from primates into humans have been made. No attempt has met with great success, but several have received some attention. In 1993, T. E. Starzl performed two liver transplants from baboons into patients suffering from liver failure. Both patients died, one after 26 days and the other after 70 days. In 1994, a pig liver was transplanted into a 26-year-old suffering from acute hepatic failure. The liver functioned only 30 hours before it was rejected by a hyperacute rejection reaction. In 1995, baboon bone marrow was infused into an HIV-infected man with the aim of boosting his weakened immune system with the baboon immune cells, which do not become infected with the virus. Although there were no complications from the transplant, the baboon bone marrow did not appear to establish itself in the recipient.

A major problem with xenotransplants is that immune rejection is often quite vigorous, even when recipients are treated with potent immunosuppressive drugs such as FK506 or rapamycin. The major response involves the action of humoral antibody and complement, leading to the development of a hyperacute rejection reaction. In addition to the problem of rejection, there is general concern that xenotransplantation has the potential of spreading pathogens from the donor to the recipient. These pathogens could potentially cause diseases, called xenozoonoses, that are fatal for humans. For example, certain viruses, including close relatives of HIV-1, found in chimpanzees, and HIV-2 and herpesvirus B, which occur in several primate species, cause limited pathogenesis in their primate hosts but can lead to deadly infections in humans. In addition, there is the fear that primate retroviruses (see Chapter 20), such as SIV, may recombine with human variants to produce new agents of disease. The possibility of introducing new viruses into humans may be greater for transplants from closely related species, such as primates, and less in the case of more distantly related species, such as pigs, because viruses are less likely to replicate in cells from unrelated species.

------------

## SUMMARY

- Graft rejection is an immunologic response displaying the attributes of specificity, memory, and self-nonself recognition. There are three major types of rejection reactions:

  - Hyperacute rejection, mediated by preexisting host antibodies to graft antigens
  - Acute graft rejection, in which $T_H$ cells and/or CTLs mediate tissue damage

  - Chronic rejection, which involves both cellular and humoral immune components

- The immune response to tissue antigens encoded within the major histocompatibility complex is the strongest force in rejection.

- The match between a recipient and potential graft donors is assessed by typing blood-group antigens and MHC class I and class II tissue antigens.

- The process of graft rejection can be divided into a sensitization stage, in which T cells are stimulated, and an effector stage, in which they attack the graft.

- In most clinical situations, graft rejection is suppressed by nonspecific immunosuppressive agents or by total lymphoid x-irradiation.

- Experimental approaches using monoclonal antibodies offer the possibility of specific immunosuppression. These antibodies may act by

  - Deleting populations of reactive cells
  - Inhibiting costimulatory signals leading to anergy in specifically reactive cells

- Certain sites in the body, including the cornea of the eye, brain, testes, and uterus, do not reject transplants despite genetic mismatch between donor and recipient.

- Specific tolerance to alloantigens is induced by exposure to them in utero or by injection of neonates.

- A major complication in bone marrow transplantation is graft-versus-host reaction, mediated by the lymphocytes contained within the donor marrow.

- The critical shortage of organs available for transplantation may be solved in the future by using organs from nonhuman species (xenotransplants).

## References

Adams, D. H. 2000. Cardiac xenotransplantation: clinical experience and future direction. *Annals of Thoracic Surgery* **70**:320.

Auchincloss, H., M. Sykes, and D. H. Sachs. 1999. Transplantation immunology. In *Fundamental Immunology,* 4th ed. W. E. Paul, ed. Lippincott-Raven, Philadelphia, p. 1175.

Claas, F. H. J., et al. 2004. A critical appraisal of HLA matching in today's renal transplantation. *Transplantation Reviews* **18**:96.

Fox, A., and L. C. Harrison. 2000. Innate immunity and graft rejection. *Immunological Reviews* **173**:141.

Halloran, P. F. 2004. Immunosuppressive drugs for kidney transplantation. *New England Journal of Medicine* **351**:26.

Harlan, D. M., and A. D. Kirk. 1999. The future of organ and tissue transplantation: can T-cell co-stimulatory pathway modifiers revolutionize the prevention of graft rejection? *Journal of the American Medical Association* **282**:1076.

Kuwaki, K., et al. 2005. Heart transplantation in baboons using alpha 1,3 galactosyltransferase gene-knockout pigs as donors: initial experience. *Nature Medicine* **11:**29.

Lenschow, D. J., et al. 1992. Long-term survival of xenogeneic pancreatic islets induced by CTLA4-Ig. *Science* **257:**789.

Obata, F., et al. 2005. Contribution of CD4$^+$ and CD8$^+$ T cells and interferon-gamma to the progress of chronic rejection of kidney allografts: the Th1 response mediates both acute and chronic rejection. *Transplant Immunology* **14:**21.

Ricordi, C., and T. B. Strom. 2004. Clinical islet transplantation: advances and immunological challenges. *Nature Reviews Immunology* **4:**259.

Sayegh, M. H., and C. B. Carpenter. 2004. Transplantation 50 years later—progress, challenges, and promises. *New England Journal of Medicine* **351:**26.

Waldmann, H., and S. Cobbold. 2004. Exploiting tolerance processes in transplantation. *Science* **305:**209.

Walsh, P. T., T. B. Strom, and L. A. Turka. 2004. Routes to transplant tolerance versus rejection: the role of cytokines. *Immunity* **20:**121.

 ## Useful Web Sites

**http://www.transweb.org**

Links to hundreds of sites giving information on all aspects of organ transplantation.

**http//www.unos.org**

United Network for Organ Sharing site has information concerning solid-organ transplantation for patients, families, doctors, and teachers.

**http://www.marrow.org**

The National Marrow Donor Program Web site contains information about all aspects of bone marrow transplantation.

 ## Study Questions

**CLINICAL FOCUS QUESTION** What features would be desirable in an ideal animal donor for xenotransplantation? How would you test your model prior to doing clinical trials in humans?

1. Indicate whether each of the following statements is true or false. If you think a statement is false, explain why.

   a. Acute rejection is mediated by preexisting host antibodies specific for antigens on the grafted tissue.

   b. Second-set rejection is a manifestation of immunologic memory.

   c. Host dendritic cells can migrate into grafted tissue and act as antigen-presenting cells.

   d. All allografts between individuals with identical HLA haplotypes will be accepted.

   e. Cytokines produced by host $T_H$ cells activated in response to alloantigens play a major role in graft rejection.

2. You are a surgeon in a major transplantation facility. One of your patients is awaiting a kidney transplant. Several cadaver organs have become available due to a serious accident. Your technician performs a microcytotoxicity test on the potential donors and gets the following results. Your patient was tested previously and was positive for the antibodies used in wells 2 and 3, and negative for the antibodies used in wells 1 and 4 for each of the HLA determinants.

|  | HLA-A | | | | HLA-B | | | | HLA-DR | | | |
|---|---|---|---|---|---|---|---|---|---|---|---|---|
|  | 1 | 2 | 3 | 4 | 1 | 2 | 3 | 4 | 1 | 2 | 3 | 4 |
| Donor 1 | ○ | ● | ○ | ● | ○ | ● | ● | ○ | ○ | ● | ● | ○ |
| Donor 2 | ○ | ● | ● | ○ | ○ | ● | ○ | ○ | ○ | ● | ○ | ○ |
| Donor 3 | ○ | ● | ● | ○ | ○ | ● | ○ | ○ | ○ | ● | ○ | ○ |
| Donor 4 | ● | ○ | ○ | ● | ● | ● | ○ | ○ | ● | ○ | ○ | ● |

   a. Which donor would be your first choice?

   b. If that organ were not usable, which one of the remaining organs would you consider using?

   c. What is the scientific basis for your decision?

   d. What test can your technician perform to confirm that the potential donor is compatible with the recipient?

3. Indicate whether a skin graft from each donor to each recipient listed in the table below would result in rejection (R) or acceptance (A). If you believe a rejection reaction would occur, indicate whether it would be a first-set rejection (FSR), occurring in 12 to 14 days, or a second-set rejection (SSR), occurring in 5 to 6 days. All the mouse strains listed have different H-2 haplotypes.

| Donor | Recipient |
|---|---|
| BALB/c | C3H |
| BALB/c | Rat |
| BALB/c | Nude mouse |
| BALB/c | C3H, had previous BALB/c graft |
| BALB/c | C3H, had previous C57BL/6 graft |
| BALB/c | BALB/c |
| BALB/c | (BALB/c × C3H)F$_1$ |
| BALB/c | (C3H × C57BL/6)F$_1$ |
| (BALB/c × C3H)F$_1$ | BALB/c |
| (BALB/c × C3H)F$_1$ | BALB/c, had previous F$_1$ graft |

4. Graft-versus-host disease (GVHD) frequently develops after certain types of transplantations.

   a. Briefly outline the mechanisms involved in GVHD.

   b. Under what conditions is GVHD likely to occur?

   c. Some researchers have found that GVHD can be diminished by prior treatment of the graft with monoclonal antibody plus complement or with monoclonal antibody conjugated with toxins. List at least two cell surface antigens to which monoclonal antibodies could be prepared and used for this purpose, and give the rationale for your choices.

5. A child who requires a kidney transplant has been offered a kidney from both parents and from five siblings.

   a. Cells from the potential donors are screened with monoclonal antibodies to the HLA-A, -B, and -C antigens in a microcytotoxicity assay. In addition, ABO blood-group typing is performed. Based on the results in the table be-

low, a kidney graft from which donor(s) is most likely to survive?

   b. Now a one-way MLR is performed using various combinations of mitomycin-treated lymphocytes. The results, expressed as counts per minute of [³H]thymidine incorporated, are shown in the table below; the stimulation index (ratio of the experimental value to the control in which identical leukocytes are mixed) is listed below in parentheses. Based on these data, a graft from which donor(s) is most likely to be accepted?

6. What is the biologic basis for attempting to use soluble CTLA-4 or anti-CD40L to block allograft rejection? Why might this be better than treating a graft recipient with CsA or FK506?

7. Immediately after transplantation, a patient is often given extra strong doses of anti-rejection drugs and then allowed to taper off as time passes. Describe the effects of the commonly used anti-rejection drugs azathioprine, cyclosporine A, FK506, and rapamycin. Why is it possible to decrease the use of some of these drugs at some point after transplantation?

## Interactive Study

www.whfreeman.com/kuby

SELF-TEST
Review OF Key Terms

| | ABO type | HLA-A type | HLA-B type | HLA-C type |
|---|---|---|---|---|
| **Recipient** | O | A1/A2 | B8/B12 | Cw3 |
| Mother | A | A1/A2 | B8/B12 | Cw1/Cw3 |
| Father | O | A2 | B12/B15 | Cw3 |
| Sibling A | O | A1/A2 | B8/B15 | Cw3 |
| Sibling B | O | A2 | B12 | Cw1/Cw3 |
| Sibling C | O | A1/A2 | B8/B12 | Cw3 |
| Sibling D | A | A1/A2 | B8/B12 | Cw3 |
| Sibling E | O | A1/A2 | B8/B15 | Cw3 |

**For use with Question 5b**

| Respondent cells | Mytomycin C–treated stimulator cells | | | | | |
|---|---|---|---|---|---|---|
| | Patient | Sibling A | Sibling B | Sibling C | Sibling D | Sibling E |
| Patient | 1,672 (1.0) | 1,800 (1.1) | 13,479 (8.1) | 5,210 (3.1) | 13,927 (8.3) | 13,808 (8.3) |
| Sibling A | 1,495 (1.6) | 933 (1.0) | 11,606 (12.4) | 8,443 (9.1) | 11,708 (12.6) | 13,430 (14.4) |
| Sibling B | 25,418 (9.9) | 26,209 (10.2) | 2,570 (1.0) | 13,170 (5.1) | 19,722 (7.7) | 4,510 (1.8) |
| Sibling C | 10,722 (6.2) | 10,714 (5.9) | 13,032 (7.5) | 1,731 (1.0) | 1,740 (1.0) | 14,365 (8.3) |
| Sibling D | 15,988 (5.1) | 13,492 (4.2) | 18,519 (5.9) | 3,300 (1.1) | 3,151 (1.0) | 18,334 (5.9) |
| Sibling E | 5,777 (6.5) | 8,053 (9.1) | 2,024 (2.3) | 6,895 (7.8) | 10,720 (12.1) | 888 (1.0) |

# chapter 18

## Immune Response to Infectious Diseases

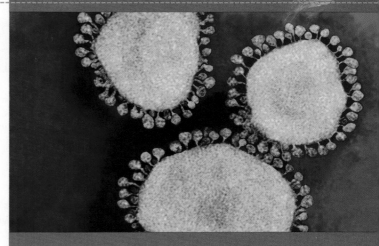

*SARS coronavirus. [Dr. Linda Stannard, UCT/Photo Researchers.]*

- Viral Infections
- Bacterial Infections
- Parasitic Diseases
- Fungal Diseases
- Emerging Infectious Diseases

I F A PATHOGEN IS TO ESTABLISH AN INFECTION IN A susceptible host, a series of coordinated events must circumvent both innate and adaptive immunity. One of the first and most important features of host innate immunity is the barrier provided by the epithelial surfaces of the skin and the lining of the gut. The difficulty of penetrating these epithelial barriers ensures that most pathogens never gain productive entry into the host. In addition to providing a physical barrier to infection, the epithelia produce chemicals that are useful in preventing infection. The secretion of gastric enzymes by specialized epithelial cells lowers the pH of the stomach and upper gastrointestinal tract, and other specialized cells in the gut produce antibacterial peptides.

A major feature of innate immunity is the presence of the normal flora in the gastrointestinal, genitourinary, and respiratory tracts. These so-called commensal organisms competitively inhibit the binding of pathogens to host cells. Innate responses can also block the establishment of infection. For example, the cell walls of some gram-positive bacteria contain a peptidoglycan that activates the alternative complement pathway, leading to opsonization and phagocytosis or possibly lysis (see Chapter 7). Some bacteria produce endotoxins such as LPS, which stimulate the production of cytokines such as TNF-α, IL-1, and IL-6 by macrophages or endothelial cells. These cytokines can activate macrophages. Phagocytosis of bacteria by neutrophils and other phagocytic cells is another highly effective line of innate defense. However, some types of bacteria that commonly grow intracellularly have developed mechanisms that allow them to resist degradation within the phagocyte.

Viruses commonly stimulate innate responses. In particular, many viruses induce the production of interferons, which can inhibit viral replication by inducing an antiviral response. Viruses are also controlled by NK cells, which frequently form the first line of defense against viral infections (see Chapter 14).

Generally, pathogens use a variety of strategies to escape destruction by the adaptive immune system. Many pathogens reduce their own antigenicity either by growing within host cells, where they are sequestered from immune attack, or by shedding their membrane antigens. Other pathogens camouflage themselves by mimicking the surfaces of host cells, either by expressing molecules with amino acid sequences similar to those of host cell membrane molecules or by acquiring a covering of host membrane molecules. Some pathogens are able to suppress the immune response selectively or to regulate it so that a branch of the immune system is activated that is ineffective against the pathogen. Continual variation in surface antigens is another strategy that enables a pathogen to elude the immune system. This antigenic variation may be due to the gradual accumulation of mutations, or it may involve an abrupt change in surface antigens.

Although both innate and adaptive immune responses to pathogens provide critical defense, infectious diseases still cause the death of millions each year. Widespread use of vaccines and drug therapy has drastically reduced mortality from infectious diseases in developed countries, but such diseases continue to be the leading cause of death in the Third World. Worldwide, an estimated 15 million deaths

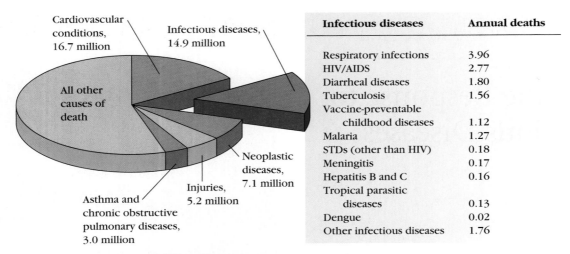

| Infectious diseases | Annual deaths |
|---|---|
| Respiratory infections | 3.96 |
| HIV/AIDS | 2.77 |
| Diarrheal diseases | 1.80 |
| Tuberculosis | 1.56 |
| Vaccine-preventable childhood diseases | 1.12 |
| Malaria | 1.27 |
| STDs (other than HIV) | 0.18 |
| Meningitis | 0.17 |
| Hepatitis B and C | 0.16 |
| Tropical parasitic diseases | 0.13 |
| Dengue | 0.02 |
| Other infectious diseases | 1.76 |

**FIGURE 18-1 Infectious diseases are among the leading causes of death worldwide.** The 14.9 million deaths attributable directly to infections are broken down by category in the table.

every year are the direct result of infectious diseases (Figure 18-1). In comparison, there are about 7.1 million cancer-related deaths and about 16.7 million deaths resulting from cardiovascular dysfunction. Adding to the infectious disease burden most heavily borne by the developing world, certain diseases are beginning to emerge or re-emerge in developed countries. New diseases such as the severe acute respiratory syndrome caused by the SARS coronavirus can spread rapidly as a result of international mobility. Influenza strains prevalent in birds adapt to cause human infection with a high rate of mortality because humans have developed little immunity to these strains. Previously rare infections by certain bacteria or fungi are increasing because of the rise in the numbers of people with impaired immunity caused by HIV infection and by the use of immunosuppressive drugs given for various purposes. Re-emerging diseases such as those caused by drug-resistant strains of *Staphylococcus aureus* and *Mycobacterium tuberculosis* are spreading at an alarming rate in developing as well as in industrialized countries. In some instances, a common bacterium may be associated with new disease, such as the necrotizing fasciitis caused by the so-called flesh-eating strain of *Streptococcus pyogenes,* an organism most commonly associated with the now rare disease scarlet fever.

We rely on our immune systems to give us protection against the common and ubiquitous pathogens as well as those emerging in the future. The survival of the human race to date indicates that this is a realistic expectation, but the large number of pathogens with rapidly evolving genetics may allow them to establish a foothold in our species. The recent history of HIV/AIDS (see Chapter 20) demonstrates the threat that infectious agents pose to public health. Pathogens sufficiently nimble to subvert immunity can cause great suffering. Although the immune system protects us, pathogenesis in a number of diseases results not from a direct assault on cells or tissues by the pathogens and their

toxins but from damage caused by the host's immune response. Such damage may result at the time of infection or appear months or more after the disease has passed. Post-infection arthritis and glomerulonephritis are examples of such aftereffects or disease sequelae. In dealing with infection, immunity is a powerful force that needs to strike a delicate balance between elimination of the pathogen and the threat of harm to the host.

In this chapter, the concepts of immunity described throughout the text are applied to selected infectious diseases caused by viruses, bacteria, parasites, and fungi—the four main types of pathogens. It is impossible in a single chapter to cover all known diseases, so we selected for discussion common diseases that affect large numbers of people and/or that illustrate general concepts, as well as some diseases that have warranted recent headlines.

## Viral Infections

Viruses are small segments of nucleic acid with a protein or lipoprotein coat. They require host resources for their replication. Typically, a virus enters a cell via a cell surface receptor for which it has affinity and preempts cell biosynthetic machinery to replicate all components of itself, including its genome. In many cases, the genomic replication process is error-prone, generating numerous mutations. Because large numbers of new viruses are produced in a replication cycle, these mutants can be selected for the ability to propagate themselves.

From a survival standpoint, the virus is more likely to thrive if it does not kill the host: sustained coexistence favors the continued survival of the virus. However, the mutability of the viral genome sometimes gives rise to lethal mutants that do not conform to this state of equilibrium with the host. If such mutants cause the death of

| TABLE 18-1 | Mechanisms of humoral and cell-mediated immune responses to viruses | |
|---|---|---|
| **Response type** | **Effector molecule or cell** | **Activity** |
| Humoral | Antibody (especially secretory IgA) | Blocks binding of virus to host cells, thus preventing infection or reinfection |
| | IgG, IgM, and IgA antibody | Blocks fusion of viral envelope with host cell's plasma membrane |
| | IgG and IgM antibody | Enhances phagocytosis of viral particles (opsonization) |
| | IgM antibody | Agglutinates viral particles |
| | Complement activated by IgG or IgM antibody | Mediates opsonization by C3b and lysis of enveloped viral particles by membrane-attack complex |
| Cell mediated | IFN-$\gamma$ secreted by $T_H$ or $T_C$ cells | Has direct antiviral activity |
| | Cytotoxic T lymphocytes (CTLs) | Kill virus-infected self cells |
| | NK cells and macrophages | Kill virus-infected cells by antibody-dependent cell-mediated cytotoxicity (ADCC) |

their host, survival of the virus requires that it spread to new hosts before replication in the original host is no longer possible. Among the survival strategies available to viruses are a long latency period before severe illness or death occurs, during which time the host may pass the virus to others. HIV employs such a strategy. Another strategy is facile transmission, in which infection is efficiently transferred during even a short acute illness. Influenza and smallpox viruses use this strategy. The life cycle of viruses pathogenic for humans may also include nonhuman hosts. The West Nile virus (WNV), for example, replicates very well in certain species of birds and is carried by mosquitoes from infected birds to so-called dead-end hosts such as horses and humans. Transmission of WNV between humans via mosquito is inefficient because the titer of virus in human blood is low and the amount of blood transferred by the insect bite is small and does not contain sufficient virus to cause infection. WNV may, however, be transferred from human to human by blood transfusion and may be passed from infected pregnant mothers to their newborns.

A number of specific immune effector mechanisms, together with nonspecific defense mechanisms, are called into play to prevent or eliminate viral infection (Table 18-1). Passage across the mucosa of the respiratory, genitourinary, and gastrointestinal tracts accounts for most viral transmission. Entrance of the virus may also occur through broken skin, usually as a result of an insect bite or puncture wound. Once inside the host, the virus is subject to an onslaught of innate and adaptive immune mechanisms bent on its destruction. The outcome of the infection depends on how effectively the host's defensive mechanisms resist the offensive tactics of the virus.

The innate immune response to viral infection primarily consists of the induction of type I interferons (IFN-$\alpha$ and IFN-$\beta$) and the activation of NK cells. Double-stranded RNA (dsRNA) molecules produced during the viral life cycle are detected by Toll-like receptors (TLRs), which induce the expression of IFN-$\alpha$ and IFN-$\beta$ by the infected cell. These cytokines can also be produced by macrophages, monocytes, and fibroblasts, but the mechanisms that induce the production of type I interferons in these cells are not completely understood. IFN-$\alpha$ and IFN-$\beta$ can induce an antiviral response or resistance to viral replication by binding to the IFN $\alpha/\beta$ receptor, thereby activating the JAK-STAT pathway (see Figure 12-10), which in turn induces the transcription of several genes. One of these genes encodes an enzyme known as 2'-5'-oligo-adenylate synthetase, which activates a ribonuclease (RNAse L) that degrades viral RNA. Other genes activated by the binding of IFN-$\alpha/\beta$ to its receptor also contribute to the inhibition of viral replication. For example, IFN-$\alpha/\beta$ binding induces a specific protein kinase called dsRNA-dependent protein kinase (abbreviated PKR, from its earlier designation, protein kinase, RNA-dependent), which leads to inactivation of protein synthesis, thus blocking viral replication in infected cells (Figure 18-2).

The binding of IFN-$\alpha$ and IFN-$\beta$ to NK cells induces lytic activity, making them very effective in killing virally infected cells. The activity of NK cells is also greatly enhanced by IL-12, a cytokine that is produced very early in the response to viral infection.

## Many viruses are neutralized by antibodies

Antibodies specific for viral surface antigens are often crucial in containing the spread of a virus during acute infection and in protecting against reinfection. Antibodies are particularly effective in protecting against infection if they are localized at the site of viral entry into the body. Most viruses express surface receptor molecules that enable them to initiate

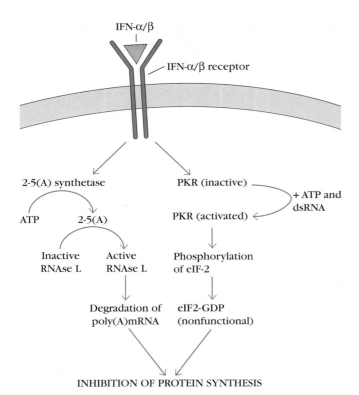

**FIGURE 18-2 Induction of antiviral activity by IFN-α and -β.** These interferons bind to the IFN receptor, which in turn induces the synthesis of both 2′-5′-oligo-adenylate synthetase (2-5(A) synthetase) and dsRNA-dependent protein kinase (PKR). The action of 2-5(A) synthetase results in the activation of RNAse L, which can degrade mRNA. PKR inactivates the translation initiation factor eIF-2 by phosphorylating it. Both pathways result in the inhibition of protein synthesis, thereby effectively blocking viral replication.

infection by binding to specific host-cell membrane molecules. For example, influenza virus binds to sialic acid residues in cell membrane glycoproteins and glycolipids, rhinovirus binds to intercellular adhesion molecules (ICAMs), and Epstein-Barr virus binds to type 2 complement receptors on B cells. If antibody to the viral receptor is produced, it can block infection by preventing the binding of viral particles to host cells. Secretory IgA in mucous secretions plays an important role in host defense against viruses by blocking viral attachment to mucosal epithelial cells. The advantage of the attenuated oral polio vaccine, considered in Chapter 19, is that it induces production of secretory IgA, which effectively blocks attachment of poliovirus along the gastrointestinal tract.

Viral neutralization by antibody sometimes involves mechanisms that operate after viral attachment to host cells. For example, antibodies may block viral penetration by binding to epitopes that are necessary to mediate fusion of the viral envelope with the plasma membrane. If the induced antibody is of a complement-activating isotype, lysis of enveloped virions can ensue. Antibody or complement can also agglutinate viral particles and function as an opsonizing agent to facilitate Fc- or C3b-receptor-mediated phagocytosis of the viral particles.

## Cell-mediated immunity is important for viral control and clearance

Although antibodies have an important role in containing the spread of a virus in the acute phases of infection, they cannot eliminate the virus if it is capable of entering a latent state in which its DNA is integrated into host chromosomal DNA. Once such an infection is established, cell-mediated immune mechanisms are most important in host defense. In general, $CD8^+$ $T_C$ cells and $CD4^+$ $T_H1$ cells are the main components of cell-mediated antiviral defense. Activated $T_H1$ cells produce a number of cytokines, including IL-2, IFN-γ, and TNF, that defend against viruses either directly or indirectly. IFN-γ acts directly by inducing an antiviral state in cells. IL-2 acts indirectly by assisting in the development of CTL precursors into an effector population. Both IL-2 and IFN-γ activate NK cells, which play an important role in host defense during the first days of many viral infections until a specific CTL response develops.

In most viral infections, specific CTL activity arises within 3 to 4 days after infection, peaks by 7 to 10 days, and then declines. Within 7 to 10 days of primary infection, most virions have been eliminated, paralleling the development of CTLs. CTLs specific for the virus eliminate virus-infected self cells and thus eliminate potential sources of new virus. The role of CTLs in defense against viruses is demonstrated by the ability of virus-specific CTLs to confer protection against that virus on nonimmune recipients by adoptive transfer. The viral specificity of the CTL can also be demonstrated with adoptive transfer: adoptive transfer of a CTL clone specific for influenza virus strain X protects mice against influenza virus X but not against influenza virus strain Y.

## Viruses can evade host defense mechanisms

Despite their restricted genome size, a number of viruses have been found to encode proteins that interfere at various levels with specific or nonspecific host defenses. Presumably, the advantage of such proteins is that they enable viruses to replicate more effectively amid host antiviral defenses. As described above, the induction of IFN-α and IFN-β is a major innate defense against viral infection, but some viruses have developed strategies to evade the action of IFN-α/β. These include hepatitis C virus, which has been shown to overcome the antiviral effect of the interferons by blocking or inhibiting the action of PKR (see Figure 18-2).

Another mechanism for evading host responses, utilized in particular by herpes simplex viruses (HSV), is inhibition of antigen presentation by infected host cells. HSV-1 and HSV-2 both express an immediate-early protein (a protein synthesized shortly after viral replication) called ICP47, which very effectively inhibits the human transporter molecule needed for antigen processing (TAP; see Figure 8-19). Inhibition of TAP blocks antigen delivery to class I MHC receptors on HSV-infected cells, thus preventing presentation of viral antigen to CD8$^+$ T cells. This results in the trapping of empty class I MHC molecules in the endoplasmic reticulum and effectively shuts down a CD8$^+$ T-cell response to HSV-infected cells.

The targeting of MHC molecules is not unique to HSV. Other viruses have been shown to down-regulate class I MHC expression shortly after infection. Two of the best-characterized examples, the adenoviruses and cytomegalovirus (CMV), use distinct molecular mechanisms to reduce the surface expression of class I MHC molecules, again inhibiting antigen presentation to CD8$^+$ T cells. Some viruses—CMV, measles virus, and HIV—have been shown to reduce levels of class II MHC molecules on the cell surface, thus blocking the function of antigen-specific antiviral helper T cells.

Antibody-mediated destruction of viruses requires complement activation, resulting either in direct lysis of the viral particle or opsonization and elimination of the virus by phagocytic cells. A number of viruses have strategies for evading complement-mediated destruction. Vaccinia virus, for example, secretes a protein that binds to the C4b complement component, inhibiting the classical complement pathway, and herpes simplex viruses have a glycoprotein component that binds to the C3b complement component, inhibiting both the classical and alternative pathways.

A number of viruses escape immune attack by constantly changing their antigens. In the influenza virus, continual antigenic variation results in the frequent emergence of new infectious strains. The absence of protective immunity to these newly emerging strains leads to repeated epidemics of influenza. Antigenic variation among rhinoviruses, the causative agent of the common cold, is responsible for our inability to produce an effective vaccine for colds. Nowhere is antigenic variation greater than in the human immunodeficiency virus (HIV), the causative agent of AIDS. Estimates suggest that HIV accumulates mutations at a rate 65 times faster than does influenza virus. Because of the importance of AIDS, a section of Chapter 20 addresses this disease in detail.

A large number of viruses evade the immune response by causing generalized immunosuppression. Among these are the paramyxoviruses that cause mumps, the measles virus, Epstein-Barr virus (EBV), cytomegalovirus, and HIV. In some cases, immunosuppression is caused by direct viral infection of lymphocytes or macrophages. The virus can then either directly destroy the immune cells by cytolytic mechanisms or alter their function. In other cases, immunosuppression is the result of a cytokine imbalance. For example, EBV produces a protein called BCRF1 that is homologous to IL-10; like IL-10, BCRF1 suppresses cytokine production by the $T_H1$ subset, resulting in decreased levels of IL-2, TNF, and IFN-$\gamma$.

## Influenza has been responsible for some of the worst pandemics in history

The influenza virus infects the upper respiratory tract and major central airways in humans, horses, birds, pigs, and even seals. In 1918–19, an influenza pandemic (worldwide epidemic) killed between 20 and 50 million people worldwide and about 675,000 in the United States. Some areas, such as Alaska and the Pacific Islands, lost more than half of their population during that pandemic. Since then, two additional pandemics have occurred, one in 1957–58, which resulted in about 70,000 deaths in the United States, and another in 1968–69, accounting for 34,000 U.S. deaths. The pandemics are caused by new influenza strains or strains that have not circulated since long before, so that most people have little immunity to them. These influenza strains differ from the subtypes of strains already in circulation, which cause seasonal outbreaks of influenza or epidemics.

### Properties of the Influenza Virus

Influenza viral particles, or virions, are roughly spherical or ovoid in shape, with an average diameter of 90 to 100 nm. The virions are surrounded by an outer envelope—a lipid bilayer acquired from the plasma membrane of the infected host cell during the process of budding. Inserted into the envelope are two glycoproteins, **hemagglutinin (HA)** and **neuraminidase (NA),** which form radiating projections that are visible in electron micrographs (Figure 18-3). The hemagglutinin projections, in the form of trimers, are responsible for the attachment of the virus to host cells. There are approximately 1000 hemagglutinin projections per influenza virion. The hemagglutinin trimer binds to sialic acid groups on host-cell glycoproteins and glycolipids by way of a conserved

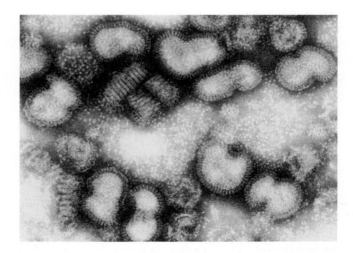

**FIGURE 18-3 Electron micrograph of influenza virus reveals roughly spherical viral particles enclosed in a lipid bilayer with protruding hemagglutinin and neuraminidase glycoprotein spikes.** [Courtesy of G. Murti, Department of Virology, St. Jude Children's Research Hospital, Memphis, Tenn.]

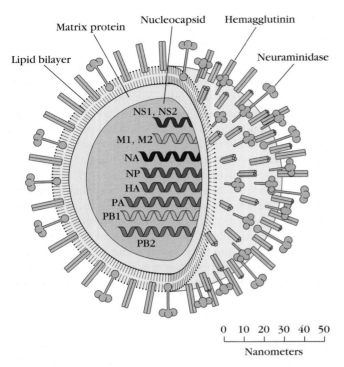

**FIGURE 18-4 Schematic representation of influenza structure.** The envelope is covered with neuraminidase and hemagglutinin spikes. Inside is an inner layer of matrix protein surrounding the nucleocapsid, which consists of eight ssRNA molecules associated with nucleoprotein. The eight RNA strands encode 10 proteins: PB1, PB2, PA, HA (hemagglutinin), NP (nucleoprotein), NA (neuraminidase), M1, M2, NS1, and NS2.

amino acid sequence that forms a small groove in the hemagglutinin molecule. Neuraminidase, as its name indicates, cleaves N-acetylneuraminic (sialic) acid from nascent viral glycoproteins and host-cell membrane glycoproteins, an activity that presumably facilitates viral budding from the infected host cell. Within the envelope, an inner layer of matrix protein surrounds the nucleocapsid, which consists of eight different strands of single-stranded RNA (ssRNA) associated with protein and RNA polymerase (Figure 18-4). Each RNA strand encodes one or more different influenza proteins.

Three basic types of influenza (A, B, and C) can be distinguished by differences in their nucleoprotein and matrix proteins. Type A, which is the most common type, is responsible for the major human pandemics. Type B causes human but not animal disease and has caused epidemics. Type C causes only mild human illness. Antigenic variation in hemagglutinin and neuraminidase distinguishes subtypes of type A influenza virus. According to the nomenclature of the World Health Organization, each virus strain is defined by its animal host of origin (specified if other than human), geographical origin, strain number, year of isolation, and antigenic description of HA and NA (Table 18-2). For example, A/Sw/Iowa/15/30 (H1N1) designates strain-A isolate 15 that arose in swine in Iowa in 1930 and has antigenic sub-

types 1 of HA and NA. Notice that the H and N proteins are antigenically distinct in these two strains. There are 13 different hemagglutinins and nine neuraminidases among the type A influenza viruses. (Neither type B nor C is classified by H and N subtypes.)

The distinguishing feature of influenza virus is its variability. The virus can change its surface antigens so completely that the immune response to infection with the virus that caused a previous epidemic gives little or no protection against the virus causing a subsequent epidemic. The antigenic variation results primarily from changes in the hemagglutinin and neuraminidase spikes protruding from the viral envelope (Figure 18-5). Two different mechanisms generate antigenic variation in HA and NA: antigenic drift and antigenic shift. **Antigenic drift** involves a series of spontaneous point mutations that occur gradually, resulting in minor changes in HA and NA. **Antigenic shift** results in the sudden emergence of a new subtype of influenza whose HA and possibly also NA are considerably different from that of the virus present in a preceding epidemic.

| TABLE 18-2 | Some influenza A strains and their hemagglutinin (H) and neuraminidase (N) subtype | |
|---|---|---|
| **Species** | **Virus strain designation** | **Antigenic subtype** |
| Human | A/Puerto Rico/8/34 | H0N1 |
| | A/Fort Monmouth/1/47 | H1N1 |
| | A/Singapore/1/57 | H2N2 |
| | A/Hong Kong/1/68 | H3N2 |
| | A/USSR/80/77 | H1N1 |
| | A/Brazil/11/78 | H1N1 |
| | A/Bangkok/1/79 | H3N2 |
| | A/Taiwan/1/86 | H1N1 |
| | A/Shanghai/16/89 | H3N2 |
| | A/Johannesburg/33/95 | H3N2 |
| | A/Wuhan/359/95 | H3N2 |
| | A/Texas/36/95 | H1N1 |
| | A/Hong Kong/156/97 | H5N1 |
| Swine | A/Sw/Iowa/15/30 | H1N1 |
| | A/Sw/Taiwan/70 | H3N2 |
| Horse (equine) | A/Eq/Prague/1/56 | H7N7 |
| | A/Eq/Miami/1/63 | H3N8* |
| Bird | A/Fowl/Dutch/27 | H7N7 |
| | A/Tern/South America/61 | H5N3 |
| | A/Turkey/Ontario/68 | H8N4 |
| | A/Chicken/Hong Kong/258/97 | H5N1[†] |

*H3N8 has recently been shown to cause flu-like illness in dogs; the species shift occurred with no reassortment of genes.

[†]As of 2006, a dangerous new H5N1 avian strain has infected approximately 175 humans with 50% mortality.

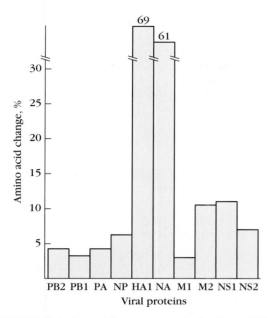

**FIGURE 18-5 Amino acid sequence variation in 10 influenza viral proteins from two H3N2 strains and one H1N1 strain.** The surface glycoproteins hemagglutinin (HA1) and neuraminidase (NA) show significant sequence variation; in contrast, the sequences of internal viral proteins, such as matrix proteins (M1 and M2) and nucleoprotein (NP), are largely conserved. [Adapted from G. G. Brownlee, 1986, in Options for the Control of Influenza, Alan R. Liss, New York.]

## Variation in Epidemic Influenza Strains

The first human influenza virus was isolated in 1934; this virus was given the subtype designation H0N1. The H0N1 subtype persisted until 1947, when a major antigenic shift generated a new subtype, H1N1, which supplanted the previous subtype and became prevalent worldwide until 1957, when H2N2 emerged. The H2N2 subtype prevailed for the next decade and was replaced in 1968 by H3N2. Antigenic shift in 1977 saw the re-emergence of H1N1. The most recent antigenic shift, in 1989, brought the re-emergence of H3N2, which remained dominant throughout the next several years. However, an H1N1 strain re-emerged in Texas in 1995, and current influenza vaccines contain both H3N2 and H1N1 strains. With each antigenic shift, hemagglutinin and neuraminidase undergo major sequence changes, resulting in major antigenic variations for which the immune system lacks memory. Thus, each antigenic shift finds the population immunologically unprepared, resulting in major outbreaks of influenza, which sometimes reach pandemic proportions.

Between pandemic-causing antigenic shifts, the influenza virus undergoes antigenic drift, generating minor antigenic variations, which account for strain differences within a subtype. The immune response contributes to the emergence of these different influenza strains. As individuals infected with a given influenza strain mount an effective immune response, the strain is eliminated. However, the accumulation of point mutations sufficiently alters the antigenicity of

some variants so that they are able to escape immune elimination (Figure 18-6a). These variants become a new strain of influenza, causing another local epidemic cycle. The role of antibody in such immunologic selection can be demonstrated in the laboratory by mixing an influenza strain with a monoclonal antibody specific for that strain and then culturing the virus in cells. The antibody neutralizes all unaltered viral particles, and only those viral particles with mutations resulting in altered antigenicity escape neutralization and are able to continue the infection. Within a short time in culture, a new influenza strain emerges.

Antigenic shift is thought to occur through genetic reassortment between influenza virions from humans and from various animals, including horses, pigs, and ducks (Figure 18-6b). The fact that influenza contains eight separate strands of ssRNA makes possible the reassortment of the RNA strands of human and animal virions within a single cell infected with both viruses. Evidence for in vivo genetic reassortment between influenza A viruses from humans and domestic pigs was obtained in 1971. After infecting a pig simultaneously with human Hong Kong influenza (H3N2) and with swine influenza (H1N1), investigators were able to recover virions expressing H3N1. In some cases, an apparent antigenic shift may represent the re-emergence of a previous strain that has remained hidden for several decades. In May 1977, a strain of influenza, A/USSR/77 (H1N1), appeared that proved to be identical to a strain that had caused an

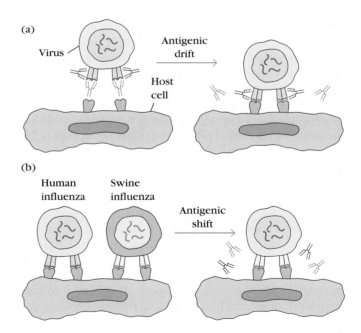

**FIGURE 18-6 Two mechanisms generate variations in influenza surface antigens.** (a) In antigenic drift, the accumulation of point mutations eventually yields a variant protein that is no longer recognized by antibody to the original antigen. (b) Antigenic shift may occur by reassortment of an entire ssRNA between human and animal virions infecting the same cell. Only four of the eight RNA strands are depicted.

epidemic 27 years earlier. The virus could have been preserved over the years in a frozen state or in an animal reservoir. When such a re-emergence occurs, the HA and NA antigens expressed are not really new; however, they will be seen as new by the immune system of anyone not previously exposed to that strain (people under the age of 27 in the 1977 epidemic, for example) because no memory cells specific for these antigenic subtypes will exist in the susceptible population. Thus, from an immunologic point of view, the re-emergence of an old influenza A strain can have the same effect as an antigenic shift that generates a new subtype. In some cases, a virus may jump from one species to another with little genetic change. Recently, the equine strain H3N8 caused an epidemic of influenza in racing dogs. The dog virus was 96% identical to the virus known to have been present in horses for about 30 years. Thus far, no human infections with H3N8 have been found, but a more careful survey of those in close contact with the infected dogs is warranted.

## The humoral response to influenza is strain specific

Humoral antibody specific for the HA molecule is produced during an influenza infection. This antibody confers protection against influenza, but its action is strain specific and is readily bypassed by antigenic drift. Antigenic drift in the HA molecule results in amino acid substitutions in several antigenic domains at the molecule's distal end (Figure 18-7). Two of these domains are on either side of the conserved sialic acid–binding cleft, which is necessary for binding of virions to target cells. Serum antibodies specific for these two regions are important in blocking initial viral infectivity. The titers of these serum antibodies peak within a few days of infection and then decrease over the next 6 months; the titers then plateau and remain fairly stable for the next several years. This antibody does not appear to be required for recovery from influenza, as patients with agammaglobulinemia recover from the disease. Instead, the serum antibody appears to play a significant role in resistance to reinfection by the same strain. When serum antibody levels are high for a particular HA molecule, both mice and humans are resistant to infection by virions expressing that HA molecule. If mice are infected with influenza virus and antibody production is experimentally suppressed, the mice recover from the infection but can be reinfected with the same viral strain. In addition to humoral responses, CTLs can play a role in immune responses to influenza.

## Avian H5N1 presents the threat of a pandemic

Since 1997, major outbreaks of influenza in poultry have been observed with occasional transmission to humans, mostly in Southeast Asia. As of March 2006, 175 human cases of infection with avian influenza strain H5N1 had been confirmed, with about 50% mortality (93 deaths). In nearly

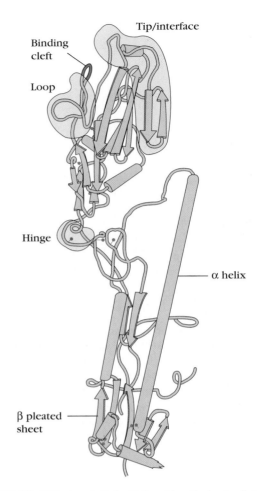

**FIGURE 18-7 Hemagglutinin.** Sialic acid on the host cells interacts with the binding cleft, which is bounded by regions—designated the loop and tip/interface—where antigenic drift is prevalent (blue areas). Antibodies to these regions are important in blocking viral infections. Continual changes in amino acid residues in these regions allow the influenza virus to evade the antibody response. Small red dots represent residues that exhibit a high degree of variation among virus strains. [Adapted from D. C. Wiley et al., 1981, Nature **289**:373.]

every case, contact with domestic or wild fowl was implicated as the source of infection, and at present, there is no evidence for facile human transmission. However, a reassortment of genes between an epidemic human strain and a lethal avian strain could cause a major pandemic. The threat of this event is heightened by the fact that the avian strains are resistant to some of the drugs commonly used to relieve the symptoms and lessen the severity of the disease.

Recent reconstruction of the virus that caused the 1918 pandemic has been accomplished using sequence data obtained from virus in tissues of victims of that outbreak, mainly recovered from bodies preserved in the permafrost in Alaska. Genomic analyses indicate that the 1918 virus derived from avian strains, and the reconstructed virus is now the object of intense study, focusing on the factors that contribute to its extreme lethality. These investigations may prove relevant to the present threat from H5N1.

# Bacterial Infections

Immunity to bacterial infections is achieved by means of antibody unless the bacterium is capable of intracellular growth, in which case delayed-type hypersensitivity has an important role. Bacteria enter the body either through a number of natural routes (e.g., the respiratory tract, the gastrointestinal tract, and the genitourinary tract) or through normally inaccessible routes opened up by breaks in mucous membranes or skin. Depending on the number of organisms entering and their virulence, different levels of host defense are enlisted. If the inoculum size and the virulence are both low, then localized tissue phagocytes may be able to eliminate the bacteria with an innate, nonspecific defense. Larger inoculums or organisms with greater virulence tend to induce an adaptive, specific immune response.

## Immune responses to extracellular and intracellular bacteria can differ

Infection by extracellular bacteria induces production of humoral antibodies, which are ordinarily secreted by plasma cells in regional lymph nodes and the submucosa of the respiratory and gastrointestinal tracts. The humoral immune response is the main protective response against extracellular bacteria. The antibodies act in several ways to protect the host from the invading organisms, including removal of the bacteria and inactivation of bacterial toxins (Figure 18-8). Extracellular bacteria can be pathogenic because they induce a localized inflammatory response or because they produce toxins. The toxins, endotoxin or exotoxin, can be cytotoxic but also may cause pathogenesis in other ways. An excellent example of this is the toxin produced by diphtheria, which exerts a toxic effect on the cell by blocking protein synthesis. Endotoxins, such as lipopolysaccharides (LPS), are generally components of bacterial cell walls, whereas exotoxins, such as diphtheria toxin, are secreted by the bacteria.

Antibody that binds to accessible antigens on the surface of a bacterium can, together with the C3b component of complement, act as an opsonin that increases phagocytosis and thus clearance of the bacterium (see Figure 18-8). In the case of some bacteria—notably, the gram-negative organisms—complement activation can lead directly to lysis of the organism. Antibody-mediated activation of the complement system can also induce localized production of immune effector molecules that help to develop an amplified and more effective inflammatory response. For example, the complement split products C3a and C5a act as anaphylatoxins, inducing local mast-cell degranulation and thus vasodilation and the extravasation of lymphocytes and neutrophils from the blood into tissue space (see Figure 18-8). Other complement split products serve as chemotactic factors for neutrophils and macrophages, thereby contributing to the buildup of phagocytic cells at the site of infection. Antibody to a bacterial toxin may bind to the toxin and neutralize it; the antibody-toxin complexes are then cleared by phagocytic cells in the same manner as any other antigen-antibody complex.

Although innate immunity is not very effective against intracellular bacterial pathogens, intracellular bacteria can activate NK cells, which in turn provide an early defense against these bacteria. Intracellular bacterial infections tend to induce a cell-mediated immune response; specifically, delayed-type hypersensitivity. In this response, cytokines secreted by CD4$^+$ T cells are important—notably IFN-$\gamma$, which activates macrophages to kill ingested pathogens more effectively (see Figure 14-15).

## Bacteria can effectively evade host defense mechanisms

There are four primary steps in bacterial infection:

- Attachment to host cells
- Proliferation
- Invasion of host tissue
- Toxin-induced damage to host cells

Host-defense mechanisms act at each of these steps, and many bacteria have evolved ways to circumvent some of them (Table 18-3).

Some bacteria have surface structures or molecules that enhance their ability to attach to host cells. A number of gram-negative bacteria, for example, have pili (long hairlike projections), which enable them to attach to the membrane of the intestinal or genitourinary tract (Figure 18-9). Other bacteria, such as *Bordetella pertussis,* secrete adhesion molecules that attach to both the bacterium and the ciliated epithelial cells of the upper respiratory tract.

Secretory IgA antibodies specific for such bacterial structures can block bacterial attachment to mucosal epithelial cells and are the main host defense against bacterial attachment. However, some bacteria (e.g., *Neisseria gonorrhoeae, Haemophilus influenzae,* and *Neisseria meningitidis*) evade the IgA response by secreting proteases that cleave secretory IgA at the hinge region; the resulting Fab and Fc fragments have a shortened half-life in mucous secretions and are not able to agglutinate microorganisms.

Some bacteria evade the IgA response of the host by changing their surface antigens. In *N. gonorrhoeae,* for example, pilin, the protein component of the pili, has a highly variable structure. Variation in the pilin amino acid sequence is generated by gene rearrangements of its coding sequence. The pilin locus consists of one or two expressed genes and 10 to 20 silent genes. Each gene is arranged into six regions called *minicassettes.* Pilin variation is generated by a process of gene conversion, in which one or more minicassettes from the silent genes replace a minicassette of the expression gene. This process generates enormous antigenic variation, which may contribute to the pathogenicity of *N. gonorrhoeae* by increasing the likelihood that expressed pili will bind firmly to epithelial cells. In addition, the continual changes in the pilin sequence allow the organism to evade neutralization by IgA.

Some bacteria possess surface structures that inhibit phagocytosis. A classic example is *Streptococcus pneumoniae,* whose

## OVERVIEW FIGURE 18-8

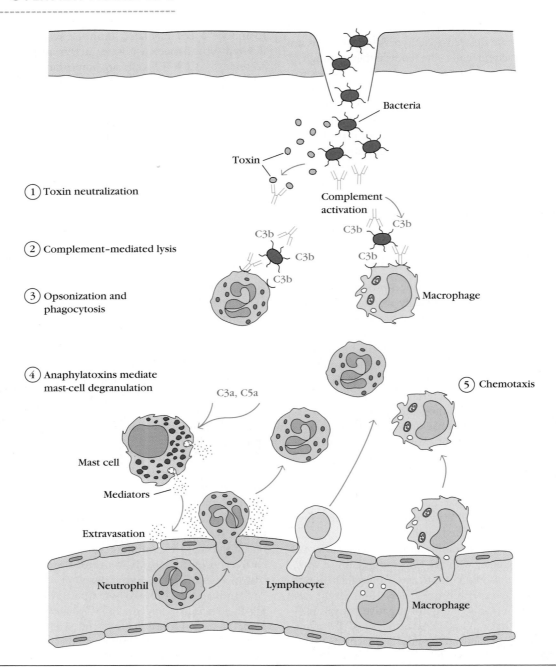

① Toxin neutralization

② Complement-mediated lysis

③ Opsonization and phagocytosis

④ Anaphylatoxins mediate mast-cell degranulation

⑤ Chemotaxis

Antibody-mediated mechanisms for combating infection by extracellular bacteria. (1) Antibody neutralizes bacterial toxins. (2) Complement activation on bacterial surfaces leads to complement-mediated lysis of bacteria. (3) Antibody and the complement split product C3b bind to bacteria, serving as opsonins to increase phagocytosis. (4) C3a and C5a, generated by antibody-initiated complement activation, induce local mast-cell degranulation, releasing substances that mediate vasodilation and extravasation of lymphocytes and neutrophils. (5) Other complement products are chemotactic for neutrophils and macrophages.

| TABLE 18-3 | Host immune responses to bacterial infection and bacterial evasion mechanisms | |
|---|---|---|
| **Infection process** | **Host defense** | **Bacterial evasion mechanisms** |
| Attachment to host cells | Blockage of attachment by secretory IgA antibodies | Secretion of proteases that cleave secretory IgA dimers (*Neisseria meningitidis, N. gonorrhoeae, Haemophilus influenzae*)<br>Antigenic variation in attachment structures (pili of *N. gonorrhoeae*) |
| Proliferation | Phagocytosis (Ab- and C3b-mediated opsonization) | Production of surface structures (polysaccharide capsule, M protein, fibrin coat) that inhibit phagocytic cells<br>Mechanisms for surviving within phagocytic cells<br>Induction of apoptosis in macrophages (*Shigella flexneri*) |
| | Complement-mediated lysis and localized inflammatory response | Generalized resistance of gram-positive bacteria to complement-mediated lysis<br>Insertion of membrane-attack complex prevented by long side chain in cell-wall LPS (some gram-negative bacteria) |
| Invasion of host tissues | Ab-mediated agglutination | Secretion of elastase that inactivates C3a and C5a (*Pseudomonas*) |
| Toxin-induced damage to host cells | Neutralization of toxin by antibody | Secretion of hyaluronidase, which enhances bacterial invasiveness |

polysaccharide capsule prevents phagocytosis very effectively. The 84 serotypes of *S. pneumoniae* differ from one another by distinct capsular polysaccharides, and during infection, the host produces antibody against the infecting serotype. This antibody protects against reinfection with the same serotype but will not protect against infection by a different serotype. In this way, *S. pneumoniae* can cause disease many times in the same individual. On other bacteria, such as *Streptococcus pyogenes,* a surface protein projection called the M protein inhibits phagocytosis. Some pathogenic staphylococci are able to assemble a protective coat from host blood proteins. These bacteria secrete a coagulase enzyme that precipitates a fibrin coat around them, shielding them from phagocytic cells.

Mechanisms for interfering with the complement system help other bacteria survive. In some gram-negative bacteria, for example, long side chains on the lipid A moiety of the cell

wall core polysaccharide help to resist complement-mediated lysis. *Pseudomonas* secretes an enzyme, elastase, that inactivates both the C3a and C5a anaphylatoxins, thereby diminishing the localized inflammatory reaction.

A number of bacteria escape host-defense mechanisms through their ability to survive within phagocytic cells. Some of these bacteria, such as *Listeria monocytogenes,* survive within phagocytic cells by escaping from the phagolysosome to the cytoplasm, a favorable environment for their growth. Other bacteria, such as *Mycobacterium avium,* block lysosomal fusion with the phagolysosome, and some mycobacteria are resistant to the oxidative attack that takes place within the phagolysosome.

## Immune responses can contribute to bacterial pathogenesis

In some bacterial infections, disease symptoms are caused not by the pathogen itself but by the immune response to the pathogen. As described in Chapter 12, pathogen-stimulated overproduction of cytokines leads to the symptoms of bacterial septic shock, food poisoning, and toxic shock syndrome. For instance, cell wall endotoxins of some gram-negative bacteria activate macrophages, resulting in release of high levels of IL-1 and TNF-$\alpha$, which can cause septic shock. In staphylococcal food poisoning and toxic shock syndrome, exotoxins produced by the pathogens function as superantigens, which can activate all T cells that express T-cell receptors with a particular $V_\beta$ domain (see Table 10-3). The resulting overproduction of cytokines by activated $T_H$ cells causes many of the symptoms of these diseases.

The ability of some bacteria to survive intracellularly within infected cells can result in chronic antigenic activation of CD4$^+$ T cells, leading to tissue destruction by a

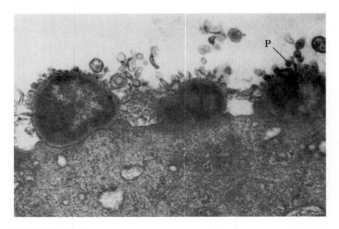

**FIGURE 18-9 Electron micrograph of *Neisseria gonorrhoeae* attaching to urethral epithelial cells.** Pili (P) extend from the gonococcal surface and mediate the attachment. *[From M. E. Ward and P. J. Watt, 1972, Journal of Infectious Disease **126**:601.]*

cell-mediated response with the characteristics of a delayed-type hypersensitivity reaction (see Chapter 15). Cytokines secreted by these activated CD4$^+$ T cells can lead to extensive accumulation and activation of macrophages, resulting in formation of a **granuloma**. The localized concentrations of lysosomal enzymes in these granulomas can cause extensive tissue necrosis. Much of the tissue damage seen with *M. tuberculosis* is due to a cell-mediated immune response.

## Diphtheria (*Corynebacterium diphtheriae*) can be controlled by immunization with inactivated toxoid

Diphtheria is the classic example of a bacterial disease caused by a secreted exotoxin to which immunity can be induced by immunization with an inactivated **toxoid**. The causative agent, a gram-positive, rodlike organism called *Corynebacterium diphtheriae*, was first described by Theodor Klebs in 1883 and was shown a year later by Friedrich Loeffler to cause diphtheria in guinea pigs and rabbits. Autopsies on the infected animals revealed that, whereas bacterial growth was limited to the site of inoculation, there was widespread damage to a variety of organs, including the heart, liver, and kidneys. This finding led Loeffler to speculate that the neurologic and cardiologic manifestations of the disease were caused by a toxic substance released by the organism.

Loeffler's hypothesis was validated in 1888, when Pierre Roux and Alexandre Yersin produced the disease in animals by injecting them with a sterile filtrate from a culture of *C. diphtheriae*. Two years later, Emil von Behring showed that an antiserum to the toxin prevented death in infected animals. He prepared a toxoid by treating the toxin with iodine trichloride and demonstrated that it could induce protective antibodies in animals. However, the toxoid prepared in this manner was still quite toxic and so unsuitable for use in humans. This problem was solved in 1923 by Gaston Ramon, who found that exposing the toxin to heat and formalin rendered it nontoxic but did not destroy its antigenicity. Clinical trials showed that formaldehyde-treated toxoid conferred a high level of protection against diphtheria.

As immunizations with the toxoid increased, the number of cases of diphtheria decreased rapidly. In the 1920s, there were approximately 200 cases of diphtheria per 100,000 population in the United States. In 2004, the Centers for Disease Control reported no diphtheria in the entire United States. Sporadic outbreaks of diphtheria have been observed in areas where vaccination coverage is allowed to lapse.

Natural infection with *C. diphtheriae* occurs only in humans. The disease is spread from one individual to another by airborne respiratory droplets. The organism colonizes the nasopharyngeal tract, remaining in the superficial layers of the respiratory mucosa. Growth of the organism itself causes little tissue damage, and only a mild inflammatory reaction develops. The virulence of the organism is due completely to its potent exotoxin. The toxin causes destruction of the underlying tissue, resulting in the formation of a tough fibrinous membrane ("pseudomembrane") composed of fibrin,

white blood cells, and dead respiratory epithelial cells. The membrane itself can cause suffocation. The exotoxin also is responsible for widespread systemic manifestations. Pronounced myocardial damage (often leading to congestive heart failure) and neurologic damage (ranging from mild muscle weakness to complete paralysis) are common.

The exotoxin that causes diphtheria symptoms is encoded by the *tox* gene carried by phage β. Within some strains of *C. diphtheriae*, phage β can exist in a state of **lysogeny,** in which the β-prophage DNA persists within the bacterial cell. Only strains that carry lysogenic phage β are able to produce the exotoxin. The diphtheria exotoxin contains two disulfide-linked chains, a binding chain and a toxin chain. The binding chain interacts with ganglioside receptors on susceptible cells, facilitating internalization of the exotoxin. Toxicity results from the inhibitory effect of the toxin chain on protein synthesis. The diphtheria exotoxin is extremely potent: a single molecule has been shown to kill a cell. Removal of the binding chain prevents the exotoxin from entering the cell, thus rendering the exotoxin nontoxic.

Today diphtheria toxoid is prepared by treating diphtheria toxin with formaldehyde. The reaction with formaldehyde cross-links the toxin, resulting in an irreversible loss in its toxicity while enhancing its antigenicity. The toxoid is administered together with tetanus toxoid and acellular *Bordetella pertussis* in a combined vaccine that is given to children beginning at 6 to 8 weeks of age (see Table 19-3). Immunization with the toxoid induces the production of antibodies (antitoxin), which can bind to the toxin and neutralize its activity. Because antitoxin levels decline slowly over time, booster doses are recommended at 10-year intervals to maintain antitoxin levels within the protective range.

## Tuberculosis (*Mycobacterium tuberculosis*) is primarily controlled by CD4$^+$ T cells

Tuberculosis is the leading cause of death in the world from a single infectious agent, killing about 1.5 million individuals every year. Roughly one-third of the world's population is infected with the causative agent *M. tuberculosis* and is at risk of developing the disease. Although tuberculosis was thought to be eliminated as a public health problem in the United States, the disease re-emerged in the early 1990s, particularly in areas where HIV-infection levels are high. In 2005, approximately 15,000 individuals were diagnosed with tuberculosis in the United States.

Although several *Mycobacterium* species can cause tuberculosis, *M. tuberculosis* is the principal causative agent. This organism is spread easily, and pulmonary infection usually results from inhalation of small droplets of respiratory secretions containing a few bacilli. The inhaled bacilli are ingested by alveolar macrophages and are able to survive and multiply intracellularly by inhibiting formation of phagolysosomes. When the infected macrophages lyse, as they eventually do, large numbers of bacilli are released. A cell-mediated response involving CD4$^+$ T cells, which is required for immunity to tuberculosis, may be responsible for

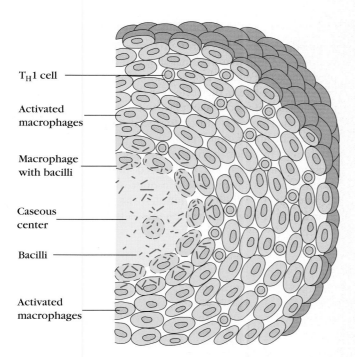

T$_H$1 cell

Activated macrophages

Macrophage with bacilli

Caseous center

Bacilli

Activated macrophages

**FIGURE 18-10 A tubercle formed in pulmonary tuberculosis.**
*[Modified from A. M. Dannenberg, 1993,* Hospital Practice *(Jan. 15):51.]*

much of the tissue damage in the disease. CD4$^+$ T-cell activity is the basis for the tuberculin skin test to the purified protein derivative (PPD) from *M. tuberculosis* (see Chapter 14).

On infection with *M. tuberculosis,* the most common clinical pattern, termed pulmonary tuberculosis, appears in about 90% of those infected. In this pattern, CD4$^+$ T cells are activated within 2 to 6 weeks after infection, inducing the infiltration of large numbers of activated macrophages. These cells wall off the organism inside a granulomatous lesion called a tubercle (Figure 18-10). A tubercle consists of a few small lymphocytes and a compact collection of activated macrophages, which sometimes differentiate into epithelioid cells or multinucleated giant cells. The massive activation of macrophages that occurs within tubercles often results in the concentrated release of lytic enzymes. These enzymes destroy nearby healthy cells, resulting in circular regions of necrotic tissue, which eventually form a lesion with a caseous (cheeselike) consistency (see Figure 18-10). As these caseous lesions heal, they become calcified and are readily visible on x-rays, where they are called Ghon complexes.

Because the activated macrophages suppress proliferation of the phagocytosed bacilli, infection is contained. Cytokines produced by CD4$^+$ T cells (T$_H$1 subset) play an important role in the response by activating macrophages so that they are able to kill the bacilli or inhibit their growth. The role of IFN-$\gamma$ in the immune response to mycobacteria has been demonstrated with knockout mice lacking IFN-$\gamma$. These mice died when they were infected with an attenuated strain of mycobacteria (BCG), whereas IFN-$\gamma^+$ normal mice survive.

Recent studies have revealed high levels of IL-12 in the pleural effusions of tuberculosis patients. The high levels of

IL-12, produced by activated macrophages, are not surprising, given the decisive role of this cytokine in stimulating T$_H$1-mediated responses (see Figure 12-12). In mouse models of tuberculosis, IL-12 has been shown to increase resistance to the disease, not only by stimulating development of T$_H$1 cells but also by inducing the production of chemokines that attract macrophages to the site of infection. When IL-12 is neutralized by antibody to IL-12, granuloma formation in tuberculous mice is blocked.

The CD4$^+$ T cell–mediated immune response mounted by the majority of people exposed to *M. tuberculosis* thus controls the infection and later protects against reinfection. However, about 10% of individuals infected with *M. tuberculosis* follow a different clinical pattern: the disease progresses to chronic pulmonary tuberculosis or to extrapulmonary tuberculosis. This progression may occur years after the primary infection. In this clinical pattern, accumulation of large concentrations of mycobacterial antigens within tubercles leads to extensive and continual chronic CD4$^+$ T-cell activation and ensuing macrophage activation. The resulting high concentrations of lytic enzymes cause the necrotic caseous lesions to liquefy, creating a rich medium that allows the tubercle bacilli to proliferate extracellularly. Eventually the lesions rupture, and the bacilli disseminate in the lung and/or are spread through the blood and lymphatic vessels to the pleural cavity, bone, urogenital system, meninges, peritoneum, or skin.

Tuberculosis is treated with several drugs used in combination, including isoniazid, rifampin, streptomycin, pyrazinamide, and ethambutol. The combination therapy of isoniazid and rifampin has been particularly effective. However, the intracellular growth of *M. tuberculosis* makes it difficult for drugs to reach the bacilli. For this reason, drug therapy must be continued for at least 9 months to eradicate the bacteria. Some patients with tuberculosis do not exhibit any clinical symptoms, and some with symptoms begin to feel better within 2 to 4 weeks after treatment begins. To avoid the side effects associated with the usual antibiotic therapy, many patients, once they feel better, stop taking the medications long before the recommended treatment period is completed. Because briefer treatment may not eradicate organisms that are somewhat resistant to the antibiotics, a multidrug-resistant strain can emerge. Noncompliance with required treatment regimes, one of the most troubling aspects of the large number of current tuberculosis cases, clearly compromises efforts to contain the spread of the disease. In settings where it is feasible, directly observed therapy, or DOT, ensures compliance with the lengthy drug regimen.

At present, the only vaccine for *M. tuberculosis* is an attenuated strain of *M. bovis* called BCG (Bacille Calmette-Guérin). The vaccine appears to provide fairly effective protection against extrapulmonary tuberculosis but has been inconsistent against pulmonary tuberculosis. In different studies, BCG has provided protection in anywhere from 0% to 80% of vaccinated individuals; in some cases, BCG vaccination has even increased the risk of infection. Moreover, after BCG vaccination, the tuberculin skin test cannot be used as an effective monitor of exposure to *M. tuberculosis.* Because of the variable effectiveness of the BCG vaccine and the inability to monitor for exposure with the skin test after vaccination, this vaccine is not used in the United States. However, the alarming increase in multidrug-resistant strains has stimulated renewed efforts to develop a more effective tuberculosis vaccine.

# Parasitic Diseases

The term *parasite* encompasses a large number of protozoan and helminthic organisms that account for an enormous disease burden, mainly in developing countries. The diversity of the parasitic universe makes it difficult to generalize, but a major difference between the types of parasites is that the protozoans are unicellular eukaryotes that usually live and multiply within host cells, whereas helminthic parasites are multicellular organisms (worms) that infest humans but have the ability to live and reproduce on their own. The majority of parasitic infections result in chronic rather than acute disease, although there are some exceptions, most notably acute malarial attacks. There is abundant evidence for evasion of the immune system by parasites, allowing them to chronically infect their host, and in some but not all cases, the mechanisms of escape are clear. Experimental systems, usually mouse models of infection, have defined how immunity to certain parasites is achieved, but the complexity of parasitic infections precludes easy generalizations.

## Protozoan diseases affect millions worldwide

Protozoans are responsible for serious diseases in humans, including amebiasis, Chagas' disease, African sleeping sickness, malaria, leishmaniasis, and toxoplasmosis. The type of immune response that develops to protozoan infection and the effectiveness of the response depend in part on the location of the parasite within the host. Many protozoans have life cycle stages in which they are free within the bloodstream, and it is during these stages that humoral antibody is most effective. Many of these same pathogens are also capable of intracellular growth; during these stages, cell-mediated immune reactions are effective in host defense. In the development of vaccines for protozoan diseases, the branch of the immune system that is most likely to confer protection must be carefully considered.

## Malaria (*Plasmodium* species) infects 600 million people worldwide

Malaria is one of the most devastating diseases in the world today, infecting nearly 10% of the world population and causing 1 to 2 million deaths every year. Malaria is caused by various species of the genus *Plasmodium*, of which *P. falciparum* is the most virulent and prevalent. The alarming development of multiple-drug resistance in *Plasmodium* and the increased resistance to pesticides of its vector, the *Anopheles* mosquito, underscore the importance of developing new strategies to hinder the spread of malaria.

### *Plasmodium* Life Cycle and Pathogenesis of Malaria

*Plasmodium* progresses through a remarkable series of developmental and maturational stages in its extremely complex life cycle. Female *Anopheles* mosquitoes, which feed on blood meals, serve as the vector for *Plasmodium,* and part of the parasite's life cycle takes place within the mosquito. (Because male *Anopheles* mosquitoes feed on plant juices, they do not transmit *Plasmodium.*)

Human infection begins when sporozoites, one of the *Plasmodium* stages, are introduced into an individual's bloodstream as an infected mosquito takes a blood meal (Figure 18-11). Within 30 minutes, the sporozoites disappear from the blood as they migrate to the liver, where they

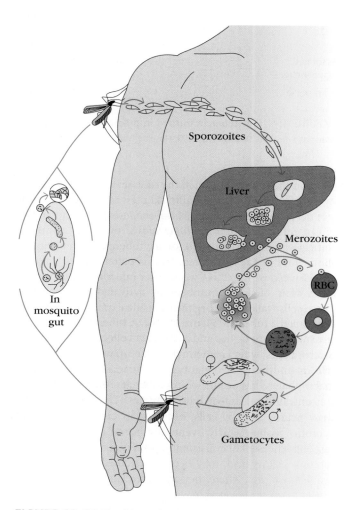

**FIGURE 18-11 The life cycle of *Plasmodium.*** Sporozoites enter the bloodstream when an infected mosquito takes a blood meal. The sporozoites migrate to the liver, where they multiply, transforming liver hepatocytes into giant multinucleate schizonts, which release thousands of merozoites into the bloodstream. The merozoites infect red blood cells, which eventually rupture, releasing more merozoites. Eventually some of the merozoites differentiate into male and female gametocytes, which are ingested by a mosquito and differentiate into gametes. The gametes fuse to form a zygote that differentiates to the sporozoite stage within the salivary gland of the mosquito.

infect hepatocytes. Sporozoites are long, slender cells that are covered by a 45-kDa protein called circumsporozoite (CS) antigen, which appears to mediate their adhesion to hepatocytes. The binding site on the CS antigen is a conserved region in the carboxyl-terminal end (called region II) that has a high degree of sequence homology with known cell adhesion molecules.

Within the liver, the sporozoites multiply extensively and undergo a complex series of transformations that culminate in the formation and release of merozoites in about a week. It has been estimated that a liver hepatocyte infected with a single sporozoite can release 5,000 to 10,000 merozoites. The released merozoites infect red blood cells, initiating the symptoms and pathology of malaria. Within a red blood cell, merozoites replicate and undergo successive differentiations; eventually the cell ruptures and releases new merozoites, which go on to infect more red blood cells. Eventually some of the merozoites differentiate into male and female gametocytes, which may be ingested by a female *Anopheles* mosquito during a blood meal. Within the mosquito's gut, the male and female gametocytes differentiate into gametes that fuse to form a zygote, which multiplies and differentiates into sporozoites within the salivary gland. The infected mosquito is now set to initiate the cycle once again.

The symptoms of malaria are recurrent chills, fever, and sweating. The symptoms peak roughly every 48 hours, when successive generations of merozoites are released from infected red blood cells. An infected individual eventually becomes weak and anemic and shows splenomegaly (an enlargement of the spleen). The large numbers of merozoites formed can block capillaries, causing intense headaches, renal failure, heart failure, or cerebral damage—often with fatal consequences. There is speculation that some of the symptoms of malaria may be caused not by *Plasmodium* itself but instead by excessive production of cytokines. This hypothesis stemmed from the observation that cancer patients treated in clinical trials with recombinant tumor necrosis factor (TNF) developed symptoms that mimicked malaria. The relation between TNF and malaria symptoms was studied by infecting mice with a mouse-specific strain of *Plasmodium,* which causes rapid death by cerebral malaria. Injection of these mice with antibodies to TNF was shown to prevent the rapid death.

## Host Response to *Plasmodium* Infection

In regions where malaria is endemic, the immune response to *Plasmodium* infection is poor. Children less than 14 years old mount the lowest immune response and consequently are most likely to develop malaria. In some regions, the childhood mortality rate for malaria reaches 50%, and worldwide, the disease kills about a million children a year. The low immune response to *Plasmodium* among children can be demonstrated by measuring serum antibody levels to the sporozoite stage. Only 22% of the children living in endemic areas have detectable antibodies to the sporozoite stage, whereas 84% of the adults have such antibodies. Even in adults, the degree of immunity is far from complete, however, and most people living in endemic regions have lifelong low-level *Plasmodium* infections.

A number of factors may contribute to the low levels of immune responsiveness to *Plasmodium*. The maturational changes from sporozoite to merozoite to gametocyte allow the organism to keep changing its surface molecules, resulting in continual changes in the antigens seen by the immune system. The intracellular phases of the life cycle in liver cells and erythrocytes also reduce the degree of immune activation generated by the pathogen and allow the organism to multiply while it is shielded from attack. Furthermore, the most accessible stage, the sporozoite, circulates in the blood for only about 30 minutes before it infects liver hepatocytes; it is unlikely that effective immune activation can occur in such a short period of time. And even when an antibody response does develop to sporozoites, *Plasmodium* has evolved a way of overcoming that response by sloughing off the surface CS-antigen coat, thus rendering the antibodies ineffective.

## Design of Malaria Vaccines

An effective vaccine for malaria should maximize the most effective immune defense mechanisms. Unfortunately, little is known of the roles that humoral and cell-mediated responses play in the development of protective immunity to this disease. Current approaches to design of malaria vaccines use a variety of approaches. An obvious target is the sporozoite stage; blocking the first invasion step could prevent infection. One experimental vaccine, for example, consists of *Plasmodium* sporozoites attenuated by x-irradiation. In one study, nine volunteers were repeatedly immunized by the bite of *P. falciparum*–infected, irradiated mosquitoes. Later challenge by the bites of mosquitoes infected with virulent *P. falciparum* revealed that six of the nine recipients were completely protected. These results are encouraging, but translating these findings into mass immunization remains problematic. Sporozoites do not grow well in cultured cells, so an enormous insectory would be required to breed mosquitoes in which to prepare enough irradiated sporozoites to vaccinate just one small village.

Other vaccine strategies are aimed at producing synthetic subunit vaccines consisting of epitopes of the blood-stage parasite that can be recognized by T cells and B cells. A more imaginative and global approach involves vaccines called transmission-blocking vaccines that target proteins expressed by the parasite in the mosquito stage, preventing that phase of the replication cycle. The female mosquito would ingest the antibodies from the vaccinated host along with the blood meal, the antibodies would prevent development of the next stage in the parasite's life cycle, and the mosquito could not pass the disease to the next bite victim, effectively blocking the transmission cycle. When a sufficient number of people were vaccinated in a given area, the transmission cycle would be broken.

To date, no effective malaria vaccine has been developed, but this is an active area of investigation, and many approaches are being tested. It is likely that a successful vaccine will utilize a combination strategy, targeting different antigens at different life stages of the malaria parasite.

## Two species of *Trypanosoma* cause African sleeping sickness

Two species of African trypanosomes, which are flagellated protozoans, can cause sleeping sickness, a chronic, debilitating disease transmitted to humans and cattle by the bite of the tsetse fly. In the bloodstream, a trypanosome differentiates into a long, slender form that continues to divide every 4 to 6 hours. The disease progresses through several stages, beginning with an early (systemic) stage in which trypanosomes multiply in the blood and progressing to a neurologic stage in which the parasite infects the central nervous system, causing meningoencephalitis and eventually the loss of consciousness.

As parasite numbers increase after infection, an effective humoral antibody response develops to the glycoprotein coat, called variant surface glycoprotein (VSG), that covers the trypanosomal surface (Figure 18-12). These antibodies eliminate most of the parasites from the bloodstream, both by complement-mediated lysis and by opsonization and subsequent phagocytosis. However, about 1% of the organisms, which bear an antigenically different VSG, escape the initial antibody response. These surviving organisms then begin to proliferate in the bloodstream, and a new wave of parasitemia is observed. The successive waves of parasitemia reflect a unique mechanism of antigenic shift by which the trypanosomes can evade the immune response to their glycoprotein antigens. This process is so effective that each new variant that arises in the course of a single infection is able to escape the humoral antibodies generated in response to the preceding variant, and so waves of parasitemia recur (Figure 18-12a).

Several unusual genetic processes generate the extensive variation in trypanosomal VSG that enables the organism to escape immunologic clearance. An individual trypanosome carries a large repertoire of VSG genes, each encoding a different VSG primary sequence. *Trypanosoma brucei*, for example, contains more than 1000 VSG genes in its genome, clustered at multiple chromosomal sites. A trypanosome expresses only a single VSG gene at a time. Activation of a VSG gene results in duplication of the gene and its transposition to a transcriptionally active expression site (ES) at the telomeric end of specific chromosomes (Figure 18-12b). Activation of a new VSG gene displaces the previous gene from the telomeric expression site. A number of chromosomes in the trypanosome have transcriptionally active expression sites at the telomeric ends, so that a number of VSG genes can potentially be expressed, but unknown control mechanisms limit expression to a single VSG expression site at a time.

There appears to be some order in the VSG variation during infection. Each new variant arises not by clonal outgrowth from a single variant cell but instead from the growth of multiple cells that have activated the same VSG gene in the current wave of parasite growth. It is not known how this process is regulated among individual trypanosomes. The continual shifts in epitopes displayed by the VSG have made the development of a vaccine for African sleeping sickness extremely difficult.

## Leishmaniasis is a useful model for demonstrating differences in host responses

The protozoan parasite *Leishmania major* provides a powerful and illustrative example of how host responses can differ between individuals. These differences can lead to either clearance of the parasite or death from the infection. *Leishmania* lives in the phagosomes of macrophages. Resistance to the infection correlates well with the production of IFN-$\gamma$ and the development of a $T_H1$ response. Elegant studies in mice have demonstrated that strains that are resistant to *Leishmania* develop a $T_H1$ response and produce IFN-$\gamma$ on infection. If these strains of mice lose either IFN-$\gamma$ or the IFN-$\gamma$ receptor, they become highly susceptible to *Leishmania*-induced fatality, underscoring the importance of IFN-$\gamma$ in containing the infection. A few strains of mice, such as BALB/c, are highly susceptible to *Leishmania* and frequently succumb to infection. These mice mount a $T_H2$-type response to *Leishmania* infection; they produce high levels of IL-4 and essentially no IFN-$\gamma$. Thus, one difference between an effective and an ineffective defense against the parasite is the development of a $T_H1$ response or $T_H2$ response. Recent studies demonstrate that a distinction between the resistant strains of mice and BALB/c is that a small restricted subset of BALB/c CD4$^+$ T cells are capable of recognizing a particular epitope on *L. major*, and this subset produces high levels of IL-4 early in the response to the parasite, skewing the response toward a predominantly $T_H2$ type. Understanding how these different T-helper responses affect the outcome of infections in general could contribute to the rational design of effective treatments and vaccines against other pathogens.

## A variety of diseases are caused by parasitic worms (helminths)

Parasitic worms, or helminths, are responsible for a wide variety of diseases in both humans and animals. Although helminths are more accessible to the immune system than protozoans, most infected individuals carry few of these parasites; for this reason, the immune system is not strongly engaged and the level of immunity generated to helminths is often very poor. More than a billion people are infected with *Ascaris*, a parasitic roundworm that infects the small intestine, and more than 300 million people are infected with *Schistosoma*, a trematode worm that causes a chronic debilitating infection. Several helminths are important pathogens of domestic animals and invade humans who ingest contaminated food. These helminths include *Taenia*, a tapeworm of cattle and pigs, and *Trichinella*, the roundworm of pigs that causes trichinosis.

Several *Schistosoma* species are responsible for the chronic, debilitating, and sometimes fatal disease **schistosomiasis** (formerly known as *bilharzia*). Three species, *S. mansoni*, *S. japonicum*, and *S. haematobium*, are the major pathogens in humans, infecting individuals in Africa, the Middle East,

## OVERVIEW FIGURE 18-12: Successive Waves of Parasitemia after Infection with *Trypanosoma* Result from Antigenic Shifts in the Parasite's Variant Surface Glycoprotein (VSG)

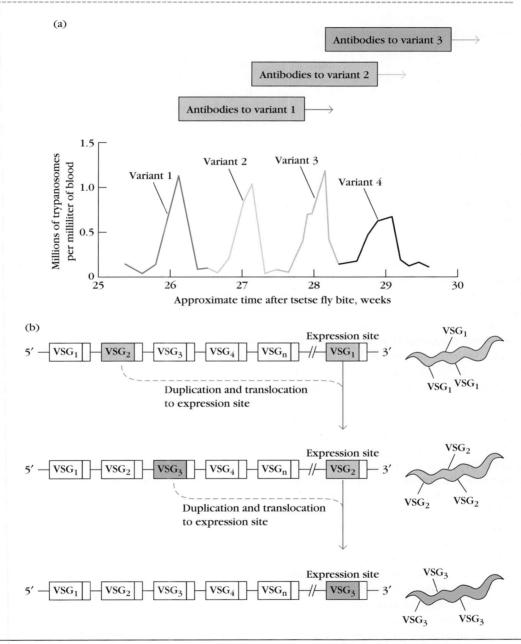

(a) Each variant that arises is unaffected by the humoral antibodies induced by the previous variant. (b) Antigenic shifts in trypanosomes occur by the duplication of gene segments encoding variant VSG molecules and their translocation to an expression site located close to the telomere. *[Part a adapted from J. Donelson, 1988, The Biology of Parasitism, Alan R. Liss, New York.]*

South America, the Caribbean, China, Southeast Asia, and the Philippines. A rise in the incidence of schistosomiasis in recent years has paralleled the increasing worldwide use of irrigation, which has expanded the habitat of the freshwater snail that serves as the intermediate host for schistosomes.

Infection occurs through contact with free-swimming infectious larvae, called cercariae, which are released from an infected snail at the rate of 300 to 3000 per day. When cercariae contact human skin, they secrete digestive enzymes that help them bore into the skin, where they shed their tails and are transformed into schistosomules. The schistosomules enter the capillaries and migrate to the lungs, then to the liver, and finally to the primary site of infection, which varies with the pathogen species. *Schistosoma mansoni* and *S. japonicum* infect the intestinal mesenteric veins; *S. haematobium* infects the veins of the urinary bladder. Once established in their final tissue site, schistosomules mature into male and female adult worms. The worms mate, and the females produce at least 300 spiny eggs a day. Unlike protozoan parasites, schistosomes and other helminths do not multiply within their hosts. The eggs produced by the female worm do not mature into adult worms in humans; instead, some of them pass into the feces or urine and are excreted to infect more snails. The number of worms in an infected individual increases only through repeated exposure to the free-swimming cercariae, and so most infected individuals carry rather low numbers of worms.

Most of the symptoms of schistosomiasis are initiated by the eggs. As many as half of the eggs produced remain in the host, where they invade the intestinal wall, liver, or bladder and cause hemorrhage. A chronic state can then develop in which the adult worms persist and the unexcreted eggs induce cell-mediated delayed-type hypersensitive reactions, resulting in large granulomas that are gradually walled off by fibrous tissue. Although the eggs are contained by the formation of the granuloma, often the granuloma itself obstructs the venous blood flow to the liver or bladder.

An immune response does develop to the schistosomes, but in most individuals, it is not sufficient to eliminate the adult worms, even though the intravascular sites of schistosome infestation should make the worm an easy target for immune attack. Instead, the worms survive for up to 20 years. The schistosomules would appear to be the forms most susceptible to attack, but because they are motile, they can evade the localized cellular buildup of immune and inflammatory cells. Adult schistosome worms also have several unique mechanisms that protect them from immune defenses. The adult worm has been shown to decrease the expression of antigens on its outer membrane and also to enclose itself in a glycolipid-and-glycoprotein coat derived from the host, masking the presence of its own antigens. Among the antigens observed on the adult worm are the host's own ABO blood-group antigens and histocompatibility antigens! The immune response is, of course, diminished by this covering made of the host's self

antigens, which must contribute to the lifelong persistence of these organisms.

The relative importance of the humoral and cell-mediated responses in protective immunity to schistosomiasis is controversial. The humoral response to infection with *S. mansoni* is characterized by high titers of antischistosome IgE antibodies, localized increases in mast cells and their subsequent degranulation, and increased numbers of eosinophils (Figure 18-13, *top*). These manifestations suggest that cytokines produced by a $T_H2$-like subset are important for the immune response: IL-4, which induces B cells to class-switch to IgE production; IL-5, which induces bone marrow precursors to differentiate into eosinophils; and IL-3, which (along with IL-4) stimulates growth of mast cells. Degranulation of mast cells releases mediators that increase the infiltration of such inflammatory cells as macrophages and eosinophils. The eosinophils express Fc receptors for IgE and IgG and bind to the antibody-coated parasite. Once bound to the parasite, an eosinophil can participate in antibody-dependent cell-mediated cytotoxicity (ADCC), releasing mediators from its granules that damage the parasite (see Figure 14-12). One eosinophil mediator, called basic protein, is particularly toxic to helminths.

Immunization studies with mice, however, suggest that this humoral IgE response may not provide protective immunity. When mice are immunized with *S. mansoni* vaccine, the protective immune response that develops is not an IgE response but rather a $T_H1$ response characterized by IFN-$\gamma$ production and macrophage accumulation (Figure 18-13, *bottom*). Furthermore, inbred strains of mice with deficiencies in mast cells or IgE develop protective immunity from vaccination, whereas inbred strains with deficiencies in cell-mediated CD4$^+$ T-cell responses fail to develop protective immunity in response to the vaccine. These studies suggest that the CD4$^+$ T-cell response may be the most important in immunity to schistosomiasis. It has been suggested that the ability to induce an ineffective $T_H2$-like response may have evolved in schistosomes as a clever defense mechanism to ensure that $T_H2$ cells produced sufficient levels of IL-10 to inhibit protective immunity mediated by the $T_H1$-like subset in the CD4$^+$ T response.

Antigens present on the membrane of cercariae and young schistosomules are promising vaccine components because these stages appear to be most susceptible to immune attack. Injecting mice and rats with monoclonal antibodies to cercariae and young schistosomules passively transferred resistance to infection with live cercariae. When these protective antibodies were used in affinity columns to purify schistosome membrane antigens from crude membrane extracts, it was found that mice immunized and boosted with these purified antigens exhibited increased resistance to a later challenge with live cercariae. Schistosome cDNA libraries were then established and screened with the protective monoclonal antibodies to identify those cDNAs encoding surface antigens. Experiments using cloned cercariae or schistosomule antigens are presently under way to

## OVERVIEW FIGURE 18-13:  Overview of the Immune Response Generated against *Schistosoma mansoni*

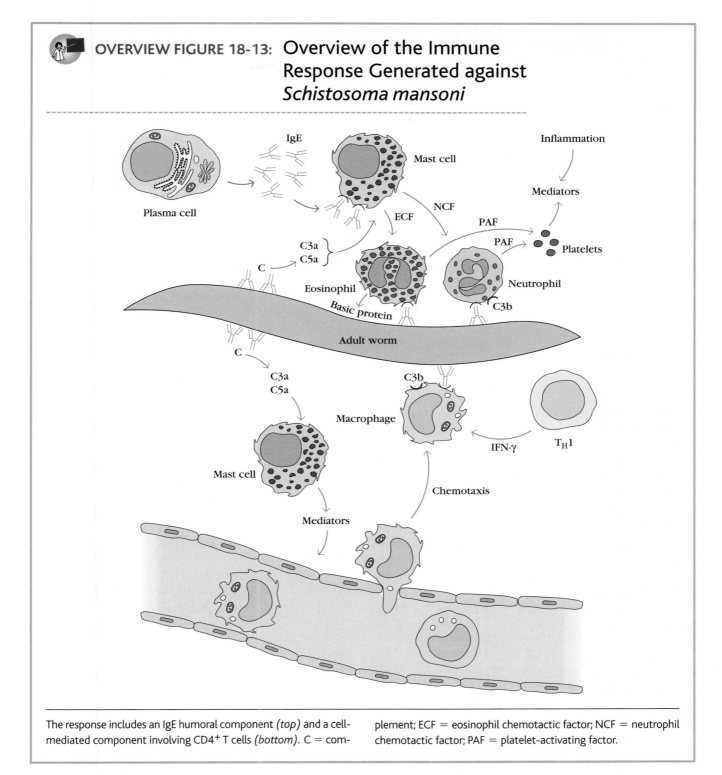

The response includes an IgE humoral component *(top)* and a cell-mediated component involving CD4+ T cells *(bottom)*. C = complement; ECF = eosinophil chemotactic factor; NCF = neutrophil chemotactic factor; PAF = platelet-activating factor.

assess their ability to induce protective immunity in animal models. However, in developing an effective vaccine for schistosomiasis, a fine line separates a beneficial immune response, which at best limits the parasite load, from a detrimental response, which in itself may become pathologic.

## Fungal Diseases

Mycologists estimate that there are as many as a million species of fungi. About 400 of these are agents of human disease. This diverse group of organisms occupies many niches

| TABLE 18-4 | Classification of fungal diseases | |
|---|---|---|
| Site of infection | Superficial | Epidermis, no inflammation |
| | Cutaneous | Skin, hair, nails |
| | Subcutaneous | Wounds, usually inflammatory |
| | Deep or systemic | Lungs, abdominal viscera, bones, CNS |
| Route of acquistion | Exogenous | Environmental, airborne, cutaneous or percutaneous |
| | Endogenous | Latent reactivation, commensal organism |
| Virulence | Primary | Inherently virulent, infects healthy host |
| | Opportunistic | Low virulence, infects immunocompromised host |

and also performs services for humans such as fermentation of bread, cheese, wine, and beer and production of penicillin, a prototypic antibiotic. Infections may result from introduction of exogenous organisms by injury or inhalation or from endogenous organisms as a result of some breach in immunity. Fungal products can be toxic, carcinogenic, or even, in the case of some mushroom products, hallucinogenic. The first human disease shown to be of fungal origin was ringworm. David Grub in 1843 isolated a fungus from a patient with ringworm and then demonstrated that it caused the disease when inoculated onto normal skin.

Fungal diseases, or mycoses, are classified by

- Site of infection—superficial, cutaneous, subcutaneous, or deep and systemic;

- Route of acquisition—exogenous or endogenous;

- Virulence—primary or opportunistic.

These categories (summarized in Table 18-4) are not mutually exclusive. For example, an infection such as coccidiomycosis may progress from a cutaneous lesion to a systemic deep infection of lungs. Cutaneous infections include attacks on skin, hair, and nails; examples are ringworm, caused by *Trichophyton tonsurans* or *Microsporum audouinii;* athlete's foot (tinea pedis); and jock itch (tinea cruris), most commonly caused by *Trichophyton mentagrophytes.* Subcutaneous infections are normally introduced by trauma and accompanied by inflammation; if inflammation is chronic, extensive tissue damage may ensue. Deep mycoses involve the lungs, the central nervous system, bones, and the abdominal viscera. Infection occurs through ingestion, inhalation, or inoculation into the bloodstream.

Virulence is divided into primary, indicating agents of high pathogenicity, and opportunistic, denoting weakly virulent agents that infect mainly individuals with compromised immunity. Most fungal infections of healthy individuals are resolved rapidly, with few clinical signs. The most commonly encountered and best-studied human fungal pathogens are *Cryptococcus neoformans, Aspergillus fumigatus, Coccidioides immitis, Histoplasma capsulatum,* and *Blastomyces dermatitidis.* Diseases caused by these fungi are named for the agent; for example, *C. neoformans* causes cryptococcosis, and *B. dermatitidis* causes blastomycosis. In each case, infection with these environmental agents is aided by predisposing conditions that include AIDS, immunosuppressive treatment, and malnutrition.

## Innate immunity controls most fungal infections

The barriers of innate immunity control most fungi. Commensal organisms also control the growth of certain potential pathogens, as demonstrated by long-term treatment with broad-spectrum antibiotics, which destroy normal mucosal bacterial flora and often lead to oral or vulvovaginal infection with *Candida albicans,* an opportunistic agent. Phagocytosis by neutrophils is a strong defense against most fungi, and so people with neutropenia (low neutrophil count) are susceptible to fungal disease.

The alternative and lectin pathways of complement activation are triggered by components present in most fungal cell walls, and resolution of infection in normal, healthy individuals is rapid. Mannose-binding protein recognizes some major fungal pathogens, including *C. albicans* and some strains of *C. neoformans* and *A. fumigatus.* Activation of complement by either the alternative or lectin pathway allows binding of components to the organism and subsequent phagocytosis and intracellular destruction by oxygen-dependent or oxygen-independent killing mechanisms. In the case of pulmonary infections, surfactant proteins present in the lung bind pathogens and enhance their phagocytosis. Cell-bound receptors, including the complement receptors CR1, CR3, and CR4 (see Chapter 7), as well as certain PRRs, bind fungal proteins and mediate phagocytosis or initiate cytokine expression that brings immune cells into play. The role of CR3, was confirmed by the fact that mortality from experimental infections of mice with *Cryptococcus* increased after an antibody to CR3 was administered. Recent studies of the role of TLRs in dealing with fungal infection show that TLR2 and TLR4 signal responses to *C. neoformans, A. fumigatus,* and *C. albicans* through MyD88 (see Chapter 3), although the exact mechanism of the response is not yet clear.

## Immunity against fungal pathogens can be acquired

A convincing demonstration of acquired immunity against fungal infection is the protection against subsequent attacks that follows infection. This protection is readily demonstrated for many bacteria and viruses but is less obvious for fungal disease because primary infection often goes unnoticed. Positive skin reactivity to fungal antigens is a good indicator of prior infection and the presence of cellular immunity. For tissue infections, a granulomatous inflammation controls spread of *C. neoformans* and *H. capsulatum,* indicating the presence of acquired cell-mediated immunity. The organism may remain in a latent state within the granuloma, reactivating only if the host becomes immunosuppressed.

The presence of antibodies usually signals that an infection was resolved and that there is lasting immunity. For example, antibodies against *C. neoformans* are commonly found in healthy subjects, indicating that infection has occurred. As discussed in Chapter 19, a vaccine based on a common cell surface polysaccharide can confer protection against several common fungi. Other studies have shown that transfer of antibodies has a similar effect.

Perhaps the most rigid test for the type of immunity most effective against fungal pathogens comes from the observation of normally rare diseases in patients with compromised immunity. Although AIDS patients suffer increased incidences of mucosal candidiasis, histoplasmosis, coccidiomycosis, and crytococcosis, there are limits to the ranges and extent of disease observed. Oropharyngeal candidiasis is frequent in women with AIDS, but vulvovaginal and systemic infection with *C. albicans* are rare. These observations in T cell–compromised AIDS patients and data showing that B cell–deficient mice have no increased susceptibility to

fungal disease indicate that other mechanisms of immunity control fungal pathogens.

The study of immunity to fungal pathogens is relatively undeveloped compared to other areas of immunology. However, the number of individuals with degrees of immunodeficiency is rising due to the spread of AIDS and the use of immunosuppressive drugs for other conditions, ensuring that research activity will increase in this area. The mechanisms by which these ubiquitous pathogens are controlled have co-evolved with their mammalian hosts, and their elucidation should open new avenues of understanding in immunology.

## Emerging Infectious Diseases

Several times a year, it seems, we hear about a new virus or bacterium arising, often in a faraway location and accompanied by severe illness or death. Newly described pathogens are referred to as *emerging pathogens*. Some emerging pathogens that have been described since the early 1970s are shown in Figure 18-14. HIV is an example of a newly emerged pathogen.

### Diseases may re-emerge for various reasons

Periodically, once-feared pathogens that have dropped from view suddenly began to infect a widening number of

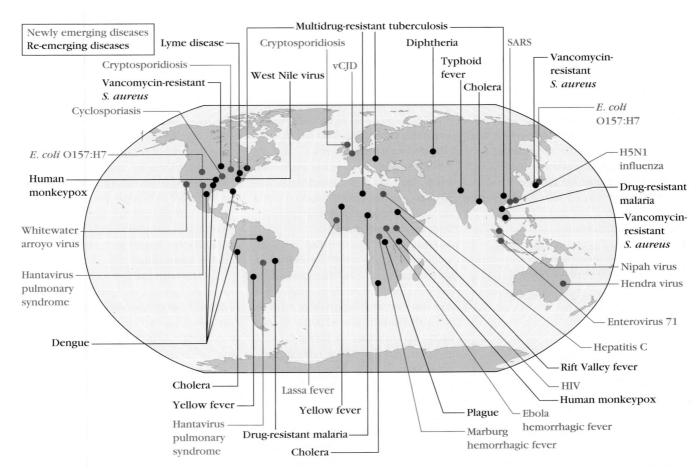

**FIGURE 18-14   Examples of emerging and re-emerging diseases showing points of origin.** Red represents newly emerging diseases, black re-emerging diseases. *[Adapted from A. S. Fauci, 2001, Infectious diseases: considerations for the 21st Century, Clinical Infectious Diseases 32:675.]*

individuals. These outbreaks are referred to as *re-emerging* infectious diseases. The re-emergence of these diseases is not surprising if we consider that bacteria can adapt to almost any environment. If they can adapt to the searing temperatures of thermal vents deep in the ocean, it is not difficult to accept that they can evolve to evade antimicrobial drugs. Examples of re-emerging diseases are also noted in Figure 18-14. The additional risk from intentionally disseminated diseases is discussed in the Clinical Focus below.

## CLINICAL FOCUS
# The Threat of Infection from Potential Agents of Bioterrorism

**The use** of human pathogens as weapons has a long history. Lord Jeffery Amherst used smallpox against native American populations before the Revolutionary War, and there are reports of attempts to spread plague and anthrax in both the distant and recent past. A few years ago, members of a dissident cult in Oregon introduced salmonella into the salad bars of several restaurants in an attempt to cause sickness and death. In 2001, anthrax spores were mailed to congressmen and news offices.

Pathogens and toxins with potential for use as weapons are called "select agents" and include bacteria, bacterial toxins, and certain viruses (see table below). The threat from such agents depends on both the severity of the disease it causes and the ease with which it can be disseminated. For example, Ebola virus causes a fulminating hemorrhagic disease, but spread of the virus requires direct contact with infected body

fluids. More worrisome are pathogens that can be spread by aerosol contact, such as anthrax, and toxins that can be added to food or water supplies, such as botulinum toxin.

It is ironic that one of the most feared bioterrorism agents is smallpox, the target of the first vaccine. Smallpox is caused by the virus *Variola major;* 30% or more of those infected with this virus die within a month of exposure, and survivors may be horribly scarred. Smallpox can spread rapidly, even before symptoms are visible. As described in Chapter 1, the vaccine for smallpox is a virus *(Vaccinia)* related to variola, which in most cases causes a localized pustule that resolves within 3 weeks. Smallpox disappeared as a consequence of widespread vaccination—the last reported case of natural infection was in 1977. As the disease was eradicated, vaccination was discontinued. In the United States, vaccination ceased in 1972. Production of the vaccine also ceased, and the remaining doses were put into storage.

Reasons for discontinuing smallpox vaccination include side effects that affect approximately 40 individuals per million vaccinees. These can be life threatening and take the form of encephalitis or disseminated skin infection. In addition, recently vaccinated individuals can spread the virus to others, especially those with compromised immunity. The occasional negative reactions to vaccinia can be treated by the administration of immunoglobulin isolated from sera of persons previously vaccinated, but this so-called Vaccinia IG, or VIG, is no longer produced and little remains available. The threat of smallpox as a bioterror-

ism agent means that vaccination must be reconsidered. It is unlikely that the vaccine produced today will be the same one used earlier. The old vaccine was produced by infecting the scarified skin of calves and scraping the infected area to collect virus. A new vaccine candidate will likely be produced under controlled conditions in a tissue-cultured cell line that is certified free of any contaminating viruses, and the actual virus used may be a more highly attenuated form of vaccinia. Stocks of VIG must be replenished before a mass vaccination effort can begin.

Most of the viruses on the select agent list are not easy to disseminate. Agents of bioterrorism prepared in a form that allows easy dispersal are referred to as *weaponized.* Although nightmare scenarios include customized viral agents engineered in the laboratory, the more probable weaponized pathogens are bacteria. An accidental release of anthrax *(Bacillus anthracis)* in Sverdlovsk in the former Soviet Union infected 79 people, of whom 68 died, underscoring the deadly potential of this organism. In late 2001, mail containing anthrax (see figure opposite) infected a number of people in multiple postal centers as the letters progressed to their destinations, giving a glimpse of how widely and rapidly a bioweapon might be spread through modern infrastructure.

*Bacillus anthracis* is a common veterinary pathogen and, like smallpox, was the subject of early vaccine efforts, in this case by Louis Pasteur. Human infection was found mainly in those working with hair or hides from animals, especially goats. Infection occurs via three different routes:

- **Inhalation** causes severe flu-like illness with high mortality unless diagnosed and treated immediately with antibiotics such as penicillin, doxycycline, or ciprofloxacin.

| Category A agents of bioterrorism |
| --- |
| Anthrax *(Bacillus anthracis)* |
| Botulism *(Clostridium botulinum toxin)* |
| Plague *(Yersinia pestis)* |
| Smallpox *(Variola major)* |
| Tularemia *(Francisella tularensis)* |
| Viral hemorrhagic fevers (filoviruses [e.g., Ebola, Marburg] and arenaviruses [e.g., Lassa, Machupo]) |

Tuberculosis is a re-emerging disease now receiving considerable attention. Fifteen years ago, public health officials were convinced that tuberculosis would soon disappear as a major health consideration in the United States. A series of events conspired to interrupt the trend, including the AIDS epidemic and an increase in immunosuppressed individuals due to other causes, allowing TB strains to gain a foothold and evolve resistance to the conventional battery of antibiotics. These individuals then passed on newly emerged, antibiotic-resistant strains of *M. tuberculosis* to others.

- **Cutaneous** exposure results in skin lesions with a characteristic black deep eschar. Cutaneous anthrax has a 20% mortality if untreated but usually responds to antibiotics.

- **Gastrointestinal** exposure results in ulcers in the ileum or cecum, bloody diarrhea, and sepsis and is nearly always fatal because of difficulty in diagnosis.

*Bacillus anthracis* is particularly deadly because the bacillus forms spores that are stable to heat, dryness, sunlight, and other factors that normally limit pathogen viability. It is relatively simple to induce spore formation, and it is spores that are used as bioweapons. Primate studies suggest that inhalation of 2500 to 55,000 spores will cause fatal disease, although the number is controversial. Victims may have flu-like symptoms; a chest x-ray will reveal a characteristic widening of the mediastinum, and blood smears will show gram-positive bacilli. Since prompt diagnosis and treatment is required for survival, it is essential that medical personnel be trained to recognize the disease.

A vaccine has been developed for anthrax, but its use has been limited to the military. The present preparation is a filtrate from cultures of a non-spore-forming strain of *B. anthracis*. Newly proposed vaccines are based on the mechanism used by the organism to infect target cells and the structure and function of anthrax-derived proteins. The major protein involved in infection is the so-called protective antigen, or PA, which pairs with either edema factor (EF) or lethal factor (LF) to cause productive infection. Antibodies that target the binding site on PA for either LF or EF are being developed as the next generation of vaccines against anthrax.

The threat from select agents of bioterrorism, like that from emerging diseases, is being addressed by careful attention to unusual infection events and by increased study of agents that lend themselves to weaponization. Research to determine the efficacy of various treatments against select agents and the windows of immunity that result from administration of antitoxins has risen to top priority in the United States following the events of September 11, 2001.

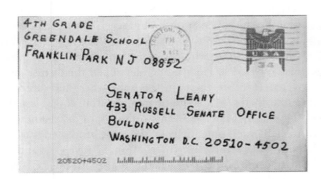

Letters to congressmen and news agencies that contained anthrax spores.
*[Courtesy of the Federal Bureau of Investigation.]*

Although the rate of infection with *M. tuberculosis* in the United States increased sharply during the early part of the 1990s, by 1995, the incidence had begun to decline again. However, the worldwide incidence of the disease is still increasing, and the World Health Organization predicts that between 1998 and 2020, 1 billion more people will become infected and over 70 million will die from this disease if preventive measures are not adopted.

Another re-emerging disease is diphtheria. In recent years, diphtheria had almost vanished from Europe because of vaccination; in 1994, however, scattered cases were reported in some of the republics of the former Soviet Union. By 1995, over 50,000 cases were reported in the same region, and thousands died from diphtheria infection. The social upheaval and instability that came with the breakup of the Soviet Union was almost certainly a major factor in the re-emergence of this disease, due to lapses in public health measures. Perhaps most important was the interruption of immunization programs. Since 1995, immunization has been re-established, and the trend has reversed, with only 13,687 cases of diphtheria reported in Russian republics in 1996, 6932 in 1998, and 1573 in 2000.

## Some fatal diseases have appeared recently

Other diseases have appeared seemingly from nowhere and, as far as we know, are caused by new pathogens. These pathogens include the widely publicized Ebola virus and *Legionella pneumophilia,* the bacterial causative agent for Legionnaires' disease. Ebola was first recognized after an outbreak in Africa in 1976 and received a great deal of attention because of the severity of disease and the rapid progression to death after the onset of symptoms. By 1977, the virus that causes this disease had been isolated and classified as a filovirus, a type of RNA virus that includes the similarly deadly Marburg virus, a close relative of Ebola. Ebola causes a particularly severe hemorrhagic fever that kills more than 50% of those infected. However, although the risk of death is very high after infection, it is fairly easy to control the spread of the virus by isolating infected individuals, a protocol that proved effective in containing the virus during the two most recent outbreaks.

Legionnaires' disease is a virulent pneumonia first reported in 221 individuals who had attended an American Legion convention in Philadelphia in 1976. Of the 221 afflicted, 34 died from the infection. The organism causing the disease was a mystery until identification of a bacterium that was named *Legionella pneumophilia*. This bacterium proliferates in cool, damp areas, including the condensing units of large commercial air-conditioning systems. Infection can spread when the air-conditioning system emits an aerosol containing the bacteria, as happened at the 1976 convention in Philadelphia. Because the hazard of such aerosols is now recognized, improved design of air-conditioning and plumbing systems has greatly reduced the danger from the disease.

In 1999, a new virus emerged in the Western Hemisphere. West Nile virus was first isolated in Uganda in 1937 and was not found outside Africa and western Asia until 1999, when it was detected in the New York City metropolitan area. By 2006, humans infected with West Nile virus had been reported in all but six states in the continental United States, a rapid spread for such a short period. West Nile virus is a flavivirus, a group of viruses usually spread by insects, often mosquitoes. The most common reservoir of the virus is birds; crows are particularly sensitive. Mosquitoes bite an infected bird and, most commonly, the virus-infected mosquito passes the virus to another bird. On occasion, the mosquito bites a human, infecting that individual with the virus. Since West Nile is not contagious between humans, it doesn't spread among human populations, and in all but a small proportion of humans, West Nile infection does not cause disease. Only in individuals with compromised immune function is the virus a health hazard. Because this virus can cross the blood-brain barrier in compromised individuals, it can cause life-threatening encephalitis or meningitis, the usual cause of death by this pathogen. Between 1999 and 2001, West Nile sickened 131 people and caused 18 deaths in the United States; in 2004, 2539 cases of West Nile had been reported to the Centers for Disease Control, with 100 deaths. These statistics indicate that West Nile is spreading and is a virus to monitor carefully. Current public health control mechanisms include education of the public about mosquito control.

## The SARS outbreak triggered a rapid international response

In November 2002, an unexplained atypical pneumonia was seen in the Guandong province of China. Within the next few months, this illness broke out in seven other provinces while proving resistant to any available treatment. In February 2003, a physician who had cared for patients ill with this disease traveled to Hong Kong, developed the disease, and infected 16 other guests in the hotel where he stayed. These guests then seeded a multinational outbreak that lasted until the end of May 2003. This outbreak of an unknown, untreatable, and potentially fatal disease prompted travel advisories, screening of international travelers, and strict quarantine of those who had contact with infected individuals. By the time the disease was contained through these extraordinary public health measures, 8096 cases had been reported, with 774 deaths. Along with the human tragedy and anxiety provoked by this disease, called severe acute respiratory syndrome (SARS), it is estimated that economic losses mounted to the tens of billions of dollars due to interruptions of travel and commerce in Asia.

A rapid response by the biomedical research community soon identified the etiologic agent of SARS as a coronavirus, so named because the spike proteins emanating from these viruses give them a crownlike appearance (Figure 18-15a). Human coronaviruses had been known for many years, but the principal human disease attributed to them was a rather mild form of the common cold. However, the newly emerged variant had not been previously seen in humans. It was soon traced to the civet cat after the earliest cases of SARS were found to be animal vendors. More recent data

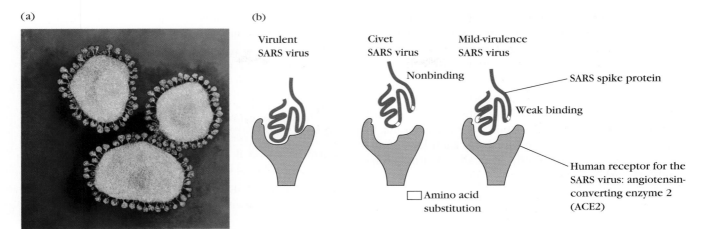

**FIGURE 18-15 The coronavirus that caused the outbreak of severe acute respiratory syndrome, or SARS.** (a) The virus is studded with spikes that in cross section give it the appearance of a crown; hence the name coronavirus. (b) The human receptor for the SARS virus is angiotensin-converting enzyme 2, or ACE2. The spike protein from the highly virulent form of the virus binds with high affinity to this receptor; the viruses found in the civet cat and in a version that causes mild human disease have two different amino acid substitutions in the spike protein. The civet cat SARS virus does not bind to the human receptor, and the version that causes mild disease in humans binds 1000 times more weakly than the most virulent form. *[Part a from Dr. Linda Stannard, UCT/Photo Researchers; part b adapted from K. V. Holmes, 2005, Science **309**:1822.]*

suggest that the bat may be the primary reservoir of the SARS coronavirus. Animal models developed for SARS virus infections showed that antibodies to the outer spike protein could thwart replication of the virus. Several vaccine candidates were proposed, and tests began. Fortunately, the feared epidemic the following winter did not occur, and a repeat of the massive public health effort was not needed.

The burning questions from the SARS outbreak are: How did this virus make the jump from animal to human, and how was it able to spread so rapidly in the human population? Answers are given by x-ray crystallographic structures of the spike protein of the SARS virus in contact with its receptor, human angiotensin-converting enzyme type 2 or ACE2. Mutations in two amino acids of the virus spike protein that contacts ACE2 converted the civet virus to one that binds 1000 times more tightly to the human protein. The importance of the binding affinity between the spike protein and ACE2 is confirmed by sequence analysis of a SARS virus from a mild case of the disease. The spike protein of this attenuated form of the virus has mutations at the key binding residues and binds only weakly to ACE2 (Figure 18-15b). The 2003 SARS epidemic provides a frightening example of the speed with which a mutation in a pathogen can move it into the human population and the potential devastation that can result.

Why are these new diseases emerging and others re-emerging? One reason suggested by public health officials is the crowding of the world's poorest populations into small areas within huge cities. Another factor is the great increase in international travel: an individual can become infected on one continent and then spread the disease on another continent all in the same day. The mass distribution and importation of food can expose large populations to potentially contaminated food. Indiscriminate use of antibiotics in humans and in veterinary applications fosters resistance in pathogens to the drugs commonly used against them. Changes in climate and weather patterns along with the extension of human populations into the previously unbroached domains of animals introduces new pathogens into human populations. Laxity in adherence to vaccination programs leads to re-emergence of diseases that were nearly eradicated. The World Health Organization and the Centers for Disease Control both actively monitor new infections and work together closely to detect and identify new infectious agents and to provide up-to-date information for travelers to parts of the world where such emerging and re-emerging infectious agents may pose a risk.

## SUMMARY

- Innate immune responses form the initial defense against pathogens. These responses include physical barriers, such as skin, and the nonspecific production of complement components, phagocytic cells, and certain cytokines in response to infection by various pathogens.

- The immune response to viral infections involves both humoral and cell-mediated components. Virus mutate rapidly and thus evade the humoral antibody response.

- The immune response to extracellular bacterial infections is generally mediated by antibody. Antibody can activate complement-mediated lysis of the bacterium, neutralize toxins, and serve as an opsonin to increase phagocytosis. Host defense against intracellular bacteria depends largely on CD4$^+$ T-cell-mediated responses.

- Both humoral and cell-mediated immune responses have been implicated in immunity to protozoan infections. In general, humoral antibody is effective against blood-borne stages of the protozoan life cycle, but once protozoans have infected host cells, cell-mediated immunity is necessary. Protozoans escape the immune response through several mechanisms.

■ Helminths are large parasites that normally do not multiply within cells. Because few of these organisms are carried by an affected individual, immune system exposure to helminths is limited; consequently, only a low level of immunity is induced. Helminths generally are attacked by antibody-mediated defenses.

■ Fungal diseases, or mycoses, are rarely severe in normal, healthy individuals but pose a greater problem for those with immunodeficiency. Both innate and acquired immunity control infection by the ubiquitous fungi.

■ Emerging and re-emerging pathogens include some newly described pathogens and others previously thought to have been controlled by public health practices. Factors leading to the emergence of such pathogens include increased travel and intense crowding of some populations.

## References

Alcami, A., and U. H. Koszinowski. 2000 Viral mechanisms of immune evasion. *Trends in Microbiology* **8:**410.

Bloom, B. R., ed. 1994. *Tuberculosis: Pathogenesis, Protection and Control.* ASM Press, Washington, DC.

Borst, P., et al. 1998. Control of VSG gene expression sites in *Trypanosoma brucei. Molecular and Biochemical Parasitology* **91:**67.

Casadevall, A. 2002. Passive antibody administration (immediate immunity) as a specific defense against biological weapons. *Emerging Infectious Diseases* **8:**833.

Doherty, P. C. 1997. Effector CD4[+] and CD8[+] T-cell mechanisms in the control of respiratory virus infections. *Immunological Reviews* **159:**105.

Kaufmann, S. H., A. Sher, and R. Ahmed, eds. 2002. *Immunology of Infectious Diseases.* ASM Press, Washington, DC.

Knodler, L. A., J. Celli, and B. B. Finlay. 2001. Pathogenic trickery: deception of host cell processes. *Nature Reviews Molecular Cell Biology* **2:**578–88.

Lamm, M. E. 1997. Interaction of antigens and antibodies at mucosal surfaces. *Annual Review of Microbiology* **51:**311.

Lane, H. C., et al. 2001. Bioterrorism: a clear and present danger. *Nature Medicine* **7:**1271.

Levitz, S. M. 2004. Interactions of Toll-like receptors with fungi. *Microbes and Infection* **6:**1351.

Li, F., et al. 2005. Structure of SARS coronavirus spike receptor-binding domain complexed with receptor. *Science* **309:**1864.

Lorenzo, M. E., H. L. Ploegh, and R. S. Tirabassi. 2001. Viral immune evasion strategies and the underlying cell biology. *Seminars in Immunology* **13:**1–9.

Merrell, D. S., and S. Falkow. 2004. Frontal and stealth attack strategies in microbial pathogenesis. *Nature* **430:**250.

Morens, D. M., G. K. Folkers, and A. S. Fauci. 2004. The challenge of emerging and re-emerging infectious diseases. *Nature* **430:**242.

Pestka, S., C. D. Krause, and M. R. Walter. 2004. Interferons, interferon-like cytokines and their receptors. *Immunological Reviews* **202:**8.

Rosenthal, S. R., et al. 2001. Developing new smallpox vaccines. *Emerging Infectious Diseases* **7:**920.

Schofield, L., and G. E. Grau. 2005. Immunological processes in malaria pathogenesis. *Nature Reviews Immunology* **5:**722.

Skowronski, D. M., et al. 2005. Severe acute respiratory syndrome. *Annual Review of Medicine* **56:**357.

Tumpey, T. M., et al. 2005. Characterization of the reconstructed 1918 Spanish influenza pandemic virus. *Science* **310:**77.

WHO Global Influenza Program Surveillance Network. 2005. Evolution of H5N1 avian influenza viruses in Asia. *Emerging Infectious Diseases* **11:**1515.

 ## Useful Web Sites

**http://www.cdc.gov/ncidod/**

The National Center for Infectious Diseases home page is a superb site for monitoring emerging diseases and a subdivision of the Centers for Disease Control (CDC). Links to the CDC are found at this site.

**http://www.niaid.nih.gov/**

National Institute of Allergy and Infectious Diseases is the NIH institute that sponsors research in infectious diseases, and its Web site provides a number of links to other relevant sites.

**http://www.who.int/**

This is the home page of the World Health Organization, the international organization that monitors infectious diseases worldwide.

**http://www.upmc-biosecurity.org/**

The University of Pittsburgh Center for Biosecurity Web site provides information about select agents and emerging diseases that may pose a security threat.

 ## Study Questions

**CLINICAL FOCUS QUESTION** VIG is used to treat individuals who display complications following administration of the smallpox vaccine. Where is VIG obtained and why is it frequently an effective treatment?

1. The effect of the MHC on the immune response to peptides of the influenza virus nucleoprotein was studied in H-2$^b$ mice that had been previously immunized with live influenza virions. The CTL activity of primed lymphocytes was determined by in vitro CML assays using H-2$^k$ fibroblasts as target cells. The target cells had been transfected with

different H-2$^b$ class I MHC genes and were infected either with live influenza or incubated with nucleoprotein synthetic peptides. The results of these assays are shown in the table below.

| Target cell (H-2$^k$ fibroblast) | Test antigen | CTL activity of influenza-primed H-2$^b$ lymphocytes (% lysis) |
|---|---|---|
| (A) Untransfected | Live influenza | 0 |
| (B) Transfected with class I $D^b$ | Live influenza | 60 |
| (C) Transfected with class I $D^b$ | Nucleoprotein peptide 365–380 | 50 |
| (D) Transfected with class I $D^b$ | Nucleoprotein peptide 50–63 | 2 |
| (E) Transfected with class I $K^b$ | Nucleoprotein peptide 365–380 | 0.5 |
| (F) Transfected with class I $K^b$ | Nucleoprotein peptide 50–63 | 1 |

a. Why was there no killing of the target cells in system A even though the target cells were infected with live influenza?

b. Why was a CTL response generated to the nucleoprotein in system C, even though it is an internal viral protein?

c. Why was there a good CTL response in system C to peptide 365–380, whereas there was no response in system D to peptide 50–63?

d. If you were going to develop a synthetic peptide vaccine for influenza in humans, how would these results obtained in mice influence your design of a vaccine?

2. Describe the nonspecific defenses that operate when a disease-producing microorganism first enters the body.

3. Describe the various specific defense mechanisms that the immune system employs to combat various pathogens.

4. What is the role of the humoral response in immunity to influenza?

5. Describe the unique mechanisms each of the following pathogens has for escaping the immune response: (a) African trypanosomes, (b) *Plasmodium* species, and (c) influenza virus.

6. M. F. Good and coworkers analyzed the effect of MHC haplotype on the antibody response to a malarial circumsporozoite (CS) peptide antigen in several recombinant congenic mouse strains. Their results are shown in the table at upper right.

a. Based on the results of this study, which MHC molecule(s) serve(s) as restriction element(s) for this peptide antigen?

b. Since antigen recognition by B cells is not MHC restricted, why is the humoral antibody response influenced by the MHC haplotype?

| Strain | H-2 alleles | | | | | Antibody response to CS peptide |
|---|---|---|---|---|---|---|
| | K | IA | IE | S | D | |
| B10.BR | k | k | k | k | k | <1 |
| B10.A (4R) | k | k | b | b | b | <1 |
| B10.HTT | s | s | k | k | d | <1 |
| B10.A (5R) | b | b | k | d | d | 67 |
| B10 | b | b | b | b | b | 73 |
| B10.MBR | b | k | k | k | q | <1 |

SOURCE: Adapted from M. F. Good et al., 1988, *Annual Review of Immunology.* **6**:633.

7. Fill in the blanks in the following statements.

a. The current vaccine for tuberculosis consists of an attenuated strain of *M. bovis* called _____.

b. Variation in influenza surface proteins is generated by _____ and _____.

c. Variation in pilin, which is expressed by many gram-negative bacteria, is generated by the process of _____.

d. The mycobacteria causing tuberculosis are walled off in granulomatous lesions called _____, which contain a small number of _____ and many _____.

e. The diphtheria vaccine is a formaldehyde-treated preparation of the exotoxin, called a _____.

f. A major contribution to nonspecific host defense against viruses is provided by _____ and _____.

g. The primary host defense against viral and bacterial attachment to epithelial surfaces is _____.

h. Two cytokines of particular importance in the response to infection with *M. tuberculosis* are _____, which stimulates development of T$_H$1 cells, and _____, which promotes activation of macrophages.

8. Despite the fact that there are no licensed vaccines for them, life-threatening fungal infections are not a problem for the general population. Why? Who may be at risk for them?

9. Discuss the factors that contribute to the emergence of new pathogens or the re-emergence of pathogens previously thought to be controlled in human populations.

10. What factors favoring new disease emergence contributed to the 2003 outbreak of SARS?

11. Which of the following are strategies used by pathogens to evade the immune system? For each correct choice, give a specific example.

a. Changing the antigens expressed on their surfaces

b. Going dormant in host cells

c. Secreting proteases to inactivate antibodies

d. Having low virulence

e. Developing resistance to complement-mediated lysis

f.  Allowing point mutations in surface epitopes, resulting in antigenic drift

g.  Increasing phagocytic activity of macrophages

12. Several diagnostic tests use ELISA technology to test patient sera for specific infections. Unfortunately, sometimes a test for Lyme disease can give a positive result when the patient is infected with *Treponema palladium* (the causative agent of syphilis) and not with *Borrelia burgdorferi*. Discuss possible explanations for such a test result.

13. Which of the following is a characteristic of the inflammatory response against extracellular bacterial infections?

a.  Increased levels of IgE

b.  Activation of self-reactive CD8$^+$ T cells

c.  Activation of complement

d.  Swelling caused by release of vasodilators

e.  Degranulation of tissue mast cells

f.  Phagocytosis by macrophages

14. Your mother may have scolded you for running around outside without shoes. This is sound advice because of the mode of transmission of the roundworm *Necator americanus*, the causative agent of hookworm.

a.  If you disobeyed your mother and caught this parasite, what cells of your immune system would fight the infection?

b.  If your doctor administered a cytokine to drive the immune response, which would be a good choice, and how would this supplement alter maturation of plasma cells to produce a more helpful class of antibody?

## Interactive Study

**www.whfreeman.com/kuby**

SELF-TEST
Review of Key Terms

MOLECULAR VISUALIZATIONS
Viral Antigens

# chapter 19

# Vaccines

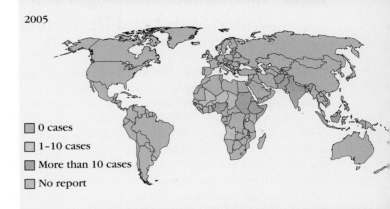

2005

- 0 cases
- 1–10 cases
- More than 10 cases
- No report

*Reported polio cases, 2005.*

- Active and Passive Immunization
- Designing Vaccines for Active Immunization
- Live, Attenuated Vaccines
- Inactivated or "Killed" Vaccines
- Subunit Vaccines
- Conjugate Vaccines
- DNA Vaccines
- Recombinant Vector Vaccines

THE DISCIPLINE OF IMMUNOLOGY HAS ITS ROOTS IN the early vaccination trials of Edward Jenner and Louis Pasteur. Since those pioneering efforts, vaccines have been developed for many diseases that were once major afflictions of mankind. The incidence of diseases such as diphtheria, measles, mumps, pertussis (whooping cough), rubella (German measles), poliomyelitis, and tetanus has declined dramatically as vaccination has become more common. Clearly, vaccination is a cost-effective weapon for disease prevention. Perhaps in no other case have the benefits of vaccination been as dramatically evident as in the eradication of smallpox, one of mankind's long-standing and most terrible scourges. Since October 1977, not a single naturally acquired smallpox case has been reported anywhere in the world. Equally encouraging is the predicted eradication of polio through large-scale vaccination programs. This program had a serious setback caused by suspension of vaccination in Nigeria and in some areas of India and Pakistan; these countries, as well as others to which the disease spread, notably Yemen and Indonesia, witnessed outbreaks of paralytic polio following this cessation of immunization. Fortunately, the program is again under way, and the introduction of a new monovalent vaccine that produces immunity against a single strain has shown efficacy in regions where only that strain of polio circulates. A new addition to the weapons against childhood disease is a vaccine against bacterial pneumonia, a major cause of infant death.

A crying need remains for vaccines against other diseases. Every year, millions throughout the world die from malaria, tuberculosis, and AIDS, diseases for which there are no effective vaccines. It is estimated by the World Health Organization (WHO) that 14,000 individuals a day become infected with HIV-1, the virus that causes AIDS. An effective vaccine could have an immense impact on the control of this tragic spread of death and disaster. In addition to the challenges presented by diseases for which no vaccines exist, the need remains to improve the safety and efficacy of present vaccines and to find ways to lower their cost and deliver them efficiently to all who need them, especially in developing countries of the world. WHO estimates that millions of infant deaths in the world are due to diseases that could be prevented by existing vaccines (see Clinical Focus on page 476).

The road to successful development of a safe and effective vaccine that can be approved for human use, manufactured at reasonable cost, and efficiently delivered to at-risk populations is long, costly, and tedious. Procedures for manufacture of materials that can be given to humans are tightly regulated, as are the protocols for testing these materials in clinical trials. Even those candidate vaccines that survive initial scrutiny and are approved for use in human trials are not guaranteed to find their way into common usage. Experience has shown that not every vaccine candidate that was successful in laboratory and animal studies prevents disease in humans. Some potential vaccines cause unacceptable side effects, and some may even worsen the disease they were meant to prevent. Live virus vaccines pose a special threat to those with primary or acquired immunodeficiency (see Chapter 20). Stringent testing is an absolute necessity, because approved vaccines will be given to large numbers of

## CLINICAL FOCUS

# Vaccination: Challenges in the United States and Developing Countries

**Many** previously common childhood diseases are seldom seen in the United States, a testament to the effectiveness of vaccination. A major barrier to similar success in the rest of the world is the difficulty of delivering vaccines to all children. However, even at home, the United States is becoming a victim of its own success. Some parents who have never encountered diseases now nearly vanquished in the United States do not consider it important to have their infants vaccinated, or they may be lax in adhering to recommended schedules of immunization. Others hold the uninformed belief that the risks associated with vaccination outweigh the risk of infection. This flawed reasoning is fueled by periodic allegations of linkage between vaccination and various disorders, such as the report circulating recently of a causal relationship between vaccination and autism, a condition of unknown etiology. Most such reports are based solely on the coincidental timing of vaccination and onset of disease or on limited sampling and poor statistical analyses. So far, no alleged associations of vaccination and disorders have withstood scrutiny that included large population samples and accepted statistical methods.

Although children in this country are protected against a variety of once-deadly diseases, this protection depends on continuation of our immunization programs. Dependency on herd immunity is dangerous for both the individual and society and may fail. For the polio vaccine, it is estimated that 95% coverage is needed to break the transmission chain of polio. Adverse reactions to vaccines must be examined thoroughly, of course, and if a vaccine causes unacceptable side reactions, the vaccination program must be reconsidered. At the same time, anecdotal reports of disease brought on by vaccines and unsupported beliefs, such as the contention that vaccines weaken the immune system, must be countered by correct information from trusted sources. To retreat from our progress in immunization by noncompliance will return us to the age when measles, mumps, whooping cough, and polio were part of the risk of growing up.

Children in the developing world suffer from a problem different from those in the United States. Examination of infant deaths worldwide shows that existing vaccines could save the lives of millions of children— there are safe, effective vaccines for five of the top 10 killers of children (Clinical Focus Table 1). Although these 10 diseases include HIV, TB, and malaria, for which no vaccines are available, administration of the vaccines that are recommended for infants in the United States could cut child mortality in the world by approximately half.

What barriers exist to achieving worldwide vaccination and complete eradication of many childhood diseases? The inability to achieve higher levels of vaccination even in the United States is an indication of the difficulty of the task. Even if suitable vaccines had been developed and compliance were universal, the ability to produce and deliver the vaccines everywhere is a major challenge. The World Health Organization (WHO) has stated that the ideal vaccine would have the following properties:

- Affordable worldwide
- Heat stable
- Effective after a single dose
- Applicable to a number of diseases
- Administered by a mucosal route
- Suitable for administration early in life

Few if any vaccines in common use today conform to all of these properties. However, the WHO goals can guide us in the pursuit of vaccines useful for worldwide application and further aid us in setting priorities, especially for development of the vaccines needed most in developing countries. For example, an HIV/AIDS vaccine that meets the WHO criteria could reach the populations most at risk and have an immediate effect on the world AIDS epidemic.

Immunization saves millions of lives, and viable vaccines are increasingly available. The challenge to the biomedical research community is to develop better, safer, cheaper, easier-to-administer forms of these vaccines so that worldwide immunization becomes a reality.

### Estimated annual deaths worldwide of children under 5 years of age, by pathogen

| Pathogen | Deaths (thousands) |
|---|---|
| *Pneumococcus** | 841 |
| Measles | 530 |
| *Haemophilus* (strains a–f)† | 945 |
| Rotavirus† | 800 |
| Malaria | 700 |
| HIV | 500 |
| RSV | 500 |
| **Pertussis** | 285 |
| **Tetanus** | 201 |
| Tuberculosis | 100 |

*Bold signifies pathogens for which an effective vaccine exists.

†A licensed vaccine is being tested for possible side effects.

SOURCE: Data derived from WHO publications.

well people. Adverse side effects, even those that occur at very low frequency, must be balanced against the potential benefit of protection by the vaccine.

Vaccine development begins with basic research. Recent advances in immunology and molecular biology have led to effective new vaccines and to promising strategies for finding new vaccine candidates. Knowledge of the differences in epitopes recognized by T cells and B cells has enabled immunologists to begin to design vaccine candidates to maximize activation of both cellular and humoral immune responses. As differences in antigen-processing pathways became evident, scientists began to design vaccines and to use adjuvants that activate innate immune processes and maximize antigen presentation with class I or class II MHC molecules. Genetic engineering techniques can be used to develop vaccines to maximize the immune response to selected epitopes and to simplify delivery of the vaccines. In this chapter we describe the vaccines now in use and describe vaccine strategies, including experimental results, that may lead to the vaccines of the future.

# Active and Passive Immunization

Immunity to infectious microorganisms can be achieved by active or passive **immunization.** In each case, immunity can be acquired either by natural processes (usually by transfer from mother to fetus or by previous infection by the organism) or by artificial means such as injection of antibodies or vaccines (Table 19-1). The agents used for inducing passive immunity include antibodies from humans or animals, whereas active immunization is achieved by inoculation with pathogens that induce immunity but do not cause dis-

| TABLE 19-1 | Acquisition of passive and active immunity |
|---|---|
| **Type** | **Acquired through** |
| Passive immunity | Natural maternal antibody |
| | Immune globulin* |
| | Humanized monoclonal antibody |
| | Antitoxin† |
| Active immunity | Natural infection |
| | Vaccines‡ |
| |    Attenuated organisms |
| |    Inactivated organisms |
| |    Purified microbial macromolecules |
| |    Cloned microbial antigens |
| |       Expressed as recombinant protein |
| |       As cloned DNA alone or in virus vectors |
| |    Multivalent complexes |
| | Toxoid§ |

*An antibody-containing solution derived from human blood, obtained by cold ethanol fractionation of large pools of plasma; available in intramuscular and intravenous preparations.

†An antibody derived from the serum of animals that have been stimulated with specific antigens.

‡A suspension of attenuated live or killed microorganisms, or antigenic portions of them, presented to a potential host to induce immunity and prevent disease.

§A bacterial toxin that has been modified to be nontoxic but retains the capacity to stimulate the formation of antitoxin.

ease or with antigenic components from the pathogens. In this section we describe current usage of passive and active immunization techniques.

## Passive immunization involves transfer of preformed antibodies

Edward Jenner and Louis Pasteur are recognized as the pioneers of vaccination, or induction of active immunity, but similar recognition is due to Emil von Behring and Hidesaburo Kitasato for their contributions to passive immunity. These investigators were the first to show that immunity elicited in one animal can be transferred to another by injecting it with serum from the first (see Clinical Focus in Chapter 4).

Passive immunization, in which preformed antibodies are transferred to a recipient, occurs naturally by transfer of maternal antibodies across the placenta to the developing fetus. Maternal antibodies to diphtheria, tetanus, streptococci, rubeola, rubella, mumps, and poliovirus all afford passively acquired protection to the developing fetus. Maternal antibodies present in colostrum and milk also provide passive immunity to the infant.

Passive immunization can also be achieved by injecting a recipient with preformed antibodies. In the past, before vaccines and antibiotics became available, passive immunization provided a major defense against various infectious diseases. Despite the risks (see Chapter 16) incurred by injecting animal sera, usually horse serum, this was the only effective therapy for otherwise fatal diseases. Currently, several conditions warrant the use of passive immunization, including the following:

- Deficiency in synthesis of antibody as a result of congenital or acquired B-cell defects, alone or together with other immunodeficiencies;

- Exposure or likely exposure to a disease that will cause complications (e.g., a child with leukemia exposed to varicella or measles) or when time does not permit adequate protection by active immunization;

- Infection by pathogens whose effects may be ameliorated by antibody. For example, if individuals who have not received up-to-date active immunization against tetanus suffer a puncture wound, they are given an injection of horse antiserum to tetanus toxin. The preformed horse antibody neutralizes any tetanus toxin produced by *Clostridium tetani* in the wound.

Passive immunization is routinely administered to individuals exposed to botulism, tetanus, diphtheria, hepatitis, measles, and rabies (Table 19-2), and it can provide immediate protection to travelers or health-care workers who will soon be exposed to an infectious organism and lack active immunity to it. Passively administered antiserum is also used to provide protection from poisonous snake and insect bites. Because passive immunization does not activate the immune system, it generates no memory response, and the protection provided is transient.

| TABLE 19-2 | Common agents used for passive immunization |
| --- | --- |
| **Disease** | **Agent** |
| Black widow spider bite | Horse antivenin |
| Botulism | Horse antitoxin |
| Cytomegalovirus | Human polyclonal Ab |
| Diphtheria | Horse antitoxin |
| Hepatitis A and B | Pooled human immunoglobulin |
| Measles | Pooled human immunoglobulin |
| Rabies | Human or horse polyclonal Ab |
| Respiratory disease | Monoclonal anti-RSV* |
| Snake bite | Horse antivenin |
| Tetanus | Pooled human immunoglobulin or horse antitoxin |
| Varicella zoster virus | Human polyclonal Ab |

*Respiratory syncytial virus

SOURCE: *Adapted from A. Casadevall, 1999, *Clinical Immunology* 93:5.

For certain diseases such as the acute respiratory failure in children caused by respiratory syncytial virus (RSV), passive immunization is the best preventative currently available. A monoclonal antibody or a combination of two monoclonal antibodies may be administered to children at risk for RSV disease. These monoclonal antibodies are prepared in mice but have been "humanized" by splicing the constant regions of human IgG to the mouse variable regions (see Chapter 5). This modification prevents many of the complications that may follow a second injection of the complete mouse antibody, which is a highly immunogenic foreign protein.

Although passive immunization may be an effective treatment, it should be used with caution because certain risks are associated with the injection of preformed antibody. If the antibody was produced in another species, such as a horse, the recipient can mount a strong response to the isotypic determinants of the foreign antibody. This anti-isotype response can cause serious complications. Some individuals, for example, produce IgE antibody specific for determinants on the injected antibody. Immune complexes of this IgE bound to the passively administered antibody can mediate systemic mast-cell degranulation, leading to systemic anaphylaxis. Other individuals produce IgG or IgM antibodies specific for the foreign antibody, which form complement-activating immune complexes. The deposition of these complexes in the tissues can lead to type III hypersensitivity reactions. Even when human gamma globulin is administered passively, the recipient can generate an anti-allotype response to the human immunoglobulin, although its intensity is usually much less than that of an anti-isotype response.

## Active immunization elicits long-term protection

Whereas the aim of passive immunization is transient protection or alleviation of an existing condition, the goal of active immunization is to elicit protective immunity and immunologic memory. When active immunization is successful, a subsequent exposure to the pathogenic agent elicits a heightened immune response that successfully eliminates the pathogen or prevents disease mediated by its products. Active immunization can be achieved by natural infection with a microorganism, or it can be acquired artificially by administration of a **vaccine** (see Table 19-1). In active immunization, as the name implies, the immune system plays an active role—proliferation of antigen-reactive T and B cells results in the formation of memory cells. Active immunization with various types of vaccines has played an important role in the reduction of deaths from infectious diseases, especially among children.

Vaccination of children is begun at about 2 months of age. The recommended program of childhood immunizations in this country, updated in 2006 by the American Academy of Pediatrics, is outlined in Table 19-3. The program includes the following vaccines:

- Hepatitis B vaccine

- Diphtheria-pertussis (acellular)-tetanus (DPaT) combined vaccine

- *Haemophilus influenzae* (Hib) vaccine

- Inactivated (Salk) polio vaccine (IPV); the oral (Sabin) vaccine is no longer recommended for use in the United States

- Measles-mumps-rubella (MMR) combined vaccine

- Varicella zoster (Var) vaccine for chickenpox

- Meningococcal vaccine (recommended at 24 months for high-risk groups; at 11 to 12 years for all children)

- Pneumococcal conjugate vaccine (PCV)

- Influenza vaccine (now recommended for children 6 to 23 months old)

- Hepatitis A vaccine

Hepatitis A vaccine is now recommended for all children between 12 and 23 months of age; previously it was given only to infants in high-risk populations. Similarly, influenza vaccines were formerly recommended for children with certain risk factors such as asthma or immunodeficiencies, but annual vaccination is now recommended for healthy children based on the increased risk of children in this age group for influenza-related hospitalization. The new list of recommendations also includes an assessment of immunity against the major childhood diseases between 11 and 12 years of age and catch-up immunizations for those requiring them.

| TABLE 19-3 | Recommended childhood immunization schedule in the United States, 2006 |
|---|---|

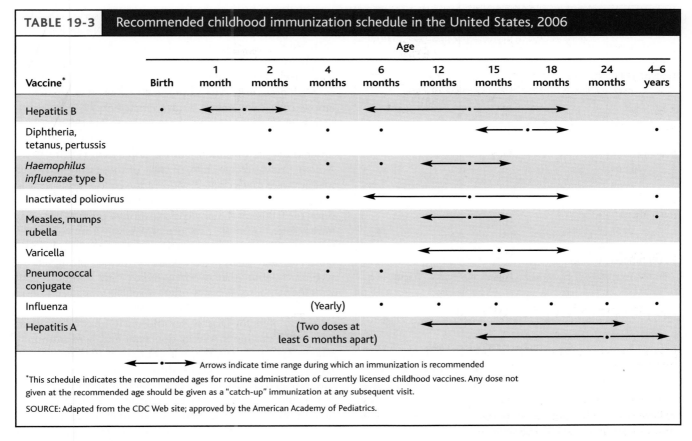

| Vacchne* | Age | | | | | | | | | |
|---|---|---|---|---|---|---|---|---|---|---|
| | Birth | 1 month | 2 months | 4 months | 6 months | 12 months | 15 months | 18 months | 24 months | 4–6 years |
| Hepatitis B | • | ←—•—→ | | | ←————————•————————→ | | | | | |
| Diphtheria, tetanus, pertussis | | | • | • | • | | ←—•—→ | | | • |
| *Haemophilus influenzae* type b | | | • | • | • | ←—•—→ | | | | |
| Inactivated poliovirus | | | • | • | ←————————•————————→ | | | | | • |
| Measles, mumps rubella | | | | | | ←—•—→ | | | | • |
| Varicella | | | | | | | ←———•———→ | | | |
| Pneumococcal conjugate | | | • | • | • | ←—•—→ | | | | |
| Influenza | | (Yearly) | | | • | • | • | • | • | • |
| Hepatitis A | | (Two doses at least 6 months apart) | | | | ←————•————————————→ | | | | |
| | | | | | | ←————————•————————→ | | | | |

←—•—→ Arrows indicate time range during which an immunization is recommended

*This schedule indicates the recommended ages for routine administration of currently licensed childhood vaccines. Any dose not given at the recommended age should be given as a "catch-up" immunization at any subsequent visit.

SOURCE: Adapted from the CDC Web site; approved by the American Academy of Pediatrics.

The introduction and spreading use of various vaccines for childhood immunization has led to a dramatic decrease in the incidence of common childhood diseases for which vaccination is currently recommended in the United States (Figure 19-1). The comparisons of disease incidence in 2004 to that reported in the peak years show dramatic drops and, in two cases, complete elimination of the disease in the United States. As long as widespread, effective immunization programs are maintained, the incidence of these childhood diseases should remain low. However, the occurrence of side reactions to a vaccine may cause a drop in its use, which can lead to re-emergence of that disease. For example, the side effects from the pertussis attenuated bacterial vaccine included seizures, encephalitis, brain damage, and even death. Decreased usage of the vaccine led to an increase in the incidence of whooping cough, with 18,957 cases in 2004. The recent development of an acellular pertussis vaccine that is as effective as the older vaccine but with none of the side effects is expected to reverse this trend.

As indicated in Table 19-3, children typically require multiple boosters (repeated inoculations) at appropriately timed intervals to achieve effective immunity. In the first months of life, the reason for this may be persistence of circulating maternal antibodies in the young infant. For example, passively acquired maternal antibodies bind to epitopes on the DPT vaccine and block adequate activation of the immune system; therefore, this vaccine must be given several times after the maternal antibody has been cleared from an infant's circula-

tion in order to achieve adequate immunity. Passively acquired maternal antibody also interferes with the effectiveness of the measles vaccine; for this reason, the MMR vaccine is not given before 12 to 15 months of age. In Third World countries, however, the measles vaccine is administered at 9 months, even though maternal antibodies are still present, because 30% to 50% of young children in these countries contract the disease before 15 months of age. Multiple immunizations with the polio vaccine are required to ensure that an adequate immune response is generated to each of the three strains of poliovirus that make up the vaccine.

Recommendations for vaccination of adults depend on the risk group. Vaccines for meningitis, pneumonia, and influenza are often given to groups living in close quarters (e.g., military recruits and incoming college students) or to individuals with reduced immunity (e.g., the elderly). Depending on their destination, international travelers are also routinely immunized against such endemic diseases as cholera, yellow fever, plague, typhoid, hepatitis, meningitis, typhus, and polio. Immunization against the deadly disease anthrax had been reserved for workers coming into close contact with infected animals or products from them. Recently, however, suspected use of anthrax spores by terrorists or in biological warfare has widened use of the vaccine to military personnel and civilians in areas at risk of attack with this deadly agent.

Vaccination is not 100% effective. With any vaccine, a small percentage of recipients will respond poorly and therefore

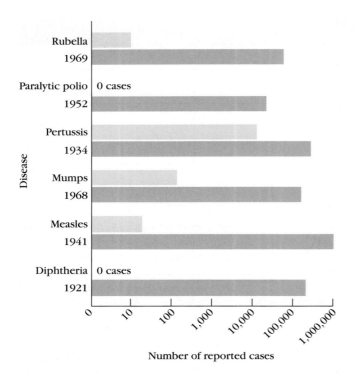

**FIGURE 19-1 Reported annual number of cases of rubella (German measles), polio, pertussis (whooping cough), mumps, measles, and diphtheria in the United States in the peak year for which data are available (orange) compared with the number of cases of each disease in 2004 (green).** Currently, vaccines are available for each of these diseases, and vaccination is recommended for all children in the United States. [Data from Centers for Disease Control.]

will not be adequately protected. This is not a serious problem if the majority of the population is immune to an infectious agent. In this case, the chance of a susceptible individual contacting an infected individual is so low that the susceptible one is not likely to become infected. This phenomenon is known as *herd immunity*. The appearance of measles epidemics among college students and unvaccinated preschool-age children in the United States during the mid- to late 1980s resulted partly from an overall decrease in vaccinations, which had lowered the herd immunity of the population (Figure 19-2). Among preschool-age children, 88% of those who developed measles were unvaccinated. Most of the college students who contracted measles had been vaccinated as children but only once; the failure of the single vaccination to protect them may have resulted from the presence of passively acquired maternal antibodies that reduced their overall response to the vaccine. The increase in the incidence of measles prompted the recommendation that children receive two immunizations with the combined measles-mumps-rubella vaccine, one at 12 to 15 months of age and the second at 4 to 6 years.

The Centers for Disease Control (CDC) has called attention to the decline in vaccination rates and herd immunity among American children. For example, a 1995 publication reported that in California nearly one-third of all infants were unvaccinated and about half of all children under the age of 2 were behind schedule on their vaccinations. Such a decrease in herd immunity portends serious consequences, as illustrated by recent events in the newly independent states of the former Soviet Union. By the mid-1990s, a diphtheria epidemic was raging in many regions of these new countries, linked to a decrease in herd immunity resulting from decreased vaccination rates after the breakup of the Soviet Union. This epidemic, which led to over 157,000 cases of diphtheria and 5000 deaths, is now controlled by mass immunization programs.

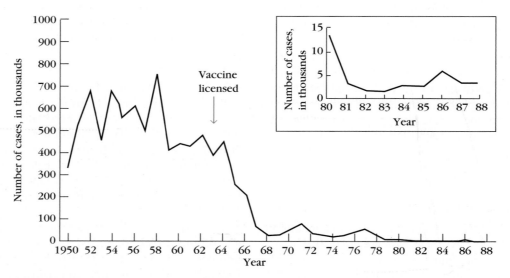

**FIGURE 19-2 Introduction of the measles vaccine in 1962 led to a dramatic decrease in the annual incidence of this disease in the United States.** Occasional outbreaks of measles in the 1980s (inset) occurred mainly among unvaccinated young children and among college students; most of the latter had been vaccinated but only once, when they were young. [Data from Centers for Disease Control.]

# Designing Vaccines for Active Immunization

The majority of infections are dealt with by nonspecific or innate immune mechanisms, which include complement, interferons, NK cells, activated phagocytes, antimicrobial compounds, and other mechanisms. The adaptive immune response provides a more flexible resistance to pathogens. Vaccination educates the adaptive immune system, preparing it to deal effectively and swiftly with pathogens not readily eliminated by innate immunity.

Several factors must be kept in mind in developing a successful vaccine. First, the development of a measurable immune response does not necessarily mean that a state of protective immunity has been achieved. Often critical is the branch of the immune system that is activated, and therefore vaccine designers must recognize the important differences between activation of the humoral and cell-mediated branches. A second factor is the development of immunologic memory. For example, a vaccine that induces a protective primary response may fail to induce the formation of memory cells, leaving the host unprotected after the primary response to the vaccine subsides.

The role of memory cells in immunity depends, in part, on the incubation period of the pathogen. In the case of influenza virus, which has a very short incubation period (1 or 2 days), disease symptoms are already under way by the time memory cells are activated. Effective protection against influenza therefore depends on maintaining high levels of neutralizing antibody by repeated immunizations; those at highest risk are immunized each year. For pathogens with a longer incubation period, maintaining detectable neutralizing antibody at the time of infection is not necessary. The poliovirus, for example, requires more than 3 days to begin to infect the central nervous system. An incubation period of this length gives the memory B cells time to respond by producing high levels of serum antibody. Thus, the vaccine for polio is designed to induce high levels of immunologic memory. After immunization with the Salk vaccine, serum antibody levels peak within 2 weeks and then decline, but the memory response continues to climb, reaching maximal levels at 6 months and persisting for years (Figure 19-3). If an immunized individual is later exposed to the poliovirus, these memory cells will respond by differentiating into plasma cells that produce high levels of serum antibody, which defend the individual from the effects of the virus.

In the remainder of this chapter, various approaches to the design of vaccines—both currently used vaccines and experimental ones—are described, with an examination of the ability of the vaccines to induce humoral and cell-mediated immunity and the production of memory cells.

As Table 19-4 indicates, the common vaccines currently in use consist of live but attenuated organisms, inactivated (killed) bacterial cells or viral particles, or protein

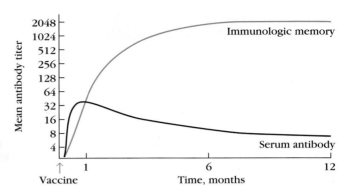

**FIGURE 19-3 Immunization with a single dose of the Salk polio vaccine induces a rapid increase in serum antibody levels, which peak by 2 weeks and then decline.** Induction of immunologic memory follows a slower time course, reaching maximal levels 6 months after vaccination. The persistence of the memory response for years after primary vaccination is responsible for immunity to poliomyelitis. *[From M. Zanetti et al., 1987,* Immunology Today *8:18.]*

or carbohydrate fragments (subunits) of the target organism. Under careful study are several new types of vaccine candidate that provide advantages in protection, production, or delivery to those in need. The primary characteristics and some advantages and disadvantages of the different types of vaccines are included in this listing. The following sections describe each of the vaccine types, indicating how they are currently used or how they may be best applied to human diseases.

# Live, Attenuated Vaccines

In some cases, microorganisms can be attenuated so that they lose their ability to cause significant disease (pathogenicity) but retain their capacity for transient growth within an inoculated host. There are examples of agents that are naturally attenuated by virtue of their inability to cause disease in a given host although they can immunize the host. The first vaccine used by Jenner is of this type; vaccinia virus (cowpox) inoculation of humans confers immunity to smallpox but does not cause smallpox. Attenuation can often be achieved by growing a pathogenic bacterium or virus for prolonged periods under abnormal culture conditions. This procedure selects mutants that are better suited to growth in the abnormal culture conditions and are therefore less capable of growth in the natural host. For example, an attenuated strain of *Mycobacterium bovis* called **Bacillus Calmette-Guérin (BCG)** was developed by growing *M. bovis* on a medium containing increasing concentrations of bile. After 13 years, this strain had adapted to growth in strong bile and had become sufficiently attenuated that it was suitable as a vaccine for tuberculosis. For reasons related to variable effectiveness and difficulties in follow-up monitoring, BLG is not used in the

| TABLE 19-4 | Classification of common vaccines for humans | | |
| --- | --- | --- | --- |
| **Vaccine type** | **Diseases** | **Advantages** | **Disadvantages** |
| Live attenuated | Measles<br>Mumps<br>Polio (Sabin vaccine)<br>Rotavirus<br>Rubella<br>Tuberculosis<br>Varicella<br>Yellow fever | Strong immune response; often lifelong immunity with few doses | Requires refrigerated storage; may mutate to virulent form |
| Inactivated or killed | Cholera<br>Influenza<br>Hepatitis A<br>Plague<br>Polio (Salk vaccine)<br>Rabies | Stable; safer than live vaccines; refrigerated storage not required | Weaker immune response than live vaccines; booster shots usually required |
| Toxoid | Diphtheria<br>Tetanus | Immune system becomes primed to recognize bacterial toxins | |
| Subunit (inactivated exotoxin) | Hepatitis B<br>Pertussis<br>Streptococcal pneumonia | Specific antigens lower the chance of adverse reactions | Difficult to develop |
| Conjugate | *Haemophilus influenzae* type B<br>Streptococcal pneumonia | Primes infant immune systems to recognize certain bacteria | |
| DNA | In clinical testing | Strong humoral and cellular immune response; relatively inexpensive to manufacture | Not yet available |
| Recombinant vector | In clinical testing | Mimics natural infection, resulting in strong immune response | Not yet available |

United States. The Sabin polio vaccine and the measles vaccine both consist of attenuated viral strains. The poliovirus used in the Sabin vaccine was attenuated by growth in monkey kidney epithelial cells. The measles vaccine contains a strain of rubella virus that was grown in duck embryo cells and later in human cell lines.

Attenuated vaccines have advantages and disadvantages. Because of their capacity for transient growth, such vaccines provide prolonged immune system exposure to the individual epitopes on the attenuated organisms, resulting in increased immunogenicity and production of memory cells. As a consequence, these vaccines often require only a single immunization, eliminating the need for repeated boosters. This property is a major advantage in Third World countries, where epidemiologic studies have shown that a significant number of individuals fail to return for subsequent boosters. The ability of many attenuated vaccines to replicate within host cells makes them particularly suitable for inducing a cell-mediated response.

The Sabin polio vaccine, consisting of three attenuated strains of poliovirus, is administered orally to children on a sugar cube or in sugar liquid. The attenuated viruses colonize the intestine and induce protective immunity to all three strains of virulent poliovirus. Sabin vaccine in the intestines induces production of secretory IgA, which serves as an important defense against naturally acquired poliovirus. The vaccine also induces IgM and IgG classes of antibody. Unlike most other attenuated vaccines, which require a single immunizing dose, the Sabin polio vaccine requires boosters, because the three strains of attenuated poliovirus in the vaccine interfere with each other's replication in the intestine. With the first immunization, one strain will predominate in its growth, inducing immunity to that strain. With the second immunization, the immunity generated by the previous immunization will limit the growth of the previously predominant strain in the vaccine, enabling one of the two remaining strains to predominate and induce immunity. Finally, with the third immunization, immunity to all three strains is achieved.

A major disadvantage of attenuated vaccines is the possibility that they will revert to a virulent form. The rate of reversion of the Sabin polio vaccine (OPV) leading to subsequent paralytic disease is about one case in 2.4 million doses of vaccine. This reversion implies that pathogenic forms of the virus are being passed by a few immunized individuals

**Reported polio cases**

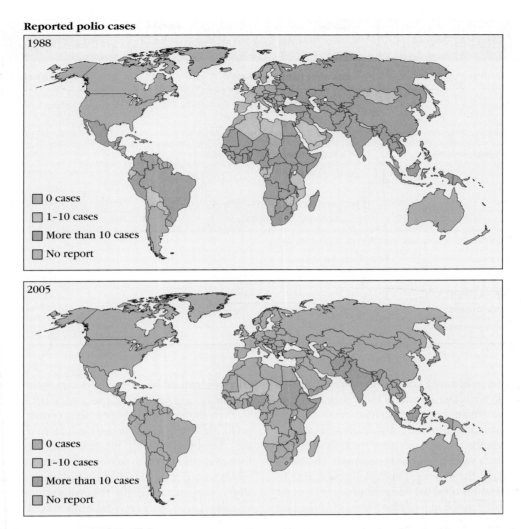

FIGURE 19-4 **Progress toward the worldwide eradication of polio.** Comparison of infection numbers for 1988 with those for 2005 show considerable progress in most parts of the world, although some areas in Africa and Asia have shown recent increases. Some experts question whether the use of live attenuated oral polio vaccine will cause reversion to pathogenic forms at a rate sufficiently high to prevent total eradication of this once-prevalent crippling disease. *[Data from WHO.]*

and can find their way into the water supply, especially in areas where sanitation standards are not rigorous or where wastewater must be recycled. This possibility has led to the exclusive use of the inactivated polio vaccine in this country (see Table 19-3). The projected eradication of paralytic polio (Figure 19-4) may be impossible as long as OPV is used anywhere in the world. The alternative inactivated Salk vaccine will likely be substituted as the number of cases decreases, although there are problems in delivering this vaccine in developing countries. Obviously, the ultimate goal of eradication is to achieve a polio-free world in which no vaccine is needed.

Attenuated vaccines also may be associated with complications similar to those seen in the natural disease. A small percentage of recipients of the measles vaccine, for example, develop postvaccine encephalitis or other complications. As shown in Table 19-5, however, the risk of vaccine-related complications is much lower than risks from infection. An independent study showed that 75 million doses of measles

vaccine were given between 1970 and 1993, with an incidence of 48 cases of vaccine-related encephalopathy. The low incidence of this side effect compared with the rate of encephalopathy associated with infection argues for the efficacy of the vaccine. A more convincing argument for vaccination is the high death rate associated with measles infection even in developed countries.

Genetic engineering techniques provide a way to attenuate a virus irreversibly by selectively removing genes that are necessary for virulence or for growth in the vaccinee. This has been done with a herpesvirus vaccine for pigs, in which the thymidine kinase gene was removed. Because thymidine kinase is required for the virus to grow in certain types of cells (e.g., neurons), removal of this gene rendered the virus incapable of causing disease.

An attenuated live vaccine against influenza was developed recently and licensed under the name FluMist. The process of attenuation involved growing the virus at lower than

| TABLE 19-5 | Risk of complications from natural measles infection compared with known risks of vaccination with a live attenuated virus in immunocompetent individuals | |
|---|---|---|
| **Complication** | **Risk after natural disease*** | **Risk after vaccination†** |
| Otitis media | 7%–9% | 0 |
| Pneumonia | 1%–6% | 0 |
| Diarrhea | 66% | 0 |
| Post-infectious encephalomyelitis | 0.5–1 per 1000 | 1 per 1,000,000 |
| SSPE | 1 per 100,000 | 0 |
| Thrombocytopenia | —‡ | 1 per 30,000§ |
| Death | 0.1–1 per 1000 (up to 5%–15% in developing countries) | 0 |

*Risk after natural measles are calculated in terms of events per number of cases.

†Risks after vaccination are calculated in terms of events per number of doses.

‡Although there have been several reports of thrombocytopenia occurring after measles, including bleeding, the risk has not been properly quantified.

§This risk has been reported after MMR vaccination and cannot be attributed only to the measles component.

MMR=measles, mumps, and rubella.

SSPE=subacute sclerosing panencephalitis.

normal temperatures until a cold-adapted strain resulted. This flu virus strain grows well at temperatures lower than 37 °C but is unable to grow at human body temperature of 37 °C. This live attenuated virus is administered intranasally and causes a transient infection in the upper respiratory tract, an infection sufficient to induce a strong immune response. The virus cannot spread beyond the upper respiratory tract because of its inability to grow at the elevated temperatures of the inner body and is therefore limited in its range. Because of the ease of administration and induction of good mucosal immunity, cold-adapted, nasally administered flu vaccines will likely soon dominate this area.

## Inactivated or "Killed" Vaccines

Another common means to achieve attenuation of a vaccine is inactivation of the pathogen by heat or chemical means so that the pathogen raises an immune response but is not capable of replication in the host. It is critically important to maintain the structure of epitopes on surface antigens during inactivation. Heat inactivation is often unsatisfactory because it causes extensive denaturation of proteins; thus, any epitopes that depend on higher orders of protein structure are likely to be altered significantly. Chemical inactivation with formaldehyde or various alkylating agents has been successful. The Salk polio vaccine is produced by formaldehyde inactivation of the polio virus.

Live attenuated vaccines generally require only one dose to induce long-lasting immunity. Killed vaccines, on the other hand, often require repeated boosters to maintain the immune status of the host. In addition, killed vaccines induce a predominantly humoral antibody response; they are less effective than attenuated vaccines in inducing cell-mediated immunity and in eliciting a secretory IgA response.

Even though the pathogens they contain are killed, inactivated whole-organism vaccines still carry certain risks. A serious complication with the first Salk vaccines arose when formaldehyde failed to kill all the virus in two vaccine lots, which caused paralytic polio in a high percentage of recipients. Risk is also encountered in the manufacture of the inactivated vaccines. Large quantities of the infectious agent must be handled prior to inactivation, and those exposed to the process are at risk of infection. There have been reports of infection of individuals involved in the manufacture of the Salk vaccine.

In general, the safety of inactivated vaccines is greater than that of live attenuated vaccines, which retain the capability to replicate and possibly revert to an active form. Inactivated vaccines in common usage include those against both viral and bacterial diseases. The classic flu vaccine is of this type, as are vaccines for hepatitis A and cholera. In addition to their relative safety, advantages of inactivated vaccines include stability and ease of storage and transport. The requirement that most inactivated vaccines be administered by injection is a drawback to their use in mass immunization campaigns.

## Subunit Vaccines

Many of the risks associated with attenuated or killed whole-organism vaccines can be avoided with vaccines that consist of specific, purified macromolecules derived from pathogens. Three general forms of vaccines that are components or subunits of the target pathogen in current use are inactivated exotoxins or toxoids, capsular polysaccharides, and recombinant protein antigens (see Table 19-4).

## Toxoids are used as vaccines

Some bacterial pathogens, including those that cause diphtheria and tetanus, produce exotoxins that account for many of the disease symptoms that result from infection. Diphtheria and tetanus vaccines have been made by purifying the bacterial exotoxin and then inactivating it with formaldehyde to form a **toxoid.** Vaccination with the toxoid induces antitoxoid antibodies, which are capable of binding to the toxin and neutralizing its effects. Conditions for the production of toxoid vaccines must be closely controlled to avoid excessive modification of the epitope structure while accomplishing complete detoxification. Obtaining sufficient quantities of the purified toxins to prepare the vaccines is achieved by cloning the exotoxin genes and expressing them in easily grown host cells.

Passive immunity to toxin can be induced by transfer of serum containing anti-toxoid antibodies. As discussed in Chapter 1, the treatment for diphtheria prior to the availability of antibiotics (or the development of an effective vaccine) entailed administration of antitoxin usually produced in horses. Similarly, disease in those exposed to tetanus can be prevented by treatment with tetanus antitoxin. If an individual who has not had a recent tetanus vaccination is exposed to tetanus, the antitoxin will be administered. Botulism is treated with horse antitoxin, but to date, no toxoid vaccine against botulism has been developed for humans.

## Bacterial polysaccharide capsules are used as vaccines

The virulence of some pathogenic bacteria depends primarily on the antiphagocytic properties of their hydrophilic polysaccharide capsule. Coating of the capsule with antibodies and/or complement greatly increases the ability of macrophages and neutrophils to phagocytose such pathogens. These findings provide the rationale for vaccines consisting of purified capsular polysaccharides.

The current vaccine for *Streptococcus pneumoniae,* which causes pneumococcal pneumonia, consists of 23 antigenically different capsular polysaccharides. The vaccine induces formation of opsonizing antibodies and is now on the list of vaccines recommended for all infants. The vaccine for *Neisseria meningitidis,* a common cause of bacterial meningitis, also consists of purified capsular polysaccharides.

## Viral glycoproteins are candidate vaccines

Although certain viral subunit glycoproteins, such as an envelope protein from HIV-1, have been tested as vaccines with little success, hope remains that these glycoproteins can protect against some diseases. A recent clinical trial of glycoprotein-D from herpes simplex virus type 2 prevented genital herpes in some vaccinees. The glycoprotein vaccine was given with an adjuvant to male and female volunteers who were first tested for serum antibodies to HSV type 1 and 2. Selected participants were negative for HSV-2 antibodies and had regular sexual partners with a history of genital herpes. The vaccine was effective in preventing genital herpes in females who were seronegative for both HSV type 1 and 2 at the onset of the study but gave little protection to those positive for HSV type 1 (which causes oral but not genital lesions). No protection of male subjects was observed whether serum antibodies to HSV were present or absent.

## Pathogen proteins are manufactured by recombinant techniques

Theoretically, the gene encoding any immunogenic protein can be cloned and expressed in bacterial, yeast, or mammalian cells using recombinant DNA technology. A number of genes encoding surface antigens from viral, bacterial, and protozoan pathogens have been successfully cloned into cellular expression systems and the expressed antigens used for vaccine development. The first such recombinant antigen vaccine approved for human use is the hepatitis B vaccine, developed by cloning the gene for the major surface antigen of hepatitis B virus (HBsAg) and expressing it in yeast cells. The recombinant yeast cells are grown in large fermenters, allowing HBsAg to accumulate in the cells. The yeast cells are harvested and disrupted, releasing the recombinant HBsAg, which is then purified by conventional biochemical techniques. Recombinant hepatitis B vaccine has been shown to induce the production of protective antibodies and holds much promise for the 250 million carriers of chronic hepatitis B worldwide.

Many of the vaccines proposed for major diseases such as malaria consist of one or more proteins from the pathogen. These proteins are biosynthesized in large quantities using appropriate cell lines and then purified using procedures that do not introduce contaminants into the product. A universal problem involves raising protective immune responses against these proteins. Certain adjuvants such as alum are approved for human use, but the more effective ones, such as Freund's complete or incomplete adjuvant, generate unacceptable side effects. A recent area of inquiry involves the search for compounds that demonstrate a strong adjuvant effect without harm to the vaccinee. Among the candidate compounds are those that activate dendritic cells and macrophages via Toll-like receptors (see Chapter 3). This activation of cells of the innate immune system would mobilize a response by components of the adaptive immune system. Among the substances being tested for adjuvant effects are oligonucleotides with the sequence $[CpG]_n$ (a common motif in bacteria) and components of the gram-negative bacterial cell wall, such as LPS.

## Use of synthetic peptides as vaccines has progressed slowly

If subunit proteins can serve as successful vaccines, it follows that we should be able to identify their most active epitopes, create synthetic peptides that mimic those epitopes, and use the peptides as vaccines. Advantages would include ease of synthesis under highly controlled conditions and virtually complete safety. Despite these very promising features, the use of synthetic peptides as vaccines has not progressed as originally hoped. Peptides are not as immunogenic as proteins, and it is difficult to elicit both humoral and cellular immunity to them. The use of conjugates and adjuvants can assist in raising protective immunity to peptides, but barriers to the widespread use of peptide vaccines remain. Balancing the disappointment of peptide vaccines has been the steady advance in techniques to produce recombinant proteins in transfected cell culture. Nonetheless, considerable theoretical interest remains in all aspects of immunity to peptides, which may yet generate insights leading to new vaccines.

## Conjugate Vaccines

One limitation of polysaccharide vaccines is their inability to activate $T_H$ cells. They activate B cells in a thymus-independent type 2 (TI-2) manner, resulting in IgM production but little class switching, no affinity maturation, and little, if any, development of memory cells. Several investigators have reported the induction of IgA-secreting plasma cells in humans receiving subcutaneous immunization with the pneumococcal polysaccharide vaccine. In this case, since $T_H$ cells are not involved in the response, the vaccine may activate IgA-specific memory B cells previously generated by naturally occurring bacterial antigens at mucosal surfaces. Because these bacteria have both polysaccharide and protein epitopes, they would activate $T_H$ cells, which in turn could mediate class switching and memory-cell formation.

One way to involve $T_H$ cells directly in the response to a polysaccharide antigen is to conjugate the antigen to some sort of protein carrier. For example, the vaccine for *Haemophilus influenzae* type b (Hib), the major cause of bacterial meningitis in children less than 5 years of age, consists of type b capsular polysaccharide covalently linked to a protein carrier, tetanus toxoid (Figure 19-5a). The polysaccharide-protein conjugate is considerably more immunogenic than the polysaccharide alone, and because it activates $T_H$ cells, it enables class switching from IgM to IgG. Although this type of vaccine can induce memory B cells, it cannot induce memory T cells specific for the pathogen. In the case of the Hib vaccine, it appears that the memory B cells can be activated to some degree in the absence of a population of memory $T_H$ cells, thus accounting for the efficacy of this vaccine. Hib infections can lead to deafness and neurologic defects; the value of this vaccine is seen in countries with broad coverage by the conjugate vaccine. The rapid decline of Hib cases in the United States is shown in Figure 19-5b.

### One polysaccharide confers protection against several fungi

A recent study of a polysaccharide component of fungi gave promising results as a conjugate vaccine in animal studies. Immunization with β-glucan isolated from brown alga and conjugated to diphtheria toxoid raised antibodies in mice and rats that protected against challenge with both

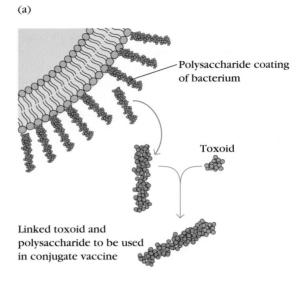

(a)

Polysaccharide coating of bacterium

Toxoid

Linked toxoid and polysaccharide to be used in conjugate vaccine

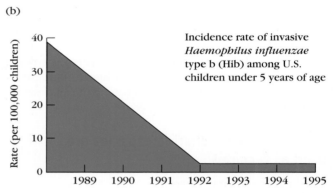

(b)

Incidence rate of invasive *Haemophilus influenzae* type b (Hib) among U.S. children under 5 years of age

**FIGURE 19-5 A conjugate vaccine protects against *Hemophilus influenzae* type b (Hib).** (a) The vaccine is prepared by conjugating the surface polysaccharide of Hib to a protein molecule. (b) Since the introduction of this vaccine in the United States, the rate of Hib disease has fallen dramatically. *[Part b, Summary of notifiable diseases, 1995, CDC Morbidity and Mortality Weekly Report.]*

*Aspergillus fumigatus* and *Candida albicans*. The protection was transferred by serum or vaginal fluid from the immunized animals, indicating that the immunity is antibody based. This finding was supported when a monoclonal antibody raised against the β-glucan also conferred protection. Infections with fungal pathogens are a serious problem for immunocompromised individuals, and the availability of immunization or antibody treatment could circumvent problems with toxicity of anti-fungal drugs while also helping to counter the problem of resistant strains emerging, an issue that is especially important in hospital settings.

### Multivalent subunit vaccines confer both cellular and humoral immunity

One of the limitations of synthetic peptide vaccines and subunit polysaccharide or protein vaccines is that they tend to be poorly immunogenic. They also tend to induce a humoral antibody response but are less likely to induce a cell-mediated

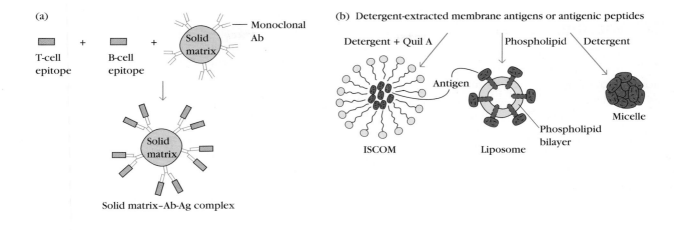

(a)

T-cell epitope + B-cell epitope + Solid matrix — Monoclonal Ab

↓

Solid matrix

Solid matrix–Ab-Ag complex

(b) Detergent-extracted membrane antigens or antigenic peptides

Detergent + Quil A    Phospholipid    Detergent

Antigen

Micelle

ISCOM    Liposome    Phospholipid bilayer

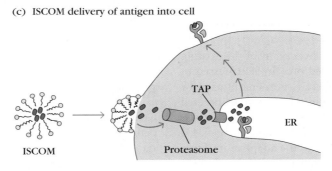

(c)  ISCOM delivery of antigen into cell

ISCOM → TAP, Proteasome, ER

**FIGURE 19-6 Multivalent subunit vaccines.** (a) Solid matrix–antibody-antigen complexes can be designed to contain synthetic peptides representing both T-cell epitopes and B-cell epitopes. (b) Protein micelles, liposomes, and immunostimulating complexes (ISCOMs) can all be prepared with extracted antigens or antigenic peptides. In micelles and liposomes, the hydrophilic residues of the antigen molecules are oriented outward. In ISCOMs, the long fatty-acid tails of the external detergent layer are adjacent to the hydrophobic residues of the centrally located antigen molecules. (c) ISCOMs and liposomes can deliver antigens inside cells, so they mimic endogenous antigens. Subsequent processing by the cytosolic pathway and presentation with class I MHC molecules induces a cell-mediated response.

response. What is needed is a method for constructing synthetic peptide vaccines that contain both immunodominant B-cell and T-cell epitopes. Furthermore, if a CTL response is desired, the vaccine must be delivered intracellularly so that the peptides can be processed and presented together with class I MHC molecules. A number of innovative techniques are being applied to develop multivalent vaccines that can present multiple copies of a given peptide or a mixture of peptides to the immune system (Figure 19-6).

One approach is to prepare solid matrix–antibody-antigen (SMAA) complexes by attaching monoclonal antibodies to particulate solid matrices and then saturating the antibody with the desired antigen. The resulting complexes are then used as vaccines. By attaching multiple monoclonal antibodies to the solid matrix, it is possible to bind a mixture of peptides or proteins, comprising immunodominant epitopes for both T cells and B cells, to the solid matrix (see Figure 19-6a). These multivalent complexes have been shown to induce vigorous humoral and cell-mediated responses. Their particulate nature contributes to their increased immunogenicity by facilitating phagocytosis by phagocytic cells.

Another means of producing a multivalent vaccine is to use detergent to incorporate protein antigens into protein micelles, lipid vesicles (called liposomes), or immunostimulating complexes (see Figure 19-6b). Mixing proteins in detergent and then removing the detergent forms micelles. The individual proteins orient themselves with their hydrophilic residues toward the aqueous environment and the hydrophobic residues at the center so as to exclude their interaction with the aqueous environment. Liposomes containing protein antigens are prepared by mixing the proteins with a suspension of phospholipids under conditions that form lipid bilayer vesicles. The proteins are incorporated into the bilayer with the hydrophilic residues exposed. Immunostimulating complexes (ISCOMs) are lipid carriers prepared by mixing protein with detergent and a glycoside called Quil A.

Membrane proteins from various pathogens, including influenza virus, measles virus, hepatitis B virus, and HIV, have been incorporated into micelles, liposomes, and ISCOMs and are currently being assessed as potential vaccines. In addition to their increased immunogenicity, liposomes and ISCOMs appear to fuse with the plasma membrane to deliver the antigen intracellularly, where it can be processed by the cytosolic pathway and thus induce a cell-mediated response (see Figure 19-6c).

# DNA Vaccines

A vaccination strategy under investigation for a number of diseases utilizes plasmid DNA encoding antigenic proteins, which is injected directly into the muscle of the recipient. Muscle cells take up the DNA, and the encoded protein antigen is expressed, leading to both a humoral antibody response and a cell-mediated response. Most surprising about this technique is that the injected DNA is taken up and expressed by the host muscle cells with much greater efficiency than in tissue culture cells. The DNA appears either to integrate into the chromosomal DNA or to be maintained for long periods in an episomal form. The viral antigen is expressed not only by the muscle cells but also by dendritic cells in the injection area. Muscle cells express low levels of class I MHC molecules and do not express costimulatory molecules, suggesting that local dendritic cells may be crucial to the development of antigenic responses to DNA vaccines (Figure 19-7).

DNA vaccines offer advantages over many of the existing vaccines. For example, the encoded protein is expressed in the host in its natural form—there is no denaturation or modification. The immune response is therefore directed to the antigen exactly as it is expressed by the pathogen. DNA vaccines also induce both humoral and cell-mediated immunity; to stimulate both arms of the adaptive immune response with non-DNA vaccines normally requires immunization with a live attenuated preparation, which introduces additional elements of risk. Finally, DNA vaccines cause prolonged expression of the antigen, which generates significant immunological memory.

The practical aspects of DNA vaccines are also very promising (see Table 19-4). Refrigeration is not required for the handling and storage of the plasmid DNA, a feature that greatly lowers the cost and complexity of delivery. The same plasmid vector can be custom tailored to make a variety of proteins, so that one manufacturing operation can produce a variety of DNA vaccines, each encoding antigen from a different pathogen. An improved method for administering these vaccines entails coating microscopic gold beads with the plasmid DNA and then delivering the coated particles through the skin into the underlying muscle with an air gun (called a *gene gun*). This will allow rapid delivery of vaccine to large populations without the requirement for huge supplies of needles and syringes.

Tests of DNA vaccines in animal models have shown that these vaccines are able to induce protective immunity against a number of pathogens, including influenza and rabies viruses. It has been further shown that the inclusion of certain DNA sequences in the vector leads to enhanced immune response. One such sequence is the CpG sequence found in pathogens; recall that this sequence is the ligand for TLR9. At present, human trials are under way with several different DNA vaccines, including those for malaria, AIDS, influenza, Ebola, and herpesvirus. A recent experimental success of a DNA vaccine involved protection of mice against challenge by the SARS coronavirus, using a DNA vector encoding the spike protein of the SARS virus. (The emergence of SARS was discussed in Chapter 18.) Immunity was transferred with serum from vaccine recipients, demonstrating that neutralizing antibodies resulted from the vaccination and were protective in the mouse SARS infection model.

Future experimental trials of DNA vaccines will mix genes for antigenic proteins with genes for cytokines or chemokines that direct the immune response to the optimum pathway. For example, the IL-12 gene may be included in a DNA vaccine; expression of IL-12 at the site of immunization will stimulate $TH_1$-type immunity induced by the vaccine.

DNA vaccines are in clinical testing and will likely be used for human immunization within the next few years. However, they are not a universal solution to the problems of vaccination. For example, only protein antigens can be encoded, and certain vaccines, such as those for pneumococcal and meningococcal infections, use polysaccharide antigens and are not candidates for delivery via DNA.

# Recombinant Vector Vaccines

Genes that encode major antigens of especially virulent pathogens can be introduced into attenuated viruses or bacteria. The attenuated organism serves as a vector, replicating within the host and expressing the gene product of the pathogen. Live attenuated virus vaccines have been prepared utilizing existing licensed vaccines and adding to them genes encoding antigens present on newly emerging pathogens. Such chimeric virus vaccines can be more quickly tested and approved than an entirely new product, thus saving valuable time. A very recent example of this type of chimera is the yellow fever vaccine that was engineered to express antigens of West Nile virus. A number of organisms have been used for vector vaccines, including vaccinia virus, the canarypox virus, attenuated poliovirus, adenoviruses, attenuated strains of *Salmonella*, the BCG strain of *Mycobacterium bovis*, and certain strains of streptococcus that normally exist in the oral cavity.

Vaccinia virus, the attenuated vaccine used to eradicate smallpox, has been widely employed as a vector vaccine. This large, complex virus, with a genome of about 200 genes, can be engineered to carry several dozen foreign genes without impairing its capacity to infect host cells and replicate. The procedure for producing a vaccinia vector that carries a foreign gene from a pathogen is outlined in Figure 19-8. The genetically engineered vaccinia expresses high levels of the inserted gene product, which can then serve as a potent immunogen in an inoculated host. Like the smallpox vaccine, genetically engineered vaccinia vector vaccines can be administered simply by scratching the skin, causing a localized infection in host cells. If the foreign gene product expressed by the vaccinia is a viral envelope protein, it is inserted into the membrane of the infected host cell, inducing development of cell-mediated immunity as well as antibody-mediated immunity.

Other attenuated vector vaccines may prove to be safer than the vaccinia vaccine. The canarypox virus has recently

Use of DNA Vaccines Raises Both Humoral and Cellular Immunity

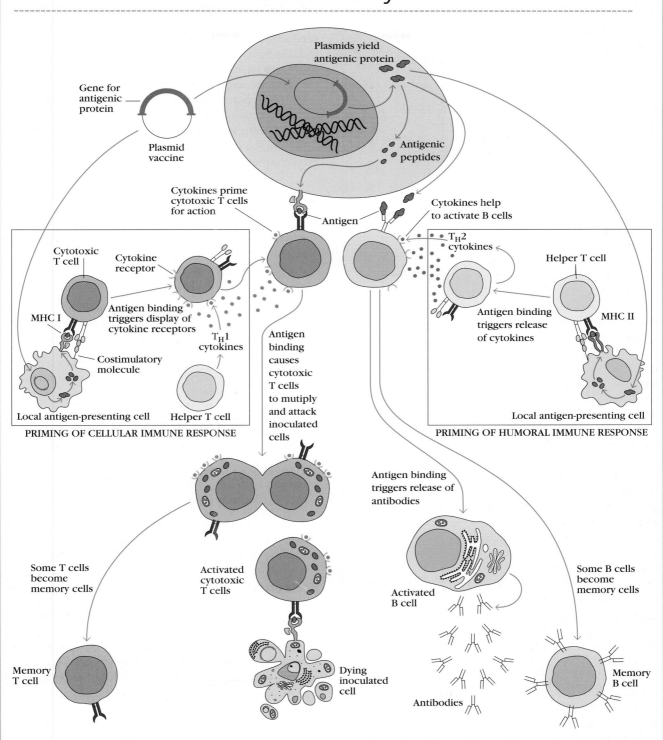

Plasmids yield antigenic protein

Gene for antigenic protein

Plasmid vaccine

Antigenic peptides

Cytokines prime cytotoxic T cells for action

Antigen

Cytokines help to activate B cells

$T_H2$ cytokines

Cytotoxic T cell

Cytokine receptor

Helper T cell

MHC I

Antigen binding triggers display of cytokine receptors

$T_H1$ cytokines

Antigen binding triggers release of cytokines

MHC II

Costimulatory molecule

Antigen binding causes cytotoxic T cells to mutiply and attack inoculated cells

Local antigen-presenting cell

Helper T cell

PRIMING OF CELLULAR IMMUNE RESPONSE

Local antigen-presenting cell

PRIMING OF HUMORAL IMMUNE RESPONSE

Antigen binding triggers release of antibodies

Some T cells become memory cells

Activated cytotoxic T cells

Activated B cell

Some B cells become memory cells

Memory T cell

Dying inoculated cell

Memory B cell

Antibodies

The injected gene is expressed in muscle cells and in nearby APCs. The peptides from the protein encoded by the DNA are expressed on the surface of both cell types after processing as an endogenous antigen by the MHC class I pathway. Cells that present the antigen in the context of class I MHC molecules stimulate development of cytotoxic T cells. The protein encoded by the injected DNA is also expressed as a soluble, secreted protein, which is taken up, processed, and presented in the context of class II MHC molecules. This pathway stimulates B-cell immunity and generates antibodies and B-cell memory against the protein. [Adapted from D. B. Weiner and R. C. Kennedy, 1999, Scientific American *281:50.*]

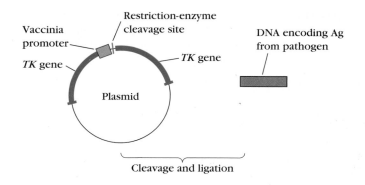

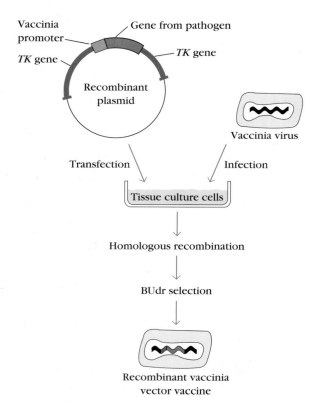

**FIGURE 19-8 Production of vaccinia vector vaccine.** The gene that encodes the desired antigen (orange) is inserted into a plasmid vector adjacent to a vaccinia promoter (pink) and flanked on either side by the vaccinia thymidine kinase (*TK*) gene (green). When tissue culture cells are incubated simultaneously with vaccinia virus and the recombinant plasmid, the antigen gene and promoter are inserted into the vaccinia virus genome by homologous recombination at the site of the nonessential TK gene, resulting in a TK⁻ recombinant virus. Cells containing the recombinant vaccinia virus are selected by addition of bromodeoxyuridine (BUdr), which kills TK⁺ cells. *[Adapted from B. Moss, 1985, Immunology Today **6**:243.]*

been tried as a vector vaccine. Like its relative vaccinia, the canarypox virus is large and easily engineered to carry multiple genes. Unlike vaccinia, the canarypox virus does not appear to be virulent even in individuals with severe immune suppression. Another possible vector is an attenuated strain of *Salmonella typhimurium*, which has been engineered with

genes from the bacterium that causes cholera. The advantage of this vector vaccine is that *Salmonella* infects cells of the mucosal lining of the gut and therefore will induce secretory IgA production. Effective immunity against a number of diseases, including cholera and gonorrhea, depends on increased production of secretory IgA at mucous membrane surfaces. Similar strategies using bacteria that are a normal part of oral flora are in development. The strategy would involve introduction of genes encoding antigens from pathogenic organisms into bacterial strains that inhabit the oral cavity or respiratory tract. Eliciting immunity at the mucosal surface could provide excellent protection at the portal used by the pathogen.

## SUMMARY

- A state of immunity can be induced by passive or active immunization. Short-term passive immunization is induced by transfer of preformed antibodies. Infection or vaccination achieves long-term active immunization.

- Three types of vaccines are currently used in humans: live attenuated (avirulent) microorganisms, inactivated (killed) microorganisms, or purified macromolecules.

- Protein components of pathogens expressed in cell culture may be effective vaccines. Polysaccharide vaccines may be conjugated to proteins to maximize immunogenicity.

- Recombinant vectors, including viruses or bacteria, engineered to carry genes from infectious microorganisms, maximize cell-mediated immunity to the encoded antigens.

- Plasmid DNA encoding protein antigens from a pathogen induces both humoral and cell-mediated immunity; DNA vaccines for several diseases are in human clinical trials.

- Realizing the optimum benefit of vaccines will require cheaper manufacture and improved delivery methods for existing vaccines.

## References

Afzal, M. F., et al. 2000. Clinical safety issues of measles, mumps, and rubella vaccines. *Bulletin of the World Health Organization* **78**:199.

Dittmann, S., 2000. Successful control of epidemic diphtheria in the states of the former Union of Soviet Socialist Republics: lessons learned. *Journal of Infectious Disease* **181** (suppl. 1):S10.

Grandi, G. 2001. Antibacterial design using genomics and proteomics. *Trends in Biotechnology* **19**:181.

Henderson, D. A. 1976. The eradication of smallpox. *Scientific American* **235**:25.

Jilek, S., et al. 2005. DNA-loaded biodegradable microparticles as vaccine delivery systems and their interaction with dendritic cells. *Advanced Drug Delivery Reviews* **57**:377.

Kew, O. M., et al. 2004. Circulating vaccine-derived polioviruses: current state of knowledge. *Bulletin of the World Health Organization* **82**:16.

Shann, F., and M. C. Steinhoff. 1999. Vaccines for children in rich and poor countries. *Lancet* **354**(suppl. II):7.

Smeeth, L., et al. 2004. MMR vaccination and pervasive developmental disorders: a case-control study. *Lancet* **364**:963.

Stanberry, L. R., et al. 2002. Glycoprotein-D-adjuvant vaccine to prevent genital herpes. *New England Journal of Medicine* **347**:1652.

Torosantucci, A., et al. 2005. A novel glyco-conjugate vaccine against fungal pathogens. *Journal of Experimental Medicine* **202**:597.

Wroe, A. L., et al. 2005. Feeling bad about immunizing our children. *Vaccine* **23**:1428.

Yang, Y-Z, et al. 2004. A DNA vaccine induces SARS coronavirus neutralization and protective immunity in mice. *Nature* **248**:561.

Zinkernagel, R. M. 2003. On natural and artificial vaccinations. *Annual Review of Immunology* **21**:515.

 ## Useful Web Sites

**http://www.VaccineAlliance.org/**

Home page of Global Alliance for Vaccines and Immunization (GAVI), a source of information about vaccines in developing countries and worldwide efforts at disease eradication. Links to major international vaccine information sites.

**http://www.ecbt.org/**

Every Child by Two offers useful information on childhood vaccination, including recommended immunization schedules.

 ## Study Questions

**CLINICAL FOCUS QUESTION** A connection between the new pneumococcus vaccine and a relatively rare form of arthritis has been reported. What data would you need to validate this report? How would you proceed to evaluate this possible connection?

1. Indicate whether each of the following statements is true or false. If you think a statement is false, explain why.

   a. Transplacental transfer of maternal IgG antibodies against measles confers short-term immunity on the fetus.

   b. Attenuated vaccines are more likely to induce cell-mediated immunity than killed vaccines are.

   c. Multivalent subunit vaccines generally induce a greater response than synthetic peptide vaccines.

   d. One disadvantage of DNA vaccines is that they don't generate significant immunologic memory.

   e. Macromolecules generally contain a large number of potential epitopes.

   f. A DNA vaccine only induces a response to a single epitope.

2. What are the advantages and disadvantages of using live attenuated organisms as vaccines?

3. A young girl who had never been immunized to tetanus stepped on a rusty nail and got a deep puncture wound. The doctor cleaned out the wound and gave the child an injection of tetanus antitoxin.

   a. Why was antitoxin given instead of a booster shot of tetanus toxoid?

   b. If the girl receives no further treatment and steps on a rusty nail again 3 years later, will she be immune to tetanus?

4. What are the advantages of the Sabin polio vaccine compared with the Salk vaccine? Why is the Sabin vaccine no longer recommended for use in the United States?

5. Why doesn't the live attenuated influenza vaccine (FluMist) cause respiratory infection?

6. In an attempt to develop a synthetic peptide vaccine, you have analyzed the amino acid sequence of a protein antigen for (a) hydrophobic peptides and (b) strongly hydrophilic peptides. How might peptides of each type be used as a vaccine to induce different immune responses?

7. Explain the phenomenon of herd immunity. How does it relate to the appearance of certain epidemics?

8. You have identified a bacterial protein antigen that confers protective immunity to a pathogenic bacterium and have cloned the gene that encodes it. The choices are either to express the protein in yeast and use this recombinant protein as a vaccine or to use the gene for the protein to prepare a DNA vaccine. Which approach would you take and why?

9. Explain the relationship between the incubation period of a pathogen and the approach needed to achieve effective active immunization.

10. List the three types of purified macromolecules that are currently used as vaccines.

11. An example of passive vaccination is when $Rh^-$ mothers carrying $Rh^+$ babies are given RhoGAM, humanized antibodies against the $Rh^+$ antigen. Why does this vaccine not cause hemagglutination in the newborn?

12. Some parents choose not to vaccinate their infants. Reasons include religion, allergic reactions, fear that the infant will develop the disease the vaccine is raised against, and recently, a fear, unsupported by research, that vaccines can cause autism. What would be the consequence if a significant proportion of the population was not vaccinated against childhood diseases such as measles or pertussis?

13. For each of the following diseases or conditions, indicate what type of vaccination is used:

   a. Polio                 1. Inactivated

   b. Chicken pox           2. Attenuated

   c. Tetanus               3. Inactivated exotoxin

   d. Hepatitis B           4. Purified macromolecule

   e. Cholera

   f. Measles

   g. Mumps

**ANALYZE THE DATA** T. W. Kim and coworkers (*J. Immunol.* **171:**2970, 2003) investigated methods to enhance the immune response against human papillomavirus (HPV)16 E7 antigen. Groups of mice were vaccinated with the following antigens incorporated in DNA vaccine constructs:

- + HPV E7 antigen
- + E7 + heat-shock protein 70 (HSP70)
- + E7 + calreticulin
- + E7 + Sorting signal of lysosome-associated membrane protein 1 (Sig/LAMP-1)

A second array of mice received the same DNA vaccines and was coadministered an additional DNA construct incorporating

(b)

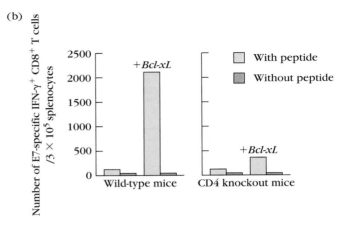

E7-specific CD8$^+$ T lymphocyte response in CD4 knockout mice vaccinated with +E7 +Sig/LAMP-1 DNA construct, with or without +*Bcl-xL* DNA construct.

the anti-apoptosis gene *Bcl-xL*. To test the efficacy of these DNA vaccine constructs in inducing a host response, spleen cells from vaccinated mice were harvested 7 days after injection, the cells were incubated in vitro with MHC class I–restricted E7 peptide (aa 49–57) overnight, and then the cells were stained for both CD8 and IFN-γ (Figure a in the left column). In another experiment Kim and group determined how effective their vaccines were if mice lacked CD4$^+$ T cells (Figure b above).

a. Which DNA vaccine(s) is (are) the most effective in inducing an immune response against papillormavirus E7 antigen? Explain your answer.

b. Propose a hypothesis to explain why expressing calreticulin in the vaccine construct was effective in inducing CD8$^+$ T cells.

c. Propose a mechanism to explain the data in Figure a.

d. If you were told that the +E7 +Sig/LAMP-l construct is the only one that targets antigen to the class II MHC processing pathway, propose a hypothesis to explain why antigen that would target MHC II molecules enhances a CD8$^+$ T-cell response. Why do you think a special signal was necessary to target antigen to MHC II?

e. What four variables contribute to the E7-specific CD8$^+$ T lymphocyte response in vitro as measured in Figure b.

(a)

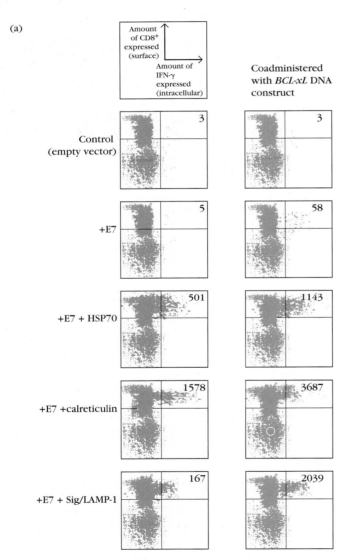

Intracellular cytokine staining followed by flow cytometry analysis to determine the E7 -specific CD8$^+$ T-cell response in mice vaccinated with DNA vaccines using intracellular targeting strategies. A DNA construct including the anti-apoptosis gene *Bcl-xL* was coadministered to the group on the right. The number at top right in each graph is the number of cells represented in the top right quadrant. [© 2003 The American Association of Immunologists, Inc.]

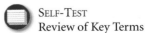 **Interactive Study**
**www.whfreeman.com/kuby**

 SELF-TEST
Review of Key Terms

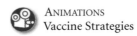 ANIMATIONS
Vaccine Strategies

# chapter 20

# AIDS and Other Immunodeficiencies

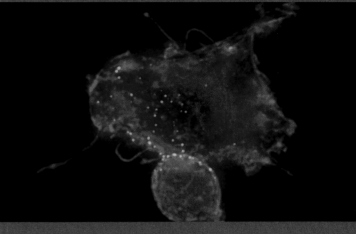

*HIV-1 (green dots) moving between dendritic cell (top) and T cell. [Courtesy of Thomas J. Hope, Northwestern University.]*

- Primary Immunodeficiencies
- AIDS and Other Acquired or Secondary Immunodeficiencies

L IKE ANY COMPLEX MULTICOMPONENT SYSTEM, THE immune system is subject to failures of some or all of its parts. These failures can have dire consequences. When the system loses its sense of self and begins to attack host cells and tissues, the result is **autoimmunity,** described in Chapter 16. When the system errs by failing to protect the host from disease-causing agents or from malignant cells, the result is **immunodeficiency,** the subject of this chapter.

A condition resulting from a genetic or developmental defect in the immune system is called a *primary immunodeficiency.* In such a condition, the defect is present at birth, although it may not manifest itself until later in life. Secondary immunodeficiency, or acquired immunodeficiency, is the loss of immune function and results from exposure to various agents. By far the most common secondary immunodeficiency is **acquired immunodeficiency syndrome,** or **AIDS,** which results from infection with human immunodeficiency virus 1 (HIV-1). In the year 2005, AIDS killed approximately 3.1 million people, including about 500,000 children under 5 years of age. HIV infection continues to spread to an estimated 14,000 people per day. AIDS patients, like other individuals with severe immunodeficiency, are at risk of infection with so-called opportunistic agents. These are microorganisms that healthy individuals can harbor with no ill consequences but that cause disease in those with impaired immune function.

The first part of this chapter describes the common primary immunodeficiencies, examines progress in identifying the genetic defects that underlie these disorders, and considers approaches to their treatment, including innovative uses of gene therapy. Animal models of primary immunodeficiency are also described. The rest of this chapter describes acquired immunodeficiency, with a strong focus on HIV infection and its effect on the immune system, AIDS, and the current status of therapeutic and prevention strategies for combating this fatal acquired immunodeficiency.

## Primary Immunodeficiencies

A primary immunodeficiency may affect either adaptive or innate immune functions. Deficiencies involving components of adaptive immunity, such as T or B cells, are thus differentiated from immunodeficiencies in which the nonspecific mediators of innate immunity, such as phagocytes or complement, are impaired. Immunodeficiencies are conveniently categorized by the type or the developmental stage of the cells involved. Figure 20-1 reviews the overall cellular development in the immune system, showing the locations of defects that give rise to primary immunodeficiencies. As we explained in Chapter 2, the two main cell lineages important to immune function are lymphoid and myeloid. Most defects that lead to immunodeficiencies affect either one or the other. The lymphoid cell disorders may affect T cells, B cells, or, in combined immunodeficiencies, both B and T cells. The myeloid cell disorders affect phagocytic function. Most of the primary immunodeficiencies are inherited, and the precise molecular variations and the genetic defects that lead to many of these dysfunctions have been determined (Table 20-1 and Figure 20-2). In addition, there are immunodeficiencies that

**OVERVIEW FIGURE 20-1:** Congenital Defects That Interrupt Hematopoiesis or Impair Functioning of Immune System Cells Result in Various Immunodeficiency Diseases

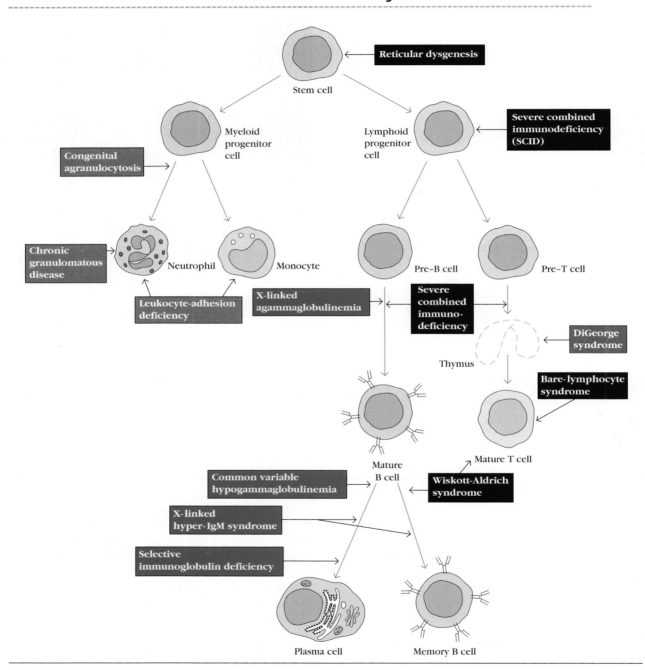

Orange boxes = phagocytic deficiencies, green = humoral deficiencies, red = cell-mediated deficiencies, and purple = combined immunodeficiencies, defects that affect more than one cell lineage.

| TABLE 20-1 | Some primary human immunodeficiency diseases and underlying genetic defects | | | |
|---|---|---|---|---|
| Immunodeficiency disease | Specific defect | Impaired function | Inheritance mode* | Chromosomal defect |
| Severe combined immunodeficiency (SCID) | *RAG-1/RAG-2* deficiency | No TCR or Ig gene rearrangement | AR | 11p13 |
| | ADA deficiency | Toxic metabolite in T and B cells | AR | 20q13 |
| | PNP deficiency | | AR | 14q13 |
| | JAK-3 deficiency | Defective signals from IL-2,-4,-7,-9,-15, | AR | 19p13 |
| | IL-2Rγ-deficiency | | XL | Xq13 |
| | ZAP-70 deficiency | Defective signal from TCR | AR | 2q12 |
| Bare-lymphocyte syndrome | Defect in MHC class II gene promoter | No class II MHC molecules | AR | 16p13 |
| Wiskott-Aldrich syndrome (WAS) | Cytoskeletal protein (CD43) | Defective T cells and platelets | XL | Xp11 |
| Interferon gamma receptor | IFN-γ–receptor defect | Impaired immunity to mycobacteria | AR | 6q23 |
| DiGeorge syndrome | Thymic aplasia | T- and B-cell development | AD | 22q11 |
| Ataxia telangiectasia | Defective cell-cycle kinase | Low IgA, IgE | AR | 11q22 |
| Gammaglobulinemias | X-linked agammaglobulinemia | Bruton's tyrosine kinase (Btk); no mature B cells | XL | Xq21 |
| | X-linked hyper-IgM syndrome | Defective CD40 ligand | XL | Xq26 |
| | Common variable immunodeficiency | Low IgG, IgA; variable IgM | Complex | |
| | Selective IgA deficiency | Low or no IgA | Complex | |
| Chronic granulomatous disease | Cyt p91$^{phox}$ | No oxidative burst for bacterial killing | XL | Xp21 |
| | Cyt p67$^{phox}$ | | AR | 1q25 |
| | Cyt p22$^{phox}$ | | AR | 16q24 |
| Chediak-Higashi syndrome | Defective intracellular transport protein (LYST) | Inability to lyse bacteria | AR | 1q42 |
| Leukocyte adhesion defect | Defective integrin β2 (CD18) | Leukocyte extravasation | AR | 21q22 |

*AR = autosomal recessive; AD = autosomal dominant; XL = X linked; "Complex" indicates conditions for which precise genetic data are not available and that may involve several interacting loci.

stem from developmental defects impairing proper function of an organ of the immune system.

The consequences of primary immunodeficiency depend on the number and type of immune system components involved. Defects in components early in the hematopoietic developmental scheme affect the entire immune system. In this category is reticular dysgenesis, a stem-cell defect that affects the maturation of all leukocytes; the resulting general failure of immunity leads to susceptibility to infection by a variety of microorganisms. Without aggressive treatment, the affected individual usually dies young from severe infection. In the more restricted case of defective phagocytic function, the major consequence is susceptibility to bacterial infection. Defects in more highly differentiated compartments of the immune system have consequences that are more specific and usually less severe. For example, an individual with selective IgA deficiency may enjoy a full life span, troubled only by a greater than normal susceptibility to infections of the respiratory and genitourinary tracts.

## Lymphoid immunodeficiencies may involve B cells, T cells, or both

The combined forms of lymphoid immunodeficiency affect both lineages and are generally lethal within the first few years of life; these arise from defects causing dysfunction in multiple cell types or in cellular interactions needed to develop immune responses. They are less common, and usually more severe, than conditions resulting from defects in more highly differentiated lymphoid cells.

B-cell immunodeficiency disorders make up a diverse spectrum of diseases ranging from the complete absence of

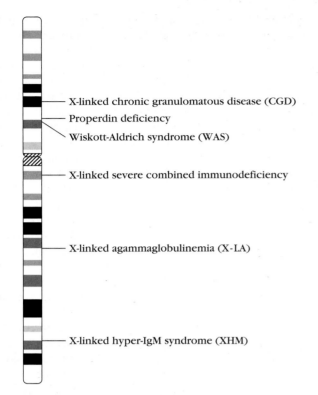

X-linked chronic granulomatous disease (CGD)
Properdin deficiency
Wiskott-Aldrich syndrome (WAS)

X-linked severe combined immunodeficiency

X-linked agammaglobulinemia (X-LA)

X-linked hyper-IgM syndrome (XHM)

**FIGURE 20-2 Several X-linked immunodeficiency diseases result from defects in loci on the X chromosome.** *[Data from the National Center for Biotechnology Information Web site.]*

mature recirculating B cells, plasma cells, and immunoglobulin to the selective absence of only certain classes of immunoglobulins. Patients with these disorders usually are subject to recurrent bacterial infections but display normal immunity to most viral and fungal infections, because the T-cell branch of the immune system is largely unaffected. Most common in patients with humoral immunodeficiencies are infections by such encapsulated bacteria as staphylococci, streptococci, and pneumococci, because antibody is critical for the opsonization and clearance of these organisms.

Because of the central role of T cells in the immune system, a T-cell deficiency can affect both the humoral and the cell-mediated responses. The impact on the cell-mediated system can be severe, with a reduction in both delayed-type hypersensitivity responses and cell-mediated cytotoxicity. Immunoglobulin deficiencies are associated primarily with recurrent infections by extracellular bacteria, but those affected have normal responses to intracellular bacteria, as well as viral and fungal infections. By contrast, defects in the cell-mediated system are associated with increased susceptibility to viral, protozoan, and fungal infections. Intracellular pathogens such as *Candida albicans, Pneumocystis carinii,* and *Mycobacteria* are often implicated, reflecting the importance of T cells in eliminating intracellular pathogens. Infections with viruses that are rarely pathogenic for the normal individ-

ual (such as cytomegalovirus or even attenuated measles vaccine) may be life threatening for those with impaired cell-mediated immunity. Defects that cause decreased T-cell counts generally also affect the humoral system, because of the requirement for $T_H$ cells in B-cell activation. Generally there is some decrease in antibody levels, particularly in the production of specific antibody after immunization.

As one might expect, combined deficiencies of the humoral and cell-mediated branches are the most serious of the immunodeficiency disorders. The onset of infections begins early in infancy, and the prognosis for these infants is early death unless therapeutic intervention reconstitutes their defective immune system. As described below, there are increasing numbers of options for the treatment of immunodeficiencies.

The immunodeficiencies that affect lymphoid function have in common the inability to mount or sustain a complete immune response against specific agents. A variety of failures can lead to such immunodeficiency. Defective intercellular communication may be rooted in deleterious mutations of genes that encode cell surface receptors or signal transduction molecules; defects in the mechanisms of gene rearrangement and other functions may prevent normal B- or T-cell responses. Figure 20-3 is an overview of the molecules involved in the more well-described interactions among T cells and B cells that give rise to specific responses, with a focus on proteins for which defects leading to immunodeficiency have been identified.

### Severe Combined Immunodeficiency (SCID)

The family of disorders termed *SCID* stems from defects in lymphoid development that affect T cells, either alone or in combination with B cells and NK cells. Depending on the underlying genetic defect, a SCID patient may have one of several lymphoid cell phenotypes. All are characterized by defective T-cell function, and dysfunction may also extend to B cells, NK cells, or both. Because T-cell defects are the common element in SCID, a recently described screening test for SCID in neonates involves assaying excised DNA circles from the TCR locus in blood cells, which are evidence of TCR gene rearrangement. Abnormal results from this assay allow early diagnosis prior to onset of life-threatening infections.

All forms of SCID have common features despite differences in the underlying genetic defects. Clinically, SCID is characterized by a very low number of circulating lymphocytes. There is a failure to mount immune responses mediated by T cells. The thymus does not develop, and the few circulating T cells in the SCID patient do not respond to stimulation by mitogens, indicating that they cannot proliferate in response to antigens. Myeloid and erythroid (red-blood-cell precursors) cells appear normal in number and function, indicating that only lymphoid cells are depleted in SCID.

SCID results in severe recurrent infections and is usually fatal in the early years of life. Although both the T and B

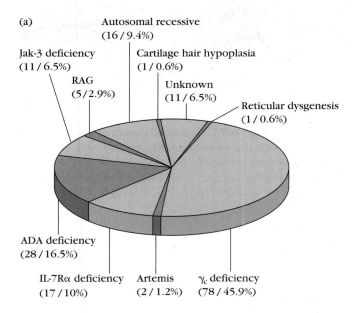

(a)

Autosomal recessive
(16 / 9.4%)

Jak-3 deficiency
(11 / 6.5%)

Cartilage hair hypoplasia
(1 / 0.6%)

RAG
(5 / 2.9%)

Unknown
(11 / 6.5%)

Reticular dysgenesis
(1 / 0.6%)

ADA deficiency
(28 / 16.5%)

IL-7Rα deficiency
(17 / 10%)

Artemis
(2 / 1.2%)

γc deficiency
(78 / 45.9%)

(b)

| Lymphocyte phenotype | | | |
|---|---|---|---|
| T | B | NK | **Type of SCID** |
| − | + | − | X-linked IL–2Rγ–chain deficiency JAK-3 deficiency CD45 deficiency |
| − | + | + | IL-7R α-chain deficiency CD3 δ-chain deficiency |
| − | − | − | Adenosine deaminase (ADA) deficiency |
| − | − | + | *RAG1* or *RAG2* deficiency Artemis deficiency |

**FIGURE 20-3  Causes of SCID.** (a) Distribution of genetic defects in 170 cases of SCID tracked over 35 years. (b) Cellular phenotypes related to different genetic defects observed in SCID. *[Part a adapted from R. H. Buckley, 2004,* Annual Review of Immunology **22**:625.*]*

lineages may be affected, the initial manifestation of SCID in infants is almost always infection by agents such as fungi or viruses that are normally dealt with by T-cell immunity. Defects in the B-cell lineage are not evident in the first few months of the affected infant's life because antibodies are passively obtained from transplacental circulation or from mother's milk. SCID infants suffer from chronic diarrhea, pneumonia, and skin, mouth, and throat lesions as well as a host of other opportunistic infections. The immune system is so compromised that even live-attenuated vaccines (such as the Sabin polio vaccine) can cause infection and disease. The life span of a SCID patient can be prolonged by preventing contact with all potentially harmful microorganisms, for example, by confinement in a sterile atmosphere.

However, extraordinary effort is required to prevent direct contact with other people and with unfiltered air; any object, including food, that comes in contact with the sequestered SCID patient must first be sterilized. Such isolation is feasible only as a temporary measure, pending treatment.

The search for defects that underlie SCID has revealed several different causes for this general failure of immunity. A recent survey of 170 patients by Rebecca Buckley at Duke University indicated that the most common cause (78 cases; see Figure 20-3) was deficiency of the common gamma chain of the IL-2 receptor (IL-2Rγ; see Figure 12-7). Defects in this chain impede signaling not only through IL-2R but also through receptors for IL-4, -7, -9, and -15 because the chain is present in receptors for all of these cytokines. Deficiency in the kinase JAK-3, which has a similar phenotype because the IL receptors signal through this molecule, accounted for 11 of the cases (see Figure 12-7c). A recent finding of an additional cytokine receptor defect found in 17 of the patients involved only the IL-7 receptor; these patients have impaired T but normal B cells and NK cells.

Another common defect is the adenosine deaminase, or ADA, deficiency, found in 28 patients in the study. Adenosine deaminase catalyzes conversion of adenosine to inosine, and its deficiency results in accumulation of adenosine, which interferes with purine metabolism and DNA synthesis. ADA deficiency results in defects in T, B, and NK cells. Defects in the recombinase activating genes (*RAG-1* and *RAG-2*), found in five patients, impair normal TCR and immunoglobulin gene rearrangements and lead to an absence of functioning T and B cells, but do not affect NK cells. As described in Chapters 5 and 9, both immunoglobulin and T-cell receptor genes require rearrangement to express the active forms of their products. The remaining cases included single instances of reticular dysgenesis and cartilage hair dysplasia or were classified as autosomal recessive defects not related to known IL-2Rγ or JAK-3 mutations. Two of the patients had defects in the gene called *Artemis*, which encodes a DNA repair and recombination enzyme. In one case, a defect in the gene for the cell surface phosphatase CD45 was found. Interestingly, this defect caused lack of αβ T cells but spared the γδ lineage and resulted in elevated B-cell numbers. Eleven of the 170 cases were of unknown origin, with no apparent genetic defect or family history of immunodeficiency.

Other known defects disrupt T-cell function and give rise to SCID-like deficiencies. One defect is characterized by depletion of CD8$^+$ T cells and involves the tyrosine kinase ZAP-70, an important element in T-cell signal transduction (see Figure 10-11). Infants with defects in ZAP-70 may have normal levels of immunoglobulin and CD4$^+$ lymphocytes, but their CD4$^+$ T cells are nonfunctional. A deficiency in the enzyme purine nucleoside phosphorylase (PNP) causes immunodeficiency by a mechanism similar to the ADA defect. A defect leading to general failure of immunity similar to SCID is failure to transcribe the genes that encode class II

MHC molecules. Without these molecules, the patient's lymphocytes cannot participate in cellular interactions with T helper cells. This type of immunodeficiency is also called the *bare-lymphocyte syndrome* (see Clinical Focus in Chapter 8). Molecular studies of a class II MHC deficiency revealed a defective interaction between a 5′ promoter sequence of the gene for the class II MHC molecule and a DNA-binding protein necessary for gene transcription. Other patients with SCID-like symptoms lack class I MHC molecules. This rare variant of immunodeficiency was ascribed to mutation in the TAP genes, which are vital to antigen processing by class I MHC molecules. This defect causes a deficit in $CD8^+$-mediated immunity, characterized by susceptibility to viral infection.

### Wiskott-Aldrich Syndrome (WAS)

The severity of this X-linked disorder increases with age and usually results in fatal infection or lymphoid malignancy. Initially, T and B lymphocytes are present in normal numbers. WAS first manifests itself by defective responses to bacterial polysaccharides and by lower-than-average IgM levels. Other responses and effector mechanisms are normal in the early stages of the syndrome. As the WAS sufferer ages, there are recurrent bacterial infections and a gradual loss of humoral and cellular responses. The syndrome includes thrombocytopenia (lowered platelet count; the existing platelets are smaller than usual and have a short half-life), which may lead to fatal bleeding. Eczema (skin rashes) in varying degrees of severity may also occur, usually beginning around 1 year of age. The defect in WAS has been mapped to the short arm of the X chromosome (see Table 20-1 and Figure 20-2) and involves a cytoskeletal glycoprotein present in lymphoid cells called sialophorin (CD43). The WAS protein is required for assembly of actin filaments required for the formation of microvesicles.

### Interferon-Gamma–Receptor Defect

An immunodeficiency that falls into the mixed-cell category involves a defect in the receptor for interferon gamma (IFN-γ; see Chapter 12). This deficiency was found in patients suffering from infection with atypical mycobacteria (intracellular organisms related to the bacteria that cause tuberculosis and leprosy). Most of those carrying this autosomal recessive trait are from families with a history of inbreeding. The susceptibility to infection with mycobacteria is selective in that those who survive these infections are not unusually susceptible to other agents, including other intracellular bacteria. This immunodeficiency points to a specific role for IFN-γ and its receptor in protection from infection with mycobacteria. Detailed analyses of patients with increased mycobacterial infections implicated defects in the IL-12 receptor as well as the IFN-γ receptor. Similar predisposition to mycobacterial infection results from defects in the NFκb pathway that preclude transcriptional activation of IFN-γ genes. A more complex defect sometimes involving tooth and skin-pigment abnormalities along with mycobacterial infection is seen in patients with defects in the IKK protein

subunit designated NEMO. Recall that IKK mediates the release of NFκb from its inhibitor, allowing it to migrate to the nucleus, where it activates immune response genes (see Figure 10-12).

Whereas SCID and the related combined immunodeficiencies affect T cells or all lymphoid cells, other primary immunodeficiencies affect B-cell function and result in the reduction or absence of some or all classes of immunoglobulins. Although the underlying defects have been identified for some of these, several of the more common deficiencies, such as common variable immunodeficiency and selective IgA deficiency, appear to involve multiple genes and a continuum of phenotypes.

### X-Linked Agammaglobulinemia

A B-cell defect called X-linked agammaglobulinemia (X-LA) or Bruton's hypogammaglobulinemia is characterized by extremely low IgG levels and by the absence of other immunoglobulin classes. Individuals with X-LA have no peripheral B cells and suffer from recurrent bacterial infections, beginning at about 9 months of age. A palliative treatment for this condition is periodic administration of immunoglobulin, but patients seldom survive past their teens. There is a defect in B-cell signal transduction in this disorder, due to a defect in a transduction molecule called Bruton's tyrosine kinase (Btk), after the investigator who described the syndrome. B cells in the X-LA patient remain in the pre–B stage with H chains rearranged but L chains in their germ-line configuration. (The Clinical Focus in Chapter 11 describes the discovery of this immunodeficiency and its underlying defect in detail.)

### X-Linked Hyper-IgM Syndrome

A peculiar immunoglobulin deficiency first thought to result from a B-cell defect has recently been shown to result instead from a defect in a T-cell surface molecule. X-linked hyper-IgM (XHM) syndrome is characterized by a deficiency of IgG, IgA, and IgE and elevated levels of IgM, sometimes as high as 10 mg/ml (normal IgM concentration is 1.5 mg/ml). Although individuals with XHM have normal numbers of B cells expressing membrane-bound IgM or IgD, they appear to lack B cells expressing membrane-bound IgG, IgA, or IgE. XHM syndrome is generally inherited as an X-linked recessive disorder (see Figure 20-2), but some forms appear to be acquired and affect both men and women. Affected individuals have high counts of IgM-secreting plasma cells in their peripheral blood and lymphoid tissue. In addition, XHM patients often have high levels of auto-antibodies to neutrophils, platelets, and red blood cells. Children with XHM suffer recurrent infections, especially respiratory infections; these are more severe than expected for a deficiency characterized by low levels of immunoglobulins.

The defect in XHM is in the gene encoding the CD40 ligand (CD40L or CD154), which maps to the X chromosome. $T_H$ cells from patients with XHM fail to express functional CD40L on their membrane. Since an interaction

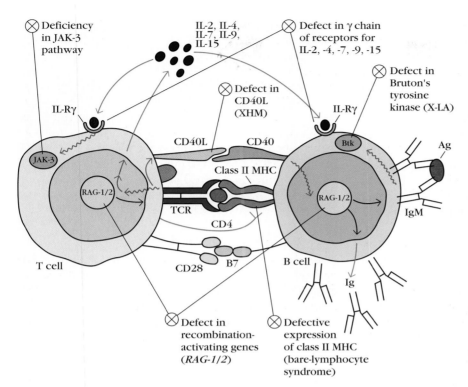

**FIGURE 20-4 Defects in cell interaction and signaling can lead to severe immunodeficiency.** The interaction of T cell and B cell is shown here with a number of the components important to the intra- and extracellular signaling pathways. A number of primary immunodeficiencies are rooted in defects in these interactions. SCID may result from defects in (1) the recombination-activating genes (*RAG-1* and *-2*) required for synthesis of the functional immunoglobulins and T-cell receptors that characterize mature B and T cells; (2) the γ chain of receptors for IL-2, -4, -7, -9, and -15 (IL-Rγ); (3) JAK-3, which transduces signals from the γ chain of the cytokine receptor; or (4) expression of the class II MHC molecule (bare-lymphocyte syndrome). X-LA results from defective transduction of activating signals from the cell surface IgM by Bruton's tyrosine kinase (Btk). XHM results from defects in CD40L that preclude normal maturation of B cells. *[Adapted from B. A. Smart and H. D. Ochs, 1997, Current Opinion in Pediatrics **9**:570.]*

between CD40 on the B cell and CD40L on the $T_H$ cell is required for B-cell activation, the absence of this costimulatory signal inhibits the B-cell response to T-dependent antigens (see Figure 11-12). The B-cell response to T-independent antigens, however, is unaffected by this defect, accounting for the production of IgM antibodies. As described in Chapter 11, class switching and formation of memory B cells both require contact with $T_H$ cells by a CD40-CD40L interaction. The absence of this interaction in XHM results in the loss of class switching to IgG, IgA, or IgE isotypes and in a failure to produce memory B cells. In addition, XHM individuals fail to produce germinal centers during a humoral response, which highlights the role of the CD40-CD40L interaction in the generation of germinal centers. Defects in cell interactions are common in lymphoid immunodeficiency, as shown in Figure 20-4.

## Common Variable Immunodeficiency (CVI)

CVI is characterized by a profound decrease in numbers of antibody-producing plasma cells, low levels of most immunoglobulin isotypes (hypogammaglobulinemia), and recurrent infections. The condition is usually manifested later in life than other deficiencies and is sometimes called late-onset hypogammaglobulinemia or, incorrectly, acquired hypogammaglobulinemia. However, CVI has a genetic component and is considered a primary immunodeficiency, although the exact pattern of inheritance is not known. Because the manifestations are very similar to those of acquired hypogammaglobulinemia, there is some confusion between the two forms (see below). Infections in CVI sufferers are most frequently bacterial and can be controlled by administration of immunoglobulin. In CVI patients, B cells fail to mature into plasma cells; however, in vitro studies show that CVI B cells are capable of maturing in response to appropriate differentiation signals. The underlying defect in CVI is not known but must involve either an in vivo blockage of the maturation of B cells to the plasma-cell stage or their inability to produce the secreted form of immunoglobulins.

## Hyper-IgE Syndrome (Job Syndrome)

A primary immunodeficiency characterized by skin abscesses, recurrent pneumonia, eczema, and elevated levels of IgE accompanies facial abnormalities and bone fragility. This multisystem disorder is autosomal dominant and has variable expressivity. The gene for hyper-IgE syndrome, or HIES, maps to chromosome 4. HIES immunologic signs include recurrent infection and eosinophilia (excessive eosinophils) in addition to elevated IgE levels.

## Selective Deficiencies of Immunoglobulin Classes

A number of immunodeficiency states are characterized by significantly lowered amounts of specific immunoglobulin isotypes. Of these, IgA deficiency is by far the most common. There are family-association data showing that IgA deficiency sometimes occurs in the same families as CVI, suggesting a relationship between these conditions. The spectrum of clinical symptoms of IgA deficiency is broad; many of those affected are asymptomatic, whereas others suffer from an assortment of serious problems. Recurrent respiratory and genitourinary tract infections resulting from lack of secreted IgA on mucosal surfaces are common. In addition, problems such as intestinal malabsorption, allergic disease, and autoimmune disorders may be associated with low IgA levels. The reasons for this variability in the clinical profile of IgA deficiency are not clear but may relate to the ability of some, but not all, patients to substitute IgM for IgA as a mucosal antibody. The defect in IgA deficiency is related to the inability of IgA B cells to undergo normal differentiation to the plasma-cell stage. The IgG subclasses IgG2 and IgG4 (see Figure 4-18) may also be deficient in IgA-deficient patients, although the surface IgA molecules on these patients' B cells appear to be expressed normally. A gene or genes outside of the immunoglobulin gene complex is suspected to be responsible for this fairly common syndrome.

Other immunoglobulin deficiencies have been reported, but these are rare. An IgM deficiency has been identified as an autosomal recessive trait. Victims of this condition are subject to severe infection by agents such as meningococcus, which causes fatal disease. IgM deficiency may be accompanied by various malignancies or by autoimmune disease. IgG deficiencies are also rare. These are often not noticed until adulthood and can be effectively treated by administration of immunoglobulin.

### Ataxia Telangiectasia

Although not classified primarily as an immunodeficiency, ataxia telangiectasia is a syndrome that includes deficiency of IgA and sometimes of IgE. The syndrome is characterized by difficulty in maintaining balance (ataxia) and by the appearance of broken capillaries (telangiectasia) in the eyes. The primary defect appears to be in a kinase involved in regulation of the cell cycle. The relationship between the immune deficiency and the other defects in ataxia telangiectasia remains obscure.

### Immune Disorders Involving the Thymus

Several immunodeficiency syndromes are grounded in failure of the thymus to undergo normal development. Thymic malfunction has a profound effect on T-cell function: all populations of T cells, including helper, cytolytic, and regulatory varieties, are affected. Immunity to viruses and fungi is especially compromised in those suffering from these conditions.

DiGeorge syndrome, or congenital thymic aplasia, in its most severe form is the complete absence of a thymus. This developmental defect, which is associated with the deletion in the embryo of a region on chromosome 22, causes immunodeficiency along with characteristic facial abnormalities, hypoparathyroidism, and congenital heart disease (Figure 20-5). The stage at which the causative developmental defect occurs has been determined, and the syndrome is sometimes called the *third and fourth pharyngeal pouch syndrome* to reflect its precise embryonic origin. The immune defect includes a profound depression of T-cell numbers and absence of T-cell responses. Although B cells are present in

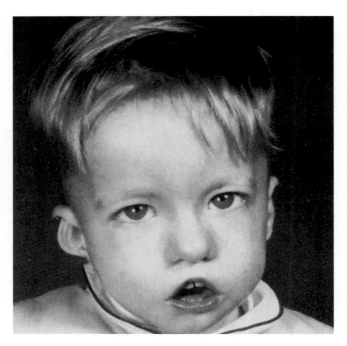

**FIGURE 20-5** A child with DiGeorge syndrome showing characteristic dysplasia of ears and mouth and abnormally long distance between the eyes. [R. Kretschmer et al., 1968, New England Journal of Medicine **279**:1295; photograph courtesy of F. S. Rosen.]

normal numbers, affected individuals do not produce antibody in response to immunization with specific antigens. Thymic transplantation is of some value for correcting the T-cell defects, but many DiGeorge patients have such severe heart disease that their chances for long-term survival are poor, even if the immune defects are corrected.

Whereas the DiGeorge syndrome results from an intrauterine or developmental anomaly, thymic hypoplasia, or the Nezelof syndrome, is an inherited disorder. The mode of inheritance for this rare disease is not known and its presentation varies, making it somewhat difficult to diagnose. As the name implies, thymic hypoplasia is a defect in which a vestigial thymus is unable to serve its function in T-cell development. In some patients, B cells are normal, whereas in others, a B-cell deficiency is secondary to the T-cell defect. Affected individuals suffer from chronic diarrhea, viral and fungal infections, and a general failure to thrive.

## Immunodeficiencies of the myeloid lineage affect innate immunity

Immunodeficiencies of the lymphoid lineage affect adaptive immunity. By contrast, defects in the myeloid-cell lineage affect the innate immune functions (see Figure 20-1). Most of these defects result in impaired phagocytic processes that are manifested by recurrent microbial infection of greater or lesser severity. The phagocytic processes may be faulty at several stages, including cell motility, adherence to and phagocytosis of organisms, and killing by macrophages.

## Reduction in Neutrophil Count

As described in Chapter 2, neutrophils are circulating granulocytes with phagocytic function. Quantitative deficiencies in neutrophils can range from an almost complete absence of cells, called agranulocytosis, to a reduction in the concentration of peripheral blood neutrophils below 1500/mm³, called granulocytopenia or neutropenia. These quantitative deficiencies may result from congenital defects or may be acquired through extrinsic factors. Acquired neutropenias are much more common than congenital ones.

Congenital neutropenia is often due to a genetic defect that affects the myeloid progenitor stem cell; it results in reduced production of neutrophils during hematopoiesis. In congenital agranulocytosis, myeloid stem cells are present in the bone marrow but rarely differentiate beyond the promyelocyte stage. As a result, children born with this condition show severe neutropenia, with counts of less than 200 neutrophils/mm³. These children suffer from frequent bacterial infections beginning as early as the first month of life; normal infants are protected at this age by maternal antibody as well as by innate immune mechanisms, including neutrophils. Experimental evidence suggests that this genetic defect results in decreased production of granulocyte colony-stimulating factor (G-CSF) and thus in a failure of the myeloid stem cell to differentiate along the granulocytic lineage (see Figure 2-1).

Neutrophils have a short life span, and their precursors must divide rapidly in the bone marrow to maintain adequate levels of these cells in the circulation. For this reason, agents such as radiation and certain drugs (e.g., chemotherapeutic drugs) that specifically damage rapidly dividing cells are likely to cause neutropenia. Occasionally, neutropenia develops in such autoimmune diseases as Sjögren's syndrome or systemic lupus erythematosus; in these conditions, auto-antibodies destroy the neutrophils. Transient neutropenia often develops after certain bacterial or viral infections, but neutrophil counts return to normal as the infection is cleared.

## Chronic Granulomatous Disease (CGD)

CGD is a genetic disease that has at least two distinct forms: an X-linked form that occurs in about 70% of patients and an autosomal recessive form found in the rest. This disease is rooted in a defect in the oxidative pathway by which phagocytes generate hydrogen peroxide and the resulting reactive products, such as hypochlorous acid (HOCL), that kill phagocytosed bacteria. CGD sufferers undergo excessive inflammatory reactions that result in gingivitis, swollen lymph nodes, and nonmalignant granulomas (lumpy subcutaneous cell masses); they are also susceptible to bacterial and fungal infection. CGD patients are not especially subject to infection by those bacteria, such as pneumococcus, that generate

their own hydrogen peroxide. In this case, the myeloperoxidase in the host cell can use the bacterial hydrogen peroxide to generate enough hypochlorous acid to thwart infection. Several related defects may lead to CGD, including a missing or defective cytochrome (cyt $b_{558}$) that functions in an oxidative pathway and defects in a protein (phagosome oxidase, or phox) that stabilizes the cytochrome. In addition to the general defect in the killer function of phagocytes, there is also a decrease in the ability of mononuclear cells to serve as APCs. Both processing and presentation of antigen are impaired. Increased amounts of antigen are required to trigger T-cell help when mononuclear cells from CGD patients are used as APCs.

The addition of IFN-γ has been shown to restore function to CGD granulocytes and monocytes in vitro. This observation prompted clinical trials of IFN-γ for CGD patients. Encouraging increases in oxidative function and restoration of cytoplasmic cytochrome have been reported in these patients. In addition, knowledge of the precise gene defects underlying CGD makes it a candidate for gene therapy, and replacement of the defective cytochrome has had promising results (see below).

## Chediak-Higashi Syndrome

This autosomal recessive disease is characterized by recurrent bacterial infections, partial oculo-cutaneous albinism (lack of skin and eye pigment), and aggressive but nonmalignant infiltration of organs by lymphoid cells. Phagocytes from patients with this immune defect contain giant granules but do not have the ability to kill bacteria. The molecular basis of the defect is a mutation in a protein (LYST) involved in the regulation of intracellular trafficking. The mutation impairs the targeting of proteins to secretory lysosomes, which makes them unable to lyse bacteria.

## Leukocyte Adhesion Deficiency (LAD)

As described in Chapter 13, cell surface molecules belonging to the integrin family of proteins function as adhesion molecules and are required to facilitate cellular interaction. Three of these, LFA-1, Mac-1, and gp150/95 (CD11a, b, and c, respectively), have a common β chain (CD18) and are variably present on different monocytic cells; CD11a is also expressed on B cells (Table 20-2). An immunodeficiency related to dysfunction of the adhesion molecules is rooted in a defect localized to the common β chain and affects expression of all three of the molecules that use this chain. This defect, called leukocyte adhesion deficiency (LAD), causes susceptibility to infection with both gram-positive and gram-negative bacteria as well as various fungi. Impairment of adhesion of leukocytes to vascular endothelium limits recruitment of cells to sites of inflammation. Viral immunity is somewhat impaired, as would be predicted from the defective T–B-cell cooperation arising from the adhesion defect. LAD varies in its severity; some affected individuals die within a few years, whereas others survive into their forties.

| TABLE 20-2 | Properties of integrin molecules that are absent in leukocyte-adhesion deficiency | | |
|---|---|---|---|
| | INTEGRIN MOLECULES* | | |
| Property | LFA-1 | CR3 | CR4 |
| CD designation | CD11a/CD18 | CD11b/CD18 | CD11c/CD18 |
| Subunit composition | $\alpha L\beta 2$ | $\alpha M\beta 2$ | $\alpha X\beta 2$ |
| Subunit molecular mass (kDa) $\alpha$ chain $\beta$ chain | 175,000 95,000 | 165,000 95,000 | 150,000 95,000 |
| Cellular expression | Lymphocytes Monocytes Macrophages Granulocytes Natural killer cells | Monocytes Macrophages Granulocytes Natural killer cells | Monocytes Macrophages Granulocytes |
| Ligand | ICAM-1 (CD 54) ICAM-2 (CD 102) | C3bi | C3bi |
| Functions inhibited with monoclonal antibody | Extravasation CTL killing T-B conjugate formation ADCC | Opsonization Granulocyte adherence, aggregation, and chemotaxis ADCC | Granulocyte adherence and aggregation |

*CR3 = type 3 complement receptor, also known as Mac-1; CR4 = type 4 complement receptor, also known as gp150/95; LFA-1, CR3, and CR4 are heterodimers containing a common $\beta$ chain but different $\alpha$ chains designated L, M, and X, respectively.

The reason for the variable disease phenotype in this disorder is not known. LAD is the subject of the Clinical Focus in Chapter 13.

## Complement defects result in immunodeficiency or immune-complex disease

Immunodeficiency diseases resulting from defects in the complement system are described in Chapter 7. Many complement deficiencies are associated with increased susceptibility to bacterial infections and/or immune-complex diseases. One of these complement disorders, a deficiency in properdin, which stabilizes the C3 convertase in the alternative complement pathway, is caused by a defect in a gene located on the X chromosome (see Figure 20-2). Defects in mannose-binding lectin (MBL) result in increased susceptibility to a variety of infections by bacterial or fungal agents. Recall from Chapter 7 that MBL is a key initiator of the complement attack on many pathogens and is an important component of the innate immune response to many organisms.

## Immunodeficiency disorders are treated by replacement of the defective element

Although there are no cures for immunodeficiency disorders, there are several treatment possibilities. In addition to the drastic option of total isolation from exposure to any microbial agent, treatment options for immunodeficiencies include the following:

- Replacement of a missing protein
- Replacement of a missing cell type or lineage
- Replacement of a missing or defective gene

For disorders that impair antibody production, the classic course of treatment is administration of the missing protein immunoglobulin. Pooled human gamma globulin given either intravenously or subcutaneously protects against recurrent infection in many types of immunodeficiency. Maintenance of reasonably high levels of serum immunoglobulin (5 mg/ml serum) will prevent most common infections in the agammaglobulinemic patient. This is generally accomplished by the administration of immunoglobulin that has been selected for antibodies directed against a particular organism. Recent advances in the preparation of human monoclonal antibodies and in the ability to genetically engineer chimeric antibodies with mouse V regions and human-derived C regions make it possible to prepare antibodies specific for important pathogens (see Chapter 5).

Advances in molecular biology make it possible to clone the genes that encode other immunologically important proteins, such as cytokines, and to express these genes in vitro, using bacterial or eukaryotic expression systems. The availability of such proteins allows new modes of therapy in which immunologically important proteins may be replaced or their concentrations increased in the patient. For example, the administration of recombinant IFN-γ has proven effective for patients with CGD, and the use of recombinant IL-2 may help to restore immune function in AIDS patients. Recombinant adenosine deaminase has been successfully administered to ADA-deficient SCID patients.

Cell replacement as therapy for immunodeficiencies has been made possible by recent progress in bone marrow transplantation (see Chapter 17). Replacement of stem cells with those from an immunocompetent donor allows development of a functional immune system (see Clinical Focus in Chapter 2). High rates of success have been reported for those who are fortunate enough to have an HLA-identical donor. Careful matching of patients with donors and the ability to enrich for stem cells by selecting for CD34+ cells continues to minimize the risk in this procedure, even when no ideal donor exists. These procedures have been highly successful with SCID infants when haploidentical (complete match of one HLA gene set or haplotype) donor marrow is used. T cells are depleted and CD34+ stem cells are enriched before introducing the donor bone marrow into the SCID infant. Because this therapy has been used only in recent years, it is not known whether transplantation cures the immunodeficiency permanently. A variation of bone marrow transplantation is the injection of paternal CD34+ cells in utero when the birth of an infant with SCID is expected.

If a single gene defect has been identified, as in adenosine deaminase deficiency or chronic granulomatous disease, replacement of the defective gene may be a treatment option. Clinical tests of such therapy are under way for SCID caused by ADA deficiency and for chronic granulomatous disease with defective p67phox, with promising initial results. Disease remission for up to 18 months was seen in the SCID patients and up to 6 months in the CGD patients. A similar procedure was used in both trials. It begins with obtaining cells (CD34+ cells are usually selected for these procedures) from the patient and transfecting them with a normal copy of the defective gene. The transfected cells are then returned to the patient. As this treatment improves, it will become applicable to a number of immunodeficiencies for which a genetic defect is well defined. As mentioned above, these include defects in genes that encode the γ chain of the IL-2 receptor, JAK-3, and ZAP-70, all of which give rise to SCID. As with all treatments, there is an element of risk in the use of gene therapy for

immunodeficiency. Two incidents in which the introduced cell populations proliferated in an uncontrolled manner, giving rise to leukemia in treated SCID patients, dictate caution in the use of gene therapy. However, the potential risk of gene therapy must be balanced against the benefit to patients who will succumb to fatal infections without treatment.

## Experimental models of immunodeficiency include genetically altered animals

Immunologists use two well-studied animal models of primary immunodeficiency for a variety of experimental purposes. One of these is the athymic, or nude, mouse; the other is the severe combined immunodeficiency, or SCID, mouse. Recent studies with genetically altered mice in which a single gene is knocked out give precise information about the role of specific genes in combating infection.

### Nude (Athymic) Mice

A genetic trait designated *nu*, which is controlled by a recessive gene on chromosome 11, was discovered in certain mice. Mice homozygous for this trait *(nu/nu)* are hairless and have a vestigial thymus (Figure 20-6). Heterozygotic, *nu/+*, litter mates have hair and a normal thymus. It is not known whether the hairlessness and the thymus defect are caused by the same gene. It is possible that two very closely linked genes control these defects, which, although unrelated, appear together in this mutant mouse. A gene that controls development may be involved, since the pathway that leads to the differential development of the thymus is related to the one that controls the skin epithelial cells. The *nu/nu* mouse cannot easily survive; under normal conditions, the mortality is 100% within 25 weeks and 50% die within the first 2 weeks after birth. Therefore, when these animals are to be used for experimental purposes, they must be maintained under conditions that protect them from infection. Precautions include use of sterilized food, water, cages, and bedding. The cages are protected from dust by placing them in a laminar flow rack or by the use of air filters fitted over the individual cages.

Nude mice lack cell-mediated immune responses, and they are unable to make antibodies to most antigens. The immunodeficiency in the nude mouse can be reversed by a thymic transplant. Because these mice can permanently

**FIGURE 20-6  A nude mouse *(nu/nu)*.** This defect leads to absence of a thymus or a vestigial thymus and cell-mediated immunodeficiency. *[Courtesy of the Jackson Laboratory, Bar Harbor, Maine.]*

tolerate both allografts and xenografts, they have a number of practical experimental uses. For example, hybridomas or solid tumors from any origin may be grown as ascites or as implanted tumors in a nude mouse. It is known that the nude mouse does not completely lack T cells; rather, it has a limited population that increases with age. The source of these T cells is not known; an intriguing possibility is that there is an extrathymic source of mature T cells. However, it is more likely that the T cells arise from the vestigial thymus. The majority of cells in the circulation of a nude mouse carry T-cell receptors of the $\gamma\delta$ type instead of the $\alpha\beta$ type that predominates in the circulation of a normal mouse.

### The SCID Mouse

In 1983, Melvin and Gayle Bosma and their colleagues described an autosomal recessive mutation in mice that gave rise to a severe deficiency in mature lymphocytes. They designated the trait SCID because of its similarity to human severe combined immunodeficiency. The SCID mouse was shown to have early B- and T-lineage cells but a virtual absence of lymphoid cells in the thymus, spleen, lymph nodes, and gut tissue, the usual locations of functional T and B cells. The precursor T and B cells in the SCID mouse appeared to be unable to differentiate into mature functional B and T lymphocytes. Inbred mouse lines carrying the SCID defect have been derived and studied in great detail. The SCID mouse can neither make antibody nor carry out delayed-type hypersensitivity (DTH) or graft rejection reactions. If the animals are not kept in an extremely clean environment, they succumb to infection early in life. Cells other than lymphocytes develop normally in the SCID mouse; red blood cells, monocytes, and granulocytes are present and functional. SCID mice may be rendered immunologically competent by transplantation of stem cells from normal mice.

The mutation in a DNA protein kinase that causes mouse SCID is a so-called leaky mutation, because a certain number of SCID mice do produce immunoglobulin. About half of these leaky SCID mice can also reject skin allografts. This finding suggests that the defective enzyme can function partly in T- and B-cell development, allowing normal differentiation of a small percentage of precursor cells. More recently, immunodeficient SCID-like mice have been developed by deletion of the recombination-activating enzymes (RAG-1 and RAG-2) responsible for the rearrangement of immunoglobulin or T-cell–receptor genes in both B- and T-cell precursors. RAG knockout mice exhibit defects in both B and T cells; neither cell type can rearrange the genes for their receptor and thus neither proceeds along a normal developmental path. Because cells with abnormal rearrangements are eliminated in vivo, both B and T cells are absent from the lymphoid organs of the RAG knockout mouse. In addition to providing a window into possible causes of combined T- and B-cell immunodeficiency, the SCID mouse has proven extremely useful in studies of cellular immunology. Because its graft rejection mechanisms do not operate, the SCID mouse can be used for studies on cells or organs from various sources. For example, immune precursor cells from human sources may be used to reestablish the SCID mouse's immune system. These human cells can develop in a normal fashion and, as a result, the SCID mouse circulation will contain immunoglobulin of human origin. In one important application, these SCID mice are infected with HIV-1. Although normal mice are not susceptible to HIV-1 infection, the SCID mouse reconstituted with human lymphoid tissue (SCID-Hu mouse) provides an animal model in which to test therapeutic or prophylactic strategies against HIV infection of the transplanted human lymphoid tissue.

# AIDS and Other Acquired or Secondary Immunodeficiencies

As described above, a variety of defects in the immune system give rise to immunodeficiency. In addition to the primary immunodeficiencies, there are also acquired, or secondary, immunodeficiencies. One that has been known for some time is called acquired hypogammaglobulinemia. (As mentioned above, this condition is sometimes confused with common variable immunodeficiency, a condition that shows genetic predisposition.) The origin of acquired hypogammaglobulinemia is unknown, and its major symptom, recurrent infection, manifests itself in young adults. The patients generally have very low but detectable levels of total immunoglobulin. T-cell numbers and function may be normal, but there are some cases with T-cell defects, which may grow more severe as the disease progresses. The disease is generally treated by immunoglobulin therapy, allowing patients to survive into their seventh and eighth decades. Unlike the similar deficiencies described above, there is no evidence for genetic transmission of this disease. Mothers with acquired hypogammaglobulinemia deliver normal infants. However, at birth the infants will be deficient in circulating immunoglobulin, because the deficiency in maternal circulation is reflected in the infant.

Another form of secondary immunodeficiency, known as agent-induced immunodeficiency, results from exposure to any of a number of chemical and biological agents that induce an immunodeficient state. Certain of these are drugs used to combat autoimmune diseases such as rheumatoid arthritis or lupus erythematosis. Corticosteroids, which are commonly used for autoimmune disorders, interfere with the immune response in order to relieve disease symptoms. Similarly, a state of immunodeficiency is deliberately induced in transplantation patients who are given immunosuppressive drugs, such as cyclosporin A, in order to blunt the attack of the immune system on transplanted organs. As described in Chapter 17, recent efforts have been made to use more specific means of inducing tolerance to allografts to circumvent the unwanted side effects of general immunosuppression. The mechanism of action of the immunosuppressive agents varies, although T cells are a common target. In addition, cytotoxic drugs or radiation treatments given to treat various forms of cancer frequently damage the dividing cells in the body, including those of the immune system, and induce a state of immunodeficiency as an unwanted consequence. Patients undergoing such therapy must be monitored closely and treated with antibiotics or immunoglobulin if infection appears.

## HIV/AIDS has claimed millions of lives worldwide

In recent years, all other forms of immunodeficiency have been overshadowed by an epidemic of severe immunodeficiency caused by the infectious agent called human immunodeficiency virus 1, or HIV-1. The disease that HIV-1 causes, acquired immunodeficiency syndrome (AIDS), was first reported in the United States in 1981 in Los Angeles, New York, and San Francisco. A group of patients displayed unusual infections, including the opportunistic fungal pathogen *Pneumocystis carinii*, which causes a pneumonia called *P. carinii* pneumonia (PCP) in people with immunodeficiency. In addition to PCP, some patients had Kaposi's sarcoma, an extremely rare skin tumor, as well as other, rarely encountered opportunistic infections. More complete evaluation of the patients showed that they had in common a marked deficiency in cellular immune responses and a significant decrease in the subpopulation of T cells that carry the CD4 marker (T helper cells). When epidemiologists examined the background of the first patients with this new syndrome, they found that the majority of those afflicted were homosexual males. As the number of AIDS cases increased and the disease was recognized throughout the world, people found to be at high risk for AIDS were homosexual males, promiscuous heterosexual individuals of either sex and their partners, intravenous drug users, people who received blood or blood products prior to 1985, and infants born to HIV-infected mothers.

Since its discovery in 1981, AIDS has increased to epidemic proportions throughout the world. As of December 2004, the cumulative total number of people in the United States reported to have died of AIDS was 524,000, and approximately a million people were living with HIV infection in 2005. Although reporting of AIDS cases is mandatory, many states do not require reporting of cases of HIV infection that have not yet progressed to AIDS, making the count of the number of HIV-infected individuals an estimate. Although the death rate from AIDS has decreased in recent years because of improved treatments, AIDS remains among the leading killers of people in the 25-to-44-year-old age range in this country (Figure 20-7). The fact that the number of yearly AIDS deaths has leveled off is encouraging but does not indicate an end to the epidemic in this country. A disturbing trend is that the epidemic is increasingly centered in women and in African Americans, who account for half of the recent AIDS deaths.

The magnitude of the AIDS epidemic in the United States is dwarfed by figures for other parts of the world. The

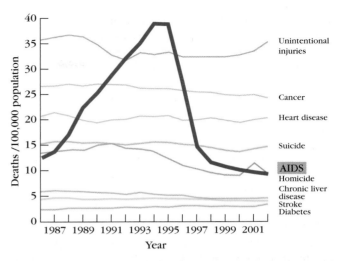

**FIGURE 20-7  Death rates for the nine leading causes of death in people ages 25 to 44 years in the United States for the years 1987 to 2002.** The red line shows that the death rate per 100,000 people caused by AIDS surpassed any other single cause of death in this age range during the period 1993 to 1995. The recent decrease in AIDS deaths in the United States is attributed to improvements in anti-HIV drug therapy, which prolongs the lives of patients. *[Data from the Centers for Disease Control Web site, www.cdc.gov/.]*

global distribution of those afflicted with AIDS is shown in Figure 20-8. In sub-Saharan Africa, an estimated 25.8 million people were living with AIDS at the end of 2005, and in South and Southeast Asia there were another 7.4 million. An estimated 40.3 million people worldwide have AIDS, including over 2.3 million children under 15 years of age. In addition, millions of children have been orphaned by the death of their parents from AIDS. Recent estimates from the World Health Organization indicate that there were 4.9 million new HIV infections in 2005, or an average of almost 13,500 people infected each day during that year. This number includes a daily infection toll of 1400 children under age 15.

The initial group of AIDS patients in the United States and Western Europe was predominantly white and male. Although this remains the group predominantly affected in these areas, more recently the distribution in the United States has shifted to include a larger proportion of women and an increasing proportion of minorities. Worldwide, the number of AIDS patients distributes more equally between males and females, and in sub-Saharan Africa, which has the highest incidence of AIDS, more than half (approximately 57%) of those afflicted are females.

## HIV-1 spreads by sexual contact, by infected blood, and from mother to infant

Although the precise mechanism by which HIV-1 infects an individual is not known, epidemiological data indicate that common means of transmission include homosexual and

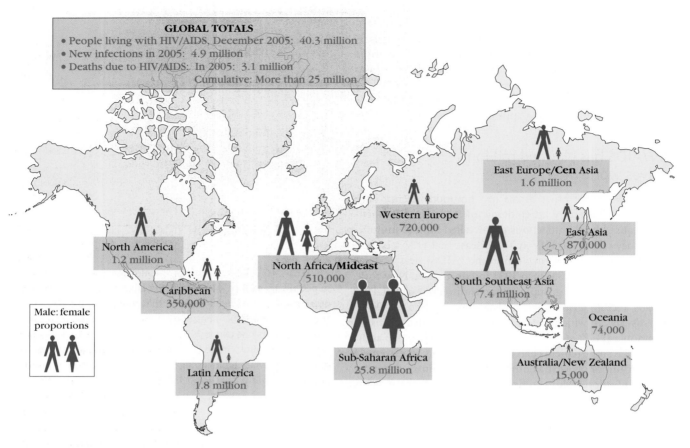

**GLOBAL TOTALS**
- People living with HIV/AIDS, December 2005: 40.3 million
- New infections in 2005: 4.9 million
- Deaths due to HIV/AIDS: In 2005: 3.1 million
  Cumulative: More than 25 million

East Europe/**Cen** Asia
1.6 million

Western Europe
720,000

East Asia
870,000

North America
1.2 million

North Africa/**Mideast**
510,000

South Southeast Asia
7.4 million

Caribbean
350,000

Oceania
74,000

Male: female
proportions

Sub-Saharan Africa
25.8 million

Australia/New Zealand
15,000

Latin America
1.8 million

**FIGURE 20-8  The global AIDS epidemic.** The estimated world-wide distribution of AIDS cases as of December 2005. Approximately 40.3 million people were living with AIDS as of December 2005; most of them were in sub-Saharan Africa and Southeast Asia. In North America and Western Europe, about 75% of those affected were men, whereas in sub-Saharan Africa about 57% of those with AIDS are women. [*Data from UNAIDS/WHO: AIDS Epidemic Update, December 2005, http://www.unaids.org/epi/2005/doc/EPIupdate2005_pdf_en/epi-update2005_en.pdf.*]

heterosexual intercourse, receipt of infected blood or blood products, and passage from mothers to infants. Before tests for HIV in the blood supply were routinely used, patients who received blood transfusions and hemophiliacs who received blood products were at risk for HIV-1 infection. Exposure to infected blood accounts for the high incidence of AIDS among intravenous drug users, who often share hypodermic needles. Infants born to mothers who are infected with HIV-1 are at high risk of infection. Unless infected mothers are treated with antiviral agents before delivery, approximately 30% of infants born to them will become infected with the virus (see Clinical Focus). Possible vehicles of passage from mother to infant include blood transferred in the birth process and milk in the nursing period. Transmission from an infected to an uninfected individual is most likely by transmission of HIV-infected cells—in particular, macrophages, dendritic cells, and lymphocytes.

In the worldwide epidemic, it is estimated that 75% of the cases of HIV transmission are attributable to heterosexual contact. Although the probability of transmission by vaginal intercourse is lower than by other means, such as IV drug use or receptive anal intercourse, the likelihood of infection is greatly enhanced by the presence of other sexually transmitted diseases (STDs). In populations where prostitution is rampant, STDs flourish and provide a powerful cofactor for the heterosexual transmission of HIV-1. Reasons for this increased infection rate include the lesions and open sores present in many STDs, which favor the transfer of HIV-infected blood during intercourse. Data from studies in India and in Uganda indicate that male circumcision significantly lowers the risk of males acquiring HIV-1 and lowers the risk of transmission to sexual partners from infected circumcised males. No similar protective effect of circumcision was observed in these studies for other sexually transmitted diseases, including herpes simplex type 2, syphilis, or gonorrhea.

Transmission of HIV-1 infection requires contact with blood, milk, semen, or vaginal fluid from an infected individual. Research workers and medical professionals who take reasonable precautions have a very low incidence of AIDS, despite repeated contact with infected materials. The risk of transmitting HIV infection can be minimized by simple precautionary measures, including the avoidance of any practice that could allow exposure of broken or abraded

## CLINICAL FOCUS
# Prevention of Infant HIV Infection by Anti-Retroviral Treatment

**It** is estimated that almost 700,000 infants became infected with HIV through mother to child transmission in 2005. The majority of these infections resulted from transmission of virus from HIV-infected mothers during childbirth or by transfer of virus from milk during breast feeding. The incidence of maternal acquired infection can be reduced by treatment of the infected mother with a course of zidovudine (AZT) for several months prior to delivery and treatment of her infant for 6 weeks after birth. This treatment regimen is widely used in the United States. However, the majority of worldwide HIV infection of infants occurs in sub-Saharan Africa and other less-developed areas, where the cost and timing of the zidovudine regimen render it an impractical solution to the problem of maternal-infant HIV transmission.

A 1999 clinical trial of the anti-retroviral nevirapine (Viramune) brought hope for a practical way to combat infant HIV infection under less-than-ideal conditions of clinical care. The trial took place at Mulago Hospital in Kampala, Uganda, and enrolled 645 mothers who tested positive for HIV infection. About half of the mothers were given a single dose of nevirapine at the onset of labor, and their infants were given a single dose 24 to 30 hours after birth. The dose and timing were dictated by the customary rapid discharge at the hospital. The control arm of the study involved a more extensive course of zidovudine, but in-country conditions did not allow exact replication of the full course administered to infected mothers in the United States. The subjects in this study were followed for at least 18 months, including 302 infants treated with zidovudine and 308 with nevirapine. Ninety-nine percent of the infants were breast-fed by the infected mothers in this study, so the test measures a considerable interval during which the infants were exposed to risk of infection.

The infants were tested for HIV-1 infection at several times after birth up to 18 months of age using an RNA PCR test and at later times for HIV-1 antibody (after maternal antibody would no longer interfere with the test). The overall rate of infection for infants born to untreated mothers is estimated to be about 37%. When the full course of zidovudine is used, the rate drops to 20%. The highly encouraging results of the Uganda study revealed infection in only 13.5% of the babies in the nevirapine group when tested at 16 weeks of age. Of those given a short course of zidovudine, 22.1% were infected at this age compared to 40.2% in a small group given placebo. In the 18-month follow-up of the test group, 15.7% of those given nevirapine were infected whereas 25.8% given zidovudine were positive for HIV-1.

From this study, it appears that the single dose of nevirapine is the most effective means found thus far to prevent maternal-infant transmission of HIV infection—even better than the more extensive and costly regimen currently used in developed countries. These results must be verified, and the possibility of unexpected side effects must be explored. However, this result gives hope for reduction of infant infection in parts of the world where access to medical care is limited.

As mentioned above, the study was designed to conform to the reality of maternal health care in Kampala; it fits this system perfectly. The use of nevirapine has other significant advantages, including stability of the drug at room temperature and reasonable cost. The dose of nevirapine administered to the mother and infant costs about 200 times less than the zidovudine regimen in current use in the United States. In fact, the treatment is sufficiently inexpensive to suggest that it may be cost-effective to treat all mothers at the time of delivery in those areas where rates of infection are high, because the nevirapine treatment costs less than the tests used to determine HIV infection. Obviously, such a strategy must be embarked on cautiously, given the danger of long-term side effects and other unexpected problems.

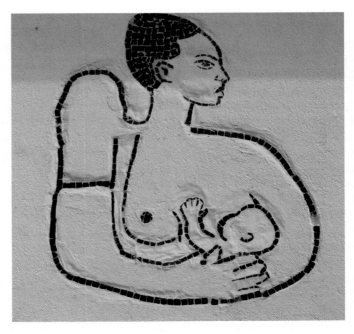

Mural showing mother and child on an outside wall of Mulago Hospital Complex in Kampala, Uganda, site of the clinical trial demonstrating that maternal-infant HIV-1 transmission was greatly reduced by nevirapine. [*Courtesy of Thomas Quinn, Johns Hopkins University.*]

## OVERVIEW FIGURE 20-9: Structure of HIV

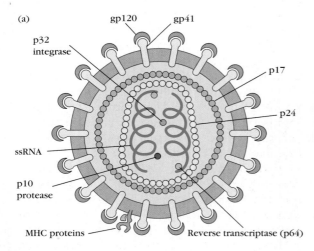

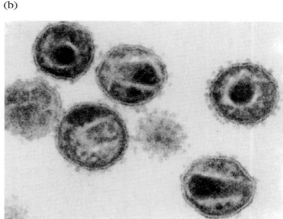

(a) Cross-sectional schematic diagram of HIV virion. Each virion expresses 72 glycoprotein projections composed of gp120 and gp41. The gp41 molecule is a transmembrane molecule that crosses the lipid bilayer of the viral envelope. Gp120 is associated with gp41 and serves as the viral receptor for CD4 on host cells. The viral envelope derives from the host cell and contains some host-cell membrane proteins, including class I and class II MHC molecules. Within the envelope is the viral core, or nucleocapsid, which includes a layer of a protein called p17 and an inner layer of a protein called p24. The HIV genome consists of two copies of single-stranded RNA, which are associated with two molecules of reverse transcriptase (p64) and nucleoid proteins p10, a protease, and p32, an integrase. (b) Electron micrograph of HIV virions magnified 200,000 times. The glycoprotein projections are faintly visible as "knobs" extending from the periphery of each virion. [Part a adapted from B. M. Peterlin and P. A. Luciw, 1988, AIDS **2**:S29; part b from a micrograph by Hans Geldenblom of the Robert Koch Institute (Berlin), in R. C. Gallo and L. Montagnier, 1988, Scientific American **259**(6):41.]

skin or any mucosal membrane to blood from a potentially infected person. The use of condoms when having sex with individuals of unknown infection status is highly recommended. One factor contributing to the spread of HIV is the long period after infection during which no clinical signs may appear but during which the infected individual may infect others. Thus, universal use of precautionary measures is important whenever and wherever infection status is uncertain.

It is a sobering thought that the epidemic of AIDS came at a time when many believed that infectious diseases no longer posed a serious threat to people in the United States and other industrialized nations. Vaccines and antibiotics controlled most serious infectious agents. The eradication of smallpox in the world had recently been celebrated, and polio was yielding to widespread vaccination efforts; these were considered milestones on the road to elimination of most infectious diseases. The outbreak of AIDS shattered this complacency and triggered a massive effort to combat the disease. In addition, the immunodeficiency that characterizes AIDS has allowed re-emergence of other infectious diseases, such as tuberculosis, which have the potential to spread into populations not infected with HIV.

## The retrovirus HIV-1 is the causative agent of AIDS

Within a few years after recognition of AIDS as an infectious disease, the causative agent was discovered and characterized by efforts in the laboratories of Luc Montagnier in Paris and Robert Gallo in Bethesda (Figure 20-9). This immunodeficiency syndrome was novel at the time in that the type of virus causing it is a **retrovirus.** Retroviruses carry their genetic information in the form of RNA. When the virus enters a cell, the RNA is reverse-transcribed to DNA by a virally encoded enzyme, reverse transcriptase (RT). As the name implies, RT reverses the normal transcription process and makes a DNA copy of the viral RNA genome. This copy, which is called a **provirus,** is integrated into the cell genome and is replicated along with the cell DNA. When the provirus is expressed to form new virions, the cell lyses. Alternatively, the provirus may remain latent in the cell until some regulatory signal starts the expression process.

Only one other human retrovirus, human T-cell lymphotropic virus I, or HTLV-I, had been described before HIV-1. This retrovirus is endemic in the southern part of Japan and in the Caribbean. Although most individuals infected with HTLV-I display no clinical signs of disease, a small percentage develop serious illness, either adult T-cell leukemia, which is aggressive and usually fatal, or a disabling progressive neurologic disorder called HTLV-I–associated myelopathy (or, in early reports, tropical spastic paraparesis). Although comparisons of their genomic sequences revealed that HIV-1 is not a close relative of HTLV-I, similarities in overall characteristics led to use of the name HTLV-III for the AIDS virus in early reports. There is also a related human virus called HIV-2, which is less pathogenic in humans than HIV-1. HIV-2 is similar to viruses isolated from monkeys; it infects certain nonhuman primates that do not become infected by HIV-1.

Viruses related to HIV-1 have been found in nonhuman primates. These viruses, variants of simian immunodeficiency virus, or SIV, cause immunodeficiency disease in certain infected monkeys. Normally, SIV strains cause no disease in their normal host but produce immunodeficiency similar to AIDS when injected into another species. For example, the virus from African green monkeys ($SIV_{agm}$) is present in a high percentage of normal healthy African green monkeys in the wild. However, when $SIV_{agm}$ is injected into macaques, it causes a severe, often lethal immunodeficiency.

A number of other animal retroviruses more or less similar to HIV-1 have been reported. These include the feline and bovine immunodeficiency viruses and the mouse leukemia virus. Study of these animal viruses has yielded information concerning the general nature of retrovirus action, but specific information about HIV-1 cannot be gained by infecting animals because HIV-1 does not replicate in them. Only the chimpanzee supports infection with HIV-1 at a level sufficient to be useful in vaccine trials, but infected chimpanzees only rarely develop AIDS, which limits the value of this model in the study of viral pathogenesis. In addition, the number of chimpanzees available for such studies is low, and both the expense and the ethical issues involved in experiments with chimpanzees preclude widespread use of this infection model. The SCID mouse (see above) reconstituted with human lymphoid tissue for infection with HIV-1 has been useful for certain studies of HIV-1 infection, especially in the development of drugs to combat viral replication.

Reasons for the limited host range of HIV-1 include not only the cell surface receptors required for entry of the virus into the host cell but dependence of the virus on host-cell factors for early events in its replication process, such as transcription and splicing of viral messages. For example, mouse cells transfected with genes that mediate expression of the human receptors for HIV-1 will not support HIV-1 replication because they lack other host factors. By contrast, cells from hamsters or rabbits transfected to express the human receptors support levels of virus replication similar to those seen in human cells. Despite some progress in understanding the factors needed for HIV-1 infection, no clear candidate for an animal model of HIV-1 infection exists. This lack of a suitable infection model hampers efforts to develop both drugs and vaccines to combat AIDS.

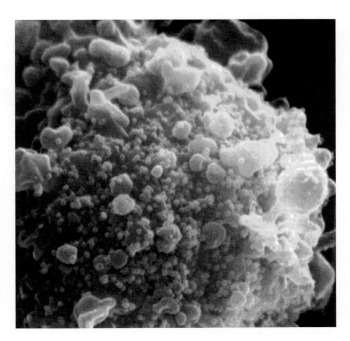

**FIGURE 20-10   Once the HIV provirus has been activated, buds representing newly formed viral particles can be observed on the surface of an infected T cell.** The extensive cell damage resulting from budding and release of virions leads to the death of infected cells. [*Courtesy of R. C. Gallo, 1988,* Journal of Acquired Immune Deficiency Syndromes *1:521.*]

## In vitro studies revealed the HIV-1 replication cycle

The AIDS virus can infect human T cells in culture, replicating itself and in many cases causing the lysis of the cell host (Figure 20-10). Much has been learned about the life cycle of HIV-1 from in vitro studies. The various proteins encoded by the viral genome have been characterized, and the functions of most of them are known (Figure 20-11).

The first step in HIV infection is viral attachment and entry into the target cell. HIV-1 infects T cells that carry the CD4 antigen on their surface; in addition, certain HIV strains will infect monocytes and other cells that have CD4 on their surface. The preference for CD4$^+$ cells is due to a high-affinity interaction between a coat (envelope, or env) protein of HIV-1 and cell surface CD4. Although the virus binds to CD4 on the cell surface, this interaction alone is not sufficient for entry and productive infection. Expression of other cell surface molecules, coreceptors present on T cells and monocytes, is required for HIV-1 infection. The infection of a T cell, depicted in Figure 20-12a, is assisted by the chemokine receptor CXCR4. An analogous receptor called CCR5 functions for the monocyte or macrophage (see Chapter 13).

After the virus has entered the cell, the RNA genome of the virus is reverse-transcribed and a cDNA copy (provirus) integrates into the host genome. The integrated provirus is transcribed and the various viral RNA messages spliced and

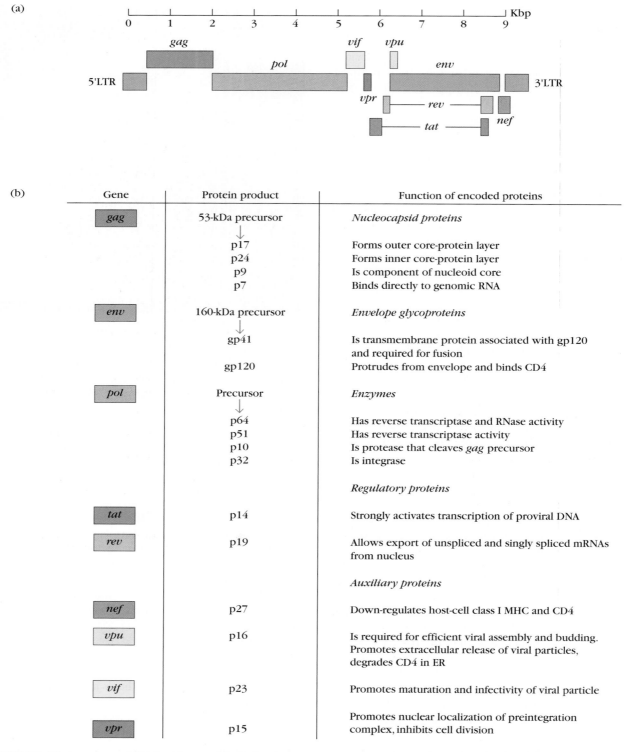

**FIGURE 20-11 Genetic organization of HIV-1 (a) and functions of encoded proteins (b).** The three major genes—*gag, pol,* and *env*—encode polyprotein precursors that are cleaved to yield the nucleocapsid core proteins, enzymes required for replication, and envelope core proteins. Of the remaining six genes, three (*tat, rev,* and *nef*) encode regulatory proteins that play a major role in controlling expression, two (*vif* and *vpu*) encode proteins required for virion maturation, and one (*vpr*) encodes a weak transcriptional activator. The 5'long terminal repeat (LTR) contains sequences to which various regulatory proteins bind. The organization of the HIV-2 and SIV genomes is very similar except that the *vpu* gene is replaced by *vpx* in both of them.

## OVERVIEW FIGURE 20-12:   HIV Infection of Target Cells and Activation of Provirus

(a) Infection of target cell

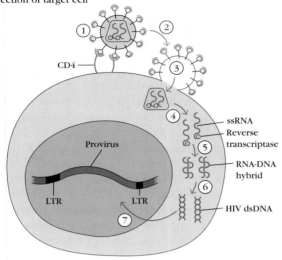

Provirus

LTR   LTR

ssRNA

Reverse transcriptase

RNA-DNA hybrid

HIV dsDNA

CD4

(b) Activation of provirus

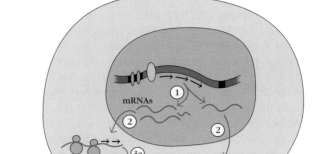

mRNAs

Precursors   Proteins

ssRNA

① HIV gp120 binds to CD4 on target cell.

② Fusogenic domain in gp41 and CXCR4, a G-protein–linked receptor in the target-cell membrane, mediate fusion.

③ Nucleocapsid containing viral genome and enzymes enters cells.

④ Viral genome and enzymes are released following removal of core proteins.

⑤ Viral reverse transcriptase catalyzes reverse transcription of ssRNA, forming RNA-DNA hybrids.

⑥ Original RNA template is partially degraded by ribonuclease H, followed by synthesis of second DNA strand to yield HIV dsDNA.

⑦ The viral dsDNA is then translocated to the nucleus and integrated into the host chromosomal DNA by the viral integrase enzyme.

① Transcription factors stimulate transcription of proviral DNA into genomic ssRNA and, after processing, several mRNAs.

② Viral RNA is exported to cytoplasm.

③a Host-cell ribosomes catalyze synthesis of viral precursor proteins.

③b Viral protease cleaves precursors into viral proteins.

④ HIV ssRNA and proteins assemble beneath the host-cell membrane, into which gp41 and gp120 are inserted.

⑤a The membrane buds out, forming the viral envelope.

⑤b Released viral particles complete maturation; incorporated precursor proteins are cleaved by viral protease present in viral particles.

(c)

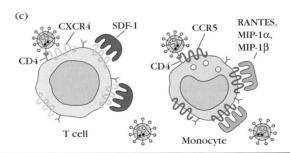

CXCR4   SDF-1

CD4

T cell

CCR5   RANTES, MIP-1α, MIP-1β

CD4

Monocyte

(a) Following entry of HIV into cells and formation of dsDNA, integration of the viral DNA into the host-cell genome creates the provirus. (b) The provirus remains latent until events in the infected cell trigger its activation, leading to formation and release of viral particles. (c) Although CD4 binds to the envelope glyco-protein of HIV-1, a second receptor is necessary for entry and infection. The T-cell–tropic strains of HIV-1 use the coreceptor CXCR4, whereas the macrophage-tropic strains use CCR5. Both are receptors for chemokines, and their normal ligands can block HIV infection of the cell.

translated into proteins, which along with a complete new copy of the RNA genome are used to form new viral particles (Figure 20-12b). The gag proteins of the virus are cleaved by the viral protease into the forms that make up the nuclear capsid (see Figure 20-10) in a mature infectious viral particle. As will be described below, different stages in this viral replication process provide targets for anti-retroviral drugs.

The discovery that CXCR4 and CCR5 serve as coreceptors for HIV-1 on T cells and macrophages, respectively, explained why some strains of HIV-1 preferentially infect T cells (T-tropic strains) whereas others prefer macrophages (M-tropic strains). A T-tropic strain uses CXCR4, whereas the M-tropic strains use CCR5. This use of different coreceptors also helped to explain the different roles of cytokines and chemokines in virus replication. It was known from in vitro studies that certain chemokines had a negative effect on virus replication whereas certain proinflammatory cytokines had a positive effect. Both of the HIV coreceptors, CCR5 and CXCR4, function as receptors for chemokines (see Table 13-2). Because the receptors cannot bind simultaneously to HIV-1 and to their chemokine ligand, there is competition for the receptor between the virus and the normal ligand (Figure 20-12c), and the chemokine can block viral entry into the host cell. Whereas the chemokines compete with HIV for usage of the coreceptor and thus inhibit viral entry, the proinflammatory cytokines induce greater expression of the chemokine receptors on the cell surface, making the cells more susceptible to viral entry.

HIV-1 infection of T cells with certain strains of virus leads to the formation of giant cells, or syncytia. These are formed by the fusion of a group of cells caused by the interaction of the viral envelope protein gp120 on the surface of infected cells with CD4 and the coreceptors on the surface of other cells, infected or not. After the initial binding, the action of other cell adhesion molecules welds the cells together in a large multinuclear mass with a characteristic fused ballooning membrane that eventually bursts. Formation of syncytia may be blocked by antibodies to some of the epitopes of the CD4 molecule, by soluble forms of the CD4 molecule (prepared by in vitro expression of a CD4 gene genetically engineered to lack the transmembrane portion), and by antibodies to cell adhesion molecules. Individual isolates of HIV-1 differ in their ability to induce syncytia formation.

Isolates of HIV-1 from different sources were formerly classified as syncytia-inducing (SI) or non-syncytium-inducing (NSI). In most cases, these differences correlated with the ability of the virus to infect T cells or macrophages: T-tropic strains were SI, whereas M-tropic strains were NSI. More recent classifications of HIV-1 are based on which coreceptor the virus uses; there is good but not absolute correlation between the use of CXCR4, which is present on T cells, and syncytia-inducing ability. The NSI strains use CCR5, which is present on monocytes. Studies of the viral envelope protein gp120 identified a region called the V3 loop, which plays a role in the choice of receptors used by the virus. A study by Mark Goldsmith and Bruce Chesebro and their colleagues indicates that a single amino acid difference in this region of gp120 may be sufficient to determine which receptor is used.

## HIV-1 infection leads to opportunistic infections

Isolation of HIV-1 and its growth in culture has allowed purification of viral proteins and the development of tests for infection with the virus. The most commonly used test is for the presence of antibodies directed against proteins of HIV-1. These generally appear in the serum of infected individuals by three months after the infection has occurred. When the antibodies appear, the individual is said to have seroconverted or to be seropositive for HIV-1. Although the precise course of HIV-1 infection and disease onset varies considerably in different patients, a general scheme for the progression of AIDS can be constructed (Figure 20-13). The course of HIV-1 infection begins with no detectable anti-HIV-1 antibodies and progresses to the full AIDS syndrome. Diagnosis of AIDS includes evidence for infection with HIV-1 (presence of antibodies or viral RNA in blood), greatly diminished numbers of CD4$^+$ T cells (<200 cells/mm$^3$), impaired or absent delayed-hypersensitivity reactions, and the occurrence of opportunistic infections (Table 20-3). Patients with AIDS generally succumb to tuberculosis, pneumonia, severe wasting diarrhea, or various malignancies. The time

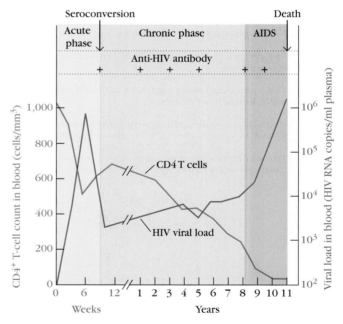

**FIGURE 20-13 Serologic profile of HIV infection showing three stages in the infection process.** Soon after infection, viral RNA is detectable in the serum. However, HIV infection is most commonly detected by the presence of anti-HIV antibodies after seroconversion, which normally occurs within a few months after infection. Clinical symptoms indicative of AIDS generally do not appear for at least 8 years after infection, but this interval is variable. The onset of clinical AIDS is usually signaled by a decrease in T-cell numbers and an increase in viral load. [Adapted from A. Fauci et al., 1996, Annals of Internal Medicine **124**:654.]

| TABLE 20-3 | Clinical diagnosis of HIV-infected individuals | | |
|---|---|---|---|

| | CLINICAL CATEGORIES* | | |
|---|---|---|---|
| CD4$^+$ T-cell count | A | B | C |
| ≥ 500/μl | A1 | B1 | **C1** |
| 200–499/μl | A2 | B2 | **C2** |
| < 200/μl | **A3** | **B3** | **C3** |

### CLASSIFICATION OF AIDS INDICATOR DISEASE

*Category A*

Asymptomatic: no symptoms at the time of HIV infection

Acute primary infection: glandular fever-like illness lasting a few weeks at the time of infection

Persistent generalized lymphadenopathy (PGL): lymph node enlargement persisting for 3 or more months with no evidence of infection

*Category B*

Bacillary angiomatosis

Candidiasis, oropharyngeal (thrush)

Candidiasis, vulvovaginal: persistent, frequent, or poorly responsive to therapy

Cervical dysplasia (moderate or severe)/cervical carcinoma in situ

Constitutional symptoms such as fever (> 38.5°C) or diarrhea lasting > 1 month

Hairy leukoplakia, oral

Herpes zoster (shingles) involving at least two distinct episodes or more than one dermatome

Idiopathic thrombocytopenic purpura

Listeriosis

Pelvic inflammatory disease, particularly by tubo-ovarian abscess

Peripheral neuropathy

*Category C*

Candidiasis of bronchi, tracheae, or lungs

Candidiasis, esophageal

Cervical cancer (invasive)

Coccidioidomycosis, disseminated or extrapulmonary

Cryptococcosis, extrapulmonary

Cryptosporidiosis, chronic intestinal (> 1 month duration)

Cytomegalovirus disease (other than liver, spleen, or nodes)

Cytomegalovirus retinitis (with loss of vision)

Encephalopathy, HIV related

Herpes simplex: chronic ulcer(s) (> 1 month duration), bronchitis, pneumonitis, or esophagitis

Histoplasmosis, disseminated or extrapulmonary

Isosporiasis, chronic intestinal (> 1 month duration)

Kaposi's sarcoma

Lymphoma, Burkitt's

Lymphoma, immunoblastic

Lymphoma, primary of brain

*Mycobacterium avium* complex or *M. Kansasii,* disseminated or extrapulmonary

*Mycobacterium tuberculosis,* any site

*Mycobacterium,* other or unidentified species, disseminated or extrapulmonary

*Pneumocystis carinii* pneumonia

Progressive multifocal leukoencephalopathy

*Salmonella* septicemia (recurrent)

Toxoplasmosis of brain

Wasting syndrome due to HIV

*All categories shown in bold type are considered AIDS. For category A diagnosis, no condition in categories B or C can be present; for category B, no category C condition can be present.

SOURCE: CDC guidelines for AIDS diagnosis, 1993 revision.

between acquisition of the virus and death from the immunodeficiency averages 9 to 12 years. In the period between infection and severe disease, there may be few symptoms. Primary infection in a minority of patients may be symptomatic with fever, lymphadenopathy (swollen lymph nodes), and a rash, but these symptoms generally do not persist more than a few weeks. Most commonly, primary infection goes unnoticed and is followed by a long chronic phase, during which the infected individual shows little or no overt sign of HIV-1 infection.

The first overt indication of AIDS may be opportunistic infection with the fungus *Candida albicans,* which causes the appearance of sores in the mouth (thrush) and, in women, a vulvovaginal yeast infection that does not respond to treatment. A persistent hacking cough caused by *P. carinii* infection of the lungs may also be an early indicator. A rise in the level of circulating HIV-1 in the plasma and concomitant drop in the number of CD4$^+$ T cells generally previews this first appearance of symptoms. Some relation between the CD4$^+$ T-cell number and the type of infection experienced by the patient has been established (see Table 20-3). Of intense interest to immunologists are the events that take place between the initial confrontation with HIV-1 and the takeover and collapse of the host immune system. Understanding how the immune system holds HIV-1 in check during this chronic phase may lead to the design of effective therapeutic and preventive strategies.

Research into the process that underlies the progression of HIV infection to AIDS has revealed a dynamic interplay between the virus and the immune system. The initial infection event causes dissemination of virus to lymphoid organs and a resultant strong immune response. A common means by which virus is disseminated from the initial point of contact to lymphoid organs, where it can contact activated T cells and begin a productive replication cycle, is probably via dendritic cells. Dendritic cells may pick up virus and bring it to the T-cell–rich areas and provide the costimulation needed to activate the T cell and render it an appropriate host for HIV-1 infection. The direct observation of HIV-1 at the site of T-cell–dendritic cell interaction supports the plausibility of this mechanism for the initial spread of HIV-1 following exposure (Figure 20-14).

Following exposure of the lymphoid tissue to virus, an immune response involving both antibody and cytotoxic CD8$^+$ T lymphocytes keeps viral replication in check; after

the initial burst of viremia (high levels of virus in the circulation), the viral level in the circulation achieves a steady state. Although the infected individual normally has no clinical signs of disease at this stage, viral replication continues and virus can be detected in circulation by sensitive PCR assays for viral RNA. These PCR-based assays, which measure **viral load** (the number of copies of viral genome in the plasma), have assumed a major role in the determination of the patient's status and prognosis. Even when the level of virus in the circulation is stable, large amounts of virus are produced in infected CD4$^+$ T cells; as many as $10^9$ virions are released every day and continually infect and destroy additional host T cells. Despite this high rate of replication, the virus is kept in check by the immune system throughout the chronic phase of infection, and the level of virus in circulation from about six months after infection is a good predictor of the course of disease. Low levels of virus in this period correlate with a longer time in which the infected individual remains free of opportunistic infection. But in the absence of treatment, the virus eventually overcomes host immune defenses, resulting in an increase in viral load above the steady-state level, a decrease in CD4$^+$ T-cell numbers, increased opportunistic infection, and eventually death of the patient.

Although the viral load in plasma remains fairly stable throughout the period of chronic HIV infection, examination of the lymph nodes and gastrointestinal tract tissue reveals a different picture. Fragments of nodes obtained by biopsy from infected subjects show high levels of infected cells at all stages of infection; in many cases, the structure of the lymph node is completely destroyed by virus long before plasma viral load increases above the steady-state level. Recent data from the laboratories of Ashley Haas and Daniel Douek indicate a dramatic depletion of CD4$^+$ T cells from the gastrointestinal tract, lymph nodes, and peripheral blood during the chronic phase of infection (Figure 20-15).

The decrease in CD4$^+$ T cells is the hallmark of AIDS. Several explanations have been advanced for the depletion of these cells in patients. In early studies, direct viral infection and destruction of CD4$^+$ T cells was discounted as the primary cause, because the large numbers of circulating HIV-infected T cells predicted by the model were not found. More recent studies indicate that the reason for the difficulty in finding the infected cells is that they are so rapidly killed by HIV; the half-life of an actively infected CD4$^+$ T cell is less than 1.5 days. Smaller numbers of CD4$^+$ T cells become infected but do not actively replicate virus. These latently infected cells persist for long periods, and the integrated proviral DNA replicates in cell division along with cell DNA. Studies in which viral load is decreased by anti-retroviral therapy show a concurrent increase in CD4$^+$ T-cell numbers in the peripheral blood. These data support a model of dynamic interaction between virus and T cells, with simultaneous high levels of viral production and rapid depletion of infected CD4$^+$ T cells. Although other mechanisms for depletion of CD4$^+$ T cells may be envisioned, infection with HIV remains the prime suspect.

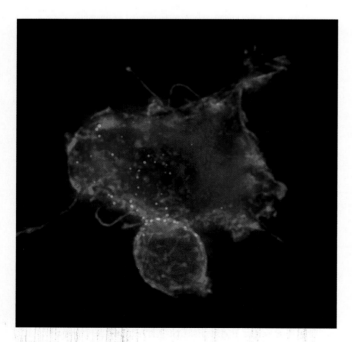

**FIGURE 20-14 Interaction between dendritic cell and T cell indicating passage of HIV-1 (green dots) between the cells.** Note that particles cluster at the interface between the large dendritic cell and the smaller T cell. *[Courtesy of Thomas J. Hope, Northwestern University.]*

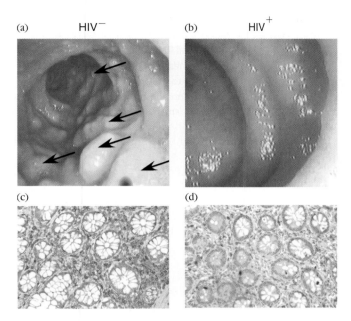

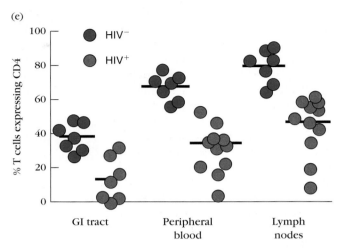

**FIGURE 20-15 Endoscopic and histologic evidence for depletion of CD4$^+$ T cells in the GI tract of AIDS patients.** Panels (a) and (b) show intestinal tract of normal individual and a stained section from a biopsy of the same area (terminal ileum) with CD4$^+$ T cells stained with antibody (rust color). Similar analysis of samples from an AIDS patient in panels (c) and (d) indicates absence of normal lymphoid tissue and sparse staining for CD4$^+$ T cells. (e) Comparison of CD4$^+$ T-cell numbers in samples from GI tract, peripheral blood (PB), and lymph nodes of AIDS-positive and -negative individuals. [*From J.M. Brenchley et al., 2004,* Journal of Experimental Medicine *200:749.*]

Not only depletion of CD4$^+$ T cells but other immunologic consequences can be measured in HIV-infected individuals during the progression to AIDS, including a decrease or absence of delayed hypersensitivity to antigens to which the individual normally reacts. Serum levels of immunoglobulins, especially IgG and IgA, show a sharp increase in the AIDS patient. This increase may be due to increased levels in HIV-infected individuals of a B-cell subpopulation with low CD21 expression and enhanced immunoglobulin secretion. This population proliferates poorly in response to B-cell mitogens. Cellular parameters of immunologic response, such as the proliferative response to mitogens, to antigens, or to alloantigens, all show a marked decrease. Generally, the HIV-infected individual loses the ability to mount T-cell responses in a predictable sequence: responses to specific antigens (for example, influenza virus) are first lost, then response to alloantigens declines, and last, the response to mitogens such as concanavalin A or phytohemagglutinin can no longer be detected. Table 20-4 lists some immune abnormalities in AIDS.

Individuals infected with HIV-1 often display dysfunction of the central and peripheral nervous systems. Specific viral DNA and RNA sequences have been detected by HIV-1 probes in the brains of children and adults with AIDS, suggesting that viral replication occurs there. Quantitative comparison of specimens from brain, lymph node, spleen, and lung of AIDS patients with progressive encephalopathy indicated that the brain was heavily infected. A frequent complication in later stages of HIV infection is AIDS dementia complex, a neurological syndrome characterized by abnormalities in cognition, motor performance, and behavior. Whether AIDS dementia and other clinical and histopathological effects observed in the central nervous systems of HIV-infected individuals are a direct effect of viral antigens on the brain, a consequence of immune responses to the virus, or a result of infection by opportunistic agents remains unknown.

## Therapeutic agents inhibit retrovirus replication

Development of a vaccine to prevent the spread of AIDS is the highest priority for immunologists, but it is also critical to develop drugs and therapies that can reverse the effects of HIV-1 in infected individuals. The number of HIV-infected people is estimated to be close to 1 million in the United States alone; for all of these individuals to develop AIDS would be an enormous tragedy. There are several strategies for development of effective antiviral drugs. The life cycle of HIV shows several susceptible points that might be blocked by pharmaceutical agents (Figure 20-16). The key to success of such therapies is that they must be specific for HIV-1 and interfere minimally with normal cell processes. Thus far, two types of antiviral agents have found their way into common usage (Table 20-5). The first success in treatment was with

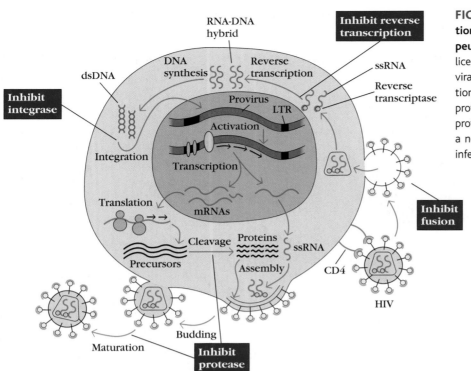

**FIGURE 20-16 Stages in the viral replication cycle that provide targets for therapeutic anti-retroviral drugs.** At present, the licensed drugs with anti-HIV activity block viral entry, block the step of reverse transcription of viral RNA to cDNA, or inhibit the viral protease necessary to cleave viral precursor proteins into the proteins needed to assemble a new virion and complete its maturation to infectious virus.

| TABLE 20-4 | Immunologic abnormalities associated with HIV infection |
|---|---|
| **Stage of infection** | **Typical abnormalities observed** |
| | LYMPH NODE STRUCTURE |
| Early | Infection and destruction of dendritic cells; some structural disruption |
| Late | Extensive damage and tissue necrosis; loss of folicular dendritic cells and germinal centers; inability to trap antigens or support activation of T and B cells |
| | T HELPER ($T_H$) CELLS |
| Early | No in vitro proliferative response to specific antigen |
| Late | Decrease in $T_H$-cell numbers and corresponding helper activities; no response to T-cell mitogens or alloantigens |
| | ANTIBODY PRODUCTION |
| Early | Enhanced nonspecific IgG and IgA production but reduced IgM synthesis |
| Late | No proliferation of B cells specific for HIV-1: no detectable anti-HIV antibodies in some patients; increased numbers of B cells with low CD21 and enhanced Ig secretion. |
| | CYTOKINE PRODUCTION |
| Early | Increased levels of some cytokines |
| Late | Shift in cytokine production from $T_H1$ subset to $T_H2$ subset |
| | DELAYED-TYPE HYPERSENSITIVITY |
| Early | Highly significant reduction in proliferative capacity of $T_H1$ cells and reduction in skin-test reactivity |
| Late | Elimination of DTH response; complete absence of skin-test reactivity |
| | T CYTOTOXIC ($T_C$) CELLS |
| Early | Normal reactivity |
| Late | Reduction but not elimination of CTL activity due to impaired ability to generate CTLs from $T_C$ cells |

| TABLE 20-5 | Some anti-HIV drugs in clinical use | |
|---|---|---|
| **Generic name (other names)** | **Typical dosage** | **Some potential side effects** |
| REVERSE TRANSCRIPTASE INHIBITORS: NUCLEOSIDE ANALOGUE | | |
| Didanosine (Videx, ddI) | 2 pills, 2 times a day on empty stomach | Nausea, diarrhea, pancreatic inflammation, peripheral neuropathy |
| Emtricitabine (Emtriva, FTC) | 1 pill, 1 time a day | Headache, diarrhea, nausea, rash |
| Lamivudine (Epivir, 3TC) | 1 pill, 2 times a day | Usually none |
| Stavudine (Zerit, d4T) | 1 pill, 2 times a day | Peripheral neuropathy |
| Zalcitabine (HIVID, ddC) | 1 pill, 3 times a day | Peripheral neuropathy, mouth inflammation, pancreatic inflammation |
| Zidovudine (Retrovir, AZT, ZDV) | 1 pill, 2 times a day | Nausea, headache, anemia, neutropenia (reduced levels of neutrophil white blood cells), weakness, insomnia |
| Pill containing lamivudine and zidovudine (Combivir) | 1 pill, 2 times a day | Same as for zidovudine |
| Abacavir (Ziagen) | 2 pills, 1 time a day | Nausea, vomiting, diarrhea, lactic acidosis (severe liver disease) |
| Tenofvir (Viread) | 1 pill, 1 time a day | Nausea, vomiting, increased risk of bone breakage |
| REVERSE TRANSCRIPTASE INHIBITORS: NONNUCLEOSIDE ANALOGUES | | |
| Delavirdine (Rescriptor) | 4 pills, 3 times a day (mixed into water); not within an hour of antacids or didanosine | Rash, headache, hepatitis |
| Nevirapine (Viramune) | 1 pill, 2 times a day | Rash, hepatitis |
| Efavirenz (Sustiva) | 1 pill, 1 time a day | Dizziness, insomnia, rash |
| PROTEASE INHIBITORS | | |
| Indinavir (Crixivan) | 2 pills, 3 times a day on empty stomach or with a low-fat snack and not within 2 hours of didanosine | Kidney stones, nausea, headache, blurred vision, dizziness, rash, metallic taste in mouth, abnormal distribution of fat, elevated triglyceride and cholesterol levels, glucose intolerance |
| Nelfinavir (Viracept) | 3 pills, 3 times a day with some food | Diarrhea, abnormal distribution of fat, elevated triglyceride and cholesterol levels, glucose intolerance |
| Ritonavir (Norvir) | 6 pills, 2 times a day (or 4 pills, 2 times a day if taken with saquinavir) with food and not within 2 hours of didanosine | Nausea, vomiting, diarrhea, abdominal pain, headache, prickling sensation in skin, hepatitis, weakness, abnormal distribution of fat, elevated triglyceride and cholesterol levels, glucose intolerance |
| Saquinavir (Invirase, a hard-gel capsule; Fortovase, a soft-gel capsule) | 6 pills, 3 times a day (or 2 pills, 2 times a day if taken with ritonavir) with a large meal | Nausea, diarrhea, headache, abnormal distribution of fat, elevated triglyceride and cholesterol levels, glucose intolerance |
| Atazanavir (Reyataz) | 2 pills, 1 time a day | Must be used with at least two other drugs |
| Fosamprenavir calcium? (Lexiva) | 2 pills, 2 times a day | Appetite loss, malaise, diarrhea, nausea, vomiting |
| FUSION INHIBITORS | | |
| Enfuvirtide (Fuzeon, T-20) | Subcutaneous injection 2 times daily | Soreness at injection site, dizziness, loss of sleep, numbness in feet and legs |

drugs that interfere with the reverse transcription of viral RNA to cDNA; several drugs in common use operate at this step. A second stage of viral replication that has proved amenable to blockade is the step at which precursor proteins are cleaved into the units needed for construction of a new mature virion. This step requires the action of a specific viral protease, which can be inhibited by chemical agents, precluding the formation of infectious viral particles. A third type of drug recently licensed for use, enfuvirtide (brand name Fuzeon), is a fusion inhibitor that prevents the virus from entering target cells.

The prototype of the drugs that interfere with reverse transcription is zidovudine, or AZT (azidothymidine). The introduction of AZT, a nucleoside analogue, into the growing cDNA chain of the retrovirus causes termination of the chain. AZT is effective in some but not all patients, and its efficacy is further limited because long-term use has several adverse side effects and because resistant viral mutants develop in treated patients. The administered AZT is used not only by the HIV-1 reverse transcriptase but also by human DNA polymerase. The incorporation of AZT into the DNA of host cells kills them. Precursors of red blood cells are especially sensitive to AZT, resulting in anemia and other side effects. A different approach to blocking reverse transcription employs drugs such as nevirapine and delaviridine, which inhibit the action of the reverse transcriptase enzyme.

A second class of drugs called protease inhibitors has proven effective when used in conjunction with AZT and/or other nucleoside analogues. Current treatment for AIDS is a combination therapy, using regimens designated HAART (*h*ighly *a*ctive *a*nti-*r*etroviral *t*herapy). In most cases, this combines the use of two nucleoside analogues and one protease inhibitor. The combination strategy appears to overcome the ability of the virus to rapidly produce mutants that are drug resistant. In many cases, HAART has lowered plasma viral load to levels that are not detectable by current methods and has improved the health of AIDS patients to the point that they can again function at a normal level. The decrease in the number of AIDS deaths in the United States in recent years (see Figure 20-7) is attributed to this advance in therapy. Despite the optimism engendered by success with HAART, present drawbacks include a strict time schedule of administration and the large number of pills to be taken every day. In addition, there may be serious side effects (see Table 20-5) that, in some patients, may be too severe to allow use of HAART.

The success of HAART in treating AIDS has opened discussion of whether it might be possible to eradicate all virus from an infected individual and thus actually cure AIDS. Most AIDS experts are not convinced that this is possible, mainly because of the persistence of latently infected CD4$^+$ T cells and macrophages, which can serve as a reservoir of infectious virus if the provirus should be activated. Even with a viral load beneath the level of detection by PCR assays, the immune system may not recover sufficiently to clear virus should it begin to replicate in response to some activation signal. In addition,

virus may persist in sites such as the brain not readily penetrated by the anti-retroviral drugs, even though the virus in circulation is undetectable. The use of immune modulators, such as recombinant IL-2, in conjunction with HAART is being examined as a strategy to help reconstitute the immune system and restore normal immune function.

In addition to the newly licensed fusion inhibitor enfuvirtide, which acts at the stage of viral attachment to the host cell, other drugs are in various stages of development. One promising class of drugs interferes with integration of the viral DNA into the host genome (see Figure 20-14). It should be stressed that the development of any drug to the point at which it can be used for patients is a long and arduous procedure. The drugs that pass the rigorous tests for safety and efficacy represent a small fraction of those that receive initial consideration.

## A vaccine may be the only way to stop the HIV/AIDS epidemic

The AIDS epidemic continues to rage despite the advances in therapeutic approaches outlined above. The present expense of HAART (as much as $15,000 per year), the strict regimen required, and the possibility of side effects precludes universal application. Even if eradication of the virus in individuals treated with combination therapy becomes possible, it will not greatly influence the epidemic in developing countries, which include the majority of AIDS victims. It is likely that effective, inexpensive, and well-tolerated drugs will be developed in the future, but at present it appears that the best option to stop the spread of AIDS is a safe, effective vaccine that prevents infection and progression to disease. Why don't we have an AIDS vaccine? The best answer to this question is to examine the special conditions that must be addressed in developing a safe, effective vaccine for this disease (Table 20-6).

*Most effective vaccines mimic the natural state of infection.* Individuals who recover from most diseases are immune from subsequent attacks. The infection by HIV-1 and progression to immunodeficiency syndrome flourishes even in the presence of circulating antibodies directed against proteins of the virus. Immunity may hold the virus in check for a time, but as mentioned above, it rarely exceeds 12 years in untreated HIV-1 infected people. In a rare subset of infected individuals called long-term nonprogressors, the period of infection without disease is longer, even indefinite. Another group for whom immunity seems to function are those who are persistently exposed but who remain seronegative. In this category are a low percentage of commercial sex workers in areas of high endemic infection, such as Nairobi, who have not become infected despite multiple daily exposures to infected individuals. Because the state of immunity (i.e., which antibodies are present and what type of cellular immunity is active) in these individuals is not clear or consistent, it is difficult to duplicate for vaccine development. Certain of the long-term nonprogressors or exposed and noninfected

| TABLE 20-6 | Why AIDS does not fit the paradigm for classic vaccine development |
|---|---|

Classic vaccines mimic natural immunity against reinfection generally seen in individuals recovered from infection; there are no recovered AIDS patients.

Most vaccines protect against disease, not against infection; HIV infection may remain latent for long periods before causing AIDS.

Most vaccines protect for years against viruses that change very little over time; HIV-1 mutates at a rapid rate and efficiently selects mutant forms that evade immunity.

Most effective vaccines are whole killed or live attenuated organisms; killed HIV-1 does not retain antigenicity, and the use of a live retrovirus vaccine raises safety issues.

Most vaccines protect against infections that are infrequently encountered; HIV may be encountered daily by individuals at high risk.

Most vaccines protect against infections through mucosal surfaces of the respiratory or gastrointestinal tract; the great majority of HIV infection is through the genital tract.

Most vaccines are tested for safety and efficacy in an animal model before trials with human volunteers; there is no suitable animal model for HIV/AIDS at present.

SOURCE: Adapted from A. S. Fauci, 1996, An HIV vaccine: breaking the paradigms, *Proceedings of the Association of American Physicians* **108**:6.

individuals have mutations and deletions in genes encoding cell coreceptors that slow the progress of viral attack on their immune system rather than an immune response that is holding HIV replication in check.

*Most vaccines prevent disease, not infection.* Polio and influenza vaccines hold the virus produced by infected cells in check so that it does not cause harm to the host, and it is then cleared. HIV-1 does not fit this model, because it integrates into the host genome and may remain latent for long periods. As described above in the context of treatment strategies, eradication of a retrovirus is not a simple matter. Clearance of a retrovirus is a difficult goal for a vaccine: every copy of the virus and every infected cell, including those latently infected, must be eradicated from the host. However, even without complete eradication, an HIV vaccine may benefit the infected individual, and a vaccine that caused a lowered viral load would help to control the spread of infection. A recent study in Uganda of sexual partners unmatched for infection showed that low viral load in the infected partner inhibited spread to the uninfected mate.

*Most vaccines prevent infection by viruses that show little variation.* The instability of its genome differentiates HIV-1 from most viruses for which successful vaccines have been developed. With the exception of influenza, for which the vaccines must be changed periodically, most viruses that can be controlled by immunization show only minor variability in structure. For comparison, consider that the rhinoviruses

that cause the common cold have more than 100 subtypes; therefore, no effective vaccine has been developed. HIV-1 shows variation in most viral antigens, and the rate of replication may be as high as $10^9$ viruses per day. This variability, along with the high rate of replication, allows the production of viruses with multiple mutations; some of these allow escape from immunity. The fact that significant differences in viral-envelope protein sequences have been seen in viral isolates taken from the same patient at different times indicates that variation occurs and that some of the variants replicate, presumably because they have learned to evade host immune defenses. Data showing that antibody from advanced AIDS patients will not neutralize virus isolated from that patient but will kill other strains of HIV-1 argues that HIV-1 does evade the immune system by mutation of proteins targeted by antibody.

*The majority of successful vaccines are live-attenuated or heat-killed organisms.* Although there are exceptions to this, notably the recombinant protein used for hepatitis B vaccine and the conjugate used for *Haemophilus influenzae* B vaccine (see Chapter 18), most of the widely used vaccines are attenuated organisms. The development of a live-attenuated retrovirus vaccine from animal viruses engineered to include HIV antigens is a possible route. However, the use of live vaccines is predicated on the supposition that the immunity raised will clear the vaccine virus from the host. This is not easily done for a retrovirus, which integrates into the host genome. A massive testing effort would be required to ensure that a live retroviral vaccine was safe and did not cause chronic host infection. On the positive side, clinical studies using other viruses such as attenuated vaccinia or canarypox as carriers for genes encoding HIV proteins have passed phase I (safety) trials and have advanced to phase II (efficacy) trials.

*For most viruses, the frequency of exposure to infection is rare or seasonal.* In many high-risk individuals, such as commercial sex workers, monogamous sexual partners of HIV-infected subjects, and intravenous drug users, the virus is encountered frequently and, potentially, in large doses. An AIDS vaccine is thus asked to prevent infection against a constant attack by the virus and/or massive doses of virus; this is not normally the case with other viruses for which immunization has proved successful.

*Most vaccines protect against respiratory or gastrointestinal infection.* In addition to the frequency of exposure to HIV, which may be extraordinarily high for some high-risk individuals, there is also the question of route. The majority of successful vaccines protect against viruses that are encountered in the respiratory and gastrointestinal tracts; the most common route of HIV-1 infection is by the genital tract. It is not known whether the immunity established by conventional vaccination procedures will protect against infection by this route. Although the lack of a completely relevant animal model precludes an in-depth test of protection, preliminary vaccine studies using rectal or vaginal challenge of immunized primates with SIV-HIV chimeric viruses (SHIV) show protection to this challenge route.

*Development of most vaccines through to clinical trials relies on animal experiments.* Testing a vaccine for safety and efficacy normally involves challenge of an animal with the virus under conditions similar to those encountered in the human. In this way, the correlates of protective immunity are established. For example, if high titers of CD8$^+$ T cells and neutralizing antibody are necessary for protection in an animal, then CD8$^+$ T-cell immunity and antibody should be measured in human trials of the vaccine. Thus far, animal studies of HIV infection and disease have yielded only a few hard facts about immune responses that are protective against infection or that prevent progression to disease. Many results involve a specific virus in a particular host and are not easily extrapolated to universal concepts, because they depend on host factors as well the relationships between the immunizing and challenge strains of virus. However, experiments have shown that passive immunization with antibodies taken from HIV-infected chimpanzees protect macaques from challenge with SHIV strains bearing HIV-envelope glycoprotein. Further indication that antibodies can prevent infection is given by studies in which monoclonal antibodies protected macaques from vaginal challenge with SHIV. In all cases, the antibodies needed to be present at the time of challenge. Post-challenge administration of antibodies was not effective in preventing infection.

Although there are no reports of great success in human HIV vaccine trials, research in this difficult area continues to be active. Despite a massive effort, progress remains slow. Numerous products are undergoing testing at early stages in human volunteers. Approximately 30 proteins, DNA vaccines, and recombinant viruses are being tested in phase I trials for safety and immunogenicity, and about five candidates have progressed to phase II trials, which involve larger populations and, in some cases, high-risk volunteers. The scale of the task to bring a vaccine candidate to a phase III trial is reflected in the evaluation of a recent phase III test of two subunit vaccines. The tests took place after 20 years of research and development and cost about $300 million; they involved 78 clinics in four countries, with 135,371 patient visits; 12,114 persons were screened and 7963 volunteers enrolled; and the study was recorded on 1,286,279 case report forms. Although no protection was observed in vaccinees, the tests were hailed by some as proof that such a large effort can be undertaken and coordinated and that lessons learned will allow successful tests of promising products.

In addition to developing a scientific rationale, behavioral and social issues influence the development and testing of candidate AIDS vaccines. Counseling concerning safe sexual practice must be part of the care given to volunteers in a vaccine trial. Will this influence the results? Would a lowering of the infection rate in all groups taking part in the trial preclude seeing a meaningful difference in the infection rate between the vaccine and the placebo groups? A further consideration is the fact that anyone successfully immunized against the AIDS virus will become seropositive and will test positive in the standard screening assays for infection. What ramifications will this have? Will the more complex viral-load assays be needed to ascertain whether an immunized individual is actually infected?

It is clear that development of an AIDS vaccine is not a simple exercise in classic vaccinology. More research is needed to understand how this viral attack against the immune system can be thwarted. Although much has been written about the subject and large-scale initiatives are proposed, the path to an effective vaccine is not obvious. It is certain only that all data must be carefully analyzed and that all possible means of creating immunity must be tested. This is one of the greatest public health challenges of our time. An intense and cooperative effort will be required to devise, test, and deliver a safe and effective vaccine for AIDS.

---

## SUMMARY

- Immunodeficiency results from the failure of one or more components of the immune system. Primary immunodeficiencies are present at birth; secondary or acquired immunodeficiencies arise from a variety of causes.

- Immunodeficiencies may be classified by the cell types involved and may affect either the lymphoid- or the myeloid-cell lineage or both.

- The gene defects that underlie primary immunodeficiency allow precise classification. Genetic defects in molecules involved in signal transduction or in cellular communication are found in many immunodeficiencies.

- Lymphoid immunodeficiencies affect T cells, B cells, NK cells, or all of these. Failure of thymic development results in severe immunodeficiency and can hinder normal development of B cells because of the lack of cellular cooperation.

- Myeloid immunodeficiency causes impaired phagocytic function. Those affected suffer from increased susceptibility to bacterial infection.

- Severe combined immunodeficiency, or SCID, always involves T-cell dysfunction and results from a number of different defects in the lymphoid lineage; it is usually fatal.

- Selective immunoglobulin deficiencies are a less-severe form of immunodeficiency and result from defects in more highly differentiated cell types.

- Immunodeficiency may be treated by replacement of the defective or missing protein, cells, or gene. Administration of human immunoglobulin is a common treatment.

- Animal models for immunodeficiency include nude and SCID mice. Gene knockout mice provide a means to study the role of specific genes on immune function.

■ Secondary immunodeficiency results from injury or infection; the most common form is HIV/AIDS, caused by a retrovirus, human immunodeficiency virus 1.

■ HIV-1 infection is spread mainly by sexual contact, by passage of blood, and from HIV-infected mother to infant.

■ Infection with HIV-1 results in severe impairment of immune function marked by depletion of CD4$^+$ T cells and death from opportunistic infection.

■ Treatment of HIV infection with anti-retroviral drugs can cause lowering of viral load and relief from infection, but no cures have been documented.

■ Efforts to develop a vaccine for HIV/AIDS have not yet been successful. The millions of new infections in the year 2005 emphasize the need for an effective vaccine.

## References

Brenchley, J. M., et al. 2004. CD4$^+$ T cell depletion during all stages of HIV disease occurs predominantly in the gastrointestinal tract. *Journal of Experimental Medicine* **200:**749.

Buckley, R. H. 2004. Molecular defects in human severe combined immunodeficiency and approaches to immune reconstitution. *Annual Review of Immunology* **22:**625.

Calvazzano-Calvo, M., et al. 2005. Gene therapy for severe combined immunodeficiency. *Annual Review of Medicine* **56:**585.

Carpenter, C. J., et al. 2000. Antiretroviral therapy in adults: updated recommendations of the International AIDS Society—USA Panel. *Journal of the American Medical Association* **283:**381.

Cohen, O. J., and A. S. Fauci. 2001. Current strategies in the treatment of HIV infection. *Advances in Internal Medicine* **46:**207.

Fauci, A. S. 2003. HIV and AIDS: 20 years of science. *Nature Medicine* **9:**839.

Francis, D. P. 2003. Candidate HIV/AIDS vaccines: lessons learned from the world's first phase III efficacy trials. *AIDS* **17:**147.

Jackson, J. B., et al. 2003. Intrapartum and neonatal single-dose nevirapine compared to zidovudine for prevention of mother-to-child transmission of HIV-1 in Kampala, Uganda: 18-month follow-up of the HIVNET 012 randomized trial. *Lancet* **362:**859.

Kinter, A., et al. 2000. Chemokines, cytokines, and HIV: a complex network of interactions that influence HIV pathogenesis. *Immunological Reviews* **177:**88.

Kohn, D. B. 2001. Gene therapy for genetic haematological disorders and immunodeficiencies. *Journal of Internal Medicine* **249:**379.

Letvin, N. L. 2005. Progress toward an HIV vaccine. *Annual Review of Medicine* **56:**213.

Li, O., et al. 2004. Functional genomic analysis of the response of HIV-1-infected lymphatic tissue to antiretroviral therapy. *Journal of Infectious Disease* **189:**572.

Moir, S., et al. 2001. HIV-1 induces phenotypic and functional perturbations of B cells in chronically infected individuals.

*Proceedings of the National Academy of Sciences U.S.A.* **98:**10362.

Moore, J. P., et al. 2004. The CCR5 and CXCR4 coreceptors—central to understanding the transmission and pathogenesis of human immunodeficiency virus type I infection. *AIDS Research and Human Retroviruses* **20:**111.

Reynolds, S. J., et al. 2004. Male circumcision and risk of HIV-1 and sexually transmitted infections in India. *Lancet* **363:**1039.

Rosenzweig, S. D., and S. M. Holland. 2005. Defects in the interferon-γ and interleukin-12 pathways. *Immunological Reviews* **203:**38.

Stebbing, J., B. Gazzard, and D. C. Douek. 2004. Where does HIV live? *New England Journal of Medicine* **350:**1872.

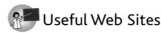 Useful Web Sites

**http://www.scid.net/**

The SCID home page contains links to periodicals and databases with information about SCID.

**http://hivinsite.ucsf.edu/**

Information about the global AIDS epidemic can be accessed from this site.

**http://hiv-web.lanl.gov/**

This Web site, maintained by Los Alamos National Laboratory, contains all available sequence data on HIV and SIV along with up-to-date reviews on topics of current interest to AIDS research.

**ftp://nlmpubs.nlm.nih.gov/aids/adatabases/drugs.txt**

This site, maintained by the National Library of Medicine, lists detailed information for several hundred drugs under development for HIV infection and opportunistic infections associated with AIDS.

**http://www.niaid.nih.gov/daids/vaccine/abtvaccines.htm**

The National Institute of Allergy and Infectious Diseases (NIAID) site includes information about AIDS vaccines and links to documents about vaccines in general.

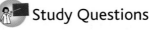 Study Questions

**CLINICAL FOCUS QUESTION** The spread of HIV/AIDS from infected mothers to infants can be reduced by single-dose regimens of the reverse transcriptase inhibitor nevirapine. What would you want to know before giving this drug to all mothers and infants (without checking infection status) at delivery in areas of high endemic infection?

1. Indicate whether each of the following statements is true or false. If you think a statement is false, explain why.

 a. DiGeorge syndrome is a congenital birth defect resulting in absence of the thymus.

 b. X-linked agammaglobulinemia (XLA) is a combined B-cell and T-cell immunodeficiency disease.

c. The hallmark of a phagocytic deficiency is increased susceptibility to viral infections.

d. In chronic granulomatous disease, the underlying defect is in a cytochrome or an associated protein.

e. Injections of immunoglobulins are given to treat individuals with X-linked agammaglobulinemia.

f. Multiple defects have been identified in human SCID.

g. Mice with the SCID defect lack functional B and T lymphocytes.

h. Mice with SCID-like phenotype can be produced by knockout of *RAG* genes.

i. Children born with SCID often manifest increased infections by encapsulated bacteria in the first months of life.

j. Failure to express class II MHC molecules in bare-lymphocyte syndrome affects cell-mediated immunity only.

2. For each of the following immunodeficiency disorders, indicate which treatment would be appropriate.

*Immunodeficiency*

a. Chronic granulomatous disease
b. ADA-deficient SCID
c. X-linked agammaglobulinemia
d. DiGeorge syndrome
e. IL-2R-deficient SCID
f. Common variable immunodeficiency

*Treatment*

1. Full bone marrow transplantation
2. Pooled human gamma globulin
3. Recombinant IFN-γ
4. Recombinant adenosine deaminase
5. Thymus transplant in an infant

3. Patients with X-linked hyper-IgM syndrome express normal genes for other antibody subtypes but fail to produce IgG, IgA, or IgE. Explain how the defect in this syndrome accounts for the lack of other antibody isotypes.

4. Patients with DiGeorge syndrome are born with either no thymus or a severely defective thymus. As such, the patient cannot develop mature helper, cytotoxic, or regulatory T cells. If an adult suffers loss of the thymus due to accident or injury, the T-cell deficiency is much less severe. Explain this phenomenon.

5. Granulocytes from patients with leukocyte adhesion deficiency (LAD) express greatly reduced amounts of three integrin molecules designated CR3, CR4, and LFA-1.

a. What is the nature of the defect that results in decreased expression or in no expression of these receptors in LAD patients?

b. What is the normal function of the integrin molecule LFA-1? Give specific examples.

6. Immunologists have studied the defect in SCID mice in an effort to understand the molecular basis for severe combined immunodeficiency in humans. In both SCID mice and humans with this disorder, mature B and T cells fail to develop.

a. In what way do rearranged Ig heavy-chain genes in SCID mice differ from those in normal mice?

b. In SCID mice, rearrangement of κ light-chain DNA is not attempted. Explain why.

c. If you introduced a rearranged, functional μ heavy-chain gene into progenitor B cells of SCID mice, would the κ light-chain DNA undergo a normal rearrangement? Explain your answer.

7. The accompanying figure outlines some of the steps in the development of immune system cells. The numbered arrows indicate the cell type whose function is defective or the developmental step that does not occur in particular immunodeficiency diseases. Identify the defective cell type or developmental step associated with each of the following diseases. Use each number only once.

a. Chronic granulomatous disease
b. Severe combined immunodeficiency disease (SCID)
c. Congenital agranulocytosis
d. Reticular dysgenesis
e. Common variable hypogammaglobulinemia
f. X-linked agammaglobulinemia
g. Leukocyte adhesion deficiency (LAD)
h. Bare-lymphocyte syndrome

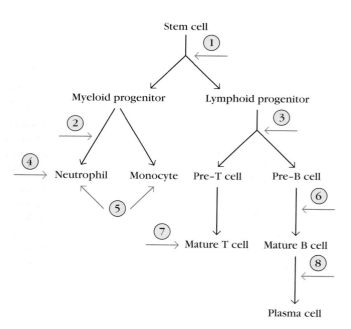

8. Indicate whether each of the following statements is true or false. If you think a statement is false, explain why.

a. HIV-1 and HIV-2 are more closely related to each other than to SIV.

b. HIV-1 causes immune suppression in both humans and chimpanzees.

c. SIV is endemic in the African green monkey.

d. The anti-HIV drugs zidovudine and indinavir both act on the same point in the viral replication cycle.

e. T-cell activation increases transcription of the HIV proviral genome.

f. Patients with advanced stages of AIDS always have detectable antibody to HIV.

g. The polymerase chain reaction is a sensitive test used to detect antibodies to HIV.

h. If HAART is successful, viral load will decrease.

9. Various mechanisms have been proposed to account for the decrease in the numbers of CD4$^+$ T cells in HIV-infected individuals. What seems to be the most likely reason for depletion of CD4$^+$ T cells?

10. Would you expect the viral load in the blood of HIV-infected individuals in the chronic phase of HIV-1 infection to vary?

11. If viral load begins to increase in the blood of an HIV-infected individual and the level of CD4$^+$ T cells decrease, what would this indicate about the infection?

12. Why do clinicians monitor the level of skin-test reactivity in HIV-infected individuals? What change might you expect to see in skin-test reactivity with progression into AIDS?

13. Certain chemokines have been shown to suppress infection of cells by HIV, and proinflammatory cytokines enhance cell infection. What is the explanation for this?

14. Treatments with combinations of anti-HIV drugs (HAART) have reduced virus levels significantly in some treated patients and delayed the onset of AIDS. If an AIDS patient becomes free of opportunistic infection and has no detectable virus in the circulation, can that person be considered cured?

15. Suppose you are a physician who has two HIV-infected patients. Patient B. W. has a fungal infection (candidiasis) in the mouth, and patient L. S. has a *Mycobacterium* infection. The CD4$^+$ T-cell counts of both patients are about 250 per mm$^3$. Would you diagnose either patient or both of them as having AIDS?

**ANALYZE THE DATA** Common variable immunodeficiency (CVID) causes low concentrations of serum Igs and leads to frequent bacterial infections in the respiratory and gastrointestinal tracts. People with CVID also have an increased prevalence of autoimmune disorders and cancers. Isgro and colleagues (*J. Immunol.* 2005, **174**:5074) examined the bone marrow of several individuals with CVID. They looked at the T cell phenotypes of CVID patients (as shown in the table below) as well as some of the cytokines made by these individuals (see figures (a) and (b) on page 524).

a. What is the impact of CVID on T helper cells?

b. Naive CTLs require IL-2 production from T helper cells in order to become activated (see Chapter 14). How might CVID impact the generation of CTLs?

c. True or false: CVID inhibits cytokine production. Explain your answer. Speculate on the physiological impact of the cytokine pattern of CVID patients.

d. Would you predict an effect of CVID on the humoral immune response?

**For use with Analyze the Data questions**

| | | Patients | | | | | | | | | | | Control ($n = 10$) |
|---|---|---|---|---|---|---|---|---|---|---|---|---|---|
| | | 1 | 2 | 5 | 6 | 7 | 8 | 9 | 10 | 11 | CVID | |
| CD4$^+$ | % | 47 | 36 | 28 | 28 | 27 | 19 | 19 | 32 | 57 | 34 | 47.5 |
| | cells/$\mu$l | 296 | 234 | 278 | 1652 | 257 | 361 | 289 | 248 | 982 | 351 | 1024 |
| Naive CD4$^+$ | % | 4 | 6.8 | 25.8 | 4 | 11.6 | 7.9 | 14 | 12 | 2 | 10 | 52 |
| | cells/$\mu$l | 12 | 16 | 72 | 66 | 30 | 29 | 40 | 30 | 20 | 31 | 519 |
| Activated CD4$^+$ | % | 2 | 5 | 22 | 3 | 9 | 8 | 12 | 9 | 1.5 | 8 | 37 |
| | cells/$\mu$l | 6 | 12 | 61 | 50 | 23 | 29 | 35 | 22 | 15 | 25 | 385 |
| CD8$^+$ | % | 31 | 38 | 30 | 57 | 44 | 56 | 47 | 45 | 21 | 39 | 20 |
| | cells/$\mu$l | 195 | 247 | 298 | 3363 | 420 | 1065 | 714 | 348 | 362 | 414 | 404 |
| Naive CD8$^+$ | % | 25 | 30 | 43.9 | 7 | 16.9 | 12.9 | 20 | 13 | 15 | 22 | 58 |
| | cells/$\mu$l | 49 | 74 | 131 | 235 | 71 | 138 | 143 | 45 | 54 | 88 | 233 |

**For use with Analyze the Data questions**

(a)

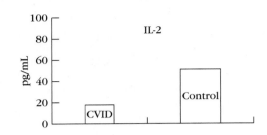

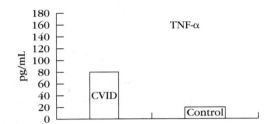

(b)

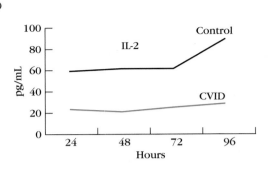

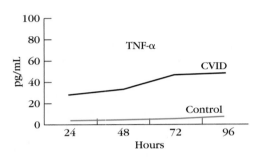

Cytokine production by bone marrow cells from patients with CVID. (a) $1 \times 10^6$ cells were cultured for 24 hours and supernatants were assayed for IL-2 and TNF-$\alpha$. (b) Supernatants were collected after 24, 48, 72, and 96 hours and assayed by ELISA.

 **Interactive Study**

**www.whfreeman.com/kuby**

 SELF-TEST
Review OF Key Terms

ANIMATIONS
Retrovirus

MOLECULAR VISUALIZATIONS
Viral Antigens
HIV-1

# chapter 21

## Cancer and the Immune System

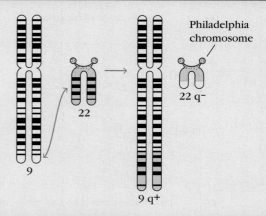

*Chromosomal translocation in chronic myelogenous leukemia.*

- Cancer: Origin and Terminology
- Malignant Transformation of Cells
- Oncogenes and Cancer Induction
- Tumors of the Immune System
- Tumor Antigens
- Tumor Evasion of the Immune System
- Cancer Immunotherapy

**A**S THE DEATH TOLL FROM INFECTIOUS DISEASE HAS declined in the Western world, cancer has become the second-ranking cause of death there, led only by heart disease. Current estimates project that one person in three in the United States will develop cancer and that one in five will die from it. From an immunologic perspective, cancer cells can be viewed as altered self cells that have escaped normal growth-regulating mechanisms. This chapter examines the unique properties of cancer cells, giving particular attention to those properties that can be recognized by the immune system. We then describe the immune responses that develop against cancer cells, as well as the methods by which cancers manage to evade those responses. The final section surveys current clinical and experimental immunotherapies for cancer.

## Cancer: Origin and Terminology

In most organs and tissues of a mature animal, a balance is maintained between cell renewal and cell death. The various types of mature cells in the body have a given life span; as these cells die, new cells are generated by the proliferation and differentiation of various types of stem cells. Under normal circumstances, the production of new cells is regulated so that the number of any particular type of cell remains constant. Occasionally, though, cells arise that no longer respond to normal growth control mechanisms. These cells give rise to clones of cells that can expand to a considerable size, producing a tumor, or **neoplasm.**

A tumor that is not capable of indefinite growth and does not invade the healthy surrounding tissue extensively is **benign.** A tumor that continues to grow and becomes progressively invasive is **malignant;** the term *cancer* refers specifically to a malignant tumor. In addition to uncontrolled growth, malignant tumors exhibit **metastasis;** in this process, small clusters of cancerous cells dislodge from a tumor, invade the blood or lymphatic vessels, and are carried to other tissues, where they continue to proliferate. In this way a primary tumor at one site can give rise to a secondary tumor at another site (Figure 21-1).

Malignant tumors or cancers are classified according to the embryonic origin of the tissue from which the tumor is derived. Most (>80%) are **carcinomas,** tumors that arise from endodermal or ectodermal tissues such as skin or the epithelial lining of internal organs and glands. The majority of cancers of the colon, breast, prostate, and lung are carcinomas. The **leukemias** and **lymphomas** are malignant tumors of hematopoietic cells of the bone marrow and account for about 9% of cancer incidence in the United States. Leukemias proliferate as single cells, whereas lymphomas tend to grow as tumor masses. **Sarcomas,** which arise less frequently (around 1% of the incidence of cancer in the United States), are derived from mesodermal connective tissues such as bone, fat, and cartilage.

## OVERVIEW FIGURE 21-1: Tumor Growth and Metastasis

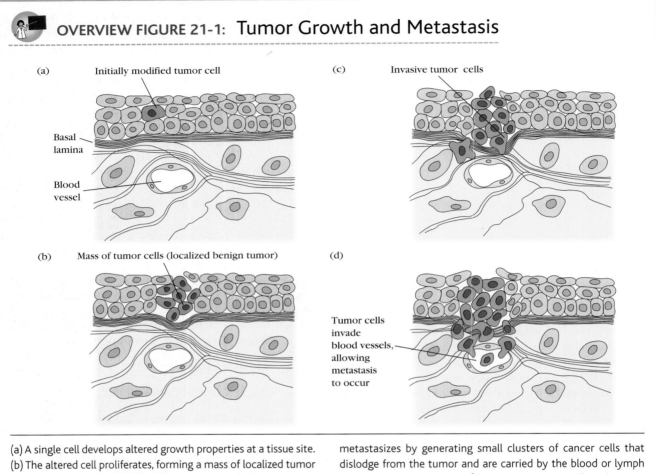

(a) A single cell develops altered growth properties at a tissue site. (b) The altered cell proliferates, forming a mass of localized tumor cells, or benign tumor. (c) The tumor cells become progressively more invasive, spreading to the underlying basal lamina. The tumor is now classified as malignant. (d) The malignant tumor metastasizes by generating small clusters of cancer cells that dislodge from the tumor and are carried by the blood or lymph to other sites in the body. [Adapted from J. Darnell et al., 1990, Molecular Cell Biology, 2nd ed., Scientific American Books.]

## Malignant Transformation of Cells

Treatment of normal cultured cells with chemical carcinogens, irradiation, and certain viruses can alter their morphology and growth properties. In some cases this process, referred to as **transformation,** makes the cells able to produce tumors when they are injected into animals. Such cells are said to have undergone malignant transformation, and they often exhibit properties in vitro similar to those of cancer cells. For example, they have decreased requirements for growth factors and serum, are no longer anchorage dependent, and grow in a density-independent fashion. Moreover, both cancer cells and transformed cells can be subcultured indefinitely; that is, for all practical purposes, they are immortal. Because of the similar properties of cancer cells and transformed cells, the process of malignant transformation has been studied extensively as a model of cancer induction.

Various chemical agents (e.g., DNA-alkylating reagents) and physical agents (e.g., ultraviolet light and ionizing radiation) that cause mutations have been shown to induce transformation. Induction of malignant transformation with chemical or physical carcinogens appears to involve multiple steps and at least two distinct phases: initiation and promotion. Initiation involves changes in the genome but does not, in itself, lead to malignant transformation. After initiation, promoters stimulate cell division and lead to malignant transformation.

The importance of mutagenesis in the induction of cancer is illustrated by diseases such as xeroderma pigmentosum. This rare disorder is caused by a defect in the gene that encodes a DNA-repair enzyme called UV-specific endonuclease. Individuals with this disease are unable to repair UV-induced mutations and consequently develop skin cancers.

Certain viruses have been linked to cancer in both experimental animals and in humans. Polyoma and SV40 viruses have been long studied in animal systems, and much is known about the role of viral proteins that induce transformation. In both cases the DNA of the viral genomes is integrated randomly into the host chromosomal DNA, including several genes that are expressed early in the course of viral replication. SV40 encodes two early proteins called large T and little T, and polyoma encodes three early proteins called large T, middle T, and little T. Each of these proteins plays a role in the malignant transformation of virus-infected cells.

Human cancers with well-established links to viral infection include

- Adult T-cell leukemia/lymphoma, which occurs in a small percentage of persons infected with human T-cell leukemia virus-1 (HTLV-1)

- Kaposi's sarcoma, which is linked to human herpesvirus-8 (HHV-8) and usually occurs in those also infected with HIV-1

- Cervical carcinoma, which is linked to infection by one of several serotypes of human papillomavirus (HPV); a vaccine for HPV is discussed in the Clinical Focus box on page 534

- Liver carcinoma that follows infection with hepatitis B virus (HBV)

- Epstein-Barr virus (EBV) infection, which is linked to Burkitt's lymphoma in African populations and nasopharyngeal carcinoma mainly in Asian populations

Of these human cancer-related viruses, EBV, HBV, and HPV are DNA viruses such as SV40 and polyoma. HTLV-1 and HHV-8 are RNA viruses. Most RNA viruses replicate in the cytosol and do not induce malignant transformation. The exceptions are retroviruses, which transcribe their RNA into DNA by means of a reverse-transcriptase enzyme and then integrate the transcript into the host's DNA. This replication cycle is similar to the replication cycles of the cytopathic retroviruses such as HIV-1 and HIV-2 and to those of the transforming retroviruses such as HTLV-1, which induce changes in the host cell that lead to malignant transformation. In some cases, retrovirus-induced transformation is related to the presence in the viral genome of **oncogenes,** or "cancer genes."

One of the best-studied transforming retroviruses is the **Rous sarcoma virus.** This virus carries an oncogene called v-*src,* which encodes a 60-kDa protein kinase (v-Src) that catalyzes the phosphorylation of tyrosine residues on proteins. The first evidence that oncogenes alone could induce malignant transformation came from studies of the v-*src* oncogene from Rous sarcoma virus. When this oncogene was cloned and transfected into normal cells in culture, the cells underwent malignant transformation.

## Oncogenes and Cancer Induction

In 1971, Howard Temin suggested that oncogenes might not be unique to transforming viruses but might also be found in normal cells; indeed, he proposed that a virus might acquire

oncogenes from the genome of an infected cell. He called these cellular genes **proto-oncogenes,** or **cellular oncogenes** (c-*onc*), to distinguish them from their viral counterparts (v-*onc*). In the mid-1970s, J. M. Bishop and H. E. Varmus identified a DNA sequence in normal chicken cells that is homologous to v-*src* from Rous sarcoma virus. This cellular oncogene was designated c-*src.* Since these early discoveries, numerous cellular oncogenes have been identified.

Sequence comparisons of viral and cellular oncogenes reveal that they are highly conserved in evolution. Although most cellular oncogenes consist of a series of exons and introns, their viral counterparts consist of uninterrupted coding sequences, suggesting that the virus might have acquired the oncogene through an intermediate RNA transcript from which the intron sequences had been removed during RNA processing. The actual coding sequences of viral oncogenes and the corresponding proto-oncogenes exhibit a high degree of homology; in some cases, a single point mutation is all that distinguishes a viral oncogene from the corresponding proto-oncogene. It has now become apparent that most, if not all, oncogenes (both viral and cellular) are derived from cellular genes that encode various growth-controlling proteins. In addition, the proteins encoded by a particular oncogene and its corresponding proto-oncogene appear to have very similar functions. As described below, the conversion of a proto-oncogene into an oncogene appears in many cases to accompany a change in the level of expression of a normal growth-controlling protein.

## Cancer-associated genes have many functions

Homeostasis in normal tissue is maintained by a highly regulated process of cellular proliferation balanced by cell death. If there is an imbalance, either at the stage of cellular proliferation or at the stage of cell death, then a cancerous state will develop. Oncogenes and tumor suppressor genes have been shown to play an important role in this process by regulating either cellular proliferation or cell death. Cancer-associated genes can be divided into three categories that reflect these different activities, summarized in Table 21-1.

### Induction of Cellular Proliferation

One category of proto-oncogenes and their oncogenic counterparts encodes proteins that induce cellular proliferation. Some of these proteins function as growth factors or growth factor receptors. Included among these are *sis,* which encodes a form of platelet-derived growth factor, and *fms, erbB,* and *neu,* which encode growth factor receptors. In normal cells, the expression of growth factors and their receptors is carefully regulated. Usually, one population of cells secretes a growth factor that acts on another population of cells that carries the receptor for the factor, thus stimulating proliferation of the second population. Inappropriate expression of either a growth factor or its receptor can result in uncontrolled proliferation.

| TABLE 21-1 | Functional classification of cancer-associated genes |
|---|---|
| **Type/name** | **Nature of gene product** |
| CATEGORY I: GENES THAT INDUCE CELLULAR PROLIFERATION | |
| Growth factors | |
| *sis* | A form of platelet-derived growth factor (PDGF) |
| Growth factor receptors | |
| *fms* | Receptor for colony-stimulating factor 1 (CSF-1) |
| *erbB* | Receptor for epidermal growth factor (EGF) |
| *neu* | Protein (HER2) related to EGF receptor |
| *erbA* | Receptor for thyroid hormone |
| Signal transducers | |
| *src* | Tyrosine kinase |
| *abl* | Tyrosine kinase |
| Ha-*ras* | GTP-binding protein with GTPase activity |
| N-*ras* | GTP-binding protein with GTPase activity |
| K-*ras* | GTP-binding protein with GTPase activity |
| Transcription factors | |
| *jun* | Component of transcription factor AP1 |
| *fos* | Component of transcription factor AP1 |
| *myc* | DNA-binding protein |
| CATEGORY II: TUMOR SUPRESSOR GENES, INHIBITORS OF CELLULAR PROLIFERATION* | |
| *Rb* | Suppressor of retinoblastoma |
| *p53* | Nuclear phosphoprotein that inhibits formation of small-cell lung cancer and colon cancers |
| *DCC* | Suppressor of colon carcinoma |
| *APC* | Suppressor of adenomatous polyposis |
| *NF1* | Suppressor of neurofibromatosis |
| *WT1* | Suppressor of Wilm's tumor |
| CATEGORY III: GENES THAT REGULATE PROGRAMMED CELL DEATH | |
| *bcl-2* | Suppressor of apoptosis |

*The activity of the normal products of the category II genes inhibits progression of the cell cycle. Loss of a gene or its inactivation by mutation in an indicated tumor-suppressor gene is associated with development of the indicated cancers.

Other oncogenes in this category encode products that function in signal transduction pathways or as transcription factors. The *src* and *abl* oncogenes encode tyrosine kinases, and the *ras* oncogene encodes a GTP-binding protein. The products of these genes act as signal transducers. The *myc*, *jun*, and *fos* oncogenes encode transcription factors. Overactivity of any of these oncogenes may result in unregulated proliferation.

## Inhibition of Cellular Proliferation

A second category of cancer-associated genes—called **tumor-suppressor genes,** or anti-oncogenes—encodes proteins that inhibit excessive cell proliferation. Inactivation of these genes results in unregulated proliferation. The prototype of this category of oncogenes is *Rb*, the retinoblastoma gene.

Hereditary retinoblastoma is a rare childhood cancer, in which tumors develop from neural precursor cells in the immature retina. The affected child has inherited a mutated *Rb* allele; somatic inactivation of the remaining *Rb* allele leads to tumor growth. Probably the single most frequent genetic abnormality in human cancer is mutation in *p53*, which encodes a nuclear phosphoprotein. Over 90% of small-cell lung cancers and over 50% of breast and colon cancers have been shown to be associated with mutations in *p53*.

## Regulation of Programmed Cell Death

A third category of cancer-associated genes regulates programmed cell death. These genes encode proteins that either block or induce apoptosis. Included in this category of oncogenes is *bcl-2*, an anti-apoptosis gene. This oncogene was originally discovered because of its association with B-cell follicular lymphoma. Since its discovery, *bcl-2* has been shown to play an important role in regulating cell survival during hematopoiesis and in the survival of selected B cells and T cells during maturation. Interestingly, the Epstein-Barr virus contains a gene that has sequence homology to *bcl-2* and may act in a similar manner to suppress apoptosis.

## Proto-oncogenes can be converted to oncogenes

In 1972, R. J. Huebner and G. J. Todaro suggested that mutations or genetic rearrangements of proto-oncogenes by carcinogens or viruses might alter the normally regulated function of these genes, converting them into potent cancer-causing oncogenes (Figure 21-2). Considerable evidence supporting this hypothesis accumulated in subsequent years. For example, some malignantly transformed cells contain multiple copies of cellular oncogenes, resulting in increased production of oncogene products. Such amplification of cellular oncogenes has been observed in cells from various types of human cancers. Several groups have identified c-*myc* oncogenes in homogeneously staining regions (HSRs) of chromosomes from cancer cells; these HSRs represent long tandem arrays of amplified genes.

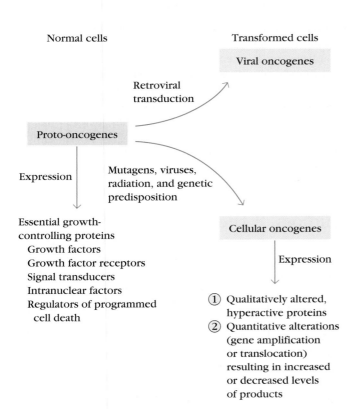

**FIGURE 21-2** Conversion of proto-oncogenes into oncogenes can involve mutation, resulting in production of qualitatively different gene products, or DNA amplification or translocation, resulting in increased or decreased expression of gene products.

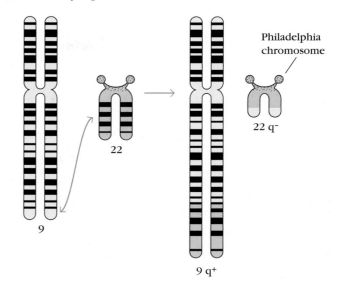

(a) Chronic myelogenous leukemia

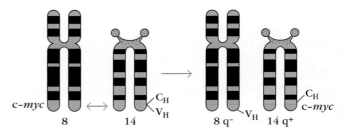

(b) Burkitt's lymphoma

**FIGURE 21-3 Chromosomal translocations in (a) chronic myelogenous leukemia (CML) and (b) Burkitt's lymphoma.** Leukemic cells from all patients with CML contain the so-called Philadelphia chromosome, which results from a translocation between chromosomes 9 and 22. Cancer cells from some patients with Burkitt's lymphoma exhibit a translocation that moves part of chromosome 8 to chromosome 14. It is now known that this translocation involves c-*myc*, a cellular oncogene. Abnormalities such as these are detected by banding analysis of metaphase chromosomes. Normal chromosomes are shown on the left and translocated chromosomes on the right.

In addition, some cancer cells exhibit chromosomal translocations, usually the movement of a proto-oncogene from one chromosomal site to another (Figure 21-3). In many cases of Burkitt's lymphoma, for example, c-*myc* is moved from its normal position on chromosome 8 to a position near the immunoglobulin heavy-chain enhancer on chromosome 14. As a result of this translocation, synthesis of the c-Myc protein, which functions as a transcription factor, increases.

Mutation in proto-oncogenes has also been associated with cellular transformation, and it may be a major mechanism by which chemical carcinogens or x-irradiation convert

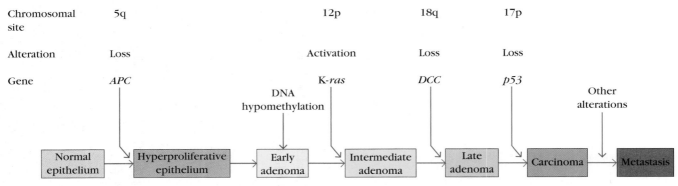

**FIGURE 21-4 Model of sequential genetic alterations leading to metastatic colon cancer.** Each of the stages indicated at the bottom is morphologically distinct, allowing researchers to determine the sequence of genetic alterations. *[Adapted from B. Vogelstein and K. W. Kinzler, 1993,* Trends in Genetics **9**:138.*]*

a proto-oncogene into a cancer-inducing oncogene. For instance, single-point mutations in c-*ras* have been detected in a significant fraction of several human cancers, including carcinomas of the bladder, colon, and lung. Some of these mutations appear to reduce the ability of Ras to associate with GTPase-stimulating proteins, thus prolonging the growth-activated state of Ras.

Viral integration into the host-cell genome may in itself serve to convert a proto-oncogene into a transforming oncogene. For example, avian leukosis virus (ALV) is a retrovirus that does not carry any viral oncogenes, yet is able to transform B cells into lymphomas. This particular retrovirus has been shown to integrate within the c-*myc* proto-oncogene, which contains three exons. Exon 1 of c-*myc* has an unknown function; exons 2 and 3 encode the Myc protein. Insertion of AVL between exon 1 and exon 2 has been shown in some cases to allow the provirus promoter to increase transcription of exons 2 and 3, resulting in increased synthesis of c-Myc.

A variety of tumors have been shown to express significantly increased levels of growth factors or growth factor receptors. Expression of the receptor for epidermal growth factor, which is encoded by c-*erbB*, has been shown to be amplified in many cancer cells. And in breast cancer, increased synthesis of the growth factor receptor encoded by c-*neu* has been linked with a poor prognosis.

## The induction of cancer is a multistep process

The development from a normal cell to a cancerous cell is usually a multistep process of clonal evolution driven by a series of somatic mutations that progressively convert the cell from normal growth to a precancerous state and finally a cancerous state.

The presence of myriad chromosomal abnormalities in precancerous and cancerous cells lends support to the role of multiple mutations in the development of cancer. This has been demonstrated in human colon cancer, which progresses in a series of well-defined morphologic stages (Figure 21-4).

Colon cancer begins as small, benign tumors called adenomas in the colorectal epithelium. These precancerous tumors grow, gradually becoming increasingly disorganized in their intracellular organization until they acquire the malignant phenotype. The morphologic stages of colon cancer have been correlated with a sequence of gene changes involving inactivation or loss of three tumor-suppressor genes (*APC*, *DCC*, and *p53*) and activation of one cellular proliferation oncogene (K-*ras*).

Studies with transgenic mice also support the role of multiple steps in the induction of cancer. Transgenic mice expressing high levels of Bcl-2 develop a population of small resting B cells, derived from secondary lymphoid follicles, that have greatly extended life spans. Gradually these transgenic mice develop lymphomas. Analysis of lymphomas from these mice has shown that approximately half have a c-*myc* translocation to the immunoglobulin H-chain locus. The synergism of Myc and Bcl-2 is highlighted in double-transgenic mice produced by mating the *bcl-2*$^+$ transgenic mice with *myc*$^+$ transgenic mice. These mice develop leukemia very rapidly.

## Tumors of the Immune System

Tumors of the immune system are classified as lymphomas or leukemias. Lymphomas proliferate as solid tumors within a lymphoid tissue such as the bone marrow, lymph nodes, or thymus; they include Hodgkin's and non-Hodgkin's lymphomas. Leukemias tend to proliferate as single cells and are detected by increased cell numbers in the blood or lymph. Leukemia can develop in lymphoid or myeloid lineages. Because the T-cell cancer caused by HTLV-1 can occur in either circulating or tissue-resident T-cell populations, the condition is called adult T-cell leukemia/lymphoma, or ATLL.

Historically, the leukemias were classified as acute or chronic according to the clinical progression of the disease. The acute leukemias appeared suddenly and progressed rapidly, whereas the chronic leukemias were much less aggressive and developed slowly as mild, barely symptomatic diseases. These

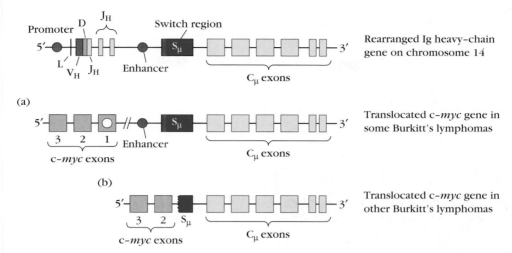

**FIGURE 21-5 In many patients with Burkitt's lymphoma, the c-*myc* gene is translocated to the immunoglobulin heavy-chain gene cluster on chromosome 14.** In some cases, the entire c-*myc* gene is inserted near the heavy-chain enhancer (a), but in other cases, only the coding exons (2 and 3) of c-*myc* are inserted at the $S_\mu$ switch site (b). Only exons 2 and 3 of c-*myc* are coding exons. Translocation may lead to overexpression of c-Myc.

clinical distinctions apply to untreated leukemias; with current treatments, the acute leukemias often have a good prognosis, and permanent remission can often be achieved. Now the major distinction between acute and chronic leukemias is the maturity of the cell involved. Acute leukemias tend to arise in less mature cells, whereas chronic leukemias arise in mature cells. The acute leukemias include **acute lymphocytic leukemia (ALL)** and **acute myelogenous leukemia (AML)**; these diseases can develop at any age and have a rapid onset. The chronic leukemias include **chronic lymphocytic leukemia (CLL)** and **chronic myelogenous leukemia (CML)**; these diseases develop slowly and are seen in adults.

A number of B- and T-cell leukemias and lymphomas involve a proto-oncogene that has been translocated into the immunoglobulin genes or T-cell receptor genes. One of the best characterized is the translocation of c-*myc* in Burkitt's lymphoma and in mouse plasmacytomas. In 75% of Burkitt's lymphoma patients, c-*myc* is translocated from chromosome 8 to the Ig heavy-chain gene cluster on chromosome 14 (see Figure 21-3b). In the remaining patients, c-*myc* remains on chromosome 8 and the κ or γ light-chain genes are translocated to a region 3′ of c-*myc*. Kappa-gene translocations from chromosome 2 to chromosome 8 occur 9% of the time, and γ-gene translocations from chromosome 22 to chromosome 8 occur 16% of the time.

Translocations of c-*myc* to the Ig heavy-chain gene cluster on chromosome 14 have been analyzed, and in some cases, the entire c-*myc* gene is translocated head to head to a region near the heavy-chain enhancer. In other cases, exons 1, 2, and 3 or exons 2 and 3 of c-*myc* are translocated head to head to the Sμ or Sα switch site (Figure 21-5). In each case, the translocation removes the *myc* coding exons from the regulatory mechanisms operating in chromosome 8 and places them in the immunoglobulin-gene region, a very active region that is expressed constitutively in these cells. The consequences of enhancer-mediated high levels of constitutive *myc* expression in lymphoid cells have been investigated in transgenic mice. In one study, mice containing a transgene consisting of all three c-*myc* exons and the immunoglobulin heavy-chain enhancer were produced. Of 15 transgenic pups born, 13 developed lymphomas of the B-cell lineage within a few months of birth.

## Tumor Antigens

The subdiscipline of tumor immunology involves the study of antigens on tumor cells and the immune response to these antigens. Two types of tumor antigens have been identified on tumor cells: **tumor-specific transplantation antigens (TSTAs)** and **tumor-associated transplantation antigens (TATAs)**. Tumor-specific antigens are unique to tumor cells and do not occur on normal cells in the body. They may result from mutations in tumor cells that generate altered cellular proteins; cytosolic processing of these proteins would give rise to novel peptides that are presented with class I MHC molecules, inducing a cell-mediated response by tumor-specific CTLs (Figure 21-6). Tumor-associated antigens, which are not unique to tumor cells, may be proteins that are expressed on normal cells during fetal development when the immune system is immature and unable to respond but that normally are not expressed in the adult. (Note that the designation "transplantation" in TSTA stems from the studies in which the antigens were discovered by transplanting them into recipient animals to measure immune responses.) Reactivation of the embryonic genes that encode these proteins in tumor cells results in their expression on the fully differentiated tumor cells. Tumor-associated antigens may also be proteins that are normally expressed at extremely low levels on normal cells but are expressed at much higher levels on tumor cells. It is now

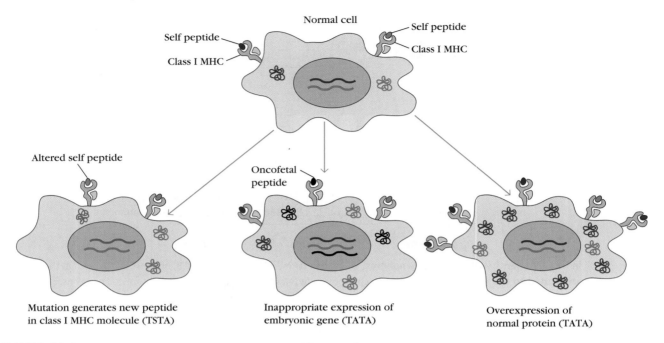

**FIGURE 21-6 Different mechanisms generate tumor-specific transplantation antigens (TSTAs) and tumor-associated transplantation antigens (TATAs).** The latter are more common.

clear that the tumor antigens recognized by human T cells fall into one of four major categories:

- Antigens encoded by genes exclusively expressed by tumors

- Antigens encoded by variant forms of normal genes that have been altered by mutation

- Antigens normally expressed only at certain stages of differentiation or only by certain differentiation lineages

- Antigens that are overexpressed in particular tumors

Many tumor antigens are cellular proteins that give rise to peptides presented with MHC molecules; typically, these antigens have been identified by their ability to induce the proliferation of antigen-specific CTLs or helper T cells.

## Some antigens are tumor specific

Tumor-specific antigens have been identified on tumors induced with chemical or physical carcinogens and on some virally induced tumors. Demonstrating the presence of tumor-specific antigens on spontaneously occurring tumors is particularly difficult because the immune response to such tumors eliminates all of the tumor cells bearing sufficient numbers of the antigens and in this way selects for cells bearing low levels of the antigens.

### Chemically or Physically Induced Tumor Antigens

Methylcholanthrene and ultraviolet light are two carcinogens that have been used extensively to generate lines of tumor cells.

When syngeneic animals are injected with killed cells from a carcinogen-induced tumor-cell line, the animals develop a specific immunologic response that can protect against later challenge by live cells of the same line but not other tumor-cell lines (Table 21-2). Even when the same chemical

| TABLE 21-2 | Immune response to methylcholanthrene (MCA) or polyoma virus (PV)* | | |
|---|---|---|---|
| **Transplanted killed tumor cells** | **Live tumor cells for challenge** | **Tumor growth** | |
| CHEMICALLY INDUCED | | | |
| MCA-induced sarcoma A | MCA-induced sarcoma A | − | |
| MCA-induced sarcoma A | MCA-induced sarcoma B | + | |
| VIRALLY INDUCED | | | |
| PV-induced sarcoma A | PV-induced sarcoma A | − | |
| PV-induced sarcoma A | PV-induced sarcoma B | − | |
| PV-induced sarcoma A | SV40-induced sarcoma C | + | |

*Tumors were induced either with MCA or PV, and killed cells from the induced tumors were injected into syngeneic animals, which were then challenged with live cells from the indicated tumor-cell lines. The absence of tumor growth after live challenge indicates that the immune response induced by tumor antigens on the killed cells provided protection against the live cells.

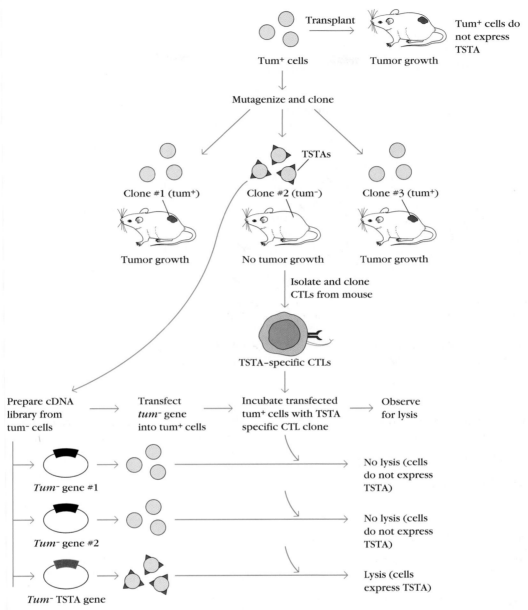

**FIGURE 21-7 One procedure for identifying genes encoding tumor-specific transplantation antigens (TSTAs).** Most TSTAs can be detected only by the cell-mediated rejection they elicit. In the first part of this procedure, a nontumorigenic (tum⁻) cell line is generated; this cell line expresses a TSTA that is recognized by syngeneic mice, which mount a cell-mediated response against it. To isolate the gene encoding the TSTA, a cosmid gene library is prepared from the tum⁻ cell line, the genes are transfected into tumorigenic tum⁻ cells, and the transfected cells are incubated with TSTA-specific CTLs.

carcinogen induces two separate tumors at different sites in the same animal, the tumor antigens are distinct and the immune response to one tumor does not protect against the other tumor.

The tumor-specific transplantation antigens of chemically induced tumors have been difficult to characterize because they cannot be identified by induced antibodies but only by their T-cell-mediated rejection. One experimental approach that has allowed identification of genes encoding some TSTAs is outlined in Figure 21-7. When a

mouse tumorigenic cell line (tum⁺), which gives rise to progressively growing tumors, is treated in vitro with a chemical mutagen, some cells are mutated so that they are no longer capable of growing into a tumor in syngeneic mice. These mutant tumor cells are designated as tum⁻ variants. Most tum⁻ variants have been shown to express TSTAs that are not expressed by the original tum⁺ tumor-cell line. When tum⁻ cells are injected into syngeneic mice, the unique TSTAs that the tum⁻ cells express are recognized by specific CTLs. The TSTA-specific CTLs destroy

## CLINICAL FOCUS
# A Vaccine That Prevents Cancer

**With** the exception of monoclonal antibodies that target specific cancer cells (see Table 21-4), most proposed immunotherapies against cancer require extremely complex protocols that entail monitoring by sophisticated clinical techniques. Consequently, application of these therapies to high-risk populations in poorer communities and in less-developed parts of the world remains a remote possibility at best. By contrast, a low-cost, easily delivered, and effective vaccine to prevent a high-frequency cancer has tremendous potential to save lives throughout the world. Recent studies on a vaccine to prevent human papillomavirus (HPV), the causative agent in cervical cancer, suggest that the goal of cancer prevention may be within reach.

Approximately 500,000 women develop cervical cancer each year, and 270,000 women die from the disease. In the United States, cervical cancer ranks about seventh among lethal cancers, whereas in the less-developed world it is second only to breast cancer as a killer of women. Periodic cervi-

cal examination (using the Papanicoloau test, or Pap smear) to detect abnormal cervical cellular infiltrates significantly reduces the risk for women, but a health-care program that includes regular Pap smears is commonly beyond the means of the less advantaged.

HPV is implicated in over 99% of cervical cancers. Multiple serotypes of HPV exist, and infection with a number of these strains causes anogenital lesions (warts) in males and females. The rate of HPV infection varies in different populations from around 7% to over 60%. A recent study of female college students at the University of Washington showed that after 5 years, more than 60% of study participants (all of whom were HPV negative when enrolled in the study) became infected. Most infections are resolved without disease; it is persistent infection leading to cervical intraepithelial neoplasia that is associated with high cancer risk. Of the dozen prevalent HPV serotypes, two, types 16 and 18, account for over 70% of the cervical

cancers and therefore offer targets for a cancer vaccine.

HPV is a DNA virus with a genome of about 8 kb. The virus encodes a surprising amount of information in its relatively small genome because it transcribes from three reading frames. Among the proteins encoded in the HPV genes are two structural proteins, L1 and L2, and a group of nonstructural proteins designated E1–E7. The contribution of these proteins to cancer development has been well documented, with the major event being the disabling of the normal apoptotic function of the infected cell by viral protein E6. Blocking apoptosis is a protective adaptation of the virus; normally the host cell would undergo apoptotic death in response to viral infection. Another viral protein, E7, activates genes that initiate the cell cycle, allowing the infected cell to proliferate. The result is an HPV-infected cell that proliferates well and has lost the ability to die by apoptosis—in other words, a potential cancer cell.

Preventing cervical cancer is therefore a matter of preventing HPV infection. Viral proteins L1 or L2 from the cancer-causing serotype(s) provide prophylactic vaccine

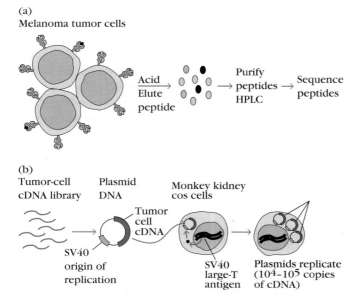

(a)
Melanoma tumor cells

Acid Elute peptide → Purify peptides HPLC → Sequence peptides

(b)
Tumor-cell cDNA library     Plasmid DNA     Monkey kidney cos cells

Tumor cell cDNA

SV40 origin of replication     SV40 large-T antigen     Plasmids replicate ($10^4$–$10^5$ copies of cDNA)

**FIGURE 21-8 Two methods used to isolate tumor antigens that induce tumor-specific CTLs.** See text for details.

the tum⁻ tumor cells, thus preventing tumor growth. To identify the genes encoding the TSTAs that are expressed on a tum⁻ cell line, a cosmid DNA library is prepared from the tum⁻ cells. Genes from the tum⁻ cells are transfected into the original tum⁺ cells, and then the transfected tum⁺ cells are tested for the expression of the tum⁻ TSTAs by their ability to activate cloned CTLs specific for the tum⁻ TSTA. A number of diverse TSTAs have been identified by this method.

In the past few years, two methods have facilitated the characterization of TSTAs (Figure 21-8). In one method, peptides bound to class I MHC molecules on the membranes of the tumor cells are eluted with acid and purified by high-pressure liquid chromatography (HPLC). In some cases, sufficient peptide is eluted to allow its sequence to be deduced by Edman degradation. In a second approach, cDNA libraries are prepared from tumor cells. These cDNA libraries are transfected transiently into COS cells, which are monkey kidney cells transfected with the gene that codes for the SV40 large-T antigen. When these cells are later transfected with plasmids containing both the tumor-cell cDNA

candidates. When L1 is expressed in transfected cell lines, it assembles into a particle that mimics the shape of the virus; these empty particles are called virus-like particles, or VLPs. Currently in testing is a vaccine composed of VLPs accompanied by adjuvant. Two candidate products are in large-scale clinical trials: one includes the VLP composed of L1 of HPV of serotypes 6, 11, 16, and 18, given with alum as adjuvant, and the other consists of VLPs from L1 of HPV 16 and 18, given with the adjuvant ASO4 (containing alum and a bacterial lipid). Serotypes 6 and 11 are included in the first vaccine mentioned in order to target the HPV types that cause anogenital warts in both men and women as well as those linked to cervical cancer in women. Inclusion of males in the vaccinated population would have the benefit of helping to break the infection cycle in populations. Although males do not benefit from vaccination against cervical cancer, prevention of the unsightly warts could provide an incentive for them to be vaccinated.

In trials with over 2000 women in the 16- to 25-year-old age group, both candidate vaccines have been highly effective at preventing genital warts and persistent HPV infection. Ongoing trials with larger groups throughout the world will determine if this vaccine is sufficiently effective to be licensed and offered to the general population. Large questions yet to be answered include which populations should be recommended for coverage if and when the vaccine clears the hurdles to licensure. The value of the HPV L1-based vaccination to those already infected is questionable, so those with no previous exposure would benefit most. The rapid spread of HPV in young sexually active cohorts, even among those using condoms, suggests that young women are the most obvious population for vaccine coverage. Although decisions about a vaccine against a sexually transmitted disease can be expected to trigger controversy, the goal of greatly reducing or eradicating lethal cervical cancer would properly carry some weight in the discussion.

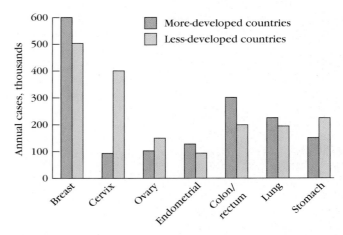

Incidence of frequently occurring cancers in women in the more- and less-developed countries. [Source: J. Cohen, 2005, High hopes and dilemmas for a cervical cancer vaccine, Science **308**:618.]

and an SV40 origin of replication, the large-T antigen stimulates plasmid replication, so that up to $10^4$ to $10^5$ plasmid copies are produced per cell. This results in high-level expression of the tumor-cell DNA.

The genes that encode some TSTAs have been shown to differ from normal cellular genes by a single point mutation. Further characterization of TSTAs has demonstrated that many of them are not cell membrane proteins; rather, as indicated already, they are short peptides derived from cytosolic proteins that have been processed and presented together with class I MHC molecules.

## Tumor antigens may be induced by viruses

In contrast to chemically induced tumors, virally induced tumors express tumor antigens shared by all tumors induced by the same virus. For example, when syngeneic mice are injected with killed cells from a particular polyoma-induced tumor, the recipients are protected against subsequent challenge with live cells from any polyoma-induced tumors (see Table 21-2). Likewise, when lymphocytes are transferred from mice with a virus-induced tumor into normal syngeneic recipients, the recipients reject subsequent transplants of all syngeneic tumors induced by the same virus. In the case of both SV40- and polyoma-induced tumors, the presence of tumor antigens is related to the neoplastic state of the cell. In humans, Burkitt's lymphoma cells have been shown to express a nuclear antigen of the Epstein-Barr virus that may indeed be a tumor-specific antigen for this type of tumor. Human papilloma virus (HPV) E6 and E7 proteins are found in more than 80% of invasive cervical cancers—the clearest example of a virally encoded tumor antigen. Consequently, there is great interest in testing as vaccine candidates the HPVs that are strongly linked to cervical cancer (see Clinical Focus).

The potential value of these virally induced tumor antigens can be seen in animal models. In one experiment, mice immunized with a preparation of genetically engineered polyoma virus tumor antigen were shown to be immune to subsequent injections of live polyoma-induced tumor cells. In another experiment, mice were immunized with a vaccinia virus vaccine engineered with the gene encoding the

(a)

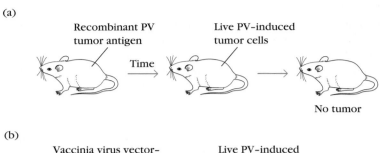

(b)

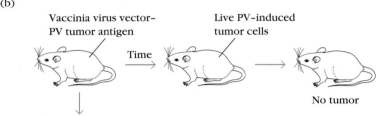

(c)                                        (d)

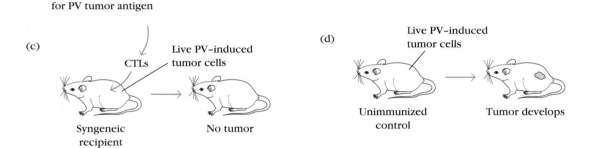

**FIGURE 21-9 Experimental induction of immunity against tumor cells induced by polyoma virus (PV).** Tumor immunity is achieved by immunizing mice with recombinant polyoma tumor antigen (a), with a vaccinia vector vaccine containing the gene encoding the PV tumor antigen (b), or with CTLs specific for the PV tumor antigen (c). Unimmunized mice (d) develop tumors when injected with live polyoma-induced tumor cells, whereas the immunized mice do not.

polyoma tumor antigen. These mice also developed immunity, rejecting later injections of live polyoma-induced tumor cells (Figure 21-9).

## Most tumor antigens are not unique to tumor cells

The majority of tumor antigens are not unique to tumor cells but are also present on normal cells. These tumor-associated transplantation antigens may be proteins usually expressed only on fetal cells but not on normal adult cells, or they may be proteins expressed at low levels by normal cells but at much higher levels by tumor cells. The latter category includes growth factors and growth factor receptors, as well as oncogene-encoded proteins.

Several growth factor receptors are expressed at significantly increased levels on tumor cells and can serve as tumor-associated antigens. For instance, a variety of tumor cells express the epidermal growth factor (EGF) receptor at levels 100 times greater than that in normal cells. An example of an overexpressed growth factor serving as a tumor-associated antigen is a transferrin growth factor, designated p97, which aids in the transport of iron into cells. Whereas normal cells express less than 8,000 molecules of p97 per cell, melanoma cells express 50,000 to 500,000 molecules of p97 per cell. The gene that encodes p97 has been cloned, and a recombinant vaccinia virus vaccine has been prepared that

carries the cloned gene. When this vaccine was injected into mice, it induced both humoral and cell-mediated immune responses, which protected the mice against live melanoma cells expressing the p97 antigen. Results such as this highlight the importance of identifying tumor antigens as potential targets of tumor immunotherapy.

### Oncofetal Tumor Antigens

**Oncofetal tumor antigens,** as the name implies, are found not only on cancerous cells but also on normal fetal cells. These antigens appear early in embryonic development, before the immune system acquires immunocompetence; if these antigens appear later on cancer cells, they are recognized as nonself and induce an immunologic response. Two well-studied oncofetal antigens are **alpha-fetoprotein (AFP)** and **carcinoembryonic antigen (CEA).**

Although the serum concentration of AFP drops from milligram levels in fetal serum to nanogram levels in normal adult serum, elevated AFP levels are found in a majority of patients with liver cancer (Table 21-3). CEA is a membrane glycoprotein found on gastrointestinal and liver cells of 2- to 6-month-old fetuses. Approximately 90% of patients with advanced colorectal cancer and 50% of patients with early colorectal cancer have increased levels of CEA in their serum; some patients with other types of cancer also exhibit increased CEA levels. However, because AFP and CEA can be found in trace amounts in some normal adults and in some

| TABLE 21-3 | Elevation of alpha-fetoprotein (AFP) and carcinoembryonic antigen (CEA) in serum of patients with various diseases | |
|---|---|---|
| Disease | No. of patients tested | % of patients with high AFP or CEA levels* |
| AFP > 400 μg/ml | | |
| Alcoholic cirrhosis | NA | 0 |
| Hepatitis | NA | 1 |
| Hepatocellular carcinoma | NA | 69 |
| Other carcinoma | NA | 0 |
| CEA > 10 mg/ml | | |
| Cancerous | | |
| Breast carcinoma | 125 | 14 |
| Colorectal carcinoma | 544 | 35 |
| Gastric carcinoma | 79 | 19 |
| Noncarcinoma malignancy | 228 | 2 |
| Pancreatic carcinoma | 55 | 35 |
| Pulmonary carcinoma | 181 | 26 |
| Noncancerous | | |
| Alcoholic cirrhosis | 120 | 2 |
| Cholecystitis | 39 | 1 |
| Nonmalignant disease | 115 | 0 |
| Pulmonary emphysema | 49 | 4 |
| Rectal polyps | 90 | 1 |
| Ulcerative colitis | 146 | 5 |

*Although trace amounts of both AFP and CEA can be found in some healthy adults, none would have levels greater than those indicated in the table.

noncancerous disease states, the presence of these oncofetal antigens is not diagnostic of tumors but rather serves to monitor tumor growth. If, for example, a patient has had surgery to remove a colorectal carcinoma, CEA levels are monitored after surgery. An increase in the CEA level is an indication of resumed tumor growth.

### Oncogene Proteins as Tumor Antigens

A number of tumors have been shown to express tumor-associated antigens encoded by cellular oncogenes. These antigens are also present in normal cells encoded by the corresponding proto-oncogene. In many cases, there is no qualitative difference between the oncogene and proto-oncogene products; instead, the increased levels of the oncogene product can be recognized by the immune system. For example, as noted earlier, human breast cancer cells exhibit elevated expression of the oncogene-encoded Neu protein, a growth factor receptor, whereas normal adult cells express only trace amounts of Neu protein. Because of this differ-

ence in the Neu level, anti-Neu monoclonal antibodies can recognize and selectively eliminate breast cancer cells without damaging normal cells.

### TATAs on Human Melanomas

Several tumor-associated transplantation antigens have been identified on human melanomas. Five of these—MAGE-1, MAGE-3, BAGE, GAGE-1, and GAGE-2—are oncofetal-type antigens. Each of these antigens is expressed on a significant proportion of human melanoma tumors, as well as on a number of other human tumors, but not on normal differentiated tissues except for the testis, where it is expressed on germ-line cells. In addition, a number of differentiation antigens expressed on normal melanocytes—including tyrosinase, gp100, Melan-A or MART-1, and gp75—are overexpressed by melanoma cells, enabling them to function as tumor-associated transplantation antigens.

Several of the human melanoma tumor antigens are shared by a number of other tumors. About 40% of human melanomas are positive for MAGE-1, and about 75% are positive for MAGE-2 or -3. In addition to melanomas, a significant percentage of glioma cell lines, breast tumors, non-small-cell lung tumors, and head or neck carcinomas express MAGE-1, -2, or -3. These shared tumor antigens could be exploited for clinical treatment. It might be possible to produce a tumor vaccine expressing the shared antigen for treatment of a number of these tumors, as described at the end of this chapter.

## Tumors can induce potent immune responses

In experimental animals, tumor antigens can be shown to induce both humoral and cell-mediated immune responses that result in the destruction of the tumor cells. In general, the cell-mediated response appears to play the major role. A number of tumors have been shown to induce tumor-specific CTLs that recognize tumor antigens presented by class I MHC on the tumor cells. However, as discussed below, expression of class I MHC molecules is decreased in a number of tumors, thereby limiting the role of specific CTLs in their destruction.

## NK cells and macrophages are important in tumor recognition

The recognition of tumor cells by NK cells is not MHC restricted. Thus, the activity of these cells is not compromised by the decreased MHC expression exhibited by some tumor cells. In some cases, Fc receptors on NK cells can bind to antibody-coated tumor cells, leading to ADCC. The importance of NK cells in tumor immunity is suggested by the mutant mouse strain called beige and by **Chediak-Higashi syndrome** in humans. In each case, a genetic defect causes marked impairment of NK cells and an associated increase in certain types of cancer.

Numerous observations indicate that activated macrophages also play a significant role in the immune response to tumors. For example, macrophages are often observed to cluster around tumors, and their presence is often correlated with tumor regression. Like NK cells, macrophages are not MHC restricted and express Fc receptors, enabling them to bind to antibody on tumor cells and mediate ADCC. The antitumor activity of activated macrophages is probably mediated by lytic enzymes and reactive oxygen and nitrogen intermediates. In addition, activated macrophages secrete a cytokine called tumor necrosis factor (TNF-$\alpha$) that has potent antitumor activity. When TNF-$\alpha$ is injected into tumor-bearing animals, it has been found to induce hemorrhage and necrosis of the tumor.

# Tumor Evasion of the Immune System

Although the immune system clearly can respond to tumor cells, the fact that so many individuals die each year from cancer suggests that the immune response to tumor cells is often ineffective. This section describes several mechanisms by which tumor cells appear to evade the immune system.

## Antitumor antibodies can enhance tumor growth

Following the discovery that antibodies could be produced against tumor-specific antigens, attempts were made to protect animals against tumor growth by active immunization with tumor antigens or by passive immunization with antitumor antibodies. Much to the surprise of the researchers, these immunizations did not protect against tumor growth; in many cases, they actually enhanced growth of the tumor.

The tumor-enhancing ability of immune sera subsequently was studied in cell-mediated lympholysis (CML) reactions in vitro. Serum taken from animals with progressive tumor growth was found to block the CML reaction, whereas serum taken from animals with regressing tumors had little or no blocking activity. K. E. Hellstrom and I. Hellstrom extended these findings by showing that children with progressive neuroblastoma had high levels of some kind of blocking factor in their sera and that children with regressive neuroblastoma did not have such factors. Since these first reports, blocking factors have been found to be associated with a number of human tumors.

In some cases, antitumor antibody itself acts as a blocking factor. Presumably the antibody binds to tumor-specific antigens and masks the antigens from cytotoxic T cells. In many cases, the blocking factors are not antibodies alone but rather antibodies complexed with tumor antigens. Although these immune complexes have been shown to block the CTL response, the mechanism of this inhibition is not known. The complexes also may inhibit ADCC by binding to Fc receptors on NK cells or macrophages and blocking their activity.

## Antibodies can modulate tumor antigens

Certain tumor-specific antigens have been observed to disappear from the surface of tumor cells in the presence of serum antibody and then to reappear after the antibody is no longer present. This phenomenon, called antigenic modulation, is readily observed when leukemic T cells are injected into mice previously immunized with a leukemic T-cell antigen (TL antigen). These mice develop high titers of anti-TL antibody, which binds to the TL antigen on the leukemic cells and induces capping, endocytosis, and/or shedding of the antigen-antibody complex. As long as antibody is present, these leukemic T cells fail to display the TL antigen and thus cannot be eliminated.

## Tumor cells frequently express low levels of class I MHC molecules

Since CD8$^+$ CTLs recognize only antigen associated with class I MHC molecules, any alteration in the expression of class I MHC molecules on tumor cells may exert a profound effect on the CTL-mediated immune response. Malignant transformation of cells is often associated with a reduction (or even a complete loss) of class I MHC molecules, and a number of tumors have been shown to express decreased levels of class I MHC molecules. As illustrated in Figure 21-10, the immune response itself may play a role in selecting for tumor cells with decreased class I MHC expression by preferentially eliminating those with normal MHC expression. The decrease in class I MHC expression is often accompanied by progressive tumor growth, and so the absence of MHC molecules on tumor cells is generally an indication of a poor prognosis.

## Tumor cells may provide poor costimulatory signals

T-cell activation requires an activating signal, triggered by recognition of a peptide-MHC molecule complex by the T-cell receptor, and a costimulatory signal, triggered by the interaction of B7 on antigen-presenting cells with CD28 on the T cells. Both signals are needed to induce IL-2 production and proliferation of T cells. The poor immunogenicity of many tumor cells may be due in large part to lack of the costimulatory molecules. Without sufficient numbers of antigen-presenting cells in the immediate vicinity of a tumor, the T cells will receive only a partial activating signal, which may lead to clonal anergy.

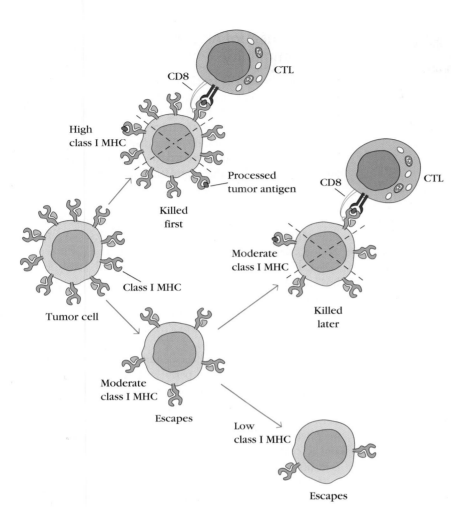

Tumor cell
Class I MHC
High class I MHC
CD8
CTL
Processed tumor antigen
Killed first
Moderate class I MHC
CD8
CTL
Killed later
Moderate class I MHC
Escapes
Low class I MHC
Escapes

**FIGURE 21-10 Down-regulation of class I MHC expression on tumor cells may allow a tumor to escape CTL-mediated recognition.** The immune response itself may play a role in selecting for tumor cells that express lower levels of class I MHC molecules by preferentially eliminating those cells expressing high levels of class I molecules. Malignant tumor cells that express fewer MHC molecules may thus escape CTL-mediated destruction.

## Cancer Immunotherapy

Immunotherapy of cancer takes several forms. The treatment may involve a general boost of the immune system through the use of an adjuvant or a cytokine or a more specific approach such as a monoclonal antibody directed against an antigen of a specific tumor type. The following sections describe immunotherapeutic agents that have been licensed for use in humans and several developmental approaches that may yield clinically useful products to fight cancer in the future.

### Manipulation of costimulatory signals can enhance immunity

Several research groups have demonstrated that tumor immunity can be enhanced by providing the costimulatory signal necessary for activation of CTL precursors (CTL-Ps; see Chapter 14). When mouse CTL-Ps are incubated with melanoma cells in vitro, antigen recognition occurs, but in the absence of a costimulatory signal, the CTL-Ps do not

proliferate and differentiate into effector CTLs. However, when the melanoma cells are transfected with the gene that encodes the B7 ligand, the CTL-Ps differentiate into effector CTLs.

These findings offer the possibility that B7-transfected tumor cells might be used to induce a CTL response in vivo. For instance, when P. Linsley, L. Chen, and their colleagues injected melanoma-bearing mice with B7+ melanoma cells, the melanomas completely regressed in more than 40% of the mice. S. Townsend and J. Allison used a similar approach to vaccinate mice against malignant melanoma. Normal mice were first immunized with irradiated, B7-transfected melanoma cells and then challenged with unaltered malignant melanoma cells. The "vaccine" was found to protect a high percentage of the mice (Figure 21-11a). It is hoped that a similar vaccine might prevent metastasis after surgical removal of a primary melanoma in human patients.

Because human melanoma antigens are shared by a number of different human tumors, it might be possible to generate a panel of B7-transfected melanoma cell lines that are

(a)

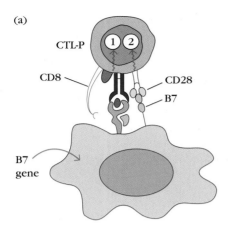

Tumor cell
transfected $\longrightarrow$ CTL activation $\longrightarrow$ Tumor destruction
with B7 gene

(b)

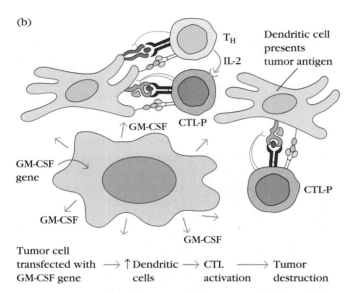

Tumor cell
transfected with $\longrightarrow \uparrow$ Dendritic $\longrightarrow$ CTL $\longrightarrow$ Tumor
GM-CSF gene      cells       activation   destruction

**FIGURE 21-11 Use of transfected tumor cells for cancer immunotherapy.** (a) Tumor cells transfected with the B7 gene express the costimulatory B7 molecule, enabling them to provide both activating signal (1) and costimulatory signal (2) to CTL-Ps. As a result of the combined signals, the CTL-Ps differentiate into effector CTLs, which can mediate tumor destruction. In effect, the transfected tumor cell acts as an antigen-presenting cell. (b) Transfection of tumor cells with the gene encoding GM-CSF allows the tumor cells to secrete high levels of GM-CSF. This cytokine will activate dendritic cells in the vicinity of the tumor, enabling the dendritic cells to present tumor antigens to both T$_H$ cells and CTL-Ps.

typed for tumor-antigen expression and for HLA expression. In this approach, the tumor antigen(s) expressed by a patient's tumor would be determined, and then the patient would be vaccinated with an irradiated B7-transfected cell line that expresses similar tumor antigen(s).

## Enhancement of APC activity can modulate tumor immunity

Mouse dendritic cells cultured in GM-CSF and incubated with tumor fragments, then reinfused into the mice, have been shown to activate both T$_H$ cells and CTLs specific for the tumor antigens. When the mice were subsequently challenged with live tumor cells, they displayed tumor immunity. These experiments have led to a number of approaches aimed at expanding the population of antigen-presenting cells, so that these cells can activate T$_H$ cells or CTLs specific for tumor antigens.

One approach that has been tried is to transfect tumor cells with the gene encoding GM-CSF. These engineered tumor cells, when reinfused into the patient, will secrete GM-CSF, enhancing the differentiation and activation of host antigen-presenting cells, especially dendritic cells. As these dendritic cells accumulate around the tumor cells, the GM-CSF secreted by the tumor cells will enhance the presentation of tumor antigens to T$_H$ cells and CTLs by the dendritic cells (Figure 21-11b).

Another way to expand the dendritic-cell population is to culture dendritic cells from peripheral-blood progenitor cells in the presence of GM-CSF, TNF-$\alpha$, and IL-4. These three cytokines induce the generation of large numbers of dendritic cells. If these dendritic cells are pulsed with tumor fragments and then reintroduced into the patient, they might activate T$_H$ and T$_C$ cells specific for the tumor antigens. Whether this hope is justified will be determined by further investigation.

A number of adjuvants, including the attenuated strains of *Mycobacterium bovis* called bacillus Calmette-Guérin (BCG) and *Corynebacterium parvuum,* have been used to boost tumor immunity. These adjuvants activate macrophages, increasing their expression of various cytokines, class II MHC molecules, and the B7 costimulatory molecule. Activated macrophages are better activators of T$_H$ cells, resulting in generalized increases in both humoral and cell-mediated responses. Thus far, adjuvants have shown only modest therapeutic results.

## Cytokine therapy can augment immune responses to tumors

The isolation and cloning of the various cytokine genes has facilitated their large-scale production. A variety of experimental and clinical approaches have been developed to use recombinant cytokines, either singly or in combination, to augment the immune response against cancer. Among the cytokines that have been evaluated in cancer immunotherapy are IFN-$\alpha$, -$\beta$, and -$\gamma$; IL-2, -4, -6, and -12; GM-CSF; and TNF. These trials produce occasional encouraging results, and IL-2 is licensed for use either alone or with other agents (such as INF-$\alpha$) for advanced kidney cancer and for metastatic melanoma.

The most notable obstacle to cytokine therapy is the complexity of the cytokine network itself, which makes it very difficult to know precisely how intervention with a given recombinant cytokine will affect the production of other cytokines. And since some cytokines act antagonistically, it is possible that intervention with a recombinant cytokine designed to enhance a particular branch of the immune response may actually lead to suppression. In addition, cytokine immunotherapy is plagued by the difficulty of administering the cytokines locally. In some cases, systemic administration of high levels of a given cytokine has been shown to lead to serious and even life-threatening consequences.

## Interferons

Large quantities of purified recombinant preparations of the interferons, IFN-$\alpha$, IFN-$\beta$, and IFN-$\gamma$, are now available. Of these, only INF-$\alpha$ is licensed for use to treat human cancer, including several types of lymphoid cancer (hairy-cell leukemia, chronic myelogenous leukemia, cutaneous T-cell lymphoma, and non-Hodgkin's lymphoma) and solid tumors such as melanoma, Kaposi's sarcoma, and renal cancer.

Interferon-mediated antitumor activity may involve several mechanisms. All three types of interferon have been shown to increase class I MHC expression on tumor cells; IFN-$\gamma$ has also been shown to increase class II MHC expression on macrophages. Given the evidence for decreased levels of class I MHC molecules on the cells of malignant tumors, the interferons may act by restoring MHC expression, thereby increasing CTL activity against tumors. In addition, the interferons have been shown to inhibit cell division of both normal and malignantly transformed cells in vitro; it is possible that some of the antitumor effects of the interferons are related to this ability to directly inhibit tumor-cell proliferation. Finally, IFN-$\gamma$ directly or indirectly increases the activity of $T_C$ cells, macrophages, and NK cells, all of which play a role in the immune response to tumor cells.

## Tumor Necrosis Factors

In some instances, the tumor necrosis factors TNF-$\alpha$ and TNF-$\beta$ have been shown to exhibit direct antitumor activity, killing some tumor cells and reducing the rate of proliferation of others while sparing normal cells (Figure 21-12). In the presence of TNF-$\alpha$ or TNF-$\beta$, a tumor undergoes visible hemorrhagic necrosis and regression. TNF-$\alpha$ has also been shown to inhibit tumor-induced vascularization (angiogenesis) by damaging the vascular endothelial cells in the vicinity of a tumor, thereby decreasing the flow of blood and oxygen that is necessary for progressive tumor growth.

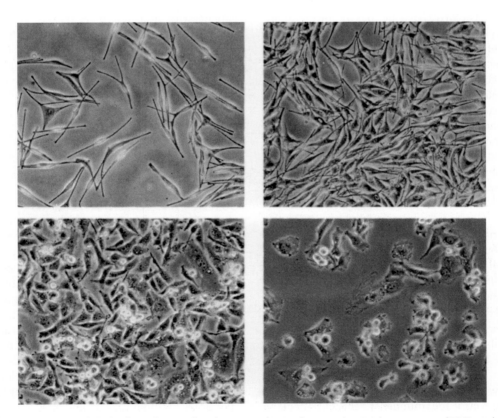

**FIGURE 21-12  Photomicrographs of cultured normal melanocytes *(top)* and cultured cancerous melanoma cells *(bottom)* in the presence *(left)* and absence *(right)* of tumor necrosis factor** (TNF-$\alpha$). Note that in the presence of TNF-$\alpha$, the cancer cells stop proliferating, whereas TNF-$\alpha$ has no inhibitory effect on proliferation of the normal cells. *[From L. J. Old, 1988, Scientific American **258**(5):59.]*

| TABLE 21-4 | Monoclonal antibodies approved for the treatment of human cancers | | |
|---|---|---|---|
| **Name** | **Trade name** | **Used to treat** | **Year approved** |
| Rituximab | Rituxan | Non-Hodgkin's lymphoma | 1997 |
| Trastuzumab | Herceptin | Breast cancer | 1998 |
| Gemtuzumab ozogamicin* | Mylotarg | Acute myelogenous leukemia (AML) | 2000 |
| Alemtuzumab | Campath | Chronic lymphocytic leukemia (CLL) | 2001 |
| Ibritumomab tiuxetan* | Zevalin | Non-Hodgkin's lymphoma | 2002 |
| Tositumomab* | Bexxar | Non-Hodgkin's lymphoma | 2003 |
| Cetuximab | Erbitux | Colorectal cancer, head and neck cancers | 2004 2006 |
| Bevacizumab | Avastin | Colorectal cancer | 2004 |

*Conjugated monoclonal antibodies

## Monoclonal antibodies are effective in treating some tumors

Monoclonal antibodies have long been used in various ways as experimental immunotherapeutic agents for cancer. At present about eight different MAbs are licensed to treat various kinds of cancer. Table 21-4 lists the monoclonal antibodies approved by the FDA and lists the cancers for which they are approved. The MAbs may be "naked," meaning unmodified, or they may be conjugated with an agent that increases their efficacy. Toxins, chemical agents, and radioactive particles are used to conjugate antibodies, which then deliver the conjugated substance to the target cell.

In one early success of monoclonal antibody treatment, R. Levy and his colleagues successfully treated a 64-year-old man with terminal B-cell lymphoma. At the time of treatment, the lymphoma had metastasized to the liver, spleen, bone marrow, and peripheral blood. Because this was a B-cell cancer, the membrane-bound antibody on all the cancerous cells had the same idiotype. By the procedure outlined in Figure 21-13, these researchers produced mouse monoclonal antibody specific for the B-lymphoma idiotype. When this mouse monoclonal anti-idiotype antibody was injected into the patient, it bound specifically to the B-lymphoma cells, because these cells expressed that particular idiotype. Since B-lymphoma cells are susceptible to complement-mediated lysis, the monoclonal antibody activated the complement system and lysed the lymphoma cells without harming other cells. After four injections with this anti-idiotype monoclonal antibody, the tumors began to shrink, and this patient entered an unusually long period of complete remission.

More recently, Levy and his colleagues have used direct immunization to recruit the immune systems of patients to an attack on their B-cell lymphoma. In a clinical trial with 41 B-cell lymphoma patients, the rearranged immunoglobulin genes of the lymphomas of each patient were isolated and used to encode the synthesis of recombinant immunoglobulin that bore the idiotype typical of the patient's tumor. Each of these Igs was coupled to keyhole limpet hemocyanin (KLH), a mollusk protein that is often used as a carrier protein because of its efficient recruitment of T-cell help. The patients were immunized with their own tumor-specific antigens, the idiotypically unique immunoglobulins produced by their own lymphomas. About 50% of the patients developed anti-idiotype antibodies against their tumors. Significantly, improved clinical outcomes were seen in the 20 patients with anti-idiotype responses but not in the others. In fact, two of the 20 experienced complete remission.

A custom approach targeting idiotypes requires a specific reagent for each lymphoma patient. This is prohibitively expensive and cannot be used as a general therapeutic approach for the thousands of patients diagnosed each year with B-cell lymphoma. A more general monoclonal antibody therapy for B-cell lymphoma is based on the fact that most B cells, whether normal or cancerous, bear lineage-distinctive antigens. For example, the monoclonal antibody Rituximab, which targets the B-cell marker CD20, is widely used to treat non-Hodgkin's lymphoma.

A variety of tumors express significantly increased levels of growth factor receptors, which are promising targets for antitumor monoclonal antibodies. For example, in 25% to 30% of women with metastatic breast cancer, a genetic alteration of the tumor cells results in the increased expression of HER2, an epidermal-growth-factor–like receptor. An anti-HER2 monoclonal antibody has been prepared and humanized and is given in large amounts (100 milligrams or more) to treat HER2-receptor-bearing breast cancers.

As mentioned above, certain of the monoclonal antibodies in clinical use are coupled with radioactive isotopes, chemotherapy drugs, or potent toxins of biological origin. In such "guided missile" therapies, the toxic agents are delivered specifically to tumor cells. This focuses the toxic effects on the tumor and spares normal tissues. Reagents known as **immunotoxins** have been constructed by coupling the inhibitor chain of a toxin (e.g., diphtheria toxin) to an antibody against a tumor-specific or tumor-associated antigen. In vitro studies have demonstrated that these "magic bullets" can kill tumor

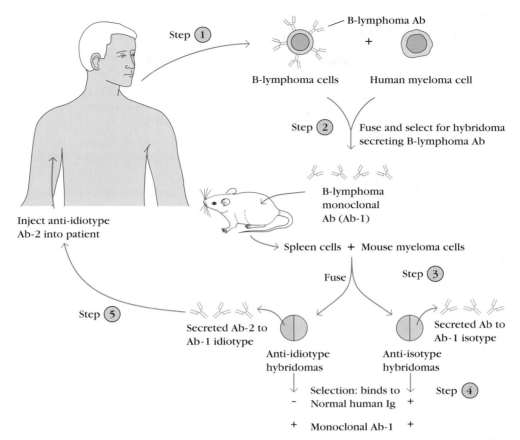

**FIGURE 21-13 Treatment of B-cell lymphoma with monoclonal antibody specific for idiotypic determinants on the cancer cells.** Because all the lymphoma cells are derived from a single transformed B cell, they all express membrane-bound antibody (Ab-1) with the same idiotype (i.e., the same antigenic specificity). In the procedure illustrated, monoclonal anti-idiotype antibody (Ab-2) against the B-lymphoma membrane-bound antibody was produced (steps 1–4). When this anti-idiotype antibody was injected into the patient (step 5), it bound selectively to B-lymphoma cells, which then were susceptible to complement-plus-antibody lysis.

cells without harming normal cells. Immunotoxins specific for tumor antigens in a variety of cancers (e.g., melanoma, colorectal carcinoma, metastatic breast carcinoma, and various lymphomas and leukemias) have been evaluated in phase I or phase II clinical trials. In a number of trials, significant numbers of leukemia and lymphoma patients exhibited partial or complete remission, and several products in use to treat leukemia are conjugated antibodies. However, the clinical responses in patients with larger tumor masses were disappointing. In some of these patients, the sheer size of the tumor may render most of its cells inaccessible to the immunotoxin.

---------------

## SUMMARY

- Tumor cells differ from normal cells by changes in the regulation of growth, allowing them to proliferate indefinitely, then invade the underlying tissue and eventually metastasize to other tissues.

- Normal cells can be transformed in vitro by chemical and physical carcinogens and by transforming viruses.

Transformed cells exhibit altered growth properties and are sometimes capable of inducing cancer when they are injected into animals.

- Proto-oncogenes encode proteins involved in control of normal cellular growth. The conversion of proto-oncogenes to oncogenes is a key step in the induction of most human cancer. This conversion may result from mutation in an oncogene, its translocation, or its amplification.

- A number of B- and T-cell leukemias and lymphomas are associated with translocated proto-oncogenes. In its new site, the translocated gene may come under the influence of enhancers or promoters that cause its transcription at higher levels than usual.

- Tumor cells display tumor-specific antigens and the more common tumor-associated antigens. In contrast to tumor antigens induced by chemicals or radiation, virally encoded tumor antigens are shared by all tumors induced by the same virus.

- The tumor antigens recognized by T cells fall into one of four major categories: antigens encoded by genes with tumor-specific expression, antigens encoded by variant forms of normal genes that have been altered by mutation, certain antigens normally expressed only at certain stages of differentiation or differentiation lineages, and antigens that are overexpressed in particular tumors.

■ The immune response to tumors includes CTL-mediated lysis, NK-cell activity, macrophage-mediated tumor destruction, and destruction mediated by ADCC. Several cytotoxic factors, including TNF-α and TNF-β, help to mediate tumor-cell killing. Tumors use several strategies to evade the immune response.

■ Experimental cancer immunotherapy is exploring a variety of approaches. Some of these are the enhancement of the costimulatory signal required for T-cell activation, genetically engineering tumor cells to secrete cytokines that may increase the intensity of the immune response against them, the therapeutic use of cytokines, and ways of increasing the activity of antigen-presenting cells.

■ Monoclonal antibodies are approved for use against several cancers. The antibodies are used in an unmodified form or are coupled with toxins, chemotherapeutic agents, or radioactive elements.

■ Key elements in the design of strategies for vaccination against cancer are the identification of significant tumor antigens, the development of strategies for the effective presentation of tumor antigens, and the generation of activated populations of helper or cytotoxic T cells.

## References

Aisenberg, A. C. 1993. Utility of gene rearrangements in lymphoid malignancies. *Annual Review of Medicine* **44:**75.

Allison, J. P., A. A. Hurwitz, and D. R. Leach. 1995. Manipulation of costimulatory signals to enhance antitumor T-cell responses. *Current Opinion in Immunology* **7:**682.

Blattman, J. N., and P. D. Greenberg. 2004. Cancer immunotherapy: a treatment for the masses. *Science* **305:**200.

Berzofsky, J. A., et al. 2004. Progress on new vaccine strategies for the immunotherapy and prevention of cancer. *Journal of Clinical Investigation* **113:**1515.

Boon, T., P. G. Coulie, and B. Van den Eynde. 1997. Tumor antigens recognized by T cells. *Immunology Today* **18:**267.

Cohen, J. 2005. High hopes and dilemmas for a cervical cancer vaccine. *Science* **308:**618.

Coulie, P. G., et al. 1994. A new gene coding for a differentiation antigen recognized by autologous cytolytic T lymphocytes on HLA-A2 melanomas. *Journal of Experimental Medicine* **180:**35.

Houghton, A. N., J. S. Gold, and N. E. Blachere. 2001. Immunity against cancer: lessons learned from melanoma. *Current Opinion in Immunology* **13:**134.

Hsu, F. J., et al. 1997. Tumor-specific idiotype vaccines in the treatment of patients with B-cell lymphoma. *Blood* **89:**3129.

Kufe, D. W. 2000. Smallpox, polio and now a cancer vaccine? *Nature Medicine* **6:**252.

Paterson, Y., and G. Ikonomidis. 1996. Recombinant *Listeria monocytogenes* cancer vaccines. *Current Opinion in Immunology* **8:**651.

Sahin, U., O. Tureci, and M. Pfreundschuh. 1997. Serological identification of human tumor antigens. *Current Opinion in Immunology* **9:**709.

Schiffman, M., and P. E. Castle. 2005. The promise of global cervical-cancer prevention. *New England Journal of Medicine* **353:**2101.

Srivastava, S. 2002. Roles of heat-shock proteins in innate and adaptive immunity. *Nature Reviews Immunology* **2:**185.

Weinberg, R. A. 1996. How cancer arises. *Scientific American* **275**(3):62.

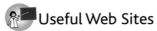 ## Useful Web Sites

**http://www.oncolink.upenn.edu/**

Oncolink offers comprehensive information about many types of cancer and is a good source of information about cancer research and advances in cancer therapy. The site is regularly updated and includes many useful links to other resources.

**http://www.cancer.org/docroot/home/index.asp**

The Web site of the American Cancer Society contains a great deal of information on the incidence, treatment, and prevention of cancer. The site also highlights significant achievements in cancer research.

**http://www.cytopathnet.org/**

A good resource for information on the cytological examination of tumors and on matters related to staining patterns that are typical of the cell populations found in a number of cancers.

 ## Study Questions

**CLINICAL FOCUS QUESTION** Why is cervical cancer a likely target for a vaccine that can prevent cancer? Can the approach being investigated for cervical cancer be applied to all types of cancer?

1. Indicate whether each of the following statements is true or false. If you think a statement is false, explain why.

   a. Hereditary retinoblastoma results from overexpression of a cellular oncogene.

   b. Translocation of c-*myc* gene is found in many patients with Burkitt's lymphoma.

   c. Multiple copies of cellular oncogenes are sometimes observed in cancer cells.

   d. Viral integration into the cellular genome may convert a proto-oncogene into a transforming oncogene.

e. All oncogenic retroviruses carry viral oncogenes.

f. The immune response against a virus-induced tumor protects against another tumor induced by the same virus.

2. You are a clinical immunologist studying acute lymphoblastic leukemia (ALL). Leukemic cells from most patients with ALL have the morphology of lymphocytes but do not express cell surface markers characteristic of mature B or T cells. You have isolated cells from ALL patients that do not express membrane Ig but do react with monoclonal antibody against a normal pre-B-cell marker (B-200). You therefore suspect that these leukemic cells are pre-B cells. How would you use genetic analysis to confirm that the leukemic cells are committed to the B-cell lineage?

3. In a recent experiment, melanoma cells were isolated from patients with early or advanced stages of malignant melanoma. At the same time, T cells specific for tetanus toxoid antigen were isolated and cloned from each patient.

   a. When early-stage melanoma cells were cultured together with tetanus-toxoid antigen and the tetanus-toxoid-specific T-cell clones, the T-cell clones were observed to proliferate. This proliferation was blocked by addition of chloroquine or by addition of monoclonal antibody to HLA-DR. Proliferation was not blocked by addition of monoclonal antibody to HLA-A, -B, -DQ, or -DP. What might these findings indicate about the early-stage melanoma cells in this experimental system?

   b. When the same experiment was repeated with advanced-stage melanoma cells, the tetanus-toxoid T-cell clones failed to proliferate in response to the tetanus-toxoid antigen. What might this indicate about advanced-stage melanoma cells?

   c. When early and advanced malignant melanoma cells were fixed with paraformaldehyde and incubated with processed tetanus toxoid, only the early-stage melanoma cells could induce proliferation of the tetanus-toxoid T-cell clones. What might this indicate about early-stage melanoma cells?

   d. How might you confirm your hypothesis experimentally?

4. What are three likely sources of tumor antigens?

5. Various cytokines have been evaluated for use in tumor immunotherapy. Describe four mechanisms by which cytokines mediate antitumor effects and the cytokines that induce each type of effect.

6. Infusion of transfected melanoma cells into cancer patients is a promising immunotherapy.

   a. Which two genes have been transfected into melanoma cells for this purpose? What is the rationale behind use of each of these genes?

   b. Why might use of such transfected melanoma cells also be effective in treating other types of cancers?

7. For each of the following descriptions, choose the most appropriate term:

| *Descriptions* | *Terms* |
|---|---|
| a. A benign or malignant tumor | 1. Sarcoma |
| b. A tumor that has arisen from endodermal tissue | 2. Carcinoma |
| c. A tumor that has arisen from mesodermal connective tissue | 3. Metastasis |
| d. A tumor that is invasive and continues to grow | 4. Neoplasm |
| e. Tumor cells that have separated from the original tumor and grow in a different part of the body | 5. Malignant |
| | 6. Leukemia |
| f. A tumor that is noninvasive | 7. Transformation |
| g. A tumor that has arisen from lymphoid cells | 8. Lymphoma |
| h. A permanent change in the genome of a cell that results in abnormal growth | 9. Benign |
| i. Cancer cells that have arisen from hematopoietic cells that do not grow as a solid tumor | |

 **Interactive Study**

**www.whfreeman.com/kuby**

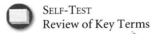 SELF-TEST
Review of Key Terms

# chapter 22

# Experimental Systems

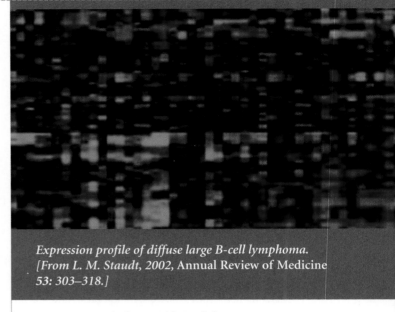

*Expression profile of diffuse large B-cell lymphoma.*
*[From L. M. Staudt, 2002,* Annual Review of Medicine
*53: 303–318.]*

E
XPERIMENTAL SYSTEMS OF VARIOUS TYPES ARE USED TO
unravel the complex cellular interactions of the immune response. In vivo systems, which involve the
whole animal, provide the most natural experimental conditions. However, in vivo systems have a myriad of unknown
and uncontrollable cellular interactions that add ambiguity
to the interpretation of data. At the other extreme are in
vitro systems, in which defined populations of lymphocytes
are studied under controlled and consequently repeatable
conditions; in vitro systems can be simplified to the extent
that individual cellular interactions can be studied effectively. Yet they have their own limitations, the most notable
of which is their artificiality. For example, providing antigen
to purified B cells in vitro does not stimulate maximal antibody production unless T cells are present. Therefore a study
of antibody production in an artificial in vitro system that
lacks T cells could lead to the incorrect conclusion that B
cells do not synthesize high levels of antibodies. One must
ask whether a cellular response observed in vitro reflects
reality or is a product of the unique conditions of the in vitro
system itself.

This chapter describes some of the experimental systems
routinely used to study the immune system and covers some
recombinant DNA techniques that have revolutionized the
study of immunology in the past two decades. Experimental
systems and techniques are also covered in other chapters.
Table 22-1 lists them and directs the reader to the appropriate
location for a description.

- Experimental Animal Models
- Cell Culture Systems
- Protein Biochemistry
- Recombinant DNA Technology
- Analysis of DNA Regulatory Sequences
- Gene Transfer into Mammalian Cells
- Microarrays—An Approach for Analyzing Patterns of Gene Expression
- Two-Photon Microscopy for In Vivo Imaging of the Immune System
- Advances in Fluorescent Technology

## Experimental Animal Models

The study of the immune system in vertebrates requires
suitable animal models. The choice of an animal depends
on its suitability for attaining a particular research goal. If
large amounts of antiserum are sought, a rabbit, goat,
sheep, or horse might be an appropriate experimental animal. If the goal is development of a protective vaccine, the
animal chosen must be susceptible to the infectious agent
so that the efficacy of the vaccine can be assessed. Mice or
rabbits can be used for vaccine development if they are

susceptible to the pathogen. But if growth of the infectious
agent is limited to humans and primates, vaccine development may require the use of monkeys, chimpanzees, or
baboons.

For most basic research in immunology, mice have been
the experimental animal of choice. They are easy to handle,
are genetically well characterized, and have a rapid breeding
cycle. The immune system of the mouse has been characterized more extensively than that of any other species. The
value of basic research in the mouse system is highlighted by
the enormous impact this research has had on clinical intervention in human disease.

| TABLE 22-1 | Immunological methods described in other chapters |
|---|---|
| **Method** | **Location** |
| Stem-cell transplantation | Ch. 2 Clinical Focus |
| Preparation of monoclonal antibodies | Fig. 4-25 |
| Genetic engineering of chimeric mouse-human monoclonal antibodies | Figure 5-24 and Ch. 5 Clinical Focus |
| Determination of antibody affinity by equilibrium dialysis | Fig. 6-2 |
| Immunodiffusion and immunoelectrophoresis | Figs. 6-6 and 6-7 |
| Hemagglutination | Fig. 6-8 |
| Radioimmunoassay (RIA) | Fig. 6-9 |
| ELISA assays | Fig. 6-10 |
| ELISPOT assay | Fig. 6-11 |
| Western blotting | Fig. 6-12 |
| Immunoprecipitation | Fig. 6-13 |
| Immunofluorescence | Fig. 6-14 |
| Flow cytometry | Fig. 6-15 |
| Mixed-lymphocyte reaction (MLR) | Figs. 14-16 and 17-4c |
| Cell-mediated lympholysis (CML) | Fig. 14-17 |
| Production of vaccinia vector vaccine | Fig. 19-8 |
| Production of multivalent subunit vaccines | Fig. 19-6 |
| HLA typing | Fig. 17-4 |

## Inbred strains can reduce experimental variation

To control experimental variation caused by differences in the genetic backgrounds of experimental animals, immunologists often work with inbred strains—that is, genetically identical animals produced by inbreeding. The rapid breeding cycle of mice makes them particularly well suited for the production of inbred strains, which are developed by repeated inbreeding between brother and sister littermates. In this way the heterozygosity of alleles that is normally found in randomly outbred mice is replaced by homozygosity at all loci. Repeated inbreeding for 20 generations usually yields an inbred strain whose progeny are homozygous at more than 98% of all loci. Around 500 different inbred strains of mice are available, each designated by a series of letters and/or numbers (Table 22-2). Most strains can be purchased by immunologists from such suppliers as the Jackson Laboratory in Bar Harbor, Maine. Inbred strains have also been produced

in rats, guinea pigs, hamsters, rabbits, and domestic fowl. Because inbred strains of animals are genetically identical (**syngeneic**) within that strain, their immune responses can be studied in the absence of variables introduced by individual genetic differences—an invaluable property. With inbred strains, lymphocyte subpopulations isolated from one animal can be injected into another animal of the same strain without eliciting a rejection reaction. This type of experimental system permitted immunologists to first demonstrate that lymphocytes from an antigen-primed animal could transfer immunity to an unprimed syngeneic recipient.

## Adoptive-transfer systems permit the in vivo examination of isolated cell populations

In some experiments, it is important to eliminate the immune responsiveness of the syngeneic host so that the response of only the transferred lymphocytes can be studied in isolation. This can be accomplished by a technique called **adoptive transfer:** first, the syngeneic host is exposed to x-rays that kill its lymphocytes; then the donor immune cells are introduced. Subjecting a mouse to high doses of x-rays (650–750 rads) can kill 99.99% of its lymphocytes, after which the activities of lymphocytes transplanted from the spleen of a syngeneic donor can be studied without interference from host lymphocytes. If the host's hematopoietic cells might influence an adoptive-transfer experiment, then higher x-ray levels (900–1000 rads) are used to eliminate the entire hematopoietic system. Mice irradiated with such doses will die unless reconstituted with bone marrow from a syngeneic donor.

The adoptive-transfer system has enabled immunologists to study the development of injected lymphoid stem cells in various organs of the recipient and has facilitated the study of various populations of lymphocytes and of the cellular interactions required to generate an immune response. Such experiments, for instance, first enabled immunologists to show that a T helper cell is necessary for B-cell activation in the humoral response. In these experiments, adoptive transfer of purified B cells or purified T cells did not produce antibody in the irradiated host. Only when both cell populations were transferred was antibody produced in response to antigen.

# Cell Culture Systems

The complexity of the cellular interactions that generate an immune response has led immunologists to rely heavily on various types of in vitro cell culture systems. A variety of cells can be cultured, including primary lymphoid cells, cloned lymphoid cell lines, and hybrid cells.

## Primary lymphoid cell cultures are derived from blood or lymphoid organs

Primary lymphoid cell cultures can be obtained by isolating lymphocytes directly from blood or lymph or from various

| TABLE 22-2 | Some inbred mouse strains commonly used in immunology | |
| --- | --- | --- |
| **Strain** | **Common substrains** | **Characteristics** |
| A | A/He<br>A/J<br>A/WySn | High incidence of mammary tumors in some substrains |
| AKR | AKR/J<br>AKR/N<br>AKR/Cum | High incidence of leukemia<br><br>*Thy 1.2* allele in AKR/Cum, and *Thy 1.1* allele in other substrains (*Thy* gene encodes a T-cell surface protein) |
| BALB/c | BALB/c*j*<br>BALB/c AnN<br>BALB/cBy | Sensitivity to radiation<br>Used in hybridoma technology<br>Many myeloma cell lines were generated in these mice |
| CBA | CBA/J<br>CBA/H<br>CBA/N | Gene *(rd)* causing retinal degeneration in CBA/J<br><br>Gene *(xid)* causing X-linked immunodeficiency in CBA/N |
| C3H | C3H/He<br>C3H/HeJ<br>C3H/HeN | Gene *(rd)* causing retinal degeneration<br>High incidence of mammary tumors in many substrains (these carry a mammary-tumor virus that is passed via maternal milk to offspring) |
| C57BL/6 | C57BL/6J<br>C57BL/6By<br>C57BL/6N | High incidence of hepatomas after irradiation<br>High complement activity |
| C57BL/10 | C57BL/10J<br>C57BL/10ScSn<br>C57BL/10N | Very close relationship to C57BL/6 but differences in at least two loci<br><br>Frequent partner in preparation of congenic mice |
| C57BR | C57BR/cd*j* | High frequency of pituitary and liver tumors<br>Very resistant to x-irradiation |
| C57L | C57L/J<br>C57L/N | Susceptibility to experimental autoimmune encephalomyelitis (EAE)<br>High frequency of pituitary and reticular cell tumors |
| C58 | C58/J<br>C58/LwN | High incidence of leukemia |
| DBA/1 | DBA/1J<br>DBA/1N | High incidence of mammary tumors |
| DBA/2 | DBA/2J<br>DBA/2N | Low immune response to some antigens<br>Low response to pneumococcal polysaccharide type III |
| HRS | HRS/J | Hairless *(hr)* gene, usually in heterozygous state |
| NZB | NZB/BINJ<br>NZB/N | High incidence of autoimmune hemolytic anemia and lupus-like nephritis<br>Autoimmune disease similar to systemic lupus erythematosus (SLE) in $F_1$ progeny from crosses with NZW |
| NZW | NZW/N | SLE-type autoimmune disease in $F_1$ progeny from crosses with NZB |
| P | P/J | High incidence of leukemia |
| SJL | SJL/J | High level of aggression and severe fighting to the point of death, especially in males<br>Tendency to develop certain autoimmune diseases, most susceptible to EAE |
| SWR | SWR/J | Tendency to develop several autoimmune diseases, especially EAE |
| 129 | 129/J<br>129/SvJ | High incidence of spontaneous teratocarcinoma |

SOURCE: Adapted from Federation of American Societies for Experimental Biology, 1979, *Biological Handbooks*, Vol. III: Inbred and Genetically Defined Strains of Laboratory Animals.

lymphoid organs by tissue dispersion. The lymphocytes can then be grown in a chemically defined basal medium (containing saline, sugars, amino acids, vitamins, trace elements, and other nutrients) to which various serum supplements are added. For some experiments, serum-free culture conditions are employed. Because in vitro culture techniques require from 10- to 100-fold fewer lymphocytes than do typical in vivo techniques, they have enabled immunologists to assess the functional properties of minor subpopulations of lymphocytes. It was by means of cell culture techniques, for example, that immunologists were first able to define the functional differences between CD4$^+$ T helper cells and CD8$^+$ T cytotoxic cells.

Cell culture techniques have also been used to identify numerous cytokines involved in the activation, growth, and differentiation of various cells involved in the immune response. Early experiments showed that media conditioned, or modified, by the growth of various lymphocytes or antigen-presenting cells would support the growth of other lymphoid cells. Conditioned media contain the secreted products from actively growing cells. Many of the individual cytokines that characterized various conditioned media have subsequently been identified and purified, and in many cases the genes that encode them have been cloned. These cytokines, which play a central role in the activation and regulation of the immune response, are described in Chapter 12 and elsewhere.

## Cloned lymphoid cell lines are important tools in immunology

A primary lymphoid cell culture comprises a heterogeneous group of cells that can be propagated only for a limited time. This heterogeneity can complicate the interpretation of experimental results. To avoid these problems, immunologists use cloned lymphoid cell lines and hybrid cells.

Normal mammalian cells generally have a finite life span in culture; that is, after a number of population doublings characteristic of the species and cell type, the cells stop dividing. In contrast, tumor cells or normal cells that have undergone **transformation** induced by chemical carcinogens or viruses can be propagated indefinitely in tissue culture; thus, they are said to be immortal. Such cells are referred to as **cell lines.**

The first cell line—the mouse fibroblast L cell—was derived in the 1940s from cultured mouse subcutaneous connective tissue by exposing the cultured cells to a chemical carcinogen, methylcholanthrene, over a 4-month period. In the 1950s, another important cell line, the HeLa cell, was derived by culturing human cervical cancer cells. Since these early studies, hundreds of cell lines have been established, each consisting of a population of genetically identical (syngeneic) cells that can be grown indefinitely in culture.

Table 22-3 lists some of the cell lines used in immunologic research and briefly describes their properties. Some of these cell lines were derived from spontaneously occurring tumors of lymphocytes, macrophages, or other accessory cells involved in the immune response. In other cases, the

| TABLE 22-3 | Cell lines commonly used in immunologic research |
|---|---|
| **Cell line** | **Description** |
| L-929 | Mouse fibroblast cell line; often used in DNA transfection studies and to assay tumor necrosis factor (TNF) |
| SP2/0 | Nonsecreting mouse myeloma; often used as a fusion partner for hybridoma secretion |
| P3X63-Ag8.653 | Nonsecreting mouse myeloma; often used as a fusion partner for hybridoma secretion |
| MPC 11 | Mouse IgG2b-secreting myeloma |
| P3X63-Ag8 | Mouse IgG1-secreting myeloma |
| MOPC 315 | Mouse IgA-secreting myeloma |
| J558 | Mouse IgA-secreting myeloma |
| 70Z/3 | Mouse pre-B-cell lymphoma; used to study early events in B-cell differentiation |
| BCL 1 | Mouse B-cell leukemia lymphoma that expresses membrane IgM and IgD and can be activated with mitogen to secrete IgM |
| CTLL-2 | Mouse T-cell line whose growth is dependent on IL-2; often used to assay IL-2 production |
| Jurkat | Human T-cell leukemia that secretes IL-2 |
| DO11.10 | Mouse T-cell hybridoma with specificity for ovalbumin |
| PU 5-1.8 | Mouse monocyte-macrophage line |
| P338 D1 | Mouse monocyte-macrophage line that secretes high levels of IL-1 |
| WEHI 265.1 | Mouse monocyte line |
| P815 | Mouse mastocytoma cells; often used as target to assess killing by cytotoxic T lymphocytes (CTLs) |
| YAC-1 | Mouse lymphoma cells; often used as target for NK cells |
| HL-60 | Human myeloid-leukemia cell line |
| COS-1 | African green monkey kidney cells transformed by SV40; often used in DNA transfection studies |

cell line was induced by transformation of normal lymphoid cells with viruses such as Abelson's murine leukemia virus (A-MLV), simian virus 40 (SV40), Epstein-Barr virus (EBV), or human T-cell leukemia virus type 1 (HTLV-1).

Lymphoid cell lines differ from primary lymphoid cell cultures in several important ways: they survive indefinitely in tissue culture, show various abnormal growth properties, and often have an abnormal number of chromosomes. Cells with more or less than the normal diploid number of chromosomes for a species are said to be aneuploid. The great advantage of cloned lymphoid cell lines is that they can be grown for extended periods in tissue culture, enabling immunologists to obtain large numbers of homogeneous cells in culture.

Until the late 1970s, immunologists did not succeed in maintaining normal T cells in tissue culture for extended periods. In 1978, a serendipitous finding led to the observation that conditioned medium containing a T-cell growth factor was required. The essential component of the conditioned medium turned out to be interleukin-2 (IL-2). By culturing normal T lymphocytes with antigen in the presence of IL-2, clones of antigen-specific T lymphocytes could be isolated. These individual clones could be propagated and studied in culture and even frozen for storage. After thawing, the clones continued to grow and express their original antigen-specific functions.

Development of cloned lymphoid cell lines has enabled immunologists to study a number of events that previously could not be examined. For example, research on the molecular events involved in activation of naive lymphocytes by antigen was hampered by the low frequency of naive B and T cells specific for a particular antigen; in a heterogeneous population of lymphocytes, the molecular changes occurring in one responding cell could not be detected against a background of $10^3$ to $10^6$ nonresponding cells. Cloned T- and B-cell lines with known antigenic specificity have provided immunologists with large homogeneous cell populations in which to study the events involved in antigen recognition. Similarly, the genetic changes corresponding to different maturational stages can be studied in cell lines that appear to be "frozen" at different stages of differentiation. Cell lines have also been useful in studying the soluble factors produced by lymphoid cells. Some cell lines secrete large quantities of various cytokines; other lines express membrane receptors for particular cytokines. These cell lines have been used by immunologists to purify various cytokines and their receptors and eventually to clone their genes.

With the advantages of lymphoid cell lines come a number of limitations. Variants arise spontaneously in the course of prolonged culture, necessitating frequent subcloning to limit the cellular heterogeneity that can develop. If variants are selected in subcloning, it is possible that two subclones derived from the same parent clone may represent different subpopulations. Moreover, any cell line derived from tumor cells or transformed cells may have unknown genetic contributions characteristic of the tumor or of the transformed state; thus, researchers must be cautious when extrapolating results obtained with cell lines to the normal situation in vivo. Nevertheless, transformed cell lines have made a major contribution to the study of the immune response, and many molecular events discovered in experiments with transformed cell lines have been shown to take place in normal lymphocytes.

## Preparing hybrid lymphoid cell lines

In somatic-cell hybridization, immunologists fuse normal B or T lymphocytes with tumor cells, obtaining hybrid cells, or heterokaryons, containing nuclei from both parent cells. Random loss of some chromosomes and subsequent cell

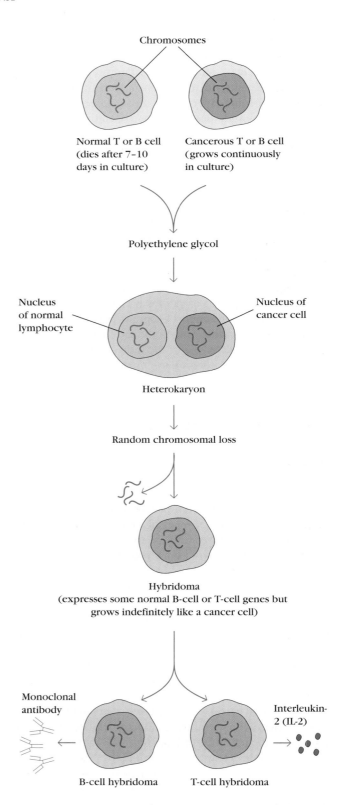

**FIGURE 22-1 Production of B-cell and T-cell hybridomas by somatic-cell hybridization.** The resulting hybridomas express some of the genes of the original normal B or T cell but also exhibit the immortal-growth properties of the tumor cell. This procedure is used to produce B-cell hybridomas that secrete monoclonal antibody and T-cell hybridomas that secrete various growth factors.

proliferation yield a clone of cells that contain a single nucleus with chromosomes from each of the fused cells; such a clone is called a **hybridoma.**

Historically, cell fusion was promoted with Sendai virus, but now it is generally done with polyethylene glycol. Normal antigen-primed B cells can be fused with cancerous plasma cells, called **myeloma cells** (Figure 22-1). The hybridoma thus formed continues to express the antibody genes of the normal B lymphocyte but is capable of unlimited growth, a characteristic of the myeloma cell. B-cell hybridomas that secrete antibody with a single antigenic specificity, called monoclonal antibody in reference to its derivation from a single clone, have revolutionized not only immunology but biomedical research and the clinical laboratory. Chapter 4 describes the production and uses of monoclonal antibodies in detail (see Figure 4-25).

T-cell hybridomas can also be obtained by fusing T lymphocytes with cancerous T-cell lymphomas, such as the mouse thymoma, BW5147. Again, the resulting hybridoma continues to express the genes of the normal T cell but acquires the immortal-growth properties of the cancerous T lymphoma cell. Immunologists have generated a number of stable hybridoma cell lines representing T-helper and T-cytotoxic lineages.

| TABLE 22-4 | Radioisotopes commonly used in immunology laboratories | | |
|---|---|---|---|
| **Isotope** | **Half-life** | **Radiation type\*** | **Autoradiography†** |
| $^{125}I$ | 60.0 days | γ | + |
| $^{131}I$ | 6.8 days | γ | + |
| $^{51}Cr$ | 27.8 days | γ | − |
| $^{32}P$ | 14.3 days | β | + |
| $^{35}S$ | 87.4 days | β | + |
| $^{14}C$ | 57.30 years | β | + |
| $^{3}H$ | 12.35 years | β | − |

\*γ (gamma) radiation may be detected in a solid scintillation counter. β (beta) radiation is detected in a liquid scintillation counter by its ability to convert energy to photons of light in a solution containing phosphorescent compounds.

†Radiation may also be detected by exposure to x-ray film. $^{35}S$ and $^{14}C$ must be placed in direct contact with film for detection. $^{3}H$ cannot be detected by normal autoradiographic techniques.

## Protein Biochemistry

The structures and functions of many important molecules of the immune system have been determined with the techniques of protein biochemistry, and many of these techniques are in constant service in experimental immunology. For example, fluorescent and radioactive labels allow immunologists to localize and visualize molecular activities, and the ability to determine such biochemical characteristics of a protein as its size, shape, and three-dimensional structure has provided essential information for understanding the functions of immunologically important molecules.

### Radiolabeling techniques allow sensitive detection of antigens or antibodies

Radioactive labels on antigen or antibody are extremely sensitive markers for detection and quantification. There are a number of ways to introduce radioactive isotopes into proteins or peptides. For example, tyrosine residues may be labeled with radioiodine by chemical or enzymatic procedures. These reactions attach an iodine atom to the phenol ring of the tyrosine molecule. One of the enzymatic iodination techniques, which uses lactoperoxidase, can label proteins on the plasma membrane of a live cell without labeling proteins in the cytoplasm, allowing the study of cell surface proteins without isolating them from other cell constituents.

A general radiolabeling of cell proteins may be carried out by growing the cells in a medium that contains one or more radiolabeled amino acids. The amino acids selected for this application are those least prone to metabolic modification during cell growth so that the radioactive label will appear in the cell protein rather than in all cell constituents. Leucine marked with $^{14}C$ or $^{3}H$, and cysteine or methionine labeled with $^{35}S$, are the most commonly used amino acids for metabolic labeling of proteins. Table 22-4 lists some properties of the radioisotopes used in immunologic research.

### Biotin labels facilitate detection of small amounts of proteins

In some instances direct labeling of proteins, especially with enzymes or other large molecules, as described in Chapter 6, may cause denaturation and loss of activity. A convenient labeling system has been developed that may be used in conjunction with the ELISA and ELISPOT assays described in Chapter 6. This labeling technique exploits the high affinity of the reaction between the vitamin biotin and avidin, a large molecule that may be labeled with radioactive isotopes, with fluorescent molecules, or with enzymes. Biotin is a small molecule (244 Da) that can be coupled to an antibody (or to any protein molecule) by a gentle chemical reaction that causes no loss of antibody activity. After the biotin-coupled antibody has reacted in the assay system, the labeled avidin is introduced and binding is measured by detecting the label on the avidin molecule (Figure 22-2). The reaction between biotin and avidin is highly specific and of such high affinity that binding between the two molecules under most assay conditions is virtually irreversible.

### Gel electrophoresis separates proteins by size and charge

When subjected to an electric field in an electrophoresis chamber, a charged molecule will move toward the oppositely charged electrode. The rate at which a charged molecule moves

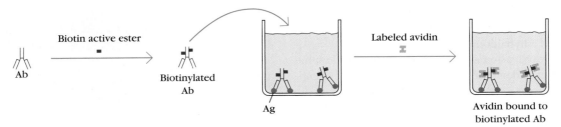

**FIGURE 22-2 Labeling of antibody with biotin.** An antibody preparation is mixed with a biotin ester, which reacts with the antibody. The biotin-labeled antibody can be used to detect antigens on a solid substrate such as the well of a microtiter plate. After unbound antibody is washed away, the bound antibody can be detected with labeled avidin. The avidin can be radioactively labeled or linked to an enzyme that catalyzes a color reaction, as in ELISA procedures (see Figure 6-10).

in a stable field (its electrophoretic mobility) depends on two factors specific to the molecule: one is the sign and magnitude of its net electric charge, and the other is its size and shape. All other factors being equal, if molecules are of equal size, the one with higher net charge will move faster in an applied electric field due to the molecular sieving properties of the solid medium. It also follows that small molecules will move faster than large ones of the same net charge. Although there are exceptions, in which the shape of a molecule may increase or decrease its frictional drag and cause atypical migration behavior, these general principles underlie all electrophoretic separations.

Most electrophoretic separations are not conducted in free solution but rather in a stable supporting medium, such as a gel. The most popular gel in research laboratories is a polymerized and cross-linked form of acrylamide. Separation on polyacrylamide gels, commonly referred to as *poly-acrylamide gel electrophoresis* (PAGE), may be used for analysis of proteins or nucleic acids (Figure 22-3a).

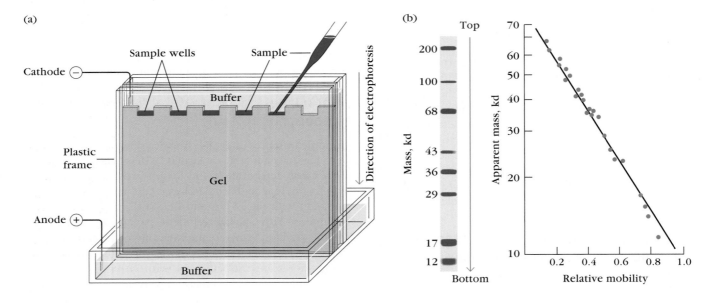

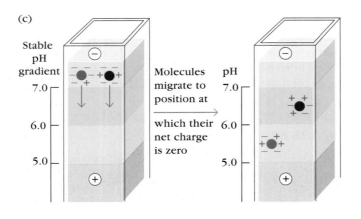

**FIGURE 22-3 Gel electrophoresis.** (a) A standard PAGE apparatus with cathode at the top and anode at the bottom. Samples are loaded on the top of the gel in sample wells, and electrophoresis is accomplished by running a current from the cathode to the anode. (b) The electrophoretic mobility, or distance traveled by a species during SDS-PAGE, is inversely proportional to the log of its molecular weight. The molecular weight of a protein is readily determined by the log of its migration distance with a standard curve that plots the migration distances of the set of standard proteins against the logs of their molecular weights. (c) Isoelectric focusing, or IEF, separates proteins solely by charge. Proteins are placed on a stable pH gradient and subjected to electrophoresis. Each protein migrates to its isoelectric point, the point at which its net charge is zero. [*Part b after K. Weber and M. Osborn, 1975, The Proteins, 3rd ed., vol. 1, Academic Press, p. 179.*]

In one common application, the electrophoresis of proteins through a polyacrylamide gel is carried out in the presence of the detergent sodium dodecyl sulfate (SDS). This method, known as SDS-PAGE, provides a relatively simple and highly effective means of separating mixtures of proteins on the basis of size. SDS is a negatively charged detergent that binds to protein in amounts proportional to the length of the protein. This binding destroys the characteristic tertiary and secondary structure of the protein, transforming it into a negatively charged rod. A protein binds so many negatively charged SDS molecules that its own intrinsic charge becomes insignificant by comparison with the net charge of the SDS molecules. Therefore, treatment of a mixture of proteins with SDS transforms them into a collection of rods whose electric charges are proportional to their molecular weights. This has two extremely useful consequences. First, it is possible to separate the components of a mixture of proteins according to molecular weight. Second, because the electrophoretic mobility, or distance traveled by a species during SDS-PAGE, is inversely proportional to the logarithm of its molecular weight, that distance is a measure of its molecular weight. To visualize the locations of the proteins, the gel is stained with a dye that reacts with protein. The migration distance of a protein in question is then compared with a plot of the distances migrated by a set of standard proteins (Figure 22-3b).

Another electrophoretic technique, isoelectric focusing (IEF), separates proteins solely on the basis of their charge. This method is based on the fact that a molecule will move in an electric field as long as it has a net positive or negative charge; molecules that bear equal numbers of positive and negative charges and therefore have a net charge of zero will not move. At most pH values, proteins (which characteristically bear a number of both positive and negative charges) have either a net negative or a net positive charge. However, for each protein there is a particular pH, called its isoelectric point (pI), at which that protein has equal numbers of positive and negative charges. Isoelectric focusing makes use of a gel containing substances, called carrier ampholytes, that arrange themselves into a continuous pH gradient when subjected to an electric field. When a mixture of proteins is applied to such a gel and subjected to electrophoresis, each protein moves until it reaches that point in the gradient where the pH of the gel is equal to its isoelectric point. It then stops moving because it has a net charge of zero. Isoelectric focusing is an extremely gentle and effective way of separating different proteins (Figure 22-3c).

A method known as two-dimensional gel electrophoresis combines the advantages of SDS-PAGE and isoelectric focusing in one of the most sensitive and discriminating ways of analyzing a mixture of proteins. In this method, one first subjects the mixture to isoelectric focusing on an IEF tube gel, which separates the molecules on the basis of their isoelectric points without regard to molecular weight. This is the first dimension. In the next step, one places the IEF gel lengthwise across the top of an SDS-polyacrylamide slab (that is, in place of the sample wells in Figure 22-3a) and

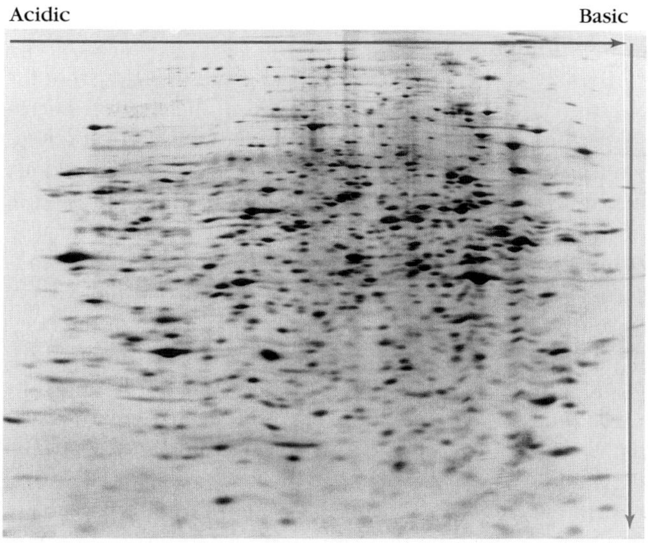

**FIGURE 22-4 Two-dimensional gel electrophoresis of $^{35}$S-methionine-labeled total cell proteins from murine thymocytes.** These proteins were first subjected to isoelectric focusing (direction of migration indicated by red arrow) and then the focused proteins were separated by SDS-PAGE (direction of migration indicated by blue arrow). The gel was exposed to x-ray film to detect the labeled proteins. *[Courtesy of B. A. Osborne.]*

runs SDS-PAGE. Preparatory to this step, all proteins have been reacted with SDS and therefore migrate out of the IEF gel and through the SDS-PAGE slab according to their molecular weights. This is the second dimension. The position of the proteins in the resulting two-dimensional gel can be visualized in a number of ways. In the least sensitive the gel is stained with a protein-binding dye (such as Coomassie blue). If the proteins have been radiolabeled, the more sensitive method of autoradiography can be used. Alternatively, silver staining is a method of great sensitivity that takes advantage of the capacity of proteins to reduce silver ions to an easily visualized deposit of metallic silver. Finally, immunoblotting—blotting of proteins onto a membrane and detection with antibody (see Figure 6-12)—can be used to locate the position of specific proteins on two-dimensional gels if an appropriate antibody is available. Figure 22-4 shows an autoradiograph of a two-dimensional gel of labeled proteins from mouse thymocytes.

## X-ray crystallography provides structural information

A great deal of information about the structure of cells, parts of cells, and even molecules has been obtained by light microscopy. The microscope uses a lens to focus radiation to form an image after it has passed through a specimen. However, a practical limitation of light microscopy is the limit of resolution. Radiation of a given wavelength cannot resolve structural features less than about half its wavelength. Since the shortest wavelength of visible light is around 400 nm, even the very best light microscopes have a theoretical limit of resolution of no less than 200 nm.

Because of the much shorter wavelength (0.004 nm) of the electron at the voltages normally used in the electron microscope, the theoretical limit of resolution of the electron microscope is about 0.002 nm. If it were possible to build an instrument that could actually approach this limit, the electron microscope could readily be used to determine the detailed atomic arrangement of biological molecules, since the constituent atoms are separated by distances of 0.1 nm to 0.2 nm. In practice, aberrations inherent in the operation of the magnetic lenses that are used to image the electron beam limit the resolution to about 0.1 nm (1 Å). This practical limit can be reached in the examination of certain specimens, particularly metals. Other considerations, however, such as specimen preparation and contrast, limit the resolution for biological materials to about 2 nm (20 Å). To determine the arrangement of a molecule's atoms, then, we must turn to x-rays, a form of electromagnetic radiation that is readily generated in wavelengths on the order of size of interatomic distances. Even though there are no microscopes with lenses that can focus x-rays into images, x-ray crystallography can reveal molecular structure at an extraordinary level of detail.

X-ray crystallography is based on the analysis of the diffraction pattern produced by the scattering of an x-ray beam as it passes though a crystal. The degree to which a particular atom scatters x-rays depends on its size. Atoms such as carbon, oxygen, or nitrogen scatter x-rays more than do hydrogen atoms, and larger atoms, such as iron, iodide, or mercury, give intense scattering. X-rays are a form of electromagnetic waves; as the scattered waves overlap, they alternately interfere with and reinforce each other. An appropriately placed detector records a pattern of spots (the diffraction pattern) whose distribution and intensities are determined by the structure of the diffracting crystal. This relationship between crystal structure and diffraction pattern is the basis of x-ray crystallographic analysis. Here is an overview of the procedures used:

*Obtaining crystals of the protein of interest.* To those who have not experienced the frustrations of crystallizing proteins, this may seem a trivial and incidental step of an otherwise highly sophisticated process. It is not. There is great variation from protein to protein in the conditions required to produce crystals that are of a size and geometrical formation appropriate for x-ray diffraction analysis. For example, myoglobin forms crystals over the course of several days at pH 7 in a 3 M solution of ammonium sulfate, but 1.5 M ammonium sulfate at pH 4 works well for a human IgG1. No set formula can be applied, and those who are consistently successful are persistent, determined, and, like great chefs, have a knack for making just the right "sauce."

*Selection and mounting.* Crystal specimens must be at least 0.1 mm in the smallest dimension and rarely exceed a few millimeters in any dimension. Once chosen, a crystal is harvested into a capillary tube along with the solution from which the crystal was grown (the "mother liquor"). This keeps the crystal from drying and maintains its solvent content, an important consideration for maintaining the internal order of the specimen. The capillary is then mounted in the diffraction apparatus.

*Generating and recording a diffraction pattern.* The precisely positioned crystal is then irradiated with x-rays of a known wavelength produced by accelerating electrons against the copper target of an x-ray tube. When the x-ray beam strikes the crystal, some of it goes straight through and some is scattered; sensitive detectors record the position and intensity of the scattered beam as a pattern of spots (Figure 22-5a, b).

*Interpreting the diffraction pattern.* The core of diffraction analysis is the mathematical deduction of the detailed structure that would produce the diffraction pattern observed. One must calculate to what extent the waves scattered by each atom have combined to reinforce or cancel each other to produce the net intensity observed for each spot in the array. A difficulty arises in the interpretation of complex diffraction patterns because the waves differ with respect to phase, the timing of the period between maxima and minima. Since the pattern observed is the net result of the interaction of many waves, information about phase is critical to calculating the distribution of electron densities that is responsible. The solution of this "phase problem" looms as a major obstacle to the derivation of a high-resolution structure of any complex molecule.

The problem is solved by derivatizing the protein—modifying it by adding heavy atoms, such as mercury, and then obtaining crystals that have the same geometry as (are isomorphous with) those of the underivatized protein. The diffraction pattern of the isomorphous crystal is obtained and compared with that of the native protein. Usually, if armed with a knowledge of the diffraction patterns of two or more isomorphous heavy-atom derivatives, a researcher can calculate the phases for the native protein by reference to the characteristic diffraction patterns generated by heavy-atom landmarks. The phases established, it is possible to move on to a calculation of the distribution of electron density. This is accomplished by Fourier synthesis, a mathematical treatment particularly suited to the analysis of periodic phenomena such as those involving waves. In this case, it is used to compute the distribution of electron density along the x, y, and z axes within a unit cell of the crystal. The deduced electron density can then be visualized on a computer (Figure 22-5c).

*Deriving the structure.* The resolution of a model depends on a number of factors. First of all, the ultimate resolution possible is set by the quality of the crystal and the internal order of the crystal. Even the highest-quality crystals have a degree of internal disorder that establishes a limit of resolution of about 2 Å. Second, a factor of paramount importance is the number of intensities fed into the Fourier synthesis. A relatively small number of spots may produce a low-resolution (6 Å) image that traces the course of the polypeptide chain but provides little additional structural

(a)

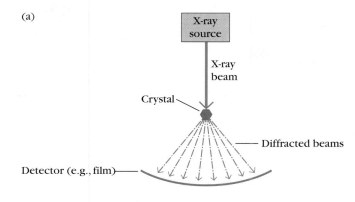

(b)

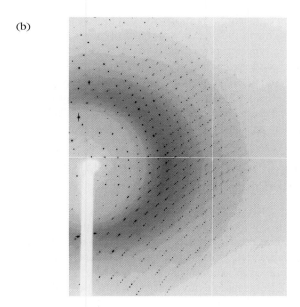

(c)

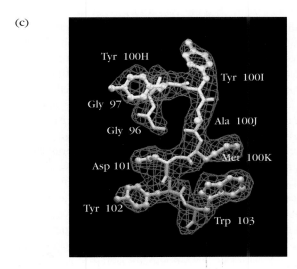

**FIGURE 22-5  X-ray crystallography.** (a) Schematic diagram of an x-ray crystallographic experiment in which an x-ray beam bombards the crystal and diffracted rays are detected. (b) Section of x-ray diffraction pattern of a crystal of murine IgG2a. (c) Section from the electron-density map of murine IgG2a. *[Part a from L. Stryer, 1995, Bio-chemistry, 4th ed.; parts b and c courtesy of A. McPherson.]*

information. On the other hand, the processing of data provided by tens of thousands of spots allows the tracing of very detailed electron-density maps. Provided one knows the amino acid sequence of the protein, such maps can guide the construction of high-resolution, three-dimensional models. Amino acid sequence data is necessary because it can be difficult, and in some cases impossible, to unambiguously distinguish among some amino acid side chains on even the most detailed electron-density maps.

Since 1960, when the first detailed structures of proteins were deduced, the structures of many thousands of proteins have been solved. These range from small and (relatively) simple proteins such as lysozyme, consisting of a single polypeptide chain, to poliovirus, an 8,500,000 dalton, stunningly complex nucleoprotein made up of RNA encased by multiple copies of four different polypeptide subunits. Of particular importance to immunologists are the large number of immunologically relevant molecules for which detailed crystal structures are now available. These include many immunoglobulins, most of the major and minor proteins involved in the MHC and T-cell-receptor complexes, and many other important immunological macromolecules, with new structures and structural variants appearing every month.

## Recombinant DNA Technology

The various techniques called recombinant DNA technology have had an impact on every area of immunologic research. Genes can be cloned, DNA can be sequenced, and recombinant proteins can be produced, supplying immunologists with defined components for study of the structure and function of the immune system at the molecular level. This section briefly describes some of the recombinant DNA techniques commonly employed in immunologic research; examples of their use have been presented throughout the book.

### Restriction enzymes cleave DNA at precise sequences

A variety of bacteria produce enzymes called *restriction endonucleases* that degrade foreign DNA (e.g., bacteriophage DNA) but spare the bacterial-cell DNA, which contains methylated residues. The discovery of these bacterial enzymes in the 1970s opened the way to a major technological advance in the field of molecular biology. Before the discovery of restriction endonucleases, double-stranded DNA (dsDNA) could be cut only with DNases. These enzymes do not recognize defined sites and therefore randomly cleave DNA into a variable series of small fragments, which are impossible to sort by size or sequence. In contrast, restriction endonucleases recognize and cleave DNA at specific sites, called restriction sites, which are short double-stranded segments of specific sequence containing four to eight nucleotides (Table 22-5).

A restriction endonuclease cuts both DNA strands at a specific point within its restriction site. Some enzymes, such as *Hpa*I, cut on the central axis and thus generate blunt-ended

| TABLE 22-5 | Some restriction enzymes and their recognition sequences | |
|---|---|---|
| Microorganism source | Abbreviation | Sequence*<br>5′ → 3′<br>3′ → 5′ |
| *Bacillus amyloliquefaciens* H | *Bam*HI | G G A T C C<br>C C T A G G |
| *Escherichia coli* RY 13 | *Eco*RI | G A A T T C<br>C T T A A G |
| *Haemophilus aegyptius* | *Hae*III | G G C C<br>C C G G |
| *Haemophilus influenzae* Rd | *Hind*III | A A G C T T<br>T T C G A A |
| *Haemophilus parainfluenzae* | *Hpa*I | G T T A A C<br>C A A T T G |
| *Nocardia otitidis-caviarum* | *Not*I | G C G G C C G C<br>C G C C G G C G |
| *Providencia stuartii* 164 | *Pst*I | G T G C A G<br>G A C G T C |
| *Staphylococcus aureus* 3A | *Sau*3A | G A T C<br>C T A G |

*Blue lines indicate locations of single-strand cuts within the restriction site. Enzymes that make off-center cuts produce fragments with short single-stranded extensions at their ends.

SOURCE: New England Biolabs, http://www.neb.com.

fragments. Other enzymes, such as *Eco*RI, cut the DNA at staggered points in the recognition site. In this case, the end of each cleaved fragment is a short segment of single-stranded DNA, called a *sticky end*. When two different DNA molecules are cut with the same restriction enzyme that makes staggered cuts, the sticky ends of the fragments are complementary; under appropriate conditions, fragments from the two molecules can be joined by base pairing to generate a recombinant DNA molecule. Several hundred different restriction endonucleases have been isolated, and many are available commercially, allowing researchers to purchase enzymes that cut DNA at defined restriction sites.

## DNA sequences are cloned into vectors

The development of DNA-cloning technology in the 1970s provided a means of amplifying a given DNA fragment to such an extent that unlimited amounts of identical DNA fragments (cloned DNA) could be produced.

## Cloning vectors are useful to replicate defined sequences of DNA

In DNA cloning, a given DNA fragment is inserted into an autonomously replicating DNA molecule, called a cloning

vector, so that the inserted DNA is replicated with the vector. A number of different viruses have been used as vectors, including bacterial viruses, insect viruses, and mammalian retroviruses. A common bacterial virus used as a vector is bacteriophage λ. If a gene is inserted into bacteriophage λ and the resulting recombinant λ phage is used to infect *E. coli,* the inserted gene will be expressed by the bacteria.

Plasmids are another common type of cloning vector. A plasmid is a small, circular, extrachromosomal DNA molecule that can replicate independently in a host cell; the most common host used in DNA cloning is *E. coli.* In general, the DNA to be cloned is inserted into a plasmid that contains an antibiotic-resistance gene. After the recombinant plasmid is incubated with bacterial cells, the cells containing the recombinant plasmid can be selected by their ability to grow in the presence of the antibiotic.

Another type of vector that is often used for cloning is called a *cosmid vector.* This type of vector is a plasmid that has been genetically engineered to contain the COS sites of λ-phage DNA, a drug-resistance gene, and a replication origin. COS sites are DNA sequences that allow any DNA up to 50 kb in length to be packaged into the λ-phage head.

## Cloning of cDNA and genomic DNA allows the isolation of defined sequences

Messenger RNA (mRNA) isolated from cells can be transcribed into complementary DNA (cDNA) with the enzyme reverse transcriptase. The cDNA can be cloned by inserting it into a plasmid vector carrying a gene that confers resistance to an antibiotic, such as ampicillin. The resulting recombinant plasmid DNA is subsequently transferred into specially treated *E. coli* cells by one of several techniques; the transfer process is called **transfection.** If the foreign DNA is incorporated into the host cell and expressed, the cell is said to be **transformed.** When the cells are cultured on agar plates containing ampicillin, only transformed cells containing the ampicillin-resistance gene will survive and grow (Figure 22-6). A collection of DNA sequences within plasmid vectors representing all the mRNA sequences derived from a cell or tissue is called a *cDNA library.* A cDNA library differs from a genomic library by virtue of the fact that it contains only the sequences derived from mRNA, the sequences that represent expressed genes.

Genomic cloning, cloning of the entire genome of an animal, requires specialized vectors. *E. coli* plasmid vectors are impractical for cloning of all the genomic DNA fragments that constitute a large genome because of the low efficiency of *E. coli* transformation and the small number of transformed colonies that can be detected on a typical petri dish. Instead, cloning vectors derived from bacteriophage λ are used to clone genomic DNA fragments obtained by cleaving chromosomal DNA with restriction enzymes (Figure 22-7). Bacteriophage λ DNA is 48.5 kb long and contains a central section of about 15 kb that is not necessary for λ replication in *E. coli* and can therefore be replaced with foreign genomic DNA. As long as the recombinant DNA does not

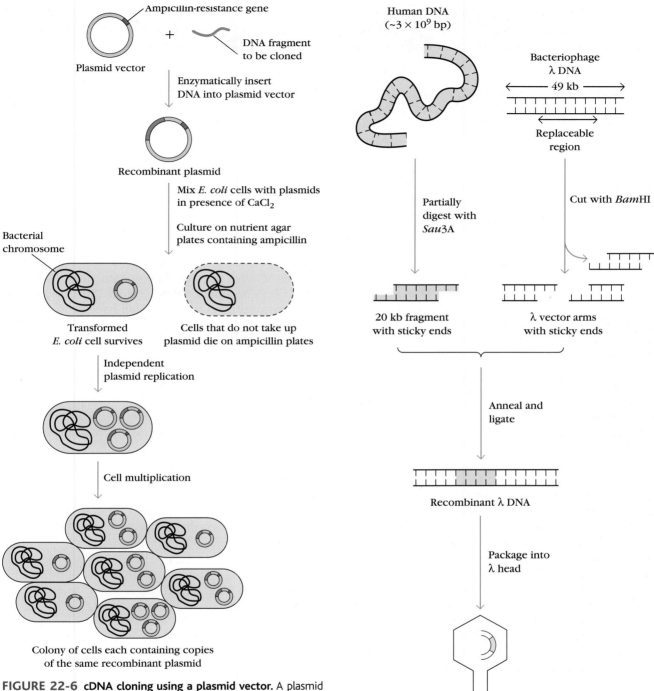

**FIGURE 22-6 cDNA cloning using a plasmid vector.** A plasmid containing a replication origin and an ampicillin-resistance gene is cut with a restriction endonuclease that produces blunt ends. After addition of a poly-C tail to the 3' ends of the cDNA and of a complementary poly-G tail to the 3' ends of the cut plasmid, the two DNAs are mixed, annealed, and joined by DNA ligase, forming the recombinant plasmid. Uptake of the recombinant plasmid into *E. coli* cells is stimulated by high concentrations of $CaCl_2$. Transformation occurs with a low frequency, but the transformed cells can be selected in the presence of ampicillin. *[Adapted from H. Lodish et al., 1995,* Molecular Cell Biology, *3rd ed., Scientific American Books.]*

**FIGURE 22-7 Genomic DNA cloning using bacteriophage λ as the vector.** Genomic DNA is partly digested with *Sau*3A, producing fragments with sticky ends. The central 15-kb region of the λ-phage DNA is cut out with *Bam*HI and discarded. These two restriction enzymes produce complementary sticky ends, so the genomic and DNA fragments can be annealed and ligated. After the resulting recombinant DNA is packaged into a λ-phage head, it can be propagated in *E. coli*.

greatly exceed the length of the original λ-phage DNA, it can be packaged into the λ-phage head and propagated in *E. coli*. This means that somewhat more than $2.0 \times 10^4$ base pairs can be cloned in one particle of λ phage. A collection of λ clones that includes all the DNA sequences of a given species is called a *genomic library*. About 1 million different recombinant λ-phage particles are needed to form a genomic DNA library representing an entire mammalian genome, which contains about $3 \times 10^9$ base pairs.

Often the 20 to 25 kb stretch of DNA that can be cloned in bacteriophage λ is not long enough to include the regulatory sequences that lie outside the 5′ and 3′ ends of the direct coding sequences of a gene. As noted already, larger genomic DNA fragments—between 30 and 50 kb in length—can be cloned in a cosmid vector. A recombinant cosmid vector, although not a fully functional bacteriophage, can infect *E. coli* and replicate as a plasmid, generating a cosmid library. Recently, a larger *E. coli* virus, called bacteriophage P1, has been used to package DNA fragments up to 100 kb long. Even larger DNA fragments, greater than 2000 kbp in length, can be cloned in yeast artificial chromosomes (YACs), which are linear DNA segments that can replicate in yeast cells (Table 22-6). The BAC, or bacterial artificial chromosome, is another useful vector. BACs can accept pieces of DNA up to 100 to 300 kb in length. Although YACs accept larger inserts of foreign DNA, BACs are much easier to propagate and are the vector of choice for many large-scale cloning efforts.

## DNA clones are selected by hybridization

Once a cDNA or genomic DNA library has been prepared, it can be screened to identify a particular DNA fragment by a technique called *in situ hybridization*. The cloned bacterial

| TABLE 22-6 | Vectors and maximum length of DNA that they can carry |
|---|---|
| Vector type | Maximum length of cloned DNA (kb) |
| Plasmid | 15 |
| Bacteriophage λ | 25 |
| Cosmid | 45 |
| Bacteriophage P1 | 100 |
| Bacterial artificial chromosome (BAC) | 100–300 |
| Yeast artificial chromosome (YAC) | >2000 |

colonies, yeast colonies, or phage plaques containing the recombinant DNA are transferred onto nitrocellulose or nylon filters by replica plating (Figure 22-8). The filter is then treated with NaOH, which both lyses the bacteria and denatures the DNA, allowing single-stranded DNA (ssDNA) to bind to the filter. The filter with bound DNA is then incubated with a radioactive probe specific for the gene of interest. The probe will hybridize with DNA in the colonies or plaques on the filter that contain the sought-after gene, and they can be identified by autoradiography. The position of the positive colonies or plaques on the filter shows where the corresponding clones can be found on the original agar plate.

Various radioactive probes can be used to screen a library. In some cases, radiolabeled mRNA or cDNA serves as the probe. If the protein encoded by the gene of interest has been purified and partly sequenced, it is possible to work backward from the amino acid sequence to deter-

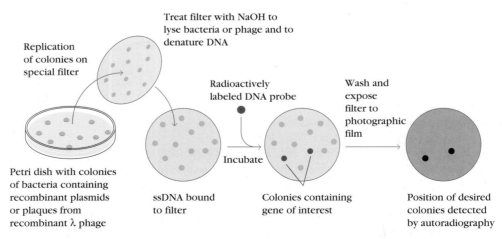

**FIGURE 22-8 Selection of specific clones from a cDNA or genomic DNA library by in situ hybridization.** A nitrocellulose or nylon filter is placed against the plate to pick up the bacterial colonies or phage plaques containing the cloned genes. After the filter is placed in a NaOH solution and heated, the denatured ssDNA becomes fixed to the filter. A radioactive probe specific for the gene of interest is incubated with the filter. The position of the colonies or plaques containing the desired gene is revealed by autoradiography.

mine the probable nucleotide sequence of the corresponding gene. A known sequence of five or six amino acid residues is all that is needed to synthesize radiolabeled oligonucleotide probes with which to screen a cDNA or genomic library for a particular gene. To cope with the degeneracy of the genetic code, peptide segments containing amino acids encoded by a limited number of codons are usually chosen. Oligonucleotides representing all possible codons for the peptide are then synthesized and used as probes to screen the DNA library.

## Southern blotting detects DNA of a given sequence

DNA fragments generated by restriction-endonuclease cleavage can be separated on the basis of length by agarose gel electrophoresis. The shorter a fragment is, the faster it moves in the gel. An elegant technique developed by E. M. Southern can be used to identify any band containing fragments with a given gene sequence (Figure 22-9). In this technique, called **Southern blotting,** DNA is cut with restriction enzymes and the fragments are separated according to size by electrophoresis on an agarose gel. Then the gel is soaked in NaOH to denature the dsDNA, and the resulting ssDNA fragments are transferred onto a nitrocellulose or nylon filter by capillary action. After transfer, the filter is incubated with an appropriate radiolabeled probe specific for the gene of interest. The probe hybridizes with the ssDNA fragment containing the gene of interest, and the position of the band containing these hybridized fragments is determined by autoradiography. Southern-blot analysis played a critical role in unraveling the mechanism by which diversity of antibodies and T-cell receptors is generated (see Figures 5-2 and 9-2).

## Northern blotting detects mRNA

**Northern blotting** (named for its similarity to Southern blotting) is used to detect the presence of specific mRNA molecules. In this procedure the mRNA is first denatured to ensure that it is in an unfolded, linear form. The mRNA molecules are then separated according to size by electrophoresis and transferred to a nitrocellulose filter, to which the mRNAs will adhere. The filter is then incubated with a labeled DNA probe and subjected to autoradiography. Northern-blot analysis is often used to determine how much of a specific mRNA is expressed in cells under different conditions. Increased levels of mRNA will bind proportionally more of the labeled DNA probe.

## The polymerase chain reaction amplifies small quantities of DNA

The polymerase chain reaction (PCR) is a powerful technique for amplifying specific DNA sequences even when they are present at extremely low levels in a complex mixture (Figure 22-10). The procedure requires that the DNA sequences that flank the target DNA sequence be known so that short oligonucleotide primers can be synthesized. The DNA mixture is denatured into single strands by a brief heat treatment. The DNA is then cooled in the presence of an excess of the oligonucleotide primers, which hybridize with the complementary ssDNA. A temperature-resistant DNA polymerase is then added, together with the four deoxyribonucleotide triphosphates, and each strand is copied. The newly synthesized DNA duplex is separated by heating and the cycle is repeated. In each cycle there is a doubling of the target DNA sequence; in only 25 cycles the desired DNA sequence can be amplified about a millionfold.

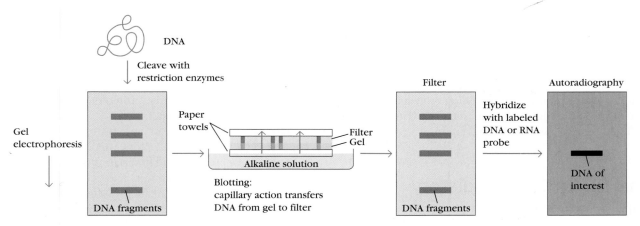

**FIGURE 22-9  The Southern-blot technique for detecting specific sequences in DNA fragments.** The DNA fragments produced by restriction-enzyme cleavage are separated by size by agarose gel electrophoresis. The agarose gel is overlaid with a nitrocellulose or nylon filter and a thick stack of paper towels. The gel is then placed in an alkaline salt solution, which denatures the DNA. As the paper towels soak up the moisture, the solution is drawn through the gel into the filter, transferring each ssDNA band to the filter. This process is called blotting. After heating, the filter is incubated with a radiolabeled probe specific for the sequence of interest; DNA fragments that hybridize with the probe are detected by autoradiography. *[Adapted from J. Darnell et al., 1990, Molecular Cell Biology, 2nd ed., Scientific American Books.]*

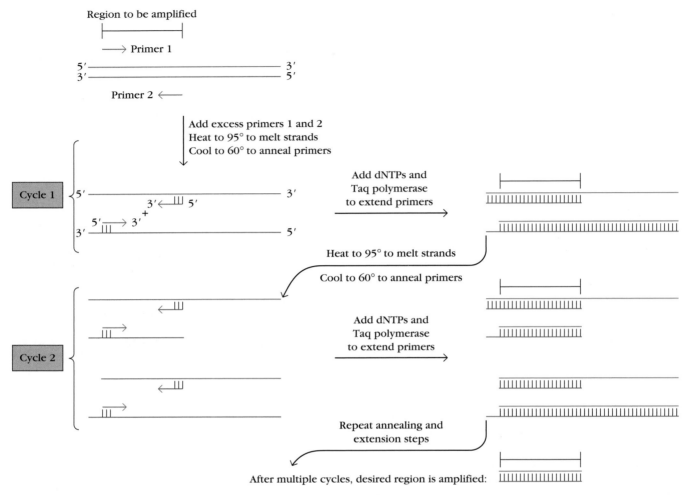

**FIGURE 22-10 The polymerase chain reaction (PCR).** DNA is denatured into single strands by a brief heat treatment and is then cooled in the presence of an excess of oligonucleotide primers complementary to the DNA sequences flanking the desired DNA segment. A heat-resistant DNA polymerase is used to copy the DNA from the 39 ends of the primers. Because all of the reaction components are heat stable, the heating and cooling cycle can be repeated many times, resulting in alternate DNA melting and synthesis and rapid amplification of a given sequence. *[Adapted from H. Lodish et al., 1995, Molecular Cell Biology, 3rd ed., Scientific American Books.]*

The DNA amplified by the PCR can be further characterized by Southern blotting, restriction-enzyme mapping, and direct DNA sequencing. The PCR technique has enabled immunologists to amplify genes encoding proteins that are important in the immune response, such as MHC molecules, the T-cell receptor, and immunoglobulins.

## Analysis of DNA Regulatory Sequences

The transcriptional activity of genes is regulated by promoter and enhancer sequences. These sequences are *cis-acting*, meaning that they regulate only genes on the same DNA molecule. The promoter sequence lies upstream from the gene it regulates and includes a TATA box, where the general transcription machinery, including RNA polymerase II, binds and begins transcription. The enhancer sequence con-

fers a high rate of transcription on the promoter. Unlike the promoter, which always lies upstream from the gene it controls, the enhancer element can be located anywhere with respect to the gene (5′ of the promoter, 3′ of the gene, or even in an intron of the gene).

The activity of enhancer and promoter sequences is controlled by transcription factors, which are DNA-binding proteins. These proteins bind to specific nucleotide sequences within promoters and enhancers and act either to enhance or suppress their activity. Enhancer and promoter sequences and their respective DNA-binding proteins have been identified by a variety of techniques, including DNA footprinting, gel-shift analysis, and the CAT assay.

### DNA footprinting identifies the sites where proteins bind DNA

The binding sites for DNA-binding proteins on enhancers and promoters can be identified by a technique called DNA

(a) DNA footprinting

(b) Gel-shift analysis

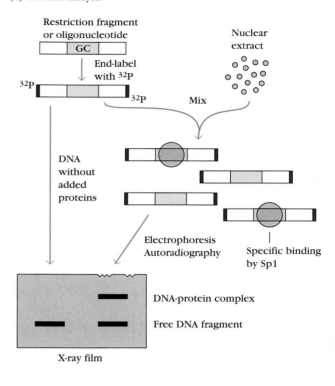

**FIGURE 22-11 Identification of DNA sequences that bind protein by DNA-footprinting and gel-shift analysis.** (a) In the footprinting technique, labeled DNA fragments containing a putative promoter or enhancer sequence are incubated in the presence and absence of a DNA-binding protein (e.g., Sp1 protein, which binds to a "GC box," a GC-rich region of DNA). After the samples are treated with DNase and the strands separated, the resulting fragments are electrophoresed; the gel then is subjected to autoradiography. A blank region (footprint) in the gel pattern indicates that protein has bound to the DNA. (b) In gel-shift analysis, a labeled DNA fragment is incubated with a cellular extract containing transcription factors. The electrophoretic mobility of the DNA-protein complex is slower than that of free DNA fragments. *[Adapted from J. D. Watson et al., 1992, Recombinant DNA, 2nd ed., W. H. Freeman and Company.]*

footprinting (Figure 22-11a). In this technique, a cloned DNA fragment containing a putative enhancer or promoter sequence is first radiolabeled at the 5′ end with $^{32}$P. The labeled DNA is then divided into two fractions: one fraction is incubated with a nuclear extract containing a DNA-binding protein; the other DNA fraction is not incubated with the extract. Both DNA samples are then digested with a nuclease or a chemical that makes random cuts in the phosphodiester bonds of the DNA, and the strands are separated. The resulting DNA fragments are run on a gel to separate fragments of different sizes. In the absence of DNA-binding proteins, a complete ladder of bands is obtained on the electrophoretic gel. When a protein that binds to a site on the DNA fragment is present, it covers some of the nucleotides, protecting that stretch of the DNA from digestion. The elec-

trophoretic pattern of such protected DNA will contain blank regions (or footprints). Each footprint represents the site within an enhancer or promoter that binds a particular DNA-binding protein.

## Gel-shift analysis identifies DNA-protein complexes

When a protein binds to a DNA fragment, forming a DNA-protein complex, the electrophoretic mobility of the DNA fragment in a gel is reduced, producing a shift in the position of the band containing that fragment. This phenomenon is the basis of gel-shift analysis. In this technique, radioactively labeled cloned DNA containing an enhancer or a promoter sequence is incubated with a nuclear extract containing a

DNA-binding protein (Figure 22-11b). The DNA-protein complex is then electrophoresed and its electrophoretic mobility is compared with that of the cloned DNA alone. A shift in the mobility indicates that a protein is bound to the DNA, retarding its migration on the electrophoretic gel.

## Luciferase assays measure transcriptional activity

One way to assess promoter activity is to engineer and clone a DNA sequence containing a *reporter gene* attached to the promoter that is being assessed. When this sequence, or construct, is introduced into eukaryotic cells, transcription will be initiated from the promoter if it is active, and the reporter gene will be transcribed and its protein product synthesized. Measuring the amount of the protein product is thus a way to determine the activity of the promoter. By introducing mutations into promoter sequences and then assaying for promoter activity with the corresponding reporter gene, conserved sequence motifs have been identified within promoters.

Most reporter genes are chosen because they encode proteins that can be easily measured, such as the firefly enzyme luciferase, an enzyme that catalyzes a reaction that emits light—the reaction that causes fireflies to glow in the dark. The more active the promoter, the more luciferase will be produced within the transfected cell. Luciferase activity is quantified by adding its substrate, luciferin, and measuring the emitted light with a luminometer. Another popular reporter gene encodes green fluorescent protein (GFP), discussed at the end of this chapter.

## Gene Transfer into Mammalian Cells

A variety of genes involved in the immune response have been isolated and cloned by use of recombinant DNA techniques. The expression and regulation of these genes has been studied by introducing them into cultured mammalian cells and, more recently, into the **germ lines** of animals.

## Cloned genes transferred into cultured cells allow in vitro analysis of gene function

Diverse techniques have been developed for transfecting genes into cells. A common technique involves the use of a retrovirus in which a viral structural gene has been replaced with the cloned gene to be transfected. The altered retrovirus is then used as a vector for introducing the cloned gene into cultured cells. Because of the properties of retroviruses, the recombinant DNA integrates into the cellular genome with a high frequency. In an alternative method, the cloned gene of interest is complexed with calcium phosphate. The calcium-phosphate–DNA complex is slowly precipitated onto the cells and the DNA is taken up by a small percentage of them. In another transfection method, called

electroporation, an electric current creates pores in cell membranes through which the cloned DNA is taken up. In both of these latter methods, the transfected DNA integrates, apparently at random sites, into the DNA of a small percentage of treated cells.

Generally, the cloned DNA being transfected is engineered to contain a selectable marker gene, such as one that confers resistance to neomycin. After transfection, the cells are cultured in the presence of neomycin. Because only the transfected cells are able to grow, the small number of transfected cells in the total cell population can be identified and selected.

Transfection of cloned genes into cells has proved to be highly effective in immunologic research. By transfecting genes involved with the immune response into cells that lack those genes, the product of a specific gene can be studied apart from interacting proteins encoded by other genes. For example, transfection of MHC genes, under the control of appropriate promoters, into a mouse fibroblast cell line (L929, or simply L cells) has enabled immunologists to study the role of MHC molecules in antigen presentation to T cells (Figure 22-12). Transfection of the gene that encodes the

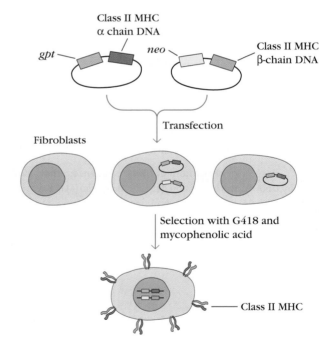

**FIGURE 22-12 Transfection of the genes encoding the class II MHC α chain and β chain into mouse fibroblast L cells, which do not normally produce these proteins.** Two constructs containing one of the MHC genes and a selectable gene were engineered: the α-chain gene with the guanine phosphoribosyl transferase gene *(gpt)*, which confers resistance to the drug G418, and the β-chain gene with a neomycin gene *(neo)*, which confers resistance to mycophenolic acid. After transfection, the cells are placed in medium containing both G418 and mycophenolic acid. Only those fibroblasts containing both the *neo* and *gpt* genes (and consequently the genes encoding the class II MHC α and β chains) will survive this selection. These fibroblasts will express both class II MHC chains on their membranes.

T-cell receptor has provided information about the antigen-MHC specificity of the T-cell receptor.

**Retroviruses,** which can infect virtually any type of mammalian cell, are a common vector used to introduce DNA into mammalian cells, particularly lymphoid cells. Retroviruses are RNA viruses that contain reverse transcriptase, an enzyme that catalyzes conversion of the viral RNA genome into DNA. The viral DNA then integrates into the host chromosomal DNA, where it is retained as a provirus, replicating along with the host chromosomal DNA at each cell division. When a retrovirus is used as a vector in research, most of the retroviral genes are removed so that the vector cannot produce viral particles; the retroviral genes that are left include a strong promoter region, located at the 5′ end of the viral genome, in a sequence called the *long terminal repeat (LTR).* If a gene is inserted into such a retroviral vector and the vector is then used to infect mammalian cells, expression of the gene will be under the control of the retroviral promoter region. Because lymphocytes are particularly resistant to many methods of transfection, the use of retroviral vectors to introduce genes into lymphoid cells has been extremely useful to immunologists.

## Cloned genes transferred into mouse embryos allow in vivo analysis of gene function

Development of techniques to introduce cloned foreign genes (called **transgenes**) into mouse embryos has permitted immunologists to study the effects of immune system genes in vivo. If the introduced gene integrates stably into the germ-line cells, it will be transmitted to the offspring.

The first step in producing transgenic mice is injection of foreign cloned DNA into a fertilized egg. In this technically demanding process, fertilized mouse eggs are held under suction at the end of a pipette and the transgene is microinjected into one of the pronuclei with a fine needle. The transgene integrates into the chromosomal DNA of the pronucleus and is passed on to the daughter cells of eggs that survive the process. The eggs then are implanted in the oviduct of "pseudopregnant" females, and transgenic pups are born after 19 or 20 days of gestation (Figure 22-13). In general, the efficiency of this procedure is low, with only one or two transgenic mice produced for every 100 fertilized egg collected.

With transgenic mice, immunologists have been able to study the expression of a given gene in a living animal. Although all the cells in a transgenic animal contain the transgene, differences in the expression of the transgene in different tissues has shed light on mechanisms of tissue-specific gene expression. By constructing a transgene with a particular promoter, researchers can control the expression of the transgene. For example, the metallothionein promoter is activated by zinc. Transgenic mice carrying a transgene linked to a metallothionein promoter express the transgene only if

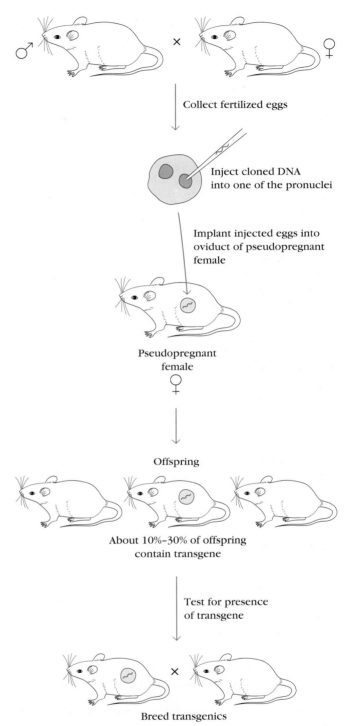

**FIGURE 22-13 General procedure for producing transgenic mice.** Fertilized eggs are collected from a pregnant female mouse. Cloned DNA (referred to as the transgene) is microinjected into one of the pronuclei of a fertilized egg. The eggs are then implanted into the oviduct of pseudopregnant foster mothers (obtained by mating normal females with a sterile male). The transgene will be incorporated into the chromosomal DNA of about 10% to 30% of the offspring and will be expressed in all of their somatic cells. If a tissue-specific promoter is linked to a transgene, then tissue-specific expression of the transgene will result.

zinc is added to their water supply. Other promoters are functional only in certain tissues; the insulin promoter, for instance, promotes transcription only in pancreatic cells. Transgenic mice carrying a transgene linked to the insulin promoter, therefore, will express the transgene in the pancreas but not in other tissues.

Because a transgene is integrated into the chromosomal DNA within the one-cell mouse embryo, it will be integrated into both somatic cells and germ-line cells. The resulting transgenic mice thus can transmit the transgene to their offspring as a Mendelian trait. In this way, it has been possible to produce lines of transgenic mice in which every member of a line contains the same transgene. A variety of such transgenic lines are currently available and are widely used in immunologic research. Included among these are lines carrying transgenes that encode immunoglobulin, T-cell receptors, class I and class II MHC molecules, various foreign antigens, and a number of cytokines. Several lines carrying oncogenes as transgenes have also been produced.

## In knockout mice, targeted genes are disrupted

One limitation of transgenic mice is that the transgene is integrated randomly within the genome. This means that some transgenes insert in regions of DNA that are not transcriptionally active, and hence the gene is not expressed. To circumvent this limitation, researchers have developed a technique in which a desired gene is targeted to specific sites within the germ line of a mouse. The primary use of this technique has been to replace a normal gene with a mutant allele or a disrupted form of the gene, thus knocking out the gene's function. Transgenic mice that carry such a disrupted gene, called *knockout mice,* have been extremely helpful to immunologists trying to understand how the removal of a particular gene product affects the immune system. Table 22-7 compares transgenic mice and knockout mice.

Production of gene-targeted knockout mice involves the following steps:

- Isolation and culture of **embryonic stem (ES) cells** from the inner cell mass of a mouse blastocyst

- Introduction of a mutant or disrupted gene into the cultured ES cells and selection of homologous recombinant cells in which the gene of interest has been knocked out (i.e., replaced by a nonfunctional form of the gene)

- Injection of homologous recombinant ES cells into a recipient mouse blastocyst and surgical implantation of the blastocyst into a pseudopregnant mouse

- Mating of chimeric offspring heterozygous for the disrupted gene to produce homozygous knockout mice

The ES cells used in this procedure are obtained by culturing the inner cell mass of a mouse blastocyst on a feeder layer of fibroblasts or in the presence of leukemia-inhibitory factor. Under these conditions, the stem cells grow but remain pluripotent and capable of differentiating later in a variety of directions, generating distinct cellular lineages (e.g., germ cells, myocardium, blood vessels, myoblasts, or nerve cells). One of the advantages of ES cells is the ease with which they can be genetically manipulated. Cloned DNA containing a desired gene can be introduced into ES cells in culture by various transfection techniques. The introduced DNA will be inserted by recombination into the chromosomal DNA of a small number of ES cells.

The insertion constructs introduced into ES cells contain three genes: the target gene of interest and two selection genes, such as *neo*$^R$, which confers neomycin resistance, and the thymidine kinase gene from herpes simplex virus ($tk^{HSV}$), which confers sensitivity to gancyclovir, a cytotoxic nucleotide analogue (Figure 22-14a). The construct often is engineered with the target gene sequence disrupted by the *neo*$^R$ gene and with the $tk^{HSV}$ gene at one end, beyond the sequence of the target gene. Most constructs will insert at random by nonhomologous recombination rather than by gene-targeted insertion through homologous recombination. As illustrated in Figure 22-14b, a two-step selection scheme is used to obtain those ES cells that have undergone homologous recombination, whereby the disrupted gene replaces the target gene.

The ES cells obtained by this procedure are heterozygous for the knockout mutation in the target gene. These cells are clonally expanded in cell culture and then injected into a mouse blastocyst, which is then implanted into a pseudopregnant female. The transgenic offspring that develop are chimeric, composed of cells derived from the genetically altered ES cells and cells derived from normal cells of the host blastocyst. When the germ-line cells are derived from the genetically altered ES cells, the genetic alteration can be passed on to the offspring. If the recombinant ES cells are

| TABLE 22-7 | Comparison of transgenic and knockout mice | |
|---|---|---|
| Characteristic | Transgenic Mice | Knockout Mice |
| Cells receiving DNA | Zygote | Embryonic stem (ES) cells |
| Means of delivery | Microinjection into zygote and implantation into foster mother | Transfer of ES cells to blastocyst and implantation into foster mother |
| Outcome | Gain of a gene | Loss of gene |

(a) Formation of recombinant ES cells

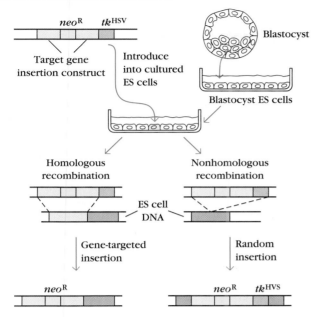

(b) Selection of ES cell carrying knockout gene

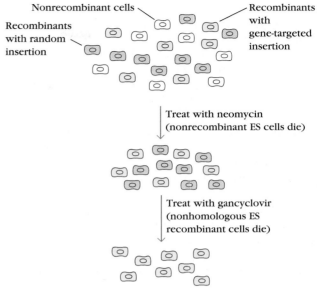

Homologous ES recombinants with targeted disruption in gene $X$ survive

**FIGURE 22-14 Formation and selection of mouse recombinant ES cells in which a particular target gene is disrupted.** (a) In the engineered insertion construct, the target gene is disrupted with the $neo^R$ gene, and the thymidine kinase $tk^{HSV}$ gene is located outside the target gene. The construct is transfected into cultured ES cells. If homologous recombination occurs, only the target gene and the $neo^R$ gene will be inserted into the chromosomal DNA of the ES cells. If nonhomologous recombination occurs, all three genes will be inserted. Recombination occurs in only about 1% of the cells, with nonhomologous recombination much more frequent than homologous recombination. (b) Selection with the neomycin-like drug G418 will kill any nonrecombinant ES cells because they lack the $neo^R$ gene. Selection with gancyclovir will kill the nonhomologous recombinants carrying the $tk^{HSV}$ gene, which confers sensitivity to gancyclovir. Only the homologous ES recombinants will survive this selection scheme. [*Adapted from H. Lodish et al., 1995,* Molecular Cell Biology, *3rd ed., Scientific American Books.*]

homozygous for black coat color (or another visible marker) and they are injected into a blastocyst homozygous for white coat color, then the chimeric progeny that carry the heterozygous knockout mutation in their germ line can be easily identified (Figure 22-15). When these are mated with each other, some of the offspring will be homozygous for the knockout mutation.

## "Knock-in" technology allows the replacement of an endogenous gene

In addition to disrupting a gene of choice, it is also possible to replace the endogenous gene with a mutated form of that gene or completely replace the endogenous gene with a DNA sequence of choice. In a recent report, for example, the CD4 gene was replaced with the gene for β-galactosidase. In these experiments, the CD4 promoter was left intact to drive the expression of β-galactosidase, which catalyzes the color change of certain reporter chemicals to blue. Because the CD4 promoter drove the expression of β-galactosidase, only those thymic cells destined to express CD4 turned blue in

the presence of the reporter chemicals. Data from these experiments were useful in tracing CD4/CD8 lineage commitment in developing T cells.

## Inducible gene targeting, the Cre/*lox* system, targets gene deletion

In addition to the deletion of genes by gene targeting, recent experimental strategies have been developed that allow the specific deletion of a gene of interest in precisely the tissue of choice. These technologies rely on the use of site-specific recombinases from bacteria or yeast. The most commonly used recombinase is Cre, isolated from bacteriophage P1. Cre recognizes a specific 34-bp site in DNA known as *loxP* and, on recognition, catalyzes a recombination. Therefore, DNA sequences that are flanked by *loxP* are recognized by Cre and the recombinational event results in the deletion of the intervening DNA sequences. In other words, animals that ubiquitously express Cre recombinase will delete all *loxP*-flanked sequences. The real innovation of this technique is that expression of the Cre recombinase gene can be controlled by

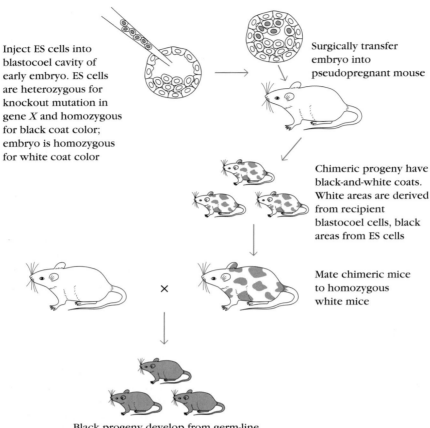

Inject ES cells into blastocoel cavity of early embryo. ES cells are heterozygous for knockout mutation in gene *X* and homozygous for black coat color; embryo is homozygous for white coat color

Surgically transfer embryo into pseudopregnant mouse

Chimeric progeny have black-and-white coats. White areas are derived from recipient blastocoel cells, black areas from ES cells

Mate chimeric mice to homozygous white mice

×

Black progeny develop from germ-line cells derived from ES cells and are heterozygous for disrupted gene *X*

**FIGURE 22-15 General procedure for producing homozygous knockout mice.** ES cells homozygous for a marker gene (e.g., black coat color) and heterozygous for a disrupted target gene (see Figure 22-16) are injected into an early embryo homozygous for an alternate marker (e.g., white coat color). The chimeric transgenic offspring, which have black-and-white coats, then are mated with homozygous white mice. The all-black progeny from this mating have ES-derived cells in their germ line, which are heterozygous for the disrupted target gene. Mating of these mice with each other produces animals homozygous for the disrupted target gene, that is, knockout mice. *[Adapted from M. R. Capecchi, 1989,* Trends in Genetics **5**:70.*]*

the use of a tissue-specific promoter. This allows tissue-specific expression of the recombinase protein and thus tissue-specific deletion of DNA flanked by *loxP*. For example, one could express Cre in B cells using the immunoglobulin promoter, and this would result in the targeted deletion of *loxP*-flanked DNA sequences only in B cells.

This technology is particularly useful when the targeted deletion of a particular gene is lethal. For example, the DNA polymerase β gene is required for embryonic development. In experiments designed to test the Cre/*lox* system, scientists flanked the mouse DNA polymerase β gene with *loxP* and mated these mice with mice carrying a Cre transgene under the control of a T-cell promoter (Figure 22-16a). The results of this mating are offspring that express the Cre recombinase specifically in T cells, allowing scientists to examine the effects of deleting the enzyme DNA polymerase β specifically in T cells. The effects of the deletion of this gene could not

be examined in a conventional gene-targeting experiment, because deletion of DNA polymerase β throughout the animal would be lethal. With the Cre/*lox* system, it is possible to examine the effects of gene deletion in a specific tissue only.

The Cre/*lox* system can also be used to turn on gene expression in a particular tissue. Just as the lack of a particular gene may be lethal during embryonic development, the expression of a gene can be toxic. To examine tissue-specific expression of such a gene, it is possible to insert a translational stop sequence flanked by *loxP* into an intron at the beginning of the gene (Figure 22-16b). Using a tissue-specific promoter driving Cre expression, the stop sequence may be deleted in the tissue of choice and the expression of the potentially toxic gene examined in this tissue. These modifications of gene-targeting technology have been very useful in determining the effects of particular genes in cells and tissues of the immune system.

(a)

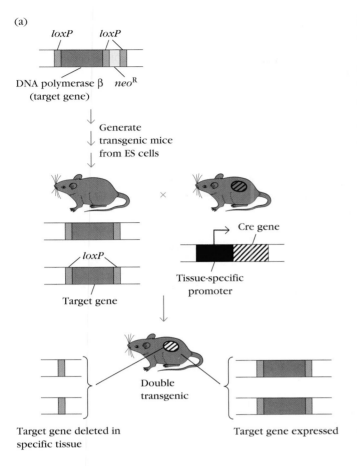

**FIGURE 22-16 Gene targeting with Cre/loxP.** (a) Conditional deletion by Cre recombinase. The targeted DNA polymerase β gene is modified by flanking the gene with *loxP* sites (for simplicity, only one allele is shown). Mice are generated from ES cells by standard procedures. Mating of the *loxP*-modified mice with a Cre transgenic will generate double transgenic mice in which the *loxP*-flanked DNA polymerase β gene will be deleted in the tissue where Cre is expressed. In this example, Cre is expressed in thymus tissue (striped), so that deletion of the *loxP*-flanked gene occurs only in the thymus (white) of the double transgenic. Other tissues and organs still express the *loxP*-flanked gene (orange). (b) Activation of gene expression using Cre/lox. A *loxP*-flanked translational STOP cassette is inserted between the promoter and the potentially toxic gene, and mice are generated from ES cells using standard procedures. These mice are mated to a transgenic line carrying the Cre gene driven by a tissue-specific promoter. In this example, Cre is expressed in the thymus, so that mating results in expression of the toxic gene (blue) solely in the thymus. Using this strategy, one can determine the effects of expression of the potentially toxic gene in a tissue-specific fashion. [*Adapted from B. Sauer, 1998, Methods **14**:381.*]

(b)

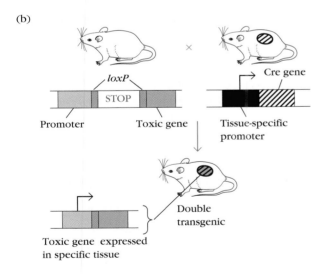

## Microarrays—An Approach for Analyzing Patterns of Gene Expression

In the past few years, a new approach has emerged designed to assess differences in gene expression between various cell types or in the same cells under different conditions. This technology, referred to as microarray technology or *expression profiling*, has the ability to rapidly and reliably scan large numbers of different mRNAs. The principle is simple and is derived from what we already know about RNA and DNA hybridization: mRNA is isolated from a given sample; then, when cDNA synthesis is initiated, the first strand of the cDNA is labeled with the tag to form a pool of target sequences.

The next step is to hybridize the labeled cDNA to nucleic acid affixed in a microarray. Many microarrays are commercially available, which fall mainly into two classes: those composed of cDNA and those composed of oligonucleotides. Microarrays of cDNAs are, as the name suggests, a collection of cDNA that has been arranged, or arrayed, on a solid substrate in defined locations (Figure 22-17a). The substrate varies but is usually a nylon membrane or glass slide. The spots of cDNA arrayed on the substrate can be as small as 5 to 10 μm in size, and more than a million spots have been arrayed on chips that are just a few centimeters across. The process of arraying the cDNA is accomplished using robotics. The cDNAs are most frequently obtained from available cDNA libraries and, in some cases, are PCR products amplified from the cDNA library using primers specific for certain known genes.

Oligonucleotide arrays are usually a collection of oligos 20 to 30 nucleotides long (Figure 22-17b), synthesized directly on the array surface using photolithography techniques that can rapidly and inexpensively synthesize different, sequence-specific oligonucleotides in each spot. The advantage of this type of array is that one needs only sequences of genes of interest; no cDNA library is needed.

Although the source of the targets used for both cDNA and oligo arrays is cDNA, the preparation of the target differs, depending on the microarray. The target preparation

## CLINICAL FOCUS
# Microarray Analysis as a Diagnostic Tool for Human Diseases

**It is almost** impossible to distinguish visually between B and T cells without molecular analysis. Similarly, it can be quite difficult to distinguish one tumor from another. Two of the best-known acute leukemias are AML, which arises from a myeloid precursor (hence the name, *acute myeloid leukemia*), and ALL (*acute lymphoid leukemia*), which arises from lymphoid precursors. Both leukemias are derived from hematopoietic stem cells, but the prognosis and treatment for the two diseases are quite different. Until recently, the two diseases could be diagnosed with some degree of confidence using a combination of surface phenotyping, karotypic analysis, and histochemical analysis, but no single test was conclusive; reliable diagnosis depended on the expertise of the clinician.

The difference between an ALL diagnosis and an AML diagnosis can mean the difference between life and death. ALL responds best to corticosteroids and chemotherapeutics such as vincristine and methotrexate. AML is usually treated with daunorubicin and cytarabine. The cure rates are dramatically diminished if the less-appropriate treatment is delivered due to misdiagnosis.

In 1999, a breakthrough in diagnosis of these two leukemias was achieved using microarray technology. Todd Golub, Eric Lander, and their colleagues isolated RNA from 38 samples of acute leukemia, labeled the RNA with biotin, and hybridized the biotinylated RNA to commercial high-

density microarrays that contained oligonucleotides corresponding to some 6817 human genes. Whenever the biotin-labeled RNA recognized a homologous oligonucleotide, hybridization occurred. Analysis revealed a group of 50 genes that were highly associated with either AML or ALL when compared with control samples. These 50 genes were then used to sample nucleic acid from 34 independent leukemias as well as samples from 24 presumed normal human bone marrow or blood samples. The result? A set of markers that clearly classified a tumor as ALL or AML.

The results of the microarray analysis further suggested that the treatments for AML and ALL can be targeted more precisely. For example, an AML expressing genes *x, y,* and *z* might respond to one treatment modality better than an AML that expresses *a, b,* and *c*. Several pharmaceutical companies have established research groups to evaluate different treatments for tumors based on the tumor's microarray profile. This designer approach to oncology is expected to produce much more effective treatments of individual tumors and, ultimately, enhanced survival rates.

Microarray analysis is likely to be very useful in the diagnosis of tumors of the immune system. Most notably, a laboratory at the National Institutes of Health (NIH) has developed a specialized DNA microarray containing more than 10,000 human cDNAs that are enriched for genes expressed in lymphocytes. Some of these cDNAs are from genes of known function;

others are unknown cDNAs derived from normal or malignantly transformed lymphocyte cDNA libraries. This specialized array is called the "Lymphochip" because the lymphocyte cDNAs are arrayed on a silicon wafer. The group at NIH asked whether they could use the Lymphochip to divide the B-cell leukemia known as diffuse large B-cell lymphoma (DLBCL) into subgroups, an important question because this type of lymphoma has a highly variable clinical course, with some patients responding well to treatment while others respond poorly. Earlier attempts to define subgroups within this group had been unsuccessful. A definition of subgroups within DLBCL could be useful in designing more effective treatments. Using the Lymphochip, the group at NCI identified two genotypically distinct subgroups of DLBCL. One group consisted of tumors expressing genes characteristic of germinal-center B cells and was called "germinal-center–B-like DLBCL (Clinical Focus Figure 1). The other group more resembled activated B cells and was termed "activated B-like DLBCL." Significantly, patients with germinal-center–B-like DLBCL had a higher survival rate than those with activated B-like DLBCL. Normally all patients with DLBCL receive multiagent chemotherapy. Patients who do not respond well to chemotherapy are then considered for bone marrow transplantation. The data obtained from this study suggest that patients with activated B-like DLBCL will not respond as well to chemotherapy and may be better served by bone marrow transplantation shortly after diagnosis. As a direct result of this work, ongoing clinical trials are evaluating how best to treat patients with activated B-like DLBCL.

Gene expression profiling is not restricted to diagnosis of cancer. This technol-

for cDNA arrays involves labeling the cDNA with different fluorescent dyes such as Cy3 and Cy5 (see Figure 22-17a). These cyanine-based dyes are easily conjugated to nucleic acids and are highly stable and emit less background fluorescence than conventional fluorescent dyes. Suppose you wish to compare two different cell types or one cell type in two different states of activation. cDNA from one population is prepared using mRNA as a template. First, strand

synthesis of the mRNA is performed using one nucleotide conjugated to Cy3. Then, using mRNA from the second cell population, cDNA is prepared using a nucleotide conjugated to Cy5. These two populations of cDNA, one marked with Cy3 and the other with Cy5, are hybridized to the microarray. If one of the targets hybridizes to a cDNA on the array, a green (Cy3) or red (Cy5) fluorescence emission is detected. If both hybridize to the cDNA, yellow fluorescence is

ogy provides us with a unique opportunity to examine differences between any distinct populations of cells. One can compare which genes are expressed in common or differentially in a naive T cell and a memory T cell. What is the difference between a normal T cell and a T cell dying by apoptosis? Comparisons like these will be a rich source of insight into differences in cell populations. The key to using this valuable information will be the development of tools to analyze the vast quantities of data that can be obtained from this new approach.

(a)

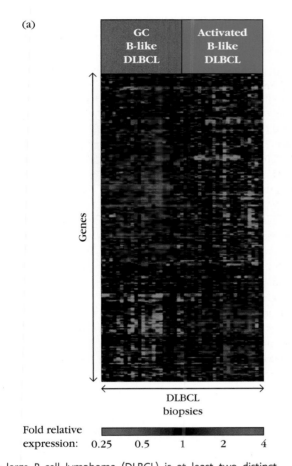

(b)

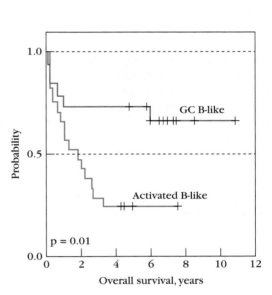

Diffuse large B-cell lymphoma (DLBCL) is at least two distinct diseases. (a) Shown are differences in gene expression between samples taken from patients with either germinal center B-like DLBCL (*left*, red) or activated B-like DLBCL (*right*, blue). Relative expression of the 100 genes (y axis) that discriminate most significantly between the two DLBCL types is depicted over a 16-fold range using the graded color scale at bottom. Note the strikingly different gene expression profiles of the two diseases. (b) Plot of overall DLBCL patient survival following chemotherapy. Gene expression profiles of tumor biopsy samples allow the assignment of patients to the correct prognostic categories and may aid in the treatment of this complex disease. *[Adapted from L. M. Staudt, 2002, Gene expression profiling of lymphoid malignancies, Annual Review of Medicine **53**:303–318.]*

detected (the combination of the red and green emissions from both dyes). The arrays are analyzed by scanning the array at two different wavelengths to distinguish between the Cy3 and Cy5 signals. The signal intensity of each dye is determined and compared, and the results are presented as a ratio between the two samples.

In the case of oligo-based microarrays, the usual approach is to label the target cDNA with a biotin-labeled nucleotide during first-strand synthesis of the mRNA. The biotin-labeled cDNA is hybridized to the oligo array and detected by the use of the fluorescent streptavidin (see Figure 22-17b). The procedure is then repeated with cDNA from the other cell type and another microarray is used. The resultant microarrays are analyzed by either phosphoimaging or fluorescent-based scanning using specialized sensors developed for scanning microarrays.

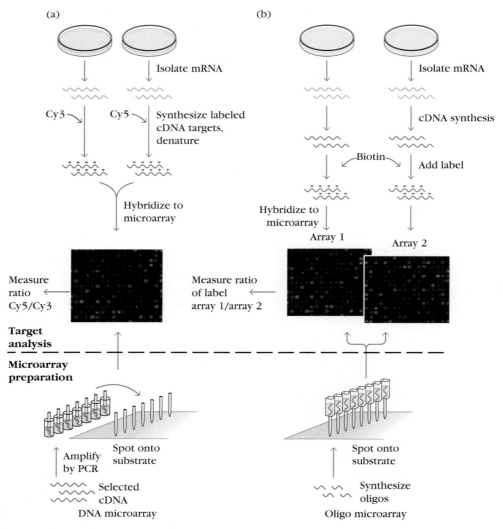

**FIGURE 22-17 DNA microarray analysis using cDNA microarrays (a) or high-density oligonucleotide microarrays (b).** As described in the text, microarray analysis relies on the isolation of RNA from the tissues or cells to be analyzed, the conversion of RNA into cDNA, and the subsequent labeling of DNA during target preparation. The labeled target sequences are hybridized to either a cDNA microarray (a) or an oligo microarray (b).

The application of microarray technology to immunology is apparent. One could easily ask, what is the difference in gene expression between T cells and B cells? Or, what is the difference between an activated T cell and a resting T cell? The list of possible comparisons is immense. To begin to answer some of the interesting immunology questions, Louis Staudt and coworkers at the NIH developed an array they called the "Lymphochip," consisting of more than 10,000 human genes and enriched in genes expressed in lymphoid cells. The Lymphochip includes genes from normal as well as transformed lymphocytes. This particular microarray has provided a great deal of useful information, including a profile of T cells compared to B cells, plasma cells compared to germinal center B cells, and gene expression patterns induced by various signaling pathways. The Lymphochip and other clinical applications of microarrays are described in the Clinical Focus on page 568.

## Two-Photon Microscopy for In Vivo Imaging of the Immune System

Until recently, microscopic examination of cells of the immune system used techniques that isolated the cells from their surrounding environment and, in many instances, fixed the cells or tissues. Data gathered from these approaches have been extremely valuable, but they represent a snapshot in time and not a dynamic view of cell-cell interactions in the body during an immune response. A recent development, two-photon laser microscopy, allows the direct real-time visualization of in vivo interactions among cells of the immune system.

The development of two-photon microscopy has its roots in confocal microscopy. Although standard fluorescent microscopy can detect labeled or tagged cells in a monolayer on a slide or culture dish, confocal microscopy has the striking advantage of "seeing" deep into tissue sections, eliminating the need to either section tissues or isolate cells on a slide. Confocal microscopy can even visualize live cells in tissue, but photobleaching and phototoxic effects are results of the laser light required. Fluorescent molecules are excited by laser light focused through a lens on a focal point, the imaging target. The emission from the fluorochrome is collected through the same lens and refocused to a spot centered in a small aperture placed in front of a detector. The energy generated along the excitation path can damage and destroy the cells in the path of this beam. In addition, the depth of detection in a tissue section is limited to only a few tens of μm.

Two-photon microscopy, like confocal microscopy, is capable of "optically sectioning" the material under examination without causing photoxic damage. Two-photon microscopy can also image tissue at depths of several hundred microns to more than a millimeter, far exceeding the reach of confocal microscopy.

The differences in the capabilities of these two techniques can be understood by comparing the technologies on which they are based. Confocal microscopy uses a single photon of light to excite the fluorochrome. In contrast, two-photon microscopy, as the name implies, excites the fluorochrome with two photons of light, each bearing half the energy of the single photon used in standard fluorescence microscopy. The fluorochromes are excited by highly focused and extremely rapid delivery of two very short pulses (approximately 100 femtoseconds) of laser light. In confocal microscopy, a fluorochrome such as fluorescein is excited by a photon of laser light at 400 nm; the same fluorochrome can be excited by two pulses of near-infrared light at 800 nm (half the energy of a 400-nm photon) if the pulses reach the fluorochrome almost simultaneously. The full intensity is only realized at the focal point, and the other portions of the tissue are not affected by the longer-wavelength photons. In short, the major advantage of this system is that excitation is confined to the focal plane, and the rest of the specimen is exposed to relatively harmless and extremely short pulses of near-infrared light. In addition, because the longer wavelengths used for excitation are not efficiently absorbed by tissue, deeper penetration of the biological sample is possible. Indeed, as we saw in Figure 11-19, it is now possible to image a lymph node in intact tissue in a living animal following the injection of antigen.

## Advances in Fluorescent Technology

In addition to the fluorochromes routinely used in microscopy and flow cytometry, such as fluorescein isothiocynate, Texas Red, rhodamine isothiocyanate, and phycoerytherin, several new fluorescent tools have become available. In particular, green fluorescent protein (GFP) and its many derivatives have been extremely useful in the analysis of both living as well as fixed cells and tissue. GFP is a naturally occurring bioluminescent protein first isolated from jellyfish. Cell biologists have taken advantage of its naturally fluorescing properties to tag the products of genes. When the gene for GFP or one of its spectral variants, such as yellow fluorescent protein (YFP), is appended to a gene of interest, the gene product will fluoresce when excited by laser light. For example, an immunoglobulin kappa light chain fused in frame to GFP would mark all cells expressing kappa light chains with a fluorescent tag. The emission from this excitation can be visualized by fluorescent microscopy or detected by flow cytometry and serves as a marker for cells expressing the tagged gene. It is also possible to produce transgenic animals that express GFP under the control of a tissue-specific promoter.

Another useful fluorochrome, 5,6-carboxyfluorescein-diacetate succinmidyl ester, or CFSE, can be loaded into cells ex vivo. For example, lymphoid cells can be removed from an animal, loaded with CFSE, and introduced into the same animal or a syngeneic member of the same strain of animals. After homing to various sites in the immune system, these cells can be visualized by fluorescence. CFSE can even be used to assess whether a cell has divided. On cell division, CFSE is distributed equally to daughter cells, with each generation becoming progressively less fluorescent. Using CFSE and flow cytometry, one can easily determine whether a population of cells has undergone cell division and even the number of times a cell has divided.

## SUMMARY

- Inbred mouse strains allow immunologists to work routinely with syngeneic, or genetically identical, animals. With these strains, aspects of the immune response can be studied uncomplicated by unknown variables arising from genetic differences between animals.

- In adoptive-transfer experiments, lymphocytes are transferred from one mouse to a syngeneic recipient mouse that has been exposed to a sublethal (or potentially lethal) dose of x-rays. The irradiation inactivates the immune cells of the recipient so that one can study the response of only the transferred cells.

- With in vitro cell culture systems, populations of lymphocytes can be studied under precisely defined conditions. Such systems include primary cultures of lymphoid cells, cloned lymphoid cell lines, and hybrid lymphoid cell lines. Unlike primary cultures, cell lines are immortal and homogeneous.

- Biochemical techniques provide tools for labeling important proteins of the immune system. Labeling antibodies with molecules such as biotin and avidin allows accurate determination of the level of antibody response. Gel electrophoresis is a convenient tool for separating and determining the molecular weight of a protein.

- The ability to identify, clone, and sequence immune system genes using recombinant DNA techniques has revolutionized the study of all aspects of the immune response. Both cDNA, which is prepared by transcribing mRNA with reverse transcriptase, and genomic DNA can be cloned. Generally, cDNA is cloned using a plasmid vector; the recombinant DNA containing the gene to be cloned is propagated in *E. coli* cells. Genomic DNA can be cloned within a bacteriophage vector or a cosmid vector, both of which are propagated in *E. coli.* Even larger genomic DNA fragments can be cloned within bacteriophage P1 vectors, which can replicate in *E. coli,* or yeast artificial chromosomes, which can replicate in yeast cells. The polymerase chain reaction (PCR) is a convenient tool for amplifying small quantities of DNA.

■ Transcription of genes is regulated by promoter and enhancer sequences; the activity of these sequences is controlled by DNA-binding proteins. Footprinting and gel-shift analysis can be used to identify DNA-binding proteins and their binding sites within the promoter or enhancer sequence. Promoter activity can be assessed by reporter genes, such as the luciferase assay and green fluorescent protein.

■ Cloned genes can be transfected (transferred) into cultured cells by several methods. Commonly, immune system genes are transfected into cells that do not normally express the gene of interest. Cloned genes also can be incorporated into the germ-line cells of mouse embryos, yielding transgenic mice, which can transmit the incorporated transgene to their offspring. Expression of a chosen gene can then be studied in a living animal. Knockout mice are transgenics in which a particular target gene has been replaced by a nonfunctional form of the gene, so the gene product is not expressed. The Cre/*lox* system provides a mechanism that allows tissue-specific expression or deletion of a particular gene.

■ Microarrays, a powerful approach for the examination of tissue-specific gene expression and comparison of gene expression in different cells, have revolutionized the study of gene regulation and gene expression.

■ Two-photon microscopy allows the tracking of cells in their biological environment over time, providing a temporal view of the behavior of lymphocytes following manipulation of the immune system.

## References

Alizadeh A. A., et al. 2000. Distinct types of diffuse large B-cell lymphoma identified by gene expression profiling. *Nature* 2000 **403**:503–11.

Bell, J. 1989. The polymerase chain reaction. *Immunology Today* **10**:351.

Betz, U. A. K., et al. 1996. Bypass of lethality with mosaic mice generated by Cre-*loxP*-mediated recombination. *Current Biology* **6**:1307.

Calahan, M. D., et al. 2002. Two-photon tissue imaging—seeing the immune system in a fresh light. *Nature Reviews Immunology* **2**:872.

Camper, S. A. 1987. Research applications of transgenic mice. *Biotechniques* **5**:638.

Capecchi, M. R. 1989. Altering the genome by homologous recombination. *Science* **244**:1288.

Denis, K. A., and O. N. Witte. 1989. Long-term lymphoid cultures in the study of B cell differentiation. In *Immunoglobulin Genes.* F. W. Alt, T. Honjo, and T. H. Rabbitts, eds. Academic Press, London, pp. 45–59.

Depamphilis, M. L., et al. 1988. Microinjecting DNA into mouse ova to study DNA replication and gene expression and to produce transgenic animals. *Biotechniques* **6**(7):622.

Golub, T. R., et al. 1999. Molecular classification of cancer: class discovery and class prediction by gene expression monitoring. *Science* **286**:531.

Koller, B. H., and O. Smithies. 1992. Altering genes in animals by gene targeting. *Annual Review of Immunology* **10**:705.

Meinl, E., et al. 1995. Immortalization of human T cells by *herpesvirus saimiri. Immunology Today* **16**:55.

Melton, D. W. 1994. Gene targeting in the mouse. *BioEssays* **16**:633.

Miller, M. J., et al. 2002. Two-photon imaging of lymphocyte motility and antigen response in intact lymph node. *Science* **296**:1869.

Sauer, B. 1998. Inducible gene targeting in mice using the Cre/*lox* system. *Methods* **14**:381.

Schlessinger, D. 1990. Yeast artificial chromosomes: tools for mapping and analysis of complex genomes. *Trends in Genetics* **6**(8):254.

Sharpe, A. H. 1995. Analysis of lymphocyte costimulation in vivo using transgenic and knockout mice. *Current Opinion in Immunology* **7**:389.

Shaffer, A. L., et al. 2001. Signatures of the immune response. *Immunity* **15**:375–85.

Schulze, A., and J. Downward. 2001. Navigating gene expression using microarrays—a technology review. *Nature Cell Biology* **3**:E190–5.

## Useful Web Sites

**http://www.jax.org/**

This is the home page for the Jackson Laboratory, the major repository of inbred mice in the world.

**http://www.neb.com/**

The home page for New England Biolabs, a molecular biology company, gives useful information concerning restriction enzymes, under Technical Reference.

**http://www.public.iastate.edu/~pedro/ research_tools.html**

This very useful site for molecular biology contains links to many informative sites and is updated regularly.

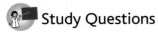

# Study Questions

**CLINICAL FOCUS QUESTION** How has microarray technology changed disease diagnosis and how is it likely to influence treatment of diseases in the future?

1. Explain why the following statements are false.

   a. The amino acid sequence of a protein can be determined from the nucleotide sequence of a genomic clone encoding the protein.

   b. Transgenic mice can be prepared by microinjection of DNA into a somatic-cell nucleus.

   c. Primary lymphoid cultures can be propagated indefinitely and are useful in studies of specific subpopulations of lymphocytes.

2. Fill in the blanks in the following statements with the most appropriate terms:

   a. In inbred mouse strains, all or nearly all genetic loci are _____; such strains are said to be_____.

   b. B-cell hybridomas are formed by fusion of _____ with _____. They are capable of _____ growth and are used to produce_____.

   c. A normal lymphoid cell that undergoes _____ can give rise to a cell line, which has an _____ life span.

3. The gene diagrammed below contains one leader (L), three exons (E), and three introns (I). Illustrate the primary transcript, mRNA, and the protein product that could be generated from such a gene.

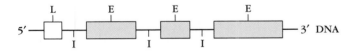

4. The term *transfection* refers to which of the following?

   a. Synthesis of mRNA from a DNA template

   b. Synthesis of protein based on an mRNA sequence

   c. Introduction of foreign DNA into a cell

   d. The process by which a normal cell becomes malignant

   e. Transfer of a signal from outside a cell to inside a cell

5. Which of the following are required to carry out the PCR?

   a. Short oligonucleotide primers

   b. Thermostable DNA polymerase

   c. Antibodies directed against the encoded protein

   d. A method for heating and cooling the reaction mixture periodically

   e. All of the above

6. Why is it necessary to include a selectable marker gene in transfection experiments?

7. What would be the result if a transgene were injected into one cell of a four-cell mouse zygote rather than into a fertilized mouse egg before it divides?

8. A circular plasmid was cleaved with *Eco*RI, producing a 5.4-kb band on a gel. A 5.4-kb band was also observed when the plasmid was cleaved with *Hind*III. Cleaving the plasmid with both enzymes simultaneously resulted in a single band 2.7 kb in size. Draw a diagram of this plasmid showing the relative location of its restriction sites. Explain your reasoning.

9. DNA footprinting is a suitable technique for identifying which of the following?

   a. Particular mRNAs in a mixture

   b. Particular tRNAs in a mixture

   c. Introns within a gene

   d. Protein-binding sites within DNA

   e. Specific DNA sites at which restriction endonucleases cleave the nucleotide chain

10. Explain briefly how you might go about cloning a gene for interleukin-2 (IL-2). Assume that you have available a monoclonal antibody specific for IL-2.

11. You have a sample of a mouse DNA-binding protein and of the mRNA that encodes it. Assuming you have a mouse genomic library available, briefly describe how you could select a clone carrying a DNA fragment that contains the gene that encodes the binding protein.

12. What are the major differences between transgenic mice and knockout mice and in the procedures for producing them?

13. How does a knock-in mouse differ from a knockout mouse?

14. How does the Cre/*lox* technology enhance knockout and knock-in strategies?

15. For each term related to recombinant DNA technology (a–i), select the most appropriate description (1–10) listed below. Each description may be used once, more than once, or not at all.

    *Terms*

    a. ____ Yeast artificial chromosome

    b. ____ Restriction endonuclease

    c. ____ cDNA

    d. ____ COS sites

    e. ____ Retrovirus

    f. ____ Plasmid

    g. ____ cDNA library

    h. ____ Sticky ends

    i. ____ Genomic library

    *Descriptions*

    (1) Cleaves mRNA at specific sites

    (2) Cleaves double-stranded DNA at specific sites

    (3) Circular genetic element that can replicate in *E. coli* cells

    (4) Used to introduce DNA into mammalian cells

    (5) Formed from action of reverse transcriptase

    (6) Collection of DNA sequences within plasmid vectors representing all of the mRNA sequences derived from a cell

(7) Produced by action of certain DNA-cleaving enzymes

(8) Used to clone very large DNA sequences

(9) Used to introduce larger-than-normal DNA fragments in λ-phage vectors

(10) Collection of λ clones that includes all the DNA sequences of a given species

16. As you investigate the current literature in immunology, you may notice that some researchers perform experiments with PBMCs (peripheral blood mononuclear cells) instead of or in addition to a cell line. The blood of patients is subjected to several isolation steps that remove the red blood cells, platelets, and granulocytes, leaving only mononuclear cells (lymphocytes and monocytes). What do you suppose are the advantages and/or disadvantages to these two cell systems?

17. Pepsin is a nonspecific protease active at acidic pH. Pepsin cleaves peptides at the C-terminal side of the inter-heavy-chain disulfide bonds, resulting in extensive degradation of the Fc fragment. Shown at right is a stained SDS-PAGE of intact (not digested) mouse IgG and mouse IgG digested with pepsin.

a. Papain is another protease that cleaves above the disulfide bond (see Figure 4.7 for activities of these proteases). What would the gel look like if the digestion was performed using papain instead?

b. What would the gel look like if β-mercaptoethanol, a reducing agent that breaks disulfide bonds, was added to the loading buffer?

c. If the gel from (a) was transferred electrophoretically onto nitrocellulose and probed with the following antibodies, what would the developed immunoblot (Western blot) look like?

(1) Anti-idiotypic mAb

(2) Polyclonal rabbit antimouse IgG

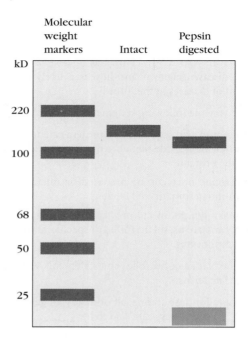

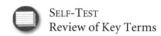

## Interactive Study

**www.whfreeman.com/kuby**

SELF-TEST
Review of Key Terms

The following table presents information about the nature, distribution, and function of the CD antigens. Because many CD antigens are known by a variety of names, synonyms are indicated in addition to the official CD designations. The mass is listed (when known) for the CD antigens that are proteins but not for those that are carbohydrates or lipids. In summarizing the expression patterns for CD antigens, we have concentrated on cells of the immune system (leukocytes). However, many of these antigens are also expressed by other cell types, and we have given some examples of these as well. The description of the functions of each CD antigen also focuses on cells of the immune system, with an occasional example of functions in other cell types. The function is recorded as "not known" if there is no immunologically relevant function known for the CD marker.

Detailed reviews of CD antigens, including cell expression patterns and functions, are available from the following online resources and are regularly updated with complete lists of references:

http://www.vetmed.wsu.edu/tkp/Search.aspx

This comprehensive, searchable database is maintained by the College of Veterinary Medicine at Washington State University.

http://www.hlda8.org/CD1toCD339.htm

This Web site provides information on the functions of known and putative CD antigens and gives links to further information on these molecules. The site is the product of a collaborative effort by numerous researchers around the world and is updated regularly through a series of international workshops known as the Human Leukocyte Differentiation Antigens (HLDA) Workshops.

http://ca.expasy.org/

This searchable database is sponsored and maintained by the Swiss Institute of Bioinformatics and provides information about the structure and sequence of proteins in addition to function and cellular distribution.

http://www.ncbi.nlm.nih.gov/PROW/

The CD index on this Web site offers a comprehensive listing of cell expression patterns and functions of these antigens in both leukocytes and other cell types. This regularly updated resource is sponsored by the National Library of Medicine.

http://www.rndsystems.com/mini_reviews.aspx

This Web site of the company R&D Systems offers updated, detailed reviews of specific CD antigens and antigen families.

Continuing research will undoubtedly identify more CD antigens and add to our understanding of their function and expression. For complete, up-to-date information about any CD antigen, consult the online resources listed above.

An additional source of information is PubMed, another Web site of the National Library of Medicine:

http://www.ncbi.nlm.nih.gov/PubMed/medline.html

This site is an online bibliographic database that contains references, many with informative abstracts, to more than 16 million articles from a comprehensive collection of medical and biological science journals. Articles from the 1950s to the present are indexed, and new references are added daily. A search engine at the Web site allows searches by key words as well as other variables, such as name of researcher or year of publication. A key word search using the name of a particular CD antigen will provide a list of the most recent journal articles that address it. For a more comprehensive search on a particular antigen, the synonyms listed in this table can be included as alternative key words.

| CD antigen. MW. Synonyms and properties. | Leukocyte expression | Function |
|---|---|---|
| **CD1a, -b, -c, -d, -e.** 43–49 kDa. MHC class I–like structure. | Dendritic cells, B cells, Langerhans cells, some thymocytes, monocytes, activated T cells, and epithelial cells | Display of nonpeptide and glycolipid antigens for T-cell responses |
| **CD2.** 45–58 kDa. LFA-2, T11, Leu5, Tp50; CD58-binding adhesion molecule, sheep red blood cell (SRBC) receptor. | T cells, thymocytes, NK cells | Adhesion molecule involved in T-cell activation |
| **CD3.** γ, 25–28 kDa; δ, 20 kDa; ε, 20 kDa. T3; composed of 3 polypeptide chains: γ, δ, ε. | Thymocytes, T cells, NKT cells | Essential role in TCR signal transduction and in cell surface expression of the TCR |

*(continued)*

| CD antigen. MW. Synonyms and properties. | Leukocyte expression | Function |
|---|---|---|
| **CD4.** 55 kDa. L3T4, T4, Leu3, Ly4, Ox38. | MHC class II–restricted T cells, some thymocytes, monocytes/macrophages, granulocytes | Coreceptor for MHC class II–restricted T-cell activation; thymic differentiation marker for T cells; receptor for HIV |
| **CD5.** 58 kDa. Leu1, T1, Ly-1, Ox19. | Mature T cells, early thymocytes; small amounts on a subset of mature B cells (B1a B cells) | Positive or negative modulation of TCR and BCR signaling, depending on the type and developmental stage of cell displaying it |
| **CD6.** 105 or 130 kDa. T12, Ox52, Tp120; member of the group B scavenger receptor cysteine-rich (SRCR) protein superfamily. | Most peripheral T cells, thymocytes, a subset of B cells | Adhesion molecule that binds developing thymocytes to thymic epithelial cells, plays a role in mature T-cell interaction with nonprofessional antigen-presenting cells |
| **CD7.** 40 kDa. gp40, Leu9, Tp41; member of the galectin-1 receptor family. | Pluripotent hematopoietic cells, T cells, thymocytes, NK cells, pre–B cells | Distinguishes primitive lymphoid progenitors from pluripotent stem cells; plays a role in regulating peripheral T-cell and NK-cell cytokine production and sensitivity to LPS-induced shock; may have costimulatory activity for T cells |
| **CD8.** T8, Leu-2, Lty-2; membrane-bound dimer of two chains, αβ heterodimer or αα homodimer. | MHC class I–restricted T cells, some thymocytes, subset of dendritic cells | Coreceptor for MHC class I–restricted T cells |
| **CD9.** 24 kDa. MRP-1, p24, DRAP-27. | Platelets, early B cells, activated B cells, activated T cells, eosinophils, basophils | Modulation of cell adhesion and migration; trigger of platelet activation |
| **CD10.** 100 kDa. Common acute lymphoblastic leukemia antigen (CALLA), EC 3.4.24.11 (neprilysin), enkephalinase, gp100, neutral endopeptidase (NEP), metalloendopeptidase. | B-cell precursors, T-cell precursors, bone marrow stromal cells, neutrophils, monocytes, keratinocytes | Membrane-bound metallopeptidase responsible for cleaving a variety of inflammatory and vasoactive peptides |
| **CD11a.** 180 kDa. α-L integrin chain, LFA-1 α (leukocyte-function–associated molecule-1 α) chain; forms LFA-1 by association with CD18. | Monocytes, macrophages, granulocytes, lymphocytes | Subunit of LFA-1, a membrane glycoprotein that provides cell-cell adhesion by interaction with ICAM-1 (intercellular adhesion molecule-1) |
| **CD11b.** 165 kDa. C3biR, CR3, Ly40, iC3b receptor, Ox42; α chain of MAC-1 integrin, forms MAC-1 by association with CD18. | Granulocytes, monocytes, NK cells, subset of T cells, subsets of B cells, myeloid dendritic cells | Subunit of MAC-1, which mediates interactions of neutrophils and monocytes with stimulated endothelium, phagocytosis of particles coated with iC3b or IgG, spreading or chemotaxis of leukocytes; ligand-binding molecule that binds C3b, ICAM-1 (CD54), ICAM-2 (CD102), and fibrinogen |
| **CD11c.** 145 kDa. α-X integrin chain, αxβ2, α chain of integrin CR4, leukocyte surface antigen, p150, 95; forms CR4 by association with CD18. | Monocytes, macrophages, NK cells, granulocytes, subset of T cell, subset of B cells, dendritic cells | Subunit of complex with CD18 that is similar to CD11b/CD18 complex, with which it acts cooperatively; the major form of CD11/CD18 on tissue macrophages |
| **CD11d.** 150 kDa. α-chain type 1 glycopeptide. | Peripheral blood leukocytes, tissue-specific macrophages | Not known; may play a role in the atherosclerotic process and macrophage differentiation |
| **CD12w.** 90–120 kDa. | Monocytes, granulocytes, platelets, and NK cells | Not known |
| **CD13.** 150 kDa. Aminopeptidase N (APN), EC 3.4.11.2, gp 150, Lap1. | Early progenitors of granulocytes and monocytes (CFU-GM), mature granulocytes and monocytes, bone marrow stromal cells, osteoclasts, a small number of large granular lymphocytes, T cells | Membrane-bound metalloprotease that catalyzes the removal of N-terminal amino acids from peptides |
| **CD14.** 53 kDa. LPS receptor (LPS-R); GPI-anchored glycosylated protein, also exists as a soluble protein. | Monocytes, macrophages, granulocytes (weak expression), B cells, dendritic cells, hepatocytes | Receptor for endotoxin (lipopolysaccharide [LPS]), activating monocytes or neutrophils to release cytokines (TNF) and up-regulate adhesion molecules when LPS is bound; binds peptidoglycan on gram-positive bacteria, Hsp60, ceramides, modified lipoproteins, and anionic phospholipids; soluble form reported to act as an acute phase protein and may regulate T-cell activation and act as B-cell mitogen |

| CD antigen. MW. Synonyms and properties. | Leukocyte expression | Function |
|---|---|---|
| **CD15.** 90 kDa. Lewis X, Lex, SSEA-1, 3-FAL, 3-FL, LNFP III (lacto-*N*-neo-fucopentose III); poly-*N*-acetyl lactosamine carbohydrate polymer of indefinite MW. | Granulocytes, monocytes, macrophages, Langerhans cells, mature bone marrow cells of myelomoncytic lineage, weakly on T cells | Suggested to be the ligand for CD61 selectins, may be important for direct carbohydrate-carbohydrate interactions |
| **CD15s.** Undefined. Sialyl Lewis X (sLex), sialylated Lewis X (SLe-x); poly-*N*-acetyl lactosamine, carbohydrate polymer of indefinite MW. | Granulocytes, monocytes, macrophages, NK cells, subsets of T cells | Strongest binding partner for E-selectin, involved with different carbohydrate-to-carbohydrate cell adhesion |
| **CD15u.** Undefined. Sulfated form of CD15. | Granulocytes, monocytes, macrophages | Carbohydrate-to-carbohydrate adhesion |
| **CD16a, -b.** 50–80 kDa. FCγRIIIA. GPI-anchored glycoprotein. | Transmembrane form in humans on NK cells, macrophages, and mast cells; GPI-anchored form expressed on neutrophils; in mice, no GPI-anchored form of CD16a has been identified and the transmembrane form is expressed on macrophages, NK cells, neutrophils, myeloid precursors, and the majority of early CD4$^+$-CD8$^+$-TCR fetal thymocytes | Subunit of low-affinity Fc receptor, involved in phagocytosis of antibody-complexed antigen and in antibody-dependent cell-mediated cytotoxicity (ADCC) |
| **CDw17.** 115-165 kDa. Lactosylceramide (LacCer); carbohydrate antigen. | Monocytes, granulocytes, basophils, platelets, subset of peripheral B cells (CD19$^+$), tonsillar dendritic cells | Not known; ability to bind bacteria for possible role in phagocytosis |
| **CD18.** 90 kDa. β2 integrin chain; β subunit that combines with CD11a, CD11b, or CD11c subunit to form integrins. | All leukocytes | Adhesion and signaling in the hematopoietic system (see entries for CD11a, CD11b, and CD11c) |
| **CD19.** >120 kDa. B4. | Follicular dendritic cells, B cells from earliest recognizable B-lineage cells during development to B-cell blasts but lost on maturation to plasma cells | Part of B-cell coreceptor with CD21 and CD81; a critical signal transduction molecule that regulates B-cell development, activation, and differentiation |
| **CD20.** 33–37 kDa. B1, Bp35; 4-transmembrane-spanning protein superfamily, type 3-, 4-span glycoprotein. | B cells | Ligation activates tyrosine kinase–dependent pathways and may have a role in B-cell proliferation and differentiation; regulates B-lymphocyte activation and proliferation via regulation of transmembrane Ca$^{2+}$ channel subunit and cell-cycle progression |
| **CD21.** 145 kDa (membrane form); 130 kDa (soluble form). C3d receptor, CR2, Epstein Barr virus receptor (EBV-R); single-chain type 1 glycoprotein. | B cells, follicular dendritic cells, subset of immature thymocytes | Receptor for EBV, C3d, C3dg, and iC3b; with CD19 and CD81, part of the B-cell coreceptor, a large signal transduction complex |
| **CD22.** 140 kDa. BL-CAM, Lyb8 in mouse; single-chain type 1 glycoprotein. | Surface of mature B cells, cytoplasm of late pro– and early pre–B cells | Adhesion molecule; signaling molecule |
| **CD23.** 45 kDa. B6, BLAST-2, FcεRII, Leu-20; low-affinity IgE receptor; C-type lectin, homotrimeric type 2 glycoprotein. | Human B cells, monocytes, weakly on other hematopoietic cells | Regulation of IgE synthesis; trigger to release TNF, IL-1, IL-6, and GM-CSF from human monocytes |
| **CD24.** 35–45 kDa. BA-1, Ly52, heat-stable antigen (HSA) in mouse. | B-cell lineage but lost at plasma-cell stage, mature granulocytes, thymocytes, normal epithelium, monocytes, Langerhans cells, erythrocytes | Mucin-like adhesion molecule shown to enhance metastatic potential of tumor cells |
| **CD25.** 55 kDa. IL-2 receptor α chain, IL-2R, Tac antigen, p55. | Activated B cells, T cells, monocytes, and oligodendrocytes, some thymocytes, macrophages, NK cells, myeloid cells, and a subset of dendritic cells | Low-affinity IL-2 receptor (IL-2Rα) and is an activation marker; induces activation and proliferation of T cells, thymocytes, NK cells, B cells, and macrophages; also plays a role in T-cell-mediated immune response. Associates with β and γ chains to form high affinity with IL-2R |

*(continued)*

| CD antigen. MW. Synonyms and properties. | Leukocyte expression | Function |
|---|---|---|
| **CD26.** 110, 120 kDa. EC 3.4.14.5, adenosine deaminase–binding protein (ADA-binding protein), dipeptidylpeptidase IV (DPP IV ectoenzyme); homodimeric type 2 glycoprotein; member of the polyoligo peptidase family. | Mature or activated T cells, B cells, NK cells, macrophages, renal proximal tubular epithelial cells, small intestinal epithelium, biliary canaliculae, splenic sinus lining cells, and prostate gland; also expressed primarily on mature thymocytes in the medulla, up-regulated on memory T cells following activation. | Proposed to function as a membrane-bound protease, a T-cell costimulatory molecule, and a cell adhesion molecule |
| **CD27.** 55 kDa. S152, T14; member of tumor necrosis factor receptor superfamily. | Mature thymocytes, activated T cells and B cells, NK cells, macrophages, most peripheral T cells, memory-type B cells | Mediation of a co-stimulatory signal for T- and B-cell activation; role in murine T-cell development |
| **CD28.** ≈ 90 kDa (homodimeric form). T44, Tp44. | Mature CD3$^+$ thymocytes, most peripheral T cells, plasma cells | Co-stimulation of T-cell proliferation and cytokine production upon binding CD80 or CD86 |
| **CD29.** 130 kDa. β1 integrin chain, GP, platelet GPIIa, VLA-β chain. | Leukocytes, weakly on granulocytes | β subunit of VLA-1 integrin, acts as a fibronectin receptor and is involved in cell adhesion and recognition in a variety of processes, including embryogenesis, hemostasis, tissue repair, immune response, metastatic diffusion of tumor cells, and development; essential to the differentiation of hematopoietic stem cells with tumor progression and metastasis/invasion |
| **CD30.** 105–120 kDa. Ber-H2 antigen, Ki-1 antigen; member of tumor necrosis factor receptor superfamily. | Activated T cells, activated B cells, activated NK cells, monocytes | Costimulates T-cell proliferation and up-regulates the expression of adhesion molecules and cytokine release; may attenuate autoreactive T-lymphocyte proliferation |
| **CD31.** 130–140 kDa. GPiia, endocam, platelet endothelial cell adhesion molecule (PECAM-1); member of immunoglobulin supergene family (IgSF). | Endothelial cells, platelets, monocytes, neutrophils, NK cells, subsets of T cells | Primarily an adhesion molecule with no enzymatic activity known; plays a role in transendothelial cell migration |
| **CD32.** 40 kDa. FCRII, FcγRII, Ly-17; multiple isoforms. | Monocytes, macrophages, Langerhans cells, neutrophils (A and C isoforms), B cells (B isoform), and platelets as well as on endothelial cells of the placenta. All isoforms are expressed on monocytes. | Receptor for Fc portion of antigen-antibody–complexed IgG molecules. Regulates B-cell function and plays a major role in immune complex–induced tissue damage; seems to have a major inhibitory role preventing the development of autoimmune disease; can trigger IgG-mediated phagocytosis and an oxidative burst in neutrophils and monocytes |
| **CD33.** 67 kDa. gp67, p67, Siglec-3; single-chain type 1 glycoprotein. | Myeloid progenitors, monocytes and macrophages, granulocyte precursors with decreasing expression following maturation and differentiation, resulting in a low level of expression on mature granulocytes | Serves as a marker for distinguishing between myeloid and lymphoid leukemias. Sialic-acid–dependent cytoadhesion molecule, a sialoadhesin; inhibits proliferation of normal and leukemic myeloid cells and might also mediate cell-cell adhesion; also plays a role in carbohydrate binding and lectin activity |
| **CD34.** 40 kDa (based on amino acid sequence). gp105–120, mucosialin; type 1 glycoprotein. | Early lymphohematopoietic stem and progenitor cells, small-vessel endothelial cells, embryonic fibroblasts, some cells in fetal and adult nervous tissue | Cell-cell adhesion molecule; also participates in inhibition of hematopoietic differentiation |
| **CD35.** Many forms: 160, 190, 220, 250, 165, 195, 225, 255 kDa. C3bR, C4bR, complement receptor type one (CR1), immune adherence receptor; member of complement component receptor superfamily (CCRSF), single-chain type 1 glycoprotein. | Erythrocytes, neutrophils, monocytes, eosinophils, B lymphocytes, and 10%–15% of T lymphocytes | Receptor for C4b/C3b-coated particles, mediating their adherence and phagocytosis; facilitator of C3b and C4b cleavage, thus limiting complement activation |

| CD antigen. MW. Synonyms and properties. | Leukocyte expression | Function |
|---|---|---|
| **CD36.** 88–113 kDa (platelet form). GPIIIb, GPIV, OKM5-antigen, PASIV; member of the scavenger receptor superfamily, single-chain type 3-, 2-span glycoprotein. | Platelets, mature monocytes/macrophages, endothelial cells; in mice, expressed on B cells | Multifunctional glycoprotein that acts as an adhesion molecule in platelet adhesion and aggregation and in platelet-monocyte or platelet-tumor cell interaction; molecule for recognition of apoptotic neutrophils and functions in the phagocytic clearance of apoptotic cells; scavenger receptor for oxidized LDL |
| **CD37.** 40–52 kDa. gp 52–50. Member of the 4-transmembrane-spanning protein superfamily (TM4SF); type 3-, 4-span glycoprotein. | B cells; low levels on T cells, neutrophils, granulocytes, and monocytes | B-cell-expressed CD37 associates noncovalently with MHC class II, CD19, CD21, and the TM4SF molecules CD53, CD81, and CD82; involved in the signal transduction pathway(s) that regulate cell development, activation, growth, and motility and may also be involved in T-cell–B-cell interactions |
| **CD38.** 45 kDa (in leukocytes), 39 kDa (soluble form). T10, ADP-ribosyl cyclase, cyclic ADP-ribose hydrolase; single-chain type 2 glycoprotein. | Variable levels on the majority of hemopoietic cells, though most strongly during early differentiation and activation and some non-hemopoietic cells; high level of expression on B cells, activated T cells, and plasma cells | NAD glycohydrolase; functions in cell adhesion; positive and negative regulator of cell activation and proliferation, depending on the cellular environment; signal transduction and calcium signaling; also acts as an ectoenzyme that participates in nucleotide metabolism |
| **CD39.** 78 kDa. Potentially a type 3-, 2-span glycoprotein. | Mantle zone and paracortical activated lymphocytes, macrophages, dendritic cells, and Langerhans cells; absent from germinal centers | May mediate B-cell homotypic adhesion and may protect activated lymphocytes through hydrolysis of extracellular ATP |
| **CD40.** 48 kDa (monomer), 85 kDa (dimer). Bp50; member of the TNF receptor superfamily (TNFRSF), single-chain type 1 glycoprotein. | All mature B-cell lineages except plasma cells, macrophages, follicular dendritic cells, activated monocytes, endothelial cells, fibroblasts, interdigitating dendritic cells, keratinocytes, CD34$^+$ hematopoietic cell progenitors | Binds to CD154. Roles in B-cell growth, differentiation, and isotype switching; apoptosis rescue signal for germinal center B cells; stimulation of cytokine production in macrophages and dendritic cells; up-regulation of adhesion molecules on dendritic cells. Plays a critical role in the regulation of cell-mediated immunity as well as antibody-mediated immunity. |
| **CD41.** αβ dimer: α, 125 kDa; β, 122 kDa. Glycoprotein IIb (GP IIb) αIIb integrin chain; forms platelet fibrinogen receptor by association with GP IIIa; single-chain type 1 glycoprotein. | Platelets, megakaryocytes | Subunit of platelet fibrinogen receptor, which mediates platelet aggregation; plays a central role in platelet activation, cohesion, coagulation, aggregation, and platelet attachment; possibly also receptor for von Willebrand factor, fibronectin, and thrombospondin; also shown to participate in cell-surface–mediated signaling |
| **CD42a, -b, -c, -d.** -a, 23 kDa; -b, 145 kDa; -c, 24 kDa; -d, 82 kDa. -a, GPIX; -b, GPIb-α; -c, GPIb-β; -d, GPV. Single-chain type 1 glycoprotein. | Platelets, megakaryocytes; CD42c also expressed on endothelium, brain, and heart tissue | CD42a–d complex serves as receptor for von Willebrand factor and thrombin. Adhesion of platelets to endothelium; amplification of platelet response to thrombin. |
| **CD43.** 95–115 kDa in naive T cells; 115–135 kDa in activated T cells and neutrophils. gpL115, leukocyte sialoglycoprotein, leukosialin, sialophorin, Ly-48. | All leukocytes except most resting B cells | Possible anti-adhesive ("barrier") molecule mediating repulsion between leukocytes and other cells |
| **CD44.** 85 kDa. ECMR III, H-CAM, HUTCH-1, Hermes, Lu (In-related), Pgp-1, gp85; multiple isoforms. | Surface of most cell types; CD44H is major isoform expressed on lymphocytes, soluble form detectable in body fluids | A recyclable receptor that binds hyaladherin (HA) and other tissue macromolecules; shown to be involved in the transduction of activation signals to the cell and functions as an adhesion molecule mediating leukocyte attachment and homing to peripheral lymphoid organs and sites of inflammation |

*(continued)*

| CD antigen. MW. Synonyms and properties. | Leukocyte expression | Function |
|---|---|---|
| **CD44R.** 85–200 kDa in epithelial cells; 85–250 kDa in lymphocytes. CD44v, CD44v9; heterogeneous group of CD44 variant isoforms designated by the variant exons that they include (v2–v10 in humans). | Epithelial cells and monocytes (constitutively), activated leukocytes | The activity of many isoforms of CD44R remains to be clarified; possible role (particularly for CD44v6) in leukocyte attachment to and rolling on endothelial cells, homing to peripheral lymphoid organs and sites of inflammation |
| **CD45.** 210–220 kDa in naive/resting T cells and NK cells; 180–200 kDa in activated/memory T cells, monocytes, granulocytes, and dendritic cells; 180 kDa in B cells. Leukocyte common antigen (LCA), T200, EC 3.1.3.4.; many different isoforms with different molecular weights, e.g., B220, CD45R, CD45RA, CD45RB, CD45RC, CD45RO. (Various isoforms of CD45 are generated by alternative splicing of 3 exons that can be inserted immediately after an N-terminal sequences of 8 aa found on all isoforms.) | All hematopoietic cells except erythrocytes; especially high on lymphocytes (10% of their surface area comprising CD45); different isoforms characteristic of differentiated subsets of various hematopoietic cells | Regulates a variety of cellular processes, including cell growth, differentiation, mitotic cycle, and oncogenic transformation; essential role in T- and B-cell antigen-receptor-mediated activation; possible role in receptor-mediated activation in other leukocytes |
| **CD46.** 52–58, 35–40 kDa. Membrane cofactor protein (MCP); member of the regulator of complement activation (RCA) family of proteins. | Peripheral-blood lymphocytes (PBL), nonhematopoietic tissues such as salivary glands and kidney ducts; widely expressed in fibroblasts and leukocytes | Binds the complement components C3b and Cb4, allowing degradation by factor 1, and is a receptor for the measles virus and for *Streptococcus pyogenes;* protective barrier against inappropriate complement activation and deposition on plasma membranes; no distinct functions have been attributed to tissue-specific isoforms and to the variable levels of expression; cross-linking leads to the down-regulation of IL-12 production via unknown mechanisms |
| **CD47.** 45–60 kDa in erthrocytes; 47–55 kDa in platelets. Rh-associated protein, gp42, integrin-associated protein (IAP), neurophilin, ovarian carcinoma antigen 3 (OA3); member of the immunoglobulin supergene family (IgSF). | Most cells; part of the Rh complex on erythrocytes and is not expressed on Rh null erythrocytes | Associates noncovalently with the CD61 β3 integrins; acts as an adhesion molecule and thrombospondin receptor; plays a role in the chemotactic and adhesive interactions of leukocytes with endothelial cells via an unclarified mechanism of action |
| **CD47R.** 120 kDa. Formally known as CDw149. MEM-133, IAP, MER6, OA3, Rh-associated protein, gp42, neurophilin. | Peripheral blood lymphocytes, monocytes, weakly on neutrophils, eosinophils, platelets | May interact with integrins and play a role in increasing intracellular calcium during cell adhesion |
| **CD48.** 45 kDa. BCM1 in mouse, Blast-1, Hu Lym3, OX-45 in rat, Sgp-60. | Widely expressed on hematopoietic cells with the exception of granulocytes and some eosinophils, platelets and erythrocytes | Recently identified as a ligand of the NK-cell inhibitory receptor CD244 (2B4); adhesion molecule via its receptor, CD2, and may participate in αβ T-cell recognition, adhesion, and costimulation and as an accessory molecule as predicted for T cell-antigen recognition |
| **CD49a.** 200 kDa. α-1 integrin chain, very late antigen 1 α chain (VLA-1 α chain). | Activated T cells, monocytes, neuronal cells, smooth muscle, IL-2-activated NK cells | Integrin that associates with CD29, binds collagen, laminin-1 |
| **CD49b.** 160 kDa. α-2 integrin chain, GPIa VLA-2 α chain. | T cells, NK cells, B cells, monocytes, platelets, megakaryocytes; neuronal, epithelial, and endothelial cells; osteoclasts | Integrin that associates with CD29, binds collagen, laminin-1; involved in the maintenance of endothelial monlayer integrity along with CD49e. Participates in adhesion and in cell-surface-mediated signaling and has a role in blood clotting and angiogenesis |
| **CD49c.** 150 kDa. α-3 integrin chain, VLA-3 α chain. | Low levels on monocytes, B and T lymphocytes | Integrin that associates with CD29 and binds to a variety of ligands, including collagen, laminin-5, fibronectin, entactin, invasin; binding affinity for each ligand depends on the cell type on which it is expressed, divalent cation concentrations, and presence of other integrin heterodimers on the cell; may have a role in cell-cell as well as cell-matrix adhesion |

| CD antigen. MW. Synonyms and properties. | Leukocyte expression | Function |
|---|---|---|
| **CD49d.** 145 kDa. α-4 integrin chain, VLA-4 α chain. | Many cell types, including T cells, B cells, monocytes, eosinophils, basophils, mast cells, thymocytes, NK cells, dendritic cells, embryonic myeloblasts and myelomonocytic cells, some melanoma cells, Kupffer cells, muscle cells, erythroblastic precursor cells, and sickle reticulocytes; not on normal red blood cells, platelets, or neutrophils | Cell adhesion molecule for attachment to cell surface ligands VCAM-1 (α4β1 and α4β7) and MAdCAM-1 (α4β7) and attachment to extracellular matrix proteins fibronectin (α4β1 and α4β7) and thrombospondin (α4β1) and for lymphocyte migration and homing; costimulatory molecule for T-cell activation |
| **CD49e.** 155 kDa. α-5 integrin chain, FNR α chain, VLA-5 α chain. | Many cell types, including T cells, very early and activated B cells, thymocytes, monocytes, platelets | Integrin that associates with CD29 and mediates binding to fibronectin providing a costimulatory signal to T cells; believed to be important for the maintenance of endothelial monolayer integrity along with CD49b; involved in adhesion, regulation of cell survival, and apoptosis; also, involved in VLA-mediated migration of monocytes into extracellular tissues |
| **CD49f.** 125 kDa. α-6 integrin chain, platelet gpI, VLA-6 α chain. | T cells, monocytes, platelets, megakaryocytes, trophoblasts | Integrin that associates with CD29 and CD104, participates in cell adhesion and migration, embryogenesis, as well as cell surface–mediated signaling; binds laminin receptor on platelets, monocytes, and T cells; provides a costimulatory signal to T cells for activation and proliferation |
| **CD50.** 115–135 kDa (neutrophils). Intercellular adhesion molecule-3, ICAM-3. | Leukocytes of all lineages; epidermal Langerhans cells and endothelial cells; constitutively expressed on resting antigen-presenting cells, including dendritic cells; released from activated lymphocytes and neutrophils and is detectable in the blood as a soluble protein | Shown to bind CD209 (DC-SIGN) as a receptor on dendritic cells and, like CD54 and CD102, is a ligand for LFA-1 CD11a/CD18 and is involved in integrin-β-dependent adhesion; acts as a costimulatory molecule in the immune response and appears to be important in the initial interaction between T and dendritic cells, leading to T-cell activation; regulates leukocyte morphology; promotes redistribution involved in cell chemotaxis and recruitment |
| **CD51.** 125 kDa. VNR-α chain, α-V integrin chain, vitronectin receptor; forms α-V integrin by association with CD61; known to form heterodimers with β1 CD29, β3 CD61, β5, β6, and β8 integrin subunits in various tissues. | Platelets, megakaryocytes, endothelial cells, certain activated leukocytes, NK cells, macrophages, and neutrophils; osteoclasts and smooth muscle cells | Subunit of α-V integrin, which binds vitronectin, von Willebrand factor, fibronectin, thrombospondin, and probably apoptotic cells; shown to mediate the binding of platelets to immobilized vitronectin without prior activation, also to interact with CD47; initiates bone resorption by mediating the adhesion of osteoclasts to osteopontin and may play a role in angiogenesis |
| **CD52.** 25–29 kDa. CAMPATH-1, HE5; single-chain GPI-anchored glycoprotein. | Highly expressed on thymocytes, lymphocytes, monocytes, macrophages, epithelial cells lining male reproductive tract | Not known |
| **CD53.** 32–42 kDa. MRC OX44; member of the 4-transmembrane-spanning protein superfamily (TM4SF). | T and B cells, monocytes, macrophages, granulocytes, dendritic cells, osteoblasts, and osteoclasts | Mediates signal transduction events involved in the regulation of cell development, activation, growth, and mortality; contributes to the transduction of C2-generated signal in T and NK cells and may play a role in growth regulation; cross-linking promotes activation of human B cells and rat macrophages |

*(continued)*

| CD antigen. MW. Synonyms and properties. | Leukocyte expression | Function |
|---|---|---|
| **CD54.** 90, 75–115 kDa. Intercellular adhesion molecule-1 (ICAM-1). | Activated endothelial cells, activated T cells, activated B cells, monocytes, endothelial cells; soluble form detectable in the blood | Ligand for CD11a/CD18 or CD11b/CD18, shown to bind to fibrinogen and hyaluronan; receptor for rhinoviruses and for RBCs infected with malarial parasites; may play a role in development and promotion of adhesion; contributes to antigen-specific T-cell activation by antigen-presenting cells, presumably by enhancing interactions between T cells and antigen-presenting cells; contributes to the extravasation of leukocytes from blood vessels, particularly in areas of inflammation; also, soluble form may inhibit the activation of CTL or NK cells by malignant cells |
| **CD55.** 80 kDa (lymphocytes), 55 kDa (erythrocytes). Decay-accelerating factor (DAF); single-chain GPI-anchored glycoprotein and secreted soluble variant, member of the regulator of complement activation (RCA) family of proteins. | Most cells; also a soluble form in plasma and body fluids | Interacts with complement components, CD97, and a 7-transmembrane domain protein that also contains three extracellular EGF domains; protective barrier against inappropriate complement activation and deposition on plasma membranes; limits formation and half-life of the C3 convertases; may be important to NK mechanisms of killing; also serves as a receptor for echovirus and coxsackie B virus |
| **CD56.** 175–220 kDa. Leu-19, NKH1, neural cell adhesion molecule (NCAM); single-chain type 1 glycoprotein with multiple isoforms; predominant isoform on NK and T cells is in the transmembrane-anchored 140 kDa glycoprotein. | Human NK cells, a subset of CD4$^+$ and CD8$^+$ T cells; also brain in the cerebellum and cortex and at neuromuscular junctions | No clear immune function although may be involved in tumor growth and spreading; regulates interactions between neurons and between neurons and muscle |
| **CD57.** 110 kDa. HNK1, Leu-7; a carbohydrate antigen (oligosaccharide), component of many glycoproteins. | NK cells; subsets of T cells, B cells | Functions in cell-cell adhesion |
| **CD58.** 55–70 kDa. Lymphocyte function-associated antigen 3 (LFA-3); single-chain type 1 glycoprotein, alternate splicing yields two forms, a transmembrane anchored form and a GPI-anchored form. | Many hematopoietic and nonhematopoietic cell types; particularly high on memory T cells and dendritic cells | Adhesion between killer T cells and target cells, antigen-presenting cells and T cells, or thymocytes and thymic epithelial cells; expressed on antigen-presenting cells and enhances T-cell antigen recognition through binding to CD2, the only known ligand for CD58 |
| **CD59.** 18–25 kDa. IF-5Ag, H19, HRF20, MACIF, MIRL, P-18 protectin. Single-chain glycoprotein, member of the Ly-6 superfamily. | Most hematopoietic and nonhematopoietic cell types | Binds to complement components C8 and C9; may bind CD2. Functions as an inhibitor of membrane attack complex (MAC) assembly, thus protecting cells from complement-mediated lysis; signaling role (as a GPI-anchored molecule) in T-cell activation; possible role in cell adhesion |
| **CD60a.** 115–125 kDa (PBMC). Carbohydrate modification found on the ganglioside GD3; previously known as CDw60. | Subset of T cells, platelets, thymic epithelium, activated keratinocytes, synovial fibroblasts, glomeruli, smooth muscle cells, astrocytes | Involved in the regulation of apoptosis and induces mitochondrial permeability transition during apoptosis; promotes B-cell help to and secretion of high levels of IL-4 from T cells |
| **CD60b.** 9-O-acetyl disialosyl ganglioside D3; carbohydrate. | T-cell subsets, activated B cells, neuroectodermal cells in the thymus epithelium and in the skin | Possible mitogen for synovial T cells |
| **CD60c.** 7-O-acetyl-GD3 (7-O-acetylated disialosyl groups); carbohydrate. | T cells | Thought to be an activating receptor for T cells |
| **CD61.** 110 kDa. CD61A, GPIIb/IIIa, β3 integrin chain; β-chain type 1 glycoprotein. | Platelets, megakaryocytes, macrophages, monocytes, mast cells, osteoclasts, endothelial cells, fibroblasts | Integrin subunit, associates with CD41 (Gp11b/11a) or CD51 (vitronectin receptor); participates in cell adhesion, cell surface–mediated signaling, and platelet aggregation |

| CD antigen. MW. Synonyms and properties. | Leukocyte expression | Function |
|---|---|---|
| **CD62E.** 97–115 kDa. E-selectin, ELAM-1, LECAM-2; C-type lectin, single-chain type 1 glycoprotein. | Acutely activated endothelium, chronic inflammatory lesions of skin and synovium | Adhesion molecule that mediates leukocyte rolling on activated endothelium at inflammatory sites; may also participate in tumor-cell adhesion during hematogenous metastasis and may play a role in angiogensis |
| **CD62L.** 74 kDa (lymphocytes), 95 kDa (neutrophils). L-selectin, LAM-1, LECAM-1, Leu-8, MEL-14, TQ-1; C-type lectin, single-chain type 1 glycoprotein. | Most peripheral-blood B cells, naive T cells, monocytes, granulocytes, subsets of NK cells and memory T cells, bone marrow lymphocytes and myeloid cells, thymocytes; soluble form detectable in blood | Adhesion molecule that mediates lymphocyte homing to high endothelial venules of peripheral lymphoid tissue and leukocyte rolling on activated endothelium at inflammatory sites |
| **CD62P.** 140 kDa. P-selectin, granule membrane protein-140 (GMP-140), platelet activation–dependent granule-external membrane protein (PADGEM); C-type lectin, single-chain type 1 glycoprotein. | Endothelium, platelets, megakaryocytes | Interaction with PSGL-1 to mediate tethering and rolling of leukocytes on the surface of activated endothelial cells, the first step in leukocyte extravasation and migration toward sites of inflammation; may contribute to inflammation-associated tissue destruction, atherogenesis, and thrombosis |
| **CD63.** 40–60 kDa. LIMP, MLA1, PTLGP40, gp55, granulophysin lysosomal membrane–associated glycoprotein 3 (LAMP-3), melanoma-associated antigen (ME491), neuroglandular antigen (NGA); lysosome-associated membrane glycoprotein, member of the transmembrane-4 superfamily subfamily 1 (TM4SF subfamily 1). | Platelets, degranulated neutrophils, monocytes, macrophages | Mediates signal transduction events that play a role in the regulation of cell development, activation, growth, and motility; may function as a transmembrane adapter protein, linking other transmembrane and signaling proteins |
| **CD64.** 72 kDa. FcγRI, FCRI; single-chain type 1 glycoprotein. | Monocytes, macrophages, blood and germinal center dendritic cells, PMN activated by IFN-γ or G-CSF, early myeloid-lineage cells | Phagocytosis: receptor-mediated endocytosis of IgG-antigen complexes; antigen capture for presentation to T cells; antibody-dependent cellular cytotoxicity; mediator of release of cytokines and reactive oxygen intermediates |
| **CD65.** Ceramide-dodecasaccharide, VIM-2; glycolipid. | Restricted to myeloid cells; with expression on most granulocytes and a proportion of monocytic cells | Not known |
| **CD65s.** Sialylated-CD65, VIM2; carbohydrate. | Granulocytes, monocytes | Possible involvement with phagocytosis and $Ca^{2+}$ influx |
| **CD66a.** 140–180 kDa. NCA-160, biliary glycoprotein (BGP); single-chain type 1 glycoprotein, member carcinoembryonic antigen (CEA) family, multiple splice variants. | Granulocytes, epithelial cells, possibly on a subset of tissue macrophages, T cells and a subpopulation of activated NK cells, a variety of epithelia and some endothelia | Mediates cell-cell adhesion by homotypic and/or heterotypic interactions with other CD66 molecules; adhesion molecule; receptor for *Neisseria gonorrheae* and *N. meningitidis;* may trigger neutrophil activation; may also contribute to the interactions of activated granulocytes with each other or with endothelium or epithelium; also thought to have tumor suppressor activity |
| **CD66b.** 95–100 kDa. CD67, CGM6, NCA-95; single-chain GPI-anchored glycoprotein, member carcinoembryonic antigen (CEA) family. | Granulocytes, hematopoietic cells of the myeloid lineage; soluble form in plasma | Similar to CD66a, receptor for *Neisseria gonorrheae* and *N. meningitides;* adhesion molecule, trigger of neutrophil activation; enhances the respiratory burst activity of neutrophils, may also regulate the adhesion activity of D11/CD18 in neutrophils; function of the soluble form is unclear |
| **CD66c.** 90 kDa. NCA, NCA-50/90; single-chain GPI-anchored glycoprotein, member carcinoembryonic antigen (CEA) family. | Granulocytes, epithelial cells, possibly on a subset of tissue macrophages, hematopoietic cells of the myeloid lineage; soluble form in plasma | Similar to CD66a and CD66b; receptor for *Neisseria gonorrheae* and *N. meningitides;* adhesion molecule, trigger of neutrophil activation, may regulate the adhesion activity of CD11/CD18 in neutrophils although direct functional evidence is lacking |

*(continued)*

| CD antigen. MW. Synonyms and properties. | Leukocyte expression | Function |
|---|---|---|
| **CD66d.** 35 kDa. CGM1; single-chain type 1 glycoprotein, member carcinoembryonic antigen (CEA) family. | Granulocytes, hematopoietic cells of the myeloid lineage, epithelial cells | Similar to CD66a–c; receptor for *Neisseria gonorrheae* and *N. meningitides*; regulates adhesion activity of CD11/CD18 in neutrophils |
| **CD66e.** 180–200 kDa. Single-chain GPI-anchored glycoprotein, member of the carcinoembryonic antigen (CEA) family. | Epithelial cells, detected in the serum | Interaction of CD66e with Kupffer cells stimulates release of various cytokines that may alter tumor-cell growth in the liver; possible adhesion molecule; also a receptor for *N. gonorrheae*; may play a role in the process of metastasis of cancer cells |
| **CD66f.** 54–72 kDa. Pregnancy-specific b1 glycoprotein, SP-1, pregnancy-specific glycoprotein (PSG). | Fetal liver; produced in placenta and released | Unclear; possible involvement in immune regulation and protection of fetus from maternal immune system; necessary for successful pregnancy since low levels in maternal blood predict spontaneous abortion |
| **CD68.** 110 kDa. gp110, macrosialin; single-chain type 1 glycoprotein, member of the scavenger receptor superfamily. | Monocytes, macrophages, dendritic cells, neutrophils, basophils, mast cells, myeloid progenitor cells, subset of CD34$^+$ hematopoietic progenitor cells, activated T cells, some peripheral-blood B cells; soluble form in blood and urine | Not known |
| **CD69.** 60 kDa. Activation-inducer molecule (AIM), EA 1, MLR3, gp34/28, very early activation (VEA); group V C-type lectin, disulfide-linked homodimeric type 2 glycoprotein. | Activated leukocytes, including T cells, thymocytes, B cells, NK cells; neutrophils, eosinophils, and Langerhans cells; subset of mature thymocytes, germinal center mantle B cells, some CD4$^+$ germinal center T cells | Involvement in early events of lymphocyte, monocyte, and platelet activation: promotes Ca$^{2+}$ influx, synthesis of cytokines and their receptors, induction of c-*myc* and c-*fos* proto-oncogene expression; role in promoting lysis mediated by activated NK cells |
| **CD70.** 75, 95 170 kDa. CD27-ligand, Ki-24 antigen; trimeric type 2 glycoprotein, member of the TNF superfamily. | Activated T and B cells and stimulated NK cells | Ligand for CD27; possible role in costimulation of B and T cells and may augment the generation of cytotoxic T cells and cytokine production |
| **CD71.** 190 kDa. T9, transferrin receptor; disulfide-linked homodimeric type 2 glycoprotein. | All proliferating cells | Iron uptake: binds ferrotransferrin at neutral pH and internalizes complex to acidic endosomal compartment, where iron is released |
| **CD72.** 39–43 kDa. Ly-19.2, Ly-32.2, Lyb-2; C-type lectin, disulfide-linked homodimeric type 2 glycoprotein. | B cells (except plasma cells), some dendritic cells, and may be expressed by tissue macrophages | May play a role in regulating the signal threshold in B cells; contains an activating tyrosine motif and possibly activates a variety of signaling pathways through the B-cell receptor and can induce MHC class II expression and B-cell proliferation |
| **CD73.** 69, 70, 72 kDa. Ecto-5′, nucleotidase; single-chain GPI-anchored glycoprotein. | Subpopulations of T cells (expression is confined to the CD28$^+$ subset) and B cells (about 75% of adult peripheral blood B cells), with expression increasing during development; follicular dendritic cells, epithelial cells, endothelial cells | Possibly regulates the availability of adenosine for interaction with the cell surface adenosine receptor by converting AMP to adenosine; can mediate costimulatory signals for T-cell activation; may play a role in mediating the interaction between B cells and follicular dendritic cells |
| **CD74.** 33, 35, 41 kDa. Class II–specific chaperone, Ii, invariant chain; homotrimers, single-chain type 2 glycoprotein. | Mostly found intracellularly in MHC class II–expressing cells, specifically, B cells, activated T cells, macrophages, activated endothelial and epithelial cells | Intracellular sorting of MHC class II molecules; function in humans is unclear |
| **CD75.** Formally known as CDw75. Carbohydrate antigen. | B cells, subpopulation of peripheral-blood T cells, erythrocytes, broad range of epithelial cells | Cell adhesion; ligand for CD22 |
| **CD75s.** Formally known as CDw76. Carbohydrate antigen. | Mature sIg$^+$ B cells, mantle zone B cells of lymphoid secondary follicles, subpopulation of T cells, mature B lineage leukemias, subsets of endothelial and epithelial cells, possibly weakly expressed on erythrocytes | Considered to be a binding partner for the B-cell-specific activation antigen CD22; exact function is unclear |

| CD antigen. MW. Synonyms and properties. | Leukocyte expression | Function |
|---|---|---|
| **CD77.** Pk blood-group antigen, Burkitt's lymphoma antigen (BLA), ceramide trihexoside (CTH), globotriaosylceramide (Gb3); sphingolipid antigen, type 2 membrane protein. | Germinal center B cells | Critical cell surface molecule able to mediate an apoptic signal; association with type 1 interferon receptor or with HIV coreceptor CXCR4 (CD184) may be essential for function; a receptor for lectins on the pili of a certain strain of *E. coli;* may be involved in the selection process within the germinal centers |
| **CD79a.** 33–45 kDa. Ig-α, MB1; type 1 glycopeptide, disulfide-linked heterodimer with CD79b. | B cells | Component of B-cell antigen receptor analogous to CD3; required for cell surface expression and signal transduction |
| **CD79b.** 37 kDa. B29, Ig-β; type 1 glycopeptide, disulfide-linked heterodimer with CD79a. | B cells | Same as CD79a |
| **CD80.** 60 kDa. B7, B7.1, BB1, Ly-53; single-chain type 1 glycoprotein. | Activated B and T cells, macrophages, low levels on resting peripheral blood monocytes and dendritic cells | Binds CD28 and CD152; costimulation of T-cell activation with CD86 when bound to CD28; inhibits T-cell activation when bound with CD152 |
| **CD81.** 26 kDa. Target for antiproliferative antigen-1 (TAPA-1). Single-chain type 3-, 4-span glycoprotein, member of the 4-transmembrane-spanning protein superfamily (TM4SF). | Broadly expressed on hematopoietic cells; expressed by endothelial and epithelial cells; absent from erythrocytes, platelets, and neutrophils | Member of CD19/CD21/Leu-13 signal transduction complex; mediates signal transduction events involved in the regulation of cell development, growth, and motility; participates in early T-cell development; binds the E2 glycoprotein of hepatitis C virus |
| **CD82.** 45–90 kDa. 4F9, C33, IA4, KAI1, R2; single-chain type 3-, 4-span glycoprotein, member of the 4-transmembrane-spanning protein superfamily (TM4SF). | Activated/differentiated hematopoietic cells, B and T cells, NK cells, monocytes, granulocytes, platelets, epithelial cells | Signal transduction: may induce T-cell spreading and pseudopod formation, modulate T-cell proliferation, and provide costimulatory signals for cytokine production; possible role in activation of monocytes |
| **CD83.** 43 kDa. HB15; single-chain type 1 glycoprotein. | Dendritic cells, B cells, Langerhans cells | May play a role in antigen presentation and/or lymphocyte activation and regulation of the immune response |
| **CD84.** 72-86 kDa. GR6. | Virtually all thymocytes, monocytes, platelets, circulating B cells | Not known |
| **CD85a.** 110 kDa. GR4. | Monocytes, macrophages, dendritic cells, granulocytes, a subpopulation of T lymphocytes | Suppression of NK-cell-mediated cytotoxicity |
| **CD86.** 80 kDa. B7.2, B70; single-chain type 1 glycoprotein. | Dendritic cells, memory B cells, germinal center B cells, monocytes | Major T-cell costimulatory molecule, interacting with CD28 (stimulatory) and CD152/CTLA4 (inhibitory) |
| **CD87.** 32–56 kDa (monocytes). Urokinase plasminogen activator receptor (uPAR); single-chain GPI-anchored glycoprotein. | T cells, NK cells, monocytes, neutrophils; nonhematopoietic cells such as vascular endothelial cells, fibroblasts, smooth muscle cells, keratinocytes, placental trophoblasts, hepatocytes | Receptor for uPA, which can convert plasminogen to plasmin; possible role in b2 integrin–dependent adherence and chemotaxis; may play a role in the process of neoplastic and inflammatory cell invasion |
| **CD88.** 43 kDa. C5a receptor, C5aR; type 3, 7-span glycoprotein, member of the 7-transmembrane-spanning protein superfamily (TM7SF). | Granulocytes, monocytes, dendritic cells, astrocytes, microglia, hepatocytes, alveolar macrophages, vascular endothelial cells | C5a-mediated inflammation, activation of granulocytes, possible function in mucosal immunity |
| **CD89.** 45–70, 55–75, 70–100, 50–65 kDa. Fcα-receptor R (Fcα-R), IgA Fc receptor, IgA receptor; single-chain type 1 glycoprotein. | Myeloid-lineage cells from promyelocytes to neutrophils and from promonocyte to monocytes; activated eosinophils, alveolar and splenic macrophages; subsets of T and B cells | Induction of phagocytosis, degranulation, respiratory burst, killing of microorganisms |
| **CD90.** 25–35 kDa. Thy-1; single-chain GPI-anchored glycoprotein. | Hematopoietic stem cells, neurons, connective tissue, thymocytes, peripheral T cells, human lymph node HEV endothelium | Possible involvement in lymphocyte costimulation; possible inhibition of proliferation and differentiation of hematopoietic stem cells |

*(continued)*

| CD antigen. MW. Synonyms and properties. | Leukocyte expression | Function |
|---|---|---|
| **CD91.** 515, 85 kDa. α-2-macroglobulin receptor (ALPHA2M-R), low-density lipoprotein receptor–related protein (LRP); single-chain type 1 glycoprotein. | Phagocytes, many nonhematopoietic cells | Endocytosis-mediating receptor expressed in coated pits that appears to play a role in the regulation of proteolytic activity and lipoprotein metabolism |
| **CD92.** 70 kDa. Formally known as CDw92; CTL1, GR9. | Monocytes, granulocytes, peripheral blood lymphocytes (PBL), mast cells | Not known; likely plays a role in signal transduction |
| **CD93.** 110 kDa. Formally known as CDw93; GR11. | Monocytes, granulocytes, endothelial cells | Not known |
| **CD94.** 70 kDa. Kp43; forms complex with NKG2 receptors, C-type lectin. | NK cells, subsets of CD8$^+$ αβ and γδ T cells | Depending on NKG2 molecule associated, may activate or inhibit NK-cell cytotoxicity and cytokine release |
| **CD95.** 45 kDa. APO-1, Fas antigen (Fas); TNF receptor superfamily, single-chain type 1 glycoprotein. | Activated T and B cells, monocytes, fibroblasts, neutrophils | Mediation of apoptosis-inducing signals |
| **CD96.** 160 kDa. T-cell activation increased late expression (TACTILE); single-chain type 1 glycoprotein. | Activated T cells, NK cells | Adhesion of activated T and NK cells during the late phase of immune response; also involved in antigen presentation and/or lymphocyte activation |
| **CD97.** 75–85 kDa (PBMC, CD97a), 28 kDa (PBMC, CD97b). BL-KDD/F12; EGF-TM7 subfamily member, type 3-, 7-span glycoprotein, three isoforms. | Activated B and T cells, monocytes, granulocytes | Binding to CD55, neutrophil migration |
| **CD98.** 80 and 45 kDa. 4F2, FRP-1, RL-388 in mouse; disulfide-linked heterodimeric type 2 glycoprotein. | Not hematopoietic specific; activated and transformed cells; lower levels on quiescent cells; high levels on monocytes | Role in regulation of cellular activation and aggregation |
| **CD99.** 32 kDa. CD99R (epitope restricted to subset of CD99 molecules), E2, *MIC2* gene product; single-chain type 1 glycoprotein. | All leukocytes; highest on thymocytes | Augments T-cell adhesion, induces apoptosis of double-positive thymocyte, participates in leukocyte migration, involved in T-cell activation and adhesion, binds to cyclophilin A |
| **CD100.** 150 kDa (PHA blasts), 120 kDa (soluble). Disulfide-linked homodimeric type 1 glycoprotein. | Most hematopoietic cells except immature bone marrow cells, RBCs, and platelets; activated T cells, germinal center B cells | Monocyte migration, T- and B-cell activation and T-B cell and T-dendritic cell interaction; shown to induce T-cell proliferation |
| **CD101.** 120 kDa. P126, V7; disulfide-linked homodimeric type 1 glycoprotein. | Monocytes, granulocytes, dendritic cells, mucosal T cells, activated peripheral blood T cells; weak on resting T, B, and NK cells | Possible costimulatory role in T-cell activation |
| **CD102.** 55–65 kDa. Intercellular adhesion molecule-2 (ICAM-2); single-chain type 1 glycoprotein. | Vascular endothelial cells, monocytes, platelets, some populations of resting lymphocytes | Like related proteins CD54 and CD50, binds CD11a/CD18 LFA-1, also reported to bind to CD11b/CD18 Mac-1; may play a role in lymphocyte recirculation. Shown to mediate adhesive interactions important for antigen-specific immune response, NK-cell-mediated clearance, lymphocyte recirculation and other cellular interactions important for immune response and surveillance; may also be involved in T-cell activation and adhesion |
| **CD103.** 150 kDa, 25 kDa. HML-1, integrin αE chain; α-chain type 1 glycopeptide. | Intraepithelial lymphocytes (in tissues such as intestine, bronchi, inflammatory skin/breast/salivary glands), many laminapropria T cells, some lymphocytes in peripheral blood and peripheral lymphoid organs | Binds to E-cadherin and integrin β7; role in the tissue-specific retention of lymphocytes at basolateral surface of intestinal epithelial cells; possible accessory molecule for activation of intraepithelial lymphocytes |

| CD antigen. MW. Synonyms and properties. | Leukocyte expression | Function |
|---|---|---|
| **CD104.** 220 kDa. β4 integrin chain, tumor-specific protein 180 antigen (TSP-1180) in mouse; β-chain type 1 glycopeptide. | CD4$^+$, CD8$^+$ thymocytes; neuronal, epithelial, and some endothelial cells; Schwann cells; trophoblasts | Integrin that associates with CD49f, binds laminins and plectin, also interacts with keratin filaments intracellularly; involved in cell-cell, cell matrix interactions, adhesion, and migration; important role in the adhesion of epithelia to basement membranes |
| **CD105.** 90 kDa. Endoglin; TGFβ type III receptor, disulfide-linked homodimeric type 1 glycoprotein. | Endothelial cells of small and large vessels; activated monocytes and tissue macrophages; stromal cells of certain tissues, including bone marrow; pre–B cells in fetal marrow; erythroid precursors in fetal and adult bone marrow; syncytiotrophoblast throughout pregnancy and cytotrophoblasts transiently during first trimester | Modulator of cellular responses to TGF-β1 |
| **CD106.** 100–110 kDa. INCAM-110, vascular-cell adhesion molecule-1 (VCAM-1); single-chain type 1 glycoprotein with multiple isoforms. | Endothelial cells, follicular and interfollicular dendritic cells, some macrophages, bone marrow stromal cells, nonvascular cell populations within joints, kidney, muscle, heart, placenta, and brain; can be induced on endothelia and other cell types in response to inflammatory cytokines | Adhesion molecule that is ligand for VLA-4; involved in leukocyte adhesion, transmigration, and costimulation of T-cell proliferation; contributes to the extravasation of lymphocytes, monocytes, basophils, and eosinophils but not neutrophils from blood vessels |
| **CD107a.** 100–120 kDa. Lysosome-associated membrane protein 1 (LAMP-1); single-chain type 1 glycoprotein. | Activated platelets, endothelial cells, tonsillar epithelium, granulocytes, T cells, macrophages, dendritic cells, lysosomal membrane, degranulated platelets, PHA-activated T cells, TNF-α–activated endothelium, FMLP-activated neutrophils | Associated with enhanced metastatic potential of tumor cells |
| **CD107b.** 100–120 kDa. Lysosome-associated membrane protein 2 (LAMP-2); single-chain type 1 glycoprotein, tissue-specific isoforms. | Granulocytes, lysosomal membrane, activated and degranulated platelets, TNF-α–activated endothelium, FMLP-activated neutrophils, tonsillar epithelium | Protection, maintenance, and adhesion of lysosomes; associated with enhanced metastatic potential of tumor cells |
| **CD108.** 80 kDa. Formally known as CDw108. John-Milton-Hagen (JMH) human-blood-group antigen; GPI-anchored glycoprotein. | Erythrocytes, circulating lymphocytes, lymphoblasts | May play a role in monocyte activation and in regulating immune cells |
| **CD109.** 175 kDa. 8A3, E123 (7D1); GPI-anchored glycoprotein. | Activated T cells, activated platelets, human umbilical vein endothelial cells | Not known |
| **CD110.** 85–92 kDa. Myeloproliferative leukemia virus oncogene (MPL), thrombopoietin receptor (TPO-R), C-MPL; cytokine receptor superfamily, single-chain type 1 glycoprotein. | Hematopoietic stem and progenitor cells, megakaryocyte progenitors, megakaryocytes, platelets | Binds thrombopoietin; main regulator of megakaryocyte and platelet formation |
| **CD111.** 75 kDa. Herpesvirus Ig-like receptor (HIgR), poliovirus receptor–related 1 (PRR1), poliovirus receptor–related 1(PVRL1), nectin 1, HevC; type 1 glycoprotein, member of the nectin family. | Expressed in multiple cell types intracellularly and in vesicle-like structures | Ca$^{2+}$-independent immunoglobulin (Ig)–like cell-cell adhesion molecules; important involvement in the formation of many types of cell-cell junctions and cell-cell contacts |
| **CD112.** 72 kDa (long isoform), 64 kDa (short isoform). Herpesvirus entry protein (HVEB), poliovirus receptor–related 2 (PRR2), poliovirus receptor–related 2(PVRL2), nectin 2; type 1 glycoprotein, member of the nectin family. | Multiple tissues and cell lineages | Hematopoietic function unclear; involved in herpesvirus entry activity and may function as a coreceptor for HSV-1, HSV-2, and pseudorabies |
| **CDw113.** Poliovirus receptor-related 3 (PRR3), poliovirus receptor-related 3 (PVRL3), nectin 3; transmembrane protein. | Epithelial cells, testis, liver, and placenta | Interacts with CD111 (nectin-1) and CD112 (nectin-2); function unclear, likely involved in adhesion |
| **CD114.** 150 kDa. CSF3R, HG-CSFR, granulocyte colony-stimulating factor receptor (G-CSFR). | All stages of granulocyte differentiation, monocytes, mature platelets, several nonhematopoietic cell types/tissues, including endothelial cells, placenta, trophoblastic cells | Specific regulator of myeloid proliferation and differentiation |

*(continued)*

| CD antigen. MW. Synonyms and properties. | Leukocyte expression | Function |
|---|---|---|
| **CD115.** 150 kDa. C-fms, colony-stimulating factor 1R (CSF-1R), macrophage colony-stimulating factor receptor (M-CSFR). | Monocytes, macrophages | Macrophage colony-stimulating factor (M-CSF) receptor |
| **CD116.** 80 kDa. GM-CSF receptor α chain; member of the cytokine receptor superfamily, α-chain type 1 glycoprotein. | Various myeloid cells, including macrophages, neutrophils, eosinophils; dendritic cells and their precursors; fibroblasts, endothelial cells | Primary binding subunit of the GM-CSF receptor |
| **CD117.** 145 kDa. c-KIT, stem-cell factor receptor (SCFR); single-chain type 1 glycoprotein. | Hematopoietic stem cells and progenitor cells, mast cells, bone marrow stromal cells | Stem-cell factor receptor, tyrosine kinase activity; early-acting hematopoietic growth factor receptor, necessary for the development of hematopoietic progenitors; capable of inducing proliferation of mast cells and is a survival factor for primordial germ cells |
| **CD118.** 190 kDa. LIF receptor, gp190; transmembrane protein belonging to the type I cytokine receptor superfamily, forms a heterodimer with gp130. | Adult and embryonic epithelial cells, monocytes, fibroblasts, embryonic stem cells, liver, placenta | High-affinity receptor (in complex with gp130) for leukemia inhibitory factor (LIF); cell differentiation, signal transduction and proliferation |
| **CD119.** 90–100 kDa. IFN-γR, IFN-γRa; class 2 cytokine receptor, type 1 glycopeptide. | Monocytes, macrophages, T and B cells, NK cells, neutrophils, fibroblasts, epithelial cells, endothelium, a wide range of tumor cells | Interferon receptor; role in host defense and the initiation and effector phases of immune responses, including macrophage activation, B- and T- cell differentiation, activation of NK cells, up-regulation of the expression of MHC class I and II antigens |
| **CD120a.** 50–60 kDa. TNFRI, p55; TNF receptor superfamily member, single-chain type 1 glycoprotein. | Constitutively on hematopoietic and nonhematopoietic cells; highest on epithelial cells | TNF and lymphotoxin-α receptor; mediates the signaling involved in proinflammatory cellular responses, programmed cell death, and antiviral activity |
| **CD120b.** 75–85 kDa. TNFRII, p75; TNF receptor superfamily member, single-chain type 1 glycoprotein. | Constitutively on hematopoietic and non-hematopoietic cells; highest on epithelial cells | TNF and lymphotoxin-α receptor; mediates the signaling involved in pro-inflammatory cellular responses, programmed cell death and anti-viral activity |
| **CD121a.** 80 kDa. IL-1R, IL-1R type 1, type 1 IL-1R; type 1 glycoprotein. | T cells, thymocytes, chondrocytes, synovial cells, hepatocytes, endothelial cells, keratinocytes; low levels on fibroblasts, lymphocytes, monocytes, macrophages, granulocytes, and dendritic, epithelial, and neural cells | Type I interleukin-1 receptor; mediates thymocyte and T-cell activation, fibroblast proliferation, induction of acute phase proteins and inflammatory reactions |
| **CDw121b.** 60–70 kDa. IL-1R type 2, type 2 IL-1R; single-chain type 1 glycoprotein. | B cells, macrophages, monocytes, neutrophils | Type I interleukin-1 receptor, likely a decoy receptor with no true function |
| **CD122.** 70–75 kDa. Interleukin-2 receptor β chain (IL-2Rβ); cytokine receptor superfamily, type 1 glycopeptide. | Activated T cells, B cells, NK cells, monocytes, macrophages, subset of resting T cells | Critical component of IL-2– and IL-15–mediated signaling; role in T-cell-mediated immune reponse; promotes proliferation and activation of T cells, thymocytes, macrophages, B cells, and NK cells |
| **CD123.** 70 kDa. IL-3 receptor α subunit (IL-3Rα). | Bone marrow stem cells, granulocytes, monocytes, megakaryocytes | IL-3 receptor chain |
| **CD124.** 140 kDa. IL-4R, IL-13R (α chain); cytokine receptor superfamily, α type 1 glycopeptide. | Mature B and T cells, hematopoietic precursors, fibroblasts, epithelial and endothelial cells, hematopoietic and nonhematopoietic cells | Receptor subunit for IL-4 and IL-13 |
| **CDw125.** 60 kDa. IL-5Rα; cytokine receptor superfamily, α type 1 glycopeptide; also exists as soluble form. | Eosinophils, activated B cells, basophils, mast cells | Low-affinity receptor for IL-5; α chain of IL-5 receptor; secreted form antagonizes IL-5-induced eosinophil activation and proliferatin |
| **CD126.** 80 kDa. Interleukin-6 receptor (IL-6R); associates with CD130, cytokine receptor superfamily, α type 1 glycopeptide; also exists as soluble form. | T cells, monocytes, activated B cells, hepatocytes, some other nonhematopoietic cells | Receptor for IL-6; soluble form is capable of binding to gp130 on cells and promoting IL-6-induced responses |

| CD antigen. MW. Synonyms and properties. | Leukocyte expression | Function |
|---|---|---|
| **CD127.** 65–75, 90 kDa. IL-7 receptor (IL-7R), IL-7 receptor α (IL-7Rα) p90; cytokine receptor superfamily, α type 1 glycopeptide. | B-cell precursors, mature resting T cells, thymocytes | Specific receptor for interleukin-7; may regulate immunoglobulin gene rearrangement |
| **CDw128b.** 58–67 kDa. CXCR2, interleukin-8 receptor B (IL-8RB) CD181; G-coupled protein receptor 1 family. | See CD181 | See CD181 |
| **CD129.** 60-65 kDa. CD129 (α chain), IL9R; α-chain type 1 glycopeptide. | Activated T-cell lines, T and B cells, both erythroid and myeloid precursors | Receptor for IL-9; promotes the growth of activated T cells, generation of erythroid and myeloid precursors |
| **CD130.** 130–140 kDa. IL-6Rβ, IL-11R, gp130; cytokine receptor superfamily, single-chain type 1 glycoprotein. | T cells, monocytes, endothelial cells; at high levels on activated and EBV-transformed B cells, plasma cells, and myelomas; lower levels on most leukocytes, epithelial cells, fibroblasts, hepatocytes, neural cells | Binding to the CD126/IL-6R complex stabilizes it, resulting in the formation of a high-affinity receptor; required for signal transduction by interleukin 6, interleukin 11, leukemia inhibitory factor, ciliary neurotrophic factor, oncostatin M, cardiotrophin-1 |
| **CD131.** 120–140 kDa. GM-CSFR, IL-3R, IL-5R (β chain); common beta subunit, cytokine receptor superfamily, β-chain type 1 glycopeptide. | Fibroblasts and endothelial cells; most myeloid cells, including early progenitors, and early B cells | Receptor subunit required for signal transduction by IL-3, GM-CSF, and IL-5 receptors |
| **CD132.** 64, 65–70 kDa. Common cytokine receptor γ chain, common γ chain, member of cytokine receptor superfamily, γ-chain type I glycopeptide. | T cells, B cells, NK cells, monocytes/macrophages, neutrophils | Subunit of IL-2, IL-4, IL-7, IL-9, and IL-15 receptors |
| **CD133.** 115–125 kDa. AC133, hematopoietic stem-cell antigen, prominin-like 1 (PROML1), prominin; pentaspan 5-transmembrane domain glycoprotein. | Hematopoietic stem cell, endothelial and epithelial cells | Stem-cell marker with no known function |
| **CD134.** 47-51 kDa. OX40; TNF receptor superfamily, single-chain type 1 glycoprotein. | Activated and regulatory T cells | Binding to OX40-ligand results in the induction of B-cell proliferation, activation, Ig production; provides necessary costimulation for T-cell proliferation, activation, adhesion, differentiation, apoptosis |
| **CD135.** 155-160 kDa. FMS-like tyrosine kinase 3 (flt3), Flk-2 in mice, STK-1; tyrosine kinase receptor, type 3, single-chain type 1 glycoprotein. | Multipotential, myelomonocytic, and primitive B-cell progenitors | Growth factor receptor for early hematopoietic progenitors, tyrosine kinase |
| **CDw136.** 150, 40 kDa. Macrophage-stimulating protein receptor (msp receptor), ron (p158-ron); tyrosine kinase receptor family, single-chain type 1 heterodimeric glycoprotein. | Macrophages; epithelial tissues, including skin, kidney, lung, liver, intestine, colon | Induction of migration, morphological change, cytokine induction, phagocytosis proliferation,and apoptosis in different target cells; may play a role in inflammation, wound healing, the mechanisms of activation in invasive growth and movement of epithelial tumors |
| **CDw137.** 39 kDa. 4-1BB, induced by lymphocyte activation (ILA), TNF receptor superfamily, single-chain type 1 glycoprotein. | T cells, B cells, monocytes, epithelial and hepatoma cells | Costimulator of T-cell proliferation via binding to 4-1BBL |
| **CD138.** 92 kDa (immature B cells), 85 kDa (plasma cells). Heparin sulfate proteoglycan, syndecan-1; type 1 glycoprotein. | Pre–B cells, immature B cells and plasma cells but not mature circulating B lymphocytes; basolateral surfaces of epithelial cells, embryonic mesenchymal cells, vascular smooth muscle cells, endothelium, neural cells, breast cancer cells | Binds to many extracellular matrix proteins and mediates cell adhesion and growth |
| **CD139.** 205-230 kDa. B-031. | B cells, monocytes, granulocytes, follicular dendritic cells, erythrocytes | Not known |

*(continued)*

| CD antigen. MW. Synonyms and properties. | Leukocyte expression | Function |
|---|---|---|
| **CD140a.** 180 kDa. PDGF receptor (PDGF-R), PDGFR, α platelet-derived growth factor receptor (PDGFRα); tyrosine kinase receptor, type 3 family. | Mesenchymal cells | Receptor for platelet-derived growth factor (PDGF); involved in cell proliferation, differentiation, and survival and in signal transduction associated with PDGFR |
| **CD140b.** 180 kDa. β platelet-derived growth factor receptor (PDGFRβ); tyrosine kinase receptor, type 3 family. | Endothelial cells, subsets of stromal cells, on mesenchymal cells | Receptor for platelet-derived growth factor (PDGF); involved in cell proliferation, differentiation, and survival and in signal transduction associated with PDGFR |
| **CD141.** 105 kDa. Fetomodulin, thrombomodulin (TM); C-type lectin, single-chain type 1 glycoprotein. | Endothelial cells, megakaryocytes, platelets, monocytes, neutrophils | Essential molecule for activation of protein C and initiation of the protein C anticoagulant pathway |
| **CD142.** 45–47 kDa. Coagulation factor III, thromboplastin, tissue factor (TF); serine protease cofactor, single-chain type 1 glycoprotein. | High levels on epidermal keratinocytes, glomerular epithelial cells, and various other epithelia; inducible on monocytes and vascular endothelial cells by various inflammatory mediators | Initiates coagulation protease cascade assembly and propagation, functions in normal hemostasis, and is a component of the cellular immune response; may play a role in tumor metastasis, breast cancer, hyperplasia, and angiogenesis |
| **CD143.** 170, 180, 90, 110 kDa. EC 3.4.15.1, angiotensin-converting enzyme (ACE), kinase II, peptidyl dipeptidase A; type 1 glycoprotein. | Endothelial cells, activated macrophages, weakly on subsets of T cells, some dendritic-cell subsets | An enzyme important in regulation of blood pressure; acts primarily as a peptidyl dipeptide hydrolase and is involved in the metabolism of two major vasoactive peptides, angiotensin II and bradykinin |
| **CD144.** 130 kDa. Cadherin-5 VE-cadherin; type 1 glycoprotein. | Endothelium, stem-cell subsets | Control of endothelial cell-cell adhesion, permeability, and migration |
| **CDw145.** 25 kDa, 90 kDa, 110 kDa. | Highly on endothelial cells | Not known |
| **CD146.** 130 kDa. A32, MCAM, MUC18, mel-CAM, S-endo; type 1 glycoprotein. | Follicular dendritic cells, endothelium, melanoma, smooth muscle, intermediate trophoblast, endothelium, a subpopulation of activated T cells | Potential adhesion molecule |
| **CD147.** 55–65 kDa. 5A11, basigin, CE9, HT7, M6, neurothelin, OX-47, extracellular-matrix metalloproteinase inducer (EMMPRIN), gp42 in mouse; type 1 glycoprotein. | All leukocytes, red blood cells, platelets, endothelial cells | Potential cell adhesion molecule and is involved in the regulation of T-cell function |
| **CD148.** 240–260, 200–250 kDa. HPTP-eta, high-cell-density–enhanced PTP 1 (DEP-1), p260; single-chain type 1 glycoprotein belonging to protein tyrosine phosphatase (PTP) family. | Granulocytes, monocytes, weakly on resting T cells and up-regulated following activation, high levels on memory T cells, dendritic cells, platelets, fibroblasts, nerve cells, Kupffer cells | Regulates signaling pathways of a variety of cellular processes, including cell growth, differentiation, mitotic cycle and oncogenic transformation, contact inhibition of cell growth |
| **CD150.** 75–95, 70 kDa. Formally known as CDw150, IPO-3, signaling lymphocyte activation molecule (SLAM); single-chain type 1 glycoprotein. | Thymocytes, subpopulation of T cells, B cells, dendritic cells, endothelial cells | B-cell and dendritic cell costimulation, T-cell activation; contributes to enhancement of immunostimulatory functions of dendritic cells |
| **CD151.** 27 kDa. PETA-3, SFA-1. Tetraspanin family, single-chain type 3-, 4- span glycoprotein. | Platelets, megakaryocytes, immature hematopoietic cells, endothelial cells | Adhesion molecule, may regulate integrin trafficking and/or function; enhances cell motility, invasion and metastasis of cancer cells |
| **CD152.** ~ 33 kDa. Cytotoxic T lymphocyte–associated protein-4 (CTLA-4) disulfide-linked homodimeric type 1 glycoprotein. | Activated T cells, perhaps some activated B cells | Negative regulator of T-cell activation |
| **CD153.** 40 kDa. CD30 ligand, CD30L; TNF superfamily, single-chain type 2 glycoprotein. | Activated T cells, activated macrophages, neutrophils, B cells | Ligand for CD30; costimulates T cells |
| **CD154.** 33 kDa. CD40 ligand (CD40L), T-BAM, TNF-related activation protein (TRAP), gp39; TNF superfamily, homotrimeric type 2 glycoprotein. | Activated CD4$^+$ T cells, small subset of CD8$^+$ T cells and γδ T cells; also activated basophils, platelets, monocytes, mast cells | Ligand for CD40, inducer of B-cell proliferation and activation, antibody class switching and germinal center formation; costimulatory molecule and a regulator of $T_H1$ generation and function; role in negative selection and peripheral tolerance |

| CD antigen. MW. Synonyms and properties. | Leukocyte expression | Function |
|---|---|---|
| **CD155.** 80–90 kDa. Polio virus receptor (PVR). | Monocytes, macrophages, thymocytes, CNS neurons | Normal function unknown; receptor for poliovirus |
| **CD156a.** 69 kDa. ADAM8, MS2 human; single-chain type 1 glycoprotein. | Neutrophils, monocytes | Involved in inflammation, cell adhesion; may play a role in muscle differentiation, signal transduction; possible involvement in extravasation of leukocytes |
| **CD156b.** 100–120 kDa. a disintegrin and metalloproteinase domain 17 (ADAM17), TNF-α converting enzyme (TACE), snake venom–like protease (cSVP); processed and unprocessed form, type 1 glycoprotein. | T cells, neutrophils, endothelial cell monocytes, dendritic cells, macrophages, polymorphonuclear leukocytes, myocytes | Primary protease that cleaves transmembrane forms of TNF-α and TGF-α to generate soluble forms |
| **CD157.** 42–45, 50 kDa. BP-3/IF-7, BST-1, Mo5; single chain, GPI-anchored protein. | Granulocytes, monocytes, macrophages, some B-cell progenitors, some T-cell progenitors | Support for growth of lymphocyte progenitors |
| **CD158a.** 58, 50 kDa. KIR2DL1, EB6, MHC class I–specific receptors, p50.1, p58.1; member of the killer-cell immunoglobulin-like receptor (KIR) family, type 1 glycoprotein. | Most NK cells, some T-cell subsets | Suppression of NK-cell- and CTL-mediated cytolytic activity on interaction with the appropriate HLA-C alleles |
| **CD158b.** 58, 50 kDa. GL183, MHC class I–specific receptors, p50.2, p58.2; member of the killer-cell immunoglobulin-like receptor (KIR) family, type 1 glycoprotein. | Most NK cells, some T-cell subsets | Suppression of NK-cell- and CTL-mediated cytolytic activity on interaction with the appropriate HLA-C alleles |
| **CD159a.** 43 kDa. NKG2, Killer-cell lectin-like receptor subfamily C, member 1 (KLRC1); type 2 glycoprotein and a member of the NKG2 family, disulfide-linked heterodimers covalently bonded to CD94. | NK cell lines, CD8$^+$ γδ cells, on some T-cell subset clones and lines | Potent negative regulator of NK cells and T-lymphocyte activation programs; implicated in activation and inhibition of NK-cell cytotoxicity and cytokine secretion |
| **CD160.** 27 kDa. BY55 antigen, NK1, NK28; expressed as a disulfide-linked multimer at cell surface, type 2 GPI-anchored glycoprotein. | Peripheral blood NK cells and CD8$^+$ T cells, IELs | Binds HLA-C, provides costimulatory signals in CD8$^+$ T lymphocytes |
| **CD161.** ~ 40 kDa. NKR-P1A, Killer-cell lectin-like receptor subfamily B member 1 (KLRB1); C-type lectin, disulfide-bonded homodimeric type 2 glycoprotein. | Most NK cells, a subset of CD4$^+$ and CD8$^+$ T cells, thymocytes | May play a role in NK-cell-mediated cytotoxicity function, induction of immature thymocyte proliferation |
| **CD162.** 110–120 kDa. PSGL-1; disulfide-linked homodimeric mucin-like type 1 glycoprotein. | Most peripheral-blood T cells, monocytes, granulocytes, B cells | Major CD62P ligand on neutrophils and T lymphocytes, mediates adhesion and leukocyte rolling and tethering on endothelial cells |
| **CD162R.** 140 kDa. Post-translational modification of PSGL-1(PEN5), PSGL1, selectin P ligand (SELPLG); poly-N-lactosame carbohydrate. | NK cells | Creates a unique binding site for L-selectin; may be a unique developmentally specific NK-cell marker |
| **CD163.** 130 kDa. GHI/61, M130; single-chain type 1 glycoprotein. | Monocytes, macrophages, myeloid cells; low level on lymphocytes, bone marrow stromal cells, subset of erythroid progenitors | Involved in hematopoietic progenitor-cell–stromal-cell interaction and endocytosis |
| **CD164.** 80–100 kDa. MUC-24, multiglycosylated core protein 24 (MGC-24v); type 1 glycoprotein. | Lymphocytes, epithelial cells, monocytes, granulocytes | Most exist intracellularly; facilitates adhesion of CD34$^+$ and plays a role in regulating hematopoietic cell proliferation |
| **CD165.** 42 kDa. AD2, gp37; membrane glycoprotein. | Peripheral lymphocytes, immature thymocytes, monocytes, most platelets; low level on thymocytes and thymic epithelial cells | Adhesive interactions, including adhesion between thymocytes and thymic epithelial cells; platelet formation |
| **CD166.** 100–105 kDa. BEN, DM-GRASP, KG-CAM, neurolin, SC-1, activated-leukocyte cell adhesion molecule (ALCAM); single-chain type 1 glycoprotein. | Activated T cells, activated monocytes, epithelium, neurons, fibroblasts, cortical and medullary thymic epithelial cells | Adhesion molecule that binds to CD6; involved in neurite extension by neurons via heterophilic and homophilic interactions; may have a role in T-cell development |

*(continued)*

| CD antigen. MW. Synonyms and properties. | Leukocyte expression | Function |
|---|---|---|
| **CD167a.** 129 kDa. α subunit 54 kDa, β subunit 63 kDa. DDR1 (discoidin domain family member 1), receptor tyrosine kinase. | Mainly normal and transformed epithelial cells, dendritic cells | Adhesion molecule and a collagen receptor |
| **CD168.** 80, 84, 80 kDa. RHAMM (receptor for hyaluronan-mediated motility); hyaluronan (HA)-binding receptor family. | Thymocytes, myelomonocytic lineage and dendritic cells; up-regulated on activated lymphocytes | Expressed on a hyaluronan-binding receptor that participates in the hyaluronan-dependent motility of thymocytes, lymphocytes, hematopoietic progenitor cells, and malignant B lymphocytes; adhesion of early thymocyte progenitors to matrix |
| **CD169.** 220 kDa. SIGLEC-1 (sialic acid binding Ig-like lectin-1); sialoadhesin, single-chain type 1 glycoprotein, soluble form results from alternate splicing. | Macrophages, dendritic cells | Mediates cell-cell interactions by binding to sialylated ligands on neutrophils, monocytes, NK cells, B cells, and a subset of CD8$^+$ T cells; may function as a pattern of self-nonself recognition receptor and mediate negative signals |
| **CD170.** 140 kDa. SIGLEC-5 (sialic acid–binding Ig-like lectin-1); type 1 glycoprotein. | Dendritic cells, macrophages, neutrophils | May function as a pattern of self-nonself recognition receptor and mediate negative signals |
| **CD171.** 200–230 kDa. L1 cell adhesion molecule; single-chain type 1 glycoprotein. | Low to intermediate expression on human lymphoid and mylelomonocytic cells, including CD4$^+$ T cells, a subset of B cells, monocytes, monocyte-derived dendritic cells, many cells of the central and peripheral nervous system | Cell adhesion molecule that plays a role in the maintenance of lymph node architecture during an immune response; kidney morphogenesis |
| **CD172a.** 110 kDa. Signal regulatory protein (SIRP α-1); single-chain type 1 glycoprotein. | CD34$^+$ stem/progenitor cells, macrophages, monocytes, granulocytes, dendritic cells; also in CNS tissue | Negative regulation of receptor tyrosine kinase–coupled signaling processes; may have a role in response to growth factors and cell adhesion |
| **CD173.** H2; blood group O antigen. | Among hematopoietic cells, present on erythrocytes and CD34$^+$ hematopoietic precursors | Glycoprotein-borne oligosaccharide; function unknown |
| **CD174.** Lewis Y blood-group antigen. | Among hematopoietic cells, present on erythrocytes and CD34$^+$ hematopoietic precursors | New hematopoietic-progenitor-cell marker; glycoprotein-borne oligosaccharide; function unknown but correlated with apoptosis; may be involved in hematopoietic-stem-cell homing |
| **CD175.** Tn; carbohydrate, tumor-specific antigen. | Variety of leukemic cells, epithelial cells, hematopoietic bone marrow cells, myeloid lineage | Not known |
| **CD175s.** Sialyl-Tn; carbohydrate, tumor-specific antigen. | Endothelial and epithelial cells and erythroblasts | Shown to bind to CD22, Siglec-3-5 and -6; function unknown |
| **CD176.** Thomson-Friedrenreich (TF) antigen: histo-blood group–related carbohydrate antigen. | Endothelial cells and erythrocytes, different types of carcinomas; CD34$^+$ hematopoietic precursor cells of the bone marrow and on various hematopoietic cell lines | May be involved in tumor metastasis such as of liver tumors and positive leukemia cells |
| **CD177.** 58–64 kDa. NB1, human neutrophil antigen-2A (HNA-2a); single-chain GPI-anchored plasma membrane glycoprotein. | Surfaces and secondary granules of neutrophils; basophils, NK cells, T-cell subsets, monocytes and endothelial cells | Role in neutrophil function unknown |
| **CD178.** Monomer of cell surface form 40 kDa; soluble forms 26–30 kDa. Fas ligand, FasL; TNF superfamily; homotrimeric type 2 glycoprotein. | Constitutive or induced expression on many cell types, including T cells, NK cells, microglial cells, neutrophils, nonhematopoietic cells such as retinal and corneal parenchymal cells | Ligand for the apoptosis-inducing receptor CD95 (Fas/APO-1); key effector of cytotoxicity and involved in Fas/FasL interaction, apoptosis, and regulation of immune responses; proposed to transduce a costimulatory signal for CD8$^+$ and naive CD4$^+$ T-cell activation |
| **CD179a.** 16–18 kDa. VpreB; polypeptide. | Selectively expressed in pro–B and early pre–B cells | Associates with CD179b to form surrogate light chain of the pre–B cell receptor; involved in early B-cell differentiation |

| CD antigen. MW. Synonyms and properties. | Leukocyte expression | Function |
|---|---|---|
| CD179b. 22 kDa. λ5; polypeptide. | Selectively expressed in pro–B and early pre–B cells | Associates with CD179a to form surrogate light chain of the pre-B-cell receptor; involved in early B-cell differentiation |
| CD180. 95–105 kDa. RP105, Ly64; leucine-rich repeat family (LRRF), type 1 glycoprotein. | Monocytes, dendritic cells, mantle zone B cells | Induces activation that leads to up-regulation of costimulatory molecules, CD80 and CD86, and an increase in cell size; promotes B-cell susceptiblity to BCR-induced cell death but not to CD95-induced apoptosis and may play a role in the transmission of a growth-promoting signal |
| CD181. 58-67 kDa. Formally known as CDw128a, IL-8R1, IL-8RA (interleukin-8 receptor A), CXCR1 (chemokine C-X-C motif receptor 1); G-protein-coupled receptor (GPCR), type 3-, 7-span glycoprotein. | Neutrophils, basophils, eosinophils, a subset of T cells, monocytes, NK and endothelial cells, keratinocytes, melanoma cells | Subunit of the IL-8 receptor; involved in neoangiogenesis |
| CD182. 58–67 kDa. Formally known as CDw128b, CXCR2, interleukin-8 receptor B (IL-8RB); G-protein-coupled receptor (GPCR), type 3-, 7-span glycoprotein. | Granulocytes, neutrophils, basophils, eosinophils, a subset of T cells, monocytes, NK and endothelial cells, keratinocytes, melanoma cells | Induces chemotaxis of neutrophils, basophils, and T lymphocytes; activates neutrophils and basophils and increases neutrophil and monocyte adhesion to endothelial cells |
| CD183. 40.6 kDa. CXCR3, IP10R, Mig-R; 7-transmembrane-spanning protein superfamily (TN7SF), type 3-, 7-span glycoprotein. | Suitably induced T cells, small subsets of B cells and NK cells, eosinophils and some dendritic cells | Signaling induces and regulates chemotactic migration of CD183-bearing leukocytes in inflammatory-associated effector T cells; may also play a role in $T_H1$-cell activation and IFN-γ production; thought to be essential for T-cell recruitment to inflammatory sites |
| CD184. ~ 40 kDa. CXCR4, fusin, LESTER (leukocyte-derived seven transmembrane domain receptor); transmembrane-spanning protein superfamily (TN7SF), type 3-, 7-span glycoprotein. | Variety of blood and tissue cells, including B and T cells, monocytes/macrophages, dendritic cells, granulocytes, megakaryocytes/platelets, lymphoid and myeloid precursor cells, endothelial and epithelial cells, astrocytes and neurons | Coreceptor for HIV; mediates blood-cell migration in response to SDF-1 and is involved in B-lympho- and myelopoiesis, cardiogenesis, blood vessel formation, and cerebellar development; costimulation of pre–B cell proliferation; induction of apoptosis |
| CD185. 45kDa. CXCR5, BLR1; G-protein-coupled receptor. | Mature B cells, Burkitt's lymphoma cells | Chemokine receptor; possible regulatory function in lymphomagenesis, B-cell differentiation, activation of mature B cells |
| CDw186. 40 kDa. CXCR6; G-protein-coupled receptor. | Subsets of B, NK, and T cells ($T_H1$) | Chemokine receptor, coreceptor for SIV, strains of HIV-2, and m-tropic HIV-1 |
| CD191. 39 kDa. CCR1, MIP-1αR, RANTES-R; G-protein-coupled receptor. | T cells, monocytes, stem-cell subsets | C-C chemokine receptors; involved in signal transduction via augmentation of intracellular calcium ion level |
| CD192. 40 kDa. CCR2, MCP-1R; G-protein-coupled receptor. | Activated NK cells and mononuclear phagocytes, T cells, B cells, endothelial cells | Chemokine receptor; binds MCP-1, -3, -4 and functions as an alternative receptor for HIV-1 infections in conjunction with CD4 |
| CD193. 45 kDa. CCR3, CKR3; G-protein-coupled receptor. | Eosinophils; weak on a subset of T cells, neutrophils, and monocytes | Chemokine receptor; binds multiple chemokines and functions as an alternative receptor for HIV-1 infections in conjunction with CD4 |
| CD195. 45 kDa. CCR5; 7-transmembrane domain G-protein-coupled receptor. | $CD4^+$ and $CD8^+$ thymocytes, Langerhans cells, peripheral blood–derived dendritic cells, hematopoietic progenitor cells and microglial cells, T lymphocytes and macrophages in both lymphoid and nonlymphoid tissues | Coreceptor for HIV; regulates innate and adaptive immune responses, chemotaxis activation, and transendothelial migration during inflammation and neutralizes HIV infection |
| CD197. 45 kDa. CCR7. | Most naive T cells and a subset of memory T cells; B cells, mature dendritic cells, NK cells, CD4 or CD8 single-positive mature thymocytes, $CD34^+$ macrophage progenitor cells | Crucial roles in naive T cells and antigen-loaded dendritic cells homing to secondary organs; involvement in T-lymphocyte adhesion and thymocyte migration |

*(continued)*

| CD antigen. MW. Synonyms and properties. | Leukocyte expression | Function |
|---|---|---|
| CDw198. 43 kDa. CCR8; G-protein-coupled receptor. | Memory T-cell subsets, thymocytes, IEL, melanoma cells, lamina propria mononuclear cells | Chemokine receptor involved in allergic inflammation; alternate coreceptor for HIV-1 in conjunction with CD4 |
| CDw198. CCR9; G-protein-coupled receptor. | Subsets of T cells, development specific | Chemokine receptor; alternate coreceptor for HIV-1 in conjunction with CD4 |
| CD200. 45–50 kDa. Ox2; single-chain type 1 glycoprotein. | Dendritic cells, thymocytes, B cells, vascular endothelium, trophoblasts, neurons, and some smooth muscle on activated T cells | May play an immunoregulatory role |
| CD201. 50 kDa. EPC-R. | Endothelial subset | Signaling of activated protein C |
| CD202b. 145 kDa. TEK, TIE2. | Endothelial and hematopoietic stem cells | Crucial role in integrity of vessels during maturation, maintenance, and remodeling |
| CD203c. 130, 150 kDa. E-NPP3 B10, PDNP3, ENPP3, bovine intestinal phosphodiesterase; ecto-nucleotide pyrophosphatase/phosphodiesterase (E-NPP) enzyme family; type 2 glycoprotein. | Basophils, mast cells, uterine tissue | May be involved in clearance of extracellular nucleotides |
| CD204. 220 kDa. Macrophage scavenger receptor (MSR); trimeric integral membrane glycoprotein. | Alveolar macrophages, Kupffer cells of the liver, splenic red pulp macrophages, sinusoidal macrophages in lymph nodes, interstitial macrophages | Role in the pathological deposition of cholesterol during atherogensis via receptor mediating uptake of low density LDL; recognition and elimination of pathogenic microorganisms; endocytosis of macromolecules |
| CD205. 205 kDa. DEC-205. | Dendritic cells, thymic epithelial cells | Function unknown |
| CD206. 162 kDa. MMR (macrophage mannose receptor), MRC1 (mannose receptor, C type lectin); pattern recognition receptors, single-chain type 1 glycoprotein. | Subset of mononuclear phagocytes (but not circulating monocytes), immature dendritic cells, hepatic and lymphatic endothelial cells | Pattern recognition receptor involved in immune responses of macrophages and immature dendritic cells; facilitates endocytosis and phagocytosis; might be important in homeostatasis |
| CD207. 40 kDa. Langerin; C-type lectin, type 2 glycoprotein. | Subset of dendritic cells, Langerhans cells | Endocytic receptor with a functional C-type lectin domain with mannose specificity that facilitates antigen recognition and uptake |
| CD208. 70–90 kDa, DC-LAMP; lysosome-associated membrane protein (LAMP) family member. | Dendritic cells | Function unknown but may participate in peptide loading onto MHC class II molecules |
| CD209. 45.7 kDa. DCSIGN (dendritic-cell-specific ICAM3-grabbing nonintegrin), HIV GP120-binding protein; C-type lectin. | Dendritic cells | Binds to HIV; role in dendritic-cell-mediated adhesion to endothelial cells and antigen endocytosis and degradation |
| CDw210. 90–110 kDa. IL-10R$\alpha$ and $\beta$; member of class 2 cytokine receptor family, single-chain type 1 glycoprotein. | Mainly on hematopoietic cells, including T and B cells, NK cells, monocytes, and macrophages | Cell-signaling and immune regulation; inhibits cytokine synthesis by activated T and NK cells, monocytes, and macrophages; blocks accessory-cell function of macrophages |
| CD212. 110, 85 kDa. IL-12R$\beta$; hematopoietin receptor family, single-chain type 1 receptor. | Activated CD4$^+$ and CD8$^+$ T cells, IL-2-activated CD56$^+$ NK cells, $\gamma\delta$ T cells; PBL, cord blood lymphocytes, subsets of monocytes | Tyrosine kinase membrane receptor for angiopoietin; involved in cell signaling and immune regulation; pleiotropic effects on NK and T cells and affects IFN-$\gamma$ production |
| CD213a1. 65 kDa. IL13RA1 (interleukin 13 receptor alpha 1) NR4 (in mouse); hemopoietic receptor family, single-chain type 1 glycoprotein. | Most human tissues, hematopoietic and nonhematopoietic | Necessary component of IL-13-and IL-4-induced signal transduction in type 2 IL-4R system but not a regulator of T-cell functions. |
| CD213a2. 60–70 kDa. IL13RA2 (interleukin 13 receptor alpha 2) NR4 (in mouse); hemopoietic receptor family, single-chain type 1 glycoprotein. | B cells, monocytes, cord PBLs, fibroblasts, immature DC | Inhibits binding of IL-13 to the IL-13 cell surface receptor |

| CD antigen. MW. Synonyms and properties. | Leukocyte expression | Function |
|---|---|---|
| CDw217. 120 kDa. IL-17R; hemopoietic receptor family, single-chain type 1 glycoprotein. | B, T, NK cells, cord blood, PBL, thymocytes; fibroblasts, epithelial cells, monocytes, macrophages, granulocytes | Not known |
| CDw218a. 70 kDa. IL-18Rα, IL-1Rrp; IL-1 receptor family. | T cells, NK cells, dendritic cells | Binds to IL-18 and induces NF-κB activation |
| CDw218b. 70 kDa. IL-18Rα, IL-18RAP; IL-1 receptor family. | T cells, NK cells, dendritic cells | Heterodimeric receptor that enhances IL-18 binding |
| CD220. 140, 70 kDa. Insulin-R. | Broad expression | Insulin signaling |
| CD221. 140, 70 kDa. IGF-1R. | Broad expression | Binds IGF with high affinity; involved in signaling, cell proliferation and differentiation |
| CD222. 280–300 kDa. Man-6p receptor (mannose-6 phosphate receptor), IGF2R (insulin-like growth factor 2 receptor). | Fibroblasts, granulocytes, lymphocytes, myocytes | Sorts newly synthesized lysosomal enzymes bearing M6P to lysosomes |
| CD223. 70 kDa. Lag-3 (lymphocyte activation gene 3); single-chain type 1 glycoprotein. | All subsets of activated T or NK cells | May promote down-regulation of TCR signaling, leading to cell inactivation, a role in down-regulating an antigen-specific response; possibly helps activate CD4$^+$ and CD8$^+$ T cells to fully activate monocytes and dendritic cells to optimize MHC class I and class II–mediated T-cell responses |
| CD224. 100 kDa. GGT (gamma glutamyl transpeptidase). | Vascular endothelium, peripheral blood macrophages, subset of B cells; lymphocytes, monocytes, granulocytes, endothelial and stem cells | Cellular detoxification, leukotriene biosynthesis, inhibition of apoptosis |
| CD225. 17 kDa. Leu13, interferon-induced protein 17 (IFI17). | Leukocytes, endothelial cells, in multiple lineages | Involved in lymphocyte activation and development; may play a role in controlling cell-to-cell interactions |
| CD226. 65 kDa. DNAM-1, PTA-1 (platelet and T-cell activation antigen 1). | NK cells, platelets, monocytes, subset of T cells | Not known |
| CD227. MUC1 (mucin 1), episialin; very large glycosylated protein with small 25 kDa subunit and larger subunits of 300–700 kDa, type I transmembrane protein. | Epithelial cells, follicular dendritic cells, monocytes, subsets of lymphocytes, B and stem cells, some myelomas; some hematopoietic cell lineages | Not known |
| CD228. 80–95 kDa. Melanotransferrin. | Stem cells, melanoma cells | Cell adhesion |
| CD229. 100–120 kDa. T-lymphocyte surface antigen, Ly-9 (lymphocyte antigen 9); CD2-subset of immunoglobulin superfamily, single-chain type 1 glycoprotein. | Predominantly in the lymph nodes, spleen, thymus, peripheral blood leukocytes | Involved in the activation of lymphocytes |
| CD230. 30–40 kDa. Prion protein, p27–30; GPI-anchored glycoprotein. | Broadly expressed on most cell types with the highest level on neurons and follicular dendritic cells; hematopoietic and nonhematopoietic cells | May inhibit apoptosis |
| CD231. 28–45 kDa. T-cell acute lymphoblastic leukemia–associated antigen 1(TALLA-1), A15, MXS1, TM4SF2, CCG-B7; tetraspan subfamily 1 member, type 3-, 4-span glycoprotein. | Strongly expressed on T-cell acute lymphoblastic leukemic cells (T-ALL), neuroblastoma cells, normal brain neurons, endothelial cells | May play a role in the regulation of cell development, activation, growth, and motility |
| CD232. 200 kDa. Virally encoded semaphorin receptor (VESPR), Plexin C1(PLXNC1); plexin family (plexin C1) of molecules. | B and NK cells, monocytes, granulocytes | Function unknown although likely to play a role in immune modulation; participates in biological activity in monocytes, some classes of dendritic cells, neutrophils, B lymphocytes |

(continued)

| CD antigen. MW. Synonyms and properties. | Leukocyte expression | Function |
|---|---|---|
| **CD233.** 95–110 kDa. Band 3, diego blood group; typically a dimer. | Erythrocytes and basolateral membrane of some cells of the distal and collecting tubules of the kidney | Maintains red-cell morphology |
| **CD234.** 35–45 kDa. FY-glycoprotein (DARC), Duffy blood group; chemokine receptor, type 3-, 7-span acidic Fy-glycoprotein. | Erythrocytes, epithelial cells of the kidney collecting duct, lung alveoli and thyroid, on neurons (Purkinje cells) in the cerebellum | Chemokine decoy receptor; binds a number of chemokines, modulates the intensity of inflammatory reactions |
| **CD235a, -b, -ab.** 35 kDa (a), 20 kDa (b). Glycophorin A, glycophorin B. | Erythrocytes | Not known |
| **CD236.** 23 kDa, 32 kDa. Glycophorin C/D. | Variety of cells and tissues both erythroid and nonerythroid in nature | Not known |
| **CD236R.** 32 kDa. Glycophorin C/D. | Variety of cells and tissues both erythroid and nonerythroid in nature | Not known |
| **CD238.** 93 kDa. Kell. | Erythrocytes and subset of stem cells | Not known |
| **CD239.** 78–85 kDa. B-CAM. | Erythrocytes and subset of stem cells | Not known |
| **CD240CE.** 30–32 kDa. Rh30CE. | Erythrocytes | Not known |
| **CD240D.** 30–32 kDa. Rh30D; type 1-, 2-transmembrane superfamily. | Erythrocytes | Not known |
| **CD241.** 50 kDa. RhAg, Rh50; type 1-, 2-transmembrane superfamily. | Erythrocytes | Forms a complex with CD47, LW, glycophorin B; unknown function |
| **CD242.** 42 kDa. ICAM-4; immunglobulin superfamily. | Erythrocytes | Cell adhesion |
| **CD243.** 170 kDa. Multidrug resistance protein 1 (MDR-1), P-gp P-glycoprotein (P-gp), ABC-B1, P170; single-chain type 3-, 12-span glycoprotein. | Stem and progenitor cells | Influences the uptake, tissue distribution, and elimination of P-gp-transported drugs and toxins; transporter in the blood-brain barrier |
| **CD243.** 170 kDa. Multidrug resistance protein 1 (MDR-1), P-gp P-glycoprotein (P-gp), ABC-B1, P170; single-chain type 3-, 12-span glycoprotein. | Stem and progenitor cells | Influences the uptake, tissue distribution, and elimination of P-gp-transported drugs and toxins; transporter in the blood-brain barrier |
| **CD244.** 63–70 kDa. 2B4, NAIL (NK-cell activation-inducing ligand); type II transmembrane (type II TM), single-chain type 1 glycoprotein. | NK cells, $\gamma\delta$ T cells, 50% of CD8$^+$ T cells, monocytes, basophils | Activation receptor for NK cells and modulates NK-cell cytokine production, cytolytic function, and extravasation; involved in NK and T-cell interactions; may be important for the development of functional CD4$^+$ T cells and possibly serves to increase cell-cell adhesion |
| **CD245.** 220–240 kDa. p220/240. | All resting peripheral blood lymphocytes | Signal transduction and costimulation of T and NK cells |
| **CD246.** 80 kDa, 200 kDa. Anaplastic lymphoma kinase (ANK); single-chain type 1 glycoprotein. | Subset of T-cell lymphomas | Receptor tyrosine kinase of unknown function |
| **CD247.** 16 kDa. Zeta chain of CD3. | NK cells during thymopoiesis and mature T cells in the periphery | T-cell activation |
| **CD248.** 175 kDa. TEM1, endosialin; C-type lectin. | Endothelial tissues, stromal fibroblasts | Tumor progression and angiogenesis |
| **CD249.** 160 kDa. Aminopeptidase A; peptidase M1 family. | Epithelial and endothelial cells | Renin-angiotensin system; immune function unknown |
| **CD252.** 34 kDa. OX-40L, gp 34; TNF superfamily. | Activated B cells, cardiac myocytes | T-cell costimulation |
| **CD253.** MW unknown. TRAIL, Apo-2L, TL2, TNFSF10; TNF superfamily. | Activated T cells; broadly expressed in tissues | Cell death |

| CD antigen. MW. Synonyms and properties. | Leukocyte expression | Function |
|---|---|---|
| CD254. 35 kDa. TRANCE, RANKL, OPGL; TNF superfamily. | Bone marrow stroma, activated T cells, lymph node | Binds to OPG and RANK, differentiation of osteoclasts, augments dendritic cells to stimulate naive T-cell proliferation |
| CD256. 16 kDa. APRIL, TALL-2; TNF superfamily. | Monocytes and macrophages | Binds TACI and BCMA; involved in B-cell proliferation |
| CD257. MW unknown. BLys, BAFF, TALL-1; TNF superfamily. | Activated monocytes; exists as a soluble form | B-cell growth factor; costimulation of Ig production |
| CD258. 28 kDa. LIGHT, HVEM-L; TNF superfamily. | Activated T cells, immature dendritic cells | Binds LTβR, involved in T-cell proliferation; receptor for HVEM |
| CD261. 57 kDa. TRAIL-R1, DR4; TNF receptor superfamily. | Activated T cells, peripheral blood leukocytes | FADD and caspase-8-mediated apoptosis |
| CD262. 60 kDa. TRAIL-R2, DR5; TNF receptor superfamily. | Widely expressed; peripheral blood leukocytes | Same as CD261 |
| CD263. 65 kDa. TRAIL-R3, DcR1, LIT; TNF receptor superfamily. | Peripheral blood leukocytes | Receptor for TRAIL but lacks a death domain |
| CD264. 35 kDa. TRAIL-R4, TRUNDO, DcR2; TNF receptor superfamily. | Peripheral blood leukocytes | Receptor for TRAIL with a truncated death domain |
| CD265. 97 kDa. RANK, TRANCE-R, ODFR; TNF receptor superfamily. | Broad | Binds TRANCE; involved in osteogenesis, T-cell–dendritic cell interactions |
| CD266. 14 kDa. TWEAK-R, FGF-inducible 14; TNF receptor superfamily. | Endothelial cell subset, placenta, kidney, heart | Cell matrix interactions, endothelial growth and migration; receptor for TWEAK |
| CD267. MW unknown. TACI, TNFR, SF13B. | B cells, activated T cells | Binds BAFF and APRIL |
| CD268. 25 kDa. BAFFR, TR13C; TNF receptor superfamily. | B cells | Binds BLyS; involved in B-cell survival |
| CD269. 20 kDa. BCMA, TNFRSF13B; TNF receptor superfamily. | Mature B cells | Receptor for APRIL and BAFF; involved in B-cell survival and proliferation |
| CD271. 45 kDa. NGFR, p75 (NTR); TNF receptor superfamily. | Neurons, bone marrow mesenchymal cells | Receptor for NGF, NT-3, NT-4; tumor suppressor, cell survival and cell death |
| CD272. 33 kDa. BTLA; glycoprotein. | $T_H1$ cells and activated B and T cells | Inhibitory receptor on T lymphocytes with similarities to CTLA-4 (CD152) and PD-1 (CD279) |
| CD273. 25 kDa. B7DC, PDL2, programmed cell death 1 ligand 2; glycoprotein. | Dendritic cells, macrophages, monocytes, activated T cells | Second ligand for PD-1 (CD279); this interaction, like that of PD-L1/PD-1, inhibits TCR-mediated proliferation and cytokine production |
| CD274. 40 kDa. B7H1, PDL1, programmed cell death 1 ligand 1. | Macrophages, epithelial and dendritic cells, NK cells, activated T cells and monocytes | Putative ligand for PD-1 (CD 279); shown to inhibit TCR-mediated proliferation and cytokine secretion; involved in costimulation and inhibition of lymphocytes |
| CD275. 40 kDa, 60 kDa. B7H2, ICOSL; inducible T-cell costimulator ligand (ICOSL). | Macrophages, dendritic cells, weak T and B cells, activated monocytes | Costimulation of T cells; reported to promote proliferation and cytokine production |
| CD276. 40–45 kDa, 110 kDa. B7H3; B7 homolog 3. | Epithelial cells, activated monocytes and T cells, dendritic-cell subsets | Stimulates T-cell proliferation and activation and IFN-γ production |
| CD277. 56 kDa. BT3.1; B7 family: butyrophilin, subfamily 3, member A1. | B cells, T cells, NK cells, dendritic cells, monocytes, endothelial cells, stem-cell subsets | May play a role in regulation of the immune response; acts as a regulator of T-cell activation and function |
| CD278. 47–57 kDa. ICOS; inducible T-cell costimulator. | $T_H2$ cells, thymocyte subsets, activated T cells | Plays a critical role in costimulating T-cell activation, development, proliferation. and cytokine production |

*(continued)*

| CD antigen. MW. Synonyms and properties. | Leukocyte expression | Function |
|---|---|---|
| **CD279.** 55 kDa. Programmed cell death 1 (PD1 or PDC1), hPD-1, SLEB2; single-chain type 1 glycoprotein. | Subset of thymocytes, activated T and B cells | Inhibits TCR-mediated proliferation and cytokine production |
| **CD280.** 180 kDa. ENDO180, uPARAP, mannose receptor, C type 2, TEM22. | Myeloid progenitor cells, fibroblasts, chondrocytes, osteoclasts, osteocytes, subsets of endothelial and macrophage cells | Not known |
| **CD281.** 90 kDa. TLR1 (Toll-like receptor 1). | Dendritic cells, keratinocytes, macrophages, monocytes, neutrophils; low level on monocytes | Recognizes pathogen-associated molecular pattern with a specificity for gram-positive bacteria; associates with and regulates TLR2 response |
| **CD282.** 87 kDa. TLR2 (Toll-like receptor 2). | Granulocytes, monocytes, macrophages, dendritic cells, keratinocytes | Interacts with microbial lipoproteins and peptidoglycans, CD14-dependent and -independent response to LPS, NF-$\kappa$B pathway |
| **CD283.** 101 kDa. TLR3 (Toll-like receptor 3). | Subsets of dendritic cells, low levels on fibroblasts, epithelial cells | Interacts with dsRNA; activates NF-$\kappa$B pathway; induces production of type I interferons; MyD88-dependent and -independent response to poly (I:C) |
| **CD284.** 85, 110, 130 kDa. TLR4 (Toll-like receptor 4). | Macrophages and endothelial cells; weak on monocytes, immature dendritic cells and neutrophils | Interacts with microbial lipoproteins; CD14-dependent response to LPS, NF-$\kappa$B pathway |
| **CD285.** TLR5 (Toll-like receptor 5). | Leukocytes | Interacts with microbial lipoproteins, NF-$\kappa$B, and responses to *Salmonella*. |
| **CD286.** TLR6 (Toll-like receptor 6). | Leukocytes | Interacts with microbial lipoproteins; regulates TLR2 response |
| **CD287.** TLR7 (Toll-like receptor 7). | Macrophages, subset of dendritic cells | Interacts with microbial nucleic acids; induces IL-12 |
| **CD288.** TLR8 (Toll-like receptor 8). | Macrophages, subset of dendritic cells | Interacts with microbial nucleic acids; induces IL-12 |
| **CD289.** 113 kDa. TLR9 (Toll-like receptor 9). | Subsets of dendritic cells, B cells, monocytes | Receptor for CpG bacterial DNA; weakly similar to TLR3, may mediate protein-protein interaction |
| **CD292.** 57 kDa. BMPR1A, ALK3; type 1 glycoprotein. | Bone progenitors | BMP2 and -4 receptor, bone development; immune function unknown |
| **CDw293.** 57 kDa. BMPR1B, ALK6; type 1 glycoprotein. | Bone progenitors | BMP receptor, bone development; immune function unknown |
| **CD294.** 55–70 kDa. CRTH2, GPR44; G-protein-coupled receptor. | T$_H$2 cells, granulocytes | Binds PGD2; provides stimulatory signals for T$_H$2 cells; involved in allergic inflammation |
| **CD295.** 132 kDa. LeptinR, LEPR; type 1 cytokine receptor. | Broad | Adipocyte metabolism |
| **CD296.** 37 kDa. ART1, RT6, ART2; ADP-ribosyltransferase. | Peripheral T cells, NK-cell subset, heart and skeletal muscle | Modifies integrins during differentiation; ADP ribosylation of target proteins |
| **CD297.** 38 kDa.ART4, dombrock blood group; ADP-ribosyltransferase. | Erythrocytes, activated monocytes | ADP ribosylation of target proteins |
| **CD298.** 52 kDa. Na$^+$/K$^+$ ATP-ase $\beta$-3 subunit. | Broad | Ion transports |
| **CD299.** 45 kDa. DC-SIGN-related, LSIGN, DC-SIGN2. | Endothelial subset | Binds ICAM-3, HIV-1 gp120; coreceptor with DC-SIGN for HIV-1 |
| **CD300a.** 60 kDa. CMRF35H, IRC1, Irp60. | Monocytes, neutrophils, T-and B-cell subsets, myeloma cells | Unknown |
| **CD300c.** MCMRF35A, LIR. | Monocytes, neutrophils, T-and B-cell subsets | Unknown |

| CD antigen. MW. Synonyms and properties. | Leukocyte expression | Function |
|---|---|---|
| **CD300e.** CMRG36L. | Unknown | Unknown |
| **CD301.** 38 kDa. MGL, HML; C-type lectin superfamily. | Immature dendritic cells | Binds Tn antigen; uptake of glycosylated antigens |
| **CD302.** 19–28 kDa. DCL1, BIMLEC; type 1 transmembrane C-type lectin receptor. | Some myeloid and Hodgkin's cell lines | Fusion protein in Hodgkin's lymphoma with DEC-205; unknown function |
| **CD303.** 38 kDa. BDCA2, HECL; C-type lectin superfamily. | Plasmacytoid dendritic cells | Inhibits IFN-α production |
| **CD304.** 130 kDa. BDCA4, neuropilin1; semaphorin family. | Neurons, subset of T cells, dendritic cells, endothelial cells, tumor cells | Coreceptor with plexin, interacts with VEGF165 and semaphorins; axonal guidance, angiogenesis, cell survival, migration |
| **CD305.** 31 kDa, 40 kDa. Leukocyte associated Ig-like receptor-1 (LAIR), p40; type 1 glycoprotein. | Expression of B cells related to maturation stage; NK and T cells, monocyte-derived dendritic cells, thymocytes, thymic precursors | Inhibitory receptor of cell function in NK cells and T and B cells; inhibits cellular activation and inflammation |
| **CD306.** LAIR2. | Unknown | Soluble form; involved in mucosal tolerance |
| **CD307.** 55–105 kDa. IRTA2; Fc receptor immunoglobulin superfamily. | B-cell subset, B-cell lymphomas | B-cell development |
| **CD309.** 230 kDa. VEGFR2, KDR; type III transmembrane tyrosine kinase. | Endothelial cells, angiogenic precursors, hemangioblast | Binds VEGF; regulates cell adhesion and signaling |
| **CD312.** 90 kDa. EMR2. | Monocytes, macrophages, myeloid dendritic cells, low on granulocytes | Cell adhesion and migration; involved in phagocytosis |
| **CD314.** NKG2D, killer-cell lectin-like receptor subfamily K, member 1; type II transmembrane protein with an extracellular C-type lectin-like domain. | NK cells primarily; also on activated macrophages and αβ CD4$^+$ T cells | Induces NK-cell activation and cytotoxicity |
| **CD315.** 135 kDa. CD9P1, SMAP6. | B-cell subsets, activated monocytes, endothelial and epithelial cells, hepatocytes, megakaryocytes | Not known |
| **CD316.** 63–78 kDa. IgSF8, KASP. | B cells, NK cells, T cells | Involved in cell migration |
| **CD317.** 29–33 kDa. BST2 (bone marrow stromal-cell antigen 2), HM1.24. | B and T cells, monocytes, NK and dendritic cells, fibroblasts, plasma and stromal cells | Not known; thought to be involved in pre-B-cell growth |
| **CD318.** 135 kDa. CDCP1, SIMA135. | Hematopoietic stem-cell subset, tumor cells | Cell adhesion with extracellular matrix |
| **CD319.** 37.5 kDa. CRACC, SLAM family member 7; multiple isoforms with possibly different functions. | NK cells, activated B-cells, NK-cell line but not in promyelocytic, B-, or T-cell lines | Mediates NK-cell activation through an SAP-independent extracellular signal-regulated ERK-mediated pathway; may play a role in lymphocyte adhesion |
| **CD320.** 30 kDa. 8D6A, 8D6. LDL receptor. | Follicular dendritic cells, germinal centers | B-cell proliferation, tumor formation |
| **CD321.** 35 kDa. JAM1, F11 receptor; immuglobulin superfamily, type 1. | Platelets, epithelial and endothelial cells | Adhesions, tight junctions |
| **CD322.** 45 kDa. Junctional adhesion molecule-2 (JAM-2), VE-JAM. | Endothelial cells, B cells, monocytes, T-cell subsets | Cell-cell adhesion; central role in the regulation of transendothelial leukocyte migration to secondary organs |
| **CD324.** 120 kDa. E-Cadherin; cadherin superfamily. | Epithelial and stem cells, erythroblasts, keratinocytes, trophoblasts, platelets | Recently shown to bind inhibitory receptors on CD8$^+$ T and NK cells; tumor suppression and cell growth and differentiation, cell adhesion |
| **CDw325.** 140 kDa. N-Cadherin, NCAD; cadherin superfamily. | Brain; skeletal and cardiac muscle | Cell adhesion, neuronal recognition |
| **CD326.** 35-40 kDa. Ep-CAM, Ly74. | Most epithelial cell membranes | Unknown |

*(continued)*

| CD antigen. MW. Synonyms and properties. | Leukocyte expression | Function |
|---|---|---|
| **CDw327.** 47 kDa (predicted). CD33L, CD33L1,OBBP1, sialic acid–binding Ig-like lectin 6 (Siglec6); leptin-binding protein. | Leukocytes, B cells | Mediates cell-cell recognition, binds leptin, possible modulator of leptin levels; likely inhibitory |
| **CDw328.** 49 kDa (predicted). AIRM1, QA79, p75, p75/AIRM1, sialic acid–binding Ig-like lectin 7 (Siglec7). | Variety of leukocytes in addition to T cells, including NK cells, monocytes, and granulocytes | Cell adhesion molecule; likely inhibitory function in NK- and T-cell activation |
| **CDw329.** 48 kDa (predicted). OBBP-like, sialic acid–binding Ig-like lectin 9 (Siglec9). | Variety of leukocytes in addition to T cells, including NK cells, monocytes, granulocytes | Cell adhesion molecule; likely inhibitory function in NK- and T-cell activation |
| **CD331.** 30 kDa. FGFR1, KAL2, N-SAM, Fms-like tyrosine kinase-2; transmembrane tyrosine kinase. | Fibroblasts, epithelial cells | High-affinity receptor for fibroblast growth factors |
| **CD332.** 115–132 kDa. FGFR3, BEK, KGFR; transmembrane tyrosine kinase. | Fibroblasts, epithelial cells | High-affinity receptor for fibroblast growth factors |
| **CD333.** 115 kDa. FGFR3, ACH, CEK2; transmembrane tyrosine kinase. | Fibroblasts, epithelial cells | High-affinity receptor for fibroblast growth factors |
| **CD334.** 110 kDa. FGF4, JTK2, TKF; transmembrane tyrosine kinase. | Fibroblasts, epithelial cells | High-affinity receptor for fibroblast growth factors |
| **CD335.** 34.5 kDa (predicted). NKp46, NCR1, Ly94, natural cytotoxicity triggering receptor 1. | NK cells | Cytotoxicity-activating receptor that may contribute to the increased efficiency with which activated NK cells mediate tumor cell lysis |
| **CD336.** 30.6 kDa (predicted). NKp44, NCR2, Ly94, natural cytotoxicity triggering receptor 2. | NK cells | Cytotoxicity-activating receptor that may contribute to the increased efficiency with which activated NK cells mediate tumor-cell lysis |
| **CD337.** 1.8 kDa (predicted). NKp30, NCR3, Ly117. | NK cells | Cytotoxicity-activating receptor that may contribute to the increased efficiency with which activated NK cells mediate tumor-cell lysis |
| **CDw338.** 73 kDa. ABCDG2, BCRP, Bcrp1, MXR; G-protein-coupled receptor, 7-transmembrane. | Stem-cell subset | Multidrug resistance transporter |
| **CD339.** 135 kDa. Jagged-1, JAG1, JAGL1, hJ1. | Stromal cells, epithelial cells, myeloma cells | Binds Notch; involved in hematopoiesis |

Data on the major biological activities and sources of many cytokines are presented. In most cases, mass is given (usually for human cytokines). In some instances, a given cytokine may have biological activities in addition to those listed here or be produced by other sources as well as the ones cited here. This list of cytokines includes most cytokines of immunological interest. However, cytokines that are not closely identified with the immune system, for example, growth hormone, are not listed. Also, with the exception of IL-8, chemokines have not been included in this compilation.

| Cytokine. MW. Synonyms. | Sources | Activity |
|---|---|---|
| Interleukin-1 (IL-1). IL-1α 17.5 kDa, IL-1β 17.3 kDa. Lymphocyte-activating factor (LAF); mononuclear cell factor (MCF); endogenous pyrogen (EP). | Many cell types, including monocytes, macrophages, dendritic cells, T and B cells, NK cells, and non-immune system cells such as vascular epithelium, fibroblasts, and some smooth muscle cells | IL-1 displays a wide variety of biological activities on many different cell types, including T cells, B cells, and monocytes. Receptors for IL-1 are also found on the other leukocytes, including eosinophils and dendritic cells, as well as on nonimmune system cells such as fibroblasts, vascular endothelial cells, and some cells of the nervous system. The in vivo effects of IL-1 include induction of fever, the acute phase response, and stimulation of neutrophil production. |
| Interleukin-2 (IL-2). 15–20 kDa. T-cell growth factor (TCGF). | T cells | Stimulates growth and differentiation of T cells, B cells, and NK cells |
| Interleukin-3 (IL-3). 15.1 kDa (monomer), 30 kDa (dimer). Multi-colony-stimulating factor (M-CSF); hematopoietic cell growth factor (HCGF); mast-cell growth factor (MCGF). | Activated T cells, mast cells, and eosinophils | Growth factor for hematopoietic cells and lymphocytes; stimulates colony formation in neutrophil, eosinophil, basophil, mast cell, erythroid, megakaryocyte, and monocytic lineages but not in lymphocytes |
| Interleukin-4 (IL-4). 15–19 kDa. B-cell-stimulating factor 1 (BSF-1). | Mast cells, T cells, and bone marrow stromal cells | Promotes growth and development of B and T cells and cells of the monocytic lineage; also affects cells outside the immune system, including endothelial cells and fibroblasts |
| Interleukin-5 (IL-5). 45 kDa. Eosinophil differentiation factor (EDF); eosinophil colony-stimulating factor (E-CSF). | Mast cells, T cells, eosinophils | Induces eosinophil formation and differentiation |
| Interleukin-6 (IL-6). 26 kDa. B-cell stimulatory factor 2 (BSF-2); hybridoma/plasmacytoma growth factor (HPGF); hepatocyte-stimulating factor (HSF). | T cells, B cells, several nonlymphoid cells, including macrophages, bone marrow stromal cells, fibroblasts, endothelial cells, and astrocytes | Regulates B and T cell functions; in vivo effects on hematopoiesis; inducer of the acute phase response |
| Interleukin-7 (IL-7). 20–28 kDa. Pre-B-cell growth factor; lymphopoietin-1 (LP-1). | Bone marrow stromal cells, thymic stromal cells, and spleen cells | Growth factor for T- and B-cell progenitors |
| Interleukin-8 (IL-8). 6–8 kDa. Neutrophil-attractant/activating protein (NAP-1); neutrophil-activating factor (NAF); granulocyte chemotactic protein (GCAP). | Many cell types, including monocytes, lymphocytes, granulocytes, and nonimmune system cells such as fibroblasts, endothelial cells, hepatocytes, and others | Chemokine that functions primarily as a chemoattractant and activator of neutrophils; also attracts basophils and some subpopulations of lymphocytes; has angiogenic activity |
| Interleukin-9 (IL-9). 32–39 kDa. P40; T-cell growth factor III. | IL-2–activated T-helper-cell populations | Stimulates proliferation of T lymphocytes and erythroid precursors |

*(continued)*

| Cytokine. MW. Synonyms. | Sources | Activity |
|---|---|---|
| Interleukin-10 (IL-10).  35–40 kDa. Cytokine synthesis inhibitory factor (CSIF). | Activated subsets of CD4$^+$ and CD8$^+$ T cells | Stimulates or enhances proliferation of B cells, thymocytes, and mast cells; in cooperation with TGF-β, stimulates IgA synthesis and secretion by human B cells; antagonizes generation of the T$_H$1 subset of helper T cells |
| Interleukin-11 (IL-11).  23 kDa. | Bone marrow stromal cells and IL-1–stimulated fibroblasts | Growth factor for plasmacytomas, megakaryocytes, and macrophage progenitor cells |
| Interleukin-12 (IL-12).  Heterodimer containing a p35 subunit of 30–33 kDa, p40 subunit of 35–44 kDa. NK cells stimulatory factor (NKSF); cytotoxic lymphocyte maturation factor (CLMF). | Macrophages and dendritic cells | Important factor in inducing differentiation of T$_H$1 subset helper T cells; also induces interferon gamma production by T cells and NK cells and enhances NK-cell activity |
| Interleukin-13 (IL-13).  10 kDa. | Activated T cells, mast cells, and NK cells | Role in T$_H$2 responses; up-regulates synthesis of IgE and suppresses inflammatory responses; involved in pathology of asthma and some allergic conditions |
| Interleukin-14 (IL-14).  60 kDa. High-molecular-weight B-cell growth factor (HMW-BCGF). | T cells | Enhances B-cell proliferation; inhibits antibody synthesis |
| Interleukin-15 (IL-15).  14–15 kDa. | Many cell types but primarily dendritic cells and cells of the monocytic lineage | Stimulates NK-cell and T-cell proliferation and development; helps to activate NK cells |
| Interleukin-16 (IL-16).  Homotetramer 60 kDa; monomer ≈17 kDa. Lymphocyte chemotactic factor (LCF). | T cells | Stimulates migration of CD4$^+$ T cells, CD4$^+$ monocytes, and eosinophils; binding of IL-16 by CD4 inhibits HIV infection of CD4$^+$ cells |
| Interleukin-17 (IL-17).  28–31 kDa. CTLA-8 (cytotoxic lymphocyte–associated antigen 8). | Primarily CD4$^+$ T cells | Supports hematopoiesis indirectly by stimulating cytokine production by epithelial, endothelial, and fibroblastic stromal cells; enhances expression of ICAM-1, thus making cells more adhesive |
| Interleukin-18 (IL-18).  18.2 kDa. Interferon gamma-inducing factor (IGIF). | Cells of the monocytic lineage and dendritic cells | Promotes differentiation of T$_H$1 subset of helper T cells; induces interferon gamma production of T cells and enhances NK-cell cytotoxicity |
| Interleukin 19 (IL-19).  Homotetramer 35–40 kDa. | LPS-stimulated monocytes and B cells | A member of the IL-10 family of cytokines, induces reactive oxygen species and proinflammatory cytokines, which eventually promotes apoptosis; shown to alter the T$_H$1/T$_H$2 balance by inhibiting IFN-γ and augmenting IL-4 and IL-13 production |
| Interleukin 20 (IL-20).  18 kDa. | Monocytes and keratinocytes | A member of the IL-10 family of cytokines; has effects in epidermal tissues; like IL-19, shown to alter the T$_H$1/T$_H$2 balance |
| Interleukin 21 (IL-21).  15 kDa. | Activated T cells | Newly discovered; enhances cytotoxic activity and IFN-γ production by activated NK cells; enhances proliferation, IFN-γ production, and cytotoxicity of CD8$^+$ T cells |
| Interleukin 22 (IL-22).  Homodimer. 25 kDa. T-cell-derived inducible factor (TIF). | Primarily CD4$^+$ T cells | A member of the IL-10 family shown to inhibit epidermal differentiation; induces acute phase responses, up-regulates pancreatitis-associated protein 1 (PAP1) and osteopontin; like IL-19 and IL-20, shown to alter the T$_H$1/T$_H$2 balance |

| Cytokine. MW. Synonyms. | Sources | Activity |
|---|---|---|
| **Interleukin 23 (IL-23).** Heterodimer of p40 subunit of IL-12 (35–40 kDa) and p19 (18.7 kDa). | Activated dendritic cells | Many of the same biological activities as IL-12 |
| **Interleukin 24 (IL-24).** 35 kDa. IL-10B; MDA7 (melanoma differentiation association protein 7). | Melanocytes, NK cells, B cells, subsets of T cells, fibroblasts | Induces TNF-α and IFN-γ and low levels of IL-1β, IL-12, and GM-CSF in human PBMC; induces selective anticancer properties in breast carcinoma cells by promoting p53 independent apoptosis; member of the IL-10 family |
| **Interleukin 25 (IL-25).** 18-20 kDa. IL-17E; stroma-derived growth factor (SF20) | Bone marrow stromal cells, subset of T cells | New member of the IL-17 family; induces production of IL-4, IL-5, IL-13, and eotaxin; in vivo introduction of IL-25 into the lung can result in airway disease, involving cytokine production, tissue reorganization, mucus secretion, and airway hyper-reactivity |
| **Interleukin 26 (IL-26).** 36 kDa homodimer. AK155. | Subset of T and NK cells | Newly identified member of the IL-10 family; may have similar functions to IL-20 |
| **Interleukin 27 (IL-27).** Heterodimer composed of the p40 subunit of IL-12 and the p28 subunit. | Produced by dendritic cells, macrophages, endothelial cells, and plasma cells | Shown to induce clonal expansion of naive CD4$^+$ T cells, to synergize with IL-12 to promote IFN-γ production from CD4$^+$ T cells, and to induce CD8 T-cell–mediated antitumor activity |
| **Interleukin 28 A/B (IL-28A/B).** 19.8 kDa. Interferon-λ2/3 (IFN-λ2/3). | Monocyte-derived dendritic cells | Newly identified interferon-like cytokine; co-expressed with IFN-β, participates in the antiviral immune response and shown to induce increased level of both MHC class I and II |
| **Interleukin 29 (IL-29).** Interferon-λ1 (IFN-λ1). | Monocyte-derived dendritic cells | Functions similarly to IL-28A/B |
| **Interleukin 30 (IL-30).** p28. | Antigen-presenting cells | Subunit of IL-27 heterodimer; functions same as IL-27 |
| **Interleukin 31 (IL-31).** | Mainly activated $T_H2$ T cells; can be induced on activated monocytes | May be involved in recruitment of polymorphonuclear cells, monocytes, and T cells to a site of skin inflammation |
| **Interleukin 32 (IL-32).** NK4. | Activated NK cells and PBMCs | Newly identified member of the IL-1 family; proinflammatory cytokine, mitogenic properties, induces TNF-α |
| **BAFF (human B-cell–activating factor).** 18 kDa. ALL-1 (TNF and apoptosis ligand-related leukocyte-expressed ligand 1); BLyS (B-lymphocyte stimulator). | T cells, cells of the monocytic lineage, and dendritic cells | Member of the TNF family, occurs in membrane bound and soluble form; supports proliferation of antigen-receptor–stimulated B cells; differentiation and survival factor for immature B cells |
| **Granulocyte colony-stimulating factor (G-CSF).** 21 kDa. | Bone marrow stromal cells and macrophages | Essential for growth and differentiation of neutrophils |
| **Granulocyte-macrophage colony-stimulating factor (GM-CSF).** 22 kDa. | T cells, macrophages, fibroblasts, and endothelial cells | Growth factor for hematopoietic progenitor cells and differentiation factor for granulocytic and monocytic cell lineages |
| **Interferon alpha (IFN-α).** 16–27 kDa. Type 1 interferon; leukocyte interferon; lymphoblast interferon. | Lymphocytes, dendritic cells, and macrophages | Induces resistance to viruses and inhibits cell proliferation; regulates expression of class I MHC molecules on nucleated cells |
| **Interferon beta (IFN-β).** 20 kDa. Type 1 interferon; fibroblast interferon. | Fibroblasts, dendritic cells, and some epithelial cells | Induces resistance to virus infection in target cells; inhibits cell proliferation and regulates expression of class I MHC molecules |

*(continued)*

| Cytokine. MW. Synonyms. | Sources | Activity |
|---|---|---|
| **Interferon gamma (IFN-γ).** Monomer 17.1 kDa; dimer 40 kDa. Type 2 interferon; immune interferon; macrophage-activating factor (MAF); T-cell interferon. | CD4$^+$ and CD8$^+$ T cells; NK cells | Affects activation, growth, and differentiation of T cells, B cells, and macrophages as well as NK cells; up-regulates MHC expression on antigen-presenting cells; signature cytokine of T$_H$1 differentiation; weak antiviral and anti-proliferative activities |
| **Interferon lambda (IFN-λ) and interferon lambda-1.** Same as IL-28 and IL-29. | | |
| **Leukemia inhibitory factor (LIF).** 45 kDa. Differentiation-inhibiting factor (DIA); differentiation-retarding factor (DRF). | Many cell types, including T cells, cells of the monocytic lineage, fibroblasts, liver, and heart | A member of the IL-6 family; major experimental application: keeps cultures of ES cells in undifferentiated state to maintain their proliferation; in vivo, in combination with other cytokines, promotes hematopoiesis, stimulates acute phase response of liver cells, increases bone resorption, enhances glucose transport and insulin resistance, alters airway contractility, and causes loss of body fat |
| **Cardiotrophin-1 (CT-1).** 21.5 kDa. | Many cell types, including cells of monocytic lineage and heart | A member of the IL-6 family shown to stimulate hepatic expression of the acute phase proteins |
| **Ciliary neurotrophic factor (CNTF).** 24 kDa. Membrane-associated neurotransmitter stimulation factor (MANS). | Schwann cells and astrocytes | A member of the IL-6 family that induces the expression of acute phase proteins in the liver and has been shown to function as an endogenous pyrogen; also shown to function in ontogenesis and may promote survival and regeneration of nerves |
| **Macrophage colony-stimulating factor (M-CSF).** Disulfide linked homodimer of 45–90 kDa. Colony-stimulating factor 1 (CSF-1). | Many cell types, including lymphocytes, monocytes, fibroblasts, epithelial cells, and others | Growth, differentiation, and survival factor for macrophage progenitors and macrophages |
| **Macrophage inhibition factor (MIF).** 12 kDa monomer is biologically active in multimeric form. | Small amounts by many cell types; major producers are activated T cells, hepatocytes, monocytes, macrophages, and epithelial cells | Activates macrophages and inhibits their migration |
| **Oncostatin M (OSM).** 28–32 kDa. Onco M; ONC. | Activated T cells, monocytes, and adherent macrophages | Many functions, including regulation of the growth and differentiation of cells during hematopoiesis, neurogenesis, and osteogenesis; shown to enhance LDL uptake and also stimulates synthesis of acute phase proteins in the liver |
| **Stem-cell factor (SCF).** 36 kDa. Kit ligand (kitL) or steel factor (SLF). | Bone marrow stromal cells, cells of other organs such as brain, kidney, lung, and placenta | Roles in development of hematopoietic gonadal and pigmental lineages; active in both membrane-bound and secreted forms |
| **Thrombopoietin (Tpo).** 60 kDa. Megakaryocyte colony-stimulating factor; thrombopoiesis stimulation factor (TSF). | Liver, kidney, and skeletal muscle | Megakaryocyte lineage-specific growth differentiation factor that regulates platelet production |
| **Transforming growth factor beta (TGF-β).** ~ 25 kDa. Differentiation-inhibiting factor. | Many nucleated cell types and found in platelets | Inhibits growth of a number of cell types; affects tissue remodeling, wound repair, development, and hematopoiesis; exerts suppressive effects on the expansion of certain immune-cell populations; switch factor for IgA |

| Cytokine. MW. Synonyms. | Sources | Activity |
|---|---|---|
| **Tumor necrosis factor alpha** (TNF-α). 52 kDa. Cachetin, TNF ligand superfamily member 2 (TNFSF2). | Monocytes, macrophages, and other cell types, including activated T cells, NK cells, neutrophils, and fibroblasts | Strong mediator of inflammatory and immune functions; known to regulate growth and differentiation of a wide variety of cell types; cytotoxic for many types of transformed cells; promotes angiogenesis, bone resorption, and thrombotic processes; suppresses lipogenetic metabolism |
| **Tumor necrosis factor beta** (TNF-β). 25 kDa. Lymphotoxin (LT); cytotoxin (CTX); differentiation-inducing factor (DIF); TNF ligand superfamily member 1 (TNFSF1). | Activated T cells, B cells, fibroblasts, astrocytes, and endothelial and epithelial cells | Inhibits osteoclasts and keratinocyte growth, anti-angiogenic, and promotes fibroblasts proliferation; induces terminal differentiation of monocytes; stimulates neutrophil production of reactive oxygen species, increases phagocytosis, and enhances adhesion |

# Glossary

**ABO blood-group antigen**   Antigenic determinants of the blood-group system defined by the agglutination of red blood cells exposed to anti-A and anti-B antibodies.

**Abzyme**   A monoclonal antibody that has catalytic activity.

**Acquired immunity**   See **adaptive immunity.**

**Acquired immunodeficiency syndrome (AIDS)**   A disease caused by human immunodeficiency virus (HIV) that is marked by significant depletion of $CD4^+$ T cells resulting in increased susceptibility to a variety of infections and cancers.

**Active immunity**   Adaptive immunity that is induced by natural exposure to a pathogen or by **vaccination.**

**Acute lymphocytic leukemia (ALL)**   A form of cancer in which there is uncontrolled proliferation of a cell of the lymphoid lineage. The proliferating cells usually are present in the blood.

**Acute myelogenous leukemia (AML)**   A form of cancer in which there is uncontrolled proliferation of a cell of the myeloid lineage. The proliferating cells usually are present in the blood.

**Acute phase protein**   One of a group of serum proteins that increase in concentration in response to inflammation. Some **complement** components and **interferons** are acute phase proteins.

**Acute phase response (APR)**   The production of certain proteins and cells that appear in the blood shortly after many infections. It is part of the host's early defense against infection and precedes the adaptive phase of the immune response (see Figure 13-12).

**Acute phase response proteins**   A class of proteins synthesized in the liver in response to inflammation; serum concentrations of these proteins increase in inflammation.

**Adaptive immunity**   Host defenses that are mediated by B and T cells following exposure to antigen and that exhibit specificity, diversity, memory, and self-nonself recognition. See also **innate immunity.**

**Adoptive transfer**   The transfer of the ability to make or participate in an immune response by the transplantation of cells of the immune system.

**Affinity**   The strength with which a ligand interacts with a binding site. It is represented quantitatively by the affinity constant $K_a$.

**Affinity constant**   The ratio of the forward ($k_1$) to the reverse ($k_{-1}$) rate constant in an antibody-antigen reaction. Equivalent to the *association constant* in biochemical terms ($K_a = k_1/k_{-1}$).

**Affinity maturation**   The increase in average antibody affinity for an antigen that occurs during the course of an immune response or in subsequent exposures to the antigen.

**Agglutination**   The aggregation or clumping of particles (e.g., latex beads) or cells (e.g., red blood cells).

**Agglutination inhibition**   The reduction of antibody-mediated clumping of particles by the addition of the soluble forms of the epitope recognized by the agglutinating antibody.

**Agglutinin**   A substance capable of mediating the clumping of cells or particles. In particular, a hemagglutinin causes clumping of red blood cells.

**Allele**   Two or more alternative forms of a gene at a particular **locus** that confer alternative characters. The presence of multiple alleles results in **polymorphism.**

**Allelic exclusion**   A process that permits expression of only one of the allelic forms of a gene. For example, a B cell expresses only one allele for an antibody heavy chain and one allele for a light chain (see Figure 5-11).

**Allergy**   A **hypersensitivity** reaction that can include hay fever, asthma, **serum sickness,** systemic **anaphylaxis,** or contact dermatitis.

**Allogeneic**   Denoting members of the same species that differ genetically.

**Allograft**   A tissue transplant between **allogeneic** individuals.

**Allotype**   A set of **allotypic determinants** characteristic of some but not all members of a species.

**Allotypic determinant**   An antigenic determinant that varies among members of a species. The constant regions of antibodies possess allotypic determinants.

**Alpha-feto protein (AFP)**   See **oncofetal tumor antigen.**

**Alternative complement pathway**   Activation of **complement** that is initiated by foreign cell surface constituents, such as microbial antigens; it is antibody independent, involves C3–C9, factors B and D, and properdin and generates the **membrane-attack complex** (see Figure 7-7).

**Alveolar macrophage**   A macrophage found in the alveoli of the lung.

**Anaphylatoxins**   The **complement** split products C3a and C5a, which mediate **degranulation** of mast cells and basophils, resulting in release of mediators that induce contraction of smooth muscle and increased vascular permeability.

**Anaphylaxis**   An immediate type I hypersensitivity reaction, which is triggered by IgE-mediated mast cell **degranulation.** Systemic anaphylaxis leads to shock and is often fatal. Localized anaphylaxis involves various types of **atopic** reactions.

**Antibody**   A protein (immunoglobulin), consisting of two identical heavy chains and two identical light chains, that recognizes a particular **epitope** on an antigen and facilitates clearance of that antigen. Membrane-bound antibody is expressed by B cells that have not encountered antigen; secreted antibody is produced by **plasma cells.** Some antibodies are multiples of the basic four-chain structure.

**Antibody-dependent cell-mediated cytotoxicity (ADCC)**   A cell-mediated reaction in which nonspecific cytotoxic cells that express **Fc receptors** (e.g., NK cells, neutrophils, macrophages) recognize bound antibody on a target cell and subsequently cause lysis of the target cell.

**Antibody molecule**   See **antibody.**

**Antigen**   Any substance (usually foreign) that binds specifically to an antibody or a T-cell receptor; often is used as a synonym for **immunogen.**

**Antigenically committed**   The state of a mature B cell displaying surface antibody specific for a single immunogen.

**Antigenic determinant**   The site on an antigen that is recognized and bound by a particular antibody, TCR-peptide-MHC complex, or TCR-ligand-CD1 complex; also called **epitope.**

**Antigenic drift**   A series of spontaneous point mutations that generate minor antigenic variations in pathogens and lead to strain differences.

**Antigenicity**   The capacity to combine specifically with antibodies or T-cell receptor/MHC.

**Antigenic peptide**   In general, a peptide capable of raising an immune response, for example, in a peptide that forms a complex with MHC that can be recognized by a T-cell receptor.

**Antigenic shift**   Sudden emergence of a new pathogen subtype possibly by genetic reassortment that has led to substantial antigenic differences.

**Antigenic specificity**   See **specificity, antigenic.**

**Antigen presentation**   See **antigen processing.**

**Antigen-presenting cell (APC)**   Any cell that can process and present antigenic peptides in association with **class II MHC molecules** and deliver a **co-stimulatory signal** necessary for T-cell activation. Macrophages, dendritic cells, and B cells constitute the professional APCs. Nonprofessional APCs, which function in antigen presentation only for short periods, include thymic epithelial cells and vascular endothelial cells.

**Antigen processing**    Degradation of antigens by one of two pathways yielding antigenic peptides that are displayed bound to MHC molecules on the surface of antigen-presenting cells or altered self cells.

**Apoptosis**    Morphologic changes associated with programmed cell death, including nuclear fragmentation, blebbing, and release of apoptotic bodies, which are phagocytosed (see Figure 2-3). In contrast to **necrosis,** it does not result in damage to surrounding cells.

**Association constant**    See **affinity constant.**

**Atopic**    Pertaining to clinical manifestations of type I (IgE-mediated) hypersensitivity, including allergic rhinitis (hay fever), eczema, asthma, and various food allergies.

**Attenuate**    To decrease virulence of a pathogen and render it incapable of causing disease. Many vaccines are attenuated bacteria or viruses that raise protective immunity without causing harmful infection.

**Autocrine**    A condition in which the cell acted on by a cytokine is the source of the cytokine.

**Autograft**    Tissue grafted from one part of the body to another in the same individual.

**Autoimmunity**    An abnormal immune response against self antigens.

**Autologous**    Denoting transplanted cells, tissues, or organs derived from the same individual.

**Avidity**    The strength of antigen-antibody binding when multiple epitopes on an antigen interact with multiple binding sites of an antibody. See also **affinity.**

**Bacillus Calmette-Guérin**    See **BCG.**

**Bacteremia**    An infection in which viable bacteria are found in the blood.

**BALT (bronchus-associated lymphoid tissue).**    See **MALT.**

**Basophil**    A nonphagocytic granulocyte that expresses Fc receptors for IgE (see Figure 2-9). Antigen-mediated cross-linkage of bound IgE induces **degranulation** of basophils.

**B cell**    See **B lymphocyte.**

**B-cell coreceptor**    A complex of three proteins (CR2, CD19, and TAPA-1) associated with the B-cell receptor. It is thought to amplify the activating signal induced by cross-linkage of the receptor.

**B-cell receptor (BCR)**    Complex comprising a membrane-bound immunoglobulin molecule and two associated signal-transducing Iga/Igb molecules.

**B-cell-specific activator protein (BSAP)**    A transcription factor encoded by the gene *Pax-5* that plays an essential role in early and later stages of B-cell development.

**BCG (Bacillus Calmette-Guérin)**    An attenuated form of *Mycobacterium bovis* used as a specific vaccine and as an adjuvant component.

**Benign**    Pertaining to a nonmalignant form of a neoplasm or a mild form of an illness.

**Bispecific antibody**    Hybrid antibody made either by chemically cross-linking two different antibodies or by fusing hybridomas that produce different monoclonal antibodies.

**Bence-Jones protein**    Protein found in high concentrations in the urine of patients with multiple myeloma; usually an Ig light chain or fragment thereof.

**$\beta_2$-microglobulin**    Invariant subunit that associates with the polymorphic $\alpha$ chain to form **class I MHC molecules;** it is not encoded by MHC genes.

**B lymphocyte**    A lymphocyte that matures in the bone marrow and expresses membrane-bound antibody. After interacting with antigen, it differentiates into antibody-secreting plasma cells and memory cells.

**Bone marrow**    The living tissue found within the hard exterior of bone.

**Booster**    Inoculation given to stimulate immunologic memory response.

**Bradykinin**    An endogenously produced peptide that produces an inflammatory response.

**Carcinoembryonic antigen (CEA)**    An oncofetal antigen (found not only on cancerous cells but also on normal cells) that can be a tumor-associated antigen.

**Carcinoma**    Tumor arising from endodermal or ectodermal tissues (e.g., skin or epithelium). Most cancers (>80%) are carcinomas.

**Carrier**    An immunogenic molecule containing antigenic determinants recognized by T cells. Conjugation of a carrier to a nonimmunogenic **hapten** renders the hapten immunogenic.

**Carrier effect**    A **secondary immune response** to a hapten depends on use of both the **hapten** and the **carrier** used in the initial immunization.

**Caspase**    A family of cysteine proteases that cleave after an aspartate residue. The term *caspase* incorporates these elements (*c*ysteine, *asp*artate, prote*ase*), which play important roles in the chain of reactions that leads to **apoptosis.**

**C (constant) gene segment**    The 3′ coding of a rearranged immunoglobulin or T-cell receptor gene. There are multiple C gene segments in germ-line DNA, but as a result of gene rearrangement and, in some cases, RNA processing, only one segment is expressed in a given protein.

**CC subgroup**    A subgroup of chemokines in which a disulfide bond links adjacent cysteines.

**CD3**    A polypeptide complex containing three dimers: a $\gamma\epsilon$ heterodimer, a $\epsilon\delta$ heterodimer, and either a $\xi\xi$ homodimer or a $\xi\eta$ heterodimer (see Figure 9-9). It is associated with the T-cell receptor and functions in signal transduction.

**CD4**    A glycoprotein that serves as a co-receptor on MHC class II–restricted T cells. Most helper T cells are CD4$^+$.

**CD8**    A dimeric protein that serves as a co-receptor on MHC class I–restricted T cells. Most cytotoxic T cells are CD8$^+$.

**Cell adhesion molecules (CAMs)**    A group of cell surface molecules that mediate intercellular adhesion. Most belong to one of four protein families: the **integrins, selectins,** mucin-like proteins, and **immunoglobulin superfamily** (see Figure 13-1).

**Cell line**    A population of cultured tumor cells or normal cells that have been subjected to chemical or viral **transformation.** Cell lines can be propagated indefinitely in culture.

**Cell-mediated immune response**    Host defenses that are mediated by antigen-specific T cells and various nonspecific cells of the immune system. It protects against intracellular bacteria, viruses, and cancer and is responsible for graft rejection. Transfer of primed T cells confers this type of immunity on the recipient. See also **humoral immune response.**

**Cell-mediated immunity**    See **cell-mediated immune response.**

**Cell-mediated lympholysis (CML)**    In vitro lysis of allogeneic cells or virus-infected syngeneic cells by T cells (see Figure 14-17); can be used as an assay for CTL activity or class I MHC activity.

**Cellular oncogene**    See **proto-oncogene.**

**Central tolerance**    Elimination of self-reactive lymphocytes in primary generative organs.

**Chediak-Higashi syndrome**    An autosomal recessive immunodeficiency caused by a defect in lysosomal granules that impairs killing by NK cells.

**Chemoattractant**    A substance that attracts leukocytes. Some chemoattractants also cause significant changes in the physiology of cells that bear receptors for them.

**Chemokine**    Any of several secreted low-molecular-weight polypeptides that mediate **chemotaxis** for different leukocytes and regulate the expression and/or adhesiveness of leukocyte **integrins.**

**Chemotactic factor**    An agent that can cause leukocytes to move up its concentration gradient.

**Chimera**    An animal or tissue composed of elements derived from genetically distinct individuals. The **SCID-human mouse** is a chimera. Also, a chimeric antibody that contains the amino acid sequence of one species in one region and the sequence of a different species in another (for example, an antibody with a human constant region and a mouse variable region).

**Chimeric antibody**    See **chimera.**

**Chromogenic substrate**    A colorless substance that is transformed into colored products by an enzymatic reaction.

**Chronic granulomatous disease**    Immunodeficiency caused by a defect in the enzyme NADPH phagosome oxidase resulting in failure to generate reactive oxygen species in neutrophils.

**Chronic lymphocytic leukemia (CLL)**    A type of leukemia in which cancerous lymphocytes are continually produced.

**Chronic myelogenous leukemia (CML)**    A type of leukemia in which cells of the myeloid lineage are continually produced.

**Cilia**    Hairlike projections on epithelial cells in the respiratory and gastrointestinal tracts; cilia function to propel microbes out of the tract.

**Classical complement pathway**    Activation of **complement** that is initiated by antigen-antibody complexes, involves C1–C9, and generates the **membrane-attack complex** (see Figure 7-5).

**Class (isotype) switching**    The change in the antibody class that a B cell produces.

**Class I MHC molecules**    Heterodimeric membrane proteins that consist of an $\alpha$ chain encoded in the **MHC** associated noncovalently with $\beta_2$-**microglobulin** (see Figure 8-3). They are expressed by nearly all nucleated cells and function to present antigen to CD8$^+$ T cells. The classical class I molecules are H-2 K, D, and L in mice and HLA-A, -B, and -C in humans.

**Class II MHC molecules**    Heterodimeric membrane proteins that consist of a noncovalently associated $\alpha$ and $\beta$ chain, both encoded in the **MHC** (see Figure 8-3). They are expressed by **antigen-presenting cells** and function to present antigen to CD4$^+$ T cells. The classical class II molecules are H-2 IA and IE in mice and HLA-DP, -DQ, and -DR in humans.

**Class III MHC molecules**    Various proteins encoded in the **MHC** but distinct from class I and class II MHC molecules. They include some complement components, two steroid 21-hydroxylases, and tumor necrosis factors $\alpha$ and $\beta$.

**Clonal anergy**    A physiological state in which cells are unable to be activated by antigen.

**Clonal deletion**    The induced death of members of a clone of lymphocytes with inappropriate receptors (those that strongly react with self during development, for example).

**Clonal selection**    The antigen-mediated activation and proliferation of members of a clone of B cells that have receptors for the antigen (or for complexes of MHC and peptides derived from the antigen, in the case of T cells).

**Clone**    Cells arising from a single progenitor cell.

**Clot**    Coagulated mass; usually refers to coagulated blood, in which conversion of fibrinogen in the plasma to fibrin has produced a jelly-like substance containing entrapped blood cells.

**Cluster of differentiation (CD)**    A collection of monoclonal antibodies that all recognize an antigen found on a particular differentiated cell type or types. Each of the antigens recognized by such a collection of antibodies is called a CD marker and assigned a unique identifying number.

**Coding joint**    The sequence at the point of union of coding sequences during V(D)J rearrangement to form rearranged antibody or T-cell receptor genes.

**Collectins**    Molecules that have surfactant qualities and can lyse bacterial cell walls.

**Complement**    A group of serum and cell membrane proteins that interact with one another and with other molecules of innate and adaptive immunity to carry out key effector functions.

**Complementarity-determining region (CDR)**    Portions of the variable regions of antibody molecules that protrude from the V domains and have the potential to contact antigens. The antigen-binding sites of antibody molecules are composed of the CDRs.

**Complement system**    See **complement.**

**Conformational determinants**    Epitopes of a protein that are composed of amino acids that are close together in the three-dimensional structure of the protein but may not be near each other in the amino acid sequence.

**Congenic**    Denoting individuals that differ genetically at a single genetic locus or region; also called *coisogenic.*

**Constant (C) region**    The nearly invariant portion of the immunoglobulin molecule that does not contain antigen-binding domains. The sequence of amino acids in the constant region determines the isotype ($\alpha$, $\gamma$, $\delta$, $\epsilon$, and $\mu$) of heavy chains and the type ($\kappa$ and $\lambda$) of light chains.

**Cortex**    The outer or peripheral layer of an organ.

**Costimulatory signal**    Additional signal that is required to induce proliferation of antigen-primed T cells and is generated by interaction of CD28 on T cells with B7 on antigen-presenting cells or altered self cells (see Figures 10-17 and 14-2). In B-cell activation, an analogous signal (competence signal 2) is provided by interaction of CD40 on B cells with CD40L on activated T$_H$ cells.

**C-reactive protein (CRP)**    An acute phase protein that participates in the complement pathway; an increased level of serum CRP is an indicator of inflammation.

**Cross-reactivity**    Ability of a particular antibody or T-cell receptor to react with two or more antigens that possess a common **epitope.**

**CXC subgroup**    A family of chemokines that contain a disulfide bridge between cysteines separated by a different amino acid residue (X).

**Cyclosporin A**    A fungal product used as a drug to suppress allograft rejection. The compound blocks T-cell activation by interfering with transcription factors and preventing gene activation.

**Cytokine**    Any of numerous secreted, low-molecular-weight proteins that regulate the intensity and duration of the immune response by exerting a variety of effects on lymphocytes and other immune cells (see Appendix II).

**Cytotoxic T lymphocyte (CTL)**    An effector T cell (usually CD8$^+$) that can mediate the lysis of target cells bearing antigenic peptides complexed with an MHC molecule. It usually arises from an antigen-activated T$_C$ cell.

**Dark zone**    A portion of the germinal center that is the site of rapid cell division by forms of B cells called centroblasts.

**Degranulation**    Discharge of the contents of cytoplasmic granules by **basophils** and **mast cells** following cross-linkage (usually by antigen) of bound IgE. It is characteristic of type **I hypersensitivity.**

**Delayed-type hypersensitivity (DTH)**    A type IV hypersensitive response mediated by sensitized T$_H$ cells, which release various **cytokines** and **chemokines** (see Figure 15-17). The response generally occurs 2–3 days after T$_H$ cells interact with antigen. It is an important part of host defense against intracellular parasites and bacteria.

**Dendritic cells (DCs)**    Bone-marrow-derived cells that descend through the myeloid and lymphoid lineages and are specialized for antigen presentation to helper T cells.

**Dermis**    Layer of skin under the epidermis that contains blood and lymph vessels, hair follicles, nerves, and nerve endings.

**Differentiation antigen**    A cell surface marker that is expressed only during a particular developmental stage or by a particular cell lineage.

**DiGeorge's syndrome**    Congenital thymic aplasia caused by deletion of a sequence on chromosome 22 during embryonic life. Consequences include immunodeficiency, facial abnormalities, and congenital heart disease.

**Direct staining**    A variation of fluorescent antibody staining in which the primary antibody is directly conjugated to the fluorescent label.

**Dissociation constant**    $K_d$, the reciprocal of the **affinity constant** ($1/K_a$).

**Double immunodiffusion**    A type of precipitation in gel analysis in which both antigen and antibody diffuse radially from wells toward each other, thereby establishing a concentration gradient. As equivalence is reached, a visible line of precipitation, a precipitin line, forms.

**Double-negative (DN) cells**    A subset of developing T cells (thymocytes) that do not express CD4 or CD8. At this early stage of T-cell development, DN cells do not express the TCR.

**Double-positive (DP) cells**    A subset of developing T cells (thymocytes) that express both CD4 and CD8. DP cells are an intermediate stage of developing thymocytes that express TCRs.

**Early B-cell factor (EBF)**    A transcription factor that is essential for early B-cell development. It is necessary for the expression of RAG.

**Edema**    Abnormal accumulation of fluid in intercellular spaces, often resulting from a failure of the lymphatic system to drain off normal leakage from the capillaries.

**Effector cell**    Any cell capable of mediating an immune function (e.g., activated $T_H$ cells, CTLs, and plasma cells).

**Effector response**    Action that occurs after recognition and binding of antigen by antibody; lysis by complement proteins is an effector response. Also called *effector function*.

**ELISA.**    See **enzyme-linked immunosorbent assay.**

**Embryonic stem cell (ES cell)**    Stem cell isolated from early embryo and grown in culture. Mouse ES cells give rise to a variety of cell types and are used to develop transgenic or knockout mouse strains.

**Endocrine**    Referring to regulatory secretions such as hormones or cytokines that pass from producer cell to target cell by the bloodstream.

**Endocytosis**    Process by which cells ingest extracellular macromolecules by enclosing them in a small portion of the plasma membrane, which invaginates and is pinched off to form an intracellular vesicle containing the ingested material.

**Endotoxins**    Certain lipopolysaccharide (LPS) components of the cell wall of gram-negative bacteria that are responsible for many of the pathogenic effects associated with these organisms. Some function as **superantigens.**

**Enzyme-linked immunosorbent assay (ELISA)**    An assay for quantitating either antibody or antigen by use of an enzyme-linked antibody and a substrate that forms a colored reaction product.

**Eosinophil**    Motile, somewhat phagocytic granulocytes that can migrate from blood to tissue spaces. They have large numbers of IgE receptors and are highly granular. They are thought to play a role in the defense against parasitic organisms such as roundworms (see Figure 2-9).

**Epidermis**    The outermost layer of the skin.

**Epitope**    The portion of an antigen that is recognized and bound by an antibody or TCR-MHC combination; also called **antigenic determinant.**

**Epitope mapping**    Localization of sites (epitopes) on an antigen molecule that are reactive with different antibodies or T-cell receptors.

**Equilibrium dialysis**    An experimental technique that can be used to determine the affinity of an antibody for antigen and its **valency** (see Figure 6-2).

**Erythroblastosis fetalis**    A type II hypersensitivity reaction in which maternal antibodies against fetal **Rh antigens** cause hemolysis of the erythrocytes of a newborn; also called *hemolytic disease of the newborn*.

**E2A**    A transcription factor that is required for the expression of the **recombination-activating genes** (RAG) as well as the expression of the

**λ5** (lambda 5)    component of the pre-B-cell receptor during B-cell development. It is essential for B-cell development.

**Exocytosis**    Process by which cells release molecules (e.g., cytokines, lytic enzymes, degradation products) contained within a membrane-bound vesicle by fusion of the vesicle with the plasma membrane (see Figure 2-8).

**Exotoxins**    Toxic proteins secreted by gram-positive and gram-negative bacteria; some function as **superantigens.** They cause food poisoning, toxic shock syndrome, and other disease states. See also **immunotoxin.**

**Extravasation**    Movement of blood cells through an unruptured vessel wall into the surrounding tissue, particularly at sites of inflammation.

**Fab fragment** (fragment antigen binding):    A monovalent antigen-binding fragment of an immunoglobulin molecule that consists of one light chain and part of one heavy chain, linked by an interchain disulfide bond. It is obtained by brief papain digestion.

**F(ab')₂ fragment**    Two Fab units linked by disulfide bridges between fragments of the heavy chain. They are obtained by digestion of antibody with pepsin.

**Fc fragment** (fragment crystallizable):    A crystallizable, non-antigen-binding fragment of an immunoglobulin molecule that consists of the carboxyl-terminal portions of both heavy chains and possesses binding sites for Fc receptors and the C1q component of complement. It is obtained by brief papain digestion.

**Fc receptor (FcR)**    Cell surface receptor specific for the Fc portion of certain classes of immunoglobulin. It is present on lymphocytes, mast cells, macrophages, and other accessory cells.

**Fibrin**    A filamentous protein produced by the action of thrombin on fibrinogen; fibrin is the main element in blood clotting.

**Fibrinopeptide**    One of two small peptides of about 20 amino acids released from fibrinogen by thrombin cleavage in the conversion to fibrin.

**Fibrosis**    A process responsible for the development of a type of scar tissue at the site of chronic inflammation.

**Fluorescent antibody**    An antibody with a **fluorochrome** conjugated to its Fc region that is used to stain cell surface molecules or tissues; the technique is called **immunofluorescence.**

**Fluorochrome**    A molecule that fluoresces when excited with appropriate wavelengths of light. See **immunofluorescence.**

**Follicular dendritic cell**    A cell with extensive dendritic extensions that is found in the follicles of lymph nodes. Although they do not express MHC class II molecules, they are richly endowed with receptors for complement and Fc receptors for antibody. They are of a lineage that is distinct from MHC class II–bearing dendritic cells.

**Fragmentin**    Enzymes present in the granules of cytotoxic lymphocytes that induce DNA fragmentation.

**Framework region (FR)**    A relatively conserved sequence of amino acids located on either side of the hypervariable regions in the variable domains of immunoglobulin heavy and light chains.

**Freund's complete adjuvant (CFA)**    A water-in-oil emulsion to which heat-killed mycobacteria have been added; antigens are administered in CFA to enhance their immunogenicity.

**Freund's incomplete adjuvant**    Freund's adjuvant lacking heat-killed mycobacteria.

**GALT**    Gut-associated lymphoid tissue. See **MALT.**

**γ (gamma)-globulin fraction**    The electrophoretic fraction of serum that contains most of the immunoglobulin classes.

**GATA-2 gene**    A gene encoding a transcription factor that is essential for the development of several hematopoietic cell lineages, including the lymphoid, erythroid, and myeloid lineages.

**Gene conversion**   Process in which portions of one gene (the recipient) are changed to those of another gene (the donor). Homologous gene conversion is a diversification mechanism used for immunoglobulin V genes in some species.

**Gene segments**   Germ-line gene sequences that are combined with others to make a complete coding sequence; Ig and TCR genes are products of V, D, J gene segments.

**Gene therapy**   General term for any measure aimed at correction of a genetic defect by introduction of a normal gene or genes.

**Genotype**   The combined genetic material inherited from both parents; also, the **alleles** present at one or more specific loci.

**Germinal center**   A region within lymph nodes and the spleen where B-cell activation, proliferation, and differentiation occurs (see Figure 11-20). Germinal centers are sites of intense B cell somatic mutation and selection.

**Germ-line theories**   Theories that explain antibody diversity by postulating that all antibodies are encoded in the host chromosomes.

**Goodpasture's syndrome**   An autoimmune disease characterized by autoantibodies specific for certain basement membrane antigens of the glomeruli of the kidneys and the alveoli of the lungs.

**Graft-versus-host disease (GVHD)**   A reaction that develops when a graft contains immunocompetent T cells that recognize and attack the recipient's cells.

**Granulocyte**   Any **leukocyte** that contains cytoplasmic granules, particularly the basophil, eosinophil, and neutrophil (see Figure 2-9).

**Granuloma**   A tumorlike mass or nodule that arises because of a chronic **inflammatory response** and contains many activated macrophages, epithelioid cells (modified macrophages), $T_H$ cells, and multinucleated giant cells formed by the fusion of macrophages.

**Granzyme**   One of a set of enzymes found in the granules of $T_C$ cells that can help to initiate apoptosis in target cells.

**Graves' disease**   An autoimmune disease in which the individual produces auto-antibodies to the receptor for thyroid-stimulating hormone TSH.

**Haplotype**   The set of **alleles** of linked genes present on one parental chromosome; commonly used in reference to the **MHC** genes.

**Hapten**   A low-molecular-weight molecule that can be made immunogenic by conjugation to a suitable carrier.

**Hapten-carrier conjugate**   A covalent combination of a small molecule (hapten) with a large carrier molecule or structure.

**Heavy (H) chain**   The larger polypeptide of an antibody molecule; it is composed of one variable domain $V_H$ and three or four constant domains ($C_H1$, $C_H2$, etc.) There are five major classes of heavy chains in humans ($\alpha$, $\gamma$, $\delta$, $\epsilon$, and $\mu$), which determine the **isotype** of an antibody (see Table 4-3).

**Hemagglutinin**   See **agglutinin.**

**Hematopoiesis**   The formation and differentiation of blood cells (see Figure 2-2).

**Hematopoietic-inducing microenvironment (HIM)**   An anatomical site that contains all of the cells and cellular factors required for the generation and development of blood cells.

**Hematopoietic stem cell (HSC)**   The cell type from which all lineages of blood cells arise.

**Hemolysis**   Alteration or destruction of red blood cells, which liberates hemoglobin.

**Heteroconjugates**   Hybrids of two different antibody molecules.

**High-endothelial venule (HEV)**   An area of a capillary venule composed of specialized cells with a plump, cuboidal ("high") shape through which lymphocytes migrate to enter various lymphoid organs (see Figure 13-7).

**Hinge region**   The segment of an immunoglobulin heavy chain between the **Fc** and **Fab** regions. It gives flexibility to the molecule and allows the two antigen-binding sites to function independently.

**Histiocyte**   An immobilized (sometimes called "tissue fixed") macrophage found in loose connective tissue.

**Histocompatible**   Denoting individuals whose major histocompatibility antigens are identical. Grafts between such individuals are generally accepted.

**Histocompatibility antigens**   Family of proteins that determines the ability of one individual to accept tissue or cell grafts from another. The major histocompatibility antigens, which are encoded by the **MHC,** function in antigen presentation.

**HLA (human leukocyte antigen) complex**   Term for the **MHC** in humans.

**Homing**   The differential migration of lymphocytes or other leukocytes to particular tissues or organs.

**Homing receptor**   A receptor that directs various populations of lymphocytes to particular lymphoid and inflammatory tissues.

**H-2 complex**   Term for the **MHC** in the mouse.

**Humanized antibody**   An antibody that contains the antigen-binding amino acid sequences of another species within the framework of a human immunoglobulin sequence.

**Humoral**   Pertaining to extracellular fluid, including the plasma, lymph, and tissue fluids.

**Humoral immune response**   Host defenses that are mediated by antibody present in the plasma, lymph, and tissue fluids. It protects against extracellular bacteria and foreign macromolecules. Transfer of antibodies confers this type of immunity on the recipient. See also **cell-mediated immune response.**

**Hybridoma**   A **clone** of hybrid cells formed by fusion of normal lymphocytes with **myeloma cells;** it retains the properties of the normal cell to produce antibodies or T-cell receptors but exhibits the immortal growth characteristic of myeloma cells. Hybridomas are used to produce **monoclonal antibody.**

**Hypersensitivity**   Exaggerated immune response that causes damage to the individual. Immediate hypersensitivity (types I, II, and III) is mediated by antibody or immune complexes, and delayed-type hypersensitivity (type IV) is mediated by $T_H$ cells.

**Hypervariable region**   One of three regions within the variable domain of each chain in immunoglobulins and T-cell receptors that exhibits the most sequence variability and contributes the most to the antigen-binding site; also called *complementarity-determining region (CDR).*

**Idiotope**   A single **antigenic determinant** in the variable domains of an antibody or T-cell receptor; also called *idiotypic determinant.* Idiotopes are generated by the unique amino acid sequence specific for each antigen.

**Idiotype**   The set of antigenic determinants (**idiotopes**) characterizing a unique antibody or T-cell receptor.

**Ikaros**   A transcription factor required for the development of all lymphoid cell lineages.

**Immediate hypersensitivity**   An exaggerated immune response mediated by antibody (type I and II) or antigen-antibody complexes (type III) that manifests within minutes to hours after exposure of a sensitized individual to antigen (see Table 15-1).

**Immunization**   The process of producing a state of immunity in a subject. See also **active immunity** and **passive immunity.**

**Immunocompetent**   Denoting a mature lymphocyte that is capable of recognizing a specific antigen and mediating an immune response.

**Immunodeficiency**   Any deficiency in the immune response. It may result from a defect in phagocytosis, the humoral response, or the cell-mediated response. Combined immunodeficiencies affect both the humoral and cell-mediated immune response (see Figure 20-1).

**Immunodominant**   Referring to epitopes that produce a more pronounced immune response than others under the same conditions.

**Immunoelectron microscopy**   A technique in which antibodies used to stain a cell or tissue are labeled with an electron-dense material and visualized with an electron microscope.

**Immunoelectrophoresis**   A technique in which an antigen mixture is first separated into its component parts by electrophoresis and then tested by **double immunodiffusion.**

**Immunofluorescence**   Technique of staining cells or tissues with **fluorescent antibody** and visualizing them under a fluorescent microscope.

**Immunogen**   A substance capable of eliciting an immune response. All immunogens are **antigens,** but some antigens (e.g., haptens) are not immunogens.

**Immunogenicity**   The capacity of a substance to induce an immune response under a given set of conditions.

**Immunoglobulin**   Protein family with antibody activity.

**Immunoglobulin fold**   Characteristic structure in immunoglobulins that consists of a domain of 100 to 110 amino acids folded into two β-pleated sheets, each containing three or four antiparallel β strands and stabilized by an intrachain disulfide bond (see Figure 4-8).

**Immunoglobulin superfamily**   Group of proteins that contain **immunoglobulin-fold** domains, or structurally related domains; it includes immunoglobulins, T-cell receptors, MHC molecules, and numerous other membrane molecules.

**Immunoreceptor tyrosine-based activation motif (ITAM)**   Amino acid sequence in the intracellular portion of signal-transducing cell surface molecules that interacts with and activates intracellular kinases after ligand binding by the surface molecule.

**Immunotoxin**   Highly cytotoxic agents produced by conjugating an antibody with a highly toxic agent, usually a protein such as ricin.

**Incomplete antibody**   Antibody that binds antigen but does not induce agglutination.

**Indirect staining**   A method of immunofluorescent staining in which the primary antibody is unlabeled and is detected with an additional fluorochrome-labeled reagent.

**Inducible nitric oxide synthetase (NOS)**   An inducible form of NOS that generates the antimicrobial compound nitric oxide from arginine.

**Inflamed**   Manifesting redness, pain, heat, and swelling.

**Inflammatory response**   A localized tissue response to injury or other trauma characterized by pain, heat, redness, and swelling. The response, which includes both localized and systemic effects, consists of altered patterns of blood flow, an influx of phagocytic and other immune cells, removal of foreign antigens, and healing of the damaged tissue.

**Innate immunity**   Nonspecific host defenses that exist prior to exposure to an antigen and involve anatomic, physiologic, endocytic and phagocytic, and inflammatory mechanisms. See also **adaptive immunity.**

**Insulin-dependent diabetes mellitus (IDDM)**   An autoimmune disease caused by an autoimmune attack on the pancreas directed against specialized insulin-producing cells, called the islets of Langerhans, scattered throughout the pancreas.

**Integrins**   A group of heterodimeric cell adhesion molecules (e.g., LFA-1, VLA-4, and Mac-1) present on various leukocytes that bind to Ig-superfamily CAMs (e.g., ICAMs, VCAM-1) on endothelium (see Figure 15-2).

**Intercellular adhesion molecules (ICAMs)**   Cellular adhesion molecules that bind to integrins. ICAMs are members of the immunoglobulin superfamily.

**Interferons (IFNs)**   Several glycoprotein **cytokines** produced and secreted by certain cells that induce an antiviral state in other cells and also help to regulate the immune response (see Table 12-1).

**Interleukins (ILs)**   A group of **cytokines** secreted by leukocytes that primarily affect the growth and differentiation of various hematopoietic and immune system cells (see Appendix II).

**Interstitial fluid**   Fluid found in the spaces between cells of an organ or tissue.

**Intraepidermal lymphocytes**   T cells found in epidermal layers.

**Intraepithelial lymphocytes (IELs)**   T cells found in the epithelial layer of organs and the gastrointestinal tract.

**Invariant (Ii) chain**   Component of the MHC class II protein that shows no genetic polymorphism. The Ii chain stabilizes the class II molecule before it has acquired an antigenic peptide.

**Isograft**   Graft between genetically identical individuals.

**Isotype**   (1) An antibody class, which is determined by the constant-region sequence of the heavy chain. The five human isotypes, designated IgA, IgD, IgE, IgG, and IgM, exhibit structural and functional differences (see Table 4-4). Also refers to the set of **isotypic determinants** that is carried by all members of a species. (2) One of the five major kinds of heavy chains in antibody molecules (α, γ, δ, ε, and μ).

**Isotype switching**   Conversion of one antibody class (**isotype**) to another resulting from the genetic rearrangement of heavy-chain constant-region genes in B cells; also called *class switching.*

**Isotypic determinant**   An **antigenic determinant** within the immunoglobulin constant regions that is characteristic of a species.

**J (joining) chain**   A polypeptide that links the heavy chains of monomeric units of polymeric IgM and di- or trimeric IgA. The linkage is by disulfide bonds between the J chain and the carboxyl-terminal cysteines of IgM or IgA heavy chains.

**J (joining) gene segment**   The part of a rearranged immunoglobulin or T-cell receptor gene that joins the variable region to the constant region and encodes part of the **hypervariable region.** There are multiple J gene segments in germ-line DNA, but gene rearrangement leaves only one in each functional rearranged gene.

**Junctional flexibility**   The diversity in antibody and T-cell receptor genes created by the imprecise joining of coding sequences during the assembly of the rearranged genes.

**Kappa (κ) chain**   One of the two types of light chains. Lambda (λ) is the other type.

**Kupffer cell**   A type of tissue-fixed macrophage found in liver.

**Lambda (λ) chain**   One of the two types of light chains. Kappa (κ) is the other type.

**λ5**   A polypeptide that associates with **Vpre-B** to form the **surrogate light chain** of the **pre-B-cell receptor.**

**Leader (L) peptide**   A short hydrophobic sequence of amino acids at the N-terminus of newly synthesized immunoglobulins; it inserts into the lipid bilayer of the vesicles that transport Ig to the cell surface. The leader is removed from the ends of mature antibody molecules by proteolysis.

**Lectin pathway**   Pathway of complement activation initiated by binding of serum protein MBL to the mannose-containing component of microbial cell walls (see Figure 7-2).

**Leukemia**   Cancer originating in any class of hematopoietic cell that tends to proliferate as single cells within the lymph or blood.

**Leukocyte**   A white blood cell. The category includes lymphocytes, granulocytes, platelets, monocytes, and macrophages.

**Leukocyte adhesion deficiency (LAD)**   An inherited disease in which the leukocytes are unable to undergo adhesion-dependent migration into sites of inflammation. Recurrent bacterial infections and impaired healing of wounds are characteristic of this disease.

**Leukocytosis**   An abnormally large number of leukocytes, usually associated with acute infection. Counts greater than 10,000/mm$^3$ may be considered leukocytosis.

**Leukotrienes**  Several lipid mediators of inflammation and type I hypersensitivity; also called *slow reactive substance of anaphylaxis (SRS-A).* They are metabolic products of arachidonic acid.

**Light (L) chains**  Immunoglobulin polypeptides of the lambda or kappa type that join with heavy-chain polypeptides to form the antibody heterodimer.

**Light zone**  A region of the germinal center that contains numerous follicular dendritic cells.

**Lipopolysaccharide (LPS)**  An oligomer of lipid and carbohydrate that constitutes the endotoxin of gram-negative bacteria. LPS acts as a polyclonal activator of murine B cells, inducing their division and differentiation into antibody-producing plasma cells.

**Locus**  The specific chromosomal location of a gene.

**LRRs (leucine-rich repeats)**  Protein structural motif containing the sequence xLxxLxLxx.

**Lymph**  Interstitial fluid derived from blood plasma that contains a variety of small and large molecules, lymphocytes, and some other cells. It circulates through the lymphatic vessels.

**Lymphatic system**  A network of vessels and nodes that conveys lymph. It returns plasma-derived interstitial fluids to the bloodstream and plays an important role in the integration of the immune system.

**Lymphatic vessels**  Thinly walled vessels through which the fluid and cells of the lymphatic system move through the lymph nodes and ultimately into the thoracic duct, where it joins the bloodstream.

**Lymph node**  A small **secondary lymphoid organ** that contains lymphocytes, macrophages, and dendritic cells and serves as a site for filtration of foreign antigen and for activation and proliferation of lymphocytes (see Figures 2-5 and 2-16). See also **germinal center.**

**Lymphoblast**  A proliferating lymphocyte.

**Lymphocyte**  A mononuclear leukocyte that mediates humoral or cell-mediated immunity. See also **B cell** and **T cell.**

**Lymphoid progenitor cell**  A cell committed to the lymphoid lineage from which all lymphocytes arise.

**Lymphokine**  A term originally applied to cytokines secreted by lymphocytes, especially by $T_H$ cells. It has been largely superseded by the more general term **cytokine.**

**Lymphoma**  A cancer of lymphoid cells that tends to proliferate as a solid tumor.

**Lysogeny**  State in which a viral genome (**provirus**) is associated with the host genome in such a way that the viral genes remain unexpressed.

**Lysosome**  A small cytoplasmic vesicle found in many types of cells that contains hydrolytic enzymes, which play an important role in the degradation of material ingested by **phagocytosis** and **endocytosis.**

**Lysozyme**  An enzyme present in tears, saliva, and mucous secretions that digests mucopeptides in bacterial cell walls and thus functions as a nonspecific antibacterial agent. It has frequently been used as a target antigen in immunological studies.

**Macrophage**  Mononuclear phagocytic leukocyte that plays roles in adaptive and innate immunity. There are many types of macrophages; some are migratory, whereas others are fixed in tissues.

**Major histocompatibility complex (MHC) molecules**  Proteins encoded by the major histocompatibility complex and classified as class I, class II, and class III MHC molecules. See also **MHC.**

**Malignant**  Refers to cancerous cells capable of uncontrolled growth.

**MALT (mucosal-associated lymphoid tissue)**  Lymphoid cells and tissues below the epithelial layer of the body's mucosal surfaces. MALT is found in the gastrointestinal tract (GALT) and bronchial tissues (BALT).

**Mannose-binding lectin (MBL)**  A serum protein that binds to mannose in microbial cell walls and initiates complement-mediated lysis.

**Marginal zone**  A diffuse region of the spleen, situated on the periphery of the **PALS (periartiolar lymphoid sheath)** between the **red pulp** and **white pulp,** that is rich in B cells and contains lymphoid follicles, which can develop into **germinal centers** (see Figure 2-17).

**Mast cell**  A bone-marrow-derived cell present in a variety of tissues that resembles peripheral blood basophils, bears **Fc receptors** for IgE, and undergoes IgE-mediated **degranulation.**

**M cells**  Specialized cells of the intestinal mucosa and other sites, such as the urogenital tract, that deliver antigen from the apical face of the cell to lymphocytes clustered in the pocket of its basolateral face.

**Medulla**  The innermost or central region of an organ.

**Membrane-attack complex (MAC)**  The complex of complement components C5–C9, which is formed in the terminal steps of either the classical or the alternative complement pathway and mediates cell lysis by creating a membrane pore in the target cell.

**Membrane-bound immunoglobulin (mIg)**  A form of antibody that is bound to a cell as a transmembrane protein. It acts as the antigen-specific receptor of B cells.

**Memory B cells**  An antigen-committed, persistent B cell. B-cell differentiation results in formation of plasma cells, which secrete antibody, and memory cells, which are involved in the **secondary responses.**

**Memory cells**  Lymphocytes generated following encounters with antigen that are characteristically long lived; they are more readily stimulated than naive lymphocytes and mediate a **secondary response** to subsequent encounters with the antigen.

**Memory, immunologic**  The attribute of the immune system mediated by **memory cells** whereby a second encounter with an antigen induces a heightened state of immune reactivity.

**Memory response**  See **memory, immunologic.**

**Mesangial cell**  A type of macrophage found in the kidney.

**Metastasis**  The colonization by tumor cells of sites distant from the primary site.

**MHC (major histocompatibility complex)**  A complex of genes encoding cell surface molecules that are required for antigen presentation to T cells and for rapid graft rejection. It is called the H-2 complex in the mouse and the HLA complex in humans.

**MHC restriction**  The characteristic of T cells that permits them to recognize antigen only after it is processed and the resulting antigenic peptide is displayed in association with either a **class I** or a **class II** MHC molecule.

**Microglial cell**  A type of macrophage found in the central nervous system.

**Minor histocompatibility loci**  Genes outside of the MHC that encode antigens contributing to graft rejection.

**Minor lymphocyte-stimulating (Mls) determinants**  Antigenic determinants encoded by endogenous retroviruses of the murine mammary tumor virus family that are displayed on the surface of certain cells.

**Mitogen**  Any substance that nonspecifically induces DNA synthesis and cell division, especially of lymphocytes. Common mitogens are concanavalin A, phytohemagglutinin, **lipopolysaccharide (LPS),** pokeweed mitogen, and various **superantigens.**

**Mixed-lymphocyte reaction (MLR)**  In vitro T-cell proliferation in response to cells expressing allogeneic MHC molecules; can be used as an assay for class II MHC activity.

**Monoclonal antibody**  Homogeneous preparation of antibody molecules, produced by a single clone of B lineage cells, often a hybridoma, all of which have the same antigenic specificity.

**Monocyte**  A mononuclear phagocytic leukocyte that circulates briefly in the bloodstream before migrating into the tissues where it becomes a **macrophage.**

**Monokine** A cytokine produced by macrophages. This term has been largely superseded by the term **cytokine.**

**Mucin** A group of serine- and threonine-rich proteins that are heavily glycosylated. They are ligands for selectins.

**Mucosal-associated lymphoid tissue (MALT)** Lymphoid tissue situated along the mucous membranes that line the digestive, respiratory, and urogenital tracts.

**Multiple myeloma** A plasma-cell cancer.

**Multiple sclerosis (MS)** An autoimmune disease that affects the central nervous system. In Western countries, it is the most common cause of neurologic disability caused by disease.

**Myasthenia gravis** An autoimmune disease mediated by antibodies that block the acetylcholine receptors on the motor end plates of muscles, resulting in progressive weakening of the skeletal muscles.

**Myeloma cell** A cancerous plasma cell.

**Myeloid progenitor cell** A cell that gives rise to cells of the myeloid lineage.

**NADPH phagosome oxidase (phox)** Enzyme that generates antimicrobial reactive oxygen species that are used by phagocytic cells to kill bacteria.

**Naive** Denoting mature B and T cells that have not encountered antigen; synonymous with *unprimed* and *virgin.*

**Natural killer (NK) cells** A class of large, granular, cytotoxic lymphocytes that do not have T- or B-cell receptors. They are antibody-independent killers of tumor cells and also can participate in **antibody-dependent cell-mediated cytotoxicity.**

**Necrosis** Morphologic changes that accompany death of individual cells or groups of cells and that release large amounts of intracellular components to the environment, leading to disruption and atrophy of tissue. See also **apoptosis.**

**Negative selection** The induction of death in lymphocytes bearing receptors that react too strongly with self antigens.

**Neonatal Fc receptor (FcR$_N$)** An MHC class I–like molecule that controls IgG half-life and transports IgG across the placenta.

**Neoplasm** Any new and abnormal growth; a benign or malignant tumor.

**Neuraminidase** An enzyme that cleaves *N*-acetylneuraminic acid (sialic acid) from glycoproteins.

**Neutrophil** A circulating, phagocytic granulocyte involved early in the **inflammatory response** (see Figure 2-9). It expresses **Fc receptors** and can participate in **antibody-dependent cell-mediated cytotoxicity.** Neutrophils are the most numerous white blood cells in the circulation.

**NK1-T cell** A lymphocyte with some of the characteristics of T cells (it has T-cell receptors) as well as those of natural killer cells.

**N-nucleotides** See **N-region nucleotides.**

**Nonproductive rearrangement** Rearrangement in which gene segments are joined out of phase so that the triplet-reading frame for translation is not preserved.

**Nonsequential epitopes** Epitopes that are far apart in the primary sequence of the polypeptide chain but close together in the tertiary structure of the molecule.

**Northern blotting** Common technique for detecting specific mRNAs, in which denatured mRNAs are separated electrophoretically and then transferred to a polymer sheet, which is incubated with a radiolabeled DNA probe specific for the mRNA of interest.

**N-region nucleotides** Nucleotides added by the enzyme terminal deoxynucleotidyl transferase (TdT) to the 3′ cut ends of V, D, and J coding segments during rearrangement.

**Nude mouse** Homozygous genetic defect (*nu/nu*) carried by an inbred mouse strain that results in the absence of the thymus and consequently a marked deficiency of T cells and cell-mediated immunity. The mice are hairless (hence the name) and can accept grafts from other species.

**Oncofetal tumor antigen** An antigen that is present during fetal development but generally is not expressed in tissues except by tumor cells. Alpha-feto protein (AFP) and carcinoembryonic antigen (CEA) are two examples that have been associated with various cancers (see Table 21-3).

**Oncogene** A gene that encodes a protein capable of inducing cellular **transformation.** Oncogenes derived from viruses are written v-*onc;* their counterparts (**proto-oncogenes**) in normal cells are written c-*onc.*

**One-turn recombination signal sequences** Immunoglobulin gene-recombination signal sequences separated by an intervening sequence of 12 base pairs.

**Opportunistic infections** Infections caused by ubiquitous microorganisms in cases of immune deficiency.

**Opsonin** A substance (e.g., an antibody or C3b) that promotes the phagocytosis of antigens by binding to them.

**Opsonization** Deposition of opsonins on an antigen, thereby promoting a stable adhesive contact with an appropriate phagocytic cell.

**Osteoclast** A bone macrophage.

**P-addition** See **P-nucleotide addition.**

**Paracortex** An area of the lymph node beneath the cortex that is populated mostly by T cells and interdigitating dendritic cells.

**Paracrine** Regulatory secretions, such as cytokines, that arrive by diffusion from a nearby cellular source.

**Passive immunity** Adaptive immunity conferred by the transfer of immune products, such as antibody or sensitized T cells, from an immune individual to a nonimmune one. See also **active immunity.**

**Passive immunization** The acquisition of immunity by receipt of preformed antibodies rather than by active production of antibodies after exposure to antigen.

**Pathogen** A disease-causing organism.

**Pathogen-associated molecular patterns (PAMPs)** Molecular patterns common to pathogens but not occurring in mammals. PAMPs are recognized by various receptors of the innate immune system.

**Pathogenesis** The means by which disease-causing organisms attack a host.

**Pattern recognition** The ability of a receptor or ligand to interact with a class of similar molecules, such as mannose-containing oligosaccharides.

**Pattern recognition receptors (PRRs)** Receptors of the innate immune system that recognize molecular patterns or motifs present on pathogens but absent in the host.

*Pax-5 gene* This gene encodes B-cell-specific activator protein (BSAP), an essential transcription factor for B-cell development.

**Pentraxins** A family of serum proteins consisting of five identical globular subunits; CRP is a pentraxin.

**Perforin** Cytolytic product of CTLs that, in the presence of $Ca^{2+}$, polymerizes to form transmembrane pores in target cells (see Figure 14-9).

**Periarteriolar lymphoid sheath (PALS)** A collar of lymphocytes encasing small arterioles of the spleen.

**Peripheral tolerance** Process by which self-reactive lymphocytes in circulation are eliminated or rendered anergic.

**Peyer's patches** Lymphoid follicles situated along the wall of the small intestine that trap antigens from the gastrointestinal tract and provide sites where B and T cells can interact with antigen.

**Phage display library** Collection of bacteriophages engineered to express specific $V_H$ and $V_L$ domains on their surface (see Figure 5-23).

**Phagocyte** A cell with the capacity to internalize and degrade microbes or particulate antigens; neutrophils and monocytes are the main phagocytes.

**Phagocytosis** The cellular uptake of particulate materials by engulfment.

**Phagolysosome** An intracellular body formed by the fusion of a phagosome with a lysosome.

**Phagosome** Intracellular vacuole containing ingested particulate materials; formed by the fusion of pseudopodia around a particle undergoing **phagocytosis.**

**P-K reaction** Prausnitz-Kustner reaction, a local skin reaction to an allergen by a normal subject at the site of injected IgE from an allergic individual. (No longer used because of risk of transmitting hepatitis or AIDS.)

**Plasma** The cell-free, fluid portion of blood, which contains all the clotting factors.

**Plasma cell** The antibody-secreting effector cell of the B lineage.

**Plasmacytoma** A plasma-cell cancer.

**Plasmapheresis** A procedure that involves the separation of blood into two components, plasma and cells. The plasma is removed and cells are returned to the individual. This procedure is done during pregnancy when the mother makes anti-Rh antibodies that react with the blood cells of the fetus.

**Plasmin** A serine protease formed by cleavage of plasminogen. Its major function is the hydrolysis of fibrin.

**P-nucleotide addition** Addition of nucleotides from cleaved hairpin loops formed by the junction of V-D or D-J gene segments during Ig or TCR gene rearrangements.

**Polyclonal antibody** A mixture of antibodies produced by a variety of B-cell clones that have recognized the same antigen. Although all of the antibodies react with the immunizing antigen, they differ from each other in amino acid sequence.

**Poly-Ig receptor** A receptor for polymeric Ig molecules (IgA or IgM) that is expressed on the basolateral surface of most mucosal epithelial cells. It transports polymeric Ig across epithelia.

**Polymorphism** Presence of multiple **alleles** at a specific genetic locus. The major histocompatibility complex is highly polymorphic.

**Positive selection** A process that permits the survival of only those T cells whose T-cell receptors recognize self MHC.

**Pre–B cell (precursor B cell)** The stage of B-cell development that follows the pro–B cell stage. Pre–B cells produce cytoplasmic μ heavy chains and most display the pre-B-cell receptor.

**Pre-B-cell receptor** A complex of the Iga/Igb heterodimer with membrane-bound Ig consisting of the μ heavy chain bound to the **surrogate light chain** Vpre-B/λ5.

**Precipitin** An antibody that aggregates a soluble antigen, forming a macromolecular complex that yields a visible precipitate.

**P-region nucleotides** See **P-nucleotide addition.**

**Pre-T-cell receptor (pre-TCR)** A complex of the CD3 group with a structure consisting of the T-cell receptor β chain complexed with a 33-kDa glycoprotein called the pre-Tα chain.

**Primary follicle** A lymphoid follicle, prior to stimulation with antigen, that contains a network of follicular dendritic cells and small resting B cells.

**Primary lymphoid organs** Organs in which lymphocyte precursors mature into antigenically committed, immunocompetent cells. In mammals, the bone marrow and thymus are the primary lymphoid organs in which B-cell and T-cell maturation occur, respectively.

**Primary response** Immune response following initial exposure to antigen; this response is characterized by short duration and low magnitude compared to response following subsequent exposures to the same antigen (**secondary response**).

**Pro–B cell (progenitor B cell)** The earliest distinct cell of the B-cell lineage.

**Productive rearrangement** The joining of V(D)J gene segments in phase to produce a VJ or V(D)J unit that can be translated in its entirety.

**Progenitor cell** A cell that has lost the capacity for self renewal and is committed to the generation of a particular cell lineage.

**Programmed cell death** An induced and ordered process in which the cell actively participates in bringing about its own death.

**Proinflammatory** Tending to cause inflammation; TNF-α and IL-1 are proinflammatory cytokines.

**Prostaglandins** A group of biologically active lipid derivatives of arachidonic acid. They mediate the inflammatory response and type I hypersensitivity reaction by inhibiting platelet aggregation, increasing vascular permeability, and inducing smooth-muscle contraction.

**Proteasome** A large multifunctional protease complex responsible for degradation of intracellular proteins.

**Protein A** An $F_C$-binding protein present on the membrane of *Staphylococcus aureus* bacteria. It is used in immunology for the detection of antigen-antibody reactions and for the purification of antibodies.

**Protein A/G** A genetically engineered $F_C$-binding protein that is a hybrid of protein A and protein G. It is used in immunology for the detection of antigen-antibody reactions and for the purification of antibodies.

**Protein G** An $F_C$-binding protein present on the membrane of *Streptococcus* bacteria. It is used in immunology for the detection of antigen-antibody reactions and for the purification of antibodies.

**Proto-oncogene** A cancer-associated gene that encodes a factor that regulates cell proliferation, survival, or death; it is required for normal cell function, but, if mutated or produced in inappropriate amounts, it becomes an oncogene, which can cause transformation of the cell (see Figure 21-2).

**Provirus** Viral DNA that is integrated into a host-cell genome in a latent state and must undergo activation before it is transcribed, leading to formation of viral particles.

**Prozone effect** The apparent decrease in an antigen-antibody reaction sometimes observed as the concentration of antigen or antibody is increased.

**Pseudogene** Nucleotide sequence that is a stable component of the genome but is incapable of being expressed. Pseudogenes are thought to have been derived by mutation of ancestral active genes.

**Pseudopodia** Membrane protrusions that extend from motile and phagocytosing cells.

**Radial immunodiffusion** A method for determining the relative concentration of an antigen. The antigen sample is placed in a well and allowed to diffuse into agar containing a suitable dilution of an antiserum. The area of the precipitin ring that forms around the well in the region of equivalence is proportional to the concentration of antigen.

**Radioimmunoassay (RIA)** A highly sensitive technique for measuring antigen or antibody that involves competitive binding of radiolabeled antigen or antibody (see Figure 6-9).

**RAG (recombination-activating genes)** Two closely linked genes, *RAG-1* and *RAG-2,* that encode proteins essential for the rearrangement of immunoglobulin gene segments in the assembly of a functional immunoglobulin gene.

**Reactive nitrogen intermediates** Highly cytotoxic antimicrobial compounds formed by the combination of oxygen and nitrogen within phagocytes such as neutrophils and macrophages.

**Reactive oxygen intermediates (ROIs)** Highly reactive compounds such as superoxide anion $O_2^{g^-}$ hydroxyl radicals (OH·)(·O⁻), and hydrogen peroxide $(H_2O_2)$ that are formed under many conditions in cells and tissues.

**Receptor editing** Process by which the T- or B-cell receptor sequence is altered to reduce affinity for self antigens.

**Recombination-activating genes** See **RAG.**

**Recombination signal sequences (RSSs)** Highly conserved heptamer and nonamer nucleotide sequences that serve as signals for the gene rearrangement process and flank each germ-line V, D, and J segment (see Figure 5-6).

**Red pulp**   Portion of the spleen consisting of a network of sinusoids populated by macrophages and erythrocytes (see Figure 2-17). It is the site where old and defective red blood cells are destroyed.

**Relative risk**   Probability that an individual with a given trait (usually, but not exclusively, a genetic trait) will acquire a disease compared with those in the same population group who lack that trait.

**Respiratory burst**   A metabolic process in activated phagocytes in which the rapid uptake of oxygen is used to produce reactive oxygen intermediates that are toxic to ingested microorganisms.

**Retrovirus**   A type of RNA virus that uses a reverse transcriptase to produce a DNA copy of its RNA genome. HIV, which causes AIDS, and HTLV, which causes adult T-cell leukemia, are both retroviruses.

**Rh antigen**   Any of a large number of antigens present on the surface of blood cells that constitute the Rh blood group. See also **erythroblastosis fetalis.**

**Rheumatoid arthritis**   A common autoimmune disorder, primarily affecting women 40 to 60 years old, causing chronic inflammation of the joints.

**Rheumatoid factor**   Autoantibody found in the serum of individuals with rheumatoid arthritis and other connective-tissue diseases.

**Rhogam**   Antibody against **Rh antigen** that is used to prevent **erythroblastosis fetalis.**

**Rocket electrophoresis**   The electrophoresis of a sample of negatively charged antigen in a gel that contains antibody. The precipitate formed between antigen and antibody has the shape of a rocket, whose height is proportional to the concentration of antigen in the sample.

**Rous sarcoma virus (RSV)**   A retrovirus that induces tumors in avian species.

**Sarcoma**   Tumor of supporting or connective tissue.

**Scatchard equation**   The equation, $(r/c) = K_a n - K_a r$, where $r$ = ratio of the concentration of bound ligand to total antibody concentration, $c$ = concentration of free ligand, and $n$ = number of binding sites per antibody molecule, allows calculation of the affinity constants of antigen with antibody.

**Schistosomiasis**   A disease caused by the parasitic worm *Schistoma.*

**SCID (severe combined immunodeficiency)**   A genetic defect in which adaptive immune responses do not occur because of a lack of T cells and possibly B and NK cells.

**SCID-human mouse**   Immunodeficient mouse into which elements of a human immune system such as bone marrow and thymic fragments have been grafted. Such mice support the differentiation of pluripotent human hematopoietic stem cells into mature immunocytes and so are valuable for studies on lymphocyte development.

**SDS-polyacrylamide gel electrophoresis (SDS-PAGE)**   An electrophoretic method for the separation of proteins. It employs SDS to denature proteins and give them negative charges; when SDS-denatured proteins are electrophoresed through polymerized-acrylamide gels, they separate according to their molecular weights.

**Secondary follicle**   A **primary follicle** after antigenic stimulation; it develops into a ring of concentrically packed B cells surrounding a germinal center.

**Secondary lymphoid organs**   Organs and tissues in which mature, immunocompetent lymphocytes encounter trapped antigens and are activated into **effector cells.** In mammals, the **lymph nodes, spleen,** and mucosal-associated lymphoid tissue (**MALT**) constitute the secondary lymphoid organs.

**Secondary response**   Immune response following exposure to previously encountered antigen; the secondary response occurs more rapidly and is greater in magnitude and duration than the **primary response.**

**Secreted immunoglobulin (sIg)**   The form of antibody that is secreted by cells of the B lineage, especially plasma cells. This form of Ig lacks a transmembrane domain.

**Secretory component**   A fragment of the **poly-Ig receptor** that remains bound to Ig after transcytosis across an epithelium and cleavage.

**Secretory IgA**   J chain–linked dimers or higher polymers of IgA that have transited epithelia and retain a bound remnant of the **poly-Ig receptor.**

**Selectin**   One of a group of monomeric **cell adhesion molecules** present on leukocytes (L-selectin) and endothelium (E- and P-selectin) that bind to mucin-like CAMs (e.g., GlyCAM, PSGL-1) (see Figure 13-1).

**Self-tolerance**   Unresponsiveness to self antigens.

**Sequential epitopes**   Two or more antigenic determinants formed by a single sequence of amino acids.

**Serum**   Fluid portion of the blood, which is free of cells and clotting factors.

**Serum sickness**   A type III hypersensitivity reaction that develops when antigen is administered intravenously, resulting in the formation of large amounts of antigen-antibody complexes and their deposition in tissues. It often develops when individuals are immunized with antiserum derived from other species.

**Severe combined immunodeficiency**   See **SCID.**

**Signaling**   Intracellular communication initiated by receptor-ligand interaction.

**Signal joint**   In V(D)J gene rearrangement, the sequence formed by the union of recombination signal sequences.

**Signal peptide**   A small sequence of amino acids, also called the *leader sequence,* that guides the heavy or light chain through the endoplasmic reticulum and is cleaved from the nascent chains before assembly of the finished immunoglobulin molecule.

**Single-positive CD4$^+$**   Thymocytes that express a CD4 coreceptor.

**Single-positive CD8$^+$**   Thymocytes that express a CD8 coreceptor.

**Slow-reacting substance of anaphylaxis (SRS-A)**   The collective term applied to leukotrienes that mediate inflammation.

**Somatic hypermutation**   The induced increase of mutation, $10^3$- to $10^6$-fold over the background rate, in the regions in and around rearranged immunoglobulin genes. In animals such as humans and mice, somatic hypermutation occurs in germinal centers.

**Somatic-variation theories**   Theories maintaining that the genome contains a small number of immunoglobulin genes from which a large number of antibody specificities are generated in somatic cells by mutation or recombination.

**Southern blotting**   A common technique for detecting specific DNA sequences in which restriction-enzyme fragments are separated electrophoretically, then denatured and transferred to a polymer sheet, which is incubated with a radioactive probe specific for the sequence of interest (see Figure 22-9).

**Sox-4**   A transcription factor that is essential for the early stages of B-cell development.

**Specificity, antigenic**   Capacity of antibody and T-cell receptor to recognize and interact with a single, unique **antigenic determinant** or **epitope.**

**Spleen**   Secondary lymphoid organ where old erythrocytes are destroyed and blood-borne antigens are trapped and presented to lymphocytes in the PALS and marginal zone (see Figure 2-17).

**STAT (signal transducer and activator of transcription)**   A family of transcription factors that bind to the phosphorylated tyrosine residues of certain receptors at their SH$_2$ domains. These molecules play an essential role in the transduction of signals from a wide variety of cytokines.

**Stem cell**   A cell from which differentiated cells derive. Stem cells are classified as totipotent, pluripotent, multipotent, or unipotent depending on the range of cell types that they can generate. See Clinical Focus in Chapter 2.

**Streptavidin**   A bacterial protein that binds to biotin with very high affinity. It is used in immunological assays to detect antibodies that have been labeled with biotin.

**Stromal cell**   A nonhematopoietic cell that supports the growth and differentiation of hematopoietic cells.

**Superantigen**   Any substance that binds to the $V_\beta$ domain of the T-cell receptor and residues in the chain of class II MHC molecules. It induces activation of all T cells that express T-cell receptors with a particular $V_\beta$ domain. It functions as potent T-cell **mitogen** and may cause food poisoning and other disorders.

**Surface plasmon resonance**   An instrumental technique for measurement of molecular interactions based on changes of reflectance properties of a sensor coated with an interactive molecule.

**Surrogate light chain**   The polypeptides **Vpre-B** and **λ5** that associate with μ heavy chains during the pre-B-cell stage of B-cell development to form the pre-B-cell receptor.

**Switch regions**   In class switching, DNA sequences located in the intron 5′ of each $C_H$ segment (except $C_\delta$).

**Syngeneic**   Denoting genetically identical individuals.

**Systemic lupus erythematosus (SLE)**   An autoimmune disease characterized by auto-antibodies to a vast array of tissue antigens such as DNA, histones, RBCs, platelets, leukocytes, and clotting factors.

**TAP (transporter associated with antigen processing)**   Heterodimeric protein present in the membrane of the rough endoplasmic reticulum (RER) that transports peptides into the lumen of the RER, where they bind to class I MHC molecules.

**T cell**   See **T lymphocyte.**

**T-cell receptor (TCR)**   Antigen-binding molecule expressed on the surface of T cells and associated with **CD3**. It is a heterodimer consisting of either an α and β chain or a γ and δ chain (see Figures 9-3 and 9-4).

**T cytotoxic ($T_C$) cells**   T cells that kill target cells in an antigen-specific manner.

**Templated somatic mutation**   See **gene conversion.**

**T helper ($T_H$) cells**   T cells that are stimulated by antigen to provide signals that promote immune responses.

**$T_H$1 subset**   Subset of T helper cells responsible for the $T_H$1 response (see Table 12-4).

**$T_H$2 subset**   Subset of T helper cells responsible for the $T_H$2 response (see Table 12-4).

**Thoracic duct**   The largest of the lymphatic vessels. It returns lymph to the circulation by emptying into the left subclavian vein near the heart.

**Thromboxane**   Lipid inflammatory mediator derived from arachidonic acid (see Figure 13-13).

**Thymocyte**   Developing T cells present in the thymus.

**Thymus**   A primary lymphoid organ, in the thoracic cavity, where T-cell maturation takes place.

**Thymus-dependent antigen**   A soluble protein that can induce antibody production only with help of $T_H$ cells; response to such antigens involves isotype switching, affinity maturation, and memory-cell production.

**Thymus-independent (TI) antigen**   B-cell responses to TI antigens do not require the help of T cells. There are two types of TI antigens: most TI-1 antigens are polyclonal B-cell activators and act without regard to the antigenic specificity of the B cell; TI-2 antigens activate B cells by extensively cross-linking the mIg receptor, and they evoke antigen-specific responses.

**Titer**   A measure of the relative strength of an antiserum. The titer is the reciprocal of the last dilution of an antiserum capable of mediating some measurable effect such as precipitation or agglutination.

**T lymphocyte**   A lymphocyte that matures in the thymus and expresses a T-cell receptor, CD3, and CD4 or CD8. Several distinct T-cell subpopulations are recognized.

**Tolerance**   A state of immunologic unresponsiveness to particular antigens or sets of antigens. Typically, an organism is unresponsive or tolerant to self antigens.

**Toll-like receptors (TLRs)**   A family of cell-surface receptors that recognize molecules from various common pathogens; for example, TLR4 recognizes bacterial lipopolysaccharides.

**Toxoid**   A toxin that has been altered to eliminate its toxicity but that still can function as an **immunogen.**

**Trafficking**   The differential migration of lymphoid cells to and from different tissues.

**Transcytosis**   The movement of antibody molecules (polymeric IgA or IgM) across epithelial layers mediated by the poly-Ig receptor.

**Transfection**   Experimental introduction of foreign DNA into cultured cells, usually followed by expression of the genes in the introduced DNA (see Figure 22-12).

**Transformation**   Change that a normal cell undergoes as it becomes malignant; also, permanent, heritable alteration in a cell resulting from the uptake and incorporation of foreign DNA into the genome.

**Transfusion reaction**   Type II hypersensitivity reaction to proteins or glycoproteins on the membrane of transfused red blood cells.

**Transgene**   A cloned foreign gene present in an animal or plant.

**T regulatory ($T_{reg}$) cell**   A type of T cell that carries CD4 on its surface and is distinguished from $T_H$ cells by surface markers, such as CD25, associated with its stage of activation.

**Tumor-associated transplantation antigens (TATAs)**   Antigens that are expressed on particular tumors or types of tumors that are not unique to tumor cells. They are generally absent from or expressed at low levels on most normal adult cells.

**Tumor-specific transplantation antigens (TSTAs)**   Antigens that are unique to tumor cells.

**Tumor-suppressor genes**   Genes that encode products that inhibit excessive cell proliferation.

**Two-turn recombination signal sequences**   Immunoglobulin gene recombination signal sequences separated by an intervening sequence of 23 base pairs.

**Vaccination**   Intentional administration of a harmless or less harmful form of a pathogen to induce a specific immune response that protects the individual against later exposure to the pathogen.

**Vaccine**   A preparation of immunogenic material used to induce immunity against pathogenic organisms.

**Valence, valency**   Numerical measure of combining capacity, generally equal to the number of binding sites. Antibody molecules are bivalent or multivalent, whereas T-cell receptors are univalent.

**V(D)J recombinase**   The set of enzymatic activities that collectively bring about the joining of gene segments into a rearranged V(D)J unit.

**Vpre-B**   A polypeptide chain that together with **λ5** forms the **surrogate light chain** of the **pre-B-cell receptor.**

**Variability**   (Antibody) variability is defined by the number of different amino acids at a given position divided by the frequency of the most common amino acid at that position (see Figure 4-12).

**V (variable) gene segment**   The 5′ coding portion of rearranged immunoglobulin and T-cell receptor genes. There are multiple V gene segments in germ-line DNA, but gene rearrangement leaves only one segment in each functional gene.

**Variable (V) region**   Amino-terminal portions of immunoglobulin and T-cell receptor chains that are highly variable and responsible for the **antigenic specificity** of these molecules.

**Vascular addressins**    Tissue-specific adhesion molecules that direct the extravasation of different populations of circulating lymphocytes into particular lymphoid organs.

**Viral load**    Concentration of virus in plasma; usually reported as copies of viral genome per unit volume of plasma.

**Western blotting**    A common technique for detecting a protein in a mixture; the proteins are separated electrophoretically and then transferred to a polymer sheet, which is flooded with radiolabeled or enzyme-conjugated antibody specific for the protein of interest.

**White pulp**    Portion of the spleen that surrounds the arteries, forming a periarteriolar lymphoid sheath (PALS) populated mainly by T cells (see Figure 2-17).

**Xenograft**    Graft or tissue transplanted from one species to another.

**X-linked hyper-IgM syndrome**    An immunodeficiency disorder in which $T_H$ cells fail to express CD40L. Patients with this disorder produce IgM but not other isotypes, they do not develop germinal centers or display somatic hypermutation, and they fail to generate memory B cells.

**X-linked severe combined immunodeficiency (XSCID)**    Immunodeficiency resulting from inherited mutations in the common γ chain of the receptor for IL-2, -4, -7, -9, and -15 that impair its ability to transmit signals from the receptor to intracellular proteins.

# Answers to Study Questions

## Chapter 1

**Clinical Focus Question:** The diet of the allergic individual must be free of tree nuts. Labels of all products eaten directly or used to prepare food for the individual should be scrutinized for nuts or nut extracts as ingredients. Children old enough to read and understand labels should be taught to do so; if not, all who offer food must be aware of the condition. Particular attention must be given to baked goods from unknown sources such as school friends' lunches. An allergist can advise as to the treatment for an acute episode; it may involve having injectable epinephrine (epipen) available. The teacher and the school nurse should be informed about the condition and should know the recommended emergency treatment.

1. Jenner's method of using cowpox infection to confer immunity to smallpox was superior to earlier methods because it carried a significantly lowered risk of serious disease. The earlier method of using material from lesions of smallpox victims conferred immunity but at the risk of acquiring the potentially lethal disease.

2. The Pasteur method for treating rabies consists of a series of inoculations with attenuated rabies virus. This process actively immunizes the recipient who then has anti-rabies to stop the progress of infection. A simple test for active immunity would be to look for anti-rabies antibodies in the recipient's blood at a time after completion of treatment when all antibodies from a passive treatment would have cleared from circulation. Alternatively, one could challenge the recipient with attenuated rabies to see whether a secondary response occurred (this treatment may be precluded by the ethical ramifications).

3. The immunized mothers would confer immunity upon their offspring because the anti-streptococcal antibodies cross the placental barrier and are present in the babies at birth. In addition the colostrums and milk from the mother would contain antibodies to protect the nursing infant from infection.

4. (a) CM. (b) H and CM. (c) H and CM. (d) H and CM. (e) CM. (f) CM. (g) H. (h) CM. (i) H. (j) H. (k) CM.

5. The four immunologic attributes are specificity, diversity, memory, and self/nonself recognition. Specificity refers to the ability of certain membrane-bound molecules on a mature lymphocyte to recognize only a single antigen (or small number of closely related antigens). Rearrangement of the immunoglobulin genes during lymphocyte maturation gives rise to antigenic specificity; it also generates a vast array of different specificities, or diversity, among mature lymphocytes. The ability of the immune system to respond to nonself-molecules, but (generally) not to self-molecules, results from the elimination during lymphocyte maturation of immature cells that recognize self-antigens. After exposure to a particular antigen, mature lymphocytes reactive with that antigen proliferate and differentiate, generating a larger population of memory cells with the same specificity; this expanded population can respond more rapidly and intensely after a subsequent exposure to the same antigen, thus displaying immunologic memory.

6. The secondary immune response involves an amplified population of memory cells. The response is more rapid and achieves higher levels than the primary response.

7. (a) Both antibodies and T-cell receptors display fine specificity for antigen; very small modifications in an antigen can prohibit its binding to its corresponding antibody or T-cell receptor. MHC molecules do not possess such fine specificity, and a variety of unrelated peptide antigens can be bound by the same MHC molecule. (b) Antibodies are expressed only by cells of the B-cell lineage; T-cell receptors are expressed by cells of the T-cell lineage; class I MHC molecules are expressed by virtually all nucleated cells; class II MHC molecules are expressed only by specialized cells that function as antigen-presenting cells (e.g., B cells, macrophages, and dendritic cells). (c) Antibodies can bind to protein or polysaccharide antigens; T-cell receptors recognize only peptides associated with MHC molecules; MHC molecules bind only processed peptides.

8. (a) Macrophages, B cells, dendritic cells. (b) Co-stimulatory; $T_H$ cells. (c) II; I. (d) Leukocyte. (e) Adaptive. (f) CD8, CD4. (g) Epitope.

9. They can recognize only antigen that is associated with class I MHC molecules.

10. Innate and adaptive immunity cooperate to give a complete protective response against pathogens. An example is the phagocytic cell, which takes up foreign material and processes it to form peptide antigens that are presented by the phagocyte. The presented antigens stimulate T cells that either provide help to B cells for production of antibody or stimulate cytotoxic T cells to provide protection against infected or cancerous cells. In addition, the phagocytic cells take part in inflammation, producing cytokines that attract T cells and B cells to the site.

11. Consequences of mild forms of immune dysfunction include sneezing, hives, and skin rashes caused by allergies. Asthma and anaphylactic reactions are more severe consequences of allergy and can result in death. Consequences of severe immune dysfunction include susceptibility to infection by a variety of microbial pathogens if the dysfunction involves immunodeficiency,

or chronic debilitating diseases, such as rheumatoid arthritis, if the dysfunction involves autoimmunity. The most common cause of immunodeficiency is infection with the retrovirus HIV-1, which leads to AIDS.

12. (a) False. (b) False. (c) False. (d) True.

13. (a) True. (b) True. (c) True. (d) True. (e) False: Antigen must be presented to T cells in the context of MHC molecules. (f) False: Peptides are added to the binding cleft of MHC class I in a vesicle. (g) True.

14. (a) 2. (b) 3. (c) 4. (d) 1.

# Chapter 2

**Clinical Focus Question:** (a) This outcome is unlikely; since the donor hematopoietic cells differentiate into T and B cells in an environment that contains antigens characteristic of both the host and donor, there is tolerance to cells and tissues of both. (b) This outcome is unlikely. T cells arising from the donor HSCs develop in the presence of the host's cells and tissues and are therefore immunologically tolerant of them. (c) This outcome is unlikely for the reasons cited in (a). (d) This outcome is a likely one because of the reasons cited in (a).

1. (a) In general, $CD4^+$ T cells that recognize class II MHC molecules function as $T_H$ cells, whereas $CD8^+$ T cells that recognize class I MHC molecules function as $T_C$ cells. However, some functional $T_H$ cells express CD8 and are class I restricted, and some functional $T_C$ cells express CD4 and are class II restricted. But these are exceptions to the general pattern. (b) The bone marrow contains few pluripotent stem cells, which constitute only about 0.05% of all bone-marrow cells. (c) $T_H$ cells recognize antigen associated with class II MHC molecules. Activation of macrophages increases their expression of class II MHC molecules. (d) Organized lymphoid follicles are also present in the tonsils, Peyer's patches, and other mucosal-associated tissue. (e) In response to infection, $T_H$ cells and macrophages are activated and secrete various cytokines that induce increased hematopoietic activity. The induced hematopoiesis expands the population of white blood cells that can fight the infection. (f) Unlike other types of dendritic cells, follicular dendritic cells do not express class II MHC molecules and thus do not function as antigen-presenting cells for $T_H$-cell activation. These cells, which are present only in lymph follicles, can trap circulating antibody-antigen complexes; this ability is thought to facilitate B-cell activation and the development of memory B cells. (g) B and T lymphocytes have antigen-binding receptors, but a small population of lymphoid cells, called null cells, does not. (h) Although many animals, such as humans and mice, generate B cells in bone marrow, others, such as some ruminants, do not. (i) So far, it has not been possible to demonstrate the presence of T or B cells in jawless fishes such as the lamprey and hagfish.

2. (a) Myeloid progenitor. (b) Granulocyte-monocyte progenitor. (c) Hematopoietic stem cell. (d) Lymphoid progenitor.

3. The primary lymphoid organs are the bone marrow (bursa of Fabricius in birds) and the thymus. These organs function as sites for B-cell and T-cell maturation, respectively.

4. The secondary lymphoid organs are the spleen, lymph nodes, and mucosal-associated lymphoid tissue (MALT) in various locations. MALT includes the tonsils, Peyer's patches, and appendix, as well as loose collections of lymphoid cells associated with the mucous membranes lining the respiratory, digestive, and urogenital tracts. All these organs trap antigen and provide sites where lymphocytes can interact with antigen and subsequently undergo clonal expansion.

5. Stem cells are capable of self-renewal and can give rise to more than one cell type, whereas progenitor cells have lost the capacity for self-renewal and are committed to a single cell lineage. Commitment of progenitor cells depends on their acquisition of responsiveness to particular growth factors.

6. The two primary roles of the thymus are the generation and the selection of a repertoire of T cells that will protect the body from infection.

7. Nude mice and people with DiGeorge's syndrome have a congenital defect that prevents development of the thymus. Both lack circulating T cells and cannot mount cell-mediated immune responses.

8. In humans, the thymus reaches maximal size during puberty. During the adult years, the thymus gradually atrophies.

9. The lethally irradiated mice serve as an assay system for multipotent HSCs, since only mice injected with these stem cells are able to reconstitute their hematopoietic systems and survive. As HSCs are successively enriched in a preparation, the total number of cells that must be injected to restore hematopoiesis and thus enable survival decreases.

10. Monocytes are the blood-borne precursors of macrophages. Monocytes have a characteristic kidney-bean-shaped nucleus and limited phagocytic and microbial killing capacity compared to macrophages. Macrophages are much larger than monocytes and undergo changes in phenotype to increase phagocytosis, antimicrobial mechanisms (oxygen-dependent and oxygen-independent), and secretion of cytokines and other immune system modulators. Tissue-specific functions are also found in tissue macrophages.

11. The bursa of Fabricius in birds is the primary site where B lymphocytes develop. Bursectomy would result in a lack of circulating B cells and humoral immunity, and it would probably be fatal.

12. After phagocytosis by macrophages, most bacteria and fungi are destroyed and broken down; the resulting antigenic peptides are displayed along with class II MHC molecules on the cell surface, where they can induce $T_H$-cell activation and subsequent antibody (humoral) response. In contrast, intracellular bacteria and fungi have various mechanisms for surviving in macrophages after phagocytosis. These bacteria thus do not induce an antibody response.

13. (a) True. (b) False: The marginal zone is rich in B cells, and PALS is rich in T cells. (c) True. (d) True. (e) False: The spleen is not supplied with afferent lymphatics. (f) False: In addition to being essential for the formation of NK cells, Ikaros is also essential for the formation of T cells and B cells. Therefore, its knockout would prevent lymph nodes as serving as sites for the generation of adaptive immune responses.

15. (a) 5. (b) 10. (c) 3. (d) 6. (e) None. (f) 4. (g) 11. (h) 14. (i) 7. (j) 12. (k) 8. (l) 16. (m) 15. (n) 2. (o) 6. (p) 5 (and all other hematopoietic cells).

# Chapter 3

**Clinical Focus Question:** Atherosclerosis is the deposit of spongy lipid-like material in the arteries that blocks normal blood flow and leads to risk of ischemia (lack of blood supply to a specific organ) and infarction (heart attack). This condition begins with deposit of macrophages on the wall of the artery and in the intima (space between wall and interior). The recruitment of macrophages is a characteristic of an inflammatory response, and more macrophages and other phagocytic cells home to the inflamed area, worsening the blockage of the vessel. Blood levels of the acute phase response protein CRP increase during an inflammatory reaction. Chronically increased CRP is indicative of chronic inflammation such as seen in atherosclerosis.

1. Cells of the innate immune system express a number of cytokines and chemokines that recruit cells of adaptive immunity to the site of infection. A key cellular bridge between innate and adaptive immunity is the dendritic cell, which may acquire antigens from a foreign agent at the site of infection, migrate back to lymphatic tissues, and present these antigens to T and B cells, giving rise to specific adaptive responses against the invading agent.

    The complement system plays a role in both types of immunity and may be activated by a number of microbial products via soluble and membrane-bound receptors or by specific antibody responses. Byproducts of complement activation stimulate cells of both the adaptive and innate immune systems and recruit them to the site of infection or injury.

2. Inflammation is characterized by redness, heat, swelling, pain, and sometimes loss of local function. The actions of leukocytes recruited to the affected area produce the observed signs of inflammation: cytokine expression can cause temperature increases and swelling due to increased capillary permeability, and distension of the affected area causes pain and redness. Increased permeability of capillaries facilitates entry of cells to the affected area, attracted by soluble molecules released during the early stages of the response. The innate immune response consists of recruitment of phagocytic cells that engulf pathogens, specific soluble molecules that attack and neutralize or kill invaders, and mobilization of adaptive responses if attackers survive the innate defenses. All of these actions rely on mustering the defenses at the site of the infection or injury, and it is this assembly of cells and chemical mediators that causes the characteristics of inflammation.

3. (a) interferon, NK. (b) iNOS, arginine, NADPH, NO. (c) NADPH phagosome oxidase, $O_2$, ROS, NO, RNS. (d) dendritic cell, class I, class II MHC molecules, $T_H$, $T_C$, dendritic cell. (e) CRP, MBL. (f) APR, CRP, MBL. (g) TLR7, TLR8, TLR9. (h) PRRs, T cell receptors, antibody. (i) PRRs, class II MHC molecules, class I MHC molecules. (j) TLR2, PRR. (k) PRRs, PAMPs. (l) TLR4, TLR2. (m) complement.

4. B. Beutler showed that *lpr* mice were resistant to endotoxin and that the genetic difference in these mice was lack of a functional TLR4 because of a single mutation in the receptor sequence. R. Medzhitov and C. Janeway demonstrated that a protein (TLR4) with homology to Toll activated the expression of immune response genes when expressed in a human cell line.

5. Disadvantages of adaptive immunity include the possibility of autoimmune responses, which may result in disease. Disadvantages seem to be overwhelmed by advantages of the combined systems. For example, innate immunity cannot respond to new pathogens that lack molecular characteristics recognized by PRRs, and there is no memory, so vaccination is not an option. The response time of adaptive immunity is slow. Given their complementary advantages and disadvantages, both systems might be deemed essential.

6. By means of antifungal proteins such as drosomycin. Humans also have a number of antimicrobial proteins that recognize molecular characteristics of groups of pathogens and are capable of destroying those pathogens.

7. (a) Blockade of integrin-ICAM interactions will inhibit adaptive immunity by impairing extravasation of T cells and macrophages and interfering with the interaction of T cells and dendritic cells. Blocking macrophages from the site of inflammation will block their activation and secretion of cytokines and chemokines that support adaptive responses. Leukocyte extravasation is inhibited because it depends on integrin-ICAM interactions; there is no effect on the acute phase response; no effect on induction of complement-mediated lysis because the components of the complement system are humoral, not cellular; inflammation is compromised because migration of cells to inflammatory sites is an important component of the inflammatory response; and finally, the intracellular pathways of ROS and RNS generation are not affected by integrin-ICAM interactions.

    (b) TLR4 is responsible for recognition of LPS by the innate immune system. In the case of infections or immunizations that do not involve LPS, none of the processes mentioned are affected. However, in the case of infections with gram-negative bacteria or exposure to LPS by other means (contaminated medications, for example), the induction of adaptive immunity, the induction of inflammation, and the induction of ROS and RNS could be inhibited.

    (c) During innate immune responses, TNF-$\alpha$ and IL-1, cytokines produced by macrophages and other cell types, mediate inflammation and affect adaptive immune responses. Failure to produce these inflammatory cytokines can result in decreased adaptive immunity and inflammation. These cytokines do not directly affect leukocyte extravasation, the acute phase response, or complement-mediated lysis.

(d) Mutations in the phox enzyme system can inhibit the generation of ROS and RNS.

(e) Knockout of class II MHC molecules will directly inhibit adaptive immunity because they are required for activation of helper T cells by antigen-presenting cells. There will be no direct effects on extravasation, the acute phase response, complement-mediated lysis, or innate stimulation of ROS and RNS.

**Analyze the Data:** (a) Yes. Part (a) of the figure demonstrates that engagement of TLRs results in significant inhibition of apoptosis, with or without the addition of the NF-κB inhibitor SN50. The inhibition of neutrophil apoptosis may offer an advantage to the host by allowing neutrophils to function longer as they fight infection. A potential disadvantage is that during chronic infection, persistent neutrophils would produce and secrete toxic substances that could result in host pathology.

(b) The data suggest that neutrophil life would not be extended. Part (b) of the figure indicates that IKK phosphorylation (and activation; see Figure 3-15) occurs in response to the TLR agonists that increase neutrophil life span. Part (a) indicates that SN50, an inhibitor of NF-κB (a product of the reaction catalyzed by IKK), decreases the effect of TLR engagement on neutrophil apoptosis. IKK phosphorylates IκB, which then dissociates from NF-κB. NF-κB then binds to DNA, where it functions as a transcriptional activator for NF-κB–dependent genes. Less NF-κB activity correlates in this experiment with less inhibition of apoptosis (part a, black bars), suggesting that defective IKK would not result in increased neutrophil half-life.

(c) From the data in (b), TLR5 and TLR7 do not appear to induce phosphorylation of IKK. These data suggest that TLR5 and TLR7 may function by alternative signal transduction pathways. Since IKK is necessary for NF-κB activation and inhibition of apoptosis, these TLR molecules may not enhance neutrophil survival.

---

# Chapter 4

**Clinical Focus Question:** (a) A large and geographically diverse population of donors will have a broader spectrum of antibody specificities than one drawn from a single region. This is because some pathogens common in one geographical area are absent from another. Consequently, people from different areas will differ in the spectrum of antibodies their immune systems have generated. Also, even in the case of those pathogens that are found in all four areas mentioned, in many cases there will be important differences in the strains of particular pathogens from continent to continent. (b) Product from company B should be administered to the population of persons who travel extensively in order to provide protection against pathogens less common in the United States. However, if they are exposed to significant numbers of visitors from foreign countries, they would benefit from the more diverse spectrum of antibodies in the company B product.

1. (a) True. (b) True. (c) False: A hapten cannot stimulate an immune response unless it is conjugated with a larger protein carrier. However, the hapten can combine with pre-formed antibody specific for the hapten. (d) True. (e) False: A T cell can recognize only peptides that have been processed and presented by MHC molecules. These epitopes tend to be internal peptides. (f) True. (g) False: Immunogens are capable of stimulating a specific immune response; antigens are capable of combining specifically with the antibodies or T-cell receptors induced during an immune response. Although all immunogens exhibit antigenicity, some antigens do not exhibit immunogenicity. (h) True: T-cell receptors do not bind free antigen. Note that this answer is correct for αβ T-cell receptors, some γδ T-cell receptors bind antigen directly. (i) False. B-cell epitopes on a protein derive not only from primary structure but also from secondary, tertiary, and even quaternary structure of a protein.

2. (a) Native BSA: Heat denaturation is likely to destroy B-cell epitopes in a globular protein, although the T-cell epitopes are generally stable to heat. (b) HEL: The immunogenicity of an antigen generally is related to its foreignness to the animal exposed. However, collagen and some other proteins highly conserved throughout evolution exhibit little foreignness across diverse species and thus often are weak antigens. (c) 150,000-MW protein: Other things being equal, larger proteins are more immunogenic than smaller ones. (d) BSA in complete Freund's adjuvant will be more immunogenic because mycobacteria included in the complete adjuvant have components that stimulate a wide range of cells types via their PRRs and thus will involve these cells in the response against the BSA. Incomplete adjuvant lacks the added mycobacteria and will not provide the broad stimulus seen with complete.

3. b, d.

4. (a) T. (b) T. (c) B. (d) B. (e) T. (f) T. (g) BT. (h) B. (i) BT.

5. (a) True. (b) True. (c) False: Multiple isotypes can appear on the surface of a single B cell. A mature B cell expresses both IgM and IgD. (d) True. (e) True. (f) True. (g) False: Secreted serum IgM is a pentamer. Because of its larger size and valency, IgM is more effective than IgG in cross-linking bacterial surface antigens, which leads to agglutination. (h) True. (i) True. (j) False: Both heavy- and light-chain variable regions contain approximately 110 amino acids.

6. (a) The molecule would have to possess the following structural features: 2 identical heavy chains and 2 identical light chains ($H_2L_2$); interchain disulfide bonds joining the heavy chains (H-H) and the heavy and light chains (H-L); a series of intrachain domains containing approximately 110 amino acids and stabilized by an intra-domain disulfide bridge forming a loop of about 60 amino acids; a single constant domain in the light chains and 3 or 4 constant domains in the heavy chains. The amino-terminal domain of both the heavy and the light chains should show sequence variation. (b) The antisera to both whole human IgG and human k chain should cross-react with the new immunoglobulin class, since both of these antisera have antibodies specific for the light chain. The new isotype would be expected to contain either kappa or lambda light

chains. (c) Reduce the interchain disulfide bonds of the new isotype with mercaptoethanol and alkylation. Separate the heavy and light chains. Immunize a rabbit with the heavy chain. After absorption on all known human isotypes, the rabbit antisera should react with the new isotype but not with any other known isotypes.

7. IgM and IgD differ in their constant-region domains, whereas antigenic specificity is determined by the variable-region domains. Molecules of IgM and IgD that have different C domains but identical $V_H$ and $V_L$ domains are found on a given B cell; thus, the cell is unispecific although it bears two isotypes.

8. Advantages of IgG compared with IgM are (1) its ability to cross the placenta and protect the developing fetus; (2) its higher serum concentration, which results in IgG antibodies binding to and neutralizing more antigen molecules and being more effective in antigen clearance; (3) its smaller size, which enables IgG to diffuse more readily into intercellular fluids. The disadvantages of IgG compared with IgM are its lower capacity to (1) agglutinate antigens and (2) activate the complement system, both of which are due to the lower valency of IgG.

9. (a) See Figures 4-6 and 4-7. The number of interchain disulfide bonds joining the heavy chains varies among IgG subclasses (see Figure 4-18). (b) Draw as a dimer containing ( heavy chains. Add J chain and secretory component (see Figure 4-19a). (c) Draw as a pentamer containing m heavy chains. The μ heavy chain contains five domains and no hinge region; the extra $C_H$ domain replaces the hinge. Add J chain. Pentameric IgM also has additional interchain disulfide bonds joining $C_H$ domains in the subunits and the J chain to two of the subunits (see Figure 4-17e).

10.

| Property | Whole IgG | H chain | L chain | Fab | F(ab′)$_2$ | Fc |
|---|---|---|---|---|---|---|
| Binds antigen | + | +/− | +/− | + | + | − |
| Bivalent antigen binding | + | − | − | − | + | − |
| Binds to Fc receptors | + | − | − | − | − | + |
| Fixes complement in presence of antigen | + | − | − | − | − | − |
| Has V domains | + | + | + | + | + | − |
| Has C domains | + | + | + | + | + | + |

11. (a) AL. (b) ID. (c) IS. (d) AL. (e) No antibodies will be formed.

12.

| | Rabbit antisera to mouse antibody component | | | | |
|---|---|---|---|---|---|
| | γ chain | κ chain | IgG Fab fragment | IgG Fc fragment | J chain |
| Mouse γ chain | Yes | No | Yes | Yes | No |
| Mouse κ chain | No | Yes | Yes | No | No |
| Mouse IgM whole | No | Yes | Yes | No | Yes |
| Mouse IgM/Fc fragment | Yes | No | No | Yes | No |

13. (a) Immunoglobulin-fold domains contain approximately 110 amino acid residues, which are arranged in two antiparallel β pleated sheets, each composed of multiple β strands separated by short loops of various lengths. Each Ig-fold domain is stabilized by an intra-domain disulfide bond between two conserved cysteine residues about 60 residues apart. (b) The Ig-fold domain structure is thought to facilitate interactions between domains; such interactions between nonhomologous domains may allow different members of the immunoglobulin superfamily to bind to each other.

14. The hypervariable regions, also called complementarity-determining regions (CDRs), are located in the loops of the immunoglobulin folds constituting the $V_H$ and $V_L$ domains. There are three hypervariable regions in each $V_H$ domain, and three in each $V_L$ domain. Residues in the hypervariable regions constitute most of the amino acids involved in antigen binding.

15. The smallest value of the variability is obtained when the same amino acid is always present at the given position and the variability is 1 (1/1 = 1). The largest value of the variability is obtained when any one of the 20 possible amino acid monomers appears at the given position with equal frequency (0.05). Variability = 20/0.05 = 400.

16. The antiserum would also contain antibodies to the kappa and lambda light chains, which are present in all isotypes. To prepare an IgG-specific antiserum, the investigator needs to reduce the interchain disulfide bonds in the mouse IgG with mercaptoethanol and isolate the heavy chain. Rabbits immunized with the mouse heavy chains alone will be specific for the IgG isotype.

17. (a) Four different antigen-binding sites (indicated in boldface) and 10 different antibody molecules: $H_sL_s/H_sL_s$, $H_mL_m/H_mL_m$, $H_sL_m/H_sL_m$, $H_mL_s/H_mL_s$, $H_sL_s/H_mL_m$, $H_sL_s/H_sL_m$, $H_sL_s/H_mL_s$, $H_mL_m/H_sL_m$, $H_mL_m/H_mL_s$, and $H_sL_m/H_mL_s$. (b) Two different antigen-binding sites and three different antibody molecules: $H_sL_s/H_sL_s$,

$H_mL_s/H_mL_s$, $H_sL_s/H_mL_s$.(c) One binding site and one antibody molecule: $H_sL_s/H_sL_s$.

18. (a) 3, 6, 10, 11. (b) 4. (c) 2, 9, 12. (d) 5. (e) 1, 3, 4, 7, 8, 11.

19. The antibody is most likely cross-linking the receptors on the cell surface, resulting in the first step of receptor activation for this type of receptor, thus mimicking the action of the ligand. Incubation of the monoclonal antibody solution with the enzyme papain will cleave the antibody in a manner that generates monovalent Fab fragments. The Fab fragments retain the ability to bind receptors but are unable to cross-link them. Using the Fab fragment in the experiment will reveal whether the antibody blocks binding of ligand to the receptor.

20. If a mother breast-feeds the infant, many immunological components are passed to the infant. Secretory IgA is the primary antibody passed in the milk. Other compounds also help the growing infant to inhibit microbial growth, including cytokines, hormones, and enzymes.

# Chapter 5

**Clinical Focus Question:** As described in the table accompanying the Clinical Focus, Zevalin is a mouse monoclonal antibody and Rituxan is a humanized monoclonal antibody. Rituxan has a long half-life in humans and Zevalin has a very short one because mouse antibodies are rapidly cleared by humans. Because Zevalin is cleared from the body within several hours of the time of its injection, it is hard to maintain therapeutically effective doses. On the other hand, a single dose of Rituxan remains in the body at therapeutically effective levels for weeks. However, if Rituxan is labeled with a radioactive isotope, the radiation is not cleared from the body but remains for long periods of time and delivers a high dose that can be injurious to the host. In contrast, mouse monoclonal antibody Zevalin can be used to bring a radioisotope to the target cell population and deliver radiation for an interval long enough to kill many tumor cells but short enough to avoid serious radiation injury to the host.

1. (a) False: $V_\kappa$ gene segments and $C_l$ are located on separate chromosomes and cannot be brought together during gene rearrangement. (b) True. (c) True. (d) True. (e) True. (f) True.

2. $V_H$ and $J_H$ gene segments cannot join because both are flanked by recombination signal sequences (RSSs) containing a 23-bp (2-turn) spacer (see Figure 5-6b). According to the one-turn/two-turn joining rule, signal sequences having a two-turn spacer can join only with signal sequences having a one-turn (12-bp) spacer.

3. Light chains: $500 \times V_L$ 3 4 $J_L$ = 2 3 $10^3$

   Heavy chains: $300 V_H \times 3$ 15 $D_H \times 3$ 4 $J_H$ = 1.8 $3 \times 10^4$

   Antibody molecules: $(2 \times 3 10^3$ LCs$) \times 3 (1.8 \times 3 10^4$ HCs$)$ = 3.6 $\times 3 10^7$

4. (a) 1, 2, 3. (b) 3. (c) 3. (d) 5. (e) 2, 3, 4. (f) 2. (g) 5.

5. Somatic mutation contributes to the variability of all three complementarity-determining regions. Additional variability is generated in the CDR3 of both heavy and light chains by junctional flexibility, which occurs during heavy-chain D-J and V-DJ rearrangements and light-chain V-J rearrangements, by P-nucleotide addition at heavy- and light-chain variable-region joints, and by N-nucleotide addition at the heavy-chain D-J and V-DJ joints. See Table 5-3.

6. (a) R: Productive rearrangement of heavy-chain allele 1 must have occurred since the cell line expresses heavy chains encoded by this allele. (b) G: Allelic exclusion forbids heavy-chain allele 2 from undergoing either productive or nonproductive rearrangement. (c) NP: The κ genes rearrange before the λ genes. Since the cell line expresses λ light chains, both κ alleles must have undergone nonproductive rearrangement, thus permitting λ-gene rearrangement to occur. (d) NP: Same reason as given in (c) above. (e) R: Productive rearrangement of λ-chain allele 1 must have occurred since the cell line expresses λ light chains encoded by this allele. (f) G: Allelic exclusion forbids λ-chain allele 2 from undergoing either productive or nonproductive rearrangement (see Figure 5-11).

7. The κ-chain DNA must have the germ-line configuration because a productive heavy-chain rearrangement must occur before the light-chain (κ) DNA can begin to rearrange.

8. (a) No. (b) Yes. (c) No. (d) No. (e) Yes.

9. Random addition of N-nucleotides at the D-J and V-DJ junctions contributes to the diversity within the CDR3 of heavy chains, but this process can result in a nonproductive rearrangement if the triplet reading frame is not preserved.

10. The joints between variable-region gene segments ($V_L$-$J_L$ and $V_H$-$D_H$-$J_H$) occur in CDR3. During formation of these joints, junctional flexibility, P-nucleotide addition, and N-nucleotide addition introduce diversity. Because these processes do not affect the rest of the variable region, CDR3 exhibits the greatest diversity.

11. Four chances (see Figure 5-11).

12. (a) 5. (b) 5, 6, 9. (c) 1. (d) 4. (e) 2, 8. (f) 2, 11. (g) 3, 7. (h) 3, 10.

13. Because mouse antibodies are rapidly cleared from the human system, therapeutic anti-idiotype antibodies derived from mice are most effective if they are humanized; i.e., genetic engineering is used to replace all mouse sequences except the CERs with human Ig sequences. It is highly unlikely that the B-cell lymphomas of different patients will have the same idiotype. Therefore, this is an approach that would have to be repeated for every B-lymphoma patient to whom it was applied.

14. (a) RNA. (b) RNA. (c) DNA. (d) DNA. (e) DNA.

15. Antibody specificity is not inherited from one's parents. It is determined at the DNA level by random arrangements of the genome coding for the antigen-binding region of the antibody. Thus, the tendency to produce IgE may be inherited from a parent, but the specificity of the IgE molecules will be determined differently in each individual.

# Chapter 6

**Clinical Focus Question:** Subpopulations of leukocytes and leukemias are usually characterized by the presence of a distinctive ensemble of cell surface markers, each marker antigenically distinct from the others. The most versatile and specific reagents for the detection and characterization of these markers are antibodies. Monoclonal antibody technology is the only general method for the preparation of antibodies specific for each the more than 200 different antigens that have been found on the surfaces of various leukocyte types, populations, and subpopulations.

1. (a) True. (b) True. (c) False: Papain digestion yields monovalent Fab fragments, which cannot crosslink antigens. (d) True. (e) True. (f) True. (g) True. (h) False: Agglutination tests are more sensitive. (i) False: these are usually performed as qualitative assays. False: the avidity of the interaction between a divalent antibody B and an antigen with multiple epitopes C could be as high as $(Ka)^2$.

2. (a) Whole bovine serum. (b) See the following figure.

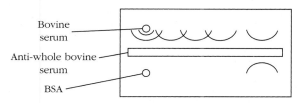

Bovine serum
Anti-whole bovine serum
BSA

3. Bottle A: H1-C1. Bottle B: H2-C2. Bottle C: H2-C1. Bottle D: H1-C2.

4. a, c, and i (chromogenic ELISA, RIA, and chemiluminescent ELISA can all be used to determine the concentration of a hapten).

5. (a) Isolate the heavy chains from the myeloma proteins of known isotype. Immunize rabbits with the isolated heavy chains to obtain antisera for each heavy-chain isotype. Absorb each of these heavy chain antisera with all of the other known heavy chain classes to remove any heavy chain isotype crossreactive antibodies. Then determine which of these anti-isotype antisera reacts with myeloma protein X using an ELISA assay. (b) Use the anti-myeloma protein antibody as a basis for the design of a quantitative ELISA assay which would allow the measurement of the level of the myeloma protein in serum.

6. (a) ELISA. (b) ELISA or RIA. (c) ELISA. (d) Immunofluoresence. (e) Agglutination. (f) ELISA. (g) Agglutination.

7. (b)

8. Affinity refers to the strength of the interaction between a *single* antigen-binding site in an antibody and the corresponding ligand. Avidity refers to the total effective strength of all the interactions between *multiple* antigen-binding sites in an antibody and multiple identical epitopes in a complex antigen. Because bacteria often have multiple copies of a particular epitope on their surface, the avidity of an antibody is a better measure than its affinity of its ability to combat bacteria.

9. (a) Antiserum #1 $K_0 = 1.2 \times 10^5$; antiserum #2 $K_0 = 4.5 \times 10^6$; antiserum #3 $K_0 = 4.5 \times 10^6$. (b) Each antibody has a valence of 2. (c) Antiserum #2. (d) The monoclonal antiserum #2 would be best because it recognizes a single epitope on the hormone and, therefore, would be less likely than the polyclonal antisera to cross-react with other serum proteins.

10. (a) Tube 1 contained $F(ab9')_2$ fragments. Because these fragments contain two antigen-binding sites, they can crosslink and eventually agglutinate SRBCs. Activation of complement by IgG requires the presence of the Fc fragment, which is missing from $F(ab9')_2$ fragments. (b) Tube 2 contained Fab fragments. Univalent Fab fragments can bind to SRBCs and inhibit subsequent agglutination by whole anti-SRBC. (c) Tube 3 contained intact antibody. (d) Tube 4 contained Fc fragments. These fragments lack antigen-binding sites and thus cannot mediate any antibody effector functions.

11. The reaction $S + L = SL$

Written according to the law of mass action, $[SL]/[S][L] = K_a$

Let $B = [SL]$ and $S = ([S_t] - [SL])$. Substituting $B/[S_t - B] F = K_a [S_t - B]$

12. (a) 2, 3, 4, 6. (b) Patient 1: 1000; patient 2: no titer; patient 3: 10; patients 1 and 3 were possibly exposed, since they have developed antibodies to Hantavirus. Patient 2 has no titer and was therefore not exposed.

13. All the Western blot strips show some reactivity, meaning that each person tested had antibodies that would bind to at least one of the flu antigens. This would indicate that the new strain is antigenically similar to existing strains and would be unlikely to cause a drastic problem if the general population reacts similarly to the tested volunteers.

14. (a) The cells need to be incubated with both primary antibodies, rinsed, and then incubated with the secondary labeled antibodies. The sample is then applied to the instrument so the fluorochrome will be excited by the laser and deflected accordingly.

    (b) The upper-left quadrant would contain cells expressing high levels of receptor, and the lower-right quadrant would contain untransfected cells or cells expressing low levels of receptor.

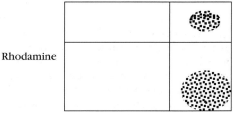

Rhodamine

Fluorescein

15. An ELISPOT uses whole live cells and can reveal how many cells in a population are expressing a soluble antigen. The relative sizes of the spots can also give an indication of the amount of antigen being produced. In a sandwich ELISA, only the amount of soluble antigen can be determined; however, the determination is quantitative rather than qualitative. In both assays, a capture antibody

is coated onto a plate. In the sandwich ELISA, the antigen solution is added without cells (e.g., culture supernatant); in the ELISPOT, the antigen-secreting cells are added directly to the plate. In both assays, the unbound antigen is washed away and enzyme-labeled secondary antibody is added. In the sandwich ELISA, a soluble chromogenic substrate is used for detection in a spectrophotometric plate reader. In the ELISPOT, the colored product must not be able to diffuse away from the spot so that the number of spots can be visually determined.

**Analyze the Data:** (a) False. The NP-related hapten NIP was inhibitory at a concentration of $2 \times 10^{-5}$ M. Therefore, B cells had a higher affinity for this hapten than NP, which required a higher concentration, $3 \times 10^{-4}$ M, to inhibit binding of labeled NP. Subtle chemical changes to antigens can improve or worsen the affinity an antibody has for it depending how modifications affect the forces that are involved in binding (e.g., hydrogen bonds, hydrophobic interactions, and van der Waals forces).

(b) Class switching most likely did occur. The idiotype specific for pNP-CGG was most likely IgM immediately after B cells completed differentiating. However, in response to the appropriate T-cell cytokines, the B cells probably class-switched, a hypothesis supported by the data in Figure 2a, which shows that the concentration of IgG went up over time.

(c) Somatic mutation occurred during the immune response to allow fine-tuning the specificity of B-cell receptors for pNP. As antigen became limiting, there was selection for higher affinity clones of B cells, resulting in affinity maturation/higher affinity.

(d) κ light-chain gene rearrangements were unsuccessful and resulted in the selection of λ light chains.

(e) Combinatorial association of the same heavy chain with a different light chain should result in the expression of an antibody with a different idiotype.

# Chapter 7

**Clinical Focus Question:** Defects in single components of the complement pathway may lead to compromised immunity to certain groups of pathogens or in some cases to immune complex disease. In most cases the outcome is not life threatening, and severe immunodeficiency is not observed because there are redundant mechanisms for dealing with common pathogens. Defects in the regulators of complement are serious because the complement system has the power to destroy cells if no checks on its activity are in place. In the case of PNH (see Clinical Focus), host red blood cells are lysed because those affected lack the cell surface molecules (DAF and MIRL) that restrict activity of complement on cells of the same species.

1. (a) True. (b) True. (c) True. (d) True. (e) False: Enveloped viruses can be lysed by complement because their outer envelope is derived from the plasma membrane of a host cell. (f) True.

2. Serum IgM is in a planar form in which the complement-binding sites in the Fc region are not accessible. Only after binding to antigen does IgM assume a conformation in which the complement-binding sites are accessible (see Figure 7-4).

3. (a) Activation of the classical pathway would not be affected. Activation of the alternate pathway would not occur. The alternate pathway relies on spontaneous cleavage of C3. (b) Clearance of immune complexes would be diminished because cleavage fragment C3b would not be available to attach immune complexes to erythrocytes via CR1 for clearance in the liver and spleen. (c) Phagocytosis would be diminished due to lack of opsonin C3b binding to bacterial cell surfaces. (d) Presentation of antigens would be diminished indirectly due to inefficient phagocytosis of bacteria by macrophages and other phagocytic antigen-presenting cells.

4. Lysis of bacteria, enveloped viruses, and cells by formation of the membrane attack complex; opsonization of invading cells to enhance phagocytosis; targeting of coated cells by immune system cells bearing complement receptors, increasing the inflammatory response at the site of infection; targeting immune complexes for clearance by offering binding sites to complement receptors on phagocytes and red blood cells.

5. (a) The classical pathway is initiated by immune complexes involving IgG or IgM; the alternative pathway generally is initiated by bacterial cell wall components; and the lectin pathway is initiated by binding of mannose-binding lectin (MBL) to microbial cell wall carbohydrates. (b) The initial reaction sequences that generate C5 convertase are the same, after the C1 step, for the classical and lectin pathways but differ between these two pathways and the alternative pathway. The terminal sequence leading from bound C5b to the MAC is the same for all three pathways. See Figure 7-2. (c) Complement activation by any of the three pathways has the *same* biological consequences because components mediating all the biological effects of complement are generated in all three.

6. (a) Innocent-bystander lysis may occur when free C5b67 binds to nearby healthy cells, including red blood cells. Binding of S protein to the free C5b67 prevents its insertion into the membrane of red blood cells and other healthy cells. In addition, red blood cells contain two membrane proteins, homologous restriction factor (HRF) and membrane inhibitor of reactive lysis (MIRL), which block the terminal sequence leading to MAC formation and thereby protect these cells from nonspecific complement-mediated lysis. (b) A defect in S protein, in the membrane-bound HRF and MIRL, or in the phospholipid membrane anchors that tether HRF and MIRL to the cell surface theoretically could lead to increased complement-mediated lysis of red blood cells. The presence of defective membrane anchors has been associated with chronic hemolytic anemia in some individuals.

7. See Table 7-2.

8. (a) 4. (b) 5. (c) 6. (d) 2. (e) 7. (f) 11. (g) 3. (h) 1. (i) 8. (j) 10. (k) 9. (l) 12.

**9.**

| | Component knocked out | | | | | | |
|---|---|---|---|---|---|---|---|
| | C1q | C4 | C3 | C5 | C6 | C9 | Factor B |
| **Complement activation** | | | | | | | |
| Formation of C3 convertase in classical pathway | A | A | NE | NE | NE | NE | NE |
| Formation of C3 convertase in alternative pathway | NE | NE | A | NE | NE | NE | A |
| Formation of C5convertase in classical pathway | A | A | A | NE | NE | NE | NE |
| Formation of C5 convertase in alternative pathway | NE | NE | A | NE | NE | NE | A |
| **Effector functions** | | | | | | | |
| C3b-mediated opsonization | D | D | A | NE | NE | NE | D |
| Neutrophil chemotaxis | D | D | D | D | D | NE | D |
| Cell lysis | D | D | A | A | A | A | D |

# Chapter 8

**Clinical Focus Question:** The TAP defiency results in lack of class I molecules on the cell surface or a type I bare-lymphocyte syndrome. This leads to partial immunodeficiency in that antigen presentation is compromised, but there are NK cells and γδ T cells to limit viral infection. Autoimmunity results from the lack of class I molecules that give negative signals through the killer-cell inhibitory receptor (KIR) molecules; interactions between KIR and class I molecules prevent the NK cells from lysing target cells. In their absence, self cells are targets of autoimmune attack on skin cells, resulting in the lesions seen in TAP-deficient patients.

The use of gene therapy to cure those affected with TAP deficiency is complicated by the fact that class I genes are expressed in nearly all nucleated cells. Because the class I product is cell bound, each deficient cell must be repaired to offset the effects of this problem. Therefore, although the replacement of the defective gene may be theoretically possible, ascertaining which cells can be repaired by transfection of the functional gene and reinfused into the host remains an obstacle.

1. (a) True. (b) True. (c) False: Class III MHC molecules are soluble proteins that do not function in antigen presentation. They include several complement components, TNF-α, TNF-β, and two heat-shock proteins. (d) False: The offspring of heterozygous parents inherit one MHC haplotype from each parent and thus will express some molecules that differ from those of each parent; for this reason, parents and offspring are histoincompatible. In contrast, siblings have a one in four chance of being histocompatible (see Figure 8-2c). (e) True. (f) False: Most nucleated cells express class I MHC molecules, but neurons, placental cells, and sperm cells at certain stages of differentiation appear to lack class I molecules. (g) True.

2. (a) Liver cells: Class I $K^d$, $K^k$, $D^d$, $D^k$, $L^d$, and $L^k$.
   (b) Macrophages: Class I $K^d$, $K^k$, $D^d$, $D^k$, $L^d$, and $L^k$. Class II $IA\alpha^k\beta^k$, $IA\alpha^d\beta^d$, $IA\alpha^k\beta^d$, $IA\alpha^d\beta^k$, $IE\alpha^k\beta^k$, $IE\alpha^d\beta^d$, $IE\alpha^d\beta^k$.

3.

| Transfected gene | MHC molecules expressed on the membrane of the transfected L cells | | | | | |
|---|---|---|---|---|---|---|
| | $D^k$ | $D^b$ | $K^k$ | $K^b$ | $IA^k$ | $IA^b$ |
| None | + | − | + | − | − | − |
| $K^b$ | + | − | + | + | − | − |
| $IA\alpha^b$ | + | − | + | − | − | − |
| $IA\beta^b$ | + | − | + | − | − | − |
| $IA\alpha^b$ and $IA\beta^b$ | + | − | + | − | − | + |

4. (a) SLJ macrophages express the following MHC molecules: $K^s$, $D^s$, $L^s$, and $IA^s$. Because of the deletion of the IEα locus, $IE^s$ is not expressed by these cells. (b) The transfected cells would express one heterologous IE molecule, $IE\alpha^k\beta^s$, and one homologous IE molecule, $IE\alpha^k\beta^k$, in addition to the molecules listed in (a).

5. See Figures 8-3, 4-6, and 4-25.

6. (a) The polymorphic residues are clustered in short stretches primarily within the membrane-distal domains of the class I and class II MHC molecules (see Figure 8-10). These regions form the peptide-binding cleft of MHC molecules. (b) MHC polymorphism is thought to arise by gene conversion of short, nearly homologous DNA sequences within unexpressed pseudogenes in the MHC to functional class I or class II genes.

7. (a) The proliferation of $T_H$ cells and IL-2 production by them is detected in assay 1, and the killing of LCM-infected target cells by cytotoxic T lymphocytes (CTLs) is detected in assay 2. (b) Assay 1 is a functional assay for class II MHC molecules, and assay 2 is a functional assay for class I molecules. (c) Class II $IA^k$ molecules are required in assay 1, and class I $D^d$ molecules are required in assay 2. (d) You could transfect L cells with the $IA^k$ gene and determine the response of the transfected cells in assay 1. Similarly, you could transfect a separate sample of L cells with the $D^d$ gene and determine the response of the transfected cells in assay 2. In each case, a positive r esponse would confirm the identity of the MHC molecules required for LCM-specific activity of the spleen cells. As a control in each case, L cells should be transfected with a different class I or class II MHC gene and assayed in the appropriate assay. (e) The immunized spleen cells express both $IA^k$ and $D^d$ molecules. Of the listed strains, only A.TL and (BALB/c × B10.A) $F_1$ express both of these MHC molecules, and thus these are the only strains from which the spleen cells could have been isolated.

8. It is not possible to predict. Since the peptide-binding cleft is identical, both MHC molecules should bind the same peptide. However, the amino acid differences outside the cleft might prevent recognition of the second MHC molecule by the T-cell receptor on the $T_C$ cells.

9. If RBCs expressed MHC molecules, then extensive tissue typing would be required before a blood transfusion, and only a few individuals would be acceptable donors for a given individual.

10. (a) No. Although those with the A99/B276 haplotype are at significantly increased relative risk, there is no absolute correlation between these alleles and the disease. (b) Nearly all of those with the disease will have the A99/B276 haplotype, but depending on the exact gene or genes responsible, this may not be a requirement for development of the disease. If the gene responsible for the disease lies between the A and B loci, then weaker associations to A99 and B276 may be observed. If the gene is located outside of the A and B regions and is linked to the haplotype only by association in a founder, then associations with other MHC genes may occur. (c) It is not possible to know how frequently the combination will occur relative to the frequency of the two individual alleles; linkage disequilibrium is difficult to predict. However, based on the data given, it may be speculated that the linkage to a disease that is fatal in individuals who have not reached reproductive years will have a negative effect on the frequency of the founder haplotype. An educated guess would be that the A99/B276 combination would be rarer than predicted on the basis of the frequency of the A99 and B276 alleles.

11. By convention, antigen-presenting cells are defined as those cells that can display antigenic peptides associated with class II MHC molecules and can deliver a costimulatory signal to CD4$^+$ $T_H$ cells. Target cells display peptides associated with class I MHC molecules to CD8$^+$ $T_C$ cells.

12. (a) Self-MHC restriction is the attribute of T cells that limits their response to antigen associated with self-MHC molecules on the membrane of antigen-presenting cells or target cells. In general, CD4$^+$ $T_H$ cells are class II MHC restricted, and CD8$^+$ $T_C$ cells are class I MHC restricted, although a few exceptions to this pattern occur. (b) Antigen processing involves the intracellular degradation of protein antigens into peptides that associate with class I or class II MHC molecules. (c) Endogenous antigens are synthesized within altered self cells (e.g., virus-infected cells or tumor cells), are processed in the cytosolic pathway, and are presented by class I MHC molecules to CD8$^+$ $T_C$ cells. (d) Exogenous antigens are internalized by antigen-presenting cells, processed in the endocytic pathway, and presented by class II MHC molecules to CD4$^+$ $T_H$ cells.

13. (a) EN: Class I molecules associate with antigenic peptides and display them on the surface of target cells to CD8$^+$ $T_C$ cells. (b) EX: Class II molecules associate with antigenic peptides and display them on surface of APCs to CD4$^+$ $T_H$ cells. (c) EX: The invariant chain interacts with the peptide-binding cleft of class II MHC molecules in the rough endoplasmic reticulum, thereby preventing binding of peptides from endogenous antigens. It also assists in folding of the class II α and β chains and in movement of class II molecules from the RER to endocytic compartments. (d) EX: Lysosomal hydrolases degrade exogenous antigens into peptides; these enzymes also degrade the invariant chain associated with class II molecules, so that the peptides and MHC molecules can associate. (e) EN: TAP, a transmembrane protein located in the RER membrane, mediates transport of antigenic peptides produced in the cytosolic pathway into the RER lumen where they can associate with class I MHC molecules. (f) B: In the endogenous pathway, vesicles containing peptide–class I MHC complexes move from the RER to the Golgi complex and then on to the cell surface. In the exogenous pathway, vesicles containing the invariant chain associated with class II MHC molecules move from the RER to the Golgi and on to endocytic compartments. (g) EN: Proteasomes are large protein complexes with multiple peptidase activity that degrade intracellular proteins within the cytosol. When associated with LMP2 and LMP7, which are encoded in the MHC region, and LMP10, which is not MHC encoded, proteasomes preferentially generate peptides that associate with class I MHC molecules. (h) EX: Antigen-presenting cells internalize exogenous (external) antigens by phagocytosis or endocytosis. (i) EN: Calnexin is a protein within the RER membrane that acts as a molecular chaperone, assisting in the folding and association of newly formed class I α chains and β$_2$-microglobulin into heterodimers. (j) EX: After degradation of the invariant chain associated with a class II MHC molecule, a small fragment called CLIP remains bound to the peptide-binding cleft, presumably preventing premature peptide loading of the MHC molecule. Eventually, CLIP is displaced by an antigenic peptide. (k) EN: Tapasin (TAP-associated protein) brings the transporter TAP into proximity with the class I MHC molecule and allows the MHC molecule to acquire an antigenic peptide (see Figure 8-20).

14. (a) Chloroquine inhibits the endocytic processing pathway, so that the APCs cannot display peptides derived from native lysozyme. The synthetic lysozyme peptide will exchange with other peptides associated with class II molecules on the APC membrane, so that it will be displayed to the $T_H$ cells and induce their activation. (b) Delay of chloroquine addition provides time for native lysozyme to be degraded in the endocytic pathway.

15. (a) Dendritic cells: constitutively express both class II MHC molecules and costimulatory signal. B cells: constitutively express class II molecules but must be activated before expressing the B7 costimulatory signal. Macrophages: must be activated before expressing either class II molecules or the B7 costimulatory signal. (b) See Table 8-3. Many nonprofessional APCs function only during sustained inflammatory responses.

16. (a) R. (b) R. (c) NR. (d) R. (e) NR. (f) R.

17. Although both HLA-DM and -DO are encoded within the MHC and both are heterodimers of α and β chains, they differ from classical class II MHC molecules in that neither is polymorphic and neither is expressed on the surface of

the cell. HLA-DM and DO differ from each other in that DM is expressed in the same cells as the classical class II molecules, whereas HLA-DO is expressed only in B cells and in the thymus. Expression of DM is up-regulated by interferon γ, which has no effect on DO expression. The function of DM is to catalyze the intracellular exchange of CLIP (class II–associated invariant chain peptide) for an antigenic peptide. The exact function of DO remains obscure, but available evidence suggests that it binds to DM and thus plays a role in down-regulation of DM activity.

18. (a) Intracellular bacteria, such as members of the *Mycobacterium* family, are a major source of nonpeptide antigens; the antigens observed in combination with CD1 are lipid and glycolipid components of the bacterial cell wall. (b) Members of the CD1 family associate with $\beta_2$-microglobulin and have structural similarity to class I MHC molecules. They are not true MHC molecules because they are not encoded within the MHC and are on a different chromosome. (c) The pathway for antigen processing taken by the CD1 molecules differs from that taken by class I MHC molecules. A major difference is that CD1 antigen processing is not inhibited in cells that are deficient in TAP, whereas class I MHC molecules cannot present antigen in TAP-deficient cells.

19. (b) The TAP1-TAP2 complex is located in the endoplasmic reticulum.

20. The offspring must have inherited HLA-A 3; HLA-B 59 and HLA-C 8 from the mother. Potential father 1 cannot be the biological father because although he shares HLA determinants with the offspring, the determinants are the same genotype inherited from the mother. Potential father 2 could be the biological father because he expresses the HLA genes expressed by the offspring that are not inherited from the mother (HLA-A 43, HLA-B 54, HLA-C 5). Potential father 3 cannot be the biological father because although he shares HLA determinants with the offspring, the determinants are the same genes inherited from the mother.

**Analyze the Data:** (a) Yes. Comparing the relative amounts of $L^d$ and $L^q$ molecules without peptides, there are about half as many open $L^q$ molecules as $L^d$ molecules. The data suggest that $L^q$ molecules form less stable peptide complexes than $L^d$ molecules.

(b) Part (a) in the figure shows that 4% of the $L^d$ molecules don't bind MCMV peptide compared to 11% of the $L^d$ after a W to R mutation. Thus, there appears to be a small decrease in peptide binding to $L^d$. It is interesting to note that nonspecific peptide binding increases severalfold after mutagenesis, based on the low amount of open-form $L^d$ W97R (mutated $L^d$) versus native $L^d$.

(c) Part (b) in the figure shows that 71% of the $L^d$ molecules don't bind tum⁻ P91A14–22 peptide after a W to R mutation, compared to 2% for native $L^d$ molecules. Thus, there is very poor binding of tum⁻ P91A14–22 peptide after a W to R mutation.

(d) You would inject a mouse that expressed $L^d$ because only 2% of the $L^d$ molecules were open forms after the addition of tum⁻ P91A14–22 peptide compared to 77% free forms when $L^q$ molecules were pulsed with peptide. Therefore,

$L^d$ would present peptide better and probably activate T cells better than $L^q$.

(e) Conserved anchor residues at the ends of the peptide bind to the MHC, allowing variability at other residues to influence which T-cell receptor engages the class I MHC–antigen complex.

# Chapter 9

**Clinical Focus Question:** Rearrangements involving immunoglobulin genes as well as those of γδ TCR genes must be tested to rule out a lymphoma involving proliferation of a B cell or γδ T cell. The cell population should be analyzed microscopically for the presence of atypical cells. If no Ig or TCR gene rearrangements are detected and the number of atypical cells is normal, it is possible that the enlargement is the result of an inflammatory response rather than a cancerous cell proliferation.

1. (a) False: The distance between CD4 and the TCR is too great for them to coprecipitate; however, CD3 and the TCR are close enough that they will coprecipitate in response to monoclonal anti-CD3. (b) True. (c) True. (d) False: The TCR variable-region genes are located on different chromosomes from the Ig variable-region genes. (e) False: All T-cell receptors have a single binding site for peptide-MHC complexes. (f) False: Because allelic exclusion is not complete for the TCR α chain, a T cell occasionally expresses two α chains resulting from rearrangement of both α-chain alleles. (g) False: The mechanisms are very similar, but T-cell receptors do not generate diversity by somatic mutation as immunoglobulins do. (h) True.

2. Functional αβ TCR genes from a $T_C$ clone specific for one hapten on an H-$2^d$ target cell were transfected into another $T_C$ clone specific for a second hapten on an H-$2^k$ target cell. Cytolysis assays revealed that the transfected $T_C$ cells killed only target cells that presented antigen associated with the original MHC restriction element.

3. See Figure 9-3.

4. (a) CD3 is a complex of three dimers containing five different polypeptide chains. It is required for the expression of the T-cell receptor and plays a role in signal transduction across the membrane. CD3 and the T-cell receptor associate to form the TCR-CD3 membrane complex (see Figure 9-9). (b) CD4 and CD8 interact with the membrane-proximal domains of class II and class I MHC molecules, respectively, thereby increasing the avidity of the interaction between T cells and peptide-MHC complexes. CD4 and CD8 also play a role in signal transduction. (c) CD2 and other accessory molecules (LFA-1, CD28, and CD45R) bind to their ligands on antigen-presenting cells or target cells. The initial contact between a T cell and antigen-presenting cell or target cell probably is mediated by these cell adhesion molecules. Subsequently, the T-cell receptor interacts with peptide-MHC complexes. These molecules also may function in signal transduction.

5. (a) TCR. (b) TCR. (c) Ig. (d) TCR/Ig. (e) TCR. (f) TCR/Ig. (g) Ig.

6. (a) The three assumptions were as follows. (1) TCR mRNA should be associated with membrane-bound polyribosomes, like the mRNAs encoding other membrane proteins; therefore, isolation of the membrane-bound polyribosomal mRNA fraction would significantly enrich the proportion of TCR mRNA in the preparations. (2) B cells and T cells would express many common genes, and the unique T-cell mRNAs would include those encoding the T-cell receptor; therefore, subtractive hybridization using B-cell mRNA would remove all the cDNAs common to both B and T cells, leaving the unique T-cell cDNA unhybridized. (3) The TCR genes undergo DNA rearrangement and therefore can be detected by Southern blotting with cDNA probes. See Figure 9-2. (b) If they wanted to identify the IL-4 gene, they should perform subtractive hybridization using a $T_H$ clone as an IL-4 producer and a $T_C$ clone as a source of mRNA lacking the message for IL-4. In addition, the gene for IL-4 would not be expected to rearrange and therefore could not be identified by Southern-blot analysis for gene rearrangement.

7.

| Gene product | cDNA source | mRNA source |
|---|---|---|
| IL-2 | A | B |
| CD8 | C | A or B |
| J chain | E | F |
| IL-1 | D | G |
| CD3 | A, B, or C | H |

8.

| Source of spleen cells from LCM-infected mice | Release of $^{51}$Cr from LCM-infected target cells | | | |
|---|---|---|---|---|
| | B10.D2 (H-2$^d$) | B10 (H-2$^b$) | B10.BR (H-2$^k$) | (BALB/C × B10)F$_1$ (H-2$^{b/d}$) |
| B10.D2 (H-2$^d$) | + | − | − | + |
| B10 (H-2$^b$) | − | + | − | + |
| BALB/c (H-2$^d$) | + | − | − | + |
| BALB/b (H-2$^b$) | − | + | − | + |

9. Although the αβ and γδ T-cell receptors have some structural differences (Figure 9-2) and functional differences in the type of antigen that is recognized, they have in common the fact that both are cell surface molecules. Immunoglobulins may be surface receptors on B cells, but their major function is to serve as soluble mediators of immunity. There is no evidence for activity of soluble forms of either the αβ or γδ T-cell receptors.

10. Two factors contribute to the high frequency of alloreactive T cells: (1) the absence of negative selection for TCR reacting with peptides contained in the foreign MHC molecules and (2) the reactivity of TCR with the exposed portions of the foreign (not previously selected against) MHC molecules.

11. The overall size of the structure determined will immediately tell whether it is TCR alone with an antigen (in which case it will be a γδ TCR) or if there is an associated complex molecule with it, indicating the presence of a class I or II MHC molecule. If the TCR contacts the MHC molecule through the peptide-binding region, this indicates αβ as opposed to γδ TCR, which makes contact with nonclassical MHC molecules via the δ chain. An important feature to be examined is the angle between the V (the portion that contacts antigen) and C domains of the TCR. This will determine if the TCR is αβ (angle of around 145 degrees) or γδ (angle closer to 110 degrees). The inclusion of either CD4 with its four globular domains or CD8 with two will indicate whether the associated MHC molecule is class II or class I. The contact region between the TCR and the antigen may further differentiate whether the antigen is presented in context of class I or class II. There are usually more contact residues between TCR and class II than class I.

12. Mice with the alpha-chain gene deleted will lack response involving the αβ TCR, but the γδ response will be intact. Loss of the CD3 chain cause a complete loss of signaling from either the αβ or the γδ TCR.

13. Both receptors use multiple gene segments, although there are more V segments that code for the BCR than for the TCR (especially for the β chain). D segments are available for the β chain of the TCR and for the heavy chain of the BCR. Multiple D segments may be included in the TCR, but only one is included in the BCR heavy chain. Both receptors use J-chain segments, but there are more J segments available for the TCR. Both receptors utilize junctional flexibility and P-nucleotide and N-nucleotide addition, although N-nucleotide addition can occur in both subunits of the TCR but only in the heavy chain of the BCR. Once the TCR is generated, no further alterations are made in the periphery. In contrast, the BCR may undergo somatic hypermutation in the periphery, enhancing antigen-binding affinity during an immune reaction.

# Chapter 10

**Clinical Focus Question:** (a) Fas-mediated signaling is important for the death-mediated regulation of lymphocyte populations in lymphocyte homeostasis. This regulation is essential because the immune response generates a sudden and often large increase in populations of responding lymphocytes. Inhibition of Fas-mediated cell death could result in progressively increasing and eventually unsustainable lymphocyte levels. The Canale-Smith syndrome demonstrates that without the culling of lymphocytes by apoptosis, severe and life-threatening disease can result.

(b) Analysis of the patient T cells reveals a large number of double negative (CD4⁻, CD⁻) T cells and almost equal numbers of CD4$^+$ and CD8$^+$ cells. A normal subject would have very few (~5%) double-negative cells and more CD4$^+$ than CD8$^+$ cells. See the Clinical Focus figure on page 266.

1. (a) The immature thymocytes express both CD4 and CD8, whereas the mature CD8$^+$ thymocytes do not express CD4. To distinguish these cells, the thymocytes are double-stained with fluorochrome-labeled anti-CD4 and anti-CD8 and analyzed in a FACS. (b) See the following table.

| Transgenic mouse | Immature thymocytes | Mature CD8$^+$ thymocytes |
|---|---|---|
| H-2$^k$ female | + | + |
| H-2$^k$ male | + | − |
| H-2$^d$ female | + | − |
| H-2$^d$ male | + | − |

(c) Because the gene encoding the H-Y antigen is on the Y chromosome, this antigen is not present in females. Thymocytes bearing the transgenic T-cell receptor, which is H-2$^k$ restricted, would undergo positive selection in both male and female H-2$^k$ transgenics. However, subsequent negative selection would eliminate thymocytes bearing the transgenic receptor, which is specific for H-Y antigen, in the male H-2$^k$ transgenics (see Figure 10-8). (d) Because the H-2$^d$ transgenics would not express the appropriate MHC molecules, T-cells bearing the transgenic T-cell receptor would not undergo positive selection.

2. Cyclosporin A blocks production of NF-ATc, one of the transcription factors necessary for proliferation of antigen-activated T$_H$ cells.

3. (a) NF-kB and NF-ATc. (b) IL-2 enhancer.

4.

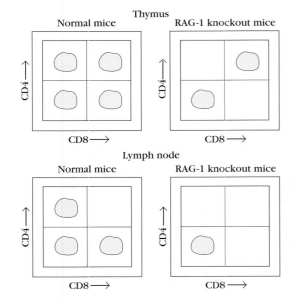

5. (a) Class I K, D, and L molecules and class II IA molecules. (b) Class I molecules only. (c) The normal H-2$^b$ mice should have both CD4$^+$ and CD8$^+$ T cells because both class I and class II MHC molecules would be present on thymic stromal cells during positive selection. H-2$^b$ mice with knockout of the *IA* gene would express no class II molecules; thus, these mice would have only CD8$^+$ cells.

6. (a) The thymus donor in experiment A was H-2$^d$ (BALB/c) and in experiment B was H-2$^b$ (C57BL/6). (b) The haplotype of the thymus donor determines the MHC restriction of the T cells in the chimeric mice. Thus, H-2$^b$ target cells were lysed in experiment B, in which the thymus donor was H-2$^b$. (c) The H-2$^k$ target cells were not lysed in either experiment because neither donor thymus expressed H-2$^k$ MHC molecules; thus, H-2$^k$-reactive T cells were not positively selected in the chimeric mice.

7. (a) Protein kinases. (b) CD45. (c) B7. (d) IL-2. (e) CD28; B7. (f) CD8. (g) Class II MHC; B7. (h) CD4. (i) Phospholipase. (j) Protein kinase. (k) CD28 stimulates and CTLA4 inhibits.

8. (a) Because the pre–T-cell receptor, which does not bind antigen, is associated with CD3, cells expressing the pre-TCR as well as the antigen-binding T-cell receptor would stain with anti-CD3. It is impossible to determine from this result how many of the CD3-staining cells are expressing complete T-cell receptors. The remaining cells are even more immature thymocytes that do not express CD3. (b) No. Because some of the CD3-staining cells express the pre-TCR or the αβ TCR instead of the complete γδ TCR, you cannot calculate the number of T$_C$ cells by simple subtraction. To determine the number of T$_C$ cells, you need fluorescent anti-CD8 antibody, which will stain only the CD8$^+$ T$_C$ cells.

9. Engagement of the T-cell receptor activates phospholipase C, which cleaves PIP$_2$, releasing diacylglycerol (DAG) and IP$_3$. DAG and IP$_3$ are potent second messengers that have wide-ranging biological effects. DAG activates protein kinase C and IP$_3$ releases Ca$^{2+}$ from intracellular stores, thus raising the intracellular Ca$^{2+}$ concentration. Phorbol esters mimic the effects of DAG, and Ca$^{2+}$ ionophores can increase intracellular concentrations of Ca$^{2+}$ by allowing extracellular Ca$^{2+}$ to enter the cell. Consequently, treatment of cells with a Ca$^{2+}$ ionophore and a phorbol ester can mimic many of the effects of T-cell receptor-mediated T-cell activation.

10. (1) Interaction between a signal and its receptor. Example: Interaction of the T-cell receptor with MHC$^+$ peptide. (2) Many signals are transduced through G proteins. Example: Signals from the TCR can activate signal transduction pathways mediated by the small G protein Ras. (3) The generation of second messengers. Example: Signals from the TCR can activate protein kinase C (PKC), which cleaves a membrane phospholipid to generate the second messengers DAG and IP$_3$. (4) Activation or inhibition of protein kinases and phosphatases. Example: DAG and IP$_3$ generated in response to signals from the TCR activate PKC. (5) Induced assembly of pathway components. Example: Engagement of MHC$^+$ peptide by the TCR promotes phosphorylation of ITAMs on subunits of the TCR complex, which creates docking sites for the recruitment of ZAP-70, a key protein tyrosine kinase of

the TCR signaling complex. (6) Signal amplification by enzyme cascade. Example: TCR-mediated initiation of the Ras pathway's MAP kinase cascade. (7) Default setting of a signal transduction pathway is off. Example: Unless the TCR is engaged by MHC$^+$ peptide, there is no TCR-mediated signaling.

11. *Signal transduction begins with the interaction between a signal and its receptor.* The interaction between the TCR and a T-cell antigen such as an MHC-peptide complex. *Many signal transduction pathways involve the signal-induced assembly of some components of the pathway; these assemblies utilize adaptor proteins.* ZAP-70 functions to phosphorylate adaptor proteins LAT and SLP-76 in TCR signaling. *Signal reception often leads to the generation within the cell of a "second messenger," a molecule or ion that can diffuse to other sites in the cell and evoke metabolic changes.* DAG and IP3 function as second messengers in TCR signaling. *Protein kinases and protein phosphatases are activated or inhibited.* ZAP-70 and p56$^{Lck}$ function as protein kinases in the first steps of the pathway. Calcineurin functions as a phosphatase, removing a phosphate from NFAT and allowing it to translocate into the nucleus, where it acts with NF-$\kappa\beta$ as a transcriptional activator. *Signals are amplified by enzyme cascades.* Several enzyme cascades function in TCR signaling. These include the PKC (protein kinase C) and RAS/MAP kinase pathways.

12. $\gamma\delta$ T cells do not have a requirement that antigen be presented by MHC. Thus, they are not limited to recognition of protein antigens. The recognition process seems to be closer to that of pattern recognition receptors found on innate immune system cells.

13. Encountering antigen can make T cells anergic if they receive a signal through the TCR (signal 1) but no signal from CD28 (signal 2). In contrast, if T cells are bound by superantigens, the TCR will be engaged regardless of antigen specificity. The interaction is tight enough to allow both signals to occur, and polyclonal activation is seen.

# Chapter 11

**Clinical Focus Question:** The HuIg purified from each donor is a sample of the antibody response mounted by the donor to whatever immunogenic stimuli were encountered during a period prior to donation. Consequently, the larger the pool of donors, the greater the diversity of antibodies present in the HuIg preparations derived from them. Antibody preparations with the broadest diversity confer protection against the greatest range of pathogens. Therefore, source C, derived from the highest number of donors, would be the best choice.

1. (a) False: The $V_H$-$D_H$-$J_H$ rearrangement occurs during the pro–B-cell stage; successful completion of heavy-chain rearrangement marks the beginning of the pre–B-cell stage, in which membrane-bound $\mu$ chains are expressed. (b) False: Immature B cells express only IgM. (c) False: TdT, which catalyzes N-nucleotide addition, is expressed only in pro–B cells. (d) True. (e) True. (f) False: Pro–B cells must interact with stromal cells to develop into pre–B cells. Progression of pre–B cells into immature B cells requires

IL-7 released from stromal cells but not direct contact. (g). True. (See Figure 11-1.)

2. (a) Progenitor B cell: no cytoplasmic or membrane staining with either reagent. (b) Pre–B cell: anti-$\mu$ staining in cytoplasm and on membrane. (c) Immature B cell: anti-$\mu$ staining in cytoplasm and on membrane. (d) Mature B cell: anti-$\mu$ and anti-$\delta$ staining in cytoplasm and on membrane. (e) Plasma cell: anti-$\mu$ staining in cytoplasm; no membrane staining but pentameric IgM is secreted.

3. The B-cell coreceptor consists of three membrane proteins: TAPA-1, CR2, and CD19; the latter belongs to the Ig superfamily. The CR2 component can bind to complement-coated antigen bound to the B-cell receptor (BCR). This binding results in the phosphorylation of CD19. Src-family tyrosine kinases (Lyn and Fyn) bind to phosphorylated CD19 and can trigger signal transduction pathways that begin with phospholipase C.

4. (a) Both single and double transgenics carried the anti-HEL transgene. Because the transgene is already rearranged, rearrangement of endogenous immunoglobulin heavy- and light-chain genes does not occur in pro–B and pre–B cells; therefore, the rearranged immunoglobulin encoded by the transgene is expressed preferentially by mature B cells. (b) Add radiolabeled HEL and use autoradiography to see whether it binds to the B-cell membrane. To determine the isotype of the membrane antibody (mIg) on these B cells, incubate the cells with fluorochrome-labeled antibodies specific for each isotype (e.g., anti-$\mu$ anti-$\delta$) and see which fluorescent antibodies stain the B cells. (c) So that the HEL transgene could be induced by adding $Zn^{2+}$ to the mouse's water supply. (d) First, B cells and T cells from the double-transgenic mice and normal syngeneic mice must be separated from each other. This can be accomplished by treating a preparation of spleen cells with a fluorochrome-labeled antibody specific for CD3, which is expressed on the membrane of T cells but not B cells, and passing the treated preparation through a FACS (see Figure 6-15). This will yield a T-cell fraction and a residual fraction containing B cells and other spleen cells. After treating the residual fraction with fluorochrome-labeled anti-IgM, pass it through a FACS to obtain a B-cell fraction. Now, mix the transgenic B cells with normal syngeneic T cells (sample A), and mix the transgenic T cells with normal syngeneic B cells (sample B). Transfer each cell mixture to lethally x-irradiated, syngeneic adoptive-transfer recipients, challenge the recipients with HEL, and determine whether they produce anti-HEL serum antibodies. If the B cells are anergic, then no antibody response would occur with sample A.

5. (a) B-cell activation by soluble protein antigens requires the involvement of TH cells. Cross-linkage of mIg on a naive B cell by thymus-dependent antigens provides competence signal 1; subsequent binding of CD40 on the B cell to CD40L on an activated $T_H$ cell provides competence signal 2. The combined effect of these signals drives the B cell from $G_0$ to the $G_1$ stage of the cell cycle and up-regulates expression of cytokine receptors on the B cell. Binding of $T_H$-derived cytokines then provides a progression signal that stimulates proliferation of the activated B cells (see Figure 11-12). (b) Binding of LPS,

a type 1 thymus-independent antigen, provides both competence signals 1 and 2 (see Figure 11-5). Efficient proliferation requires a cytokine-mediated progression signal.

6. (a) Light zone. (b) Paracortex. (c) Centroblasts; dark zone. (d) Centrocytes; follicular dendritic cells. (e) Light zone; $T_H$ cells. (f) Light zone. (g) Medulla. (h) Centroblasts.

7. (a) See Table 11-2. (b) TI antigens.

8. (a) Each mIg molecule is associated with one molecule of a heterodimer called Ig-α/Ig-β, forming the B-cell receptor (BCR). Both Ig-α and Ig-β have long cytoplasmic tails, which are capable of mediating signal transduction to the cell interior (see Figure 11-8). (b) B-cell activating and differentiating signals—provided by antigen binding, interaction with $T_H$ cells, or cytokine binding—trigger intracellular signal transduction pathways that ultimately generate active transcription factors. These then translocate to the nucleus, where they stimulate or inhibit transcription of specific genes (see Figure 11-10).

9. (a) The recipient mice had a haplotype other than H-2$^b$. (b) The purpose of these experiments was to see the effect of a self antigen, represented by the transgene-encoded K$^b$ that is expressed only in the periphery. Linking the K$^b$ transgene to a liver-specific promoter ensured that the K$^b$ class I molecule would be expressed in the periphery but not in the bone marrow, where immature B cells could interact with it. (c) These results suggest that exposure to self antigens in the periphery can lead to negative selection and apoptosis (clonal deletion) in some cases (see Figure 11-15).

10. (a) True. (b) True. (c) True. (d) False: Antigenic competition will reduce the response to SRBC.

11. (a) The early stages of the response will be dominated by IgM. Although the average affinity of the antibody produced will be lower than that expected during a secondary response after challenge with the same antigen, it will be higher than that obtained during a primary response to a high dose of the antigen. (b) The antibody will tend to be largely IgG and of higher affinity than that obtained during a primary response to a low dose of the antigen. (c) The average affinity of the antibody will be of lower affinity than that obtained during the primary response to a low dose of the antigen. Most of the antibody made during the early stages of the response will be IgM. (d) The average affinity of the antibody will be higher than that of antibody made during a primary response to a high dose of the antigen, but not necessarily lower than that to antibody made in a primary response to low doses. It will be of lower average affinity than the antibody produced during a secondary response to a low dose of the antigen. It will be mostly IgG.

12. (a) Sample B yields a single restriction fragment with each digest characteristic of unrearranged germ-line DNA. Thus, this sample is from liver cells. (b) The two fragments obtained with the *Bam*HI digest of sample C indicate that both heavy-chain alleles are rearranged, whereas no κ-chain rearrangement has occurred. Thus, this sample is from pre–B lymphoma cells. (c) The blots obtained with sample A indicate that one heavy-chain allele is rearranged and one light-chain allele is rearranged. This pattern indicates that this sample is from IgM-secreting myeloma cells.

13. (a) True. (b) False: AP-1 is a transcription factor that when activated by a signal transduction cascade will translocate to the nucleus from the cytosol. (c) True. (d) False: Class switching is a feature of the response to thymus-dependent (TD) antigens, arising from interactions with $T_H$ cells. Thymus-independent antigens can activate B cells without assistance from T helper cells, including the interactions that lead to class switching. (e) False: B cells mature into effector cells in the lymph node. (f) False: As B cells develop into plasma cells, production of membrane immunoglobulin ceases in favor of secreted antibody, rendering plasma cells unresponsive to the presence of antigen. (g) True.

14. (a) Class switching. (b) IL-4. (c) VpreB; λ5. (d) Plasma cells.

# Chapter 12

**Clinical Focus Question:** All the agents mentioned in this question along with a description of their nature appear in the table that concludes the Clinical Focus section of this chapter. Open a Web browser and search using the name of the agent to find the manufacturer's information about the drug, including its side effects.

1. (a) False: The high-affinity IL-2 receptor comprises three subunits—the α, β, and γ chains—all of which are transmembrane proteins. (b) False: Anti-TAC binds to the 55-kDa α chain of the IL-2 receptor. (c) False: Although all class I and class II cytokine receptors contain two or three subunits, the receptors for IL-1, IL-8, TNF-α, TNF-β, and some other cytokines have only one chain. (d) False: Low levels of the IL-2R β chain are expressed in resting T cells, although expression is increased greatly after activation. The α chain is expressed only by activated T cells. (e) False: The cytosolic domains of class I and class II cytokine receptors appear to be closely associated with intracellular tyrosine kinases but do not themselves possess tyrosine kinase activity. (f) True.

2. Only antigen-activated T cells will proliferate, because they express the high-affinity IL-2 receptor, whereas resting T cells do not and therefore cannot respond to IL-2.

3. Cytokines, growth factors, and hormones are all secreted proteins that bind to receptors on target cells, eliciting various biological effects. Cytokines tend to be produced by a variety of cells, although their production is carefully regulated, and they exert their effects on several cell types; most cytokines also act in an autocrine or paracrine fashion. Growth factors, unlike cytokines, are often produced constitutively. Hormones, unlike cytokines, generally act over long distances (endocrine effect) on one or a few types of target cells.

4. (a) γ chain and β chain (low level). (b) α, β, and γ chains. (c) γ and β chain (low level); cyclosporin A prevents the gene activation that leads to increased expression of the β and α chains. (d) γ chain and β chain (low level). (e) α, β, and γ chains. (f) β and γ chains.

5. (a) Superantigens bind to class II MHC molecules outside of the normal peptide-binding cleft; unlike normal antigens, they are not internalized and processed by antigen-presenting cells but bind directly to class II molecules. Superantigens also bind to regions of the $V_\beta$ domain of the T-cell receptor that are not involved in binding normal antigenic peptides. Superantigens exhibit specificity for one or a few $V_\beta$ domains; thus, a given superantigen can activate all T cells that express the $V_\beta$ domain(s) for which it is specific regardless of the antigenic specificity of the T cells. (b) A given superantigen can activate 5%–25% of $T_H$ cells, leading to excessive production of cytokines. The high levels of cytokines are thought to cause the symptoms associated with food poisoning and toxic shock syndrome. (c) Yes. To exert their effect, superantigens must form a ternary complex with a class II MHC molecule and T-cell receptor.

6. The receptors for IL-3, IL-5, and GM-CSF contain a common signal-transducing β chain. Cytokine binding to each of these receptors probably triggers a similar signal transduction pathway.

7. (a) The $T_H1$ subset is responsible for classical cell-mediated functions (e.g., delayed-type hypersensitivity and activation of $T_C$ cells). Viral infections and intracellular pathogens are likely to induce a $T_H1$ response. (b) The $T_H2$ subset functions primarily as a helper for B-cell activation. This subset may be best suited to respond to free-living bacteria and helminthic parasites and may mediate allergic reactions, since IL-4 and IL-5 are known to induce IgE production and eosinophil activation, respectively.

8. (a) False: Whereas Il-6 does increase the production of acute phase proteins, these proteins are proinflammatory and do not shut down the immune response. (b) False: Cytokines can act in an autocrine fashion (on the cells secreting them), and they act in paracrine and endocrine fashions also. (c) True. (d) False: $T_H1$ cells secrete IFNγ, IL-2, and TNF-β, which lead to activation of macrophages and increased production of IgG in B cells. (e) True. (f) False: Activation of T cells results in production of IL-2 and its receptor, not IL-1.

9. The upper-right quadrant. The intermediate affinity receptor is expressed on cells in the upper-left quadrant, and the low-affinity receptor is expressed on cells in the lower-right quadrant. The bottom-left quadrant contains cells that do not express the IL-2 receptor.

10. (a) 2, 7. (b) 4, 8. (c) 9. (d) 5. (e) 6. (f) 1, 3. (g) 3.

11. (a) IL-4. (b) Chemokines. (c) G proteins. (d) IL-7. (e) No. (f) IFNα.

12. (d).

13. (a), (b), (c), (d).

# Chapter 13

**Clinical Focus Question:** The inflammatory process is essential for providing an effective defense against pathogens. A key element of this process is the migration of leukocytes into sites of inflammation. The mobilization of leukocyte populations from the circulation depends on integrin-mediated cell adhesion. CD18 is an integrin subunit, and defects in this molecule result in a failure of intergrin function. See Figure 13-5 for a presentation of the cell biology of leukocyte migration.

1. (a) False: Various chemokines are chemotactic for all types of leukocytes. (b) False: Integrins are expressed by various leukocytes but not by endothelial cells. (c) True. (d) True. (e) True. (f) False: Systemic effects, called the acute phase response, are induced by cytokines generated in the localized acute inflammatory response. (g) True. (h) False: Granulomas may form in sites of chronic infection but are unlikely during an acute inflammatory response.

2. The increased expression of ICAMs on vascular endothelial cells near an inflammatory site facilitates adherence of leukocytes to the blood vessel wall, resulting in increased migration of leukocytes into the area.

3. (a) Rolling, activation, arrest/adhesion, and transendothelial migration (see Figures 13-5a and 13-8). (b) Neutrophils generally extravasate at sites of inflammation because they bind to the cell adhesion molecules that are induced on vascular endothelium early in the inflammatory response. (c) Different lymphocyte subsets express homing receptors that bind to tissue-specific adhesion molecules (vascular addressins) on HEVs in different lymphoid organs, on inflamed endothelium, or on venules in tertiary sites. The homing of particular lymphocytes to certain sites is facilitated by chemokines that preferentially attract different lymphocytes. Thus, differences in (1) vascular addressins, (2) homing receptors, and (3) chemokines and their receptors determine the recirculation pattern of particular lymphocyte subsets.

4. IL-1, IL-6, and TNF-α (see Table 13-3).

5. TNF-α released by activated tissue macrophages during a localized acute inflammatory response acts on vascular endothelial cells and macrophages, inducing secretion of colony-stimulating factors (CSFs), which stimulate hematopoiesis in the bone marrow. This is part of the systemic acute phase response.

6. (a) N. (b) 1. (c) N. (d) 3. (e) N. (f) 2. (g) 3 (see Figure 13-5).

7. IFN-γ stimulates activation of macrophages, resulting in increased expression of class II MHC molecules, increased microbicidal activity, and cytokine production. The accumulation of large numbers of activated macrophages is responsible for much of the tissue damage associated with chronic inflammation. TNF-α secreted by activated macrophages also contributes to the tissue wasting common in chronic inflammation. These two cytokines act synergistically to facilitate migration of large numbers of cells to sites of chronic inflammation.

8. Binding of all these cytokines to their receptors on hepatocytes induces formation of the same transcription factor, NF-16, that stimulates transcription of acute phase proteins.

9. (a) 7. (b) 1, 6. (c) 10, 11. (d) 2. (e) 8. (f) 4. (g) 10. (h) 9.

10. (a) A defect in lymphocyte homing to mucosal tissues. (b) A defect in rolling of leukocytes. (c) Leukocyte adhesion deficiency, increased bacterial infections.

11. (a) and (d) are correct matches.

12. Bone marrow transplant, using stem cells isolated from the patient, in which the defective gene is replaced by a functional copy. Alternately, bone marrow transplant from a compatible donor.

13. Neutrophils can bind to the rapidly expressed P-selectin and low levels of ICAM-1 constitutively expressed on endothelial cells, whereas monocytes bind to E-selectin, ICAM-1, and VCAM at higher levels, which arrive at the endothelial cell surface after a delay due to the time required for new protein synthesis.

14. The kinin system produces bradykinin, which increases vascular permeability, pain, and smooth muscle contraction. Bradykinin also induces formation of the C5a and C5b components of the complement cascade. The clotting system also produces mediators that increase vascular permeability and proinflammatory cytokines. Increased vascular permeability allows fluid and cells to pass more easily into the tissue, leading to the characteristic signs of an inflammatory response (pain, swelling, redness, heat). It is advantageous for phagocytic cells to be called to a site of tissue damage due to the likelihood of exposure to invading bacteria.

**Analyze the Data:** (a) Gram-negative bacterial infection. LPS is found in the outer membrane of gram-negative bacteria. (b) Based on the increase in MPO activity in wild-type mice injected with LPS, it is evident that LPS induces the recruitment of neutrophils into the lung. The inflammatory response is diminished when LPS is injected into $5\text{-}LOX^{-/-}$ mice, suggesting the involvement of leukotrienes (a product of the lipoxygenase pathway; see Figure 13-12) in the recruitment of neutrophils. (c) Leukotriene $B_4$. (d) The 5′-lipoxygenase knockout mice had higher levels of inflammation in response to LPS than untreated controls and lower levels than wild-type mice treated with LPS. Therefore, neutrophils are recruited by factors in addition to products of the lipoxygenase pathway, which could include the activation of complement by LPS and the production of C5a, the induction of IL-8, or the induction of other chemokines by LPS.

---

# Chapter 14

**Clinical Focus Question:** Arthritis is characterized by inflammation of a joint leading to damaged tissue, swelling, and pain. Psoriatic arthritis is accompanied by skin lesions caused by immune attack (psoriasis). Association of KIR/MHC combinations with susceptibility to arthritic disease would likely stem from a deficiency of inhibitory signals (for example, absence of MHC alleles that produce inhibitory ligands for specific KIR molecules), leading to damage of host cells and tissues. Diabetes is another autoimmune disease; exhibiting destruction of the host pancreatic islet cells that produce insulin, the same mechanisms predicted for arthritis could operate in diabetes. In both cases, absence of inhibitory NK signals could lead to damage inflicted by NK cells, directly or through their recruitment of other effector cells.

1. (a) True. (b) True. (c) True. (d) False: There are two pathways by which cytotoxic T cells kill target cells. One pathway is perforin dependent, and the other uses FAS ligand displayed by the CTL to induce death in FAS-expressing target cells.

2. The monoclonal antibody to LFA-1 should block formation of the CTL–target-cell conjugate. This should inhibit killing of the target cell and, therefore, should result in diminished $^{51}$Cr release in the CML assay.

3.

| Population 1 | Population 2 | Proliferation |
|---|---|---|
| C57BL/6 (H-2$^b$) | CBA (H-2$^k$) | 1 and 2 |
| C57BL/6 (H-2$^b$) | CBA (H-2$^k$) mitomycin C-treated | 1 |
| C57BL/6 (H-2$^b$) | (CBA × C57BL/6) F$_1$ (H-2$^{k/b}$) | 1 |
| C57BL/6 (H-2$^b$) | C57L (H-2$^b$) | Neither |

4. (a) CD4$^+$ T$_H$ cells. (b) To demonstrate the identity of the proliferating cells, you could incubate them with fluorescein-labeled anti-CD4 monoclonal antibody and rhodamine-labeled anti-CD8 monoclonal antibody. The proliferating cells will be stained only with the anti-CD4 reagent. (c) As CD4$^+$ T$_H$ cells recognize allogeneic class II MHC molecules on the stimulator cells, they are activated and begin to secrete IL-2, which then autostimulates T$_H$-cell proliferation. Thus, the extent of proliferation is directly related to the level of IL-2 produced.

5. (a) Neither. (b) Both. (c) CTL. (d) CTL. (e) T$_H$ cell. (f) CTL. (g) T$_H$ cell. (h) CTL. (i) T$_H$ cell. (j) T$_H$ cell. (k) Both. (l) Both. (m) Both. (n) Both. (o) T$_H$ cell. (p) Neither. (q) CTL. (r) T$_H$ cell.

6.

| Source of primed spleen cells | $^{51}$Cr release from LCM-infected target cells | | | |
|---|---|---|---|---|
| | B10.D2 (H-2$^d$) | B10 (H-2$^b$) | B10.BR (H-2$^k$ | (BALB/c × B10) F$_1$ (H-2$^{b/d}$) |
| B10.D2 (H-2$^d$) | + | – | – | + |
| B10 (H-2$^b$) | – | + | – | + |
| BALB/c (H-2$^d$) | + | – | – | + |
| (BALB/c × B10) | | | | |
| (H-2$^{b/d}$) | + | + | – | + |

7. To determine T$_C$ activity specific for influenza, perform a CML reaction by incubating spleen cells from the infected mouse with influenza-infected syngeneic target cells. To determine T$_H$ activity, incubate the spleen cells from the infected mouse with syngeneic APCs presenting influenza peptides; measure IL-2 production.

8. NK cells mediate lysis of tumor cells and virus-infected cells by perforin-induced pore formation, as in the mechanism employed by CTLs. NK cells are able to distinguish normal cells from infected cells because normal cells express relatively high levels of class I MHC molecules, which appears to protect them from NK cells. The balance between positive signals generated by the engagement of activating receptors (NKR-P1 and others) on NK cells and negative signals induced by recognition of self MHC molecules by inhibitory receptors, such as CD94/NKG2 and the KIR family on NK cells, regulates NK-cell-mediated killing.

9. (a–e) The outcome in all cases is no lysis mediated by cytotoxic T cells. There are two pathways by which cytotoxic T cells kill target cells, one that is perforin dependent and one that uses FasL to induce death in Fas-expressing target cells. Because both of these pathways are inoperative in the double-knockout mice, there will be no T-cell-mediated cytotoxicity.

10. If the HLA type is known, MHC tetramers bound to the peptide generated from gp120 and labeled with a fluorescent tag can be used to specifically label all of the CD8$^+$ T cells in a sample that have T-cell receptors capable of recognizing this complex of HLA$^+$ peptide.

11. (a) True. (b) True. (c) True. (d) False: The signal cascade from Fas recruits FADD adapter proteins, not G proteins. (e) True. (f) False: FasL is expressed on the cell inducing apoptosis; the target cell expresses Fas.

12. (a) NK cells are thought to utilize an opposing-signal mechanism to detect infected cells. If a potential target cell expresses normal levels of class I MHC molecules, receptors on the NK cell (KIR, CD94, NKG2) induce a signal transduction cascade indicating that the target cell should not be killed. This overrides pro-killing signals generated via ligands binding to activating receptors on the NK cell. In virally infected cells with decreased levels of class I MHC, the activating receptor signal predominates.

   (b) The mechanism of NK killing is similar to that of killing by CTL cells. NK cells express FasL and contain granules containing perforin and granzyme, equipping them to induce apoptosis in the target cell. (c) NK cells are thought to arise from the same progenitor cell as T cells in the bone marrow but do not appear to develop in the thymus. The exact mechanisms of NK-cell differentiation are not known.

13. In ADCC, cells expressing Fc receptors recognize antibody bound to the cell surface. Ideally, the Ab-targeted cell is a tumor cell or is infected with a virus. If the antigen bound by the antibody is expressed on normal cells, uninfected cells will also be targeted. Neutrophils, eosinophils, and macrophages will bind to cells expressing the antigen and release lytic enzymes, NK cells and eosiniphils will release perforin, and macrophages and NK cells will release TNF. The result is cell death and tissue damage, possibly leading to an autoimmune disorder or type II hypersensitivity.

**Analyze the Data:** (a) Epitopes 2, 12, and 18 generated high CTL activity, and epitopes 5 and 21 generated medium activity. (b) It is possible that different peptides use distinct anchor residues, which would make this prediction more difficult.

However, if we make the assumption that the same amino acids would be bound by HLA-A2, it appears that a leucine (L) on the amino terminus side separated by four amino acids from a threonine (T) is the only common motif for the five most immunogenic peptides (2, 12, 18, 5, and 21). All of the peptides that generate high CTL activity have two consecutive leucines on the amino terminal side as well. The problem with the threonines serving as anchors is that peptide 2 has four amino acids at the carboxyl side of the T, which seems to result in the end of the peptide extending out of the binding pocket. This would be a very unusual configuration. Peptide 12 may have a less dramatic but similar problem. Therefore, a leucine in pocket 2 of HLA-A2 would be consistent with the data generated by Matsumura in 1992 (*Science* **257**:927). Thus, the main anchor may be a leucine residue at the amino end of the binding pocket, with possible contribution by T at the carboxyl end under some circumstances. (c) It is possible that there are no T cells specific for those peptides, even if they are presented. Therefore, you would not see a CTL response. (d) CTLs only recognize antigen in the context of self-MHC molecules. Therefore, in order to assess CTL activity, the T2 cells also had to express HLA-A2. (e) Class I MHC molecules typically bind peptides containing eight to 10 residues (see Figure 8-7). Peptide 2 is 11 residues long, suggesting that it bulges in the middle when bound. Since it appears to be a major epitope for CTL killing, bulging does not seem to interfere with CTL interaction and may contribute.

---

# Chapter 15

**Clinical Focus Question:** The FceRIβ is the high-affinity IgE receptor. It is found on mast cells, where it binds IgE. When an allergen cross-links IgE on the surface of a mast cell, the mast cell becomes activated and releases histamine. Therefore, mutations that enhance the FcεRIβ interaction with IgE might be expected to result in more sensitive mast cells that are degranulated more easily. IL-4 is at least partially responsible for Ig class switching to IgE, suggesting that mutations in IL-4 that more effectively drive class switching to IgE could also participate in a genetic predisposition to asthma.

1. (a) False: IgE is increased. (b) False: IL-4 increases IgE production. (c) True. (d) False: Antihistamines are helpful in type I hypersensitivity, which involves release of histamine by mast-cell degranulation. Because type III hypersensitivity primarily involves immune-complex deposition, antihistamines are ineffective. (e) False: Most pollen allergens contain multiple allergenic components. (f) False: Unlike IgG, IgE cannot pass through the placenta. (g) True. (h) True. (i) True. (j) False: $T_H1$ cells are active in type IV hypersensitivity and produce cytokines that activate macrophages. (k) True. (l) True. (m) True. (n) False: Repeated injections of allergens (hyposensitization) is thought to cause a shift to a $T_H1$ response so that IgG is produced instead of IgE.

2. (a) The complete antibodies would cross-link FcεRI molecules on the membrane of mast cells and basophils, resulting in their activation and degranulation. The released mediators would induce vasodilation, smooth muscle contraction, and a local wheal-and-flare reaction.

Because the Fab fragment is monovalent, it cannot cross-link FcεRI molecules and thus cannot induce degranulation. However, this type of antireceptor antibody could bind to FcεRI and might thereby block binding of IgE to the receptors. (b) The response induced by complete anti-FcεRI antibodies does not depend on allergen-specific IgE and thus would be similar in allergic and nonallergic mice. Injection of Fab fragments of anti-FcεRI might prevent allergic mice from reacting to an allergen if these fragments block binding of IgE to mast cells and basophils.

3. Engineer chimeric monoclonal antibodies to snake venom that contain mouse variable regions but human heavy- and light-chain constant regions (see Figure 5-23).

4. (a) Type I hypersensitivity: localized atopic reaction that results from allergen cross-linkage of fixed IgE on skin mast cells, inducing degranulation and mediator release. (b) Type III hypersensitivity: immune complexes of antibody and insect antigens form and are deposited locally, causing an Arthus-type reaction resulting from complement activation and complement split products. (c) Type IV hypersensitivity: sensitized $T_H$ cells release their mediators, inducing macrophage accumulation and activation. Tissue damage results from lysosomal enzymes released by the macrophages.

5. (a) IV. (b) All four types. (c) I and III. (d) IV. (e) I. (f) II. (g) I. (h) I. (i) II. (j) II. (k) I.

6. Answer shown in table below.

| Immunologic event | Hypersensitivity | | | |
|---|---|---|---|---|
| | Type I | Type II | Type III | Type IV |
| IgE-mediated degranulation of mast cells | + | | | |
| Lysis of antibody-coated blood cells by complement | | + | | |
| Tissue destruction in response to poison oak | | | | + |
| C3a- and C5a-mediated mast-cell degranulation | | (some) | + | |
| Chemotaxis of neutrophils | | | + | |
| Chemotaxis of eosinophils | + | | | |
| Activation of macrophages by IFN-γ | | | | + |
| Deposition of antigen-antibody complexes on basement membranes of capillaries | | | + | |
| Sudden death due to vascular collapse (shock) shortly after injection or ingestion of antigen | + | | | |

7. Memory B cells generated during a previous pregnancy (sensitization) become activated to secrete anti-Rh IgG, which crosses the placenta and binds to fetal red blood cells. Complement can be activated on the surface of the red blood cells, resulting in red-cell lysis.

8. Immune complexes tend to lodge in capillaries, causing inflammation and subsequent tissue damage.

9. (a)1. (b) 5. (c) 2. (d) 3. (e) 7. (f) 6. (g) 4.

10. The early phase of asthma is characterized by degranulation of mast cells and release of inflammatory mediators such as histamine, leukotriene C4, and prostaglandin D2. The consequences are a localized type I hypersensitivity in the lung, including increased mucus secretion, vasodilation, and bronchoconstriction. The late-phase response is characterized by infiltration of other leukocytes, characteristic of chronic inflammation. These include the presence of eosinophils, lymphocytes, and neutrophils in the airway space and release of mediators characteristic of a $T_H2$-type response, including PAF, IL-4, IL-5, eosinophil chemotactic factor, neutrophil chemotactic factor, TNF-α, and $LTC_4$. Consequences in the lung are localized inflammation, damage and loss of epithelial cells, fibrosis of the basement membrane, and decreased lung function.

11. RhoGam; erythroblastosis fetalis.

**Analyze the Data:** (a) This is a type I hypersensitivity response mediated by IgE. Helminth infections are frequently associated with type I hypersensitivity responses. (b) Histamine is a major component of human mast cells and basophil granules and is released during basophil degranulation. Lacking an assay for histamine, you can measure other primary mediators such as heparin, serotonin, or various proteases or secondary mediators such as leukotrienes, platelet-activating factor, or TNF. (c) IgE. This isotype binds to the surface of basophils and mast cells, inducing degranulation and cellular activation. (d) Basophils and mast cells bind IgE. When antigen is introduced, in this case *Brugia malayi* antigen, the basophils cross-link FcRI and induce degranulation. Cross-linking induces a drop in cAMP, calcium flux from both extracellular and intracellular pools, activation of tyrosine kinases, phosphorylation of phospholipase C, and a series of biochemical reactions that result in cellular activation. (e) Eosinophils would be expected in the late-phase response, recruited to the region by the secretion of platelet activating factor, leukotrienes, eosinophil chemotactic factor, and other mediators. (f) It induces an IgE response. IgE binds to both mast cells and basophils to make the person "sensitive" to the *Brugia malayi* antigen. (g) This would be a $T_H2$ response. IL-4 induces Stat6, which promotes B cells to class-switch to transcribe IgE. Stat6 activation has negative feedback on the activation of transcription of type 1 cytokines such as IFN-γ. Therefore, IFN-γ responses may be predicted to go down.

# Chapter 16

**Clinical Focus Question:** The observations that women mount more robust immune responses than men and that these responses tend to be more $T_H1$-like may, in part, explain gender differences in susceptibility to autoimmunity. Since the $T_H1$ type of response is proinflammatory, the development of autoimmunity may be enhanced.

1.  The process called central tolerance eliminates lymphocytes with receptors displaying affinity for self antigens in the thymus or in the bone marrow. A self-reactive lymphocyte may escape elimination in these primary lymphoid organs if the self antigen is not encountered there or if the affinity for the self antigen is below that needed to trigger the induction of apoptotic death. Self-reactive lymphocytes escaping central tolerance elimination are kept from harming the host by peripheral tolerance, which involves three major strategies: induction of cell death or apoptosis, induction of anergy (a state of nonresponsiveness), or induction of an antigen-specific population of regulatory T cells that keeps the self-reactive cells in check.

2.  Tolerance is necessary to remove self-reactive B and T lymphocytes. Without tolerance, which can be defined as unresponsiveness to an antigen, massive autoimmunity or self-reactivity would result.

3.  Receptor editing is a process by which B cells (but not T cells) exchange the potentially autoreactive V region of the immunoglobulin with another V gene, thus changing antigen specificity and avoiding self-reactivity.

4.  (a) 6. (b) 8. (c) 10. (d) 9. (e) 12. (f) 7. (g) 3. (h) 11. (i) 2. (j) 1. (k) 5. (l) 4.

5.  (a) EAE is induced by injecting mice or rats with myelin basic protein in complete Freund's adjuvant. (b) The animals that recover from EAE are now resistant to EAE. If they are given a second injection of myelin basic protein in complete Freund's adjuvant, they no longer develop EAE. (c) If T cells from mice with EAE are transferred to normal syngeneic mice, the mice will develop EAE. See Figure 16-11.

6.  A number of viruses have been shown to possess proteins that share sequences with myelin basic protein (MBP). Since the encephalitogenic peptides of MBP are known, it is possible to test these peptides to see whether they bear sequence homology to known viral protein sequences. Computer analysis has revealed a number of viral peptides that bear sequence homology to encephalitogenic peptides of MBP. By immunizing rabbits with these viral sequences, it was possible to induce EAE. The studies on the encephalitogenic peptides of MBP also showed that different peptides induced EAE in different strains. Thus, the MHC haplotype will determine which cross-reacting viral peptides will be presented and, therefore, will influence the development of EAE.

7.  (1) A virus might express an antigenic determinant that cross-reacts with a self-component. (2) A viral infection might induce localized concentrations of IFN-γ. The IFN-γ might then induce inappropriate expression of class II MHC molecules on non-antigen-presenting cells, enabling self peptides presented together with the class II MHC molecules on these cells to activate $T_H$ cells. (3) A virus might damage an organ, resulting in release of antigens that are normally sequestered from the immune system.

8.  (a) So that the IFN-γ transgene would be expressed only by pancreatic beta cells. (b) There was a cellular infiltration of lymphocytes and macrophages similar to that seen in insulin-dependent diabetes mellitus. (c) The IFN-γ transgene induced the pancreatic beta cells to express class II MHC molecules. (d) A localized viral infection in the pancreas might result in the localized production of IFN-γ by activated T cells. The IFN-γ might then induce the inappropriate expression of class II MHC molecules by pancreatic beta cells as well as the production of other cytokines such as IL-1 or TNF. If self peptides are presented by the class II MHC molecules, then IL-1 might provide the necessary costimulatory signal to activate T cells against the self peptides. Alternatively, the TNF might also cause localized cellular damage. See Figure 16-13.

9.  Anti-CD4 monoclonal antibodies have been used to block $T_H$ activity. Monoclonal antibodies (anti-TAC) specific for the high-affinity IL-2 receptor have been tried to block activated $T_H$ cells. The association of some autoimmune diseases with restricted expression of the T-cell receptor has prompted researchers to use monoclonal antibody specific for T-cell receptors carrying particular V domains. Finally, antibodies against specific MHC alleles associated with increased risk for autoimmunity have been tested in EAE models.

10. (a) True. (b) False: IL-12, which promotes the development of $T_H1$ cells, increases the autoimmune response to MBP plus adjuvant. (c) False: The presence of HLA B27 is strongly associated with susceptibility to ankylosing spondylitis but is not the only factor required for development of the disease. (d) True. (e) True.

11. (a) 5; 1. (b) 1; 3; 4. (c) 1; 3. (d) 1; 2; 3.

12. Activation of early complement components leads to the formation of C3b, which binds to CR1 in RBCs to clear immune complexes from the bloodstream. The immune complexes are carried to the liver and spleen, where they are stripped from the RBCs and degraded. Complement activation in patients lacking C1, C4, or CR1 may be induced via the alternate or MBL pathway.

13. (a) Polyclonal B cell activation can occur as result of infection with gram-negative bacteria, cytomegalovirus, or EBV, which induce nonspecific proliferation of B cells; some self-reactive B cells can be stimulated in this process. (b) If normally sequestered antigens are exposed, self-reactive T cells may be stimulated. (c) The immune response against a virus may cross-react with normal cellular antigens, as in the case of molecular mimicry. (d) Increased expression of TCR molecules should not lead to autoimmunity; however, if the expression is not regulated in the thymus, self-reactive cells could be produced. (e) Increased expression of class II MHC molecules has been seen in IDDM and Grave's disease, suggesting that inappropriate antigen presentation may stimulate self-reactive T cells.

14. (a) 1. False. 2. False. 3. False. 4. True. (b) 100. (c) All three could be used to verify this experiment.

# Chapter 17

**Clinical Focus Question:** The ideal animal for producing organs for xenotransplantation would have body size roughly equivalent to that of humans and could be genetically altered to eliminate any antigens that cause acute rejection. It should be free of any disease that can be passed to humans. The test of the organs must include transplantation to nonhuman primates and observation periods that are sufficiently long to ascertain that the organ remains fully functional in the new host and that no disease is transmitted.

1. (a) False: Acute rejection is cell mediated and probably involves the first-set rejection mechanism (see Figures 17-1b and 17-7). (b) True. (c) False: Passenger leukocytes are donor dendritic cells that express class I MHC molecules and high levels of class II MHC molecules. They migrate from the grafted tissue to regional lymph nodes of the recipient, where host immune cells respond to alloantigens on them. (d) False: A graft that is matched for the major histocompatibility antigens, encoded in the HLA, may be rejected because of differences in the minor histocompatibility antigens encoded at other loci. (e) True.

2. (a) Dark wells indicate that the cells have taken up the dye and are positive for that antigen. Donor 2 is a perfect match with the recipient on all antigens tested and would therefore be the first choice. (b) Donor 1 would make an acceptable second choice, given that only one antigen is a mismatch. (c) Donor 4 is a mismatch for both HLA-A and HLA-DR antigens tested and is a match with only one antigen of HLA-B tested. Donors 1 and 3 are both mismatched at one antigen; however, donor 3 is mismatched with an HLA-DR antigen, which is a class II MHC molecule. Mismatches in class II MHC are more likely to lead to rejection than mismatches in class I MHC. Therefore, donor 1 is less likely to be rejected. (d) A one-way mixed lymphocyte reaction should be done to determine compatibility.

3. Answer to Question 3 below.

as TNF may mediate direct cytolytic damage to the host cells. (b) GVHD develops when the donated organ or tissue contains immunocompetent lymphocytes and when the host is immune suppressed. (c) The donated organ or tissue could be treated with monoclonal antibodies to CD3, CD4, or the high-affinity IL-2 receptor to deplete donor $T_H$ cells. The rationale behind this approach is to diminish $T_H$-cell activation in response to the alloantigens of the host. The use of anti-CD3 will deplete all T cells; the use of anti-CD4 will deplete all $T_H$ cells; the use of anti-IL-2R will deplete only the activated $T_H$ cells.

5. (a) Sibling C is the best potential donor based on the match of ABO blood type and all class I MHC antigens. (b) Sibling A is the best donor as indicated by the low proliferative response in a one-way MLR with the recipient. This result indicates that class II antigens are matched with those of the recipient. By these tests, sibling A would be selected to be the donor, because the class II antigens are more important in determining potential rejection.

6. The use of soluble CTLA4 or anti-CD40 ligand to promote acceptance of allografts is based on the requirement of a T cell for a costimulatory signal when its receptor is bound. Even if the recipient T cell recognizes the graft as foreign, the presence of CTLA4 or anti-CD40L will prevent the T cell from becoming activated because it does not receive a second signal through the CD40 or CD28 receptor (see Figure 17-9). Instead of becoming activated, T cells stimulated in the presence of these blocking molecules become anergic. The advantage of using soluble CTLA4 or anti-CD40L is that these molecules affect only those T cells involved in the reaction against the allograft. These allograft-specific T cells will become anergic, but the general population of T cells will remain normal. More general immunosuppressive measures, such as the use of

| Donor | Recipient | Response | Type of rejection |
|---|---|---|---|
| BALB/c | C3H | R | FSR |
| BALB/c | Rat | R | FSR |
| BALB/c | Nude mouse | A | |
| BALB/c | C3H, had previous BALB/c graft | R | SSR |
| BALB/c | C3H, had previous C57BL/6 graft | R | FSR |
| BALB/c | BALB/c | A | |
| BALB/c | (BALB/c × C3H) $F_1$ | A | |
| BALB/c | (C3H × C57BL/6) $F_1$ | R | FSR |
| (BALB/c × C3H) $F_1$ | BALB/c | R | FSR |
| (BALB/c × C3H) $F_1$ | BALB/c, had previous $F_1$ graft | R | SSR |

4. (a) Graft-versus-host disease (GVHD) develops as donor T cells recognize alloantigens on cells of an immune-suppressed host. The response develops as donor $T_H$ cells are activated in response to recipient peptide-MHC complexes displayed on antigen-presenting cells. Cytokines elaborated by these $T_H$ cells activate a variety of effector cells, including NK cells, CTLs, and macrophages, which damage the host tissue. In addition, cytokines such

CsA or FK506, cause immunodeficiency and subsequent susceptibility to infection.

7. Azathioprine is a mitotic inhibitor used to block proliferation of graft-specific T cells. Cyclosporin A and FK506 block activation of resting T cells by interfering with a signal transduction pathway that leads to assembly of the transcription factor NFAT. Rapamycin inhibits the activation of $T_H$ cells by arresting the cell cycle. Ideally, if

early rejection is inhibited by preventing a response by specific T cells, these cells may be rendered tolerant of the graft over time. Lowering the dosage of the drugs is desirable because of decreased side effects in the long term.

# Chapter 18

**Clinical Focus Question:** VIG, or vaccinia-specific immunoglobulin, is obtained from individuals who have been immunized with vaccinia, the smallpox vaccine. In past years, when smallpox vaccination was common, VIG was obtained by collecting serum from immunized individuals; due to discontinuation of smallpox vaccinations, VIG is currently in short supply. VIG is very effective because vaccinia is cleared quite efficiently by antibody.

1. (a) Because the infected target cells expressed $H-2^k$ MHC molecules but the primed T cells were $H-2^b$ restricted. (b) Because the influenza nucleoprotein is processed by the endogenous processing pathway and the resulting peptides are presented by class I MHC molecules. (c) Probably because the transfected class I $D^b$ molecule is only able to present peptide 365–380 and not peptide 50–63. Alternatively, peptide 50–63 may not be a T-cell epitope. (d) These results suggest that a cocktail of several immunogenic peptides would be more likely to be presented by different MHC haplotypes in humans and would provide the best vaccines for humans.

2. Nonspecific host defenses include ciliated epithelial cells, bactericidal substances in mucous secretions, complement split products activated by the alternative pathway that serve both as opsonins and as chemotactic factors, and phagocytic cells.

3. Specific host defenses include secretory IgA in the mucous secretions, IgG and IgM in the tissue fluids, the classical complement pathway, complement split products, the opsonins (IgM, IgG, and C3b), and phagocytic cells. Cytokines produced during the specific immune response, including IFN-$\gamma$, TNF, IL-1, and IL-6, contribute to the overall intensity of the inflammatory response.

4. Humoral antibody peaks within a few days of infection and binds to the influenza HA glycoprotein, blocking viral infection of host epithelial cells. Because the antibody is strain specific, its major role is in protecting against re-infection with the same strain of influenza.

5. (a) African trypanosomes are capable of antigenic shifts in the variant surface glycoprotein (VSG). The antigenic shifts are accomplished as gene segments encoding parts of the VSG are duplicated and translocated to transcriptionally active expression sites. (b) *Plasmodium* evades the immune system by continually undergoing maturational changes from sporozoite to merozoite to gametocyte, allowing the organism to continually change its surface molecules. In addition, the intracellular phases of its life cycle reduce the level of immune activation. Finally, the organism is able to slough off its circumsporozoite coat after antibody binds to it. (c) Influenza is able to evade the immune response through frequent antigenic changes in its hemagglutinin and neuraminidase glycoproteins. The antigenic changes

are accomplished by the accumulation of small point mutations (antigenic drift) or through genetic reassortment of RNA between influenza virions from humans and animals (antigenic shift).

6. (a) $IA^b$. (b) Because antigen-specific, MHC-restricted $T_H$ cells participate in B-cell activation.

7. (a) BCG (Bacille Calmette-Guérin). (b) Antigenic shift; antigenic drift. (c) Gene conversion. (d) Tubercles; $T_H$ cells; activated macrophages. (e) Toxoid. (f) Interferon $\alpha$; interferon $\gamma$. (g) Secretory IgA. (h) IL-12; IFN-$\gamma$.

8. Most fungal infections prevalent in the general population do not lead to severe disease and are dealt with by innate immune mechanisms. Problematic fungal infections are more commonly seen in those with some form of immunodeficiency, such as patients with HIV/AIDS or those with immunosuppression caused by therapeutic measures.

9. One possible reason for the emergence of new pathogens is the crowding of the world's poorest populations into very small places within huge cities, because population density increases the spread of disease. Another factor is the increase in international travel. Other features of modern life that may contribute include mass distribution of food, which exposes large populations to potentially contaminated food, and unhygienic food preparation.

10. The major factor contributing to the rapid spread of the SARS virus is the prevalence of international travel. The disease was carried by a physician from mainland China to Hong Kong, where a number of residents in the same hotel acquired the disease and carried it to their next destination.

11. (a) Influenza virus changes surface expression of neuraminidase and hemagglutinin. (b) Herpesviruses remains dormant in nerve cells. Later emergence can cause outbreaks of cold sores or shingles (chicken-pox virus). (c) *Neisseria* secretes proteases that cleave IgA. (d) False. (e) Several gram-positive bacteria resist complement-mediated lysis. (f) Influenza virus accumulates mutations from year to year. (g) False.

12. Surface epitopes of *Treponema* are structurally similar to epitopes on *Borrelia*, leading to the cross-reaction of antibodies and false-positive results. Newer tests have been developed with antibodies against epitopes that are specific for *Borrelia*.

13. (a) No, IgE is raised against allergens, not infectious diseases. (b) No, autoreactive T cells are activated only by intracellular infections. The statements in (c) through (f) are correct.

14. (a) Large, granular cells such as mast cells and eosinophils. Neutrophils and macrophages will also be involved. (b) Cytokines such as IL-4, IL-5, and IL-3 would help drive the $T_H1$ response already present; however, theraputic cytokines to drive the response toward a $T_H2$ response may be more beneficial to longer-term immunity.

# Chapter 19

**Clinical Focus Question:** Any connection between vaccination and a subsequent adverse reaction must be evaluated by valid clinical trials involving sufficient numbers of subjects in the control group (those given a placebo) and experimental group (those receiving the vaccine). This is needed to give a statistically correct assessment of the effects of the vaccine versus other possible causes for the adverse event. Such clinical studies must be carried out in a double-blind manner; that is, neither the subject nor the caregiver should know who received the vaccine and who received the placebo until the end of the observation period.

In the example cited, it is possible that the adverse event (increased incidence of arthritis) was caused by an infection occurring at the same time the new vaccine was administered. Determining the precise cause of this side effect may not be possible, but ascertaining whether it is caused by the vaccine is possible by appropriate studies of the vaccinated and control populations.

1. (a) True. (b) True. (c) True. (d) False: Because DNA vaccines allow prolonged exposure to antigen, they are likely to generate immunologic memory. (e) True. (f) False: A DNA vaccine contains the gene encoding an entire protein antigen, which most likely contains multiple epitopes.

2. Because attenuated organisms are capable of limited growth within host cells, they are processed by the cytosolic pathway and presented on the membrane of infected host cells together with class I MHC molecules. These vaccines, therefore, usually can induce a cell-mediated immune response. The limited growth of attenuated organisms within the host often eliminates the need for booster doses of the vaccine. Also, if the attenuated organism is able to grow along mucous membranes, then the vaccine will be able to induce the production of secretory IgA. The major disadvantage of attenuated whole-organism vaccines is that they may revert to a virulent form. They also are more unstable than other types of vaccines, requiring refrigeration to maintain their activity.

3. (a) The antitoxin was given to inactivate any toxin that might be produced if *Clostridium tetani* infected the wound. The antitoxin was necessary because the girl had not been previously immunized and, therefore, did not have circulating antibody to tetanus toxin or memory B cells specific for tetanus toxin. (b) Because of the treatment with antitoxin, the girl would not develop immunity to tetanus as a result of the first injury. Therefore, after the second injury, 3 years later, she will require another dose of antitoxin. To develop long-term immunity, she should be vaccinated with tetanus toxoid.

4. The Sabin polio vaccine is attenuated, whereas the Salk vaccine is inactivated. The Sabin vaccine thus has the usual advantages of an attenuated vaccine compared with an inactivated one (see Answer 2). Moreover, since the Sabin vaccine is capable of limited growth along the gastrointestinal tract, it induces production of secretory IgA. The attenuated Sabin vaccine can cause life-threatening infection in individuals, such as children with AIDS, whose immune systems are severely suppressed.

5. The virus strains used for the nasally administered vaccines are temperature-sensitive mutants that cannot grow at human body temperature (37° C). The live attenuated virus can grow in the upper respiratory tract to induce immunity but will not grow in the lower respiratory to cause influenza.

6. T-cell epitopes generally are internal peptides, which commonly contain a high proportion of hydrophobic residues. In contrast, B-cell epitopes are located on an antigen's surface, where they are accessible to antibody, and contain a high proportion of hydrophilic residues. Thus, synthetic hydrophobic peptides are most likely to represent T-cell epitopes and induce cell-mediated response, whereas synthetic hydrophilic peptides are most likely to represent accessible B-cell epitopes and induce an antibody response.

7. When the majority of a population is immune to a particular pathogen—that is, there is herd immunity—then the probability of the few susceptible members of the population contacting an infected individual is very low. Thus, susceptible individuals are not likely to become infected with the pathogen. If the number of immunized individuals decreases sufficiently, most commonly because of reduction in vaccination rates, then herd immunity no longer operates to protect susceptible individuals and infection may spread rapidly in a population, leading to an epidemic.

8. In this hypothetical situation, the gene can be cloned into an expression system and the protein expressed and purified in order to test it as a recombinant protein vaccine. Alternatively, the gene can be cloned into a plasmid vector that can be injected directly and tested as a DNA vaccine. Use of the cloned gene as a DNA vaccine is more efficient, because it eliminates the steps required for preparation of the protein and its purification. However, the plasmid containing the gene for the protective antigen must be suitably purified for use in human trials. DNA vaccines have a greater ability to stimulate both the humoral and cellular arms of the immune system than protein vaccines do and thus may confer more complete immunity. The choice must also take into consideration the fact that recombinant protein vaccines are in widespread use but DNA vaccines for human use are still in early test phases.

9. Pathogens with a short incubation period (e.g., influenza virus) cause disease symptoms before a memory-cell response can be induced. Protection against such pathogens is achieved by repeated reimmunizations to maintain high levels of neutralizing antibody. For pathogens with a longer incubation period (e.g., polio virus), the memory-cell response is rapid enough to prevent development of symptoms, and high levels of neutralizing antibody at the time of infection are unnecessary.

10. Bacterial capsular polysaccharides, inactivated bacterial exotoxins (toxoids), and surface protein antigens. The latter two commonly are produced by recombinant DNA technology. In addition, the use of DNA molecules to direct synthesis of antigens on immunization is being evaluated.

11. The passive vaccination antibodies bind to Fc receptors on the maternal B cells, which keeps the $Rh^+$ B cells from becoming activated and producing antibodies against the developing fetus. In this way, the vaccination protects the newborn against attack from the maternal immune system.

12. A possible loss of herd immunity in the population. Even in a vaccinated population of children, a small percentage

may have diminished immunity to the diseased target due to differences among MHC molecule expression in a population, providing a reservoir for the disease. In addition, most vaccinated individuals, if exposed to the disease, will develop mild illness. Exposure of unvaccinated individuals to either source of disease would put them at risk for serious illness. Epidemics within adult populations would have more serious consequences, and infant mortality due to these diseases would increase.

13. (a) 1 or 2. (b) 2. (c) 3. (d) 4. (e) 1, (f) 2, (g) 2.

**Analyze the Data:** (a) The pSG5DNA-Bcl-xL targeting calreticulin (CRT) and LAMP-1 are the most effective vaccines at inducing $CD8^+$ T cells to make IFN-$\gamma$. The pSG5DNA-Bcl-xL targeting HSP70 also activated $CD8^+$ T cells. However, the pSG5 construct without the anti-apoptosis gene targeting CRT also induced a good $CD8^+$ T-cell response. (b) Calreticulin is a chaperone protein associated with partially folded class I MHC molecules in the endoplasmic reticulum. Associating E7 antigen with the chaperone may enhance loading of class I MHC molecules with E7, making the antigen more available to T cells once it is expressed on cells. (c) The DNA vaccines co-injected with pSG5DNA-Bcl-xL were effective in inducing $CD8^+$ T cells, possibly because the expression of anti-apoptotic genes in dendritic cells allowed those cells to survive longer and present antigen to T cells for a longer time. The longer they presented antigen, the longer the host would respond to produce antigen-specific T cells. (d) The data in Figure b indicate that in the absence of $CD4^+$ helper T cells (in CD4 knockout mice), there is ineffective activation of $CD8^+$ T cells. Therefore, T-cell help is necessary to activate the CD8 response; by targeting antigen to MHC II, you more efficiently activate helper T cells. The Sig/E7/LAMP-1 construct was necessary because most antigens presented by MHC II molecules are processed by the endocytic pathway, and the Sig/E7/LAMP-1 construct targets antigen to the Golgi, where the E7 peptides can be exchanged with CLIP and inserted into MHC II. (e) Helper T cells (poor response in the CD4 knockout mice), long-lived dendritic cells (immunization with the pSG5DNA-Bcl-xL improves the response), antigen (the absence of peptide failed to induce a response), and the targeting of antigen to MHC II (immunizing with the Sig/E7/LAMP-1 construct is the only one that induces a $CD8^+$ T-cell response).

# Chapter 20

**Clinical Focus Question:** Before Nevirapine can be universally administered to all mothers at delivery, it is necessary to know that the benefit of this treatment outweighs the risk. The benefit of the drug, as learned from studies already conducted, is that it reduces significantly the transmission of HIV to infants born of infected mothers. A study of the risk of Nevirapine administration to normal mothers and their infants must be carried out, and only if there are minimal or no side effects can universal treatment be recommended.

1. (a) True. (b) False: X-linked agammaglobulinemia is characterized by a reduction in B cells and an absence of immunoglobulins. (c) False: Phagocytic defects result in recurrent bacterial and fungal infections. (d) True. (e) True. (f) True. (g) True. (h) True. (i) False: These children are usually able to eliminate common encapsulated bacteria with antibody plus complement but are susceptible to viral, protozoan, fungal, and intracellular bacterial pathogens, which are eliminated by the cell-mediated branch of the immune system. (j) False: Humoral immunity also is affected because class II–restricted $T_H$ cells must be activated for an antibody response to occur.

2. (a) 3. (b) 4. (c) 2. (d) 5. (e) 1. (f) 2.

3. The defect in X-linked hyper-IgM syndrome is in CD40L expressed on B cells. CD40L mediates binding of B cells to T cells and sends costimulatory signals to the B cell for class switching. Without CD40L, class switching does not occur and the B cells do not express other antibody isotypes.

4. As discussed in Chapter 10, the thymus is the location of differentiation and maturation of helper and cytotoxic T cells. Positive and negative selection occur in this organ as well. Thus, the thymocytes produced in the bone marrow of patients with DiGeorge syndrome do not have the ability to mature into effector cell types. In Chapter 2, we noted that the thymus decreases in size and function with age. In the adult, effector-cell populations have already been produced (peak thymus size occurs during puberty); therefore, a defect after this stage would cause less severe T-cell deficiency.

5. (a) Leukocyte adhesion deficiency results from biosynthesis of a defective $\beta$ chain in LFA-1, CR3, and CR4, which all contain the same $\beta$ chain. (b) LFA-1 plays a role in cell adhesion by binding to ICAM-1 expressed on various types of cells. Binding of LFA-1 to ICAM-1 is involved in the interactions between $T_H$ cells and B cells, between CTLs and target cells, and between circulating leukocytes and vascular endothelial cells.

6. (a) The rearranged heavy-chain genes in SCID mice lack the D and/or J gene segments. (b) According to the model of allelic exclusion discussed in Chapter 5, a productive heavy-chain gene rearrangement must occur before $\kappa$-chain genes are rearranged. Since SCID mice lack productive heavy-chain rearrangement, they do not attempt $\kappa$ light-chain rearrangement. (c) Yes: The rearrange heavy-chain gene would be transcribed to yield a functional $\mu$ heavy chain. The presence of the heavy chain then would induce rearrangement of the $\mu$-chain gene (see Figure 5-11).

7. (a) 4. (b) 3. (c) 2. (d) 1. (e) 8. (f) 6. (g) 5. (h) 7.

8. (a) False: HIV-2 and SIV are more closely related. (b) False: HIV-1 infects chimpanzees but does not cause immune suppression. (c) True. (d) False: Zidovudine (AZT) acts at the level of reverse transcription of the viral genome, whereas indinavir is an inhibitor of the viral protease. (e) True. (f) False: Patients with advanced AIDS sometimes have no detectable serum antibody to HIV. (g) False: The PCR detects HIV proviral DNA in latently infected cells. (h) True.

9. The most likely reason for T-cell depletion in AIDS is the cytopathic effects of HIV infection. If the amount of virus in circulation is decreased by the use of antiviral agents, the number of T cells will increase.

10. No. In the chronic phase of HIV infection, viral replication and the $CD4^+$ T-cell number are in a dynamic equilibrium, and the level of virus remains relatively constant.

11. An increase in the viral load and a decrease in CD4$^+$ T-cell levels indicates that HIV infection is progressing from the chronic phase into AIDS. Often infection by an opportunistic agent occurs when the viral load increases and the CD4$^+$ T-cell level drops. The disease AIDS in an HIV-1 infected individual is defined by the presence of certain opportunistic infections or by a CD4$^+$ T-cell count of less than 200 cells/mm$^3$ (see Table 20-3).

12. Skin-test reactivity is monitored to indicate the functional activity of T$_H$ cells. As AIDS progresses and CD4$^+$ T cells decline, there is a decline in skin-test reactivity to common antigens.

13. Receptors for certain chemokines, such as CXCR4 and CCR5, also function as coreceptors for HIV-1. The chemokine that is the normal ligand for the receptor competes with the virus for binding to the receptor and can thus inhibit cell infection by blocking attachment of the virus. Cytokines that activate T cells stimulate infection because they increase expression of receptors used by the virus (see Figure 20-12c).

14. No. There are latently infected cells that reside in lymph nodes or in other sites. These cells can be activated and begin producing virus, thus causing a relapse of the disease. In a true cure for AIDS, the patient must be free of all cells containing HIV DNA.

15. Patient LS fits the definition of AIDS, whereas patient BW does not. The clinical diagnosis of AIDS among HIV-infected individuals depends on both the T-cell count and the presence of various indicator diseases. See Table 20-3.

**Analyze the Data:** (a) CVID lowers the number and percentage of T helper cells. (b) Table 1 indicates that there are fewer naive CD8$^+$ T cells in CVID patients. This could mean chronic activation in CVID has made these T cells less dependent on IL-2. Alternatively, Figure 2 shows that bone marrow cells from CVID patients make less IL-2. If this is also true of T$_H$ cells or there are fewer T$_H$ cells (as seen in Table 1), then the generation of CTLs may be impaired. This example of the complexity of the immune system demonstrates that specific responses are not always easily predictable. (c) False. The data in Figure b show that the kinetics and overall production of TNF-$\alpha$ is greater in CVID than in normal patients, but IL-2 is lower. Thus, there is no consistent impact. Based on what we know about TNF-$\alpha$ activity, it might be responsible for increased pathology or cell death, perhaps explaining the loss of both CD4$^+$ and CD8$^+$ cells in CVID patients. (d) According to Chapter 20, CVID lowers IgG, IgA, and IgM. However, based on the data presented in Table 1, one might predict that in the absence of T-cell help, there will be a significant impact on class switching.

# Chapter 21

**Clinical Focus Question:** Because cervical cancer is linked to human papilloma virus infection, a preventive cancer vaccine may be possible—if HPV infection is prevented, then cervical cancer will be prevented. Other cancers that may be targets for such prevention include adult T-cell leukemia/lymphoma and the liver cancer that is linked to hepatitis B infection. Most cancers have not been clearly linked to an agent of infection, and therefore a preventive vaccine is not an option.

1. (a) False: Hereditary retinoblastoma results from inactivation of both alleles of *Rb*, a tumor suppressor gene. (b) True. (c) True. (d) True. (e) False: Some oncogenic retroviruses do not have viral oncogenes. (f) True.

2. Cells of the pre-B-cell lineage have rearranged the heavy-chain genes and express the $\mu$ heavy chain in their cytoplasm (see Figure 11-3). You could perform Southern-blot analysis with a C$_\mu$ probe to see whether the heavy-chain genes have rearranged. You could also perform fluorescent staining with antibody specific for the cytoplasmic $\mu$ heavy chain.

3. (a) Early-stage melanoma cells appear to be functioning as antigen-presenting cells, processing the antigen by the exogenous pathway and presenting the tetanus toxoid antigen together with the class II MHC DR molecule. (b) Advanced-stage melanoma cells might have a reduction in the expression of class II MHC molecules, or they might not be able to internalize and process the antigen by the exogenous route. (c) Since the paraformaldehyde-fixed early melanoma cells could present processed tetanus toxoid, they must express class II MHC molecules on their surface. (d) Stain the early and advanced melanoma cells with fluorescent monoclonal antibody specific for class II MHC molecules.

4. Tumor antigens may be encoded by genes expressed only by tumors, may be products of genes overexpressed by the tumor or genes normally expressed only at certain stages of differentiation, or may be products of normal genes that are altered by mutation. In some cases viral products may be tumor antigens.

5. IFN-$\alpha$, IFN-$\beta$, and IFN-$\gamma$ enhance the expression of class I MHC molecules on tumor cells, thereby increasing the CTL response to tumors. IFN-$\gamma$ also increases the activity of CTLs, macrophages, and NK cells, each of which plays a role in the immune response to tumors. TNF-$\alpha$ and TNF-$\beta$ have direct antitumor activity, inducing hemorrhagic necrosis and tumor regression. IL-2 activates LAK and TIL cells, which both have antitumor activity.

6. (a) Melanoma cells transfected with the B7 gene are able to deliver the costimulatory signal necessary for activation of CTL precursors to effector CTLs, which then can destroy tumor cells. Melanoma cells transfected with the GM-CSF gene secrete this cytokine, which stimulates the activation and differentiation of antigen-presenting cells, especially dendritic cells, in the vicinity. The dendritic cells then can present tumor antigens to T$_H$ cells and CTL precursors, enhancing the generation of effector CTLs (see Figure 21-11). (b) Because some of the tumor antigens on human melanomas are expressed by other types of cancers, transfected melanoma cells might be effective against other tumors carrying identical antigens.

7. (a) 4, (b) 2. (c) 1. (d) 5. (e) 3. (f) 9. (g) 8. (h) 7. (i) 6.

# Chapter 22

**Clinical Focus Question:** Microarrays provide the clinician with the ability to more accurately diagnose a disease. For example, if gene expression patterns vary in two phenotypically similar tumors, it may be possible to redefine the disease as two distinct diseases with concomitantly distinct treatments.

1. (a) The genomic clone contains intervening sequences (introns) that are removed during processing of the primary transcript and therefore do not specify any amino acid residues in the protein product. (b) DNA must be microinjected into a fertilized egg so that the DNA (transgene) will be passed on to all daughter cells. (c) Primary lymphoid cultures have a finite life span and contain a heterogeneous population of cells.

2. (a) Homozygous; syngeneic. (b) Normal antigen-primed B cells; cancerous plasma cells (myeloma cells); unlimited; monoclonal antibodies. (c) Transformation; indefinite.

3.

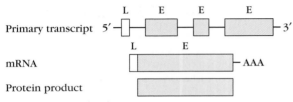

4. c.

5. a, b, d.

6. Transfected DNA integrates only into a small percentage of cells. If a selectable marker gene is also present, the small number of transfected cells will grow in the appropriate medium, whereas the much larger number of nontransfected cells will die.

7. The mouse would become a mosaic in which the transgene would have incorporated into some of the somatic cells but not all.

8. Since cleavage with each restriction enzyme alone produced a single band, the plasmid, which is circular, must contain one restriction site for each enzyme. Since simultaneous cleavage of the 5.4-kb plasmid with both enzymes yielded a single band, the fragments from each cleavage must have the same length. Thus, the restriction sites must be equidistant from each other in both directions on the plasmid, as shown in the following diagram.

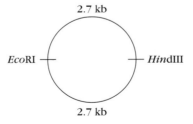

9. d.

10. Isolate mRNA from activated T cells and transcribe it into cDNA using reverse transcriptase. Insert cDNA into a suitable expression vector, such as plasmid DNA carrying an ampicillin selection gene. Transfer the recombinant plasmid DNA into *E. coli* and grow in the presence of ampicillin to select for bacteria containing the plasmid DNA. Test the bacterial culture supernatant for the presence of IL-2 by seeing whether the monoclonal antibody to IL-2 reacts with the culture supernatant. Once a bacterial culture is identified that is secreting IL-2, the cDNA can be cloned.

11. Radiolabel the mRNA and use the labeled mRNA as a probe to screen the genomic library by in situ hybridization. Replate and culture the selected clone, and isolate the cloned DNA from the host DNA.

12. In production of transgenic mice, a functional foreign gene is added to the genome; the process involves introducing a cloned natural gene or cDNA into a fertilized egg, which then is implanted into a pseudopregnant female. In production of knockout mice, a nonfunctional form of a mouse gene replaces the functional gene, so that the animal does not express the encoded product; the process involves selecting transfected embryonic stem (ES) cells carrying the mutated gene and injecting the ES cells into a blastocyst, which then is implanted into a pseudopregnant female.

13. Knockout mice are animals in which a particular gene or DNA sequence has been deleted, whereas knock-in mice are animals in which a piece of DNA has been exchanged for the endogenous DNA sequence.

14. The Cre/*lox* system makes it possible to specifically delete or insert a DNA sequence of choice into the tissue of choice at a particular time during development by using tissue-specific promoters to drive the change in the DNA sequence. This technology thus enhances knockout and knock-in strategies by making them more selective. Also, sometimes restricting the change in sequence to a particular tissue will allow the consequences of a particular deletion or insertion to be studied in that one tissue even though it would be lethal to the organism if present in all tissues.

15. (a) 8. (b) 2. (c) 5. (d) 9. (e) 4. (f) 3. (g) 6. (h) 7. (i) 10.

16. Cells taken directly from patients or normal donors more closely represent the in vivo activity of these cell types, which is an advantage to using PBMCs. However, they are a mixed cell population, and it is not always clear how cell-cell interactions may affect experimental results. Using cell lines eliminates the effects of cell-cell interactions and clearly defines which specific cell type is responsible for an effect. Also, experiments performed with PBMCs will exhibit donor-to-donor variation, whereas cell lines will produce more uniform results due to the genetic homogeneity of the cell population. Finally, cell lines are often transformed or are derived from cancer cells and may deviate in significant ways from the in vivo cell types they represent.

**17.**

(a)

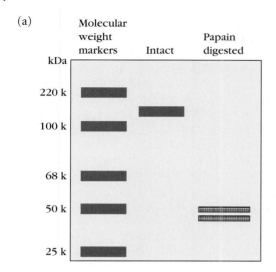

(b)

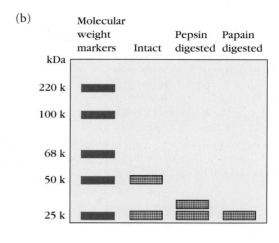

(c)

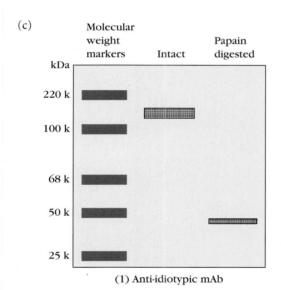

(1) Anti-idiotypic mAb

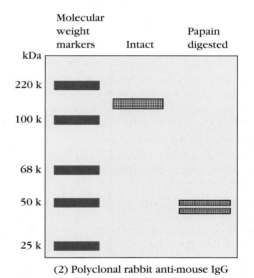

(2) Polyclonal rabbit anti-mouse IgG

# Index

Page numbers followed by f indicate figures; those followed by t indicate tables. Page numbers preceded by A or B refer to appendices.